Birkhäuser

Probability and Its Applications

More information about this series at http://www.springer.com/series/4893

Andrei N. Borodin • Paavo Salminen

Handbook of Brownian Motion - Facts and Formulae

Second Edition

Birkhäuser

Andrei N. Borodin
Steklov Mathematical Institute
St. Petersburg, Russia
and
Department of Mathematics and Mechanics
St. Petersburg State University
St. Petersburg, Russia

Paavo Salminen
Åbo Akademi University
Faculty of Science and Engineering
Mathematics and Statistics
Turku, Finland

2015 corrected reprint of the second edition

ISSN 2297-0371 ISSN 2297-0398 (electronic)
Probability and Its Applications
ISBN 978-3-7643-6705-3 ISBN 978-3-0348-8163-0 (eBook)
DOI 10.1007/978-3-0348-8163-0

Library of Congress Control Number: 2015936694

Mathematics Subject Classification (2010): 60J65, 60J60, 60H05, 60H10, 60J55

Springer Basel Heidelberg New York Dordrecht London

Printed on acid-free paper

Springer Basel AG is part of Springer Science+Business Media (www.birkhauser-science.com)

CONTENTS

PREFACE TO THE FIRST EDITION

There are two parts in this book. The first part is devoted mainly to the properties of linear diffusions in general and Brownian motion in particular. The second part consists of tables of distributions of functionals of Brownian motion and related processes.

The primary aim of this book is to give an easy reference to a large number of facts and formulae associated to Brownian motion. We have tried to do this in a "handbook–style". By this we mean that results are given without proofs but are equipped with a reference, where a proof or a derivation can be found. It is our belief and experience that such a material would be very much welcome by students and people working with applications of diffusions and Brownian motion. In discussions with many of our colleagues we have found that they share this point of view.

Our original plan included more things than we were able to realize. It turned out very soon when trying to put the plan into practice that the material would be too wide to be published under one cover. Excursion theory, which most of the recent results concerning linear Brownian motion and diffusions can be classified as, is only touched upon slightly here, not to mention Brownian motion in several dimensions which enters only through the discussion of Bessel processes. On the other hand, much attention is given to the theory of local time. The Markov property of local time, the so called Ray–Knight theorems, are discussed in a chapter of their own, and one can find a large number of distributions associated with local times in the tables. The concept of local time has proved to be of importance in many applications of diffusions; for instance, in economics, mathematical finance, and optimal control. We hope that the material presented in this book will be useful in these and other applications.

In the tables there are more than 1550 numbered formulae associated directly or via some transformations to Brownian motion. The systematic structure of the tables is explained in Introduction on p. 120[1]. Some of the formulae are certainly obvious, tautological, and perhaps too complicated to be of any real use. In spite of this a decision was made to keep the structure and not to expand the tables with more "exotic" formulae. The main reason for this decision was that nothing like these tables has been published before, and so we think that one should start from the "bottom". We also believe that the structure of the tables makes them easy to

[1]On p. 151 in this edition.

use, especially for people who do not deal with Brownian motion or diffusions on a daily basis. It is, further, our hope that the presentation has some pedagogical merits when used in teaching and by students.

As mentioned above results in this book are usually given without proofs. There are many superb research monographs and text books, where diffusions and Brownian motion play the central role, and where proofs of the theorems in our book are given. In particular, we have been using books by Dynkin (1965), Blumenthal and Getoor (1968), McKean (1969), Ito and McKean (1974, 2nd ed.), Gikhman and Skorokhod (1974, 1975, 1979), Williams (1979), Ikeda and Watanabe (1981), Knight (1981), Chung and Williams (1983), Doob (1984), Rogers and Williams (1987), Revuz and Yor (1991), and Karatzas and Shreve (1991, 2nd ed.). Many of these, e.g., Doob (1984), Rogers and Williams (1987), Revuz and Yor (1991), Karatzas and Shreve (1991, 2nd ed.), give, besides the proofs, also historical comments and notes concerning the origin and development of the theorems. Therefore, whenever possible, a reference is given to a monograph. We want to stress heavily (the obvious thing) that the presentation in Part I is not everywhere meant to be mathematically complete. To achieve this one must in several places study the different sources given in the text and via them, perhaps, also the original research papers.

Paul McGill, Jim Pitman and Marc Yor read parts of the material of the first and latter versions. We thank them for their comments, criticism and encouragement. We are also grateful to Göran Högnas and Juha Vuolle–Apiala for their comments to a more final version.

During the course of this work we have received financial support from the Academy of Finland, the Research Institute of the Åbo Akademi Foundation, the Foundation of the University of Turku, and the Russian Foundation of Fundamental Research, grant No. 93-01-01454. This support is gratefully acknowledged. It is also our pleasure to thank the mathematical departments at the Steklov Institute, St. Petersburg's branch, the University of Turku, and the Åbo Akademi University for their hospitality.

Our special thanks goes to P.E. Baskakova for helping us with the complicated typing work.

Finally, we want to thank the staff of Birkhauser for their patience when this project was dragging in time, and for their excellent work when preparing the real thing. It is also our pleasure to thank the copy–editor whose careful work made the text more readable.

St. Petersburg, Russia, and Åbo (Turku), Finland, March 1996

A.N. Borodin P. Salminen

PREFACE TO THE SECOND EDITION

Compared with the first edition published 1996 this second edition is revised and considerably expanded. The number of new formulae in the tables is more than 1000 and some of the old formulae in the tables and especially in Appendix 1 have found a new equivalent form.

It is our feeling that the first edition has been used, in some extent, for problems arising in financial mathematics. To meet better requirements from this field we have included in the theory part a new subsection on some recent results on geometric Brownian motion and in the formula part there is also a new section with ca. 200 formulae devoted to geometric Brownian motion.

Section V on Ray–Knight theorems has undergone much revision. Aim has been to make it more readable and also to present more detailed explanations. We have also included summarizing tables for ease of reference.

There is a new subsection in Section VI on Feynman–Kac formula concerning functionals stopped at the first range time. This is the random time when the maximum minus minimum of a process reaches a given level. We think that the first range time is an important stopping time with many applications on different fields. A number of formulae associated to the first range time of Brownian motion are displayed in the formula part of the book.

New processes are added to Appendix 1 "Briefly on some diffusions". In particular, we have studied radial Ornstein–Uhlenbeck processes and their squares. These processes are stationary with interesting equilibrium probability distributions; like gamma-, Rayleigh-, Maxwell- and χ^2-distributions. Radial Ornstein–Uhlenbeck processes appear also in the formula part with ca. 160 new formulae.

We have also expanded the introduction to the tables with new references and some explicit examples. We hope that these examples make it easier to use and understand the results in Section VI. There are also three new appendices and the old appendices have been expanded. The first new appendix is on formulae for inverse Laplace transforms, the next one on the second order ordinary differential equations and their fundamental solutions, and the third one on formulae for n-fold differentiation of the product and composition of two functions. The appendix on differential equations is needed when applying the Feynman–Kac formula, and the appendix on differentiation is useful when computing, e.g., moments of a functional of a random process by differentiating a Laplace transform.

This second edition has about 200 pages more than the first edition. However, the added material covers more than that. To keep the book in a reasonable size we have not included some formulae from the first edition to the second one. From our point of view the excluded formulae are simple consequences of others and can be constructed easily, if needed.

We apologize sincerely for the misprints in the first edition and hope that the positive experiences with the book have weighted more than the negative ones. We have worked hard to correct all the misprints and other errors. However, due to the nature of the book and the fast modern word processing technology it would be too optimistic to claim that this second edition does not have any defects. We therefore appreciate to have feedback from our readers also on this edition. Our e-mail addresses are printed on the information page of this book.

We express our warm thanks to Marc Yor for encouragement, many concrete proposals for improvements, pointing out misprints, and giving new references. We are also grateful to Jim Pitman for comments and interest in this work. Andrei Sudakov helped us with the secrets of the AMS-TEXpackage and we thank him for his help.

Financial support from the Research Institute of the Åbo Akademi Foundation, the Magnus Ehrnrooth's Foundation (Finland), Vaisala Foundation (Finland), and the Russian Foundation of Fundamental Research, grant No. 99-01-0112 are gratefully acknowledged.

St. Petersburg, Russia, and Åbo (Turku), Finland, October 2001

A.N. Borodin P. Salminen

PREFACE TO THE CORRECTED SECOND PRINTING OF THE SECOND EDITION

Our main aim with this printing has been to correct some typos and flaws still remaining in the second edition. For a list of the corrections, see our home pages. In this process it was very tempting to digress from the aim, include new material and steer toward a third edition. Although the immense mathematical richness of Brownian motion and many important applications would have made this possible and, perhaps, desirable we feel that it is time to close the project. However, there are small bonuses with the new printing, e.g., a discussion on perpetual integral functionals of Brownian motion with drift in Chapter 4 No.'s 34 and 35 and a presentation of CEV processes in Appendix 1 No. 27.

Financial support from the Magnus Ehrnrooth's Foundation (Finland) and from Svenska kulturfonden via Stiftelsernas professorspool (Finland) are gratefully acknowledged.

St. Petersburg, Russia, and Åbo (Turku), Finland, June 2014

A.N. Borodin P. Salminen

NOTATION

Chapters are referred to with Roman numerals. Inside the chapters we use running Arabic numbering for the items. For example, V.6 is item 6 in the fifth chapter, A1.1 is item 1 in the first appendix which is at the end of Part 1, and A2.2 is item 2 in the second appendix at the end of Part 2. When referring to an item in the same chapter the Roman numeral is omitted and we write, e.g., No. 6.

The symbol
$:=$ means "is defined to be equal to",

$\sim$ means "is identical in law to".

$(\Omega, \mathcal{F}, \mathbf{P})$ – probability space.

ω – element in Ω.

$t \mapsto \omega(t)$ – continuous function in the canonical framework, see I.12.

$(\Omega, \mathcal{F}, \{\mathcal{F}_t\}, \mathbf{P})$ – filtered probability space, see I.3.

$\{\mathcal{F}_t : t \geq 0\}$, $\{\mathcal{F}_t\}$ – filtration, see I.3.

$\{\mathcal{F}_t^c : t \geq 0\}$, $\{\mathcal{F}_t^c\}$ – canonical filtration, see I.12.

$\{\mathcal{F}_t^o : t \geq 0\}$, $\{\mathcal{F}_t^o\}$ – natural filtration, see I.4.

$\mathcal{B}(E)$ – Borel σ-algebra of subsets of E.

$\mathcal{C}(E)$ – continuous functions from E to $\mathbb{R}$.

$\mathcal{C}_b(E)$ – continuous and bounded functions from E to $\mathbb{R}$.

$\mathcal{C}_0(E)$ – continuous and bounded functions from E to $\mathbb{R}$ vanishing at infinity.

$\mathcal{C}(\mathbb{R}_+, \mathbb{R}^d)$ – continuous functions from $\mathbb{R}_+$ to $\mathbb{R}^d$.

$\mathrm{Exp}(\alpha)$ – exponential distribution with parameter α.

$\tau \sim \mathrm{Exp}(\alpha)$ – random variable τ is exponentially distributed with parameter α.

I – interval on $\mathbb{R}$.

l, r – left and right, respectively, endpoint of I.

int I – interior of I.

$\mathcal{C}^1(I)$ – continuously differentiable functions from I to $\mathbb{R}$.

$\mathcal{C}^2(I)$ – two times continuously differentiable functions from I to $\mathbb{R}$.

$\mathcal{C}_b^2(I)$ – bounded two times continuously differentiable functions from I to $\mathbb{R}$.

$B_{\mathcal{B}}(I)$ – bounded Borel-measurable functions from I to $\mathbb{R}$.

Leb – Lebesgue measure.

$\varepsilon_{\{x\}}$ – Dirac measure at x.

$\delta_0(\cdot)$ – Dirac δ-function.

$\mathbf{P}_x$ – probability measure associated with a process started at x.

$\mathbf{E}_x$ – expectation operator associated with a process started at x.

θ_t – shift operator, see I.13.

$\mathbb{I}_A(\omega)$ – indicator function.

$\mathbf{E}_x\big(\xi(\omega); A\big) := \mathbf{E}_x\big(\xi(\omega)\mathbb{I}_A(\omega)\big)$.

$\mathbf{E}_x\big(\xi(\omega); \eta(\omega) \in dy\big) := \dfrac{\partial}{\partial y}\mathbf{E}_x\big(\xi(\omega); \eta(\omega) < y\big)dy$.

$\mathcal{G}$ – infinitesimal generator, see I.8, II.7.

m – speed measure, see II.4.

k – killing measure, see II.4.

s – scale function, see II.4.

$p(t; x, y)$ – transition density with respect to the speed measure, see II.4.

$G_\alpha(x, y),\ G(x, y) := G_0(x, y)$ – Green function, see II.11.

ψ_α – fundamental increasing solution, see II.10.

φ_α – fundamental decreasing solution, see II.10.

f^+ – right derivative of f with respect to the scale function.

f^- – left derivative of f with respect to the scale function.

$\ell(t, x)$ – local time with respect to the Lebesgue measure.

$\varrho(t, z) = \inf\{s :\ \ell(s, z) > t\}$ – inverse local time.

$L(t, x)$ – local time with respect to the speed measure, see II.13.

$a \vee b := \max\{a, b\}$.

$a \wedge b := \min\{a, b\}$.

$\check{M}_t := \sup\{X_s : 0 \le s \le t\}$ – maximum of the process X.

$\hat{M}_t := \inf\{X_s : 0 \le s \le t\}$ – minimum of the process X.

$\check{H}(t) := \inf\{s < t : X_s = \check{M}_t\}$ – first location of the maximum before time t.

$\hat{H}(t) := \inf\{s < t : X_s = \hat{M}_t\}$ – first location of the minimum before time t.

$H_z := \inf\{t : X_t = z\}$ – first hitting time of z.

$H_{a,b} := \inf\{t : X_s \notin (a, b)\}$ – first exit time from (a, b).

$\mathcal{L}_\gamma^{-1}$ – inverse Laplace transform with respect to γ.

$f(t) * g(t) := \int_0^t f(s) g(t-s)\, ds$ – convolution of the functions f and g.

sh, ch, th, cth – hyperbolic functions, see A2.1.

Γ – Gamma function, see A2.2.

$J_\nu,\ Y_\nu$ – Bessel functions, see A2.3.

$I_\nu,\ K_\nu$ – modified Bessel functions, see A2.4.

Ai – Airy function, see A2.5.

He_n – Hermite polynomials, see A2.6.

Erf, Erfc, Erfi, Erfid – error functions, see A2.8.

D_ν – parabolic cylinder functions, see A2.9.

$M(a, b, \cdot),\ U(a, b, \cdot)$ – Kummer functions, see A2.10.

$M_{n,m},\ W_{n,m}$ – Whittaker functions, see A2.10.

$F(\alpha, \beta, \gamma, \cdot),\ G(\alpha, \beta, \gamma, \cdot)$ – hypergeometric functions, see A2.11.

$P_p^q,\ \widetilde{Q}_p^q$ – Legendre functions, see A2.12.

For further notations associated to special functions, see A2.

Part I

THEORY

CHAPTER I

STOCHASTIC PROCESSES IN GENERAL

1. Basic definitions

1. Let $(\Omega, \mathcal{F}, \mathbf{P})$ be a probability space, $(E, \mathcal{E})$ a measurable space, and Σ an arbitrary set. A *stochastic process* is a family $X = \{X_t : t \in \Sigma\}$ of random variables $X_t : \Omega \mapsto E$. The space $(E, \mathcal{E})$ is referred to as the *state space* of X. Given a point ω in the sample space Ω, the mapping $t \mapsto X_t(\omega)$ is called a *trajectory* or *realization* or *sample path* of the process X. For the general theory of stochastic processes see Dellacherie and Meyer (1976, 1980, 1983, 1987) and Gihman and Skorohod (1974, 1975, 1979).

Two stochastic processes X and Y defined on the same probability space are said to be *modifications* of each other if $\mathbf{P}(X_t = Y_t) = 1$ for all $t \in \Sigma$.

Two stochastic processes X and Y taking values in the same state space $(E, \mathcal{E})$ but not necessarily defined on the same probability space are said to be *identical in law* or *versions* of each other if they have the same *finite dimensional distributions*. In other words, for all $t_i \in \Sigma$, and $A_i \in \mathcal{E}$, $i = 1, 2, \dots, n$,

$$\mathbf{P}(X_{t_1} \in A_1, \dots, X_{t_n} \in A_n) = \mathbf{P}(Y_{t_1} \in A_1, \dots, Y_{t_n} \in A_n).$$

Two stochastic processes X and Y defined on the same probability space are said to be *indistinguishable* or *equivalent* if there exists a set $N \in \mathcal{F}$ such that $\mathbf{P}(N) = 0$ and $X_\cdot(\omega) = Y_\cdot(\omega)$ for all $\omega \in \Omega \setminus N$.

A stochastic process with $\Sigma = \mathbb{R}$ or $\mathbb{R}_+$ is called *stationary* if the finite dimensional distributions are invariant under translations of the time parameter, that is,

$$\mathbf{P}(X_{t_1} \in A_1, \dots, X_{t_n} \in A_n) = \mathbf{P}(X_{t_1+h} \in A_1, \dots, X_{t_n+h} \in A_n)$$

for all $h \in \Sigma$, $t_i \in \Sigma$, and $A_i \in \mathcal{E}$, $i = 1, 2, \dots, n$.

In the general theory of stochastic processes Σ is usually a subset of $\mathbb{R}$ and E is a complete separable metrizable space (*Polish*) or a locally compact space with countable base (*LCCB*). In this framework, the σ-algebra $\mathcal{E}$ consists of the Borel subsets of E, denoted by $\mathcal{B}(E)$.

We assume – if nothing else is said – that $\Sigma = [0, \infty)$.

2. Let X be a stochastic process taking values in $\mathbb{R}^d$. Assume that $\mathbb{R}^d$ is equipped with the usual topology. Then X is said to be *a.s. continuous* or to have *a.s.*

continuous sample paths if there exists a set $N \in \mathcal{F}$ such that $\mathbf{P}(N) = 0$ and the mapping $t \mapsto X_t(\omega)$ is continuous for all $\omega \in \Omega \setminus N$. An analogous statement is used to define, e.g., a process which is *a.s. right continuous and has left limits* (for short, r.c.l.l.).

The process X is said to be *stochastically continuous* or *continuous in probability* if for every $t > 0$ and $\varepsilon > 0$,

$$\lim_{s \to t} \mathbf{P}(|X_s - X_t| > \varepsilon) = 0,$$

and *continuous in mean square* if for every $t \geq 0$,

$$\lim_{s \to t} \mathbf{E}|X_t - X_s|^2 = 0.$$

Kolmogorov's criterion for continuity. Let the time parameter of X be n-dimensional, that is $t = (t_1, \dots, t_n)$, $t_i \geq 0$. Assume that there exist positive constants α, β, and M_k, $k = 1, 2, \dots$, such that

(a) $$\mathbf{E}|X_t - X_s|^\alpha \leq M_k |t - s|^{n+\beta}$$

for every s and t with $s_i, t_i \leq k, i = 1, \dots, n$. Then X has a continuous modification. For a proof in the one-dimensional case, see Ikeda and Watanabe (1981) p. 20. The n-dimensional case is proved similarly. See also Meyer (1981).

3. A family of σ-algebras $\{\mathcal{F}_t : t \geq 0\}$ (for short, $\{\mathcal{F}_t\}$) on $(\Omega, \mathcal{F})$ is called a *filtration* if

$$\mathcal{F}_s \subseteq \mathcal{F}_t \subseteq \mathcal{F} \text{ for every } s \leq t.$$

A filtration is said to be *right continuous* if for every s,

$$\mathcal{F}_s = \mathcal{F}_{s+} := \bigcap_{t>s} \mathcal{F}_t.$$

The collection $(\Omega, \mathcal{F}, \{\mathcal{F}_t\}, \mathbf{P})$ is called a *filtered probability space.* It is said to satisfy the *usual conditions* if

(a) $\mathcal{F}$ is $\mathbf{P}$-*complete*,

(b) $\mathcal{F}_0$ contains all $\mathbf{P}$-null sets of $\mathcal{F}$,

(c) $\{\mathcal{F}_t\}$ is right continuous.

The meaning of (a) is that a set $A \subseteq \Omega$ belongs to $\mathcal{F}$ if there exist A_1 and A_2 in $\mathcal{F}$ such that $A_1 \subseteq A \subseteq A_2$ and $\mathbf{P}(A_1) = \mathbf{P}(A_2)$.

The *usual augmentation* of a filtered probability space $(\Omega, \mathcal{F}, \{\mathcal{F}_t\}, \mathbf{P})$ is a filtered probability space $(\Omega, \mathcal{G}, \{\mathcal{G}_t\}, \mathbf{P})$, where

($\star$) $\mathcal{G}$ is the $\mathbf{P}$-*completion* of $\mathcal{F}$,

($\star$) $\mathcal{G}_t$ is for every t the σ-algebra generated by $\mathcal{F}_{t+}$ and the $\mathbf{P}$-null sets of $\mathcal{G}$.

By definition, $A \subseteq \Omega$ belongs to $\mathcal{G}$ if there exist A_1 and A_2 in $\mathcal{F}$ such that $A_1 \subseteq A \subseteq A_2$ and $\mathbf{P}(A_1) = \mathbf{P}(A_2)$.

It can be proved (Rogers and Williams (1994) p. 172) that $(\Omega, \mathcal{G}, \{\mathcal{G}_t\}, \mathbf{P})$ satisfies the usual conditions.

4. A stochastic process X defined on a filtered probability space $(\Omega, \mathcal{F}, \{\mathcal{F}_t\}, \mathbf{P})$ is said to be *progressively measurable* if for every t the mapping $(s, \omega) \mapsto X_s(\omega)$ from $[0,t] \times \Omega$ to E is $\mathcal{B}([0,t]) \times \mathcal{F}_t$-measurable. We say that X is *adapted* (to the filtration $\{\mathcal{F}_t\}$) if for every t the random variable X_t is $\mathcal{F}_t$-measurable. Notice that X is always adapted to its *natural filtration* $\{\mathcal{F}_t^o := \sigma\{X_s : s \leq t\}\}$. An adapted process with right or left continuous paths is progressively measurable (see Revuz and Yor (2001) p. 44).

In general, a stochastic process X is called *measurable* if the mapping $(t, \omega) \mapsto X_t(\omega)$ is $\mathcal{B}([0, \infty)) \times \mathcal{F}$-measurable.

5. A *stopping time* with respect to a filtration $\{\mathcal{F}_t\}$ is a mapping $T : \Omega \mapsto [0, \infty) \cup \{\infty\}$ such that $\{T \leq t\} \in \mathcal{F}_t$ for every t.

In the case $\{\mathcal{F}_t\}$ is right continuous, T is a stopping time if and only if $\{T < t\} \in \mathcal{F}_t$ for every t.

The σ-algebra $\mathcal{F}_T$ which, informally, consists of all events occurring before or at a given stopping time T is defined by

$$A \in \mathcal{F}_T \quad \Leftrightarrow \quad A \in \mathcal{F} \text{ and } A \cap \{T \leq t\} \in \mathcal{F}_t \quad \text{for all } t.$$

2. Markov processes, transition functions, resolvents, and generators

6. An adapted stochastic process $X = \{X_t : t \geq 0\}$ defined on a filtered probability space $(\Omega, \mathcal{F}, \{\mathcal{F}_t\}, \mathbf{P})$ and taking values in a general state space $(E, \mathcal{E})$ is called a *time-homogeneous Markov process* (Revuz and Yor (2001) p. 81) if for every t there exists a *kernel* $P_t : E \times \mathcal{E} \mapsto [0, 1]$, called a *transition function*, such that

(a) for all $A \in \mathcal{E}$ and $t, s \geq 0$,

$$\mathbf{P}(X_{t+s} \in A | \mathcal{F}_t) = P_s(X_t, A) \qquad \text{a.s.},$$

(b) for all $A \in \mathcal{E}$ and t the mapping $x \mapsto P_t(x, A)$ is measurable,

(c) for all $x \in E$ and t the mapping $A \mapsto P_t(x, A)$ is a measure,

(d) for all $x \in E$ and $A \in \mathcal{E}$,

$$P_0(x, A) = \varepsilon_{\{x\}}(A) := \begin{cases} 1, & x \in A, \\ 0, & x \notin A, \end{cases}$$

(e) the *Chapman–Kolmogorov equation* holds:

$$P_{t+s}(x, A) = \int_E P_t(x, dy) P_s(y, A), \quad x \in E, \ A \in \mathcal{E}, \ t, s \geq 0.$$

The time-homogeneity and the Markov property are expressed in (a). Intuitively, the Markov property means that the behavior of X in the future depends only on the present state of X and not on the behavior in the past. In other words, the future and the past are independent given the present.

We remark that there are also more general definitions of a Markov process, and definitions in which the approach is not directly via transition functions. For these see, e.g., Dynkin (1965), Blumenthal and Getoor (1968), and Ethier and Kurtz (1986) p. 156.

Properties (b) and (c) are the defining properties of a kernel. In general, a mapping $K : E \times \mathcal{E} \mapsto [0, \infty)$ is called a *kernel* if

($\star$) for every $A \in \mathcal{E}$ the mapping $x \mapsto K(x, A)$ is measurable,

($\star$) for every $x \in E$ the mapping $A \mapsto K(x, A)$ is a measure.

A Markov process is called *normal* if $\lim_{t \to 0} P_t(x, E) = 1$ for all $x \in E$, and *conservative* (or *honest*) if $P_t(x, E) = 1$ for all t and $x \in E$ (see Dynkin (1965) Vol I p. 47 and 85). A non-conservative process is often made conservative by adjoining an extra state ∂ (called *cemetery* or *coffin state*) to E. Let $E^\partial := E \cup \{\partial\}$ and $\mathcal{E}^\partial := \sigma\{\mathcal{E}, \partial\}$. Then the conservative transition function $\{P_t^\partial\}$ is introduced in the state space $(E^\partial, \mathcal{E}^\partial)$ as follows:

($\star$) $P_t^\partial(x, A) := P_t(x, A) \quad x \in E, A \in \mathcal{E}$,

($\star$) $P_t^\partial(x, \partial) := 1 - P_t(x, E), \quad x \in E$,

($\star$) $P_t^\partial(\partial, A) := \varepsilon_{\{\partial\}}(A), \quad A \in \mathcal{E}^\partial$.

7. A transition function P is called *jointly measurable* if for all $A \in \mathcal{E}$ the mapping $(t, x) \mapsto P_t(x, A)$ is $\mathcal{B}([0, \infty)) \times \mathcal{E}$-measurable. In this case we may construct a family of operators $\{R_\lambda : \lambda > 0\}$ from $B_\mathcal{E}(E)$ to $B_\mathcal{E}(E)$ by setting

$$R_\lambda f(x) := \int_0^\infty e^{-\lambda t} P_t f(x) dt,$$

where

$$P_t f(x) := \int_E P_t(x, dy) f(y),$$

and $B_\mathcal{E}(E)$ is the set of bounded $\mathcal{E}$-measurable functions from E to $\mathbb{R}$.

The family $\{R_\lambda\}$ is called the *resolvent* of $\{P_t\}$. It has the following properties (see Gihman and Skorohod (1975) vol. II p. 92, Rogers and Williams (1994) p. 233):

($\star$) the *resolvent equation*

$$R_\alpha - R_\beta = (\beta - \alpha) R_\beta R_\alpha, \quad \alpha, \beta > 0,$$

and, consequently, $R_\alpha R_\beta = R_\beta R_\alpha$.

($\star$) $\frac{d^n}{d\lambda^n} R_\lambda f = (-1)^n n! R_\lambda^{n+1} f$,

($\star$) $R_\lambda f = \sum_{n=0}^{\infty} (\lambda_o - \lambda)^n R_{\lambda_o}^{n+1} f$ for all λ such that $|\lambda - \lambda_o| < \lambda_o$ with $\lambda_o > 0$,

($\star$) $\|R_\lambda\| \le 1/\lambda$, where $\|R_\lambda\| := \sup\{\|R_\lambda f\| : \|f\| = 1\}$ and $\|f\| := \sup |f(x)|$,

($\star$) if $f_n \downarrow 0$ (pointwise on E) then $R_\lambda f_n \downarrow 0$ (pointwise on E) for all $\lambda > 0$.

8. The *(strong) infinitesimal generator* $\mathcal{G}$ of a Markov process X (or its transition function P) is defined as

$$\mathcal{G}f := \lim_{t \downarrow 0} \frac{P_t f - f}{t}$$

for $f \in B_{\mathcal{E}}(E)$ such that the limit exists in the sup norm $\|\cdot\|$. The set of all these f is called the *domain* of $\mathcal{G}$.

3. Feller processes, Feller–Dynkin processes, and the strong Markov property

9. The definition of a Markov process in No. 6 does not give enough structure to build a rich theory. In particular, more assumptions are needed to be able to prove regularity properties of sample paths such as right continuity and existence of limits from the left (r.c.l.l.). Regularity is of importance, e.g., when extending the Markov property to be valid at stopping times. Also in stochastic calculus regularity of sample paths is paramount. In the general theory of Markov processes one usually works with processes having a r.c.l.l. modification (see, e.g., the definition of a *standard process* in Blumenthal and Getoor (1968) p. 45 or Sharpe (1988) p. 220).

In this section, E is a LCCB space and $\mathcal{E}$ its Borel σ-algebra. The notation $\mathcal{C}_b(E)$ is used for the set of bounded continuous $\mathcal{E}$-measurable functions from E to $\mathbb{R}$, and $\mathcal{C}_0(E)$ is the subset of $\mathcal{C}_b(E)$ consisting of functions vanishing at infinity.

10. A *Feller semigroup* is a family $\{P_t\}$ of positive linear operators mapping $\mathcal{C}_b(E)$ into $\mathcal{C}_b(E)$ such that

(a) $P_0 = I$ (the identity operator),

(b) $0 \le P_t f \le 1$ for all $f \in \mathcal{C}_b(E)$ such that $0 \le f \le 1$,

(c) $P_{s+t} = P_s P_t$,

(d) $\lim_{t \downarrow 0} \|P_t f - f\| = 0$ for all $f \in \mathcal{C}_b(E)$.

A Markov process whose transition function is a Feller semigroup is called a *Feller process.* A Feller process does not, in general, have a r.c.l.l. modification. (Dynkin (1965) Theorem 3.6, p. 92 gives sufficient conditions for this to hold).

11. A *Feller–Dynkin semigroup* is a family $\{P_t\}$ of positive linear operators mapping $\mathcal{C}_0(E)$ into $\mathcal{C}_0(E)$ such that (a)–(d) in No. 10 hold when $\mathcal{C}_b(E)$ is replaced by $\mathcal{C}_0(E)$. Property (d) in No. 10, the *strong continuity* condition, follows (Rogers and Williams (1994) p. 241) in this case from the *pointwise continuity* condition:

$$\lim_{t\downarrow 0} P_t f(x) = f(x) \text{ for all } f \in \mathcal{C}_0(E) \text{ and } x \in E.$$

A Markov process whose transition function is a Feller–Dynkin semigroup is called a *Feller–Dynkin process.* A Feller–Dynkin process (in a LCCB space) always has a r.c.l.l. modification (Rogers and Williams (1994) p. 243, Revuz and Yor (2001) p. 91, Blumenthal and Getoor (1968) p. 46, and Dynkin (1965) p. 92).

12. Let X be a Feller process or a Feller–Dynkin process, and consider it in the state space $(E^\partial, \mathcal{E}^\partial)$ and assume, therefore, that X is conservative (see No. 6). Let $P^\partial = \{P_t^\partial\}$ be its transition function and $\nu(\cdot) = \mathbf{P}(X_0 \in \cdot)$ its *initial distribution.* A *canonical realization* of X (or of the pair (ν, P^∂)) is a process Y in the probability space $(E^{\partial,\infty}, \mathcal{F}^c_\infty, \mathbf{P}_\nu)$, where $E^{\partial,\infty}$ is the set of functions $\omega : [0,\infty) \to E^\partial$ such that $Y_t(\omega) := \omega(t)$, $\mathcal{F}^c_\infty := \sigma\{\omega(t) : t \geq 0\}$, and $\mathbf{P}_\nu$ is the unique probability measure (Rogers and Williams (1994) p. 244) on $(E^{\partial,\infty}, \mathcal{F}^c_\infty)$ such that for all $0 \leq t_1 \leq t_2 \leq \cdots \leq t_n$, and $A_i \in \mathcal{E}^\partial$,

$$\begin{aligned}\mathbf{P}_\nu(&Y_0 \in A_0, \dots, Y_{t_n} \in A_n)\\ &= \int_{A_0} \cdots \int_{A_n} \nu(dx_0) P^\partial_{t_1}(x_0, dx_1) \cdot \ldots \cdot P^\partial_{t_n - t_{n-1}}(x_{n-1}, dx_n).\end{aligned}$$

It can be proved (Revuz and Yor (2001) p. 81) that Y is a Markov process in the filtered probability space $(E^{\partial,\infty}, \mathcal{F}^c_\infty, \{\mathcal{F}^c_t := \sigma\{\omega(s) : s \leq t\}\}, \mathbf{P}_\nu)$ for any initial distribution ν. We let $\mathbf{P}_x$ denote $\mathbf{P}_\nu$ when $\nu = \varepsilon_{\{x\}}$.

In the case X is a Feller–Dynkin process the space $E^{\partial,\infty}$ can be replaced by the space D, the so-called *Skorohod space,* which consists of all r.c.l.l. functions $\omega : [0,\infty) \mapsto E^\partial$ such that $\omega(t) = \partial$ for all $t > s$ if $\omega(s) = \partial$ or $\lim_{u\uparrow s} \omega(u) = \partial$. Let ω_∂ denote the function which is identically equal to ∂. Functions in D are often also extended to $[0,\infty]$ by setting $\omega(\infty) = \partial$. Moreover, a usual convention is that $\eta(\omega_\partial) = 0$ for any $\mathcal{F}^c_\infty$-measurable random variable η.

The *lifetime* ζ of a path $\omega \in D$ is defined by $\zeta(\omega) := \inf\{t : \omega_t = \partial\}$. Observe that ζ is an $\mathcal{F}^c_{t+}$-stopping time.

13. The *Markov property* of a r.c.l.l. Feller process or a Feller–Dynkin process X in the canonical framework can be expressed as follows: Let ξ and η be, respectively, bounded $\mathcal{F}^c_t$- and $\mathcal{F}^c_\infty$-measurable random variables. Then for any initial distribution ν,

$$\text{(a)}, \qquad \mathbf{E}_\nu(\xi\ \eta \circ \theta_t) = \mathbf{E}_\nu(\xi\ \mathbf{E}_{X_t}\eta),$$

where θ_t is the so-called *shift operator*, i.e., a mapping defined for every t via

$$\theta_t(\omega) = \omega', \quad \omega'(s) := \omega(t+s) \quad \text{for every } s \geq 0,$$

and $\eta \circ \theta_t(\omega) := \eta(\theta_t(\omega))$. Note that in (a) it is necessary that $x \mapsto \mathbf{E}_x(\eta)$ is $\mathcal{E}$-measurable. This follows from (b) in No. 6 by the monotone class theorem.

The process X has also the *strong Markov property.* To state this let T be an $\mathcal{F}^c_{t+}$-stopping time, and introduce the σ-algebra $\mathcal{F}^c_{T+}$ (see. No. 5) by setting

$$A \in \mathcal{F}^c_{T+} \quad \Leftrightarrow \quad A \in \mathcal{F}^c_\infty \text{ and } A \cap \{T < t\} \in \mathcal{F}^c_t \quad \text{for all } t.$$

Then the value of X at time T, denoted X_T ($= \partial$ if $T = \infty$), is $\mathcal{F}^c_{T+}$-measurable (Rogers and Williams (1994) p. 183). Let ξ and η be, respectively, $\mathcal{F}^c_{T+}$- and $\mathcal{F}^c_\infty$-measurable bounded random variables. The strong Markov property says that for any initial distribution ν,

$$\mathbf{E}_\nu(\xi\ \eta \circ \theta_T) = \mathbf{E}_\nu(\xi\ \mathbf{E}_{X_T}\eta),$$

where θ_T stands for the shift operator evaluated at T with the convention $\theta_T(\omega) = \omega_\partial$ if $T = \infty$.

14. It is an important fact – especially when working with discontinuous processes (Rogers and Williams (1994) p. 248) – that the above presented Markov properties can be extended to be valid with respect to appropriately completed filtrations.

To present this, let $\mathcal{F}^\nu_\infty$ be the $\mathbf{P}_\nu$-completion of $\mathcal{F}^c_\infty$ and for every t let $\mathcal{F}^\nu_t$ be the σ-algebra containing $\mathcal{F}^c_t$ and the $\mathbf{P}_\nu$-null sets from $\mathcal{F}^\nu_\infty$. We also introduce the *universal completions*

$$\mathcal{F}_t := \bigcap_\nu \mathcal{F}^\nu_t, \quad \mathcal{F}_\infty := \bigcap_\nu \mathcal{F}^\nu_\infty,$$

where ν runs through the set of all probability measures on $(E^\partial, \mathcal{E}^\partial)$.

Assume that X is a r.c.l.l. Feller process or a Feller–Dynkin process. Then it can be proved (Revuz and Yor (2001) p. 93 for Feller–Dynkin processes) that $\{\mathcal{F}^\nu_t\}$ and $\{\mathcal{F}_t\}$ are right continuous. Therefore, $(E^{\partial,\infty}, \mathcal{F}^\nu_\infty, \{\mathcal{F}^\nu_t\}, \mathbf{P}_\nu)$ is the usual augmentation of $(E^{\partial,\infty}, \mathcal{F}^c_\infty, \{\mathcal{F}^c_t\}, \mathbf{P}_\nu)$. Let T be a stopping time with respect to the filtration $\{\mathcal{F}_t\}$. Then for any initial distribution ν (Dynkin (1965) p. 99, p. 104 or Rogers and Williams (1994) p. 250) X has the strong Markov property

$$\mathbf{E}_\nu(\xi\ \eta \circ \theta_T) = \mathbf{E}_\nu(\xi\ \mathbf{E}_{X_T}(\eta)),$$

where ξ and η are, respectively, $\mathcal{F}_T$- and $\mathcal{F}_\infty$-measurable bounded random variables, and $\mathcal{F}_T$ is defined as in No. 5 with respect to $\{\mathcal{F}_t\}$. An equivalent formulation is

$$\mathbf{E}_\nu(\eta \circ \theta_T | \mathcal{F}_T) = \mathbf{E}_{X_T}(\eta) \quad \mathbf{P}_\nu\text{-a.s. on } \{T < \infty\}.$$

4. Martingales

15. In this section it is assumed that $(\Omega, \mathcal{F}, \{\mathcal{F}_t\}, \mathbf{P})$ is a filtered probability space satisfying the usual conditions (see No. 3). A real valued stochastic process $M = \{M_t : t \geq 0\}$ defined on this space is called a *martingale* (or $(\mathbf{P}, \mathcal{F}_t)$-*martingale* or *martingale with respect to* $\{\mathcal{F}_t\}$) if for every $t \geq 0$,

(a) M is adapted, that is, M_t is $\mathcal{F}_t$-measurable,

(b) $\mathbf{E}|M_t| < \infty$,

(c) $\mathbf{E}(M_t|\mathcal{F}_s) = M_s$ a.s. for every $s \le t$.

The process M is called a *supermartingale* if it has properties (a), (b), and

(d) $\mathbf{E}(M_t|\mathcal{F}_s) \le M_s$ a.s. for every $s \le t$.

The process M is called a *submartingale* if it has properties (a), (b), and

(e) $\mathbf{E}(M_t|\mathcal{F}_s) \ge M_s$ a.s. for every $s \le t$.

By the definition of conditional expectation the martingale property (c) (or (d) or (e)) is equivalent to

$$\mathbf{E}(M_t \mathbb{I}_A) = \mathbf{E}(M_s \mathbb{I}_A) \text{ (or } \mathbf{E}(M_t \mathbb{I}_A) \le \mathbf{E}(M_s \mathbb{I}_A) \text{ or } \mathbf{E}(M_t \mathbb{I}_A) \ge \mathbf{E}(M_s \mathbb{I}_A)),$$

respectively, which should hold for all s and t such that $s \le t$ and for all $A \in \mathcal{F}_s$.

Clearly, if M is a submartingale, then $-M$ is a supermartingale. A martingale is both a super- and a submartingale.

Under the assumption that $(\Omega, \mathcal{F}, \{\mathcal{F}_t\}, \mathbf{P})$ satisfies the usual conditions it can be proved (Rogers and Williams (1994) p. 173) that a sub- or supermartingale M has a r.c.l.l. modification if and only if the mapping

$$t \mapsto \mathbf{E}(M_t)$$

is right continuous. In particular, a martingale has in this case always a r.c.l.l. modification.

For the rest of this section all of the sub- and supermartigales we consider are assumed to be r.c.l.l.

16. Convergence. Let M be a sub- or supermartingale. Assume that M is bounded in $\mathbf{L}^1$, that is

$$\sup_t \mathbf{E}|M_t| < \infty.$$

Then

(a) $$\lim_{t\to\infty} M_t =: M_\infty \quad \text{exists a.s. and} \quad \mathbf{E}|M_\infty| < \infty.$$

This is due to J.L. Doob, and is based on the so-called upcrossing lemma. See Doob (1953) p. 319 and 316, where the result is formulated for martingales in discrete time. For more recent reproductions see Ikeda and Watanabe (1981) p. 30, and Rogers and Williams (1994) p. 176. We refer also to Chung (1998) for an article on Doob's fundamental contributions in probability theory.

Next recall that a stochastic process X is called *uniformly integrable* if for every $\varepsilon > 0$ there exists a number $c = c(\varepsilon) > 0$ such that for every $t \ge 0$,

$$\mathbf{E}(|X_t|\,;\,|X_t| > c) < \varepsilon.$$

Notice that a uniformly integrable stochastic process is bounded in $\mathbf{L}^1$.

Now assume additionally that M is uniformly integrable. Then $M_t \to M_\infty$ also in $\mathbf{L}^1$, that is

(b) $$\lim_{t\to\infty} \mathbf{E}|M_t - M_\infty| = 0.$$

Moreover, if M is a martingale,

(c) $$M_t = \mathbf{E}(M_\infty|\mathcal{F}_t) \quad \text{a.s.}$$

Observe that a non-negative supermartingale is always convergent (but not necessarily uniformly integrable).

A martingale which is bounded in $\mathbf{L}^2$ is uniformly integrable, and also bounded in $\mathbf{L}^1$. Therefore, (a), (b), and (c) hold for martingales bounded in $\mathbf{L}^2$.

17. Consider an arbitrary real valued stochastic process $M = \{M_t : t \geq 0\}$ on the probability space $(\Omega, \mathcal{F}, \{\mathcal{F}_t\}, \mathbf{P})$. Assume that M_0 is $\mathcal{F}_0$-measurable. Then M is called a *local martingale* if there exists an increasing sequence of stopping times $\{T_n\}$ such that $T_n \uparrow \infty$ and each *stopped process* $\{M_{T_n\wedge t} - M_0 : t \geq 0\}$ is a martingale, where

$$M_{T_n\wedge t} = \begin{cases} M_{T_n}, & T_n < t, \\ M_t, & T_n \geq t. \end{cases}$$

A martingale is always a local martingale. In fact, we have the following general result. Let M be a sub- or supermartingale and T a stopping time. Then $\{M_{T\wedge t} : t \geq 0\}$ is also a sub- or supermartingale, respectively, with respect to $\{\mathcal{F}_t\}$. For the proof see, e.g., Revuz and Yor (2001) p. 71, or Doob (1984) p. 471. In particular, we have, respectively,

$$\mathbf{E}M_{T\wedge t} = \mathbf{E}M_0, \qquad \mathbf{E}M_{T\wedge t} \geq \mathbf{E}M_0, \qquad \mathbf{E}M_{T\wedge t} \leq \mathbf{E}M_0.$$

18. **Optional stopping.** Let M be a uniformly integrable or non-negative supermartingale and $M_\infty := \lim_{t\to\infty} M_t$. Then for all stopping times S and T such that $S \leq T$ we have

(a) $\mathbf{E}(M_T|\mathcal{F}_S) \leq M_S$ a.s.

In particular,

(b) $\mathbf{E}M_T \leq \mathbf{E}M_S \leq \mathbf{E}M_0$.

In (a) and (b) it is understood that $M_T = M_\infty$ on the set $\{T = \infty\}$. For a proof in discrete time, see Doob (1984) p. 440, and for continuous time Gihman and Skorohod (1979) p. 11, and Rogers and Williams (1994) p. 189.

If there exists a constant C such that $S \leq T \leq C$ a.s. then (a) holds because M restricted to a finite time interval is uniformly integrable. Furthermore, if M is bounded it is uniformly integrable and hence (a) holds.

For a uniformly integrable martingale the inequalities above can be replaced by equalities.

19. Maximal inequalities. Let $t > 0$ and $\alpha > 0$ be given. The following inequalities hold for a submartingale M :

$$\alpha \mathbf{P}\Big(\sup_{s\le t} M_s \ge \alpha\Big) \le \mathbf{E}\Big(M_t;\ \sup_{s\le t} M_s \ge \alpha\Big)$$
$$\le \mathbf{E}(M_t \vee 0) \le \mathbf{E}|M_t|,$$

$$\alpha \mathbf{P}\Big(\inf_{s\le t} M_s \le -\alpha\Big) \le \mathbf{E}\Big(M_t;\ \inf_{s\le t} M_s > -\alpha\Big) - \mathbf{E}(M_0)$$
$$\le \mathbf{E}(M_t \vee 0) - \mathbf{E}(M_0) \le \mathbf{E}|M_t| + \mathbf{E}|M_0|.$$

If M is a supermartingale, then $-M$ is a submartingale and, therefore,

$$\alpha \mathbf{P}\Big(\inf_{s\le t} M_s \le -\alpha\Big) \le -\mathbf{E}(M_t \wedge 0) \le \mathbf{E}|M_t|,$$

$$\alpha \mathbf{P}\Big(\sup_{s\le t} M_s \ge \alpha\Big) \le -\mathbf{E}(M_t \wedge 0) + \mathbf{E}(M_0) \le \mathbf{E}|M_t| + \mathbf{E}|M_0|.$$

Especially, if M is a positive supermartingale we have the inequality

$$\alpha \mathbf{P}\Big(\sup_{s\le t} M_s \ge \alpha\Big) \le \mathbf{E}(M_0).$$

For proofs, see Doob (1984) p. 442, and Ikeda and Watanabe (1981) p. 27.

20. $\mathbf{L}^p$-inequalities. Let M be a positive submartingale. Assume that $\mathbf{E}M_t^p < \infty$ for a given $t > 0$ and for some $p \ge 1$. Then for $\alpha > 0$,

$$\alpha^p \mathbf{P}\Big(\sup_{s\le t} M_s \ge \alpha\Big) \le \mathbf{E}\Big(M_t^p\,;\ \sup_{s\le t} M_s \ge \alpha\Big) \le \mathbf{E}(M_t^p),$$

and if $p > 1$, then

(a) $$\mathbf{E}\Big(\sup_{s\le t} M_s^p\Big) \le \Big(\frac{p}{p-1}\Big)^p \mathbf{E}M_t^p.$$

The inequality (a) for $p = 2$ is called the *Doob–Kolmogorov inequality.* Notice that if N is a martingale, then $|N|$ is a positive submartingale. From (a) it follows (for $p > 1$) that $\sup_{s\le t} M_s$ is in $\mathbf{L}^p$ if and only if M_t is in $\mathbf{L}^p$.

For proofs, see Chung and Williams (1990) p. 15 or Doob (1984) p. 444.

21. Doob–Meyer decomposition. A stochastic process X defined on the probability space $(\Omega, \mathcal{F}, \{\mathcal{F}_t\}, \mathbf{P})$ is said to be of *class* (D) if the family

$$\{X_T :\ T \text{ a finite stopping time }\}$$

is uniformly integrable. The Doob–Meyer decomposition theorem says that a r.c.l.l. adapted process Z is a submartingale belonging to the class (D) if and only if there exist a uniformly integrable martingale M, $M_0 = 0$, and a predictable (see III.2) integrable increasing process A, $A_0 = 0$, such that a.s. for all t,

$$Z_t = Z_0 + M_t + A_t.$$

Moreover, for a given submartingale Z in the class (D) this decomposition is unique, and A is called the *compensator* of Z. We remark that uniformly integrable martingales belong to the class (D). In particular, martingales bounded in $\mathbf{L}^p, p > 1$, are uniformly integrable, and, hence, in (D).

We refer to Rogers and Williams (2000) p. 367 for a proof and further developments. For a definition of a smaller class (DL), and the Doob–Meyer decomposition for submartingales in this class, see Ikeda and Watanabe (1981) p. 35. Moreover, the Doob–Meyer decomposition can be formulated for *local submartingales*; see Rogers and Williams (2000) p. 375.

22. The Burkholder–Davis–Gundy inequality. Let M, $M_0 = 0$, be a continuous local martingale. Then M^2 is a local submartingale. Let $\langle M\rangle$ be the predictable increasing process associated to M^2 by the local form of the Doob–Meyer decomposition. The process $\langle M\rangle$ is continuous in this case. Let $M_t^* := \sup_{s\le t}|M_s|$. The Burkholder–Davis–Gundy inequality says that for every $p > 0$ there exist constants c_p and C_p such that for all stopping times T,

$$c_p\mathbf{E}(\langle M\rangle_T^p) \le \mathbf{E}((M_T^*)^{2p}) \le C_p\mathbf{E}(\langle M\rangle_T^p).$$

For a direct proof of the existence of $\langle M\rangle$ see Revuz and Yor (2001) p. 124. For a proof and more general formulations of the Burkholder–Davis–Gundy inequality see Lenglart, Lépingle and Pratelli (1980) and Revuz and Yor (2001) p. 160.

23. An adapted r.c.l.l. process X defined on the probability space $(\Omega, \mathcal{F}, \{\mathcal{F}_t\}, \mathbf{P})$ is said to be a *semimartingale* (see Rogers and Williams (2000) p. 23, Revuz and Yor (2001) p. 127) if there exist a local martingale M, $M_0 = 0$, and an adapted process A, $A_0 = 0$, of finite variation such that a.s. for all t,

$$X_t = X_0 + M_t + A_t.$$

Because A could also be a (discontinuous) martingale the semimartingale representation is not in general unique. However, if the semimartingale representation holds with A predictable, then A is the unique predictable process for which the representation is valid. Also if X is a continuous semimartingale and M in its representation is continuous, then A is unique.

CHAPTER II

LINEAR DIFFUSIONS

1. Basic facts

1. We adopt here the general approach to linear diffusions as presented in the book by K. Itô and H.P. McKean which appeared in 1965. References are made to the second printing from 1974.

Let I be an interval with left endpoint $l \geq -\infty$ and right endpoint $r \leq \infty$. Let X be a time-homogeneous Markov process taking values in I (see I.6) and $\mathcal{F}^c$ its canonical filtration (see I.12). Further, let $\mathbf{P}_x$ denote the probability measure associated to X when started at $x \in I$. Then X is called a *linear* (or *one-dimensional*) *diffusion* if for all $x \in I$,

(a) $t \mapsto X_t(\omega)$ is continuous on $[0, \zeta)$ $\quad \mathbf{P}_x$-a.s.,

(b) $\mathbf{E}_x(\eta \circ \theta_T | \mathcal{F}^c_{T+}) = \mathbf{E}_{X_T}(\eta)$ $\quad \mathbf{P}_x$-a.s.,

where ζ is the lifetime of X, η is a bounded $\mathcal{F}^c_\infty$-measurable random variable, T is an $\mathcal{F}^c_{t+}$-stopping time, and $\theta_\cdot$ is the shift operator (see I.13). The σ-algebra $\mathcal{F}^c_{T+}$ is constructed as in I.13. Recall also the conventions in I.12 concerning the cemetery state ∂. In particular, if $T = \infty$ both the left- and the right-hand side in (b) are taken to be equal to zero.

For the definition above we refer to Itô and McKean (1974) p. 84, Freedman (1971) p. 103, and Rogers and Williams (2000) p. 271. For an approach based on extended generators, martingale problems and stochastic differential equations, see Stroock and Varadhan (1979) p. 136, and Revuz and Yor (2001) p. 294, Øksendal (1998) p. 108.

2. Roughly speaking, a linear diffusion is a strong Markov process with continuous paths taking values on an interval. Notice, or see Itô and McKean (1974) p. 84, that if $\tau \sim \text{Exp}(1)$, then our strong Markov property excludes processes of the type

$$X_t = \begin{cases} 0, & t \leq \tau, \\ t - \tau, & t > \tau. \end{cases}$$

The strong Markov property also implies that for all $x \in I$,

(a) $\mathbf{P}_x(\exists\, \varepsilon(\omega) > 0 \text{ such that } X_t(\omega) \geq x \;\forall\, t \in (0, \varepsilon(\omega))) = 0$ or 1,

(b) $\mathbf{P}_x(\exists\, \varepsilon(\omega) > 0 \text{ such that } X_t(\omega) \leq x \;\forall\, t \in (0, \varepsilon(\omega))) = 0$ or 1.

In fact, the *Blumenthal 0–1 law* (Itô and McKean(1974) p. 85) holds:

$$\mathbf{P}_x(A) = 0 \text{ or } 1, \quad \text{for all } x \in I \text{ and } A \in \mathcal{F}^c_{0+}.$$

3. Unless otherwise stated, we consider only *regular diffusions* (Dynkin (1965) Vol II p. 121); letting $H_y := \inf\{t : X_t = y\}$, then X is a regular diffusion if for every $x, y \in I$,

(a) $$\mathbf{P}_x(H_y < \infty) > 0.$$

It can be proved (Rogers and Williams (2000) p. 273) that if X is regular, then the probabilities in (a) and (b) in No. 2 equal 0 for every $x \in \text{int}I$. It holds also for every $x \in I$ that

$$\mathbf{P}_x\big(\exists\, \varepsilon(\omega) > 0 \text{ such that } X_t(\omega) \in I \ \forall\, t \in (0, \varepsilon(\omega))\big) = 1,$$

that is, the lifetime ζ is positive a.s. Moreover, if $l \in I$, then

$$\mathbf{P}_l\big(\exists \varepsilon(\omega) > 0 \, \text{such that } X_t(\omega) > l \ \forall t \in (0, \varepsilon(\omega))\big) = 0,$$

and an analogous equality holds for r.

Notice also that (a) means that the state space of a regular diffusion consists of only one *communicating* class (Itô and McKean (1974) p. 92). Using terminology from the theory of Markov chains, regularity here means *irreducibility.*

We refer to Blumenthal and Getoor (1968) p. 61 for a general definition of a *regular point.*

We stress that our definition of regularity does not allow absorbing boundaries and is hence slightly more restrictive than the definition in Freedman (1971) p. 106 and Rogers and Williams (2000) p. 271, or in Karlin and Taylor (1981) p. 158.

4. Every diffusion has three basic characteristics: speed measure, scale function, and killing measure.

The *speed measure* m is a measure on $\mathcal{B}(I)$ such that $0 < m((a, b)) < \infty$, $l < a < b < r$. For every $t > 0$ and $x \in I$ the measure $A \mapsto P_t(x, A)$, $A \in \mathcal{B}(I)$, (cf. I.6) is absolutely continuous with respect to m :

$$P_t(x, A) = \int_A p(t; x, y) m(dy).$$

The density p may be taken to be positive, jointly continuous in all variables, and symmetric, that is $p(t; x, y) = p(t; y, x)$. For a proof see Itô and McKean (1974) p. 149. Due to this symmetry X is in duality with itself relative to m (for a definition of duality see Blumenthal and Getoor (1968) p. 253).

The *killing measure* k is a measure on $\mathcal{B}(I)$ such that $k((a,b)) < \infty$, $l < a < b < r$. It is associated to the distribution of the location of the process at its lifetime $\zeta := \inf\{t : X_t \notin I\}$:

$$\mathbf{P}_x(X_{\zeta-} \in A;\ \zeta < t) = \int_0^t ds \int_A k(dy)\, p(s; x, y), \quad A \in \mathcal{B}(I).$$

The *scale function* s is an increasing continuous function from I to $\mathbb{R}$. It is connected to the drift of the diffusion. In particular, if $k((a,b)) = 0$ for an interval $(a,b) \subset I$, then for $a \le x \le b$,

$$\mathbf{P}_x(H_a < H_b) = 1 - \mathbf{P}_x(H_b < H_a) = \frac{s(b) - s(x)}{s(b) - s(a)}.$$

We say that X is in *natural scale* if $s(x) = x$. In this case and if, e.g., $k \equiv 0$ and $I = \mathbb{R}$, the diffusion X is a local martingale and can be constructed as a random time change of a standard Brownian motion (see No. 16).

In No. 9 we relate m, k, and s – in the absolutely continuous case – with the *infinitesimal parameters* of X.

5. The transition semigroup $\{P_t\}$ of a diffusion maps $\mathcal{C}_b(I)$ into $\mathcal{C}_b(I)$ (Breiman (1968) p. 358, Freedman (1971) p. 116, Rogers and Williams (2000) p. 291) and is strongly continuous (Dynkin (1965) Vol. II p. 137 (I closed), p. 162 (I open)). It follows that a linear diffusion is a Feller process (cf. I.10).

The semigroup $\{P_t\}$ does not in general map $\mathcal{C}_0(I)$ into $\mathcal{C}_0(I)$, and hence a diffusion is not in general a Feller–Dynkin process (cf. I.11) (see Rogers and Williams (2000) p. 292 for an example).

A linear diffusion is not in general a semimartingale (cf. I.23). An example of such a diffusion (Yor (1978)) is $X := \{\sqrt{|W_t|} : t \ge 0\}$, where W is standard Brownian motion. Notice that X is a diffusion because $|W|$ is a diffusion and $x \mapsto \sqrt{x}$, $x \ge 0$, is a one-to-one mapping.

A linear diffusion can be viewed as a standard process in the sense of the definition (9.2) in Blumenthal and Getoor (1968) p. 45.

6. Boundary classification. The behavior of X in the vicinity of the endpoints of I is determined by the properties of the basic characteristics m, s, and k. To explain this let z be such that $l < z < r$. Then the following terminology is used (Itô and Mckean (1974) p. 108, Ethier and Kurtz (1986) p. 366):

The left-hand endpoint l is called *exit* if

$$\int_{(l,z)} (m((a,z)) + k((a,z)))s(da) < \infty,$$

and *entrance* if

$$\int_{(l,z)} (s(z) - s(a))(m(da) + k(da)) < \infty.$$

The right-hand endpoint r is called *exit* if

$$\int_{(z,r)} (m((z,a)) + k((z,a)))s(da) < \infty,$$

and *entrance* if

$$\int_{(z,r)} (s(a) - s(z))(m(da) + k(da)) < \infty.$$

A boundary point which is both entrance and exit is called *non-singular.* A diffusion reaches its non-singular boundaries with positive probability, and it is, a priori, possible to start a diffusion from a non-singular boundary. In this case the basic characteristics alone do not determine the process uniquely and the description of the process must be completed by giving a boundary condition at each non-singular boundary point. This is discussed in the next section.

A boundary point which is neither entrance nor exit is called *natural.* A natural boundary point cannot be reached in finite time, and hence does not belong to the state space of X.

Consider the endpoint l and suppose that it is natural. Then l is said to be *attractive* if $\lim_{a\downarrow l} s(a) > -\infty$. In this case and if, e.g., $k \equiv 0$, we have $\lim_{t\uparrow\infty} X_t = l$ with positive probability.

A boundary point which is entrance but not exit, for short *entrance-not-exit,* cannot be reached from an interior point of I, and hence does not belong to the state space of a regular diffusion. The difference between an entrance-not-exit and a natural boundary is that it is possible in the former case, but not in the latter, to start the process from the boundary. Analytically this means that for every $t > 0$ and entrance boundary l, the measure $P_t(x, \cdot)$ converges (weakly) to a (sub)probability measure $P_t(l, \cdot) \not\equiv 0$ as x tends to l.

A boundary point which is exit but not entrance, for short *exit-not-entrance,* is reached from an interior point of I with positive probability. It is not, however, possible to start the process from an exit-not-entrance boundary point.

7. The *(weak) infinitesimal generator* of X is the operator $\mathcal{G}^\bullet$ defined by

$$\mathcal{G}^\bullet f := \lim_{t\downarrow 0} \frac{P_t f - f}{t}$$

applied to $f \in \mathcal{C}_b(I)$ for which the limit exists pointwise, is in $\mathcal{C}_b(I)$, and

$$\sup_{t>0} \left\| \frac{P_t f - f}{t} \right\| < \infty.$$

Let $\mathcal{D}(\mathcal{G}^\bullet)$ denote this set of functions.

Assume that both l and r are non-singular and belong to I. Let f^+ and f^- denote the right and left derivative of a function f with respect to the scale function s:

$$f^+(x) := \lim_{h\downarrow 0} \frac{f(x+h) - f(x)}{s(x+h) - s(x)}, \quad f^-(x) := \lim_{h\downarrow 0} \frac{f(x) - f(x-h)}{s(x) - s(x-h)}.$$

We define a set of functions $\mathcal{D}(\mathcal{G})$ by saying that a function $f \in \mathcal{C}_b(I)$ belongs to $\mathcal{D}(\mathcal{G})$ if there exists a function $g \in \mathcal{C}_b(I)$ such that for every $l < a < b < r$,

(a) $\int_{[a,b)} g(x)m(dx) = f^-(b) - f^-(a) - \int_{[a,b)} f(x)k(dx),$

(b) $\int_{(a,b]} g(x)m(dx) = f^+(b) - f^+(a) - \int_{(a,b]} f(x)k(dx),$

(c) $g(l)m(\{l\}) = f^+(l+) - f(l)k(\{l\}),$ if $m(\{l\}) < \infty,\ k(\{l\}) < \infty,$

(d) $g(r)m(\{r\}) = -f^-(r-) - f(r)k(\{r\}),$ if $m(\{r\}) < \infty,\ k(\{r\}) < \infty.$

Let $\mathcal{G}f := g$ for $f \in \mathcal{D}(\mathcal{G})$. The existence of f^+ and f^- is part of the definition of $\mathcal{D}(\mathcal{G})$. From (a) and (b) it is seen that for $f \in \mathcal{D}(\mathcal{G})$ the derivatives f^+ and f^- are right and left continuous, respectively. The result (Itô and McKean (1974) p. 100, 117, and 135, Mandl (1968), Freedman (1971) p. 131, Ethier and Kurtz (1986)) is

$$\mathcal{G} = \mathcal{G}^\bullet, \quad \mathcal{D}(\mathcal{G}) = \mathcal{D}(\mathcal{G}^\bullet).$$

The following terminology concerning non-singular boundaries is often used. Consider the left endpoint l. It is called

($\star$) *reflecting,* if $m(\{l\}) = k(\{l\}) = 0,$

($\star$) *sticky,* if $m(\{l\}) > 0,\ k(\{l\}) = 0,$

($\star$) *elastic,* if $m(\{l\}) = 0,\ k(\{l\}) > 0.$

A diffusion "spends no time" and "does not die" at a reflecting boundary point. Supposing that l is reflecting, then the first statement means that for all $t \geq 0$ and $x \in I$,

$$\mathbf{P}_x(\text{Leb}(\{s \leq t :\ X_s = l\}) = 0) = 1,$$

and the meaning of the second one is

$$\mathbf{P}_x(X_{\zeta-} = l) = 0.$$

If l is sticky, then X does not die at l but spends a positive amount of time there, i.e.,

$$\mathbf{P}_x(\text{Leb}(\{s \leq t :\ X_s = l\}) > 0) > 0.$$

If l is elastic, then X does not spend any time at l but

$$\mathbf{P}_x(X_{\zeta-} = l) > 0.$$

The condition in (c) can be given meaning also when $m(\{l\}) = \infty$ and/or $k(\{l\}) = \infty$. Indeed, dividing in (c) by $m(\{l\})$ and letting $m(\{l\})$ tend to infinity gives the condition

(e) $g(l) = \mathcal{G}f(l) = 0.$

In this case l is called *absorbing,* and we may define $\mathbf{P}_l(X_t = l$ for all $t) = 1$. Now X is not regular.

Dividing in (c) by $k(\{l\})$ and letting $k(\{l\})$ tend to infinity gives the condition

(f) $f(l) = 0.$

In this case l is called a *killing* boundary, meaning that the diffusing particle is – if it hits l – immediately sent to the cemetery state ∂. Here $l \notin I$.

The conditions (e) and (f) can also be combined

$$g(l) = -\gamma(l)f(l), \quad \gamma(l) > 0.$$

The picture here is that the particle – if it hits l – stays there a random amount of time which is exponentially distributed with parameter $\gamma(l)$ and independent of the motion prior to hitting of l. After that time the particle is sent to the cemetery state ∂. We call this kind of boundary point a *trap*, and remark that in Itô and McKean (1974) p. 91 a more general meaning is given to this term.

We refer also to Itô and McKean (1974) p. 186 and Rogers and Williams (2000) p. 439, where general boundary conditions are discussed in the context of Brownian motion (so-called *Feller's Brownian motions*).

8. If the boundary l is not non-singular, the infinitesimal generator is still given as in (a) and (b) in No. 7 but the condition (c) must be replaced by:

($\star$) $f^+(l+) = 0$ if l is entrance-not-exit,

($\star$) $f(l+) = 0$ if l is exit-not-entrance,

($\star$) no condition is needed if l is natural.

9. We consider here the special case in which the basic characteristics are absolutely continuous with respect to the Lebesgue measure and have smooth derivatives. In other words,

$$m(dx) = m(x)dx, \quad k(dx) = k(x)dx, \quad s(x) = \int^x s'(y)dy,$$

where the functions m and s' are continuous and positive and k is continuous and non-negative. Moreover, if s'' is continuous, then the infinitesimal generator $\mathcal{G} : \mathcal{D}(\mathcal{G}) \mapsto \mathcal{C}_b(I)$ is a second order differential operator

$$\mathcal{G}f(x) = \frac{1}{2}a^2(x)\frac{d^2}{dx^2}f(x) + b(x)\frac{d}{dx}f(x) - c(x)f(x).$$

The functions a, b and c – the *infinitesimal parameters* of X – are related to m, k and s via the formulae

$$m(x) = 2\,a^{-2}(x)e^{B(x)}, \quad s'(x) = e^{-B(x)}, \quad k(x) = 2\,a^{-2}(x)\,c(x)e^{B(x)}$$

with $B(x) := \int^x 2\,a^{-2}(y)\,b(y)dy$.

The domain of $\mathcal{G}$ consists of all functions in $\mathcal{C}_b(I)$ such that $\mathcal{G}f \in \mathcal{C}_b(I)$ together with the appropriate boundary conditions as explained in No. 7 and 8. Because of absolute continuity, the boundary condition at a non-singular boundary can – strictly speaking – only be reflection or killing.

The function a^2 is called the *infinitesimal variance*, b the *infinitesimal mean*, and c the *infinitesimal killing rate* because, in many cases,

$$\lim_{t\to 0}\frac{1}{t}\,\mathbf{E}_x(X_t-x)^2=a^2(x),$$

$$\lim_{t\to 0}\frac{1}{t}\,\mathbf{E}_x(X_t-x)=b(x),$$

and

$$\lim_{t\to 0}\frac{1}{t}\Big(\mathbf{P}_x(\zeta>t)-1\Big)=-c(x),$$

see, e.g., Lamperti (1977) p. 130, and Karlin and Taylor (1981) p. 159. Notice that the scale function s has the property

$$\frac{1}{2}a^2(x)\frac{d^2}{dx^2}s(x)+b(x)\frac{d}{dx}s(x)=0,$$

and the function m – the density of the speed measure – has the adjoint property

$$\frac{1}{2}\frac{d^2}{dx^2}(a^2(x)m(x))-\frac{d}{dx}(b(x)m(x))=0.$$

10. Let $H_z:=\inf\{t:\ X_t=z\}$ be the *first hitting time* of $z\in I$. Then (Itô and McKean (1974) p. 128) for $\alpha>0$,

$$\mathbf{E}_x(e^{-\alpha H_z})=\begin{cases}\dfrac{\psi_\alpha(x)}{\psi_\alpha(z)}, & x\le z,\\[2mm] \dfrac{\varphi_\alpha(x)}{\varphi_\alpha(z)}, & x\ge z,\end{cases}$$

where ψ_α and φ_α are continuous solutions of the generalized differential equation

(a) $$\mathcal{G}u=\alpha u.$$

This equation should be read as follows: u is a function which satisfies

$$\alpha\int_{[a,b)}u(x)m(dx)=u^-(b)-u^-(a)-\int_{[a,b)}u(x)k(dx)$$

for all a and b such that $l<a<b<r$. In the absolute continuous case u is a solution of the ODE

$$\frac{1}{2}a^2(x)\frac{d^2}{dx^2}u(x)+b(x)\frac{d}{dx}u(x)-c(x)u(x)=\alpha u(x).$$

The functions ψ_α and φ_α can be characterized as the unique (up to a multiplicative constant) positive solutions of (a) by firstly demanding that ψ_α is increasing and φ_α is decreasing, and secondly posing conditions at the non-singular boundary points. For ψ_α a boundary condition is only needed at l:

if $l \in I$,

$$\alpha\psi_\alpha(l)m(\{l\}) = \psi_\alpha^+(l) - \psi_\alpha(l)k(\{l\}),$$

if $l \notin I$,

$$\psi_\alpha(l+) = 0.$$

In the latter case l is a killing boundary. For φ_α we have, similarly,

if $r \in I$,

$$\alpha\varphi_\alpha(r)m(\{r\}) = -\varphi_\alpha^-(r) - \varphi_\alpha(r)k(\{r\}),$$

if $r \notin I$,

$$\varphi_\alpha(r-) = 0.$$

The functions ψ_α and φ_α also have the following properties at l (and analogous properties at r):

if l is entrance-not-exit,

$$\psi_\alpha(l+) > 0,\ \psi_\alpha^+(l+) = 0,\ \varphi_\alpha(l+) = +\infty,\ \varphi_\alpha^+(l+) > -\infty,$$

if l is exit-not-entrance,

$$\psi_\alpha(l+) = 0,\ \psi_\alpha^+(l+) > 0,\ \varphi_\alpha(l+) < +\infty,\ \varphi_\alpha^+(l+) = -\infty,$$

if l is natural,

$$\psi_\alpha(l+) = 0,\ \psi_\alpha^+(l+) = 0,\ \varphi_\alpha(l+) = +\infty,\ \varphi_\alpha^+(l+) = -\infty.$$

11. The functions ψ_α and φ_α are called the *fundamental solutions* of (a) in No. 10. They are linearly independent and all solutions can be expressed as their linear combinations. Moreover, the so-called *Wronskian*

$$w_\alpha := \psi_\alpha^+(x)\varphi_\alpha(x) - \psi_\alpha(x)\varphi_\alpha^+(x) = \psi_\alpha^-(x)\varphi_\alpha(x) - \psi_\alpha(x)\varphi_\alpha^-(x)$$

is independent of x.

Let p be the transition density of X with respect to the speed measure (see No. 4) and introduce the *Green function*

$$G_\alpha(x, y) := \int_0^\infty e^{-\alpha t} p(t; x, y)dt.$$

Then

$$G_\alpha(x, y) = \begin{cases} w_\alpha^{-1}\psi_\alpha(x)\varphi_\alpha(y), & x \le y, \\ w_\alpha^{-1}\psi_\alpha(y)\varphi_\alpha(x), & x \ge y. \end{cases}$$

If the killing measure $k \equiv 0$ and non-singular boundaries are killing boundaries, then

$$G_0(x,y) := \int_0^\infty p(t;x,y)dt = \lim_{a\downarrow l, b\uparrow r} \frac{(s(x)-s(a))(s(b)-s(y))}{s(b)-s(a)}, \quad x \le y.$$

Under the same assumptions but with l reflecting, we have $\psi_0 \equiv 1$ and hence

$$G_0(x,y) = \lim_{b\uparrow r}(s(b)-s(y)), \quad x \le y.$$

It can be proved (Itô and McKean (1974) p. 149, Lamperti (1977) p. 127) that $(t,x) \mapsto p(t;x,\cdot)$ solves for $t > 0$ and $x \in \operatorname{int} I$ the generalized parabolic differential equation

(a) $$\frac{\partial}{\partial t} u(t,x) = \mathcal{G}u(t,x).$$

The boundary conditions for $p(t;x,\cdot)$ at l and r can be deduced from the corresponding conditions for ψ_α and φ_α given in No. 10.

The equation (a) is the so-called *backward Kolmogorov equation* for X introduced in Kolmogorov (1931) and further studied in Feller (1936). See Lamperti (1977) p. 127 for a short historical view of Kolmogorov's equation, and Shiryayev (1989) for an article on Kolmogorov's life and activities. We refer also to Jacobsen (1996) who discusses Laplace's early work on an urn model leading to a partial differential equation which is the forward Kolmogorov equation for an Ornstein–Uhlenbeck process.

12. A diffusion X is said to be *recurrent* if $\mathbf{P}_x(H_y < \infty) = 1$ for all $x, y \in I$. Notice that

$$\mathbf{P}_x(H_y < \infty) = \lim_{\alpha\to 0} \mathbf{E}_x(e^{-\alpha H_y}).$$

A diffusion which is not recurrent is called *transient*. It can be proved (Itô and McKean (1974) p. 159 for the case $x = y$) that X is transient if and only if

$$\lim_{\alpha\downarrow 0} G_\alpha(x,y) < \infty$$

for some $x, y \in I$, and hence for all $x, y \in I$. From the resolvent equation it follows that this limit always exists or equals $+\infty$. Therefore X is recurrent if and only if

$$\lim_{\alpha\downarrow 0} G_\alpha(x,y) = \infty.$$

A recurrent diffusion is called *null recurrent* if $\mathbf{E}_x(H_y) = \infty$ for all $x, y \in I$ and *positively recurrent* if $\mathbf{E}_x(H_y) < \infty$ for all $x, y \in I$. It can be proved that a recurrent diffusion is positively recurrent if and only if $m(I) < \infty$. In the recurrent case we have (see Salminen (1993))

(a) $$\lim_{\alpha\downarrow 0} \alpha G_\alpha(x,x) = \frac{1}{m(I)}.$$

The speed measure m is also easily seen to be a *stationary* (or *equilibrium* or *invariant*) *measure* of a recurrent diffusion, that is, it satisfies for all t and $A \in \mathcal{B}(I)$,

$$mP_t(A) := \int_I m(dx)\, P_t(x,A) = m(A).$$

2. Local time

13. For a regular diffusion X there exists (see Itô and McKean (1974) p. 174 and p. 183, Freedman (1971) p. 159 and Rogers and Williams (2000) p. 289) a family of random variables

$$\{L(t,x):\ x\in I,\ t\geq 0\},$$

called the *local time* of X, such that

(a) $\int_0^t \mathbb{I}_A(X_s)\,ds = \int_A L(t,x)\,m(dx)$ a.s., $A\in\mathcal{B}(I)$,

(b) $L(t,x)=\lim_{\epsilon\downarrow 0}\dfrac{\text{Leb}(\{s<t:\ x-\epsilon<X_s<x+\epsilon\})}{m((x-\epsilon,x+\epsilon))}$ a.s.,

(c) $L(t,x,\omega)=L(s,x,\omega)+L(t-s,x,\theta_s(\omega))$ a.s.,

where in (c) $s<t$, and $\theta_.$ is the shift operator (see I.13). If $m(\{x\})>0$, then $L(t,x)$ is proportional to the amount of time X spends at x up to t.

We have the following connection between the local time and the transition density:

(d) $$\mathbf{E}_x(L(t,y))=\int_0^t p(s;x,y)\,ds.$$

In the transient case, $\mathbf{E}_x(L(\zeta,y))=\mathbf{E}_x(L(\infty,y))=G_0(x,y)$. In fact, (see No. 27)

$$\mathbf{P}_x(L(\infty,x)>\alpha)=\exp\Big(-\frac{\alpha}{G_0(x,x)}\Big).$$

For a fixed x the process $L^x=\{L(t,x):\ t\geq 0\}$ is called the *local time process* of X at the point x. Using the definition in No. 21 it is seen that L^x is a *continuous additive functional* of X. Because L^x is non-decreasing it induces a (random) measure on the time axis. The support of this measure, that is, the complement of the largest open set of measure zero, is a.s. equal to the set $\{t:\ X_t=x\}$. This fact is often expressed by saying that L^x increases only when X is at x, and can be proved using the random time change technique (see No. 16) from the corresponding property of the Brownian local time (see IV.7).

Regularity of a point and existence of local time at that point are equivalent properties for a large class of stochastic processes; see Blumenthal and Getoor (1968) p. 216. For a survey article on local times in general see Geman and Horowitz (1980).

14. Assume that X is started at x, and introduce the *inverse local time process* $\rho^x=\{\rho(t,x):\ t\geq 0\}$ at x by setting

$$\rho(t,x):=\inf\{s:\ L(s,x)>t\}.$$

Then ρ^x has the properties:

(a) $\rho(0,x)=0$ a.s.,

(b) $t \mapsto \rho(t,x)$ is right-continuous and non-decreasing,

(c) $\rho(t+s,x) - \rho(t,x) \sim \rho(s,x)$ for all t, $s \geq 0$,

(d) for $0 \leq t_1 < \cdots < t_n$ the random variables

$$\rho(t_1,x),\ \rho(t_2,x) - \rho(t_1,x), \dots,\ \rho(t_n,x) - \rho(t_{n-1},x)$$

are independent.

In general, a stochastic process having the properties (b)–(d) is called a *subordinator*, or a *non-decreasing Lévy process.* Due to (c) and (d) we say that ρ^x has *stationary* and *independent increments.* Subordinators play prominent roles, e.g., in the excursion theory of Markov processes.

To give the *Lévy–Khintchine representation* of ρ^x let $n^x_\pm$ be the measures defined by

$$\int_0^\infty (1 - e^{-\alpha l}) n^x_\pm(dl) = \lim_{y \to x\pm} \frac{\mathbf{P}_y(H_x < \infty) - \mathbf{E}_y(e^{-\alpha H_x})}{\pm(s(y) - s(x))},$$

and

$$n^x_\pm(\infty) := \lim_{y \to x\pm} \frac{1 - \mathbf{P}_y(H_x < \infty)}{\pm(s(y) - s(x))}.$$

Then the Lévy–Khintchine representation is

$$\mathbf{E}_x(e^{-\alpha\rho(t,x)}) = \exp\left(-t\left(m(\{x\})\alpha + \int_0^\infty (1 - e^{-\alpha l}) n^x(dl) + n^x(\infty)\right)\right),$$

where

$$n^x := n^x_- + n^x_+, \qquad n^x(\infty) := n^x_+(\infty) + n^x_-(\infty).$$

Note that

$$\mathbf{P}_x(\rho(t,x) < \infty) = \exp(-t\, n^x(\infty)).$$

It follows that X is transient if and only if $n^x(\infty) > 0$ for some x. Computing the limits above (see Itô and McKean (1974) p. 214, where the recurrent case is studied, see also Pitman and Yor (1999a)) shows that

$$\mathbf{E}_x(e^{-\alpha\rho(t,x)}) = \exp\Big(-\frac{t}{G_\alpha(x,x)}\Big).$$

Notice that, if X is recurrent, it follows from (a) in No. 12 that for all $x \in I$,

$$\mathbf{E}_x(\rho(t,x)) = m(I)\, t.$$

15. Let X be recurrent and $0 \in I$. Let $\rho := \rho^0$ be the inverse local time process at 0 and consider for a given $x \in I$ the process

$$L(\rho) = \{L(\rho(t), x) : t \geq 0\}.$$

Using the strong Markov property it is seen that $L(\rho)$ is a subordinator. Its Lévy–Khintchine representation is

$$\mathbf{E}_0(e^{-\alpha L(\rho(t),x)}) = \exp\Big(-\frac{\alpha\, t}{1+\alpha|s(x)-s(0)|}\Big)$$
$$= \exp\left(-t\int_0^\infty (1-e^{-\alpha l})n(dl)\right),$$

where

$$n(dl) = \beta^2\, e^{-\beta l}\, dl, \quad \beta := |s(x)-s(0)|^{-1}.$$

For the proof when X is a Brownian motion see Itô and McKean (1974) p. 74. General recurrent diffusions are considered in Csáki and Salminen (1996). It is also possible to have analogous representations for transient diffusions. In this case, however, the process ρ is exploding.

Notice that the representation above does not depend on the speed measure and because $n([0,\infty)) < \infty$ the process $L(\rho)$ is a compound Poisson process. Straightforward computations show that

$$\mathbf{E}_0(L(\rho(t),x)) = t, \qquad \mathbf{E}_0(L^2(\rho(t),x)) = t^2 + 2t|s(x)-s(0)|.$$

16. Consider a diffusion X in natural scale, i.e., when $s(x) = x$, and assume that $k \equiv 0$. This imposes some restrictions on the boundary behavior. For instance, a finite boundary point cannot be entrance-not-exit and an infinite boundary point cannot be exit. As usual, let m denote the speed measure of X. Further, let W be a standard Brownian motion on $\mathbb{R}$ and assume that $X_0 = W_0$. We explain here how to obtain a version of X from W via a *random time change* (or a *random time substitution.*) This construction makes the fundamental role of local time in the theory of diffusions apparent.

Let $\{L(t,x) : \ x \in \mathbb{R},\ t \geq 0\}$ be the local time of W and extend m to $\mathcal{B}(\mathbb{R})$ by setting $m(A) = 0$ if $A \cap I = \emptyset$. Consider the functional

$$L^{(m)}(t) := \int_{\mathbb{R}} L(t,y)\, m(dy).$$

In the case $m(dx) = 2m(x)\, dx$ we have

$$L^{(m)}(t) = \int_0^t m(W_u)\, du.$$

It can be proved (Itô and McKean (1974) p. 168) that $t \mapsto L^{(m)}(t)$ is non-decreasing and continuous (when finite). This also follows from No. 21 because $L^{(m)}$ constitutes an additive functional of W. It may happen that $L^{(m)}(t) = \infty$ with positive probability for some t. In fact, we force this to happen in an important special case. Indeed, assume that l is non-singular for X. Then it is impossible on the basis of m alone to say whether l is reflecting or killing. In order to distinguish between the two classes, it is assumed that in the latter case $m(\{l\}) = \infty$, i.e., l is regarded as an absorbing state. This implies that $L^{(m)}(t) = \infty$ when $t > H_l := \inf\{t : W_t = l\}$.

Now let

$$\rho^{(m)}(t) := \inf\{u : \ L^{(m)}(u) > t\}$$

be the right continuous inverse of $L^{(m)}$. If l is absorbing, then $\rho^{(m)}(t) = H_l$ for all $t \geq L^{(m)}(H_l-)$. The random time change of W based on $L^{(m)}$ is the process

$$W^{(m)} := \{W_{\rho^{(m)}(t)} : \ t \geq 0\}.$$

Then, when killing boundaries are changed to absorbing ones,

$$X \sim W^{(m)}.$$

Further, letting $\{L_X(t,x) : \ x \in I, \ t \geq 0\}$ be the local time of X with respect to the speed measure m, then

$$\{L_X(t,x) \ : \ t \geq 0, \ x \in I\} \sim \{L(\rho^{(m)}(t), x) \ : \ t \geq 0, \ x \in I\}.$$

If X is not in natural scale the process defined by $\widetilde{X}_t := s(X_t)$, where s is the scale function of X, is a diffusion in natural scale for which the results above apply. In particular, if $\widetilde{m}$ is the speed measure of $\widetilde{X}$, then

$$\{X_t : t \geq 0\} \sim \{s^{(-1)}(W_{\rho^{(\widetilde{m})}(t)}) : t \geq 0\},$$

where $s^{(-1)}(x) := \inf\{y : s(y) > x\}$ is the inverse of s.

These results are due to Itô and McKean (1974) p. 167–174 and Volkonskij (1958). For proofs see also Rogers and Williams (2000) p. 277–283 and Freedman (1971) p. 151–159.

From the definition of $L^{(m)}$ it is evident that if a Brownian particle is moving at time t on the region, where m is "big", then $L^{(m)}$ is increasing rapidly. Consequently, $\rho^{(m)}(t+h) \simeq \rho^{(m)}(t)$ for "large" $h > 0$, and hence the increment $X_{t+h} - X_t$ is "small". Due to this property m has been given the name *speed measure*. An extreme case is obtained at the points with positive m-mass (so-called *sticky points*, cf. the definition of a sticky boundary in No. 7) in the vicinity of which X will "spend much time".

3. Passage times

17. Assume that $X_0 = x < r$ and let $S_a^+ := \inf\{t : X_t > a\}$, $a > x$. The *first passage process*

$$S^+ := \{S_a^+ : \ a \geq x\}$$

is non-decreasing and, due to the strong Markov property, has independent increments. Using No. 10 and 14 it is seen (Itô and McKean (1974) p. 144) that S^+ has the Lévy–Khintchine representation

$$\mathbf{E}_x(e^{-\alpha S_a^+}) = \exp\left(-\int_x^a s(dy)\left(\int_0^\infty (1-e^{-\alpha l})n_-^y(dl) + n_-^y(\infty)\right)\right).$$

Analogously, for $S_a^- := \inf\{t : X(t) < a\}$, $a < x$, we have

$$\mathbf{E}_x(e^{-\alpha S_a^-}) = \exp\left(-\int_a^x s(dy)\left(\int_0^\infty (1-e^{-\alpha l})n_+^y(dl) + n_+^y(\infty)\right)\right).$$

For the definitions of $n_\pm^y$, see No. 14.

18. Assume that X is recurrent and $0 \in I$. Using the strong Markov property it is seen that the process

$$L(H) := \{L(H_b, 0) : b \geq 0\}$$

has independent increments. To find its Lévy–Khintchine representation observe that for $0 \leq a < b$,

$$\mathbf{E}_0(e^{-\alpha(L(H_b,0)-L(H_a,0))}) = \mathbf{P}_a(H_b < H_0) + \mathbf{P}_a(H_0 < H_b)\mathbf{E}_0(e^{-\alpha L(H_b,0)}).$$

But (see No. 27)

$$\mathbf{E}_0(e^{-\alpha L(H_b,0)}) = \frac{1}{\alpha(s(b)-s(0))+1},$$

giving

$$\begin{aligned}\mathbf{E}_0(e^{-\alpha(L(H_b,0)-L(H_a,0))}) &= \frac{\alpha(s(a)-s(0))+1}{\alpha(s(b)-s(0))+1}\\ &= \exp\left(-\int_0^\infty (1-e^{-\alpha l})n(dl)\right),\end{aligned}$$

where

$$n(dl) := l^{-1}(e^{-\beta(b)\,l} - e^{-\beta(a)\,l})\,dl, \quad \beta(x) := (s(x)-s(0))^{-1}.$$

For Brownian motion with $a = 0$ this result is proved in Itô and McKean (1974) p. 72. The proof therein is based on the scaling property.

19. Here we consider the joint distribution of the *maximum up to time t and its location.* Assume that the diffusion X is conservative and for $t > 0$ let

$$\check{M}_t := \sup\{X_s : s \leq t\}.$$

Then for $x < z$,

$$\mathbf{P}_x(\check{M}_t \geq z) = \mathbf{P}_x(H_z < t) = \int_0^t n_x(z,s)ds,$$

where $n_x(z,\cdot)$ is the $\mathbf{P}_x$-density of H_z (Itô and McKean (1974) p. 154). Taking the Laplace transform with respect to t gives

$$\int_0^\infty e^{-\alpha t}\mathbf{P}_x(\check{M}_t \geq z)dt = \frac{1}{\alpha}\frac{\psi_\alpha(x)}{\psi_\alpha(z)}.$$

Notice that if $\tau \sim \text{Exp}(\alpha)$ and is independent of X, then

$$\mathbf{P}_x(\check{M}_\tau \geq z) = \mathbf{P}_x(H_z < \tau) = \mathbf{E}_x(\exp(-\alpha H_z)).$$

In particular, if X does not hit r it follows that

$$\mathbf{E}_x(\check{M}_\tau) = x + \int_x^r \frac{\psi_\alpha(x)}{\psi_\alpha(z)}dz.$$

For $z < r$ we have

$$\int_0^\infty e^{-\alpha t}\mathbf{P}_x(\check{M}_t \in dz)dt = \frac{\psi_\alpha(x)}{\psi_\alpha^2(z)}\, s(dz) \int_{[l,z)\cap I} \psi_\alpha(y)m(dy),$$

and

$$\int_0^\infty e^{-\alpha t}\mathbf{P}_x(\check{M}_t \in dz, X_t \in dy)dt = \frac{\psi_\alpha(x)\psi_\alpha(y)}{\psi_\alpha^2(z)}\, s(dz)\, m(dy).$$

Let

$$\check{H}(t) := \inf\{s < t :\ X_s = \check{M}_t\}$$

be the first time point that X attains its maximum before t. (In fact, the time point of the maximum is a.s. unique in the case $\check{M}_t < r$.) Then for $x \vee y < z$ and $v < t$,

$$\mathbf{P}_x(\check{M}_t \in dz, X_t \in dy, \check{H}(t) \in dv) = n_x(z,v)n_y(z,t-v)\, s(dz)\, m(dy)\, dv.$$

This formula is valid for all (regular) diffusions, not just for conservative ones. For a proof, see Csáki, Földes and Salminen (1987). Notice, however, that the distribution might have an atom at r :

$$\begin{aligned}\mathbf{P}_x(\check{M}_t = r, X_t \in dy, \check{H}(t) \in dv) &= \mathbf{P}_x(H_r \in dv, X_t \in dy)\\ &= n_x(r,v)p(t-v;r,y)\, m(dy)\, dv.\end{aligned}$$

Observe that the distribution of the global maximum $\check{M} := \sup X_t$ is (cf. No. 10)

$$\mathbf{P}_x(\check{M} > z) = \mathbf{P}_x(H_z < \infty) = \lim_{\alpha\to 0}\mathbf{E}_x(e^{-\alpha H_z}) = \lim_{\alpha\to 0}\frac{\psi_\alpha(x)}{\psi_\alpha(z)}.$$

20. Let $y \in I$ be given and define the *last passage* or *last exit time* at y by

$$\lambda_y := \sup\{t :\ X_t = y\}.$$

If X does not visit y at all we define $\lambda_y = 0$. Therefore, $\{\lambda_y = 0\} = \{H_y = \infty\}$. If X is recurrent, then $\lambda_y = \infty$ a.s. for all $y \in I$. In the transient case $\lambda_y < \infty$ a.s. for all $y \in I$, we have (see Pitman and Yor (1981), Salminen (1984))

(a)
$$\mathbf{P}_x(0 < \lambda_y \leq t) = \int_0^t \frac{p(u; x, y)}{G_0(y, y)} du.$$

Further

$$\mathbf{P}_x(\lambda_y = 0) = \mathbf{P}_x(H_y = \infty) = 1 - \mathbf{P}_x(0 < \lambda_y < \infty) = 1 - \frac{G_0(x, y)}{G_0(y, y)}.$$

Now let t be fixed and define $\lambda_y^t := \sup\{u < t : X_u = y\}$. To find the distribution of λ_y^t let τ be an exponentially distributed random variable with parameter α, independent of X, and let $\lambda_y^\tau := \sup\{u < \tau : X_u = y\}$. From (a),

$$\mathbf{P}_x(\lambda_x^\tau \in du) = \frac{e^{-\alpha u} p(u; x, x)}{G_\alpha(x, x)} du.$$

Let $f(\cdot, x)$ be the inverse of the Laplace transform $1/\alpha G_\alpha(x, x)$. Then

$$\mathbf{P}_x(\lambda_x^t \in du) = f(t - u, x) \, p(u; x, x) du, \quad t > u.$$

Notice also that
$$\mathbf{E}_x(\exp(-\beta \lambda_x^\tau)) = \frac{G_{\alpha+\beta}(x, x)}{G_\alpha(x, x)}.$$

We refer to Chung and Getoor (1977), Chung (1982b), and (1995) for interpretations and applications of last exit times in potential theory.

4. Additive functionals and killing

21. Let $\mathbf{P}_\nu$ denote the probability measure associated to a diffusion X when X is started with the initial distribution ν. Let $\mathcal{F}_t$ and $\mathcal{F}_\infty$ be the universal completions of the canonical filtration $\mathcal{F}_t^c$ and the σ-algebra $\mathcal{F}_\infty^c$, respectively, (cf. I.14). Following, e.g., Blumenthal and Getoor (1968) p. 148 or Revuz and Yor (2001) p. 401, a stochastic process $A = \{A_t : t \geq 0\}$, $A_0 = 0$, taking values in $[0, \infty]$ is called an *additive functional* of X if it has the following properties for every $x \in I$:

(a) A_t is $\mathcal{F}_t$-measurable for every $t \geq 0$,

(b) $t \mapsto A_t$ is $\mathbf{P}_x$-a.s. non-decreasing, right-continuous, and

$$A_t = A_{\zeta-}, \quad t \geq \zeta,$$

(c)   for every s and t,

$$A_{s+t} = A_t + \mathbb{1}_{[0,\zeta)}(t) A_s \circ \theta_t, \quad \mathbf{P}_x\text{-a.s.}$$

It can be proved that every additive functional of a regular diffusion is a.s. continuous. Indeed, we notice first that all additive functionals of a diffusion are *natural* (Blumenthal and Getoor (1968) p. 153). Combining this with the fact that all α-excessive functions are continuous (see No. 29) the claim follows using Exercise (4.17) in Blumenthal and Getoor (1968) p. 293. The continuity at $t = \zeta$ is clear from (b).

An additive functional A of a diffusion is said to have the *strong Markov property* if

(c')   for every $\mathcal{F}_t$-stopping time T and non-negative random variable S

$$A_{T(\omega)+S(\omega)}(\omega) = A_{T(\omega)}(\omega) + \mathbb{1}_{[0,\zeta(\omega))}(T(\omega))\, A_{S(\omega)}(\theta_T(\omega)) \quad \mathbf{P}_x\text{-a.s.}$$

From Proposition (1.13) in Blumenthal and Getoor (1968) p. 152 it follows that every additive functional of a diffusion has the strong Markov property.

22. Let $\hat{X}$ be a diffusion having a killing measure $k \not\equiv 0$. We describe here how a version of $\hat{X}$ can be constructed from a diffusion X having the same speed and scale as $\hat{X}$ but for which the killing measure equals zero. It is assumed that both $\hat{X}$ and X attain values on the same interval I.

Let $\{L(x,t) : x \in I,\ t \geq 0\}$ be the local time of X, and define

$$L^{(k)}(t) := \int_I L(t,y)\, k(dy).$$

From the properties of local time in No. 13 it follows that $L^{(k)}$ is an additive functional. Let $\tau \sim \text{Exp}(1)$ independent of X, and introduce

$$X_t^\bullet := \begin{cases} X_t, & L^{(k)}(t) < \tau, \\ \partial, & L^{(k)}(t) \geq \tau. \end{cases}$$

Then $\hat{X}$ and $X^\bullet$ are identical in law. We say that X is killed according to the additive functional $L^{(k)}$. For a proof, see Itô and McKean (1974) p. 179–183.

23. Let A be an arbitrary additive functional of a diffusion X. We assume, for simplicity, that the killing measure of X equals zero. Let τ be as in No. 22 and define

$$\hat{X}_t := \begin{cases} X_t, & A_t < \tau, \\ \partial, & A_t \geq \tau. \end{cases}$$

Then $\hat{X}$ is a Markov process with continuous paths. Moreover, because A has the strong Markov property it is seen that $\hat{X}$ is a diffusion. Letting k denote the killing measure of $\hat{X}$ it follows from No. 22 that for every t a.s.

$$A_t = \int_I L(t,x)\, k(dx).$$

Due to this property k is also called the *representing measure* of the additive functional A.

24. Let X and $L^{(k)}$ be as in No. 22. The distribution of $L^{(k)}$ can be characterized by choosing $f \equiv 1$ on I in the following double Laplace transform:

$$\int_0^\infty e^{-\gamma t} \mathbf{E}_x \Big(e^{-\eta\, L^{(k)}(t)}\, f(X_t) \Big) dt = \int_I \tilde{G}_\gamma(x,y)\, f(y)\, m(dy),$$

where $\tilde{G}$ is the Green function of a diffusion with the characteristics m, s and $\tilde{k} := \eta\, k$. The equality above is a compact form of the famous *Feynman–Kac formula* for diffusions. In Chapter VI this formula is studied for Brownian motion in a number of special cases. Recall also that if $k(dx) = k(x)m(dx)$, then the occupation time formula (a) in No. 13 gives

$$L^{(k)}(t) = \int_I L(t,y)k(y)m(dy) = \int_0^t k(X_s)\, ds.$$

In particular, to find the distribution of $L(t,z)$ for given t and z (under the assumption $m(\{z\}) = 0$), let the killing measure in No. 22 be the Dirac measure at z. Then functions ψ and φ in terms of which the corresponding Green function $\tilde{G}$ is expressed are solutions of the generalized differential equation (see No's. 7 and 10)

$$\mathcal{G}u = \gamma u$$

such that

$$u^+(z+) - u^+(z-) = \eta u(z),$$

or, equivalently,

$$u^-(z+) - u^-(z-) = \eta u(z).$$

In the absolutely continuous case (cf. No. 9) we may consider the local time with respect to the Lebesgue measure. Letting ℓ denote this local time we have

$$\ell(t,x) = m(x)L(t,x).$$

Now φ and ψ, associated to the distribution of ℓ via the Feynman–Kac formula, are solutions of the ODE

$$\frac{1}{2}a^2(x)\, u''(x) + b(x)\, u'(x) = \gamma u(x)$$

such that

$$u'(z+) - u'(z-) = \frac{2\eta}{a^2(z)} u(z).$$

The distribution of $L(t,x)$ is obtained more easily, perhaps, by choosing ζ in (a) in No. 27 to be exponentially distributed and independent of X. Then the result (a) in No. 27 gives the inversion with respect to η of the double Laplace transform computed above.

25. Let A be an additive functional of X and k_o the measure associated to it as explained in No. 23. Then by (d) in No. 13,

(a) $$\mathbf{E}_x A_t = \int_I k_o(dy) \int_0^t ds\, p(s; x, y).$$

To find the second moment notice that

$$\begin{aligned}\mathbf{E}_x A_t^2 &= \mathbf{E}_x\Big(\int_0^t dA_{t_1} \int_0^t dA_{t_2}\Big) = 2\,\mathbf{E}_x\Big(\int_0^t dA_{t_1} \int_{t_1}^t dA_{t_2}\Big)\\ &= 2 \int_I k_o(dy) \int_0^t dt_1\, p(t_1; x, y) \mathbf{E}_y(A_{t-t_1}).\end{aligned}$$

We may now proceed recursively and obtain

$$\begin{aligned}\mathbf{E}_x A_t^n &= \mathbf{E}_x\Big(\int_0^t dA_{t_1}\Big(\int_0^t dA_{t_2}\Big)^{n-1}\Big) = n\,\mathbf{E}_x\Big(\int_0^t dA_{t_1}\Big(\int_{t_1}^t dA_{t_2}\Big)^{n-1}\Big)\\ &= n \int_I k_o(dy) \int_0^t dt_1\, p(t_1; x, y) \mathbf{E}_y(A_{t-t_1}^{n-1}).\end{aligned}$$

This formula is called the *Kac moment formula,* see Kac (1949, 1951), where a special case is treated. Here and in the next two sections our main reference is Fitzsimmons and Pitman (1999) in which more general processes are considered.

Now let $A^{(1)}$ and $A^{(2)}$ be two additive functionals with representing measures k_1 and k_2, respectively. Then

$$\begin{aligned}\mathbf{E}_x(A_t^{(1)} A_t^{(2)}) &= \mathbf{E}_x\Big(\int_0^t dA_{t_1}^{(1)} \int_0^t dA_{t_2}^{(2)}\Big)\\ &= \mathbf{E}_x\Big(\int_0^t dA_{t_1}^{(1)} \int_{t_1}^t dA_{t_2}^{(2)}\Big) + \mathbf{E}_x\Big(\int_0^t dA_{t_2}^{(2)} \int_{t_2}^t dA_{t_1}^{(1)}\Big)\\ &= \int_I k_1(dy) \int_0^t dt_1 p(t_1; x, y) \mathbf{E}_y(A_{t-t_1}^{(2)})\\ &\qquad + \int_I k_2(dy) \int_0^t dt_2 p(t_2; x, y) \mathbf{E}_y(A_{t-t_2}^{(1)}).\end{aligned}$$

In particular, for local times we have

$$\mathbf{E}_x(L(t, y)^n) = n \int_0^t dt_1\, p(t_1; x, y) \mathbf{E}_y(L(t - t_1, y)^{n-1})$$

and

$$\begin{aligned}\mathbf{E}_x(L(t, y)L(t, z)) = \int_0^t dt_1\, p(t_1; x, y) \mathbf{E}_y(L(t - t_1, z))&\\ + \int_0^t dt_1\, p(t_1; x, z) \mathbf{E}_z(L(t - t_1, y)).&\end{aligned}$$

Further, for $f \in B_{\mathcal{B}}(I)$,

$$\mathbf{E}_x(A_t\, f(X_t)) = \int_I k_o(dy) \int_0^t ds\, p(s; x, y)\mathbf{E}_y(f(X_{t-s})).$$

Choosing here $f \equiv 1$ on I gives $\mathbf{E}_x(A_t^n;\, t < \zeta)$. In general,

$$\mathbf{E}_x(A_t^n\, f(X_t)) = n \int_I k_o(dy) \int_0^t ds\, p(s; x, y)\mathbf{E}_y(A_{t-s}^{n-1} f(X_{t-s})).$$

26. We present here some formulae associated to the location of X at its lifetime ζ. First, recall the formula displayed for indicators in No. 4:

$$\text{(a)} \qquad \mathbf{E}_x(f(X_{\zeta-});\ \zeta \le t) = \int_0^t ds \int_I k(dy)\, p(s; x, y) f(y), \quad f \in B_{\mathcal{B}}(I).$$

Because $f(\partial) = 0$ this does not, in general, give the distribution of ζ. On the other hand,

$$\mathbf{P}_x(\zeta > t) = \int_I p(t; x, y) m(dy).$$

From (a) we obtain

$$\mathbf{E}_x(e^{-\gamma\zeta} f(X_{\zeta-})) = \int_I G_\gamma(x, y) f(y) k(dy).$$

Using the properties of ψ_0 and φ_0 given in No. 10 it follows that for $l < a < b \le x \le c < d < r$,

$$\mathbf{P}_x(a \le X_{\zeta-} < b) = \varphi_0(x)(\psi_0^-(b) - \psi_0^-(a)),$$
$$\mathbf{P}_x(c < X_{\zeta-} \le d) = \psi_0(x)(\varphi_0^+(d) - \varphi_0^+(c)),$$

where ψ_0 and φ_0 are normalized so that

$$G_0(x, y) = \psi_0(y)\varphi_0(x), \quad x \ge y.$$

Furthermore, in the case $\zeta < \infty$ and $X_{\zeta-} \in I$ a.s., we have

$$\int_I G_\gamma(x, y)(k(dy) + \gamma\, m(dy)) = 1.$$

27. Let A be an additive functional of X and k_o the measure associated to it as explained in No. 23. Then from (a) in No. 25,

$$\mathbf{E}_x A_\zeta = \int_I G_0(x, y) k_o(dy)$$

and

$$\mathbf{E}_x(A_\zeta^n) = n\int_I G_0(x,y)\mathbf{E}_y(A_\zeta^{n-1})k_o(dy).$$

In particular,

$$\mathbf{E}_x(L(\zeta,x)^n) = n!\,G_0(x,x)^n,$$

which leads to

(a) $$\mathbf{P}_x(L(\zeta,x) > \alpha) = \exp(-\frac{\alpha}{G_0(x,x)}).$$

Further,

(b) $$\mathbf{E}_x(A_\zeta^n f(X_{\zeta-})) = n\int_I G_0(x,y)\mathbf{E}_y(A_\zeta^{n-1} f(X_{\zeta-}))k_o(dy).$$

Choosing here $f \equiv 1$ on I gives $\mathbf{E}_x(A_\zeta^n;\ X_{\zeta-} \in I)$.

5. Excessive functions

28. A non-negative measurable function $h : I \mapsto \mathbb{R} \cup \{\infty\}$ is called *α-excessive,* $\alpha \geq 0$, for X if it has the following two properties:

($\star$) $e^{-\alpha t}\mathbf{E}_x(h(X_t)) \leq h(x)$ for all $x \in I$ and $t \geq 0$,

($\star$) $e^{-\alpha t}\mathbf{E}_x(h(X_t)) \to h(x)$ for all $x \in I$ as $t \downarrow 0$.

An α-excessive function h is called *α-invariant* if for all $x \in I$ and $t \geq 0$,

$$e^{-\alpha t}\mathbf{E}_x(h(X_t)) = h(x).$$

A 0-excessive function is simply called *excessive* and, similarly, a 0-invariant function is called *invariant.* Recall the convention that a function f defined on I is extended to $I^\partial := I \cup \{\partial\}$ by setting $f(\partial) = 0$

29. A regular diffusion is *standard* in the sense of Dynkin (1965) Vol I p. 104 and Blumenthal and Getoor (1968). Consequently, using Dynkin (1965) Vol II Theorems 15.5 p. 134 and 12.4 p. 7 (see also Blumenthal and Getoor (1968) p. 93) it is seen that a non-negative function h is α-excessive if and only if h is continuous and satisfies for all $x \in I$,

(a) $$\mathbf{E}_x(e^{-\alpha T}h(X_T)) \leq h(x),$$

where $T := \inf\{t\ :\ X_t \in \Gamma\}$ and Γ is an arbitrary compact subset of I (recall the convention $h(X_T) = 0$ if $T = \infty$). Because $X_{H_z} = z$ a.s. when $H_z < \infty$, we obtain from No. 10 and (a) that an α-excessive function h satisfies the inequality

$$\frac{h(x)}{h(z)} \geq \begin{cases} \dfrac{\psi_\alpha(x)}{\psi_\alpha(z)}, & x \leq z, \\[2ex] \dfrac{\varphi_\alpha(x)}{\varphi_\alpha(z)}, & x \geq z. \end{cases}$$

From (a) and because X is assumed to be regular it follows that, if h is α-excessive and $h(x) = 0$ for some $x \in I$, then $h \equiv 0$. Further, if h is α-excessive and $h(x) = \infty$ for some $x \in I$, then $h \equiv \infty$ on I. These two functions are called *trivial* α-excessive functions.

If X is recurrent (see No. 12), then a non-negative measurable function h is excessive if and only if it is constant on I. The following converse is also true: if X is a diffusion such that every excessive function for X is constant on I, then X is recurrent. See Getoor (1980) for a discussion of recurrence and transience for general Markov processes.

We remark also that if h is excessive (invariant), then, by the Markov property, $\{h(X_t) \ : \ t \geq 0\}$ is a positive supermartingale (martingale) with respect to the natural filtration of X. In particular, it follows from the martingale convergence theorem that $\lim_{t\to\infty} h(X_t)$ exists a.s., and from the optional stopping theorem

$$\mathbf{E}_x(h(X_T) \leq \mathbf{E}_x(h(X_S))$$

for all stopping times S and T such that $S \leq T$.

30. Using (a) in No. 29 it is seen that for every $y \in (l, r)$ the function $x \mapsto G_\alpha(x, y)$ is α-excessive, as are $x \mapsto \psi_\alpha(x)$ and $x \mapsto \varphi_\alpha(x)$.

These functions are *minimal* in the sense that an arbitrary non-trivial α-excessive function h can be represented as a linear combination of them. To give the exact statement let $x_o \in I$ be given and consider a non-negative measurable function h on I such that $h(x_o) = 1$. Then h is α-excessive if and only if there exists a probability measure μ on $[l, r]$ such that for all $x \in I$,

$$\text{(a)} \qquad h(x) = \int_{(l,r)} \frac{G_\alpha(x,y)}{G_\alpha(x_o,y)}\mu(dy) + \frac{\varphi_\alpha(x)}{\varphi_\alpha(x_o)}\mu(\{l\}) + \frac{\psi_\alpha(x)}{\psi_\alpha(x_o)}\mu(\{r\}).$$

The measure μ is called the *representing measure* of h.

From (a) it can be deduced (see Salminen (1985)) that α-excessive functions of regular diffusions are continuous. Moreover, if φ_α and ψ_α are differentiable at a given point $z \in I$ then for any α-excessive function h it holds that h^+ and h^- exist and satisfy

$$h^-(z) \geq h^+(z).$$

Furthermore, h is differentiable with respect to the scale function if and only if the representing measure of h does not charge z.

In general, the idea to represent α-excessive functions in terms of the minimal ones comes from classical potential theory. The corresponding problem therein is to represent positive harmonic functions on Greenian subsets of $\mathbb{R}^n$ in terms of the minimal harmonic functions. See Doob (1984) Chapter XII: The Martin Boundary, p. 195–225, Bass (1995), and Chung and Walsh (2005) Chapter 14. One-dimensional diffusions are especially treated in, e.g., Lai (1973) and Salminen (1984, 1985)).

31. Let h, x_o, and μ be as in No. 30. Recall that the sample space of X is the canonical space $(I^{\partial,\infty}, \mathcal{F}^c_\infty)$. The elements of this space are denoted by ω and they are functions mapping $[0,\infty)$ to I^∂ (see I.12). Using the probability measure $\mathbf{P}$ associated to X we construct a new probability measure $\mathbf{P}^h$ by setting, for $A \in \mathcal{F}^c_t \cap \{\zeta > t\}$,

$$\mathbf{P}^h_x(A) := e^{-\alpha t}\mathbf{E}_x\Big(\frac{h(\omega(t))}{h(x)};\, A\Big).$$

Let T be an $\mathcal{F}^c_{t+}$-stopping time and recall the definition of $\mathcal{F}^c_{T+}$ in No. 1. Then for $A \in \mathcal{F}^c_{T+} \cap \{\zeta > T\}$,

$$\mathbf{P}^h_x(A) = \mathbf{E}_x\Big(e^{-\alpha T}\frac{h(\omega(T))}{h(x)};\, A\Big).$$

The co-ordinate process under the measure $\mathbf{P}^h$ is called *Doob's h-transform* or an *α-excessive transform* of X. It is a regular diffusion and we let X^h denote this process.

Let m be the speed measure of X and p the transition density with respect to m. Then the semigroup associated to X^h is

$$\begin{aligned} P^h_t f(x) &:= \mathbf{E}^h_x(f(\omega(t))) = e^{-\alpha t}\mathbf{E}_x\Big(f(\omega(t))\frac{h(\omega(t))}{h(x)}\Big) \\ &= \int_I e^{-\alpha t} p(t;x,y) f(y)\,\frac{h(y)}{h(x)}\,m(dy), \quad f \in \mathcal{C}_b(I). \end{aligned}$$

In many cases, we can readily check that the basic characteristics of X^h are given as follows:

(a) speed measure $m^h(dy) = h^2(y)m(dy)$,

(b) scale measure $s^h(dy) = h^{-2}(y)s(dy)$,

(c) killing measure $k^h(dy) = (G^h(x_o,y))^{-1}\mu(dy)$, $y \in I$, $G^h(x,y) := \dfrac{G_\alpha(x,y)}{h(x)h(y)}$.

The transition density p^h of X^h with respect to m^h is given by

$$p^h(t;x,y) := \frac{e^{-\alpha t}p(t;x,y)}{h(x)h(y)},$$

and, hence, G^h as defined above is the Green function of X^h. We refer to Jacobsen (1974), Williams (1974), and Salminen (1984), where h-transforms based on the scale function and the Green kernel are appearing and analysed. For the connection between the representing measure of an excessive function and the terminal distribution of the corresponding h-transform in a general Markovian setting, see Kunita and Watanabe (1965). For diffusions the terminal distribution can be expressed, on the other hand, in terms of the killing measure as explained in No. 26. Combining these two approaches results to the formula for the klling measure of the h-transform presented in (c). See also Langer and Schenk (1990).

We stress furthermore that the boundary behaviors of X and X^h can be very different. For example, an entrance boundary of X might turn into a non-singular killing boundary for X^h.

32. Constructing an excessive transform of a diffusion, or more generally of a Markov process, can be viewed as conditioning the process to behave in some desired fashion.

For example, assume that the diffusion X is transient but $k \equiv 0$. Then

$$\mathbf{P}_x\Big(\lim_{t\to\zeta} X_t = l \text{ or } r\Big) = 1.$$

Now let $(X^+, \mathbf{P}^+_\cdot)$ be the excessive transform of X obtained using the function ψ_0. It can be proved that this ψ_0-transform has the property

$$\mathbf{P}^+_x\Big(\lim_{t\to\zeta} X^+_t = r\Big) = 1.$$

If $\mathbf{P}_x(\lim_{t\to\zeta} X_t = r) > 0$, the function $\psi_0(x)$ is a constant multiple of this probability, and consequently, X^+ is identical in law to X given that $\lim_{t\to\zeta} X(t) = r$.

An illustrative case is when X is a Brownian motion with drift $\mu < 0$. The ψ_0-transform is a Brownian motion with drift $-\mu > 0$. For examples of excessive transforms of other types see IV.23. See also Doob (1984) Chapter X.

33. Excessive transforms also arise naturally when we are considering *time reversed* processes.

Let X be as in No. 32. Fix two points $a, b \in I$ and let $h_a := G_0(\cdot, a)$ and $h_b := G_0(\cdot, b)$. Furthermore, let $(X^a, \mathbf{P}^a_\cdot)$ and $(X^b, \mathbf{P}^b_\cdot)$ denote the h_a- and h_b-transforms of X, respectively. Then

$$\{X^a_t,\ 0 < t < \zeta;\ \mathbf{P}^a_b\} \ \sim\ \{X^b_{\zeta-t},\ 0 < t < \zeta;\ \mathbf{P}^b_a\}.$$

We remark also that

$$\{X^a_t,\ 0 \le t < \zeta;\ \mathbf{P}^a_a\} \ \sim\ \{X_t,\ 0 \le t < \lambda_a;\ \mathbf{P}_a\},$$

where $\lambda_a := \sup\{s:\ X_s = a\}$ is the last exit time at a (cf. No. 20).

An important special case is obtained when we take X to be a Brownian motion living in $\mathbb{R}_+$ and killed at the first hitting time to zero. Then, letting $a = 0$ and $b > 0$, we have $h_0 \equiv 1$ and $h_b(x) = x \wedge b$. Hence, the time reversal of X, we assume that $X_0 = b$, is identical in law to the h_b-transform of X (started at 0). Simple computations show further that this transform is identical in law to a 3-dimensional Bessel process killed at the last exit time λ_b. More generally, for $\nu > 0$ let $R^{(\nu)}$ be a Bessel process of order ν (see IV.39) starting at 0 and $R^{(-\nu)}$ a Bessel process of order $-\nu$ starting at a point $a > 0$; if $-1 < -\nu$ we let 0 be a killing boundary for $R^{(-\nu)}$. Then

$$\{R^{(\nu)}_{\lambda_a - t}\ :\ 0 \le t \le \lambda_a\} \quad \sim \quad \{R^{(-\nu)}_t\ :\ 0 \le t \le H_0\}.$$

We refer to Williams (1974) for additional information, see also Meyer, Smythe and Walsh (1971), Jacobsen (1974), Pitman and Yor (1981), Sharpe (1980), Salminen (1984). These results can also be obtained in the general time reversal framework of Nagasawa (1964).

6. Ergodic results

34. Let X be a recurrent diffusion and $\{L(t,x): x \in I, t \geq 0\}$ its local time (with respect to the speed measure). Let μ_1 and μ_2 be two measures on $\mathcal{B}(I)$ and introduce, for $i = 1, 2$,

$$L^{(i)}(t) := \int_I L(t,x)\,\mu_i(dx).$$

Then in the case $0 < \mu_2(I) < \infty$ we have (Itô and McKean (1974) p. 228–229, Revuz and Yor (2001) p. 427, Rogers and Williams (2000) p. 300) a.s.

$$\lim_{t\to\infty} \frac{L^{(1)}(t)}{L^{(2)}(t)} = \frac{\mu_1(I)}{\mu_2(I)}.$$

35. We present here a number of special cases of the general result in No. 34. All limits below exist a.s.

Consider first the case in which $\mu_i(dx) = f_i(x)m(dx)$, where $f_i \in B_{\mathcal{B}}(I)$, $f_i \geq 0$, and m is the speed measure of X. Due to the occupation time formula (a) in No. 13 we have

$$L^{(i)}(t) = \int_0^t f_i(X_s)ds.$$

Consequently, if $\int_I f_2(x)\,m(dx) < \infty$,

(a) $$\lim_{t\to\infty} \frac{\int_0^t f_1(X_s)ds}{\int_0^t f_2(X_s)ds} = \frac{\int_I f_1(x)m(dx)}{\int_I f_2(x)m(dx)}.$$

Letting $A_i \in \mathcal{B}(I)$ and taking $f_i = \mathbb{1}_{A_i}$ in (a) gives in the case $0 < m(A_2) < \infty$,

$$\lim_{t\to\infty} \frac{\text{Leb}\{s<t:\ X_s \in A_1\}}{\text{Leb}\{s<t:\ X_s \in A_2\}} = \frac{m(A_1)}{m(A_2)}.$$

Let $x \in I$ be given and choose $\mu_2 = \varepsilon_{\{x\}}$, i.e., μ_2 is the Dirac measure at x. Then $L^{(2)}(t) = L(t,x)$ and we have

(b) $$\lim_{t\to\infty} \frac{L^{(1)}(t)}{L(t,x)} = \mu_1(I).$$

Let μ_2 be the Dirac measure at x as in (b) and μ_1 the Dirac measure at another point $y \in I$. Then

$$\lim_{t\to\infty} \frac{L(t,y)}{L(t,x)} = 1.$$

Letting $\mu_2 = m$, the speed measure of X, and using the occupation time formula give $L^{(2)}(t) = t$. Consequently, in the case $\mu_1(I) < \infty$, we have

$$\lim_{t\to\infty} \frac{L^{(1)}(t)}{t} = \frac{\mu_1(I)}{m(I)},$$

where the right-hand side is defined to be zero when $m(I) = \infty$. Taking here μ_1 to be the Dirac measure at y gives

(c) $$\lim_{t\to\infty} \frac{L(t,x)}{t} = \frac{1}{m(I)}.$$

36. Let X be as in No. 34, and let ν_1 and ν_2 be two initial probability distributions. Introduce for $i = 1, 2$ and $A \in \mathcal{B}(I)$,

$$\nu_i(t, A) := \int_I \mathbf{P}_x(X_t \in A)\nu_i(dx), \quad \nu(t, A) := \nu_1(t, A) - \nu_2(t, A).$$

Consider the total variation of the signed measure $\nu(t, \cdot)$:

$$\begin{aligned} \|\nu(t)\| &:= \sup\{|\nu(t, A)| + |\nu(t, I \setminus A)| : \ A \in \mathcal{B}(I)\} \\ &= 2\sup\{|\nu(t, A)| : \ A \in \mathcal{B}(I)\}. \end{aligned}$$

Then

(a) $$\lim_{t\to\infty} \|\nu(t)\| = 0.$$

For a proof, see Lindvall (1983), where rates of convergence are also derived. See also Rogers and Williams (2000) p. 301. We remark that for (a) to hold for a general regular diffusion it is necessary and sufficient that the tail σ-field of the diffusion is trivial. This is the case, e.g., for recurrent diffusions, see Rösler (1979).

Assume now that X is positively recurrent. Then, $m(I) < \infty$ and $\pi(\cdot) := m(\cdot)/m(I)$ is the stationary probability measure of X. Let $x \in I$ be given and choose $\nu_1 = \varepsilon_{\{x\}}$ and $\nu_2 = \pi$. Then the result above gives

$$\lim_{t\to\infty} \|P_t(x, \cdot) - \pi\| = 0.$$

In particular, we have for every $f \in B_{\mathcal{B}}(I)$,

(b) $$\lim_{t\to\infty} \mathbf{E}_x(f(X_t)) = \int_I f(y)\,\pi(dy).$$

37. Let X be positively recurrent. Then using (b) in No. 36 and l'Hopital's rule we have

(a) $$\lim_{t\to\infty} \frac{1}{t}\,\mathbf{E}_x\Big(\int_0^t f(X_s)\,ds\Big) = \int_I f(y)\,\pi(dy), \quad f \in B_{\mathcal{B}}(I).$$

Combining (a) with (a) in No. 35, where we take $f_1 = f$ and $f_2 \equiv 1$, shows that the convergence

$$\lim_{t\to\infty} \frac{1}{t} \int_0^t f(X_s)\,ds = \int_I f(y)\,\pi(dy)$$

takes place also in $\mathbf{L}^1$. We remark that (a) can also be deduced using the Chacon–Ornstein ergodic theorem, see Revuz and Yor (2001) (3.18) Exercise p. 429. From this approach it also follows that

$$\lim_{t\to\infty} \frac{1}{t}\, \mathbf{E}(L(t,x)) = \lim_{t\to\infty} \frac{1}{t} \int_0^t p(s;x,y)\, ds = \frac{1}{m(I)};$$

hence, (c) in No. 35 holds also in $\mathbf{L}^1$. Moreover, it can be proved using spectral representations (see Salminen (1996)) that

$$\lim_{t\to\infty} p(t;x,y) = \frac{1}{m(I)} \quad \forall x,y \in I.$$

CHAPTER III

STOCHASTIC CALCULUS

1. Stochastic integration with respect to Brownian motion

1. We present here very briefly the basic facts of the theory of stochastic integration in the case, where the integrator is a Brownian motion. Let $(\Omega, \mathcal{F}, \{\mathcal{F}_t\}, \mathbf{P})$ be a probability space satisfying the usual conditions (see I.3) and W an $\mathcal{F}_t$-Brownian motion (see IV.1) in this space. The goal is to give some meaning to stochastic integrals of the type

$$\text{(a)}, \qquad \int_0^t f(W_s)\, dW_s,$$

where f is a "smooth" function. Because $t \mapsto W_t$ is of infinite variation on any interval, it is not possible to define such integrals by using classical approaches from the theory of integration. This difficulty was overcome by Kiyosi Itô. His far-reaching basic idea presented in Itô (1944) is that stochastic integrals of the type (a) can be defined via an isometry. The notion to which this approach leads us is called the *Itô integral* and the theory is called the *Itô calculus.* In particular, using the differentiation rule in the Itô calculus (see No. 8), it is seen that for all $t \geq 0$,

$$\int_0^t 2\, W_s\, dW_s = W_t^2 - t \quad \text{a.s.}$$

In this chapter the words "a.s." in the statements of the above kind are usually omitted.

A classical textbook on Itô's theory is McKean (1969). Our discussion of stochastic integrals follows mainly Chung and Williams (1990).

2. We begin by introducing the so-called *predictable* (or *previsible*) processes, a subset of which appears as the set of integrands in the general theory of stochastic integration. To consider only predictable integrands instead of progressively measurable (see I.4) is not a loss of generality because stochastic integrals with progressively measurable integrands are indistinguishable from integrals with predictable integrands (see Ikeda and Watanabe (1981) p. 45, Revuz and Yor (2001) p. 172).

The *predictable σ-field* $\mathcal{P}$ is the smallest σ-field which contains all sets of the form $\{0\} \times A$ and $(a, b] \times B$, where $0 < a < b$, $A \in \mathcal{F}_0$, and $B \in \mathcal{F}_a$. A function $F : [0, \infty) \times \Omega \mapsto \mathbb{R}$ is called *predictable* if it is $\mathcal{P}$-measurable. Analogously, a stochastic process $X = \{X_t : t \geq 0\}$ is predictable if the mapping $(t,\omega) \mapsto X_t(\omega)$ is predictable. By the monotone class theorem, predictable processes are adapted (and progressively measurable).

We remark that $\mathcal{P}$ is also the smallest σ-field on $[0,\infty) \times \Omega$ making all adapted and (left) continuous processes measurable (Revuz and Yor (2001) p. 171). The smallest σ-field on $[0,\infty)\times\Omega$ making all adapted and right continuous processes measurable is called the *optional* σ-field. Using for this the notation $\mathcal{O}$, we have

$$\mathcal{P} \subset \mathcal{O} \subset Prog,$$

where $Prog$ stands for the σ-field generated by the progressively measurable processes.

Let $(a_i, b_i] \times A_i$, $i = 1, 2, \ldots, n$, be predictable such that $0 < a_1 < b_1 \leq a_2 < b_2 \leq \cdots \leq a_n < b_n$, and define the predictable process $C = \{C_t : t \geq 0\}$ by setting

$$C_t(\omega) = \sum_{i=1}^{n} \lambda_i \mathbb{I}_{(a_i,b_i]}(t)\mathbb{I}_{A_i}(\omega) + \lambda_0 \mathbb{I}_0(t)\mathbb{I}_{A_0}(\omega), \quad \lambda_i \in \mathbb{R}.$$

Processes of this type are called *simple predictable* processes.

3. The stochastic integral of C with respect to W is defined to be

$$\int C_s\, dW_s := \sum_{i=1}^{n} \lambda_i \mathbb{I}_{A_i}(\omega)(W_{b_i}(\omega) - W_{a_i}(\omega)).$$

Because of the simple predictable structure of C we have

$$\mathbf{E}\left(\int C_s\, dW_s\right)^2 = \mathbf{E}\int_0^\infty C_s^2 ds.$$

This formula is of key importance for further development.

4. Let $\mathbf{L}^2$ be the space of square integrable random variables. Then $\mathbf{L}^2$ is a Hilbert space when equipped with the norm

$$\| X \|_2 := (\mathbf{E}X^2)^{1/2}, \quad X \in \mathbf{L}^2.$$

Further, let $\mathcal{L}^2$ be the space of predictable processes such that

(a)
$$\mathbf{E}\int_0^\infty X_s^2\, ds < \infty.$$

Then $\mathcal{L}^2$ is a Hilbert space when equipped with the norm

$$\| X \|_2^\circ := \left(\mathbf{E}\int_0^\infty X_s^2\, ds\right)^{1/2}, \quad X \in \mathcal{L}^2.$$

In particular, if C is a simple predictable process, then $C \in \mathcal{L}^2$ and the mapping

$$\text{(b)} \qquad C \mapsto \int C_s \, dW_s$$

is an *isometry* from a subset of $\mathcal{L}^2$ into $\mathbf{L}^2$. The set of simple predictable processes is dense in $(\mathcal{L}^2, \| \cdot \|_2^\circ)$ (Chung and Williams (1990) Lemma 2.4 p. 37). Consequently, the linear isometry in (b) can be extended uniquely to a linear isometry from the whole of $\mathcal{L}^2$ into $\mathbf{L}^2$, thus defining the stochastic integral of $X \in \mathcal{L}^2$ with respect to the Brownian motion W.

We remark that if $\{C^{(n)}\}$ is a sequence of simple predictable processes converging in norm $\| \cdot \|_2^\circ$ to a predictable process $X \in \mathcal{L}^2$, then the corresponding sequence of stochastic integrals converges in $\mathbf{L}^2$ to the stochastic integral of X.

5. Let $\mathcal{L}_\ell^2$ be the set of predictable processes X such that $Y^{(t)} := X \, \mathbb{1}_{[0,t]} \in \mathcal{L}^2$ for each $t \geq 0$. The stochastic integral of $Y^{(t)}$ is well defined for each t and we denote it by

$$\int_0^t X_s \, dW_s := \int Y_s^{(t)} \, dW_s.$$

By the isometry property we have

$$\int_0^0 X_s \, dW_s := \int Y_s^{(0)} \, dW_s = 0.$$

The set $\mathcal{L}_\ell^2$ of integrands can be further extended by a localization argument. Indeed, let $\mathcal{L}_{loc}^2$ be the set of predictable processes X for which there exists a sequence of stopping times $\{T_k\}$ such that $T_k \uparrow \infty$ and

$$\mathbf{E} \int_0^\infty X_s^2 \, \mathbb{1}_{[0,T_k]} \, ds < \infty,$$

where

$$[0, T_k] := \{(t, \omega) \, : \, 0 \leq t \leq T_k(\omega), \; \omega \in \Omega\}.$$

Then viewing W as a local martingale and recalling that $[0, T_k]$ is predictable (see Chung and Williams (1990) p. 28) the stochastic integral of $X \in \mathcal{L}_{loc}^2$ with respect to W can be introduced on $[0, T_k]$ for every k, similarly as is done above for deterministic t. These different constructions are consistent, and hence lead to the stochastic integral of X on the whole time axis. See, e.g., Chung and Williams (1990) p. 44 or Revuz and Yor (2001) p. 140. We remark (see Rogers and Williams (2000) p. 55) that $\mathcal{L}_{loc}^2$ consists precisely of predictable processes such that for all t a.s.

$$\int_0^t X_s^2 \, ds < \infty.$$

6. We give here some properties of the stochastic integral.

Additivity I. For $X \in \mathcal{L}^2_{loc}$

$$\int_0^{t+s} X_u\, dW_u = \int_0^t X_u\, dW_u + \int_t^{t+s} X_u\, dW_u.$$

Additivity II. Let A and B be a.s. finite $\mathcal{F}_0$−measurable random variables and $X, Y \in \mathcal{L}^2_{loc}$. Then

$$\int_0^t (A\,X_s + B\,Y_s)\, dW_s = A\int_0^t X_s\, dW_s + B\int_0^t Y_s\, dW_s.$$

Covariance. For $X, Y \in \mathcal{L}^2_\ell$

$$\mathbf{E}\left(\int_0^t X_s\, dW_s \int_0^t Y_s\, dW_s\right) = \mathbf{E}\left(\int_0^t X_s Y_s\, ds\right).$$

Martingale property and continuity. For $X \in \mathcal{L}^2_{loc}$

$$X \cdot W := \left\{\int_0^t X_s\, dW_s\, :\, t \geq 0\right\}$$

defines a continuous local martingale. Furthermore, if $X \in \mathcal{L}^2_\ell$, then $X \cdot W$ is a continuous $\mathbf{L}^2$−martingale (Chung and Williams (1990) p. 38).

Riemann sums. Let $X \in \mathcal{L}^2_{loc}$ be left continuous and $\{t_i^{(n)}\, :\, i = 1, 2, \ldots, n\}$ a sequence of subdivisions of the interval $[0, t]$ such that

$$\sup_i (t_i^{(n)} - t_{i-1}^{(n)}) \to 0 \quad \text{as } n \to \infty.$$

Then (Revuz and Yor (2001) p. 142)

$$\int_0^t X_s\, dW_s = \lim_{n\to\infty} \sum_{i=1}^n X_{t_i^{(n)}}(W_{t_{i+1}^{(n)}} - W_{t_i^{(n)}}),$$

where the limit is in probability.

7. Let $\{\mathcal{G}_t^o\}$be the natural completed filtration of Brownian motion W, and let $\mathcal{L}^2(W)$ be the class of predictable processes X with respect to $\{\mathcal{G}_t^o\}$ such that (a) in No. 4 holds. Then the *Itô representation theorem* states (see, e.g., Revuz and Yor (2001) p. 199 or Øksendal (1998) p. 51) that for any square integrable $\mathcal{G}_\infty^o$-measurable random variable F there exists a unique $H \in \mathcal{L}^2(W)$ such that

$$F = \mathbf{E}(F) + \int_0^\infty H_s\, dW_s.$$

Next, let M be a square integrable $\mathcal{G}_t^o$-martingale and $\mathcal{L}^2_\ell(W)$ the class of predictable processes X such that $X\,\mathbb{I}_{[0,t]} \in \mathcal{L}^2(W)$ for all $t \geq 0$. Then there exists a unique $H \in \mathcal{L}^2_\ell(W)$ such that

$$M_t = \mathbf{E}(M_0) + \int_0^t H_s\, dW_s.$$

This result, *the martingale representation theorem,* can also be formulated for local martingales, see Revuz and Yor (2001) p. 199. An important consequence is that all local $\mathcal{G}_t^o$-martingales have continuous versions.

2. The Itô and Tanaka formulae

8. It is often of interest to study the properties of the process $f(W) = \{f(W_t) : t \geq 0\}$, where f is a given smooth function. In the case, where f is twice continuously differentiable, the celebrated *Itô formula* says that

$$f(W_t) - f(W_0) = \int_0^t f'(W_s)\,dW_s + \frac{1}{2}\int_0^t f''(W_s)\,ds.$$

Notice that this is also the semimartingale decomposition of $f(W)$. The stochastic integral is well defined because $f'(W) \in \mathcal{L}^2_{loc}$ due to continuity.

9. For a more general form of the Itô formula consider a real valued function $(x,y) \mapsto f(x,y)$ such that f'_x, f''_{xx}, and f'_y are continuous. Let V be a continuous $\mathcal{F}_t$-adapted process of finite variation on finite intervals. Then

$$\begin{aligned} &f(W_t, V_t) - f(W_0, V_0) \\ &\quad = \int_0^t f'_x(W_s, V_s)\,dW_s + \int_0^t f'_y(W_s, V_s)\,dV_s + \frac{1}{2}\int_0^t f''_{xx}(W_s, V_s)\,ds, \end{aligned}$$

where the second integral on the right-hand side is a Lebesgue–Stieltjes integral.

10. Consider a family of continuous semimartingales $\{Z^{(i)};\ i = 1, 2, \ldots, n\}$ given by

$$Z_t^{(i)} := Z_0^{(i)} + \int_0^t X_s^{(i)}\,dW_s + \int_0^t Y_s^{(i)}\,ds,$$

where $Z_0^{(i)}$ are $\mathcal{F}_0$-measurable random variables, $X^{(i)} \in \mathcal{L}^2_{loc}$, and $Y^{(i)}$ are $\mathcal{F}_t$-adapted and $\mathcal{B}[0,\infty) \times \mathcal{F}$-measurable processes such that for all $t > 0$,

$$\int_0^t |Y_s^{(i)}|\,ds < \infty.$$

Let $f : \mathbb{R}_+ \times \mathbb{R}^n \mapsto \mathbb{R}$ be a continuous function of $(t, x_1, \ldots, x_n)$ such that the partial derivatives

$$f'_0 := \frac{\partial f}{\partial t} \quad f'_i := \frac{\partial f}{\partial x_i} \qquad f''_{ij} := \frac{\partial^2 f}{\partial x_i \partial x_j}$$

exist and are continuous. Then setting $\vec{Z}_t := (Z_t^{(1)}, Z_t^{(2)}, \ldots, Z_t^{(n)})$ we have

$$\begin{aligned} f(t,\vec{Z}_t) - f(0,\vec{Z}_0) &= \int_0^t f'_0(s,\vec{Z}_s)\,ds \\ &+ \sum_{i=1}^n \int_0^t f'_i(s,\vec{Z}_s)(X_s^{(i)}\,dW_s + Y_s^{(i)}\,ds) + \frac{1}{2}\sum_{i,j=1}^n \int_0^t f''_{ij}(s,\vec{Z}_s) X_s^{(i)} X_s^{(j)}\,ds. \end{aligned}$$

11. Let $\vec{W} := (W^{(1)}, W^{(2)}, \ldots, W^{(d)})$ be a d-dimensional Brownian motion, that is, $W^{(i)}$ are independent linear Brownian motions. For a continuous function $f : \mathbb{R}^d \mapsto \mathbb{R}$ such that f_i' and f_{ii}'' exist and are continuous for every i, we extend the Itô formula to $\mathbb{R}^d$:

$$\begin{aligned} f(\vec{W}_t) - f(\vec{W}_0) &= \sum_{i=1}^{d} \int_0^t f_i'(\vec{W}_s)\, dW_s^{(i)} + \frac{1}{2} \sum_{i=1}^{d} \int_0^t f_{ii}''(\vec{W}_s)\, ds \\ &= \int_0^t \nabla f(\vec{W}_s)\, d\vec{W}_s + \frac{1}{2} \int_0^t \Delta f(\vec{W}_s)\, ds. \end{aligned}$$

12. Let $\ell(\cdot, x)$ denote the Brownian local time at the point $x \in \mathbb{R}$ such that

$$\ell(t, x) = \lim_{\epsilon \downarrow 0} \frac{1}{2\epsilon} \int_0^t \mathbb{1}_{(x-\epsilon, x+\epsilon)}(W_s)\, ds,$$

where the limit exists a.s. and in $\mathbf{L}^2$ (cf. IV.8). We may and do take ℓ to be jointly continuous and $\mathcal{F}_t$-adapted. Setting $\operatorname{sgn}(y) := \mathbb{1}_{(0,+\infty)}(y) - \mathbb{1}_{(-\infty,0)}(y)$, the *Tanaka formula* says that

$$\text{(a)} \qquad |W_t - x| - |W_0 - x| = \int_0^t \operatorname{sgn}(W_s - x)\, dW_s + \ell(t, x).$$

There is also a one-sided version

$$(W_t - x) \vee 0 - (W_0 - x) \vee 0 = \int_0^t \mathbb{1}_{(0,\infty)}(W_s - x)\, dW_s + \frac{1}{2} \ell(t, x).$$

13. To generalize (a) in No. 12, let F be a linear combination of convex functions. Recall (Roberts and Varberg (1973) p. 23) that this is the case if and only if there exists a function f of finite variation such that

$$F(x) = F(0) + \int_0^x f(y) dy.$$

Then letting $f(dy)$ be the signed measure associated to f via its representation as a difference of two increasing functions, we have

$$F(W_t) - F(W_0) = \int_0^t f(W_s)\, dW_s + \frac{1}{2} \int_{-\infty}^{\infty} \ell(t, y)\, f(dy).$$

Next, consider a function F given by

$$\text{(a)} \qquad F(x) := \int_0^x f(y) dy,$$

where f is locally bounded and measurable. Then (see Bouleau and Yor (1981))

$$F(W_t) = \int_0^t f(W_s)\,dW_s - \frac{1}{2}\int_{-\infty}^{\infty} f(y)\,\ell(t,dy).$$

Finally, letting F be as in (a) with f locally square integrable, it is proved in Föllmer, Protter and Shiryayev (1995) that

$$F(W_t) = \int_0^t f(W_s)\,dW_s + \frac{1}{2}[f(W),W]_t,$$

where

$$[f(W),W]_t := \lim_{n\to\infty}\sum_{i=1}^{n}(f(W_{t_{i+1}^{(n)}}) - f(W_{t_i^{(n)}}))(W_{t_{i+1}^{(n)}} - W_{t_i^{(n)}})$$

and the limit exists in probability along a sequence of subdivisions $\{t_i^{(n)}\}$ as introduced in No. 6. In Föllmer and Protter (2000) this result is generalised for multidimensional Brownian motion.

3. Stochastic differential equations – strong solutions

14. Let $(\Omega, \mathcal{F}, \mathbf{P})$ be a probability space on which a d–dimensional Brownian motion $W = \{W_t : \ t \geq 0\}$, and a finite-valued random variable $\xi \in \mathbb{R}^r$ with law μ are given. Assume that W is started from the origin and that ξ and W are independent. Introduce for $t \geq 0$,

$$\mathcal{F}_t := \sigma\{\xi,\ \{W_s : \ s \leq t\}\},$$

and let $(\Omega, \mathcal{G}, \{\mathcal{G}_t\}, \mathbf{P})$ be the usual augmentation of $(\Omega, \mathcal{F}, \{\mathcal{F}_t\}, \mathbf{P})$. Then, following Rogers and Williams (2000) p. 125, the object $(\Omega, \mathcal{G}, \{\mathcal{G}_t\}, \mathbf{P}, W, \xi)$ is called a *minimal set-up*.

Let $a : \mathbb{R}_+ \times \mathbb{R}^r \mapsto \mathbb{R}^r \times \mathbb{R}^d$ and $b : \mathbb{R}_+ \times \mathbb{R}^r \mapsto \mathbb{R}^r$ be Borel-measurable functions. A process $X,\ X_0 = \xi$, on the set-up $(\Omega, \mathcal{G}, \{\mathcal{G}_t\}, \mathbf{P})$ taking values in $\mathbb{R}^r$ is said to be a *strong non-exploding solution* with the initial distribution μ of the stochastic differential equation (SDE)

$$dX_t = a(t,X_t)dW_t + b(t,X_t)dt \qquad (\dagger)$$

if

(a) X is $\mathcal{G}_t$–adapted,

(b) $\int_0^t (|a(s,X_s)|^2 + |b(s,X_s)|)\,ds < \infty$ for all t a.s.,

(c) $X_t = \xi + \int_0^t a(s,X_s)\,dW_s + \int_0^t b(s,X_s)\,ds$ for all t a.s.

Note that due to (b) the integrals in (c) are well defined. It is also clear from (c) that X is a continuous semimartingale – in fact, a (non-time-homogeneous) diffusion. Non-exploding means that X is well defined for all t. The notation $|\cdot|$ stands for the usual Euclidian norm, i.e., if $A = (A_{i,j})$ is an $r \times d$ matrix, then

$$|A|^2 := \sum_{i,j} |A_{i,j}|^2.$$

Strong uniqueness is said to hold for the equation (†) if any two strong solutions in any given minimal set-up are indistinguishable or equivalently, because of the continuity, modifications of each other. Observe that to verify strong uniqueness we have to consider all minimal set-ups.

Remark. Introducing the notion of strong uniqueness we follow the approach in Karatzas and Shreve (1991) Chapter 5. It is also common to work with the more general notion of pathwise uniqueness (see No. 18).

15. Assume that the functions a and b satisfy the "global" Lipschitz condition

(a) there exists a constant C such that for all t and $x, y \in \mathbb{R}^r$

$$|a(t,x) - a(t,y)| + |b(t,x) - b(t,y)| \le C|x - y|,$$

and the boundedness condition

(b) for all $T > 0$ there exists a constant C_T such that for all $t \le T$

$$|a(t,0)| + |b(t,0)| \le C_T.$$

Under these conditions the SDE (†) has a unique strong non-exploding solution, see Itô (1951) and, e.g., Rogers and Williams (2000) p. 128, where the SDE (†) is in this case said to be pathwise exact.

We remark also that in this case the strong solution X can be expressed a.s. as

$$X = F(\xi, W),$$

where F is a measurable mapping from $\mathbb{R}^r \times \mathcal{C}(\mathbb{R}_+, \mathbb{R}^d)$ to $\mathcal{C}(\mathbb{R}_+, \mathbb{R}^r)$ (for a proof see Rogers and Williams (2000) p. 136; for the definition of measurability in this context see ibid. p. 127).

16. Assume that the functions a and b satisfy the "local" Lipschitz condition

(a) for every $N > 0$ there exists a constant C_N such that for all t and $|x|, |y| \le N$,

$$|a(t,x) - a(t,y)| + |b(t,x) - b(t,y)| \le C_N|x - y|,$$

and the growth condition

(b) for all $T > 0$ there exists a constant C_T such that for all $t \le T$ and $x \in \mathbb{R}^r$,

$$|a(t,x)| + |b(t,x)| \le C_T(1 + |x|).$$

Under these conditions the SDE (†) has a unique strong non-exploding solution (Rogers and Williams (2000) p. 132). The condition (b) is needed to guarantee the solution to be non-exploding. Without (b) the equation still has a unique strong solution up to the eventual explosion time.

4. Stochastic differential equations – weak solutions

17. We say that the SDE (†) in No. 14 has a weak solution (X, W) with initial distribution μ if there exists a filtered probability space $(\Omega, \mathcal{F}, \{\mathcal{F}_t\}, \mathbf{P})$ satisfying the usual conditions and carrying a Brownian motion W, a continuous semimartingale X, and a random variable ξ having the law μ such that

($\star$) X and W are $\mathcal{F}_t$–adapted,

($\star$) $\int_0^t (|a(s, X_s)|^2 + |b(s, X_s)|)\, ds < \infty$ for all t a.s.,

($\star$) $X_t = \xi + \int_0^t a(s, X_s)\, dW_s + \int_0^t b(s, X_s)\, ds$ for all t a.s.

It is clear that a strong solution is always also a weak solution. See Ikeda and Watanabe (1981) p. 152, Barlow (1982), (1988) for examples showing that the converse is not true.

18. Let (X, W) and (X', W') be any two weak solutions of the SDE (†) defined on the same filtered probability space such that $W = W'$ and $X_0 = X_0'$ a.s. Then we say that *pathwise uniqueness* holds for (†) if X and X' are indistinguishable.

Let (X, W) and (X', W') be any two weak solutions of (†) with initial distribution μ. Then we say that *uniqueness in law* holds for (†) if X and X' are versions of each other, i.e., have the same finite dimensional distributions.

It can be proved that pathwise uniqueness implies uniqueness in law. Moreover, given that a weak solution exists for every initial distribution μ and that pathwise uniqueness holds, then there also exists a unique strong solution. These results are due to Yamada and Watanabe (1971); for proofs see Ikeda and Watanabe (1981) p. 149 and 152, and Rogers and Williams (2000) p. 151.

19. Assume that the functions a and b in the SDE (†) are jointly continuous and bounded in all variables (in the Euclidian norm $|\cdot|$), and that μ has a compact support. Then the equation (†) has a non-exploding weak solution (X, W) with initial distribution μ (Ikeda and Watanabe (1981) p. 155).

20. Assume that a and b in the SDE (†) do not depend explicitly on t. Let $a^\star$ be the transpose of the (matrix) function a, and suppose that $a \cdot a^\star$ is bounded, continuous and uniformly positive definite, i.e., there exists $\alpha > 0$ such that for all $x, \lambda \in \mathbb{R}^r$,

$$\lambda^\star \cdot a(x) \cdot a^\star(x) \cdot \lambda \geq \alpha\, \lambda^\star \cdot \lambda\ .$$

Assume further that the function b is bounded. Then uniqueness in law holds for the equation (†). This is due to Stroock and Varadhan (1969); see Ikeda and Watanabe (1981) p. 171.

5. One-dimensional stochastic differential equations

21. Let us take $d = r = 1$ in the SDE (†), and assume that a and b therein are bounded. Then pathwise uniqueness holds if the following conditions are fulfilled:

(a) there exists an increasing function $\rho : [0,\infty) \mapsto [0,\infty)$ such that $\rho(0) = 0$ and for all $\epsilon > 0$,

$$\int_0^\epsilon \rho^{-2}(u)\,du = \infty,$$

and for every t, and $x, y \in \mathbb{R}$,

$$|a(t,x) - a(t,y)| \le \rho(|x-y|),$$

(b) there exists a non-decreasing concave function $\kappa : [0,\infty) \mapsto [0,\infty)$ such that $\kappa(0) = 0$ and for all $\epsilon > 0$,

$$\int_0^\epsilon \kappa^{-1}(u)du = \infty,$$

and for every t, and $x, y \in \mathbb{R}$,

$$|b(t,x) - b(t,y)| \le \kappa(|x-y|).$$

If, moreover, a and b are continuous, the equation (†) has a unique strong solution (cf. No. 18 and 19). For a proof see Yamada and Watanabe (1971). In Ikeda and Watanabe (1981) p. 168 a proof in the time-homogeneous case is given, and in Karatzas and Shreve (1991) p. 291 the result is discussed under the more restrictive assumption that b is globally Lipschitz in the sense of No. 15 (a).

We remark that pathwise uniqueness still holds without (b) if (a) holds and if there exists $\varepsilon > 0$ such that $|a(t,x)| > \varepsilon$ for all t and x (LeGall (1983)). See also Perkins (1982).

22. Assume, as in No. 21, that a and b are bounded. Then pathwise uniqueness holds if a satisfies the following conditions:

($\star$) there exists $\varepsilon > 0$ such that $a(t,x) > \varepsilon$ for all t and x,

($\star$) there exists a bounded increasing function f such that for all t, x, and y

$$|a(t,x) - a(t,y)|^2 \le |f(x) - f(y)|.$$

See LeGall (1983) or Revuz and Yor (2001) p. 390, where the time-homogeneous case is discussed.

23. Here we assume that a and b are independent of t. In other words, (†) has the form

(a) $$dX_t = a(X_t)dW_t + b(X_t)dt.$$

If b is bounded and measurable and a is Lipschitz continuous such that $a(x) > \varepsilon$ for some $\varepsilon > 0$ and all $x \in \mathbb{R}$ then (a) has a unique strong solution. This due to Zvonkin (1974), see also Rogers and Wiiliams (2000) p. 179.

24. If a and b are continuous and bounded in No. 23 (a) and if $a^2(x) > \varepsilon$ for all x for some $\varepsilon > 0$, then (a) has a weak solution which is unique in law (see No. 19 and 20). This solution is non-exploding and, in fact, a linear diffusion (in the sense of the definition in II.1). The infinitesimal generator is the second order differential operator

$$\mathcal{G}f(x) := \frac{1}{2}a^2(x)\frac{d^2}{dx^2}f(x) + b(x)\frac{d}{dx}f(x)$$

acting on the domain $\mathcal{D} := \{f : f, \mathcal{G}f \in \mathcal{C}_b(\mathbb{R})\}$. Due to the assumptions on a and b the boundary points $\pm\infty$ are natural (cf. II.6). This is an alternative way to express the fact that the solution is non-exploding.

25. Let a be a measurable function and consider the equation

(a) $$dX_t = a(X_t)dW_t.$$

Every solution of (a) is non-exploding (in the case $a \in \mathcal{C}^1(\mathbb{R})$, see McKean (1969) p. 57, and, in general, Karatzas and Shreve (1991) p. 332). Recall that a solution X such that $\mathbf{P}(X_t = \xi \text{ for all } t) = 1$ is called *trivial* and otherwise *non-trivial.* By Engelbert and Schmidt (1985) the SDE (a) has for every initial distribution μ a non-trivial weak solution if and only if $x \mapsto a^{-2}(x)$ is locally integrable. Moreover, uniqueness in law holds if and only if $a^2(x) > 0$ for all x.

26. Assume now that a and b in No. 23 (a) are measurable and satisfy the following conditions:

($\star$) $a^2(x) > 0$ for all $x \in \mathbb{R}$,

($\star$) $x \mapsto a^{-2}(x)$ and $x \mapsto |b(x)|/a^2(x)$ are locally integrable.

Then using the method of *transformation of drift* (cf. No. 27) on the SDE (a) in No. 25 it can be shown that for every initial distribution μ the equation (a) in No. 24 has a weak solution defined up to an explosion time and the solution is unique in law (Engelbert and Schmidt (1984, 1991) or Karatzas and Shreve (1991) p. 341).

6. The Cameron–Martin–Girsanov transformation of measure

27. We consider here W up to a given time T in its canonical framework. For this let $\mathcal{C}_T$ be the space of continuous functions $\omega : [0,T] \mapsto \mathbb{R}$ and $\mathcal{F}_T^c := \sigma\{\omega(s); 0 \le s \le T\}$ the smallest σ-field making all co-ordinate mappings measurable. Introduce also the filtration $\{\mathcal{F}_t^c\}$, $\mathcal{F}_t^c := \sigma\{\omega(s); s \le t\}$. Let $\mathbf{P}$ be a probability measure on $(\mathcal{C}_T, \mathcal{F}_T^c, \{\mathcal{F}_t^c\})$ such that the process W defined for $0 \le t \le T$ by

$$W_t(\omega) := \omega(t), \quad W_0(\omega) := \omega(0) = 0,$$

is a standard Brownian motion. Further, let $(\mathcal{C}_T, \mathcal{F}_T, \{\mathcal{F}_t\}, \mathbf{P})$ be the usual augmentation of $(\mathcal{C}_T, \mathcal{F}_T^c, \{\mathcal{F}_t^c\}, \mathbf{P})$. Then W is a (stopped) Brownian motion also in this larger space.

A probability measure $\mathbf{Q}$ on $(\mathcal{C}_T, \mathcal{F}_T)$ is said to be *absolutely continuous* with respect to $\mathbf{P}$ if for all $A \in \mathcal{F}_T$ such that $\mathbf{P}(A) = 0$ it is also the case that $\mathbf{Q}(A) = 0$. We are interested in the structure of a measure $\mathbf{Q}$, when restricted to $\mathcal{F}_t$, which has this property.

Let $\mu = \{\mu(t) : 0 \le t \le T\}$ be in $\mathcal{L}^2_{loc}$ and introduce

$$Z_t := \exp\left(\int_0^t \mu(s)\, d\omega(s) - \frac{1}{2}\int_0^t \mu^2(s)\, ds\right).$$

Assume that μ is such that $\{Z_t : 0 \le t \le T\}$ is a martingale. This is the case if the Novikov condition

$$\mathbf{E}\Big(\exp\Big(\frac{1}{2}\int_0^T \mu^2(s)ds\Big)\Big) < +\infty$$

holds (see Ikeda and Watanabe (1981) p.142, and Liptser and Shiryaev (1977) Section 6.2 which contains also many illuminating examples).

The *Cameron–Martin–Girsanov theorem* states that there exists a unique measure $\mathbf{Q}$ on $(\mathcal{C}_T, \mathcal{F}_T)$ such that it is absolutely continuous with respect to $\mathbf{P}$ and Z_t is a version of its Radon–Nikodym derivative when the measures are restricted to $\mathcal{F}_t$, i.e.,

$$\frac{d\mathbf{Q}}{d\mathbf{P}}\Big|_{\mathcal{F}_t} = Z_t \qquad \mathbf{P} - \text{a.s.}$$

Further, under $\mathbf{Q}$, the process $\{\omega(t) - \int_0^t \mu(s)\, ds : 0 \le t \le T\}$ is a (stopped) standard Brownian motion.

For the proof see Rogers and Williams (2000) p. 82, where uncompleted filtrations are used and the result is formulated in the case $T = \infty$. In Revuz and Yor (2001) Chapter VIII one can find a generalization to the semi-martingale framework.

28. Consider in $\mathbb{R}$ the SDEs

$$dX_t = a(t, X_t)dW_t + b_1(t, X_t)dt,$$
$$dY_t = a(t, Y_t)dW_t + b_2(t, Y_t)dt,$$

where $0 \le t \le T$ as in No. 27 and $X_0 = Y_0 = 0$. We assume (for simplicity) that the functions a, b_1, and b_2 satisfy the "global" conditions in No. 15. Then both equations have non-exploding strong unique solutions. Let $\mathbf{P}^X$ and $\mathbf{P}^Y$ be the probability measures in $(\mathcal{C}_T, \mathcal{F}_T)$ associated to the solutions X and Y, respectively.

Assume further that

($\star$) $a(t,x) \neq 0$ for all $0 \leq t \leq T$ and $x \in \mathbb{R}$,

($\star$) the function $\alpha(t,\omega(t)) := a^{-1}(t,\omega(t))\big(b_1(t,\omega(t)) - b_2(t,\omega(t))\big)$ is bounded $\mathbf{P}^X$-a.s. for all t,

($\star$) $\{Z_t : 0 \leq t \leq T\}$ is a $\mathbf{P}^X$-martingale, where

$$Z_t := \exp\Big(-\int_0^t a^{-2}(s,\omega(s))\big(b_1(s,\omega(s)) - b_2(s,\omega(s))\big)d\omega(s) + \frac{1}{2}\int_0^t a^{-2}(s,\omega(s)))\big(b_1^2(s,\omega(s)) - b_2^2(s,\omega(s))\big)ds\Big),$$

($\star$) for $i = 1,2$,

$$\mathbf{P}^X\Big(\int_0^T a^{-2}(s,\omega(s))\, b_i^2(s,\omega(s))\, ds < \infty\Big) = 1.$$

Then for every $0 \leq t \leq T$ the measure $\mathbf{P}^Y$ is absolutely continuous with respect to $\mathbf{P}^X$ when restricted to $\mathcal{F}_t$ with the Radon–Nikodym derivative Z_t :

$$\frac{d\mathbf{P}^Y}{d\mathbf{P}^X}\Big|_{\mathcal{F}_t} = Z_t \qquad \mathbf{P}^X - \text{a.s.}$$

For a proof, see Liptser and Shiryaev (1977) p. 275, where also the derivative of $\mathbf{P}^X$ with respect to $\mathbf{P}^Y$ is given. We refer also to Gihman and Skorohod (1979) p. 261.

Notice that the process Z can also be expressed as

$$Z_t = \exp\Big(-\int_0^t \alpha(s,\omega(s))\, d\beta_s - \frac{1}{2}\int_0^t \alpha(s,\omega(s))^2\, ds\Big),$$

where β is a Brownian motion in $(\mathcal{C}_T, \mathcal{F}_T, \mathbf{P}^X)$.

29. To formulate an important special case of the result in No. 28 let X be a diffusion on $\mathbb{R}$ having the generator

$$\frac{1}{2}\frac{d^2}{dx^2} + b(x)\frac{d}{dx},$$

where b is continuous (but not necessarily bounded). Assume also that both $-\infty$ and $+\infty$ are not-exit boundaries, i.e., letting m be the speed measure and s the scale function (see II.9) we have (see II.6) for any $z \in \mathbf{R}$,

(a) $$\int_{(z,+\infty)} m((z,a))s(da) = \int_{(-\infty,z)} m((a,z))s(da) = \infty.$$

This is the one-dimensional version of Khasminskii's test for explosion (see Rogers and Williams (2000) p. 297). Assuming (a) we have for all $t \geq 0$,

$$\mathbf{P}^X\Big(\int_0^t b^2(\omega(s))ds < \infty\Big) = \mathbf{P}\Big(\int_0^t b^2(\omega(s))ds < \infty\Big) = 1,$$

where $\mathbf{P}^X$ and $\mathbf{P}$ are the measures associated to X and the Wiener process, respectively. Using Theorem 7.6 in Liptser and Shiryayev (1977) p. 246 it is seen that $\mathbf{P}^X$ is absolutely continuous with respect to $\mathbf{P}$ when restricted to $\mathcal{F}_t$ with the Radon–Nikodym derivative

$$\left.\frac{d\mathbf{P}^X}{d\mathbf{P}}\right|_{\mathcal{F}_t} = \exp\left(\int_0^t b(\omega(s))\,d\omega(s) - \frac{1}{2}\int_0^t b^2(\omega(s))\,ds\right) \quad \mathbf{P}-\text{a.s.}$$

CHAPTER IV

BROWNIAN MOTION

1. Definition and basic properties

1. Having defined linear diffusions (cf. II.1), *standard one-dimensional Brownian motion* $W = \{W_t : t \geq 0\}$ can be defined as a linear diffusion on $\mathbb{R}$ with the speed measure $m(dx) = 2\,dx$ and the scale function $s(x) = x$. This is equivalent to saying that the infinitesimal generator of W is the second order differential operator

$$\mathcal{G}f(x) := \frac{1}{2}\frac{d^2}{dx^2}f(x)$$

acting on the domain $\mathcal{D}(\mathcal{G}) = \{f \in \mathcal{C}_b^2(\mathbb{R}) : f'' \in \mathcal{C}_b(\mathbb{R})\}$. For the basic data of W see A1.1.

From the early days of Brownian motion we mention the papers by Bachelier (1900) and Einstein (1905). See Nelson (1967) for a discussion on Robert Brown's and Albert Einstein's works, and Taqqu (2001) for an article on Louis Bachelier and his times. There is an English translation of Bachelier's dissertation in Cootner (1964). It was, however, Norbert Wiener who in his paper from 1923 "put the Brownian motion on a firm mathematical foundation" (Itô and McKean (1974) p. viii). We refer to Doob (1966) and Kac (1966) for articles concerning Wiener's contributions to Brownian motion and related topics. Much of the modern analysis of Brownian motion is based on the ideas of Paul Lévy; see Loeve (1973) and Schwartz (1988) for memorial articles on Lévy.

There are now several different constructions of Brownian motion, in other words, proofs for its existence; see, e.g., Freedman (1971), Knight (1981), and Karatzas and Shreve (1991). In these the starting point is usually the following definition:

Standard one-dimensional Brownian motion initiated at x on a probability space $(\Omega, \mathcal{F}, \mathbf{P})$ is a stochastic process W such that

(a) $W_0 = x$ a.s.,

(b) $s \mapsto W_s$ is continuous a.s.,

(c) for all $0 = t_0 < t_1 < \cdots < t_n$ the increments

$$W_{t_n} - W_{t_{n-1}}, W_{t_{n-1}} - W_{t_{n-2}}, \ldots, W_{t_1} - W_{t_0}$$

are normally distributed with

$$\mathbf{E}(W_{t_i} - W_{t_{i-1}}) = 0, \qquad \mathbf{E}(W_{t_i} - W_{t_{i-1}})^2 = t_i - t_{i-1}.$$

Because uncorrelated normally distributed random variables are independent it follows from (c) that the increments of Brownian motion are independent.

When W is to be defined on a filtered probability space $(\Omega, \mathcal{F}, \{\mathcal{F}_t\}, \mathbf{P})$ it is stated that W is $\mathcal{F}_t$-adapted, has the properties (a), (b), and (c), where (c) can be reformulated in the following compressed form:

(c') $$\mathbf{E}(\exp(\mathrm{i}\,\mu\,(W_t - W_s))|\mathcal{F}_s) = \exp\Big(-\frac{\mu^2}{2}(t-s)\Big) \quad \text{for all } \mu \in \mathbb{R}.$$

In this setting we often say that W is an $\mathcal{F}_t$-Brownian motion. Of course, W is always an $\mathcal{F}_t^o$-Brownian motion, where $\{\mathcal{F}_t^o\}$ is the natural filtration generated by W. An example (see Yor (1992a) p. 3) of a Brownian motion which is not an $\mathcal{F}_t^o$-Brownian motion is given by

$$\tilde{W}_t := W_t - \int_0^t \frac{W_s}{s}\,ds, \ t > 0, \quad \tilde{W}_0 := 0.$$

Notice that $t \mapsto \tilde{W}_t$ is continuous (use, e.g., the law of iterated logarithm, see No. 5). The fact that $\tilde{W}$ is not an $\mathcal{F}_t^o$-Brownian motion follows from No. 3.

In the sequel we consider only standard Brownian motions, and the word "standard" will be omitted. It is also assumed, if nothing else is said, that W is given in its canonical framework (cf. II.1).

2. Here is a list of basic distributional properties of W. Assume that $W_0 = 0$.

Spatial homogeneity. For every $x \in \mathbb{R}$ the process $x + W$ is a Brownian motion initiated at x.

Symmetry. $-W$ is a Brownian motion.

Reflection principle. Let T be a finite stopping time with respect to $\{\mathcal{F}_{t+}^c\}$. Then the process given by

$$Y_t := \begin{cases} W_t, & t \le T, \\ 2W_T - W_t, & t \ge T, \end{cases}$$

is a Brownian motion.

Scaling. For every $c > 0$ the process $\{\sqrt{c}\,W_{t/c} : \ t \ge 0\}$ is a Brownian motion.

Time inversion. The process given by

$$Z_t := \begin{cases} 0, & t = 0, \\ t\,W_{1/t}, & t > 0, \end{cases}$$

is a Brownian motion.

Time reversibility. For a given $t > 0$,

$$\{W_s : \ 0 \le s \le t\} \ \sim \ \{W_{t-s} - W_t : \ 0 \le s \le t\}.$$

3. Lévy's martingale characterization. A continuous real-valued process X in a filtered probability space $(\Omega, \mathcal{F}, \{\mathcal{F}_t\}, \mathbf{P})$ is an $\mathcal{F}_t$-Brownian motion if and only if both X itself and $\{X_t^2 - t : t \geq 0\}$ are $\mathcal{F}_t$-martingales (Revuz and Yor (2001) p. 150).

4. Strong law of large numbers.

$$\lim_{t \to \infty} \frac{W_t}{t} = 0 \quad \text{a.s.}$$

5. Laws of the iterated logarithm (Revuz and Yor (2001) p. 53).

$$\limsup_{t \downarrow 0} \frac{W_t}{\sqrt{2t \ln \ln(1/t)}} = 1 \quad \text{a.s.}$$

$$\liminf_{t \downarrow 0} \frac{W_t}{\sqrt{2t \ln \ln(1/t)}} = -1 \quad \text{a.s.}$$

$$\limsup_{t \to \infty} \frac{W_t}{\sqrt{2t \ln \ln t}} = 1 \quad \text{a.s.}$$

$$\liminf_{t \to \infty} \frac{W_t}{\sqrt{2t \ln \ln t}} = -1 \quad \text{a.s.}$$

6. Local properties of sample paths. For proofs see, e.g., Freedman (1971), Karatzas and Shreve (1991), Knight (1981), Revuz and Yor (2001).

Continuity. The first property, already stated above as a defining property, is that Brownian paths $t \mapsto W_t$ are continuous a.s.

Hölder continuity. In fact, $t \mapsto W_t$ is locally Hölder continuous of order α for every $\alpha < 1/2$. In other words, for all $T > 0$, $0 < \alpha < 1/2$ and almost all ω there exists a constant $C_{T,\alpha}(\omega)$ such that for all $t, s < T$,

$$|W_t(\omega) - W_s(\omega)| \leq C_{T,\alpha}(\omega)|t - s|^{\alpha}.$$

Modulus of continuity. Recall that for a continuous function $f : [0, \infty) \mapsto \mathbb{R}$ its modulus of continuity on the interval $[0, T]$ is a function $\delta \mapsto \varepsilon(\delta)$ such that

$$|f(t_2) - f(t_1)| < \varepsilon(\delta) \quad \text{for all } t_1, t_2 \in \Delta_\delta := \{(s, t) : 0 < |t - s| < \delta,\ s, t < T\}.$$

For Brownian paths

$$\limsup_{\delta \to 0} \sup_{\Delta_\delta} \frac{|W_{t_2} - W_{t_1}|}{\sqrt{2\delta \ln(1/\delta)}} = 1 \quad \text{a.s.}$$

Nowhere differentiability. Brownian paths are a.s. nowhere locally Hölder continuous of order $\alpha > 1/2$. In particular, Brownian paths are nowhere differentiable.

Variation. Brownian paths are of infinite variation on intervals, that is, for every $s \le t$ a.s.

$$\sup \sum |W_{t_i} - W_{t_{i-1}}| = \infty,$$

where the supremum is over all subdivisions $s \le t_1 \le \cdots \le t_n \le t$ of the interval (s,t). On the other hand, let $\Delta_n := \{t_i^{(n)}\}$ be a sequence of subdivisions of $[s,t]$ such that $\Delta_n \subseteq \Delta_{n+1}$ and

$$\lim_{n\to\infty} \max_i |t_i^{(n)} - t_{i-1}^{(n)}| = 0.$$

Then a.s. and in $\mathbf{L}^2$,

$$\lim_{n\to\infty} \sum |W_{t_i^{(n)}} - W_{t_{i-1}^{(n)}}|^2 = t - s.$$

This fact is expressed by saying that the *quadratic variation* of W over $[s,t]$ is $t-s$.

Local maxima and minima. Recall that for a continuous function $f : [0,\infty) \to \mathbb{R}$ a point t is called a point of local (strict) maximum if there exists $\varepsilon > 0$ such that for all $s \in (t-\varepsilon, t+\varepsilon)$ we have $f(s) \le f(t)$ ($f(s) < f(t)$, $s \ne t$). A point of local minimum is defined analogously. Then for almost every $\omega \in \Omega$ the set of points of local maxima for the Brownian path $W_\cdot(\omega)$ is countable and dense in $[0,\infty)$, each local maximum is strict, and there is no interval on which the path is monotone.

Points of increase and decrease. Recall that for a continuous function $f : [0,\infty) \to \mathbb{R}$ a point t is called a point of increase if there exists $\varepsilon > 0$ such that for all $s \in (0,\varepsilon)$ we have $f(t-s) \le f(t) \le f(t+s)$. A point of decrease is defined analogously. Then for almost every $\omega \in \Omega$ the Brownian path $W_\cdot(\omega)$ has no points of increase or decrease.

Level sets. For a given ω and $x \in \mathbb{R}$ let $\mathcal{Z}_x(\omega) := \{t : W_t(\omega) = x\}$. Then a.s. the random set $\mathcal{Z}_x(\omega)$ is unbounded and of the Lebesgue measure 0. It is closed and has no isolated points, i.e., is dense in itself. A set with these properties is called *perfect.* The Hausdorff dimension of $\mathcal{Z}_x$ is $1/2$ (Itô and McKean (1974) p. 50). Its complement $\mathcal{Z}_x^c$ is a countable union of open intervals; so-called excursion intervals.

2. Brownian local time

7. For $t > 0$ let $\mathcal{Z}_0(t) := \mathcal{Z}_0 \cap [0,t]$, where $\mathcal{Z}_0$ is the level set of 0. As explained in No. 6, the Lebesgue measure is not an appropriate tool to measure $\mathcal{Z}_0(t)$. Paul Lévy's ingenious observation is that there exists, however, a nondecreasing (random) function determined by $\mathcal{Z}_0$. This function, called the *Brownian local time at* 0, is an additive functional of Brownian motion and as such unique up to a normalizing constant. Every point in $\mathcal{Z}_0$ is a (left and/or right) strict increase point for the local time and it is constant on every open interval in $\mathcal{Z}_0^c$.

Therefore, the local time at 0 measures the sets $\mathcal{Z}_0(t)$. In fact, it is the exact Hausdorff measure of the level sets (Taylor and Wendel (1966), Perkins (1981)).

To motivate briefly the existence of the local time consider the process

$$W^+ = \{|W_t| : \ t \geq 0\}.$$

This is called *reflecting Brownian motion.* It is a linear diffusion; see A1.2 for its basic characteristics. Let

$$W^\circ = \{\check{M}_t - W_t : \ t \geq 0\},$$

where $\check{M}_t := \sup\{W_s : \ s \leq t\}$. Then by Lévy (for a proof see Itô and McKean (1974) p. 41)

(a) $$W^+ \sim W^\circ.$$

Consider now the set $\mathcal{Z}_0^\circ := \{t : \ \check{M}_t - W_t = 0\}$. If there exists a function determined by $\mathcal{Z}_0^\circ$ with the desired properties, then there exists by (a) also a function determined by $\mathcal{Z}_0 = \{t : \ |W_t| = 0\}$ having the same properties. The claim is that $\check{M} = \{\check{M}_t : \ t \geq 0\}$ is what we want. Because the asserted properties of $\check{M}$ are obvious the real problem is to construct for every t the value of $\check{M}_t$ using only the set $\mathcal{Z}_0^\circ \cap [0,t]$. For this see Itô and McKean (1974) p. 43, Balkema and Chung (1991), and Karatzas and Shreve (1991) Section 6.2.

8. Let $t \mapsto \ell(t,0)$ denote the local time at zero constructed above. It is clear that we can find in a similar way the local time at an arbitrary point x. Lévy also proved that a.s.

(a) $$\ell(t,x) = \lim_{\varepsilon\downarrow 0} \frac{1}{2\varepsilon} \int_0^t \mathbb{I}_{(x-\varepsilon,x+\varepsilon)}(W_s)ds.$$

This gives the occupation time formula

$$\int_0^t \mathbb{I}_A(W_s)\, ds = \int_A \ell(t,x)\, dx, \quad A \in \mathcal{B}(\mathbb{R}).$$

Further, $(t,x) \mapsto \ell(t,x)$ is continuous a.s. This is due to Trotter (1958) (Itô and McKean (1974) p. 63, Ikeda and Watanabe (1981) p. 113). Convergence in (a) holds also in $\mathbf{L}^2$ (Chung and Williams (1990) p. 145).

We restate using our new notation the basic distributional property explained in No. 7,

$$\{(|W_t|, \ell(t,0)) : \ t \geq 0\} \ \sim \ \{(\check{M}_t - W_t, \check{M}_t) : \ t \geq 0\}.$$

Notice also that under the usual conditions $\{\ell(t,x) : \ t \geq 0\}$ can by (a) be taken to be $\mathcal{F}_t$-adapted. For survey articles on Brownian local time, see McKean (1975) and Borodin (1989a).

9. We describe two additional constructions of $\ell(t,0)$ due to Lévy. Assume that $W_0 = 0$.

(A) For $\varepsilon > 0$ let $T(\varepsilon, t)$ be the real number obtained by adding the lengths smaller than ε of all open intervals in the complement of $\mathcal{Z}_0(t)$. Then (Itô and McKean (1974) p. 43 or Karatzas and Shreve (1991) p. 414)

$$\lim_{\varepsilon \downarrow 0} \sqrt{\frac{\pi}{2\varepsilon}}\, T(\varepsilon, t) = \ell(t,0) \qquad \text{a.s.}$$

(B) Introduce for $\varepsilon > 0$ and $n = 2, 3, \dots,$

$$H_\varepsilon^{(1)} := \inf\{s : W_s = \varepsilon\}, \quad H_0^{(1)} := \inf\{s > H_\varepsilon^{(1)} : W_s = 0\},$$

$$H_\varepsilon^{(n)} := \inf\{s > H_0^{(n-1)} : W_s = \varepsilon\}, \quad H_0^{(n)} := \inf\{s > H_\varepsilon^{(n)} : W_s = 0\}.$$

Then $D(\varepsilon, t) := \sup\{n : H_0^{(n)} \le t\}$ is called the number of downcrossings of $(0, \varepsilon)$ up to time t and we have (Itô and McKean (1974) p. 48 and p. 222, Karatzas and Shreve (1991) p. 416; see also Williams (1977) and Walsh (1978))

$$\lim_{\varepsilon \downarrow 0} \varepsilon\, D(\varepsilon, t) = \frac{1}{2}\,\ell(t,0) \qquad \text{a.s.}$$

10. We conclude this subsection with some miscellaneous properties of ℓ.

Semimartingale property. For every $t > 0$ the process $x \mapsto \ell(t,x)$ is a semimartingale, see Perkins (1982).

Scaling. For $c > 0$,

$$\{\sqrt{c}\,\ell(t/c, x/\sqrt{c}) : 0 \le t < \infty, x \in \mathbb{R}\} \sim \{\ell(t,x) : 0 \le t < \infty, x \in \mathbb{R}\}.$$

Laws of the iterated logarithm.

$$\limsup_{t \to \infty} \frac{\ell(t,0)}{\sqrt{2t \ln\ln t}} = 1 \quad \text{a.s.}$$

$$\limsup_{t \to \infty} \frac{\sup_{y \in \mathbb{R}} \ell(t,y)}{\sqrt{2t \ln\ln t}} = 1 \quad \text{a.s.}$$

$$\liminf_{t \to \infty} \frac{\sqrt{\ln\ln t}}{\sqrt{t}} \sup_{y \in \mathbb{R}} \ell(t,y) = \gamma \quad \text{a.s.}$$

For proofs see Kesten (1965). In Csáki and Földes (1986) the value of the constant γ is found to be $j_1\sqrt{2}$, where j_1 is the smallest positive zero of the Bessel function J_0.

Modulus of continuity. The exact modulus of continuity uniform in x of Brownian local time $\ell(t,x)$ with respect to t is (Perkins (1981)):

$$\limsup_{h \downarrow 0} \sup_{(t,x) \in [0,T] \times \mathbb{R}} \frac{|\ell(t+h,x) - \ell(t,x)|}{\sqrt{2h \ln(1/h)}} = 1 \quad \text{a.s.}$$

The exact modulus of continuity uniform in t on intervals $[0,T]$ of Brownian local time $\ell(t,x)$ with respect to x is (Borodin (1985)):

$$\limsup_{\delta \downarrow 0} \sup_{x \in \mathbb{R}} \frac{\sup_{0<t<T} |\ell(t,x+\delta) - \ell(t,x)|}{2((\ell(T,x) + \ell(T,x+\delta))\delta \ln(1/\delta))^{1/2}} = 1 \quad \text{a.s.}$$

3. Excursions

11. In this section we study excursions of Brownian motion W which start and end at zero. Our basic reference is Revuz and Yor (2001) Chapter XII. We refer also to Blumenthal (1992) and Yor (1995).

Let E be the set which consists of continuous functions $f : [0, \infty) \mapsto \mathbb{R},\ f(0) = 0$, with the property

$$\forall f \in E\ \exists \zeta = \zeta(f) < \infty : \ f(s) = 0 \text{ for all } s \geq \zeta(f) \\ \text{and } f(s) \neq 0 \text{ for all } 0 < s < \zeta(f).$$

Let δ denote the function which is identically equal to 0. It is assumed that $\delta \in E$. The elements in E are called excursions and the function $\zeta : E \mapsto [0, \infty)$ is called the *excursion length.* Let $\mathcal{E}$ be the smallest σ-algebra on E making all co-ordinate mappings measurable. The measurable space $(E, \mathcal{E})$ is called the *excursion space.*

Let ℓ be the Brownian local time at 0 with the properties given in No. 8 and

$$\varrho_t := \inf\{s : \ \ell(s, 0) > t\}$$

its right continuous inverse. The *excursion process* $\Xi = \{\Xi_s : \ s \geq 0\}$ for the excursions of W from 0 to 0 is a stochastic process taking values in $(E, \mathcal{E})$ and defined on the sample space of W as follows

(a) if $l_s := \varrho_s - \varrho_{s-} > 0$, then Ξ_s is the mapping $r \mapsto \mathbb{I}_{[0,l_s]}(r) W(\varrho_{s-} + r)$,

(b) if $l_s = 0$ or if the mapping in (a) is not in E, then $\Xi_s(\omega) = \delta$.

Notice that ϱ_t is for every t an $\{\mathcal{F}_s\}$-stopping time, and let $\mathcal{F}_{\varrho_t}$ be the σ-algebra as defined in I.5 using ϱ_t as T. Then the famous result due to Itô (1970) is that Ξ is a stationary $\mathcal{F}_{\varrho_t}$-*Poisson point process.* The proof is based on the strong Markov property, see, e.g., Revuz and Yor (2001) p. 481.

12. A time-homogeneous Poisson point process is completely characterized by its mean measure. The mean measure of Ξ is called the *Itô measure* (or *characteristic measure*) and we denote it by $\mathbf{n}$. For $A \in \mathcal{E}$ and $s < t$ let

$$N(A; s, t) := |\{(u, \Xi_u) : \ u \in (s, t],\ \Xi_u \in A\}|,$$

that is, N is the number of points in the set in the braces. Then

$$\mathbf{E}(N(A; s, t)) = (t - s)\, \mathbf{n}(A)$$

(see Ikeda and Watanabe (1981) p. 43). The Itô measure $\mathbf{n}$ is an (infinite) σ-finite measure. To characterize it let $\hat{p}$ be the transition density of the diffusion obtained from W by killing at the first hitting time H_0 :

$$\hat{p}(t; x, y) = \begin{cases} \dfrac{1}{\sqrt{2\pi t}} \Big(\exp\Big(-\dfrac{(x - y)^2}{2t} \Big) - \exp\Big(-\dfrac{(x + y)^2}{2t} \Big) \Big), & xy > 0, \\ 0, & xy \leq 0. \end{cases}$$

Recall also that

$$n_x(0,t) := \mathbf{P}_x(H_0 \in dt)/dt = \frac{|x|}{\sqrt{2\pi t^3}} \exp\Big(-\frac{x^2}{2t}\Big).$$

Then for $0 < t_1 < \cdots < t_k$, $A_i \in \mathcal{B}(\mathbb{R})$, $i = 1, 2, \ldots, k$, and

$$\Gamma := \{f \in E : \ f(t_1) \in A_1, \ldots, f(t_k) \in A_k\} \in \mathcal{E}$$

we have

$$\text{(a)} \qquad \mathbf{n}(\Gamma) = \int_{A_1} dx_1 \cdots \int_{A_k} dx_k \, n_{x_1}(0, t_1)\, \hat{p}(t_2 - t_1; x_1, x_2) \cdot \ldots \cdot \hat{p}(t_k - t_{k-1}; x_{k-1}, x_k),$$

see Revuz and Yor (2001) p. 494.

13. We remark also that from a realization of Ξ we can construct the corresponding realization of W. This is achieved by first setting

$$\varrho_t := \sum_{s \le t} \zeta(\Xi_s), \quad \varrho_{t-} := \sum_{s < t} \zeta(\Xi_s).$$

Next let $\Xi_s(r)$ be the value of Ξ_s at time $0 \le r < \infty$. Then for $t \ge 0$

$$W_t := \Xi_s(t - \varrho_{s-}), \quad \varrho_{s-} \le t \le \varrho_s,$$

defines a Brownian motion and the continuous inverse of ϱ gives its local time at 0 (see Revuz and Yor (2001) p. 482).

14. In No. 15 and 16 below two explicit characterizations of $\mathbf{n}$ are discussed in more detail. These are called Itô's and Williams' descriptions, respectively. There is also a description due to Bismut (1985); this is presented in Revuz and Yor (2001) p. 502. We also refer to Biane and Yor (1987) for further developments and as a source for many formulae associated to the excursion theory of Brownian motion.

It is convenient to study separately positive and negative excursions. Therefore introduce

$$E^+ := \{f \in E : \ f(s) \ge 0 \text{ for all } s\} \cup \{\delta\},$$

$$E^- := \{f \in E : \ f(s) \le 0 \text{ for all } s\} \cup \{\delta\},$$

and let $\mathbf{n}^+ := \mathbf{n}(\cdot \cap E^+)$ and $\mathbf{n}^- := \mathbf{n}(\cdot \cap E^-)$ be the restrictions of $\mathbf{n}$ on E^+ and E^-, respectively.

15. Here we take a closer look at the Itô measure $\mathbf{n}$ from the point of view of the lengths of the excursions. We call this Itô's description. It is also in the spirit of Lévy (1948), pp. 225–237, see also Itô and McKean (1974) pp. 75–79 and Revuz and Yor (2001).

In (a) in No. 12 let $n = 1$, $A_1 = \mathbb{R}$ and integrate over x_1. This gives

$$\mathbf{n}(\zeta > t) = \int_{\mathbb{R}} n_x(0,t)\,dx = \frac{2}{\sqrt{2\pi t}}.$$

Because excursions take place during the constant periods of ℓ this formula is also obtained directly from the Lévy measure in the Lévy–Khintchine representation of ϱ_t (cf. II.14; notice, however, that the local time is here taken with respect to the Lebesgue measure):

$$\begin{aligned}\mathbf{E}_0(\exp(-\lambda\varrho_t)) &= \exp\Big(-t\int_0^\infty (1-e^{-\lambda l})\frac{1}{\sqrt{2\pi l^3}}dl\Big)\\ &= \exp(-t\sqrt{2\lambda}).\end{aligned}$$

Due to the spatial symmetry of W we have

$$\text{(a)} \qquad \mathbf{n}^+(\zeta > t) = \mathbf{n}^-(\zeta > t) = \frac{1}{\sqrt{2\pi t}}.$$

To proceed we need the notion of *excursion bridge.* Let $\hat{W}$ be a Brownian motion on $(0,\infty)$ killed when it hits 0. Let $\mathbf{P}_{y,l,0}$ be the probability measure associated to $\hat{W}$ when started at y and conditioned (in the sense of Doob's h-transforms) to hit 0 at time l. In other words with canonical notations, letting for $0 < t_1 < \cdots < t_k < l$ and $A_i \in \mathcal{B}((0,\infty))$,

$$\Gamma_l := \{\omega :\ \omega_{t_1} \in A_1, \dots, \omega_{t_k} \in A_k\},$$

we have

$$\begin{aligned}\mathbf{P}_{y,l,0}(\Gamma_l) = \int_{A_1} dx_1 \cdots \int_{A_k} dx_k\, \hat{p}(t_1; y, x_1)&\\ \cdot \dots \cdot \hat{p}(t_k - t_{k-1}; x_{k-1}, x_k)\, \frac{n_{x_k}(0, l-t_k)}{n_y(0,l)}&.\end{aligned}$$

The process governed by $\mathbf{P}_{y,l,0}$ is a non-time-homogeneous strong Markov process. The family of measures $\{\mathbf{P}_{y,l,0} :\ y > 0\}$ can be extended by defining

$$\mathbf{P}^+_{0,l,0} := \lim_{y\downarrow 0} \mathbf{P}_{y,l,0}.$$

The limit is in the sense of the convergence of finite dimensional distributions. The process governed by $\mathbf{P}^+_{0,l,0}$ is called the *positive excursion bridge of length l* for the excursions from 0. It has the finite dimensional distributions

$$\begin{aligned}\mathbf{P}^+_{0,l,0}(\Gamma_l) = \int_{A_1} dx_1 \cdots \int_{A_k} dx_k\, n_{x_1}(0,t_1)\, \hat{p}(t_2 - t_1; x_1, x_2)&\\ \cdot \dots \cdot \hat{p}(t_k - t_{k-1}; x_{k-1}, x_k)\, \frac{n_{x_k}(0, l-t_k)}{n^0_+(l)}&\end{aligned}$$

with

$$n^0_+(l) := \mathbf{n}^+(\zeta \in dl)/dl = \frac{1}{2\sqrt{2\pi l^3}}.$$

Then for $\Gamma^+ := \{f \in E^+ : f(t_1) \in A_1, \dots, f(t_k) \in A_k\}$ Itô's description of $\mathbf{n}^+$ is

$$\begin{aligned}\mathbf{n}^+(\Gamma^+) &= \int_{t_k}^{\infty} \mathbf{n}^+(\Gamma^+; \zeta \in dl)\\ &= \int_{t_k}^{\infty} \mathbf{P}^+_{0,l,0}(\Gamma^+)\, \mathbf{n}^+(\zeta \in dl).\end{aligned}$$

An analogous description, with the obvious modifications, is valid for $\mathbf{n}^-$.

Now let $R^{(3)}$, $R^{(3)}_0 = 0$, be a 3-dimensional Bessel process, $\mathbf{Q}_0$ the probability measure associated to it, and r its transition density with respect to the speed measure (see A1.22). Introduce the function

$$h(z,t;l,0) := \begin{cases} \dfrac{r(l-t;z,0)}{r(l;0,0)}, & t < l,\\ 0, & t \geq l.\end{cases}$$

Let $\mathbf{Q}_{0,l,0}$ denote the law of the h-transform of $R^{(3)}$. This is constructed similarly to the Brownian bridge in No. 23 and called the 3-dimensional Bessel bridge of length l from 0 to 0. It is a straightforward computation to check that the laws $\mathbf{Q}_{0,l,0}$ and $\mathbf{P}^+_{0,l,0}$ are the same.

16. We turn now to *Williams' description* of $\mathbf{n}$ (see Williams (1979) p. 98). For $f \in E^+$ let

$$m(f) := \sup\{f(t) : t \geq 0\}.$$

The starting point in Williams' approach is the $\mathbf{n}^+$- distribution of $m = m(f)$

(a) $$\mathbf{n}^+(m > z) = \frac{1}{2z}.$$

For proofs see Rogers and Williams (1987) p. 425 or Revuz and Yor (2001) p. 492. Still another way to prove this is to compute the distribution of the maximum of a 3-dimensional Bessel bridge and to use Itô's description of $\mathbf{n}^+$. Indeed, using absolute continuity and the strong Markov property it is seen that

(b), $$\mathbf{Q}_{0,l,0}(\check{M} > z) = \frac{\sqrt{2\pi l^3}}{z} \int_0^l r_0(z,t)\, n_z(0, l-t)\, dt,$$

where $\check{M}$ is the maximum and

$$r_0(z,t) := \mathbf{Q}_0(H_z \in dt)/dt.$$

Putting this into Itô's description and integrating over l gives the desired formula. We remark that in Chung (1976) (see also Kennedy (1976), and Durrett and Iglehart (1977)) it is proved that

$$\mathbf{Q}_{0,l,0}(\check{M} > z) = 2\sum_{k=1}^{\infty}\Big(\frac{4k^2z^2}{l} - 1\Big)\exp\Big(-\frac{2k^2z^2}{l}\Big).$$

For further results on the maximum of a Bessel bridge see Pitman and Yor (1999b).

To proceed with Williams' description we notice that $\hat{W}$ on $(0,\infty)$ conditioned in the sense of Doob's h-transforms to drift to ∞ is identical in law to a 3-dimensional Bessel process. The excessive function used is simply the scale function $s(x)=x$ of W. This observation makes the appearance of 3-dimensional Bessel processes in Williams' description plausible. Now let $R^{(3,1)}$ and $R^{(3,2)}$ be two independent 3-dimensional Bessel processes starting from 0, and for a given $z>0$ let

$$H_z^{(i)} := \inf\{t:\ R_t^{(3,i)} = z\}, \quad i=1,2.$$

Introduce the process

$$Z_t^{(z)} := \begin{cases} R_t^{(3,1)}, & 0 \le t \le H_z^{(1)}, \\ z - R^{(3,2)}_{t-H_z^{(1)}}, & H_z^{(1)} \le t \le H_z^{(1)} + H_z^{(2)}, \\ 0, & H_z^{(1)} + H_z^{(2)} \le t, \end{cases}$$

and let $K(z,\Gamma^+)$, $\Gamma^+ \in E^+$, be the kernel (see I.6) associated to $Z^{(z)}$. Then Williams' description for positive excursions is

$$\mathbf{n}^+(\Gamma^+) = \int_0^\infty K(z,\Gamma^+)\,\mathbf{n}^+(m\in dz).$$

An analogous description is valid for negative excursions.

17. Here we give a few additional formulae derived from No. 15 (a) and 16 (a),(b). First,

$$\mathbf{n}^+(m>z,\ \zeta\in dl) = \frac{dl}{2z}\int_0^l r_0(z,t)\,n_z(0,l-t)\,dt.$$

Next recall that the Laplace transform of the $\mathbf{Q}_0$-distribution of the first hitting time for a 3-dimensional Bessel process is given by (see formula 5.2.0.1)

$$\mathbf{Q}_0(\exp(-\lambda H_x)) = \frac{\sqrt{2\lambda}\,x}{\operatorname{sh}(\sqrt{2\lambda}\,x)}.$$

From Williams' description we obtain

$$\mathbf{n}^+(e^{-\lambda\zeta};\ m\in dz) = \Big(\frac{\sqrt{2\lambda}\,z}{\operatorname{sh}(\sqrt{2\lambda}\,z)}\Big)^2\,\frac{1}{2z^2}\,dz,$$

and, consequently,

$$\mathbf{n}^+(m\in dz,\ \zeta\in dl) = \frac{dl\,dz}{2z^2}\int_0^l r_0(z,t)\,r_0(z,l-t)\,dt.$$

18. Excursions straddling a fixed time. Let $t>0$ be given, and introduce the random variables

$$G_t := \sup\{s<t:\ W_s=0\}, \quad D_t := \inf\{s>t:\ W_s=0\}.$$

Then

$$U_s := \begin{cases} W_{s+G_t}, & 0 \le s < D_t - G_t, \\ \partial, & s \ge D_t - G_t, \end{cases}$$

defines a process U called the *Brownian excursion process straddling* t.

We characterize the law of U under the measure $\mathbf{P}_0$. For proofs see Chung (1976), and, in a more general framework, Getoor (1979b) and Getoor and Sharpe (1982).

To start with, G_t has the arcsine distribution (cf. II.20), that is

$$\mathbf{P}_0(G_t \in du) = \frac{du}{\pi\sqrt{u(t-u)}} \quad (\text{or } \mathbf{P}_0(G_t \le u) = \frac{2}{\pi}\arcsin\sqrt{\frac{u}{t}}\,).$$

Using the Markov property it is seen that for $v > t$,

$$\begin{aligned}\mathbf{P}_0(D_t \in dv) &= \int_{\mathbb{R}} \mathbf{P}_0(W_t \in dy)\mathbf{P}_y(H_0 \in dv - t) \\ &= \frac{1}{\pi v}\sqrt{\frac{t}{v-t}}\,dv.\end{aligned}$$

The joint distribution is

$$\mathbf{P}_0(G_t \in du,\, D_t \in dv) = \frac{du\,dv}{2\pi\sqrt{u\,(v-u)^3}}, \quad 0 < u < t < v,$$

giving

$$\mathbf{P}_0(G_t \in du,\, D_t - G_t \in dl) = \frac{du\,dl}{2\pi\sqrt{u\,l^3}}, \quad (t-l) \vee 0 < u < t,$$

and

$$\mathbf{P}_0(D_t - G_t \in dl) = \frac{\sqrt{t} - \sqrt{(t-l)\vee 0}}{\pi\sqrt{l^3}}\,dl.$$

Recall the definitions of $\hat{p}(t;x,y)$ and $n_x(0,t)$ in No. 12. Then for $0 < u_1 < \cdots < u_k < l, , u < t,\ u+l > t$, and $y_1, y_2, \dots, y_k$ either all positive or all negative we have

$$\begin{aligned}&\mathbf{P}_0(G_t \in du,\, U_{u_1} \in dy_1,\, U_{u_2} \in dy_2,\, \dots, U_{u_k} \in dy_k,\, D_t - G_t \in dl) \\ &\quad = \frac{1}{\sqrt{2\pi u}} du\, n_{y_1}(0,u_1)\,dy_1\,\hat{p}(u_2 - u_1; y_1, y_2)\,dy_2 \\ &\qquad\qquad \cdot \ldots \cdot \hat{p}(u_k - u_{k-1}; y_{k-1}, y_k)\,dy_k\, n_{y_k}(0, l - u_n)\,dl.\end{aligned}$$

This leads to

$$\begin{aligned}&\mathbf{P}_0(U_{u_1} \in dy_1,\, U_{u_2} \in dy_2,\, \dots, U_{u_k} \in dy_k |\, G_t = u,\, D_t - G_t = l) \\ &\quad = \sqrt{2\pi l^3}\; n_{y_1}(0,u_1)\,dy_1\,\hat{p}(u_2 - u_1; y_1, y_2)\,dy_2 \\ &\qquad\qquad \cdot \ldots \cdot\ \hat{p}(u_k - u_{k-1}; y_{k-1}, y_k)\,dy_k\, n_{y_k}(0, l - u_k) \\ &\quad = \frac{1}{2}\,\mathbf{Q}_{0,l,0}(\omega(u_1) \in dy_1,\, \omega(u_2) \in dy_2,\, \dots, \omega(u_k) \in dy_k),\end{aligned}$$

where $\mathbf{Q}_{0,l,0}$ is the law (in the canonical setting) of the 3-dimensional Bessel bridge as introduced in No. 15.

19. Here we take a closer look at W prior to time t given that $G_t = u$.

First, given $G_t = u$ the process $\{W_s : 0 \le s \le G_t\}$ is independent of U and has the law of Brownian bridge from 0 to 0 with length u (Revuz and Yor (2001) p. 492).

Second, let

$$V_s := \begin{cases} W_{s+G_t}, & 0 \le s < t - G_t, \\ \partial, & s \ge t - G_t. \end{cases}$$

This defines a process V called the *Brownian meandering process ending at t*. To analyze V we have for $0 < u_1 < \cdots < u_k < t - u$ the joint distributions

$$\begin{aligned} &\mathbf{P}_0(G_t \in du,\ V_{u_1} \in dy_1,\ V_{u_2} \in dy_2,\ \ldots,\ V_{u_k} \in dy_k,\ W_t \in dy) \\ &\quad = \frac{1}{\sqrt{2\pi u}} du\, n_{y_1}(0, u_1)\, dy_1\, \hat{p}(u_2 - u_1; y_1, y_2)\, dy_2 \\ &\qquad \cdot \ldots \cdot \hat{p}(u_k - u_{k-1}; y_{k-1}, y_k)\, dy_k \hat{p}(t - u - u_k; y_k, y)\, dy. \end{aligned}$$

In particular,

$$\mathbf{P}_0(G_t \in du,\ W_t \in dy) = \frac{|y|}{2\pi\sqrt{u(t-u)^3}} \exp\Big(-\frac{y^2}{2(t-u)}\Big)\, du\, dy, \qquad 0 < u < t,$$

leading for $y > 0$ to the conditional law

$$\begin{aligned} &\mathbf{P}_0(V_{u_1} \in dy_1, V_{u_2} \in dy_2, \ldots, V_{u_k} \in dy_k \mid G_t = u,\ W_t = y) \\ &\qquad = \mathbf{Q}_{0,t-u,y}\big(\omega(u_1) \in dy_1,\ \omega(u_2) \in dy_2,\ \ldots, \omega(u_k) \in dy_k\big), \end{aligned}$$

where $\mathbf{Q}_{0,l,y}$ is the measure (in the canonical setting) associated to the 3-dimensional Bessel bridge from 0 to y having length l. By symmetry, this also characterizes the case $y < 0$. Further, we have the following absolute continuity relation (see Imhof (1984))

$$\begin{aligned} &\mathbf{P}_0(V_{u_1} \in dy_1,\ V_{u_2} \in dy_2,\ \ldots, V_{u_k} \in dy_k, W_t \in dy \mid G_t = u) \\ \text{(a)}\quad &\qquad = \frac{c}{y}\, \mathbf{Q}_0\big(\omega(u_1) \in dy_1,\ \omega(u_2) \in dy_2,\ \ldots, \omega(u_k) \in dy_k, \omega(t-u) \in dy\big), \end{aligned}$$

where $\mathbf{Q}_0$ is the measure (in the canonical setting) associated to the 3-dimensional Bessel process starting at 0 and $c = (\pi(t-u)/2)^{1/2}$.

A Brownian meandering process ending at t is often taken to be non-negative and normalized to have length 1 and, hence, defined as

$$\Big\{\frac{1}{\sqrt{t - G_t}} |W_{G_t + u(t-G_t)}| : 0 \le u \le 1\Big\},$$

and called simply *Brownian meander* (see Yor (1992a) p. 41). The absolute continuity property displayed in (a) takes in this setting the form

$$\frac{d\mathbf{Q}^m}{d\mathbf{Q}_0}\bigg|_{\mathcal{F}^c_{v+}} = \sqrt{\frac{\pi}{2}}\,\frac{1}{\omega(1)} \qquad \mathbf{Q}_0\text{-a.s.},$$

where $v \le 1$, $\mathbf{Q}^m$ is the measure associated to the Brownian meander, and $\{\mathcal{F}^c_v\}$ is the canonical filtration in the space of continuous functions.

For results concerning Brownian meander see, e.g., Chung (1976, 1982a), Imhof (1984), Biane and Yor (1987), (1988), Yor (1992a), Bertoin and Pitman (1994), and Pitman and Yor (1996).

4. Brownian bridge

20. Let $x, y \in \mathbb{R}$ and $l > 0$ be given. A continuous Gaussian process $X = \{X_t : 0 \le t \le l\}$, $X_0 = x$, such that

$$\mathbf{E}\,X_t = x + (y-x)\,\frac{t}{l}, \quad \mathbf{Cov}(X_t, X_s) = (s \wedge t) - \frac{s\,t}{l}$$

is called *a Brownian bridge from x to y of length l.*

To show that there indeed exists a continuous modification of a Gaussian process having the stated covariance Kolmogorov's criterion in I.2 can be used (see also Gihman and Skorohod (1974) p. 193).

We indicate the parameters of X by writing $X^{x,l,y}$. The law is denoted by $\mathbf{P}_{x,l,y}$. Observe that $\mathbf{E}X_l = y$ and $\mathbf{Cov}(X_t, X_s) = 0$ if $s = l$ or $t = l$. Therefore, a.s. $X_l = y$. Because of this property X is called a Brownian bridge. Other names are *pinned Brownian motion* and *tied-down Brownian motion.*

21. Let W be a standard Brownian motion started at 0. Then we have the following two descriptions of a Brownian bridge:

(a)
$$X^{x,l,x} \sim \{x + W_t - \tfrac{t}{l}\,W_l : 0 \le t \le l\},$$

(b)
$$X^{x,l,x} \sim \Big\{x + \frac{l-t}{l}\,W\Big(\frac{l\,t}{l-t}\Big) : 0 \le t \le l\Big\}.$$

In (b) the process in braces is defined to be equal to x when $t = l$.

Moreover, $X^{x,l,y}$ can be viewed as $X^{x,l,x}$ with a drift. The exact statement is

$$X^{x,l,y} \sim \{X^{x,l,x}_t + (y-x)\tfrac{t}{l} : 0 \le t \le l\}.$$

22. A Brownian bridge of length l from x to y can also be characterized as the pathwise unique solution of the SDE (Ikeda and Watanabe (1981) p. 229)

$$\begin{cases} dX_t = \dfrac{y - X_t}{l-t}\,dt + dW_t, & 0 \le t < l, \\ X_0 = x. \end{cases}$$

The solution is

$$X_t = x + (y - x)\,\frac{t}{l} + (l - t)\int_0^t \frac{dW_s}{l-s}.$$

We refer to Yor (1997a) p. 34 for an approach to the Brownian bridge via enlargement of filtrations, see also Jeulin and Yor (1979), and Jeulin (1980) p. 119.

23. One more approach to the Brownian bridge is via Doob's h-transform. To explain this let p be the transition density of W and notice that the function

$$(x,t) \mapsto h(x,t;y,l) = \begin{cases} p(l-t;x,y), & 0 \le t < l, \\ 0, & t \ge l, \end{cases}$$

is for every $y \in \mathbb{R}$ and $l > 0$ space-time excessive for W. Indeed, letting $\bar{x} = (s,x)$ be the starting state,

$$\mathbf{E}_{\bar{x}}(h(W_t,t;y,l)) = \begin{cases} h(x,s;y,l), & 0 \le s \le t < l, \\ 0, & \text{otherwise.} \end{cases}$$

Using the function h we build the h-transform of W. This is a non-time-homogeneous Markov process $W^h = \{W_t^h : 0 \le t < l\}$ having the finite dimensional distributions

$$\mathbf{P}_x^h(W_{t_1}^h \in dx_1,\, W_{t_2}^h \in dx_2,\, \dots, W_{t_n}^h \in dx_n) \\ = \prod_{i=1}^n p(t_i - t_{i-1};\, x_{i-1}, x_i)\,\frac{h(x_n,t_n;y,l)}{h(x,0;y,l)},$$

where $0 = t_0 < t_1 < \cdots < t_n < l$, $x_0 := x$ and $x_i \in \mathbb{R}$. It is a straightforward computation to check that the measures $\mathbf{P}_x^h$ and $\mathbf{P}_{x,l,y}$ are the same.

Intuitively, a Brownian bridge $X^{x,l,y}$ is a Brownian motion W started at x and conditioned to be at y at time l. The h-transform techniques makes this kind of conditioning with respect to a set of probability 0 plausible.

From the h-transform approach we also get the absolute continuity result

$$\left.\frac{d\mathbf{P}_{x,l,y}}{d\mathbf{P}_x}\right|_{\mathcal{F}_{t+}^c} = \frac{h(\omega(t),t;y,l)}{h(x,0;y,l)} \qquad \mathbf{P}_x\text{-a.s.},$$

where $t < l$ and $\{\mathcal{F}_t^c\}$ is the canonical filtration in the space of continuous functions.

24. Vervaat's path transformation. Let X denote $X^{0,1,0}$, and introduce

$$\hat{M} := \inf\{X_t : 0 \le t \le 1\} \quad \text{and} \quad \hat{H} := \inf\{t : X_t = \hat{M}\}.$$

Then $\hat{H} \sim U(0,1)$, that is, $\hat{H}$ is uniformly distributed on $(0,1)$. Define next the process Y by

$$Y_t := \begin{cases} X_{\hat{H}+t} - \hat{M}, & 0 \le t \le 1 - \hat{H}, \\ X_{t-(1-\hat{H})} - \hat{M}, & 1 - \hat{H} \le t \le 1. \end{cases}$$

Let R denote the 3-dimensional Bessel bridge from 0 to 0 of length 1 (see No. 15). A result due to Vervaat (1979) is that $\hat{H}$ and Y are independent and

$$Y \quad \sim \quad R.$$

Conversely, let $\xi \sim U(0,1)$ and independent of R, and set

$$Z_t := \begin{cases} R_{\xi+t} - R_\xi, & 0 \le t \le 1 - \xi, \\ R_{t-(1-\xi)} - R_\xi, & 1 - \xi \le t \le 1. \end{cases}$$

Then (Biane (1986))

$$Z \quad \sim \quad X.$$

An immediate implication of Vervaat's result is that the range of X and the maximum of R have the same laws. For the law of the maximum see No. 16, and for the distribution of the range the formula 1.1.15.8 (2), see also Bass and Koshnevisan (1995).

Verwaat's path transformation can be used (see Chaumont, Hobson, and Yor (2001)) to explain the following puzzling identity presented in Donati-Martin, Matsumoto and Yor (2002)

$$\mathbf{E}\left(\left(\int_0^1 \exp(\alpha X_u)\,du\right)^{-1}\right) = 1 \quad \text{for all } \alpha \in \mathbb{R},$$

see also Donati-Martin, Matsumoto and Yor (2000b).

25. Let X and R be as in No. 24, and let $\ell^R(x)$ be the total local time at x of R such that

$$\int_0^1 \mathbb{1}_B(R_s)\,ds = \int_B \ell^R(x)\,2dx, \quad B \in \mathcal{B}(\mathbb{R}).$$

Let, further,

$$A_x^R := \int_0^1 \mathbb{1}_{(-\infty,x)}(R_s)\,ds$$

be the time R spends below the level x and introduce for $0 \le t \le 1$ its inverse

$$\alpha_t^R := \inf\{x : A_x^R \ge t\}.$$

Then (Jeulin (1985) p. 264)

$$\{\ell^R(\alpha_t^R) : 0 \le t \le 1\} \sim R.$$

In particular,

$$\sup\{\ell^R(x):\ x \geq 0\} \ \sim\ \sup\{R_t:\ 0 \leq t \leq 1\}.$$

Now let $\ell^X(x)$ be the total local time at x of X taken with respect to $2dx$. As above, let A_x^X be the time X spends below the level x and α_t^X its inverse. Then the result is also as above (Biane (1986)), that is,

$$\{\ell^X(\alpha_t^X):\ 0 \leq t \leq 1\} \ \sim\ R.$$

In particular,

$$\sup\{\ell^X(x):\ x \in \mathbb{R}\} \ \sim\ \sup\{R_t:\ 0 \leq t \leq 1\}.$$

For further results concerning the Brownian bridge and related notions see Biane and Yor (1987), Pitman and Yor (1996), and Pitman (1999).

26. We have not explicitly displayed formulae for Brownian bridges in Part II of the book. Instead, we give the joint distributions of functionals up to time t and the location of W at time t. The distributions of the corresponding functionals of a Brownian bridge can easily be obtained from these. For example, by the formula 1.1.1.8 from Part II for $x, z \leq b$,

$$\mathbf{P}_x\Big(\sup_{0\leq t\leq l} W_t < b,\, W_l \in dz\Big) = \frac{1}{\sqrt{2\pi\, l}}\Big(\exp\Big(-\frac{(z-x)^2}{2l}\Big) - \exp\Big(-\frac{(z+x-2b)^2}{2l}\Big)\Big)\, dz.$$

Therefore, dividing above by $\mathbf{P}_x(W_t \in dz)$ gives

$$\mathbf{P}_{x,l,z}\Big(\sup_{0\leq t\leq l} X_t^{x,l,z} < b\Big) = 1 - \exp\Big(-\frac{(z+x-2b)^2}{2l} + \frac{(z-x)^2}{2l}\Big).$$

Another approach is to use absolute continuity, namely,

$$\mathbf{P}_{x,l,z}\Big(\sup_{0\leq t\leq l} X_t^{x,l,z} > b\Big) = \int_0^l \mathbf{E}_x\Big(\frac{p(l-t;b,z)}{p(l;x,z)};\ H_b \in dt\Big) = \frac{1}{p(l;x,z)}\int_0^l p(l-t;b,z)\,\mathbf{P}_x(H_b \in dt).$$

The integral can be computed, e.g., by taking the Laplace transform.

5. Brownian motion with drift

27. Let $\mu \in \mathbb{R}$ be given. A linear diffusion on $\mathbb{R}$ having the operator

$$\mathcal{G} = \frac{1}{2}\frac{d^2}{dx^2} + \mu\frac{d}{dx}$$

as the infinitesimal generator acting on the domain

$$\mathcal{D}(\mathcal{G}) = \{f : \ f, \mathcal{G}f \in \mathcal{C}_b(\mathbb{R})\}$$

is called a *Brownian motion with drift* μ. We let $W^{(\mu)} = \{W_t^{(\mu)} : \ t \geq 0\}$ denote this process. For the basic data of $W^{(\mu)}$ see A1.14.

28. Brownian motion with drift can be constructed from W by adding a drift term

$$W^{(\mu)} \ \sim \ \{W_t + \mu t : \ t \geq 0\}.$$

It can also be obtained as an h-transform of W (in space-time). Indeed, the function

$$(x,t) \mapsto h(x,t;\mu) := \exp\Bigl(\mu x - \frac{\mu^2 t}{2}\Bigr)$$

is space-time invariant for W, i.e.,

$$\mathbf{E}_{s,x}(h(W_t, t; \mu)) = h(x, s; \mu).$$

This is seen immediately from (c') in No. 1. Using the function h we build the h-transform of W. It turns out that this is a time-homogeneous Markov process identical in law to $W^{(\mu)}$.

Notice also that if W is an $\mathcal{F}_t$-Brownian motion, then the process

$$\{h(W_t, t; \mu) : \ t \geq 0\}$$

is a $(\mathbf{P}, \mathcal{F}_t)$-martingale.

29. The following properties of $W^{(\mu)}$ can be deduced from the corresponding properties of W :

Independent increments. For all $0 = t_0 < t_1 < \cdots < t_n$ the increments

$$W_{t_n}^{(\mu)} - W_{t_{n-1}}^{(\mu)}, W_{t_{n-1}}^{(\mu)} - W_{t_{n-2}}^{(\mu)}, \ldots, W_{t_1}^{(\mu)} - W_{t_0}^{(\mu)}$$

are independent, and normally distributed with

$$\mathbf{E}(W_{t_i}^{(\mu)} - W_{t_{i-1}}^{(\mu)}) = \mu(t_i - t_{i-1}), \quad \mathbf{Var}(W_{t_i}^{(\mu)} - W_{t_{i-1}}^{(\mu)}) = t_i - t_{i-1}.$$

Spatial homogeneity. Let $x \in \mathbb{R}$ be given and assume that $W_0^{(\mu)} = 0$. Then $x + W^{(\mu)}$ is a Brownian motion with drift μ started at x.

Symmetry. $-W^{(\mu)}$ is a Brownian motion with drift $-\mu$.

Scaling. For every $c > 0$ the process $\{\sqrt{c}\,W_{t/c}^{(\sqrt{c}\mu)} : t \geq 0\}$ started at 0 is identical in law to $W^{(\mu)}$ started at 0.

Time inversion. The process defined by

$$Z_t := \begin{cases} \mu, & t = 0, \\ tW_{1/t}^{(\mu)}, & t > 0, \end{cases}$$

is a Brownian motion started at μ.

Time reversibility. For a given $t > 0$,

$$\{W_s^{(\mu)} - W_0^{(\mu)} : 0 \leq s \leq t\} \sim \{W_{t-s}^{(-\mu)} - W_t^{(-\mu)} : 0 \leq s \leq t\}.$$

Strong law of large numbers.

$$\lim_{t\to\infty} \frac{W_t^{(\mu)}}{t} = \mu \quad \text{a.s.}$$

Laws of the iterated logarithm.

$$\limsup_{t\downarrow 0} \frac{W_t^{(\mu)}}{\sqrt{2t\ln\ln(1/t)}} = 1 \quad \text{a.s.}$$

$$\liminf_{t\downarrow 0} \frac{W_t^{(\mu)}}{\sqrt{2t\ln\ln(1/t)}} = -1 \quad \text{a.s.}$$

All *local properties of the sample paths* as presented in No. 6 hold except that the level sets $\mathcal{Z}_x(\omega) := \{t : W_t^{(\mu)}(\omega) = x\}$ are bounded a.s. when $\mu \neq 0$.

30. The h-transform approach leads to the following absolute continuity result (cf. the Cameron–Martin–Girsanov transformation of measure in III.26–28):

$$\frac{d\mathbf{P}_x^{(\mu)}}{d\mathbf{P}_x}\bigg|_{\mathcal{F}_{t+}^c} = \frac{h(\omega(t), t; \mu)}{h(x, 0; \mu)} = \exp\Big(\mu(\omega(t) - x) - \frac{\mu^2 t}{2}\Big) \qquad \mathbf{P}_x\text{-a.s.},$$

where $\mathbf{P}_x^{(\mu)}$ and $\mathbf{P}_x$ are the measures associated to $W^{(\mu)}$ and W, respectively.

31. Consider the last exit time $\lambda_y := \sup\{t : W_t^{(\mu)} = y\}$. By the general results in II.20 (for the Green function G_0 and the transition density p, see A1.14),

$$\begin{aligned}\mathbf{P}_x(\lambda_y \in dt) &= \frac{p(t; x, y)}{G_0(y, y)}\, dt \\ &= \frac{|\mu|}{\sqrt{2\pi\, t}} \exp\left(\mu(y - x) - \frac{\mu^2\, t}{2} - \frac{(x-y)^2}{2t}\right) dt,\end{aligned}$$

and

$$\mathbf{P}_x(\lambda_y = 0) = \mathbf{P}_x(H_y = \infty) = \begin{cases} 1 - e^{-2\mu\,(x-y)}, & \mu(x-y) \geq 0, \\ 0, & \mu(x-y) \leq 0. \end{cases}$$

32. Suppose that $\mu > 0$ and $W^{(\mu)}$ is started at 0. Let

$$\hat{M} := \inf\{W_t^{(\mu)} : \ t \geq 0\} \quad \text{and} \quad \hat{H} := \inf\{t : \ W_t^{(\mu)} = \hat{M}\}.$$

We find here the joint distribution of $\hat{M}$ and $\hat{H}$. Let

$$\hat{M}_t := \inf\{W_s^{(\mu)} : \ 0 \leq s \leq t\} \quad \text{and} \quad \hat{H}(t) := \inf\{s \leq t : \ W_s^{(\mu)} = \hat{M}_t\}.$$

Then $\hat{M}_t \to \hat{M}$ and $\hat{H}_t \to \hat{H}$ a.s. as $t \to \infty$. Using II.19 it is seen that for $\alpha, \beta, \gamma > 0$,

$$\begin{aligned}
&\int_0^\infty e^{-\gamma\, t} \mathbf{E}_0(\exp(\alpha \hat{M}_t - \beta \hat{H}_t))\, dt \\
&\qquad = \int_{-\infty}^0 s(dz)\, e^{\alpha\, z} \int_z^\infty m(dy)\, \frac{\varphi_\gamma(y)}{\varphi_{\beta+\gamma}(z)\, \varphi_\gamma(z)} \\
&\qquad = \frac{1}{\gamma}\, \frac{\sqrt{2\gamma + \mu^2} + \mu}{\sqrt{2(\beta+\gamma) + \mu^2} + \alpha + \mu} =: F(\gamma).
\end{aligned}$$

Consequently,

$$\begin{aligned}
\mathbf{E}_0(\exp(\alpha \hat{M} - \beta \hat{H})) &= \lim_{t\to\infty} \mathbf{E}_0(\exp(\alpha \hat{M}_t - \beta \hat{H}_t)) \\
&= \lim_{\gamma \to 0} \gamma\, F(\gamma) = \frac{2\,\mu}{\sqrt{2\,\beta + \mu^2} + \alpha + \mu}.
\end{aligned}$$

Because $\hat{M} \sim \operatorname{Exp}(2\,\mu)$ on $(-\infty, 0)$ it is seen that $\hat{H}$ given $\hat{M} = z$ is distributed as H_z for the process $W^{(-\mu)}$.

33. The result in No. 32 can be expanded considerably. This leads to the *path decomposition* of $W^{(\mu)}$ due to David Williams (see Rogers and Williams (2000) p. 436).

First, we construct from $W^{(\mu)}$ a new process living on $(0, \infty)$ by conditioning $W^{(\mu)}$ to not hit 0. For this let

$$h(x) := \mathbf{P}_x(H_0 = \infty) = 1 - e^{-2\mu\, x}, \quad x > 0.$$

Then the desired conditioning is achieved by building the excessive transform of $W^{(\mu)}$ using the function h. The transform is a time-homogeneous diffusion on $(0, \infty)$ with the generator

$$\mathcal{G}^{+} = \frac{1}{2}\frac{d^2}{dx^2} + \mu\, \operatorname{cth}(\mu\, x) \frac{d}{dx}.$$

Straightforward computations show that for this process 0 is an entrance-not-exit boundary.

Let W^+ denote the diffusion having the generator $\mathcal{G}^+$ and which is started at 0. Further, let $S \sim \text{Exp}(2\,\mu)$ on $(-\infty, 0)$. Assume that W^+, S and $W^{(-\mu)}$ are defined on the same probability space and that they are independent. Let $H_S := \inf\{t;\ W_t^{(-\mu)} = S\}$ and define

$$X_t^{(\mu)} := \begin{cases} W_t^{(-\mu)}, & 0 \le t < H_S, \\ S + W_{t-H_S}^+, & t \ge H_S. \end{cases}$$

Then, if $X_0^{(\mu)} = W_0^{(\mu)} = 0$, the Williams path decomposition theorem says that

$$X^{(\mu)} \sim W^{(\mu)}.$$

This result can be viewed as a canonical example of the various path decomposition theorems. It is possible to formulate similar explicit results for all transient diffusions. We refer to Rogers and Williams (2000) for references, and, especially, for the path decomposition of killed Brownian motion. See also Williams (1974), Biane (1993), and Pitman and Yor (1996).

34. Consider the *perpetual integral functional*

$$I_\infty(f) := \int_0^\infty f(W_t^{(\mu)})\,dt,$$

where f is a non-negative, locally bounded, and Borel measurable function and $\mu > 0$. It holds that

(a) $$I_\infty(f) < \infty \ \text{ a.s.} \ \Leftrightarrow \ \int^\infty f(x)dx < \infty,$$

(b) $\int^\infty f(x)\,dx < \infty$ and $\exists m > 0 \ \ f(x) = O(|x|^m)_{x\to-\infty}$

$$\Rightarrow \ \forall n \in \mathbb{N},\ \forall x \quad \mathbf{E}_x\big(I_\infty(f)^n\big) < \infty,$$

(c) $\int^\infty f(x)\,dx < \infty$ and $f(x) = O(1)_{x\to-\infty}$

$$\Rightarrow \ \exists \gamma > 0,\ \forall x\ \mathbf{E}_x\big(\exp(\gamma I_\infty(f))\big) < \infty,$$

(d) $$\sup_x \mathbf{E}_x\big(I_\infty(f)\big) < \infty \ \Leftrightarrow \ \int_{-\infty}^\infty f(x)\,dx < \infty.$$

For proofs based on Ray-Knight theorems and *Jeulin's lemma*, see Salminen and Yor (2005a); for (a), see also Engelbert and Senf (1991).

35. It is interesting that perpetual integral functionals of $W^{(\mu)}$ with $\mu > 0$ can be viewed as first hitting times for a properly scaled and time changed diffusions. To explain this, let f be a monotone, twice continuously differentiable function such that $r := \lim_{x\to\infty} f(x)$ exists and introduce the additive functional

$$J_s := \int_0^s (f'(W_u^{(\mu)}))^2\, du.$$

Assume further that $f'(x) \neq 0$ and

$$\int^\infty (f'(x))^2 dx < \infty.$$

Let Z be a diffusion given by

$$Z_t := f(W_{\alpha_t}^{(\mu)}),$$

for $t \geq 0$ such that

$$\alpha_t := \inf\{s : J_s > t\} < \infty.$$

Then a.s.

$$\int_0^\infty (f'(W_s^{(\mu)}))^2\, ds = \inf\{t : Z_t = r\}.$$

Moreover, Z is a solution of the SDE

$$dZ_t = d\beta_t + G \circ f^{-1}(Z_t)\, dt, \quad Z_0 = f(0),$$

where β is a Brownian motion and

$$G(x) := \frac{1}{(f'(x))^2}\Big(\frac{1}{2} f''(x) + \mu\, f'(x)\Big).$$

For a proof, see Salminen and Yor (2005b). In Khosnevisan et al. (2006) the result is generalized for fairly general diffusions. In Salminen and Wallin (2007) numerical metods for calculating the distributions of perpetual functionals are studied. For other related works, see Salminen and Yor (2003), Decamps et al. (2005), and Borodin and Salminen (2006). These papers contain also many examples, in particular Salminen and Yor (2005b). A famous example – Dufresne's identity – is presented in No. 50.

6. Bessel processes

36. Let $\{W^{(i)} : i = 1, 2, \dots, n\}$ be a family of independent Brownian motions. The process $R^{(n)}$ defined by

$$R_t^{(n)} := \sqrt{(W_t^{(1)})^2 + (W_t^{(2)})^2 + \cdots + (W_t^{(n)})^2}, \quad t \geq 0,$$

is called an *n-dimensional Bessel process* or a *Bessel process of order* $\nu := n/2-1$. Clearly, $R^{(n)}$ is the *radial part* of an n-dimensional Brownian motion. Special properties of Brownian motion imply (Itô and McKean (1974) p. 60) that $R^{(n)}$ is a linear diffusion. Its generator is

$$\mathcal{G}^{(n)} = \frac{1}{2}\frac{d^2}{dx^2} + \frac{n-1}{2x}\frac{d}{dx}$$

with the domain $D^{(n)} = \{f : \ f, \mathcal{G}^{(n)} f \in \mathcal{C}_b[0,\infty)\}$ for $n \geq 2$ and $D^{(1)} = \{f : f \in \mathcal{C}_b^2[0,\infty), \ f'(0+) = 0\}$.

37. Notice that $R^{(1)}$ is identical in law to a reflected Brownian motion and 0 is in this case a non-singular boundary point. For $n \geq 2$ the boundary point 0 is entrance-not-exit. This means that the n-dimensional Brownian motion (for $n \geq 2$) started from the origin does not hit the origin after time 0. However, because $R^{(2)}$ is recurrent the 2-dimensional Brownian motion when started outside the origin hits every ball with center at the origin. For $n > 2$ the process $R^{(n)}$ is transient and drifts to ∞. It follows that n-dimensional Brownian motion (for $n > 2$) is transient.

38. To justify the name "Bessel process" we remark that the fundamental solutions φ_α and ψ_α (see II.10 and 11) are given in terms of the modified Bessel functions I and K (for these and other basic characteristics of $R^{(n)}$ see A1.21). In fact, if the ODE

$$\mathcal{G}^{(n)} u = \alpha u$$

is multiplied by x^2 we obtain a form of the (modified) Bessel differential equation. See Abramowitz and Stegun (1972) Section 9.6 for details concerning Bessel's ODE and its solutions. See also A2.4 and A4.10, A4.11.

39. The property of n-dimensional Brownian motion that its radial part is a linear diffusion is a special one. We take a closer look at this in general.

First, notice that if X is a linear diffusion on $\mathbb{R}$ such that $X \sim -X$, then X^2 is also a diffusion. In particular, W^2 is a diffusion with the generator

$$2x\frac{d^2}{dx^2} + \frac{d}{dx}.$$

Secondly, we remark that the requirement that the sum of two independent non-negative diffusions X_1 and X_2 is also a diffusion is very exclusive. It has been proved by Shiga and Watanabe (1973) that this is the case if and only if the generators of X_1 and X_2 are of the corresponding forms

$$ax\frac{d^2}{dx^2} + (bx + c\alpha_i)\frac{d}{dx}, \quad i = 1, 2,$$

where $a > 0, \ b \in \mathbb{R}, \ c \geq 0$ and $\alpha_i \geq 0, \ i = 1, 2$. The boundary point 0 is

($\star$) exit-not-entrance if $c\alpha_i = 0$,

($\star$) non-singular if $0 < c\alpha_i < a$,

($\star$) entrance-not-exit if $c\alpha_i \geq a$.

In the non-singular case we let the boundary condition be reflection. The generator of $X_1 + X_2$ is, then

$$\text{(a)} \qquad ax\frac{d^2}{dx^2} + (bx + c(\alpha_1 + \alpha_2))\frac{d}{dx}.$$

Thirdly, if X is a non-negative diffusion, then $\sqrt{X}$ is also a diffusion because $x \mapsto \sqrt{x}$ is increasing. Moreover, if the generator of X is of the form in (a), then the generator of $\sqrt{X}$ is

$$\text{(b)} \qquad \frac{a}{4}\frac{d^2}{dx^2} + \left(\frac{bx}{2} + \frac{2c(\alpha_1 + \alpha_2) - a}{4x}\right)\frac{d}{dx}.$$

Diffusions associated to the generator in (b) can be viewed as radial Ornstein–Uhlenbeck processes; see A1.25 for their basic characteristics. Diffusions associated to the generator in (a) – squared radial Ornstein–Uhlenbeck processes, see A1.26 – play a prominent role in the Ray–Knight theorems; see Chapter 5. In mathematical finance they are often called CIR-processes, due to the work of Cox, Ingersoll and Ross (1985).

40. Let $\check{M}_t := \sup\{W_s : s \leq t\}$, where W is initiated at 0. Then, see No. 7, a result due to Lévy is that

$$R^{(1)} \sim |W| \sim \{\check{M}_t - W_t : t \geq 0\}.$$

In Pitman (1975) (see also Williams (1979) p. 96) a similar construction is given for $R^{(3)}$:

$$R^{(3)} \sim \{2\check{M}_t - W_t : t \geq 0\}.$$

The filtration generated by $\{\check{M}_t - W_t : t \geq 0\}$ is equal to the Brownian filtration, but the filtration generated by $\{2\check{M}_t - W_t : t \geq 0\}$ is strictly smaller than the Brownian filtration (see Revuz and Yor (2001) p. 253). We remark also that if $c \notin \{0, 1, 2\}$, then the process $\{c\check{M}_t - W_t : t \geq 0\}$ is not Markovian.

There are extensions of these results when W is replaced with $W^{(\mu)}$. We have that

$$\{\check{M}_t^{(\mu)} - W_t^{(\mu)} : t \geq 0\}$$

is identical in law with a reflecting Brownian motion on $\mathbb{R}^+$ with drift $-\mu$ (see, e.g., Harrison (1985)). For the basic characteristics of this process see A1.16. Further (see Rogers and Pitman (1981)), the process

$$\{2\check{M}_t^{(\mu)} - W_t^{(\mu)} : t \geq 0\}$$

is a linear diffusion with the generator (cf. No. 33)

$$\frac{1}{2}\frac{d^2}{dx^2} + |\mu| \operatorname{cth}(|\mu|\, x)\frac{d}{dx}.$$

For further extensions of Pitman's theorem we refer to Rogers (1981), Salminen (1983), Saisho and Tanemura (1990), Bertoin (1992), Rauscher (1997) and Matsumoto and Yor (2000),(2001) (for the last one see also No. 53).

41. Let $n = 2\nu + 2$ for $\nu \in \mathbb{R}$. Then a linear diffusion having the generator

$$\mathcal{G}^{(n)} = \frac{1}{2}\frac{d^2}{dx^2} + \frac{2\nu+1}{2x}\frac{d}{dx},$$

is called a *Bessel process of order* ν. Clearly, the Bessel processes in No. 35 are obtained from these when $\nu = -1/2, 0, 1/2, 1, \ldots$. With a slight abuse of notation we let $R^{(\nu)}$ below denote a diffusion with the generator $\mathcal{G}^{(n)}$.

For the basic characteristics of Bessel processes see A1.21. Recall that the boundary point 0 is:

($\star$) exit-not-entrance if $\nu \le -1$,

($\star$) non-singular if $-1 < \nu < 0$,

($\star$) entrance-not-exit if $\nu \ge 0$.

42. Assume that $R_0^{(\nu)} = 0$ and $\nu > -1$. For $-1 < \nu < 0$ let 0 be reflecting. Then $R^{(\nu)}$ has the following properties.

Scaling. For every $c > 0$,

$$R^{(\nu)} \sim \{\sqrt{c}\, R_{t/c}^{(\nu)} : t \ge 0\}.$$

Time inversion. Letting

$$Z_t := \begin{cases} 0, & t = 0, \\ t\, R_{1/t}^{(\nu)}, & t > 0, \end{cases}$$

it holds (see Shiga and Watanabe (1973)) that

$$R^{(\nu)} \sim \{Z_t : t \ge 0\}.$$

From this we obtain

$$\lim_{t\to\infty} \frac{R_t^{(\nu)}}{t} = 0 \quad \text{a.s.}$$

43. Laws of the iterated logarithm. Let $\nu \ge 0$; then

(a) $$\limsup_{t\downarrow 0} \frac{R_t^{(\nu)}}{\sqrt{2t \ln\ln(1/t)}} = 1 \quad \text{a.s.},$$

(b) $$\limsup_{t\to\infty} \frac{R_t^{(\nu)}}{\sqrt{2t \ln\ln t}} = 1 \quad \text{a.s.}$$

For $-1/2 \le \nu < 0$, (a) and (b) hold if the boundary condition is reflection. For a proof of (a) for $\nu = (n-2)/2$, $n = 1, 2, \dots$, see Itô and McKean (1974) p. 61. For arbitrary $\nu \ge -1/2$, a proof of (a) is indicated in Revuz and Yor (2001) p. 450. Observe that (b) follows from (a) by time inversion.

44. For $\nu \ne -1/2$ the drift term in the generator of a Bessel process is (only) locally Lipschitz. Therefore, we can a priori expect that $R^{(\nu)}$ is characterizable via the stochastic differential equation

$$\text{(a)} \qquad \begin{cases} dR_t^{(\nu)} = dW_t + \dfrac{2\nu+1}{2R_t^{(\nu)}}\, dt, & t > 0, \\ R_0^{(\nu)} = x > 0, \end{cases}$$

only up to the "explosion" time $H_0 := \inf\{t : \ R_t^{(\nu)} = 0\}$.

In the case $\nu \ge 0$ the boundary point 0 is entrance-not-exit, and so $H_0 = \infty$ a.s. It can be proved, in this case, that $R^{(\nu)}$ is the unique strong solution of (a) (see Chung and Williams (1990) p. 252).

We remark that the description via the SDE in (a) is also sufficient to characterize $R^{(\nu)}$ in the case $\nu \le -1$ because then 0 is exit-not-entrance.

When $-1 < \nu < 0$ a boundary condition at 0 is involved. For instance, if $\nu = -1/2$ and the condition at 0 is reflection, then

$$R^{(\nu)} \sim |W|,$$

and there does not exist a SDE in the classical sense for $R^{(\nu)}$. See *Skorokhod's equation,* e.g., in Ikeda and Watanabe (1981) p. 120. We refer also to Bertoin (1990) and Yor (1997a) p. 6.

45. The scale function of $R^{(\nu)}$, $\nu > 0$, is $s(x) = -x^{-2\nu}/(2\nu)$. Hence, the function $h(x) := s(\infty) - s(x) = x^{-2\nu}/(2\nu)$ is excessive for $R^{(\nu)}$. Standard computations show that the corresponding h-transform (cf. II.31) of $R^{(\nu)}$ is $R^{(-\nu)}$. If $0 < \nu < 1$, in this setting, the boundary condition at 0 for $R^{(-\nu)}$ is killing.

Let $\mathbf{P}^{(\nu)}$ and $\mathbf{P}^{(-\nu)}$ be the measures associated to $R^{(\nu)}$ and $R^{(-\nu)}$, respectively. From the h-transform property we get the absolute continuity property

$$\mathbf{P}_x^{(-\nu)}(A;\ H_0 > t) = \mathbf{E}_x^{(\nu)}\Big(\frac{h(\omega(t))}{h(x)};\ A\Big), \qquad A \in \mathcal{F}_{t+}^c.$$

46. The speed measure of $R^{(\nu)}$, $\nu > 0$, is $m(dx) = 2\,x^{2\nu+1}\,dx$, and the transition density with respect to m is (see A1.21)

$$p^{(\nu)}(t; x, y) = \frac{1}{2t}(xy)^{-\nu} \exp\Big(-\frac{x^2+y^2}{2t}\Big) I_\nu\big(\tfrac{xy}{t}\big).$$

Because for small values of z,

$$I_\nu(z) \simeq \frac{1}{\Gamma(\nu+1)}\Big(\frac{z}{2}\Big)^\nu$$

we have

$$p^{(\nu)}(t;0,y) = \lim_{x\to 0} p^{(\nu)}(t;x,y) = \frac{t^{-\nu-1}}{2^{\nu+1}\Gamma(\nu+1)} \exp\Bigl(-\frac{y^2}{2t}\Bigr).$$

From II.20 we obtain the distribution of the last exit time

$$\mathbf{P}_x^{(\nu)}(\lambda_y \in dt) = \frac{p^{(\nu)}(t;x,y)}{-s(y)}\, dt.$$

The time reversal property in II.33 gives further (see also Getoor and Sharpe (1979) and Getoor (1979a))

$$\mathbf{P}_y^{(-\nu)}(H_0 \in dt) = \mathbf{P}_0^{(\nu)}(\lambda_y \in dt).$$

47. We conclude this chapter by presenting a special case of a formula due to Pitman and Yor (1982) (see also Yor (1992a)). For $\nu \ge -1$ let $S^{(\nu)} := (R^{(\nu)})^2$ be the squared Bessel process of order ν, see A1.23. When $-1 < \nu < 0$ the boundary condition at 0 is reflection. Then

$$\begin{aligned}
&\mathbf{E}_x^{(\nu)}\Bigl(\exp\bigl(-\alpha S_t^{(\nu)} - \frac{\beta^2}{2}\int_0^t S_s^{(\nu)}\, ds\bigr)\Bigr) \\
&\quad = \bigl(\operatorname{ch}(\beta t) + 2\alpha\beta^{-1}\operatorname{sh}(\beta t)\bigr)^{-\nu-1} \exp\Bigl(-\frac{x\beta}{2}\frac{(1+2\alpha\beta^{-1}\operatorname{cth}(\beta t))}{(\operatorname{cth}(\beta t)+2\alpha\beta^{-1})}\Bigr)
\end{aligned}$$

and

$$\begin{aligned}
&\mathbf{E}_x^{(\nu)}\Bigl(\exp\bigl(-\frac{\beta^2}{2}\int_0^t S_s^{(\nu)}\, ds\bigr) \,\big|\, S_t^{(\nu)} = y\Bigr) \\
&\qquad = \frac{\beta t}{\operatorname{sh}(\beta t)} \exp\Bigl(\frac{x+y}{2t}(1-\beta t\operatorname{cth}(\beta t))\Bigr) \frac{I_\nu\Bigl(\frac{\beta\sqrt{xy}}{\operatorname{sh}(\beta t)}\Bigr)}{I_\nu\Bigl(\frac{\sqrt{xy}}{t}\Bigr)}.
\end{aligned}$$

Choosing here $\nu = -1/2$ gives the classical Cameron–Martin formula for Brownian motion (see also 1.1.9.3):

$$\mathbf{E}_0\Bigl(\exp\bigl(-\frac{\beta^2}{2}\int_0^t W_s^2\, ds\bigr)\Bigr) = (\operatorname{ch}(\beta t))^{-1/2}.$$

When $\nu = 0$ we obtain Lévy's area formula for 2-dimensional Brownian motion, i.e., letting $W^{(1)}$ and $W^{(2)}$ be two independent Brownian motions started at 0, then

$$\begin{aligned}
&\mathbf{E}\Bigl(\exp\bigl(i\beta\int_0^1 (W_s^{(1)}\, dW_s^{(2)} - W_s^{(2)}\, dW_s^{(1)})\bigr) \,\Big|\, W_1^{(1)} = x,\ W_1^{(2)} = y\Bigr) \\
&\qquad = \mathbf{E}_0^{(0)}\Bigl(\exp\bigl(-\frac{\beta^2}{2}\int_0^1 S_s^{(0)}\, ds\bigr) \,\big|\, S_1^{(0)} = x^2+y^2\Bigr) \\
&\qquad = \frac{\beta}{\operatorname{sh}\beta} \exp\Bigl(\frac{x^2+y^2}{2}(1-\beta\operatorname{cth}\beta)\Bigr).
\end{aligned}$$

7. Geometric Brownian motion

48. For $\sigma \neq 0, \mu \in \mathbb{R}$, and $x > 0$ consider the linear SDE

$$dX_t = \sigma\, X_t\, dW_t + \mu\, X_t\, dt, \quad X_0 = x,$$

where W is a standard Brownian started at 0. By III.15 this equation has a strong unique non-exploding solution, and it is a simple application of Itô's formula to show that the solution is given by

$$V_t := x \exp\Big(\sigma\, W_t + \Big(\mu - \frac{\sigma^2}{2}\Big)t\Big), \quad t \geq 0.$$

The process $V = \{V_t \ : \ t \geq 0\}$ is called a *geometric* or *exponential Brownian motion.* It is a linear diffusion with the generator

$$\frac{1}{2}\sigma^2\, x^2\, \frac{d^2}{dx^2} + \mu\, x\, \frac{d}{dx}.$$

Geometric Brownian motion is of primary interest in mathematical finance and insurance, where it is used as an (unrealistic but computional) model for various economical processes. Many recent results on geometric Brownian motion originate from this setting, see, e.g., Geman and Yor (1993), Dufresne (1990), Paulsen (1993), Shepp and Shiryaev (1993), and Yor (1992a).

For the basic characteristics of geometric Brownian motion, see A1.20. Part II Section 9 contains formulae associated to geometric Brownian motion; therein the drift coefficient μ is, for typographical reasons, taken to be $\sigma^2\nu + \dfrac{\sigma^2}{2}$.

49. We consider here transformations of geometric Brownian motion inspired by Lévy's construction of reflected Brownian motion and Pitman's construction of 3-dimensional Bessel process (see No. 40).

For $\mu \in \mathbb{R}$ let $W_t^{(\mu)} := W_t + \mu\, t$ and define the process $Y^{(1,\mu)}$ by setting

$$Y_t^{(1,\mu)} := \exp\big(-W_t^{(\mu)}\big) \sup_{s \leq t}\big\{\exp\big(W_s^{(\mu)}\big)\big\}.$$

Then, using the fact that the process

$$\Big\{\sup_{s \leq t} W_s^{(\mu)} - W_t^{(\mu)} \ : \ t \geq 0\Big\}$$

is identical in law to a reflecting Brownian motion with drift $-\mu$ (cf. No. 40), it is seen that $Y^{(1,\mu)}$ is a reflecting geometric Brownian motion with the generator

$$\mathcal{G} := \frac{1}{2}\, x^2\, \frac{d^2}{dx^2} + \big(\tfrac{1}{2} - \mu\big)\, x\, \frac{d}{dx}$$

acting on the domain

$$\mathcal{D} = \{f : \ f,\ \mathcal{G}f \in \mathcal{C}_b([1,\infty)),\ f'(1+) = 0\}.$$

The filtrations generated by W and $Y^{(1,\mu)}$ are the same. The process $Y^{(1,\mu)}$ is used in Shepp and Shiryaev (1994) when analyzing "Russian" options, and is associated to the notion of "dual martingale measure".

To formulate a similar extension of Pitman's theorem introduce $Y^{(2,\mu)}$ by setting

$$Y_t^{(2,\mu)} := \exp\bigl(-W_t^{(\mu)}\bigr) \sup_{s\le t}\bigl\{\exp\bigl(2\,W_s^{(\mu)}\bigr)\bigr\}.$$

Then, using Itô's formula, it is seen that $Y^{(2,\mu)}$ is a linear diffusion on $[1,\infty)$ starting at 1 (which is an entrance-not-exit boundary) with the generator

$$\frac{1}{2}\,x^2\,\frac{d^2}{dx^2} + \bigl(\tfrac{1}{2} + |\mu|\,\mathrm{cth}(|\mu|\ln x)\bigr)\,x\,\frac{d}{dx}.$$

The filtration generated by $Y^{(2,\mu)}$ is strictly smaller than the filtration generated by W.

50. Dufresne's identity. For $a \neq 0$, $b > 0$ we have (see Dufresne (1990))

$$\int_0^\infty e^{a\,W_s - b\,s}\,ds \quad \sim \quad \frac{2}{a^2}\,Z^{-1},$$

where $W_0 = 0$ and Z is a $\Gamma(\alpha,1)$-distributed random variable with $\alpha = 2b/a^2$, i.e.,

$$\mathbf{P}(Z \in dz) = \frac{1}{\Gamma(\alpha)}\,z^{\alpha-1}\,e^{-z}\,dz$$

(cf. 9.1.8.4.(1)). The proof of Dufresne's identity in Yor (1992c) is based on the representation of geometric Brownian motion as a random time change of a Bessel process (the so-called *Lamperti relation*). To explain this (cf. No.35), let R be a Bessel process of order $-\nu < 0$, started at 1 and independent of W. Then

(a) $$\bigl\{e^{W_t - \nu\,t}\ :\ t \ge 0\bigr\} \quad \sim \quad \bigl\{R_{A_t^{(-\nu)}}\ :\ t \ge 0\bigr\},$$

where

$$A_t^{(-\nu)} := \int_0^t e^{2(W_s - \nu s)}\,ds.$$

As $t \to \infty$

$$e^{W_t - \nu\,t} \ \to\ 0 \quad \text{a.s.}$$

and it follows from (a) that

$$A_\infty^{(-\nu)} \quad \sim \quad H_0^{(-\nu)} := \inf\{t\ :\ R_t^{(-\nu)} = 0\}.$$

The distribution of $H_0^{(-\nu)}$ can be computed using standard methods (cf. II.10), and this leads to Dufresne's identity. Another, perhaps shorter, approach is by time reversal using last exit times as explained in No. 45 and 46.

51. The law of the additive functional

$$A_t^{(\nu)} := \int_0^t e^{2\,(W_s + \nu\, s)}\, ds$$

is of interest in mathematical finance (especially in studies concerning Asian options), in mathematical physics (in studies on one-dimensional disorder systems, see Comtet, Monthus, and Yor (1998) and references therein), and in the theory of hyperbolic Brownian motion (see Yor (1997b)). We notice that the distribution of $A_t^{(\nu)}$ is given in 9.1.8.4. and the joint distribution of $A_t^{(\nu)}$ and $W_t^{(\nu)}$ in 9.1.8.8. In particular, for $\nu = 0$ we have ($A_t := A_t^{(0)}$)

$$\text{(a)} \qquad \mathbf{P}_0(A_t \in du,\ W_t \in dx) = \exp\Bigl(-\frac{1+e^{2x}}{2u}\Bigr)\theta_{e^x/u}(t)\frac{du\, dx}{u}.$$

The function $t \mapsto \theta_r(t)$ in (a) is the (unnormalized) Hartman–Watson density characterized via its Laplace transform:

$$\int_0^\infty \exp(-\tfrac{1}{2}\beta^2 t)\, \theta_r(t)\, dt = I_\beta(r),$$

where I_β is the usual modified Bessel function. We have also the integral representation due to Yor (1980),

$$\theta_r(t) = \frac{r}{\sqrt{2\pi^3 t}} \int_0^\infty \exp\Bigl(\frac{\pi^2 - y^2}{2t} - r\,\mathrm{ch}(y)\Bigr)\, \mathrm{sh}(y)\, \sin\bigl(\tfrac{\pi y}{t}\bigr)\, dy.$$

In A2.14 and in the tables, $\mathrm{i}_y(z)$ denotes this function and the connection between the different notations is

$$\mathrm{i}_y(z) = 2\,\theta_z(2y).$$

For further results (especially for positive and negative moments of A_t) see Yor (1980), (1992b), Dufresne (1999), and Donati-Martin, Matsumoto and Yor (2000a), (2000b).

52. Bougerol's identity. Consider the functional $t \mapsto A_t$ as introduced in No. 51. Then, letting W° be another standard Brownian motion started at 0 independent of W, we have (see Bougerol (1983)) for every fixed $t \ge 0$,

$$\text{(a)} \qquad W^\circ_{A_t} \quad \sim \quad \mathrm{sh}\,(W_t).$$

An equivalent formulation of (a) is

$$\int_0^t e^{W_s}\, dW_s^\circ \quad \sim \quad \mathrm{sh}\,(W_t).$$

In Alili, Dufresne and Yor (1997) it is proved that (a) can be extended to an equivalence in law for processes:

$$\Bigl\{e^{W_t}\int_0^t e^{-W_s}\, dW_s^\circ \ :\ t \ge 0\Bigr\} \quad \sim \quad \bigl\{\mathrm{sh}(W_t)\ :\ t \ge 0\bigr\}.$$

These results can be generalized for Brownian motion with drift. To give a flavor of this, we quote from Alili, Dufresne and Yor (1997) the result

$$\int_0^t e^{W_s+s}\, dW_s^\circ \quad \sim \quad \mathrm{sh}\,(W_t + \varepsilon\, t),$$

where ε is a random variable independent of W such that

$$\mathbf{P}(\varepsilon = +1) = \mathbf{P}(\varepsilon = -1) = 1/2.$$

We refer to Yor (1992b), Alili, Dufresne and Yor (1997), Alili and Gruet (1997), Comtet, Monthus and Yor (1998), and Matsumoto and Yor (1998) for proofs, further results and generalizations (especially for Brownian motion with drift) concerning Bougerol's identity.

53. We present here two additional extensions of Lévy's and Pitman's theorems (cf. No. 40 and No. 49). For the first one define the process $Z^{(1,\mu)}$ by setting

$$Z_t^{(1,\mu)} := \exp\bigl(-W_t^{(\mu)}\bigr) \int_0^t \exp\bigl(W_s^{(\mu)}\bigr) ds, \quad \mu \in \mathbb{R}.$$

Then (see Kramkov and Mordecki (1994) and Matsumoto and Yor (2000)) $Z^{(1,\mu)}$ is a linear diffusion with the generator

$$\frac{1}{2}\, x^2\, \frac{d^2}{dx^2} + \Bigl(\bigl(\tfrac{1}{2} - \mu\bigr)\, x + 1\Bigr)\, \frac{d}{dx}.$$

Moreover, the filtration generated by $Z^{(1,\mu)}$ is the same as the filtration generated by W.

In Kramkov and Mordecki (1994) $Z^{(1,\mu)}$ is used as a tool to solve the optimal stopping problem associated to the pricing of an American perpetual integral option.

For the second one introduce

$$Z_t^{(2,\mu)} := \exp\bigl(-W_t^{(\mu)}\bigr) \int_0^t \exp\bigl(2\, W_s^{(\mu)}\bigr) ds, \quad \mu \in \mathbb{R}.$$

Then (see Matsumoto and Yor (1999), (2000), (2001)) the process $Z^{(2,\mu)}$ is a linear diffusion with the generator

$$\text{(a)} \qquad \frac{1}{2}\, x^2\, \frac{d^2}{dx^2} + \left(\bigl(\tfrac{1}{2} - \mu\bigr)\, x + \frac{K_{1+\mu}(1/x)}{K_\mu(1/x)}\right) \frac{d}{dx},$$

where K_μ is the modified Bessel function of order μ. Using the relationship

$$K_{1+\mu}(x) = \mu\, x^{-1}\, K_\mu(x) - K_\mu'(x)$$

the generator in (a) takes the form

$$\frac{1}{2}\, x^2\, \frac{d^2}{dx^2} + \left(\frac{1}{2}\, x - \frac{K_\mu'(1/x)}{K_\mu(1/x)}\right) \frac{d}{dx}.$$

It is also proved in Matsumoto and Yor (2000) that the filtration generated by $Z^{(2,\mu)}$ is strictly smaller than the filtration generated by W; in analog with Pitman's theorem, see No. 40.

CHAPTER V

LOCAL TIME AS A MARKOV PROCESS

1. Diffusion local time

1. Let X be a (regular) diffusion on an interval I. It is assumed that the scale function s is in $\mathcal{C}^2(I)$ and that the infinitesimal generator (see II.7) is given by the ordinary differential operator

$$\frac{1}{2}a^2(x)\frac{d^2}{dx^2} + b(x)\frac{d}{dx}$$

with continuous a and b. For simplicity we also assume that $a^2(x) > 0$ for all $x \in I$. Then $s' > 0$ in the interior of I, and the measure $m^\star$ given by

$$m^\star(A) = \int_A a^{-2}(x)\,dx, \quad A \in \mathcal{B}(I)$$

is well defined and finite on compact subsets of I. From the existence of the local time L (see II.13) it follows that X has a local time $\{\ell(t,x) :\ t \geq 0,\ x \in I\}$ with respect to the measure $m^\star$ such that a.s. for all $t \geq 0$ and $A \in \mathcal{B}(I)$,

$$\int_0^t \mathbb{I}_A(X_s)\,ds = \int_A \ell(t,x)\,m^\star(dx).$$

In fact, letting B be as in II.9

$$\ell(t,x) = 2\,e^{B(x)}L(t,x).$$

Below we give results concerning the Markovian character of the process

$$\{\ell(\zeta,x) :\ x \in I\},$$

where ζ refers to the lifetime of X killed in an appropriate way. These kinds of results are usually called *Ray–Knight theorems,* due to the fundamental contributions of Ray (1963) and Knight (1963). Our basic reference, however, is Section 4 in Walsh (1978). For a different approach based on the representation theorems in the excursion filtrations see Jeulin and Yor (1979), McGill (1982) and Jeulin (1985).

The results in No. 2 and 3 are obtained from No. 4 – which is Walsh's theorem 4.1 – by an appropriate limiting procedure. The local time used by Walsh is constructed from downcrossings, as explained in the case of Brownian motion in IV.9 (see also Itô and McKean (1974) p. 222). Notice that our local time ℓ is twice Walsh's local time. The infinitesimal generators in this chapter are taken in the extended sense, see, e.g., Revuz and Yor (2001) p. 285.

2. We assume first that X is transient and $X_{\zeta-} = r$ a.s. Recall that r denotes the right endpoint of I. It is also assumed that r is not elastic (the case with elastic r is covered in No. 4). Let $X_0 = x$ and $\hat{M} := \inf\{X_t : t \geq 0\}$. Then the process

$$\{\ell(\zeta, y) : y \geq \hat{M}\}$$

is a non-time-homogeneous diffusion absorbed at 0 and having the space-time infinitesimal generator:

(a) for $\hat{M} < y \leq x$,

$$2z\frac{\partial^2}{\partial z^2} + \left(4 - \left(\frac{s''(y)}{s'(y)} + \frac{2\,s'(y)}{s(r) - s(y)}\right) z\right)\frac{\partial}{\partial z} + \frac{\partial}{\partial y},$$

(b) for $x < y < r$,

$$2z\frac{\partial^2}{\partial z^2} + \left(2 - \left(\frac{s''(y)}{s'(y)} + \frac{2\,s'(y)}{s(r) - s(y)}\right) z\right)\frac{\partial}{\partial z} + \frac{\partial}{\partial y}.$$

The process in (b) is started from the position of the process in (a) at time x but is otherwise independent of it. The distribution of $\hat{M}$ is

$$\mathbf{P}_x(\hat{M} \leq y) = \frac{s(r) - s(x)}{s(r) - s(y)}.$$

If l is non-singular, then due to the assumption $X_{\zeta-} = r$ a.s., the boundary condition at l is reflection. To complete the description in this case the $\mathbf{P}_x$-distribution of $\ell(\zeta, l)$ is needed. This can be computed as explained in II.27 (a). It can also be obtained by considering $-X$ and using the result in No. 3.

3. Let X be as in No. 2 but now let $X_{\zeta-} = l$ a.s., where l is the left endpoint of I. Then the process

$$\{\ell(\zeta, y) : y > l\}$$

is a non-time-homogeneous diffusion absorbed at 0 having the space-time infinitesimal generator:

(a) for $l < y \leq x$,

$$2z\frac{\partial^2}{\partial z^2} + \left(2 - \frac{s''(y)}{s'(y)}\, z\right)\frac{\partial}{\partial z} + \frac{\partial}{\partial y},$$

(b) for $x \leq y < r$,

$$2z\frac{\partial^2}{\partial z^2} - \frac{s''(y)}{s'(y)}\, z\,\frac{\partial}{\partial z} + \frac{\partial}{\partial y}.$$

The process in (b) is started from the position of the process in (a) at time x but is otherwise independent of it. In case l is elastic the $\mathbf{P}_x$-distribution of $\ell(\zeta, l)$ is needed to complete the description (cf. II.27 (a)).

4. Next assume that X, $X_0 = x$, is recurrent. Let L be the local time of X with respect to the speed measure and $\tau \sim \mathrm{Exp}(c)$, $c > 0$, independent of X. For a given $z_o \in I$, a new diffusion $\hat{X}$ is constructed by killing X at time

$$\zeta = \inf\{t : \ L(t, z_o) \geq \tau\}.$$

The killing measure of $\hat{X}$ is $c\,\varepsilon_{\{z_o\}}$. We normalize the scale function s so that $s(z_o) = 0$. Consider for $x \leq z_o$,

$$\hat{\mathbf{E}}_{z_o}(e^{-\alpha H_x}) = \frac{\varphi_\alpha(z_o)}{\varphi_\alpha(x)},$$

where φ_α is the fundamental decreasing solution associated to $\hat{X}$. Let $\alpha \to 0$ to obtain

$$\hat{\mathbf{P}}_{z_o}(H_x < \infty) = \frac{1}{\varphi_0(x)} = \frac{1}{1 - c\,s(x)}, \qquad x \leq z_o,$$

(see II.24; we may take $\varphi_0 \equiv 1$ for $x \geq z_o$). Following Walsh (1978), we introduce

$$p(x) := \begin{cases} 1, & x \geq z_o, \\ \dfrac{1}{1 - c\,s(x)}, & x \leq z_o. \end{cases}$$

Then

$$\{\ell(\zeta, y) : \ y > \hat{M}\}$$

is a non-time-homogeneous diffusion absorbed at 0 having the space-time infinitesimal generator:

(a) for $\hat{M} < y \leq x \wedge z_o$,

$$2z\frac{\partial^2}{\partial z^2} + \left(4 - \left(\frac{s''(y)}{s'(y)} + \frac{2c\,s'(y)}{1 - c\,s(y)}\right) z\right)\frac{\partial}{\partial z} + \frac{\partial}{\partial y},$$

(b) for $x \wedge z_o < y \leq x \vee z_o$,

$$2z\frac{\partial^2}{\partial z^2} + \left(2 - \frac{(p^2(y)\,s'(y))'}{p^2(y)\,s'(y)}\, z\right)\frac{\partial}{\partial z} + \frac{\partial}{\partial y},$$

(c) for $x \vee z_o < y < r$,

$$2z\frac{\partial^2}{\partial z^2} - \frac{s''(y)}{s'(y)}\, z\,\frac{\partial}{\partial z} + \frac{\partial}{\partial y}.$$

The processes in (a), (b), and (c) are knotted together in the same manner as in No's. 2 and 3. Observe also that if we study $-\hat{X}$ instead of $\hat{X}$ the result above characterizes the local time process (or any part of it) in the reversed y-direction. This is practical in some computations. Notice that the coefficient of z in the drift terms can in every case be written as in (b).

5. As can be seen from above, the squared Bessel processes of dimensions 0, 2, and 4 (or of orders −1, 0, and 1, respectively) have important roles in the Ray–Knight theorems. We let $(Z^{(\delta)}, \mathbf{Q}^{(\delta)})$, $\delta = 0,2,4$, respectively, denote these processes. The corresponding generators are

$$2z\frac{d^2}{dz^2} + \delta\frac{d}{dz}, \qquad \delta = 0,2,4.$$

These processes appear in No's. 6, 10 and 17 below. For basic characteristics of squared Bessel processes see A1.23.

For $\theta \in \mathbb{R}$ we let $(Z^{(\delta,\theta)}, \mathbf{Q}^{(\delta,\theta)})$, $\delta = 0,2,4$, denote diffusions with generators

$$2z\frac{d^2}{dz^2} + (\delta - \theta z)\frac{d}{dz}, \qquad \delta = 0,2,4,$$

respectively. These are particular cases of squared radial Ornstein–Uhlenbeck processes (see A1.26 for their basic characteristics, and also IV.37), and are used below in No's. 9, 11, 15 and 16 when treating Brownian motion with drift.

To construct non-time-homogeneous diffusions with generators of the form displayed, e.g., in (a) and (b) in No. 3, a deterministic space-time transformation can be applied. Indeed, let g be an increasing function in $\mathcal{C}^2(I)$ such that $g' > 0$. Then for $a \in I$ the process

(a) $$\Big\{\frac{1}{g'(y)} Z^{(\delta)}_{g(y)-g(a)} \,:\, y \geq a\Big\}$$

has the extended generator

$$2z\frac{\partial^2}{\partial z^2} + \Big(\delta - \frac{g''(y)}{g'(y)}\, z\Big)\frac{\partial}{\partial z} + \frac{\partial}{\partial y}$$

(cf. McGill (1982)). To give an example, notice that

$$Z^{(\delta,\theta)} \;\sim\; \Big\{e^{-\theta t} Z^{(\delta)}_{(e^{\theta t}-1)/\theta} \,:\, t \geq 0\Big\}.$$

Deterministic space-time transformations as in (a) are used to compute distributions of the transformed "complicated" process in terms of the corresponding known distributions of the original "simple" process (see the remarks in No's. 6, 8, 17, 18 and 19).

2. Local time of Brownian motion

6. Local time stopped at the first hitting time. Let $b \in \mathbb{R}$ be given and suppose that $W_0 = x \leq b$. We are interested in characterizing the law of the process

$$\ell_b := \{\ell(H_b, y) :\; y \leq b\}.$$

Recall that the scale function of W is $s(x) = x$ and the measure $m^\star$ is the Lebesgue measure. In the case (A) below the y-direction is from right to left, i.e, from b to $-\infty$. In the case (B) it is the opposite.

(A) From No. 3 by considering $-W$ we obtain

$$\{\ell(H_b, b-y) : 0 \le y \le b-x\} \sim \{Z_y^{(2)} : 0 \le y \le b-x\},$$

$$\{\ell(H_b, x-y) : y \ge 0\} \sim \{Z_y^{(0)} : y \ge 0\}.$$

The squared Bessel process $Z^{(2)}$ is started at its entrance boundary 0 and $Z^{(0)}$ is started from the position of $Z^{(2)}$ at time $b-x$; otherwise $Z^{(0)}$ and $Z^{(2)}$ are independent. The process $Z^{(0)}$ is run until it hits its exit boundary 0.

(B) Given that $\hat{M} := \inf\{W_t : t \le H_b\} = m$ we have from No. 2,

$$\{\ell(H_b, m+y) : 0 \le y \le x-m\} \sim \{Y_y^{(4,b-m)} : 0 \le y \le x-m\},$$

$$\{\ell(H_b, x+y) : 0 \le y < b-x\} \sim \{Y_y^{(2,b-x)} : 0 \le y < b-x\},$$

where $Y_0^{(4,b-m)} := 0$ and the generators of $Y^{(4,b-m)}$ and $Y^{(2,b-x)}$ are

$$2z\frac{\partial^2}{\partial z^2} + \left(4 - \frac{2z}{b-m-y}\right)\frac{\partial}{\partial z} + \frac{\partial}{\partial y},$$

$$2z\frac{\partial^2}{\partial z^2} + \left(2 - \frac{2z}{b-x-y}\right)\frac{\partial}{\partial z} + \frac{\partial}{\partial y},$$

respectively. The process $Y^{(2,b-x)}$ is started from the position of $Y^{(4,b-m)}$ at time $x-m$ but otherwise $Y^{(2,b-x)}$ and $Y^{(4,b-m)}$ are independent.

Remark 1. The distribution of $Z_{b-x}^{(2)}$ can be found in A1.23. Hence (see also II.27 (a)),

$$\mu((v,\infty)) := \mathbf{P}_x(\ell(H_b,x) > v) = \mathbf{Q}_0^{(2)}(Z_{b-x}^{(2)} > v) = \exp\Bigl(-\frac{v}{2(b-x)}\Bigr).$$

Using, e.g., time reversal as explained in the case of Bessel processes in IV.46 we have

(a) $$\mathbf{Q}_y^{(0)}(H_0 \in dt) = \frac{y}{2t^2}e^{-y/2t}\,dt.$$

Let $\hat{M} := \inf\{W_t : t \le H_b\}$. Then $\ell(H_b, \hat{M}) = 0$, and it can be checked using (a) that

$$\mathbf{P}_x(\hat{M} < m) = \frac{b-x}{b-m} = \int_0^\infty \mu(dy)\mathbf{Q}_y^{(0)}(H_0 \ge x-m), \quad m \le x.$$

Remark 2. Let for $a > 0$ and $0 \le y < a$,

$$y \mapsto g(y,a) := \frac{1}{a-y} - \frac{1}{a}.$$

From No. 5,

$$Y^{(4,b-m)} \sim \Bigl\{\frac{1}{g'(y,b-m)}\,Z_{g(y,b-m)}^{(4)} : 0 \le y \le x-m\Bigr\},$$

with $Z_0^{(4)} = 0$, and

$$Y^{(2,b-x)} \sim \Big\{\frac{1}{g'(y,b-x)}\, Z^{(2)}_{g(y,b-x)} \,:\, 0 \le y < b-x\Big\},$$

where $\dfrac{1}{g'(0,b-x)} Z_0^{(2)}$ has the law μ_m given in (b) below. It follows from the SLLN (cf. IV.4) that $\lim_{y\to b-x} Y_y^{(2,b-x)} = 0$ a.s.

Remark 3. Using the explicit expressions for the transition density of $Z^{(4)}$ it is seen that $Y^{(4,b-m)}$, when considered on the time axis $[0, b-m)$, is identical in law to $Z^{(4)}$ started at 0 and conditioned to hit 0 at time $b-m$. The construction is similar to the construction of the 3-dimensional Bessel bridge (see IV.15) and the Brownian bridge (see IV.23). From this description of $Y^{(4,b-m)}$ it follows that

(b) $$\mu_m(dv) := \mathbf{P}_x(\ell(H_b, x) \in dv | \hat{M} = m) = \lambda^2 v\, e^{-\lambda v}\, dv$$

with $\lambda = (b-m)/2(x-m)(b-x)$. Moreover, it is seen that $Y^{(2,b-x)}$ is identical in law to $Z^{(2)}$ started with the initial distribution μ_m and conditioned to hit 0 at time $b-x$.

7. **Local time stopped at the first range time.** Assume that W is started at x. For $r > 0$ define the first range time (of the level r) via

$$\theta_r := \inf\Big\{t : \sup_{0\le s\le t} W_s - \inf_{0\le s\le t} W_s = r\Big\}.$$

Given $W_{\theta_r} = z > x$ the process

$$\ell_\theta := \{\ell(\theta_r, y) :\; z - r \le y \le z\}.$$

is identical in law to the process

$$\{\ell(H_z, y) :\; z - r \le y \le z\}$$

given $\hat{M} := \inf\{W_t :\; t \le H_z\} = z - r$. Clearly, we have a similar description in the case $W_{\theta_r} = z < x$. Therefore, the law of ℓ_θ can be deduced from (B) in No. 6.

Remark 1. From Remark 3 in No. 6, using symmetry,

$$\mathbf{P}_x(\ell(\theta_r, x) \in dv | W_{\theta_r} = z) = \eta^2 v\, e^{-\eta v}\, dv$$

with $\eta = r/2(r - |x-z|)|z-x|$. Further, from VI.22,

$$\mathbf{P}_x(W_{\theta_r} \in dz) = \frac{|z-x|}{r^2}\, dz, \qquad |z-x| \le r.$$

Remark 2. Because $Y^{(4,b-m)}$, as introduced in No. 6, is a diffusion bridge starting from 0 and ending at 0 at time $b-m$ it has the time reversal property (see Salminen (1997))

(a) $$\{Y_y^{(4,b-m)} : 0 \le y \le b-m\} \sim \{Y_{b-m-y}^{(4,b-m)} : 0 \le y \le b-m\}.$$

Applying (a) with $b = z$ and $m = z - r$ we obtain, given $W_{\theta_r} = z > x$,

$$\{\ell(\theta_r, x-y) : 0 \le y \le x-z+r\} \sim \{Y_{z-x+y}^{(4,r)} : 0 \le y \le x-z+r\}.$$

This identity in law is used in the computations in VI.23.

8. Local time stopped at the first exit time from a finite interval. Let $a < b$ be given and suppose that $W_0 = x \in (a,b)$. Consider

$$\ell_{a,b} := \{\ell(H_{a,b}, y) : a \le y \le b\},$$

given $W_{H_{a,b}} = b$. Under this condition $\{W_t : 0 \le t < H_{a,b}\}$ is a linear diffusion R which is killed when it hits b. The generator of R is

$$\frac{1}{2}\frac{d^2}{dx^2} + \frac{1}{x-a}\frac{d}{dx}, \qquad x > a,$$

the scale function is $s(x) = -(x-a)^{-1}$, and the measure $m^\star$ is the Lebesgue measure. Therefore, the local time of R up to H_b (with respect to $m^\star$) is identical in law to $\ell_{a,b}$ given $W_{H_{a,b}} = b$ and the results in No's. 2 and 3 are applicable for $\ell_{a,b}$. Notice also that the boundary point a is entrance-not-exit for the diffusion R.

(A) From No. 3, conditioning on $W_{H_{a,b}} = b$,

$$\{\ell(H_{a,b}, b-y) : 0 \le y \le b-x\} \sim \{Y_y^{(2,b-a)} : 0 \le y \le b-x\},$$

$$\{\ell(H_{a,b}, x-y) : 0 \le y < x-a\} \sim \{Y_y^{(0,x-a)} : 0 \le y < x-a\},$$

where $Y_0^{(2,b-a)} := 0$ and the generators of $Y^{(2,b-a)}$ and $Y^{(0,x-a)}$ are

$$2z\frac{\partial^2}{\partial z^2} + \Big(2 - \frac{2z}{b-a-y}\Big)\frac{\partial}{\partial z} + \frac{\partial}{\partial y},$$

$$2z\frac{\partial^2}{\partial z^2} - \frac{2z}{x-a-y}\frac{\partial}{\partial z} + \frac{\partial}{\partial y},$$

respectively. The process $Y^{(0,x-a)}$ is started from the position of $Y^{(2,b-a)}$ at time $b-x$ but otherwise $Y^{(0,x-a)}$ and $Y^{(2,b-a)}$ are independent.

(B) Here we consider the same process as in (A) but with the y-direction reversed. Using No. 2 and (B) in No. 6 it is seen that given $W_{H_{a,b}} = b$ and $\hat{M}_{a,b} := \inf\{W_t : t \le H_{a,b}\} = m$,

$$\{\ell(H_{a,b}, m+y) : 0 \le y \le b-m\}$$

is identical in law to

$$\{\ell(H_b, m+y) : \ 0 \le y \le b-m\}$$

given $\hat{M}_b := \inf\{W_t : \ t \le H_b\} = m$. Of course, the (conditional) distributions of $\hat{M}_b$ and $\hat{M}_{a,b}$ are different.

Remark 1. Comparing the generator of $Y^{(2,b-a)}$ with the generator of $Y^{(2,b-x)}$ in (B) in No. 6 reveals that $Y^{(2,b-a)}$ can be described similarly as $Y^{(2,b-x)}$. Indeed, $Y^{(2,b-a)}$ is identical in law to $Z^{(2)}$ started from 0 and conditioned to hit 0 at time $b-a$. This gives

$$\mu((v,\infty)) := \mathbf{P}_x(\ell(H_{a,b},x) > v \mid W_{H_{a,b}} = b) = \exp\Big(-\frac{(b-a)v}{2(b-x)(x-a)}\Big).$$

An alternative way to find this distribution is to use II.27 (a) (see also 1.3.3.6).

Remark 2. We show that $Y^{(0,x-a)}$ is identical in law to $Z^{(0)}$ started with the initial distribution μ and conditioned to hit 0 before time $x-a$. To realize the conditioning we use Doob's h-transform techniques. For this, assume first that $Z^{(0)}$ is started at x_o and let $n_{x_o}(0,\cdot)$ be the $\mathbf{Q}^{(0)}_{x_o}$-density of H_0, see (a) in No. 6. Define for $\xi > 0$,

$$h^\star(z,t;0,\xi) := \begin{cases} \dfrac{n_z(0,\xi-t)}{n_{x_o}(0,\xi)}, & 0 \le t < \xi, \\ 0, & t \ge \xi, \end{cases}$$

and introduce for $\xi < x-a$ the measure

$$\kappa(d\xi) := \mathbf{Q}^{(0)}_{x_o}(H_0 \in d\xi | H_0 < x-a) = \frac{x_o}{2\xi^2}\exp\Big(-\frac{x_o}{2\xi} + \frac{x_o}{2(x-a)}\Big)d\xi.$$

Keeping z and t fixed we integrate the function $\xi \mapsto h^\star(z,t;0,\xi)$ with respect to κ to obtain

$$h(z,t) := \int_0^{x-a} h^\star(z,t;0,\xi)\kappa(d\xi) = \exp\Big(-\frac{z}{2(x-a-t)} + \frac{x_o}{2(x-a)}\Big).$$

The function h is proportional to the probability that starting $Z^{(0)}$ from the space-time point (z,t), $t < x-a$, the hitting time H_0 will be less than $x-a$. The desired conditioning can now be realized using this function h. Because

$$-h'_t(z,t) = 2z\, h''_{zz}(z,t)$$

it is seen that the h-transform of $Z^{(0)}$ has the same extended generator as $Y^{(0,x-a)}$.

9. Local time stopped at an exponential time. Let τ be an exponentially distributed random variable (with parameter $\lambda > 0$) independent of W. By No. 4

we obtain (see the remark below) that under the conditions $W_0 = x$, $W_\tau = z_o \geq x$, and $\hat{M}_\tau := \inf\{W_t :\ t \leq \tau\} = m$,

$$\{\ell(\tau, m+y) :\ 0 \leq y \leq x-m\} \sim \{Z_y^{(4,\theta)} :\ 0 \leq y \leq x-m\},$$
$$\{\ell(\tau, x+y) :\ 0 \leq y \leq z_o - x\} \sim \{Z_y^{(2,\theta)} :\ 0 \leq y \leq z_o - x\},$$
$$\{\ell(\tau, z_o+y) :\ y \geq 0\} \sim \{Z_y^{(0,\theta)} :\ y \geq 0\},$$

where the processes $Z^{(\delta,\theta)}$, $\delta = 0, 2, 4$, are as in No. 5 with $\theta = 2\sqrt{2\lambda}$.

Remark. To show that No. 4 can be applied let $\hat{W}$ be W killed at time τ. Then $\hat{W}$ conditioned on $W_\tau = z_o$ is a linear diffusion $\bar{W}$ obtained from $\hat{W}$ by constructing the h-transform with $h = 2\sqrt{2\lambda}\, G_\lambda(\cdot, z_o)$, where G is the Green function of W :

$$G_\lambda(x,y) = \frac{1}{2\sqrt{2\lambda}} \exp(-\sqrt{2\lambda}|x-y|).$$

From II.31 it is seen that

$$m^h(dy) = \exp(-2\sqrt{2\lambda}|y - z_o|)\, 2dy, \quad s^h(dy) = \exp(2\sqrt{2\lambda}|y - z_o|)\, dy,$$
$$k^h(dy) = 2\sqrt{2\lambda}\, \varepsilon_{z_o}(dy)$$

can be taken to be the speed, scale, and killing measure, respectively, of the h-transform $\bar{W}$. Hence, $\bar{W}$ satisfies the assumptions in No. 4. Consequently, letting ζ be the lifetime of $\bar{W}$ and ℓ^o the local time of $\bar{W}$ with respect to the Lebesgue measure, we have conditionally on $W_\tau = z_o$,

$$\{\ell(\tau, y) :\ y \in \mathbb{R}\} \sim \{\ell^o(\zeta, y) :\ y \in \mathbb{R}\}.$$

Next, we remark that for all $x \leq y \leq z_o$,

$$\mathbf{P}_x(\ell(\tau, y) \geq v \mid W_\tau = z_o) = \exp(-\sqrt{2\lambda}\, v).$$

Finally, observe by considering $-W$ that $Z^{(0,\theta)}$ and $Z^{(4,\theta)}$ are, in a sense (cf. II.33), time reversals of each other.

10. Local time stopped at the inverse local time. For $z_o \in \mathbb{R}$ let

$$\varrho := \varrho(t, z_o) := \inf\{s :\ \ell(s, z_o) > t\}$$

be the inverse Brownian local time evaluated at $t \geq 0$. Assume that $W_0 = x < z_o$. Then

$$\{\ell(\varrho, z_o - y) :\ 0 \leq y \leq z_o - x\} \sim \{Z_y^{(2)} :\ 0 \leq y \leq z_o - x\},$$
$$\{\ell(\varrho, x - y) :\ y \geq 0\} \sim \{Z_y^{(0\,|\,1)} :\ y \geq 0\},$$
$$\{\ell(\varrho, z_o + y) :\ y \geq 0\} \sim \{Z_y^{(0\,|\,2)} :\ y \geq 0\},$$

where $Z^{(0\,|\,1)}$ and $Z^{(0\,|\,2)}$ are 0-dimensional squared Bessel processes. The processes $Z^{(2)}$ and $Z^{(0\,|\,2)}$ are started at t and $Z^{(0\,|\,1)}$ from the position of $Z^{(2)}$ at time $z_o - x$ but otherwise $Z^{(0\,|\,1)}$, $Z^{(0\,|\,2)}$ and $Z^{(2)}$ are independent.

Remark. The statement above cannot be proved directly using the general results in No's. 2, 3, and 4, and we refer to Knight (1981) p. 137 for a proof. However, the Laplace transforms of the finite dimensional distributions can be obtained by "inverting" in No. 9 with respect to λ, see also VI.16.

3. Local time of Brownian motion with drift

11. Total local time. Let $W^{(\mu)}$ be a Brownian motion with drift $\mu > 0$ started at x and $\ell^{(\mu)}$ its local time with respect to the Lebesgue measure. Because $W^{(\mu)}$ is transient its total local time $\ell^{(\mu)}(\infty, y)$ is a.s. finite for all y. Here we characterize the law of

$$\{\ell^{(\mu)}(\infty, y) : \ y \in \mathbb{R}\}.$$

The processes $Z^{(\delta,\theta)}$, $\delta = 0, 2, 4$, appearing below are as in No. 5 with $\theta = 2\mu$.

(A) By No. 3,

$$\{\ell^{(\mu)}(\infty, -y) : \ -\infty < y \le -x\} \ \sim \ \{Z_y^{(2,\theta)} : \ -\infty < y \le -x\},$$

$$\{\ell^{(\mu)}(\infty, x - y) : \ y \ge 0\} \ \sim \ \{Z_y^{(0,\theta)} : \ y \ge 0\}.$$

(B) Next, by No. 2, given $\hat{M} := \inf\{W_t^{(\mu)} : \ t \ge 0\} = m$

$$\{\ell^{(\mu)}(\infty, m + y) : \ 0 \le y \le x - m\} \ \sim \ \{Z_y^{(4,\theta)} : \ 0 \le y \le x - m\},$$

$$\{\ell^{(\mu)}(\infty, x + y) : \ y \ge 0\} \ \sim \ \{Z_y^{(2,\theta)} : \ y \ge 0\}.$$

Remark. Notice that $Z^{(2,\theta)}$ is positively recurrent and its stationary probability distribution is an exponential distribution with parameter μ. Because $Z^{(2,\theta)}$ in (A) is considered on the time axis $(-\infty, -x)$, it is, in fact, considered in its stationary state. This means that the finite dimensional distributions of $Z^{(2,\theta)}$ are given for $t_1 < t_2 < \cdots < t_n < -x$ by

$$\begin{aligned} &\mathbf{Q}_{\hat{m}}^{(2,\theta)}(Z_{t_1}^{(2,\theta)} \in dz_1, \, Z_{t_2}^{(2,\theta)} \in dz_2, \, \dots, \, Z_{t_n}^{(2,\theta)} \in dz_n) \\ &\quad = \hat{m}(dz_1)\, \hat{r}(t_2 - t_1; z_1, z_2)\, m(dz_2) \cdot \ldots \cdot \hat{r}(t_n - t_{n-1}; z_{n-1}, z_n)\, m(dz_n), \end{aligned}$$

where $\hat{m}$, m, and $\hat{r}$ are the stationary probability measure, speed measure, and transition density of $Z^{(2,\theta)}$, respectively. In particular,

$$\begin{aligned} \mathbf{P}_x(\ell^{(\mu)}(\infty, y) \ge v) &= \mathbf{P}_x(\ell^{(\mu)}(\infty, y) \ge v, \ H_y < \infty) \\ &= \mathbf{P}_x(H_y < \infty)\mathbf{P}_y(\ell^{(\mu)}(\infty, y) \ge v) \\ &= \begin{cases} e^{-\mu v}, & y \ge x, \\ e^{-2\mu(x-y) - \mu v}, & y \le x, \end{cases} \end{aligned}$$

(cf. II.27 (a) for $y = x$).

12. Local time stopped at the first hitting time, part 1. The next Ray–Knight theorem for $W^{(\mu)}$, $\mu > 0$, gives the law of

$$\{\ell^{(\mu)}(H_b, y) : \ y \le b\}$$

when $W^{(\mu)}$ is initiated at $x < b$. The processes $Z^{(\delta,\theta)}$, $\delta = 0, 2, 4$, are as in No. 11.

(A) Considering $-W^{(\mu)}$ it is seen from No. 3 that

$$\{\ell^{(\mu)}(H_b, b-y):\ 0 \le y \le b-x\} \sim \{Z_y^{(2,\theta)}:\ 0 \le y \le b-x\},$$

$$\{\ell^{(\mu)}(H_b, x-y):\ y \ge 0\} \sim \{Z_y^{(0,\theta)}:\ y \ge 0\}.$$

(B) Given that $\hat{M} := \inf\{W_t^{(\mu)}:\ t \le H_b\} = m$,

$$\{\ell^{(\mu)}(H_b, m+y):\ 0 \le y \le x-m\} \sim \{Y_y^{(4,\mu,b-m)}:\ 0 \le y \le x-m\},$$

$$\{\ell^{(\mu)}(H_b, x+y):\ 0 \le y < b-x\} \sim \{Y_y^{(2,\mu,b-x)}:\ 0 \le y < b-x\},$$

where $Y_0^{(4,\mu,b-m)} := 0$ and the generators of $Y^{(4,\mu,b-m)}$ and $Y^{(2,\mu,b-x)}$ are

$$2z\frac{\partial^2}{\partial z^2} + (4 - 2\mu\, z \operatorname{cth}(\mu(b-m-y)))\frac{\partial}{\partial z} + \frac{\partial}{\partial y},$$

$$2z\frac{\partial^2}{\partial z^2} + (2 - 2\mu\, z \operatorname{cth}(\mu\,(b-x-y)))\frac{\partial}{\partial z} + \frac{\partial}{\partial y},$$

respectively. The process $Y^{(2,\mu,b-x)}$ is started from the position of $Y^{(4,\mu,b-m)}$ at time $x-m$ but otherwise $Y^{(2,\mu,b-x)}$ and $Y^{(4,\mu,b-m)}$ are independent.

Remark 1. In particular,

(a) $$\mathbf{P}_x(\ell^{(\mu)}(H_b, x) \ge v) = \exp\Bigl(-\frac{\mu}{1-\exp(-2\mu\,(b-x))}v\Bigr)$$

(see also 2.2.3.4).

Remark 2. It can be proved that $Y^{(2,\mu,b-x)}$ is identical in law to $Z^{(2,2\mu)}$ started with the exponential distribution given in (a) and conditioned to hit 0 at time $b-x$.

13. Local time stopped at the first hitting time, part 2. Let $W^{(\mu)}$, $\mu > 0$, be started at $x > a$ and killed if it hits a. For this process we study the total local time

$$\ell_\infty^{(\mu)} := \{\ell^{(\mu)}(\infty, y):\ y \ge a\}.$$

Conditioned on $H_a < \infty$, the process $W^{(\mu)}$ is identical in law to $W^{(-\mu)}$ killed when it hits a. It follows that the law of $\ell_\infty^{(\mu)}$ is as described in No. 12. On the other hand, given $H_a = \infty$ and $\hat{M} := \inf\{W_t:\ t \ge 0\} = m$, we obtain from No. 2,

$$\{\ell^{(\mu)}(\infty, m+y):\ 0 \le y \le x-m\} \sim \{Z_y^{(4,\theta)}:\ 0 \le y \le x-m\},$$

$$\{\ell^{(\mu)}(\infty, x+y):\ y \ge 0\} \sim \{Z_y^{(2,\theta)}:\ y \ge 0\},$$

where $Z^{(\delta,\theta)}$, $\delta = 0, 2, 4$, are as in No's. 11 and 12.

Remark 1. The process $W^{(\mu)}$ given $H_a = \infty$ is identical in law (cf. IV.33) to the diffusion X^+ associated to the generator

$$\frac{1}{2}\frac{d^2}{dx^2} + \mu \operatorname{cth}(\mu\,(x-a))\frac{d}{dx}.$$

The scale function of X^+ is $s(x) = 1 - \operatorname{cth}(\mu\,(x-a))$. Notice that $s(a) = -\infty$ and $s(+\infty) = 0$. Let ℓ^+ denote the total local time of X^+. Then $\ell^{(\mu)}_\infty$ under the condition $H_a = \infty$ is identical in law to ℓ^+, for which No. 2 can be applied. Using the scale function of X^+ it is seen that

$$\mathbf{P}_x(\hat{M} \in dm \mid H_a = \infty) = \mu\, e^{2\mu\,(m-a)}\,(\operatorname{cth}(\mu\,(x-a)) - 1)\,dm,$$

and so, evoking the explicit form of the transition density of $Z^{(4,\theta)}$ (see A1.26) and integrating, we obtain

(a) $$\mathbf{P}_x(\ell^{(\mu)}(\infty, x) \geq v \mid H_a = \infty) = \exp\Bigl(-\frac{\mu \exp(\mu\,(x-a))}{2\operatorname{sh}(\mu\,(x-a))}v\Bigr).$$

An alternative and more direct way to compute this distribution is to use II.27 (a) (see also 2.2.3.2.(1)).

Remark 2. We leave the complete description in the other y-direction to the reader. From the point of view of explicit computation it is, however, good to know that

$$\{\ell^{(\mu)}(\infty, x-y) :\ 0 \leq y < x-a\} \ \sim\ \{Y_y^{(0,\mu,x-a)} :\ 0 \leq y < x-a\},$$

where the generator of $Y^{(0,\mu,x-a)}$ is

$$2z\frac{\partial^2}{\partial z^2} - 2\mu\, z \operatorname{cth}(\mu(x-a-y))\frac{\partial}{\partial z} + \frac{\partial}{\partial y}.$$

The process $Y^{(0,\mu,x-a)}$ is started at time 0 with the exponential distribution given in (a). We remark that $Y^{(0,\mu,x-a)}$ is identical in law to $Z^{(0,2\mu)}$ conditioned to hit 0 before time $x-a$.

14. Local time stopped at the first exit time from a finite interval. Consider $W^{(\mu)}$ started at $x \in (a,b)$ and killed when it exits (a,b) (cf. No. 7). Conditioned on $W^{(\mu)}_{H_{a,b}} = a$, this process is a diffusion with the generator (cf. No. 13)

$$\frac{1}{2}\frac{d^2}{dx^2} - |\mu| \operatorname{cth}(|\mu|\,(b-x))\frac{d}{dx},$$

and having a as a killing boundary. Therefore, the Ray–Knight theorem here is very similar to the one in No. 12. For the y-direction from left to right we obtain from No. 3 that, given $W^{(\mu)}_{H_{a,b}} = a$,

$$\{\ell^{(\mu)}(H_{a,b}, a+y) :\ 0 \leq y \leq x-a\} \ \sim\ \{Y_y^{(2,|\mu|,b-a)} :\ 0 \leq y \leq x-a\},$$

$$\{\ell^{(\mu)}(H_{a,b},x+y):\ 0\le y<b-x\}\ \sim\ \{Y_y^{(0,|\mu|,b-x)}:\ 0\le y<b-x\},$$

where $Y_0^{(2,|\mu|,b-a)}:=0$ and the generators of $Y^{(2,|\mu|,b-a)}$ and $Y^{(0,|\mu|,b-x)}$ are

$$2z\frac{\partial^2}{\partial z^2}+\big(2-2|\mu|\,z\,\mathrm{cth}(|\mu|(b-a-y))\big)\frac{\partial}{\partial z}+\frac{\partial}{\partial y},$$

$$2z\frac{\partial^2}{\partial z^2}-2|\mu|\,z\,\mathrm{cth}(|\mu|(b-x-y))\frac{\partial}{\partial z}+\frac{\partial}{\partial y},$$

respectively.

Remark 1. The process $Y^{(2,|\mu|,b-a)}$ is identical in law to $Z^{(2,2|\mu|)}$ started at 0 and conditioned to hit 0 at time $b-a$. This gives, e.g.,

(a) $$\mathbf{P}_x(\ell^{(\mu)}(H_{a,b},x)\ge v\mid W^{(\mu)}_{H_{a,b}}=a)=\exp\Big(-\frac{|\mu|\,\mathrm{sh}(|\mu|(b-a))}{2\,\mathrm{sh}(|\mu|(x-a))\,\mathrm{sh}(|\mu|(b-x))}v\Big).$$

The process $Y^{(0,|\mu|,b-x)}$ is identical in law to $Z^{(0,2|\mu|)}$ started with the exponential distribution given in (a) and conditioned to hit 0 before time $b-x$.

Remark 2. We leave the complete description in the other y-direction also here to the interested reader. Notice, however, that from the characterization of $Y^{(2,|\mu|,b-a)}$ (or from No. 2) it follows that

$$\{\ell^{(\mu)}(H_{a,b},x-y):\ 0\le y<x-a\}\ \sim\ \{Y_y^{(2,|\mu|,x-a)}:\ 0\le y<x-a\},$$

where $Y^{(2,|\mu|,x-a)}$ is started at time 0 with the distribution given in (a) and the generator of $Y^{(2,|\mu|,x-a)}$ is

$$2z\frac{\partial^2}{\partial z^2}+\big(2-2|\mu|\,z\ \mathrm{cth}(|\mu|(x-a-y))\big)\frac{\partial}{\partial z}+\frac{\partial}{\partial y}.$$

Hence $Y^{(2,|\mu|,x-a)}$ is identical in law to $Z^{(2,2|\mu|)}$ conditioned to hit 0 at time $x-a$.

15. Local time stopped at an exponential time. Let $\tau\sim\mathrm{Exp}(\lambda)$, independent of $W^{(\mu)}$, and let $\hat W^{(\mu)}$ be $W^{(\mu)}$ killed at time τ. Recall (see A1.14) that the Green function of $W^{(\mu)}$ is

$$G_\alpha(x,y)=\frac{1}{2\sqrt{2\alpha+\mu^2}}e^{-(\sqrt{2\alpha+\mu^2}+\mu)\,x}e^{(\sqrt{2\alpha+\mu^2}-\mu)\,y},\quad x\ge y.$$

Then, $\hat W^{(\mu)}$ under the condition $W_\tau^{(\mu)}=z_o$ is a diffusion which can be constructed as an h-transform of $\hat W^{(\mu)}$ using

$$h(y)=\frac{G_\lambda(y,z_o)}{G_\lambda(z_o,z_o)}.$$

Computing the basic characteristics of this h-process and comparing these with the corresponding characteristics in No. 9 it is seen that the conditioned processes are very similar in both cases. Essentially, all we have to do is to change the parameter λ in No. 9 to $\lambda + \mu^2/2$. Once this change is made we obtain the desired description for

$$\{\ell^{(\mu)}(\tau, y) : \ y \in \mathbb{R}\}.$$

16. Local time stopped at the inverse local time. For $z_o \in \mathbb{R}$ let

$$\varrho := \varrho^{(\mu)}(t, z_o) := \inf\{s : \ \ell^{(\mu)}(s, z_o) > t\}$$

be the inverse local time evaluated at $t \geq 0$. Assume that $W_0^{(\mu)} = x < z_o$, and let $Z^{(0,\theta)}$ and $Z^{(2,\theta)}$ be as in No. 5 with $\theta = 2|\mu|$. Then given $\varrho < \infty$,

$$\{\ell^{(\mu)}(\varrho, z_o - y) : \ 0 \leq y \leq z_o - x\} \ \sim \ \{Z_y^{(2,\theta)} : \ 0 \leq y \leq z_o - x\},$$

$$\{\ell^{(\mu)}(\varrho, x - y) : \ y \geq 0\} \ \sim \ \{Z_y^{(0,\theta\,|\,1)} : \ y \geq 0\},$$

$$\{\ell^{(\mu)}(\varrho, z_o + y) : \ y \geq 0\} \ \sim \ \{Z_y^{(0,\theta\,|\,2)} : \ y \geq 0\},$$

where $Z^{(0,\theta\,|\,1)}$ and $Z^{(0,\theta\,|\,2)}$ are independent copies of $Z^{(0,\theta)}$. The processes $Z^{(2,\theta)}$ and $Z^{(0,\theta\,|\,2)}$ are started at t, and $Z^{(0,\theta\,|\,1)}$ is initiated from the position of $Z^{(2,\theta)}$ at time $z_o - x$.

4. Local time of Bessel process

17. Total local time when $\nu > 0$. Let $R^{(\nu)}$ denote a Bessel process of order ν and $\ell^{(\nu)}$ its local time with respect to the Lebesgue measure. For $\nu > 0$, $R^{(\nu)}$ is transient and we may consider its total local time $\ell^{(\nu)}(\infty, y)$, $y > 0$. We assume here (and below) $R_0^{(\nu)} = x > 0$, and leave the descriptions in the case $x = 0$ to the reader.

(A) Applying No. 3 for $-R^{(\nu)}$ gives

$$\{\ell^{(\nu)}(\infty, -y) : \ -\infty < y \leq -x\} \ \sim \ \{\hat{X}_y^{(2,0)} : \ -\infty < y \leq -x\},$$

$$\{\ell^{(\nu)}(\infty, x - y) : \ 0 \leq y < x\} \ \sim \ \{\hat{X}_y^{(0,x)} : \ 0 \leq y < x\},$$

where the extended generators of $\hat{X}^{(2,0)}$ and $\hat{X}^{(0,x)}$ are

$$2z\,\frac{\partial^2}{\partial z^2} + \left(2 - \frac{(2\nu + 1)\,z}{-y}\right)\frac{\partial}{\partial z} + \frac{\partial}{\partial y},$$

$$2z\,\frac{\partial^2}{\partial z^2} - \frac{(2\nu + 1)\,z}{x - y}\,\frac{\partial}{\partial z} + \frac{\partial}{\partial y},$$

respectively.

(B) Given $\hat{M} := \inf\{R_t^{(\nu)} : \ t \geq 0\} = m$ we obtain from No. 2,

$$\{\ell^{(\nu)}(\infty, m + y) : \ 0 \leq y \leq x - m\} \ \sim \ \{\tilde{X}_y^{(4,m)} : \ 0 \leq y \leq x - m\},$$

$$\{\ell^{(\nu)}(\infty, x + y) : \ y \geq 0\} \ \sim \ \{\tilde{X}_y^{(2,x)} : \ y \geq 0\},$$

where $\tilde{X}_0^{(4,m)} := 0$ and the generators of $\tilde{X}^{(4,m)}$ and $\tilde{X}^{(2,x)}$ are

$$2z\,\frac{\partial^2}{\partial z^2} + \Big(4 - \frac{(2\nu-1)\,z}{m+y}\Big)\,\frac{\partial}{\partial z} + \frac{\partial}{\partial y},$$

$$2z\,\frac{\partial^2}{\partial z^2} + \Big(2 - \frac{(2\nu-1)\,z}{x+y}\Big)\,\frac{\partial}{\partial z} + \frac{\partial}{\partial y},$$

respectively. In the 3-dimensional case, that is, $\nu = 1/2$,

$$\tilde{X}^{(4,m)} \sim Z^{(4)}, \quad \tilde{X}^{(2,x)} \sim Z^{(2)}.$$

Remark 1. For the description in (A) it is essential to have the $\mathbf{P}_x^{(\nu)}$-distribution of $\ell^{(\nu)}(\infty, y),\ y \ge x$, which can be found using II.27 (a):

(a) $$\mathbf{P}_x^{(\nu)}(\ell^{(\nu)}(\infty,y) \ge v) = \mathbf{P}_y^{(\nu)}(\ell^{(\nu)}(\infty,y) \ge v) = \exp(-\frac{\nu}{y}v).$$

Using this we can write down the finite-dimensional distributions of $\hat{X}^{(2,0)}$ similarly as explained for Brownian motion with drift in the Remark in No. 11.

Remark 2. Let for $a \ge 0$ and $0 < y < a$

$$y \mapsto h(y,a) := (a-y)^{-2\nu} - a^{-2\nu}.$$

Then for all $b > x$, assuming that the initial distributions coincide, we have from No. 5,

$$\{\hat{X}^{(2,0)}_{-b+y} : 0 \le y \le b-x\} \sim \Big\{\frac{1}{h'(y,b)} Z^{(2)}_{h(y,b)} : 0 \le y \le b-x\Big\},$$

$$\{\hat{X}^{(0,x)}_y : 0 \le y < x\} \sim \Big\{\frac{1}{h'(y,x)} Z^{(0)}_{h(y,x)} : 0 \le y < x\Big\}.$$

In the 3-dimensional case $\hat{X}^{(0,x)}$ is identical in law to $Z^{(0)}$ conditioned to hit 0 before time x (cf. No. 8).

Remark 3. Similarly as in Remark 2, introduce for $a > 0$ and $y \ge 0$ the function

$$y \mapsto g(y,a) := (a+y)^{2\nu} - a^{2\nu}.$$

Then from No. 5 assuming that the initial distributions coincide,

$$\{\tilde{X}^{(4,m)}_y : 0 \le y \le x-m\} \sim \Big\{\frac{1}{g'(y,m)} Z^{(4)}_{g(y,m)} : 0 \le y \le x-m\Big\},$$

$$\{\tilde{X}^{(2,x)}_y : y \ge 0\} \sim \Big\{\frac{1}{g'(y,x)} Z^{(2)}_{g(y,x)} : y \ge 0\Big\}.$$

18. Total local time when $\nu < 0$. Let $\ell^{(\nu)}$ be as in No. 17 with $\nu < 0$. If $\nu > -1$ then 0 is non-singular and we pose on it the boundary condition of killing. The reflecting case is considered in No's. 19 and 20.

Using the result in No. 3 (we leave the other y-direction to the reader) it is seen that

$$\{\ell^{(\nu)}(\infty,y):\ 0\le y\le x\}\ \sim\ \{\bar X^{(2,0)}_y:\ 0\le y\le x\},$$

$$\{\ell^{(\nu)}(\infty,x+y):\ y\ge 0\}\ \sim\ \{\bar X^{(0,x)}_y:\ y\ge 0\},$$

where the generators of $\bar X^{(2,0)}$ and $\bar X^{(0,x)}$ are

$$2z\,\frac{\partial^2}{\partial z^2}+\Big(2+\frac{(2\nu+1)\,z}{y}\Big)\,\frac{\partial}{\partial z}+\frac{\partial}{\partial y},$$

$$2z\,\frac{\partial^2}{\partial z^2}+\frac{(2\nu+1)\,z}{x+y}\,\frac{\partial}{\partial z}+\frac{\partial}{\partial y},$$

respectively. For $\nu=-1/2$ the Bessel process $R^{(\nu)}$ is a Brownian motion on $(0,\infty)$ killed when it hits 0. The above result, in this case, has already been presented in No. 6.

Remark 1. To express $\bar X^{(2,0)}$ and $\bar X^{(0,x)}$ in terms of $Z^{(2)}$ and $Z^{(0)}$, introduce for $a\ge 0$ and $y\ge 0$,

$$y\mapsto g(y,a):=(a+y)^{-2\nu}-a^{-2\nu}.$$

Then

$$\Big\{\bar X^{(2,0)}_y:\ 0<y\le x\Big\}\ \sim\ \Big\{\frac{1}{g'(y,0)}\,Z^{(2)}_{g(y,0)}:\ 0<y\le x\Big\},$$

$$\Big\{\bar X^{(0,x)}_y:\ y\ge 0\Big\}\ \sim\ \Big\{\frac{1}{g'(y,x)}\,Z^{(0)}_{g(y,x)}:\ y\ge 0\Big\},$$

where $Z^{(2)}$ is started at 0.

Remark 2. Notice that

$$\mathbf{P}^{(\nu)}_x(\ell^{(\nu)}(\infty,x)\ge v)=\exp\big(-\frac{|\nu|}{x}v\big)$$

which coincides with the corresponding distribution in the case $\nu>0$, see (a) in No. 17. This can be explained by the h-transform property of Bessel processes (cf. IV.43) and the last exit decompositions.

19. Local time stopped at the first hitting time. Next we study the process

$$\{\ell^{(\nu)}(H_b,y):\ 0\le y\le b\}$$

under the condition $R^{(\nu)}_0=x<b$. We consider only the case when the y-direction is from right to left.

Assume first that $\nu>0$ or $-1<\nu<0$ with 0 reflecting. Then from No. 3, by considering $-R^{(\nu)}$,

$$\{\ell^{(\nu)}(H_b,b-y):\ 0\le y\le b-x\}\ \sim\ \{\hat X^{(2,b)}_y:\ 0\le y\le b-x\},$$

$$\{\ell^{(\nu)}(H_b,x-y):\ 0\le y\le x\}\ \sim\ \{\hat X^{(0,x)}_y:\ 0\le y\le x\},$$

where the generators of $\hat{X}^{(2,b)}$ and $\hat{X}^{(0,x)}$ (cf. (A) in No. 17) are

$$2z\,\frac{\partial^2}{\partial z^2} + \Big(2 - \frac{(2\nu+1)\,z}{b-y}\Big)\,\frac{\partial}{\partial z} + \frac{\partial}{\partial y},$$

$$2z\,\frac{\partial^2}{\partial z^2} - \frac{(2\nu+1)\,z}{x-y}\,\frac{\partial}{\partial z} + \frac{\partial}{\partial y},$$

respectively. Notice that for $\nu = -1/2$ we have the Ray–Knight theorem for reflecting Brownian motion.

Assume next $\nu \le -1$ or $-1 < \nu < 0$ with 0 a killing boundary. It is now possible that $H_b = \infty$; this case is discussed in No. 20. However, conditioning on $H_b < \infty$,

$$\{R_t^{(\nu)} : t < H_b\} \sim \{R_t^{(-\nu)} : t < H_b\}$$

(see IV.43), and hence, under the condition $H_b < \infty$,

$$\{\ell^{(\nu)}(H_b, y) : 0 \le y \le b\} \sim \{\ell^{(-\nu)}(H_b, y) : 0 \le y \le b\}.$$

For $\nu = -1/2$, i.e., $R^{(\nu)}$ is a killed Brownian motion, these results are already given in No. 6.

Remark. Using the characterization of $\hat{X}^{(2,b)}$ in terms of $Z^{(2)}$ (see Remark 2 in No. 17) we obtain for $\nu > 0$ and $x < b$

$$\mathbf{P}_x^{(\nu)}(\ell^{(\nu)}(H_b, x) \ge v) = \exp\Big(-\frac{h'(b-x,b)}{2\,h(b-x,b)}v\Big),$$

where

$$y \mapsto h(y,b) := (b-y)^{-2\nu} - b^{-2\nu}.$$

By II.27 (a) the parameter of this exponential distribution can also be expressed as $1/m(x)G(x,x)$, where m is the speed measure and $G := G_0$ is the Green function of $R^{(\nu)}$ killed at H_b.

20. Local time stopped at the first exit time from a finite interval. Let $0 < a < b$ and $x \in (a,b)$ be given and consider the process

$$\ell^{(\nu)}(H_{a,b}) := \{\ell^{(\nu)}(H_{a,b}, y) : a \le y \le b\}.$$

We are interested in characterizing the law of $\ell^{(\nu)}(H_{a,b})$ under the condition $H_{a,b} = H_a$. The complete result is not given but we indicate the facts needed to be able to apply No. 3. For $\nu > 0$ let

$$g(x) := \frac{s(b) - s(x)}{s(b) - s(a)} = \mathbf{P}_x^{(\nu)}(H_{a,b} = H_a),$$

where $s(x) = -x^{-2\nu}/2\nu$ is the scale function of $R^{(\nu)}$ (see A1.21). Use g to build the excessive transform of $R^{(\nu)}$. This transform is identical in law to $R^{(\nu)}$

conditioned by $H_{a,b} = H_a$. Standard computations show that the generator of the transform is

$$\frac{1}{2}\frac{d^2}{dx^2} + \left(\frac{2\nu+1}{2x} + \frac{g'(x)}{g(x)}\right)\frac{d}{dx}.$$

Observe that this makes sense also when $a = 0$ if we take $g(x) = s(b) - s(x)$. The scale function of the h-transform can be obtained, e.g., from the general formula in II.31. The diffusion characterization of $\ell^{(\nu)}(H_{a,b})$ given $H_{a,b} = H_a$ follows by applying No. 3 to the excessive transform.

Similar computations when $\nu < 0$ show that the results above do not depend on the sign of ν, that is, conditioned on the event $H_{a,b} = H_a$,

$$\{\ell^{(\nu)}(H_{a,b}, y) : \ a \le y \le b\} \ \sim \ \{\ell^{(-\nu)}(H_{a,b}, y) : \ a \le y \le b\}.$$

When $\nu = -1/2$ the Bessel process $R^{(\nu)}$ is a Brownian motion on $\mathbb{R}_+$ killed when it hits 0. Therefore, the above result in the case $\nu = \pm 1/2$ has already been discussed in No. 8.

21. Local time stopped at an exponential time. The discussion here follows that of No. 9 (see also No. 15). Assume that $\nu > 0$, and let $\tau \sim \text{Exp}(\lambda)$ independent of $R^{(\nu)}$. First, $R^{(\nu)}$ started at x and conditioned on $R^{(\nu)}_\tau = z_o > x$, is the excessive transform of $R^{(\nu)}$ obtained by using

$$h(x) := \frac{G_\lambda(x, z_o)}{G_\lambda(z_o, z_o)}.$$

Let $\bar{R}^{(\nu)}$ denote this h-transform and let ζ and $\bar{\ell}^{(\nu)}$ be its lifetime and local time, respectively. Then under the condition $R^{(\nu)}_\tau = z_o$,

$$\{\ell^{(\nu)}(\tau, y) : \ y \ge 0\} \ \sim \ \{\bar{\ell}^{(\nu)}(\zeta, y) : \ y \ge 0\}.$$

The latter process can be characterized as explained in No. 4. A properly normalized scale function is

$$s^h(x) = \int_{z_o}^{x} \left(\frac{G_\lambda(y, z_o)}{G_\lambda(z_o, z_o)}\right)^{-2} s(dy)$$

$$= \begin{cases} \varphi_\lambda^2(z_o)\Big(\dfrac{\psi_\lambda(x)}{\varphi_\lambda(x)} - \dfrac{\psi_\lambda(z_o)}{\varphi_\lambda(z_o)}\Big), & x \ge z_o, \\[2ex] \psi_\lambda^2(z_o)\Big(\dfrac{\varphi_\lambda(z_o)}{\psi_\lambda(z_o)} - \dfrac{\varphi_\lambda(x)}{\psi_\lambda(x)}\Big), & x \le z_o, \end{cases}$$

where $s(dy) = y^{-2\nu-1}\,dy$ is the scale measure (cf. A1.21), and

$$\varphi_\lambda(x) = x^{-\nu} K_\nu(x\sqrt{2\lambda}), \quad \psi_\lambda(x) = x^{-\nu} I_\nu(x\sqrt{2\lambda}).$$

The function p appearing in No. 4 takes the form

$$p(x) = \begin{cases} 1, & x \ge z_o, \\[1ex] \dfrac{\varphi_\lambda(z_o)\,\psi_\lambda(x)}{\varphi_\lambda(x)\,\psi_\lambda(z_o)}, & x \le z_o, \end{cases}$$

and hence, for all x,

$$\frac{(p^2(x)\,(s^h(x))')'}{p^2(x)\,(s^h(x))'} = \frac{s''(x)}{s'(x)} - 2\,\frac{\varphi_\lambda'(x)}{\varphi_\lambda(x)} = -\frac{1}{x} - \sqrt{2\lambda}\frac{K_\nu'(x\sqrt{2\lambda})}{K_\nu(x\sqrt{2\lambda})}.$$

Putting this into the expressions in No. 4 gives us the Ray–Knight theorem for $\bar{R}^{(\nu)}$, leading to the desired characterization. In particular, for $y \in [x, z_o]$,

$$\mathbf{P}_x^{(\nu)}(\ell^{(\nu)}(\tau, y) \geq v | R^{(\nu)}(\tau) = z_o) = \exp\Big(-\frac{v}{m(y)\,G_\lambda(y,y)}\Big),$$

where m is the density of the speed measure.

Now let $\nu < 0$. Straightforward computations show that $\bar{R}^{(\nu)}$ is identical in law to $\bar{R}^{(-\nu)}$. Therefore, from the results above we also obtain results for $R^{(\nu)}$, $\nu < 0$, by doing some obvious changes. The case $\nu = 0$ is left to the reader.

We conclude by considering the case $\nu = 1/2$. Now (see A1.22)

$$\frac{(p^2(x)\,(s^h(x))')'}{p^2(x)\,(s^h(x))'} = 2\sqrt{2\lambda}.$$

Consequently, the description is very similar to the one in No. 9. Notice, however, that the process

$$\{\ell^{(1/2)}(\tau, x-y) :\ 0 \leq y \leq x\}$$

must be different. In fact, it is identical in law to a non-time-homogeneous diffusion having the generator

$$2v\frac{\partial^2}{\partial z^2} - 2\sqrt{2\lambda}\,z\ \mathrm{cth}(\sqrt{2\lambda}(x-y))\frac{\partial}{\partial z} + \frac{\partial}{\partial y}.$$

It can be proved that this is also the generator of $Z^{(0,\theta)}$ with $\theta = 2\sqrt{2\lambda}$ conditioned to hit 0 before time x (cf. Remark 1 in No. 14).

5. Summarizing tables

Below we review in a table form some of the Ray–Knight results presented above. For Bessel processes of positive order we have included only three cases, and Bessel processes of negative order do not appear at all. For these (as well as for reflecting Brownian motion) we refer to No's. 17–21. In the following tables references are given to the sections, where the result is discussed in more detail and where one can also find the unexplained notations explained. For typographical reasons the partial differential operator $\partial/\partial y$ associated to the time-parameter y (in non-time-homogeneous cases) is not displayed. Recall that H_a is the first hitting time of a, $H_{a,b}$ is the first exit time from the interval (a, b), $\tau \sim \mathrm{Exp}(\lambda)$ independent of the considered process, ϱ is the inverse local time at z_o, the starting state of the process is denoted by x, and m is the value of the infimum of the process up to the considered stopping time.

Brownian motion

Local time process	Generator
No. **6** (A)	
$\{\ell(H_b, b-y):\ 0 \le y \le b-x\}$	$2z\frac{d^2}{dz^2} + 2\frac{d}{dz}$
$\{\ell(H_b, x-y):\ y \ge 0\}$	$2z\frac{d^2}{dz^2}$
No. **6** (B)	
$\{\ell(H_b, m+y):\ 0 \le y \le x-m\}$	$2z\frac{\partial^2}{\partial z^2} + \left(4 - \frac{2z}{b-m-y}\right)\frac{\partial}{\partial z}$
$\{\ell(H_b, x+y):\ 0 \le y < b-x\}$	$2z\frac{\partial^2}{\partial z^2} + \left(2 - \frac{2z}{b-x-y}\right)\frac{\partial}{\partial z}$
No. **8** Given $H_{a,b} = H_b$,	
$\{\ell(H_{a,b}, b-y):\ 0 \le y \le b-x\}$	$2z\frac{\partial^2}{\partial z^2} + \left(2 - \frac{2z}{b-a-y}\right)\frac{\partial}{\partial z}$
$\{\ell(H_{a,b}, x-y):\ 0 \le y < x-a\}$	$2z\frac{\partial^2}{\partial z^2} - \frac{2z}{x-a-y}\frac{\partial}{\partial z}$
No. **9** Given $W_\tau = z_o$,	
$\{\ell(\tau, m+y):\ 0 \le y \le x-m\}$	$2z\frac{d^2}{dz^2} + (4 - 2z\sqrt{2\lambda})\frac{d}{dz}$
$\{\ell(\tau, x+y):\ 0 \le y \le z_o-x\}$	$2z\frac{d^2}{dz^2} + (2 - 2z\sqrt{2\lambda})\frac{d}{dz}$
$\{\ell(\tau, z_o+y):\ y \ge 0\}$	$2z\frac{d^2}{dz^2} - 2z\sqrt{2\lambda}\frac{d}{dz}$
No. **10**	
$\{\ell(\varrho, z_o-y):\ 0 \le y \le z_o-x\}$	$2z\frac{d^2}{dz^2} + 2\frac{d}{dz}$
$\{\ell(\varrho, x-y):\ y \ge 0\}$	$2z\frac{d^2}{dz^2}$
$\{\ell(\varrho, z_o+y):\ y \ge 0\}$	$2z\frac{d^2}{dz^2}$

Brownian motion with drift $\mu > 0$

Local time process	Generator
No. **11** (A) $\{\ell^{(\mu)}(\infty, -y) : -\infty < y \le -x\}$ $\{\ell^{(\mu)}(\infty, x-y) : y \ge 0\}$	$2z\frac{d^2}{dz^2} + (2 - 2\mu z)\frac{d}{dz}$ $2z\frac{d^2}{dz^2} - 2\mu z\frac{d}{dz}$
No. **11** (B) $\{\ell^{(\mu)}(\infty, m+y) : 0 \le y \le x-m\}$ $\{\ell^{(\mu)}(\infty, x+y) : y \ge 0\}$	$2z\frac{d^2}{dz^2} + (4 - 2\mu z)\frac{d}{dz}$ $2z\frac{d^2}{dz^2} + (2 - 2\mu z)\frac{d}{dz}$
No. **12** (A) $\{\ell^{(\mu)}(H_b, b-y) : 0 \le y \le b-x\}$ $\{\ell^{(\mu)}(H_b, x-y) : y \ge 0\}$	$2z\frac{d^2}{dz^2} + (2 - 2\mu z)\frac{d}{dz}$ $2z\frac{d^2}{dz^2} - 2\mu z\frac{d}{dz}$
No. **12** (B) $\{\ell^{(\mu)}(H_b, m+y) : 0 \le y \le x-m\}$ $\{\ell^{(\mu)}(H_b, x+y) : 0 \le y < b-x\}$	$2z\frac{\partial^2}{\partial z^2} + (4 - 2\mu\, z\,\mathrm{cth}(\mu\, y_1))\frac{\partial}{\partial z}$ $2z\frac{\partial^2}{\partial z^2} + (2 - 2\mu\, z\,\mathrm{cth}(\mu\, y_2))\frac{\partial}{\partial z}$ $y_1 := b-m-y,\ y_2 := b-x-y$
No. **13** Given $H_a = \infty$, $\{\ell^{(\mu)}(\infty, m+y) : 0 \le y \le x-m\}$ $\{\ell^{(\mu)}(\infty, x+y) : y \ge 0\}$	$2z\frac{d^2}{dz^2} + (4 - 2\mu z)\frac{d}{dz}$ $2z\frac{d^2}{dz^2} + (2 - 2\mu z)\frac{d}{dz}$
No. **14** Given $H_{a,b} = H_a$, $\{\ell^{(\mu)}(H_{a,b}, a+y) : 0 \le y \le x-a\}$ $\{\ell^{(\mu)}(H_{a,b}, x+y) : 0 \le y < b-x\}$	$2z\frac{\partial^2}{\partial z^2} + \big(2 - 2\|\mu\|\, z\,\mathrm{cth}(\|\mu\| y_1)\big)\frac{\partial}{\partial z}$ $2z\frac{\partial^2}{\partial z^2} - 2\|\mu\|\, z\,\mathrm{cth}(\|\mu\| y_2)\frac{\partial}{\partial z}$ $y_1 := b-a-y,\ y_2 := b-x-y$

Brownian motion with drift $\mu > 0$ (cont'd)

Local time process	Generator
No. **15** Given $W_\tau^{(\mu)} = z_o$,	
$\{\ell^{(\mu)}(\tau, m+y) : 0 \le y \le x-m\}$	$2z\frac{d^2}{dz^2} + (4 - 2z\sqrt{2\lambda+\mu^2})\frac{d}{dz}$
$\{\ell^{(\mu)}(\tau, x+y) : 0 \le y \le z_o - x\}$	$2z\frac{d^2}{dz^2} + (2 - 2z\sqrt{2\lambda+\mu^2})\frac{d}{dz}$
$\{\ell^{(\mu)}(\tau, z_o+y) : y \ge 0\}$	$2z\frac{d^2}{dz^2} - 2z\sqrt{2\lambda+\mu^2}\frac{d}{dz}$
No. **16** Given $\varrho < \infty$,	
$\{\ell^{(\mu)}(\varrho, z_o-y) : 0 \le y \le z_o - x\}$	$2z\frac{d^2}{dz^2} + (2 - 2\lvert\mu\rvert z)\frac{d}{dz}$
$\{\ell^{(\mu)}(\varrho, x-y) : y \ge 0\}$	$2z\frac{d^2}{dz^2} - 2\lvert\mu\rvert z\frac{d}{dz}$
$\{\ell^{(\mu)}(\varrho, z_o+y) : y \ge 0\}$	$2z\frac{d^2}{dz^2} - 2\lvert\mu\rvert z\frac{d}{dz}$

Bessel process of order $\nu > 0$

Local time process	Generator
No. **17** (A)	
$\{\ell^{(\nu)}(\infty, -y) : -\infty < y \le -x\}$	$2z\,\frac{\partial^2}{\partial z^2} + \left(2 - \frac{(2\nu+1)\,z}{-y}\right)\frac{\partial}{\partial z}$
$\{\ell^{(\mu)}(\infty, x-y) : 0 \le y < x\}$	$2z\,\frac{\partial^2}{\partial z^2} - \frac{(2\nu+1)\,z}{x-y}\,\frac{\partial}{\partial z}$
No. **17** (B)	
$\{\ell^{(\nu)}(\infty, m+y) : 0 \le y \le x-m\}$	$2z\,\frac{\partial^2}{\partial z^2} + \left(4 - \frac{(2\nu-1)\,z}{m+y}\right)\frac{\partial}{\partial z}$
$\{\ell^{(\nu)}(\infty, x+y) : y \ge 0\}$	$2z\,\frac{\partial^2}{\partial z^2} + \left(2 - \frac{(2\nu-1)\,z}{x+y}\right)\frac{\partial}{\partial z}$
No. **19**	
$\{\ell^{(\nu)}(H_b, b-y) : 0 \le y \le b-x\}$	$2z\,\frac{\partial^2}{\partial z^2} + \left(2 - \frac{(2\nu+1)\,z}{b-y}\right)\frac{\partial}{\partial z}$
$\{\ell^{(\nu)}(H_b, x-y) : 0 \le y \le x\}$	$2z\,\frac{\partial^2}{\partial z^2} - \frac{(2\nu+1)\,z}{x-y}\,\frac{\partial}{\partial z}$

CHAPTER VI

DIFFERENTIAL SYSTEMS ASSOCIATED TO BROWNIAN MOTION

1. The Feynman–Kac formula

1. For $T > 0$ let $F : \mathbb{R} \mapsto \mathbb{R}$, $f : [0,T] \times \mathbb{R} \mapsto \mathbb{R}_+$ and $g : [0,T] \times \mathbb{R} \mapsto \mathbb{R}$ be given functions. Assume that g is bounded and that F is bounded and Hölder continuous. If, moreover, g and f are Hölder continuous locally in t, that is, e.g. for g, there exist for any $N < \infty$ constants $C > 0$ and $\alpha > 0$ such that

$$|g(t,x) - g(t,y)| < C\,|x-y|^{\alpha}$$

for all $t < N$, then the *Feynman–Kac formula* (cf. II.24) says that

$$\begin{aligned} v(t,x) := \mathbf{E}\Big(F(W_T)\exp\Big(-\int_t^T f(s,W_s)\,ds\Big) \\ + \int_t^T g(s,W_s)\exp\Big(-\int_t^s f(v,W_v)\,dv\Big)\,ds\,\Big|\,W_t = x\Big) \end{aligned}$$

is the unique solution of the problem

$$\begin{aligned} -u_t'(t,x) &= \frac{1}{2}u_{xx}''(t,x) - f(t,x)u(t,x) + g(t,x), \quad (t,x) \in (0,T)\times\mathbb{R}, \\ u(T,x) &= F(x), \quad x \in \mathbb{R}. \end{aligned}$$

For a proof (for more general diffusions than Brownian motion) see, e.g., Karatzas and Shreve (1991) p. 366–368. The Hölder conditions above are needed to guarantee enough smoothness on v; see Durrett (1996) p. 130–141. We also remark that the boundedness assumption can be relaxed to a polynomial growth condition.

Another much used form of the Feynman–Kac formula says that

$$\begin{aligned} (t,x) \mapsto \mathbf{E}_x\Big(F(W_t)\exp\Big(-\int_0^t f(t-s,W_s)\,ds\Big) \\ + \int_0^t g(t-s,W_s)\exp\Big(-\int_0^s f(t-v,W_v)\,dv\Big)\,ds\Big) \end{aligned}$$

is the unique solution of the problem

$$\begin{aligned} u_t'(t,x) &= \frac{1}{2}u_{xx}''(t,x) - f(t,x)u(t,x) + g(t,x), \quad t>0,\ x \in \mathbb{R}, \\ u(0,x) &= F(x), \quad x \in \mathbb{R}. \end{aligned}$$

For proofs we refer to Durrett (1996) p. 130–141, and Karatzas and Shreve (1991) p. 268–270.

The issue 14(7)(1986) of Annals of Probability is dedicated to the memory of Mark Kac. In particular, Kesten (1986) discusses the influence of Mark Kac on probability theory.

2. In this chapter we present a number of differential systems for finding distributions of various functionals associated to W. Our basic tool hereby is the Feynman–Kac formula. We are mainly working in the time-homogeneous case, and exploit the Feynman–Kac formula as presented for general diffusions in II.24. This approach leads to ordinary differential problems which are more suitable for analytical solutions than parabolic systems.

For a vector $\vec{x} = (x_1, x_2, \dots, x_k)$ and a non-negative piecewise continuous function f consider the random vector

$$\vec{R} := \vec{R}(f, \vec{x}) := \Big(\int_0^t f(W_s) ds, \ell(t, x_1), \ell(t, x_2), \dots, \ell(t, x_k) \Big),$$

where $\ell(\cdot, x_i)$ is the Brownian local time (with respect to Lebesgue measure) at the point x_i. Our aim is to characterize the law of $\vec{R}$ at certain stopping times. For this we introduce the additive functional

$$t \mapsto A(f, \vec{x}, t) := \int_0^t f(W_s) ds + \sum_{i=1}^k \gamma_i \ell(t, x_i), \qquad \gamma_i \geq 0.$$

The functional A is considered under the side condition

$$\begin{aligned} C(a, b, t) :=& \{a < \inf_{s \leq t} W_s < \sup_{s \leq t} W_s < b\} \\ =& \{H_{a,b} > t\}. \end{aligned}$$

It is possible to generalize the treatment of A presented in the sections below and to recursively use the Markov property to characterize the distribution of non-time-homogeneous functionals of the type

$$A(\vec{f}_n, \vec{x}, \vec{t}_n) := \sum_{j=1}^n \Big(\int_{t_{j-1}}^{t_j} f_j(W_s) ds + \sum_{i=1}^k \gamma_{i,j} (\ell(t_j, x_i) - \ell(t_{j-1}, x_i)) \Big),$$

where $f_j : \mathbb{R} \mapsto \mathbb{R}_+$ are non-negative piecewise continuous functions, $\gamma_{i,j} \geq 0$ and $0 = t_0 < t_1 \leq \dots \leq t_n$. However, this is fairly straightforward and is omitted.

3. We also comment below how the corresponding differential systems for a Brownian motion with drift are obtained. There are two standard general approaches: the first one is based on absolute continuity or the Cameron–Martin–Girsanov formula (see III.26–28 and IV.30), and in the second $W^{(\mu)}$ is considered as a linear diffusion and the Laplace operator appearing in the differential equations for Brownian motion is replaced by

$$\mathcal{G}^{(\mu)} = \frac{1}{2}\frac{d^2}{dx^2} + \mu\frac{d}{dx}.$$

To demonstrate these two approaches in the Feynman–Kac formula consider the time-homogeneous case and assume also that $g \equiv 0$. Then by absolute continuity

$$\begin{aligned}
v_\mu(t,x) :&= \mathbf{E}_x\Big(F(W_t^{(\mu)})\exp\Big(-\int_0^t f(W_s^{(\mu)})\,ds\Big)\Big)\\
&= \mathbf{E}_x\Big(F(W_t)\exp\Big(\mu(W_t-x)-\mu^2t/2-\int_0^t f(W_s)\,ds\Big)\Big)\\
&= e^{-\mu x}\,\mathbf{E}_x\Big(e^{\mu W_t}F(W_t)\exp\Big(-\int_0^t (f(W_s)+\mu^2/2)\,ds\Big)\Big)\\
&=: e^{-\mu x}\,v(t,x).
\end{aligned}$$

Therefore, v is the solution of the problem

$$\begin{aligned}
&u_t'(t,x) = \frac{1}{2}u_{xx}''(t,x) - \Big(\frac{\mu^2}{2}+f(x)\Big)u(t,x), \quad t>0,\ x\in\mathbb{R},\\
&u(0,x) = e^{\mu x}\,F(x), \quad x\in\mathbb{R}.
\end{aligned}$$

On the other hand, using the differential operator $\mathcal{G}^{(\mu)}$ it is seen that v_μ is the solution of the problem

$$\begin{aligned}
&u_t'(t,x) = \frac{1}{2}u_{xx}''(t,x) + \mu\,u_x'(t,x) - f(x)u(t,x), \quad t>0,\ x\in\mathbb{R},\\
&u(0,x) = F(x), \quad x\in\mathbb{R}.
\end{aligned}$$

We emphasize that the local time of $W^{(\mu)}$ is taken with respect to Lebesgue measure. As will be seen, this leads to seemingly simpler conditions than some other normalizations.

2. Exponential stopping

In this section $\tau \sim \mathrm{Exp}(\lambda)$, independent of W, and $\hat{W}$ denotes W killed when it exits the interval (a,b).

4. Let F be a piecewise continuous bounded function and let $A := A(f,\vec{x},\cdot)$ be as in No. 2. Then

(a) $$v(x) := \mathbf{E}_x\big(F(W_\tau)\exp(-A(f,\vec{x},\tau));\ C(a,b,\tau)\big),$$

is the solution of the problem

(b) $$\frac{1}{2}u''(x) - (\lambda+f(x))u(x) = -\lambda F(x), \qquad x\in(a,b)\setminus\{x_1,\dots,x_k\},$$

(c) $$u'(x_i+) - u'(x_i-) = 2\gamma_i u(x_i), \qquad i=1,\dots,k,$$

(d) $$u(a+) = u(b-) = 0.$$

Remark. When we say (here and below in this chapter) that "v is the solution of the problem" we mean that v is the unique continuous bounded function which satisfies the ODE (b) and the conditions (c) and (d). If f and/or F have discontinuity points, then (b) holds on every interval of continuity of f and F. At the discontinuity points v'' does not in general exist but v' exists and is continuous.

To ensure that the function v given in (a) is the correct one, let $\tilde{p}$ and $\tilde{G}_\lambda$ be the transition density and the Green function, respectively, of the process $\tilde{W}$ obtained from $\hat{W}$ by killing it (additionally) according to the additive functional A (see II.22 and 23). The Green function $\tilde{G}_\lambda$ is given (see II.11) in terms of the functions $\tilde{\psi}_\lambda$ and $\tilde{\varphi}_\lambda$. These satisfy

(b')
$$\frac{1}{2}u''(x) - (\lambda + f(x))u(x) = 0 \qquad x \in (a,b) \setminus \{x_1, \dots, x_k\},$$

and (c) and are such that $\tilde{\psi}_\lambda(a) = 0$ and $\tilde{\varphi}_\lambda(b) = 0$ (see II.24). Notice that

$$v(x) = \int_0^\infty dt\, \lambda e^{-\lambda t} \int_a^b 2dy\, \tilde{p}(t; x, y)\, F(y) = \lambda \int_a^b \tilde{G}_\lambda(x,y) F(y)\, 2dy,$$

and it can be proved by straightforward differentiation that v has the claimed properties.

Next, replace W with $W^{(\mu)}$ in (a), and let v_μ be the corresponding function. Then

(e)
$$v_\mu(x) = \lambda \int_a^b \tilde{G}_\lambda^{(\mu)}(x,y) F(y)\, m(y)\, dy,$$

where $m(y) = 2\, e^{2\mu y}$ is the density of the speed measure. The Green function is computed in the usual way. Notice that in this case the fundamental solutions also satisfy (c) due to the normalization of $\ell^{(\mu)}$.

For $a = -\infty$, say, the condition "$u(a+) = 0$" must be replaced by the condition "u is bounded in a neighborhood of $-\infty$".

5. Especially, choose in No. 4

$$F(x) = \mathbb{1}_{(-\infty, z)}(x), \quad z \in (a,b).$$

Then

$$\begin{aligned} v_z'(x,z) := v_z'(x) &= \frac{\partial}{\partial z} \mathbf{E}_x(\exp(-A(f, \vec{x}, \tau));\ C(a,b,\tau),\ W_\tau \le z) \\ &= 2\lambda \tilde{G}_\lambda(x,z). \end{aligned}$$

For a Brownian motion with drift the factor 2 above must be replaced by $m(z) = 2e^{2\mu z}$ (see (e) in No. 4). Recalling the definition of the Wronskian it is seen that for both W and $W^{(\mu)}$ and for $z \notin \{x_1, \dots, x_k\}$

$$\frac{\partial}{\partial x} v_z'(z+, z) - \frac{\partial}{\partial x} v_z'(z-, z) = -2\lambda.$$

6. Here we characterize the first two moments of the functional A as solutions of a differential system. For this let (cf. 1.3.0.1)

$$g(x) := \mathbf{P}_x(\tau \le H_{a,b}) = 1 - \mathbf{E}_x\big(\exp(-\lambda H_{a,b})\big)$$
$$= 1 - \frac{\operatorname{ch}((b+a-2x)\sqrt{\lambda/2})}{\operatorname{ch}((b-a)\sqrt{\lambda/2})}.$$

Then

$$v_1(x) := \mathbf{E}_x\big(A(f,\vec{x},\tau);\ C(a,b,\tau)\big)$$

is the solution of the problem (see Remark in No. 4)

$$\begin{aligned}
&\frac{1}{2}u''(x) - \lambda u(x) = -f(x)g(x), \quad x \in (a,b)\setminus\{x_1,\dots,x_k\},\\
&u'(x_i+) - u'(x_i-) = -2\gamma_i g(x_i), \quad i = 1,2,\dots,k,\\
&u(a+) = u(b-) = 0.
\end{aligned}$$

To briefly motivate this, let $\hat{G}$ be the Green function of $\hat{W}$. Then (see also (b) in II.27) by the Markov property

$$\mathbf{E}_x\Big(\int_0^\tau f(W_s)\,ds;\ \tau < H_{a,b}\Big) = \int_a^b \hat{G}_\lambda(x,y)f(y)g(y)\,2dy,$$

and

$$\mathbf{E}_x\big(\ell(\tau,x_i);\ \tau < H_{a,b}\big) = 2\hat{G}_\lambda(x,x_i)g(x_i),$$

explaining the above condition involving first derivatives.

The second moment is obtained by iteration. The result is that

$$v_2(x) := \mathbf{E}_x\big(A^2(f,\vec{x},\tau);\ C(a,b,\tau)\big)$$

is the solution of the problem

$$\begin{aligned}
&\frac{1}{2}u''(x) - \lambda u(x) = -2f(x)v_1(x), \quad x \in (a,b)\setminus\{x_1,\dots,x_k\},\\
&u'(x_i+) - u'(x_i-) = -4\gamma_i v_1(x_i), \quad i = 1,2,\dots,k,\\
&u(a+) = u(b-) = 0.
\end{aligned}$$

In the case $a = -\infty$ and/or $b = \infty$, some condition must be posed on f, e.g., that it does not grow faster than exponentially, to guarantee existence of the expectations. Notice also that if $a = -\infty$, say, then g above must be replaced by its limit as $a \to -\infty$:

$$\lim_{a\to-\infty} \mathbf{P}_x(\tau \le H_{a,b}) = \mathbf{P}_x(\tau \le H_b).$$

7. Combining No's. 5 and 6 shows that

$$v_1(x) := \frac{\partial}{\partial z}\mathbf{E}_x\big(A(f,\vec{x},\tau);\ C(a,b,\tau),\ W_\tau \le z\big)$$

is the solution of the problem (see Remark in No. 4)

$$\begin{aligned}&\frac{1}{2}u''(x)-\lambda u(x)=-2f(x)\hat{G}_\lambda(x,z), \quad x\in(a,b)\setminus\{x_1,\dots,x_k\},\\&u'(x_i+)-u'(x_i-)=-4\gamma_i\hat{G}_\lambda(x_i,z), \quad i=1,2,\dots,k,\\&u(a+)=u(b-)=0.\end{aligned}$$

Furthermore,

$$v_2(x):=\frac{\partial}{\partial z}\mathbf{E}_x\big(A^2(f,\vec{x},\tau);\ C(a,b,\tau),\ W_\tau\le z\big)$$

is the solution of the problem

$$\begin{aligned}&\frac{1}{2}u''(x)-\lambda u(x)=-2f(x)v_1(x), \quad x\in(a,b)\setminus\{x_1,\dots,x_k\},\\&u'(x_i+)-u'(x_i-)=-4\gamma_i v_1(x_i), \quad i=1,2,\dots,k,\\&u(a+)=u(b-)=0.\end{aligned}$$

We remark that

$$v_1(x)=\int_a^b \hat{G}_\lambda(x,y)\ f(y)\ 2\hat{G}_\lambda(y,z)\ 2dy+\sum_{i=1}^k 4\gamma_i\hat{G}_\lambda(x,x_i)\hat{G}_\lambda(x_i,z).$$

8. Let $f:\mathbb{R}_+\mapsto\mathbb{R}_+,\ f(0)=0$, be piecewise continuous and introduce

$$V(f,\tau):=\int_{-\infty}^{+\infty} f(\ell(\tau,y))\,dy.$$

Then using the Ray–Knight theorem in V.9 it can be proved (Borodin (1989b)) that for $h>0$,

$$\mathbf{E}_x\Big(\exp(-V(f,\tau));\ \sup_y \ell(\tau,y)<h\Big)=\int_0^h \sqrt{2\lambda}e^{-\sqrt{2\lambda}l}\,v_0(l)\,v_2(l)\,dl,$$

where v_0 is the solution of the problem

$$\begin{aligned}&2x\,u''(x)-2\sqrt{2\lambda}\,x\,u'(x)-f(x)u(x)=0, \qquad x\in(0,h),\\&u(0)=1, \quad u(h)=0,\end{aligned}$$

and v_2 is the solution of the problem

$$\begin{aligned}&2xu''(x)+(2-2\sqrt{2\lambda}\,x)u'(x)\\&\qquad -(\sqrt{2\lambda}+f(x))u(x)=-\sqrt{2\lambda}v_0(x), \qquad x\in(0,h),\\&u(h)=0.\end{aligned}$$

Choosing $f \equiv 0$ here gives the distribution $\sup_y \ell(\tau, y)$ (see 1.1.11.2 and Borodin (1989b)).

For Brownian motion with drift use V.15 to obtain ($\gamma := \sqrt{2\lambda + \mu^2}$)

$$\mathbf{E}_x\Big(\exp(-V(f,\tau));\ \sup_y \ell^{(\mu)}(\tau, y) < h\Big)$$
$$= \frac{\lambda}{\gamma - \mu} \int_0^h e^{-\gamma l}\ v_0(l) v_2^-(l)\ dl + \frac{\lambda}{\gamma + \mu} \int_0^h e^{-\gamma l}\ v_0(l) v_2^+(l)\ dl,$$

where v_0 is the solution of the problem

$$2x\, u''(x) - 2\gamma x\, u'(x) - f(x) u(x) = 0, \quad x \in (0, h),$$
$$u(0) = 1, \quad u(h) = 0.$$

The functions v_2^+ and v_2^- are the solutions of the problems

$$2xu''(x) + (2 - 2\gamma x) u'(x)$$
$$- (\gamma \pm \mu + f(x)) u(x) = -(\gamma \pm \mu) v_0(x), \quad x \in (0, h),$$
$$u(h) = 0,$$

respectively.

3. Stopping at first exit time

9. The function

$$v(x) := \mathbf{E}_x\big(\exp(-A(f, \vec{x}, H_{a,b}))\big)$$

is the solution of the problem

$$\frac{1}{2} u''(x) - f(x) u(x) = 0, \qquad x \in (a,b) \setminus \{x_1, \dots, x_k\},$$
$$u'(x_i+) - u'(x_i-) = 2\gamma_i u(x_i), \qquad i = 1, \dots, k,$$
$$u(a+) = u(b-) = 1.$$

To explain this let $\tilde{W}$, $\tilde{\varphi}_0$, and $\tilde{\psi}_0$ be as in No. 4. Then

$$v(x) = \mathbf{P}_x(\tilde{W}_{\zeta-} = a \text{ or } b) = \frac{\tilde{\psi}_0(x)}{\tilde{\psi}_0(b)} + \frac{\tilde{\varphi}_0(x)}{\tilde{\varphi}_0(a)},$$

and the result follows. From this it is also obvious that

$$x \mapsto \mathbf{E}_x\big(\exp(-A(f, \vec{x}, H_{a,b}));\ W_{H_{a,b}} = b\big) = \frac{\tilde{\psi}_0(x)}{\tilde{\psi}_0(b)}$$

is the solution of the problem

$$\begin{aligned}&\frac{1}{2}u''(x) - f(x)u(x) = 0, \qquad x \in (a,b) \setminus \{x_1, \dots, x_k\},\\&u'(x_i+) - u'(x_i-) = 2\gamma_i u(x_i), \qquad i = 1, \dots, k,\\&u(a+) = 0, \ u(b-) = 1.\end{aligned}$$

A similar statement holds for the function

$$x \mapsto \mathbf{E}_x\big(\exp(-A(f, \vec{x}, H_{a,b})); \ W_{H_{a,b}} = a\big) = \frac{\tilde{\varphi}_0(x)}{\tilde{\varphi}_0(a)}.$$

10. It follows from No. 9 that for $c \in (a,b)$,

$$x \mapsto \mathbf{E}_x\big(\exp(-A(f, \vec{x}, H_c)); \ C(a, b, H_c)\big)$$

is the solution of the problem

$$\begin{aligned}&\frac{1}{2}u''(x) - f(x)u(x) = 0, \qquad x \in (a,b) \setminus \{c, x_1, \dots, x_k\},\\&u'(x_i+) - u'(x_i-) = 2\gamma_i u(x_i), \qquad i = 1, \dots, k,\\&u(a+) = u(b-) = 0, \ u(c) = 1.\end{aligned}$$

11. We characterize the first two moments of $A(f, \vec{x}, H_{a,b})$ as solutions to differential problems. The first moment

$$v(x) := \mathbf{E}_x\big(A(f, \vec{x}, H_{a,b})\big)$$

is the solution of the problem

$$\begin{aligned}&\frac{1}{2}u''(x) = -f(x), \qquad x \in (a,b) \setminus \{x_1, \dots, x_k\},\\&u'(x_i+) - u'(x_i-) = -2\gamma_i, \qquad i = 1, \dots, k,\\&u(a+) = u(b-) = 0.\end{aligned}$$

The second moment is obtained by iteration. The result is that

$$x \mapsto \mathbf{E}_x\big(A^2(f, \vec{x}, H_{a,b})\big)$$

is the solution of the problem

$$\begin{aligned}&\frac{1}{2}u''(x) = -2f(x)\,v(x), \quad x \in (a,b) \setminus \{x_1, \dots, x_k\},\\&u'(x_i+) - u'(x_i-) = -4\gamma_i v(x_i), \quad i = 1, 2, \dots, k,\\&u(a+) = u(b-) = 0.\end{aligned}$$

For an explanation of these, see II.27 and recall (cf. II.11) that the Green kernel is given by

$$G_0(x,y) := \frac{(b-x)(y-a)}{b-a}, \quad x \ge y.$$

12. Next we consider the first two moments of the functional $A(f, \vec{x}, H_{a,b})$ under the side condition $W_{H_{a,b}} = b$. Recall that

$$g(x) := \mathbf{P}_x(W_{H_{a,b}} = b) = \mathbf{P}_x\big(H_b < H_a\big) = \frac{x-a}{b-a}.$$

Then

$$v(x) := \mathbf{E}_x\big(A(f, \vec{x}, H_{a,b});\ W_{H_{a,b}} = b\big)$$

is the solution of

$$\begin{aligned}
&\frac{1}{2}u''(x) = -f(x)g(x), \qquad x \in (a,b) \setminus \{x_1, \dots, x_k\},\\
&u'(x_i+) - u'(x_i-) = -2\gamma_i g(x_i), \qquad i = 1, \dots, k,\\
&u(a+) = u(b-) = 0.
\end{aligned}$$

Further,

$$x \mapsto \mathbf{E}_x\big(A^2(f, \vec{x}, H_{a,b});\ W_{H_{a,b}} = b\big)$$

is the solution of the problem

$$\begin{aligned}
&\frac{1}{2}u''(x) = -2f(x)\,v(x), \qquad x \in (a,b) \setminus \{x_1, \dots, x_k\},\\
&u'(x_i+) - u'(x_i-) = -4\gamma_i v(x_i), \qquad i = 1, \dots, k,\\
&u(a+) = u(b-) = 0.
\end{aligned}$$

13. Let f be as in No. 8 and introduce

$$V(f, H_z) := \int_{-\infty}^{+\infty} f(\ell(H_z, y))\,dy.$$

For $h > 0$ and $x < z$ consider the function

$$v(x,z) := \mathbf{E}_x\Big(\exp(-V(f, H_z));\ \sup_y \ell(H_z, y) < h\Big).$$

By the spatial homogeneity of Brownian motion v depends on x and z only through the difference $z - x$. Let $v(u) := v(x,z)$, where $u = z - x$. Then using the Ray–Knight theorem in V.6 it can be proved (Borodin (1989b), see also Salminen and Vallois (2005) and cf. No. 23) that

(a) $$\int_0^\infty e^{-\alpha u} v(u)\,du = \int_0^h \tilde{G}_\alpha(0,y) v_0(y)\,\frac{1}{2}dy,$$

where

$$v_0(y) := \mathbf{E}_y\Big(\exp\Big(-\int_0^\infty f(Z_s^{(0)})ds\Big);\ \sup_{s\ge 0} Z_s^{(0)} < h\Big)$$

and $Z^{(0)}$ is a squared 0-dimensional Bessel process. Therefore, v_0 is the solution of the problem

$$\begin{aligned} &2x\,u''(x) - f(x)u(x) = 0, \qquad x \in (0,h),\\ &u(0) = 1, \quad u(h) = 0. \end{aligned}$$

The Green function $\tilde{G}$ appearing in (a) is associated to the process $\tilde{Z}^{(2)}$ obtained from the squared 2-dimensional Bessel process $Z^{(2)}$ (recall that $\frac{dx}{2}$ is the speed measure of $Z^{(2)}$) killed when it hits h and which is also killed according to the additive functional

$$t \mapsto A(f,t) := \int_0^t f(Z_s^{(2)})\,ds.$$

In other words, $\tilde{Z}^{(2)}$ is defined via

$$\tilde{Z}_t^{(2)} = \begin{cases} Z_t^{(2)}, & t < \zeta \wedge H_h,\\ \partial, & t \ge \zeta \wedge H_h, \end{cases}$$

where

$$H_h := \inf\{t\ :\ Z_t^{(2)} = h\} \quad \text{and} \quad \zeta := \inf\{t\ :\ A(f,t) > \tau\}$$

with $\tau \sim \text{Exp}(1)$, independent of $Z^{(2)}$.

14. Consider for $h > 0$ and $a < 0 < b$ the function

$$v_b(a,b) := \mathbf{E}_0\Big(\exp(-V(f,H_{a,b}));\ \sup_y \ell(H_{a,b},y) < h,\ W_{H_{a,b}} = b\Big),$$

where f is as in No. 8 and

$$V(f,H_{a,b}) := \int_{-\infty}^{+\infty} f(\ell(H_{a,b},y))\,dy.$$

Using the Ray–Knight theorem in V.8 it can be proved that (Borodin 1989b)

$$\int_0^\infty db\,e^{-\alpha b}\int_{-\infty}^0 da\,e^{\beta a} v_b(a,b) = \frac{1}{2\beta}\int_0^h \tilde{G}_\alpha(0,y)v_0(y)\,dy,$$

where $\tilde{G}$ is as in No. 13 and v_0 is the solution of the problem

$$\begin{aligned} &2x\,u''(x) - (\beta + f(x))u(x) = 0, \qquad x \in (0,h),\\ &u(0) = 1, \quad u(h) = 0. \end{aligned}$$

The function

$$v_a(a,b) := \mathbf{E}_0\Big(\exp(-V(f,H_{a,b}));\ \sup_y \ell(H_{a,b},y) < h,\ W_{H_{a,b}} = a\Big)$$

has a similar representation with the roles of α and β changed.

4. Stopping at inverse additive functional

15. Let g be a non-negative piecewise continuous function and introduce

$$A(g,t) := \int_0^t g(W_s)ds.$$

For $t > 0$ we consider the stopping time $\nu(t) := \inf\{s : A(g,s) > t\}$. Let $\tau \sim \text{Exp}(\lambda)$ independent of W. Then the function

$$v(x) := \mathbf{E}_x\big(\exp(-A(f,\vec{x},\nu(\tau)));\ C(a,b,\nu(\tau))\big)$$

is the solution of the problem

$$\begin{aligned}
&\frac{1}{2}u''(x) - (\lambda g(x) + f(x))u(x) = -\lambda g(x), \qquad x \in (a,b)\setminus\{x_1,\dots,x_k\},\\
&u'(x_i+) - u'(x_i-) = 2\gamma_i u(x_i), \qquad i = 1,\dots,k,\\
&u(a+) = u(b-) = 0.
\end{aligned}$$

To briefly motivate this, let $\hat{W}$ be as in No. 4 and construct $\bar{W}$ from $\hat{W}$ by killing it (see II.22) according to the additive functional

$$t \mapsto A(f,\vec{x},t) + \lambda A(g,t).$$

Hence

$$k(dy) = 2(f(y) + \lambda g(y))dy + \sum_{i=1}^{k} 2\gamma_i \varepsilon_{\{x_i\}}(dy)$$

is the killing measure of $\bar{W}$. Let $\bar{G}$ and $\bar{\zeta}$ be the Green function (with respect to $2dx$) and the lifetime, respectively, of $\bar{W}$. Then (see II.26)

$$v(x) = \mathbf{P}_x(\bar{\zeta} = \nu(\tau)) = \lambda \int_a^b \bar{G}_0(x,z)g(z)2dz,$$

and the result follows from No. 4.

16. For $t \geq 0$ and $z \in (a,b)$ let $\varrho(t,z) := \inf\{s : \ell(s,z) > t\}$ be the inverse local time and let τ be as in No. 15. Then

$$v(x) := \mathbf{E}_x\big(\exp(-A(f,\vec{x},\varrho(\tau,z)));\ C(a,b,\varrho(\tau,z))\big)$$

is the solution of the problem

$$\begin{aligned}
&\frac{1}{2}u''(x) - f(x)u(x) = 0, \qquad x \in (a,b)\setminus\{z,x_1,\dots,x_k\},\\
&u'(z+) - u'(z-) = 2\lambda u(z) - 2\lambda,\\
&u'(x_i+) - u'(x_i-) = 2\gamma_i u(x_i), \qquad i = 1,\dots,k,\\
&u(a+) = u(b-) = 0,
\end{aligned}$$

where it is assumed that $z \neq x_i,\ i = 1, 2, \dots, k$. If $z = x_i$ for some i, then

$$u'(z+) - u'(z-) = 2(\lambda + \gamma_i)u(z) - 2\lambda.$$

This can be ensured similarly as the result in No. 15.

17. Consider the functional in No. 16 for Brownian motion with drift $W^{(\mu)}$. Due to transience of $W^{(\mu)}$ it may happen that $\varrho^{(\mu)}(\tau, z) = \infty$, and if $a = -\infty$ and $\mu < 0$, for instance, this appears in the formula. Indeed, set $C(b, \cdot) := C(-\infty, b, \cdot)$ and $\varrho^{(\mu)} := \varrho^{(\mu)}(\tau, z)$. Then

$$\begin{aligned} v(x) :=& \mathbf{E}_x\big(\exp(-A^{(\mu)}(f, \vec{x}, \varrho^{(\mu)}));\ C(b, \varrho^{(\mu)})\big) \\ =& \mathbf{E}_x\big(\exp(-A^{(\mu)}(f, \vec{x}, \varrho^{(\mu)}));\ C(b, \varrho^{(\mu)}),\ \varrho^{(\mu)} < \infty\big) \\ &\qquad + \mathbf{E}_x\big(\exp(-A^{(\mu)}(f, \vec{x}, \infty));\ C(b, \infty),\ \varrho^{(\mu)} = \infty\big) \\ =:& \ v_1(x) + v_2(x). \end{aligned}$$

Let $\bar{W}^{(\mu)}$ be the process obtained from $W^{(\mu)}$ by killing at H_b and also according to the additive functional

$$t \mapsto A^{(\mu)}(f, \vec{x}, t) + \lambda \ell^{(\mu)}(t, z).$$

Then if $z \neq x_i,\ i = 1, \dots, k$,

$$v_1(x) = \mathbf{P}_x(\bar{W}^{(\mu)}_{\zeta-} = z) = \lambda m(z) \bar{G}^{(\mu)}_0(x, z),$$

and

$$v_2(x) = \mathbf{P}_x(\lim_{t\to\infty} \bar{W}^{(\mu)}_t = -\infty) = \lim_{y\to-\infty} \frac{\bar{\varphi}_0(x)}{\bar{\varphi}_0(y)},$$

where $m(z) = 2e^{2\mu z}$ is the density of the speed measure and $\bar{\varphi}_0$ is the non-increasing fundamental solution of the system

$$\begin{aligned} &\frac{1}{2}u''(x) + \mu u'(x) - f(x)u(x) = 0, \qquad x \in (-\infty, b) \setminus \{z, x_1, \dots, x_k\}, \\ &u'(z+) - u'(z-) = 2\lambda u(z), \\ &u'(x_i+) - u'(x_i-) = 2\gamma_i u(x_i), \qquad i = 1, \dots, k, \\ &u(b) = 0. \end{aligned}$$

18. The Laplace transform v in No. 16 can be inverted explicitly. Analytically this is due to the fact that λ appears in a simple form only in one of the side conditions. However, there is also a more probabilistic explanation. For this let $\tilde{\varrho}(\cdot, z)$ be the inverse local time at z of the process $\tilde{W}$ obtained from $\hat{W}$ by killing it according to the additive functional $A(f, \vec{x}, \cdot)$ (cf. No. 4). Then, letting $\tilde{\zeta}$ be the lifetime of $\tilde{W}$, we have

$$\begin{aligned} \mathbf{P}_z(\tilde{\varrho}(t, z) < \infty) &= \mathbf{P}_z(\tilde{\varrho}(t, z) < \tilde{\zeta}) \\ &= \mathbf{E}_z\big(\exp(-A(f, \vec{x}, \varrho(t, z)));\ C(a, b, \varrho(t, z))\big). \end{aligned}$$

From II.14 we have

$$\mathbf{P}_z(\tilde{\varrho}(t,z) < \infty) = \exp\Big(-\frac{t}{2\tilde{G}_0(z,z)}\Big),$$

where $\tilde{G}$ is the Green function of $\tilde{W}$. Let $\tilde{H}$ be the first hitting time functional for $\tilde{W}$. Then, by the strong Markov property,

$$\begin{aligned}\mathbf{P}_x(\tilde{\varrho}(t,z) < \infty) &= \mathbf{P}_x(\tilde{H}_z < \infty,\ \tilde{\varrho}(t,z) < \infty)\\ &= \mathbf{P}_x(\tilde{H}_z < \infty)\mathbf{P}_z(\tilde{\varrho}(t,z) < \infty).\end{aligned}$$

Because

$$\mathbf{P}_x(\tilde{H}_z < \infty) = \begin{cases} \dfrac{\tilde{\psi}_0(x)}{\tilde{\psi}_0(z)}, & x \le z,\\[2ex] \dfrac{\tilde{\varphi}_0(x)}{\tilde{\varphi}_0(z)}, & x \ge z,\end{cases}$$

we have done the promised inversion. Here, $\tilde{\psi}_0$ and $\tilde{\varphi}_0$ are the fundamental solutions of the differential system associated to $\tilde{W}$.

The description is very similar for $W^{(\mu)}$. In this case we have

$$\mathbf{P}_z(\tilde{\varrho}^{(\mu)}(t,z) < \infty) = \exp\Big(-\frac{t}{m(z)\tilde{G}_0^{(\mu)}(z,z)}\Big),$$

where $\tilde{G}^{(\mu)}$ is the Green function of the diffusion obtained from $W^{(\mu)}$ similarly as $\tilde{W}$ is constructed from W.

Observe that $\{\varrho^{(\mu)}(\tau,z) = \infty\} = \{\ell^{(\mu)}(\infty,z) < \tau\}$ and so the function v_2 in No. 17 can be written as

$$v_2(x) = \mathbf{E}_x\big(\exp(-A^{(\mu)}(f,\vec{x},\infty) - \lambda\ell^{(\mu)}(\infty,z));\ C(b,\infty)\big).$$

In some cases this can be inverted to find an expression for

$$\mathbf{E}_x\big(\exp(-A^{(\mu)}(f,\vec{x},\infty));\ C(b,\infty),\ \varrho^{(\mu)}(t,z) = \infty\big).$$

19. The result in No. 16 can be generalized to stopping times of the type

$$\hat{\varrho}(t_1,t_2,\dots,t_n) := \varrho(t_1,z_1)\wedge\varrho(t_2,z_2)\wedge\cdots\wedge\varrho(t_n,z_n).$$

Consider the case $n = 2$, and let τ_1 and τ_2 be two independent exponentially distributed random variables (with parameters λ_1 and λ_2, respectively,) independent of W. Then in the case $z_1, z_2 \notin \{x_1,\dots,x_k\}$ and $z_1 \ne z_2$,

$$\begin{aligned}v(x) :=&\mathbf{E}_x\big(\exp(-A(f,\vec{x},\hat{\varrho}(\tau_1,\tau_2)));\ C(a,b,\hat{\varrho}(\tau_1,\tau_2))\big)\\ =&\mathbf{P}_x(\bar{W}_{\zeta-} = z_1 \ \text{ or } \ z_2)\\ =&2\lambda_1\bar{G}_0(x,z_1) + 2\lambda_2\bar{G}_0(x,z_2),\end{aligned}$$

where $\bar{G}$ is the Green function of the process $\bar{W}$ obtained from $\hat{W}$ by killing according to the additive functional

$$t \mapsto A(f, \vec{x}, t) + \lambda_1 \ell(t, z_1) + \lambda_2 \ell(t, z_2).$$

20. Analogously to the generalization in No. 19, we can generalize the result in No. 15. Consider the case $n = 2$ and let $\hat{\nu}(t_1, t_2) := \nu_1(t_1) \wedge \nu_2(t_2)$, where

$$\nu_j(t) := \inf\Big\{ s : \int_0^s g_j(W_v)dv > t \Big\}, \quad j = 1, 2,$$

and g_j, $j = 1, 2$, are non-negative piecewise continuous functions. Then

$$v(x) := \mathbf{E}_x\big(\exp(-A(f, \vec{x}, \hat{\nu}(\tau_1, \tau_2))); \ C(a, b, \hat{\nu}(\tau_1, \tau_2))\big)$$

is the solution of the problem

$$\begin{aligned}
&\frac{1}{2}u''(x) - \Big(\lambda_1 g_1(x) + \lambda_2 g_2(x) + f(x)\Big)u(x) \\
&\qquad\qquad\qquad = -\lambda_1 g_1(x) - \lambda_2 g_2(x), \qquad x \in (a, b) \setminus \{x_1, \dots, x_k\}, \\
&u'(x_i+) - u'(x_i-) = 2\gamma_i u(x_i), \qquad i = 1, \dots, k, \\
&u(a+) = u(b-) = 0.
\end{aligned}$$

21. Consider for $h > 0$ the function

$$v(x) := \mathbf{E}_x\Big(\exp(-V(f, \varrho(t, z))); \ \sup_y \ell(\varrho(t, z), y) < h\Big),$$

where f is as in No. 8 and

$$V(f, \varrho(t, z)) := \int_{-\infty}^{+\infty} f(\ell(\varrho(t, z), y)) \, dy.$$

This generalizes in a sense the result in No. 13 and we use the notation therein. By the Ray–Knight theorem in V.10 we have

$$v(x) = v_0(t) \, v_2(t, |z - x|), \quad t < h,$$

where v_0 is as in No. 13 and

$$v_2(t, y) := \mathbf{E}_t\Big(v_0(Z_y^{(2)}) \exp\Big(-\int_0^y f(Z_u^{(2)})du\Big); \ \sup_{u \le y} Z_u^{(2)} < h\Big).$$

Then for $\tilde{G}$ as in No. 13,

$$\int_0^\infty e^{-\alpha y} v_2(t, y) \, dy = \int_0^h \tilde{G}_\alpha(t, u) \, v_0(u) \, \frac{1}{2} du.$$

In the case $z = x$ we simply have

$$\mathbf{E}_x\Big(\exp(-V(f, \varrho(t, x))); \ \sup_y \ell(\varrho(t, x), y) < h\Big) = v_0^2(t), \quad t < h.$$

5. Stopping at first range time

22. Let $r > 0$ and consider the first range time

$$\theta_r := \inf\Big\{t : \sup_{0\le s\le t} W_s - \inf_{0\le s\le t} W_s = r\Big\}.$$

For results on first range time see Feller (1951), Imhof (1985), Vallois (1993), (1995), Borodin (1999), and Chong, Cowan and Holst (2000), and Salminen and Vallois (2005). We have the following fundamental identity which holds for $z \le x \le z + r$,

$$\text{(a)}\qquad \frac{\partial}{\partial z}\mathbf{P}_x(\theta_r \in dt, W_{\theta_r} \le z) = \frac{\partial}{\partial r}\mathbf{P}_x\Big(H_z \in dt, \sup_{0\le s\le t} W_s - \inf_{0\le s\le t} W_s \le r\Big).$$

Clearly,

$$\mathbf{P}_x\Big(H_z \in dt, \sup_{0\le s\le t} W_s - \inf_{0\le s\le t} W_s \le r\Big) = \mathbf{P}_x(H_z \in dt, H_{z+r} > t).$$

Using this in (a) and integrating over t gives for $x - r \le z \le x$,

$$\frac{\partial}{\partial z}\mathbf{P}_x(W_{\theta_r} \le z) = \frac{\partial}{\partial r}\mathbf{P}_x(H_z < H_{z+r}) = \frac{\partial}{\partial r}\frac{z + r - x}{r} = \frac{x - z}{r^2},$$

and, further, by symmetry

$$\mathbf{P}_x(W_{\theta_r} \in dz) = \frac{|z - x|}{r^2}\,dz, \qquad |z - x| \le r.$$

The identity (a) can be generalized to (see Borodin (1999))

$$\begin{aligned}&\frac{\partial}{\partial z}\mathbf{E}_x\big(\exp(-A(f, \vec{x}, \theta_r)); \theta_r \in dt, W_{\theta_r} < z\big)\\ &\quad = \frac{\partial}{\partial r}\mathbf{E}_x\big(\exp(-A(f, \vec{x}, H_z)); H_z \in dt, H_{z+r} > t\big),\end{aligned}$$

where $z < x < z + r$. Integrating over t and using the results and the notation from No. 9 (with $a = z$ and $b = z + r$) lead to

$$\text{(b)}\qquad \frac{\partial}{\partial z}\mathbf{E}_x\big(\exp(-A(f, \vec{x}, \theta_r)); W_{\theta_r} < z\big) = \frac{\partial}{\partial r}\frac{\tilde{\varphi}_0(x)}{\tilde{\varphi}_0(z)}.$$

Analogously, for $z - r < x < z$ (now $a = z - r$ and $b = z$ in No. 9)

$$\text{(c)}\qquad \frac{\partial}{\partial z}\mathbf{E}_x\big(\exp(-A(f, \vec{x}, \theta_r)); W_{\theta_r} < z\big) = \frac{\partial}{\partial r}\frac{\tilde{\psi}_0(x)}{\tilde{\psi}_0(z)}.$$

Notice that the functions $\tilde{\varphi}_0$ and $\tilde{\psi}_0$ depend on r via the boundary conditions. It is possible to write the right-hand sides of (b) and (c) in a form, where the dependence on r is more explicit. To explain this assume that $f \not\equiv 0$ and/or $\gamma_i \neq 0$ for some i. Then the functions $\tilde{\varphi}_0$ and $\tilde{\psi}_0$ can be expressed as

$$\begin{aligned}\tilde{\varphi}_0(x) &= \psi(z+r)\varphi(x) - \psi(x)\varphi(z+r),\\ \tilde{\psi}_0(x) &= \psi(x)\varphi(z-r) - \psi(z-r)\varphi(x),\end{aligned}$$

where $\varphi := \varphi_0$ and $\psi := \psi_0$ are fundamental solutions of the system

$$\begin{aligned}&\frac{1}{2}u''(x) - f(x)u(x) = 0, \quad x \in \mathbb{R} \setminus \{x_1, \dots, x_k\},\\ &u'(x_i+) - u'(x_i-) = 2\gamma_i u(x_i), \quad i = 1, \dots, k.\end{aligned}$$

Using these expressions we can, in fact, compute the derivatives in (b) and (c). To do this, define

$$\rho(x,y) := \psi(x)\varphi(y) - \psi(y)\varphi(x).$$

Then for any a, b and c,

$$\rho(b,c)\,\frac{\partial}{\partial b}\rho(a,b) - \rho(a,b)\,\frac{\partial}{\partial b}\rho(b,c) = -\omega\rho(a,c),$$

where

$$\omega := \psi'(x)\varphi(x) - \psi(x)\varphi'(x)$$

is a constant (Wronskian). Consequently, for $z < x < z+r$ (the case $z-r < x < z$ is left to the reader),

$$\frac{\partial}{\partial r}\frac{\tilde{\varphi}_0(x)}{\tilde{\varphi}_0(z)} = \frac{\partial}{\partial r}\frac{\rho(z+r,x)}{\rho(z+r,z)} = -\frac{\partial}{\partial r}\frac{\rho(x,z+r)}{\rho(z+r,z)} = \frac{\omega\rho(x,z)}{\rho^2(z+r,z)},$$

and we have the formulae

$$\frac{\partial}{\partial z}\mathbf{E}_x\big(\exp(-A(f,\vec{x},\theta_r)); W_{\theta_r} < z\big) = \begin{cases} \dfrac{\omega\rho(x,z)}{\rho^2(z+r,z)}, & x-r < z \le x,\\[2mm] \dfrac{\omega\rho(z,x)}{\rho^2(z,z-r)}, & x \le z < x+r,\end{cases}$$

and

$$\mathbf{E}_x\big(F(W_{\theta_r})\exp(-A(f,\vec{x},\theta_r))\big) = \omega\int_x^{x+r}\frac{F(z)\rho(z,x) + F(z-r)\rho(x,z-r)}{\rho^2(z,z-r)}\,dz,$$

where F is measurable and bounded. The corresponding formulae for Brownian motion with drift can be derived using absolute continuity.

23. Let $f : \mathbb{R}_+ \mapsto \mathbb{R}_+$, $f(0) = 0$, be piecewise continuous and consider

$$V(f,\theta_r) := \int_{-\infty}^{+\infty} f(\ell(\theta_r,y))\,dy.$$

Using the Ray–Knight theorem for Brownian motion stopped at θ_r (see V.7) we obtain (see Borodin (1999) – also for Brownian motion with drift – and Salminen and Vallois (2005))

$$\int_0^\infty e^{-\alpha r}\, \mathbf{E}_x\Big(\exp(-V(f,\theta_r));\ \sup_y \ell(\theta_r,y) < h\Big)\, dr = \int_0^h y\, \tilde{G}_\alpha(0,y)\, \hat{G}_\alpha(0,y)\, dy,$$

where $\tilde{G}_\alpha$ is as in No. 13 and the Green function $\hat{G}_\alpha$ is associated to the process $\tilde{Z}^{(4)}$ obtained from the squared Bessel process $Z^{(4)}$ killed when it hits h and which is also killed according to the additive functional

$$t \mapsto \int_0^t f(Z_s^{(4)})\, ds.$$

APPENDIX 1

BRIEFLY ON SOME DIFFUSIONS

In this appendix we give basic characteristics and other facts and formulae associated with some diffusions appearing frequently in the literature – and in our own text. This collection may also serve as a sample of the examples highlighting different aspects in the general theory of linear diffusions.

Diffusions below are considered in the canonical framework. In particular, sample paths are continuous functions $t \mapsto \omega(t)$ taking values in $\mathbb{R}$ up to the lifetime. See No. I.12.

Transition densities are given with respect to speed measures. This normalization gives us symmetric densities (cf. No. II.4). To get the density with respect to Lebesgue measure our densities must be multiplied by the speed density (with respect to Lebesgue measure). See also No. II.9.

The Green functions are obtained from the transition densities, and vice versa. We present below the Green functions in the form

$$G_\alpha(x,y) = w_\alpha^{-1}\varphi_\alpha(x)\psi_\alpha(y), \quad x \geq y.$$

By symmetry, this description is sufficient. Notice also that with this notation the decreasing fundamental solution is always a function of x, and the increasing one is always a function of y. Recall from No. II.10 the connection with the hitting times and the fundamental solutions.

For properties of special functions, Laplace transforms, and inverse Laplace transforms we refer to Abramowitz and Stegun (1970), and Erdelyi and et al. (1953), (1954). At the end of the formula part of this book there is a list of properties of some special functions (Appendix 2) appearing in the tables and in this appendix and also there is a collection of the most used formulae for the inverse Laplace transforms (Appendix 3).

1. Brownian motion

Speed measure: $m(dx) = 2\,dx$.

Scale function: $s(x) = x$.

Generator: $\mathcal{G}f = \dfrac{1}{2}\dfrac{d^2 f}{dx^2}$.

Domain: $\mathcal{D} = \{f :\ f,\ \mathcal{G}f \in \mathcal{C}_b(\mathbb{R})\}$.

Green function: $G_\alpha(x,y) = w_\alpha^{-1}\, e^{-x\sqrt{2\alpha}}\, e^{y\sqrt{2\alpha}}, \quad x \ge y\,.$

Wronskian: $w_\alpha = 2\sqrt{2\alpha}$.

Transition density w.r.t. m: $p(t;x,y) = \frac{1}{2\sqrt{2\pi t}} \exp\Bigl(-\frac{(x-y)^2}{2t}\Bigr)$.

Spectral representation of the transition density:

$$\begin{aligned} p(t;x,y) &= \frac{1}{4\pi}\int_{-\infty}^{\infty} e^{-\lambda^2 t/2}\,(\cos(\lambda x)\cos(\lambda y) + \sin(\lambda x)\sin(\lambda y))\,d\lambda \\ &= \frac{1}{2\pi}\int_0^{\infty} e^{-\lambda^2 t/2}\,\cos(\lambda(x-y))\,d\lambda. \end{aligned}$$

2. Brownian motion on $[a,\infty)$ reflected at a

Speed measure: $m(dx) = 2\,dx$.

Scale function: $s(x) = x$.

Generator: $\mathcal{G}f = \frac{1}{2}\frac{d^2 f}{dx^2}, \quad x > a, \quad \mathcal{G}f(a) = \frac{1}{2} f''(a+)$.

Domain: $\mathcal{D} = \{f:\ f,\ \mathcal{G}f \in \mathcal{C}_b([a,\infty)),\ f'(a+) = 0\}$.

Green function: $G_\alpha(x,y) = w_\alpha^{-1}\, e^{-(x-a)\sqrt{2\alpha}}\, \mathrm{ch}((y-a)\sqrt{2\alpha})\,, \quad x \ge y \ge a\,.$

Wronskian: $w_\alpha = \sqrt{2\alpha}$.

Transition density w.r.t. m:

$$p(t;x,y) = \frac{1}{2\sqrt{2\pi t}}\Bigl(\exp\Bigl(-\frac{(x-y)^2}{2t}\Bigr) + \exp\Bigl(-\frac{(x+y-2a)^2}{2t}\Bigr)\Bigr).$$

Spectral representation of the transition density:

$$p(t;x,y) = \frac{1}{\pi}\int_0^{\infty} e^{-\lambda^2 t/2} \cos(\lambda(x-a))\,\cos(\lambda(y-a))\,d\lambda.$$

Let W be a Brownian motion process. Then $\{|W_t - a| + a:\ t \ge 0\}$ is the Brownian motion on $[a,\infty)$ reflected at a.

3. Brownian motion on (a,∞) killed at a

Speed measure: $m(dx) = 2\,dx$.

Scale function: $s(x) = x$.

Generator: $\mathcal{G}f = \frac{1}{2}\frac{d^2 f}{dx^2}, \quad x > a$.

Domain: $\mathcal{D} = \{f:\ f,\ \mathcal{G}f \in \mathcal{C}_b((a,\infty)),\ f(a+) = 0\}$.

Green function: $G_\alpha(x,y) = w_\alpha^{-1} e^{-(x-a)\sqrt{2\alpha}} \operatorname{sh}((y-a)\sqrt{2\alpha}), \quad x \ge y > a\,.$

Wronskian: $w_\alpha = \sqrt{2\alpha}$.

Transition density w.r.t. m:

$$p(t;x,y) = \frac{1}{2\sqrt{2\pi t}}\Big(\exp\Big(-\frac{(x-y)^2}{2t}\Big) - \exp\Big(-\frac{(x+y-2a)^2}{2t}\Big)\Big).$$

Spectral representation of the transition density:

$$p(t;x,y) = \frac{1}{\pi}\int_0^\infty e^{-\lambda^2 t/2} \sin(\lambda(x-a))\,\sin(\lambda(y-a))\,d\lambda.$$

4. Brownian motion on $[a,\infty)$ absorbed at a

Speed measure: $m(dx) = 2\,dx$.

Scale function: $s(x) = x$.

Generator: $\mathcal{G}f = \frac{1}{2}\frac{d^2 f}{dx^2}, \quad x > a, \quad \mathcal{G}f(a) = \frac{1}{2}f''(a+)$.

Domain: $\mathcal{D} = \{f:\ f,\ \mathcal{G}f \in \mathcal{C}_b([a,\infty)),\ f''(a+) = 0\}$.

Green function: $G_\alpha(x,y) = w_\alpha^{-1} e^{-(x-a)\sqrt{2\alpha}} \operatorname{sh}((y-a)\sqrt{2\alpha}), \quad x \ge y > a\,.$

Wronskian: $w_\alpha = \sqrt{2\alpha}$.

Resolvent at a: $R_\alpha f(a) = \frac{1}{\alpha} f(a)$.

Transition density w.r.t. m and absorption probability:

$$p(t;x,y) = \frac{1}{2\sqrt{2\pi t}}\Big(\exp\Big(-\frac{(x-y)^2}{2t}\Big) - \exp\Big(-\frac{(x+y-2a)^2}{2t}\Big)\Big), \quad x > a,\ y > a,$$

$$\mathbf{P}_x(\omega(t) = a) = \begin{cases} \operatorname{Erfc}\Big(\frac{x-a}{\sqrt{2t}}\Big), & x > a,\\ 1, & x = a. \end{cases}$$

5. Brownian motion on $[a,b]$ reflected at a and b

Speed measure: $m(dx) = 2\,dx$.

Scale function: $s(x) = x$.

Generator: $\mathcal{G}f = \frac{1}{2}\frac{d^2}{dx^2}, \quad a < x < b,$

$\mathcal{G}f(a) = \frac{1}{2}f''(a+), \quad \mathcal{G}f(b) = \frac{1}{2}f''(b-)$.

Domain: $\mathcal{D} = \{f:\ f,\ \mathcal{G}f \in \mathcal{C}_b([a,b]),\ f'(a+) = f'(b-) = 0\}$.

Green function:

$$G_\alpha(x,y) = w_\alpha^{-1}\,\mathrm{ch}((b-x)\sqrt{2\alpha})\,\mathrm{ch}((y-a)\sqrt{2\alpha}), \quad b \ge x \ge y \ge a.$$

Wronskian: $w_\alpha = \sqrt{2\alpha}\,\mathrm{sh}((b-a)\sqrt{2\alpha})$.

Transition density w.r.t. m:

$$p(t;x,y) = \frac{1}{2\sqrt{2\pi t}} \sum_{n=-\infty}^{\infty} \Big(\exp\Big(-\frac{(x-y+2n(b-a))^2}{2t}\Big) + \exp\Big(-\frac{(x+y-2a+2n(b-a))^2}{2t}\Big)\Big).$$

Spectral representation of the transition density:

$$p(t;x,y) = \frac{1}{b-a}\Big(\frac{1}{2} + \sum_{n=1}^{\infty} \exp\Big(-\frac{n^2\pi^2}{2(b-a)^2}t\Big)\cos\Big(\frac{n\pi}{b-a}(x-a)\Big)\cos\Big(\frac{n\pi}{b-a}(y-a)\Big)\Big).$$

6. Brownian motion on (a,b) killed at a or b

Speed measure: $m(dx) = 2\,dx$.

Scale function: $s(x) = x$.

Generator: $\mathcal{G}f = \frac{1}{2}\frac{d^2f}{dx^2}, \quad a < x < b$.

Domain: $\mathcal{D} = \{f:\ f,\ \mathcal{G}f \in \mathcal{C}_b((a,b)),\ f(a+) = f(b-) = 0\}$.

Green function:

$$G_\alpha(x,y) = w_\alpha^{-1}\,\mathrm{sh}((b-x)\sqrt{2\alpha})\,\mathrm{sh}((y-a)\sqrt{2\alpha}), \quad b > x \ge y > a\ .$$

Wronskian: $w_\alpha = \sqrt{2\alpha}\,\mathrm{sh}((b-a)\sqrt{2\alpha})$.

Transition density w.r.t. m:

$$p(t;x,y) = \frac{1}{2\sqrt{2\pi t}} \sum_{n=-\infty}^{\infty} \Big(\exp\Big(-\frac{(x-y+2n(b-a))^2}{2t}\Big) - \exp\Big(-\frac{(x+y-2a+2n(b-a))^2}{2t}\Big)\Big).$$

Spectral representation of the transition density:

$$p(t;x,y) = \frac{1}{b-a}\sum_{n=1}^{\infty} \exp\Big(-\frac{n^2\pi^2}{2(b-a)^2}t\Big)\sin\Big(\frac{n\pi}{b-a}(x-a)\Big)\sin\Big(\frac{n\pi}{b-a}(y-a)\Big).$$

7. Brownian motion killed elastically at 0

Speed measure: $m(dx) = 2\,dx$.

Scale function: $s(x) = x$.

Killing measure: $\kappa(dx) = \gamma\,\varepsilon_{\{0\}}(dx),\ \gamma > 0$.

Generator: $\mathcal{G}f = \frac{1}{2}\frac{d^2 f}{dx^2},\quad x \neq 0,\quad \mathcal{G}f(0) = \frac{1}{2}f''(0+) = \frac{1}{2}f''(0-)$.

Domain: $\mathcal{D} = \{f:\ f,\ \mathcal{G}f \in \mathcal{C}_b(\mathbb{R}),\ f'(0+) - f'(0-) = \gamma f(0)\}$.

Green function:

$$G_\alpha(x,y) = \begin{cases} w_\alpha^{-1}\Big(e^{-x\sqrt{2\alpha}} - \dfrac{\gamma}{\sqrt{2\alpha}}\operatorname{sh}(x\sqrt{2\alpha})\Big)\,e^{y\sqrt{2\alpha}}, & 0 \geq x \geq y, \\ w_\alpha^{-1}\,e^{-x\sqrt{2\alpha}}\,e^{y\sqrt{2\alpha}}, & x \geq 0 \geq y, \\ w_\alpha^{-1}\,e^{-x\sqrt{2\alpha}}\Big(\dfrac{\gamma}{\sqrt{2\alpha}}\operatorname{sh}(y\sqrt{2\alpha}) + e^{y\sqrt{2\alpha}}\Big), & x \geq y \geq 0, \end{cases}$$

$$= \frac{1}{2\sqrt{2\alpha}}\Big(e^{-|x-y|\sqrt{2\alpha}} - \frac{\gamma}{w_\alpha}e^{-(|x|+|y|)\sqrt{2\alpha}}\Big).$$

Wronskian: $w_\alpha = 2\sqrt{2\alpha} + \gamma$.

Transition density w.r.t. m:

$$p(t;x,y) = \frac{1}{2\sqrt{2\pi t}}\exp\Big(-\frac{(x-y)^2}{2t}\Big) - \frac{\gamma}{8}\exp\Big(\frac{\gamma|x| + \gamma|y|}{2} + \frac{\gamma^2 t}{8}\Big)\operatorname{Erfc}\Big(\frac{2|x| + 2|y| + \gamma t}{2\sqrt{2t}}\Big).$$

8. Brownian motion sticky at 0

Speed measure: $m(dx) = 2\,dx + 2c\,\varepsilon_{\{0\}}(dx),\quad c > 0$.

Scale function: $s(x) = x$.

Generator: $\mathcal{G}f = \frac{1}{2}\frac{d^2 f}{dx^2},\quad x \neq 0,\quad \mathcal{G}f(0) = \frac{1}{2}f''(0+) = \frac{1}{2}f''(0-)$.

Domain: $\mathcal{D} = \{f:\ f,\ \mathcal{G}f \in \mathcal{C}_b(\mathbb{R}),\ c\,f''(0+) = f'(0+) - f'(0-)\}$.

Green function:

$$G_\alpha(x,y) = \begin{cases} w_\alpha^{-1}\Big(e^{-x\sqrt{2\alpha}} - c\sqrt{2\alpha}\operatorname{sh}(x\sqrt{2\alpha})\Big)\,e^{y\sqrt{2\alpha}}, & 0 \geq x \geq y, \\ w_\alpha^{-1}\,e^{-x\sqrt{2\alpha}}\,e^{y\sqrt{2\alpha}}, & x \geq 0 \geq y, \\ w_\alpha^{-1}\,e^{-x\sqrt{2\alpha}}\Big(c\sqrt{2\alpha}\operatorname{sh}(y\sqrt{2\alpha}) + e^{y\sqrt{2\alpha}}\Big), & x \geq y \geq 0, \end{cases}$$

$$= \frac{1}{2\sqrt{2\alpha}}\Big(e^{-|x-y|\sqrt{2\alpha}} - \frac{2\alpha c}{w_\alpha}e^{-(|x|+|y|)\sqrt{2\alpha}}\Big).$$

Wronskian: $w_\alpha = 2\sqrt{2\alpha} + 2\alpha c$.

Transition density w.r.t. m:

$$p(t;x,y) = \frac{1}{2\sqrt{2\pi t}}\Big(\exp\Big(-\frac{(x-y)^2}{2t}\Big) - \exp\Big(-\frac{(|x|+|y|)^2}{2t}\Big)\Big) + \frac{1}{2c}\exp\Big(\frac{2|x|+2|y|}{c} + \frac{2t}{c^2}\Big)\ \mathrm{Erfc}\Big(\frac{|x|+|y|}{\sqrt{2t}} + \frac{\sqrt{2t}}{c}\Big).$$

For results on sticky Brownian motion see Harrison and Lemoine (1981), Amir (1991), Chitashvili (1997), and Warren (1997).

9. Brownian motion sticky at 0 and killed elastically at 0

Speed measure: $m(dx) = 2\,dx + 2c\,\varepsilon_{\{0\}}(dx), \quad c > 0$.

Scale function: $s(x) = x$.

Killing measure: $\kappa(dx) = \gamma\,\varepsilon_{\{0\}}(dx),\ \gamma > 0$.

Generator: $\mathcal{G}f = \frac{1}{2}\frac{d^2 f}{dx^2}, \quad x \neq 0, \quad \mathcal{G}f(0) = \frac{1}{2}f''(0+) = \frac{1}{2}f''(0-)$.

Domain: $\mathcal{D} = \{f :\ f,\ \mathcal{G}f \in \mathcal{C}_b(\mathbb{R}),\ c\,f''(0+) = f'(0+) - f'(0-) - \gamma f(0)\}$.

Green function:

$$G_\alpha(x,y) = \begin{cases} w_\alpha^{-1}\Big(e^{-x\sqrt{2\alpha}} - \dfrac{2\alpha c+\gamma}{\sqrt{2\alpha}}\,\mathrm{sh}(x\sqrt{2\alpha})\Big)\,e^{y\sqrt{2\alpha}}, & 0 \ge x \ge y,\\ w_\alpha^{-1}\,e^{-x\sqrt{2\alpha}}\,e^{y\sqrt{2\alpha}}, & x \ge 0 \ge y,\\ w_\alpha^{-1}\,e^{-x\sqrt{2\alpha}}\Big(\dfrac{2\alpha c+\gamma}{\sqrt{2\alpha}}\,\mathrm{sh}(y\sqrt{2\alpha}) + e^{y\sqrt{2\alpha}}\Big), & x \ge y \ge 0, \end{cases}$$

$$= \frac{1}{2\sqrt{2\alpha}}\Big(e^{-|x-y|\sqrt{2\alpha}} - e^{-(|x|+|y|)\sqrt{2\alpha}}\Big) + \frac{1}{w_\alpha}e^{-(|x|+|y|)\sqrt{2\alpha}}.$$

Wronskian: $w_\alpha = \gamma + 2\sqrt{2\alpha} + 2\alpha c$.

Transition density w.r.t. m:

$$\begin{aligned} p(t;x,y) = {} & \frac{1}{2\sqrt{2\pi t}}\Big(\exp\Big(-\frac{(x-y)^2}{2t}\Big) - \exp\Big(-\frac{(|x|+|y|)^2}{2t}\Big)\Big)\\ & + \frac{\Upsilon+1}{4c\Upsilon}\exp\Big(\frac{(|x|+|y|)\gamma}{1-\Upsilon} + \frac{(\Upsilon+1)^2 t}{2c^2}\Big)\ \mathrm{Erfc}\Big(\frac{|x|+|y|}{\sqrt{2t}} + \frac{(1+\Upsilon)\sqrt{t}}{c\sqrt{2}}\Big)\\ & + \frac{\Upsilon-1}{4c\Upsilon}\exp\Big(\frac{(|x|+|y|)\gamma}{1+\Upsilon} + \frac{(\Upsilon-1)^2 t}{2c^2}\Big)\ \mathrm{Erfc}\Big(\frac{|x|+|y|}{\sqrt{2t}} + \frac{(1-\Upsilon)\sqrt{t}}{c\sqrt{2}}\Big), \end{aligned}$$

where $\Upsilon = \sqrt{1-\gamma c}$. The epression for the density is valid for all non-negative values on c and γ; also when $\gamma c > 1$. We remark that the densities in No. 7 and No. 8 can be obtained from the density above by letting $c \to 0$ and $\gamma \to 0$, respectively.

10. Reflecting Brownian motion on $[0,\infty)$ killed elastically at 0

Speed measure: $m(dx) = 2\,dx$.

Scale function: $s(x) = x$.

Killing measure: $k(dx) = \gamma\,\varepsilon_{\{0\}}(dx),\ \gamma > 0$.

Generator: $\mathcal{G}f = \frac{1}{2}\frac{d^2 f}{dx^2}, \quad x > 0, \quad \mathcal{G}f(0) = \frac{1}{2}f''(0+)$.

Domain: $\mathcal{D} = \{f:\ f,\ \mathcal{G}f \in \mathcal{C}_b([0,\infty)),\ f'(0+) = \gamma f(0)\}$.

Green function:

$$G_\alpha(x,y) = w_\alpha^{-1}\, e^{-x\sqrt{2\alpha}} \Big(\mathrm{ch}(y\sqrt{2\alpha}) + \frac{\gamma}{\sqrt{2\alpha}}\,\mathrm{sh}(y\sqrt{2\alpha})\Big), \quad x \ge y \ge 0.$$

Wronskian: $w_\alpha = \sqrt{2\alpha} + \gamma$.

Transition density w.r.t. m:

$$\begin{aligned} p(t;x,y) = {} & \frac{1}{2\sqrt{2\pi t}}\Big(\exp\Big(-\frac{(x-y)^2}{2t}\Big) + \exp\Big(-\frac{(x+y)^2}{2t}\Big)\Big) \\ & - \frac{\gamma}{2}\exp\Big(\gamma(x+y) + \frac{\gamma^2 t}{2}\Big)\,\mathrm{Erfc}\Big(\frac{x+y+\gamma t}{\sqrt{2t}}\Big), \qquad x \ge 0,\ y \ge 0. \end{aligned}$$

11. Reflecting Brownian motion on $[0,\infty)$ sticky at 0 and killed elastically at 0

Speed measure: $m(dx) = 2\,dx + 2c\,\varepsilon_{\{0\}}(dx), \quad c > 0$.

Scale function: $s(x) = x$.

Killing measure: $\kappa(dx) = \gamma\,\varepsilon_{\{0\}}(dx),\ \gamma > 0$.

Generator: $\mathcal{G}f = \frac{1}{2}\frac{d^2 f}{dx^2}, \quad x > 0, \quad \mathcal{G}f(0) = \frac{1}{2}f''(0+)$.

Domain: $\mathcal{D} = \{f:\ f,\ \mathcal{G}f \in \mathcal{C}_b([0,\infty)),\ c\,f''(0+) = f'(0+) - \gamma f(0)\}$.

Green function:

$$G_\alpha(x,y) = w_\alpha^{-1} e^{-x\sqrt{2\alpha}} \Big(\mathrm{ch}(y\sqrt{2\alpha}) + \frac{2\alpha c + \gamma}{\sqrt{2\alpha}}\,\mathrm{sh}(y\sqrt{2\alpha})\Big), \qquad x \ge y \ge 0.$$

Wronskian: $w_\alpha = \gamma + \sqrt{2\alpha} + 2\alpha c$.

Transition density w.r.t. m:

$$\begin{aligned} p(t;x,y) = {} & \frac{1}{2\sqrt{2\pi t}}\Big(\exp\Big(-\frac{(x-y)^2}{2t}\Big) - \exp\Big(-\frac{(x+y)^2}{2t}\Big)\Big) \\ & + \frac{\Upsilon+1}{4c\Upsilon}\exp\Big(\frac{(x+y)2\gamma}{1-\Upsilon} + \frac{(\Upsilon+1)^2 t}{8c^2}\Big)\,\mathrm{Erfc}\Big(\frac{x+y}{\sqrt{2t}} + \frac{(1+\Upsilon)\sqrt{t}}{2c\sqrt{2}}\Big) \\ & + \frac{\Upsilon-1}{4c\Upsilon}\exp\Big(\frac{(x+y)2\gamma}{1+\Upsilon} + \frac{(\Upsilon-1)^2 t}{8c^2}\Big)\,\mathrm{Erfc}\Big(\frac{x+y}{\sqrt{2t}} + \frac{(1-\Upsilon)\sqrt{t}}{2c\sqrt{2}}\Big), \end{aligned}$$

where $\Upsilon = \sqrt{1-4\gamma c}$ and $x \wedge y \ge 0$. The epression for the density is valid for all non-negative values on c and γ; also when $4\gamma c > 1$.

12. Brownian motion skew at zero

This process is usually indexed by one parameter $\beta \in (0,1)$ (called the skewness parameter of the process). Skew Brownian motion behaves like ordinary Brownian motion when off the origin but if, e.g., $\beta > 1/2$, it has more tendency to move up than down from the origin. In particular, letting $\{\omega(t) : t \geq 0\}$ denote a skew Brownian motion starting at 0 we have for every $t > 0$ the following property

$$\mathbf{P}_0(\omega(t) \geq 0) = 1 - \mathbf{P}_0(\omega(t) \leq 0) = \beta.$$

Speed measure:

$$m(dx) = \begin{cases} 4\beta\, dx, & x > 0, \\ 4(1-\beta)\, dx, & x < 0. \end{cases}$$

Scale function:

$$s(x) = \begin{cases} \dfrac{x}{2\beta}, & x \geq 0, \\ \dfrac{x}{2(1-\beta)}, & x \leq 0. \end{cases}$$

Generator: $\mathcal{G}f = \frac{1}{2}\frac{d^2 f}{dx^2}, \quad x \neq 0, \quad \mathcal{G}f(0) = \frac{1}{2}f''(0+) = \frac{1}{2}f''(0-).$

Domain: $\mathcal{D} = \{f : f, \mathcal{G}f \in \mathcal{C}_b(\mathbb{R}), (1-\beta)f'(0-) = \beta f'(0+)\}.$

Green function:

$$G_\alpha(x,y) = \begin{cases} w_\alpha^{-1}\left(\dfrac{1-2\beta}{1-\beta}\operatorname{sh}(x\sqrt{2\alpha}) + e^{-x\sqrt{2\alpha}}\right)e^{y\sqrt{2\alpha}}, & 0 \geq x \geq y, \\ w_\alpha^{-1}\, e^{-x\sqrt{2\alpha}}\, e^{y\sqrt{2\alpha}}, & x \geq 0 \geq y, \\ w_\alpha^{-1}\, e^{-x\sqrt{2\alpha}}\left(\dfrac{1-2\beta}{\beta}\operatorname{sh}(y\sqrt{2\alpha}) + e^{y\sqrt{2\alpha}}\right), & x \geq y \geq 0, \end{cases}$$

$$= \frac{e^{-|x-y|\sqrt{2\alpha}} - e^{-(|x|+|y|)\sqrt{2\alpha}}}{2\sqrt{2\alpha}(1+(2\beta-1)\operatorname{sgn}(x\wedge y))} + \frac{e^{-(|x|+|y|)\sqrt{2\alpha}}}{2\sqrt{2\alpha}}.$$

Wronskian: $w_\alpha = 2\sqrt{2\alpha}.$

Transition density w.r.t. m:

$$p(t;x,y) = \frac{e^{-(x-y)^2/2t} - e^{-(|x|+|y|)^2/2t}}{2\sqrt{2\pi t}(1+(2\beta-1)\operatorname{sgn}(x\wedge y))} + \frac{e^{-(|x|+|y|)^2/2t}}{2\sqrt{2\pi t}}.$$

For further results concerning skew Brownian motion see Walsh (1978), Harrison and Shepp (1981), and Barlow (1988).

13. Brownian motion skew at 0, sticky at 0 and killed elastically at 0

Speed measure: $m(dx) = 2(1+(2\beta-1)\operatorname{sgn} x)\,dx + 2c\,\varepsilon_{\{0\}}(dx), \quad c>0.$

Scale function: $s(x) = \dfrac{x}{1+(2\beta-1)\operatorname{sgn} x}.$

Killing measure: $\kappa(dx) = \gamma\,\varepsilon_{\{0\}}(dx),\ \gamma>0.$

Generator: $\mathcal{G}f = \dfrac{1}{2}\dfrac{d^2 f}{dx^2}, \quad x\neq 0, \quad \mathcal{G}f(0) = \frac{1}{2}f''(0+) = \frac{1}{2}f''(0-).$

Domain:

$$\mathcal{D} = \{f:\ f,\ \mathcal{G}f \in \mathcal{C}_b(\mathbb{R}),\ c\,f''(0+) = 2\beta f'(0+) - 2(1-\beta)f'(0-) - \gamma f(0)\}.$$

Green function:

$$G_\alpha(x,y) = \begin{cases} w_\alpha^{-1}\Big(\Big(2 - \dfrac{w_\alpha}{2(1-\beta)\sqrt{2\alpha}}\Big)\operatorname{sh}(x\sqrt{2\alpha}) + e^{-x\sqrt{2\alpha}}\Big)\,e^{y\sqrt{2\alpha}}, & 0\ge x\ge y,\\ w_\alpha^{-1}\,e^{-x\sqrt{2\alpha}}\,e^{y\sqrt{2\alpha}}, & x\ge 0\ge y,\\ w_\alpha^{-1}e^{-x\sqrt{2\alpha}}\Big(\Big(\dfrac{w_\alpha}{2\beta\sqrt{2\alpha}} - 2\Big)\operatorname{sh}(y\sqrt{2\alpha}) + e^{y\sqrt{2\alpha}}\Big), & x\ge y\ge 0,\end{cases}$$

$$= \frac{e^{-|x-y|\sqrt{2\alpha}} - e^{-(|x|+|y|)\sqrt{2\alpha}}}{2\sqrt{2\alpha}(1+(2\beta-1)\operatorname{sgn}(x\wedge y))} + \frac{e^{-(|x|+|y|)\sqrt{2\alpha}}}{w_\alpha}.$$

Wronskian: $w_\alpha = \gamma + 2\sqrt{2\alpha} + 2\alpha c.$

Transition density w.r.t. m:

$$\begin{aligned} p(t;x,y) &= \frac{1}{2\sqrt{2\pi t}(1+(2\beta-1)\operatorname{sgn}(x\wedge y))}\Big(e^{-(x-y)^2/2t} - e^{-(|x|+|y|)^2/2t}\Big)\\ &+ \frac{\Upsilon+1}{4c\Upsilon}\exp\Big(\frac{(|x|+|y|)\gamma}{1-\Upsilon} + \frac{(\Upsilon+1)^2 t}{2c^2}\Big)\operatorname{Erfc}\Big(\frac{|x|+|y|}{\sqrt{2t}} + \frac{(1+\Upsilon)\sqrt{t}}{c\sqrt{2}}\Big)\\ &+ \frac{\Upsilon-1}{4c\Upsilon}\exp\Big(\frac{(|x|+|y|)\gamma}{1+\Upsilon} + \frac{(\Upsilon-1)^2 t}{2c^2}\Big)\operatorname{Erfc}\Big(\frac{|x|+|y|}{\sqrt{2t}} + \frac{(1-\Upsilon)\sqrt{t}}{c\sqrt{2}}\Big),\end{aligned}$$

where $\Upsilon = \sqrt{1-\gamma c}$. The epression for the density is valid for all non-negative values on c and γ; also when $\gamma c > 1$.

14. Brownian motion with drift

Speed measure: $m(dx) = 2e^{2\mu x}dx.$

Scale function: $s(x) = \dfrac{1}{2\mu}(1-e^{-2\mu x}).$

Generator: $\mathcal{G}f = \dfrac{1}{2}\dfrac{d^2 f}{dx^2} + \mu\dfrac{d f}{dx}.$

Domain: $\mathcal{D} = \{f: \ f, \ \mathcal{G}f \in \mathcal{C}_b(\mathbb{R})\}$.

Green function:

$$G_\alpha(x,y) = w_\alpha^{-1}\, e^{-(\sqrt{2\alpha+\mu^2}+\mu)x} e^{(\sqrt{2\alpha+\mu^2}-\mu)y}, \quad x \ge y.$$

Wronskian: $w_\alpha = 2\sqrt{2\alpha+\mu^2}$.

Transition density w.r.t. m:

$$p(t;x,y) = \frac{1}{2\sqrt{2\pi t}} \exp\Bigl(-\mu(x+y) - \frac{\mu^2 t}{2} - \frac{(x-y)^2}{2t}\Bigr).$$

Spectral representation of the transition density:

$$p(t;x,y) = \frac{1}{2\pi} e^{-\mu(y+x)} \int_0^\infty e^{-(\mu^2+\lambda^2)t/2} \cos(\lambda(x-y))\, d\lambda \ .$$

Absolute continuity:

$$\frac{d\mathbf{P}_x^{(\mu)}}{d\mathbf{P}_x}\Big|_{\mathcal{F}_{t+}^c} = \exp\Bigl(\omega(t) - x) - \frac{\mu^2 t}{2}\Bigr) \qquad \mathbf{P}_x\text{-a.s.},$$

where $\mathbf{P}_x^{(\mu)}$ and $\mathbf{P}_x$ are the measures associated to Brownian motion with drift and Brownian motion, respectively.

15. Brownian motion with alternating drift

Speed measure: $m(dx) = 2e^{2\mu|x|}dx$.

Scale function: $s(x) = \dfrac{1}{2\mu \operatorname{sgn} x}(1 - e^{-2\mu|x|})$.

Generator:

$$\mathcal{G}f = \frac{1}{2}\frac{d^2 f}{dx^2} + \mu \operatorname{sgn} x \frac{df}{dx}, \quad x \ne 0, \quad \mathcal{G}f(0) = \frac{1}{2}f''(0+) = \frac{1}{2}f''(0-).$$

Domain: $\mathcal{D} = \{f: \ f, \ \mathcal{G}f \in \mathcal{C}_b(\mathbb{R})\}$.

Positive recurrence: if $\mu < 0$ the process is positively recurrent. The stationary probability measure is

$$\hat{m}(dx) = |\mu| e^{2\mu|x|} dx.$$

Green function:

$$G_\alpha(x,y) = \begin{cases} w_\alpha^{-1}\Bigl(e^{-(\Upsilon-\mu)x} + \frac{\mu}{\Upsilon}\bigl(e^{-(\Upsilon-\mu)x} - e^{(\Upsilon+\mu)x}\bigr)\Bigr)e^{(\Upsilon+\mu)y}, & 0 \ge x \ge y, \\ w_\alpha^{-1}\, e^{-(\Upsilon+\mu)x} e^{(\Upsilon+\mu)y}, & x \ge 0 \ge y, \\ w_\alpha^{-1}\, e^{-(\Upsilon+\mu)x}\Bigl(e^{(\Upsilon-\mu)y} + \frac{\mu}{\Upsilon}\bigl(e^{(\Upsilon-\mu)y} - e^{-(\Upsilon+\mu)y}\bigr)\Bigr), & x \ge y \ge 0, \end{cases}$$

$$= \frac{1}{2\sqrt{2\alpha+\mu^2}} e^{-\mu(|x|+|y|)}\Bigl(e^{-|x-y|\sqrt{2\alpha+\mu^2}} - \frac{2\mu}{w_\alpha} e^{-(|x|+|y|)\sqrt{2\alpha+\mu^2}}\Bigr),$$

where $\Upsilon = \sqrt{2\alpha + \mu^2}$.

Wronskian: $w_\alpha = 2(\sqrt{2\alpha + \mu^2} + \mu)$.

Transition density w.r.t. m:

$$p(t;x,y) = \frac{1}{2\sqrt{2\pi t}} \exp\Big(-\mu(|x|+|y|) - \frac{\mu^2}{2}t - \frac{(x-y)^2}{2t}\Big) - \frac{\mu}{4} \operatorname{Erfc}\Big(\frac{|x|+|y|+\mu t}{\sqrt{2t}}\Big).$$

Absolute continuity:

$$\frac{d\mathbf{P}_x^{(\pm)}}{d\mathbf{P}_x}\Big|_{\mathcal{F}_{t+}^c} = \exp\Big(\mu(|\omega(t)| - |x|) - \mu\,\ell(t,0) - \frac{\mu^2\,t}{2}\Big) \qquad \mathbf{P}_x\text{-a.s.},$$

where $\mathbf{P}_x^{(\pm)}$ and $\mathbf{P}_x$ are the measures associated to Brownian motion with alternating drift and Brownian motion, respectively, and $\ell(t,0)$ is the local time in the canonical framework with respect to Lebesgue measure, that is,

$$\ell(t,0) := \lim_{\varepsilon \downarrow 0} \frac{1}{2\varepsilon} \int_0^t \mathbb{1}_{(-\varepsilon,\varepsilon)}(\omega_s)ds.$$

Notice that if $\{\omega(t) : t \geq 0\}$ is a Brownian motion with alternating drift (also called a bang-bang Brownian motion) then $\{|\omega(t)| : t \geq 0\}$ is a reflected Brownian motion with drift (see No. 16), see also Graversen and Shiryaev (2000).

16. Brownian motion on $[0,\infty)$ with drift and reflected at 0

Speed measure: $m(dx) = 2e^{2\mu x}dx$.

Scale function: $s(x) = \frac{1}{2\mu}(1 - e^{-2\mu x})$.

Generator: $\mathcal{G}f = \frac{1}{2}\frac{d^2 f}{dx^2} + \mu\frac{d f}{dx}, \quad x > 0, \quad \mathcal{G}f(0) = \frac{1}{2}f''(0+) + \mu f'(0+)$.

Domain: $\mathcal{D} = \{f : f,\ \mathcal{G}f \in \mathcal{C}_b([0,\infty)),\ f'(0+) = 0\}$.

Positive recurrence: if $\mu < 0$ the process is positively recurrent. The stationary probability measure is

$$\hat{m}(dx) = 2|\mu|e^{2\mu x}dx.$$

Green function:

$$G_\alpha(x,y) = w_\alpha^{-1}\, e^{-(\Upsilon+\mu)x}\Big(\frac{\Upsilon+\mu}{2\Upsilon}e^{(\Upsilon-\mu)y} + \frac{\Upsilon-\mu}{2\Upsilon}e^{-(\Upsilon+\mu)y}\Big), \quad x \geq y,$$

where $\Upsilon = \sqrt{2\alpha + \mu^2}$.

Wronskian: $w_\alpha = \sqrt{2\alpha + \mu^2} + \mu$.

Transition density w.r.t. m:

$$p(t;x,y) = \frac{1}{2\sqrt{2\pi t}} \exp\Big(-\mu(x+y) - \frac{\mu^2}{2}t\Big)\Big(\exp\Big(-\frac{(x-y)^2}{2t}\Big) + \exp\Big(-\frac{(x+y)^2}{2t}\Big)\Big)$$
$$- \frac{\mu}{2}\operatorname{Erfc}\Big(\frac{x+y+\mu t}{\sqrt{2t}}\Big), \qquad x \ge 0,\ y \ge 0.$$

Absolute continuity (Kinkladze (1982)):

$$\frac{d\mathbf{P}_x^{(\mu)}}{d\mathbf{P}_x}\Big|_{\mathcal{F}_{t+}^c} = \exp\Big(\mu(\omega(t) - x) - \frac{\mu\,\ell(t,0)}{2} - \frac{\mu^2 t}{2}\Big) \qquad \mathbf{P}_x\text{-a.s.},$$

where $\mathbf{P}_x^{(\mu)}$ and $\mathbf{P}_x$ are the measures associated to reflected Brownian motion with drift and reflected Brownian motion, respectively, and $\ell(t,0)$ is the local time in the canonical framework with respect to Lebesgue measure, that is,

$$\ell(t,0) := \lim_{\varepsilon \downarrow 0} \frac{1}{\varepsilon} \int_0^t \mathbb{I}_{[0,\varepsilon)}(\omega_s)\,ds.$$

17. Brownian motion on $(0,\infty)$ with drift and killed at 0

Speed measure: $m(dx) = 2e^{2\mu x}dx$.

Scale function: $s(x) = \dfrac{1}{2\mu}(1 - e^{-2\mu x})$.

Generator: $\mathcal{G}f = \dfrac{1}{2}\dfrac{d^2 f}{dx^2} + \mu\dfrac{df}{dx}, \quad x > 0$.

Domain: $\mathcal{D} = \{f:\ f,\ \mathcal{G}f \in \mathcal{C}_b((0,\infty)),\ f(0+) = 0\}$.

Probability that the process drifts to ∞:

$$\mathbf{P}_x\Big(\lim_{t\to\infty}\omega(t) = \infty\Big) = \begin{cases} 1 - e^{-2\mu x}, & \mu \ge 0, \\ 0, & \mu \le 0. \end{cases}$$

Green function:

$$G_\alpha(x,y) = w_\alpha^{-1}\, e^{-(\sqrt{2\alpha+\mu^2}+\mu)x}\Big(e^{(\sqrt{2\alpha+\mu^2}-\mu)y} - e^{-(\sqrt{2\alpha+\mu^2}+\mu)y}\Big), \quad x \ge y.$$

Wronskian: $w_\alpha = 2\sqrt{2\alpha + \mu^2}$.

Transition density w.r.t. m:

$$p(t;x,y) = \frac{1}{2\sqrt{2\pi t}} \exp\Big(-\mu(x+y) - \frac{\mu^2}{2}t\Big)\Big(\exp\Big(-\frac{(x-y)^2}{2t}\Big) - \exp\Big(-\frac{(x+y)^2}{2t}\Big)\Big).$$

Spectral representation of the transition density:

$$p(t;x,y) = \frac{1}{\pi}e^{-\mu(x+y)} \int_0^\infty e^{-(\mu^2+\lambda^2)t/2} \sin(\lambda x)\sin(\lambda y)\,d\lambda.$$

18. Brownian motion with drift and killed elastically at 0

Speed measure: $m(dx) = 2e^{2\mu x}dx$.

Scale function: $s(x) = \frac{1}{2\mu}(1 - e^{-2\mu x})$.

Killing measure: $k(dx) = \gamma\varepsilon_{\{0\}}(dx),\ \gamma > 0$.

Generator: $\mathcal{G}f = \frac{1}{2}\frac{d^2 f}{dx^2} + \mu\frac{d f}{dx}, \quad x \neq 0$,

$\mathcal{G}f(0) = \frac{1}{2}f''(0+) + \mu f'(0+) = \frac{1}{2}f''(0-) + \mu f'(0-)$.

Domain: $\mathcal{D} = \{f:\ f,\ \mathcal{G}f \in \mathcal{C}_b(\mathbb{R}),\ f'(0+) - f'(0-) = \gamma f(0)\}$.

Green function:

$$G_\alpha(x,y) = \begin{cases} w_\alpha^{-1}\left(\left(1 + \frac{\gamma}{2\Upsilon}\right)e^{-(\Upsilon+\mu)x} - \frac{\gamma}{2\Upsilon}e^{(\Upsilon-\mu)x}\right)e^{(\Upsilon-\mu)y}, & 0 \geq x \geq y, \\ w_\alpha^{-1}\, e^{-(\Upsilon+\mu)x}\, e^{(\Upsilon-\mu)y}, & x \geq 0 \geq y, \\ w_\alpha^{-1}\, e^{-(\Upsilon+\mu)x}\left(\left(1 + \frac{\gamma}{2\Upsilon}\right)e^{(\Upsilon-\mu)y} - \frac{\gamma}{2\Upsilon}e^{-(\Upsilon+\mu)y}\right), & x \geq y \geq 0, \end{cases}$$

$$= \frac{1}{2\sqrt{2\alpha + \mu^2}}e^{-\mu(x+y)}\left(e^{-|x-y|\sqrt{2\alpha+\mu^2}} - \frac{\gamma}{w_\alpha}e^{-(|x|+|y|)\sqrt{2\alpha+\mu^2}}\right),$$

where $\Upsilon = \sqrt{2\alpha + \mu^2}$.

Wronskian: $w_\alpha = 2\sqrt{2\alpha + \mu^2} + \gamma$.

Transition density w.r.t. m:

$$p(t;x,y) = \exp\Big(-\mu(x+y) - \frac{\mu^2 t}{2}\Big)\Big(\frac{1}{2\sqrt{2\pi t}}\exp\Big(-\frac{(x-y)^2}{2t}\Big) - \frac{\gamma}{8}\exp\Big(\frac{\gamma|x| + \gamma|y|}{2} + \frac{\gamma^2 t}{8}\Big)\operatorname{Erfc}\Big(\frac{2|x| + 2|y| + \gamma t}{2\sqrt{2t}}\Big)\Big).$$

19. Brownian motion on $[0,\infty)$ with drift and killed elastically at 0

Speed measure: $m(dx) = 2e^{2\mu x}dx$.

Scale function: $s(x) = \frac{1}{2\mu}(1 - e^{-2\mu x})$.

Killing measure: $k(dx) = \gamma\varepsilon_{\{0\}}(dx),\ \gamma > 0$.

Generator: $\mathcal{G}f = \frac{1}{2}\frac{d^2 f}{dx^2} + \mu\frac{d f}{dx}, \quad x > 0$,

$\mathcal{G}f(0) = \frac{1}{2}f''(0+) + \mu f'(0+)$.

Domain: $\mathcal{D} = \{f:\ f,\ \mathcal{G}f \in \mathcal{C}_b([0,\infty)),\ f'(0+) = \gamma f(0)\}$.

Green function:

$$G_\alpha(x,y) = w_\alpha^{-1}\, e^{-(\Upsilon+\mu)x}$$
$$\times\Big(\Big(\frac{1}{2}+\frac{\gamma+\mu}{2\Upsilon}\Big)e^{(\Upsilon-\mu)y}+\Big(\frac{1}{2}-\frac{\gamma+\mu}{2\Upsilon}\Big)e^{-(\Upsilon+\mu)y}\Big), \quad x\ge y\ge 0,$$

where $\Upsilon = \sqrt{2\alpha+\mu^2}$.

Wronskian: $w_\alpha = \sqrt{2\alpha+\mu^2}+\gamma+\mu$.

Transition density w.r.t. m:

$$p(t;x,y) = \frac{1}{2\sqrt{2\pi t}}\exp\Big(-\mu(y+x)-\frac{\mu^2}{2}t\Big)\Big(\exp\Big(-\frac{(x-y)^2}{2t}\Big)+\exp\Big(-\frac{(x+y)^2}{2t}\Big)\Big)$$
$$-\frac{\mu+\gamma}{2}\exp\Big(\gamma(x+y+\mu t)+\frac{\gamma^2 t}{2}\Big)\operatorname{Erfc}\Big(\frac{x+y+\mu t+\gamma t}{\sqrt{2t}}\Big), \quad x\ge 0, y\ge 0.$$

20. Geometric (or exponential) Brownian motions

These diffusions are usually indexed by two parameters $\sigma^2>0$ and $\mu\in\mathbb{R}$. Set $\nu = \frac{\mu}{\sigma^2}-\frac{1}{2}$.

Speed measure: $m(dx) = \frac{2}{\sigma^2}x^{2\nu-1}dx$.

Scale function:

$$s(x) = \begin{cases} -\dfrac{x^{-2\nu}}{2\nu}, & \nu\ne 0,\\ \ln x, & \nu = 0.\end{cases}$$

Boundary classification: for all values of ν both boundary points 0 and ∞ are natural. Moreover

if $\nu<0$, then $\lim_{t\to\infty}\omega(t) = 0$ a.s.,

if $\nu>0$, then $\lim_{t\to\infty}\omega(t) = \infty$ a.s.,

if $\nu=0$, then $\liminf_{t\to\infty}\omega(t)=0$, $\limsup_{t\to\infty}\omega(t) = +\infty$ a.s.

Generator: $\mathcal{G}f = \frac{1}{2}\sigma^2x^2\frac{d^2f}{dx^2}+\mu x\frac{df}{dx}, \quad x>0.$

Domain: $\mathcal{D} = \{f:\ f,\ \mathcal{G}f\in\mathcal{C}_b((0,\infty))\}$.

Green function:

$$G_\alpha(x,y) = w_\alpha^{-1}\,x^{-\sqrt{\nu^2+2\alpha/\sigma^2}-\nu}\,y^{\sqrt{\nu^2+2\alpha/\sigma^2}-\nu}, \quad x\ge y>0.$$

Wronskian: $w_\alpha = 2\sqrt{\nu^2+2\alpha/\sigma^2}$.

Transition density w.r.t. m:

$$p(t;x,y) = \frac{|\sigma|}{2\sqrt{2\pi t}}(xy)^{-\nu}\exp\Big(-\frac{\sigma^2\nu^2 t}{2}-\frac{(\ln y-\ln x)^2}{2\sigma^2 t}\Big).$$

Let $\{\omega(t)\ :\ t\ge 0\}$ be a Brownian motion started at 0. Then for $x>0$

$$\{x\exp((\mu-\sigma^2/2)t+\sigma\,\omega(t)) :\ t\ge 0\}$$

is a geometric Brownian motion started at x and having the parameters μ, σ^2.

21. Bessel processes

Bessel processes are indexed by a parameter $\nu \in \mathbb{R}$. Recall from No. IV.34 that $\nu = n/2 - 1$, where n is the "dimension" parameter, i.e, $n = 2\nu + 2$. The state space of a Bessel process is $[0, \infty)$ or $(0, \infty)$ depending on the value of ν and the boundary condition at 0. The transition density of a Bessel process is computed in Molchanov (1967).

Speed measure:

$$m_\nu(dx) = 2x^{2\nu+1}dx.$$

Scale function:

$$s_\nu(x) = \begin{cases} -\dfrac{x^{-2\nu}}{2\nu}, & \nu \neq 0, \\ \log x, & \nu = 0. \end{cases}$$

Boundary classification: for all values of ν, the boundary point ∞ is natural. The nature of the boundary point 0 depends on the value of ν as follows:

if $\nu \geq 0$, then 0 is entrance-not-exit,

if $-1 < \nu < 0$, then 0 is non-singular,

if $\nu \leq -1$, then 0 is exit-not-entrance.

In the non-singular case the boundary condition at 0 is usually reflection or killing.

Generator: $\mathcal{G}f = \dfrac{1}{2}\dfrac{d^2 f}{dx^2} + \dfrac{2\nu + 1}{2x}\dfrac{d f}{dx}, \quad x > 0.$

Domain: if $\nu \geq 0$ or if $-1 < \nu < 0$ and 0 is reflecting,

$$\mathcal{D} = \{f : \ f, \ \mathcal{G}f \in \mathcal{C}_b([0,\infty)), \ f^+(0+) = 0\};$$

if $\nu \leq -1$ or if $-1 < \nu < 0$ and 0 is killing,

$$\mathcal{D} = \{f : \ f, \ \mathcal{G}f \in \mathcal{C}_b([0,\infty)), \ f(0) = 0\}.$$

Green function:

if $\nu \geq 0$ or if $-1 < \nu < 0$ and 0 is reflecting,

$$G_\alpha(x,y) = w_\alpha^{-1}\, x^{-\nu} K_\nu(x\sqrt{2\alpha})\, y^{-\nu} I_\nu(y\sqrt{2\alpha}), \quad x \geq y > 0,$$
$$G_\alpha(0,x) = w_\alpha^{-1} \left(\frac{\sqrt{2\alpha}}{2}\right)^\nu \frac{1}{\Gamma(\nu+1)} x^{-\nu} K_\nu(x\sqrt{2\alpha}), \quad x > 0;$$

if $\nu \leq -1$ or if $-1 < \nu < 0$ and 0 is killing,

$$G_\alpha(x,y) = w_\alpha^{-1}\, x^{-\nu} K_\nu(x\sqrt{2\alpha})\, y^{-\nu} I_{-\nu}(y\sqrt{2\alpha}), \quad x \geq y > 0.$$

Wronskian: $w_\alpha = 1$.

Transition density w.r.t. m:

if $\nu \geq 0$ or if $-1 < \nu < 0$ and 0 is reflecting,

$$p(t;x,y) = \frac{1}{2t}\,(xy)^{-\nu}\exp\Bigl(-\frac{x^2+y^2}{2t}\Bigr)\,I_\nu\Bigl(\frac{xy}{t}\Bigr),$$
$$p(t;0,y) = \frac{1}{2^{\nu+1}\,t^{\nu+1}\,\Gamma(\nu+1)}\exp\Bigl(-\frac{y^2}{2t}\Bigr);$$

if $\nu \leq -1$ or if $-1 < \nu < 0$ and 0 is killing,

$$p(t;x,y) = \frac{1}{2t}\,(xy)^{-\nu}\exp\Bigl(-\frac{x^2+y^2}{2t}\Bigr)\,I_{|\nu|}\Bigl(\frac{xy}{t}\Bigr).$$

Absolute continuity:

$$\frac{d\mathbf{P}_x^{(\mu)}}{d\mathbf{P}_x^{(\nu)}}\Big|_{\mathcal{F}_{t+}^c} = \frac{(\omega(t))^{\mu-\nu}}{x^{\mu-\nu}}\exp\Bigl(-\frac{\mu^2-\nu^2}{2}\int_0^t\frac{ds}{\omega^2(s)}\Bigr) \qquad \mathbf{P}_x^{(\nu)}\text{-a.s. on } \{H_0 > t\},$$

where $x > 0$ and $\mathbf{P}^{(\nu)}$ is the measure associated to a Bessel process of order ν. Notice that if $\nu \geq 0$ the condition "on $\{H_0 > t\}$" can be omitted. Putting here $\nu = -1/2$ gives the density w.r.t. the Wiener measure. For a proof in the case $\mu \geq 0$ and $\nu \geq 0$ see Yor (1980). The general case is obtained from No. IV.43 using the chain rule.

22. Special case of No. 21: 3-dimensional Bessel process, (i.e., $\nu = 1/2$)

Speed measure: $m(dx) = 2x^2dx$.

Scale function: $s(x) = -\dfrac{1}{x}$.

Boundary classification: ∞ is natural and 0 is entrance-not-exit.

Generator: $\mathcal{G}f = \dfrac{1}{2}\dfrac{d^2f}{dx^2} + \dfrac{1}{x}\dfrac{df}{dx}, \quad x > 0.$

Domain: $\mathcal{D} = \{f:\ f,\ \mathcal{G}f \in \mathcal{C}_b([0,\infty)),\ f^+(0+) = 0\}$.

Green function:

$$G_\alpha(x,y) = w_\alpha^{-1}\,\frac{1}{x}e^{-x\sqrt{2\alpha}}\,\frac{1}{y}\bigl(e^{y\sqrt{2\alpha}} - e^{-y\sqrt{2\alpha}}\bigr), \qquad x \geq y > 0,$$
$$G_\alpha(0,x) = \frac{1}{x}\,e^{-\sqrt{2\alpha}x}, \qquad x > 0.$$

Wronskian: $w_\alpha = 2\sqrt{2\alpha}$.

Transition density w.r.t. m:

$$p(t;x,y) = \frac{1}{2\sqrt{2\pi t}}\,\frac{1}{xy}\Bigl(\exp\Bigl(-\frac{(x-y)^2}{2t}\Bigr) - \exp\Bigl(-\frac{(x+y)^2}{2t}\Bigr)\Bigr),$$
$$p(t;0,y) = \frac{1}{\sqrt{2\pi t^3}}\exp\Bigl(-\frac{y^2}{2t}\Bigr).$$

Absolute continuity:

$$\frac{d\mathbf{P}_x^{(1/2)}}{d\mathbf{P}_x}\Big|_{\mathcal{F}_{t+}^c} = \frac{\omega(t)}{x} \qquad \mathbf{P}_x\text{-a.s. on } \{H_0 > t\},$$

where $x > 0$, $\mathbf{P}^{(1/2)}$ is the measure associated to a three-dimensional Bessel process, and $\mathbf{P}$ is the measure associated to a Brownian motion.

23. Squared Bessel processes

Squared Bessel processes are indexed by a parameter $\nu \in \mathbb{R}$. The state space of these diffusions is $[0,\infty)$ or $(0,\infty)$ depending on the value of ν and the boundary condition at 0.

Let $\{\omega(t) : t \geq 0\}$ be a Bessel process of index (or order) ν. Then $\{\omega(t)^2 : t \geq 0\}$ is a squared Bessel process of order ν.

Speed measure:

$$m_\nu(dx) = \frac{x^\nu}{2}dx.$$

Scale function:

$$s_\nu(x) = \begin{cases} -\dfrac{x^{-\nu}}{\nu}, & \nu \neq 0, \\ \log x, & \nu = 0. \end{cases}$$

Boundary classification: for all values of ν, the boundary point ∞ is natural. The nature of the boundary point 0 depends on the value of ν as follows:

if $\nu \geq 0$, then 0 is entrance-not-exit,

if $-1 < \nu < 0$, then 0 is non-singular,

if $\nu \leq -1$, then 0 is exit-not-entrance.

Generator: $\mathcal{G}f = 2x\dfrac{d^2 f}{dx^2} + (2\nu+2)\dfrac{d f}{dx}, \quad x > 0\,.$

Domain: if $\nu \geq 0$ or if $-1 < \nu < 0$ and 0 is reflecting,

$$\mathcal{D} = \{f :\ f,\ \mathcal{G}f \in \mathcal{C}_b([0,\infty)),\ f^+(0+) = 0\},$$

if $\nu \leq -1$ or if $-1 < \nu < 0$ and 0 is killing,

$$\mathcal{D} = \{f :\ f,\ \mathcal{G}f \in \mathcal{C}_b([0,\infty)),\ f(0) = 0\}\,.$$

Green function:

if $\nu \geq 0$ or if $-1 < \nu < 0$ and 0 is reflecting,

$$G_\alpha(x,y) = w_\alpha^{-1} x^{-\nu/2} K_\nu(\sqrt{2\alpha x})\, y^{-\nu/2} I_\nu(\sqrt{2\alpha y}), \quad x \geq y > 0,$$
$$G_\alpha(0,x) = w_\alpha^{-1}\Big(\frac{\sqrt{2\alpha}}{2}\Big)^\nu \frac{1}{\Gamma(\nu+1)} x^{-\nu/2} K_\nu(\sqrt{2\alpha x}), \quad x > 0;$$

if $\nu \le -1$ or if $-1 < \nu < 0$ and 0 is killing,

$$G_\alpha(x,y) = w_\alpha^{-1} x^{-\nu/2} K_\nu(\sqrt{2\alpha x})\, y^{-\nu/2} I_{-\nu}(\sqrt{2\alpha y}), \quad x \ge y > 0.$$

Wronskian: $w_\alpha = 1/2$.

Transition density w.r.t. m:

if $\nu \ge 0$ or if $-1 < \nu < 0$ and 0 reflecting,

$$p(t;x,y) = \frac{1}{t}\,(xy)^{-\nu/2} \exp\Big(-\frac{x+y}{2t}\Big)\, I_\nu\Big(\frac{\sqrt{xy}}{t}\Big),$$
$$p(t;0,y) = \frac{1}{2^\nu\, t^{\nu+1}\Gamma(\nu+1)} \exp\Big(-\frac{y}{2t}\Big);$$

if $\nu \le -1$ or if $-1 < \nu < 0$ and 0 is killing,

$$p(t;x,y) = \frac{1}{t}\,(xy)^{-\nu/2} \exp\Big(-\frac{x+y}{2t}\Big)\, I_{|\nu|}\Big(\frac{\sqrt{xy}}{t}\Big).$$

24. Ornstein–Uhlenbeck process

Ornstein–Uhlenbeck processes are usually indexed by one parameter $\gamma \in \mathbb{R}$. For $\gamma = 0$ we have Brownian motion. The behavior of Ornstein–Uhlenbeck processes is very different depending on whether $\gamma > 0$ or $\gamma < 0$.

Speed measure: $m_\gamma(dx) = 2\exp(-\gamma x^2)\, dx$.

Scale function: $s_\gamma(x) = \int_0^x \exp(\gamma y^2)\, dy$.

Generator: $\mathcal{G} f = \frac{1}{2}\frac{d^2 f}{dx^2} - \gamma x \frac{d f}{dx}$.

Domain: $\mathcal{D} = \{f :\ f,\ \mathcal{G} f \in \mathcal{C}_b(\mathbb{R})\}$.

Doob's transformation: Let $\mathbf{P}_x$ denote the measure associated to Brownian motion, and for $\gamma \ne 0$ introduce

$$\tau_t := \frac{e^{2\gamma t} - 1}{2\gamma}.$$

Then, under the measure $\mathbf{P}_x$, the process

$$\{e^{-\gamma t}\, \omega(\tau_t) :\ t \ge 0\}$$

is identical in law to an Ornstein–Uhlenbeck process of index γ started at x.

Absolute continuity:

$$\frac{d\mathbf{P}_x^{(\gamma)}}{d\mathbf{P}_x}\Big|_{\mathcal{F}_{t+}^c} = \exp\Big(-\frac{\gamma}{2}(\omega^2(t) - x^2) + \frac{\gamma t}{2} - \frac{\gamma^2}{2}\int_0^t \omega^2(s)\, ds\Big) \qquad \mathbf{P}_x\text{-a.s.},$$

where $\mathbf{P}^{(\gamma)}$ is the measure associated to an Ornstein–Uhlenbeck process of index γ and $\mathbf{P}$ is the Wiener measure. By the chain rule,

$$\frac{d\mathbf{P}_x^{(-\gamma)}}{d\mathbf{P}_x^{(\gamma)}}\bigg|_{\mathcal{F}_{t+}^c} = \exp\bigl(\gamma(\omega^2(t) - x^2) - \gamma t\bigr) \qquad \mathbf{P}_x^{(\gamma)}\text{-a.s.}$$

(A) The case $\gamma > 0$.

The process is positively recurrent. Its stationary probability measure is

$$\hat{m}(dx) = \frac{\sqrt{\gamma}}{\sqrt{\pi}} \exp\bigl(-\gamma x^2\bigr)\, dx.$$

Green function:

$$G_\alpha(x,y) = w_\alpha^{-1} \exp\Bigl(\frac{\gamma x^2}{2}\Bigr) D_{-\alpha/\gamma}(x\sqrt{2\gamma}) \times \exp\Bigl(\frac{\gamma y^2}{2}\Bigr) D_{-\alpha/\gamma}(-y\sqrt{2\gamma}), \qquad x \geq y.$$

Wronskian: $w_\alpha = \dfrac{2\sqrt{\gamma\pi}}{\Gamma(\alpha/\gamma)}$.

Transition density w.r.t. m :

$$p(t;x,y) = \frac{1}{2}\frac{\sqrt{\gamma}}{\sqrt{2\pi\,\mathrm{sh}(\gamma t)}} \exp\Bigl(\frac{\gamma t}{2} + \frac{\gamma(x^2+y^2)}{2} - \frac{(x^2+y^2)\gamma\,\mathrm{ch}(\gamma t) - 2xy\gamma}{2\,\mathrm{sh}(\gamma t)}\Bigr).$$

(B) The case $\gamma < 0$.

Green function:

$$G_\alpha(x,y) = w_\alpha^{-1} \exp\Bigl(-\frac{|\gamma| x^2}{2}\Bigr) D_{-\alpha/|\gamma|-1}(x\sqrt{2|\gamma|}) \times \exp\Bigl(-\frac{|\gamma| y^2}{2}\Bigr) D_{-\alpha/|\gamma|-1}(-y\sqrt{2|\gamma|}), \qquad x \geq y.$$

Wronskian: $w_\alpha = \dfrac{2\sqrt{|\gamma|\pi}}{\Gamma(\alpha/|\gamma|+1)}$.

The transition density w.r.t. m has the same form as for $\gamma > 0$:

$$p(t;x,y) = \frac{1}{2}\frac{\sqrt{|\gamma|}}{\sqrt{2\pi\,\mathrm{sh}(|\gamma| t)}} \exp\Bigl(-\frac{|\gamma| t}{2} - \frac{|\gamma|(x^2+y^2)}{2} - \frac{(x^2+y^2)|\gamma|\,\mathrm{ch}(|\gamma| t) - 2xy|\gamma|}{2\,\mathrm{sh}(|\gamma| t)}\Bigr).$$

For $\gamma < 0$ the Ornstein–Uhlenbeck process is transient and, by symmetry,

$$\mathbf{P}_0\Bigl(\lim_{t\to\infty} \omega(t) = +\infty\Bigr) = \mathbf{P}_0\Bigl(\lim_{t\to\infty} \omega(t) = -\infty\Bigr) = 1/2.$$

In Part II Section 7 we consider the Ornstein–Uhlenbeck process indexed by two parameters $\sigma > 0$ and $\theta > 0$ with the generator

$$\mathcal{G}f = \sigma^2\theta\frac{d^2 f}{dx^2} - \theta x\frac{d f}{dx}, \qquad \theta > 0\,.$$

The one parameter model in the case $\gamma > 0$ is obtained from the two parameter model by setting $\sigma = 1/\sqrt{2\gamma}$ and $\theta = \gamma$.

25. Radial Ornstein–Uhlenbeck process

Let $\{\omega^{(i)}(t) : t \geq 0\}$, $i = 1, 2, \ldots n$, be independent Ornstein–Uhlenbeck processes with the same parameter γ. Then the process

$$\Big\{\Big(\sum_{k=1}^{n}\big(\omega^{(k)}(t)\big)^2\Big)^{1/2} \ : \ t \geq 0\Big\}$$

is called radial Ornstein-Uhlenbeck process with parameters $\nu = n/2 - 1$ and γ. For $\gamma = 0$ we have a Bessel process of order ν. Below we let, as in the case of Bessel processes, ν be an arbitrary real number.

Speed measure: $m(dx) = 2x^{2\nu+1}\exp\big(-\gamma x^2\big)\,dx$.

Scale function: $s(x) = \int^x y^{-2\nu-1}\exp(\gamma y^2)\,dy$.

Doob's transformation: let $\mathbf{P}_x$ denote the measure associated to a Bessel process of order ν started at x, and for $\gamma \neq 0$ introduce

$$\tau_t := \frac{e^{2\gamma t} - 1}{2\gamma}.$$

Then, under the measure $\mathbf{P}_x$, the process

$$\{e^{-\gamma t}\,\omega(\tau_t) : \ t \geq 0\}$$

is identical in law to a radial Ornstein–Uhlenbeck process with the parameters ν and γ started at x.

Boundary classification: for all values of ν, the boundary point ∞ is natural. The nature of the boundary point 0 depends on the value of ν as follows:

if $\nu \geq 0$, then 0 is entrance-not-exit,

if $-1 < \nu < 0$, then 0 is non-singular,

if $\nu \leq -1$, then 0 is exit-not-entrance.

In the non-singular case we take below the boundary condition at 0 to be reflection or killing.

Generator: $\mathcal{G}f = \dfrac{1}{2}\dfrac{d^2 f}{dx^2} + \Big(\dfrac{2\nu+1}{2x} - \gamma x\Big)\dfrac{df}{dx}, \quad x > 0.$

Domain:

if $\nu \geq 0$ or if $-1 < \nu < 0$ and 0 is reflecting,

$$\mathcal{D} = \{f : \ f,\ \mathcal{G}f \in \mathcal{C}_b([0,\infty)),\ f^{+}(0+) = 0\};$$

if $\nu \leq -1$ or if $-1 < \nu < 0$ and 0 is killing,

$$\mathcal{D} = \{f : \ f,\ \mathcal{G}f \in \mathcal{C}_b([0,\infty)),\ f(0) = 0\}.$$

Absolute continuity: for $\mu \geq 0$, $\nu \in \mathbb{R}$, and $x > 0$

$$\frac{d\mathbf{P}_x^{(\mu)}}{d\mathbf{P}_x^{(\nu)}}\Big|_{\mathcal{F}_{t+}^c} = \frac{(\omega(t))^{\mu-\nu}}{x^{\mu-\nu}} \exp\Big(\gamma(\mu-\nu)t - \frac{\mu^2-\nu^2}{2}\int_0^t \frac{ds}{\omega^2(s)}\Big) \quad \mathbf{P}_x^{(\nu)}\text{-a.s. on } \{H_0 > t\},$$

where $\mathbf{P}_x^{(\mu)}$ and $\mathbf{P}_x^{(\nu)}$ are the measures associated to the radial Ornstein–Uhlenbeck processes of order μ and ν, respectively. Notice that if $\nu \geq 0$ the condition "on $\{H_0 > t\}$" can be omitted. Putting here $\nu = -1/2$ gives the density w.r.t. the Ornstein–Uhlenbeck measure.

(A) The case $\gamma > 0$.

For $\nu \geq 0$, or if $-1 < \nu < 0$ and 0 is a reflecting boundary the process is positively recurrent, and its stationary probability measure is

$$\hat{m}(dx) = \frac{2\gamma^{\nu+1}x^{2\nu+1}}{\Gamma(\nu+1)} \exp\Big(-\gamma x^2\Big)\,dx.$$

For $\nu = 0$ and $\gamma = 1/2$ this is the so-called Rayleigh distribution, for $\nu = 1/2$ and $\gamma = 1/2$ we have the Maxwell distribution and, in general, for $\nu = n/2 - 1$, $n = 1, 2, \ldots$, and $\gamma = 1/2$ the distribution of the square root of the χ^2-distributed random variable with n degrees of freedom.

Green function:

if $\nu \geq 0$ or if $-1 < \nu < 0$ and 0 is reflecting,

$$G_\alpha(x,y) = w_\alpha^{-1} x^{-\nu-1} \exp\Big(\frac{\gamma x^2}{2}\Big) W_{-\frac{\alpha}{2\gamma}+\frac{\nu+1}{2},\frac{\nu}{2}}(\gamma x^2)$$
$$\times\, y^{-\nu-1} \exp\Big(\frac{\gamma y^2}{2}\Big) M_{-\frac{\alpha}{2\gamma}+\frac{\nu+1}{2},\frac{\nu}{2}}(\gamma y^2), \quad x \geq y.$$

Wronskian: $w_\alpha = \dfrac{2\gamma\Gamma(1+\nu)}{\Gamma(\alpha/2\gamma)}$.

If $\nu \leq -1$ or if $-1 < \nu < 0$ and 0 is killing,

$$G_\alpha(x,y) = w_\alpha^{-1} x^{-\nu-1} \exp\Big(\frac{\gamma x^2}{2}\Big) W_{-\frac{\alpha}{2\gamma}+\frac{\nu+1}{2},-\frac{\nu}{2}}(\gamma x^2)$$
$$\times\, y^{-\nu-1} \exp\Big(\frac{\gamma y^2}{2}\Big) M_{-\frac{\alpha}{2\gamma}+\frac{\nu+1}{2},-\frac{\nu}{2}}(\gamma y^2), \quad x \geq y.$$

Wronskian: $w_\alpha = \dfrac{2\gamma\Gamma(1-\nu)}{\Gamma(\frac{\alpha}{2\gamma}-\nu)}$.

Transition density w.r.t. m:

if $\nu \geq 0$ or if $-1 < \nu < 0$ and 0 is reflecting,

$$p(t;x,y) = \frac{\gamma e^{\gamma(\nu+1)t}}{2x^\nu y^\nu \operatorname{sh}(\gamma t)} \exp\Big(-\frac{\gamma e^{-\gamma t}(x^2+y^2)}{2\operatorname{sh}(\gamma t)}\Big) I_\nu\Big(\frac{\gamma xy}{\operatorname{sh}(\gamma t)}\Big),$$

$$p(t;0,y) = \frac{\gamma^{\nu+1}e^{\gamma(\nu+1)t}}{2^{\nu+1}\Gamma(\nu+1)(\operatorname{sh}(\gamma t))^{\nu+1}} \exp\Big(-\frac{\gamma e^{-\gamma t}y^2}{2\operatorname{sh}(\gamma t)}\Big);$$

if $\nu \le -1$ or if $-1 < \nu < 0$ and 0 is killing,

$$p(t; x, y) = \frac{\gamma e^{\gamma(\nu+1)t}}{2x^\nu y^\nu \operatorname{sh}(\gamma t)} \exp\Big(-\frac{\gamma e^{-\gamma t}(x^2+y^2)}{2\operatorname{sh}(\gamma t)}\Big) I_{-\nu}\Big(\frac{\gamma x y}{\operatorname{sh}(\gamma t)}\Big).$$

(B) The case $\gamma < 0$.

Green function:

if $\nu \ge 0$ or if $-1 < \nu < 0$ and 0 is reflecting,

$$G_\alpha(x, y) = w_\alpha^{-1} x^{-\nu-1} \exp\Big(-\frac{|\gamma|\, x^2}{2}\Big) W_{-\frac{\alpha}{2|\gamma|}-\frac{\nu+1}{2}, \frac{\nu}{2}}\big(|\gamma| x^2\big)$$
$$\times\, y^{-\nu-1} \exp\Big(-\frac{|\gamma|\, y^2}{2}\Big) M_{-\frac{\alpha}{2|\gamma|}-\frac{\nu+1}{2}, \frac{\nu}{2}}\big(|\gamma| y^2\big), \quad x \ge y.$$

Wronskian: $w_\alpha = \dfrac{2|\gamma|\Gamma(1+\nu)}{\Gamma(\frac{\alpha}{2|\gamma|}+\nu+1)}$.

If $\nu \le -1$ or if $-1 < \nu < 0$ and 0 is killing,

$$G_\alpha(x, y) = w_\alpha^{-1} x^{-\nu-1} \exp\Big(-\frac{|\gamma|\, x^2}{2}\Big) W_{-\frac{\alpha}{2|\gamma|}-\frac{\nu+1}{2}, \frac{|\nu|}{2}}\big(|\gamma| x^2\big)$$
$$\times\, y^{-\nu-1} \exp\Big(-\frac{|\gamma|\, y^2}{2}\Big) M_{-\frac{\alpha}{2|\gamma|}-\frac{\nu+1}{2}, \frac{|\nu|}{2}}\big(|\gamma| y^2\big), \quad x \ge y.$$

Wronskian: $w_\alpha = \dfrac{2|\gamma|\Gamma(1-\nu)}{\Gamma(\frac{\alpha}{2|\gamma|}+1)}$.

Transition density w.r.t. m has the same form as for $\gamma > 0$.

In Part II Section 8 the radial Ornstein–Uhlenbeck process is parameterized to have the generator

$$\mathcal{G}f = \sigma^2\theta \frac{d^2 f}{dx^2} + \Big(\frac{\sigma^2\theta(2\nu+1)}{x} - \theta x\Big)\frac{df}{dx}, \quad \theta > 0.$$

26. Squared radial Ornstein–Uhlenbeck process

Let $\{\omega(t) : t \ge 0\}$ be a radial Ornstein–Uhlenbeck process with the parameters ν and γ. Then the process

$$\big\{\omega^2(t) : t \ge 0\big\}$$

is called squared radial Ornstein–Uhlenbeck process with the parameters ν and γ. For $\gamma = 0$ we have a squared Bessel process of order ν.

Speed measure: $m(dx) = \frac{1}{2} x^\nu \exp(-\gamma x)\, dx$.

Scale function: $s(x) = \int^x y^{-\nu-1} \exp(\gamma y)\, dy$.

Time transformation: let $\mathbf{P}_x$ denote the measure associated to a squared Bessel process of order ν started at x, and for $\gamma \neq 0$ introduce

$$\tau_t := \frac{e^{\gamma t} - 1}{\gamma}.$$

Then, under the measure $\mathbf{P}_x$, the process

$$\{e^{-\gamma t}\,\omega(\tau_t) : \ t \geq 0\}$$

is identical in law to a squared radial Ornstein–Uhlenbeck process with the parameters ν and γ started at x.

Boundary classification: For all values of ν and γ, the boundary point ∞ is natural. The nature of the boundary point 0 depends on the value of ν as follows:

if $\nu \geq 0$, then 0 is entrance-not-exit,

if $-1 < \nu < 0$, then 0 is non-singular,

if $\nu \leq -1$, then 0 is exit-not-entrance.

In the non-singular case we take below the boundary condition at 0 to be reflection or killing.

Generator: $\mathcal{G}f = 2x\dfrac{d^2 f}{dx^2} + \Big(2\nu + 2 - 2\gamma x\Big)\dfrac{df}{dx}$.

Domain:

if $\nu \geq 0$ or if $-1 < \nu < 0$ and 0 is reflecting,

$$\mathcal{D} = \{f : \ f,\ \mathcal{G}f \in \mathcal{C}_b([0,\infty)),\ f^+(0+) = 0\};$$

if $\nu \leq -1$ or if $-1 < \nu < 0$ and 0 is killing,

$$\mathcal{D} = \{f : \ f,\ \mathcal{G}f \in \mathcal{C}_b([0,\infty)),\ f(0) = 0\}.$$

(A) The case $\gamma > 0$.

For $\nu \geq 0$, or if $-1 < \nu < 0$ and 0 is a reflecting boundary the process is positively recurrent, and its stationary probability measure is

$$\hat{m}(dx) = \frac{\gamma^{\nu+1} x^{\nu}}{\Gamma(\nu+1)} \exp\Big(-\gamma\, x\Big)\, dx,$$

i.e., the gamma–distribution with the parameters γ and $\nu + 1$.

Green function:

if $\nu \geq 0$ or if $-1 < \nu < 0$ and 0 is reflecting,

$$G_\alpha(x,y) = w_\alpha^{-1} x^{-(\nu+1)/2} \exp\Big(\frac{\gamma\, x}{2}\Big)\, W_{-\frac{\alpha}{2\gamma}+\frac{\nu+1}{2},\frac{\nu}{2}}(\gamma x)$$
$$\times\, y^{-(\nu+1)/2} \exp\Big(\frac{\gamma\, y}{2}\Big)\, M_{-\frac{\alpha}{2\gamma}+\frac{\nu+1}{2},\frac{\nu}{2}}(\gamma y), \quad x \geq y.$$

Wronskian: $w_\alpha = \dfrac{\gamma\Gamma(1+\nu)}{\Gamma(\alpha/2\gamma)}$.

If $\nu \le -1$ or if $-1 < \nu < 0$ and 0 is killing,

$$G_\alpha(x,y) = w_\alpha^{-1} x^{-(\nu+1)/2} \exp\Big(\frac{\gamma x}{2}\Big) W_{-\frac{\alpha}{2\gamma}+\frac{\nu+1}{2},-\frac{\nu}{2}}(\gamma x) \times y^{-(\nu+1)/2} \exp\Big(\frac{\gamma y}{2}\Big) M_{-\frac{\alpha}{2\gamma}+\frac{\nu+1}{2},-\frac{\nu}{2}}(\gamma y), \quad x \ge y.$$

Wronskian: $w_\alpha = \dfrac{\gamma\Gamma(1-\nu)}{\Gamma(\frac{\alpha}{2\gamma}-\nu)}$.

Transition density w.r.t. m:

if $\nu \ge 0$ or if $-1 < \nu < 0$ and 0 is reflecting,

$$p(t;x,y) = \frac{\gamma e^{\gamma(\nu+1)t}}{x^{\nu/2} y^{\nu/2} \operatorname{sh}(\gamma t)} \exp\Big(-\frac{\gamma e^{-\gamma t}(x+y)}{2\operatorname{sh}(\gamma t)}\Big) I_\nu\Big(\frac{\gamma\sqrt{xy}}{\operatorname{sh}(\gamma t)}\Big),$$

$$p(t;0,y) = \frac{\gamma^{\nu+1} e^{\gamma(\nu+1)t}}{2^\nu \Gamma(\nu+1)(\operatorname{sh}(\gamma t))^{\nu+1}} \exp\Big(-\frac{\gamma e^{-\gamma t} y}{2\operatorname{sh}(\gamma t)}\Big);$$

if $\nu \le -1$ or if $-1 < \nu < 0$ and 0 is killing,

$$p(t;x,y) = \frac{\gamma e^{\gamma(\nu+1)t}}{x^{\nu/2} y^{\nu/2} \operatorname{sh}(\gamma t)} \exp\Big(-\frac{\gamma e^{-\gamma t}(x^2+y^2)}{2\operatorname{sh}(\gamma t)}\Big) I_{-\nu}\Big(\frac{\gamma\sqrt{xy}}{\operatorname{sh}(\gamma t)}\Big).$$

B) The case $\gamma < 0$.

Green function:

if $\nu \ge 0$ or if $-1 < \nu < 0$ and 0 is reflecting,

$$G_\alpha(x,y) = w_\alpha^{-1} x^{-(\nu+1)/2} \exp\Big(\frac{\gamma x}{2}\Big) W_{-\frac{\alpha}{2|\gamma|}-\frac{\nu+1}{2},\frac{\nu}{2}}(-\gamma x) \times y^{-(\nu+1)/2} \exp\Big(\frac{\gamma y}{2}\Big) M_{-\frac{\alpha}{2|\gamma|}-\frac{\nu+1}{2},\frac{\nu}{2}}(-\gamma y), \quad x \ge y.$$

Wronskian: $w_\alpha = \dfrac{|\gamma|\Gamma(1+\nu)}{\Gamma(\frac{\alpha}{2|\gamma|}+\nu+1)}$.

If $\nu \le -1$ or if $-1 < \nu < 0$ and 0 is killing,

$$G_\alpha(x,y) = w_\alpha^{-1} x^{-(\nu+1)/2} \exp\Big(\frac{\gamma x}{2}\Big) W_{-\frac{\alpha}{2|\gamma|}-\frac{\nu+1}{2},\frac{|\nu|}{2}}(-\gamma x) \times y^{-(\nu+1)/2} \exp\Big(\frac{\gamma y}{2}\Big) M_{-\frac{\alpha}{2|\gamma|}-\frac{\nu+1}{2},\frac{|\nu|}{2}}(-\gamma y), \quad x \ge y.$$

Wronskian: $w_\alpha = \dfrac{|\gamma|\Gamma(1-\nu)}{\Gamma(\frac{\alpha}{2|\gamma|}+1)}$.

Transition density w.r.t. m has the same form as for $\gamma > 0$.

27. Constant elasticity of variance (CEV) process

The CEV process is a weak solution of the SDE

$$dS_t = \mu S_t\,dt + \sigma S_t^{1+\beta}\,dW_t, \qquad S_0 = x > 0.$$

considered on $(0, +\infty)$ up to $H_0 := \inf\{t : S_t = 0\}$ (if finite). The parameters β and μ are arbitrary real numbers. For $\beta = 0$ we recover the geometric Brownian motion treated in No. 20 and for $\beta = -1/2$ the squared radial Ornstein-Uhlenbeck process with the parameter $\nu = -1$, see No. 26. We assume now that $\beta \neq 0$.

Speed measure: $m(dx) = \frac{2}{\sigma^2}\,x^{-2(1+\beta)}\exp\Big(-\frac{\mu}{\sigma^2\beta}\,x^{-2\beta}\Big)\,dx.$

Scale function: $s(x) = \int^x \exp\Big(\frac{\mu}{\sigma^2\beta}\,y^{-2\beta}\Big)\,dy.$

Boundary classification: The nature of the boundary point ∞ depends on the value of β as follows:

if $\beta < 0$, then ∞ is natural,

if $\beta > 0$, then ∞ is entrance-not-exit.

The nature of the boundary point 0 depends on the values of β as follows:

if $\beta < -1/2$, then 0 is non-singular,

if $-1/2 \le \beta < 0$, then 0 is exit-not-entrance,

if $\beta > 0$, then 0 is natural.

In the non-singular case, 0 is taken to be a killing boundary.

Generator: $\mathcal{G}f = \frac{1}{2}\,\sigma^2 x^{2+2\beta}\,\frac{d^2 f}{dx^2} + \mu x\,\frac{d\,f}{dx}, \quad x > 0.$

Domain:

if $\beta > 0$,

$$\mathcal{D} = \{f :\ f,\ \mathcal{G}f \in \mathcal{C}_b((0,\infty)),\ \lim_{x\to\infty} f'(x) = 0\},$$

if $\beta < -1/2$ and 0 is a killing boundary, or $-1/2 \le \beta < 0$,

$$\mathcal{D} = \{f :\ f,\ \mathcal{G}f \in \mathcal{C}_b((0,\infty)),\ f(0+) = 0\}.$$

Transience and recurrence:

if $\beta < 0$ and $\mu \le 0$, then $H_0 < \infty$ a.s.,

if $\beta < 0$ and $\mu > 0$, then $\lim_{t\to\infty} S_t = +\infty$ with positive probability,

if $\beta > 0$ and $\mu \le 0$, then $\lim_{t\to\infty} S_t = 0$ a.s.,

if $\beta > 0$ and $\mu > 0$, then the process is positively recurrent.

Green function:

If $\beta < 0$, $\mu \neq 0$,

$$G_\alpha(x,y) = w_\alpha^{-1} \exp\Big(\frac{\epsilon|\mu|(x^{-2\beta}+y^{-2\beta})}{2\sigma^2|\beta|}\Big)(xy)^{\beta+1/2} \times W_{n,m}\Big(\frac{|\mu|x^{-2\beta}}{\sigma^2|\beta|}\Big) M_{n,m}\Big(\frac{|\mu|y^{-2\beta}}{\sigma^2|\beta|}\Big), \qquad x \geq y,$$

where (and also below)

$$\epsilon = \operatorname{sign}(\mu\beta), \quad n = \epsilon\Big(\frac{1}{2}+\frac{1}{4\beta}\Big) - \frac{\alpha}{2|\mu\beta|}, \quad m = \frac{1}{4|\beta|}.$$

Wronskian: $w_\alpha = \dfrac{2|\mu|\Gamma(2m+1)}{\sigma^2\Gamma(m-n+1/2)}$.

If $\beta > 0$, $\mu \neq 0$,

$$G_\alpha(x,y) = w_\alpha^{-1} \exp\Big(\frac{\epsilon|\mu|(x^{-2\beta}+y^{-2\beta})}{2\sigma^2|\beta|}\Big)(xy)^{\beta+1/2} \times M_{n,m}\Big(\frac{|\mu|x^{-2\beta}}{\sigma^2|\beta|}\Big) W_{n,m}\Big(\frac{|\mu|y^{-2\beta}}{\sigma^2|\beta|}\Big), \qquad x \geq y.$$

Wronskian: $w_\alpha = \dfrac{2|\mu|\Gamma(2m+1)}{\sigma^2\Gamma(m-n+1/2)}$.

If $\beta < 0$, $\mu = 0$,

$$G_\alpha(x,y) = w_\alpha^{-1}\sqrt{xy}\, K_{1/2|\beta|}\Big(\frac{x^{-\beta}}{\sigma|\beta|}\sqrt{2\alpha}\Big) I_{1/2|\beta|}\Big(\frac{y^{-\beta}}{\sigma|\beta|}\sqrt{2\alpha}\Big), \qquad x \geq y.$$

Wronskian: $w_\alpha = |\beta|$.

If $\beta > 0$, $\mu = 0$,

$$G_\alpha(x,y) = w_\alpha^{-1}\sqrt{xy}\, I_{1/2|\beta|}\Big(\frac{x^{-\beta}}{\sigma|\beta|}\sqrt{2\alpha}\Big) K_{1/2|\beta|}\Big(\frac{y^{-\beta}}{\sigma|\beta|}\sqrt{2\alpha}\Big), \qquad x \geq y.$$

Wronskian: $w_\alpha = |\beta|$.

Transition density w.r.t. m:

if $\beta \neq 0$, $\mu \neq 0$,

$$p(t;x,y) = \frac{|\mu|\, e^{\mu(\beta+1/2)t}}{2\,\mathrm{sh}(|\mu\beta|t)}\sqrt{xy}\exp\Big(-\frac{\mu e^{-\mu\beta t}(x^{-2\beta}+y^{-2\beta})}{2\sigma^2\beta\,\mathrm{sh}(\mu\beta t)}\Big) I_{1/2|\beta|}\Big(\frac{\mu x^{-\beta}y^{-\beta}}{\sigma^2\beta\,\mathrm{sh}(\mu\beta t)}\Big),$$

if $\beta \neq 0$, $\mu = 0$,

$$p(t;x,y) = \frac{\sqrt{xy}}{2|\beta|t}\exp\Big(-\frac{(x^{-2\beta}+y^{-2\beta})}{2\sigma^2\beta^2 t}\Big) I_{1/2|\beta|}\Big(\frac{x^{-\beta}y^{-\beta}}{\sigma^2\beta^2 t}\Big).$$

CEV processes were introduced by Cox (1975). We refer also to the papers Emanuel and MacBeth (1982), Goldenberg (1991) and Davydov and Linetsky (2001).

Part II

TABLES OF DISTRIBUTIONS OF FUNCTIONALS OF BROWNIAN MOTION AND RELATED PROCESSES

INTRODUCTION

The tables are divided into nine sections:

Let X denote any of these processes. In the tables we consider functionals of X up to the following stopping times:

1. τ — exponentially distributed random "time" independent of X.
2. $H_z := \inf\{s : X_s = z\}$ — first hitting time of the point z.
3. $H_{a,b} := \inf\{s : X_s \notin (a,b)\}$ — first exit time from the interval (a,b).
4. $\varrho(t,z) := \inf\{s : \ell(s,z) = t\}$ — inverse local time at the point z
5. $\theta_v := \min\{t : \sup_{0\le s\le t} X_s - \inf_{0\le s\le t} X_s = v\}$ — first range time at the level v.

Recall that considering functionals of X stopped at the independent exponential time τ corresponds to taking the Laplace transform of the functional with respect to the time parameter. Hence, inverting the transform gives the distribution at a fixed time t. All of the above mentioned stopping times are not treated for every process (see the list of contents of the tables on p. vii). In particular, formulae for the first range time are computed only for Brownian motion.

The aim has been to give a "systematic" presentation. The structure of the tables is displayed inside the individual sections via a triple numbering system. The first number characterizes the stopping time. The second indicates a functional or a group of functionals. There is a list of these below the reference numbers indicated. The third number refers to the Laplace transform or to the distribution itself; usually several special cases are considered.

Roughly speaking, Brownian motion is the most computationally tractable diffusion. The structure of the tables is therefore determined by the first section, which also contains more formulae than the other sections. In particular, this means that, e.g., formula 1.3.2 for a Radial Ornstein–Uhlenbeck process in Section 8 has as its counterpart formula 1.3.2 for a Brownian motion in Section 1. On the other hand, formula 1.3.3 in Section 1 has no counterpart in Section 8. This strive toward a systematic presentation has also lead to, e.g., there not being a Subsection 3 in Section 3. In Section 6 there is no Subsection 1 because the formulae in Subsection 1 in Section 4 (when taking the parameter ν equal to 0) are also valid in this case. In Section 1 some notation is also introduced that will be used in Sections 2 and 5.

Many formulae contain so called special functions. Concerning these there is Appendix 2 at the end of the tables. For more complete information on special functions we refer to Erdelyi et al. (1953), Abramowitz and Stegun (1970), and Lebedev (1972) .

It is important to notice that the local times in the tables are densities of the occupation time measure with respect to Lebesgue measure:

$$\ell(t,x) = \lim_{\varepsilon \to 0} \frac{1}{\varepsilon} \int_0^t \mathbb{I}_{[x,x+\varepsilon)}(X_s)\, ds,$$

where X is any of the diffusions in the tables. Letting $L(t,x)$ be the local time with respect to the speed measure of X and $m(x)$ the density of the speed measure with respect to Lebesgue measure, then (see No. II.24)

(a) $$\ell(t,x) = m(x)\, L(t,x).$$

1. List of functionals

The number of the functional on this list corresponds to the second number in the triple numbering system used in the tables. Here X is any of the processes and T is any of the stopping times considered.

1. $\sup_{0 \le s \le T} X_s$

2. $\inf_{0 \le s \le T} X_s$

3. $\ell(T, r)$

4. $\int_0^T \mathbb{I}_{[r,\infty)}(X_s) ds$

5. $\int_0^T \mathbb{I}_{(-\infty,r]}(X_s)ds$

6. $\int_0^T \mathbb{I}_{(-\infty,r)}(X_s)ds, \quad \int_0^T \mathbb{I}_{[r,\infty)}(X_s)ds$

7. $\int_0^T \mathbb{I}_{[r,u]}(X_s)ds$

8. $\int_0^T X_s ds$ or $\int_0^T |X_s| ds$

9. $\int_0^T X_s^2 ds$

10. $\int_0^T e^{2\beta X_s} ds$

11. $\sup_{y\in\mathbb{R}} \ell(T,y)$

12. $\check{H}(T):=\inf\{s<T : X_s=\sup_{0\le s\le T} X_s\}$ or $\hat{H}(T):=\inf\{s<T : X_s=\inf_{0\le s\le T} X_s\}$

13. $\check{H}(T), \quad \sup_{0\le s\le T} X_s$

14. $\hat{H}(T), \quad \inf_{0\le s\le T} X_s$

15. $\inf_{0\le s\le T} X_s, \quad \sup_{0\le s\le T} X_s$

16. $\ell(T,r), \quad \sup_{0\le s\le T} X_s$

17. $\ell(T,r), \quad \inf_{0\le s\le T} X_s$

18. $\ell(T,r), \quad \ell(T,u)$

19. $\int_0^T \frac{ds}{X_s}$

20. $\int_0^T \frac{ds}{X_s^2}$

21. $\int_0^T \frac{ds}{X_s^2}, \quad \int_0^T X_s^2 ds$

22. $\displaystyle\int_0^T \frac{ds}{X_s^2}, \quad \int_0^T \frac{ds}{X_s}$

23. $\displaystyle\int_0^T (p\mathbb{1}_{(-\infty,r)}(X_s) + q\mathbb{1}_{[r,\infty)}(X_s))ds, \quad \sup_{0\le s\le T} X_s$

24. $\displaystyle\int_0^T (p\mathbb{1}_{(-\infty,r)}(X_s) + q\mathbb{1}_{[r,\infty)}(X_s))ds, \quad \inf_{0\le s\le T} X_s$

25. $\ell(T,r), \quad \inf_{0\le s\le T} X_s, \quad \sup_{0\le s\le T} X_s$

26. $\displaystyle\int_0^T (p\mathbb{1}_{(-\infty,r)}(X_s) + q\mathbb{1}_{[r,\infty)}(X_s))ds, \quad \inf_{0\le s\le T} X_s, \quad \sup_{0\le s\le T} X_s$

27. $\displaystyle\int_0^T \mathbb{1}_{(-\infty,r)}(X_s)ds, \quad \int_0^T \mathbb{1}_{[r,\infty)}(X_s)ds, \quad \ell(T,r)$

28. $\hat{H}(T), \quad \check{H}(T), \quad \inf_{0\le s\le T} X_s, \quad \sup_{0\le s\le T} X_s$

29. $\displaystyle\int_0^T \mathbb{1}_{(-\infty,r)}(X_s)ds, \quad \int_0^T \mathbb{1}_{[r,\infty)}(X_s)ds, \quad \inf_{0\le s\le T} X_s, \quad \sup_{0\le s\le T} X_s, \quad \ell(T,r)$

30. $\displaystyle\int_0^T e^{2\beta X_s}ds, \quad \int_0^T e^{\beta X_s}ds$

31. $\sup_{0\le s\le T} \big(\ell(s,u) - \alpha^{-1}\ell(s,r)\big), \ \alpha > 0$

2. Comments and references

Many formulae in the tables have a general formulation which often can be found in Part I, especially in Chapters II, IV, V, and VI. In particular, we have included formulae based on the results in No.'s II.19, 25, 26, and IV.44. However, the main tool has been the Feynman–Kac formula which for Brownian motion and Brownian motion with drift is discussed in Chapter VI; see also No. II.24. Although Chapter VI is devoted to Brownian motion, the solutions therein are given in terms of the Green functions, indicating the form of the solution for an arbitrary diffusion. It is explained in No. II.11 how to compute the Green function of a given diffusion.

It should be possible for the reader to check a great number of the formulae in Sections 1 and 2 by solving the corresponding differential problems in Chapter VI or, whichever is easier, by checking that the solution is as claimed. In the "Exponential stopping" subsections one must also have some familiarity with inverting Laplace transforms. Hereby, Kamke (1943) and Erdelyi and et al. (1954) are good references. In Appendix 3 we have collected a number of "key" Laplace transforms mainly taken from Erdelyi and et al. (1954), and Appendix 4 concerning second order differential equations is extracted mainly from Kamke (1943).

Below we give more detailed comments concerning the particular functionals in the list above. In the papers referred to one can usually find a special case of the considered functional or further developments.

Functionals 1 and 2. Formulae associated to these can be derived using standard diffusion theory. Notice that, e.g.,

$$\mathbf{P}_x\big(\inf_{0\le t\le H_c} X_t < a\big) = \mathbf{P}_x(H_a < H_c),$$

$$\mathbf{P}_x\big(\inf_{0\le t\le \tau} X_t < a\big) = \mathbf{P}_x(H_a < \tau) = \mathbf{E}_x\big(e^{-\lambda H_a}\big),$$

and see No. II.4 and 10 for these expressions for general regular diffusions. We refer also to No. VI.10 and 4, respectively, which when choosing $b = +\infty$, $f \equiv 0$, $\gamma_i = 0, i = 1, 2, \dots k$ and $F \equiv 1$ lead to the same solutions. A general form of

$$\mathbf{P}_x\big(\inf_{0\le s\le \varrho(t,z)} X_s > a\big)$$

is given in No. VI.18.

Functional 3. Recall the formula (a) in No. II.27:

$$\mathbf{P}_x(L(\zeta, x) > \alpha) = \exp\big(-\frac{\alpha}{G_0(x,x)}\big),$$

where ζ is the life time and G_0 is the Green function of the considered diffusion. Using this, the strong Markov property and the relation $\ell(t,x) = m(x)\,L(t,x)$ give

$$\text{(b)}\qquad \mathbf{P}_z(\ell(\zeta, x) > \alpha) = \mathbf{P}_z(H_x < \infty)\exp\big(-\frac{\alpha}{m(x)G_0(x,x)}\big),$$

where m is the density of the speed measure with respect to the Lebesgue measure (when it exists). Choosing in (b) $\zeta = \tau$ or H_c or $H_{a,b}$ and computing the corresponding Green function give many of the formulae displayed in the tables. In this way we may also find the distribution of local time of a diffusion stopped at $\varrho(\tau, x)$ and this can be inverted with respect to λ. For the approach via differential equations see No. VI.4, 9, 16, 18, and 22.

Functionals 4–7. Deriving distributions of these is a typical application of the Feynman–Kac formula. Also here it is possible to formulate general results for a large class of diffusions. Indeed, consider, e.g., the problem of finding the joint distribution of the occupation times

$$\text{(c)}\qquad A_t^- := \int_0^t \mathbb{I}_{(-\infty,0)}(X_s)ds, \quad A_t^+ := \int_0^\tau \mathbb{I}_{(0,\infty)}(X_s)ds,$$

where X is a regular diffusion on the interval $I = (l, r)$ containing 0. We assume that the boundaries l and r are entrance-not-exit or natural and that there is no killing. Moreover suppose that the infinitesimal generator is given by a smooth ordinary differential operator, denoted $\mathcal{G}$, as in No. II.9. From No. VI.4 we have

$$\text{(d)}\quad \begin{aligned}\mathbf{E}_0\Big(\exp(-\alpha_1 A^+_\tau - \alpha_2 A^-_\tau)\Big) &= \lambda \int_l^r \tilde{G}_\lambda(0,y)\, m(dy)\\ &= \frac{\lambda}{\tilde{w}_\lambda}\Big(\tilde{\varphi}_\lambda(0)\int_l^0 \tilde{\psi}_\lambda(y)\, m(dy) + \tilde{\psi}_\lambda(0)\int_0^r \tilde{\varphi}_\lambda(y)\, m(dy)\Big).\end{aligned}$$

Due to the assumptions on the boundary behavior, the functions $\tilde{\varphi}_\lambda$ and $\tilde{\psi}_\lambda$ can be expressed (cf. No. II.10) in terms of the fundamental solutions φ_λ and ψ_λ associated to X as follows

$$\text{(e)}\quad \begin{aligned}\tilde{\varphi}_\lambda(x) &= \begin{cases} c_1\,\varphi_{\lambda+\alpha_2}(x) + c_2\,\psi_{\lambda+\alpha_2}(x), & x \le 0,\\ \varphi_{\lambda+\alpha_1}(x), & x \ge 0,\end{cases}\\ \tilde{\psi}_\lambda(x) &= \begin{cases} \psi_{\lambda+\alpha_2}(x), & x \le 0,\\ c_3\,\varphi_{\lambda+\alpha_1}(x) + c_4\,\psi_{\lambda+\alpha_1}(x), & x \ge 0,\end{cases}\end{aligned}$$

where the constants c_1, c_2, c_3 and c_4 are determined by the demand that $\tilde{\varphi}_\lambda$ and $\tilde{\psi}_\lambda$ are in $\mathcal{C}^1(I)$. Using the fact that φ_λ and ψ_λ are solutions of the ODE $\mathcal{G}u = \lambda u$ and computing the Wronskian $\tilde{w}_\lambda$ we obtain

$$\mathbf{E}_0\Big(\exp(-\alpha_1 A^+_\tau - \alpha_2 A^-_\tau)\Big) = \Big(\frac{\lambda}{\lambda+\alpha_2}F(0) - \frac{\lambda}{\lambda+\alpha_1}G(0)\Big)/(F(0) - G(0)),$$

where

$$F(0) := \psi'_{\lambda+\alpha_2}(0)/\psi_{\lambda+\alpha_2}(0) \quad \text{and} \quad G(0) := \varphi'_{\lambda+\alpha_1}(0)/\varphi_{\lambda+\alpha_1}(0).$$

For this formula (with a proof based on excursion theory) see Truman and Williams (1990). To find the corresponding formula when τ is replaced with H_z, $z > 0$ we can proceed as explained in No. VI.9. This gives using the notation above

$$\mathbf{E}_0\Big(\exp(-\alpha_1 A^+_{H_z} - \alpha_2 A^-_{H_z})\Big) = \tilde{\psi}_0(0)/\tilde{\psi}_0(z).$$

We refer to Pitman and Yor (1999) for an alternative derivation of this formula and connections to excursion theory. See Knight (1969), (1978), and Jeulin and Yor (1981) for formulae for Brownian motion. Further, we remark that the arcsine law of Lévy (1939) is included in 1.1.4.4. and the formula 2.1.4.4. is due to Takàcs (1996).

Functional 8. To analyze this functional using the Feynman–Kac formula one should be able to solve the ODE

$$\mathcal{G}u(x) - \alpha\,|x|u(x) = \lambda u(x),$$

where $\alpha > 0$, $\lambda \ge 0$, and $\mathcal{G}$ is the differential operator associated to the considered diffusion. In Appendix 4 the fundamental solutions are given for Brownian motion in Eq. 2, for Brownian motion with drift in Eq. 9, and for Ornstein–Uhlenbeck processes in Eq. 18. Notice also that Eq. 15.a when multiplied by x^2 gives the solutions for geometric Brownian motion, and Eq. 15 when multiplied by x gives the ODE associated to squared Bessel processes. Especially, for the formula 1.1.8.5 we refer to Shepp (1982), where the functional is considered for Brownian bridge; see also Rice (1982) and Johnson and Killeen (1983). The formula 9.1.8.4(1) is first presented in Dufresne (1990), see also Yor (1992c).

Functional 9. As in the previous case one can find the needed differential equations and their solutions in Appendix 4. The Cameron–Martin formula is 1.1.9.3; see Cameron and Martin (1945). Special cases of the general formula in No. IV.44 due to Pitman and Yor (1982) are displayed here at many spots. See also Wenocur (1990). Furthermore, we refer to Neveu (1968) for additional references and results on quadratic functionals. For more recent works, see Ikeda, Kusuoka and Manabe (1995) and Jeanblanc, Pitman and Yor (1997).

Functional 10. Formulae associated to this functional are only displayed for Brownian motion, Brownian motion with drift and three-dimensional Bessel process. In these cases the computations can be done studing the functional 8 for geometric Brownian motion. For further results, see Yor (1992a,b).

Functional 11. For Brownian motion, see No. VI.8, 13, 14, 21, and 23. These results are due to Borodin (1982, 1989), see also Csáki, Földes and Salminen (1987) and Eisenbaum (1990). For 1.1.11.4 see Csáki and Földes (1986).

Functionals 12–14. These are special cases of the general formula in No. II.19, see Csáki, Földes and Salminen (1987). See also Lévy (1948), Vincze (1957), Shepp (1979), Imhof (1984), Louchard (1984), and Pitman and Yor (1996).

Functional 15. The formula 1.1.15.4 is due to Lévy. For 1.1.15.8 see Feller (1951) and for 1.1.15.8(2) Kac, Kiefer and Wolfowitz (1955). From the latter one we can deduce the distribution of the range of a Brownian bridge. For an explanation that this distribution is the same as the distribution of the maximum of a three-dimensional Bessel bridge, see No. IV.24. We refer further to Doob (1949) for the probability that Brownian motion stays between two linear boundaries.

Functionals 16–17. See Chapter VI.

Functional 18. Joint distributions of (stopped) local times can be calculated using the Ray–Knight theorems; see Chapter V.

Functionals 19–27, 29, 30. These are typical applications of the Feynman–Kac formula with a side condition, see Chapter VI. In particular, we have for the functional 27 with the notations (and assumptions) as in (c) and (d)

$$\mathbf{E}_0\Big(\exp\big(-\alpha_1 A^+_\tau - \alpha_2 A^-_\tau - \gamma\,\ell(\tau,0)\big)\Big) = \lambda \int_l^r \tilde{G}_\lambda(0,y)\, m(dy),$$

and the Green function $\tilde{G}_\lambda$ is expressed in terms of $\tilde{\varphi}_\lambda$ and $\tilde{\psi}_\lambda$ satisfying (e). The constants c_1, c_2, c_3 and c_4 are determined by the demand that $\tilde{\varphi}_\lambda$ and $\tilde{\psi}_\lambda$ are continuous and satisfy

$$\tilde{\varphi}'_\lambda(0+) - \tilde{\varphi}'_\lambda(0-) = \frac{2\gamma}{a^2(0)}\tilde{\varphi}_\lambda(0),$$

and

$$\tilde{\psi}'_\lambda(0+) - \tilde{\psi}'_\lambda(0-) = \frac{2\gamma}{a^2(0)}\tilde{\psi}_\lambda(0),$$

respectively, where a^2 is the infinitesimal variance of X. See Karatzas and Shreve (1984), where the case with Brownian motion is discussed. A general formula can also be derived when X is stopped at the first hitting time H_z, cf. Pitman and Yor (1999).

Functional 28. The joint distributions of the functionals 12–14 can be obtained from the formula in No. II.19 by conditioning.

Functional 31. For this, see Borodin (2000).

1. BROWNIAN MOTION

1. Exponential stopping

1.0.1 $\mathbf{E}e^{-\alpha\tau} = \dfrac{\lambda}{\lambda+\alpha}$

1.0.2 $\mathbf{P}(\tau > t) = e^{-\lambda t}$

1.0.3 $\mathbf{E}_x e^{i\beta W_\tau} = \dfrac{2\lambda}{2\lambda+\beta^2} e^{i\beta x}$

1.0.4 $\mathbf{E}_x e^{i\beta W_t} = e^{i\beta x - \beta^2 t/2}$

1.0.5 $\mathbf{P}_x(W_\tau \in dz) = \dfrac{\sqrt{\lambda}}{\sqrt{2}} e^{-|z-x|\sqrt{2\lambda}} dz$

1.0.6 $\mathbf{P}_x\big(W_t \in dz\big) = \dfrac{1}{\sqrt{2\pi t}} e^{-(z-x)^2/2t} dz$

1.1.1 $\mathbf{E}_x \exp\big(-\gamma \sup_{0\le s\le\tau} W_s\big) = \dfrac{\sqrt{2\lambda}}{\gamma+\sqrt{2\lambda}} e^{-\gamma x}$

1.1.2 $x \le y$ $\mathbf{P}_x\big(\sup_{0\le s\le\tau} W_s \ge y\big) = e^{(x-y)\sqrt{2\lambda}}$

1.1.3 $\mathbf{E}_x \exp\big(-\gamma \sup_{0\le s\le t} W_s\big) = e^{-\gamma x+\gamma^2 t/2}\,\text{Erfc}(\gamma\sqrt{t/2})$

1.1.4 $x \le y$ $\mathbf{P}_x\big(\sup_{0\le s\le t} W_s \ge y\big) = \text{Erfc}\Big(\dfrac{y-x}{\sqrt{2t}}\Big)$

1.1.5 $\mathbf{E}_x\big\{\exp(-\gamma \sup_{0\le s\le\tau} W_s); W_\tau \in dz\big\} = \dfrac{2\lambda}{\gamma+2\sqrt{2\lambda}} e^{-\gamma(x\vee z)-|z-x|\sqrt{2\lambda}} dz$

1.1.6 $\mathbf{P}_x\big(\sup_{0\le s\le\tau} W_s \ge y, W_\tau \in dz\big) = \dfrac{\sqrt{\lambda}}{\sqrt{2}} e^{(z+x-2y)\sqrt{2\lambda}} dz, \quad x\vee z \le y$

1	W $\qquad\qquad\qquad\qquad$ $\tau \sim \mathrm{Exp}(\lambda)$, independent of W
1.1.7	$\mathbf{E}_x\{\exp(-\gamma \sup_{0\le s\le t} W_s); W_t\in dz\}$
	$= e^{-\gamma(x\vee z)}\Big(\frac{1}{\sqrt{2\pi t}}e^{-(z-x)^2/2t} - \frac{\gamma}{4}e^{\gamma\|z-x\|/2+\gamma^2t/8}\,\mathrm{Erfc}\Big(\frac{\|z-x\|}{\sqrt{2t}}+\frac{\gamma\sqrt{t}}{2\sqrt{2}}\Big)\Big)dz$
1.1.8 $x\le y$	$\mathbf{P}_x\big(\sup_{0\le s\le t} W_s\ge y, W_t\in dz\big) = \frac{1}{\sqrt{2\pi t}}e^{-(\|z-y\|+y-x)^2/2t}dz$
1.2.1	$\mathbf{E}_x\exp(\gamma \inf_{0\le s\le \tau} W_s) = \frac{\sqrt{2\lambda}}{\gamma+\sqrt{2\lambda}}e^{\gamma x}$
1.2.2 $y\le x$	$\mathbf{P}_x\big(\inf_{0\le s\le \tau} W_s\le y\big) = e^{(y-x)\sqrt{2\lambda}}$
1.2.3	$\mathbf{E}_x\exp(\gamma \inf_{0\le s\le t} W_s) = e^{\gamma x+\gamma^2t/2}\,\mathrm{Erfc}(\gamma\sqrt{t/2})$
1.2.4 $y\le x$	$\mathbf{P}_x\big(\inf_{0\le s\le t} W_s\le y\big) = \mathrm{Erfc}\Big(\frac{x-y}{\sqrt{2t}}\Big)$
1.2.5	$\mathbf{E}_x\{\exp(\gamma \inf_{0\le s\le \tau} W_s); W_\tau\in dz\} = \frac{2\lambda}{\gamma+2\sqrt{2\lambda}}e^{\gamma(x\wedge z)-\|z-x\|\sqrt{2\lambda}}dz$
1.2.6	$\mathbf{P}_x\big(\inf_{0\le s\le \tau} W_s\le y, W_\tau\in dz\big) = \frac{\sqrt{\lambda}}{\sqrt{2}}e^{(2y-z-x)\sqrt{2\lambda}}dz, \quad y\le x\wedge z$
1.2.7	$\mathbf{E}_x\{\exp(\gamma \inf_{0\le s\le t} W_s); W_t\in dz\}$
	$= e^{\gamma(x\wedge z)}\Big(\frac{1}{\sqrt{2\pi t}}e^{-(z-x)^2/2t} - \frac{\gamma}{4}e^{\gamma\|z-x\|/2+\gamma^2t/8}\,\mathrm{Erfc}\Big(\frac{\|z-x\|}{\sqrt{2t}}+\frac{\gamma\sqrt{t}}{2\sqrt{2}}\Big)\Big)dz$
1.2.8 $y\le x$	$\mathbf{P}_x\big(\inf_{0\le s\le t} W_s\le y, W_t\in dz\big) = \frac{1}{\sqrt{2\pi t}}e^{-(\|z-y\|-y+x)^2/2t}dz$
1.3.1	$\mathbf{E}_x e^{-\gamma\ell(\tau,r)} = 1-\frac{\gamma}{\gamma+\sqrt{2\lambda}}e^{-\|r-x\|\sqrt{2\lambda}}$

1 W	$\tau \sim \mathrm{Exp}(\lambda)$, independent of W
1.3.2 $0<y$	$\mathbf{P}_x\big(\ell(\tau,r)\in dy\big)=\sqrt{2\lambda}e^{-(y+\lvert r-x\rvert)\sqrt{2\lambda}}dy$
(1)	$\mathbf{P}_x\big(\ell(\tau,r)=0\big)=1-e^{-\lvert r-x\rvert\sqrt{2\lambda}}$
1.3.3	$\mathbf{E}_x e^{-\gamma\ell(t,r)}=1-\operatorname{Erfc}\Big(\frac{\lvert r-x\rvert}{\sqrt{2t}}\Big)+e^{\lvert r-x\rvert\gamma+\gamma^2t/2}\operatorname{Erfc}\Big(\frac{\lvert r-x\rvert}{\sqrt{2t}}+\frac{\gamma\sqrt{t}}{\sqrt{2}}\Big)$
1.3.4 $0<y$	$\mathbf{P}_x\big(\ell(t,r)\in dy\big)=\frac{\sqrt{2}}{\sqrt{\pi t}}e^{-(y+\lvert r-x\rvert)^2/2t}dy$
(1)	$\mathbf{P}_x\big(\ell(t,r)=0\big)=1-\operatorname{Erfc}\Big(\frac{\lvert r-x\rvert}{\sqrt{2t}}\Big)$
1.3.5	$\mathbf{E}_x\big\{e^{-\gamma\ell(\tau,r)};W_\tau\in dz\big\}=\frac{\sqrt{\lambda}}{\sqrt{2}}e^{-\lvert z-x\rvert\sqrt{2\lambda}}dz-\frac{\gamma\sqrt{\lambda/2}}{\gamma+\sqrt{2\lambda}}e^{-(\lvert z-r\rvert+\lvert r-x\rvert)\sqrt{2\lambda}}dz$
1.3.6 $0<y$	$\mathbf{P}_x\big(\ell(\tau,r)\in dy,W_\tau\in dz\big)=\lambda e^{-(\lvert z-r\rvert+\lvert r-x\rvert+y)\sqrt{2\lambda}}dydz$
(1)	$\mathbf{P}_x\big(\ell(\tau,r)=0,W_\tau\in dz\big)=\frac{\sqrt{\lambda}}{\sqrt{2}}e^{-\lvert z-x\rvert\sqrt{2\lambda}}dz-\frac{\sqrt{\lambda}}{\sqrt{2}}e^{-(\lvert z-r\rvert+\lvert r-x\rvert)\sqrt{2\lambda}}dz$
1.3.7	$\mathbf{E}_x\big\{e^{-\gamma\ell(t,r)};W_t\in dz\big\}=\frac{1}{\sqrt{2\pi t}}e^{-(z-x)^2/2t}dz$ $-\frac{\gamma}{2}\exp\big((\lvert z-r\rvert+\lvert r-x\rvert)\gamma+\gamma^2t/2\big)\operatorname{Erfc}\Big(\frac{\gamma\sqrt{t}}{\sqrt{2}}+\frac{\lvert z-r\rvert+\lvert r-x\rvert}{\sqrt{2t}}\Big)dz$
1.3.8 $0<y$	$\mathbf{P}_x\big(\ell(t,r)\in dy,W_t\in dz\big)$ $=\frac{1}{t\sqrt{2\pi t}}\big(y+\lvert z-r\rvert+\lvert r-x\rvert\big)e^{-(\lvert z-r\rvert+\lvert r-x\rvert+y)^2/2t}dydz$
(1)	$\mathbf{P}_x\big(\ell(t,r)=0,W_t\in dz\big)=\frac{1}{\sqrt{2\pi t}}e^{-(z-x)^2/2t}dz-\frac{1}{\sqrt{2\pi t}}e^{-(\lvert z-r\rvert+\lvert r-x\rvert)^2/2t}dz$

1 W	$\tau \sim \mathrm{Exp}(\lambda)$, independent of W
1.4.1	$\mathbf{E}_x \exp\Big(-\gamma \int_0^\tau \mathbb{I}_{[r,\infty)}(W_s)ds\Big)$
$x \le r$	$= 1 - \Big(1 - \frac{\sqrt{\lambda}}{\sqrt{\lambda+\gamma}}\Big)e^{(x-r)\sqrt{2\lambda}}$
$r \le x$	$= \frac{\lambda}{\lambda+\gamma} - \Big(\frac{\lambda}{\lambda+\gamma} - \frac{\sqrt{\lambda}}{\sqrt{\lambda+\gamma}}\Big)e^{(r-x)\sqrt{2\lambda+2\gamma}}$
1.4.2	$\mathbf{P}_x\Big(\int_0^\tau \mathbb{I}_{[r,\infty)}(W_s)ds \in dy\Big)$
$x \le r$	$= \frac{\sqrt{\lambda}}{\sqrt{\pi y}} e^{(x-r)\sqrt{2\lambda}-\lambda y}dy$
$r \le x$	$= \lambda e^{-\lambda y}\,\mathrm{Erf}\Big(\frac{x-r}{\sqrt{2y}}\Big)dy + \frac{\sqrt{\lambda}}{\sqrt{\pi y}} e^{-\lambda y-(x-r)^2/2y}dy$
(1) $x \le r$	$\mathbf{P}_x\Big(\int_0^\tau \mathbb{I}_{[r,\infty)}(W_s)ds = 0\Big) = 1 - e^{(x-r)\sqrt{2\lambda}}$
1.4.3	$\mathbf{E}_x \exp\Big(-\gamma \int_0^t \mathbb{I}_{[r,\infty)}(W_s)ds\Big)$
$x \le r$	$= \mathrm{Erf}\Big(\frac{r-x}{\sqrt{2t}}\Big) + \frac{1}{\pi}e^{-\gamma t}\int_0^t \frac{1}{\sqrt{s(t-s)}} e^{\gamma s-(r-x)^2/2s}ds$
$r \le x$	$= e^{-\gamma t}\,\mathrm{Erf}\Big(\frac{x-r}{\sqrt{2t}}\Big) + \frac{1}{\pi}\int_0^t \frac{1}{\sqrt{s(t-s)}} e^{-\gamma s-(x-r)^2/2s}ds$
$x = r$	$= e^{-\gamma t/2} I_0\Big(\frac{\gamma t}{2}\Big)$
1.4.4	$\mathbf{P}_x\Big(\int_0^t \mathbb{I}_{[r,\infty)}(W_s)ds \in dv\Big) = \begin{cases} \frac{\mathbb{I}_{[0,t]}(v)}{\pi\sqrt{v(t-v)}} e^{-(r-x)^2/2(t-v)}dv, & x \le r \\ \frac{\mathbb{I}_{[0,t]}(v)}{\pi\sqrt{v(t-v)}} e^{-(x-r)^2/2v}dv, & r \le x \end{cases}$
(1) $x \le r$	$\mathbf{P}_x\Big(\int_0^t \mathbb{I}_{[r,\infty)}(W_s)ds = 0\Big) = 1 - \mathrm{Erfc}\Big(\frac{r-x}{\sqrt{2t}}\Big)$

1	W $\qquad\qquad\qquad$ $\tau \sim \operatorname{Exp}(\lambda)$, independent of W
(2) $r \le x$	$\mathbf{P}_x\Big(\int_0^t \mathbb{I}_{[r,\infty)}(W_s)ds = t\Big) = 1 - \operatorname{Erfc}\Big(\frac{x-r}{\sqrt{2t}}\Big)$
1.4.5	$\mathbf{E}_x\Big\{\exp\Big(-\gamma\int_0^\tau \mathbb{I}_{[r,\infty)}(W_s)ds\Big); W_\tau \in dz\Big\} =: \lambda Q_\lambda^{(4)}(z)dz$
$x \le r$ $z \le r$	$= \frac{\lambda}{\sqrt{2\lambda}} e^{-\lvert z-x\rvert\sqrt{2\lambda}}dz + \lambda\Big(\frac{2}{\sqrt{2\lambda}+\sqrt{2\lambda+2\gamma}} - \frac{1}{\sqrt{2\lambda}}\Big)e^{(z+x-2r)\sqrt{2\lambda}}dz$
$x \le r$ $r \le z$	$= \frac{2\lambda}{\sqrt{2\lambda}+\sqrt{2\lambda+2\gamma}} e^{(x-r)\sqrt{2\lambda}+(r-z)\sqrt{2\lambda+2\gamma}}dz$
$r \le x$ $z \le r$	$= \frac{2\lambda}{\sqrt{2\lambda}+\sqrt{2\lambda+2\gamma}} e^{(r-x)\sqrt{2\lambda+2\gamma}+(z-r)\sqrt{2\lambda}}dz$
$r \le x$ $r \le z$	$= \frac{\lambda}{\sqrt{2\lambda+2\gamma}}\Big(e^{-\lvert z-x\rvert\sqrt{2\lambda+2\gamma}} + \Big(1 - \frac{2\sqrt{2\lambda}}{\sqrt{2\lambda+2\gamma}+\sqrt{2\lambda}}\Big)e^{(2r-x-z)\sqrt{2\lambda+2\gamma}}\Big)dz$
1.4.6	$\mathbf{P}_x\Big(\int_0^\tau \mathbb{I}_{[r,\infty)}(W_s)ds \in dy, W_\tau \in dz\Big) =: \lambda B_\lambda^{(4)}(y,z)dydz$
$x \le r$ $z \le r$	$= \lambda\Big(\frac{\sqrt{2}}{\sqrt{\pi y}}e^{-\lambda y} - \sqrt{2\lambda}\operatorname{Erfc}(\sqrt{\lambda y})\Big)e^{(z+x-2r)\sqrt{2\lambda}}dydz$
$x \le r$ $r \le z$	$= \Big(\frac{\lambda\sqrt{2}}{\sqrt{\pi y}}e^{(x-r)\sqrt{2\lambda}-\lambda y-(z-r)^2/2y} - \sqrt{2\lambda^3}e^{(z+x-2r)\sqrt{2\lambda}}\operatorname{Erfc}\Big(\frac{z-r}{\sqrt{2y}}+\sqrt{\lambda y}\Big)\Big)dydz$
$r \le x$ $z \le r$	$= \Big(\frac{\lambda\sqrt{2}}{\sqrt{\pi y}}e^{(z-r)\sqrt{2\lambda}-\lambda y-(x-r)^2/2y} - \sqrt{2\lambda^3}e^{(z+x-2r)\sqrt{2\lambda}}\operatorname{Erfc}\Big(\frac{x-r}{\sqrt{2y}}+\sqrt{\lambda y}\Big)\Big)dydz$
$r \le x$ $r \le z$	$= \lambda\Big(\frac{1}{\sqrt{2\pi y}}e^{-\lambda y}\big(e^{-(z-x)^2/2y} + e^{-(2r-z-x)^2/2y}\big)$ $\qquad - \sqrt{2\lambda}e^{(z+x-2r)\sqrt{2\lambda}}\operatorname{Erfc}\Big(\frac{z+x-2r}{\sqrt{2y}} + \sqrt{\lambda y}\Big)\Big)dydz$
(1) $x \le r$ $z \le r$	$\mathbf{P}_x\Big(\int_0^\tau \mathbb{I}_{[r,\infty)}(W_s)ds = 0,\ W_\tau \in dz\Big) = \frac{\sqrt{\lambda}}{\sqrt{2}}\big(e^{-\lvert z-x\rvert\sqrt{2\lambda}} - e^{(z+x-2r)\sqrt{2\lambda}}\big)dz$

1 W $\tau \sim \operatorname{Exp}(\lambda)$, independent of W

1.4.7 $\mathbf{E}_x\Big\{\exp\Big(-\gamma\int_0^t \mathbb{I}_{[r,\infty)}(W_s)ds\Big); W_t\in dz\Big\} =: V_x^{(4)}(t,z)dz$

$x<r$, $z<r$

$$= \frac{1}{\sqrt{2\pi t}}\big(e^{-(z-x)^2/2t} - e^{-(2r-x-z)^2/2t}\big)dz + \frac{2r-x-z}{2\pi}\Big(\frac{1-e^{-\gamma t}}{\gamma t^{3/2}}\Big) * \Big(\frac{1}{t^{3/2}}e^{-(2r-x-z)^2/2t}\Big)dz$$

$x<r$, $r<z$

$$= \frac{(z-r)(r-x)}{(2\pi)^{3/2}}\Big(\frac{1-e^{-\gamma t}}{\gamma t^{3/2}}\Big) * \Big(\frac{e^{-\gamma t}}{t^{3/2}}e^{-(z-r)^2/2t}\Big) * \Big(\frac{1}{t^{3/2}}e^{-(r-x)^2/2t}\Big)dz$$

$r<x$, $z<r$

$$= \frac{(x-r)(r-z)}{(2\pi)^{3/2}}\Big(\frac{1-e^{-\gamma t}}{\gamma t^{3/2}}\Big) * \Big(\frac{e^{-\gamma t}}{t^{3/2}}e^{-(x-r)^2/2t}\Big) * \Big(\frac{1}{t^{3/2}}e^{-(r-z)^2/2t}\Big)dz$$

$r<x$, $r<z$

$$= \frac{e^{-\gamma t}}{\sqrt{2\pi t}}\big(e^{-(z-x)^2/2t} - e^{-(x+z-2r)^2/2t}\big)dz + \frac{x+z-2r}{2\pi}\Big(\frac{1-e^{-\gamma t}}{\gamma t^{3/2}}\Big) * \Big(\frac{e^{-\gamma t}}{t^{3/2}}e^{-(x+z-2r)^2/2t}\Big)dz$$

$x=r$, $z=r$

$$= \frac{1-e^{-\gamma t}}{(2\pi)^{1/2}\gamma t^{3/2}}dz$$

$x=r$, $r<z$

$$= \frac{z-r}{2\pi}\Big(\frac{1-e^{-\gamma t}}{\gamma t^{3/2}}\Big) * \Big(\frac{e^{-\gamma t}}{t^{3/2}}e^{-(z-r)^2/2t}\Big)dz$$

$x=r$, $z<r$

$$= \frac{r-z}{2\pi}\Big(\frac{1-e^{-\gamma t}}{\gamma t^{3/2}}\Big) * \Big(\frac{1}{t^{3/2}}e^{-(r-z)^2/2t}\Big)dz$$

1.4.8 $v<t$

$\mathbf{P}_x\Big(\int_0^t \mathbb{I}_{[r,\infty)}(W_s)ds\in dv, W_t\in dz\Big) =: F_x^{(4)}(t,v,z)dvdz$

$$= \int_0^\infty B_x^{(27)}(t-v,v,y,z)dy\,dvdz \qquad \text{for } B_x^{(27)}(u,v,y,z) \text{ see 1.27.6}$$

$x\le r$, $z\le r$

$$= \frac{(2r-x-z)\sqrt{t-v}}{\pi t^2\sqrt{v}}\exp\Big(-\frac{(2r-x-z)^2}{2(t-v)}\Big)dvdz + \Big(\frac{1}{\sqrt{2\pi}t^{3/2}} - \frac{(2r-x-z)^2}{\sqrt{2\pi}t^{5/2}}\Big)e^{-(2r-x-z)^2/2t}\operatorname{Erfc}\Big(\frac{(2r-x-z)\sqrt{v}}{\sqrt{2t(t-v)}}\Big)dvdz$$

1	W $\qquad\qquad\qquad \tau \sim \operatorname{Exp}(\lambda)$, independent of W
$x \le r$ $r \le z$	$$= \Big(\frac{(z-r)\sqrt{v}}{\pi t^2\sqrt{t-v}} + \frac{(r-x)\sqrt{t-v}}{\pi t^2\sqrt{v}}\Big)\Big) \exp\Big(-\frac{(z-r)^2}{2v} - \frac{(r-x)^2}{2(t-v)}\Big)dvdz$$ $$+\Big(\frac{1}{\sqrt{2\pi}t^{3/2}} - \frac{(2r-x-z)^2}{\sqrt{2\pi}t^{5/2}}\Big)e^{-(2r-x-z)^2/2t} \operatorname{Erfc}\Big(\frac{(z-r)\sqrt{t-v}}{\sqrt{2tv}} + \frac{(r-x)\sqrt{v}}{\sqrt{2t(t-v)}}\Big)dvdz$$
$r \le x$ $z \le r$	$$= \Big(\frac{(x-r)\sqrt{v}}{\pi t^2\sqrt{t-v}} + \frac{(r-z)\sqrt{t-v}}{\pi t^2\sqrt{v}}\Big)\Big) \exp\Big(-\frac{(x-r)^2}{2v} - \frac{(r-z)^2}{2(t-v)}\Big)dvdz$$ $$+\Big(\frac{1}{\sqrt{2\pi}t^{3/2}} - \frac{(2r-x-z)^2}{\sqrt{2\pi}t^{5/2}}\Big)e^{-(2r-x-z)^2/2t} \operatorname{Erfc}\Big(\frac{(x-r)\sqrt{t-v}}{\sqrt{2tv}} + \frac{(r-z)\sqrt{v}}{\sqrt{2t(t-v)}}\Big)dvdz$$
$r \le x$ $r \le z$	$$= \frac{(x+z-2r)\sqrt{v}}{\pi t^2\sqrt{t-v}} \exp\Big(-\frac{(x+z-2r)^2}{2v}\Big)dvdz$$ $$+\Big(\frac{1}{\sqrt{2\pi}t^{3/2}} - \frac{(x+z-2r)^2}{\sqrt{2\pi}t^{5/2}}\Big)e^{-(x+z-2r)^2/2t} \operatorname{Erfc}\Big(\frac{(x+z-2r)\sqrt{t-v}}{\sqrt{2tv}}\Big)dvdz$$
(1) $x \le r$ $z \le r$	$$\mathbf{P}_x\Big(\int_0^t \mathbb{I}_{[r,\infty)}(W_s)ds = 0, W_t \in dz\Big) = \frac{1}{\sqrt{2\pi t}}\Big(e^{-(z-x)^2/2t} - e^{-(2r-x-z)^2/2t}\Big)dz$$
(2) $r \le x$ $r \le z$	$$\mathbf{P}_x\Big(\int_0^t \mathbb{I}_{[r,\infty)}(W_s)ds = t, W_t \in dz\Big) = \frac{1}{\sqrt{2\pi t}}\Big(e^{-(z-x)^2/2t} - e^{-(x+z-2r)^2/2t}\Big)dz$$
(3)	$$\mathbf{P}_x\Big(\int_0^t \mathbb{I}_{[x,\infty)}(W_s)ds \in dv \Big\vert W_t = x\Big) = \frac{1}{t}\mathbb{I}_{[0,t]}(v)dv$$
1.5.1	$$\mathbf{E}_x \exp\Big(-\gamma\int_0^\tau \mathbb{I}_{(-\infty,r]}(W_s)ds\Big)$$
$x \le r$	$$= \frac{\lambda}{\lambda+\gamma} - \Big(\frac{\lambda}{\lambda+\gamma} - \frac{\sqrt{\lambda}}{\sqrt{\lambda+\gamma}}\Big)e^{(x-r)\sqrt{2\lambda+2\gamma}}$$
$r \le x$	$$= 1 - \Big(1 - \frac{\sqrt{\lambda}}{\sqrt{\lambda+\gamma}}\Big)e^{(r-x)\sqrt{2\lambda}}$$
1.5.2	$$\mathbf{P}_x\Big(\int_0^\tau \mathbb{I}_{(-\infty,r]}(W_s)ds \in dy\Big)$$
$x \le r$	$$= \lambda e^{-\lambda y} \operatorname{Erf}\Big(\frac{r-x}{\sqrt{2y}}\Big)dy + \frac{\sqrt{\lambda}}{\sqrt{\pi y}}e^{-\lambda y-(r-x)^2/2y}dy$$
$r \le x$	$$= \frac{\sqrt{\lambda}}{\sqrt{\pi y}}e^{(r-x)\sqrt{2\lambda}-\lambda y}dy$$

1 W	$\tau \sim \mathrm{Exp}(\lambda)$, independent of W
(1) $r \le x$	$\mathbf{P}_x\Big(\int_0^\tau \mathbb{I}_{(-\infty,r]}(W_s)ds = 0\Big) = 1 - e^{(r-x)\sqrt{2\lambda}}$
1.5.3	$\mathbf{E}_x \exp\Big(-\gamma \int_0^t \mathbb{I}_{(-\infty,r]}(W_s)ds\Big)$
$x \le r$	$= e^{-\gamma t}\,\mathrm{Erf}\Big(\frac{r-x}{\sqrt{2t}}\Big) + \frac{1}{\pi}\int_0^t \frac{1}{\sqrt{s(t-s)}} e^{-\gamma s-(r-x)^2/2s} ds$
$r \le x$	$= \mathrm{Erf}\Big(\frac{x-r}{\sqrt{2t}}\Big) + \frac{1}{\pi} e^{-\gamma t} \int_0^t \frac{1}{\sqrt{s(t-s)}} e^{\gamma s-(x-r)^2/2s} ds$
$x = r$	$= e^{-\gamma t/2} I_0\big(\frac{\gamma t}{2}\big)$
1.5.4	$\mathbf{P}_x\Big(\int_0^t \mathbb{I}_{(-\infty,r]}(W_s)ds \in dv\Big) = \begin{cases} \frac{\mathbb{I}_{[0,t]}(v)}{\pi\sqrt{v(t-v)}} e^{-(r-x)^2/2v} dv, & x \le r \\ \frac{\mathbb{I}_{[0,t]}(v)}{\pi\sqrt{v(t-v)}} e^{-(x-r)^2/2(t-v)} dv, & r \le x \end{cases}$
(1) $r \le x$	$\mathbf{P}_x\Big(\int_0^t \mathbb{I}_{(-\infty,r]}(W_s)ds = 0\Big) = 1 - \mathrm{Erfc}\Big(\frac{x-r}{\sqrt{2t}}\Big)$
(2) $x \le r$	$\mathbf{P}_x\Big(\int_0^t \mathbb{I}_{(-\infty,r]}(W_s)ds = t\Big) = 1 - \mathrm{Erfc}\Big(\frac{r-x}{\sqrt{2t}}\Big)$
1.5.5	$\mathbf{E}_x\Big\{\exp\Big(-\gamma \int_0^\tau \mathbb{I}_{(-\infty,r]}(W_s)ds\Big); W_\tau \in dz\Big\} =: \lambda Q_\lambda^{(5)}(z)dz$
$x \le r$, $z \le r$	$= \frac{\lambda}{\sqrt{2\lambda+2\gamma}}\Big(e^{-\lvert z-x\rvert\sqrt{2\lambda+2\gamma}} + \Big(1 - \frac{2\sqrt{2\lambda}}{\sqrt{2\lambda+2\gamma}+\sqrt{2\lambda}}\Big) e^{(z+x-2r)\sqrt{2\lambda+2\gamma}}\Big) dz$
$x \le r$, $r \le z$	$= \frac{2\lambda}{\sqrt{2\lambda}+\sqrt{2\lambda+2\gamma}} e^{(x-r)\sqrt{2\lambda+2\gamma}+(r-z)\sqrt{2\lambda}} dz$
$r \le x$, $z \le r$	$= \frac{2\lambda}{\sqrt{2\lambda}+\sqrt{2\lambda+2\gamma}} e^{(r-x)\sqrt{2\lambda}+(z-r)\sqrt{2\lambda+2\gamma}} dz$
$r \le x$, $r \le z$	$= \frac{\lambda}{\sqrt{2\lambda}} e^{-\sqrt{2\lambda}\lvert z-x\rvert} dz + \lambda\Big(\frac{2}{\sqrt{2\lambda}+\sqrt{2\lambda+2\gamma}} - \frac{1}{\sqrt{2\lambda}}\Big) e^{(2r-x-z)\sqrt{2\lambda}} dz$

1 W $\qquad\qquad$ $\tau \sim \operatorname{Exp}(\lambda)$, independent of W

1.5.6

$$\mathbf{P}_x\Big(\int_0^\tau \mathbb{I}_{(-\infty,r]}(W_s)ds \in dy,\ W_\tau \in dz\Big) =: \lambda B_\lambda^{(5)}(y,z)dydz$$

$x \le r$, $z \le r$

$$= \lambda\Big(\frac{1}{\sqrt{2\pi y}}e^{-\lambda y}\big(e^{-(z-x)^2/2y} + e^{-(2r-x-z)^2/2y}\big) - \sqrt{2\lambda}e^{(2r-x-z)\sqrt{2\lambda}}\operatorname{Erfc}\Big(\frac{2r-x-z}{\sqrt{2y}} + \sqrt{\lambda y}\Big)\Big)dydz$$

$x \le r$, $r \le z$

$$= \Big(\frac{\lambda\sqrt{2}}{\sqrt{\pi y}}e^{(r-z)\sqrt{2\lambda}-\lambda y-(r-x)^2/2y} - \sqrt{2\lambda^3}e^{(2r-x-z)\sqrt{2\lambda}}\operatorname{Erfc}\Big(\frac{r-x}{\sqrt{2y}} + \sqrt{\lambda y}\Big)\Big)dydz$$

$r \le x$, $z \le r$

$$= \Big(\frac{\lambda\sqrt{2}}{\sqrt{\pi y}}e^{(r-x)\sqrt{2\lambda}-\lambda y-(z-r)^2/2y} - \sqrt{2\lambda^3}e^{(2r-x-z)\sqrt{2\lambda}}\operatorname{Erfc}\Big(\frac{r-z}{\sqrt{2y}} + \sqrt{\lambda y}\Big)\Big)dydz$$

$r \le x$, $r \le z$

$$= \lambda\Big(\frac{\sqrt{2}}{\sqrt{\pi y}}e^{-\lambda y} - \sqrt{2\lambda}\operatorname{Erfc}(\sqrt{\lambda y})\Big)e^{(2r-x-z)\sqrt{2\lambda}}dydz$$

(1) $r \le x$, $r \le z$

$$\mathbf{P}_x\Big(\int_0^\tau \mathbb{I}_{(-\infty,r]}(W_s)ds = 0,\ W_\tau \in dz\Big) = \frac{\sqrt{\lambda}}{\sqrt{2}}\big(e^{-|z-x|\sqrt{2\lambda}} - e^{(2r-x-z)\sqrt{2\lambda}}\big)dz$$

1.5.7

$$\mathbf{E}_x\Big\{\exp\Big(-\gamma\int_0^t \mathbb{I}_{(-\infty,r]}(W_s)ds\Big); W_t \in dz\Big\} =: V_x^{(5)}(t,z)dz$$

$x < r$, $z < r$

$$= \frac{1}{\sqrt{2\pi t}}e^{-\gamma t}\big(e^{-(x-z)^2/2t} - e^{-(2r-x-z)^2/2t}\big)dz + \frac{2r-x-z}{2\pi}\Big(\frac{1-e^{-\gamma t}}{\gamma t^{3/2}}\Big) * \Big(\frac{e^{-\gamma t}}{t^{3/2}}e^{-(2r-x-z)^2/2t}\Big)dz$$

$x < r$, $r < z$

$$= \frac{(z-r)(r-x)}{(2\pi)^{3/2}}\Big(\frac{1-e^{-\gamma t}}{\gamma t^{3/2}}\Big) * \Big(\frac{e^{-\gamma t}}{t^{3/2}}e^{-(r-x)^2/2t}\Big) * \Big(\frac{1}{t^{3/2}}e^{-(z-r)^2/2t}\Big)dz$$

$r < x$, $z < r$

$$= \frac{(x-r)(r-z)}{(2\pi)^{3/2}}\Big(\frac{1-e^{-\gamma t}}{\gamma t^{3/2}}\Big) * \Big(\frac{e^{-\gamma t}}{t^{3/2}}e^{-(r-z)^2/2t}\Big) * \Big(\frac{1}{t^{3/2}}e^{-(x-r)^2/2t}\Big)dz$$

$r < x$, $r < z$

$$= \frac{1}{\sqrt{2\pi t}}\Big(e^{-(x-z)^2/2t} - e^{-(x+z-2r)^2/2t}\Big)dz$$

1	W $\qquad\qquad\qquad\qquad \tau \sim \text{Exp}(\lambda)$, independent of W
	$+\frac{x+z-2r}{2\pi}\Big(\frac{1-e^{-\gamma t}}{\gamma t^{3/2}}\Big) * \Big(\frac{1}{t^{3/2}} e^{-(x+z-2r)^2/2t}\Big)dz$
$x=r$ $z=r$	$=\frac{1-e^{-\gamma t}}{(2\pi)^{1/2}\gamma t^{3/2}}dz$
$x=r$ $r<z$	$=\frac{z-r}{2\pi}\Big(\frac{1-e^{-\gamma t}}{\gamma t^{3/2}}\Big) * \Big(\frac{1}{t^{3/2}} e^{-(z-r)^2/2t}\Big)dz$
$x=r$ $z<r$	$=\frac{r-z}{2\pi}\Big(\frac{1-e^{-\gamma t}}{\gamma t^{3/2}}\Big) * \Big(\frac{e^{-\gamma t}}{t^{3/2}} e^{-(r-z)^2/2t}\Big)dz$
1.5.8 $v<t$	$\mathbf{P}_x\Big(\int_0^t \mathbb{I}_{(-\infty,r]}(W_s)ds \in dv, W_t \in dz\Big) =: F_x^{(5)}(t,v,z)dvdz$
	$=\int_0^\infty B_x^{(27)}(v,t-v,y,z)dy\,dvdz$ for $B_x^{(27)}(u,v,y,z)$ see 1.27.6
$x\le r$ $z\le r$	$=\frac{(2r-x-z)\sqrt{v}}{\pi t^2\sqrt{t-v}}\exp\Big(-\frac{(2r-x-z)^2}{2v}\Big)dvdz$ $+\Big(\frac{1}{\sqrt{2\pi}t^{3/2}}-\frac{(2r-x-z)^2}{\sqrt{2\pi}t^{5/2}}\Big)e^{-(2r-x-z)^2/2t}\,\text{Erfc}\Big(\frac{(2r-x-z)\sqrt{t-v}}{\sqrt{2tv}}\Big)dvdz$
$x\le r$ $r\le z$	$=\Big(\frac{(z-r)\sqrt{t-v}}{\pi t^2\sqrt{v}}+\frac{(r-x)\sqrt{v}}{\pi t^2\sqrt{t-v}}\Big)\exp\Big(-\frac{(z-r)^2}{2(t-v)}-\frac{(r-x)^2}{2v}\Big)dvdz$ $+\Big(\frac{1}{\sqrt{2\pi}t^{3/2}}-\frac{(2r-x-z)^2}{\sqrt{2\pi}t^{5/2}}\Big)e^{-(2r-x-z)^2/2t}\,\text{Erfc}\Big(\frac{(z-r)\sqrt{v}}{\sqrt{2t(t-v)}}+\frac{(r-x)\sqrt{t-v}}{\sqrt{2tv}}\Big)dvdz$
$r\le x$ $z\le r$	$=\Big(\frac{(x-r)\sqrt{t-v}}{\pi t^2\sqrt{v}}+\frac{(r-z)\sqrt{v}}{\pi t^2\sqrt{t-v}}\Big)\exp\Big(-\frac{(x-r)^2}{2(t-v)}-\frac{(r-z)^2}{2v}\Big)dvdz$ $+\Big(\frac{1}{\sqrt{2\pi}t^{3/2}}-\frac{(2r-x-z)^2}{\sqrt{2\pi}t^{5/2}}\Big)e^{-(2r-x-z)^2/2t}\,\text{Erfc}\Big(\frac{(x-r)\sqrt{v}}{\sqrt{2t(t-v)}}+\frac{(r-z)\sqrt{t-v}}{\sqrt{2tv}}\Big)dvdz$
$r\le x$ $r\le z$	$=\frac{(x+z-2r)\sqrt{t-v}}{\pi t^2\sqrt{v}}\exp\Big(-\frac{(x+z-2r)^2}{2(t-v)}\Big)dvdz$ $+\Big(\frac{1}{\sqrt{2\pi}t^{3/2}}-\frac{(x+z-2r)^2}{\sqrt{2\pi}t^{5/2}}\Big)e^{-(x+z-2r)^2/2t}\,\text{Erfc}\Big(\frac{(x+z-2r)\sqrt{v}}{\sqrt{2t(t-v)}}\Big)dvdz$

1	W $\qquad\qquad\qquad\qquad$ $\tau \sim \operatorname{Exp}(\lambda)$, independent of W
(1) $r \le x$ $r \le z$	$\mathbf{P}_x\Big(\int_0^t \mathbb{I}_{(-\infty,r]}(W_s)ds = 0, W_t \in dz\Big) = \frac{1}{\sqrt{2\pi t}}\Big(e^{-(z-x)^2/2t} - e^{-(x+z-2r)^2/2t}\Big)dz$
(2) $x \le r$ $z \le r$	$\mathbf{P}_x\Big(\int_0^t \mathbb{I}_{(-\infty,r]}(W_s)ds = t, W_t \in dz\Big) = \frac{1}{\sqrt{2\pi t}}\Big(e^{-(z-x)^2/2t} - e^{-(2r-x-z)^2/2t}\Big)dz$
(3)	$\mathbf{P}_x\Big(\int_0^t \mathbb{I}_{(-\infty,x]}(W_s)ds \in dv \Big\vert W_t = x\Big) = \frac{1}{t}\mathbb{I}_{[0,t]}(v)dv$
1.6.1	$\mathbf{E}_x \exp\Big(-\int_0^\tau \big(p\mathbb{I}_{(-\infty,r)}(W_s) + q\mathbb{I}_{[r,\infty)}(W_s)\big)ds\Big)$
$x \le r$	$= \frac{\lambda}{\lambda+p} - \Big(\frac{\lambda}{\lambda+p} - \frac{\lambda}{\sqrt{(\lambda+p)(\lambda+q)}}\Big)e^{(x-r)\sqrt{2\lambda+2p}}$
$r \le x$	$= \frac{\lambda}{\lambda+q} - \Big(\frac{\lambda}{\lambda+q} - \frac{\lambda}{\sqrt{(\lambda+p)(\lambda+q)}}\Big)e^{(r-x)\sqrt{2\lambda+2q}}$
1.6.2	$\mathbf{P}_x\Big(\int_0^\tau \mathbb{I}_{(-\infty,r)}(W_s)ds \in du,\ \int_0^\tau \mathbb{I}_{[r,\infty)}(W_s)ds \in dv\Big)$
	$= \begin{cases} \frac{\lambda}{\pi\sqrt{uv}}e^{-\lambda(u+v)-(r-x)^2/2u}dudv, & x \le r \\ \frac{\lambda}{\pi\sqrt{uv}}e^{-\lambda(u+v)-(x-r)^2/2v}dudv, & r \le x \end{cases}$
(1) $x \le r$	$\mathbf{P}_x\Big(\int_0^\tau \mathbb{I}_{(-\infty,r)}(W_s)ds \in du, \int_0^\tau \mathbb{I}_{[r,\infty)}(W_s)ds = 0\Big) = \lambda e^{-\lambda u}\operatorname{Erf}\Big(\frac{r-x}{\sqrt{2u}}\Big)du$
(2) $r \le x$	$\mathbf{P}_x\Big(\int_0^\tau \mathbb{I}_{(-\infty,r)}(W_s)ds = 0, \int_0^\tau \mathbb{I}_{[r,\infty)}(W_s)ds \in dv\Big) = \lambda e^{-\lambda v}\operatorname{Erf}\Big(\frac{x-r}{\sqrt{2v}}\Big)dv$
1.6.3	$\mathbf{E}_x \exp\Big(-\int_0^t \big(p\mathbb{I}_{(-\infty,r)}(W_s) + q\mathbb{I}_{[r,\infty)}(W_s)\big)ds\Big)$
$x \le r$	$= e^{-pt}\operatorname{Erf}\Big(\frac{r-x}{\sqrt{2t}}\Big) + \frac{1}{\pi}e^{-qt}\int_0^t \frac{1}{\sqrt{s(t-s)}}e^{-(p-q)s-(r-x)^2/2s}ds$
$r \le x$	$= e^{-qt}\operatorname{Erf}\Big(\frac{x-r}{\sqrt{2t}}\Big) + \frac{1}{\pi}e^{-pt}\int_0^t \frac{1}{\sqrt{s(t-s)}}e^{-(q-p)s-(x-r)^2/2s}ds$

1 W $\tau \sim \operatorname{Exp}(\lambda)$, independent of W

$x=r$	$= e^{-(p+q)t/2} I_0\Big(\frac{\lvert p-q\rvert t}{2}\Big)$
1.6.4	$\mathbf{P}_x\Big(\int_0^t \big(p\mathbb{I}_{(-\infty,r)}(W_s) + q\mathbb{I}_{[r,\infty)}(W_s)\big)ds \in dv\Big)$
$x \le r$	$= \frac{\mathbb{I}_{((q\wedge p)t,(q\vee p)t)}(v)}{\pi\sqrt{\lvert pt-v\rvert\lvert v-qt\rvert}} e^{-(r-x)^2\lvert p-q\rvert/2\lvert v-qt\rvert} dv$
$r \le x$	$= \frac{\mathbb{I}_{((q\wedge p)t,(q\vee p)t)}(v)}{\pi\sqrt{\lvert pt-v\rvert\lvert v-qt\rvert}} e^{-(x-r)^2\lvert p-q\rvert/2\lvert pt-v\rvert} dv$
(1) $x \le r$	$\mathbf{P}_x\Big(\int_0^t \big(p\mathbb{I}_{(-\infty,r)}(W_s) + q\mathbb{I}_{[r,\infty)}(W_s)\big)ds = pt\Big) = 1 - \operatorname{Erfc}\Big(\frac{r-x}{\sqrt{2t}}\Big)$
(2) $r \le x$	$\mathbf{P}_x\Big(\int_0^t \big(p\mathbb{I}_{(-\infty,r)}(W_s) + q\mathbb{I}_{[r,\infty)}(W_s)\big)ds = qt\Big) = 1 - \operatorname{Erfc}\Big(\frac{x-r}{\sqrt{2t}}\Big)$
1.6.5	$\mathbf{E}_x\Big\{\exp\Big(-\int_0^\tau \big(p\mathbb{I}_{(-\infty,r)}(W_s) + q\mathbb{I}_{[r,\infty)}(W_s)\big)ds\Big); W_\tau \in dz\Big\} =: \lambda Q_\lambda^{(6)}(z)dz$
$x \le r$ $z \le r$	$= \frac{\lambda}{\sqrt{2\lambda+2p}} e^{-\lvert z-x\rvert\sqrt{2\lambda+2p}} dz$ $+\lambda\Big(\frac{2}{\sqrt{2\lambda+2p}+\sqrt{2\lambda+2q}} - \frac{1}{\sqrt{2\lambda+2p}}\Big) e^{(z+x-2r)\sqrt{2\lambda+2p}} dz$
$x \le r$ $r \le z$	$= \frac{2\lambda}{\sqrt{2\lambda+2p}+\sqrt{2\lambda+2q}} e^{(x-r)\sqrt{2\lambda+2p}+(r-z)\sqrt{2\lambda+2q}} dz$
$r \le x$ $z \le r$	$= \frac{2\lambda}{\sqrt{2\lambda+2p}+\sqrt{2\lambda+2q}} e^{(r-x)\sqrt{2\lambda+2q}+(z-r)\sqrt{2\lambda+2p}} dz$
$r \le x$ $r \le z$	$= \frac{\lambda}{\sqrt{2\lambda+2q}} e^{-\lvert z-x\rvert\sqrt{2\lambda+2q}} dz$ $+\lambda\Big(\frac{2}{\sqrt{2\lambda+2p}+\sqrt{2\lambda+2q}} - \frac{1}{\sqrt{2\lambda+2q}}\Big) e^{(2r-x-z)\sqrt{2\lambda+2q}} dz$
1.6.6	$\mathbf{P}_x\Big(\int_0^\tau \mathbb{I}_{(-\infty,r)}(W_s)ds \in du, \int_0^\tau \mathbb{I}_{[r,\infty)}(W_s)ds \in dv, W_\tau \in dz\Big)$

1 W $\tau \sim \text{Exp}(\lambda)$, independent of W

$$= \lambda e^{-\lambda(u+v)} F_x^{(4)}(u+v,v,z)\,du\,dv\,dz \qquad \text{for } F_x^{(4)}(t,v,z) \text{ see 1.4.8}$$

(1) $$\mathbf{P}_x\Big(\int_0^\tau \mathbb{I}_{(-\infty,r)}(W_s)ds \in du,\ \int_0^\tau \mathbb{I}_{[r,\infty)}(W_s)ds = 0,\ W_\tau \in dz\Big)$$

$x \le r$, $z \le r$
$$= \lambda e^{-\lambda u}\Big(\frac{1}{\sqrt{2\pi u}} e^{-(z-x)^2/2u} - \frac{1}{\sqrt{2\pi u}} e^{-(2r-x-z)^2/2u}\Big) du\,dz$$

(2) $$\mathbf{P}_x\Big(\int_0^\tau \mathbb{I}_{(-\infty,r)}(W_s)ds = 0,\ \int_0^\tau \mathbb{I}_{[r,\infty)}(W_s)ds \in dv,\ W_\tau \in dz\Big)$$

$r \le x$, $r \le z$
$$= \lambda e^{-\lambda v}\Big(\frac{1}{\sqrt{2\pi v}} e^{-(z-x)^2/2v} - \frac{1}{\sqrt{2\pi v}} e^{-(x+z-2r)^2/2v}\Big) dv\,dz$$

1.6.7 $$\mathbf{E}_x\Big\{\exp\Big(-\int_0^t (p\mathbb{I}_{(-\infty,r)}(W_s) + q\mathbb{I}_{[r,\infty)}(W_s))ds\Big); W_t \in dz\Big\} =: V_x^{(6)}(t,z)dz$$

$x < r$, $z < r$
$$= \frac{e^{-pt}}{\sqrt{2\pi t}}\big(e^{-(z-x)^2/2t} - e^{-(2r-x-z)^2/2t}\big)dz$$
$$+\frac{2r-x-z}{2\pi}\Big(\frac{|e^{-pt}-e^{-qt}|}{|p-q|t^{3/2}}\Big) * \Big(\frac{e^{-pt}}{t^{3/2}} e^{-(2r-x-z)^2/2t}\Big)dz$$

$x < r$, $r < z$
$$= \frac{(z-r)(r-x)}{(2\pi)^{3/2}}\Big(\frac{|e^{-pt}-e^{-qt}|}{|p-q|t^{3/2}}\Big) * \Big(\frac{e^{-qt}}{t^{3/2}} e^{-(z-r)^2/2t}\Big) * \Big(\frac{e^{-pt}}{t^{3/2}} e^{-(r-x)^2/2t}\Big)dz$$

$r < x$, $z < r$
$$= \frac{(x-r)(r-z)}{(2\pi)^{3/2}}\Big(\frac{|e^{-pt}-e^{-qt}|}{|p-q|t^{3/2}}\Big) * \Big(\frac{e^{-qt}}{t^{3/2}} e^{-(x-r)^2/2t}\Big) * \Big(\frac{e^{-pt}}{t^{3/2}} e^{-(r-z)^2/2t}\Big)dz$$

$r < x$, $r < z$
$$= \frac{e^{-qt}}{\sqrt{2\pi t}}\big(e^{-(z-x)^2/2t} - e^{-(x+z-2r)^2/2t}\big)dz$$
$$+\frac{x+z-2r}{2\pi}\Big(\frac{|e^{-pt}-e^{-qt}|}{|p-q|t^{3/2}}\Big) * \Big(\frac{e^{-qt}}{t^{3/2}} e^{-(x+z-2r)^2/2t}\Big)dz$$

$x = r$, $z = r$
$$= \frac{|e^{-pt}-e^{-qt}|}{(2\pi)^{1/2}|p-q|t^{3/2}}dz$$

$x = r$, $r < z$
$$= \frac{z-r}{2\pi}\Big(\frac{|e^{-pt}-e^{-qt}|}{|p-q|t^{3/2}}\Big) * \Big(\frac{e^{-qt}}{t^{3/2}} e^{-(z-r)^2/2t}\Big)dz$$

1 W	$\tau \sim \operatorname{Exp}(\lambda)$, independent of W
$x=r$ $z<r$	$= \frac{r-z}{2\pi}\Big(\frac{\lvert e^{-pt}-e^{-qt}\rvert}{\lvert p-q\rvert t^{3/2}}\Big) * \Big(\frac{e^{-pt}}{t^{3/2}} e^{-(r-z)^2/2t}\Big)$
1.6.8	$\mathbf{P}_x\Big(\int_0^t \big(p\mathbb{I}_{(-\infty,r)}(W_s)+q\mathbb{I}_{[r,\infty)}(W_s)\big)ds \in dv, W_t\in dz\Big)$
	$= \begin{cases} \frac{1}{\lvert p-q\rvert}F_x^{(4)}\Big(t,\frac{v-pt}{q-p},z\Big)dvdz, & pt<v<qt, \quad \text{for } F_x^{(4)}(t,v,z) \text{ see 1.4.8} \\ \frac{1}{\lvert p-q\rvert}F_x^{(5)}\Big(t,\frac{v-qt}{p-q},z\Big)dvdz, & qt<v<pt, \quad \text{for } F_x^{(5)}(t,v,z) \text{ see 1.5.8} \\ 0, & \text{otherwise} \end{cases}$
(1) $x\le r$	$\mathbf{P}_x\Big(\int_0^t \big(p\mathbb{I}_{(-\infty,r)}(W_s)+q\mathbb{I}_{[r,\infty)}(W_s)\big)ds = pt, W_t\in dz\Big)$
$z\le r$	$= \frac{1}{\sqrt{2\pi t}}e^{-(z-x)^2/2t}dz - \frac{1}{\sqrt{2\pi t}}e^{-(2r-x-z)^2/2t}dz$
(2) $r\le x$	$\mathbf{P}_x\Big(\int_0^t \big(p\mathbb{I}_{(-\infty,r)}(W_s)+q\mathbb{I}_{[r,\infty)}(W_s)\big)ds = qt, W_t\in dz\Big)$
$r\le z$	$= \frac{1}{\sqrt{2\pi t}}e^{-(z-x)^2/2t}dz - \frac{1}{\sqrt{2\pi t}}e^{-(x+z-2r)^2/2t}dz$
1.7.1	$\mathbf{E}_x \exp\Big(-\gamma\int_0^\tau \mathbb{I}_{[r,u]}(W_s)ds\Big)$
$x\le r$	$= 1 - \frac{\gamma \operatorname{sh}((u-r)\sqrt{(\lambda+\gamma)/2})e^{(x-r)\sqrt{2\lambda}}}{\sqrt{\lambda+\gamma}(\sqrt{\lambda+\gamma}\operatorname{sh}((u-r)\sqrt{(\lambda+\gamma)/2})+\sqrt{\lambda}\operatorname{ch}((u-r)\sqrt{(\lambda+\gamma)/2}))}$
$r\le x$ $x\le u$	$= \frac{\lambda}{\lambda+\gamma} + \frac{\gamma\sqrt{\lambda}\operatorname{ch}((2x-u-r)\sqrt{(\lambda+\gamma)/2})}{(\lambda+\gamma)(\sqrt{\lambda+\gamma}\operatorname{sh}((u-r)\sqrt{(\lambda+\gamma)/2})+\sqrt{\lambda}\operatorname{ch}((u-r)\sqrt{(\lambda+\gamma)/2}))}$
$u\le x$	$= 1 - \frac{\gamma \operatorname{sh}((u-r)\sqrt{(\lambda+\gamma)/2})e^{(u-x)\sqrt{2\lambda}}}{\sqrt{\lambda+\gamma}(\sqrt{\lambda+\gamma}\operatorname{sh}((u-r)\sqrt{(\lambda+\gamma)/2})+\sqrt{\lambda}\operatorname{ch}((u-r)\sqrt{(\lambda+\gamma)/2}))}$
1.7.4	$\mathbf{P}_x\Big(\int_0^t \mathbb{I}_{[r,u]}(W_s)ds\in dv\Big)$

1 W	$\tau \sim \text{Exp}(\lambda)$, independent of W
$x \le r$	$= 2\int_0^\infty \mathrm{h}_{t-v}(0, y+r-x)\, \mathrm{ec}_v\,\big(0, 0, \frac{u-r}{2}, 0, y\big) dy\, dv$
	$+2\int_0^\infty \mathrm{h}_{t-v}(1, y+r-x)\, \mathrm{esc}_v\,\big(\frac{u-r}{2}, \frac{u-r}{2}, 0, y\big) dy\, dv$
$r \le x$ $x \le u$	$= 2\int_0^\infty \mathrm{h}_{t-v}(0, y)\, \mathrm{ecc}_v\,\big(\frac{u+r-2x}{2}, \frac{u-r}{2}, 0, y\big) dy\, dv$
	$+2\int_0^\infty \mathrm{h}_{t-v}(1, y)\, \mathrm{ecsc}_v\,\big(\frac{u+r-2x}{2}, \frac{u-r}{2}, \frac{u-r}{2}, y\big) dy\, dv$
$u \le x$	$= 2\int_0^\infty \mathrm{h}_{t-v}(0, y+x-u)\, \mathrm{ec}_v\,\big(0, 0, \frac{u-r}{2}, 0, y\big) dy\, dv$
	$+2\int_0^\infty \mathrm{h}_{t-v}(1, y+x-u)\, \mathrm{esc}_v\,\big(\frac{u-r}{2}, \frac{u-r}{2}, 0, y\big) dy\, dv$
(1)	$\mathbf{P}_x\Big(\int_0^t \mathbb{I}_{[r,u]}(W_s) ds = 0\Big) = \begin{cases} 1 - \text{Erfc}\Big(\frac{r-x}{\sqrt{2t}}\Big), & x \le r \\ 0, & r \le x \le u \\ 1 - \text{Erfc}\Big(\frac{x-u}{\sqrt{2t}}\Big), & u \le x \end{cases}$
(2)	$\mathbf{P}_x\Big(\int_0^t \mathbb{I}_{[r,u]}(W_s) ds = t\Big) = 1 - \widetilde{\mathrm{cc}}_t\big(\frac{u+r-2x}{2}, \frac{u-r}{2}\big)$ $\qquad r \le x \le u$
1.8.3	$\mathbf{E}_x \exp\Big(-\gamma\int_0^t W_s ds\Big) = \exp\Big(-\gamma x t + \frac{\gamma^2 t^3}{6}\Big)$
1.8.4	$\mathbf{P}_x\Big(\int_0^t W_s ds \in dy\Big) = \frac{\sqrt{3}}{\sqrt{2\pi t^3}} \exp\Big(-\frac{3(y-xt)^2}{2t^3}\Big) dy$
1.8.5	$\mathbf{E}_0\Big\{\exp\Big(-\gamma\int_0^\tau \lvert W_s\rvert ds\Big);\ W_\tau \in dz\Big\} = -\frac{\lambda\,\text{Ai}(2^{1/3}\gamma^{-2/3}(\lambda+\gamma\lvert z\rvert))}{(2\gamma)^{1/3}\,\text{Ai}'(2^{1/3}\gamma^{-2/3}\lambda)} dz$
1.8.7	$\mathbf{E}_x\Big\{\exp\Big(-\gamma\int_0^t W_s ds\Big); W_t \in dz\Big\} = \frac{1}{\sqrt{2\pi t}} \exp\Big(-\frac{(z-x)^2}{2t} - \frac{\gamma(z+x)t}{2} + \frac{\gamma^2 t^3}{24}\Big) dz$

1 W	$\tau \sim \operatorname{Exp}(\lambda)$, independent of W
(1)	$$\mathbf{E}_0\Big\{\exp\Big(-\gamma\int_0^t |W_s|ds\Big); W_t\in dz\Big\} = -\sum_{k=1}^{\infty}\frac{\operatorname{Ai}(\alpha_k'+(2\gamma)^{1/3}|z|)e^{(\gamma^2/2)^{1/3}t\alpha_k'}}{2^{2/3}\gamma^{-1/3}\alpha_k'\operatorname{Ai}(\alpha_k')}dz$$
1.8.8	$$\mathbf{E}_x\Big(\int_0^t W_s ds\in dy,\ W_t\in dz\Big) = \frac{\sqrt{3}}{\pi t^2}\exp\Big(-\frac{(z-x)^2}{2t}-\frac{3(2y-(z+x)t)^2}{2t^3}\Big)dydz$$
1.9.3	$$\mathbf{E}_x\exp\Big(-\frac{\gamma^2}{2}\int_0^t W_s^2ds\Big) = \frac{1}{\sqrt{\operatorname{ch}(t\gamma)}}\exp\Big(-\frac{x^2\gamma\operatorname{sh}(t\gamma)}{2\operatorname{ch}(t\gamma)}\Big)$$
(1)	$$\mathbf{E}_x\exp\Big(-\beta W_t^2-\frac{\gamma^2}{2}\int_0^t W_s^2ds\Big)$$ $$=\frac{1}{\sqrt{\operatorname{ch}(t\gamma)+2\beta\gamma^{-1}\operatorname{sh}(t\gamma)}}\exp\Big(-\frac{x^2(\gamma\operatorname{sh}(t\gamma)+2\beta\operatorname{ch}(t\gamma))}{2(\operatorname{ch}(t\gamma)+2\beta\gamma^{-1}\operatorname{sh}(t\gamma))}\Big)$$
(2) $0<x$	$$\mathbf{E}_x\Big\{\exp\Big(-\frac{\gamma^2}{2}\int_0^t W_s^2ds\Big);\ 0<\inf_{0\le s\le t}W_s\Big\}$$ $$=\frac{1}{\sqrt{\operatorname{ch}(t\gamma)}}\exp\Big(-\frac{x^2\gamma\operatorname{sh}(t\gamma)}{2\operatorname{ch}(t\gamma)}\Big)\operatorname{Erf}\Big(\frac{x\sqrt{\gamma}}{\sqrt{\operatorname{sh}(2t\gamma)}}\Big)$$
1.9.4	$$\mathbf{P}_x\Big(\int_0^t W_s^2ds\in dy\Big) = \operatorname{ec}_y(0,1/2,t,0,x^2/2)dy$$
(1)	$$\mathbf{P}_0\Big(\int_0^t W_s^2ds<y\Big) = \frac{\sqrt{2}}{\sqrt{\pi}}\sum_{k=0}^{\infty}\frac{(-1)^k\Gamma(k+1/2)}{k!}\operatorname{Erfc}\Big(\frac{(4k+1)t}{2\sqrt{2y}}\Big)$$
1.9.5	$$\mathbf{E}_x\Big\{\exp\Big(-\frac{\gamma^2}{2}\int_0^\tau W_s^2ds\Big);\ W_\tau\in dz\Big\}$$ $$=\begin{cases}\dfrac{\lambda\Gamma(1/2+\lambda/\gamma)}{\sqrt{\pi\gamma}}D_{-1/2-\lambda/\gamma}(-x\sqrt{2\gamma})D_{-1/2-\lambda/\gamma}(z\sqrt{2\gamma})dz, & x\le z\\ \dfrac{\lambda\Gamma(1/2+\lambda/\gamma)}{\sqrt{\pi\gamma}}D_{-1/2-\lambda/\gamma}(x\sqrt{2\gamma})D_{-1/2-\lambda/\gamma}(-z\sqrt{2\gamma})dz, & z\le x\end{cases}$$
1.9.7	$$\mathbf{E}_x\Big\{\exp\Big(-\frac{\gamma^2}{2}\int_0^t W_s^2ds\Big);\ W_t\in dz\Big\}$$ $$=\frac{\sqrt{\gamma}}{\sqrt{2\pi\operatorname{sh}(t\gamma)}}\exp\Big(-\frac{(x^2+z^2)\gamma\operatorname{ch}(t\gamma)-2xz\gamma}{2\operatorname{sh}(t\gamma)}\Big)dz$$

1	W $\qquad\qquad\qquad\qquad$ $\tau \sim \operatorname{Exp}(\lambda)$, independent of W
1.9.8	$\mathbf{P}_x\Big(\int_0^t W_s^2 ds \in dy,\ W_t \in dz\Big) = (2\pi)^{-1/2}\,\mathrm{ee}_y(1/2, t, (x^2+z^2)/2, -xz)dydz$
(1)	$\mathbf{P}_0\Big(\int_0^t W_s^2 ds < y \Big\vert W_t = 0\Big)$
	$= \sum_{k=0}^{\infty} \frac{t\Gamma(k+1/2)\sqrt{4k+1}}{\sqrt{y}\pi^{3/2}k!} e^{-(4k+1)^2t^2/16y} K_{1/4}\Big(\frac{(4k+1)^2t^2}{16y}\Big)$
1.10.2	$\mathbf{P}_x\Big(\int_0^\tau e^{2\beta W_s} ds \in dy\Big)$
$\beta \neq 0$	$= \frac{\lambda\Gamma(\sqrt{\lambda}/\vert\beta\vert\sqrt{2})}{\vert\beta\vert\sqrt{2y}\Gamma(1+\sqrt{2\lambda}/\vert\beta\vert)} \exp\Big(-\beta x - \frac{e^{2\beta x}}{4\beta^2 y}\Big) M_{1/2,\sqrt{\lambda}/\vert\beta\vert\sqrt{2}}\Big(\frac{e^{2\beta x}}{2\beta^2 y}\Big)dy$
1.10.4 $\beta \neq 0$	$\mathbf{P}_x\Big(\int_0^t e^{2\beta W_s} ds \in dy\Big) = \frac{\sqrt{2}\vert\beta\vert}{\sqrt{y}} \exp\Big(-\beta x - \frac{e^{2\beta x}}{4\beta^2 y}\Big)\ \mathrm{m}_{2\beta^2 t}\Big(-\frac{1}{2}, \frac{e^{2\beta x}}{4\beta^2 y}\Big)dy$
1.10.5	$\mathbf{E}_x\Big\{\exp\Big(-\gamma \int_0^\tau e^{2\beta W_s} ds\Big);\ W_\tau \in dz\Big\}$
$0 < \beta$	$= \begin{cases} \frac{2\lambda}{\beta} I_{\sqrt{2\lambda}/\beta}\Big(\frac{\sqrt{2\gamma}}{\beta}e^{\beta x}\Big) K_{\sqrt{2\lambda}/\beta}\Big(\frac{\sqrt{2\gamma}}{\beta}e^{\beta z}\Big)dz, & x \le z \\ \frac{2\lambda}{\beta} K_{\sqrt{2\lambda}/\beta}\Big(\frac{\sqrt{2\gamma}}{\beta}e^{\beta x}\Big) I_{\sqrt{2\lambda}/\beta}\Big(\frac{\sqrt{2\gamma}}{\beta}e^{\beta z}\Big)dz, & z \le x \end{cases}$
$\beta < 0$	$= \begin{cases} \frac{2\lambda}{\vert\beta\vert} K_{\sqrt{2\lambda}/\vert\beta\vert}\Big(\frac{\sqrt{2\gamma}}{\vert\beta\vert}e^{\beta x}\Big) I_{\sqrt{2\lambda}/\vert\beta\vert}\Big(\frac{\sqrt{2\gamma}}{\vert\beta\vert}e^{\beta z}\Big)dz, & x \le z \\ \frac{2\lambda}{\vert\beta\vert} I_{\sqrt{2\lambda}/\vert\beta\vert}\Big(\frac{\sqrt{2\gamma}}{\vert\beta\vert}e^{\beta x}\Big) K_{\sqrt{2\lambda}/\vert\beta\vert}\Big(\frac{\sqrt{2\gamma}}{\vert\beta\vert}e^{\beta z}\Big)dz, & z \le x \end{cases}$
1.10.6 $\beta \neq 0$	$\mathbf{P}_x\Big(\int_0^\tau e^{2\beta W_s} ds \in dy, W_\tau \in dz\Big) = \frac{\lambda}{\vert\beta\vert y} \exp\Big(-\frac{e^{2\beta x}+e^{2\beta z}}{2\beta^2 y}\Big) I_{\sqrt{2\lambda}/\vert\beta\vert}\Big(\frac{e^{\beta(x+z)}}{\beta^2 y}\Big)dydz$
1.10.7 $\beta \neq 0$	$\mathbf{E}_x\Big\{\exp\Big(-\gamma \int_0^t e^{2\beta W_s} ds\Big);\ W_t \in dz\Big\} = \vert\beta\vert\ \mathrm{ki}_{\beta^2 t/2}\Big(\frac{\sqrt{2\gamma}}{\vert\beta\vert}e^{\beta x}, \frac{\sqrt{2\gamma}}{\vert\beta\vert}e^{\beta z}\Big)dz$
1.10.8 $\beta \neq 0$	$\mathbf{P}_x\Big(\int_0^t e^{2\beta W_s} ds \in dy, W_t \in dz\Big) = \frac{\vert\beta\vert}{2y} \exp\Big(-\frac{e^{2\beta x}+e^{2\beta z}}{2\beta^2 y}\Big)\ \mathrm{i}_{\beta^2 t/2}\Big(\frac{e^{\beta(x+z)}}{\beta^2 y}\Big)dydz$

1 W $\tau \sim \operatorname{Exp}(\lambda)$, independent of W

1.11.2 $$\mathbf{P}_x\big(\sup_{y\in\mathbb{R}} \ell(\tau,y) \ge h\big) = \frac{h\sqrt{2\lambda}I_1(h\sqrt{\lambda/2})}{\operatorname{sh}^2(h\sqrt{\lambda/2})I_0(h\sqrt{\lambda/2})}$$

1.11.4 $$\mathbf{P}\big(\sup_{y\in\mathbb{R}} \ell(t,y) < h\big) = 4\sum_{k=1}^{\infty} \frac{1}{\sin^2 j_{0,k}} e^{-2j_{0,k}^2 t/h^2}$$
$$+4\sum_{k=1}^{\infty}\Big(\frac{4t\pi k J_1(\pi k)}{h^2 J_0(\pi k)} + \frac{J_1(\pi k)}{\pi k J_0(\pi k)} - \frac{J_1^2(\pi k)}{J_0^2(\pi k)} - 1\Big) e^{-2\pi^2 k^2 t/h^2}$$

1.12.1 $$\mathbf{E}_x e^{-\gamma \check{H}(\tau)} = \mathbf{E}_x e^{-\gamma \hat{H}(\tau)} = \frac{\sqrt{\lambda}}{\sqrt{\lambda+\gamma}}$$

1.12.2 $$\mathbf{P}_x\big(\check{H}(\tau)\in dv\big) = \mathbf{P}_x\big(\hat{H}(\tau)\in dv\big) = \frac{\sqrt{\lambda}e^{-\lambda v}}{\sqrt{\pi v}}dv$$

1.12.3 $$\mathbf{E}_x e^{-\gamma \check{H}(t)} = \mathbf{E}_x e^{-\gamma \hat{H}(t)} = e^{-\gamma t/2} I_0\Big(\frac{\gamma t}{2}\Big)$$

1.12.4 $$\mathbf{P}_x\big(\check{H}(t) < v\big) = \mathbf{P}_x\big(\hat{H}(t) < v\big) = \frac{2}{\pi}\operatorname{arctg}\Big(\frac{\sqrt{v}}{\sqrt{t-v}}\Big), \quad 0\le v\le t$$

1.12.5 $$\mathbf{E}_x\{e^{-\gamma \check{H}(\tau)}; W_\tau \in dz\} = \begin{cases} \dfrac{\lambda\sqrt{2}}{\sqrt{\lambda+\gamma}+\sqrt{\lambda}} e^{(x-z)\sqrt{2\lambda+2\gamma}}dz, & x\le z\\[2ex] \dfrac{\lambda\sqrt{2}}{\sqrt{\lambda+\gamma}+\sqrt{\lambda}} e^{(z-x)\sqrt{2\lambda}}dz, & z\le x \end{cases}$$

(1) $$\mathbf{E}_x\{e^{-\gamma \hat{H}(\tau)}; W_\tau \in dz\} = \begin{cases} \dfrac{\lambda\sqrt{2}}{\sqrt{\lambda+\gamma}+\sqrt{\lambda}} e^{(x-z)\sqrt{2\lambda}}dz, & x\le z\\[2ex] \dfrac{\lambda\sqrt{2}}{\sqrt{\lambda+\gamma}+\sqrt{\lambda}} e^{(z-x)\sqrt{2\lambda+2\gamma}}dz, & z\le x \end{cases}$$

1.12.6 $$\mathbf{P}_x\big(\check{H}(\tau)\in dv, W_\tau\in dz\big)$$

$x\le z$ $$= \Big(\frac{\lambda\sqrt{2}}{\sqrt{\pi v}} e^{-\lambda v-(z-x)^2/2v} - \lambda^{3/2}\sqrt{2}e^{(z-x)\sqrt{2\lambda}}\operatorname{Erfc}\Big(\frac{z-x}{\sqrt{2v}}+\sqrt{\lambda v}\Big)\Big)dvdz$$

$z\le x$ $$= \Big(\frac{\lambda\sqrt{2}}{\sqrt{\pi v}} e^{-\lambda v-(x-z)\sqrt{2\lambda}} - \lambda^{3/2}\sqrt{2}e^{-(x-z)\sqrt{2\lambda}}\operatorname{Erfc}\big(\sqrt{\lambda v}\big)\Big)dvdz$$

(1) $$\mathbf{P}_x\big(\hat{H}(\tau)\in dv, W_\tau\in dz\big)$$

1	W — $\tau \sim \operatorname{Exp}(\lambda)$, independent of W
$x \le z$	$= \Big(\frac{\lambda\sqrt{2}}{\sqrt{\pi v}} e^{-\lambda v-(z-x)\sqrt{2\lambda}} - \lambda^{3/2}\sqrt{2}e^{-(z-x)\sqrt{2\lambda}}\operatorname{Erfc}\big(\sqrt{\lambda v}\big)\Big)dvdz$
$z \le x$	$= \Big(\frac{\lambda\sqrt{2}}{\sqrt{\pi v}} e^{-\lambda v-(x-z)^2/2v} - \lambda^{3/2}\sqrt{2}e^{(x-z)\sqrt{2\lambda}}\operatorname{Erfc}\Big(\frac{x-z}{\sqrt{2v}} + \sqrt{\lambda v}\Big)\Big)dvdz$
1.12.7	$\mathbf{E}_x\big\{e^{-\gamma\check{H}(t)}; W_t \in dz\big\}$
$x \le z$	$= \Big(\frac{1-e^{-\gamma t}}{\sqrt{2\pi}\gamma t^{3/2}}\Big) * \Big(\frac{z-x}{\sqrt{2\pi}t^{3/2}} e^{-\gamma t-(z-x)^2/2t}\Big)dz$
$z \le x$	$= \Big(\frac{1-e^{-\gamma t}}{\sqrt{2\pi}\gamma t^{3/2}}\Big) * \Big(\frac{x-z}{\sqrt{2\pi}t^{3/2}} e^{-(x-z)^2/2t}\Big)dz$
(1)	$\mathbf{E}_x\big\{e^{-\gamma\hat{H}(t)}; W_t \in dz\big\}$
$x \le z$	$= \Big(\frac{1-e^{-\gamma t}}{\sqrt{2\pi}\gamma t^{3/2}}\Big) * \Big(\frac{z-x}{\sqrt{2\pi}t^{3/2}} e^{-(z-x)^2/2t}\Big)dz$
$z \le x$	$= \Big(\frac{1-e^{-\gamma t}}{\sqrt{2\pi}\gamma t^{3/2}}\Big) * \Big(\frac{x-z}{\sqrt{2\pi}t^{3/2}} e^{-\gamma t-(x-z)^2/2t}\Big)dz$
1.12.8 $v < t$	$\mathbf{P}_x\big(\check{H}(t)\in dv, W_t\in dz\big) = \int_{x\vee z}^{\infty} \frac{(y-x)(y-z)}{\pi(v(t-v))^{3/2}} \exp\Big(-\frac{(y-x)^2}{2v} - \frac{(y-z)^2}{2(t-v)}\Big)dy\, dvdz$
(1) $v < t$	$\mathbf{P}_x\big(\hat{H}(t)\in dv, W_t\in dz\big) = \int_{-\infty}^{x\wedge z} \frac{(x-y)(z-y)}{\pi(v(t-v))^{3/2}} \exp\Big(-\frac{(x-y)^2}{2v} - \frac{(z-y)^2}{2(t-v)}\Big)dy\, dvdz$
1.13.1 $x < y$	$\mathbf{E}_x\big\{e^{-\gamma\check{H}(\tau)}; \sup_{0\le s\le\tau} W_s \in dy\big\} = \sqrt{2\lambda}e^{-(y-x)\sqrt{2\lambda+2\gamma}}dy$
1.13.2 $x < y$	$\mathbf{P}_x\big(\check{H}(\tau) \in dv, \sup_{0\le s\le\tau} W_s \in dy\big) = \frac{\sqrt{2}\lambda(y-x)}{\sqrt{2\pi}v^{3/2}} \exp\Big(-\lambda v - \frac{(y-x)^2}{2v}\Big)dvdy$
1.13.3 $x < y$	$\mathbf{E}_x\big\{e^{-\gamma\check{H}(t)}; \sup_{0\le s\le t} W_s \in dy\big\} = \Big(\frac{y-x}{\pi t^{3/2}} \exp\Big(-\gamma t - \frac{(y-x)^2}{2t}\Big)\Big) * \frac{1}{\sqrt{t}}dy$
1.13.4 $x < y$	$\mathbf{P}_x\big(\check{H}(t) \in dv, \sup_{0\le s\le t} W_s \in dy\big) = \frac{\mathbb{I}_{[0,t]}(v)(y-x)}{\pi v^{3/2}(t-v)^{1/2}} \exp\Big(-\frac{(y-x)^2}{2v}\Big)dvdy$

1	W $\qquad\qquad\qquad\qquad$ $\tau \sim \text{Exp}(\lambda)$, independent of W
1.13.5 $x<y$ $z<y$	$\mathbf{E}_x\{e^{-\gamma\check{H}(\tau)}; \sup_{0\le s\le\tau} W_s \in dy, W_\tau \in dz\} = 2\lambda e^{(x-y)\sqrt{2\lambda+2\gamma}} e^{(z-y)\sqrt{2\lambda}} dydz$
1.13.6 $x<y$ $z<y$	$\mathbf{P}_x\big(\check{H}(\tau) \in dv, \sup_{0\le s\le\tau} W_s \in dy, W_\tau \in dz\big)$ $= \frac{2\lambda(y-x)}{\sqrt{2\pi}v^{3/2}} \exp\Big(-\lambda v - \frac{(y-x)^2}{2v}\Big) e^{-(y-z)\sqrt{2\lambda}} dvdydz$
1.13.7 $x<y$ $z<y$	$\mathbf{E}_x\{e^{-\gamma\check{H}(t)}; \sup_{0\le s\le t} W_s \in dy,\ W_t \in dz\}$ $= \frac{(y-x)(y-z)}{\pi}\Big(\frac{1}{t^{3/2}} \exp\Big(-\gamma t - \frac{(y-x)^2}{2t}\Big)\Big) * \Big(\frac{1}{t^{3/2}} \exp\Big(-\frac{(y-z)^2}{2t}\Big)\Big) dydz$
1.13.8 $v<t$	$\mathbf{P}_x\big(\check{H}(t) \in dv, \sup_{0\le s\le t} W_s \in dy, W_t \in dz\big)$ $= \frac{(y-x)(y-z)}{\pi(v(t-v))^{3/2}} \exp\Big(-\frac{(y-x)^2}{2v} - \frac{(y-z)^2}{2(t-v)}\Big) dy\, dvdz, \qquad x\vee z<y$
1.14.1 $y<x$	$\mathbf{E}_x\{e^{-\gamma\hat{H}(\tau)}; \inf_{0\le s\le\tau} W_s \in dy\} = \sqrt{2\lambda} e^{-(x-y)\sqrt{2\lambda+2\gamma}} dy$
1.14.2 $y<x$	$\mathbf{P}_x\big(\hat{H}(\tau) \in dv, \inf_{0\le s\le\tau} W_s \in dy\big) = \frac{\sqrt{2\lambda}(x-y)}{\sqrt{2\pi}v^{3/2}} \exp\Big(-\lambda v - \frac{(x-y)^2}{2v}\Big) dvdy$
1.14.3 $y<x$	$\mathbf{E}_x\{e^{-\gamma\hat{H}(t)}; \inf_{0\le s\le t} W_s \in dy\} = \Big(\frac{x-y}{\pi t^{3/2}} \exp\Big(-\gamma t - \frac{(x-y)^2}{2t}\Big)\Big) * \frac{1}{\sqrt{t}} dy$
1.14.4 $y<x$	$\mathbf{P}_x\big(\hat{H}(t) \in dv, \inf_{0\le s\le t} W_s \in dy\big) = \frac{\mathbb{I}_{[0,t]}(v)(x-y)}{\pi v^{3/2}(t-v)^{1/2}} \exp\Big(-\frac{(x-y)^2}{2v}\Big) dvdy$
1.14.5 $y<x$ $y<z$	$\mathbf{E}_x\{e^{-\gamma\hat{H}(\tau)}; \inf_{0\le s\le\tau} W_s \in dy, W_\tau \in dz\} = 2\lambda e^{(y-x)\sqrt{2\lambda+2\gamma}} e^{(y-z)\sqrt{2\lambda}} dydz$

1	W $\tau \sim \mathrm{Exp}(\lambda)$, independent of W
1.14.6 $y < x$ $y < z$	$\mathbf{P}_x\big(\hat{H}(\tau) \in dv, \inf_{0\le s\le\tau} W_s \in dy, W_\tau \in dz\big)$ $= \frac{2\lambda(x-y)}{\sqrt{2\pi}v^{3/2}} \exp\Big(-\lambda v - \frac{(x-y)^2}{2v}\Big) e^{-(z-y)\sqrt{2\lambda}} dv dy dz$
1.14.7 $y < x$ $y < z$	$\mathbf{E}_x\{e^{-\gamma\hat{H}(t)}; \inf_{0\le s\le t} W_s \in dy,\ W_t \in dz\}$ $= \frac{(x-y)(z-y)}{\pi}\Big(\frac{1}{t^{3/2}}\exp\Big(-\gamma t - \frac{(x-y)^2}{2t}\Big)\Big) * \Big(\frac{1}{t^{3/2}}\exp\Big(-\frac{(z-y)^2}{2t}\Big)\Big) dy dz$
1.14.8 $v < t$	$\mathbf{P}_x\big(\hat{H}(t) \in dv, \inf_{0\le s\le t} W_s \in dy, W_t \in dz\big)$ $= \frac{(x-y)(z-y)}{\pi(v(t-v))^{3/2}} \exp\Big(-\frac{(x-y)^2}{2v} - \frac{(z-y)^2}{2(t-v)}\Big) dy\, dv dz, \qquad y < x \wedge z$
1.15.2	$\mathbf{P}_x\big(a < \inf_{0\le s\le\tau} W_s, \sup_{0\le s\le\tau} W_s < b\big) = 1 - \frac{\mathrm{ch}((b+a-2x)\sqrt{\lambda/2})}{\mathrm{ch}((b-a)\sqrt{\lambda/2})}$
(1)	$\mathbf{P}_x\big(\sup_{0\le s\le\tau} W_s - \inf_{0\le s\le\tau} W_s < y\big) = \frac{\mathrm{sh}^2(y\sqrt{\lambda/2})}{\mathrm{ch}^2(y\sqrt{\lambda/2})}$
1.15.4	$\mathbf{P}_x\big(a < \inf_{0\le s\le t} W_s, \sup_{0\le s\le t} W_s < b\big)$ $= \frac{1}{\sqrt{2\pi t}} \sum_{k=-\infty}^{\infty} \int_a^b \big(e^{-(z-x+2k(b-a))^2/2t} - e^{-(z+x-2a+2k(b-a))^2/2t}\big) dz$
(1)	$\mathbf{P}_x\big(\sup_{0\le s\le t} W_s - \inf_{0\le s\le t} W_s < y\big) = 1 + 4\sum_{k=1}^{\infty}(-1)^k k\, \mathrm{Erfc}\Big(\frac{ky}{\sqrt{2t}}\Big)$
1.15.6	$\mathbf{P}_x\big(a < \inf_{0\le s\le\tau} W_s, \sup_{0\le s\le\tau} W_s < b, W_\tau \in dz\big)$ $= \frac{\sqrt{\lambda}(\mathrm{ch}((b-a-\lvert z-x\rvert)\sqrt{2\lambda}) - \mathrm{ch}((b+a-z-x)\sqrt{2\lambda}))}{\sqrt{2}\,\mathrm{sh}((b-a)\sqrt{2\lambda})} dz, \quad a < x \wedge z,\ x \vee z < b$
(1)	$\mathbf{P}_x\big(\sup_{0\le s\le\tau} W_s - \inf_{0\le s\le\tau} W_s < y, W_\tau \in dz\big)$ for $\lvert z-x\rvert < y$

1	W	$\tau \sim \text{Exp}(\lambda)$, independent of W

$$= \frac{\lambda}{\operatorname{sh}^2(y\sqrt{2\lambda})}\Big(\frac{\operatorname{sh}\big((y-|z-x|)\sqrt{2\lambda}\big)\operatorname{ch}(y\sqrt{2\lambda})}{\sqrt{2\lambda}} - (y-|z-x|)\operatorname{ch}\big(|z-x|\sqrt{2\lambda}\big)\Big)dz$$

1.15.8 $\mathbf{P}_x\big(a < \inf_{0\le s\le t} W_s,\ \sup_{0\le s\le t} W_s < b\,, W_t \in dz\big)$

$$= \frac{1}{\sqrt{2\pi t}}\sum_{k=-\infty}^{\infty}\big(e^{-(z-x+2k(b-a))^2/2t} - e^{-(z+x-2a+2k(b-a))^2/2t}\big)dz, \quad a < x\wedge z,\ x\vee z < b$$

(1) $\mathbf{P}_x\big(\sup_{0\le s\le t}|W_s| \ge b, W_t \in dz\big)$

$$= \frac{1}{\sqrt{2\pi t}}\sum_{k=-\infty}^{\infty}\big(e^{-(z-x+4kb)^2/2t} - e^{-(z+x+2b+4kb)^2/2t}\big)dz, \qquad -b < x\wedge z,\ x\vee z < b$$

(2) $\mathbf{P}_x\big(\sup_{0\le s\le t} W_s - \inf_{0\le s\le t} W_s < y, W_t \in dz\big)$ for $|z-x| < y$

$$= \frac{1}{\sqrt{2\pi t}}\sum_{k=-\infty}^{\infty}\Big(2k+1-\frac{2k}{t}(y-|z-x|)(|z-x|+2ky)\Big)e^{-(|z-x|+2ky)^2/2t}dz$$

1.16.1, $r < b$: $\mathbf{E}_x\big\{e^{-\gamma\ell(\tau,r)};\ \sup_{0\le s\le\tau} W_s < b\big\} = 1 - e^{-(b-x)\sqrt{2\lambda}}$

$$-\frac{\gamma(1-e^{-(b-r)\sqrt{2\lambda}})}{\sqrt{2\lambda}+\gamma(1-e^{-2(b-r)\sqrt{2\lambda}})}\big(e^{-|x-r|\sqrt{2\lambda}} - e^{-(2b-r-x)\sqrt{2\lambda}}\big)$$

1.16.2, $r < b$: $\mathbf{P}_x\big(\ell(\tau,r) \in dy,\ \sup_{0\le s\le\tau} W_s < b\big)$

$0 < y$:

$$= \frac{\sqrt{2\lambda}(e^{-|x-r|\sqrt{2\lambda}} - e^{-(2b-r-x)\sqrt{2\lambda}})}{(1+e^{-(b-r)\sqrt{2\lambda}})(1-e^{-2(b-r)\sqrt{2\lambda}})}\exp\Big(-\frac{y\sqrt{2\lambda}}{1-e^{-2(b-r)\sqrt{2\lambda}}}\Big)dy$$

(1) $\mathbf{P}_x\big(\ell(\tau,r) = 0,\ \sup_{0\le s\le\tau} W_s < b\big) = 1 - e^{-(b-x)\sqrt{2\lambda}} - \dfrac{e^{-|x-r|\sqrt{2\lambda}} - e^{-(2b-r-x)\sqrt{2\lambda}}}{1+e^{-(b-r)\sqrt{2\lambda}}}$

1 W $\tau \sim \text{Exp}(\lambda)$, independent of W

1.16.4 $\mathbf{P}_x\Big(\frac{1}{2}\ell(t,r) \in dy,\ \sup_{0\le s\le t} W_s < b\Big)$ $\qquad r \vee x \le b$

$0 < y$

$$= (\mathrm{es}_t(-1,2,b-r,y+|x-r|-2b+2r,y) + \mathrm{es}_t(-1,2,b-r,y+b-x,y)$$
$$- \mathrm{es}_t(-1,2,b-r,y+|x-r|-b+r,y) - \mathrm{es}_t(-1,2,b-r,y+r-x,y))dy$$

(1) $\mathbf{P}_x\big(\ell(t,r) = 0,\ \sup_{0\le s\le t} W_s < b\big)$ $\qquad r \le x \le b$

$$= \frac{1}{\sqrt{2\pi t}} \sum_{k=-\infty}^{\infty} \int_r^b \big(e^{-(z-x+2k(b-r))^2/2t} - e^{-(z+x-2r+2k(b-r))^2/2t}\big)dz$$

1.16.5 $r < b$ $z < b$

$$\mathbf{E}_x\{e^{-\gamma\ell(\tau,r)};\ \sup_{0\le s\le \tau} W_s < b, W_\tau \in dz\} = \frac{\sqrt{\lambda}}{\sqrt{2}}\Big\{e^{-|z-x|\sqrt{2\lambda}} - e^{(x+z-2b)\sqrt{2\lambda}}$$
$$- \frac{\gamma(e^{-|x-r|\sqrt{2\lambda}} - e^{-(2b-x-r)\sqrt{2\lambda}})}{\sqrt{2\lambda} + \gamma(1 - e^{-2(b-r)\sqrt{2\lambda}})}\big(e^{-|z-r|\sqrt{2\lambda}} - e^{-(2b-z-r)\sqrt{2\lambda}}\big)\Big\}dz$$

1.16.6 $r < b$ $z < b$ $0 < y$

$$\mathbf{P}_x\big(\ell(\tau,r) \in dy,\ \sup_{0\le s\le \tau} W_s < b, W_\tau \in dz\big) = \exp\Big(-\frac{y\sqrt{2\lambda}}{1 - e^{-2(b-r)\sqrt{2\lambda}}}\Big)$$
$$\times \frac{\lambda(e^{-|x-r|\sqrt{2\lambda}} - e^{-(2b-r-x)\sqrt{2\lambda}})(e^{-|z-r|\sqrt{2\lambda}} - e^{-(2b-z-r)\sqrt{2\lambda}})}{(1 - e^{-2(b-r)\sqrt{2\lambda}})^2}dydz$$

(1)

$$\mathbf{P}_x\big(\ell(\tau,r) = 0,\ \sup_{0\le s\le \tau} W_s < b, W_\tau \in dz\big) = \frac{\sqrt{\lambda}}{\sqrt{2}}\Big\{e^{-|z-x|\sqrt{2\lambda}} - e^{(x+z-2b)\sqrt{2\lambda}}$$
$$- \frac{e^{-|x-r|\sqrt{2\lambda}} - e^{-(2b-x-r)\sqrt{2\lambda}}}{1 - e^{-2(b-r)\sqrt{2\lambda}}}\big(e^{-|z-r|\sqrt{2\lambda}} - e^{-(2b-z-r)\sqrt{2\lambda}}\big)\Big\}dz$$

1.16.8 $\mathbf{P}_x\Big(\frac{1}{2}\ell(t,r) \in dy,\ \sup_{0\le s\le t} W_s < b, W_t \in dz\Big) =: F_x^{(16)}(t,y,z)dydz$

for $r \vee x \vee z < b$

$0 < y$

$$= \frac{1}{2}(\mathrm{es}_t(0,2,b-r,y+|x-r|+|z-r|-2b+2r,y)$$
$$+ \mathrm{es}_t(0,2,b-r,y+2b-x-z,y) - \mathrm{es}_t(0,2,b-r,y+|x-r|+r-z,y)$$
$$- \mathrm{es}_t(0,2,b-r,y+|z-r|+r-x,y))dydz$$

1 W $\tau \sim \operatorname{Exp}(\lambda)$, independent of W

(1) $\mathbf{P}_x\big(\ell(t,r)=0,\ \sup_{0\le s\le t} W_s<b, W_t\in dz\big)$ $r<x\wedge z,\ x\vee z<b$

$$=\frac{1}{\sqrt{2\pi t}}\sum_{k=-\infty}^{\infty}\big(e^{-(z-x+2k(b-r))^2/2t}-e^{-(z+x-2r+2k(b-r))^2/2t}\big)dz$$

1.17.1 $a<r$

$$\mathbf{E}_x\big\{e^{-\gamma\ell(\tau,r)};a<\inf_{0\le s\le\tau}W_s\big\}=1-e^{-(x-a)\sqrt{2\lambda}}$$

$$-\frac{\gamma(1-e^{-(r-a)\sqrt{2\lambda}})}{\sqrt{2\lambda}+\gamma(1-e^{-2(r-a)\sqrt{2\lambda}})}\big(e^{-|x-r|\sqrt{2\lambda}}-e^{-(x+r-2a)\sqrt{2\lambda}}\big)$$

1.17.2 $a<r$

$$\mathbf{P}_x\big(\ell(\tau,r)\in dy, a<\inf_{0\le s\le\tau}W_s\big)$$

$0<y$

$$=\frac{\sqrt{2\lambda}(e^{-|x-r|\sqrt{2\lambda}}-e^{-(x+r-2a)\sqrt{2\lambda}})}{(1+e^{-(r-a)\sqrt{2\lambda}})(1-e^{-2(r-a)\sqrt{2\lambda}})}\exp\Big(-\frac{y\sqrt{2\lambda}}{1-e^{-2(r-a)\sqrt{2\lambda}}}\Big)dy$$

(1)

$$\mathbf{P}_x\big(\ell(\tau,r)=0,a<\inf_{0\le s\le\tau}W_s\big)=1-e^{-(x-a)\sqrt{2\lambda}}-\frac{e^{-|x-r|\sqrt{2\lambda}}-e^{-(x+r-2a)\sqrt{2\lambda}}}{1+e^{-(r-a)\sqrt{2\lambda}}}$$

1.17.4 $\mathbf{P}_x\Big(\frac{1}{2}\ell(t,r)\in dy, a<\inf_{0\le s\le t}W_s\Big)$ $a<x\wedge z\wedge r$

$0<y$

$$=(\mathrm{es}_t(-1,2,r-a,y+|x-r|-2r+2a,y)+\mathrm{es}_t(-1,2,r-a,y+x-a,y)$$
$$-\mathrm{es}_t(-1,2,r-a,y+|x-r|-r+a,y)-\mathrm{es}_t(-1,2,r-a,y+x-r,y))dy$$

(1) $\mathbf{P}_x\big(\ell(t,r)=0,a<\inf_{0\le s\le t}W_s\big)$ $a\le x\le r$

$$=\frac{1}{\sqrt{2\pi t}}\sum_{k=-\infty}^{\infty}\int_a^r\big(e^{-(z-x+2k(r-a))^2/2t}-e^{-(z+x-2a+2k(r-a))^2/2t}\big)dz$$

1.17.5 $a<r$ $a<z$

$$\mathbf{E}_x\big\{e^{-\gamma\ell(\tau,r)};a<\inf_{0\le s\le\tau}W_s,W_\tau\in dz\big)=\frac{\sqrt{\lambda}}{\sqrt{2}}\Big\{e^{-|z-x|\sqrt{2\lambda}}-e^{(2a-x-z)\sqrt{2\lambda}}$$

$$-\frac{\gamma(e^{-|x-r|\sqrt{2\lambda}}-e^{-(x+r-2a)\sqrt{2\lambda}})}{\sqrt{2\lambda}+\gamma(1-e^{-2(r-a)\sqrt{2\lambda}})}\big(e^{-|z-r|\sqrt{2\lambda}}-e^{-(z+r-2a)\sqrt{2\lambda}}\big)\Big\}dz$$

1 W $\tau \sim \text{Exp}(\lambda)$, independent of W

1.17.6 ($a < r$, $a < z$, $0 < y$)

$$\mathbf{P}_x\big(\ell(\tau,r) \in dy, a < \inf_{0\le s\le\tau} W_s, W_\tau \in dz\big) = \exp\Big(-\frac{y\sqrt{2\lambda}}{1-e^{-2(r-a)\sqrt{2\lambda}}}\Big)$$

$$\times \frac{\lambda(e^{-|x-r|\sqrt{2\lambda}} - e^{-(x+r-2a)\sqrt{2\lambda}})(e^{-|z-r|\sqrt{2\lambda}} - e^{-(z+r-2a)\sqrt{2\lambda}})}{(1-e^{-2(r-a)\sqrt{2\lambda}})^2} dydz$$

(1)

$$\mathbf{P}_x\big(\ell(\tau,r) = 0, a < \inf_{0\le s\le\tau} W_s, W_\tau \in dz\big) = \frac{\sqrt{\lambda}}{\sqrt{2}}\Big\{e^{-|z-x|\sqrt{2\lambda}} - e^{(2a-x-z)\sqrt{2\lambda}}$$

$$-\frac{e^{-|x-r|\sqrt{2\lambda}} - e^{-(x+r-2a)\sqrt{2\lambda}}}{1-e^{-2(r-a)\sqrt{2\lambda}}}\big(e^{-|z-r|\sqrt{2\lambda}} - e^{-(z+r-2a)\sqrt{2\lambda}}\big)\Big\}dz$$

1.17.8

$$\mathbf{P}_x\Big(\frac{1}{2}\ell(t,r) \in dy, a < \inf_{0\le s\le t} W_s, W_t \in dz\Big) =: F^{(17)}_{a,x}(t,y,z)dydz$$

for $a < x \wedge z \wedge r$

$0 < y$

$$= \frac{1}{2}(\text{es}_t(0,2,r-a,y+|x-r|+|z-r|-2r+2a,y)$$

$$+\text{es}_t(0,2,r-a,y+x+z-2a,y) - \text{es}_t(0,2,r-a,y+|x-r|+z-r,y)$$

$$-\text{es}_t(0,2,r-a,y+|z-r|+x-r,y))dydz$$

(1) $a < x \wedge z,\ x \vee z < r$

$$\mathbf{P}_x\big(\ell(t,r) = 0, a < \inf_{0\le s\le t} W_s, W_t \in dz\big)$$

$$= \frac{1}{\sqrt{2\pi t}} \sum_{k=-\infty}^{\infty} \big(e^{-(z-x+2k(r-a))^2/2t} - e^{-(z+x-2a+2k(r-a))^2/2t}\big)dz$$

1.18.1

$$\mathbf{E}_x e^{-\gamma\ell(\tau,r)-\eta\ell(\tau,u)} = 1$$

$$-\frac{\gamma(\sqrt{2\lambda}+\eta(1-e^{-|u-r|\sqrt{2\lambda}}))e^{-|r-x|\sqrt{2\lambda}} + \eta(\sqrt{2\lambda}+\gamma(1-e^{-|u-r|\sqrt{2\lambda}}))e^{-|u-x|\sqrt{2\lambda}}}{(\sqrt{2\lambda}+\gamma)(\sqrt{2\lambda}+\eta) - \gamma\eta e^{-2|u-r|\sqrt{2\lambda}}}$$

1.18.5

$$\mathbf{E}_x\big\{e^{-\gamma\ell(\tau,r)-\eta\ell(\tau,u)} \mid W_\tau = z\big\}$$

$$= 1 - \frac{\gamma((\sqrt{2\lambda}+\eta)e^{-|r-x|\sqrt{2\lambda}} - \eta e^{-(|u-x|+|u-r|)\sqrt{2\lambda}})e^{-(|z-r|-|z-x|)\sqrt{2\lambda}}}{(\sqrt{2\lambda}+\gamma)(\sqrt{2\lambda}+\eta) - \gamma\eta e^{-2|u-r|\sqrt{2\lambda}}}$$

1 W $\tau \sim \mathrm{Exp}(\lambda)$, independent of W

$$-\frac{\eta\big((\sqrt{2\lambda}+\gamma)e^{-|u-x|\sqrt{2\lambda}}-\gamma e^{-(|r-x|+|u-r|)\sqrt{2\lambda}}\big)e^{-(|z-u|-|z-x|)\sqrt{2\lambda}}}{(\sqrt{2\lambda}+\gamma)(\sqrt{2\lambda}+\eta)-\gamma\eta e^{-2|u-r|\sqrt{2\lambda}}}$$

(1) $\mathbf{E}_x\{e^{-\gamma\ell(\tau,r)} \mid \ell(\tau,u)=g,\, W_\tau=z\}$

for $r<u\le x\wedge z$ or $x\vee z\le u<r$

$$=\exp\Big(-g\gamma e^{-2|r-u|\sqrt{2\lambda}}\Big(1+\frac{\gamma}{\sqrt{2\lambda}}\big(1-e^{-2|r-u|\sqrt{2\lambda}}\big)\Big)^{-1}\Big),$$

for $x\wedge z\le r<u\le x\vee z$ or $x\wedge z\le u<r\le x\vee z$

$$=\Big(1+\frac{\gamma}{\sqrt{2\lambda}}\big(1-e^{-2|r-u|\sqrt{2\lambda}}\big)\Big)^{-1}$$
$$\times\exp\Big(-g\gamma e^{-2|r-u|\sqrt{2\lambda}}\Big(1+\frac{\gamma}{\sqrt{2\lambda}}\big(1-e^{-2|r-u|\sqrt{2\lambda}}\big)\Big)^{-1}\Big)$$

1.18.6 $\mathbf{P}_x\big(\ell(\tau,r)\in dy \mid \ell(\tau,u)=g,\, W_\tau=z\big)$

for $r<u\le x\wedge z$ or $x\vee z\le u<r$

$$=\frac{\sqrt{\lambda g}}{\sqrt{2y}\,\mathrm{sh}(|r-u|\sqrt{2\lambda})}\exp\Big(-\frac{\sqrt{\lambda}(ye^{|r-u|\sqrt{2\lambda}}+ge^{-|r-u|\sqrt{2\lambda}})}{\sqrt{2}\,\mathrm{sh}(|r-u|\sqrt{2\lambda})}\Big)I_1\Big(\frac{\sqrt{2\lambda gy}}{\mathrm{sh}(|r-u|\sqrt{2\lambda})}\Big)dy$$

for $x\wedge z\le r<u\le x\vee z$ or $x\wedge z\le u<r\le x\vee z$

$$=\frac{\sqrt{\lambda}e^{|r-u|\sqrt{2\lambda}}}{\sqrt{2}\,\mathrm{sh}(|r-u|\sqrt{2\lambda})}\exp\Big(-\frac{\sqrt{\lambda}(ye^{|r-u|\sqrt{2\lambda}}+ge^{-|r-u|\sqrt{2\lambda}})}{\sqrt{2}\,\mathrm{sh}(|r-u|\sqrt{2\lambda})}\Big)I_0\Big(\frac{\sqrt{2\lambda gy}}{\mathrm{sh}(|r-u|\sqrt{2\lambda})}\Big)dy$$

(1) $\mathbf{P}_x\{\ell(\tau,r)=0 \mid \ell(\tau,u)=g,\, W_\tau=z\}=\exp\Big(-\dfrac{g\sqrt{2\lambda}}{e^{2|r-u|\sqrt{2\lambda}}-1}\Big)$

for $r<u\le x\wedge z$ or $x\vee z\le u<r$

1.18.8 $\mathbf{P}_x\big(\ell(t,r)\in dy,\, \ell(t,u)\in dg,\, W_t\in dz\big)=:F^{(18)}_{x,t}(y,g,z)dydgdz$

for $r<u\le x\wedge z$ or $x\vee z\le u<r$

$$=\frac{\sqrt{g}}{2\sqrt{y}}\,\mathrm{is}_t(1,|r-u|,|z+x-2u|+(y+g)/2,(y+g)/2,\sqrt{gy}/2)dydgdz$$

1 W $\tau \sim \operatorname{Exp}(\lambda)$, independent of W

for $x \wedge z \le r < u \le x \vee z$ or $x \wedge z \le u < r \le x \vee z$

$$= \frac{1}{2}\operatorname{is}_t(0,|r-u|,|z-x|-|r-u|+(y+g)/2,(y+g)/2,\sqrt{gy}/2)dydgdz$$

1.19.2 $0<x$

$$\mathbf{P}_x\Big(\int_0^\tau \frac{ds}{W_s}\in dy,\ 0<\inf_{0\le s\le\tau} W_s\Big)=\frac{\lambda x}{\operatorname{ch}^2(y\sqrt{\lambda/2})}\exp\Big(-\frac{x\sqrt{2\lambda}\operatorname{sh}(y\sqrt{\lambda/2})}{\operatorname{ch}(y\sqrt{\lambda/2})}\Big)dy$$

1.19.4 $0<x$

$$\mathbf{P}_x\Big(\int_0^t \frac{ds}{W_s}\in dy,\ 0<\inf_{0\le s\le t} W_s\Big)=x\operatorname{ec}_t(0,2,y/2,0,x)dy$$

1.19.5 $0<x$

$$\mathbf{E}_x\Big\{\exp\Big(-\gamma\int_0^\tau\frac{ds}{W_s}\Big);\ 0<\inf_{0\le s\le\tau}W_s, W_\tau\in dz\Big\}=:\lambda Q_\lambda^{(19)}(z)dz$$

$x\le z$

$$=\frac{\sqrt{\lambda}}{\sqrt{2}}\Gamma(1+\gamma/\sqrt{2\lambda})M_{-\gamma/\sqrt{2\lambda},1/2}(2x\sqrt{2\lambda})W_{-\gamma/\sqrt{2\lambda},1/2}(2z\sqrt{2\lambda})dz$$

$z\le x$

$$=\frac{\sqrt{\lambda}}{\sqrt{2}}\Gamma(1+\gamma/\sqrt{2\lambda})W_{-\gamma/\sqrt{2\lambda},1/2}(2x\sqrt{2\lambda})M_{-\gamma/\sqrt{2\lambda},1/2}(2z\sqrt{2\lambda})dz$$

1.19.6 $0<x$

$$\mathbf{P}_x\Big(\int_0^\tau \frac{ds}{W_s}\in dy,\ 0<\inf_{0\le s\le\tau} W_s,\ W_\tau\in dz\Big)$$

$$=\frac{\lambda\sqrt{2\lambda xz}}{\operatorname{sh}(y\sqrt{\lambda/2})}\exp\Big(-\frac{(x+z)\sqrt{2\lambda}\operatorname{ch}(y\sqrt{\lambda/2})}{\operatorname{sh}(y\sqrt{\lambda/2})}\Big)I_1\Big(\frac{2\sqrt{2\lambda xz}}{\operatorname{sh}(y\sqrt{\lambda/2})}\Big)dydz$$

1.19.8 $0<x$

$$\mathbf{P}_x\Big(\int_0^t \frac{ds}{W_s}\in dy, 0<\inf_{0\le s\le t} W_s, W_t\in dz\Big)=\sqrt{xz}\operatorname{is}_t(1,y/2,0,x+z,\sqrt{xz})dydz$$

1.20.3 $0<x$

$$\mathbf{E}_x\Big\{\exp\Big(-\frac{\gamma^2}{2}\int_0^t\frac{ds}{W_s^2}\Big);\ 0<\inf_{0\le s\le t}W_s\Big\}$$

$$=\frac{(2t)^{1/4}\Gamma(3/4+\sqrt{1+4\gamma^2}/4)}{\sqrt{x}\Gamma(\sqrt{\gamma^2+1/4}+1)}e^{-x^2/4t}M_{-1/4,\sqrt{4\gamma^2+1}/4}\Big(\frac{x^2}{2t}\Big)$$

1.20.4 $0<x$

$$\mathbf{P}_x\Big(\int_0^t \frac{ds}{W_s^2}\in dy,\ 0<\inf_{0\le s\le t} W_s\Big)=\frac{2(2t)^{1/4}}{\sqrt{x}}\,e^{-y/8-x^2/4t}\ \operatorname{m}_{2y}\Big(\frac{1}{4},\frac{x^2}{4t}\Big)dy$$

1	W $\qquad\qquad\qquad\qquad \tau \sim \mathrm{Exp}(\lambda)$, independent of W
1.20.5 $0<x$	$\mathbf{E}_x\Big\{\exp\Big(-\frac{\gamma^2}{2}\int_0^\tau \frac{ds}{W_s^2}\Big);\ 0<\inf_{0\le s\le\tau} W_s,\ W_\tau\in dz\Big\}$ $=\begin{cases} 2\lambda\sqrt{xz}\,I_{\sqrt{\gamma^2+1/4}}(x\sqrt{2\lambda})K_{\sqrt{\gamma^2+1/4}}(z\sqrt{2\lambda})dz, & 0<x\le z\\ 2\lambda\sqrt{xz}\,K_{\sqrt{\gamma^2+1/4}}(x\sqrt{2\lambda})I_{\sqrt{\gamma^2+1/4}}(z\sqrt{2\lambda})dz, & z\le x\end{cases}$
1.20.6 $0<x$	$\mathbf{P}_x\Big(\int_0^\tau \frac{ds}{W_s^2}\in dy,\ 0<\inf_{0\le s\le\tau} W_s,\ W_\tau\in dz\Big)$ $=\lambda\sqrt{xz}e^{-y/8}\,\mathrm{ki}_{y/2}(x\sqrt{2\lambda},z\sqrt{2\lambda})dydz$
1.20.7 $0<x$	$\mathbf{E}_x\Big\{\exp\Big(-\frac{\gamma^2}{2}\int_0^t \frac{ds}{W_s^2}\Big);\ 0<\inf_{0\le s\le t} W_s,\ W_t\in dz\Big\}$ $=\frac{\sqrt{xz}}{t}e^{-(x^2+z^2)/2t}I_{\sqrt{\gamma^2+1/4}}\Big(\frac{xz}{t}\Big)dz$
1.20.8 $0<x$	$\mathbf{P}_x\Big(\int_0^t \frac{ds}{W_s^2}\in dy,\ 0<\inf_{0\le s\le t} W_s, W_t\in dz\Big)$ $=\frac{\sqrt{xz}}{2t}e^{-(x^2+z^2)/2t-y/8}\ \mathrm{i}_{y/2}\Big(\frac{xz}{t}\Big)dydz$
1.21.3 $0<x$	$\mathbf{E}_x\Big\{\exp\Big(-\int_0^t\Big(\frac{p^2}{2W_s^2}+\frac{q^2W_s^2}{2}\Big)ds\Big);\ 0<\inf_{0\le s\le t} W_s\Big\}$ $=\frac{(2\,\mathrm{sh}(tq))^{1/4}\Gamma(3/4+\sqrt{4p^2+1}/4)}{\sqrt{x}(q\,\mathrm{ch}(tq))^{1/4}\Gamma(1+\sqrt{p^2+1/4})}\exp\Big(-\frac{x^2q\,\mathrm{ch}(2tq)}{2\,\mathrm{sh}(2tq)}\Big)M_{-\frac14,\sqrt{4p^2+1}/4}\Big(\frac{x^2q}{\mathrm{sh}(2tq)}\Big)$
(1)	$\mathbf{E}_x\Big\{\exp\Big(-\frac{q^2}{2}\int_0^t W_s^2ds\Big);\ \int_0^t\frac{ds}{W_s^2}\in dy,\ 0<\inf_{0\le s\le t} W_s\Big\}$
$0<x$	$=\frac{2(2\,\mathrm{sh}(tq))^{1/4}}{\sqrt{x}(q\,\mathrm{ch}(tq))^{1/4}}\exp\Big(-\frac{y}{8}-\frac{x^2q\,\mathrm{ch}(2tq)}{2\,\mathrm{sh}(2tq)}\Big)\,\mathrm{m}_{2y}\Big(\frac14,\frac{x^2q}{2\,\mathrm{sh}(2tq)}\Big)dy$

1	W $\qquad\qquad\qquad\qquad\qquad\qquad$ $\tau \sim \mathrm{Exp}(\lambda)$, independent of W
1.21.5 $0<x$	$\mathbf{E}_x\Big\{\exp\Big(-\int_0^\tau\big(\frac{p^2}{2W_s^2}+\frac{q^2W_s^2}{2}\big)ds\Big);\ 0<\inf_{0\le s\le\tau}W_s, W_\tau\in dz\Big\} =: \lambda Q_\lambda^{(21)}(z)dz$
$x\le z$	$=\frac{\lambda\Gamma((1+\sqrt{p^2+1/4}+\lambda/q)/2)}{q\Gamma(1+\sqrt{p^2+1/4})\sqrt{xz}}M_{-\lambda/2q,\sqrt{4p^2+1}/4}(qx^2)W_{-\lambda/2q,\sqrt{4p^2+1}/4}(qz^2)dz$
$z\le x$	$=\frac{\lambda\Gamma((1+\sqrt{p^2+1/4}+\lambda/q)/2)}{q\Gamma(1+\sqrt{p^2+1/4})\sqrt{xz}}W_{-\lambda/2q,\sqrt{4p^2+1}/4}(qx^2)M_{-\lambda/2q,\sqrt{4p^2+1}/4}(qz^2)dz$
1.21.7 $0<x$	$\mathbf{E}_x\Big\{\exp\Big(-\int_0^t\big(\frac{p^2}{2W_s^2}+\frac{q^2W_s^2}{2}\big)ds\Big);\ 0<\inf_{0\le s\le t}W_s, W_t\in dz\Big\}$
	$=\frac{q\sqrt{xz}}{\mathrm{sh}(tq)}\exp\Big(-\frac{(x^2+z^2)q\,\mathrm{ch}(tq)}{2\,\mathrm{sh}(tq)}\Big)I_{\sqrt{p^2+1/4}}\Big(\frac{xzq}{\mathrm{sh}(tq)}\Big)dz$
(1)	$\mathbf{E}_x\Big\{\exp\Big(-\frac{q^2}{2}\int_0^t W_s^2ds\Big);\ \int_0^t\frac{ds}{W_s^2}\in dy,\ 0<\inf_{0\le s\le t}W_s,\ W_t\in dz\Big\}$
	$=\frac{q\sqrt{xz}}{2\,\mathrm{sh}(tq)}\exp\Big(-\frac{y}{8}-\frac{(x^2+z^2)q\,\mathrm{ch}(tq)}{2\,\mathrm{sh}(tq)}\Big)\,\mathrm{i}_{y/2}\Big(\frac{xzq}{\mathrm{sh}(tq)}\Big)dydz$
(2)	$\mathbf{E}_x\Big\{\exp\Big(-\frac{p^2}{2}\int_0^t\frac{ds}{W_s^2}\Big);\ \int_0^t W_s^2ds\in dy,\ 0<\inf_{0\le s\le t}W_s,\ W_t\in dz\Big\}$
	$=\sqrt{xz}\,\mathrm{is}_y(\sqrt{p^2+1/4},t,0,(x^2+z^2)/2,xz/2)dydz$
1.21.8 $0<x$	$\mathbf{P}_x\Big(\int_0^t W_s^2ds\in dg,\ \int_0^t\frac{ds}{W_s^2}\in dy,\ 0<\inf_{0\le s\le t}W_s,\ W_t\in dz\Big)$
	$=2^{-1}\sqrt{xz}e^{-y/8}\,\mathrm{ei}_g(y/2,t,(x^2+z^2)/2,xz)dgdydz$
1.22.1 (1)	$\mathbf{E}_x\Big\{\exp\Big(-\frac{p^2}{2}\int_0^\tau\frac{ds}{W_s^2}\Big);\ \int_0^\tau\frac{ds}{W_s}\in dy,\ 0<\inf_{0\le s\le\tau}W_s\Big\}=\frac{\sqrt{\lambda}\,\mathrm{sh}(y\sqrt{\lambda/2})}{\sqrt{2}\,\mathrm{ch}(y\sqrt{\lambda/2})}$
$0<x$	$\times\frac{\Gamma(3/2+\sqrt{p^2+1/4})}{\Gamma(1+\sqrt{4p^2+1})}\exp\Big(-\frac{x\sqrt{2\lambda}\,\mathrm{ch}(y\sqrt{2\lambda})}{\mathrm{sh}(y\sqrt{2\lambda})}\Big)M_{-1,\sqrt{p^2+1/4}}\Big(\frac{2x\sqrt{2\lambda}}{\mathrm{sh}(y\sqrt{2\lambda})}\Big)dy$

1 W	$\tau \sim \operatorname{Exp}(\lambda)$, independent of W
1.22.2 $0<x$	$\mathbf{P}_x\Big(\int_0^\tau \frac{ds}{W_s^2}\in dg,\ \int_0^\tau \frac{ds}{W_s}\in dy,\ 0<\inf_{0\le s\le\tau} W_s\Big)$
	$=\frac{\sqrt{\lambda}\,\mathrm{sh}(y\sqrt{\lambda/2})}{2\sqrt{2}\,\mathrm{ch}(y\sqrt{\lambda/2})}\exp\Big(-\frac{g}{8}-\frac{x\sqrt{2\lambda}\,\mathrm{ch}(y\sqrt{2\lambda})}{\mathrm{sh}(y\sqrt{2\lambda})}\Big)\,\mathrm{m}_{g/2}\Big(1,\frac{x\sqrt{2\lambda}}{\mathrm{sh}(y\sqrt{2\lambda})}\Big)dgdy$
1.22.5 $0<x$	$\mathbf{E}_x\Big\{\exp\Big(-\int_0^\tau\big(\frac{p^2}{2W_s^2}+\frac{q}{W_s}\big)ds\Big);\ 0<\inf_{0\le s\le\tau} W_s, W_\tau\in dz\Big\}=:\lambda Q_\lambda^{(22)}(z)dz$
	$=\frac{\sqrt{\lambda}\Gamma(\sqrt{p^2+1/4}+1/2+q/\sqrt{2\lambda})}{\sqrt{2}\Gamma(2\sqrt{p^2+1/4}+1)}$
	$\times\begin{cases} M_{-q/\sqrt{2\lambda},\sqrt{p^2+1/4}}(2x\sqrt{2\lambda})W_{-q/\sqrt{2\lambda},\sqrt{p^2+1/4}}(2z\sqrt{2\lambda})dz, & x\le z\\ W_{-q/\sqrt{2\lambda},\sqrt{p^2+1/4}}(2x\sqrt{2\lambda})M_{-q/\sqrt{2\lambda},\sqrt{p^2+1/4}}(2z\sqrt{2\lambda})dz, & z\le x\end{cases}$
(1)	$\mathbf{E}_x\Big\{\exp\Big(-\frac{p^2}{2}\int_0^\tau\frac{ds}{W_s^2}\Big);\ \int_0^\tau\frac{ds}{W_s}\in dy,\ 0<\inf_{0\le s\le\tau}W_s,\ W_\tau\in dz\Big\}$
	$=\frac{\lambda\sqrt{2\lambda xz}}{\mathrm{sh}(y\sqrt{\lambda/2})}\exp\Big(-\frac{(x+z)\sqrt{2\lambda}\,\mathrm{ch}(y\sqrt{\lambda/2})}{\mathrm{sh}(y\sqrt{\lambda/2})}\Big)I_{\sqrt{4p^2+1}}\Big(\frac{2\sqrt{2\lambda xz}}{\mathrm{sh}(y\sqrt{\lambda/2})}\Big)dydz$
1.22.6 $0<x$	$\mathbf{P}_x\Big(\int_0^\tau\frac{ds}{W_s^2}\in dg,\ \int_0^\tau\frac{ds}{W_s}\in dy,\ 0<\inf_{0\le s\le\tau}W_s,\ W_\tau\in dz\Big)$
	$=\frac{\lambda\sqrt{2\lambda xz}}{8\,\mathrm{sh}(y\sqrt{\lambda/2})}\exp\Big(-\frac{g}{8}-\frac{(x+z)\sqrt{2\lambda}\,\mathrm{ch}(y\sqrt{\lambda/2})}{\mathrm{sh}(y\sqrt{\lambda/2})}\Big)\,\mathrm{i}_{g/8}\Big(\frac{2\sqrt{2\lambda xz}}{\mathrm{sh}(y\sqrt{\lambda/2})}\Big)dgdydz$
1.22.7 (1)	$\mathbf{E}_x\Big\{\exp\Big(-\frac{p^2}{2}\int_0^t\frac{ds}{W_s^2}\Big);\ \int_0^t\frac{ds}{W_s}\in dy,\ 0<\inf_{0\le s\le t}W_s,W_t\in dz\Big\}$
$0<x$	$=\sqrt{xz}\ \mathrm{is}_t(\sqrt{4p^2+1},y/2,0,x+z,\sqrt{xz})dydz$
1.22.8 $0<x$	$\mathbf{P}_x\Big(\int_0^t\frac{ds}{W_s^2}\in dg,\ \int_0^t\frac{ds}{W_s}\in dy,\ 0<\inf_{0\le s\le t}W_s,\ W_t\in dz\Big)$
	$=8^{-1}\sqrt{xz}e^{-g/8}\,\mathrm{ei}_t(g/8,y/2,x+z,2\sqrt{xz})dgdydz$

1 W $\tau \sim \text{Exp}(\lambda)$, independent of W

1.23.1 $r < b$

$$\mathbf{E}_x\Big\{\exp\Big(-\int_0^\tau \big(p\mathbb{I}_{(-\infty,r)}(W_s) + q\mathbb{I}_{[r,\infty)}(W_s)\big)ds\Big);\ \sup_{0\le s\le\tau} W_s < b\Big\}$$

$x \le r$

$$= \frac{\lambda}{\lambda+p} - \frac{\lambda(\Upsilon_p^2 + 2(q-p)\operatorname{ch}((b-r)\Upsilon_q))e^{(x-r)\Upsilon_p}}{(\lambda+p)\Upsilon_q(\Upsilon_p\operatorname{sh}((b-r)\Upsilon_q) + \Upsilon_q\operatorname{ch}((b-r)\Upsilon_q))}$$

$r \le x$, $x \le b$

$$= \frac{\lambda}{\lambda+q}\big(1 - e^{(x-b)\Upsilon_q}\big) + \frac{\lambda(2(q-p) + (\Upsilon_p^2 - \Upsilon_p\Upsilon_q)e^{(r-b)\Upsilon_q})\operatorname{sh}((b-x)\Upsilon_q)}{(\lambda+q)\Upsilon_p(\Upsilon_p\operatorname{sh}((b-r)\Upsilon_q) + \Upsilon_q\operatorname{ch}((b-r)\Upsilon_q))}$$

1.23.5 $r < b$

$$\mathbf{E}_x\Big\{\exp\Big(-\int_0^\tau \big(p\mathbb{I}_{(-\infty,r)}(W_s) + q\mathbb{I}_{[r,\infty)}(W_s)\big)ds\Big);\ \sup_{0\le s\le\tau} W_s < b,\ W_\tau \in dz\Big\}$$

$z < b$

$$=: \lambda Q_\lambda^{(23)}(z)dz$$

$x \le r$, $z \le r$

$$= \frac{\lambda}{\Upsilon_p}e^{-|z-x|\Upsilon_p}dz + \frac{\lambda(\Upsilon_p\operatorname{sh}((b-r)\Upsilon_q) - \Upsilon_q\operatorname{ch}((b-r)\Upsilon_q))e^{(z+x-2r)\Upsilon_p}}{\Upsilon_p(\Upsilon_p\operatorname{sh}((b-r)\Upsilon_q) + \Upsilon_q\operatorname{ch}((b-r)\Upsilon_q))}dz$$

$x \le r$, $r \le z$

$$= \frac{2\lambda\operatorname{sh}((b-z)\Upsilon_q)e^{(x-r)\Upsilon_p}}{\Upsilon_p\operatorname{sh}((b-r)\Upsilon_q) + \Upsilon_q\operatorname{ch}((b-r)\Upsilon_q)}dz$$

$r \le x$, $z \le r$

$$= \frac{2\lambda\operatorname{sh}((b-x)\Upsilon_q)e^{(z-r)\Upsilon_p}}{\Upsilon_p\operatorname{sh}((b-r)\Upsilon_q) + \Upsilon_q\operatorname{ch}((b-r)\Upsilon_q)}dz$$

$r \le x$, $r \le z$

$$= \frac{\lambda}{\Upsilon_q}\Big(e^{-|z-x|\Upsilon_q} - e^{(x+z-2b)\Upsilon_q} + \frac{2(\Upsilon_q - \Upsilon_p)\operatorname{sh}((b-z)\Upsilon_q)\operatorname{sh}((b-x)\Upsilon_q)e^{(r-b)\Upsilon_q}}{\Upsilon_p\operatorname{sh}((b-r)\Upsilon_q) + \Upsilon_q\operatorname{ch}((b-r)\Upsilon_q)}\Big)dz$$

1.24.1 $a < r$

$$\mathbf{E}_x\Big\{\exp\Big(-\int_0^\tau \big(p\mathbb{I}_{(-\infty,r)}(W_s) + q\mathbb{I}_{[r,\infty)}(W_s)\big)ds\Big);\ a < \inf_{0\le s\le\tau} W_s\Big\}$$

$a \le x$, $x \le r$

$$= \frac{\lambda}{\lambda+p}\big(1 - e^{(a-x)\Upsilon_p}\big) - \frac{\lambda(2(q-p) + (\Upsilon_p\Upsilon_q - \Upsilon_q^2)e^{(a-r)\Upsilon_p})\operatorname{sh}((x-a)\Upsilon_p)}{(\lambda+p)\Upsilon_q(\Upsilon_q\operatorname{sh}((r-a)\Upsilon_p) + \Upsilon_p\operatorname{ch}((r-a)\Upsilon_p))}$$

$$\Upsilon_s := \sqrt{2\lambda + 2s}$$

1 W	$\tau \sim \mathrm{Exp}(\lambda)$, independent of W
$r \le x$	$= \dfrac{\lambda}{\lambda+q} - \dfrac{\lambda(\Upsilon_q^2 - 2(q-p)\operatorname{ch}((r-a)\Upsilon_p))e^{(r-x)\Upsilon_q}}{\Upsilon_p(\lambda+q)(\Upsilon_q \operatorname{sh}((r-a)\Upsilon_p) + \Upsilon_p \operatorname{ch}((r-a)\Upsilon_p))}$
1.24.5 $a < r$	$\mathbf{E}_x\Big\{\exp\Big(-\int_0^\tau \big(p\mathbb{I}_{(-\infty,r)}(W_s) + q\mathbb{I}_{[r,\infty)}(W_s)\big)ds\Big); a < \inf_{0\le s\le\tau} W_s,\ W_\tau \in dz\Big\}$
$a < z$	$=: \lambda Q^{(24)}_{a,\lambda}(z)dz$
$x \le r$ $z \le r$	$= \dfrac{\lambda}{\Upsilon_p}\Big(e^{-\lvert z-x\rvert\Upsilon_p} - e^{(2a-z-x)\Upsilon_p} + \dfrac{2(\Upsilon_p - \Upsilon_q)\operatorname{sh}((z-a)\Upsilon_p)\operatorname{sh}((x-a)\Upsilon_p)e^{(a-r)\Upsilon_p}}{\Upsilon_q \operatorname{sh}((r-a)\Upsilon_p) + \Upsilon_p \operatorname{ch}((r-a)\Upsilon_p)}\Big)dz$
$x \le r$ $r \le z$	$= \dfrac{2\lambda \operatorname{sh}((x-a)\Upsilon_p)e^{(r-z)\Upsilon_q}}{\Upsilon_q \operatorname{sh}((r-a)\Upsilon_p) + \Upsilon_p \operatorname{ch}((r-a)\Upsilon_p)}dz$
$r \le x$ $z \le r$	$= \dfrac{2\lambda \operatorname{sh}((z-a)\Upsilon_p)e^{(r-x)\Upsilon_q}}{\Upsilon_q \operatorname{sh}((r-a)\Upsilon_p) + \Upsilon_p \operatorname{ch}((r-a)\Upsilon_p)}dz$
$r \le x$ $r \le z$	$= \dfrac{\lambda}{\Upsilon_q}e^{-\lvert z-x\rvert\Upsilon_q}dz + \dfrac{\lambda(\Upsilon_q \operatorname{sh}((r-a)\Upsilon_p) - \Upsilon_p \operatorname{ch}((r-a)\Upsilon_p))e^{(2r-x-z)\Upsilon_q}}{\Upsilon_q(\Upsilon_q \operatorname{sh}((r-a)\Upsilon_p) + \Upsilon_p \operatorname{ch}((r-a)\Upsilon_p))}dz$
1.25.1 $r < b$	$\mathbf{E}_x\{e^{-\gamma\ell(\tau,r)}; a < \inf_{0\le s\le\tau} W_s,\ \sup_{0\le s\le\tau} W_s < b\}$
$a \le x$ $x \le r$	$= 1 - \dfrac{\operatorname{ch}((r+a-2x)\sqrt{\lambda/2})}{\operatorname{ch}((r-a)\sqrt{\lambda/2})}$ $+ \dfrac{\sqrt{2\lambda}\operatorname{sh}((b-a)\sqrt{\lambda/2})\operatorname{sh}((b-r)\sqrt{\lambda/2})\operatorname{sh}((x-a)\sqrt{2\lambda})}{(\gamma \operatorname{sh}((b-r)\sqrt{2\lambda})\operatorname{sh}((r-a)\sqrt{2\lambda}) + \sqrt{\lambda/2}\operatorname{sh}((b-a)\sqrt{2\lambda}))\operatorname{ch}((r-a)\sqrt{\lambda/2})}$
$r \le x$ $x \le b$	$= 1 - \dfrac{\operatorname{ch}((b+r-2x)\sqrt{\lambda/2})}{\operatorname{ch}((b-r)\sqrt{\lambda/2})}$ $+ \dfrac{\sqrt{2\lambda}\operatorname{sh}((b-a)\sqrt{\lambda/2})\operatorname{sh}((r-a)\sqrt{\lambda/2})\operatorname{sh}((b-x)\sqrt{2\lambda})}{(\gamma \operatorname{sh}((b-r)\sqrt{2\lambda})\operatorname{sh}((r-a)\sqrt{2\lambda}) + \sqrt{\lambda/2}\operatorname{sh}((b-a)\sqrt{2\lambda}))\operatorname{ch}((b-r)\sqrt{\lambda/2})}$
	$\Upsilon_s := \sqrt{2\lambda + 2s}$

1.25.2 $r<b$

$$\mathbf{P}_x\Big(\ell(\tau,r)\in dy, a<\inf_{0\le s\le\tau}W_s, \sup_{0\le s\le\tau}W_s<b\Big)$$

$$=\exp\Big(-\frac{y\sqrt{\lambda/2}\,\operatorname{sh}((b-a)\sqrt{2\lambda})}{\operatorname{sh}((b-r)\sqrt{2\lambda})\operatorname{sh}((r-a)\sqrt{2\lambda})}\Big)$$

$$\times\begin{cases}\dfrac{\sqrt{\lambda/2}\,\operatorname{sh}((b-a)\sqrt{\lambda/2})\operatorname{sh}((x-a)\sqrt{2\lambda})}{\operatorname{ch}((b-r)\sqrt{\lambda/2})\operatorname{ch}((r-a)\sqrt{\lambda/2})\operatorname{sh}((r-a)\sqrt{2\lambda})}dy, & a\le x\le r\\[2ex] \dfrac{\sqrt{\lambda/2}\,\operatorname{sh}((b-a)\sqrt{\lambda/2})\operatorname{sh}((b-x)\sqrt{2\lambda})}{\operatorname{sh}((b-r)\sqrt{2\lambda})\operatorname{ch}((b-r)\sqrt{\lambda/2})\operatorname{ch}((r-a)\sqrt{\lambda/2})}dy, & r\le x\le b\end{cases}$$

(1)

$$\mathbf{P}_x\Big(\ell(\tau,r)=0, \sup_{0\le s\le\tau}W_s<b\Big)$$

$$=\begin{cases}1-\dfrac{\operatorname{ch}((r+a-2x)\sqrt{\lambda/2})}{\operatorname{ch}((r-a)\sqrt{\lambda/2})}, & a\le x\le r\\[2ex] 1-\dfrac{\operatorname{ch}((b+r-2x)\sqrt{\lambda/2})}{\operatorname{ch}((b-r)\sqrt{\lambda/2})}, & r\le x\le b\end{cases}$$

1.25.5 $r<b$ $z<b$

$$\mathbf{E}_x\Big\{e^{-\gamma\ell(\tau,r)}; a<\inf_{0\le s\le\tau}W_s, \sup_{0\le s\le\tau}W_s<b, W_\tau\in dz\Big\}=:\lambda Q^{(25)}_{a,\lambda}(z)dz$$

$z\le r$, $x\le r$

$$=\frac{\sqrt{\lambda}\big(\operatorname{ch}((r-a-|z-x|)\sqrt{2\lambda})-\operatorname{ch}((r+a-x-z)\sqrt{2\lambda})\big)}{\sqrt{2}\operatorname{sh}((r-a)\sqrt{2\lambda})}dz$$

$$+\frac{\lambda\operatorname{sh}((b-r)\sqrt{2\lambda})\operatorname{sh}((z-a)\sqrt{2\lambda})\operatorname{sh}((x-a)\sqrt{2\lambda})}{(\gamma\operatorname{sh}((b-r)\sqrt{2\lambda})\operatorname{sh}((r-a)\sqrt{2\lambda})+\sqrt{\lambda/2}\operatorname{sh}((b-a)\sqrt{2\lambda}))\operatorname{sh}((r-a)\sqrt{2\lambda})}dz$$

$z\le r$, $r\le x$

$$=\frac{\lambda\operatorname{sh}((b-x)\sqrt{2\lambda})\operatorname{sh}((z-a)\sqrt{2\lambda})}{(\gamma\operatorname{sh}((b-r)\sqrt{2\lambda})\operatorname{sh}((r-a)\sqrt{2\lambda})+\sqrt{\lambda/2}\operatorname{sh}((b-a)\sqrt{2\lambda}))}dz$$

$r\le z$, $x\le r$

$$=\frac{\lambda\operatorname{sh}((b-z)\sqrt{2\lambda})\operatorname{sh}((x-a)\sqrt{2\lambda})}{(\gamma\operatorname{sh}((b-r)\sqrt{2\lambda})\operatorname{sh}((r-a)\sqrt{2\lambda})+\sqrt{\lambda/2}\operatorname{sh}((b-a)\sqrt{2\lambda}))}dz$$

$r\le z$, $r\le x$

$$=\frac{\sqrt{\lambda}\big(\operatorname{ch}((b-r-|z-x|)\sqrt{2\lambda})-\operatorname{ch}((b+r-z-x)\sqrt{2\lambda})\big)}{\sqrt{2}\operatorname{sh}((b-r)\sqrt{2\lambda})}dz$$

$$+\frac{\lambda\operatorname{sh}((b-z)\sqrt{2\lambda})\operatorname{sh}((r-a)\sqrt{2\lambda})\operatorname{sh}((b-x)\sqrt{2\lambda})}{(\gamma\operatorname{sh}((b-r)\sqrt{2\lambda})\operatorname{sh}((r-a)\sqrt{2\lambda})+\sqrt{\lambda/2}\operatorname{sh}((b-a)\sqrt{2\lambda}))\operatorname{sh}((b-r)\sqrt{2\lambda})}dz$$

1 W	$\tau \sim \operatorname{Exp}(\lambda)$, independent of W
1.25.6 $r<b$ $z<b$	$\mathbf{P}_x\big(\ell(\tau,r)\in dy, a<\inf\limits_{0\le s\le\tau} W_s, \sup\limits_{0\le s\le\tau} W_s<b, W_\tau\in dz\big) =: \lambda B^{(25)}_{a,\lambda}(y,z)dydz$
$z\le r$ $x\le r$	$= \dfrac{\lambda\operatorname{sh}((z-a)\sqrt{2\lambda})\operatorname{sh}((x-a)\sqrt{2\lambda})}{\operatorname{sh}^2((r-a)\sqrt{2\lambda})}\exp\Big(-\dfrac{y\sqrt{\lambda/2}\operatorname{sh}((b-a)\sqrt{2\lambda})}{\operatorname{sh}((b-r)\sqrt{2\lambda})\operatorname{sh}((r-a)\sqrt{2\lambda})}\Big)dydz$
$z\le r$ $r\le x$	$= \dfrac{\lambda\operatorname{sh}((b-x)\sqrt{2\lambda})\operatorname{sh}((z-a)\sqrt{2\lambda})}{\operatorname{sh}((b-r)\sqrt{2\lambda})\operatorname{sh}((r-a)\sqrt{2\lambda})}\exp\Big(-\dfrac{y\sqrt{\lambda/2}\operatorname{sh}((b-a)\sqrt{2\lambda})}{\operatorname{sh}((b-r)\sqrt{2\lambda})\operatorname{sh}((r-a)\sqrt{2\lambda})}\Big)dydz$
$r\le z$ $x\le r$	$= \dfrac{\lambda\operatorname{sh}((b-z)\sqrt{2\lambda})\operatorname{sh}((x-a)\sqrt{2\lambda})}{\operatorname{sh}((b-r)\sqrt{2\lambda})\operatorname{sh}((r-a)\sqrt{2\lambda})}\exp\Big(-\dfrac{y\sqrt{\lambda/2}\operatorname{sh}((b-a)\sqrt{2\lambda})}{\operatorname{sh}((b-r)\sqrt{2\lambda})\operatorname{sh}((r-a)\sqrt{2\lambda})}\Big)dydz$
$r\le z$ $r\le x$	$= \dfrac{\lambda\operatorname{sh}((b-z)\sqrt{2\lambda})\operatorname{sh}((b-x)\sqrt{2\lambda})}{\operatorname{sh}^2((b-r)\sqrt{2\lambda})}\exp\Big(-\dfrac{y\sqrt{\lambda/2}\operatorname{sh}((b-a)\sqrt{2\lambda})}{\operatorname{sh}((b-r)\sqrt{2\lambda})\operatorname{sh}((r-a)\sqrt{2\lambda})}\Big)dydz$
(1)	$\mathbf{P}_x\big(\ell(\tau,r)=0, a<\inf\limits_{0\le s\le\tau} W_s, \sup\limits_{0\le s\le\tau} W_s<b, W_\tau\in dz\big)$
$z\le r$ $x\le r$	$= \dfrac{\sqrt{\lambda}\big(\operatorname{ch}((r-a-\lvert z-x\rvert)\sqrt{2\lambda})-\operatorname{ch}((r+a-x-z)\sqrt{2\lambda})\big)}{\sqrt{2}\operatorname{sh}((r-a)\sqrt{2\lambda})}dz$
$r\le z$ $r\le x$	$= \dfrac{\sqrt{\lambda}\big(\operatorname{ch}((b-r-\lvert z-x\rvert)\sqrt{2\lambda})-\operatorname{ch}((b+r-z-x)\sqrt{2\lambda})\big)}{\sqrt{2}\operatorname{sh}((b-r)\sqrt{2\lambda})}dz$
1.25.8	$\mathbf{P}_x\big(\ell(t,r)\in dy, a<\inf\limits_{0\le s\le t} W_s, \sup\limits_{0\le s\le t} W_s<b, W_t\in dz\big) =: F^{(25)}_{a,x}(t,y,z)dydz$
	for $a<r<b,\quad a<z<b$
$z\le r$ $x\le r$	$= \operatorname{es}_t(0,0,b-r,0,y/2) * \operatorname{esss}_t(z-a,x-a,r-a,y/2)dydz$
$z\le r$ $r\le x$	$= \operatorname{ess}_t(b-x,b-r,0,y/2) * \operatorname{ess}_t(z-a,r-a,0,y/2)dydz$
$r\le z$ $x\le r$	$= \operatorname{ess}_t(b-z,b-r,0,y/2) * \operatorname{ess}_t(x-a,r-a,0,y/2)dydz$
$r\le z$ $r\le x$	$= \operatorname{esss}_t(b-z,b-x,b-r,y/2) * \operatorname{es}_t(0,0,r-a,0,y/2)dydz$

1 W $\tau \sim \operatorname{Exp}(\lambda)$, independent of W

1.26.1 $r < b$, $a < r$

$$\mathbf{E}_x\Big\{\exp\Big(-\int_0^\tau (p\mathbb{I}_{(-\infty,r)}(W_s) + q\mathbb{I}_{[r,\infty)}(W_s))ds\Big);\ a < \inf_{0\le s\le\tau} W_s,\ \sup_{0\le s\le\tau} W_s < b\Big\}$$

$a \le x$, $x \le r$

$$= \frac{\lambda}{\lambda+p}\Big(1 - \frac{\operatorname{ch}((r+a-2x)\Upsilon_p/2)}{\operatorname{ch}((r-a)\Upsilon_p/2)}\Big) + \frac{2\lambda(\Upsilon_p \operatorname{th}((b-r)\Upsilon_q/2) + \Upsilon_q \operatorname{th}((r-a)\Upsilon_p/2))\operatorname{sh}((x-a)\Upsilon_p)}{\Upsilon_p\Upsilon_q(\Upsilon_q \operatorname{cth}((b-r)\Upsilon_q) + \Upsilon_p \operatorname{cth}((r-a)\Upsilon_p))\operatorname{sh}((r-a)\Upsilon_p)}$$

$r \le x$, $x \le b$

$$= \frac{\lambda}{\lambda+q}\Big(1 - \frac{\operatorname{ch}((b+r-2x)\Upsilon_q/2)}{\operatorname{ch}((b-r)\Upsilon_q/2)}\Big) + \frac{2\lambda(\Upsilon_p \operatorname{th}((b-r)\Upsilon_q/2) + \Upsilon_q \operatorname{th}((r-a)\Upsilon_p/2))\operatorname{sh}((b-x)\Upsilon_q)}{\Upsilon_p\Upsilon_q(\Upsilon_q \operatorname{cth}((b-r)\Upsilon_q) + \Upsilon_p \operatorname{cth}((r-a)\Upsilon_p))\operatorname{sh}((b-r)\Upsilon_q)}$$

1.26.5 $r < b$, $a < r$

$$\mathbf{E}_x\Big\{\exp\Big(-\int_0^\tau (p\mathbb{I}_{(-\infty,r)}(W_s) + q\mathbb{I}_{[r,\infty)}(W_s))ds\Big);\ a < \inf_{0\le s\le\tau} W_s,\ \sup_{0\le s\le\tau} W_s < b,\ W_\tau \in dz\Big\} =: \lambda Q^{(26)}_{a,\lambda}(z)dz$$

$z \le r$, $x \le r$

$$= \frac{\lambda}{\Upsilon_p}\big(e^{-|z-x|\Upsilon_p} - e^{-(z+x-2a)\Upsilon_p}\big)dz + \frac{2\lambda(\Upsilon_p \operatorname{th}((b-r)\Upsilon_q) - \Upsilon_q)\operatorname{sh}((z-a)\Upsilon_p)\operatorname{sh}((x-a)\Upsilon_p)e^{(a-r)\Upsilon_p}}{\Upsilon_p(\Upsilon_p \operatorname{th}((b-r)\Upsilon_q)\operatorname{ch}((r-a)\Upsilon_p) + \Upsilon_q \operatorname{sh}((r-a)\Upsilon_p))}dz$$

$z \le r$, $r \le x$

$$= \frac{2\lambda \operatorname{sh}((z-a)\Upsilon_p)\operatorname{sh}((b-x)\Upsilon_q)}{\Upsilon_p \operatorname{sh}((b-r)\Upsilon_q)\operatorname{ch}((r-a)\Upsilon_p) + \Upsilon_q \operatorname{ch}((b-r)\Upsilon_q)\operatorname{sh}((r-a)\Upsilon_p)}dz$$

$r \le z$, $x \le r$

$$= \frac{2\lambda \operatorname{sh}((b-z)\Upsilon_q)\operatorname{sh}((x-a)\Upsilon_p)}{\Upsilon_p \operatorname{sh}((b-r)\Upsilon_q)\operatorname{ch}((r-a)\Upsilon_p) + \Upsilon_q \operatorname{ch}((b-r)\Upsilon_q)\operatorname{sh}((r-a)\Upsilon_p)}dz$$

$r \le z$, $r \le x$

$$= \frac{\lambda}{\Upsilon_q}\big(e^{-|z-x|\Upsilon_q} - e^{(x+z-2b)\Upsilon_q}\big)dz + \frac{2\lambda(\Upsilon_q \operatorname{th}((r-a)\Upsilon_p) - \Upsilon_p)\operatorname{sh}((b-z)\Upsilon_q)\operatorname{sh}((b-x)\Upsilon_q)e^{(r-b)\Upsilon_q}}{\Upsilon_q(\Upsilon_p \operatorname{sh}((b-r)\Upsilon_q) + \Upsilon_q \operatorname{ch}((b-r)\Upsilon_q)\operatorname{th}((r-a)\Upsilon_p))}dz$$

$$\Upsilon_s := \sqrt{2\lambda + 2s}$$

1 W $\tau \sim \mathrm{Exp}(\lambda)$, independent of W

1.27.1 $\mathbf{E}_x\Big\{\exp\Big(-\int_0^\tau (p\mathbb{I}_{(-\infty,r)}(W_s)+q\mathbb{I}_{[r,\infty)}(W_s))ds\Big);\ \frac12\ell(\tau,r)\in dy\Big\}$

$x\le r$ $$=\Big(\frac{2\lambda}{\sqrt{2\lambda+2p}}+\frac{2\lambda}{\sqrt{2\lambda+2q}}\Big)e^{(x-r)\sqrt{2\lambda+2p}}e^{-y(\sqrt{2\lambda+2p}+\sqrt{2\lambda+2q})}dy$$

$r\le x$ $$=\Big(\frac{2\lambda}{\sqrt{2\lambda+2p}}+\frac{2\lambda}{\sqrt{2\lambda+2q}}\Big)e^{(r-x)\sqrt{2\lambda+2q}}e^{-y(\sqrt{2\lambda+2p}+\sqrt{2\lambda+2q})}dy$$

(1) $\mathbf{E}_x\Big\{\exp\Big(-\int_0^\tau (p\mathbb{I}_{(-\infty,r)}(W_s)+q\mathbb{I}_{[r,\infty)}(W_s))ds\Big);\ \ell(\tau,r)=0\Big\}$

$x\le r$ $$=\frac{\lambda}{\lambda+p}\big(1-e^{(x-r)\sqrt{2\lambda+2p}}\big)$$

$r\le x$ $$=\frac{\lambda}{\lambda+q}\big(1-e^{(r-x)\sqrt{2\lambda+2q}}\big)$$

1.27.2 $\mathbf{P}_x\Big(\int_0^\tau \mathbb{I}_{(-\infty,r)}(W_s)ds\in du,\ \int_0^\tau \mathbb{I}_{[r,\infty)}(W_s)ds\in dv,\ \frac12\ell(\tau,r)\in dy\Big)/dy$

$x\le r$ $$=\frac{\lambda}{\pi\sqrt{uv}}\Big(\frac{y+r-x}{u}+\frac{y}{v}\Big)e^{-\lambda(u+v)-(y+r-x)^2/2u-y^2/2v}dudv$$

or $$=-\frac{\lambda e^{-\lambda(u+v)}}{\pi\sqrt{uv}}\frac{d}{dy}e^{-(y+r-x)^2/2u-y^2/2v}\,dudv$$

$r\le x$ $$=\frac{\lambda}{\pi\sqrt{uv}}\Big(\frac{y}{u}+\frac{y+x-r}{v}\Big)e^{-\lambda(u+v)-y^2/2u-(y+x-r)^2/2v}dudv$$

or $$=-\frac{\lambda e^{-\lambda(u+v)}}{\pi\sqrt{uv}}\frac{d}{dy}e^{-y^2/2u-(y+x-r)^2/2v}\,dudv$$

(1) $x\le r$ $\mathbf{P}_x\Big(\int_0^\tau \mathbb{I}_{(-\infty,r)}(W_s)ds\in du,\ \int_0^\tau \mathbb{I}_{[r,\infty)}(W_s)ds=0, \ell(\tau,r)=0\Big)$

$$=\lambda e^{-\lambda u}\Big(1-\mathrm{Erfc}\Big(\frac{r-x}{\sqrt{2u}}\Big)\Big)du$$

(2) $x\le r$ $\mathbf{P}_x\Big(\int_0^\tau \mathbb{I}_{(-\infty,r)}(W_s)ds=0,\ \int_0^\tau \mathbb{I}_{[r,\infty)}(W_s)ds\in dv, \ell(\tau,r)=0\Big)$

$$=\lambda e^{-\lambda v}\Big(1-\mathrm{Erfc}\Big(\frac{x-r}{\sqrt{2v}}\Big)\Big)dv$$

1 W $\qquad\tau \sim \mathrm{Exp}(\lambda)$, independent of W

1.27.4
$$\mathbf{P}_x\Big(\int_0^t \big(p\mathbb{1}_{(-\infty,r)}(W_s)+q\mathbb{1}_{[r,\infty)}(W_s)\big)ds\in dv,\ \frac{1}{2}\ell(t,r)\in dy\Big)/dy$$

for $(p\wedge q)t\le v\le (p\vee q)t$

$x\le r$
$$=\frac{|p-q|((y+r-x)|pt-v|+y|v-qt|)}{\pi|pt-v|^{3/2}|v-qt|^{3/2}}\exp\Big(-\frac{y^2|p-q|}{2|pt-v|}-\frac{(y+r-x)^2|p-q|}{2|v-qt|}\Big)\,dv$$

or
$$=-\frac{1}{\pi|pt-v|^{1/2}|v-qt|^{1/2}}\frac{d}{dy}\exp\Big(-\frac{y^2|p-q|}{2|pt-v|}-\frac{(y+r-x)^2|p-q|}{2|v-qt|}\Big)\,dv$$

$r\le x$
$$=\frac{|p-q|(y|pt-v|+(y+x-r)|v-qt|)}{\pi|pt-v|^{3/2}|v-qt|^{3/2}}\exp\Big(-\frac{(y+x-r)^2|p-q|}{2|pt-v|}-\frac{y^2|p-q|}{2|v-qt|}\Big)\,dv$$

or
$$=-\frac{1}{\pi|pt-v|^{1/2}|v-qt|^{1/2}}\frac{d}{dy}\exp\Big(-\frac{(y+x-r)^2|p-q|}{2|pt-v|}-\frac{y^2|p-q|}{2|v-qt|}\Big)\,dv$$

(1) $x\le r$
$$\mathbf{P}_x\Big(\int_0^t\big(p\mathbb{1}_{(-\infty,r)}(W_s)+q\mathbb{1}_{[r,\infty)}(W_s)\big)ds=pt,\ \ell(t,r)=0\Big)=1-\mathrm{Erfc}\Big(\frac{r-x}{\sqrt{2t}}\Big)$$

(2) $r\le x$
$$\mathbf{P}_x\Big(\int_0^t\big(p\mathbb{1}_{(-\infty,r)}(W_s)+q\mathbb{1}_{[r,\infty)}(W_s)\big)ds=qt,\ \ell(t,r)=0\Big)=1-\mathrm{Erfc}\Big(\frac{x-r}{\sqrt{2t}}\Big)$$

1.27.5
$$\mathbf{E}_x\Big\{\exp\Big(-\int_0^\tau\big(p\mathbb{1}_{(-\infty,r)}(W_s)+q\mathbb{1}_{[r,\infty)}(W_s)\big)ds\Big);\ \frac{1}{2}\ell(\tau,r)\in dy,\ W_\tau\in dz\Big\}$$

$x\le r$, $z\le r$
$$=2\lambda e^{-y\sqrt{2\lambda+2q}}e^{-(y+2r-x-z)\sqrt{2\lambda+2p}}\,dydz$$

$x\le r$, $r\le z$
$$=2\lambda e^{-(y+z-r)\sqrt{2\lambda+2q}}e^{-(y+r-x)\sqrt{2\lambda+2p}}\,dydz$$

$r\le x$, $z\le r$
$$=2\lambda e^{-(y+x-r)\sqrt{2\lambda+2q}}e^{-(y+r-z)\sqrt{2\lambda+2p}}\,dydz$$

$r\le x$, $r\le z$
$$=2\lambda e^{-(y+x+z-2r)\sqrt{2\lambda+2q}}e^{-y\sqrt{2\lambda+2p}}\,dydz$$

(1)
$$\mathbf{E}_x\Big\{\exp\Big(-\int_0^\tau\big(p\mathbb{1}_{(-\infty,r)}(W_s)+q\mathbb{1}_{[r,\infty)}(W_s)\big)ds\Big);\ \ell(\tau,r)=0,\ W_\tau\in dz\Big\}$$

1	W $\qquad\qquad$ $\tau \sim \operatorname{Exp}(\lambda)$, independent of W
$x \le r$ $z \le r$	$= \dfrac{\lambda}{\sqrt{2\lambda+2p}}\Big(e^{-\lvert z-x\rvert\sqrt{2\lambda+2p}} - e^{(z+x-2r)\sqrt{2\lambda+2p}}\Big)dz$
$r \le x$ $r \le z$	$= \dfrac{\lambda}{\sqrt{2\lambda+2q}}\Big(e^{-\lvert z-x\rvert\sqrt{2\lambda+2q}} - e^{(2r-x-z)\sqrt{2\lambda+2q}}\Big)dz$
1.27.6	$\mathbf{P}_x\Big(\int_0^\tau \mathbb{I}_{(-\infty,r)}(W_s)ds \in du, \int_0^\tau \mathbb{I}_{[r,\infty)}(W_s)ds \in dv, \frac{1}{2}\ell(\tau,r)\in dy, W_\tau \in dz\Big)$ $=: \lambda e^{-\lambda(u+v)}B_x^{(27)}(u,v,y,z)dudvdydz$
$x \le r$ $z \le r$	$= \lambda e^{-\lambda(u+v)}\dfrac{(y+2r-x-z)y}{\pi u^{3/2}v^{3/2}}\exp\Big(-\dfrac{(y+2r-x-z)^2}{2u} - \dfrac{y^2}{2v}\Big)dudvdydz$
$x \le r$ $r \le z$	$= \lambda e^{-\lambda(u+v)}\dfrac{(y+r-x)(y+z-r)}{\pi u^{3/2}v^{3/2}}\exp\Big(-\dfrac{(y+r-x)^2}{2u} - \dfrac{(y+z-r)^2}{2v}\Big)dudvdydz$
$r \le x$ $z \le r$	$= \lambda e^{-\lambda(u+v)}\dfrac{(y+r-z)(y+x-r)}{\pi u^{3/2}v^{3/2}}\exp\Big(-\dfrac{(y+r-z)^2}{2u} - \dfrac{(y+x-r)^2}{2v}\Big)dudvdydz$
$r \le x$ $r \le z$	$= \lambda e^{-\lambda(u+v)}\dfrac{y(y+x+z-2r)}{\pi u^{3/2}v^{3/2}}\exp\Big(-\dfrac{y^2}{2u} - \dfrac{(y+x+z-2r)^2}{2v}\Big)dudvdydz$
(1)	$\mathbf{P}_x\Big(\int_0^\tau \mathbb{I}_{(-\infty,r)}(W_s)ds \in du,\ \int_0^\tau \mathbb{I}_{[r,\infty)}(W_s)ds = 0,\ \ell(\tau,r)=0, W_\tau \in dz\Big)$
$x \le r$ $z \le r$	$= \lambda e^{-\lambda u}\Big(\dfrac{1}{\sqrt{2\pi u}}e^{-(z-x)^2/2u} - \dfrac{1}{\sqrt{2\pi u}}e^{-(2r-x-z)^2/2u}\Big)dudz$
(2)	$\mathbf{P}_x\Big(\int_0^\tau \mathbb{I}_{(-\infty,r)}(W_s)ds = 0,\ \int_0^\tau \mathbb{I}_{[r,\infty)}(W_s)ds \in dv,\ \ell(\tau,r)=0, W_\tau \in dz\Big)$
$r \le x$ $r \le z$	$= \lambda e^{-\lambda v}\Big(\dfrac{1}{\sqrt{2\pi v}}e^{-(z-x)^2/2v} - \dfrac{1}{\sqrt{2\pi v}}e^{-(x+z-2r)^2/2v}\Big)dvdz$
1.27.8	$\mathbf{P}_x\Big(\int_0^t (p\mathbb{I}_{(-\infty,r)}(W_s) + q\mathbb{I}_{[r,\infty)}(W_s))ds \in dv, \frac{1}{2}\ell(t,r)\in dy, W_t \in dz\Big)$ $= \dfrac{1}{\lvert p-q\rvert}B_x^{(27)}\Big(\dfrac{\lvert qt-v\rvert}{\lvert p-q\rvert}, \dfrac{\lvert pt-v\rvert}{\lvert p-q\rvert}, y, z\Big)\mathbb{I}_{((q\wedge p)t,(q\vee p)t)}(v)\,dvdydz$

1 W $\tau \sim \text{Exp}(\lambda)$, independent of W

for $B_x^{(27)}(u,v,y,z)$ see 1.27.6

(1) $x \le r$ $z \le r$

$$\mathbf{P}_x\Big(\int_0^t \big(p\mathbb{I}_{(-\infty,r)}(W_s)+q\mathbb{I}_{[r,\infty)}(W_s)\big)ds = pt, \ell(t,r)=0, W_t\in dz\Big)$$

$$= \frac{1}{\sqrt{2\pi t}}e^{-(z-x)^2/2t}dz - \frac{1}{\sqrt{2\pi t}}e^{-(2r-x-z)^2/2t}dz$$

(2) $r \le x$ $r \le z$

$$\mathbf{P}_x\Big(\int_0^t \big(p\mathbb{I}_{(-\infty,r)}(W_s)+q\mathbb{I}_{[r,\infty)}(W_s)\big)ds = qt, \ell(t,r)=0, W_t\in dz\Big)$$

$$= \frac{1}{\sqrt{2\pi t}}e^{-(z-x)^2/2t}dz - \frac{1}{\sqrt{2\pi t}}e^{-(x+z-2r)^2/2t}dz$$

1.28.1 $a < x$ $x < b$

$$\mathbf{E}_x\Big\{e^{-\gamma\hat H(\tau)-\eta\check H(\tau)}; \inf_{0\le s\le\tau} W_s\in da, \sup_{0\le s\le\tau} W_s\in db\Big\}$$

$$=\Big\{\frac{\sqrt{2\lambda(2\lambda+2\eta)}\,\text{sh}((b-x)\sqrt{2\lambda+2\gamma+2\eta})}{\text{sh}((b-a)\sqrt{2\lambda+2\gamma+2\eta})\,\text{sh}((b-a)\sqrt{2\lambda+2\eta})}$$

$$+\frac{\sqrt{2\lambda(2\lambda+2\gamma)}\,\text{sh}((x-a)\sqrt{2\lambda+2\gamma+2\eta})}{\text{sh}((b-a)\sqrt{2\lambda+2\gamma+2\eta})\,\text{sh}((b-a)\sqrt{2\lambda+2\gamma})}\Big\}\frac{\text{sh}((b-a)\sqrt{\lambda/2})}{\text{ch}((b-a)\sqrt{\lambda/2})}dadb$$

1.28.2

$$\mathbf{P}_x\big(\hat H(\tau)\in du, \check H(\tau)\in dv, \inf_{0\le s\le\tau} W_s\in da, \sup_{0\le s\le\tau} W_s\in db\big)$$

$u < v$

$$=\sqrt{2\lambda}e^{-\lambda v}\,\text{th}((b-a)\sqrt{\lambda/2})\,\text{ss}_u(b-x,b-a)\,\text{s}_{v-u}(b-a)\,dudvdadb$$

$v < u$

$$=\sqrt{2\lambda}e^{-\lambda u}\,\text{th}((b-a)\sqrt{\lambda/2})\,\text{ss}_v(x-a,b-a)\,\text{s}_{u-v}(b-a)\,dudvdadb$$

1.28.4

$$\mathbf{P}_x\big(\hat H(t)\in du, \check H(t)\in dv, \inf_{0\le s\le t} W_s\in da, \sup_{0\le s\le t} W_s\in db\big)$$

for $0 < u\wedge v < u\vee v < t, \qquad a < x < b$

$u < v$

$$=2\,\text{ss}_u(b-x,b-a)\,\text{s}_{v-u}(b-a)\,\text{sc}_{t-v}((b-a)/2,(b-a)/2)dudvdadb$$

$v < u$

$$=2\,\text{ss}_v(x-a,b-a)\,\text{s}_{u-v}(b-a)\,\text{sc}_{t-u}((b-a)/2,(b-a)/2)dudvdadb$$

1 W $\tau \sim \mathrm{Exp}(\lambda)$, independent of W

1.28.5 $\mathbf{E}_x\{e^{-\gamma\hat{H}(\tau)-\eta\check{H}(\tau)};\ \inf_{0\le s\le\tau} W_s\in da,\ \sup_{0\le s\le\tau} W_s\in db,\ W_\tau\in dz\}$

$a<x$

$x<b$

$$=\frac{2\lambda\sqrt{2\lambda+2\eta}\,\mathrm{sh}((b-x)\sqrt{2\lambda+2\gamma+2\eta})\,\mathrm{sh}((z-a)\sqrt{2\lambda})}{\mathrm{sh}((b-a)\sqrt{2\lambda+2\gamma+2\eta})\,\mathrm{sh}((b-a)\sqrt{2\lambda+2\eta})\,\mathrm{sh}((b-a)\sqrt{2\lambda})}dadbdz$$

$$+\frac{2\lambda\sqrt{2\lambda+2\gamma}\,\mathrm{sh}((x-a)\sqrt{2\lambda+2\gamma+2\eta})\,\mathrm{sh}((b-z)\sqrt{2\lambda})}{\mathrm{sh}((b-a)\sqrt{2\lambda+2\gamma+2\eta})\,\mathrm{sh}((b-a)\sqrt{2\lambda+2\gamma})\,\mathrm{sh}((b-a)\sqrt{2\lambda})}dadbdz$$

1.28.6 $\mathbf{P}_x\big(\hat{H}(\tau)\in du, \check{H}(\tau)\in dv, \inf_{0\le s\le\tau} W_s\in da, \sup_{0\le s\le\tau} W_s\in db, W_\tau\in dz\big)$

$u<v$ $$=2\lambda e^{-\lambda v}\frac{\mathrm{sh}((z-a)\sqrt{2\lambda})}{\mathrm{sh}((b-a)\sqrt{2\lambda})}\,\mathrm{ss}_u(b-x,b-a)\,\mathrm{s}_{v-u}(b-a)\,dudvdadbdz$$

$v<u$ $$=2\lambda e^{-\lambda u}\frac{\mathrm{sh}((b-z)\sqrt{2\lambda})}{\mathrm{sh}((b-a)\sqrt{2\lambda})}\,\mathrm{ss}_v(x-a,b-a)\,\mathrm{s}_{u-v}(b-a)\,dudvdadbdz$$

1.28.8 $\mathbf{P}_x\big(\hat{H}(t)\in du, \check{H}(t)\in dv, \inf_{0\le s\le t} W_s\in da, \sup_{0\le s\le t} W_s\in db, W_t\in dz\big)$

for $\quad 0<u\wedge v<u\vee v<t, \qquad a<x\wedge z<x\vee z<b$

$u<v$ $$=2\,\mathrm{ss}_u(b-x,b-a)\,\mathrm{s}_{v-u}(b-a)\,\mathrm{ss}_{t-v}(z-a,b-a)dudvdadbdz$$

$v<u$ $$=2\,\mathrm{ss}_v(x-a,b-a)\,\mathrm{s}_{u-v}(b-a)\,\mathrm{ss}_{t-u}(b-z,b-a)dudvdadbdz$$

1.29.1 $\mathbf{E}_x\Big\{\exp\Big(-\int_0^\tau\big(p\mathbb{I}_{(-\infty,r)}(W_s)+q\mathbb{I}_{[r,\infty)}(W_s)\big)ds\Big); a<\inf_{0\le s\le\tau} W_s,$

$\sup_{0\le s\le\tau} W_s<b, \frac{1}{2}\ell(\tau,r)\in dy\Big\}$

$$=2\lambda\exp\Big(-\frac{y\Upsilon_q\,\mathrm{ch}((b-r)\Upsilon_q)}{\mathrm{sh}((b-r)\Upsilon_q)}-\frac{y\Upsilon_p\,\mathrm{ch}((r-a)\Upsilon_p)}{\mathrm{sh}((r-a)\Upsilon_p)}\Big)$$

$$\times\Big(\frac{\mathrm{sh}((b-r)\Upsilon_q/2)}{\Upsilon_q\,\mathrm{ch}((b-r)\Upsilon_q/2)}+\frac{\mathrm{sh}((r-a)\Upsilon_p/2)}{\Upsilon_p\,\mathrm{ch}((r-a)\Upsilon_p/2)}\Big)\begin{cases}\dfrac{\mathrm{sh}((x-a)\Upsilon_p)}{\mathrm{sh}((r-a)\Upsilon_p)}dy, & a\le x\le r\\[2mm] \dfrac{\mathrm{sh}((b-x)\Upsilon_q)}{\mathrm{sh}((b-r)\Upsilon_q)}dy, & r\le x\le b\end{cases}$$

1 W $\tau \sim \operatorname{Exp}(\lambda)$, independent of W

(1) $$\mathbf{E}_x\Big\{\exp\Big(-\int_0^\tau \big(p\mathbb{I}_{(-\infty,r)}(W_s)+q\mathbb{I}_{[r,\infty)}(W_s)\big)ds\Big);a<\inf_{0\le s\le\tau}W_s,\ \sup_{0\le s\le\tau}W_s<b,\ell(\tau,r)=0\Big\}$$

$$=\begin{cases}\dfrac{\lambda}{\lambda+p}\Big(1-\dfrac{\operatorname{ch}((2x-r-a)\Upsilon_p/2)}{\operatorname{ch}((r-a)\Upsilon_p/2)}\Big), & a\le x\le r\\ \dfrac{\lambda}{\lambda+q}\Big(1-\dfrac{\operatorname{ch}((b+r-2x)\Upsilon_q/2)}{\operatorname{ch}((b-r)\Upsilon_q/2)}\Big), & r\le x\le b\end{cases}$$

1.29.5 $$\mathbf{E}_x\Big\{\exp\Big(-\int_0^\tau \big(p\mathbb{I}_{(-\infty,r)}(W_s)+q\mathbb{I}_{[r,\infty)}(W_s)\big)ds\Big);a<\inf_{0\le s\le\tau}W_s,\ \sup_{0\le s\le\tau}W_s<b,\frac{1}{2}\ell(\tau,r)\in dy,W_\tau\in dz\Big\}=:\lambda Q_\lambda^{(29)}(y,z)dydz$$

$$=2\lambda\exp\Big(-\frac{y\Upsilon_q\operatorname{ch}((b-r)\Upsilon_q)}{\operatorname{sh}((b-r)\Upsilon_q)}-\frac{y\Upsilon_p\operatorname{ch}((r-a)\Upsilon_p)}{\operatorname{sh}((r-a)\Upsilon_p)}\Big)$$

$$\times\begin{cases}\dfrac{\operatorname{sh}((x-a)\Upsilon_p)\operatorname{sh}((z-a)\Upsilon_p)}{\operatorname{sh}^2((r-a)\Upsilon_p)}dydz, & a\le x\le r,\ a\le z\le r\\ \dfrac{\operatorname{sh}((b-x)\Upsilon_q)\operatorname{sh}((z-a)\Upsilon_p)}{\operatorname{sh}((b-r)\Upsilon_q)\operatorname{sh}((r-a)\Upsilon_p)}dydz, & a\le z\le r\le x\le b\\ \dfrac{\operatorname{sh}((b-z)\Upsilon_q)\operatorname{sh}((x-a)\Upsilon_p)}{\operatorname{sh}((b-r)\Upsilon_q)\operatorname{sh}((r-a)\Upsilon_p)}dydz, & a\le x\le r\le z\le b\\ \dfrac{\operatorname{sh}((b-x)\Upsilon_q)\operatorname{sh}((b-z)\Upsilon_q)}{\operatorname{sh}^2((b-r)\Upsilon_q)}dydz, & r\le x\le b,\ r\le z\le b\end{cases}$$

(1) $$\mathbf{E}_x\Big\{\exp\Big(-\int_0^\tau \big(p\mathbb{I}_{(-\infty,r)}(W_s)+q\mathbb{I}_{[r,\infty)}(W_s)\big)ds\Big);a<\inf_{0\le s\le\tau}W_s,\ \sup_{0\le s\le\tau}W_s<b,\ell(\tau,r)=0,W_\tau\in dz\Big\}$$

$x\le r$, $z\le r$ $$=\frac{\lambda(\operatorname{ch}((r-a-|z-x|)\Upsilon_p)-\operatorname{ch}((r+a-z-x)\Upsilon_p))}{\Upsilon_p\operatorname{sh}((r-a)\Upsilon_p)}dz$$

$r\le x$, $r\le z$ $$=\frac{\lambda(\operatorname{ch}((b-r-|z-x|)\Upsilon_q)-\operatorname{ch}((b+r-z-x)\Upsilon_q))}{\Upsilon_q\operatorname{sh}((b-r)\Upsilon_q)}dz$$

$$\Upsilon_s:=\sqrt{2\lambda+2s}$$

1 W $\qquad$ $\tau \sim \text{Exp}(\lambda)$, independent of W

1.29.6
$$\mathbf{P}_x\Big(\int_0^\tau 1\!\!1_{(-\infty,r)}(W_s)ds \in du, \int_0^\tau 1\!\!1_{[r,\infty)}(W_s)ds \in dv, a < \inf_{0\le s\le\tau} W_s, \sup_{0\le s\le\tau} W_s < b, \frac{1}{2}\ell(\tau,r) \in dy, W_\tau \in dz\Big)$$

$$=: \lambda e^{-\lambda(u+v)} B_x^{(29)}(u,v,y,z)dudvdydz$$

$z \le r$, $x \le r$
$$= \lambda e^{-\lambda(u+v)}\, 2\,\text{esss}_u(x-a, z-a, r-a, y)\,\text{es}_v(0,0,b-r,0,y)dudvdydz$$

$z \le r$, $r \le x$
$$= \lambda e^{-\lambda(u+v)}\, 2\,\text{ess}_u(z-a, r-a, 0, y)\,\text{ess}_v(b-x, b-r, 0, y)dudvdydz$$

$r \le z$, $x \le r$
$$= \lambda e^{-\lambda(u+v)}\, 2\,\text{ess}_u(x-a, r-a, 0, y)\,\text{ess}_v(b-z, b-r, 0, y)dudvdydz$$

$r \le z$, $r \le x$
$$= \lambda e^{-\lambda(u+v)}\, 2\,\text{es}_u(0,0,r-a,0,y)\,\text{esss}_v(b-z, b-x, b-r, y)dudvdydz$$

(1)
$$\mathbf{P}_x\Big(\int_0^\tau 1\!\!1_{(-\infty,r)}(W_s)ds \in du,\ \int_0^\tau 1\!\!1_{[r,\infty)}(W_s)ds = 0,\ a < \inf_{0\le s\le\tau} W_s, \sup_{0\le s\le\tau} W_s < b, \ell(\tau,r) = 0, W_\tau \in dz\Big)$$

for $a < x \wedge z \le x \vee z < r$

$$= \lambda e^{-\lambda u}(\text{cs}_u(r-a-|z-x|, r-a) - \text{cs}_u(r+a-z-x, r-a))dudz$$

(2)
$$\mathbf{P}_x\Big(\int_0^\tau 1\!\!1_{(-\infty,r)}(W_s)ds = 0,\ \int_0^\tau 1\!\!1_{[r,\infty)}(W_s)ds \in dv,\ a < \inf_{0\le s\le\tau} W_s, \sup_{0\le s\le\tau} W_s < b, \ell(\tau,r) = 0, W_\tau \in dz\Big)$$

for $r < x \wedge z \le x \vee z < b$

$$= \lambda e^{-\lambda v}(\text{cs}_v(b-r-|z-x|, b-r) - \text{cs}_v(b+r-z-x, b-r))dvdz$$

1 W $\tau \sim \mathrm{Exp}(\lambda)$, independent of W

1.29.8 $$\mathbf{P}_x\Big(\int_0^t \big(p\mathbb{I}_{(-\infty,r)}(W_s)+q\mathbb{I}_{[r,\infty)}(W_s)\big)ds\in dv,\, a<\inf_{0\le s\le t}W_s,\ \sup_{0\le s\le t}W_s<b,\ \frac{1}{2}\ell(t,r)\in dy,\, W_t\in dz\Big)$$

$$=\frac{1}{|p-q|}B_x^{(29)}\Big(\frac{|qt-v|}{|p-q|},\frac{|pt-v|}{|p-q|},y,z\Big)\mathbb{I}_{((q\wedge p)t,(q\vee p)t)}(v)\,dvdydz$$

for $B_x^{(29)}(u,v,y,z)$ see 1.29.6

(1) $$\mathbf{P}_x\Big(\int_0^t \mathbb{I}_{(-\infty,r)}(W_s)ds=t,\ \int_0^t \mathbb{I}_{[r,\infty)}(W_s)ds=0,\ a<\inf_{0\le s\le t}W_s,\ \sup_{0\le s\le t}W_s<b,\ell(t,r)=0,W_t\in dz\Big)$$

for $a<x\wedge z\le x\vee z<r$

$$=\mathrm{cs}_t(r-a-|z-x|,r-a)dz-\mathrm{cs}_t(r+a-z-x,r-a)dz$$

(2) $$\mathbf{P}_x\Big(\int_0^t \mathbb{I}_{(-\infty,r)}(W_s)ds=0,\ \int_0^t \mathbb{I}_{[r,\infty)}(W_s)ds=t,\ a<\inf_{0\le s\le t}W_s,\ \sup_{0\le s\le t}W_s<b,\ell(t,r)=0,W_t\in dz\Big)$$

for $r<x\wedge z\le x\vee z<b$

$$=\mathrm{cs}_t(b-r-|z-x|,b-r)dz-\mathrm{cs}_t(b+r-z-x,b-r)dz$$

1.30.1 (1) $$\mathbf{E}_x\Big\{\exp\Big(-\frac{p^2}{2}\int_0^\tau e^{2\beta W_s}ds\Big);\int_0^\tau e^{\beta W_s}ds\in dy\Big\}=\frac{\lambda\sqrt{p\,\mathrm{ch}(yp|\beta|/2)}\Gamma(\sqrt{2\lambda}/|\beta|)}{\sqrt{|\beta|\,\mathrm{sh}(yp|\beta|/2)}\Gamma(1+\sqrt{8\lambda}/|\beta|)}$$

$\beta\ne 0$ $$\times\exp\Big(-\frac{\beta x}{2}-\frac{pe^{\beta x}\,\mathrm{ch}(yp|\beta|)}{|\beta|\,\mathrm{sh}(yp|\beta|)}\Big)M_{1/2,\sqrt{2\lambda}/|\beta|}\Big(\frac{2pe^{\beta x}}{|\beta|\,\mathrm{sh}(yp|\beta|)}\Big)dy$$

1 W $\tau \sim \text{Exp}(\lambda)$, independent of W

1.30.3 (1) $\mathbf{E}_x\Big\{\exp\Big(-\frac{p^2}{2}\int_0^t e^{2\beta W_s}ds\Big);\ \int_0^t e^{\beta W_s}ds\in dy\Big\}$

$\beta\neq 0$ $= \frac{\beta^2\sqrt{p\,\text{ch}(yp|\beta|/2)}}{2\sqrt{|\beta|\,\text{sh}(yp|\beta|/2)}}\exp\Big(-\frac{\beta x}{2}-\frac{pe^{\beta x}\,\text{ch}(yp|\beta|)}{|\beta|\,\text{sh}(yp|\beta|)}\Big)\,\text{m}_{\beta^2t/2}\Big(-\frac{1}{2},\frac{pe^{\beta x}}{|\beta|\,\text{sh}(yp|\beta|)}\Big)dy$

1.30.5 $\beta\neq 0$ $\mathbf{E}_x\Big\{\exp\Big(-\int_0^\tau\Big(\frac{p^2}{2}e^{2\beta W_s}+qe^{\beta W_s}\Big)ds\Big);\ W_\tau\in dz\Big\}=:\lambda Q_\lambda^{(30)}(z)dz$

$$= e^{-\beta(x+z)/2}\frac{\lambda\Gamma(\sqrt{2\lambda}/|\beta|+1/2+q/p|\beta|)}{p\Gamma(2\sqrt{2\lambda}/|\beta|+1)}$$

$$\times\begin{cases} M_{-q/p|\beta|,\sqrt{2\lambda}/|\beta|}\big(\frac{2p}{|\beta|}e^{\beta x}\big)W_{-q/p|\beta|,\sqrt{2\lambda}/|\beta|}\big(\frac{2p}{|\beta|}e^{\beta z}\big)dz, & \beta x\le\beta z\\ W_{-q/p|\beta|,\sqrt{2\lambda}/|\beta|}\big(\frac{2p}{|\beta|}e^{\beta x}\big)M_{-q/p|\beta|,\sqrt{2\lambda}/|\beta|}\big(\frac{2p}{|\beta|}e^{\beta z}\big)dz, & \beta z\le\beta x\end{cases}$$

(1) $\mathbf{E}_x\Big\{\exp\Big(-\frac{p^2}{2}\int_0^\tau e^{2\beta W_s}ds\Big);\int_0^\tau e^{\beta W_s}ds\in dy, W_\tau\in dz\Big\}=:\lambda B_\lambda^{(30)}(y,z)dydz$

$$=\frac{\lambda p}{\text{sh}(yp|\beta|/2)}\exp\Big(-\frac{p(e^{\beta x}+e^{\beta z})\,\text{ch}(yp|\beta|/2)}{|\beta|\,\text{sh}(yp|\beta|/2)}\Big)I_{2\sqrt{2\lambda}/|\beta|}\Big(\frac{2pe^{\beta(x+z)/2}}{|\beta|\,\text{sh}(yp|\beta|/2)}\Big)dydz$$

1.30.6 $\mathbf{P}_x\Big(\int_0^\tau e^{2\beta W_s}ds\in dg,\ \int_0^\tau e^{\beta W_s}ds\in dy,\ W_\tau\in dz\Big)$

$\beta\neq 0$ $=\lambda\,\text{is}_g\Big(\frac{2\sqrt{2\lambda}}{|\beta|},\frac{y|\beta|}{2},0,\frac{1}{|\beta|}\big(e^{\beta x}+e^{\beta z}\big),\frac{1}{|\beta|}e^{\beta(x+z)/2}\Big)dgdydz$

1.30.7 (1) $\mathbf{E}_x\Big\{\exp\Big(-\frac{p^2}{2}\int_0^t e^{2\beta W_s}ds\Big);\ \int_0^t e^{\beta W_s}ds\in dy,\ W_t\in dz\Big\}=:F_{x,t}^{(30)}(y,z)dydz$

$\beta\neq 0$ $=\frac{\beta^2p}{8\,\text{sh}(yp|\beta|/2)}\exp\Big(-\frac{p(e^{\beta x}+e^{\beta z})\,\text{ch}(yp|\beta|/2)}{|\beta|\,\text{sh}(yp|\beta|/2)}\Big)\,\text{i}_{\beta^2t/8}\Big(\frac{2pe^{\beta(x+z)/2}}{|\beta|\,\text{sh}(yp|\beta|/2)}\Big)dydz$

1 W $\qquad \tau \sim \operatorname{Exp}(\lambda)$, independent of W

1.30.8 $\mathbf{P}_x\Big(\int_0^t e^{2\beta W_s}ds \in dg,\ \int_0^t e^{\beta W_s}ds \in dy,\ W_t \in dz\Big)$

$\beta \neq 0$ $\quad = \frac{\beta^2}{8}\operatorname{ei}_g\Big(\frac{\beta^2 t}{8}, \frac{y|\beta|}{2}, \frac{1}{|\beta|}\big(e^{\beta x}+e^{\beta z}\big), \frac{2}{|\beta|}e^{\beta(x+z)/2}\Big)dg\,dy\,dz$

1.31.1 $\mathbf{E}_{x/\sqrt{2\lambda}}\exp\big(-\gamma\sqrt{2\lambda}\sup_{0\le s\le\tau}\big(\ell(s,u/\sqrt{2\lambda})-\alpha^{-1}\ell(s,r/\sqrt{2\lambda})\big)\big)$

$0<\alpha$

$$= 1-\frac{\gamma(2-2e^{-2|u-r|}-\Upsilon)e^{|u-r|-|x-r|}+\gamma\Upsilon e^{-|u-x|}}{2\gamma(1-e^{-2|u-r|})+\Upsilon}$$

1.31.2 $\mathbf{P}_{x/\sqrt{2\lambda}}\Big(\sup_{0\le s\le\tau}\big(\ell(s,u/\sqrt{2\lambda})-\alpha^{-1}\ell(s,r/\sqrt{2\lambda})\big) > h/\sqrt{2\lambda}\Big) = \Big[e^{|u-r|-|x-r|}$

$0<\alpha$

$$-\frac{\Upsilon}{2(1-e^{-2|u-r|})}\big(e^{|u-r|-|x-r|}-e^{-|u-x|}\big)\Big]\exp\Big(-\frac{\Upsilon h}{2(1-e^{-2|u-r|})}\Big)$$

1.31.5 $\mathbf{E}_{x/\sqrt{2\lambda}}\Big\{\exp\big(-\gamma\sqrt{2\lambda}\sup_{0\le s\le\tau}\big(\ell(s,u/\sqrt{2\lambda})-\alpha^{-1}\ell(s,r/\sqrt{2\lambda})\big)\big); W_\tau \in dz/\sqrt{2\lambda}\Big\}$

$0<\alpha$

$$= \frac{1}{2}e^{-|z-x|}dz - \frac{\gamma e^{-|z-u|}\big[(2-2e^{-2|u-r|}-\Upsilon)e^{|u-r|-|x-r|}+\Upsilon e^{-|u-x|}\big]}{2\big(2\gamma(1-e^{-2|u-r|})+\Upsilon\big)}dz$$

1.31.6 $\mathbf{P}_{x/\sqrt{2\lambda}}\Big(\sup_{0\le s\le\tau}\big(\ell(s,u/\sqrt{2\lambda})-\alpha^{-1}\ell(s,r/\sqrt{2\lambda})\big) > h/\sqrt{2\lambda},\ W_\tau \in dz/\sqrt{2\lambda}\Big)$

$0<\alpha$

$$= \frac{1}{2}\Big[e^{|u-r|-|x-r|}-\frac{\Upsilon}{2(1-e^{-2|u-r|})}\big(e^{|u-r|-|x-r|}-e^{-|u-x|}\big)\Big]$$

$$\times\exp\Big(-|z-u|-\frac{\Upsilon h}{2(1-e^{-2|u-r|})}\Big)dz$$

$$\Upsilon := \sqrt{(\alpha-1)^2+4\alpha(1-e^{-2|u-r|})}+1-\alpha$$

1 W $H_z = \min\{s : W_s = z\}$

2. Stopping at first hitting time

2.0.1 $\mathbf{E}_x e^{-\alpha H_z} = e^{-|x-z|\sqrt{2\alpha}}$

2.0.2 $\mathbf{P}_x\big(H_z \in dt\big) = \frac{|x-z|}{\sqrt{2\pi}t^{3/2}} \exp\Big(-\frac{(x-z)^2}{2t}\Big)dt$

2.1.1 $z \le x$ $\mathbf{E}_x \exp\big(-\gamma \sup_{0\le s\le H_z} W_s\big) = e^{-\gamma x} - \gamma(x-z)e^{-\gamma z}\int_{x-z}^{\infty}\frac{1}{y}e^{-\gamma y}dy$

2.1.2 $\mathbf{P}_x\big(\sup_{0\le s\le H_z} W_s < y\big) = \begin{cases} 1, & x \le z \\ \frac{y-x}{y-z}, & z \le x \le y \end{cases}$

2.1.4 $\mathbf{E}_x\big\{e^{-\alpha H_z}; \sup_{0\le s\le H_z} W_s < y\big\} = \begin{cases} e^{-(z-x)\sqrt{2\alpha}}, & x \le z \le y \\ \frac{\mathrm{sh}((y-x)\sqrt{2\alpha})}{\mathrm{sh}((y-z)\sqrt{2\alpha})}, & z \le x \le y \end{cases}$

(1) $\mathbf{P}_x\big(\sup_{0\le s\le H_z} W_s < y, H_z \in dt\big) = \begin{cases} \frac{|x-z|}{\sqrt{2\pi}t^{3/2}} \exp\Big(-\frac{(x-z)^2}{2t}\Big)dt, & x \le z \le y \\ \mathrm{ss}_t(y-x, y-z)dt, & z \le x \le y \end{cases}$

2.2.1 $x \le z$ $\mathbf{E}_x \exp\big(\gamma \inf_{0\le s\le H_z} W_s\big) = e^{\gamma x} - \gamma(z-x)e^{\gamma z}\int_{z-x}^{\infty}\frac{1}{y}e^{-\gamma y}dy$

2.2.2 $\mathbf{P}_x\big(\inf_{0\le s\le H_z} W_s > y\big) = \begin{cases} \frac{x-y}{z-y}, & y \le x \le z \\ 1, & y \le z \le x \end{cases}$

2.2.4 $\mathbf{E}_x\big\{e^{-\alpha H_z}; \inf_{0\le s\le H_z} W_s > y\big\} = \begin{cases} \frac{\mathrm{sh}((x-y)\sqrt{2\alpha})}{\mathrm{sh}((z-y)\sqrt{2\alpha})}, & y \le x \le z \\ e^{-(x-z)\sqrt{2\alpha}}, & y \le z \le x \end{cases}$

(1) $\mathbf{P}_x\big(\inf_{0\le s\le H_z} W_s > y, H_z \in dt\big) = \begin{cases} \mathrm{ss}_t(x-y, z-y)dt, & y \le x \le z \\ \frac{|x-z|}{\sqrt{2\pi}t^{3/2}} \exp\Big(-\frac{(x-z)^2}{2t}\Big)dt, & y \le z \le x \end{cases}$

1 W $H_z = \min\{s : W_s = z\}$

2.3.1 $$\mathbf{E}_x e^{-\gamma\ell(H_z,r)} = \begin{cases} 1, & x \le z \le r, \quad r \le z \le x \\ \dfrac{1+2\gamma|r-x|}{1+2\gamma|z-r|}, & r \wedge z \le x \le r \vee z \\ \dfrac{1}{1+2\gamma|z-r|}, & z \le r \le x, \quad x \le r \le z \end{cases}$$

2.3.2 $$\mathbf{P}_x\big(\ell(H_z,r) \in dy\big) = \begin{cases} 0, & x \le z \le r, \quad r \le z \le x \\ \dfrac{|x-z|}{2(r-z)^2}\exp\Big(-\dfrac{y}{2|r-z|}\Big)dy, & z \wedge r \le x \le r \vee z \\ \dfrac{1}{2|r-z|}\exp\Big(-\dfrac{y}{2|r-z|}\Big)dy, & z \le r \le x, \quad x \le r \le z \end{cases}$$

(1) $$\mathbf{P}_x\big(\ell(H_z,r) = 0\big) = \begin{cases} 1, & x \le z \le r, \quad r \le z \le x \\ \dfrac{|r-x|}{|r-z|}, & z \wedge r \le x \le r \vee z \\ 0, & z \le r \le x, \quad x \le r \le z \end{cases}$$

2.3.3 $$\mathbf{E}_x e^{-\alpha H_z - \gamma\ell(H_z,r)}$$

$$= e^{-|z-x|\sqrt{2\alpha}} \begin{cases} 1, & x \le z \le r, \quad r \le z \le x \\ \dfrac{\sqrt{2\alpha}+\gamma(1-e^{-2|x-r|\sqrt{2\alpha}})}{\sqrt{2\alpha}+\gamma(1-e^{-2|z-r|\sqrt{2\alpha}})}, & r \wedge z \le x \le r \vee z \\ \dfrac{\sqrt{2\alpha}}{\sqrt{2\alpha}+\gamma(1-e^{-2|z-r|\sqrt{2\alpha}})}, & z \le r \le x, \quad x \le r \le z \end{cases}$$

2.3.4 $$\mathbf{E}_x\big\{e^{-\alpha H_z}; \ell(H_z,r) \in dy\big\} = \exp\Big(-\frac{y\sqrt{2\alpha}}{1-e^{-2|r-z|\sqrt{2\alpha}}}\Big)$$

$$\times \begin{cases} 0, & x \le z \le r, \quad r \le z \le x \\ \dfrac{\sqrt{2\alpha}\,\mathrm{sh}(|x-z|\sqrt{2\alpha})}{2\,\mathrm{sh}^2(|r-z|\sqrt{2\alpha})}dy, & z \wedge r \le x \le r \vee z \\ \dfrac{\sqrt{2\alpha}}{2\,\mathrm{sh}(|r-z|\sqrt{2\alpha})}e^{-|r-x|\sqrt{2\alpha}}dy, & z \le r \le x, \quad x \le r \le z \end{cases}$$

(1) $$\mathbf{E}_x\big\{e^{-\alpha H_z}; \ell(H_z,r) = 0\big\} = \begin{cases} e^{-|z-x|\sqrt{2\alpha}}, & x \le z \le r, \quad r \le z \le x \\ \dfrac{\mathrm{sh}(|x-r|\sqrt{2\alpha})}{\mathrm{sh}(|z-r|\sqrt{2\alpha})}, & r \wedge z \le x \le r \vee z \\ 0, & z \le r \le x, \quad x \le r \le z \end{cases}$$

1 W $H_z = \min\{s : W_s = z\}$

(2) $$\mathbf{P}_x\Big(H_z \in dt,\ \frac{1}{2}\ell(H_z, r) \in dy\Big)$$

for $z \wedge r \le x \le r \vee z$

$$= \frac{1}{2}(\mathrm{es}_t(1, 2, |r-z|, y - |x-z|, y) - \mathrm{es}_t(1, 2, |r-z|, y + |x-z|, y))dtdy$$

for $z \le r \le x, \quad x \le r \le z$

$$= \mathrm{es}_t(1, 1, |r-z|, y + |x-r|, y)dtdy$$

2.4.1 $$\mathbf{E}_x \exp\Big(-\gamma \int_0^{H_z} 𝟙_{[r,\infty)}(W_s)ds\Big)$$

$z \le r$ $$= \begin{cases} 1, & x \le z \\ \dfrac{1 + (r-x)\sqrt{2\gamma}}{1 + (r-z)\sqrt{2\gamma}}, & z \le x \le r \\ \dfrac{1}{1 + (r-z)\sqrt{2\gamma}} e^{-(x-r)\sqrt{2\gamma}}, & r \le x \end{cases}$$

$r \le z$ $$= \begin{cases} \dfrac{1}{\mathrm{ch}((z-r)\sqrt{2\gamma})}, & x \le r \\ \dfrac{\mathrm{ch}((x-r)\sqrt{2\gamma})}{\mathrm{ch}((z-r)\sqrt{2\gamma})}, & r \le x \le z \\ e^{-(x-z)\sqrt{2\gamma}}, & z \le x \end{cases}$$

2.4.2 $$\mathbf{P}_x\Big(\int_0^{H_z} 𝟙_{[r,\infty)}(W_s)ds \in dy\Big)$$

$z < r$ $$= \begin{cases} 0, & x \le z \\ \dfrac{x-z}{(r-z)^2}\Big(\dfrac{1}{\sqrt{2\pi y}} - \dfrac{1}{2(r-z)} e^{y/2(r-z)^2} \operatorname{Erfc}\Big(\dfrac{\sqrt{y}}{(r-z)\sqrt{2}}\Big)\Big)dy, & z \le x \le r \\ \dfrac{1}{(r-z)\sqrt{2\pi y}} e^{-(x-r)^2/2y} dy - & \\ -\dfrac{1}{2(r-z)^2} \exp\Big(\dfrac{x-r}{r-z} + \dfrac{y}{2(r-z)^2}\Big) \operatorname{Erfc}\Big(\dfrac{x-r}{\sqrt{2y}} + \dfrac{\sqrt{y}}{(r-z)\sqrt{2}}\Big)dy, & r \le x \end{cases}$$

$r < z$ $$= \begin{cases} \mathrm{cc}_y(0, z-r)dy, & x \le r \\ \mathrm{cc}_y(x-r, z-r)dy, & r \le x \le z \\ \dfrac{x-z}{\sqrt{2\pi} y^{3/2}} e^{-(x-z)^2/2y} dy, & z \le x \end{cases}$$

1 $\quad W \qquad\qquad H_z = \min\{s : W_s = z\}$

(1)
$z \le r$

$$\mathbf{P}_x\Big(\int_0^{H_z} \mathbb{I}_{[r,\infty)}(W_s)ds = 0\Big) = \begin{cases} 1, & x \le z \\ \dfrac{r-x}{r-z}, & z \le x \le r \\ 0, & r \le x \end{cases}$$

2.4.4

$$\mathbf{P}_x\Big(\int_0^{H_z} \mathbb{I}_{[r,\infty)}(W_s)ds \in dv,\ H_z \in dt\Big) = \int_0^\infty B_z^{(27)}(t-v,v,y)dy\,dvdt$$

for $B_z^{(27)}(u,v,y)$ see 2.27.2

(1)

$$\mathbf{P}_x\Big(\int_0^{H_z} \mathbb{I}_{[r,\infty)}(W_s)ds = 0,\ H_z \in dt\Big) = \begin{cases} \mathrm{h}_t(1, z-x)dt, & x \le z \\ \mathrm{ss}_t(r-x, r-z)dt, & z \le x \le r \end{cases}$$

2.5.1

$$\mathbf{E}_x \exp\Big(-\gamma \int_0^{H_z} \mathbb{I}_{(-\infty,r]}(W_s)ds\Big)$$

$z \le r$

$$= \begin{cases} e^{-(z-x)\sqrt{2\gamma}}, & x \le z \\ \dfrac{\mathrm{ch}((r-x)\sqrt{2\gamma})}{\mathrm{ch}((r-z)\sqrt{2\gamma})}, & z \le x \le r \\ \dfrac{1}{\mathrm{ch}((r-z)\sqrt{2\gamma})}, & r \le x \end{cases}$$

$r \le z$

$$= \begin{cases} \dfrac{1}{1+(z-r)\sqrt{2\gamma}} e^{-(r-x)\sqrt{2\gamma}}, & x \le r \\ \dfrac{1+(x-r)\sqrt{2\gamma}}{1+(z-r)\sqrt{2\gamma}}, & r \le x \le z \\ 1, & z \le x \end{cases}$$

2.5.2

$$\mathbf{P}_x\Big(\int_0^{H_z} \mathbb{I}_{(-\infty,r]}(W_s)ds \in dy\Big)$$

$z < r$

$$= \begin{cases} \dfrac{z-x}{\sqrt{2\pi}y^{3/2}} e^{-(z-x)^2/2y}dy, & x \le z \\ \mathrm{cc}_y(r-x, r-z)dy, & z \le x \le r \\ \mathrm{cc}_y(0, r-z)dy, & r \le x \end{cases}$$

$r < z$

$$= \begin{cases} \dfrac{1}{(z-r)\sqrt{2\pi y}} e^{-(r-x)^2/2y}dy - \\ \quad -\dfrac{1}{2(z-r)^2}\exp\Big(\dfrac{r-x}{z-r} + \dfrac{y}{2(z-r)^2}\Big)\operatorname{Erfc}\Big(\dfrac{r-x}{\sqrt{2y}} + \dfrac{\sqrt{y}}{(z-r)\sqrt{2}}\Big)dy, & x \le r \\ \dfrac{z-x}{(z-r)^2}\Big(\dfrac{1}{\sqrt{2\pi y}} - \dfrac{1}{2(z-r)} e^{y/2(z-r)^2}\operatorname{Erfc}\Big(\dfrac{\sqrt{y}}{(z-r)\sqrt{2}}\Big)\Big)dy, & r \le x \le z \\ 0, & z \le x \end{cases}$$

1 W $H_z = \min\{s : W_s = z\}$

(1) $r \le z$

$$\mathbf{P}_x\Big(\int_0^{H_z} \mathbb{1}_{(-\infty,r]}(W_s)ds = 0\Big) = \begin{cases} 0, & x \le r \\ \dfrac{x-r}{z-r}, & r \le x \le z \\ 1, & z \le x \end{cases}$$

2.5.4

$$\mathbf{P}_x\Big(\int_0^{H_z} \mathbb{1}_{(-\infty,r)}(W_s)ds \in dv,\ H_z \in dt\Big) = \int_0^{\infty} B_z^{(27)}(v, t-v, y)dy\,dvdt$$

for $B_z^{(27)}(u,v,y)$ see 2.27.2

(1)

$$\mathbf{P}_x\Big(\int_0^{H_z} \mathbb{1}_{(-\infty,r)}(W_s)ds = 0,\ H_z \in dt\Big) = \begin{cases} \mathrm{ss}_t(x-r, z-r)dt, & r \le x \le z \\ \mathrm{h}_t(1, x-z)dt, & z \le x \end{cases}$$

2.6.1

$$\mathbf{E}_x \exp\Big(-\int_0^{H_z} (p\mathbb{1}_{(-\infty,r)}(W_s) + q\mathbb{1}_{[r,\infty)}(W_s))ds\Big) =: Q_z^{(6)}(2p, 2q)$$

$z \le r$

$$= \begin{cases} e^{-(z-x)\sqrt{2p}}, & x \le z \\ \dfrac{\sqrt{p}\,\mathrm{ch}((r-x)\sqrt{2p}) + \sqrt{q}\,\mathrm{sh}((r-x)\sqrt{2p})}{\sqrt{p}\,\mathrm{ch}((r-z)\sqrt{2p}) + \sqrt{q}\,\mathrm{sh}((r-z)\sqrt{2p})}, & z \le x \le r \\ \dfrac{\sqrt{p}e^{-(x-r)\sqrt{2q}}}{\sqrt{p}\,\mathrm{ch}((r-z)\sqrt{2p}) + \sqrt{q}\,\mathrm{sh}((r-z)\sqrt{2p})}, & r \le x \end{cases}$$

$r \le z$

$$= \begin{cases} \dfrac{\sqrt{q}e^{-(r-x)\sqrt{2p}}}{\sqrt{q}\,\mathrm{ch}((z-r)\sqrt{2q}) + \sqrt{p}\,\mathrm{sh}((z-r)\sqrt{2q})}, & x \le r \\ \dfrac{\sqrt{q}\,\mathrm{ch}((x-r)\sqrt{2q}) + \sqrt{p}\,\mathrm{sh}((x-r)\sqrt{2q})}{\sqrt{q}\,\mathrm{ch}((z-r)\sqrt{2q}) + \sqrt{p}\,\mathrm{sh}((z-r)\sqrt{2q})}, & r \le x \le z \\ e^{-(x-z)\sqrt{2q}}, & z \le x \end{cases}$$

2.6.2

$$\mathbf{P}_x\Big(\int_0^{H_z} \mathbb{1}_{(-\infty,r)}(W_s)ds \in du,\ \int_0^{H_z} \mathbb{1}_{[r,\infty)}(W_s)ds \in dv\Big)$$

$$= \int_0^{\infty} B_z^{(27)}(u,v,y)dy\,dudv$$ for $B_z^{(27)}(u,v,y)$ see 2.27.2

(1)

$$\mathbf{P}_x\Big(\int_0^{H_z} \mathbb{1}_{(-\infty,r)}(W_s)ds \in du,\ \int_0^{H_z} \mathbb{1}_{[r,\infty)}(W_s)ds = 0\Big)$$

$$= \begin{cases} \mathrm{h}_u(1, z-x)du, & x \le z \\ \mathrm{ss}_u(r-x, r-z)du, & z \le x \le r \end{cases}$$

1 W $H_z = \min\{s : W_s = z\}$

(2) $$\mathbf{P}_x\Big(\int_0^{H_z} \mathbb{I}_{(-\infty,r)}(W_s)ds = 0,\ \int_0^{H_z} \mathbb{I}_{[r,\infty)}(W_s)ds \in dv\Big)$$

$$= \begin{cases} \mathrm{ss}_v(x-r, z-r)dv, & r \le x \le z \\ \mathrm{h}_v(1, x-z)dv, & z \le x \end{cases}$$

2.7.1
$r < u$

$$\mathbf{E}_x \exp\Big(-\gamma \int_0^{H_z} \mathbb{I}_{[r,u]}(W_s)ds\Big)$$

$z \le r$

$$= \begin{cases} 1, & x \le z \\ \dfrac{(r-x)\sqrt{2\gamma}\,\mathrm{sh}((u-r)\sqrt{2\gamma}) + \mathrm{ch}((u-r)\sqrt{2\gamma})}{(r-z)\sqrt{2\gamma}\,\mathrm{sh}((u-r)\sqrt{2\gamma}) + \mathrm{ch}((u-r)\sqrt{2\gamma})}, & z \le x \le r \\ \dfrac{\mathrm{ch}((u-x)\sqrt{2\gamma})}{(r-z)\sqrt{2\gamma}\,\mathrm{sh}((u-r)\sqrt{2\gamma}) + \mathrm{ch}((u-r)\sqrt{2\gamma})}, & r \le x \le u \\ \dfrac{1}{(r-z)\sqrt{2\gamma}\,\mathrm{sh}((u-r)\sqrt{2\gamma}) + \mathrm{ch}((u-r)\sqrt{2\gamma})}, & u \le x \end{cases}$$

$u \le z$

$$= \begin{cases} \dfrac{1}{(z-u)\sqrt{2\gamma}\,\mathrm{sh}((u-r)\sqrt{2\gamma}) + \mathrm{ch}((u-r)\sqrt{2\gamma})}, & x \le r \\ \dfrac{\mathrm{ch}((x-r)\sqrt{2\gamma})}{(z-u)\sqrt{2\gamma}\,\mathrm{sh}((u-r)\sqrt{2\gamma}) + \mathrm{ch}((u-r)\sqrt{2\gamma})}, & r \le x \le u \\ \dfrac{(x-u)\sqrt{2\gamma}\,\mathrm{sh}((u-r)\sqrt{2\gamma}) + \mathrm{ch}((u-r)\sqrt{2\gamma})}{(z-u)\sqrt{2\gamma}\,\mathrm{sh}((u-r)\sqrt{2\gamma}) + \mathrm{ch}((u-r)\sqrt{2\gamma})}, & u \le x \le z \\ 1, & z \le x \end{cases}$$

2.7.2
$r < u$

$$\mathbf{P}_x\Big(\int_0^{H_z} \mathbb{I}_{[r,u]}(W_s)ds \in dy\Big) =: B_z^{(7)}(y)dy$$

$$= \begin{cases} \dfrac{(x-z)}{(r-z)}\,\mathrm{rs}_y(0, u-r, u-r, r-z)dy, & z \le x \le r \\ \mathrm{rs}_y(0, u-r, u-x, r-z))dy, & r \le x \le u \\ \mathrm{rs}_y(0, u-r, 0, r-z)dy, & u \le x \end{cases}$$

$$= \begin{cases} \mathrm{rs}_y(0, u-r, 0, z-u)dy, & x \le r \\ \mathrm{rs}_y(0, u-r, x-r, z-u)dy, & r \le x \le u \\ \dfrac{(z-x)}{(z-u)}\,\mathrm{rs}_y(0, u-r, u-r, z-u)dy, & u \le x \le z \end{cases}$$

1 W $H_z = \min\{s : W_s = z\}$

2.8.1 (1) $$\mathbf{E}_x \exp\Big(-\gamma \int_0^{H_0} |W_s| ds\Big) = 3^{2/3}\Gamma(2/3)\,\mathrm{Ai}\big((2\gamma)^{1/3}|x|\big)$$

2.8.2 (1) $$\mathbf{P}_x\Big(\int_0^{H_0} |W_s| ds \in dy\Big) = \Big(\frac{2}{9y^4}\Big)^{1/3} \frac{|x|}{\Gamma(1/3)} e^{-2|x|^3/9y} dy$$

2.9.3 $$\mathbf{E}_x \exp\Big(-\alpha H_z - \frac{\gamma^2}{2}\int_0^{H_z} W_s^2 ds\Big) = \begin{cases} \dfrac{D_{-1/2-\alpha/\gamma}(-x\sqrt{2\gamma})}{D_{-1/2-\alpha/\gamma}(-z\sqrt{2\gamma})}, & x \le z \\ \dfrac{D_{-1/2-\alpha/\gamma}(x\sqrt{2\gamma})}{D_{-1/2-\alpha/\gamma}(z\sqrt{2\gamma})}, & z \le x \end{cases}$$

2.9.4 (1) $$\mathbf{E}_x\Big\{\exp\Big(-\frac{\gamma^2}{2}\int_0^{H_0} W_s^2 ds\Big); H_0 \in dt\Big\} = \frac{|x|\gamma^{3/2}}{\sqrt{2\pi}\,\mathrm{sh}^{3/2}(t\gamma)} \exp\Big(-\frac{x^2\gamma\,\mathrm{ch}(t\gamma)}{2\,\mathrm{sh}(t\gamma)}\Big) dt$$

(2) $$\mathbf{P}_x\Big(\int_0^{H_0} W_s^2 ds \in dy, H_0 \in dt\Big) = \frac{|x|}{\sqrt{2\pi}}\,\mathrm{es}_y\big(\frac{3}{2}, \frac{3}{2}, t, 0, \frac{x^2}{2}\big) dy dt$$

2.10.3 $$\mathbf{E}_x \exp\Big(-\alpha H_z - \gamma\int_0^{H_z} e^{2\beta W_s} ds\Big) = \begin{cases} \dfrac{I_{\sqrt{2\alpha}/|\beta|}\Big(\frac{\sqrt{2\gamma}}{|\beta|}e^{\beta x}\Big)}{I_{\sqrt{2\alpha}/|\beta|}\Big(\frac{\sqrt{2\gamma}}{|\beta|}e^{\beta z}\Big)}, & (z-x)\beta \ge 0 \\ \dfrac{K_{\sqrt{2\alpha}/|\beta|}\Big(\frac{\sqrt{2\gamma}}{|\beta|}e^{\beta x}\Big)}{K_{\sqrt{2\alpha}/|\beta|}\Big(\frac{\sqrt{2\gamma}}{|\beta|}e^{\beta z}\Big)}, & (z-x)\beta \le 0 \end{cases}$$

2.11.2 $$\lambda\int_0^\infty e^{-\lambda z}\mathbf{P}_0\big(\sup_{y\in\mathbb{R}} \ell(H_z, y) < h\big) dz = 1 + \frac{2}{\lambda h}\Big(\frac{1}{I_0(\sqrt{2\lambda h})} - 1\Big)$$

(1) $z \ge 0$ $$\mathbf{P}_0\big(\sup_{y\in\mathbb{R}} \ell(H_z, y) < h\big) = 8\sum_{k=1}^\infty \frac{\exp(-z j_{0,k}^2/2h)}{j_{0,k}^3 J_1(j_{0,k})}$$

2.12.1 $$\mathbf{E}_x e^{-\gamma \check{H}(H_z)} = \mathbf{E}_x e^{-\gamma \hat{H}(H_z)} = \mathrm{sh}(|x-z|\sqrt{2\gamma}) \int_{|x-z|}^\infty \frac{1}{h\,\mathrm{sh}(h\sqrt{2\gamma})} dh$$

2.12.2 $$\mathbf{P}_x\big(\check{H}(H_z) \in du\big) = \mathbf{P}_x\big(\hat{H}(H_z) \in du\big) = \int_{|x-z|}^\infty h^{-1}\,\mathrm{ss}_u(|x-z|, h) dh\, du$$

1 W	$H_z = \min\{s : W_s = z\}$
2.12.4 $u<t$	$\mathbf{P}_x\big(\check H(H_z)\in du,\ H_z\in dt\big)=\mathbf{P}_x\big(\hat H(H_z)\in du,\ H_z\in dt\big)$ $=\int_{\lvert x-z\rvert}^{\infty}\mathrm{ss}_u(\lvert x-z\rvert,h)\,\mathrm{s}_{t-u}(h)dh\,dudt$
2.13.1 $z<x$ $x<y$	$\mathbf{E}_x\big\{e^{-\gamma\check H(H_z)};\ \sup_{0\le s\le H_z} W_s\in dy\big\}=\frac{\mathrm{sh}((x-z)\sqrt{2\gamma})}{(y-z)\,\mathrm{sh}((y-z)\sqrt{2\gamma})}dy$
2.13.2 $z<x$ $x<y$	$\mathbf{P}_x\big(\check H(H_z)\in du,\ \sup_{0\le s\le H_z} W_s\in dy\big)=(y-z)^{-1}\,\mathrm{ss}_u(x-z,y-z)dudy$
2.13.4 $z<x$ $x<y$	$\mathbf{E}_x\big\{e^{-\gamma\check H(H_z)-\alpha H_z};\ \sup_{0\le s\le H_z} W_s\in dy\big\}=\frac{\sqrt{2\alpha}\,\mathrm{sh}((x-z)\sqrt{2\gamma+2\alpha})}{\mathrm{sh}((y-z)\sqrt{2\alpha})\,\mathrm{sh}((y-z)\sqrt{2\gamma+2\alpha})}dy$
(1) $u<t$	$\mathbf{P}_x\big(\check H(H_z)\in du,\ H_z\in dt,\ \sup_{0\le s\le H_z} W_s\in dy\big)$ $=\mathrm{ss}_u(x-z,y-z)\,\mathrm{s}_{t-u}(y-z)dudtdy$
2.14.1 $x<z$ $y<x$	$\mathbf{E}_x\big\{e^{-\gamma\hat H(H_z)};\ \inf_{0\le s\le H_z} W_s\in dy\big\}=\frac{\mathrm{sh}((z-x)\sqrt{2\gamma})}{(z-y)\,\mathrm{sh}((z-y)\sqrt{2\gamma})}dy$
2.14.2 $x<z$ $y<x$	$\mathbf{P}_x\big(\hat H(H_z)\in du,\ \inf_{0\le s\le H_z} W_s\in dy\big)=(z-y)^{-1}\,\mathrm{ss}_u(z-x,z-y)dudy$
2.14.4 $x<z$ $y<x$	$\mathbf{E}_x\big\{e^{-\gamma\hat H(H_z)-\alpha H_z};\ \inf_{0\le s\le H_z} W_s\in dy\big\}=\frac{\sqrt{2\alpha}\,\mathrm{sh}((z-x)\sqrt{2\gamma+2\alpha})}{\mathrm{sh}((z-y)\sqrt{2\alpha})\,\mathrm{sh}((z-y)\sqrt{2\gamma+2\alpha})}dy$
(1) $u<t$	$\mathbf{P}_x\big(\hat H(H_z)\in du,\ H_z\in dt,\ \inf_{0\le s\le H_z} W_s\in dy\big)$ $=\mathrm{ss}_u(z-x,z-y)\,\mathrm{s}_{t-u}(z-y)dudtdy$

1 W $H_z = \min\{s : W_s = z\}$

2.15.2 $$\mathbf{P}_x\big(a < \inf_{0\le s\le H_z} W_s,\ \sup_{0\le s\le H_z} W_s < b\big) = \begin{cases} \dfrac{x-a}{z-a}, & a \le x \le z \\ \dfrac{b-x}{b-z}, & z \le x \le b \end{cases}$$

2.15.4 $$\mathbf{E}_x\{e^{-\alpha H_z}; a < \inf_{0\le s\le H_z} W_s,\ \sup_{0\le s\le H_z} W_s < b\} = \begin{cases} \dfrac{\operatorname{sh}((x-a)\sqrt{2\alpha})}{\operatorname{sh}((z-a)\sqrt{2\alpha})}, & a \le x \le z \\ \dfrac{\operatorname{sh}((b-x)\sqrt{2\alpha})}{\operatorname{sh}((b-z)\sqrt{2\alpha})}, & z \le x \le b \end{cases}$$

(1) $$\mathbf{P}_x\big(H_z \in dt, a < \inf_{0\le s\le t} W_s,\ \sup_{0\le s\le t} W_s < b\big) = \begin{cases} \operatorname{ss}_t(x-a, z-a)dt, & a \le x \le z \\ \operatorname{ss}_t(b-x, b-z)dt, & z \le x \le b \end{cases}$$

2.16.1 $$\mathbf{E}_x\{e^{-\gamma\ell(H_z,r)};\ \sup_{0\le s\le H_z} W_s < b\}$$
$r < b$
$z < r$

$$= \begin{cases} \dfrac{b-x}{b-z+2\gamma(b-r)(r-z)}, & r \le x \le b \\ \dfrac{r-x}{r-z} + \dfrac{(b-r)(x-z)}{(r-z)(b-z+2\gamma(b-r)(r-z))}, & z \le x \le r \end{cases}$$

2.16.2 $$\mathbf{P}_x\big(\ell(H_z,r) \in dy,\ \sup_{0\le s\le H_z} W_s < b\big)$$
$r < b$
$z < r$

$$= \begin{cases} \dfrac{b-x}{2(b-r)(r-z)} \exp\Big(-\dfrac{(b-z)y}{2(b-r)(r-z)}\Big)dy, & r \le x \le b \\ \dfrac{x-z}{2(r-z)^2} \exp\Big(-\dfrac{(b-z)y}{2(b-r)(r-z)}\Big)dy, & z \le x \le r \end{cases}$$

(1) $$\mathbf{P}_x\big(\ell(H_z,r) = 0,\ \sup_{0\le s\le H_z} W_s < b\big) = \begin{cases} 0, & r \le x \le b \\ \dfrac{r-x}{r-z}, & z \le x \le r \end{cases}$$

2.17.1 $$\mathbf{E}_x\{e^{-\gamma\ell(H_z,r)};\ a < \inf_{0\le s\le H_z} W_s\}$$
$a < r$
$r < z$

$$= \begin{cases} \dfrac{x-r}{z-r} + \dfrac{(z-x)(r-a)}{(z-r)(z-a+2\gamma(z-r)(r-a))}, & r \le x \le z \\ \dfrac{x-a}{z-a+2\gamma(z-r)(r-a)}, & a \le x \le r \end{cases}$$

1 W	$H_z = \min\{s : W_s = z\}$

2.17.2 $a < r$, $r < z$

$$\mathbf{P}_x\big(\ell(H_z,r) \in dy, a < \inf_{0\le s\le H_z} W_s\big)$$

$$= \begin{cases} \dfrac{z-x}{2(z-r)^2}\exp\Big(-\dfrac{(z-a)y}{2(z-r)(r-a)}\Big)dy, & r \le x \le z \\ \dfrac{x-a}{2(z-r)(r-a)}\exp\Big(-\dfrac{(z-a)y}{2(z-r)(r-a)}\Big)dy, & a \le x \le r \end{cases}$$

(1) $$\mathbf{P}_x\big(\ell(H_z,r) = 0, a < \inf_{0\le s\le H_z} W_s\big) = \begin{cases} \dfrac{x-r}{z-r}, & r \le x \le z \\ 0, & a \le x \le r \end{cases}$$

2.18.1 $r < u$

$$\mathbf{E}_x \exp\big(-\gamma\ell(H_z,r) - \eta\ell(H_z,u)\big)$$

$z < r$

$$= \begin{cases} 1, & x \le z \\ \dfrac{1+2\gamma(r-x)+2\eta(u-x)+4\gamma\eta(u-r)(r-x)}{1+2\gamma(r-z)+2\eta(u-z)+4\gamma\eta(u-r)(r-z)}, & z \le x \le r \\ \dfrac{1+2\eta(u-x)}{1+2\gamma(r-z)+2\eta(u-z)+4\gamma\eta(u-r)(r-z)}, & r \le x \le u \\ \dfrac{1}{1+2\gamma(r-z)+2\eta(u-z)+4\gamma\eta(u-r)(r-z)}, & u \le x \end{cases}$$

$u < z$

$$= \begin{cases} \dfrac{1}{1+2\gamma(z-r)+2\eta(z-u)+4\gamma\eta(z-u)(u-r)}, & x \le r \\ \dfrac{1+2\gamma(x-r)}{1+2\gamma(z-r)+2\eta(z-u)+4\gamma\eta(z-u)(u-r)}, & r \le x \le u \\ \dfrac{1+2\gamma(x-r)+2\eta(x-u)+4\gamma\eta(x-u)(u-r)}{1+2\gamma(z-r)+2\eta(z-u)+4\gamma\eta(z-u)(u-r)}, & u \le x \le z \\ 1, & z \le x \end{cases}$$

2.18.2 $r < u$

$$\mathbf{P}_x\big(\ell(H_z,r) \in dy,\ \ell(H_z,u) \in dg\big) =: B^{(18)}_{x,z}(y,g)dydg$$

$z < r$

$$= \begin{cases} \dfrac{x-z}{r-z}p_r dydg, & z \le x \le r \\ \dfrac{u-x}{u-r}p_r dydg + \dfrac{x-r}{u-r}p_u dydg, & r \le x \le u \\ p_u dydg, & u \le x \end{cases}$$

$u < z$

$$= \begin{cases} q_r dydg, & x \le r \\ \dfrac{u-x}{u-r}q_r dydg + \dfrac{x-r}{u-r}q_u dydg, & r \le x \le u \\ \dfrac{z-x}{z-u}q_u dydg, & u \le x \le z \end{cases}$$

1 W $H_z = \min\{s : W_s = z\}$

where

$$p_r := \frac{\sqrt{y}}{4(u-r)(r-z)\sqrt{g}} \exp\Big(-\frac{y+g}{2(u-r)} - \frac{y}{2(r-z)}\Big) I_1\Big(\frac{\sqrt{yg}}{u-r}\Big)$$

$$q_r := \frac{1}{4(z-u)(u-r)} \exp\Big(-\frac{y+g}{2(u-r)} - \frac{g}{2(z-u)}\Big) I_0\Big(\frac{\sqrt{yg}}{u-r}\Big)$$

$$p_u := \frac{1}{4(u-r)(r-z)} \exp\Big(-\frac{y+g}{2(u-r)} - \frac{y}{2(r-z)}\Big) I_0\Big(\frac{\sqrt{yg}}{u-r}\Big)$$

$$q_u := \frac{\sqrt{g}}{4(z-u)(u-r)\sqrt{y}} \exp\Big(-\frac{y+g}{2(u-r)} - \frac{g}{2(z-u)}\Big) I_1\Big(\frac{\sqrt{yg}}{u-r}\Big)$$

(1) $$\mathbf{P}_x\big(\ell(H_z,r) \in dy,\ \ell(H_z,u) = 0\big)$$
$$= \begin{cases} \dfrac{x-z}{2(r-z)^2} \exp\Big(-\dfrac{y(u-z)}{2(u-r)(r-z)}\Big) dy, & z \le x \le r \\ \dfrac{u-x}{2(u-r)(r-z)} \exp\Big(-\dfrac{y(u-z)}{2(u-r)(r-z)}\Big) dy, & r \le x \le u \end{cases}$$

(2) $$\mathbf{P}_x\big(\ell(H_z,r) = 0,\ \ell(H_z,u) \in dg\big)$$
$$= \begin{cases} \dfrac{x-r}{2(z-u)(u-r)} \exp\Big(-\dfrac{g(z-r)}{2(z-u)(u-r)}\Big) dg, & r \le x \le u \\ \dfrac{z-x}{2(z-u)^2} \exp\Big(-\dfrac{g(z-r)}{2(z-u)(u-r)}\Big) dg, & u \le x \le z \end{cases}$$

(3) $$\mathbf{P}_x\big(\ell(H_z,r) = 0,\ \ell(H_z,u) = 0\big) = \begin{cases} \dfrac{r-x}{r-z}, & z \le x \le r \\ \dfrac{x-u}{z-u}, & u \le x \le z \end{cases}$$

2.19.1 $$\mathbf{E}_x\Big\{\exp\Big(-\gamma\int_0^{H_z} \frac{ds}{W_s}\Big);\ 0 < \inf_{0 \le s \le H_z} W_s\Big\} = \begin{cases} \dfrac{x^{1/2} I_1(2\sqrt{2\gamma x})}{z^{1/2} I_1(2\sqrt{2\gamma z})}, & 0 < x \le z \\ \dfrac{x^{1/2} K_1(2\sqrt{2\gamma x})}{z^{1/2} K_1(2\sqrt{2\gamma z})}, & z \le x \end{cases}$$

2.19.1 (1) $$\mathbf{E}_x \exp\Big(-\gamma\int_0^{H_0} \frac{ds}{W_s}\Big) = 2\sqrt{2\gamma x} K_1(2\sqrt{2\gamma x})$$ $0 \le x$

2.19.2 (1) $$\mathbf{P}_x\Big(\int_0^{H_0} \frac{ds}{W_s} \in dy\Big) = \frac{2x}{y^2} e^{-2x/y} dy$$ $0 \le x$

1 W $H_z = \min\{s : W_s = z\}$

2.20.1 $$\mathbf{E}_x\Big\{\exp\Big(-\frac{\gamma^2}{2}\int_0^{H_z}\frac{ds}{W_s^2}\Big); 0 < \inf_{0\le s\le H_z} W_s\Big\} = \begin{cases} \Big(\dfrac{x}{z}\Big)^{1/2+\sqrt{1/4+\gamma^2}}, & 0 < x \le z \\ \Big(\dfrac{x}{z}\Big)^{1/2-\sqrt{1/4+\gamma^2}}, & z \le x \end{cases}$$

2.20.4 $$\mathbf{P}_x\Big(\int_0^{H_z}\frac{ds}{W_s^2} \in dy,\ 0 < \inf_{0\le s\le H_z} W_s\Big)$$

$$= \begin{cases} \dfrac{x^{1/2}\ln(z/x)}{z^{1/2}\sqrt{2\pi}y^{3/2}} e^{-y/8} e^{-\ln^2(z/x)/2y} dy, & 0 < x \le z \\ \dfrac{x^{1/2}\ln(x/z)}{z^{1/2}\sqrt{2\pi}y^{3/2}} e^{-y/8} e^{-\ln^2(x/z)/2y} dy, & z \le x \end{cases}$$

2.21.1 $$\mathbf{E}_x\Big\{\exp\Big(-\int_0^{H_z}\Big(\frac{p^2}{2W_s^2} + \frac{q^2W_s^2}{2}\Big)ds\Big);\ 0 < \inf_{0\le s\le H_z} W_s\Big\}$$

$$= \begin{cases} \dfrac{x^{1/2} I_{\sqrt{4p^2+1}/4}(qx^2/2)}{z^{1/2} I_{\sqrt{4p^2+1}/4}(qz^2/2)}, & 0 < x \le z \\ \dfrac{x^{1/2} K_{\sqrt{4p^2+1}/4}(qx^2/2)}{z^{1/2} K_{\sqrt{4p^2+1}/4}(qz^2/2)}, & z \le x \end{cases}$$

2.22.1 $$\mathbf{E}_x\Big\{\exp\Big(-\int_0^{H_z}\Big(\frac{p^2}{2W_s^2} + \frac{q}{W_s}\Big)ds\Big);\ 0 < \inf_{0\le s\le H_z} W_s\Big\}$$

$$= \begin{cases} \dfrac{x^{1/2} I_{\sqrt{4p^2+1}}(2\sqrt{2qx})}{z^{1/2} I_{\sqrt{4p^2+1}}(2\sqrt{2qz})}, & 0 < x \le z \\ \dfrac{x^{1/2} K_{\sqrt{4p^2+1}}(2\sqrt{2qx})}{z^{1/2} K_{\sqrt{4p^2+1}}(2\sqrt{2qz})}, & z \le x \end{cases}$$

2.27.1 $$\mathbf{E}_x\Big\{\exp\Big(-\int_0^{H_z}\big(p\mathbb{I}_{(-\infty,r)}(W_s) + q\mathbb{I}_{[r,\infty)}(W_s)\big)ds\Big); \tfrac{1}{2}\ell(H_z, r) \in dy\Big\}$$

$$=: Q_z^{(27)}(2p, 2q, y)dy$$

$z \le r$, $x \le r$
$$= \frac{\sqrt{2p}\,\mathrm{sh}((x-z)\sqrt{2p})}{\mathrm{sh}^2((r-z)\sqrt{2p})}\exp\big(-y\sqrt{2p}\,\mathrm{cth}((r-z)\sqrt{2p}) - y\sqrt{2q}\big)dy$$

$z \le r$, $r \le x$
$$= \frac{\sqrt{2p}}{\mathrm{sh}((r-z)\sqrt{2p})}\exp\big(-y\sqrt{2p}\,\mathrm{cth}((r-z)\sqrt{2p}) - (y+x-r)\sqrt{2q}\big)dy$$

1 W $H_z = \min\{s : W_s = z\}$

$r \le z$ $x \le r$	$= \frac{\sqrt{2q}}{\operatorname{sh}((z-r)\sqrt{2q})} \exp\big(-y\sqrt{2q}\operatorname{cth}((z-r)\sqrt{2q}) - (y+r-x)\sqrt{2p}\big)dy$
$r \le z$ $r \le x$	$= \frac{\sqrt{2q}\operatorname{sh}((z-x)\sqrt{2q})}{\operatorname{sh}^2((z-r)\sqrt{2q})} \exp\big(-y\sqrt{2q}\operatorname{cth}((z-r)\sqrt{2q}) - y\sqrt{2p}\big)dy$
(1)	$\mathbf{E}_x\Big\{\exp\Big(-\int_0^{H_z}(p\mathbb{I}_{(-\infty,r)}(W_s) + q\mathbb{I}_{[r,\infty)}(W_s))ds\Big); \ell(H_z,r)=0\Big\}$
	$= \begin{cases} e^{-(z-x)\sqrt{2p}}, & x \le z \\ \frac{\operatorname{sh}((r-x)\sqrt{2p})}{\operatorname{sh}((r-z)\sqrt{2p})}, & z \le x \le r \\ \frac{\operatorname{sh}((x-r)\sqrt{2q})}{\operatorname{sh}((z-r)\sqrt{2q})}, & r \le x \le z \\ e^{-(x-z)\sqrt{2q}}, & z \le x \end{cases}$
2.27.2	$\mathbf{P}_x\Big(\int_0^{H_z}\mathbb{I}_{(-\infty,r)}(W_s)ds \in du, \int_0^{H_z}\mathbb{I}_{[r,\infty)}(W_s)ds \in dv, \frac{1}{2}\ell(H_z,r) \in dy\Big)$
	$=: B_z^{(27)}(u,v,y)dudvdy$
$z \le r$ $x \le r$	$= \operatorname{escs}_u(x-z, o, r-z, y)\operatorname{h}_v(1,y)dudvdy$
$z \le r$ $r \le x$	$= \operatorname{es}_u(1,1,r-z,0,y)\operatorname{h}_v(1,y+x-r)dudvdy$
$r \le z$ $x \le r$	$= \operatorname{h}_u(1,y+r-x)\operatorname{es}_v(1,1,z-r,0,y)dudvdy$
$r \le z$ $r \le x$	$= \operatorname{h}_u(1,y)\operatorname{escs}_v(z-x,0,z-r,y)dudvdy$
(1)	$\mathbf{P}_x\Big(\int_0^{H_z}\mathbb{I}_{(-\infty,r)}(W_s)ds \in du, \int_0^{H_z}\mathbb{I}_{[r,\infty)}(W_s)ds = 0, \ell(H_z,r)=0\Big)$
	$= \begin{cases} \operatorname{h}_u(1,z-x)du, & x \le z \\ \operatorname{ss}_u(r-x,r-z)du, & z \le x \le r \end{cases}$
(2)	$\mathbf{P}_x\Big(\int_0^{H_z}\mathbb{I}_{(-\infty,r)}(W_s)ds = 0, \int_0^{H_z}\mathbb{I}_{[r,\infty)}(W_s)ds \in dv, \ell(H_z,r)=0\Big)$

1 W	$H_z = \min\{s : W_s = z\}$
	$$= \begin{cases} \mathrm{ss}_v(x-r, z-r)dv, & r \le x \le z \\ \mathrm{h}_v(1, x-z)dv, & z \le x \end{cases}$$
2.30.1	$$\mathbf{E}_x \exp\Bigl(-\int_0^{H_z} (\frac{p^2}{2} e^{2\beta W_s} + q e^{\beta W_s}) ds\Bigr)$$ $$= \begin{cases} \dfrac{e^{-\beta x/2} M_{-q/p\lvert\beta\rvert,0}\bigl(\frac{2p}{\lvert\beta\rvert} e^{\beta x}\bigr)}{e^{-\beta z/2} M_{-q/p\lvert\beta\rvert,0}\bigl(\frac{2p}{\lvert\beta\rvert} e^{\beta z}\bigr)}, & (z-x)\beta \ge 0 \\ \dfrac{e^{-\beta x/2} W_{-q/p\lvert\beta\rvert,0}\bigl(\frac{2p}{\lvert\beta\rvert} e^{\beta x}\bigr)}{e^{-\beta z/2} W_{-q/p\lvert\beta\rvert,0}\bigl(\frac{2p}{\lvert\beta\rvert} e^{\beta z}\bigr)}, & (z-x)\beta \le 0 \end{cases}$$
2.31.1 $0 < \alpha$	$$\mathbf{E}_x \exp\Bigl(-\frac{\gamma}{2} \sup_{0 \le s \le H_z} \bigl(\ell(s,u) - \alpha^{-1}\ell(s,r)\bigr)\Bigr)$$ for $z \le x \wedge r \wedge u$ or $x \vee r \vee u \le z$ $$= 1 - \frac{\gamma(\Upsilon + \lvert r-z\rvert - \alpha\lvert u-z\rvert)(\lvert u-z\rvert \wedge \lvert x-z\rvert)}{2\gamma(\lvert u-z\rvert \wedge \lvert r-z\rvert)\lvert u-r\rvert + \Upsilon + \lvert r-z\rvert - \alpha\lvert u-z\rvert}$$ $$- \frac{\gamma(2(\lvert u-z\rvert \wedge \lvert r-z\rvert)\lvert u-r\rvert - (\Upsilon + \lvert r-z\rvert - \alpha\lvert u-z\rvert)\lvert u-z\rvert)(\lvert x-z\rvert \wedge \lvert r-z\rvert)}{(2\gamma(\lvert u-z\rvert \wedge \lvert r-z\rvert)\lvert u-r\rvert + \Upsilon + \lvert r-z\rvert - \alpha\lvert u-z\rvert)(\lvert u-z\rvert \wedge \lvert r-z\rvert)}$$
2.31.2 $0 < \alpha$	$$\mathbf{P}_x\Bigl(\sup_{0 \le s \le H_z} \bigl(\ell(s,u) - \alpha^{-1}\ell(s,r)\bigr) > h\Bigr)$$ for $z \le x \wedge r \wedge u$ or $x \vee r \vee u \le z$ $$= \exp\Bigl(-\frac{(\Upsilon + \lvert r-z\rvert - \alpha\lvert u-z\rvert)h}{4(\lvert u-z\rvert \wedge \lvert r-z\rvert)\lvert u-r\rvert}\Bigr)\Bigl[\frac{\lvert x-z\rvert \wedge \lvert r-z\rvert}{\lvert u-z\rvert \wedge \lvert r-z\rvert}$$ $$- \frac{\Upsilon + \lvert r-z\rvert - \alpha\lvert u-z\rvert}{2(\lvert u-z\rvert \wedge \lvert r-z\rvert)\lvert u-r\rvert}\Bigl(\frac{(\lvert x-z\rvert \wedge \lvert r-z\rvert)\lvert u-z\rvert}{\lvert u-z\rvert \wedge \lvert r-z\rvert} - \lvert u-z\rvert \wedge \lvert x-z\rvert\Bigr)\Bigr]$$
	$$\Upsilon := \sqrt{(\lvert r-z\rvert - \alpha\lvert u-z\rvert)^2 + 4\alpha(\lvert u-z\rvert \wedge \lvert r-z\rvert)\lvert u-r\rvert}$$

1 W $H = H_{a,b} = \min\{s : W_s \notin (a,b)\}$

3. Stopping at first exit time

3.0.1 $$\mathbf{E}_x e^{-\alpha H} = \frac{\operatorname{sh}\big((b-x)\sqrt{2\alpha}\big) + \operatorname{sh}\big((x-a)\sqrt{2\alpha}\big)}{\operatorname{sh}\big((b-a)\sqrt{2\alpha}\big)} = \frac{\operatorname{ch}\big((b+a-2x)\sqrt{\alpha/2}\big)}{\operatorname{ch}\big((b-a)\sqrt{\alpha/2}\big)}$$

3.0.2 $$\mathbf{P}_x(H \in dt) = \operatorname{ss}_t(b-x, b-a)dt + \operatorname{ss}_t(x-a, b-a)dt = \operatorname{cc}_t\Big(\frac{b+a-2x}{2}, \frac{b-a}{2}\Big)dt$$

3.0.3 $$\mathbf{E}_x e^{i\beta W_H} = \frac{b-x}{b-a}e^{i\beta a} + \frac{x-a}{b-a}e^{i\beta b}$$

3.0.4 (a) $$\mathbf{P}_x\big(W_H = a\big) = \frac{b-x}{b-a}$$

3.0.4 (b) $$\mathbf{P}_x\big(W_H = b\big) = \frac{x-a}{b-a}$$

3.0.5 (a) $$\mathbf{E}_x\big\{e^{-\alpha H};\ W_H = a\big\} = \frac{\operatorname{sh}\big((b-x)\sqrt{2\alpha}\big)}{\operatorname{sh}\big((b-a)\sqrt{2\alpha}\big)}$$

3.0.5 (b) $$\mathbf{E}_x\big\{e^{-\alpha H};\ W_H = b\big\} = \frac{\operatorname{sh}\big((x-a)\sqrt{2\alpha}\big)}{\operatorname{sh}\big((b-a)\sqrt{2\alpha}\big)}$$

3.0.6 (a) $$\mathbf{P}_x\big(H \in dt, W_H = a\big) = \operatorname{ss}_t(b-x, b-a)dt$$

3.0.6 (b) $$\mathbf{P}_x\big(H \in dt, W_H = b\big) = \operatorname{ss}_t(x-a, b-a)dt$$

3.1.6 $$\mathbf{P}_x\Big(\sup_{0\le s\le H} W_s \ge y, W_H = a\Big) = \frac{(b-y)(x-a)}{(b-a)(y-a)}, \quad a \le x \le y \le b$$

3.1.8 (1) $$\mathbf{E}_x\big\{e^{-\alpha H};\ \sup_{0\le s\le H} W_s \ge y, W_H = a\big\}$$

1 W $\qquad H = H_{a,b} = \min\{s : W_s \notin (a,b)\}$

	$= \dfrac{\operatorname{sh}((b-y)\sqrt{2\alpha})\operatorname{sh}((x-a)\sqrt{2\alpha})}{\operatorname{sh}((b-a)\sqrt{2\alpha})\operatorname{sh}((y-a)\sqrt{2\alpha})}, \quad a \le x \le y \le b$
3.1.8 (2)	$\mathbf{P}_x\big(\sup_{0\le s\le H} W_s \ge y, W_H = a, H \in dt\big)$ $= \operatorname{ss}_t(b-y, b-a) * \operatorname{ss}_t(x-a, y-a)dt, \quad a \le x \le y \le b$
3.2.6	$\mathbf{P}_x\big(\inf_{0\le s\le H} W_s \le y, W_H = b\big) = \dfrac{(b-x)(y-a)}{(b-y)(b-a)}, \quad a \le y \le x \le b$
3.2.8 (1)	$\mathbf{E}_x\{e^{-\alpha H}; \inf_{0\le s\le H} W_s \le y, W_H = b\}$ $= \dfrac{\operatorname{sh}((b-x)\sqrt{2\alpha})\operatorname{sh}((y-a)\sqrt{2\alpha})}{\operatorname{sh}((b-y)\sqrt{2\alpha})\operatorname{sh}((b-a)\sqrt{2\alpha})}, \quad a \le y \le x \le b$
3.2.8 (2)	$\mathbf{P}_x\big(\inf_{0\le s\le H} W_s \le y, W_H = b, H \in dt\big)$ $= \operatorname{ss}_t(b-x, b-y) * \operatorname{ss}_t(y-a, b-a)dt, \quad a \le y \le x \le b$
3.3.1	$\mathbf{E}_x e^{-\gamma\ell(H,r)} = \begin{cases} \dfrac{b-a+2\gamma(x-r)(r-a)}{b-a+2\gamma(b-r)(r-a)}, & r \le x \le b \\ \dfrac{b-a+2\gamma(b-r)(r-x)}{b-a+2\gamma(b-r)(r-a)}, & a \le x \le r \end{cases}$
3.3.2	$\mathbf{P}_x\big(\ell(H,r) \in dy\big) = \begin{cases} \dfrac{(b-x)(b-a)}{2(b-r)^2(r-a)} \exp\Big(-\dfrac{(b-a)y}{2(b-r)(r-a)}\Big)dy, & r \le x \le b \\ \dfrac{(x-a)(b-a)}{2(b-r)(r-a)^2} \exp\Big(-\dfrac{(b-a)y}{2(b-r)(r-a)}\Big)dy, & a \le x \le r \end{cases}$
(1)	$\mathbf{P}_x\big(\ell(H,r) = 0\big) = \begin{cases} \dfrac{x-r}{b-r}, & r \le x \le b \\ \dfrac{r-x}{r-a}, & a \le x \le r \end{cases}$
3.3.3	$\mathbf{E}_r e^{-\alpha H-\gamma\ell(H,r)} = \dfrac{\sqrt{2\alpha}\operatorname{sh}((b-r)\sqrt{2\alpha}) + \sqrt{2\alpha}\operatorname{sh}((r-a)\sqrt{2\alpha})}{\sqrt{2\alpha}\operatorname{sh}((b-a)\sqrt{2\alpha}) + 2\gamma\operatorname{sh}((b-r)\sqrt{2\alpha})\operatorname{sh}((r-a)\sqrt{2\alpha})} =: L$

1 W $H = H_{a,b} = \min\{s : W_s \notin (a,b)\}$

$$\mathbf{E}_x e^{-\alpha H - \gamma \ell(H,r)} = \begin{cases} \dfrac{\operatorname{sh}((x-r)\sqrt{2\alpha})}{\operatorname{sh}((b-r)\sqrt{2\alpha})} + \dfrac{\operatorname{sh}((b-x)\sqrt{2\alpha})}{\operatorname{sh}((b-r)\sqrt{2\alpha})} L, & r \le x \le b \\ \dfrac{\operatorname{sh}((r-x)\sqrt{2\alpha})}{\operatorname{sh}((r-a)\sqrt{2\alpha})} + \dfrac{\operatorname{sh}((x-a)\sqrt{2\alpha})}{\operatorname{sh}((r-a)\sqrt{2\alpha})} L, & a \le x \le r \end{cases}$$

3.3.4 $$\mathbf{E}_x\{e^{-\alpha H};\ \ell(H,r) \in dy\} = \exp\Big(-\frac{\sqrt{2\alpha}\operatorname{sh}((b-a)\sqrt{2\alpha})y}{2\operatorname{sh}((b-r)\sqrt{2\alpha})\operatorname{sh}((r-a)\sqrt{2\alpha})}\Big)$$

$$\times \begin{cases} \dfrac{\sqrt{2\alpha}\operatorname{sh}((b-x)\sqrt{2\alpha})[\operatorname{sh}((b-r)\sqrt{2\alpha}) + \operatorname{sh}((r-a)\sqrt{2\alpha})]}{2\operatorname{sh}^2((b-r)\sqrt{2\alpha})\operatorname{sh}((r-a)\sqrt{2\alpha})} dy, & r \le x \le b \\ \dfrac{\sqrt{2\alpha}\operatorname{sh}((x-a)\sqrt{2\alpha})[\operatorname{sh}((b-r)\sqrt{2\alpha}) + \operatorname{sh}((r-a)\sqrt{2\alpha})]}{2\operatorname{sh}((b-r)\sqrt{2\alpha})\operatorname{sh}^2((r-a)\sqrt{2\alpha})} dy, & a \le x \le r \end{cases}$$

(1) $$\mathbf{E}_x\{e^{-\alpha H};\ \ell(H,r) = 0\} = \begin{cases} \dfrac{\operatorname{sh}((x-r)\sqrt{2\alpha})}{\operatorname{sh}((b-r)\sqrt{2\alpha})}, & r \le x \le b \\ \dfrac{\operatorname{sh}((r-x)\sqrt{2\alpha})}{\operatorname{sh}((r-a)\sqrt{2\alpha})}, & a \le x \le r \end{cases}$$

3.3.5 (a) $$\mathbf{E}_x\{e^{-\gamma\ell(H,r)}; W_H = a\} =: Q_a^{(3)}(2\gamma) = \begin{cases} \dfrac{b-x}{b-a+2\gamma(b-r)(r-a)}, & r \le x \le b \\ \dfrac{b-x+2\gamma(b-r)(r-x)}{b-a+2\gamma(b-r)(r-a)}, & a \le x \le r \end{cases}$$

3.3.5 (b) $$\mathbf{E}_x\{e^{-\gamma\ell(H,r)}; W_H = b\} =: Q_b^{(3)}(2\gamma) = \begin{cases} \dfrac{x-a+2\gamma(x-r)(r-a)}{b-a+2\gamma(b-r)(r-a)}, & r \le x \le b \\ \dfrac{x-a}{b-a+2\gamma(b-r)(r-a)}, & a \le x \le r \end{cases}$$

3.3.6 (a) $$\mathbf{P}_x\big(\ell(H,r) \in dy, W_H = a\big) =: B_a^{(3)}(y)dy$$

$$= \begin{cases} \dfrac{b-x}{2(b-r)(r-a)} \exp\Big(-\dfrac{(b-a)y}{2(b-r)(r-a)}\Big) dy, & r \le x \le b \\ \dfrac{x-a}{2(r-a)^2} \exp\Big(-\dfrac{(b-a)y}{2(b-r)(r-a)}\Big) dy, & a \le x \le r \end{cases}$$

(1) $$\mathbf{P}_x\big(\ell(H,r) = 0, W_H = a\big) = \begin{cases} 0, & r \le x \le b \\ \dfrac{r-x}{r-a}, & a \le x \le r \end{cases}$$

1 W $H = H_{a,b} = \min\{s : W_s \notin (a,b)\}$

3.3.6 (b) $\mathbf{P}_x\big(\ell(H,r) \in dy, W_H = b\big) =: B_b^{(3)}(y)dy$

$$= \begin{cases} \dfrac{b-x}{2(b-r)^2} \exp\Big(-\dfrac{(b-a)y}{2(b-r)(r-a)}\Big)dy, & r \le x \le b \\ \dfrac{x-a}{2(b-r)(r-a)} \exp\Big(-\dfrac{(b-a)y}{2(b-r)(r-a)}\Big)dy, & a \le x \le r \end{cases}$$

(1) $\mathbf{P}_x\big(\ell(H,r) = 0, W_H = b\big) = \begin{cases} \dfrac{x-r}{b-r}, & r \le x \le b \\ 0, & a \le x \le r \end{cases}$

3.3.7 (a) $\mathbf{E}_x\{e^{-\alpha H - \gamma \ell(H,r)};\ W_H = a\} =: V_a^{(3)}(2\alpha)$

$$= \begin{cases} \dfrac{\sqrt{2\alpha}\,\mathrm{sh}((b-x)\sqrt{2\alpha})}{\sqrt{2\alpha}\,\mathrm{sh}((b-a)\sqrt{2\alpha}) + 2\gamma\,\mathrm{sh}((b-r)\sqrt{2\alpha})\,\mathrm{sh}((r-a)\sqrt{2\alpha})}, & r \le x \le b \\ \dfrac{\sqrt{2\alpha}\,\mathrm{sh}((b-x)\sqrt{2\alpha}) + 2\gamma\,\mathrm{sh}((b-r)\sqrt{2\alpha})\,\mathrm{sh}((r-x)\sqrt{2\alpha})}{\sqrt{2\alpha}\,\mathrm{sh}((b-a)\sqrt{2\alpha}) + 2\gamma\,\mathrm{sh}((b-r)\sqrt{2\alpha})\,\mathrm{sh}((r-a)\sqrt{2\alpha})}, & a \le x \le r \end{cases}$$

3.3.7 (b) $\mathbf{E}_x\{e^{-\alpha H - \gamma \ell(H,r)};\ W_H = b\} =: V_b^{(3)}(2\alpha)$

$$= \begin{cases} \dfrac{\sqrt{2\alpha}\,\mathrm{sh}((x-a)\sqrt{2\alpha}) + 2\gamma\,\mathrm{sh}((x-r)\sqrt{2\alpha})\,\mathrm{sh}((r-a)\sqrt{2\alpha})}{\sqrt{2\alpha}\,\mathrm{sh}((b-a)\sqrt{2\alpha}) + 2\gamma\,\mathrm{sh}((b-r)\sqrt{2\alpha})\,\mathrm{sh}((r-a)\sqrt{2\alpha})}, & r \le x \le b \\ \dfrac{\sqrt{2\alpha}\,\mathrm{sh}((x-a)\sqrt{2\alpha})}{\sqrt{2\alpha}\,\mathrm{sh}((b-a)\sqrt{2\alpha}) + 2\gamma\,\mathrm{sh}((b-r)\sqrt{2\alpha})\,\mathrm{sh}((r-a)\sqrt{2\alpha})}, & a \le x \le r \end{cases}$$

3.3.8 (a) $\mathbf{E}_x\{e^{-\alpha H};\ \ell(H,r) \in dy, W_H = a\} = \exp\Big(-\dfrac{\sqrt{2\alpha}\,\mathrm{sh}((b-a)\sqrt{2\alpha})y}{2\,\mathrm{sh}((b-r)\sqrt{2\alpha})\,\mathrm{sh}((r-a)\sqrt{2\alpha})}\Big)$

$$\times \begin{cases} \dfrac{\sqrt{2\alpha}\,\mathrm{sh}((b-x)\sqrt{2\alpha})}{2\,\mathrm{sh}((b-r)\sqrt{2\alpha})\,\mathrm{sh}((r-a)\sqrt{2\alpha})}dy, & r \le x \le b \\ \dfrac{\sqrt{2\alpha}\,\mathrm{sh}((x-a)\sqrt{2\alpha})}{2\,\mathrm{sh}^2((r-a)\sqrt{2\alpha})}dy, & a \le x \le r \end{cases}$$

(1) $\mathbf{E}_x\{e^{-\alpha H};\ \ell(H,r) = 0, W_H = a\} = \begin{cases} 0, & r \le x \le b \\ \dfrac{\mathrm{sh}((r-x)\sqrt{2\alpha})}{\mathrm{sh}((r-a)\sqrt{2\alpha})}, & a \le x \le r \end{cases}$

1 W $H = H_{a,b} = \min\{s : W_s \notin (a,b)\}$

3.3.8 (b)

$$\mathbf{E}_x\{e^{-\alpha H};\ \ell(H,r)\in dy, W_H=b\}=\exp\Big(-\frac{\sqrt{2\alpha}\,\mathrm{sh}((b-a)\sqrt{2\alpha})y}{2\,\mathrm{sh}((b-r)\sqrt{2\alpha})\,\mathrm{sh}((r-a)\sqrt{2\alpha})}\Big)$$

$$\times\begin{cases}\dfrac{\sqrt{2\alpha}\,\mathrm{sh}((b-x)\sqrt{2\alpha})}{2\,\mathrm{sh}^2((b-r)\sqrt{2\alpha})}dy, & r\le x\le b\\[2mm] \dfrac{\sqrt{2\alpha}\,\mathrm{sh}((x-a)\sqrt{2\alpha})}{2\,\mathrm{sh}((b-r)\sqrt{2\alpha})\,\mathrm{sh}((r-a)\sqrt{2\alpha})}dy, & a\le x\le r\end{cases}$$

(1)

$$\mathbf{E}_x\{e^{-\alpha H};\ \ell(H,r)=0, W_H=b\}=\begin{cases}\dfrac{\mathrm{sh}((x-r)\sqrt{2\alpha})}{\mathrm{sh}((b-r)\sqrt{2\alpha})}, & r\le x\le b\\ 0, & a\le x\le r\end{cases}$$

3.4.1

$$\mathbf{E}_x\exp\Big(-\gamma\int_0^H \mathbb{1}_{[r,b]}(W_s)ds\Big)$$

$$=\begin{cases}\dfrac{r-x}{r-a}+\dfrac{(x-a)(\mathrm{sh}((b-r)\sqrt{2\gamma})+\sqrt{2\gamma}(r-a))}{(r-a)(\mathrm{sh}((b-r)\sqrt{2\gamma})+\sqrt{2\gamma}(r-a)\,\mathrm{ch}((b-r)\sqrt{2\gamma}))}, & a\le x\le r\\[2mm] \dfrac{\mathrm{sh}((x-r)\sqrt{2\gamma})}{\mathrm{sh}((b-r)\sqrt{2\gamma})} & \\ \quad+\dfrac{\mathrm{sh}((b-x)\sqrt{2\gamma})(\mathrm{sh}((b-r)\sqrt{2\gamma})+\sqrt{2\gamma}(r-a))}{\mathrm{sh}((b-r)\sqrt{2\gamma})(\mathrm{sh}((b-r)\sqrt{2\gamma})+\sqrt{2\gamma}(r-a)\,\mathrm{ch}((b-r)\sqrt{2\gamma}))}, & r\le x\le b\end{cases}$$

3.4.5 (a)

$$\mathbf{E}_x\Big\{\exp\Big(-\gamma\int_0^H \mathbb{1}_{[r,b]}(W_s)ds\Big);\ W_H=a\Big\}=:Q_a^{(4)}(2\gamma)$$

$$=\begin{cases}\dfrac{\mathrm{sh}((b-r)\sqrt{2\gamma})+(r-x)\sqrt{2\gamma}\,\mathrm{ch}((b-r)\sqrt{2\gamma})}{\mathrm{sh}((b-r)\sqrt{2\gamma})+(r-a)\sqrt{2\gamma}\,\mathrm{ch}((b-r)\sqrt{2\gamma})}, & a\le x\le r\\[2mm] \dfrac{\mathrm{sh}((b-x)\sqrt{2\gamma})}{\mathrm{sh}((b-r)\sqrt{2\gamma})+(r-a)\sqrt{2\gamma}\,\mathrm{ch}((b-r)\sqrt{2\gamma})}, & r\le x\le b\end{cases}$$

3.4.5 (b)

$$\mathbf{E}_x\Big\{\exp\Big(-\gamma\int_0^H \mathbb{1}_{[r,b]}(W_s)ds\Big);\ W_H=b\Big\}=:Q_b^{(4)}(2\gamma)$$

$$=\begin{cases}\dfrac{(x-a)\sqrt{2\gamma}}{\mathrm{sh}((b-r)\sqrt{2\gamma})+(r-a)\sqrt{2\gamma}\,\mathrm{ch}((b-r)\sqrt{2\gamma})}, & a\le x\le r\\[2mm] \dfrac{\mathrm{sh}((x-r)\sqrt{2\gamma})+(r-a)\sqrt{2\gamma}\,\mathrm{ch}((x-r)\sqrt{2\gamma})}{\mathrm{sh}((b-r)\sqrt{2\gamma})+(r-a)\sqrt{2\gamma}\,\mathrm{ch}((b-r)\sqrt{2\gamma})}, & r\le x\le b\end{cases}$$

1 W $H = H_{a,b} = \min\{s : W_s \notin (a,b)\}$

3.4.6 (a)

$$\mathbf{P}_x\Big(\int_0^H \mathbb{I}_{[r,b]}(W_s)ds \in dy, W_H = a\Big) =: B_a^{(4)}(y)dy$$

$$= \begin{cases} \dfrac{x-a}{r-a}\,\mathrm{rc}_y(0, b-r, b-r, r-a)dy, & a \le x \le r \\ \mathrm{rc}_y(0, b-r, b-x, r-a)dy, & r \le x \le b \end{cases}$$

(1)

$$\mathbf{P}_x\Big(\int_0^H \mathbb{I}_{[r,b]}(W_s)ds = 0, W_H = a\Big) = \begin{cases} \dfrac{r-x}{r-a}, & a \le x \le r \\ 0, & r \le x \le b \end{cases}$$

3.4.6 (b)

$$\mathbf{P}_x\Big(\int_0^H \mathbb{I}_{[r,b]}(W_s)ds \in dy, W_H = b\Big) =: B_b^{(4)}(y)dy$$

$$= \begin{cases} (x-a)\widetilde{\mathrm{rc}}_y(1, b-r, 0, r-a)dy, & a \le x \le r \\ \mathrm{ss}_y(x-r, b-r)dy & \\ \quad +(r-a)\,\mathrm{ss}_y(b-x, b-r) * \widetilde{\mathrm{rc}}_y(1, b-r, 0, r-a)dy, & r \le x \le b \end{cases}$$

(1)

$$\mathbf{P}_x\Big(\int_0^H \mathbb{I}_{[r,b]}(W_s)ds = 0, W_H = b\Big) = 0$$

3.4.8 (a)

$$\mathbf{P}_x\Big(\int_0^H \mathbb{I}_{[r,b]}(W_s)ds \in dv,\, H \in dt, W_H = a\Big) = \int_0^\infty B_a^{(27)}(t-v, v, y)dy\,dvdt$$

for $B_a^{(27)}(u,v,y)$ see 3.27.6(a)

3.4.8 (b)

$$\mathbf{P}_x\Big(\int_0^H \mathbb{I}_{[r,b]}(W_s)ds \in dv,\, H \in dt, W_H = b\Big) = \int_0^\infty B_b^{(27)}(t-v, v, y)dy\,dvdt$$

for $B_b^{(27)}(u,v,y)$ see 3.27.6(b)

3.5.1

$$\mathbf{E}_x \exp\Big(-\gamma\int_0^H \mathbb{I}_{[a,r]}(W_s)ds\Big)$$

$$= \begin{cases} \dfrac{\mathrm{sh}((r-x)\sqrt{2\gamma})}{\mathrm{sh}((r-a)\sqrt{2\gamma})} & \\ +\dfrac{\mathrm{sh}((x-a)\sqrt{2\gamma})(\mathrm{sh}((r-a)\sqrt{2\gamma}) + (b-r)\sqrt{2\gamma})}{\mathrm{sh}((r-a)\sqrt{2\gamma})(\mathrm{sh}((r-a)\sqrt{2\gamma}) + (b-r)\sqrt{2\gamma}\,\mathrm{ch}((r-a)\sqrt{2\gamma}))}, & a \le x \le r \\ \dfrac{x-r}{b-r} + \dfrac{(b-x)(\mathrm{sh}((r-a)\sqrt{2\gamma}) + \sqrt{2\gamma}(b-r))}{(b-r)(\mathrm{sh}((r-a)\sqrt{2\gamma}) + (b-r)\sqrt{2\gamma}\,\mathrm{ch}((r-a)\sqrt{2\gamma}))}, & r \le x \le b \end{cases}$$

1 W $H = H_{a,b} = \min\{s : W_s \notin (a,b)\}$

3.5.5 (a)

$$\mathbf{E}_x\Big\{\exp\Big(-\gamma\int_0^H \mathbb{I}_{[a,r]}(W_s)ds\Big), W_H = a\Big\} =: Q_a^{(5)}(2\gamma)$$

$$= \begin{cases} \dfrac{\operatorname{sh}((r-x)\sqrt{2\gamma}) + \sqrt{2\gamma}(b-r)\operatorname{ch}((r-x)\sqrt{2\gamma})}{\operatorname{sh}((r-a)\sqrt{2\gamma}) + \sqrt{2\gamma}(b-r)\operatorname{ch}((r-a)\sqrt{2\gamma})}, & a \le x \le r \\ \dfrac{(b-x)\sqrt{2\gamma}}{\operatorname{sh}((r-a)\sqrt{2\gamma}) + (b-r)\sqrt{2\gamma}\operatorname{ch}((r-a)\sqrt{2\gamma})}, & r \le x \le b \end{cases}$$

3.5.5 (b)

$$\mathbf{E}_x\Big\{\exp\Big(-\gamma\int_0^H \mathbb{I}_{[a,r]}(W_s)ds\Big);\ W_H = b\Big\} =: Q_b^{(5)}(2\gamma)$$

$$= \begin{cases} \dfrac{\operatorname{sh}((x-a)\sqrt{2\gamma})}{\operatorname{sh}((r-a)\sqrt{2\gamma}) + (b-r)\sqrt{2\gamma}\operatorname{ch}((r-a)\sqrt{2\gamma})}, & a \le x \le r \\ \dfrac{\operatorname{sh}((r-a)\sqrt{2\gamma}) + (x-r)\sqrt{2\gamma}\operatorname{ch}((r-a)\sqrt{2\gamma})}{\operatorname{sh}((r-a)\sqrt{2\gamma}) + (b-r)\sqrt{2\gamma}\operatorname{ch}((r-a)\sqrt{2\gamma})}, & r \le x \le b \end{cases}$$

3.5.6 (a)

$$\mathbf{P}_x\Big(\int_0^H \mathbb{I}_{[a,r]}(W_s)ds \in dy, W_H = a\Big) =: B_a^{(5)}(y)dy$$

$$= \begin{cases} \operatorname{ss}_y(r-x, r-a)dy \\ \quad +(b-r)\operatorname{ss}_y(x-a, r-a) * \widetilde{\operatorname{rc}}_y(1, r-a, 0, b-r)dy, & a \le x \le r \\ (b-x)\widetilde{\operatorname{rc}}_y(1, r-a, 0, b-r)dy, & r \le x \le b \end{cases}$$

(1)

$$\mathbf{P}_x\Big(\int_0^H \mathbb{I}_{[a,r]}(W_s)ds = 0, W_H = a\Big) = 0$$

3.5.6 (b)

$$\mathbf{P}_x\Big(\int_0^H \mathbb{I}_{[a,r]}(W_s)ds \in dy, W_H = b\Big) =: B_b^{(5)}(y)dy$$

$$= \begin{cases} \operatorname{rc}_y(0, r-a, x-a, b-r)dy, & a \le x \le r \\ \dfrac{b-x}{b-r}\operatorname{rc}_y(0, r-a, r-a, b-r)dy, & r \le x \le b \end{cases}$$

(1)

$$\mathbf{P}_x\Big(\int_0^H \mathbb{I}_{[a,r]}(W_s)ds = 0, W_H = b\Big) = \begin{cases} 0, & a \le x \le r \\ \dfrac{x-r}{b-r}, & r \le x \le b \end{cases}$$

1 W $H = H_{a,b} = \min\{s : W_s \notin (a,b)\}$

3.5.8 (a)

$$\mathbf{P}_x\Big(\int_0^H \mathbb{I}_{[a,r]}(W_s)ds \in dv, H \in dt, W_H = a\Big) = \int_0^\infty B_a^{(27)}(v, t-v, y)dy\,dvdt$$

for $B_a^{(27)}(u,v,y)$ see 3.27.6(a)

3.5.8 (b)

$$\mathbf{P}_x\Big(\int_0^H \mathbb{I}_{[a,r]}(W_s)ds \in dv, H \in dt, W_H = b\Big) = \int_0^\infty B_b^{(27)}(v, t-v, y)dy\,dvdt$$

for $B_b^{(27)}(u,v,y)$ see 3.27.6(b)

3.6.1

$$\mathbf{E}_x \exp\Big(-\int_0^H (p\mathbb{I}_{[a,r]}(W_s) + q\mathbb{I}_{[r,b]}(W_s))ds\Big)$$

$$= \begin{cases} \dfrac{\operatorname{sh}((x-r)\sqrt{2q})}{\operatorname{sh}((b-r)\sqrt{2q})} + \dfrac{\operatorname{sh}((b-x)\sqrt{2q})}{\operatorname{sh}((b-r)\sqrt{2q})}Q, & r \le x \le b \\ \dfrac{\operatorname{sh}((r-x)\sqrt{2p})}{\operatorname{sh}((r-a)\sqrt{2p})} + \dfrac{\operatorname{sh}((x-a)\sqrt{2p})}{\operatorname{sh}((r-a)\sqrt{2p})}Q, & a \le x \le r \end{cases}$$

where

$$Q := \frac{\sqrt{p}\operatorname{sh}((b-r)\sqrt{2q}) + \sqrt{q}\operatorname{sh}((r-a)\sqrt{2p})}{\sqrt{q}\operatorname{ch}((b-r)\sqrt{2q})\operatorname{sh}((r-a)\sqrt{2p}) + \sqrt{p}\operatorname{sh}((b-r)\sqrt{2q})\operatorname{ch}((r-a)\sqrt{2p})}$$

3.6.5 (a)

$$\mathbf{E}_x\Big\{\exp\Big(-\int_0^H (p\mathbb{I}_{[a,r]}(W_s) + q\mathbb{I}_{[r,b]}(W_s))ds\Big);\ W_H = a\Big\} =: Q_a^{(6)}(2p,2q)$$

$a \le x$, $x \le r$

$$= \frac{\sqrt{q}\operatorname{ch}((b-r)\sqrt{2q})\operatorname{sh}((r-x)\sqrt{2p}) + \sqrt{p}\operatorname{sh}((b-r)\sqrt{2q})\operatorname{ch}((r-x)\sqrt{2p})}{\sqrt{q}\operatorname{ch}((b-r)\sqrt{2q})\operatorname{sh}((r-a)\sqrt{2p}) + \sqrt{p}\operatorname{sh}((b-r)\sqrt{2q})\operatorname{ch}((r-a)\sqrt{2p})}$$

$r \le x$, $x \le b$

$$= \frac{\sqrt{p}\operatorname{sh}((b-x)\sqrt{2q})}{\sqrt{q}\operatorname{ch}((b-r)\sqrt{2q})\operatorname{sh}((r-a)\sqrt{2p}) + \sqrt{p}\operatorname{sh}((b-r)\sqrt{2q})\operatorname{ch}((r-a)\sqrt{2p})}$$

3.6.5 (b)

$$\mathbf{E}_x\Big\{\exp\Big(-\int_0^H (p\mathbb{I}_{[a,r]}(W_s) + q\mathbb{I}_{[r,b]}(W_s))ds\Big);\ W_H = b\Big\} =: Q_b^{(6)}(2p,2q)$$

$a \le x$, $x \le r$

$$= \frac{\sqrt{q}\operatorname{sh}((x-a)\sqrt{2p})}{\sqrt{q}\operatorname{ch}((b-r)\sqrt{2q})\operatorname{sh}((r-a)\sqrt{2p}) + \sqrt{p}\operatorname{sh}((b-r)\sqrt{2q})\operatorname{ch}((r-a)\sqrt{2p})}$$

1 W $H = H_{a,b} = \min\{s : W_s \notin (a,b)\}$

$r \le x$
$x \le b$

$$= \frac{\sqrt{q}\,\mathrm{ch}((x-r)\sqrt{2q})\,\mathrm{sh}((r-a)\sqrt{2p}) + \sqrt{p}\,\mathrm{sh}((x-r)\sqrt{2q})\,\mathrm{ch}((r-a)\sqrt{2p})}{\sqrt{q}\,\mathrm{ch}((b-r)\sqrt{2q})\,\mathrm{sh}((r-a)\sqrt{2p}) + \sqrt{p}\,\mathrm{sh}((b-r)\sqrt{2q})\,\mathrm{ch}((r-a)\sqrt{2p})}$$

3.6.6 (a)

$$\mathbf{P}_x\Big(\int_0^H \mathbb{I}_{[a,r)}(W_s)ds \in du,\ \int_0^H \mathbb{I}_{[r,b]}(W_s)ds \in dv,\ W_H = a\Big)$$

$$= \int_0^\infty B_a^{(27)}(u,v,y)dy\,dudv$$

for $B_a^{(27)}(u,v,y)$ see 3.27.6(a)

3.6.6 (b)

$$\mathbf{P}_x\Big(\int_0^H \mathbb{I}_{[a,r)}(W_s)ds \in du,\ \int_0^H \mathbb{I}_{[r,b]}(W_s)ds \in dv,\ W_H = b\Big)$$

$$= \int_0^\infty B_b^{(27)}(u,v,y)dy\,dudv$$

for $B_b^{(27)}(u,v,y)$ see 3.27.6(b)

3.7.1
$a \le r$
$u \le b$

$$\mathbf{E}_r \exp\Big(-\gamma\int_0^H \mathbb{I}_{[r,u]}(W_s)ds\Big) =: Q_r$$

$$= \frac{\mathrm{sh}((u-r)\sqrt{2\gamma}) + (b-u)\sqrt{2\gamma}\,\mathrm{ch}((u-r)\sqrt{2\gamma}) + (r-a)\sqrt{2\gamma}}{(1+(b-u)(r-a)2\gamma)\,\mathrm{sh}((u-r)\sqrt{2\gamma}) + (b-u+r-a)\sqrt{2\gamma}\,\mathrm{ch}((u-r)\sqrt{2\gamma})}$$

$$\mathbf{E}_u \exp\Big(-\gamma\int_0^H \mathbb{I}_{[r,u]}(W_s)ds\Big) =: Q_u$$

$$= \frac{\mathrm{sh}((u-r)\sqrt{2\gamma}) + (b-u)\sqrt{2\gamma} + (r-a)\sqrt{2\gamma}\,\mathrm{ch}((u-r)\sqrt{2\gamma})}{(1+(b-u)(r-a)2\gamma)\,\mathrm{sh}((u-r)\sqrt{2\gamma}) + (b-u+r-a)\sqrt{2\gamma}\,\mathrm{ch}((u-r)\sqrt{2\gamma})}$$

$$\mathbf{E}_x \exp\Big(-\gamma\int_0^H \mathbb{I}_{[r,u]}(W_s)ds\Big)$$

$$= \begin{cases} \dfrac{r-x}{r-a} + \dfrac{x-a}{r-a}Q_r, & a \le x \le r \\ \dfrac{\mathrm{sh}((u-x)\sqrt{2\gamma})}{\mathrm{sh}((u-r)\sqrt{2\gamma})}Q_r + \dfrac{\mathrm{sh}((x-r)\sqrt{2\gamma})}{\mathrm{sh}((u-r)\sqrt{2\gamma})}Q_u, & r \le x \le u \\ \dfrac{x-u}{b-u} + \dfrac{b-x}{b-u}Q_u, & u \le x \le b \end{cases}$$

1	W $\qquad\qquad H = H_{a,b} = \min\{s : W_s \notin (a,b)\}$
3.7.5 (a)	$\mathbf{E}_x\Big\{\exp\Big(-\gamma\int_0^H \mathbb{I}_{[r,u]}(W_s)ds\Big);\ W_H = a\Big\} =: Q_a^{(7)}(2\gamma)$
$a \le x$ $x \le r$	$= \dfrac{(1+(b-u)(r-x)2\gamma)\operatorname{sh}((u-r)\sqrt{2\gamma}) + (b-u+r-x)\sqrt{2\gamma}\operatorname{ch}((u-r)\sqrt{2\gamma})}{(1+(b-u)(r-a)2\gamma)\operatorname{sh}((u-r)\sqrt{2\gamma}) + (b-u+r-a)\sqrt{2\gamma}\operatorname{ch}((u-r)\sqrt{2\gamma})}$
$r \le x$ $x \le u$	$= \dfrac{\operatorname{sh}((u-x)\sqrt{2\gamma}) + (b-u)\sqrt{2\gamma}\operatorname{ch}((u-x)\sqrt{2\gamma})}{(1+(b-u)(r-a)2\gamma)\operatorname{sh}((u-r)\sqrt{2\gamma}) + (b-u+r-a)\sqrt{2\gamma}\operatorname{ch}((u-r)\sqrt{2\gamma})}$
$u \le x$ $x \le b$	$= \dfrac{(b-x)\sqrt{2\gamma}}{(1+(b-u)(r-a)2\gamma)\operatorname{sh}((u-r)\sqrt{2\gamma}) + (b-u+r-a)\sqrt{2\gamma}\operatorname{ch}((u-r)\sqrt{2\gamma})}$
3.7.5 (b)	$\mathbf{E}_x\Big\{\exp\Big(-\gamma\int_0^H \mathbb{I}_{[r,u]}(W_s)ds\Big);\ W_H = b\Big\} =: Q_b^{(7)}(2\gamma)$
$a \le x$ $x \le r$	$= \dfrac{(x-a)\sqrt{2\gamma}}{(1+(b-u)(r-a)2\gamma)\operatorname{sh}((u-r)\sqrt{2\gamma}) + (b-u+r-a)\sqrt{2\gamma}\operatorname{ch}((u-r)\sqrt{2\gamma})}$
$r \le x$ $x \le u$	$= \dfrac{\operatorname{sh}((x-r)\sqrt{2\gamma}) + (r-a)\sqrt{2\gamma}\operatorname{ch}((x-r)\sqrt{2\gamma})}{(1+(b-u)(r-a)2\gamma)\operatorname{sh}((u-r)\sqrt{2\gamma}) + (b-u+r-a)\sqrt{2\gamma}\operatorname{ch}((u-r)\sqrt{2\gamma})}$
$u \le x$ $x \le b$	$= \dfrac{(1+(x-u)(r-a)2\gamma)\operatorname{sh}((u-r)\sqrt{2\gamma}) + (x-u+r-a)\sqrt{2\gamma}\operatorname{ch}((u-r)\sqrt{2\gamma})}{(1+(b-u)(r-a)2\gamma)\operatorname{sh}((u-r)\sqrt{2\gamma}) + (b-u+r-a)\sqrt{2\gamma}\operatorname{ch}((u-r)\sqrt{2\gamma})}$
3.8.1	$\mathbf{E}_x \exp\Big(-\gamma\int_0^H W_s ds\Big)$
	$= \dfrac{xbS_{1/3}(\frac{2}{3}b^{3/2}\sqrt{2\gamma}, \frac{2}{3}x^{3/2}\sqrt{2\gamma}) + xaS_{1/3}(\frac{2}{3}x^{3/2}\sqrt{2\gamma}, \frac{2}{3}a^{3/2}\sqrt{2\gamma})}{abS_{1/3}(\frac{2}{3}b^{3/2}\sqrt{2\gamma}, \frac{2}{3}a^{3/2}\sqrt{2\gamma})}$
3.8.5 (a)	$\mathbf{E}_x\Big\{\exp\Big(-\gamma\int_0^H W_s ds\Big); W_H = a\Big\} = \dfrac{xS_{1/3}(\frac{2}{3}b^{3/2}\sqrt{2\gamma}, \frac{2}{3}x^{3/2}\sqrt{2\gamma})}{aS_{1/3}(\frac{2}{3}b^{3/2}\sqrt{2\gamma}, \frac{2}{3}a^{3/2}\sqrt{2\gamma})}$
3.8.5 (b)	$\mathbf{E}_x\Big\{\exp\Big(-\gamma\int_0^H W_s ds\Big); W_H = b\Big\} = \dfrac{xS_{1/3}(\frac{2}{3}x^{3/2}\sqrt{2\gamma}, \frac{2}{3}a^{3/2}\sqrt{2\gamma})}{bS_{1/3}(\frac{2}{3}b^{3/2}\sqrt{2\gamma}, \frac{2}{3}a^{3/2}\sqrt{2\gamma})}$

1 W $H = H_{a,b} = \min\{s : W_s \notin (a,b)\}$

3.9.3	$\mathbf{E}_x \exp\Bigl(-\alpha H - \frac{\gamma^2}{2}\int_0^H W_s^2 ds\Bigr)$ $= \frac{e^{(a^2-x^2)\gamma/2} S(\frac12+\frac{\alpha}{\gamma}, b\sqrt{2\gamma}, x\sqrt{2\gamma}) + e^{(b^2-x^2)\gamma/2} S(\frac12+\frac{\alpha}{\gamma}, x\sqrt{2\gamma}, a\sqrt{2\gamma})}{S(\frac12+\frac{\alpha}{\gamma}, b\sqrt{2\gamma}, a\sqrt{2\gamma})}$
3.9.7 (a)	$\mathbf{E}_x\Bigl\{\exp\Bigl(-\alpha H - \frac{\gamma^2}{2}\int_0^H W_s^2 ds\Bigr); W_H = a\Bigr\} = \frac{e^{a^2\gamma/2} S(\frac12+\frac{\alpha}{\gamma}, b\sqrt{2\gamma}, x\sqrt{2\gamma})}{e^{x^2\gamma/2} S(\frac12+\frac{\alpha}{\gamma}, b\sqrt{2\gamma}, a\sqrt{2\gamma})}$
3.9.7 (b)	$\mathbf{E}_x\Bigl\{\exp\Bigl(-\alpha H - \frac{\gamma^2}{2}\int_0^H W_s^2 ds\Bigr); W_H = b\Bigr\} = \frac{e^{b^2\gamma/2} S(\frac12+\frac{\alpha}{\gamma}, x\sqrt{2\gamma}, a\sqrt{2\gamma})}{e^{x^2\gamma/2} S(\frac12+\frac{\alpha}{\gamma}, b\sqrt{2\gamma}, a\sqrt{2\gamma})}$
3.10.1	$\mathbf{E}_x \exp\Bigl(-\gamma\int_0^H e^{2\beta W_s} ds\Bigr) = \frac{S_0\bigl(\frac{\sqrt{2\gamma}}{\lvert\beta\rvert}e^{\beta b}, \frac{\sqrt{2\gamma}}{\lvert\beta\rvert}e^{\beta x}\bigr) + S_0\bigl(\frac{\sqrt{2\gamma}}{\lvert\beta\rvert}e^{\beta x}, \frac{\sqrt{2\gamma}}{\lvert\beta\rvert}e^{\beta a}\bigr)}{S_0\bigl(\frac{\sqrt{2\gamma}}{\lvert\beta\rvert}e^{\beta b}, \frac{\sqrt{2\gamma}}{\lvert\beta\rvert}e^{\beta a}\bigr)}$
3.10.7 (a)	$\mathbf{E}_x\Bigl\{\exp\Bigl(-\alpha H - \gamma\int_0^H e^{2\beta W_s} ds\Bigr);\ W_H = a\Bigr\} = \frac{S_{\sqrt{2\alpha}/\lvert\beta\rvert}\bigl(\frac{\sqrt{2\gamma}}{\lvert\beta\rvert}e^{\beta b}, \frac{\sqrt{2\gamma}}{\lvert\beta\rvert}e^{\beta x}\bigr)}{S_{\sqrt{2\alpha}/\lvert\beta\rvert}\bigl(\frac{\sqrt{2\gamma}}{\lvert\beta\rvert}e^{\beta b}, \frac{\sqrt{2\gamma}}{\lvert\beta\rvert}e^{\beta a}\bigr)}$
3.10.7 (b)	$\mathbf{E}_x\Bigl\{\exp\Bigl(-\alpha H - \gamma\int_0^H e^{2\beta W_s} ds\Bigr);\ W_H = b\Bigr\} = \frac{S_{\sqrt{2\alpha}/\lvert\beta\rvert}\bigl(\frac{\sqrt{2\gamma}}{\lvert\beta\rvert}e^{\beta x}, \frac{\sqrt{2\gamma}}{\lvert\beta\rvert}e^{\beta a}\bigr)}{S_{\sqrt{2\alpha}/\lvert\beta\rvert}\bigl(\frac{\sqrt{2\gamma}}{\lvert\beta\rvert}e^{\beta b}, \frac{\sqrt{2\gamma}}{\lvert\beta\rvert}e^{\beta a}\bigr)}$
3.12.5 (a)	$\mathbf{E}_x\{e^{-\gamma\check{H}(H)};\ W_H = a\} = \int_{x-a}^{b-a} \frac{\operatorname{sh}((x-a)\sqrt{2\gamma})}{y\operatorname{sh}(y\sqrt{2\gamma})} dy$
3.12.5 (b)	$\mathbf{E}_x\{e^{-\gamma\hat{H}(H)};\ W_H = b\} = \int_{b-x}^{b-a} \frac{\operatorname{sh}((b-x)\sqrt{2\gamma})}{y\operatorname{sh}(y\sqrt{2\gamma})} dy$
3.12.6 (a)	$\mathbf{P}_x\bigl(\check{H}(H) \in du,\ W_H = a\bigr) = \int_{x-a}^{b-a} y^{-1}\operatorname{ss}_u(x-a, y) dy\, du$
3.12.6 (b)	$\mathbf{P}_x\bigl(\hat{H}(H) \in du,\ W_H = b\bigr) = \int_{b-x}^{b-a} y^{-1}\operatorname{ss}_u(b-x, y) dy\, du$
3.12.7 (a)	$\mathbf{E}_x\{e^{-\alpha H-\gamma\check{H}(H)};\ W_H = a\} = \int_{x-a}^{b-a} \frac{\sqrt{2\alpha}\operatorname{sh}((x-a)\sqrt{2\gamma+2\alpha})}{\operatorname{sh}(y\sqrt{2\alpha})\operatorname{sh}(y\sqrt{2\gamma+2\alpha})} dy$

1	W $\qquad\qquad H = H_{a,b} = \min\{s : W_s \notin (a,b)\}$
3.12.7 (b)	$\mathbf{E}_x\{e^{-\alpha H - \gamma \hat{H}(H)};\ W_H = b\} = \int_{b-x}^{b-a} \frac{\sqrt{2\alpha}\,\mathrm{sh}((b-x)\sqrt{2\gamma+2\alpha})}{\mathrm{sh}(y\sqrt{2\alpha})\,\mathrm{sh}(y\sqrt{2\gamma+2\alpha})} dy$
3.12.8 (a) $u < t$	$\mathbf{P}_x\big(\check{H}(H) \in du,\ H \in dt,\ W_H = a\big) = \int_{x-a}^{b-a} \mathrm{ss}_u(x-a, y)\,\mathrm{s}_{t-u}(y) dy\, dudt$
3.12.8 (b) $u < t$	$\mathbf{P}_x\big(\hat{H}(H) \in du,\ H \in dt,\ W_H = b\big) = \int_{b-x}^{b-a} \mathrm{ss}_u(b-x, y)\,\mathrm{s}_{t-u}(y) dy\, dudt$
3.18.1 $r < u$	$\mathbf{E}_x \exp\big(-\gamma \ell(H,r) - \eta \ell(H,u)\big)$
$a \le x$ $x \le r$	$= \frac{b-a+2\gamma(b-r)(r-x)+2\eta(b-u)(u-x)+4\gamma\eta(b-u)(u-r)(r-x)}{b-a+2\gamma(b-r)(r-a)+2\eta(b-u)(u-a)+4\gamma\eta(b-u)(u-r)(r-a)}$
$r \le x$ $x \le u$	$= \frac{b-a+2\gamma(x-r)(r-a)+2\eta(b-u)(u-x)}{b-a+2\gamma(b-r)(r-a)+2\eta(b-u)(u-a)+4\gamma\eta(b-u)(u-r)(r-a)}$
$u \le x$ $x \le b$	$= \frac{b-a+2\gamma(x-r)(r-a)+2\eta(x-u)(u-a)+4\gamma\eta(x-u)(u-r)(r-a)}{b-a+2\gamma(b-r)(r-a)+2\eta(b-u)(u-a)+4\gamma\eta(b-u)(u-r)(r-a)}$
3.18.5 (a)	$\mathbf{E}_x\big\{\exp\big(-\gamma \ell(H,r) - \eta \ell(H,u)\big);\ W_H = a\big\} =: Q_a^{(18)}(2\gamma, 2\eta)$
$a \le x$ $x \le r$	$= \frac{b-x+2\gamma(b-r)(r-x)+2\eta(b-u)(u-x)+4\gamma\eta(b-u)(u-r)(r-x)}{b-a+2\gamma(b-r)(r-a)+2\eta(b-u)(u-a)+4\gamma\eta(b-u)(u-r)(r-a)}$
$r \le x$ $x \le u$	$= \frac{b-x+2\eta(b-u)(u-x)}{b-a+2\gamma(b-r)(r-a)+2\eta(b-u)(u-a)+4\gamma\eta(b-u)(u-r)(r-a)}$
$u \le x$ $x \le b$	$= \frac{b-x}{b-a+2\gamma(b-r)(r-a)+2\eta(b-u)(u-a)+4\gamma\eta(b-u)(u-r)(r-a)}$
3.18.5 (b)	$\mathbf{E}_x\big\{\exp\big(-\gamma \ell(H,r) - \eta \ell(H,u)\big);\ W_H = b\big\} =: Q_b^{(18)}(2\gamma, 2\eta)$
$a \le x$ $x \le r$	$= \frac{x-a}{b-a+2\gamma(b-r)(r-a)+2\eta(b-u)(u-a)+4\gamma\eta(b-u)(u-r)(r-a)}$
$r \le x$ $x \le u$	$= \frac{x-a+2\gamma(x-r)(r-a)}{b-a+2\gamma(b-r)(r-a)+2\eta(b-u)(u-a)+4\gamma\eta(b-u)(u-r)(r-a)}$

1 W	$H = H_{a,b} = \min\{s : W_s \notin (a,b)\}$
$u \le x$ $x \le b$	$$= \frac{x-a+2\gamma(x-r)(r-a)+2\eta(x-u)(u-a)+4\gamma\eta(x-u)(u-r)(r-a)}{b-a+2\gamma(b-r)(r-a)+2\eta(b-u)(u-a)+4\gamma\eta(b-u)(u-r)(r-a)}$$
3.18.6 (a)	$\mathbf{P}_x\big(\ell(H,r)\in dy,\ \ell(H,u)\in dg,\ W_H=a\big) =: B^{(18)}_{x,a}(y,g)dydg$ $$= \begin{cases} \frac{x-a}{r-a}p_r dydg, & a\le x\le r \\ \frac{u-x}{u-r}p_r dydg + \frac{x-r}{u-r}p_u dydg, & r\le x\le u \\ \frac{b-x}{b-u}p_u dydg, & u\le x \end{cases}$$ where $$p_r := \frac{\sqrt{y}}{4(u-r)(r-a)\sqrt{g}}\exp\Big(-\frac{y+g}{2(u-r)}-\frac{y}{2(r-a)}-\frac{g}{2(b-u)}\Big)I_1\Big(\frac{\sqrt{yg}}{u-r}\Big)$$ $$p_u := \frac{1}{4(u-r)(r-a)}\exp\Big(-\frac{y+g}{2(u-r)}-\frac{y}{2(r-a)}-\frac{g}{2(b-u)}\Big)I_0\Big(\frac{\sqrt{yg}}{u-r}\Big)$$
(1)	$\mathbf{P}_x\big(\ell(H,r)\in dy,\ \ell(H,u)=0,\ W_H=a\big)$ $$= \begin{cases} \frac{x-a}{2(r-a)^2}\exp\Big(-\frac{y(u-a)}{2(u-r)(r-a)}\Big)dy, & a\le x\le r \\ \frac{u-x}{2(u-r)(r-a)}\exp\Big(-\frac{y(u-a)}{2(u-r)(r-a)}\Big)dy, & r\le x\le u \end{cases}$$
(2)	$\mathbf{P}_x\big(\ell(H,r)=0,\ \ell(H,u)=0,\ W_H=a\big) = \frac{r-x}{r-a}, \qquad a\le x\le r$
3.18.6 (b)	$\mathbf{P}_x\big(\ell(H,r)\in dy,\ \ell(H,u)\in dg,\ W_H=b\big) =: B^{(18)}_{x,b}(y,g)dydg$
$u<b$	$$= \begin{cases} \frac{x-a}{r-a}q_r dydg, & x\le r \\ \frac{u-x}{u-r}q_r dydg + \frac{x-r}{u-r}q_u dydg, & r\le x\le u \\ \frac{b-x}{b-u}q_u dydg, & u\le x\le b \end{cases}$$ where $$q_r := \frac{1}{4(b-u)(u-r)}\exp\Big(-\frac{y+g}{2(u-r)}-\frac{y}{2(r-a)}-\frac{g}{2(b-u)}\Big)I_0\Big(\frac{\sqrt{yg}}{u-r}\Big)$$ $$q_u := \frac{\sqrt{g}}{4(b-u)(u-r)\sqrt{y}}\exp\Big(-\frac{y+g}{2(u-r)}-\frac{y}{2(r-a)}-\frac{g}{2(b-u)}\Big)I_1\Big(\frac{\sqrt{yg}}{u-r}\Big)$$

1 W $\qquad H = H_{a,b} = \min\{s : W_s \notin (a,b)\}$

(1)
$$\mathbf{P}_x\big(\ell(H,r)=0,\ \ell(H,u)\in dg,\ W_H=b\big) = \begin{cases} \dfrac{x-r}{2(b-u)(u-r)}\exp\Big(-\dfrac{g(b-r)}{2(b-u)(u-r)}\Big)dg, & r\le x\le u \\ \dfrac{b-x}{2(b-u)^2}\exp\Big(-\dfrac{g(b-r)}{2(b-u)(u-r)}\Big)dg, & u\le x\le b \end{cases}$$

(2)
$$\mathbf{P}_x\big(\ell(H,r)=0,\ \ell(H,u)=0,\ W_H=b\big) = \frac{x-u}{b-u}, \qquad u\le x\le b$$

3.19.5 (a)
$$\mathbf{E}_x\Big\{\exp\Big(-\gamma\int_0^H \frac{ds}{W_s}\Big);\ W_H=a\Big\} = \frac{xS_1(2\sqrt{2\gamma b},2\sqrt{2\gamma x})}{aS_1(2\sqrt{2\gamma b},2\sqrt{2\gamma a})} \qquad 0<a\le x\le b$$

3.19.5 (b)
$$\mathbf{E}_x\Big\{\exp\Big(-\gamma\int_0^H \frac{ds}{W_s}\Big);\ W_H=b\Big\} = \frac{xS_1(2\sqrt{2\gamma x},2\sqrt{2\gamma a})}{bS_1(2\sqrt{2\gamma b},2\sqrt{2\gamma a})} \qquad 0<a\le x\le b$$

3.20.5 (a) $0<a$
$$\mathbf{E}_x\Big\{\exp\Big(-\frac{\gamma^2}{2}\int_0^H \frac{ds}{W_s^2}\Big);\ W_H=a\Big\} = \frac{x^{1/2}\big((b/x)^{\sqrt{1/4+\gamma^2}}-(x/b)^{\sqrt{1/4+\gamma^2}}\big)}{a^{1/2}\big((b/a)^{\sqrt{1/4+\gamma^2}}-(a/b)^{\sqrt{1/4+\gamma^2}}\big)}$$

3.20.5 (b) $0<a$
$$\mathbf{E}_x\Big\{\exp\Big(-\frac{\gamma^2}{2}\int_0^H \frac{ds}{W_s^2}\Big);\ W_H=b\Big\} = \frac{x^{1/2}\big((x/a)^{\sqrt{1/4+\gamma^2}}-(a/x)^{\sqrt{1/4+\gamma^2}}\big)}{b^{1/2}\big((b/a)^{\sqrt{1/4+\gamma^2}}-(a/b)^{\sqrt{1/4+\gamma^2}}\big)}$$

3.21.5 (a)
$$\mathbf{E}_x\Big\{\exp\Big(-\int_0^H \big(\frac{p^2}{2W_s^2}+\frac{q^2W_s^2}{2}\big)ds\Big);\ W_H=a\Big\}$$
$$= \frac{x^{\sqrt{p^2+1/4}+1/2}S_{\sqrt{4p^2+1}/4}(qb^2/2,qx^2/2)}{a^{\sqrt{p^2+1/4}+1/2}S_{\sqrt{4p^2+1}/4}(qb^2/2,qa^2/2)}, \qquad 0<a\le x\le b$$

3.21.5 (b)
$$\mathbf{E}_x\Big\{\exp\Big(-\int_0^H \big(\frac{p^2}{2W_s^2}+\frac{q^2W_s^2}{2}\big)ds\Big);\ W_H=b\Big\}$$
$$= \frac{x^{\sqrt{p^2+1/4}+1/2}S_{\sqrt{4p^2+1}/4}(qx^2/2,qa^2/2)}{b^{\sqrt{p^2+1/4}+1/2}S_{\sqrt{4p^2+1}/4}(qb^2/2,qa^2/2)}, \qquad 0<a\le x\le b$$

3.22.5 (a)
$$\mathbf{E}_x\Big\{\exp\Big(-\int_0^H \big(\frac{p^2}{2W_s^2}+\frac{q}{W_s}\big)ds\Big);\ W_H=a\Big\}$$
$$= \frac{x^{\sqrt{p^2+1/4}+1/2}S_{\sqrt{4p^2+1}}(2\sqrt{2qb},2\sqrt{2qx})}{a^{\sqrt{p^2+1/4}+1/2}S_{\sqrt{4p^2+1}}(2\sqrt{2qb},2\sqrt{2qa})}, \qquad 0<a\le x\le b$$

1 W $H = H_{a,b} = \min\{s : W_s \notin (a,b)\}$

3.22.5 (b)

$$\mathbf{E}_x\Big\{\exp\Big(-\int_0^H \big(\frac{p^2}{2W_s^2} + \frac{q}{W_s}\big)ds\Big);\ W_H = b\Big\}$$

$$= \frac{x^{\sqrt{p^2+1/4}+1/2} S_{\sqrt{4p^2+1}}(2\sqrt{2qx}, 2\sqrt{2qa})}{b^{\sqrt{p^2+1/4}+1/2} S_{\sqrt{4p^2+1}}(2\sqrt{2qb}, 2\sqrt{2qa})}, \qquad 0 < a \le x \le b$$

3.27.1

$$\mathbf{E}_x\Big\{\exp\Big(-\int_0^H (p\mathbb{1}_{[a,r)}(W_s) + q\mathbb{1}_{[r,b]}(W_s))ds\Big); \tfrac{1}{2}\ell(H,r) \in dy\Big\}$$

$$= \exp\big(-y(\sqrt{2q}\,\mathrm{cth}((b-r)\sqrt{2q}) + \sqrt{2p}\,\mathrm{cth}((r-a)\sqrt{2p}))\big)$$

$$\times \begin{cases} \dfrac{\sqrt{2}\,\mathrm{sh}((b-x)\sqrt{2q})[\sqrt{p}\,\mathrm{sh}((b-r)\sqrt{2q}) + \sqrt{q}\,\mathrm{sh}((r-a)\sqrt{2p})]}{\mathrm{sh}^2((b-r)\sqrt{2q})\,\mathrm{sh}((r-a)\sqrt{2p})}dy, & r \le x \le b \\ \dfrac{\sqrt{2}\,\mathrm{sh}((x-a)\sqrt{2p})[\sqrt{p}\,\mathrm{sh}((b-r)\sqrt{2q}) + \sqrt{q}\,\mathrm{sh}((r-a)\sqrt{2p})]}{\mathrm{sh}((b-r)\sqrt{2q})\,\mathrm{sh}^2((r-a)\sqrt{2p})}dy, & a \le x \le r \end{cases}$$

3.27.5 (a)

$$\mathbf{E}_x\Big\{\exp\Big(-\int_0^H (p\mathbb{1}_{[a,r)}(W_s) + q\mathbb{1}_{[r,b]}(W_s))ds\Big); \tfrac{1}{2}\ell(H,r) \in dy, W_H = a\Big\}$$

$$= \exp\big(-y(\sqrt{2q}\,\mathrm{cth}((b-r)\sqrt{2q}) + \sqrt{2p}\,\mathrm{cth}((r-a)\sqrt{2p}))\big)$$

$$\times \begin{cases} \dfrac{\sqrt{2p}\,\mathrm{sh}((b-x)\sqrt{2q})}{\mathrm{sh}((b-r)\sqrt{2q})\,\mathrm{sh}((r-a)\sqrt{2p})}dy, & r \le x \le b \\ \dfrac{\sqrt{2p}\,\mathrm{sh}((x-a)\sqrt{2p})}{\mathrm{sh}^2((r-a)\sqrt{2p})}dy, & a \le x \le r \end{cases}$$

(1)

$$\mathbf{E}_x\Big\{\exp\Big(-\int_0^H (p\mathbb{1}_{[a,r)}(W_s) + q\mathbb{1}_{[r,b]}(W_s))ds\Big); \ell(H_z,r) = 0, W_H = a\Big\}$$

$$= \frac{\mathrm{sh}((r-x)\sqrt{2p})}{\mathrm{sh}((r-a)\sqrt{2p})}, \qquad a \le x \le r$$

3.27.5 (b)

$$\mathbf{E}_x\Big\{\exp\Big(-\int_0^H (p\mathbb{1}_{[a,r)}(W_s) + q\mathbb{1}_{[r,b]}(W_s))ds\Big); \tfrac{1}{2}\ell(H_z,r) \in dy, W_H = b\Big\}$$

$$= \exp\big(-y(\sqrt{2q}\,\mathrm{cth}((b-r)\sqrt{2q}) + \sqrt{2p}\,\mathrm{cth}((r-a)\sqrt{2p}))\big)$$

$$\times \begin{cases} \dfrac{\sqrt{2q}\,\mathrm{sh}((b-x)\sqrt{2q})}{\mathrm{sh}^2((b-r)\sqrt{2q})}dy, & r \le x \le b \\ \dfrac{\sqrt{2q}\,\mathrm{sh}((x-a)\sqrt{2p})}{\mathrm{sh}((b-r)\sqrt{2q})\,\mathrm{sh}((r-a)\sqrt{2p})}dy, & a \le x \le r \end{cases}$$

1 W $\qquad H = H_{a,b} = \min\{s : W_s \notin (a,b)\}$

(1) $$\mathbf{E}_x\Big\{\exp\Big(-\int_0^H \big(p\mathbb{I}_{[a,r)}(W_s) + q\mathbb{I}_{[r,b]}(W_s)\big)ds\Big); \ell(H,r)=0, W_H = b\Big\}$$

$$= \frac{\operatorname{sh}((x-r)\sqrt{2q})}{\operatorname{sh}((b-r)\sqrt{2q})}, \qquad r \le x \le b$$

3.27.6 (a) $$\mathbf{P}_x\Big(\int_0^H \mathbb{I}_{[a,r)}(W_s)ds \in du,\ \int_0^H \mathbb{I}_{[r,b]}(W_s)ds \in dv,\ \frac{1}{2}\ell(H,r) \in dy, W_H = a\Big)$$

$$=: B_a^{(27)}(u,v,y)dudvdy$$

$$= \begin{cases} \operatorname{escs}_u(x-a, 0, r-a, y)dudvdy, & a \le x \le r \\ \operatorname{es}_u(1,1,r-a,0,y)\operatorname{ess}_v(b-x, b-r, 0, y)dudvdy, & r \le x \le b \end{cases}$$

3.27.6 (b) $$\mathbf{P}_x\Big(\int_0^H \mathbb{I}_{[a,r)}(W_s)ds \in du,\ \int_0^H \mathbb{I}_{[r,b]}(W_s)ds \in dv,\ \frac{1}{2}\ell(H,r) \in dy, W_H = b\Big)$$

$$=: B_b^{(27)}(u,v,y)dudvdy$$

$$= \begin{cases} \operatorname{ess}_u(x-a, r-a, 0, y)\operatorname{es}_v(1,1,b-r,0,y)dudvdy, & a \le x \le r \\ \operatorname{es}_u(0,0,r-a,0,y)\operatorname{escs}_v(b-x, 0, b-r, y)dudvdy, & r \le x \le b \end{cases}$$

3.30.5 (a) $$\mathbf{E}_x\Big\{\exp\Big(-\int_0^H \Big(\frac{p^2}{2}e^{2\beta W_s} + qe^{\beta W_s}\Big)ds\Big);\ W_H = a\Big\}$$

$$= \frac{\exp\Big(-\frac{p}{|\beta|}e^{\beta x}\Big)S\Big(\frac{q}{p|\beta|}+\frac{1}{2}, 1, \frac{2p}{|\beta|}e^{\beta b}, \frac{2p}{|\beta|}e^{\beta x}\Big)}{\exp\Big(-\frac{p}{|\beta|}e^{\beta a}\Big)S\Big(\frac{q}{p|\beta|}+\frac{1}{2}, 1, \frac{2p}{|\beta|}e^{\beta b}, \frac{2p}{|\beta|}e^{\beta a}\Big)}$$

3.30.5 (b) $$\mathbf{E}_x\Big\{\exp\Big(-\int_0^H \Big(\frac{p^2}{2}e^{2\beta W_s} + qe^{\beta W_s}\Big)ds\Big);\ W_H = b\Big\}$$

$$= \frac{\exp\Big(-\frac{p}{|\beta|}e^{\beta x}\Big)S\Big(\frac{q}{p|\beta|}+\frac{1}{2}, 1, \frac{2p}{|\beta|}e^{\beta x}, \frac{2p}{|\beta|}e^{\beta a}\Big)}{\exp\Big(-\frac{p}{|\beta|}e^{\beta b}\Big)S\Big(\frac{q}{p|\beta|}+\frac{1}{2}, 1, \frac{2p}{|\beta|}e^{\beta b}, \frac{2p}{|\beta|}e^{\beta a}\Big)}$$

1 W $H = H_{a,b} = \min\{s : W_s \notin (a,b)\}$

3.31.5 (a)

$$\mathbf{E}_{x(b-a)+a}\Big\{\exp\Big(-\frac{\gamma}{2(b-a)}\sup_{0\le s\le H}\big(\ell(s,u(b-a)+a)-\alpha^{-1}\ell(s,r(b-a)+a)\big)\Big);$$

$0<\alpha$

$$W_H=a\Big\} = 1-x-\frac{\gamma(1-u)(\Upsilon+r-r^2-\alpha u+\alpha u^2)(u\wedge x-ux)}{2\gamma(u\wedge r-ur)|u-r|+\Upsilon+r-r^2-\alpha u+\alpha u^2}$$

$$-\frac{\gamma(1-u)(2(u\wedge r-ur)|u-r|-(\Upsilon+r-r^2-\alpha u+\alpha u^2)(u-u^2))(x\wedge r-xr)}{(2\gamma(u\wedge r-ur)|u-r|+\Upsilon+r-r^2-\alpha u+\alpha u^2)(u\wedge r-ur)}$$

3.31.5 (b)

$$\mathbf{E}_{x(b-a)+a}\Big\{\exp\Big(-\frac{\gamma}{2(b-a)}\sup_{0\le s\le H}\big(\ell(s,u(b-a)+a)-\alpha^{-1}\ell(s,r(b-a)+a)\big)\Big);$$

$0<\alpha$

$$W_H=b\Big\} = x-\frac{\gamma u(\Upsilon+r-r^2-\alpha u+\alpha u^2)(u\wedge x-ux)}{2\gamma(u\wedge r-ur)|u-r|+\Upsilon+r-r^2-\alpha u+\alpha u^2}$$

$$-\frac{\gamma u(2(u\wedge r-ur)|u-r|-(\Upsilon+r-r^2-\alpha u+\alpha u^2)(u-u^2))(x\wedge r-xr)}{(2\gamma(u\wedge r-ur)|u-r|+\Upsilon+r-r^2-\alpha u+\alpha u^2)(u\wedge r-ur)}$$

3.31.6 (a)

$$\mathbf{P}_{x(b-a)+a}\Big(\sup_{0\le s\le H}\big(\ell(s,u(b-a)+a)-\alpha^{-1}\ell(s,r(b-a)+a)\big)>(b-a)h,$$

$0<\alpha$

$$W_H=a\Big) = (1-u)\exp\Big(-\frac{(\Upsilon+r-r^2-\alpha u+\alpha u^2)h}{4(u\wedge r-ur)|u-r|}\Big)\Big[\frac{x\wedge r-xr}{u\wedge r-ur}$$

$$-\frac{\Upsilon+r-r^2-\alpha u+\alpha u^2}{2(u\wedge r-ur)|u-r|}\Big(\frac{(x\wedge r-xr)(u-u^2)}{u\wedge r-ur}-u\wedge x+ux\Big)\Big]$$

3.31.6 (b)

$$\mathbf{P}_{x(b-a)+a}\Big(\sup_{0\le s\le H}\big(\ell(s,u(b-a)+a)-\alpha^{-1}\ell(s,r(b-a)+a)\big)>(b-a)h,$$

$0<\alpha$

$$W_H=b\Big) = u\exp\Big(-\frac{(\Upsilon+r-r^2-\alpha u+\alpha u^2)h}{4(u\wedge r-ur)|u-r|}\Big)\Big[\frac{x\wedge r-xr}{u\wedge r-ur}$$

$$-\frac{\Upsilon+r-r^2-\alpha u+\alpha u^2}{2(u\wedge r-ur)|u-r|}\Big(\frac{(x\wedge r-xr)(u-u^2)}{u\wedge r-ur}-u\wedge x+ux\Big)\Big]$$

$$\Upsilon := \sqrt{(r-r^2-\alpha u+\alpha u^2)^2+4\alpha(u\wedge r-ur)|u-r|}$$

1 W $\varrho = \varrho(v,z) = \min\{s : \ell(s,z) = v\}$

4. Stopping at inverse local time

4.0.1 $\mathbf{E}_x e^{-\alpha\varrho} = e^{-(|z-x|+v)\sqrt{2\alpha}}$

4.0.2 $\mathbf{P}_x\big(\varrho \in dt\big) = \frac{|z-x|+v}{\sqrt{2\pi}t^{3/2}} \exp\big(-\frac{(|z-x|+v)^2}{2t}\big)dt$

4.1.1 $\mathbf{E}_x \exp\big(-\gamma \sup_{0\le s\le\varrho} W_s\big) = e^{-\gamma z}\sqrt{2v\gamma}K_1\big(\sqrt{2v\gamma}\big)$

$x \le z$

4.1.2 $$\mathbf{P}_x\big(\sup_{0\le s\le\varrho} W_s < y\big) = \begin{cases} \exp\Big(-\dfrac{v}{2(y-z)}\Big), & x \le z \\ \dfrac{y-x}{y-z}\exp\Big(-\dfrac{v}{2(y-z)}\Big), & z \le x \le y \end{cases}$$

4.1.4 $\mathbf{E}_x\big\{e^{-\alpha\varrho}; \sup_{0\le s\le\varrho} W_s < y\big\}$

$$= \begin{cases} \exp\Big(-(z-x)\sqrt{2\alpha} - \dfrac{v\sqrt{\alpha}e^{(y-z)\sqrt{2\alpha}}}{\sqrt{2}\,\mathrm{sh}((y-z)\sqrt{2\alpha})}\Big), & x \le z \le y \\ \dfrac{\mathrm{sh}((y-x)\sqrt{2\alpha})}{\mathrm{sh}((y-z)\sqrt{2\alpha})}\exp\Big(-\dfrac{v\sqrt{\alpha}e^{(y-z)\sqrt{2\alpha}}}{\sqrt{2}\,\mathrm{sh}((y-z)\sqrt{2\alpha})}\Big), & z \le x \le y \end{cases}$$

(1) $$\mathbf{P}_x\big(\sup_{0\le s\le\varrho} W_s < y, \varrho \in dt\big) = \begin{cases} \mathrm{es}_t\big(0,0,y-z,z-x+\frac{v}{2},\frac{v}{2}\big)dt, & x \le z \le y \\ \mathrm{ess}_t\big(y-x,y-z,\frac{v}{2},\frac{v}{2}\big)dt, & z \le x \le y \end{cases}$$

4.2.1 $\mathbf{E}_x \exp\big(\gamma \inf_{0\le s\le\varrho} W_s\big) = e^{\gamma z}\sqrt{2v\gamma}K_1\big(\sqrt{2v\gamma}\big)$

$z \le x$

4.2.2 $$\mathbf{P}_x\big(\inf_{0\le s\le\varrho} W_s > y\big) = \begin{cases} \dfrac{x-y}{z-y}\exp\Big(-\dfrac{v}{2(z-y)}\Big), & y \le x \le z \\ \exp\Big(-\dfrac{v}{2(z-y)}\Big), & y \le z \le x \end{cases}$$

4.2.4 $\mathbf{E}_x\big\{e^{-\alpha\varrho}; \inf_{0\le s\le\varrho} W_s > y\big\}$

1 W $\qquad \varrho=\varrho(v,z)=\min\{s:\ell(s,z)=v\}$

$$=\begin{cases}\dfrac{\operatorname{sh}((x-y)\sqrt{2\alpha})}{\operatorname{sh}((z-y)\sqrt{2\alpha})}\exp\Big(-\dfrac{v\sqrt{\alpha}e^{(z-y)\sqrt{2\alpha}}}{\sqrt{2}\operatorname{sh}((z-y)\sqrt{2\alpha})}\Big), & y\le x\le z\\ \exp\Big(-(x-z)\sqrt{2\alpha}-\dfrac{v\sqrt{\alpha}e^{(z-y)\sqrt{2\alpha}}}{\sqrt{2}\operatorname{sh}((z-y)\sqrt{2\alpha})}\Big), & y\le z\le x\end{cases}$$

(1) $$\mathbf{P}_x\Big(\inf_{0\le s\le\varrho}W_s>y,\varrho\in dt\Big)=\begin{cases}\operatorname{ess}_t\big(x-y,z-y,\frac{v}{2},\frac{v}{2}\big)dt, & y\le x\le z\\ \operatorname{es}_t\big(0,0,z-y,x-z+\frac{v}{2},\frac{v}{2}\big)dt, & y\le z\le x\end{cases}$$

4.3.1 $$\mathbf{E}_x e^{-\gamma\ell(\varrho,r)}=\begin{cases}\exp\Big(-\dfrac{\gamma v}{1+2\gamma|r-z|}\Big), & x\le z\le r,\quad r\le z\le x\\ \dfrac{1+2\gamma|r-x|}{1+2\gamma|r-z|}\exp\Big(-\dfrac{\gamma v}{1+2\gamma|r-z|}\Big), & r\wedge z\le x\le r\vee z\\ \dfrac{1}{1+2\gamma|r-z|}\exp\Big(-\dfrac{\gamma v}{1+2\gamma|r-z|}\Big), & z\le r\le x,\quad x\le r\le z\end{cases}$$

4.3.2 $\mathbf{P}_x\big(\ell(\varrho,r)\in dy\big)$

$$=\begin{cases}\dfrac{1}{2|r-z|}\dfrac{\sqrt{v}}{\sqrt{y}}\exp\Big(-\dfrac{(v+y)}{2|r-z|}\Big)I_1\Big(\dfrac{\sqrt{vy}}{|r-z|}\Big)dy, & x\le z\le r,\quad r\le z\le x\\ \dfrac{1}{2(r-z)^2}\exp\Big(-\dfrac{(v+y)}{2|r-z|}\Big)\Big[|r-x|\dfrac{\sqrt{v}}{\sqrt{y}}I_1\Big(\dfrac{\sqrt{vy}}{|r-z|}\Big) & \\ \qquad\qquad +|z-x|I_0\Big(\dfrac{\sqrt{vy}}{|r-z|}\Big)\Big]dy, & z\wedge r\le x\le r\vee z\\ \dfrac{1}{2|r-z|}\exp\Big(-\dfrac{(v+y)}{2|r-z|}\Big)I_0\Big(\dfrac{\sqrt{vy}}{|r-z|}\Big)dy, & z\le r\le x,\quad x\le r\le z\end{cases}$$

(1) $$\mathbf{P}_x\big(\ell(\varrho,r)=0\big)=\begin{cases}\exp\Big(-\dfrac{v}{2|r-z|}\Big), & x\le z\le r,\quad r\le z\le x\\ \dfrac{|r-x|}{|r-z|}\exp\Big(-\dfrac{v}{2|r-z|}\Big), & z\wedge r\le x\le r\vee z\\ 0, & z\le r\le x,\quad x\le r\le z\end{cases}$$

4.3.3 $$\mathbf{E}_x e^{-\alpha\varrho-\gamma\ell(\varrho,r)}=:V_z^{(3)}(2\alpha)=\exp\Big(-|z-x|\sqrt{2\alpha}-\frac{v\sqrt{2\alpha}(\gamma+\sqrt{2\alpha})}{\sqrt{2\alpha}+\gamma(1-e^{-2|r-z|\sqrt{2\alpha}})}\Big)$$

$$\times\begin{cases}1, & x\le z\le r,\quad r\le z\le x\\ \dfrac{\sqrt{2\alpha}+\gamma(1-e^{-2|r-x|\sqrt{2\alpha}})}{\sqrt{2\alpha}+\gamma(1-e^{-2|r-z|\sqrt{2\alpha}})}, & r\wedge z\le x\le r\vee z\\ \dfrac{\sqrt{2\alpha}}{\sqrt{2\alpha}+\gamma(1-e^{-2|r-z|\sqrt{2\alpha}})}, & z\le r\le x,\quad x\le r\le z\end{cases}$$

1 W $\qquad \varrho = \varrho(v,z) = \min\{s : \ell(s,z) = v\}$

4.3.4 $\mathbf{E}_x\{e^{-\alpha\varrho}; \ell(\varrho,r) \in dy\} =: F_z^{(3)}(2\alpha, y)dy = \exp\Big(-\frac{(v+y)\sqrt{2\alpha}}{1-e^{-2|r-z|\sqrt{2\alpha}}}\Big)$

$$\times \begin{cases} \frac{\sqrt{\alpha}e^{-|z-x|\sqrt{2\alpha}}}{\sqrt{2}\,\mathrm{sh}(|r-z|\sqrt{2\alpha})}\frac{\sqrt{v}}{\sqrt{y}}I_1\Big(\frac{\sqrt{2\alpha vy}}{\mathrm{sh}(|r-z|\sqrt{2\alpha})}\Big)dy, & x \le z \le r, \quad r \le z \le x \\ \frac{\sqrt{\alpha}}{\sqrt{2}\,\mathrm{sh}^2(|r-z|\sqrt{2\alpha})}\Big[\mathrm{sh}(|r-x|\sqrt{2\alpha})\frac{\sqrt{v}}{\sqrt{y}}I_1\Big(\frac{\sqrt{2\alpha vy}}{\mathrm{sh}(|r-z|\sqrt{2\alpha})}\Big) & \\ \qquad +\,\mathrm{sh}(|z-x|\sqrt{2\alpha})I_0\Big(\frac{\sqrt{2\alpha vy}}{\mathrm{sh}(|r-z|\sqrt{2\alpha})}\Big)\Big]dy, & z\wedge r \le x \le r \vee z \\ \frac{\sqrt{\alpha}e^{-|r-x|\sqrt{2\alpha}}}{\sqrt{2}\,\mathrm{sh}(|r-z|\sqrt{2\alpha})}I_0\Big(\frac{\sqrt{2\alpha vy}}{\mathrm{sh}(|r-z|\sqrt{2\alpha})}\Big)dy, & z \le r \le x, \quad x \le r \le z \end{cases}$$

(1) $\mathbf{E}_x\{e^{-\alpha\varrho}; \ell(\varrho,r) = 0\}$

$$= \exp\Big(-\frac{v\sqrt{2\alpha}}{1-e^{-2|r-z|\sqrt{2\alpha}}}\Big)\begin{cases} e^{-|z-x|\sqrt{2\alpha}}, & x \le z \le r, \quad r \le z \le x \\ \frac{\mathrm{sh}(|r-x|\sqrt{2\alpha})}{\mathrm{sh}(|r-z|\sqrt{2\alpha})}, & r\wedge z \le x \le r \vee z \\ 0, & z \le r \le x, \quad x \le r \le z \end{cases}$$

(2) $\mathbf{P}_x\big(\ell(\varrho,r) \in dy,\ \varrho \in dt\big) =: F_z^{(3)}(y,t)dydt$

for $x \le z \le r$ or $r \le z \le x$

$$= \frac{\sqrt{v}}{2\sqrt{y}}\,\mathrm{is}_t(1, |r-z|, |z-x| + (y+v)/2, (y+v)/2, \sqrt{vy}/2)dydt$$

for $z \le r \le x$ or $x \le r \le z$

$$= \frac{1}{2}\,\mathrm{is}_t(0, |r-z|, |r-x| + (y+v)/2, (y+v)/2, \sqrt{vy}/2)dydt$$

4.4.1 $\mathbf{E}_x \exp\Big(-\gamma\int_0^{\varrho} \mathbb{I}_{[r,\infty)}(W_s)ds\Big)$

$z \le r$
$$= \begin{cases} \exp\Big(-\frac{v\sqrt{2\gamma}}{2(1+(r-z)\sqrt{2\gamma})}\Big), & x \le z \\ \frac{1+(r-x)\sqrt{2\gamma}}{1+(r-z)\sqrt{2\gamma}}\exp\Big(-\frac{v\sqrt{2\gamma}}{2(1+(r-z)\sqrt{2\gamma})}\Big), & z \le x \le r \\ \frac{1}{1+(r-z)\sqrt{2\gamma}}\exp\Big((r-x)\sqrt{2\gamma} - \frac{v\sqrt{2\gamma}}{2(1+(r-z)\sqrt{2\gamma})}\Big), & r \le x \end{cases}$$

1 W	$\varrho = \varrho(v,z) = \min\{s : \ell(s,z) = v\}$
$r \le z$	$$= \begin{cases} \dfrac{1}{\operatorname{ch}((z-r)\sqrt{2\gamma})} \exp\Big(-\dfrac{v\sqrt{\gamma}e^{(z-r)\sqrt{2\gamma}}}{\sqrt{2}\operatorname{ch}((z-r)\sqrt{2\gamma})}\Big), & x \le r \\ \dfrac{\operatorname{ch}((x-r)\sqrt{2\gamma})}{\operatorname{ch}((z-r)\sqrt{2\gamma})} \exp\Big(-\dfrac{v\sqrt{\gamma}e^{(z-r)\sqrt{2\gamma}}}{\sqrt{2}\operatorname{ch}((z-r)\sqrt{2\gamma})}\Big), & r \le x \le z \\ \exp\Big((z-x)\sqrt{2\gamma} - \dfrac{v\sqrt{\gamma}e^{(z-r)\sqrt{2\gamma}}}{\sqrt{2}\operatorname{ch}((z-r)\sqrt{2\gamma})}\Big), & z \le x \end{cases}$$
4.4.2	$\mathbf{P}_x\Big(\int_0^{\varrho} \mathbb{I}_{[r,\infty)}(W_s)ds \in dy\Big)$
$z \le r$ $x \le z$	$$= \frac{\sqrt{v}}{2\sqrt{\pi}(r-z)y^{3/2}} \exp\Big(-\frac{v}{2(r-z)}\Big) \int_0^{\infty} \sqrt{u} \exp\Big(-\frac{u^2}{2y} - \frac{u}{r-z}\Big) I_1\Big(\frac{\sqrt{2vu}}{r-z}\Big) du\, dy$$
$z \le r$ $r \le x$	$$= \frac{1}{\sqrt{2\pi}(r-z)y^{3/2}} \exp\Big(-\frac{v-2x+2r}{2(r-z)}\Big) \int_{x-r}^{\infty} u \exp\Big(-\frac{u^2}{2y} - \frac{u}{r-z}\Big) I_0\Big(\frac{\sqrt{2vu}}{r-z}\Big) du\, dy$$
$r \le z$ $x \le r$	$= \operatorname{ec}_y(0, 1, z-r, v/2, v/2)dy$
$r \le x$ $x \le z$	$= \operatorname{ecc}_y(x-r, z-r, v/2, v/2)dy$
$r \le z$ $z \le x$	$= \operatorname{ec}_y(0, 0, z-r, v/2+x-z, v/2)dy$
(1) $z \le r$	$$\mathbf{P}_x\Big(\int_0^{\varrho} \mathbb{I}_{[r,\infty)}(W_s)ds = 0\Big) = \begin{cases} \exp\Big(-\dfrac{v}{2(r-z)}\Big), & x \le z \\ \dfrac{r-x}{r-z} \exp\Big(-\dfrac{v}{2(r-z)}\Big), & z \le x \le r \\ 0, & r \le x \end{cases}$$
4.5.1	$\mathbf{E}_x \exp\Big(-\gamma \int_0^{\varrho} \mathbb{I}_{(-\infty,r]}(W_s)ds\Big)$
$z \le r$	$$= \begin{cases} \exp\Big((x-z)\sqrt{2\gamma} - \dfrac{v\sqrt{\gamma}e^{(r-z)\sqrt{2\gamma}}}{\sqrt{2}\operatorname{ch}((r-z)\sqrt{2\gamma})}\Big), & x \le z \\ \dfrac{\operatorname{ch}((r-x)\sqrt{2\gamma})}{\operatorname{ch}((r-z)\sqrt{2\gamma})} \exp\Big(-\dfrac{v\sqrt{\gamma}e^{(r-z)\sqrt{2\gamma}}}{\sqrt{2}\operatorname{ch}((r-z)\sqrt{2\gamma})}\Big), & z \le x \le r \\ \dfrac{1}{\operatorname{ch}((r-z)\sqrt{2\gamma})} \exp\Big(-\dfrac{v\sqrt{\gamma}e^{(r-z)\sqrt{2\gamma}}}{\sqrt{2}\operatorname{ch}((r-z)\sqrt{2\gamma})}\Big), & r \le x \end{cases}$$

1 W	$\varrho=\varrho(v,z)=\min\{s:\ell(s,z)=v\}$
$r\le z$	$=\begin{cases}\dfrac{1}{1+(z-r)\sqrt{2\gamma}}\exp\Bigl((x-r)\sqrt{2\gamma}-\dfrac{v\sqrt{2\gamma}}{2(1+(z-r)\sqrt{2\gamma})}\Bigr), & x\le r\\ \dfrac{1+(x-r)\sqrt{2\gamma}}{1+(z-r)\sqrt{2\gamma}}\exp\Bigl(-\dfrac{v\sqrt{2\gamma}}{2(1+(z-r)\sqrt{2\gamma})}\Bigr), & r\le x\le z\\ \exp\Bigl(-\dfrac{v\sqrt{2\gamma}}{2(1+(z-r)\sqrt{2\gamma})}\Bigr), & z\le x\end{cases}$
4.5.2	$\mathbf{P}_x\Bigl(\int_0^{\varrho}\mathbb{I}_{(-\infty,r]}(W_s)ds\in dy\Bigr)$
$z\le r$ $x\le z$	$=\operatorname{ec}_y(0,0,r-z,v/2+z-x,v/2)dy$
$z\le x$ $x\le r$	$=\operatorname{ecc}_y(r-x,r-z,v/2,v/2)dy$
$z\le r$ $r\le x$	$=\operatorname{ec}_y(0,1,r-z,v/2,v/2)dy$
$r\le z$ $x\le r$	$=\dfrac{1}{\sqrt{2\pi}(z-r)y^{3/2}}\exp\Bigl(-\dfrac{v-2r+2x}{2(z-r)}\Bigr)\int_{r-x}^{\infty}u\exp\Bigl(-\dfrac{u^2}{2y}-\dfrac{u}{z-r}\Bigr)I_0\Bigl(\dfrac{\sqrt{2vu}}{z-r}\Bigr)du\,dy$
$r\le z$ $z\le x$	$=\dfrac{\sqrt{v}}{2\sqrt{\pi}(z-r)y^{3/2}}\exp\Bigl(-\dfrac{v}{2(z-r)}\Bigr)\int_0^{\infty}\sqrt{u}\exp\Bigl(-\dfrac{u^2}{2y}-\dfrac{u}{z-r}\Bigr)I_1\Bigl(\dfrac{\sqrt{2vu}}{z-r}\Bigr)du\,dy$
(1) $r\le z$	$\mathbf{P}_x\Bigl(\int_0^{\varrho}\mathbb{I}_{(-\infty,r]}(W_s)ds=0\Bigr)=\begin{cases}0, & x\le r\\ \dfrac{x-r}{z-r}\exp\Bigl(-\dfrac{v}{2(z-r)}\Bigr), & r\le x\le z\\ \exp\Bigl(-\dfrac{v}{2(z-r)}\Bigr), & z\le x\end{cases}$
4.6.1	$\mathbf{E}_x\exp\Bigl(-\int_0^{\varrho}\bigl(p\mathbb{I}_{(-\infty,r)}(W_s)+q\mathbb{I}_{[r,\infty)}(W_s)\bigr)ds\Bigr)=:Q_z^{(6)}(2p,2q)$
$z\le r$	$=\exp\Bigl(-\dfrac{v\sqrt{p}(\sqrt{p}+\sqrt{q})e^{(r-z)\sqrt{2p}}}{\sqrt{2}(\sqrt{p}\operatorname{ch}((r-z)\sqrt{2p})+\sqrt{q}\operatorname{sh}((r-z)\sqrt{2p}))}\Bigr)$ $\times\begin{cases}e^{-(z-x)\sqrt{2p}}, & x\le z\\ \dfrac{\sqrt{p}\operatorname{ch}((r-x)\sqrt{2p})+\sqrt{q}\operatorname{sh}((r-x)\sqrt{2p})}{\sqrt{p}\operatorname{ch}((r-z)\sqrt{2p})+\sqrt{q}\operatorname{sh}((r-z)\sqrt{2p})}, & z\le x\le r\\ \dfrac{\sqrt{p}e^{-(x-r)\sqrt{2q}}}{\sqrt{p}\operatorname{ch}((r-z)\sqrt{2p})+\sqrt{q}\operatorname{sh}((r-z)\sqrt{2p})}, & r\le x\end{cases}$
$r\le z$	$=\exp\Bigl(-\dfrac{v\sqrt{q}(\sqrt{p}+\sqrt{q})e^{(z-r)\sqrt{2q}}}{\sqrt{2}(\sqrt{q}\operatorname{ch}((z-r)\sqrt{2q})+\sqrt{p}\operatorname{sh}((z-r)\sqrt{2q}))}\Bigr)$

1 W $\varrho = \varrho(v,z) = \min\{s : \ell(s,z) = v\}$

$$\times\begin{cases} \dfrac{\sqrt{q}e^{-(r-x)\sqrt{2p}}}{\sqrt{q}\,\mathrm{ch}((z-r)\sqrt{2q}) + \sqrt{p}\,\mathrm{sh}((z-r)\sqrt{2q})}, & x \le r \\ \dfrac{\sqrt{q}\,\mathrm{ch}((x-r)\sqrt{2q}) + \sqrt{p}\,\mathrm{sh}((x-r)\sqrt{2q})}{\sqrt{q}\,\mathrm{ch}((z-r)\sqrt{2q}) + \sqrt{p}\,\mathrm{sh}((z-r)\sqrt{2q})}, & r \le x \le z \\ e^{-(x-z)\sqrt{2q}}, & z \le x \end{cases}$$

4.6.2 $\mathbf{P}_x\Big(\int_0^\varrho \mathbb{I}_{(-\infty,r)}(W_s)ds \in du,\ \int_0^\varrho \mathbb{I}_{[r,\infty)}(W_s)ds \in dg\Big)$

$$= \int_0^\infty B_z^{(27)}(u,g,y)dy\,dudg \qquad \text{for } B_z^{(27)}(u,g,y) \quad \text{see 4.27.2}$$

4.7.1 $\mathbf{E}_x \exp\Big(-\gamma\int_0^\varrho \mathbb{I}_{[r,u]}(W_s), ds\Big)$

$z \le r$
$$= \exp\Big(-\frac{v\sqrt{\gamma}\,\mathrm{sh}((u-r)\sqrt{2\gamma})}{\sqrt{2}((r-z)\sqrt{2\gamma}\,\mathrm{sh}((u-r)\sqrt{2\gamma}) + \mathrm{ch}((u-r)\sqrt{2\gamma}))}\Big)$$

$$\times\begin{cases} \dfrac{(r-x)\sqrt{2\gamma}\,\mathrm{sh}((u-r)\sqrt{2\gamma}) + \mathrm{ch}((u-r)\sqrt{2\gamma})}{(r-z)\sqrt{2\gamma}\,\mathrm{sh}((u-r)\sqrt{2\gamma}) + \mathrm{ch}((u-r)\sqrt{2\gamma})}, & z \le x \le r \\ \dfrac{\mathrm{ch}((u-x)\sqrt{2\gamma})}{(r-z)\sqrt{2\gamma}\,\mathrm{sh}((u-r)\sqrt{2\gamma}) + \mathrm{ch}((u-r)\sqrt{2\gamma})}, & r \le x \le u \\ \dfrac{1}{(r-z)\sqrt{2\gamma}\,\mathrm{sh}((u-r)\sqrt{2\gamma}) + \mathrm{ch}((u-r)\sqrt{2\gamma})}, & u \le x \end{cases}$$

$u \le z$
$$= \exp\Big(-\frac{v\sqrt{\gamma}\,\mathrm{sh}((u-r)\sqrt{2\gamma})}{\sqrt{2}((z-u)\sqrt{2\gamma}\,\mathrm{sh}((u-r)\sqrt{2\gamma}) + \mathrm{ch}((u-r)\sqrt{2\gamma}))}\Big)$$

$$\times\begin{cases} \dfrac{1}{(z-u)\sqrt{2\gamma}\,\mathrm{sh}((u-r)\sqrt{2\gamma}) + \mathrm{ch}((u-r)\sqrt{2\gamma})}, & x \le r \\ \dfrac{\mathrm{ch}((x-r)\sqrt{2\gamma})}{(z-u)\sqrt{2\gamma}\,\mathrm{sh}((u-r)\sqrt{2\gamma}) + \mathrm{ch}((u-r)\sqrt{2\gamma})}, & r \le x \le u \\ \dfrac{(x-u)\sqrt{2\gamma}\,\mathrm{sh}((u-r)\sqrt{2\gamma}) + \mathrm{ch}((u-r)\sqrt{2\gamma})}{(z-u)\sqrt{2\gamma}\,\mathrm{sh}((u-r)\sqrt{2\gamma}) + \mathrm{ch}((u-r)\sqrt{2\gamma})}, & u \le x \le z \end{cases}$$

4.8.1 $\mathbf{E}_x \exp\Big(-\gamma\int_0^{\varrho(v,0)} |W_s|ds\Big) = 3^{2/3}\Gamma(2/3)\,\mathrm{Ai}\big((2\gamma)^{1/3}|x|\big)\exp\Big(-\frac{(6\gamma)^{1/3}\Gamma(2/3)v}{\Gamma(1/3)}\Big)$

4.9.3 $\mathbf{E}_x \exp\Big(-\alpha\varrho - \frac{\gamma^2}{2}\int_0^\varrho W_s^2 ds\Big)$

$$= \begin{cases} \dfrac{D_{-\frac12-\frac{\alpha}{\gamma}}(-x\sqrt{2\gamma})}{D_{-\frac12-\frac{\alpha}{\gamma}}(-z\sqrt{2\gamma})}\exp\Big(-\dfrac{v\sqrt{\pi\gamma}}{\Gamma(\frac12+\frac{\alpha}{\gamma})D_{-\frac12-\frac{\alpha}{\gamma}}(-z\sqrt{2\gamma})D_{-\frac12-\frac{\alpha}{\gamma}}(z\sqrt{2\gamma})}\Big), & x \le z \\ \dfrac{D_{-\frac12-\frac{\alpha}{\gamma}}(x\sqrt{2\gamma})}{D_{-\frac12-\frac{\alpha}{\gamma}}(z\sqrt{2\gamma})}\exp\Big(-\dfrac{v\sqrt{\pi\gamma}}{\Gamma(\frac12+\frac{\alpha}{\gamma})D_{-\frac12-\frac{\alpha}{\gamma}}(-z\sqrt{2\gamma})D_{-\frac12-\frac{\alpha}{\gamma}}(z\sqrt{2\gamma})}\Big), & z \le x \end{cases}$$

1	W $\qquad\qquad\qquad\qquad \varrho = \varrho(v,z) = \min\{s : \ell(s,z) = v\}$
4.10.3	$$\mathbf{E}_x \exp\Big(-\alpha\varrho - \gamma\int_0^{\varrho} e^{2\beta W_s}ds\Big) = \exp\Big(-\frac{v\lvert\beta\rvert}{2I_{\sqrt{2\alpha}/\lvert\beta\rvert}\Big(\frac{\sqrt{2\gamma}}{\lvert\beta\rvert}e^{\beta z}\Big)K_{\sqrt{2\alpha}/\lvert\beta\rvert}\Big(\frac{\sqrt{2\gamma}}{\lvert\beta\rvert}e^{\beta z}\Big)}\Big) \times\begin{cases}\dfrac{I_{\sqrt{2\alpha}/\lvert\beta\rvert}\Big(\frac{\sqrt{2\gamma}}{\lvert\beta\rvert}e^{\beta x}\Big)}{I_{\sqrt{2\alpha}/\lvert\beta\rvert}\Big(\frac{\sqrt{2\gamma}}{\lvert\beta\rvert}e^{\beta z}\Big)}, & (z-x)\beta\ge 0\\ \dfrac{K_{\sqrt{2\alpha}/\lvert\beta\rvert}\Big(\frac{\sqrt{2\gamma}}{\lvert\beta\rvert}e^{\beta x}\Big)}{K_{\sqrt{2\alpha}/\lvert\beta\rvert}\Big(\frac{\sqrt{2\gamma}}{\lvert\beta\rvert}e^{\beta z}\Big)}, & (z-x)\beta\le 0\end{cases}$$
4.11.2	$$\lambda\int_0^{\infty} e^{-\lambda z}\mathbf{P}_0\big(\sup_{y\in\mathbb{R}}\ell(\varrho,y) < h\big)dz = \big(1-\frac{v}{h}\big)\Big[1-\frac{v}{h}+\frac{2}{\lambda h}\Big(\frac{I_0(\sqrt{2\lambda v})}{I_0(\sqrt{2\lambda h})}-1\Big)\Big]\mathbb{I}_{[v,\infty)}(h)$$
(1) $z \ge 0$	$$\mathbf{P}_0\big(\sup_{y\in\mathbb{R}}\ell(\varrho,y) < h\big) = 8\big(1-\frac{v}{h}\big)\sum_{k=1}^{\infty}\frac{\exp(-zj_{0,k}^2/2h)J_0(j_{0,k}\sqrt{v/h})}{j_{0,k}^3 J_1(j_{0,k})}\mathbb{I}_{[0,h]}(v)$$
(2)	$$\mathbf{P}_0\big(\sup_{y\in\mathbb{R}}\ell(\varrho(v,0),y) < h\big) = \big(1-\frac{v}{h}\big)^2\mathbb{I}_{[0,h]}(v)$$
4.13.1	$$\mathbf{E}_x\{e^{-\gamma\check{H}(\varrho)}; \sup_{0\le s\le\varrho} W_s \in dy\} = \frac{\operatorname{sh}((x-z)^+\sqrt{2\gamma})}{(y-z)\operatorname{sh}((y-z)\sqrt{2\gamma})}\exp\Big(-\frac{v}{2(y-z)}\Big)dy + \frac{2\sqrt{2\gamma}e^{-(y-z)\sqrt{2\gamma}}}{1-2(y-z)\sqrt{2\gamma}-e^{2(z-y)\sqrt{2\gamma}}}\Big(\exp\Big(-\frac{v\sqrt{2\gamma}}{1-e^{2(z-y)\sqrt{2\gamma}}}\Big)-\exp\Big(-\frac{v}{2(y-z)}\Big)\Big) \times\begin{cases} e^{-(z-x)\sqrt{2\gamma}}dy, & x\le z\\ \dfrac{\operatorname{sh}((y-x)\sqrt{2\gamma})}{\operatorname{sh}((y-z)\sqrt{2\gamma})}dy, & z\le x\le y\end{cases}$$
4.14.1	$$\mathbf{E}_x\{e^{-\gamma\hat{H}(\varrho)}; \inf_{0\le s\le\varrho} W_s \in dy\} = \frac{\operatorname{sh}((z-x)^+\sqrt{2\gamma})}{(z-y)\operatorname{sh}((z-y)\sqrt{2\gamma})}\exp\Big(-\frac{v}{2(z-y)}\Big)dy + \frac{2\sqrt{2\gamma}e^{-(z-y)\sqrt{2\gamma}}}{1-2(z-y)\sqrt{2\gamma}-e^{2(y-z)\sqrt{2\gamma}}}\Big(\exp\Big(-\frac{v\sqrt{2\gamma}}{1-e^{2(y-z)\sqrt{2\gamma}}}\Big)-\exp\Big(-\frac{v}{2(z-y)}\Big)\Big) \times\begin{cases}\dfrac{\operatorname{sh}((x-y)\sqrt{2\gamma})}{\operatorname{sh}((z-y)\sqrt{2\gamma})}dy, & y\le x\le z\\ e^{-(x-z)\sqrt{2\gamma}}dy, & z\le x\end{cases}$$

1 W $\varrho = \varrho(v,z) = \min\{s : \ell(s,z) = v\}$

4.15.2 $\mathbf{P}_x\big(a < \inf_{0\le s\le\varrho} W_s,\ \sup_{0\le s\le\varrho} W_s < b\big)$

$$= \begin{cases} \frac{x-a}{z-a}\exp\Big(-\frac{(b-a)v}{2(z-a)(b-z)}\Big), & a \le x \le z \\ \frac{b-x}{b-z}\exp\Big(-\frac{(b-a)v}{2(z-a)(b-z)}\Big), & z \le x \le b \end{cases}$$

(1) $$\mathbf{P}_x\big(\sup_{0\le s\le\varrho} |W_s| \le b\big) = \begin{cases} \frac{b+x}{b+z}\exp\Big(-\frac{bv}{(b+z)(b-z)}\Big), & -b \le x \le z \\ \frac{b-x}{b-z}\exp\Big(-\frac{bv}{(b+z)(b-z)}\Big), & z \le x \le b \end{cases}$$

4.15.4 $\mathbf{E}_x\big\{e^{-\alpha\varrho}; a < \inf_{0\le s\le\varrho} W_s,\ \sup_{0\le s\le\varrho} W_s < b\big\}$

$$= \begin{cases} \frac{\operatorname{sh}((x-a)\sqrt{2\alpha})}{\operatorname{sh}((z-a)\sqrt{2\alpha})}\exp\Big(-\frac{v\sqrt{2\alpha}\operatorname{sh}((b-a)\sqrt{2\alpha})}{2\operatorname{sh}((b-z)\sqrt{2\alpha})\operatorname{sh}((z-a)\sqrt{2\alpha})}\Big), & a \le x \le z \\ \frac{\operatorname{sh}((b-x)\sqrt{2\alpha})}{\operatorname{sh}((b-z)\sqrt{2\alpha})}\exp\Big(-\frac{v\sqrt{2\alpha}\operatorname{sh}((b-a)\sqrt{2\alpha})}{2\operatorname{sh}((b-z)\sqrt{2\alpha})\operatorname{sh}((z-a)\sqrt{2\alpha})}\Big), & z \le x \le b \end{cases}$$

(1) $\mathbf{E}_x\big\{e^{-\alpha\varrho}; \sup_{0\le s\le\varrho} |W_s| \le b\big\}$

$$= \begin{cases} \frac{\operatorname{sh}((b+x)\sqrt{2\alpha})}{\operatorname{sh}((b+z)\sqrt{2\alpha})}\exp\Big(-\frac{v\sqrt{2\alpha}\operatorname{sh}((2b\sqrt{2\alpha})}{\operatorname{ch}(2b\sqrt{2\alpha})-\operatorname{ch}(2z\sqrt{2\alpha})}\Big), & -b \le x \le z \\ \frac{\operatorname{sh}((b-x)\sqrt{2\alpha})}{\operatorname{sh}((b-z)\sqrt{2\alpha})}\exp\Big(-\frac{v\sqrt{2\alpha}\operatorname{sh}(2b\sqrt{2\alpha})}{\operatorname{ch}(2b\sqrt{2\alpha})-\operatorname{ch}(2z\sqrt{2\alpha})}\Big), & z \le x \le b \end{cases}$$

4.16.1
$r < b$
$z < r$

$\mathbf{E}_x\big\{e^{-\gamma\ell(\varrho,r)}; \sup_{0\le s\le\varrho} W_s < b\big\}$

$$= \exp\Big(-\frac{v(1+2\gamma(b-r))}{2(b-z+2\gamma(b-r)(r-z))}\Big)\begin{cases} \frac{b-x}{b-z+2\gamma(b-r)(r-z)}, & r \le x \le b \\ \frac{b-x+2\gamma(b-r)(r-x)}{b-z+2\gamma(b-r)(r-z)}, & z \le x \le r \end{cases}$$

4.16.2
$r < b$
$z < r$

$$\mathbf{P}_x\big(\ell(\varrho,r)\in dy,\ \sup_{0\le s\le\varrho} W_s < b\big) = \exp\Big(-\frac{v}{2(r-z)} - \frac{(b-z)y}{2(b-r)(r-z)}\Big)$$

1 W $\varrho = \varrho(v,z) = \min\{s : \ell(s,z) = v\}$

$$\times \begin{cases} \dfrac{b-x}{2(b-r)(r-z)} I_0\Big(\dfrac{\sqrt{vy}}{r-z}\Big)dy, & r \le x \le b \\ \dfrac{1}{2(r-z)^2}\Big[(r-x)\dfrac{\sqrt{v}}{\sqrt{y}} I_1\Big(\dfrac{\sqrt{vy}}{r-z}\Big) + (x-z) I_0\Big(\dfrac{\sqrt{vy}}{r-z}\Big)\Big]dy, & z \le x \le r \end{cases}$$

(1) $$\mathbf{P}_x\big(\ell(\varrho,r) = 0, \sup_{0 \le s \le \varrho} W_s < b\big) = \begin{cases} 0, & r \le x \le b \\ \dfrac{r-x}{r-z} \exp\Big(-\dfrac{v}{2(r-z)}\Big), & z \le x \le r \end{cases}$$

4.17.1 $a < r$ $r < z$ $$\mathbf{E}_x\big\{e^{-\gamma \ell(\varrho,r)}; a < \inf_{0 \le s \le \varrho} W_s\big\}$$

$$= \exp\Big(-\frac{v(1+2\gamma(r-a))}{2(z-a+2\gamma(z-r)(r-a))}\Big) \begin{cases} \dfrac{x-a+2\gamma(x-r)(r-a)}{z-a+2\gamma(z-r)(r-a)}, & r \le x \le z \\ \dfrac{x-a}{z-a+2\gamma(z-r)(r-a)}, & a \le x \le r \end{cases}$$

4.17.2 $a < r$ $r < z$ $$\mathbf{P}_x\big(\ell(\varrho,r) \in dy, a < \inf_{0 \le s \le \varrho} W_s\big) = \exp\Big(-\frac{v}{2(z-r)} - \frac{(z-a)y}{2(z-r)(r-a)}\Big)$$

$$\times \begin{cases} \dfrac{1}{2(z-r)^2}\Big[(x-r)\dfrac{\sqrt{v}}{\sqrt{y}} I_1\Big(\dfrac{\sqrt{vy}}{z-r}\Big) + (z-x) I_0\Big(\dfrac{\sqrt{vy}}{z-r}\Big)\Big]dy, & r \le x \le z \\ \dfrac{x-a}{2(z-r)(r-a)} I_0\Big(\dfrac{\sqrt{vy}}{z-r}\Big)dy, & a \le x \le r \end{cases}$$

(1) $$\mathbf{P}_x\big(\ell(\varrho,r) = 0, a < \inf_{0 \le s \le \varrho} W_s\big) = \begin{cases} \dfrac{x-r}{z-r} \exp\Big(-\dfrac{v}{2(z-r)}\Big), & r \le x \le z \\ 0, & a \le x \le r \end{cases}$$

4.18.1 $r < u$ $$\mathbf{E}_x \exp\big(-\gamma \ell(\varrho,r) - \eta \ell(\varrho,u)\big)$$

$z < r$ $$= \exp\Big(-\frac{v(\gamma+\eta+2\gamma\eta(u-r))}{1+2\gamma(r-z)+2\eta(u-z)+4\gamma\eta(u-r)(r-z)}\Big)$$

$$\times \begin{cases} \dfrac{1+2\gamma(r-x)+2\eta(u-x)+4\gamma\eta(u-r)(r-x)}{1+2\gamma(r-z)+2\eta(u-z)+4\gamma\eta(u-r)(r-z)}, & z \le x \le r \\ \dfrac{1+2\eta(u-x)}{1+2\gamma(r-z)+2\eta(u-z)+4\gamma\eta(u-r)(r-z)}, & r \le x \le u \\ \dfrac{1}{1+2\gamma(r-z)+2\eta(u-z)+4\gamma\eta(u-r)(r-z)}, & u < x \end{cases}$$

1 W $\varrho = \varrho(v,z) = \min\{s : \ell(s,z) = v\}$

$u < z$

$$= \exp\Big(-\frac{v(\gamma+\eta+2\gamma\eta(u-r))}{1+2\gamma(z-r)+2\eta(z-u)+4\gamma\eta(z-u)(u-r)}\Big)$$

$$\times\begin{cases} \dfrac{1}{1+2\gamma(z-r)+2\eta(z-u)+4\gamma\eta(z-u)(u-r)}, & x < r \\ \dfrac{1+2\gamma(x-r)}{1+2\gamma(z-r)+2\eta(z-u)+4\gamma\eta(z-u)(u-r)}, & r \le x \le u \\ \dfrac{1+2\gamma(x-r)+2\eta(x-u)+4\gamma\eta(x-u)(u-r)}{1+2\gamma(z-r)+2\eta(z-u)+4\gamma\eta(z-u)(u-r)}, & u \le x \le z \end{cases}$$

4.19.1 $\mathbf{E}_x\Big\{\exp\Big(-\gamma\int_0^\varrho \frac{ds}{W_s}\Big);\ 0 < \inf_{0\le s\le\varrho} W_s\Big\}$

$$= \begin{cases} \dfrac{x^{1/2}I_1(2\sqrt{2\gamma x})}{z^{1/2}I_1(2\sqrt{2\gamma z})}\exp\Big(-\dfrac{v}{4zI_1(2\sqrt{2\gamma z})K_1(2\sqrt{2\gamma z})}\Big), & 0 < x \le z \\ \dfrac{x^{1/2}K_1(2\sqrt{2\gamma x})}{z^{1/2}K_1(2\sqrt{2\gamma z})}\exp\Big(-\dfrac{v}{4zI_1(2\sqrt{2\gamma z})K_1(2\sqrt{2\gamma z})}\Big), & z \le x \end{cases}$$

4.20.1 $\mathbf{E}_x\Big\{\exp\Big(-\frac{\gamma^2}{2}\int_0^\varrho \frac{ds}{W_s^2}\Big);\ 0 < \inf_{0\le s\le\varrho} W_s\Big\}$

$$= \begin{cases} \Big(\dfrac{x}{z}\Big)^{1/2+\sqrt{\gamma^2+1/4}}\exp\Big(-\dfrac{v\sqrt{\gamma^2+1/4}}{z}\Big), & 0 < x \le z \\ \Big(\dfrac{x}{z}\Big)^{1/2-\sqrt{\gamma^2+1/4}}\exp\Big(-\dfrac{v\sqrt{\gamma^2+1/4}}{z}\Big), & z \le x \end{cases}$$

4.20.4
$0 < y$ $\mathbf{P}_x\Big(\int_0^\varrho \frac{ds}{W_s^2} \in dy,\ 0 < \inf_{0\le s\le\varrho} W_s\Big)$

$$= \begin{cases} \dfrac{x^{1/2}(\ln(z/x)+v/z)}{z^{1/2}\sqrt{2\pi}y^{3/2}}e^{-y/8}e^{-(\ln(z/x)+v/z)^2/2y}dy, & 0 < x \le z \\ \dfrac{x^{1/2}(\ln(x/z)+v/z)}{z^{1/2}\sqrt{2\pi}y^{3/2}}e^{-y/8}e^{-(\ln(x/z)+v/z)^2/2y}dy, & z \le x \end{cases}$$

4.21.1
$0 < x$ $\mathbf{E}_x\Big\{\exp\Big(-\int_0^\varrho\big(\frac{p^2}{2W_s^2}+\frac{q^2W_s^2}{2}\big)ds\Big);\ 0 < \inf_{0\le s\le\varrho} W_s\Big\}$

$$= \begin{cases} \dfrac{\sqrt{x}I_{\sqrt{4p^2+1}/4}(qx^2/2)}{\sqrt{z}I_{\sqrt{4p^2+1}/4}(qz^2/2)}\exp\Big(-\dfrac{v}{zI_{\sqrt{4p^2+1}/4}(qz^2/2)K_{\sqrt{4p^2+1}/4}(qz^2/2)}\Big), & x \le z \\ \dfrac{\sqrt{x}K_{\sqrt{4p^2+1}/4}(qx^2/2)}{\sqrt{z}K_{\sqrt{4p^2+1}/4}(qz^2/2)}\exp\Big(-\dfrac{v}{zI_{\sqrt{4p^2+1}/4}(qz^2/2)K_{\sqrt{4p^2+1}/4}(qz^2/2)}\Big), & z \le x \end{cases}$$

1 W $\qquad \varrho=\varrho(v,z)=\min\{s:\ell(s,z)=v\}$

4.22.1
$0<x$

$$\mathbf{E}_x\Big\{\exp\Big(-\int_0^\varrho\big(\frac{p^2}{2W_s^2}+\frac{q}{W_s}\big)ds\Big);\ 0<\inf_{0\le s\le\varrho}W_s\Big\}$$

$$=\begin{cases}\dfrac{\sqrt{x}I_{\sqrt{4p^2+1}}(2\sqrt{2qx})}{\sqrt{z}I_{\sqrt{4p^2+1}}(2\sqrt{2qz})}\exp\Big(-\dfrac{v}{4zI_{\sqrt{4p^2+1}}(2\sqrt{2qz})K_{\sqrt{4p^2+1}}(2\sqrt{2qz})}\Big), & x\le z\\ \dfrac{\sqrt{x}K_{\sqrt{4p^2+1}}(2\sqrt{2qx})}{\sqrt{z}K_{\sqrt{4p^2+1}}(2\sqrt{2qz})}\exp\Big(-\dfrac{v}{4zI_{\sqrt{4p^2+1}}(2\sqrt{2qz})K_{\sqrt{4p^2+1}}(2\sqrt{2qz})}\Big), & z\le x\end{cases}$$

4.23.1
$r<b$
$z<r$

$$\mathbf{E}_x\Big\{\exp\Big(-\int_0^\varrho\big(p\mathbb{I}_{(-\infty,r)}(W_s)+q\mathbb{I}_{[r,b]}(W_s)\big)ds\Big);\ \sup_{0\le s\le\varrho}W_s<b\Big\}$$

$$=\exp\Big(-\frac{v\sqrt{p}(\sqrt{p}\,\mathrm{th}((b-r)\sqrt{2q})+\sqrt{q})e^{(r-z)\sqrt{2p}}}{\sqrt{2}(\sqrt{p}\,\mathrm{th}((b-r)\sqrt{2q})\,\mathrm{ch}((r-z)\sqrt{2p})+\sqrt{q}\,\mathrm{sh}((r-z)\sqrt{2p}))}\Big)$$

$$\times\begin{cases}\dfrac{\sqrt{p}\,\mathrm{sh}((b-x)\sqrt{2q})}{\sqrt{p}\,\mathrm{sh}((b-r)\sqrt{2q})\,\mathrm{ch}((r-z)\sqrt{2p})+\sqrt{q}\,\mathrm{ch}((b-r)\sqrt{2q})\,\mathrm{sh}((r-z)\sqrt{2p})}, & r\le x,\ x\le b\\ \dfrac{\sqrt{p}\,\mathrm{sh}((b-r)\sqrt{2q})\,\mathrm{ch}((r-x)\sqrt{2p})+\sqrt{q}\,\mathrm{ch}((b-r)\sqrt{2q})\,\mathrm{sh}((r-x)\sqrt{2p})}{\sqrt{p}\,\mathrm{sh}((b-r)\sqrt{2q})\,\mathrm{ch}((r-z)\sqrt{2p})+\sqrt{q}\,\mathrm{ch}((b-r)\sqrt{2q})\,\mathrm{sh}((r-z)\sqrt{2p})}, & z\le x,\ x\le r\\ e^{-(z-x)\sqrt{2p}}, & x<z\end{cases}$$

4.24.1
$a<r$
$r<z$

$$\mathbf{E}_x\Big\{\exp\Big(-\int_0^\varrho\big(p\mathbb{I}_{[a,r)}(W_s)+q\mathbb{I}_{[r,\infty)}(W_s)\big)ds\Big);a<\inf_{0\le s\le\varrho}W_s\Big\}$$

$$=\exp\Big(-\frac{v\sqrt{q}(\sqrt{p}+\sqrt{q}\,\mathrm{th}((r-a)\sqrt{2p}))e^{(z-r)\sqrt{2q}}}{\sqrt{2}(\sqrt{p}\,\mathrm{sh}((z-r)\sqrt{2q})+\sqrt{q}\,\mathrm{ch}((z-r)\sqrt{2q})\,\mathrm{th}((r-a)\sqrt{2p}))}\Big)$$

$$\times\begin{cases}e^{-(z-x)\sqrt{2p}}, & z<x\\ \dfrac{\sqrt{p}\,\mathrm{sh}((x-r)\sqrt{2q})\,\mathrm{ch}((r-a)\sqrt{2p})+\sqrt{q}\,\mathrm{ch}((x-r)\sqrt{2q})\,\mathrm{sh}((r-a)\sqrt{2p})}{\sqrt{p}\,\mathrm{sh}((z-r)\sqrt{2q})\,\mathrm{ch}((r-a)\sqrt{2p})+\sqrt{q}\,\mathrm{ch}((z-r)\sqrt{2q})\,\mathrm{sh}((r-a)\sqrt{2p})}, & r\le x,\ x\le z\\ \dfrac{\sqrt{q}\,\mathrm{sh}((x-a)\sqrt{2p})}{\sqrt{p}\,\mathrm{sh}((z-r)\sqrt{2q})\,\mathrm{ch}((r-a)\sqrt{2p})+\sqrt{q}\,\mathrm{ch}((z-r)\sqrt{2q})\,\mathrm{sh}((r-a)\sqrt{2p})}, & a\le x,\ x\le r\end{cases}$$

1 W $\varrho=\varrho(v,z)=\min\{s:\ell(s,z)=v\}$

4.27.1 $\mathbf{E}_x\Big\{\exp\Big(-\int_0^{\varrho}\big(p\mathbb{I}_{(-\infty,r)}(W_s)+q\mathbb{I}_{[r,\infty)}(W_s)\big)ds\Big);\ell(\varrho,r)\in dy\Big\}$

$$=: Q_z^{(27)}(2p,2q,y)dy$$

$z\le r$

$$=\exp\Big(-\frac{v\sqrt{p}}{\sqrt2}-\frac{y\sqrt{q}}{\sqrt2}-\frac{(v+y)\sqrt{p}\,\mathrm{ch}((r-z)\sqrt{2p})}{\sqrt2\,\mathrm{sh}((r-z)\sqrt{2p})}\Big)$$

$$\times\begin{cases}\dfrac{\sqrt{pv}e^{-(z-x)\sqrt{2p}}}{\sqrt2\,\mathrm{sh}((r-z)\sqrt{2p})\sqrt{y}}I_1\Big(\dfrac{\sqrt{2pvy}}{\mathrm{sh}((r-z)\sqrt{2p})}\Big)dy, & x\le z\\ \dfrac{\sqrt{p}}{\sqrt2\,\mathrm{sh}^2((r-z)\sqrt{2p})}\Big[\mathrm{sh}((r-x)\sqrt{2p})\dfrac{\sqrt{v}}{\sqrt{y}}I_1\Big(\dfrac{\sqrt{2pvy}}{\mathrm{sh}((r-z)\sqrt{2p})}\Big) & \\ \qquad+\mathrm{sh}((x-z)\sqrt{2p})I_0\Big(\dfrac{\sqrt{2pvy}}{\mathrm{sh}((r-z)\sqrt{2p})}\Big)\Big]dy, & z\le x\le r\\ \dfrac{\sqrt{p}e^{-(x-r)\sqrt{2q}}}{\sqrt2\,\mathrm{sh}((r-z)\sqrt{2p})}I_0\Big(\dfrac{\sqrt{2pvy}}{\mathrm{sh}((r-z)\sqrt{2p})}\Big)dy, & r\le x\end{cases}$$

$r\le z$

$$=\exp\Big(-\frac{v\sqrt{q}}{\sqrt2}-\frac{y\sqrt{p}}{\sqrt2}-\frac{(v+y)\sqrt{q}\,\mathrm{ch}((z-r)\sqrt{2q})}{\sqrt2\,\mathrm{sh}((z-r)\sqrt{2q})}\Big)$$

$$\times\begin{cases}\dfrac{\sqrt{q}e^{-(r-x)\sqrt{2p}}}{\sqrt2\,\mathrm{sh}((z-r)\sqrt{2q})}I_0\Big(\dfrac{\sqrt{2qvy}}{\mathrm{sh}((z-r)\sqrt{2q})}\Big)dy, & x\le r\\ \dfrac{\sqrt{q}}{\sqrt2\,\mathrm{sh}^2((z-r)\sqrt{2q})}\Big[\mathrm{sh}((x-r)\sqrt{2q})\dfrac{\sqrt{v}}{\sqrt{y}}I_1\Big(\dfrac{\sqrt{2qvy}}{\mathrm{sh}((z-r)\sqrt{2q})}\Big) & \\ \qquad+\mathrm{sh}((z-x)\sqrt{2q})I_0\Big(\dfrac{\sqrt{2qvy}}{\mathrm{sh}((z-r)\sqrt{2q})}\Big)\Big]dy, & r\le x\le z\\ \dfrac{\sqrt{qv}e^{-(x-z)\sqrt{2q}}}{\sqrt2\,\mathrm{sh}((z-r)\sqrt{2q})\sqrt{y}}I_1\Big(\dfrac{\sqrt{2qvy}}{\mathrm{sh}((z-r)\sqrt{2q})}\Big)dy, & z\le x\end{cases}$$

(1) $\mathbf{E}_x\Big\{\exp\Big(-\int_0^{\varrho}\big(p\mathbb{I}_{(-\infty,r)}(W_s)+q\mathbb{I}_{[r,\infty)}(W_s)\big)ds\Big);\ell(\varrho,r)=0\Big\}$

$z\le r$

$$=\exp\Big(-\frac{v\sqrt{p}e^{(r-z)\sqrt{2p}}}{\sqrt2\,\mathrm{sh}((r-z)\sqrt{2p})}\Big)\begin{cases}e^{-(z-x)\sqrt{2p}}, & x\le z\\ \dfrac{\mathrm{sh}((r-x)\sqrt{2p})}{\mathrm{sh}((r-z)\sqrt{2p})}, & z\le x\le r\\ 0, & r\le x\end{cases}$$

$r\le z$

$$=\exp\Big(-\frac{v\sqrt{q}e^{(z-r)\sqrt{2q}}}{\sqrt2\,\mathrm{sh}((z-r)\sqrt{2q})}\Big)\begin{cases}0, & x\le r\\ \dfrac{\mathrm{sh}((x-r)\sqrt{2q})}{\mathrm{sh}((z-r)\sqrt{2q})}, & r\le x\le z\\ e^{-(x-z)\sqrt{2q}}, & z\le x\end{cases}$$

1 W $\varrho = \varrho(v,z) = \min\{s : \ell(s,z) = v\}$

4.27.2 $\mathbf{P}_x\Big(\int_0^\varrho \mathbb{I}_{(-\infty,r)}(W_s)ds \in du,\ \int_0^\varrho \mathbb{I}_{[r,\infty)}(W_s)ds \in dg,\ \ell(\varrho,r) \in dy\Big)$

$$=: B_z^{(27)}(u,g,y)dudgdy$$

$x \le z$, $z \le r$:
$$= \frac{\sqrt{v}}{2\sqrt{y}}\,\mathrm{is}_u(1, r-z, z-x+v/2, (y+v)/2, \sqrt{vy}/2)\,\mathrm{h}_g(1, y/2)dudgdy$$

$z \le r$, $r \le x$:
$$= \frac{1}{2}\,\mathrm{is}_u(0, r-z, v/2, (y+v)/2, \sqrt{vy}/2)\,\mathrm{h}_g(1, x-r+y/2)dudgdy$$

$x \le r$, $r \le z$:
$$= \frac{1}{2}\,\mathrm{h}_u(1, r-x+y/2)\,\mathrm{is}_g(0, z-r, v/2, (y+v)/2, \sqrt{vy}/2)dudgdy$$

$r \le z$, $z \le x$:
$$= \frac{\sqrt{v}}{2\sqrt{y}}\,\mathrm{h}_u(1, y/2)\,\mathrm{is}_g(1, z-r, x-z+v/2, (y+v)/2, \sqrt{vy}/2)dudgdy$$

(1) $\mathbf{P}_x\Big(\int_0^\varrho \mathbb{I}_{(-\infty,r)}(W_s)ds \in du,\ \int_0^\varrho \mathbb{I}_{[r,\infty)}(W_s)ds = 0,\ \ell(\varrho,r) = 0\Big)$

$$= \begin{cases} \mathrm{es}_u(0, 0, r-z, z-x+v/2, v/2)du, & x \le z \\ \mathrm{ess}_u(r-x, r-z, v/2, v/2)du, & z \le x \le r \end{cases}$$

(2) $\mathbf{P}_x\Big(\int_0^\varrho \mathbb{I}_{(-\infty,r)}(W_s)ds = 0,\ \int_0^\varrho \mathbb{I}_{[r,\infty)}(W_s)ds \in dg,\ \ell(\varrho,r) = 0\Big)$

$$= \begin{cases} \mathrm{ess}_g(x-r, z-r, v/2, v/2)dg, & r \le x \le z \\ \mathrm{es}_g(0, 0, z-r, x-z+v/2, v/2)dg, & z \le x \end{cases}$$

4.30.1 $\mathbf{E}_x \exp\Big(-\int_0^\varrho \Big(\frac{p^2}{2}e^{2\beta W_s} + qe^{\beta W_s}\Big)ds\Big)$

$$= \exp\Big(-\frac{pve^{\beta z}}{\Gamma(1/2 + q/p|\beta|)M_{-q/p|\beta|,0}\big(\frac{2p}{|\beta|}e^{\beta z}\big)W_{-q/p|\beta|,0}\big(\frac{2p}{|\beta|}e^{\beta z}\big)}\Big)$$

$$\times \begin{cases} \dfrac{e^{-\beta x/2}M_{-q/p|\beta|,0}\big(\frac{2p}{|\beta|}e^{\beta x}\big)}{e^{-\beta z/2}M_{-q/p|\beta|,0}\big(\frac{2p}{|\beta|}e^{\beta z}\big)}, & (z-x)\beta \ge 0 \\ \dfrac{e^{-\beta x/2}W_{-q/p|\beta|,0}\big(\frac{2p}{|\beta|}e^{\beta x}\big)}{e^{-\beta z/2}W_{-q/p|\beta|,0}\big(\frac{2p}{|\beta|}e^{\beta z}\big)}, & (z-x)\beta \le 0 \end{cases}$$

1 W $\theta_v = \min\{t : \sup_{0\le s\le t} W_s - \inf_{0\le s\le t} W_s = v\}$

5. Stopping at first range time

5.0.1 $$\mathbf{E}_x e^{-\alpha\theta_v} = \frac{1}{\operatorname{ch}^2(v\sqrt{\alpha/2})}$$

5.0.2 $$\mathbf{P}_x(\theta_v \in dt) = \frac{4v}{t\sqrt{2\pi t}}\sum_{k=1}^{\infty}(-1)^{k-1}k^2 e^{-k^2v^2/2t}\,dt$$

5.0.3 $$\mathbf{E}_x e^{i\beta W_{\theta_v}} = \frac{2}{\beta^2 v^2}e^{i\beta x}(\beta v\sin(\beta v) + \cos(\beta v) - 1)$$

5.0.4 $$\mathbf{P}_x\big(W_{\theta_v} \in dz\big) = \frac{|z-x|}{v^2}dz, \qquad |z-x| \le v$$

5.0.5 $$\mathbf{E}_x\big\{e^{-\alpha\theta_v};\ W_{\theta_v} \in dz\big\} = \frac{\sqrt{2\alpha}\operatorname{sh}(|z-x|\sqrt{2\alpha})}{\operatorname{sh}^2(v\sqrt{2\alpha})}dz, \qquad |z-x| \le v$$

5.0.6 $$\mathbf{P}_x\big(\theta_v \in dt, W_{\theta_v} \in dz\big) \qquad |z-x| \le v$$
$$= \frac{\sqrt{2}}{t\sqrt{\pi t}}\sum_{k=-\infty}^{\infty} k\big(1 - t^{-1}(|x-z| + 2kv)^2\big)e^{-(|x-z|+2kv)^2/2t}\,dtdz$$

5.3.5 $$\mathbf{E}_x\big\{e^{-\gamma\ell(\theta_v,r)};\ W_{\theta_v} \in dz\big\} \qquad |z-x| \le v$$
$$= \begin{cases} \dfrac{|z-x|}{(v+2\gamma(v-|z-r|)|r-z|)^2}dz, & z\wedge r \le x \le z\vee r \\ \dfrac{|z-x| + 2\gamma|z-r||r-x|}{(v+2\gamma(v-|z-r|)|r-z|)^2}dz, & z\wedge x \le r \le z\vee x \end{cases}$$

5.3.6 $$\mathbf{P}_x\big(\ell(\theta_v,r) \in dy, W_{\theta_v} \in dz\big) = \exp\Big(-\frac{vy}{2(v-|z-r|)|z-r|}\Big)$$
$$\times\begin{cases} \dfrac{|z-x|y}{4(v-|z-r|)^2(z-r)^2}dydz, & z\wedge r \le x \le z\vee r \\ \dfrac{2(v-|z-r|)|x-r| + (v-|z-x|)y}{4(v-|z-r|)^3|z-r|}dydz, & z\wedge x \le r \le z\vee x \end{cases}$$

5.3.7 $$\mathbf{E}_x\big\{e^{-\alpha\theta_v-\gamma\ell(\theta_v,r)};\ W_{\theta_v} \in dz\big\} \qquad |z-x| \le v$$
$$= \begin{cases} \dfrac{\sqrt{2\alpha}\operatorname{sh}(|z-x|\sqrt{2\alpha})}{\big(\operatorname{sh}(v\sqrt{2\alpha}) + \frac{\gamma\sqrt{2}}{\sqrt{\alpha}}\operatorname{sh}((v-|z-r|)\sqrt{2\alpha})\operatorname{sh}(|r-z|\sqrt{2\alpha})\big)^2}dz, & z\wedge r \le x \le z\vee r \\ \dfrac{\sqrt{2\alpha}\operatorname{sh}(|z-x|\sqrt{2\alpha}) + 2\gamma\operatorname{sh}(|z-r|\sqrt{2\alpha})\operatorname{sh}(|r-x|\sqrt{2\alpha})}{\big(\operatorname{sh}(v\sqrt{2\alpha}) + \frac{\gamma\sqrt{2}}{\sqrt{\alpha}}\operatorname{sh}((v-|z-r|)\sqrt{2\alpha})\operatorname{sh}(|r-z|\sqrt{2\alpha})\big)^2}dz, & z\wedge x \le r \le z\vee x \end{cases}$$

1 W $\theta_v = \min\{t : \sup_{0\le s\le t} W_s - \inf_{0\le s\le t} W_s = v\}$

5.3.8 $\mathbf{E}_x\{e^{-\alpha\theta_v};\ \ell(\theta_v,r)\in dy, W_{\theta_v}\in dz\}$

$$= \exp\Big(-\frac{\sqrt{2\alpha}\,\mathrm{sh}(v\sqrt{2\alpha})y}{2\,\mathrm{sh}((v-|z-r|)\sqrt{2\alpha})\,\mathrm{sh}(|z-r|\sqrt{2\alpha})}\Big)$$

$$\times\begin{cases}\dfrac{\alpha\sqrt{2\alpha}\,\mathrm{sh}(|z-x|\sqrt{2\alpha})y}{2\,\mathrm{sh}^2((v-|z-r|)\sqrt{2\alpha})\,\mathrm{sh}^2(|z-r|\sqrt{2\alpha})}dydz, & z\wedge r\le x\le z\vee r\\ \dfrac{\alpha\,\mathrm{sh}(|x-r|\sqrt{2\alpha})}{\mathrm{sh}^2((v-|z-r|)\sqrt{2\alpha})\,\mathrm{sh}(|z-r|\sqrt{2\alpha})}dydz & \\ \qquad+\dfrac{\alpha\sqrt{2\alpha}\,\mathrm{sh}((v-|z-x|)\sqrt{2\alpha})y}{2\,\mathrm{sh}^3((v-|z-r|)\sqrt{2\alpha})\,\mathrm{sh}(|z-r|\sqrt{2\alpha})}dydz, & z\wedge x\le r\le z\vee x\end{cases}$$

5.4.5 $\mathbf{E}_x\Big\{\exp\Big(-\gamma\int_0^{\theta_v}\mathbb{1}_{[r,\infty)}(W_s)ds\Big);\ W_{\theta_v}\in dz\Big\}$

$$=\begin{cases}\dfrac{2\gamma(x-z)}{(\mathrm{sh}((v+z-r)\sqrt{2\gamma})+(r-z)\sqrt{2\gamma}\,\mathrm{ch}((v+z-r)\sqrt{2\gamma}))^2}dz, & z\le x\le r\\ \dfrac{\sqrt{2\gamma}\,\mathrm{sh}((x-r)\sqrt{2\gamma})+2\gamma(r-z)\,\mathrm{ch}((x-r)\sqrt{2\gamma})}{(\mathrm{sh}((v+z-r)\sqrt{2\gamma})+(r-z)\sqrt{2\gamma}\,\mathrm{ch}((v+z-r)\sqrt{2\gamma}))^2}dz, & r\le x\le z+v\end{cases}$$

$$=\begin{cases}\dfrac{\sqrt{2\gamma}\,\mathrm{sh}((z-r)\sqrt{2\gamma})+2\gamma(r-x)\,\mathrm{ch}((z-r)\sqrt{2\gamma})}{(\mathrm{sh}((z-r)\sqrt{2\gamma})+(v-z+r)\sqrt{2\gamma}\,\mathrm{ch}((z-r)\sqrt{2\gamma}))^2}dz, & z-v\le x\le r\\ \dfrac{\sqrt{2\gamma}\,\mathrm{sh}((z-x)\sqrt{2\gamma})}{(\mathrm{sh}((z-r)\sqrt{2\gamma})+(v-z+r)\sqrt{2\gamma}\,\mathrm{ch}((z-r)\sqrt{2\gamma}))^2}dz, & r\le x\le z+v\end{cases}$$

5.5.5 $\mathbf{E}_x\Big\{\exp\Big(-\gamma\int_0^{\theta_v}\mathbb{1}_{(-\infty,r]}(W_s)ds\Big), W_{\theta_v}\in dz\Big\}$

$$=\begin{cases}\dfrac{\sqrt{2\gamma}\,\mathrm{sh}((x-z)\sqrt{2\gamma})}{(\mathrm{sh}((r-z)\sqrt{2\gamma})+(v+z-r)\sqrt{2\gamma}\,\mathrm{ch}((r-z)\sqrt{2\gamma}))^2}dz, & z\le x\le r\\ \dfrac{\sqrt{2\gamma}\,\mathrm{sh}((r-z)\sqrt{2\gamma})+2\gamma(x-r)\,\mathrm{ch}((r-z)\sqrt{2\gamma})}{(\mathrm{sh}((r-z)\sqrt{2\gamma})+(v+z-r)\sqrt{2\gamma}\,\mathrm{ch}((r-z)\sqrt{2\gamma}))^2}dz, & r\le x\le z+v\end{cases}$$

$$=\begin{cases}\dfrac{\sqrt{2\gamma}\,\mathrm{sh}((r-x)\sqrt{2\gamma})+2\gamma(z-r)\,\mathrm{ch}((r-x)\sqrt{2\gamma})}{(\mathrm{sh}((v-z+r)\sqrt{2\gamma})+(z-r)\sqrt{2\gamma}\,\mathrm{ch}((v-z+r)\sqrt{2\gamma}))^2}dz, & z-v\le x\le r\\ \dfrac{2\gamma(z-x)}{(\mathrm{sh}((v-z+r)\sqrt{2\gamma})+(z-r)\sqrt{2\gamma}\,\mathrm{ch}((v-z+r)\sqrt{2\gamma}))^2}dz, & r\le x\le z\end{cases}$$

5.6.5 $\mathbf{E}_x\Big\{\exp\Big(-\int_0^{\theta_v}(p\mathbb{1}_{(-\infty,r]}(W_s)+q\mathbb{1}_{[r,\infty)}(W_s))ds\Big);\ W_{\theta_v}\in dz\Big\}$

1 W $\theta_v = \min\{t : \sup_{0\le s\le t} W_s - \inf_{0\le s\le t} W_s = v\}$

$z \le x$, $x \le r$
$$= \frac{q\sqrt{2p}\,\mathrm{sh}((x-z)\sqrt{2p})}{(\sqrt{q}\,\mathrm{ch}((v+z-r)\sqrt{2q})\,\mathrm{sh}((r-z)\sqrt{2p}) + \sqrt{p}\,\mathrm{sh}((v+z-r)\sqrt{2q})\,\mathrm{ch}((r-z)\sqrt{2p}))^2}dz$$

$r \le x$, $x \le b$
$$= \frac{q\sqrt{2p}\,\mathrm{ch}((x-r)\sqrt{2q})\,\mathrm{sh}((r-z)\sqrt{2p}) + p\sqrt{2q}\,\mathrm{sh}((x-r)\sqrt{2q})\,\mathrm{ch}((r-z)\sqrt{2p})}{(\sqrt{q}\,\mathrm{ch}((v+z-r)\sqrt{2q})\,\mathrm{sh}((r-z)\sqrt{2p}) + \sqrt{p}\,\mathrm{sh}((v+z-r)\sqrt{2q})\,\mathrm{ch}((r-z)\sqrt{2p}))^2}dz$$

where $b = z+v,\ a = z-v$

$a \le x$, $x \le r$
$$= \frac{q\sqrt{2p}\,\mathrm{ch}((z-r)\sqrt{2q})\,\mathrm{sh}((r-x)\sqrt{2p}) + p\sqrt{2q}\,\mathrm{sh}((z-r)\sqrt{2q})\,\mathrm{ch}((r-x)\sqrt{2p})}{(\sqrt{q}\,\mathrm{ch}((z-r)\sqrt{2q})\,\mathrm{sh}((v-z+r)\sqrt{2p}) + \sqrt{p}\,\mathrm{sh}((z-r)\sqrt{2q})\,\mathrm{ch}((v-z+r)\sqrt{2p}))^2}dz$$

$r \le x$, $x \le z$
$$= \frac{p\sqrt{2q}\,\mathrm{sh}((z-x)\sqrt{2q})}{(\sqrt{q}\,\mathrm{ch}((z-r)\sqrt{2q})\,\mathrm{sh}((v-z+r)\sqrt{2p}) + \sqrt{p}\,\mathrm{sh}((z-r)\sqrt{2q})\,\mathrm{ch}((v-z+r)\sqrt{2p}))^2}dz$$

5.7.5 $\mathbf{E}_x\Big\{\exp\Big(-\gamma\int_0^{\theta_v} \mathbb{1}_{[r,u]}(W_s)ds\Big);\ W_{\theta_v}\in dz\Big\}$

$z \le x$, $x \le r$
$$= \frac{(x-z)2\gamma}{((1+(v+z-u)(r-z)2\gamma)\,\mathrm{sh}((u-r)\sqrt{2\gamma}) + (v-u+r)\sqrt{2\gamma}\,\mathrm{ch}((u-r)\sqrt{2\gamma}))^2}dz$$

$r \le x$, $x \le u$
$$= \frac{\sqrt{2\gamma}\,\mathrm{sh}((x-r)\sqrt{2\gamma}) + (r-z)2\gamma\,\mathrm{ch}((x-r)\sqrt{2\gamma})}{((1+(v+z-u)(r-z)2\gamma)\,\mathrm{sh}((u-r)\sqrt{2\gamma}) + (v-u+r)\sqrt{2\gamma}\,\mathrm{ch}((u-r)\sqrt{2\gamma}))^2}dz$$

$u \le x$, $x \le b$
$$= \frac{(1+(x-u)(r-z)2\gamma)\sqrt{2\gamma}\,\mathrm{sh}((u-r)\sqrt{2\gamma}) + (x-u+r-z)2\gamma\,\mathrm{ch}((u-r)\sqrt{2\gamma})}{((1+(v+z-u)(r-z)2\gamma)\,\mathrm{sh}((u-r)\sqrt{2\gamma}) + (v-u+r)\sqrt{2\gamma}\,\mathrm{ch}((u-r)\sqrt{2\gamma}))^2}dz$$

where $b = z+v,\ a = z-v$

$a \le x$, $x \le r$
$$= \frac{(1+(z-u)(r-x)2\gamma)\sqrt{2\gamma}\,\mathrm{sh}((u-r)\sqrt{2\gamma}) + (z-u+r-x)2\gamma\,\mathrm{ch}((u-r)\sqrt{2\gamma})}{((1+(z-u)(v-z+r)2\gamma)\,\mathrm{sh}((u-r)\sqrt{2\gamma}) + (v-u+r)\sqrt{2\gamma}\,\mathrm{ch}((u-r)\sqrt{2\gamma}))^2}dz$$

$r \le x$, $x \le u$
$$= \frac{\sqrt{2\gamma}\,\mathrm{sh}((u-x)\sqrt{2\gamma}) + (z-u)2\gamma\,\mathrm{ch}((u-x)\sqrt{2\gamma})}{((1+(z-u)(v-z+r)2\gamma)\,\mathrm{sh}((u-r)\sqrt{2\gamma}) + (v-u+r)\sqrt{2\gamma}\,\mathrm{ch}((u-r)\sqrt{2\gamma}))^2}dz$$

$u \le x$, $x \le z$
$$= \frac{(z-x)2\gamma}{((1+(z-u)(v-z+r)2\gamma)\,\mathrm{sh}((u-r)\sqrt{2\gamma}) + (v-u+r)\sqrt{2\gamma}\,\mathrm{ch}((u-r)\sqrt{2\gamma}))^2}dz$$

1	W $\qquad\theta_v = \min\{t : \sup_{0\le s\le t} W_s - \inf_{0\le s\le t} W_s = v\}$
5.8.5	$\mathbf{E}_x\Big\{\exp\Big(-\gamma\int_0^{\theta_v} W_s ds\Big); W_{\theta_v}\in dz\Big\}$ $= \begin{cases} \dfrac{3^{5/3}xS_{1/3}(\frac23 x^{3/2}\sqrt{2\gamma},\frac23 z^{3/2}\sqrt{2\gamma})}{4z(z+v)^2S^2_{1/3}(\frac23(z+v)^{3/2}\sqrt{2\gamma},\frac23 z^{3/2}\sqrt{2\gamma})}dz, & z\le x\le z+v\\ \dfrac{3^{5/3}xS_{1/3}(\frac23 z^{3/2}\sqrt{2\gamma},\frac23 x^{3/2}\sqrt{2\gamma})}{4z(z-v)^2S^2_{1/3}(\frac23 z^{3/2}\sqrt{2\gamma},\frac23(z-v)^{3/2}\sqrt{2\gamma})}dz, & z-v\le x\le z\end{cases}$
5.9.5	$\mathbf{E}_x\Big\{\exp\Big(-2\gamma\int_0^{\theta_v} W_s^2 ds\Big); W_{\theta_v}\in dz\Big\}$ $= \begin{cases} \dfrac{2xS_{1/4}(x^2\sqrt{\gamma},z^2\sqrt{\gamma})}{z(z+v)^2\gamma^{1/4}S^2_{1/4}((z+v)^2\sqrt{\gamma},z^2\sqrt{\gamma})}dz, & z\le x\le z+v\\ \dfrac{2xS_{1/4}(z^2\sqrt{\gamma},x^2\sqrt{\gamma})}{z(z-v)^2\gamma^{1/4}S^2_{1/4}(z^2\sqrt{\gamma},(z-v)^2\sqrt{\gamma})}dz, & z-v\le x\le z\end{cases}$
5.10.5	$\mathbf{E}_x\Big\{\exp\Big(-\gamma\int_0^{\theta_v} e^{2\beta W_s}ds\Big);\ W_{\theta_v}\in dz\Big\}$ $= \begin{cases} \dfrac{\beta S_0\Big(\frac{\sqrt{2\gamma}}{\lvert\beta\rvert}e^{\beta x},\frac{\sqrt{2\gamma}}{\lvert\beta\rvert}e^{\beta z}\Big)}{S_0^2\Big(\frac{\sqrt{2\gamma}}{\lvert\beta\rvert}e^{\beta(z+v)},\frac{\sqrt{2\gamma}}{\lvert\beta\rvert}e^{\beta z}\Big)}dz, & z\le x\le z+v\\ \dfrac{\beta S_0\Big(\frac{\sqrt{2\gamma}}{\lvert\beta\rvert}e^{\beta z},\frac{\sqrt{2\gamma}}{\lvert\beta\rvert}e^{\beta x}\Big)}{S_0^2\Big(\frac{\sqrt{2\gamma}}{\lvert\beta\rvert}e^{\beta z},\frac{\sqrt{2\gamma}}{\lvert\beta\rvert}e^{\beta(z-v)}\Big)}dz, & z-v\le x\le z\end{cases}$
5.11.2	$\lambda\int_0^\infty e^{-\lambda v}\mathbf{P}_x\big(\sup_{y\in\mathbb{R}}\ell(\theta_v,y)\ge h\big)dv = \dfrac{\sqrt{\lambda h}}{\sqrt2 I_0(\sqrt{2\lambda h})I_1(\sqrt{2\lambda h})}$
5.11.4	$\mathbf{P}\big(\sup_{y\in\mathbb{R}}\ell(\theta_v,y)<h\big) = \sum_{k=1}^\infty\Big(\frac{1}{J_1^2(j_{0,k})}e^{-j^2_{0,k}v/2h} - \frac{1}{J_0^2(j_{1,k})}e^{-j^2_{1,k}v/2h}\Big)$
5.13.1	$\mathbf{E}_x\big\{e^{-\gamma\check H(\theta_v)}; W_{\theta_v}<x\big\} = \dfrac{\operatorname{sh}(v\sqrt{\gamma/2})}{v\sqrt{2\gamma}\operatorname{ch}(v\sqrt{\gamma/2})}$
5.13.2	$\mathbf{P}_x\big(\check H(\theta_v)\in du,\ W_{\theta_v}<x\big) = v^{-1}\operatorname{sc}_u(v/2,v/2)du$
5.13.3	$\mathbf{E}_x\big\{e^{-\alpha\theta_v-\gamma\check H(\theta_v)}; W_{\theta_v}<x\big\} = \dfrac{\sqrt{\alpha}\operatorname{sh}(v\sqrt{(\alpha+\gamma)/2})}{\sqrt{\gamma+\alpha}\operatorname{sh}(v\sqrt{2\alpha})\operatorname{ch}(v\sqrt{(\alpha+\gamma)/2})}$
5.13.4 $u<t$	$\mathbf{P}_x\big(\check H(\theta_v)\in du,\ \theta_v\in dt,\ W_{\theta_v}<x\big) = \operatorname{sc}_u(v/2,v/2)\operatorname{s}_{t-u}(v)dudt$

1	W $\qquad\qquad\qquad\qquad \theta_v = \min\{t : \sup_{0\le s\le t} W_s - \inf_{0\le s\le t} W_s = v\}$
5.13.5 $z < x$	$\mathbf{E}_x\{e^{-\gamma\check{H}(\theta_v)}; W_{\theta_v} \in dz\} = \frac{\operatorname{sh}((x-z)\sqrt{2\gamma})}{v\operatorname{sh}(v\sqrt{2\gamma})}dz$
5.13.6 $z < x$	$\mathbf{P}_x\big(\check{H}(\theta_v) \in du,\ W_{\theta_v} \in dz\big) = v^{-1}\operatorname{ss}_u(x-z, v)dudz$
5.13.7 $z < x$	$\mathbf{E}_x\{e^{-\alpha\theta_v-\gamma\check{H}(\theta_v)}; W_{\theta_v} \in dz\} = \frac{\sqrt{2\alpha}\operatorname{sh}((x-z)\sqrt{2\alpha+2\gamma})}{\operatorname{sh}(v\sqrt{2\alpha})\operatorname{sh}(v\sqrt{2\alpha+2\gamma})}dz$
5.13.8 $u < t$	$\mathbf{P}_x\big(\check{H}(\theta_v) \in du,\ \theta_v \in dt,\ W_{\theta_v} \in dz\big) = \operatorname{ss}_u(x-z, v)\operatorname{s}_{t-u}(v)dudtdz$
5.14.1	$\mathbf{E}_x\{e^{-\gamma\hat{H}(\theta_v)}; W_{\theta_v} > x\} = \frac{\operatorname{sh}(v\sqrt{\gamma/2})}{v\sqrt{2\gamma}\operatorname{ch}(v\sqrt{\gamma/2})}$
5.14.2	$\mathbf{P}_x\big(\hat{H}(\theta_v) \in du,\ W_{\theta_v} > x\big) = v^{-1}\operatorname{sc}_u(v/2, v/2)du$
5.14.3	$\mathbf{E}_x\{e^{-\alpha\theta_v-\gamma\hat{H}(\theta_v)}; W_{\theta_v} > x\} = \frac{\sqrt{\alpha}\operatorname{sh}(v\sqrt{(\alpha+\gamma)/2})}{\sqrt{\gamma+\alpha}\operatorname{sh}(v\sqrt{2\alpha})\operatorname{ch}(v\sqrt{(\alpha+\gamma)/2})}$
5.14.4 $u < t$	$\mathbf{P}_x\big(\hat{H}(\theta_v) \in du,\ \theta_v \in dt,\ W_{\theta_v} > x\big) = \operatorname{sc}_u(v/2, v/2)\operatorname{s}_{t-u}(v)dudt$
5.14.5 $x < z$	$\mathbf{E}_x\{e^{-\gamma\hat{H}(\theta_v)}; W_{\theta_v} \in dz\} = \frac{\operatorname{sh}((z-x)\sqrt{2\gamma})}{v\operatorname{sh}(v\sqrt{2\gamma})}dz$
5.14.6 $x < z$	$\mathbf{P}_x\big(\hat{H}(\theta_v) \in du,\ W_{\theta_v} \in dz\big) = v^{-1}\operatorname{ss}_u(z-x, v)dudz$
5.14.7 $x < z$	$\mathbf{E}_x\{e^{-\alpha\theta_v-\gamma\hat{H}(\theta_v)}; W_{\theta_v} \in dz\} = \frac{\sqrt{2\alpha}\operatorname{sh}((z-x)\sqrt{2\alpha+2\gamma})}{\operatorname{sh}(v\sqrt{2\alpha})\operatorname{sh}(v\sqrt{2\alpha+2\gamma})}dz$
5.14.8 $u < t$	$\mathbf{P}_x\big(\hat{H}(\theta_v) \in du,\ \theta_v \in dt,\ W_{\theta_v} \in dz\big) = \operatorname{ss}_u(z-x, v)\operatorname{s}_{t-u}(v)dudtdz$
5.18.5	$\mathbf{E}_x\{\exp\big(-\gamma\ell(\theta_v, r) - \eta\ell(\theta_v, u)\big);\ W_{\theta_v} \in dz\}$
$z \le x$ $x \le r$	$= \frac{x-z}{(v+2\gamma(v+z-r)(r-z)+2\eta(v+z-u)(u-z)+4\gamma\eta(v+z-u)(u-r)(r-z))^2}dz$
$r \le x$ $x \le u$	$= \frac{x-z+2\gamma(x-r)(r-z)}{(v+2\gamma(v+z-r)(r-z)+2\eta(v+z-u)(u-z)+4\gamma\eta(v+z-u)(u-r)(r-z))^2}dz$

1 W $\theta_v = \min\{t : \sup_{0\le s\le t} W_s - \inf_{0\le s\le t} W_s = v\}$

$u \le x$, $x \le b$

$$= \frac{x-z+2\gamma(x-r)(r-z)+2\eta(x-u)(u-z)+4\gamma\eta(x-u)(u-r)(r-z)}{(v+2\gamma(v+z-r)(r-z)+2\eta(v+z-u)(u-z)+4\gamma\eta(v+z-u)(u-r)(r-z))^2}dz$$

where $b = z + v,\ a = z - v$

$a \le x$, $x \le r$

$$= \frac{z-x+2\gamma(z-r)(r-x)+2\eta(z-u)(u-x)+4\gamma\eta(z-u)(u-r)(r-x)}{(v+2\gamma(z-r)(v+r-z)+2\eta(z-u)(v+u-z)+4\gamma\eta(z-u)(u-r)(v+r-z))^2}dz$$

$r \le x$, $x \le u$

$$= \frac{z-x+2\eta(z-u)(u-x)}{(v+2\gamma(z-r)(v+r-z)+2\eta(z-u)(v+u-z)+4\gamma\eta(z-u)(u-r)(v+r-z))^2}dz$$

$u \le x$, $x \le z$

$$= \frac{z-x}{(v+2\gamma(z-r)(v+r-z)+2\eta(z-u)(v+u-z)+4\gamma\eta(z-u)(u-r)(v+r-z))^2}dz$$

5.18.6 $z \le x$

$$\mathbf{P}_x\big(\ell(\theta_v, r) \in dy,\ \ell(\theta_v, u) \in dg,\ W_{\theta_v} \in dz\big)$$

$$= \begin{cases} \dfrac{x-z}{r-z}\, p_r dydgdz, & z \le x \le r \\ \dfrac{u-x}{u-r}\, p_r dydgdz + \dfrac{x-r}{u-r}\, p_u dydgdz, & r \le x \le u \\ \Big(\dfrac{z+v-x}{z+v-u} + \dfrac{2(x-u)}{g}\Big) p_u dydgdz, & u \le x \le z+v \end{cases}$$

where

$$p_r := \frac{\sqrt{yg}}{8(z+v-u)^2(u-r)(r-z)} \exp\Big(-\frac{y+g}{2(u-r)} - \frac{y}{2(r-z)} - \frac{g}{2(z+v-u)}\Big) I_1\Big(\frac{\sqrt{yg}}{u-r}\Big)$$

$$p_u := \frac{g}{8(z+v-u)^2(u-r)(r-z)} \exp\Big(-\frac{y+g}{2(u-r)} - \frac{y}{2(r-z)} - \frac{g}{2(z+v-u)}\Big) I_0\Big(\frac{\sqrt{yg}}{u-r}\Big)$$

$x \le z$

$$\mathbf{P}_x\big(\ell(\theta_v, r) \in dy,\ \ell(\theta_v, u) \in dg,\ W_{\theta_v} \in dz\big)$$

$$= \begin{cases} \Big(\dfrac{2(r-x)}{y} + \dfrac{x+v-z}{r+v-z}\Big) q_r dydgdz, & z-v \le x \le r \\ \dfrac{u-x}{u-r}\, q_r dydgdz + \dfrac{x-r}{u-r}\, q_u dydgdz, & r \le x \le u \\ \dfrac{z-x}{z-u}\, q_u dydgdz, & u \le x \le z \end{cases}$$

where

1 W $\theta_v = \min\{t : \sup_{0\le s\le t} W_s - \inf_{0\le s\le t} W_s = v\}$

$$q_r := \frac{y}{8(z-u)(u-r)(v+r-z)^2}\exp\Big(-\frac{y+g}{2(u-r)} - \frac{y}{2(r+v-z)} - \frac{g}{2(z-u)}\Big)I_0\Big(\frac{\sqrt{yg}}{u-r}\Big)$$

$$q_u := \frac{\sqrt{yg}}{8(z-u)(u-r)(v+r-z)^2}\exp\Big(-\frac{y+g}{2(u-r)} - \frac{y}{2(r+v-z)} - \frac{g}{2(z-u)}\Big)I_1\Big(\frac{\sqrt{yg}}{u-r}\Big)$$

5.19.5 $\mathbf{E}_x\Big\{\exp\Big(-\gamma\int_0^{\theta_v}\frac{ds}{W_s}\Big);\ W_{\theta_v}\in dz\Big\}$

$$= \begin{cases} \dfrac{xS_1(2\sqrt{2\gamma x}, 2\sqrt{2\gamma z})}{16z(z+v)^2\gamma S_1^2(2\sqrt{2\gamma(z+v)}, 2\sqrt{2\gamma z})}dz, & 0<z\le x\le z+v \\ \dfrac{xS_1(2\sqrt{2\gamma z}, 2\sqrt{2\gamma x})}{16z(z-v)^2\gamma S_1^2(2\sqrt{2\gamma z}, 2\sqrt{2\gamma(z-v)})}dz, & 0<z-v\le x\le z \end{cases}$$

5.20.5 $\mathbf{E}_x\Big\{\exp\Big(-\frac{\gamma^2}{2}\int_0^{\theta_v}\frac{ds}{W_s^2}\Big);\ W_{\theta_v}\in dz\Big\}$

$$= \begin{cases} \dfrac{2\sqrt{1/4+\gamma^2}\sqrt{x}\big((x/z)^{\sqrt{1/4+\gamma^2}} - (z/x)^{\sqrt{1/4+\gamma^2}}\big)}{\sqrt{z}(z+v)\big(((z+v)/z)^{\sqrt{1/4+\gamma^2}} - (z/(z+v))^{\sqrt{1/4+\gamma^2}}\big)^2}dz, & z\le x\le z+v \\ \dfrac{2\sqrt{1/4+\gamma^2}\sqrt{x}\big((z/x)^{\sqrt{1/4+\gamma^2}} - (x/z)^{\sqrt{1/4+\gamma^2}}\big)}{\sqrt{z}(z-v)\big((z/(z-v))^{\sqrt{1/4+\gamma^2}} - ((z-v)/z)^{\sqrt{1/4+\gamma^2}}\big)^2}dz, & z-v\le x\le z \end{cases}$$

5.21.5 $\mathbf{E}_x\Big\{\exp\Big(-\int_0^{\theta_v}\big(\frac{p^2}{2W_s^2} + \frac{q^2W_s^2}{2}\big)ds\Big);\ W_{\theta_v}\in dz\Big\}$

$$= \begin{cases} \dfrac{2(2x/qz)^{\sqrt{1/4+p^2}}(x/z)^{1/2}S_{\sqrt{4p^2+1}/4}(qx^2/2, qz^2/2)}{(z+v)^{2\sqrt{1/4+p^2}+1}S^2_{\sqrt{4p^2+1}/4}(q(z+v)^2/2, qz^2/2)}dz, & 0<z\le x\le z+v \\ \dfrac{2(2x/qz)^{\sqrt{1/4+p^2}}(x/z)^{1/2}S_{\sqrt{4p^2+1}/4}(qz^2/2, qx^2/2)}{(z-v)^{2\sqrt{1/4+p^2}+1}S^2_{\sqrt{4p^2+1}/4}(qz^2/2, q(z-v)^2/2)}dz, & 0<z-v\le x\le z \end{cases}$$

5.22.5 $\mathbf{E}_x\Big\{\exp\Big(-\int_0^{\theta_v}\big(\frac{p^2}{2W_s^2} + \frac{q}{W_s}\big)ds\Big);\ W_{\theta_v}\in dz\Big\}$

$$= \begin{cases} \dfrac{2^{-3\sqrt{4p^2+1}}(x/z)^{\sqrt{1/4+p^2}+1/2}S_{\sqrt{4p^2+1}}(2\sqrt{2qx}, 2\sqrt{2qz})}{2q^{\sqrt{4p^2+1}}(z+v)^{\sqrt{4p^2+1}+1}S^2_{\sqrt{4p^2+1}}(2\sqrt{2q(z+v)}, 2\sqrt{2qz})}dz, & z\le x\le z+v \\ \dfrac{2^{-3\sqrt{4p^2+1}}(x/z)^{\sqrt{1/4+p^2}+1/2}S_{\sqrt{4p^2+1}}(2\sqrt{2qz}, 2\sqrt{2qx})}{2q^{\sqrt{4p^2+1}}(z-v)^{\sqrt{4p^2+1}+1}S^2_{\sqrt{4p^2+1}}(2\sqrt{2qz}, 2\sqrt{2q(z-v)})}dz, & z-v\le x\le z \end{cases}$$

1 W $\theta_v = \min\{t : \sup_{0\le s\le t} W_s - \inf_{0\le s\le t} W_s = v\}$

5.27.5 $\mathbf{E}_x\Big\{\exp\Big(-\int_0^{\theta_v}\big(p\mathbb{I}_{(-\infty,r)}(W_s)+q\mathbb{I}_{[r,\infty)}(W_s)\big)ds\Big);\ \frac{1}{2}\ell(\theta_v,r)\in dy, W_{\theta_v}\in dz\Big\}$

$z\le x$
$$= \exp\big(-y\sqrt{2q}\,\mathrm{cth}((v+z-r)\sqrt{2q}) - y\sqrt{2p}\,\mathrm{cth}((r-z)\sqrt{2p})\big)$$
$$\times\begin{cases} \dfrac{2\sqrt{pq}\,\mathrm{sh}((x-r)\sqrt{2q})}{\mathrm{sh}^2((v+z-r)\sqrt{2q})\,\mathrm{sh}((r-z)\sqrt{2p})}dydz \\ \qquad + \dfrac{2yq\sqrt{2p}\,\mathrm{sh}((v+z-x)\sqrt{2q})}{\mathrm{sh}^3((v+z-r)\sqrt{2q})\,\mathrm{sh}((r-z)\sqrt{2p})}dydz, & r\le x\le z+v \\ \dfrac{2yq\sqrt{2p}\,\mathrm{sh}((x-z)\sqrt{2p})}{\mathrm{sh}^2((v+z-r)\sqrt{2q})\,\mathrm{sh}^2((r-z)\sqrt{2p})}dydz, & z\le x\le r \end{cases}$$

$x\le z$
$$= \exp\big(-y\sqrt{2q}\,\mathrm{cth}((z-r)\sqrt{2q}) - y\sqrt{2p}\,\mathrm{cth}((v+r-z)\sqrt{2p})\big)$$
$$\times\begin{cases} \dfrac{2yp\sqrt{2q}\,\mathrm{sh}((z-x)\sqrt{2q})}{\mathrm{sh}^2((z-r)\sqrt{2q})\,\mathrm{sh}^2((v+r-z)\sqrt{2p})}dydz, & r\le x\le z \\ \dfrac{2\sqrt{pq}\,\mathrm{sh}((r-x)\sqrt{2p})}{\mathrm{sh}((z-r)\sqrt{2q})\,\mathrm{sh}^2((v+r-z)\sqrt{2p})}dydz \\ \qquad + \dfrac{2yp\sqrt{2q}\,\mathrm{sh}((v+x-z)\sqrt{2p})}{\mathrm{sh}((z-r)\sqrt{2q})\,\mathrm{sh}^3((v+r-z)\sqrt{2p})}dydz, & z-v\le x\le r \end{cases}$$

5.27.6 $\mathbf{P}_x\Big(\int_0^{\theta_v}\mathbb{I}_{(-\infty,r)}(W_s)ds\in du, \int_0^{\theta_v}\mathbb{I}_{[r,\infty)}(W_s)ds\in dg, \frac{1}{2}\ell(\theta_v,r)\in dy, W_{\theta_v}\in dz\Big)$

$z\le x$
$$= \begin{cases} \mathrm{es}_u(1,1,r-z,0,y)\big(\mathrm{escs}_g(x-r,o,v+z-r,y) \\ \qquad +\frac{y}{2}\,\mathrm{es}_g(2,3,v+z-r,x-z-v,y) \\ \qquad -\frac{y}{2}\,\mathrm{es}_g(2,3,v+z-r,v+z-x,y)\big)dudgdydz, & r\le x\le z+v \\ y\,\mathrm{es}_g(2,2,v+z-r,0,y)\,\mathrm{escs}_u(x-z,0,r-z,y)dudgdydz, & z\le x\le r \end{cases}$$

$x\le z$
$$= \begin{cases} y\,\mathrm{es}_u(2,2,v+r-z,0,y)\,\mathrm{escs}_g(z-x,0,z-r,y)dudgdydz, & r\le x\le z \\ \mathrm{es}_g(1,1,z-r,0,y)\big(\mathrm{escs}_u(r-x,0,v+r-z,y) \\ \qquad +\frac{y}{2}\,\mathrm{es}_u(2,3,v+r-z,z-x-v,y) \\ \qquad -\frac{y}{2}\,\mathrm{es}_u(2,3,v+r-z,v+x-z,y)\big)dudgdydz, & z-v\le x\le r \end{cases}$$

2. BROWNIAN MOTION WITH DRIFT

1. Exponential stopping

1.0.3 $$\mathbf{E}_x e^{i\beta W_\tau^{(\mu)}} = \frac{2\lambda}{2\lambda - 2i\beta\mu + \beta^2} e^{i\beta x}$$

1.0.4 $$\mathbf{E}_x e^{i\beta W_t^{(\mu)}} = e^{i\beta(x+\mu t) - \beta^2 t/2}$$

1.0.5 $$\mathbf{P}_x(W_\tau^{(\mu)} \in dz) = \frac{\lambda}{\sqrt{2\lambda+\mu^2}} e^{\mu(z-x) - |z-x|\sqrt{2\lambda+\mu^2}} dz$$

1.0.6 $$\mathbf{P}_x\big(W_t^{(\mu)} \in dz\big) = \frac{1}{\sqrt{2\pi t}} e^{-(z-\mu t-x)^2/2t} dz$$

1.1.1 $$\mathbf{E}_x \exp\big(-\gamma \sup_{0\le s\le\tau} W_s^{(\mu)}\big) = \frac{\sqrt{2\lambda+\mu^2}-\mu}{\gamma + \sqrt{2\lambda+\mu^2}-\mu} e^{-\gamma x}$$

1.1.2
$x \le y$ $$\mathbf{P}_x\big(\sup_{0\le s\le\tau} W_s^{(\mu)} \ge y\big) = e^{\mu(y-x) + (x-y)\sqrt{2\lambda+\mu^2}}$$

1.1.3 $$\mathbf{E}_x \exp\big(-\gamma \sup_{0\le s\le t} W_s^{(\mu)}\big)$$

$$= \frac{\gamma-\mu}{\gamma-2\mu} e^{-\gamma x + \gamma(\gamma-2\mu)t/2} \operatorname{Erfc}\Big(\frac{(\gamma-\mu)\sqrt{t}}{\sqrt{2}}\Big) - \frac{\mu}{\gamma-2\mu} e^{-\gamma x} \operatorname{Erfc}\Big(\frac{\mu\sqrt{t}}{\sqrt{2}}\Big)$$

(1) $$\mathbf{E}_x \exp\big(-\gamma \sup_{0\le s<\infty} W_s^{(\mu)}\big) = \frac{|\mu|-\mu}{\gamma+|\mu|-\mu} e^{-\gamma x}$$

1.1.4
$x \le y$ $$\mathbf{P}_x\big(\sup_{0\le s\le t} W_s^{(\mu)} \ge y\big) = \frac{1}{2}\operatorname{Erfc}\Big(\frac{y-x}{\sqrt{2t}} - \frac{\mu\sqrt{t}}{\sqrt{2}}\Big) + \frac{1}{2} e^{2\mu(y-x)} \operatorname{Erfc}\Big(\frac{y-x}{\sqrt{2t}} + \frac{\mu\sqrt{t}}{\sqrt{2}}\Big)$$

2 $W_s^{(\mu)} = \mu s + W_s$ $\tau \sim \text{Exp}(\lambda)$, independent of $W^{(\mu)}$

(1) $x \le y$ $\mu < 0$

$$\mathbf{P}_x\Big(\sup_{0\le s<\infty} W_s^{(\mu)} \ge y\Big) = e^{2\mu(y-x)}, \qquad x \le y$$

1.1.5

$$\mathbf{E}_x\Big\{\exp\Big(-\gamma \sup_{0\le s\le\tau} W_s^{(\mu)}\Big); W_\tau^{(\mu)} \in dz\Big\} = \frac{2\lambda}{\gamma + 2\sqrt{2\lambda+\mu^2}} e^{\mu(z-x)-\gamma(x\vee z)-|z-x|\sqrt{2\lambda+\mu^2}}dz$$

1.1.6 $x \le y$ $z \le y$

$$\mathbf{P}_x\Big(\sup_{0\le s\le\tau} W_s^{(\mu)} \ge y, W_\tau^{(\mu)} \in dz\Big) = \frac{\lambda}{\sqrt{2\lambda+\mu^2}} e^{\mu(z-x)+(z+x-2y)\sqrt{2\lambda+\mu^2}}dz$$

1.1.7

$$\mathbf{E}_x\Big\{\exp\Big(-\gamma \sup_{0\le s\le t} W_s^{(\mu)}\Big); W_t^{(\mu)} \in dz\Big\} = e^{\mu(z-x)-\mu^2 t/2-\gamma(x\vee z)} \times\Big(\frac{1}{\sqrt{2\pi t}} e^{-(z-x)^2/2t} - \frac{\gamma}{4} e^{\gamma|z-x|/2+\gamma^2 t/8} \operatorname{Erfc}\Big(\frac{|z-x|}{\sqrt{2t}} + \frac{\gamma\sqrt{t}}{2\sqrt{2}}\Big)\Big)dz$$

1.1.8 $x \le y$

$$\mathbf{P}_x\Big(\sup_{0\le s\le t} W_s^{(\mu)} \ge y, W_t^{(\mu)} \in dz\Big) = \frac{1}{\sqrt{2\pi t}} e^{\mu(z-x)-\mu^2 t/2-(|z-y|+y-x)^2/2t}dz$$

1.2.1

$$\mathbf{E}_x \exp\Big(\gamma \inf_{0\le s\le\tau} W_s^{(\mu)}\Big) = \frac{\mu+\sqrt{2\lambda+\mu^2}}{\gamma+\mu+\sqrt{2\lambda+\mu^2}} e^{\gamma x}$$

1.2.2 $y \le x$

$$\mathbf{P}_x\Big(\inf_{0\le s\le\tau} W_s^{(\mu)} \le y\Big) = e^{\mu(y-x)+(y-x)\sqrt{2\lambda+\mu^2}}$$

1.2.3

$$\mathbf{E}_x \exp\Big(\gamma \inf_{0\le s\le t} W_s^{(\mu)}\Big) = \frac{\gamma+\mu}{\gamma+2\mu} e^{\gamma x+\gamma(\gamma+2\mu)t/2} \operatorname{Erfc}\Big(\frac{(\gamma+\mu)\sqrt{t}}{\sqrt{2}}\Big) + \frac{\mu}{\gamma+2\mu} e^{\gamma x} \operatorname{Erfc}\Big(-\frac{\mu\sqrt{t}}{\sqrt{2}}\Big)$$

(1)

$$\mathbf{E}_x \exp\Big(\gamma \inf_{0\le s<\infty} W_s^{(\mu)}\Big) = \frac{\mu+|\mu|}{\gamma+\mu+|\mu|} e^{\gamma x}$$

1.2.4 $y \le x$

$$\mathbf{P}_x\Big(\inf_{0\le s\le t} W_s^{(\mu)} \le y\Big) = \frac{1}{2} \operatorname{Erfc}\Big(\frac{x-y}{\sqrt{2t}} + \frac{\mu\sqrt{t}}{\sqrt{2}}\Big) + \frac{1}{2} e^{2\mu(y-x)} \operatorname{Erfc}\Big(\frac{x-y}{\sqrt{2t}} - \frac{\mu\sqrt{t}}{\sqrt{2}}\Big)$$

2 $W_s^{(\mu)} = \mu s + W_s$ $\tau \sim \text{Exp}(\lambda)$, independent of $W^{(\mu)}$

(1) $y \le x$, $\mu > 0$

$$\mathbf{P}_x\Big(\inf_{0\le s<\infty} W_s^{(\mu)} \le y\Big) = e^{2\mu(y-x)}$$

1.2.5

$$\mathbf{E}_x\big\{\exp(\gamma \inf_{0\le s\le\tau} W_s^{(\mu)}); W_\tau^{(\mu)} \in dz\big\} = \frac{2\lambda}{\gamma + 2\sqrt{2\lambda+\mu^2}} e^{\mu(z-x)+\gamma(x\wedge z)-|z-x|\sqrt{2\lambda+\mu^2}}dz$$

1.2.6 $y \le x$, $y \le z$

$$\mathbf{P}_x\Big(\inf_{0\le s\le\tau} W_s^{(\mu)} \le y, W_\tau^{(\mu)} \in dz\Big) = \frac{\lambda}{\sqrt{2\lambda+\mu^2}} e^{\mu(z-x)+(2y-z-x)\sqrt{2\lambda+\mu^2}}dz$$

1.2.7

$$\mathbf{E}_x\big\{\exp(\gamma \inf_{0\le s\le t} W_s^{(\mu)}); W_t^{(\mu)} \in dz\big\} = e^{\mu(z-x)-\mu^2 t/2+\gamma(x\wedge z)}$$

$$\times\Big(\frac{1}{\sqrt{2\pi t}}e^{-(z-x)^2/2t} - \frac{\gamma}{4}e^{\gamma|z-x|/2+\gamma^2 t/8}\,\text{Erfc}\Big(\frac{|z-x|}{\sqrt{2t}} + \frac{\gamma\sqrt{t}}{2\sqrt{2}}\Big)\Big)dz$$

1.2.8 $y \le x$

$$\mathbf{P}_x\Big(\inf_{0\le s\le t} W_s^{(\mu)} \le y, W_t^{(\mu)} \in dz\Big) = \frac{1}{\sqrt{2\pi t}} e^{\mu(z-x)-\mu^2 t/2-(|z-y|+x-y)^2/2t}dz$$

1.3.1

$$\mathbf{E}_x e^{-\gamma\ell(\tau,r)} = 1 - \frac{\gamma}{\gamma+\sqrt{2\lambda+\mu^2}} e^{\mu(r-x)-|r-x|\sqrt{2\lambda+\mu^2}}$$

1.3.2 $0 < y$

$$\mathbf{P}_x\big(\ell(\tau,r) \in dy\big) = \sqrt{2\lambda+\mu^2}\,e^{\mu(r-x)-(y+|r-x|)\sqrt{2\lambda+\mu^2}}dy$$

(1)

$$\mathbf{P}_x\big(\ell(\tau,r) = 0\big) = 1 - e^{\mu(r-x)-|r-x|\sqrt{2\lambda+\mu^2}}$$

1.3.3

$$\mathbf{E}_x e^{-\gamma\ell(t,r)} = 1 - \frac{\gamma}{2}e^{\mu(r-x)}\Big[\frac{1}{\gamma-|\mu|}e^{|\mu||r-x|}\,\text{Erfc}\Big(\frac{|r-x|}{\sqrt{2t}} + \frac{|\mu|\sqrt{t}}{\sqrt{2}}\Big)$$

$$+\frac{1}{\gamma+|\mu|}e^{-|\mu||r-x|}\,\text{Erfc}\Big(\frac{|r-x|}{\sqrt{2t}} - \frac{|\mu|\sqrt{t}}{\sqrt{2}}\Big)\Big]$$

$$+\frac{\gamma^2}{\gamma^2-\mu^2}e^{\mu(r-x)+|r-x|\gamma+(\gamma^2-\mu^2)t/2}\,\text{Erfc}\Big(\frac{|r-x|}{\sqrt{2t}} + \frac{\gamma\sqrt{t}}{\sqrt{2}}\Big)$$

2 $W_s^{(\mu)} = \mu s + W_s$ $\qquad \tau \sim \operatorname{Exp}(\lambda)$, independent of $W^{(\mu)}$

(1) $$\mathbf{E}_x e^{-\gamma\ell(\infty,r)} = 1 - \frac{\gamma}{\gamma+|\mu|} e^{\mu(r-x)-|\mu||r-x|}$$

1.3.4 $0<y$ $$\mathbf{P}_x\big(\ell(t,r)\in dy\big) = \frac{\sqrt{2}}{\sqrt{\pi t}} e^{\mu(r-x)-\mu^2 t/2-(y+|r-x|)^2/2t} dy$$
$$+\frac{|\mu|}{2} e^{\mu(r-x)}\Big[e^{-|\mu|(y+|r-x|)}\operatorname{Erfc}\Big(\frac{y+|r-x|}{\sqrt{2t}} - \frac{|\mu|\sqrt{t}}{\sqrt{2}}\Big)$$
$$-e^{|\mu|(y+|r-x|)}\operatorname{Erfc}\Big(\frac{y+|r-x|}{\sqrt{2t}} + \frac{|\mu|\sqrt{t}}{\sqrt{2}}\Big)\Big] dy$$

(1) $$\mathbf{P}_x\big(\ell(t,r)=0\big) = 1 - \frac{1}{2} e^{\mu(r-x)}\Big[e^{|\mu||r-x|}\operatorname{Erfc}\Big(\frac{|r-x|}{\sqrt{2t}} + \frac{|\mu|\sqrt{t}}{\sqrt{2}}\Big)$$
$$+e^{-|\mu||r-x|}\operatorname{Erfc}\Big(\frac{|r-x|}{\sqrt{2t}} - \frac{|\mu|\sqrt{t}}{\sqrt{2}}\Big)\Big]$$

(2) $0<y$ $$\mathbf{P}_x\big(\ell(\infty,r)\in dy\big) = |\mu| e^{\mu(r-x)-|\mu||r-x|-|\mu|y} dy$$

(3) $$\mathbf{P}_x\big(\ell(\infty,r)=0\big) = 1 - e^{\mu(r-x)-|\mu||r-x|}$$

1.3.5 $$\mathbf{E}_x\{e^{-\gamma\ell(\tau,r)}; W_\tau^{(\mu)}\in dz\} = \frac{\lambda}{\sqrt{2\lambda+\mu^2}} e^{\mu(z-x)-|z-x|\sqrt{2\lambda+\mu^2}} dz$$
$$-\frac{\gamma\lambda}{\sqrt{2\lambda+\mu^2}(\gamma+\sqrt{2\lambda+\mu^2})} e^{\mu(z-x)-(|z-r|+|r-x|)\sqrt{2\lambda+\mu^2}} dz$$

1.3.6 $0<y$ $$\mathbf{P}_x\big(\ell(\tau,r)\in dy, W_\tau^{(\mu)}\in dz\big) = \lambda e^{\mu(z-x)-(|z-r|+|r-x|+y)\sqrt{2\lambda+\mu^2}} dy dz$$

(1) $$\mathbf{P}_x\big(\ell(\tau,r)=0, W_\tau^{(\mu)}\in dz\big) = \frac{\lambda}{\sqrt{2\lambda+\mu^2}} e^{\mu(z-x)-|z-x|\sqrt{2\lambda+\mu^2}} dz$$
$$-\frac{\lambda}{\sqrt{2\lambda+\mu^2}} e^{\mu(z-x)-(|z-r|+|r-x|)\sqrt{2\lambda+\mu^2}} dz$$

2	$W_s^{(\mu)} = \mu s + W_s$ $\qquad\qquad$ $\tau \sim \operatorname{Exp}(\lambda)$, independent of $W^{(\mu)}$
1.3.7	$\mathbf{E}_x\{e^{-\gamma\ell(t,r)}; W_t^{(\mu)} \in dz\} = \frac{1}{\sqrt{2\pi t}} e^{-(z-\mu t-x)^2/2t} dz$ $-\frac{\gamma}{2} e^{\mu(z-x)+(\lvert z-r\rvert+\lvert r-x\rvert)\gamma+(\gamma^2-\mu^2)t/2} \operatorname{Erfc}\Big(\frac{\gamma\sqrt{t}}{\sqrt{2}} + \frac{\lvert z-r\rvert+\lvert r-x\rvert}{\sqrt{2t}}\Big) dz$
1.3.8 $0 < y$	$\mathbf{P}_x\big(\ell(t,r) \in dy, W_t^{(\mu)} \in dz\big)$ $= \frac{1}{t\sqrt{2\pi t}}\big(y + \lvert z-r\rvert + \lvert r-x\rvert\big) e^{\mu(z-x)-\mu^2 t/2-(\lvert z-r\rvert+\lvert r-x\rvert+y)^2/2t} dydz$
(1)	$\mathbf{P}_x\big(\ell(t,r) = 0, W_t^{(\mu)} \in dz\big)$ $= \frac{1}{\sqrt{2\pi t}} e^{-(z-\mu t-x)^2/2t} dz - \frac{1}{\sqrt{2\pi t}} e^{\mu(z-x)-\mu^2 t/2-(\lvert z-r\rvert+\lvert r-x\rvert)^2/2t} dz$
1.4.1	$\mathbf{E}_x \exp\Big(-\gamma \int_0^\tau \mathbb{I}_{[r,\infty)}\big(W_s^{(\mu)}\big) ds\Big)$
$x \le r$	$= 1 - \frac{\gamma(\sqrt{2\lambda+2\gamma+\mu^2}+\mu)}{(\lambda+\gamma)(\sqrt{2\lambda+2\gamma+\mu^2}+\sqrt{2\lambda+\mu^2})} e^{(x-r)(\sqrt{2\lambda+\mu^2}-\mu)}$
$r \le x$	$= \frac{\lambda}{\lambda+\gamma} + \frac{\gamma(\sqrt{2\lambda+\mu^2}-\mu)}{(\lambda+\gamma)(\sqrt{2\lambda+2\gamma+\mu^2}+\sqrt{2\lambda+\mu^2})} e^{(r-x)(\sqrt{2\lambda+\mu^2+2\gamma}+\mu)}$
1.4.2	$\mathbf{P}_x\Big(\int_0^\tau \mathbb{I}_{[r,\infty)}\big(W_s^{(\mu)}\big) ds \in dy\Big)$
$x \le r$	$= \frac{\lambda e^{-\lambda y-(r-x)(\sqrt{2\lambda+\mu^2}-\mu)}}{\sqrt{\lambda+\mu^2/2}+\mu/\sqrt{2}} \Big(\frac{e^{-\mu^2 y/2}}{\sqrt{\pi y}} + \frac{\mu}{\sqrt{2}} \operatorname{Erfc}\Big(-\frac{\mu\sqrt{y}}{\sqrt{2}}\Big)\Big) dy$
$r \le x$	$= \lambda e^{-\lambda y}\Big(1 - \frac{1}{2}\operatorname{Erfc}\Big(\frac{x-r}{\sqrt{2y}} + \frac{\mu\sqrt{y}}{\sqrt{2}}\Big) - \frac{1}{2} e^{2\mu(r-x)} \operatorname{Erfc}\Big(\frac{x-r}{\sqrt{2y}} - \frac{\mu\sqrt{y}}{\sqrt{2}}\Big)\Big) dy$ $+ \frac{\lambda e^{-\lambda y}}{\sqrt{\lambda+\mu^2/2}+\mu/\sqrt{2}} \Big(\frac{e^{-(x-r+\mu y)^2/2y}}{\sqrt{\pi y}} + \frac{\mu}{\sqrt{2}} e^{2\mu(r-x)} \operatorname{Erfc}\Big(\frac{x-r}{\sqrt{2y}} - \frac{\mu\sqrt{y}}{\sqrt{2}}\Big)\Big) dy$
(1) $x \le r$	$\mathbf{P}_x\Big(\int_0^\tau \mathbb{I}_{[r,\infty)}\big(W_s^{(\mu)}\big) ds = 0\Big) = 1 - e^{(x-r)(\sqrt{2\lambda+\mu^2}-\mu)}$

2 $W_s^{(\mu)} = \mu s + W_s$ $\tau \sim \mathrm{Exp}(\lambda)$, independent of $W^{(\mu)}$

1.4.3 (1) $\mu < 0$

$$\mathbf{E}_x \exp\Big(-\gamma \int_0^\infty \mathbb{I}_{[r,\infty)}\big(W_s^{(\mu)}\big)ds\Big) = \begin{cases} 1 - \dfrac{\sqrt{2\gamma+\mu^2}+\mu}{\sqrt{2\gamma+\mu^2}-\mu} e^{-2\mu(x-r)}, & x \le r \\ -\dfrac{2\mu e^{(r-x)(\sqrt{2\gamma+\mu^2}+\mu)}}{\sqrt{2\gamma+\mu^2}-\mu}, & r \le x \end{cases}$$

1.4.4

$$\mathbf{P}_x\Big(\int_0^t \mathbb{I}_{[r,\infty)}\big(W_s^{(\mu)}\big)ds \in dy\Big) = B_x^{(6)}(t-y,y)\mathbb{I}_{(0,t)}(y)dy$$

for $B_x^{(6)}(u,v)$ see 1.6.2

(1) $x \le r$

$$\mathbf{P}_x\Big(\int_0^t \mathbb{I}_{[r,\infty)}\big(W_s^{(\mu)}\big)ds = 0\Big)$$

$$= 1 - \frac{1}{2}\operatorname{Erfc}\Big(\frac{r-x}{\sqrt{2t}} - \frac{\mu\sqrt{t}}{\sqrt{2}}\Big) - \frac{1}{2}e^{2\mu(r-x)}\operatorname{Erfc}\Big(\frac{r-x}{\sqrt{2t}} + \frac{\mu\sqrt{t}}{\sqrt{2}}\Big)$$

(2) $r \le x$

$$\mathbf{P}_x\Big(\int_0^t \mathbb{I}_{[r,\infty)}\big(W_s^{(\mu)}\big)ds = t\Big)$$

$$= 1 - \frac{1}{2}\operatorname{Erfc}\Big(\frac{x-r}{\sqrt{2t}} + \frac{\mu\sqrt{t}}{\sqrt{2}}\Big) - \frac{1}{2}e^{2\mu(r-x)}\operatorname{Erfc}\Big(\frac{x-r}{\sqrt{2t}} - \frac{\mu\sqrt{t}}{\sqrt{2}}\Big)$$

(3) $\mu < 0$

$$\mathbf{P}_x\Big(\int_0^\infty \mathbb{I}_{[r,\infty)}\big(W_s^{(\mu)}\big)ds \in dy\Big)$$

$x \le r$

$$= -\frac{\mu\sqrt{2}}{\sqrt{\pi y}}e^{2(r-x)\mu-\mu^2 y/2}dy - \mu^2 e^{2(r-x)\mu}\operatorname{Erfc}\Big(-\frac{\mu\sqrt{y}}{\sqrt{2}}\Big)dy$$

$r \le x$

$$= -\frac{\mu\sqrt{2}}{\sqrt{\pi y}}e^{-(x-r+\mu y)^2/2y}dy - \mu^2 e^{2(r-x)\mu}\operatorname{Erfc}\Big(\frac{x-r}{\sqrt{2y}} - \frac{\mu\sqrt{y}}{\sqrt{2}}\Big)dy$$

(4) $x \le r$

$$\mathbf{P}_x\Big(\int_0^\infty \mathbb{I}_{[r,\infty)}\big(W_s^{(\mu)}\big)ds = 0\Big) = 1 - e^{(x-r)(|\mu|-\mu)}$$

1.4.5

$$\mathbf{E}_x\Big\{\exp\Big(-\gamma\int_0^\tau \mathbb{I}_{[r,\infty)}\big(W_s^{(\mu)}\big)ds\Big); W_\tau^{(\mu)} \in dz\Big\} = \lambda e^{\mu(z-x)}Q^{(4)}_{\lambda+\mu^2/2}(z)dz$$

for $Q_\lambda^{(4)}(z)$ see 1.1.4.5

2 $W_s^{(\mu)} = \mu s + W_s$ $\tau \sim \operatorname{Exp}(\lambda)$, independent of $W^{(\mu)}$

1.4.6 $$\mathbf{P}_x\Big(\int_0^\tau \mathbb{I}_{[r,\infty)}(W_s^{(\mu)})ds \in dy, W_\tau^{(\mu)} \in dz\Big) = \lambda e^{\mu(z-x)} B^{(4)}_{\lambda+\mu^2/2}(y,z)dydz$$

for $B^{(4)}_\lambda(y,z)$ see 1.1.4.6

(1) $x \le r$, $z \le r$
$$\mathbf{P}_x\Big(\int_0^\tau \mathbb{I}_{[r,\infty)}(W_s^{(\mu)})ds = 0,\ W_\tau^{(\mu)} \in dz\Big)$$
$$= \frac{\lambda}{\sqrt{2\lambda+\mu^2}}\big(e^{\mu(z-x)-|z-x|\sqrt{2\lambda+\mu^2}} - e^{\mu(z-x)+(z+x-2r)\sqrt{2\lambda+\mu^2}}\big)dz$$

1.4.7 $$\mathbf{E}_x\Big\{\exp\Big(-\gamma\int_0^t \mathbb{I}_{[r,\infty)}(W_s^{(\mu)})ds\Big); W_t^{(\mu)} \in dz\Big\} = e^{\mu(z-x)-\mu^2 t/2} V_x^{(4)}(t,z)dz$$

for $V_x^{(4)}(t,z)$ see 1.1.4.7

1.4.8 $v < t$
$$\mathbf{P}_x\Big(\int_0^t \mathbb{I}_{[r,\infty)}(W_s^{(\mu)})ds \in dv, W_t^{(\mu)} \in dz\Big) = e^{\mu(z-x)-\mu^2 t/2} F_x^{(4)}(t,v,z)dvdz$$

for $F_x^{(4)}(t,v,z)$ see 1.1.4.8

(1) $x \le r$, $z \le r$
$$\mathbf{P}_x\Big(\int_0^t \mathbb{I}_{[r,\infty)}(W_s^{(\mu)})ds = 0, W_t^{(\mu)} \in dz\Big)$$
$$= \frac{1}{\sqrt{2\pi t}}e^{-(z-\mu t-x)^2/2t}dz - \frac{1}{\sqrt{2\pi t}}e^{\mu(z-x)-\mu^2 t/2-(2r-x-z)^2/2t}dz$$

(2) $r \le x$, $r \le z$
$$\mathbf{P}_x\Big(\int_0^t \mathbb{I}_{[r,\infty)}(W_s^{(\mu)})ds = t, W_t^{(\mu)} \in dz\Big)$$
$$= \frac{1}{\sqrt{2\pi t}}e^{-(z-\mu t-x)^2/2t}dz - \frac{1}{\sqrt{2\pi t}}e^{\mu(z-x)-\mu^2 t/2-(x+z-2r)^2/2t}dz$$

1.5.1 $$\mathbf{E}_x \exp\Big(-\gamma\int_0^\tau \mathbb{I}_{(-\infty,r]}(W_s^{(\mu)})ds\Big)$$

$x \le r$
$$= \frac{\lambda}{\lambda+\gamma} + \frac{\gamma(\sqrt{2\lambda+\mu^2}+\mu)}{(\lambda+\gamma)(\sqrt{2\lambda+2\gamma+\mu^2}+\sqrt{2\lambda+\mu^2})}e^{(x-r)(\sqrt{2\lambda+\mu^2+2\gamma}-\mu)}$$

$r \le x$
$$= 1 - \frac{\gamma(\sqrt{2\lambda+2\gamma+\mu^2}-\mu)}{(\lambda+\gamma)(\sqrt{2\lambda+2\gamma+\mu^2}+\sqrt{2\lambda+\mu^2})}e^{(r-x)(\sqrt{2\lambda+\mu^2}+\mu)}$$

2 $W_s^{(\mu)} = \mu s + W_s$ $\tau \sim \text{Exp}(\lambda)$, independent of $W^{(\mu)}$

1.5.2 $\mathbf{P}_x\Big(\int_0^\tau \mathbb{1}_{(-\infty,r]}\big(W_s^{(\mu)}\big)ds \in dy\Big)$

$x \le r$
$$= \lambda e^{-\lambda y}\Big(1 - \frac{1}{2}\,\text{Erfc}\Big(\frac{r-x}{\sqrt{2y}} - \frac{\mu\sqrt{y}}{\sqrt{2}}\Big) - \frac{1}{2}e^{2\mu(r-x)}\,\text{Erfc}\Big(\frac{r-x}{\sqrt{2y}} + \frac{\mu\sqrt{y}}{\sqrt{2}}\Big)\Big)dy$$
$$+ \frac{\lambda e^{-\lambda y}}{\sqrt{\lambda+\mu^2/2} - \mu/\sqrt{2}}\Big(\frac{e^{-(r-x-\mu y)^2/2y}}{\sqrt{\pi y}} - \frac{\mu}{\sqrt{2}}e^{2\mu(r-x)}\,\text{Erfc}\Big(\frac{r-x}{\sqrt{2y}} + \frac{\mu\sqrt{y}}{\sqrt{2}}\Big)\Big)dy$$

$r \le x$
$$= \frac{\lambda e^{-\lambda y-(x-r)(\sqrt{2\lambda+\mu^2}+\mu)}}{\sqrt{\lambda+\mu^2/2} - \mu/\sqrt{2}}\Big(\frac{e^{-\mu^2 y/2}}{\sqrt{\pi y}} - \frac{\mu}{\sqrt{2}}\,\text{Erfc}\Big(\frac{\mu\sqrt{y}}{\sqrt{2}}\Big)\Big)dy$$

(1)
$r \le x$
$$\mathbf{P}_x\Big(\int_0^\tau \mathbb{1}_{(-\infty,r]}\big(W_s^{(\mu)}\big)ds = 0\Big) = 1 - e^{(r-x)(\sqrt{2\lambda+\mu^2}+\mu)}$$

1.5.3
(1)
$\mu > 0$
$$\mathbf{E}_x \exp\Big(-\gamma\int_0^\infty \mathbb{1}_{(-\infty,r]}\big(W_s^{(\mu)}\big)ds\Big) = \begin{cases} \dfrac{2\mu e^{(x-r)(\sqrt{2\gamma+\mu^2}-\mu)}}{\sqrt{2\gamma+\mu^2}+\mu}, & x \le r \\[2ex] 1 - \dfrac{\sqrt{2\gamma+\mu^2}-\mu}{\sqrt{2\gamma+\mu^2}+\mu}e^{2\mu(r-x)}, & r \le x \end{cases}$$

1.5.4
$$\mathbf{P}_x\Big(\int_0^t \mathbb{1}_{(-\infty,r]}\big(W_s^{(\mu)}\big)ds \in dy\Big) = B_x^{(6)}(y, t-y)\mathbb{1}_{(0,t)}(y)dy$$

for $B_x^{(6)}(u,v)$ see 1.6.2

(1)
$r \le x$
$$\mathbf{P}_x\Big(\int_0^t \mathbb{1}_{(-\infty,r]}\big(W_s^{(\mu)}\big)ds = 0\Big)$$
$$= 1 - \frac{1}{2}\,\text{Erfc}\Big(\frac{x-r}{\sqrt{2t}} + \frac{\mu\sqrt{t}}{\sqrt{2}}\Big) - \frac{1}{2}e^{2\mu(r-x)}\,\text{Erfc}\Big(\frac{x-r}{\sqrt{2t}} - \frac{\mu\sqrt{t}}{\sqrt{2}}\Big)$$

(2)
$x \le r$
$$\mathbf{P}_x\Big(\int_0^t \mathbb{1}_{(-\infty,r]}\big(W_s^{(\mu)}\big)ds = t\Big)$$
$$= 1 - \frac{1}{2}\,\text{Erfc}\Big(\frac{r-x}{\sqrt{2t}} - \frac{\mu\sqrt{t}}{\sqrt{2}}\Big) - \frac{1}{2}e^{2\mu(r-x)}\,\text{Erfc}\Big(\frac{r-x}{\sqrt{2t}} + \frac{\mu\sqrt{t}}{\sqrt{2}}\Big)$$

(3)
$\mu > 0$
$$\mathbf{P}_x\Big(\int_0^\infty \mathbb{1}_{(-\infty,r]}\big(W_s^{(\mu)}\big)ds \in dy\Big)$$

$x \le r$
$$= \frac{\mu\sqrt{2}}{\sqrt{\pi y}}e^{-(r-x-\mu y)^2/2y}dy - \mu^2 e^{2(r-x)\mu}\,\text{Erfc}\Big(\frac{r-x}{\sqrt{2y}} + \frac{\mu\sqrt{y}}{\sqrt{2}}\Big)dy$$

2 $W_s^{(\mu)} = \mu s + W_s$ $\qquad\tau \sim \mathrm{Exp}(\lambda)$, independent of $W^{(\mu)}$

$r \le x$	$= \dfrac{\mu\sqrt{2}}{\sqrt{\pi y}} e^{2(r-x)\mu - \mu^2 y/2} dy - \mu^2 e^{2(r-x)\mu} \operatorname{Erfc}\Big(\dfrac{\mu\sqrt{y}}{\sqrt{2}}\Big) dy$
1.5.5	$\mathbf{E}_x\Big\{\exp\Big(-\gamma\int_0^\tau \mathbb{I}_{(-\infty,r]}\big(W_s^{(\mu)}\big)ds\Big); W_\tau^{(\mu)} \in dz\Big\} = \lambda e^{\mu(z-x)} Q^{(5)}_{\lambda+\mu^2/2}(z)dz$
	for $Q^{(5)}_\lambda(z)$ see 1.1.5.5
1.5.6	$\mathbf{P}_x\Big(\int_0^\tau \mathbb{I}_{(-\infty,r]}\big(W_s^{(\mu)}\big)ds \in dy,\ W_\tau^{(\mu)} \in dz\Big) = \lambda e^{\mu(z-x)} B^{(5)}_{\lambda+\mu^2/2}(y,z)dydz$
	for $B^{(5)}_\lambda(y,z)$ see 1.1.5.6
(1) $r < x$	$\mathbf{P}_x\Big(\int_0^\tau \mathbb{I}_{(-\infty,r]}\big(W_s^{(\mu)}\big)ds = 0,\ W_\tau^{(\mu)} \in dz\Big)$
$r \le z$	$= \dfrac{\lambda}{\sqrt{2\lambda+\mu^2}}\big(e^{\mu(z-x)-\lvert z-x\rvert\sqrt{2\lambda+\mu^2}} - e^{\mu(z-x)+(2r-x-z)\sqrt{2\lambda+\mu^2}}\big)dz$
1.5.7	$\mathbf{E}_x\Big\{\exp\Big(-\gamma\int_0^t \mathbb{I}_{(-\infty,r]}\big(W_s^{(\mu)}\big)ds\Big); W_t^{(\mu)} \in dz\Big\} = e^{\mu(z-x)-\mu^2 t/2} V_x^{(5)}(t,z)dz$
	for $V_x^{(5)}(t,z)$ see 1.1.5.7
1.5.8 $v < t$	$\mathbf{P}_x\Big(\int_0^t \mathbb{I}_{(-\infty,r]}\big(W_s^{(\mu)}\big)ds \in dv, W_t^{(\mu)} \in dz\Big) = e^{\mu(z-x)-\mu^2 t/2} F_x^{(5)}(t,v,z)dvdz$
	for $F_x^{(5)}(t,v,z)$ see 1.1.5.8
(1) $r \le x$	$\mathbf{P}_x\Big(\int_0^t \mathbb{I}_{(-\infty,r]}\big(W_s^{(\mu)}\big)ds = 0, W_t^{(\mu)} \in dz\Big)$
$z \le r$	$= \dfrac{1}{\sqrt{2\pi t}} e^{-(z-\mu t-x)^2/2t}dz - \dfrac{1}{\sqrt{2\pi t}} e^{\mu(z-x)-\mu^2 t/2-(2r-x-z)^2/2t}dz$
(2) $x \le r$	$\mathbf{P}_x\Big(\int_0^t \mathbb{I}_{(-\infty,r]}\big(W_s^{(\mu)}\big)ds = t, W_t^{(\mu)} \in dz\Big)$
$r \le z$	$= \dfrac{1}{\sqrt{2\pi t}} e^{-(z-\mu t-x)^2/2t}dz - \dfrac{1}{\sqrt{2\pi t}} e^{\mu(z-x)-\mu^2 t/2-(x+z-2r)^2/2t}dz$
1.6.1	$\mathbf{E}_x \exp\Big(-\int_0^\tau \big(p\mathbb{I}_{(-\infty,r)}\big(W_s^{(\mu)}\big) + q\mathbb{I}_{[r,\infty)}\big(W_s^{(\mu)}\big)\big)ds\Big)$

2 $W_s^{(\mu)} = \mu s + W_s$ $\tau \sim \text{Exp}(\lambda)$, independent of $W^{(\mu)}$

$x \le r$ $$= \frac{\lambda}{\lambda+p}\Big(1 - e^{-(r-x)(\sqrt{2\lambda+2p+\mu^2}-\mu)}\Big) + \frac{2\lambda e^{-(r-x)(\sqrt{2\lambda+2p+\mu^2}-\mu)}}{(\sqrt{2\lambda+2q+\mu^2}-\mu)(\sqrt{2\lambda+2p+\mu^2}+\mu)}$$

$r \le x$ $$= \frac{\lambda}{\lambda+q}\Big(1 - e^{-(x-r)(\sqrt{2\lambda+2q+\mu^2}+\mu)}\Big) + \frac{2\lambda e^{-(x-r)(\sqrt{2\lambda+2q+\mu^2}+\mu)}}{(\sqrt{2\lambda+2q+\mu^2}-\mu)(\sqrt{2\lambda+2p+\mu^2}+\mu)}$$

1.6.2 $$\mathbf{P}_x\Big(\int_0^\tau \mathbb{I}_{(-\infty,r)}(W_s^{(\mu)})ds \in du,\ \int_0^\tau \mathbb{I}_{[r,\infty)}(W_s^{(\mu)})ds \in dv\Big)$$

$$=: \lambda e^{-\lambda(u+v)} B_x^{(6)}(u,v)dudv$$

$x \le r$ $$= \lambda e^{-\lambda(u+v)}\Big(\frac{e^{-(r-x-\mu u)^2/2u}}{\sqrt{\pi u}} - \frac{\mu}{\sqrt{2}}e^{2\mu(r-x)}\operatorname{Erfc}\Big(\frac{r-x}{\sqrt{2u}} + \frac{\mu\sqrt{u}}{\sqrt{2}}\Big)\Big)$$

$$\times\Big(\frac{e^{-\mu^2 v/2}}{\sqrt{\pi v}} + \frac{\mu}{\sqrt{2}}\operatorname{Erfc}\Big(-\frac{\mu\sqrt{v}}{\sqrt{2}}\Big)\Big)dudv$$

$r \le x$ $$= \lambda e^{-\lambda(u+v)}\Big(\frac{e^{-\mu^2 u/2}}{\sqrt{\pi u}} - \frac{\mu}{\sqrt{2}}\operatorname{Erfc}\Big(\frac{\mu\sqrt{u}}{\sqrt{2}}\Big)\Big)$$

$$\times\Big(\frac{e^{-(x-r+\mu v)^2/2v}}{\sqrt{\pi v}} + \frac{\mu}{\sqrt{2}}e^{2\mu(r-x)}\operatorname{Erfc}\Big(\frac{x-r}{\sqrt{2v}} - \frac{\mu\sqrt{v}}{\sqrt{2}}\Big)\Big)dudv$$

(1) $x \le r$ $$\mathbf{P}_x\Big(\int_0^\tau \mathbb{I}_{(-\infty,r)}(W_s^{(\mu)})ds \in du,\ \int_0^\tau \mathbb{I}_{[r,\infty)}(W_s^{(\mu)})ds = 0\Big)$$

$$= \lambda e^{-\lambda u}\Big(1 - \frac{1}{2}\operatorname{Erfc}\Big(\frac{r-x}{\sqrt{2u}} - \frac{\mu\sqrt{u}}{\sqrt{2}}\Big) - \frac{1}{2}e^{2\mu(r-x)}\operatorname{Erfc}\Big(\frac{r-x}{\sqrt{2u}} + \frac{\mu\sqrt{u}}{\sqrt{2}}\Big)\Big)du$$

(2) $r \le x$ $$\mathbf{P}_x\Big(\int_0^\tau \mathbb{I}_{(-\infty,r)}(W_s^{(\mu)})ds = 0,\ \int_0^\tau \mathbb{I}_{[r,\infty)}(W_s^{(\mu)})ds \in dv\Big)$$

$$= \lambda e^{-\lambda v}\Big(1 - \frac{1}{2}\operatorname{Erfc}\Big(\frac{x-r}{\sqrt{2v}} + \frac{\mu\sqrt{v}}{\sqrt{2}}\Big) - \frac{1}{2}e^{2\mu(r-x)}\operatorname{Erfc}\Big(\frac{x-r}{\sqrt{2v}} - \frac{\mu\sqrt{v}}{\sqrt{2}}\Big)\Big)dv$$

1.6.4 $$\mathbf{P}_x\Big(\int_0^t (p\mathbb{I}_{(-\infty,r)}(W_s^{(\mu)}) + q\mathbb{I}_{[r,\infty)}(W_s^{(\mu)}))ds \in dv\Big)$$

$$= \frac{1}{|p-q|}B_x^{(6)}\Big(\frac{|qt-v|}{|p-q|}, \frac{|pt-v|}{|p-q|}\Big)\mathbb{I}_{((q\wedge p)t,(q\vee p)t)}(v)\,dv \quad \text{for } B_x^{(6)}(u,v)$$ see 1.6.2

2 $W_s^{(\mu)} = \mu s + W_s$ $\qquad \tau \sim \text{Exp}(\lambda)$, independent of $W^{(\mu)}$

(1) $x \le r$

$$\mathbf{P}_x\Big(\int_0^t \big(p\mathbb{I}_{(-\infty,r)}\big(W_s^{(\mu)}\big) + q\mathbb{I}_{[r,\infty)}\big(W_s^{(\mu)}\big)\big)ds = pt\Big)$$

$$= 1 - \frac{1}{2}\,\text{Erfc}\Big(\frac{r-x}{\sqrt{2t}} - \frac{\mu\sqrt{t}}{\sqrt{2}}\Big) - \frac{1}{2}e^{2\mu(r-x)}\,\text{Erfc}\Big(\frac{r-x}{\sqrt{2t}} + \frac{\mu\sqrt{t}}{\sqrt{2}}\Big)$$

(2) $r \le x$

$$\mathbf{P}_x\Big(\int_0^t \big(p\mathbb{I}_{(-\infty,r)}\big(W_s^{(\mu)}\big) + q\mathbb{I}_{[r,\infty)}\big(W_s^{(\mu)}\big)\big)ds = qt\Big)$$

$$= 1 - \frac{1}{2}\,\text{Erfc}\Big(\frac{x-r}{\sqrt{2t}} + \frac{\mu\sqrt{t}}{\sqrt{2}}\Big) - \frac{1}{2}e^{2\mu(r-x)}\,\text{Erfc}\Big(\frac{x-r}{\sqrt{2t}} - \frac{\mu\sqrt{t}}{\sqrt{2}}\Big)$$

1.6.5

$$\mathbf{E}_x\Big\{\exp\Big(-\int_0^\tau \big(p\mathbb{I}_{(-\infty,r)}\big(W_s^{(\mu)}\big) + q\mathbb{I}_{[r,\infty)}\big(W_s^{(\mu)}\big)\big)ds\Big); W_\tau^{(\mu)} \in dz\Big\}$$

$$= \lambda e^{\mu(z-x)} Q^{(6)}_{\lambda+\mu^2/2}(z)dz, \qquad \text{for } Q_\lambda^{(6)}(z) \text{ see 1.1.6.5}$$

1.6.6

$$\mathbf{P}_x\Big(\int_0^\tau \mathbb{I}_{(-\infty,r)}\big(W_s^{(\mu)}\big)ds \in du, \int_0^\tau \mathbb{I}_{[r,\infty)}\big(W_s^{(\mu)}\big)ds \in dv, W_\tau^{(\mu)} \in dz\Big)$$

$$= \lambda e^{\mu(z-x)} e^{-(\lambda+\mu^2/2)(u+v)} F_x^{(4)}(u+v, v, z)dudvdz$$

for $F_x^{(4)}(t,v,z)$ see 1.1.4.8

1.6.7

$$\mathbf{E}_x\Big\{\exp\Big(-\int_0^t \big(p\mathbb{I}_{(-\infty,r)}\big(W_s^{(\mu)}\big) + q\mathbb{I}_{[r,\infty)}\big(W_s^{(\mu)}\big)\big)ds\Big); W_t^{(\mu)} \in dz\Big\}$$

$$= e^{\mu(z-x)-\mu^2 t/2} V_x^{(6)}(t,z)dz, \qquad \text{for } V_x^{(6)}(t,z) \text{ see 1.1.6.7}$$

1.6.8

$$\mathbf{P}_x\Big(\int_0^t \big(p\mathbb{I}_{(-\infty,r)}\big(W_s^{(\mu)}\big) + q\mathbb{I}_{[r,\infty)}\big(W_s^{(\mu)}\big)\big)ds \in dv, W_t^{(\mu)} \in dz\Big)$$

$$= \frac{1}{|p-q|}\, e^{\mu(z-x)-\mu^2 t/2} \begin{cases} F_x^{(4)}\Big(t, \dfrac{v-pt}{q-p}, z\Big)dvdz, & pt < v < qt \\ F_x^{(5)}\Big(t, \dfrac{v-qt}{p-q}, z\Big)dvdz, & qt < v < pt \\ 0, & \text{otherwise} \end{cases}$$

for $F_x^{(4)}(t,v,z)$ see 1.1.4.8, $\qquad$ for $F_x^{(5)}(t,v,z)$ see 1.1.5.8

2 $W_s^{(\mu)} = \mu s + W_s$ $\qquad \tau \sim \text{Exp}(\lambda)$, independent of $W^{(\mu)}$

(1) $x \le r$

$$\mathbf{P}_x\Big(\int_0^t \big(p\mathbb{I}_{(-\infty,r)}\big(W_s^{(\mu)}\big) + q\mathbb{I}_{[r,\infty)}\big(W_s^{(\mu)}\big)\big)ds = pt, W_t^{(\mu)} \in dz\Big)$$

$z \le r$

$$= \frac{1}{\sqrt{2\pi t}} e^{-(z-\mu t-x)^2/2t} dz - \frac{1}{\sqrt{2\pi t}} e^{\mu(z-x)-\mu^2 t/2-(2r-x-z)^2/2t} dz$$

(2) $r \le x$

$$\mathbf{P}_x\Big(\int_0^t \big(p\mathbb{I}_{(-\infty,r)}\big(W_s^{(\mu)}\big) + q\mathbb{I}_{[r,\infty)}\big(W_s^{(\mu)}\big)\big)ds = qt, W_t^{(\mu)} \in dz\Big)$$

$r \le z$

$$= \frac{1}{\sqrt{2\pi t}} e^{-(z-\mu t-x)^2/2t} dz - \frac{1}{\sqrt{2\pi t}} e^{\mu(z-x)-\mu^2 t/2-(x+z-2r)^2/2t} dz$$

1.7.1

$$\mathbf{E}_x \exp\Big(-\gamma \int_0^\tau \mathbb{I}_{[r,u]}\big(W_s^{(\mu)}\big)ds\Big)$$

$x \le r$

$$= 1 - \frac{\gamma S_r}{\lambda+\gamma} e^{(x-r)(\sqrt{2\lambda+\mu^2}-\mu)}$$

$r \le x$, $x \le u$

$$= \frac{\lambda}{\lambda+\gamma} + \frac{\gamma(1-S_r)}{\lambda+\gamma} \frac{\mathrm{sh}((u-x)\Upsilon_\lambda)e^{(r-x)\mu}}{\mathrm{sh}((u-r)\Upsilon_\lambda)} + \frac{\gamma(1-S_u)}{\lambda+\gamma} \frac{\mathrm{sh}((x-r)\Upsilon_\lambda)e^{(u-x)\mu}}{\mathrm{sh}((u-r)\Upsilon_\lambda)}$$

$u \le x$

$$= 1 - \frac{\gamma S_u}{\lambda+\gamma} e^{(u-x)(\sqrt{2\lambda+\mu^2}+\mu)}$$

where

$$S_r := \frac{(\Upsilon_\lambda - \mu)(\Upsilon_\lambda - \sqrt{2\lambda+\mu^2})\,\mathrm{sh}((u-r)\Upsilon_\lambda) + (\sqrt{2\lambda+\mu^2}+\mu)\Upsilon_\lambda(e^{(u-r)\Upsilon_\lambda} - e^{(u-r)\mu})}{2(2\lambda+\mu^2+\gamma)\,\mathrm{sh}((u-r)\Upsilon_\lambda) + 2\Upsilon_\lambda\sqrt{2\lambda+\mu^2}\,\mathrm{ch}((u-r)\Upsilon_\lambda)}$$

$$S_u := \frac{(\Upsilon_\lambda + \mu)(\Upsilon_\lambda - \sqrt{2\lambda+\mu^2})\,\mathrm{sh}((u-r)\Upsilon_\lambda) + (\sqrt{2\lambda+\mu^2}-\mu)\Upsilon_\lambda(e^{(u-r)\Upsilon_\lambda} - e^{(r-u)\mu})}{2(2\lambda+\mu^2+\gamma)\,\mathrm{sh}((u-r)\Upsilon_\lambda) + 2\Upsilon_\lambda\sqrt{2\lambda+\mu^2}\,\mathrm{ch}((u-r)\Upsilon_\lambda)}$$

1.7.2 (1)

$$\mathbf{P}_x\Big(\int_0^\tau \mathbb{I}_{[r,u]}\big(W_s^{(\mu)}\big)ds = 0\Big) = \begin{cases} 1 - e^{(x-r)(\sqrt{2\lambda+\mu^2}-\mu)}, & x \le r \\ 1 - e^{(u-x)(\sqrt{2\lambda+\mu^2}+\mu)}, & u \le x \end{cases}$$

$$\Upsilon_s := \sqrt{2s + 2\gamma + \mu^2}$$

2 $W_s^{(\mu)} = \mu s + W_s$ $\tau \sim \operatorname{Exp}(\lambda)$, independent of $W^{(\mu)}$

1.7.3 (1) $\mathbf{E}_x \exp\Big(-\gamma \int_0^\infty \mathbb{I}_{[r,u]}\big(W_s^{(\mu)}\big) ds\Big)$

$x \le r$
$$= 1 - e^{-(r-x)(|\mu|-\mu)} + Q_r e^{-(r-x)(|\mu|-\mu)}$$

$r \le x$, $x \le u$
$$= Q_r e^{(r-x)\mu} \frac{\operatorname{sh}((u-x)\sqrt{2\gamma+\mu^2})}{\operatorname{sh}((u-r)\sqrt{2\gamma+\mu^2})} + Q_u e^{(u-x)\mu} \frac{\operatorname{sh}((x-r)\sqrt{2\gamma+\mu^2})}{\operatorname{sh}((u-r)\sqrt{2\gamma+\mu^2})}$$

$u \le x$
$$= 1 - e^{-(x-u)(|\mu|+\mu)} + Q_u e^{-(x-u)(|\mu|+\mu)}$$

where

$$Q_r := \frac{(|\mu|-\mu)\big(\sqrt{2\gamma+\mu^2}\operatorname{cth}((u-r)\sqrt{2\gamma+\mu^2}) + |\mu|\big) + \frac{(|\mu|+\mu)\sqrt{2\gamma+\mu^2}\, e^{(u-r)\mu}}{\operatorname{sh}((u-r)\sqrt{2\gamma+\mu^2})}}{2(\gamma+\mu^2) + 2|\mu|\sqrt{2\gamma+\mu^2}\operatorname{cth}((u-r)\sqrt{2\gamma+\mu^2})}$$

$$Q_u := \frac{(|\mu|+\mu)\big(\sqrt{2\gamma+\mu^2}\operatorname{cth}((u-r)\sqrt{2\gamma+\mu^2}) + |\mu|\big) + \frac{(|\mu|-\mu)\sqrt{2\gamma+\mu^2}\, e^{(r-u)\mu}}{\operatorname{sh}((u-r)\sqrt{2\gamma+\mu^2})}}{2(\gamma+\mu^2) + 2|\mu|\sqrt{2\gamma+\mu^2}\operatorname{cth}((u-r)\sqrt{2\gamma+\mu^2})}$$

1.7.4 (1) $\mu > 0$ $\mathbf{P}_x\Big(\int_0^\infty \mathbb{I}_{[r,u]}\big(W_s^{(\mu)}\big) ds \in dv\Big)$

$x \le r$
$$= \mu e^{(u-r)\mu - \mu^2 v/2} \int_0^v e^{-\mu^2 y/2} \operatorname{es}_{v-y}(1,1,u-r,0,\mu y) dy\, dv$$

$r \le x$, $x \le u$
$$= \mu e^{(u-x)\mu - \mu^2 v/2} \int_0^v e^{-\mu^2 y/2} \big(\operatorname{escs}_{v-y}(u-x,0,u-r,\mu y)$$
$$+ \mu \operatorname{ess}_{v-y}(x-r,u-r,0,\mu y) + \operatorname{escs}_{v-y}(x-r,u-r,u-r,\mu y)\big) dy\, dv$$

$u \le x$
$$= \mu e^{2(u-x)\mu - \mu^2 v/2} \int_0^v e^{-\mu^2 y/2} \big(\mu \operatorname{es}_{v-y}(0,0,u-r,0,\mu y)$$
$$+ \operatorname{escs}_{v-y}(u-r,u-r,u-r,\mu y)\big) dy\, dv$$

(2)
$$\mathbf{P}_x\Big(\int_0^\infty \mathbb{I}_{[r,u]}\big(W_s^{(\mu)}\big) ds = 0\Big) = \begin{cases} 1 - e^{-(r-x)(|\mu|-\mu)}, & x \le r \\ 1 - e^{-(x-u)(|\mu|+\mu)}, & u \le x \end{cases}$$

1.8.3
$$\mathbf{E}_x \exp\Big(-\gamma \int_0^t W_s^{(\mu)} ds\Big) = \exp\Big(-\gamma x t - \gamma\mu \frac{t^2}{2} + \frac{\gamma^2 t^3}{6}\Big)$$

2 $W_s^{(\mu)} = \mu s + W_s$	$\tau \sim \operatorname{Exp}(\lambda)$, independent of $W^{(\mu)}$
1.8.4	$\mathbf{P}_x\Big(\int_0^t W_s^{(\mu)} ds \in dy\Big) = \frac{\sqrt{3}}{\sqrt{2\pi t^3}} \exp\Big(-\frac{3(y - xt - \mu t^2/2)^2}{2t^3}\Big) dy$
1.8.5	$\mathbf{E}_0\Big\{\exp\Big(-\gamma\int_0^\tau \lvert W_s^{(\mu)}\rvert ds\Big); W_\tau^{(\mu)} \in dz\Big\} = -\frac{\lambda e^{\mu z} \operatorname{Ai}(2^{1/3}\gamma^{-2/3}(\lambda + \mu^2/2 + \gamma\lvert z\rvert))}{(2\gamma)^{1/3} \operatorname{Ai}'(2^{1/3}\gamma^{-2/3}(\lambda + \mu^2/2))} dz$
1.8.7	$\mathbf{E}_x\Big\{\exp\Big(-\gamma \int_0^t W_s^{(\mu)} ds\Big);\ W_t^{(\mu)} \in dz\Big\}$
	$= \frac{1}{\sqrt{2\pi t}} \exp\Big(-\frac{(z - \mu t - x)^2}{2t} - \frac{\gamma(z + x)t}{2} + \frac{\gamma^2 t^3}{24}\Big) dz$
(1)	$\mathbf{E}_0\Big\{\exp\Big(-\gamma \int_0^t \lvert W_s^{(\mu)}\rvert ds\Big); W_t^{(\mu)} \in dz\Big\}$
	$= -e^{\mu z - \mu^2 t/2} \sum_{k=1}^{\infty} \gamma^{1/3} \exp\big(2^{-1/3}\gamma^{2/3} t\alpha_k'\big) \frac{\operatorname{Ai}(\alpha_k' + 2^{1/3}\gamma^{1/3}\lvert z\rvert)}{2^{2/3}\alpha_k' \operatorname{Ai}(\alpha_k')} dz$
1.8.8	$\mathbf{E}_x\Big(\int_0^t W_s^{(\mu)} ds \in dy,\ W_t^{(\mu)} \in dz\Big)$
	$= \frac{\sqrt{3}}{\pi t^2} \exp\Big(-\frac{(z - \mu t - x)^2}{2t} - \frac{3(2y - (z + x)t)^2}{2t^3}\Big) dy dz$
(1)	$\mathbf{P}_0\Big(\int_0^t \lvert W_s^{(\mu)}\rvert ds < y \,\Big\vert\, W_t^{(\mu)} = 0\Big) = \sum_{k=1}^{\infty} \frac{(9/y)^{1/3}\sqrt{\pi t}}{2^{1/6}\lvert\alpha_k'\rvert} \exp\Big(\frac{(\alpha_k')^3 t^3}{27y^2}\Big) \operatorname{Ai}\Big(\frac{(\alpha_k')^2 t^2}{(3y\sqrt{2})^{4/3}}\Big)$
1.9.3	$\mathbf{E}_x \exp\Big(-\frac{\gamma^2}{2}\int_0^t (W_s^{(\mu)})^2 ds\Big) = \frac{e^{-\mu x}}{\sqrt{\operatorname{ch}(t\gamma)}} \exp\Big(-\frac{\mu^2 t}{2} - \frac{(x^2\gamma^2 - \mu^2)\operatorname{sh}(t\gamma) - 2\mu x\gamma}{2\gamma \operatorname{ch}(t\gamma)}\Big)$
1.9.5	$\mathbf{E}_x\Big\{\exp\Big(-\frac{\gamma^2}{2}\int_0^\tau (W_s^{(\mu)})^2 ds\Big);\ W_\tau^{(\mu)} \in dz\Big\}$
$x \le z$	$= \frac{\lambda\Gamma(\frac{1}{2} + \frac{2\lambda + \mu^2}{2\gamma})}{\sqrt{\pi\gamma}} e^{\mu(z-x)} D_{-\frac{1}{2} - \frac{2\lambda+\mu^2}{2\gamma}}(-x\sqrt{2\gamma}) D_{-\frac{1}{2} - \frac{2\lambda+\mu^2}{2\gamma}}(z\sqrt{2\gamma}) dz$
$z \le x$	$= \frac{\lambda\Gamma(\frac{1}{2} + \frac{2\lambda + \mu^2}{2\gamma})}{\sqrt{\pi\gamma}} e^{\mu(z-x)} D_{-\frac{1}{2} - \frac{2\lambda+\mu^2}{2\gamma}}(x\sqrt{2\gamma}) D_{-\frac{1}{2} - \frac{2\lambda+\mu^2}{2\gamma}}(-z\sqrt{2\gamma}) dz$

2 $W_s^{(\mu)} = \mu s + W_s$ $\qquad \tau \sim \operatorname{Exp}(\lambda)$, independent of $W^{(\mu)}$

1.9.7 $$\mathbf{E}_x\Big\{\exp\Big(-\frac{\gamma^2}{2}\int_0^t (W_s^{(\mu)})^2 ds\Big);\ W_t^{(\mu)} \in dz\Big\}$$

$$= \frac{\sqrt{\gamma}}{\sqrt{2\pi \operatorname{sh}(t\gamma)}} \exp\Big(\mu(z-x) - \frac{\mu^2 t}{2} - \frac{(x^2+z^2)\gamma \operatorname{ch}(t\gamma) - 2xz\gamma}{2\operatorname{sh}(t\gamma)}\Big)dz$$

1.9.8 $$\mathbf{P}_x\Big(\int_0^t (W_s^{(\mu)})^2 ds \in dy,\ W_t^{(\mu)} \in dz\Big)$$

$$= (2\pi)^{-1/2} e^{\mu(z-x)-\mu^2 t/2}\, \operatorname{ee}_y(1/2, t, (x^2+z^2)/2, -xz)\,dydz$$

1.10.2 $$\mathbf{P}_x\Big(\int_0^\tau e^{2\beta W_s^{(\mu)}} ds \in dy\Big) = \frac{\lambda(2\beta^2 y)^{\mu/2\beta}\Gamma(\mu/2\beta + \sqrt{2\lambda+\mu^2}/2|\beta|)}{|\beta|\sqrt{2y}\Gamma(1+\sqrt{2\lambda+\mu^2}/|\beta|)}$$

$\beta \neq 0$ $$\times \exp\Big(-(\beta+\mu)x - \frac{e^{2\beta x}}{4\beta^2 y}\Big) M_{1/2-\mu/2\beta,\sqrt{2\lambda+\mu^2}/2|\beta|}\Big(\frac{e^{2\beta x}}{2\beta^2 y}\Big)dy$$

1.10.3 $\frac{\mu}{\beta} < 0$ $$\mathbf{E}_x \exp\Big(-\gamma\int_0^\infty e^{2\beta W_s^{(\mu)}} ds\Big) = \frac{2(\sqrt{2\gamma}/2|\beta|)^{|\mu|/|\beta|}}{\Gamma(|\mu|/|\beta|)} e^{-\mu x} K_{|\mu|/|\beta|}\Big(\frac{\sqrt{2\gamma}}{|\beta|}e^{\beta x}\Big)$$

1.10.4 $$\mathbf{P}_x\Big(\int_0^t e^{2\beta W_s^{(\mu)}} ds \in dy\Big) \qquad \mu/2\beta > -1$$

$\beta \neq 0$ $$= \frac{\sqrt{2}|\beta|(2\beta^2 y)^{\mu/2\beta}}{\sqrt{y}} \exp\Big(-(\beta+\mu)x - \frac{\mu^2 t}{2} - \frac{e^{2\beta x}}{4\beta^2 y}\Big)\, \operatorname{m}_{2\beta^2 t}\Big(\frac{\mu}{2\beta} - \frac{1}{2}, \frac{e^{2\beta x}}{4\beta^2 y}\Big)dy$$

(1) $\frac{\mu}{\beta} < 0$ $$\mathbf{P}_x\Big(\int_0^\infty e^{2\beta W_s^{(\mu)}} ds \in dy\Big) = \frac{(2\beta^2 y)^{-|\mu|/|\beta|}}{y\Gamma(|\mu|/|\beta|)} \exp\Big(-2\mu x - \frac{e^{2\beta x}}{2\beta^2 y}\Big)dy$$

1.10.5 $$\mathbf{E}_x\Big\{\exp\Big(-\gamma\int_0^\tau e^{2\beta W_s^{(\mu)}} ds\Big);\ W_\tau^{(\mu)} \in dz\Big\}$$

$0 < \beta$ $$= e^{\mu(z-x)} \begin{cases} \dfrac{2\lambda}{\beta} I_{\sqrt{2\lambda+\mu^2}/\beta}\Big(\dfrac{\sqrt{2\gamma}}{\beta}e^{\beta x}\Big) K_{\sqrt{2\lambda+\mu^2}/\beta}\Big(\dfrac{\sqrt{2\gamma}}{\beta}e^{\beta z}\Big)dz, & x \le z \\ \dfrac{2\lambda}{\beta} K_{\sqrt{2\lambda+\mu^2}/\beta}\Big(\dfrac{\sqrt{2\gamma}}{\beta}e^{\beta x}\Big) I_{\sqrt{2\lambda+\mu^2}/\beta}\Big(\dfrac{\sqrt{2\gamma}}{\beta}e^{\beta z}\Big)dz, & z \le x \end{cases}$$

2 $W_s^{(\mu)} = \mu s + W_s$ $\qquad \tau \sim \operatorname{Exp}(\lambda)$, independent of $W^{(\mu)}$

$\beta < 0$	$= e^{\mu(z-x)} \begin{cases} \frac{2\lambda}{\lvert\beta\rvert} K_{\sqrt{2\lambda+\mu^2}/\lvert\beta\rvert}\Big(\frac{\sqrt{2\gamma}}{\lvert\beta\rvert}e^{\beta x}\Big) I_{\sqrt{2\lambda+\mu^2}/\lvert\beta\rvert}\Big(\frac{\sqrt{2\gamma}}{\lvert\beta\rvert}e^{\beta z}\Big)dz, & x \le z \\ \frac{2\lambda}{\lvert\beta\rvert} I_{\sqrt{2\lambda+\mu^2}/\lvert\beta\rvert}\Big(\frac{\sqrt{2\gamma}}{\lvert\beta\rvert}e^{\beta x}\Big) K_{\sqrt{2\lambda+\mu^2}/\lvert\beta\rvert}\Big(\frac{\sqrt{2\gamma}}{\lvert\beta\rvert}e^{\beta z}\Big)dz, & z \le x \end{cases}$
1.10.6 $\beta \ne 0$	$\mathbf{P}_x\Big(\int_0^\tau e^{2\beta W_s^{(\mu)}}ds \in dy, W_\tau^{(\mu)} \in dz\Big)$
	$= \frac{\lambda}{\lvert\beta\rvert y}\exp\Big(\mu(z-x) - \frac{e^{2\beta x}+e^{2\beta z}}{2\beta^2 y}\Big) I_{\sqrt{2\lambda+\mu^2}/\lvert\beta\rvert}\Big(\frac{e^{\beta(x+z)}}{\beta^2 y}\Big)dydz$
1.10.7	$\mathbf{E}_x\Big\{\exp\Big(-\gamma\int_0^t e^{2\beta W_s^{(\mu)}}ds\Big);\ W_t^{(\mu)} \in dz\Big\}$
$\beta \ne 0$	$= \lvert\beta\rvert\, e^{\mu(z-x)-\mu^2 t/2}\, \operatorname{ki}_{\beta^2 t/2}\Big(\frac{\sqrt{2\gamma}}{\lvert\beta\rvert}e^{\beta x}, \frac{\sqrt{2\gamma}}{\lvert\beta\rvert}e^{\beta z}\Big)dz$
1.10.8	$\mathbf{P}_x\Big(\int_0^t e^{2\beta W_s^{(\mu)}}ds \in dy,\ W_t^{(\mu)} \in dz\Big)$
$\beta \ne 0$	$= \frac{\lvert\beta\rvert}{2y}\exp\Big(\mu(z-x) - \frac{\mu^2 t}{2} - \frac{e^{2\beta x}+e^{2\beta z}}{2\beta^2 y}\Big)\, \operatorname{i}_{\beta^2 t/2}\Big(\frac{e^{\beta(x+z)}}{\beta^2 y}\Big)dydz$
1.11.2	$\mathbf{P}_x\big(\sup_{y\in\mathbb{R}} \ell(\tau,y) \ge h\big) = \frac{h^2(2\lambda+\mu^2)}{8\operatorname{sh}^2(h\sqrt{2\lambda+\mu^2}/2)}$
	$\times\Bigg(\frac{M(\frac{3}{2}+\frac{\mu}{2\sqrt{2\lambda+\mu^2}}, 3, h\sqrt{2\lambda+\mu^2})}{M(\frac{1}{2}+\frac{\mu}{2\sqrt{2\lambda+\mu^2}}, 1, h\sqrt{2\lambda+\mu^2})} + \frac{M(\frac{3}{2}-\frac{\mu}{2\sqrt{2\lambda+\mu^2}}, 3, h\sqrt{2\lambda+\mu^2})}{M(\frac{1}{2}-\frac{\mu}{2\sqrt{2\lambda+\mu^2}}, 1, h\sqrt{2\lambda+\mu^2})}\Bigg)$
1.12.1	$\mathbf{E}_x e^{-\gamma \check{H}_\mu(\tau)} = \frac{\sqrt{2\lambda+\mu^2}-\mu}{\sqrt{2\lambda+2\gamma+\mu^2}-\mu}$
(1)	$\mathbf{E}_x e^{-\gamma \hat{H}_\mu(\tau)} = \frac{\sqrt{2\lambda+\mu^2}+\mu}{\sqrt{2\lambda+2\gamma+\mu^2}+\mu}$
1.12.2	$\mathbf{P}_x\big(\check{H}_\mu(\tau) \in dv\big) = (\sqrt{2\lambda+\mu^2}-\mu)e^{-\lambda v}\Big(\frac{e^{-\mu^2 v/2}}{\sqrt{2\pi v}} + \frac{\mu}{2}\operatorname{Erfc}\Big(-\frac{\mu\sqrt{v}}{\sqrt{2}}\Big)\Big)dv$
(1)	$\mathbf{P}_x\big(\hat{H}_\mu(\tau) \in dv\big) = (\sqrt{2\lambda+\mu^2}+\mu)e^{-\lambda v}\Big(\frac{e^{-\mu^2 v/2}}{\sqrt{2\pi v}} - \frac{\mu}{2}\operatorname{Erfc}\Big(\frac{\mu\sqrt{v}}{\sqrt{2}}\Big)\Big)dv$

2 $W_s^{(\mu)} = \mu s + W_s$ $\tau \sim \operatorname{Exp}(\lambda)$, independent of $W^{(\mu)}$

1.12.3 $$\mathbf{E}_x e^{-\gamma \check{H}_\mu(t)} = \Big(\frac{e^{-\gamma t-\mu^2 t/2}}{\sqrt{\pi t}} + \frac{\mu e^{-\gamma t}}{\sqrt{2}} \operatorname{Erfc}\Big(-\frac{\mu\sqrt{t}}{\sqrt{2}}\Big)\Big) * \Big(\frac{e^{-\mu^2 t/2}}{\sqrt{\pi t}} - \frac{\mu}{\sqrt{2}} \operatorname{Erfc}\Big(\frac{\mu\sqrt{t}}{\sqrt{2}}\Big)\Big)$$

(1) $$\mathbf{E}_x e^{-\gamma \hat{H}_\mu(t)} = \Big(\frac{e^{-\gamma t-\mu^2 t/2}}{\sqrt{\pi t}} - \frac{\mu e^{-\gamma t}}{\sqrt{2}} \operatorname{Erfc}\Big(\frac{\mu\sqrt{t}}{\sqrt{2}}\Big)\Big) * \Big(\frac{e^{-\mu^2 t/2}}{\sqrt{\pi t}} + \frac{\mu}{\sqrt{2}} \operatorname{Erfc}\Big(-\frac{\mu\sqrt{t}}{\sqrt{2}}\Big)\Big)$$

(2) $$\mathbf{E}_x e^{-\gamma \check{H}_\mu(\infty)} = \frac{|\mu| - \mu}{\sqrt{2\gamma + \mu^2} - \mu}$$

(3) $$\mathbf{E}_x e^{-\gamma \hat{H}_\mu(\infty)} = \frac{|\mu| + \mu}{\sqrt{2\gamma + \mu^2} + \mu}$$

1.12.4 $$\mathbf{P}_x\big(\check{H}_\mu(t) \in dv\big) \qquad 0 \le v \le t$$

$$= \Big(\frac{e^{-\mu^2 v/2}}{\sqrt{\pi v}} + \frac{\mu}{\sqrt{2}} \operatorname{Erfc}\Big(-\frac{\mu\sqrt{v}}{\sqrt{2}}\Big)\Big)\Big(\frac{e^{-\mu^2 (t-v)/2}}{\sqrt{\pi (t-v)}} - \frac{\mu}{\sqrt{2}} \operatorname{Erfc}\Big(\frac{\mu\sqrt{t-v}}{\sqrt{2}}\Big)\Big) dv$$

(1) $$\mathbf{P}_x\big(\hat{H}_\mu(t) \in dv\big) \qquad 0 \le v \le t$$

$$= \Big(\frac{e^{-\mu^2 v/2}}{\sqrt{\pi v}} - \frac{\mu}{\sqrt{2}} \operatorname{Erfc}\Big(\frac{\mu\sqrt{v}}{\sqrt{2}}\Big)\Big)\Big(\frac{e^{-\mu^2 (t-v)/2}}{\sqrt{\pi (t-v)}} + \frac{\mu}{\sqrt{2}} \operatorname{Erfc}\Big(-\frac{\mu\sqrt{t-v}}{\sqrt{2}}\Big)\Big) dv$$

(2) $$\mathbf{P}_x\big(\check{H}_\mu(\infty) \in dv\big) = (|\mu| - \mu)\Big(\frac{e^{-\mu^2 v/2}}{\sqrt{2\pi v}} + \frac{\mu}{2} \operatorname{Erfc}\Big(-\frac{\mu\sqrt{v}}{\sqrt{2}}\Big)\Big) dv$$

(3) $$\mathbf{P}_x\big(\hat{H}_\mu(\infty) \in dv\big) = (|\mu| + \mu)\Big(\frac{e^{-\mu^2 v/2}}{\sqrt{2\pi v}} - \frac{\mu}{2} \operatorname{Erfc}\Big(\frac{\mu\sqrt{v}}{\sqrt{2}}\Big)\Big) dv$$

1.12.5 $$\mathbf{E}_x\big\{e^{-\gamma \check{H}_\mu(\tau)}; W_\tau^{(\mu)} \in dz\big\} = \begin{cases} \dfrac{2\lambda e^{(x-z)(\sqrt{2\lambda+\mu^2+2\gamma}-\mu)}}{\sqrt{2\lambda+\mu^2+2\gamma} + \sqrt{2\lambda+\mu^2}} dz & x \le z \\ \dfrac{2\lambda e^{(z-x)(\sqrt{2\lambda+\mu^2}+\mu)}}{\sqrt{2\lambda+\mu^2+2\gamma} + \sqrt{2\lambda+\mu^2}} dz & z \le x \end{cases}$$

(1) $$\mathbf{E}_x\big\{e^{-\gamma \hat{H}_\mu(\tau)}; W_\tau^{(\mu)} \in dz\big\} = \begin{cases} \dfrac{2\lambda e^{(x-z)(\sqrt{2\lambda+\mu^2}-\mu)}}{\sqrt{2\lambda+\mu^2+2\gamma} + \sqrt{2\lambda+\mu^2}} dz & x \le z \\ \dfrac{2\lambda e^{(z-x)(\sqrt{2\lambda+\mu^2+2\gamma}+\mu)}}{\sqrt{2\lambda+\mu^2+2\gamma} + \sqrt{2\lambda+\mu^2}} dz & z \le x \end{cases}$$

2 $W_s^{(\mu)} = \mu s + W_s$ $\tau \sim \text{Exp}(\lambda)$, independent of $W^{(\mu)}$

1.12.7 $\mathbf{E}_x\{e^{-\gamma \check{H}_\mu(t)}; W_t^{(\mu)} \in dz\}$

$x \le z$ $$= e^{\mu(z-x)-\mu^2 t/2}\Big(\frac{1-e^{-\gamma t}}{\sqrt{2\pi}\gamma t^{3/2}}\Big) * \Big(\frac{z-x}{\sqrt{2\pi}t^{3/2}} e^{-\gamma t-(z-x)^2/2t}\Big)dz$$

$z \le x$ $$= e^{\mu(z-x)-\mu^2 t/2}\Big(\frac{1-e^{-\gamma t}}{\sqrt{2\pi}\gamma t^{3/2}}\Big) * \Big(\frac{x-z}{\sqrt{2\pi}t^{3/2}} e^{-(x-z)^2/2t}\Big)dz$$

(1) $\mathbf{E}_x\{e^{-\gamma \hat{H}_\mu(t)}; W_t^{(\mu)} \in dz\}$

$x \le z$ $$= e^{\mu(z-x)-\mu^2 t/2}\Big(\frac{1-e^{-\gamma t}}{\sqrt{2\pi}\gamma t^{3/2}}\Big) * \Big(\frac{z-x}{\sqrt{2\pi}t^{3/2}} e^{-(z-x)^2/2t}\Big)dz$$

$z \le x$ $$= e^{\mu(z-x)-\mu^2 t/2}\Big(\frac{1-e^{-\gamma t}}{\sqrt{2\pi}\gamma t^{3/2}}\Big) * \Big(\frac{x-z}{\sqrt{2\pi}t^{3/2}} e^{-\gamma t-(x-z)^2/2t}\Big)dz$$

1.12.8 $\mathbf{P}_x\big(\check{H}_\mu(t) \in dv, W_t^{(\mu)} \in dz\big)$
$v < t$

$$= e^{\mu(z-x)-\mu^2 t/2}\int_{x\vee z}^{\infty} \frac{(y-x)(y-z)}{\pi(v(t-v))^{3/2}} \exp\Big(-\frac{(y-x)^2}{2v} - \frac{(y-z)^2}{2(t-v)}\Big)dy\, dvdz$$

(1) $\mathbf{P}_x\big(\hat{H}_\mu(t) \in dv, W_t^{(\mu)} \in dz\big)$
$v < t$

$$= e^{\mu(z-x)-\mu^2 t/2}\int_{-\infty}^{x\wedge z} \frac{(x-y)(z-y)}{\pi(v(t-v))^{3/2}} \exp\Big(-\frac{(x-y)^2}{2v} - \frac{(z-y)^2}{2(t-v)}\Big)dy\, dvdz$$

1.13.1 $\mathbf{E}_x\{e^{-\gamma \check{H}_\mu(\tau)}; \sup_{0\le s\le\tau} W_s^{(\mu)} \in dy\} = \dfrac{2\lambda}{\mu+\sqrt{2\lambda+\mu^2}} e^{-(y-x)(\sqrt{2\lambda+2\gamma+\mu^2}-\mu)}dy$
$x < y$

1.13.2 $\mathbf{P}_x\big(\check{H}_\mu(\tau) \in dv, \sup_{0\le s\le\tau} W_s^{(\mu)} \in dy\big)$
$x < y$

$$= \frac{2\lambda(y-x)}{\sqrt{2\pi}(\sqrt{2\lambda+\mu^2}+\mu)v^{3/2}} \exp\Big(\mu(y-x) - (\lambda+\frac{\mu^2}{2})v - \frac{(y-x)^2}{2v}\Big)dvdy$$

1.13.3 $\mathbf{E}_x\{e^{-\gamma \check{H}_\mu(t)}; \sup_{0\le s\le t} W_s^{(\mu)} \in dy\}$
$x < y$

2 $W_s^{(\mu)} = \mu s + W_s$ $\quad\tau \sim \mathrm{Exp}(\lambda)$, independent of $W^{(\mu)}$

$$= \Big(\frac{y-x}{\sqrt{\pi}t^{3/2}} e^{\mu(y-x)-\gamma t-\mu^2 t/2} \exp\Big(-\frac{(y-x)^2}{2t}\Big)\Big) * \Big(\frac{e^{-\mu^2 t/2}}{\sqrt{t}} - \frac{\mu}{\sqrt{2}} \operatorname{Erfc}\Big(\frac{\mu\sqrt{t}}{\sqrt{2}}\Big)\Big) dy$$

(1) $x < y$

$$\mathbf{E}_x\big\{e^{-\gamma \check{H}_\mu(\infty)};\ \sup_{0\le s<\infty} W_s^{(\mu)} \in dy\big\} = (|\mu|-\mu) e^{-(y-x)(\sqrt{2\gamma+\mu^2}-\mu)} dy$$

1.13.4 $x < y$

$$\mathbf{P}_x\big(\check{H}_\mu(t) \in dv,\ \sup_{0\le s\le t} W_s^{(\mu)} \in dy\big) \qquad 0 \le v \le t$$

$$= \frac{y-x}{\sqrt{\pi}v^{3/2}} \exp\Big(-\frac{(y-x-\mu v)^2}{2v}\Big)\Big(\frac{e^{-\mu^2(t-v)/2}}{\sqrt{\pi(t-v)}} - \frac{\mu}{\sqrt{2}} \operatorname{Erfc}\Big(\frac{\mu\sqrt{t-v}}{\sqrt{2}}\Big)\Big) dv dy$$

(1) $x < y$

$$\mathbf{P}_x\big(\check{H}_\mu(\infty) \in dv,\ \sup_{0\le s<\infty} W_s^{(\mu)} \in dy\big) = \frac{(|\mu|-\mu)(y-x)}{\sqrt{2\pi}v^{3/2}} \exp\Big(-\frac{(y-x-\mu v)^2}{2v}\Big) dv dy$$

1.13.5

$$\mathbf{E}_x\big\{e^{-\gamma \check{H}_\mu(\tau)};\ \sup_{0\le s\le \tau} W_s^{(\mu)} \in dy, W_\tau^{(\mu)} \in dz\big\} \qquad x \vee z < y$$

$$= 2\lambda e^{\mu(z-x)} e^{(x-y)\sqrt{2\lambda+2\gamma+\mu^2}} e^{(z-y)\sqrt{2\lambda+\mu^2}} dy dz$$

1.13.6

$$\mathbf{P}_x\big(\check{H}_\mu(\tau) \in dv,\ \sup_{0\le s\le \tau} W_s^{(\mu)} \in dy, W_\tau^{(\mu)} \in dz\big) \qquad x \vee z < y$$

$$= \frac{2\lambda(y-x)}{\sqrt{2\pi}v^{3/2}} \exp\Big(\mu(z-x) - \frac{(y-x)^2}{2v} - \big(\lambda + \frac{\mu^2}{2}\big)v - (y-z)\sqrt{2\lambda+\mu^2}\Big) dv dy dz$$

1.13.7

$$\mathbf{E}_x\big\{e^{-\gamma \check{H}_\mu(t)};\ \sup_{0\le s\le t} W_s^{(\mu)} \in dy,\ W_t^{(\mu)} \in dz\big\} \qquad x \vee z < y$$

$$= \Big(\frac{y-x}{\pi t^{3/2}} \exp\Big(-\gamma t - \frac{(y-x-\mu t)^2}{2t}\Big)\Big) * \Big(\frac{y-z}{t^{3/2}} \exp\Big(-\frac{(y-z+\mu t)^2}{2t}\Big)\Big) dy dz$$

1.13.8 $v < t$

$$\mathbf{P}_x\big(\check{H}_\mu(t) \in dv,\ \sup_{0\le s\le t} W_s^{(\mu)} \in dy, W_t^{(\mu)} \in dz\big) \qquad x \vee z < y$$

$$= \frac{(y-x)(y-z)}{\pi(v(t-v))^{3/2}} \exp\Big(-\frac{(y-x-\mu v)^2}{2v} - \frac{(y-z+\mu(t-v))^2}{2(t-v)}\Big) dy\, dv dz$$

2 $W_s^{(\mu)} = \mu s + W_s$ $\tau \sim \operatorname{Exp}(\lambda)$, independent of $W^{(\mu)}$

1.14.1 ($y < x$)

$$\mathbf{E}_x\{e^{-\gamma \hat{H}_\mu(\tau)}; \inf_{0\le s\le\tau} W_s^{(\mu)} \in dy\} = \frac{2\lambda}{\sqrt{2\lambda+\mu^2}-\mu} e^{-(x-y)(\sqrt{2\lambda+2\gamma+\mu^2}+\mu)} dy$$

1.14.2 ($y < x$)

$$\mathbf{P}_x\big(\hat{H}_\mu(\tau) \in dv, \inf_{0\le s\le\tau} W_s^{(\mu)} \in dy\big)$$

$$= \frac{2\lambda(x-y)}{\sqrt{2\pi}(\sqrt{2\lambda+\mu^2}-\mu)v^{3/2}} \exp\Big(\mu(y-x) - (\lambda+\frac{\mu^2}{2})v - \frac{(x-y)^2}{2v}\Big) dv dy$$

1.14.3 ($y < x$)

$$\mathbf{E}_x\{e^{-\gamma \hat{H}_\mu(t)}; \inf_{0\le s\le t} W_s^{(\mu)} \in dy\}$$

$$= \Big(\frac{x-y}{\sqrt{\pi}t^{3/2}} e^{\mu(y-x)-\gamma t-\mu^2 t/2} \exp\Big(-\frac{(x-y)^2}{2t}\Big)\Big) * \Big(\frac{e^{-\mu^2 t/2}}{\sqrt{t}} + \frac{\mu}{\sqrt{2}} \operatorname{Erfc}\Big(-\frac{\mu\sqrt{t}}{\sqrt{2}}\Big)\Big) dy$$

(1) ($y < x$)

$$\mathbf{E}_x\{e^{-\gamma \hat{H}_\mu(\infty)}; \inf_{0\le s<\infty} W_s^{(\mu)} \in dy\} = (|\mu|+\mu) e^{-(x-y)(\sqrt{2\gamma+\mu^2}+\mu)} dy$$

1.14.4 ($y < x$) $0 \le v \le t$

$$\mathbf{P}_x\big(\hat{H}_\mu(t) \in dv, \inf_{0\le s\le t} W_s^{(\mu)} \in dy\big)$$

$$= \frac{x-y}{\sqrt{\pi}v^{3/2}} \exp\Big(-\frac{(x-y+\mu v)^2}{2v}\Big)\Big(\frac{e^{-\mu^2(t-v)/2}}{\sqrt{\pi(t-v)}} + \frac{\mu}{\sqrt{2}} \operatorname{Erfc}\Big(-\frac{\mu\sqrt{t-v}}{\sqrt{2}}\Big)\Big) dv dy$$

(1) ($y < x$)

$$\mathbf{P}_x\big(\hat{H}_\mu(\infty) \in dv, \inf_{0\le s<\infty} W_s^{(\mu)} \in dy\big) = \frac{(|\mu|+\mu)(x-y)}{\sqrt{2\pi}v^{3/2}} \exp\Big(-\frac{(x-y+\mu v)^2}{2v}\Big) dv dy$$

1.14.5 $y < x \wedge z$

$$\mathbf{E}_x\{e^{-\gamma \hat{H}_\mu(\tau)}; \inf_{0\le s\le\tau} W_s^{(\mu)} \in dy, W_\tau^{(\mu)} \in dz\}$$

$$= 2\lambda e^{\mu(z-x)} e^{(y-x)\sqrt{2\lambda+2\gamma+\mu^2}} e^{(y-z)\sqrt{2\lambda+\mu^2}} dy dz$$

1.14.6 $y < x \wedge z$

$$\mathbf{P}_x\big(\hat{H}_\mu(\tau) \in dv, \inf_{0\le s\le\tau} W_s^{(\mu)} \in dy, W_\tau^{(\mu)} \in dz\big)$$

$$= \frac{2\lambda(x-y)}{\sqrt{2\pi}v^{3/2}} \exp\Big(\mu(z-x) - \frac{(x-y)^2}{2v} - (\lambda+\frac{\mu^2}{2})v - (z-y)\sqrt{2\lambda+\mu^2}\Big) dv dy dz$$

2 $W_s^{(\mu)} = \mu s + W_s$ $\tau \sim \operatorname{Exp}(\lambda)$, independent of $W^{(\mu)}$

1.14.7 $\mathbf{E}_x\{e^{-\gamma \hat{H}_\mu(t)}; \inf_{0\le s\le t} W_s^{(\mu)} \in dy,\ W_t^{(\mu)} \in dz\}$ $y < x \wedge z$

$$= \Big(\frac{x-y}{\pi t^{3/2}} \exp\Big(-\gamma t - \frac{(x-y+\mu t)^2}{2t}\Big)\Big) * \Big(\frac{z-y}{t^{3/2}} \exp\Big(-\frac{(z-y-\mu t)^2}{2t}\Big)\Big) dydz$$

1.14.8 $v < t$ $\mathbf{P}_x\big(\hat{H}_\mu(t) \in dv, \inf_{0\le s\le t} W_s^{(\mu)} \in dy, W_t^{(\mu)} \in dz\big)$ $y < x \wedge z$

$$= \frac{(x-y)(z-y)}{\pi(v(t-v))^{3/2}} \exp\Big(-\frac{(x-y+\mu v)^2}{2v} - \frac{(z-y-\mu(t-v))^2}{2(t-v)}\Big) dy\, dvdz$$

1.15.2 $\mathbf{P}_x\big(a < \inf_{0\le s\le \tau} W_s^{(\mu)};\ \sup_{0\le s\le \tau} W_s^{(\mu)} < b\big)$

$$= 1 - \frac{e^{\mu(a-x)} \operatorname{sh}\big((b-x)\sqrt{2\lambda+\mu^2}\big) + e^{\mu(b-x)} \operatorname{sh}\big((x-a)\sqrt{2\lambda+\mu^2}\big)}{\operatorname{sh}\big((b-a)\sqrt{2\lambda+\mu^2}\big)}$$

(1) $\mathbf{P}_x\big(\sup_{0\le s\le \tau} W_s^{(\mu)} - \inf_{0\le s\le \tau} W_s^{(\mu)} < y\big)$

$$= 1 - \frac{2\sqrt{2\lambda+\mu^2}}{\operatorname{sh}^2(y\sqrt{2\lambda+\mu^2})} \Big(\frac{\operatorname{sh}^2(y(\sqrt{2\lambda+\mu^2}+\mu)/2)}{\sqrt{2\lambda+\mu^2}+\mu} + \frac{\operatorname{sh}^2(y(\sqrt{2\lambda+\mu^2}-\mu)/2)}{\sqrt{2\lambda+\mu^2}-\mu}\Big)$$

1.15.4 $\mathbf{P}_x\big(a < \inf_{0\le s\le t} W_s^{(\mu)};\ \sup_{0\le s\le t} W_s^{(\mu)} < b\big)$

$$= \frac{e^{-\mu^2 t/2}}{\sqrt{2\pi t}} \sum_{k=-\infty}^{\infty} \int_a^b e^{\mu(z-x)} \big(e^{-(z-x+2k(b-a))^2/2t} - e^{-(z+x-2a+2k(b-a))^2/2t}\big) dz$$

(1) $\mathbf{P}_x\big(\sup_{0\le s\le t} W_s^{(\mu)} - \inf_{0\le s\le t} W_s^{(\mu)} < y\big)$

$$= \sum_{k=-\infty}^{\infty} e^{-2k\mu y}\Big[\big(\tfrac{1}{2} - k\mu y\big)\Big(\operatorname{Erf}\Big(\frac{(2k+1)y}{\sqrt{2t}} - \frac{\mu\sqrt{t}}{\sqrt{2}}\Big) - \operatorname{Erf}\Big(\frac{(2k-1)y}{\sqrt{2t}} - \frac{\mu\sqrt{t}}{\sqrt{2}}\Big)\Big)$$

$$+k\big(2 - 2k\mu y + \mu^2 t\big)\Big\{\operatorname{Erf}\Big(\frac{(2k+1)y}{\sqrt{2t}} - \frac{\mu\sqrt{t}}{\sqrt{2}}\Big) + \operatorname{Erf}\Big(\frac{(2k-1)y}{\sqrt{2t}} - \frac{\mu\sqrt{t}}{\sqrt{2}}\Big)$$

2 $W_s^{(\mu)} = \mu s + W_s$ $\tau \sim \mathrm{Exp}(\lambda)$, independent of $W^{(\mu)}$

$$-2\,\mathrm{Erf}\Big(\frac{2ky}{\sqrt{2t}} - \frac{\mu\sqrt{t}}{\sqrt{2}}\Big)\Big\} + \frac{k\mu\sqrt{2t}}{\sqrt{\pi}}\Big\{2\exp\Big(-\Big(\frac{2ky}{\sqrt{2t}} - \frac{\mu\sqrt{t}}{\sqrt{2}}\Big)^2\Big)$$

$$-\exp\Big(-\Big(\frac{(2k+1)y}{\sqrt{2t}} - \frac{\mu\sqrt{t}}{\sqrt{2}}\Big)^2\Big) - \exp\Big(-\Big(\frac{(2k-1)y}{\sqrt{2t}} - \frac{\mu\sqrt{t}}{\sqrt{2}}\Big)^2\Big)\Big\}\Big]$$

1.15.6 $\mathbf{P}_x\big(a < \inf_{0\le s\le\tau} W_s^{(\mu)};\ \sup_{0\le s\le\tau} W_s^{(\mu)} < b, W_\tau^{(\mu)} \in dz\big)$

$$= \frac{\lambda\big(\mathrm{ch}\big((b-a-|z-x|)\sqrt{2\lambda+\mu^2}\big) - \mathrm{ch}\big((b+a-z-x)\sqrt{2\lambda+\mu^2}\big)\big)}{e^{\mu(x-z)}\sqrt{2\lambda+\mu^2}\,\mathrm{sh}\big((b-a)\sqrt{2\lambda+\mu^2}\big)}dz, \qquad a < x\wedge z,\ x\vee z < b$$

(1) $\mathbf{P}_x\big(\sup_{0\le s\le\tau} W_s^{(\mu)} - \inf_{0\le s\le\tau} W_s^{(\mu)} < y, W_\tau^{(\mu)} \in dz\big)$ for $|z-x| < y$

$$= \frac{\lambda e^{\mu(z-x)}}{\mathrm{sh}^2(y\sqrt{2\lambda+\mu^2})}\Big(\mathrm{sh}\big((y-|z-x|)\sqrt{2\lambda+\mu^2}\big)\frac{\mathrm{ch}(y\sqrt{2\lambda+\mu^2})}{\sqrt{2\lambda+\mu^2}} - (y-|z-x|)\,\mathrm{ch}\big(|z-x|\sqrt{2\lambda+\mu^2}\big)\Big)dz$$

1.15.8 $\mathbf{P}_x\big(a < \inf_{0\le s\le t} W_s^{(\mu)};\ \sup_{0\le s\le t} W_s^{(\mu)} < b\,, W_t^{(\mu)} \in dz\big)$

for $a < x\wedge z \le x\vee z < b$

$$= \frac{1}{\sqrt{2\pi t}}e^{\mu(z-x)-\mu^2t/2}\sum_{k=-\infty}^{\infty}\big(e^{-(z-x+2k(b-a))^2/2t} - e^{-(z+x-2a+2k(b-a))^2/2t}\big)dz$$

(1) $\mathbf{P}_x\big(\sup_{0\le s\le t} W_s^{(\mu)} - \inf_{0\le s\le t} W_s^{(\mu)} < y, W_t^{(\mu)} \in dz\big)$ for $|z-x| < y$

$$= \frac{e^{\mu(z-x)-\mu^2t/2}}{\sqrt{2\pi t}}\sum_{k=-\infty}^{\infty}\Big(2k+1 - \frac{2k(y-|z-x|)(|z-x|+2ky)}{t}\Big)e^{-(|z-x|+2ky)^2/2t}dz$$

1.16.1 $r < b$ $\mathbf{E}_x\big\{e^{-\gamma\ell(\tau,r)};\ \sup_{0\le s\le\tau} W_s^{(\mu)} < b\big\} = 1 - e^{(b-x)(\mu-\sqrt{2\lambda+\mu^2})}$

$$-\frac{\gamma\big(e^{\mu(r-x)} - e^{\mu(b-x)-(b-r)\sqrt{2\lambda+\mu^2}}\big)}{\sqrt{2\lambda+\mu^2}+\gamma\big(1-e^{-2(b-r)\sqrt{2\lambda+\mu^2}}\big)}\big(e^{-|x-r|\sqrt{2\lambda+\mu^2}} - e^{-(2b-r-x)\sqrt{2\lambda+\mu^2}}\big)$$

2 $W_s^{(\mu)} = \mu s + W_s$ $\tau \sim \operatorname{Exp}(\lambda)$, independent of $W^{(\mu)}$

1.16.2 $r < b$ $0 < y$

$$\mathbf{P}_x\Big(\ell(\tau,r)\in dy,\ \sup_{0\le s\le\tau} W_s^{(\mu)} < b\Big) = \exp\Big(-\frac{y\sqrt{2\lambda+\mu^2}}{1-e^{-2(b-r)\sqrt{2\lambda+\mu^2}}}\Big)$$

$$\times\frac{\sqrt{2\lambda+\mu^2}\,(e^{\mu(r-x)} - e^{\mu(b-x)-(b-r)\sqrt{2\lambda+\mu^2}})(e^{-|x-r|\sqrt{2\lambda+\mu^2}} - e^{-(2b-r-x)\sqrt{2\lambda+\mu^2}})}{(1-e^{-2(b-r)\sqrt{2\lambda+\mu^2}})^2}dy$$

(1)

$$\mathbf{P}_x\Big(\ell(\tau,r)=0,\ \sup_{0\le s\le\tau} W_s^{(\mu)} < b\Big) = 1 - e^{(b-x)(\mu-\sqrt{2\lambda+\mu^2})}$$

$$-\frac{(e^{\mu(r-x)} - e^{\mu(b-x)-(b-r)\sqrt{2\lambda+\mu^2}})(e^{-|x-r|\sqrt{2\lambda+\mu^2}} - e^{-(2b-r-x)\sqrt{2\lambda+\mu^2}})}{1-e^{-2(b-r)\sqrt{2\lambda+\mu^2}}}$$

1.16.3 (1) $\mu < 0$

$$\mathbf{E}_x\Big\{e^{-\gamma\ell(\infty,r)};\ \sup_{0\le s<\infty} W_s^{(\mu)} < b\Big\} \qquad r\vee x\le b$$

$$= 1 - e^{\mu(|x-r|+r-x)} - \frac{\mu(e^{\mu(|x-r|+r-x)} - e^{2\mu(b-x)})}{\gamma(1-e^{2\mu(b-r)})-\mu}$$

1.16.4 (1) $\mu < 0$

$$\mathbf{P}_x\Big(\ell(\infty,r)\in dy,\ \sup_{0\le s<\infty} W_s^{(\mu)} < b\Big) \qquad r\vee x\le b$$

$$= \frac{|\mu|(e^{\mu(|x-r|+r-x)} - e^{2\mu(b-x)})}{1-e^{2\mu(b-r)}}\exp\Big(-\frac{y|\mu|}{1-e^{2\mu(b-r)}}\Big)dy$$

(2)

$$\mathbf{P}_x\Big(\ell(\infty,r)=0,\ \sup_{0\le s<\infty} W_s^{(\mu)} < b\Big) = 1 - e^{\mu(|r-x|+r-x)}$$

1.16.5

$$\mathbf{E}_x\Big\{e^{-\gamma\ell(\tau,r)};\ \sup_{0\le s\le\tau} W_s^{(\mu)} < b, W_\tau^{(\mu)}\in dz\Big\}$$

$$= \frac{\lambda}{\sqrt{2\lambda+\mu^2}}e^{\mu(z-x)}\Big\{e^{-|z-x|\sqrt{2\lambda+\mu^2}} - e^{-(2b-z-x)\sqrt{2\lambda+\mu^2}}$$

$$-\frac{\gamma(e^{-|x-r|\sqrt{2\lambda+\mu^2}} - e^{-(2b-x-r)\sqrt{2\lambda+\mu^2}})(e^{-|z-r|\sqrt{2\lambda+\mu^2}} - e^{-(2b-z-r)\sqrt{2\lambda+\mu^2}})}{\sqrt{2\lambda+\mu^2}+\gamma(1-e^{-2(b-r)\sqrt{2\lambda+\mu^2}})}\Big\}dz$$

1.16.6

$$\mathbf{P}_x\Big(\ell(\tau,r)\in dy,\ \sup_{0\le s\le\tau} W_s^{(\mu)} < b, W_\tau^{(\mu)}\in dz\Big)$$

2 $W_s^{(\mu)} = \mu s + W_s$ $\tau \sim \operatorname{Exp}(\lambda)$, independent of $W^{(\mu)}$

$$= \lambda \exp\Big(\mu(z-x) - \frac{y\sqrt{2\lambda+\mu^2}}{1-e^{-2(b-r)\sqrt{2\lambda+\mu^2}}}\Big)$$

$$\times \frac{(e^{-|x-r|\sqrt{2\lambda+\mu^2}} - e^{-(2b-r-x)\sqrt{2\lambda+\mu^2}})(e^{-|z-r|\sqrt{2\lambda+\mu^2}} - e^{-(2b-z-r)\sqrt{2\lambda+\mu^2}})}{(1-e^{-2(b-r)\sqrt{2\lambda+\mu^2}})^2} dydz$$

1.16.8 $\mathbf{P}_x\Big(\frac{1}{2}\ell(t,r) \in dy, \sup_{0\le s\le t} W_s^{(\mu)} < b, W_t^{(\mu)} \in dz\Big)$

$$= e^{\mu(z-x)-\mu^2 t/2} F_x^{(16)}(t,y,z)dydz, \qquad \text{for } F_x^{(16)}(t,y,z) \text{ see 1.1.16.8}$$

(1) $\mathbf{P}_x\big(\ell(t,r) = 0, \sup_{0\le s\le t} W_s^{(\mu)} < b, W_t^{(\mu)} \in dz\big)$ $r < x \wedge z,\ x \vee z < b$

$$= e^{\mu(z-x)-\mu^2 t/2}\frac{1}{\sqrt{2\pi t}} \sum_{k=-\infty}^{\infty} \big(e^{-(z-x+2k(b-r))^2/2t} - e^{-(z+x-2r+2k(b-r))^2/2t}\big)dz$$

1.17.1 $a < r$ $\mathbf{E}_x\big\{e^{-\gamma\ell(\tau,r)}, a < \inf_{0\le s\le\tau} W_s^{(\mu)}\big\} = 1 - e^{(a-x)(\mu+\sqrt{2\lambda+\mu^2})}$

$$- \frac{\gamma(e^{\mu(r-x)} - e^{\mu(a-x)-(r-a)\sqrt{2\lambda+\mu^2}})}{\sqrt{2\lambda+\mu^2} + \gamma(1-e^{-2(r-a)\sqrt{2\lambda+\mu^2}})}\big(e^{-|x-r|\sqrt{2\lambda+\mu^2}} - e^{-(x+r-2a)\sqrt{2\lambda+\mu^2}}\big)$$

1.17.2 $a < r$ $\mathbf{P}_x\big(\ell(\tau,r) \in dy, a < \inf_{0\le s\le\tau} W_s^{(\mu)}\big) = \exp\Big(-\frac{y\sqrt{2\lambda+\mu^2}}{1-e^{-2(r-a)\sqrt{2\lambda+\mu^2}}}\Big)$

$$\times \frac{\sqrt{2\lambda+\mu^2}(e^{\mu(r-x)} - e^{\mu(a-x)-(r-a)\sqrt{2\lambda+\mu^2}})(e^{-|x-r|\sqrt{2\lambda+\mu^2}} - e^{-(x+r-2a)\sqrt{2\lambda+\mu^2}})}{(1-e^{-2(r-a)\sqrt{2\lambda+\mu^2}})^2} dy$$

(1) $a < r$ $\mathbf{P}_x\big(\ell(\tau,r) = 0, a < \inf_{0\le s\le\tau} W_s^{(\mu)}\big) = 1 - e^{(a-x)(\mu+\sqrt{2\lambda+\mu^2})}$

$$- \frac{(e^{\mu(r-x)} - e^{\mu(a-x)-(r-a)\sqrt{2\lambda+\mu^2}})}{1-e^{-2(r-a)\sqrt{2\lambda+\mu^2}}}\big(e^{-|x-r|\sqrt{2\lambda+\mu^2}} - e^{-(x+r-2a)\sqrt{2\lambda+\mu^2}}\big)$$

1.17.3 (1) $\mathbf{E}_x\big\{e^{-\gamma\ell(\infty,r)}; a < \inf_{0\le s<\infty} W_s^{(\mu)}\big\}$ $a \le r \wedge x$

$\mu > 0$
$$= 1 - e^{-\mu(|x-r|+x-r)} + \frac{\mu(e^{-\mu(|x-r|+x-r)} - e^{-2\mu(x-a)})}{\gamma(1-e^{-2\mu(r-a)}) + \mu}$$

2 $W_s^{(\mu)} = \mu s + W_s$ $\tau \sim \operatorname{Exp}(\lambda)$, independent of $W^{(\mu)}$

1.17.4 (1) $\mathbf{P}_x\big(\ell(\infty,r)\in dy, a<\inf_{0\le s<\infty} W_s^{(\mu)}\big)$ $a\le r\wedge x$

$\mu>0$ $$= \frac{\mu(e^{-\mu(|x-r|+x-r)}-e^{-2\mu(x-a)})}{1-e^{-2\mu(r-a)}}\exp\Big(-\frac{y\mu}{1-e^{-2\mu(r-a)}}\Big)dy$$

(2) $a<r$ $$\mathbf{P}_x\big(\ell(\infty,r)=0, a<\inf_{0\le s<\infty} W_s^{(\mu)}\big)=1-e^{-\mu(|x-r|+x-r)}$$

1.17.5 $$\mathbf{E}_x\big\{e^{-\gamma\ell(\tau,r)}; a<\inf_{0\le s\le\tau} W_s^{(\mu)}, W_\tau^{(\mu)}\in dz\big\}$$

$$= \frac{\lambda}{\sqrt{2\lambda+\mu^2}}e^{\mu(z-x)}\Big\{e^{-|z-x|\sqrt{2\lambda+\mu^2}}-e^{-(z+x-2a)\sqrt{2\lambda+\mu^2}}$$

$$-\frac{\gamma(e^{-|x-r|\sqrt{2\lambda+\mu^2}}-e^{-(x+r-2a)\sqrt{2\lambda+\mu^2}})(e^{-|z-r|\sqrt{2\lambda+\mu^2}}-e^{-(z+r-2a)\sqrt{2\lambda+\mu^2}})}{\sqrt{2\lambda+\mu^2}+\gamma(1-e^{-2(r-a)\sqrt{2\lambda+\mu^2}})}\Big\}dz$$

1.17.6 $$\mathbf{P}_x\big(\ell(\tau,r)\in dy, a<\inf_{0\le s\le\tau} W_s^{(\mu)}, W_\tau^{(\mu)}\in dz\big)$$

$$= \lambda\exp\Big(\mu(z-x)-\frac{y\sqrt{2\lambda+\mu^2}}{1-e^{-2(r-a)\sqrt{2\lambda+\mu^2}}}\Big)$$

$$\times\frac{(e^{-|x-r|\sqrt{2\lambda+\mu^2}}-e^{-(x+r-2a)\sqrt{2\lambda+\mu^2}})(e^{-|z-r|\sqrt{2\lambda+\mu^2}}-e^{-(z+r-2a)\sqrt{2\lambda+\mu^2}})}{(1-e^{-2(r-a)\sqrt{2\lambda+\mu^2}})^2}dydz$$

1.17.8 $$\mathbf{P}_x\Big(\frac{1}{2}\ell(t,r)\in dy, a<\inf_{0\le s\le t} W_s^{(\mu)}, W_t^{(\mu)}\in dz\Big)$$

$$= e^{\mu(z-x)-\mu^2t/2}F_{a,x}^{(17)}(t,y,z)dydz,$$ for $F_{a,x}^{(17)}(t,y,z)$ see 1.1.17.8

(1) $\mathbf{P}_x\big(\ell(t,r)=0, a<\inf_{0\le s\le t} W_s^{(\mu)}, W_t^{(\mu)}\in dz\big)$ $a<x\wedge z,\ x\vee z<r$

$$= e^{\mu(z-x)-\mu^2t/2}\frac{1}{\sqrt{2\pi t}}\sum_{k=-\infty}^{\infty}\big(e^{-(z-x+2k(r-a))^2/2t}-e^{-(z+x-2a+2k(r-a))^2/2t}\big)dz$$

1.18.1 $$\mathbf{E}_x e^{-\gamma\ell(\tau,r)-\eta\ell(\tau,u)} = 1-\frac{\gamma\big(e^{\mu(r-x)}(\Upsilon_0+\eta)-\eta e^{\mu(u-x)-|u-r|\Upsilon_0}\big)e^{-|r-x|\Upsilon_0}}{(\Upsilon_0+\gamma)(\Upsilon_0+\eta)-\gamma\eta e^{-2|u-r|\Upsilon_0}}$$

2 $W_s^{(\mu)} = \mu s + W_s$ $\tau \sim \text{Exp}(\lambda)$, independent of $W^{(\mu)}$

$$-\frac{\eta\big(e^{\mu(u-x)}(\Upsilon_0+\gamma) - \gamma e^{\mu(r-x)-|u-r|\Upsilon_0}\big)e^{-|u-x|\Upsilon_0}}{(\Upsilon_0+\gamma)(\Upsilon_0+\eta) - \gamma\eta e^{-2|u-r|\Upsilon_0}}$$

1.18.3 (1)

$$\mathbf{E}_x e^{-\gamma\ell(\infty,r)-\eta\ell(\infty,u)} = 1 - \frac{\gamma\big(e^{\mu(r-x)}(|\mu|+\eta) - \eta e^{\mu(u-x)-|u-r||\mu|}\big)e^{-|r-x||\mu|}}{(|\mu|+\gamma)(|\mu|+\eta) - \gamma\eta e^{-2|u-r||\mu|}}$$

$$-\frac{\eta\big(e^{\mu(u-x)}(|\mu|+\gamma) - \gamma e^{\mu(r-x)-|u-r||\mu|}\big)e^{-|u-x||\mu|}}{(|\mu|+\gamma)(|\mu|+\eta) - \gamma\eta e^{-2|u-r||\mu|}}$$

1.18.4 (1)

$$\mathbf{P}_x\big(\ell(\infty,r)\in dy,\ \ell(\infty,u)\in dg\big)$$

for $x \le r < u,\ \mu > 0$ or $u < r \le x,\ \mu < 0$

$$= \frac{\mu^2 e^{|u-r||\mu|}}{2\,\text{sh}(|u-r||\mu|)} \exp\Big(-\frac{|\mu|(y+g)}{1-e^{-2|u-r||\mu|}}\Big) I_0\Big(\frac{|\mu|\sqrt{yg}}{\text{sh}(|u-r||\mu|)}\Big) dydg$$

for $r < u \le x, \mu > 0$ or $x \le u < r, \mu < 0$

$$= \frac{\mu^2 e^{-2|u-x||\mu|}\sqrt{g}}{2\,\text{sh}(|u-r||\mu|)\sqrt{y}} \exp\Big(-\frac{|\mu|(y+g)}{1-e^{-2|u-r||\mu|}}\Big) I_1\Big(\frac{|\mu|\sqrt{yg}}{\text{sh}(|u-r||\mu|)}\Big) dydg$$

for $r < x < u, \mu > 0$ or $u < x < r, \mu < 0$

$$= \frac{\mu^2 e^{|u-x||\mu|}}{2\,\text{sh}(|u-r||\mu|)} \exp\Big(-\frac{|\mu|(y+g)}{1-e^{-2|u-r||\mu|}}\Big) \Big[\text{sh}(|u-x||\mu|) I_0\Big(\frac{|\mu|\sqrt{yg}}{\text{sh}(|u-r||\mu|)}\Big)$$

$$+ \text{sh}(|x-r||\mu|)\sqrt{g/y}\, I_1\Big(\frac{|\mu|\sqrt{yg}}{\text{sh}(|u-r||\mu|)}\Big)\Big] dydg$$

(2)

$$\mathbf{P}_x\big(\ell(\infty,r)=0,\ \ell(\infty,u)\in dg\big)$$

for $r < x < u, \mu > 0$ or $u < x < r, \mu < 0$

$$= |\mu| e^{|u-x||\mu|} \frac{\text{sh}(|x-r||\mu|)}{\text{sh}(|u-r||\mu|)} \exp\Big(-\frac{g|\mu|}{1-e^{-2|\mu||u-r|}}\Big) dg$$

for $r < u \le x, \mu > 0$ or $x \le u < r, \mu < 0$

$$= |\mu| e^{-2|\mu||u-x|} \exp\Big(-\frac{g|\mu|}{1-e^{-2|\mu||u-r|}}\Big) dg$$

2 $W_s^{(\mu)} = \mu s + W_s$ $\tau \sim \text{Exp}(\lambda)$, independent of $W^{(\mu)}$

for $r < u \le x,\ \mu > 0$ or $x \le u < r,\ \mu < 0$

(3) $\mathbf{P}_x\big(\ell(\infty, r) = 0,\ \ell(\infty, u) = 0\big) = 1 - e^{-2|\mu||u-x|}$

1.18.5 $\mathbf{E}_x\{e^{-\gamma\ell(\tau,r)-\eta\ell(\tau,u)} \mid W_\tau^{(\mu)} = z\}$

$$= 1 - \frac{\gamma\big((\Upsilon_0+\eta)e^{-|r-x|\Upsilon_0} - \eta e^{-(|u-x|+|u-r|)\Upsilon_0}\big)e^{-(|z-r|-|z-x|)\Upsilon_0}}{(\Upsilon_0+\gamma)(\Upsilon_0+\eta) - \gamma\eta e^{-2|u-r|\Upsilon_0}}$$
$$- \frac{\eta\big((\Upsilon_0+\gamma)e^{-|u-x|\Upsilon_0} - \gamma e^{-(|r-x|+|u-r|)\Upsilon_0}\big)e^{-(|z-u|-|z-x|)\Upsilon_0}}{(\Upsilon_0+\gamma)(\Upsilon_0+\eta) - \gamma\eta e^{-2|u-r|\Upsilon_0}}$$

(1) $\mathbf{E}_x\{e^{-\gamma\ell(\tau,r)} \mid \ell(\tau,u) = g,\ W_\tau^{(\mu)} = z\}$

for $r < u \le x \wedge z$ or $x \vee z \le u < r$

$$= \exp\Big(-\frac{g\gamma\Upsilon_0 e^{-2|r-u|\Upsilon_0}}{\Upsilon_0 + \gamma(1 - e^{-2|r-u|\Upsilon_0})}\Big),$$

for $x \wedge z \le r < u \le x \vee z$ or $x \wedge z \le u < r \le x \vee z$

$$= \frac{\Upsilon_0}{\Upsilon_0 + \gamma(1 - e^{-2|r-u|\Upsilon_0})}\exp\Big(-\frac{g\gamma\Upsilon_0 e^{-2|r-u|\Upsilon_0}}{\Upsilon_0 + \gamma(1 - e^{-2|r-u|\Upsilon_0})}\Big)$$

1.18.6 $\mathbf{P}_x\big(\ell(\tau,r) \in dy \mid \ell(\tau,u) = g,\ W_\tau^{(\mu)} = z\big)$

for $r < u \le x \wedge z$ or $x \vee z \le u < r$

$$= \frac{\Upsilon_0\sqrt{g}}{2\sqrt{y}\,\text{sh}(|r-u|\Upsilon_0)}\exp\Big(-\frac{\Upsilon_0(ye^{|r-u|\Upsilon_0} + ge^{-|r-u|\Upsilon_0})}{2\,\text{sh}(|r-u|\Upsilon_0)}\Big) I_1\Big(\frac{\Upsilon_0\sqrt{gy}}{\text{sh}(|r-u|\Upsilon_0)}\Big)dy$$

for $x \wedge z \le r < u \le x \vee z$ or $x \wedge z \le u < r \le x \vee z$

$$= \frac{\Upsilon_0 e^{|r-u|\Upsilon_0}}{2\,\text{sh}(|r-u|\Upsilon_0)}\exp\Big(-\frac{\Upsilon_0(ye^{|r-u|\Upsilon_0} + ge^{-|r-u|\Upsilon_0})}{2\,\text{sh}(|r-u|\Upsilon_0)}\Big) I_0\Big(\frac{\Upsilon_0\sqrt{gy}}{\text{sh}(|r-u|\Upsilon_0)}\Big)dy$$

(1) $\mathbf{P}_x\big(\ell(\tau,r) = 0 \mid \ell(\tau,u) = g,\ W_\tau^{(\mu)} = z\big) = \exp\Big(-\frac{g\Upsilon_0}{e^{2|r-u|\Upsilon_0} - 1}\Big)$

for $r < u \le x \wedge z$ or $x \vee z \le u < r$

$\Upsilon_0 := \sqrt{2\lambda + \mu^2}$

2 $W_s^{(\mu)} = \mu s + W_s$ $\tau \sim \operatorname{Exp}(\lambda)$, independent of $W^{(\mu)}$

1.18.8 $\mathbf{P}_x\big(\ell(t,r) \in dy, \ell(t,u) \in dg, W_t^{(\mu)} \in dz\big) = e^{\mu(z-x)-\mu^2 t/2} F_{x,t}^{(18)}(y,g,z)\,dy\,dg\,dz$

for $F_{x,t}^{(18)}(y,g,z)$ see 1.1.18.8

1.19.2, $0 < x$:

$$\mathbf{P}_x\Big(\int_0^\tau \frac{ds}{W_s^{(\mu)}} \in dy,\ 0 < \inf_{0 \le s \le \tau} W_s^{(\mu)}\Big) = \frac{(2\lambda+\mu^2)}{\operatorname{sh}^2(y\sqrt{2\lambda+\mu^2}/2)}$$

$$\times \frac{\lambda x}{(\sqrt{2\lambda+\mu^2}\operatorname{cth}(y\sqrt{2\lambda+\mu^2}/2)-\mu)^2}\exp\Big(-\frac{2\lambda x}{\sqrt{2\lambda+\mu^2}\operatorname{cth}(y\sqrt{2\lambda+\mu^2}/2)-\mu}\Big)dy$$

1.19.5, $0 < x$:

$$\mathbf{E}_x\Big\{\exp\Big(-\gamma\int_0^\tau \frac{ds}{W_s^{(\mu)}}\Big);\ 0 < \inf_{0 \le s \le \tau} W_s^{(\mu)}, W_\tau^{(\mu)} \in dz\Big\}$$

$$= \lambda e^{\mu(z-x)} Q_{\lambda+\mu^2/2}^{(19)}(z)\,dz, \qquad \text{for } Q_\lambda^{(19)}(z) \text{ see 1.1.19.5}$$

1.19.6, $0 < x$:

$$\mathbf{P}_x\Big(\int_0^\tau \frac{ds}{W_s^{(\mu)}} \in dy,\ 0 < \inf_{0 \le s \le \tau} W_s^{(\mu)},\ W_\tau^{(\mu)} \in dz\Big) = \frac{\lambda\sqrt{(2\lambda+\mu^2)xz}}{\operatorname{sh}(y\sqrt{2\lambda+\mu^2}/2)}$$

$$\times \exp\Big(\mu(z-x) - \frac{(x+z)\sqrt{2\lambda+\mu^2}\operatorname{ch}(y\sqrt{2\lambda+\mu^2}/2)}{\operatorname{sh}(y\sqrt{2\lambda+\mu^2}/2)}\Big) I_1\Big(\frac{2\sqrt{(2\lambda+\mu^2)xz}}{\operatorname{sh}(y\sqrt{2\lambda+\mu^2}/2)}\Big)dy\,dz$$

1.19.8, $0 < x$:

$$\mathbf{P}_x\Big(\int_0^t \frac{ds}{W_s^{(\mu)}} \in dy,\ 0 < \inf_{0 \le s \le t} W_s^{(\mu)},\ W_t^{(\mu)} \in dz\Big)$$

$$= e^{\mu(z-x)-\mu^2 t/2}\sqrt{xz}\,\operatorname{is}_t(1, y/2, 0, x+z, \sqrt{xz})\,dy\,dz$$

1.20.5, $0 < x$:

$$\mathbf{E}_x\Big\{\exp\Big(-\frac{\gamma^2}{2}\int_0^\tau \frac{ds}{(W_s^{(\mu)})^2}\Big);\ 0 < \inf_{0 \le s \le \tau} W_s^{(\mu)},\ W_\tau^{(\mu)} \in dz\Big\}$$

$x \le z$: $$= 2\lambda e^{\mu(z-x)}\sqrt{xz}\, I_{\sqrt{\gamma^2+1/4}}\big(x\sqrt{2\lambda+\mu^2}\big) K_{\sqrt{\gamma^2+1/4}}\big(z\sqrt{2\lambda+\mu^2}\big)dz$$

$z \le x$: $$= 2\lambda e^{\mu(z-x)}\sqrt{xz}\, K_{\sqrt{\gamma^2+1/4}}\big(x\sqrt{2\lambda+\mu^2}\big) I_{\sqrt{\gamma^2+1/4}}\big(z\sqrt{2\lambda+\mu^2}\big)dz$$

1.20.6, $0 < x$:

$$\mathbf{P}_x\Big(\int_0^\tau \frac{ds}{(W_s^{(\mu)})^2} \in dy,\ 0 < \inf_{0 \le s \le \tau} W_s^{(\mu)},\ W_\tau^{(\mu)} \in dz\Big)$$

$$= \lambda\sqrt{xz}\, e^{\mu(z-x)-y/8}\operatorname{ki}_{y/2}(x\sqrt{2\lambda+\mu^2}, z\sqrt{2\lambda+\mu^2})\,dy\,dz$$

2 $W_s^{(\mu)} = \mu s + W_s$ $\tau \sim \text{Exp}(\lambda)$, independent of $W^{(\mu)}$

1.20.7
$0 < x$

$$\mathbf{E}_x\Big\{\exp\Big(-\frac{\gamma^2}{2}\int_0^t \frac{ds}{(W_s^{(\mu)})^2}\Big);\ 0 < \inf_{0\le s\le t} W_s^{(\mu)},\ W_t^{(\mu)} \in dz\Big\}$$

$$= \frac{\sqrt{xz}}{t} e^{\mu(z-x)-\mu^2 t/2} e^{-(x^2+z^2)/2t} I_{\sqrt{\gamma^2+1/4}}\Big(\frac{xz}{t}\Big)dz$$

1.20.8
$0 < x$

$$\mathbf{P}_x\Big(\int_0^t \frac{ds}{(W_s^{(\mu)})^2} \in dy,\ 0 < \inf_{0\le s\le t} W_s^{(\mu)},\ W_t^{(\mu)} \in dz\Big)$$

$$= \frac{\sqrt{xz}}{t} e^{\mu(z-x)-\mu^2 t/2-(x^2+z^2)/2t-y/8}\, \mathrm{i}_{y/2}\Big(\frac{xz}{t}\Big)dydz$$

1.21.5
$0 < x$

$$\mathbf{E}_x\Big\{\exp\Big(-\int_0^\tau\Big(\frac{p^2}{2(W_s^{(\mu)})^2} + \frac{q^2(W_s^{(\mu)})^2}{2}\Big)ds\Big); 0 < \inf_{0\le s\le \tau} W_s^{(\mu)}, W_\tau^{(\mu)} \in dz\Big\}$$

$$= \lambda e^{\mu(z-x)} Q^{(21)}_{\lambda+\mu^2/2}(z)dz, \qquad \text{for } Q^{(21)}_\lambda(z) \text{ see 1.1.21.5}$$

1.21.7
$0 < x$

$$\mathbf{E}_x\Big\{\exp\Big(-\int_0^t\Big(\frac{p^2}{2(W_s^{(\mu)})^2} + \frac{q^2(W_s^{(\mu)})^2}{2}\Big)ds\Big);\ 0 < \inf_{0\le s\le t} W_s^{(\mu)}, W_t^{(\mu)} \in dz\Big\}$$

$$= \frac{q\sqrt{xz}}{\mathrm{sh}(tq)} \exp\Big(\mu(z-x) - \frac{\mu^2 t}{2} - \frac{(x^2+z^2)q\,\mathrm{ch}(tq)}{2\,\mathrm{sh}(tq)}\Big) I_{\sqrt{p^2+1/4}}\Big(\frac{xzq}{\mathrm{sh}(tq)}\Big)dz$$

1.21.8
$0 < x$

$$\mathbf{P}_x\Big(\int_0^t (W_s^{(\mu)})^2 ds \in dg, \int_0^t \frac{ds}{(W_s^{(\mu)})^2} \in dy,\ 0 < \inf_{0\le s\le t} W_s^{(\mu)}, W_t^{(\mu)} \in dz\Big)$$

$$= 2^{-1}\sqrt{xz} e^{\mu(z-x)-\mu^2 t/2-y/8}\, \mathrm{ei}_g(y/2, t, (x^2+z^2)/2, xz)dgdydz$$

1.22.5
$0 < x$

$$\mathbf{E}_x\Big\{\exp\Big(-\int_0^\tau\Big(\frac{p^2}{2(W_s^{(\mu)})^2} + \frac{q}{W_s^{(\mu)}}\Big)ds\Big);\ 0 < \inf_{0\le s\le \tau} W_s^{(\mu)}, W_\tau^{(\mu)} \in dz\Big\}$$

$$= \lambda e^{\mu(z-x)} Q^{(22)}_{\lambda+\mu^2/2}(z)dz, \qquad \text{for } Q^{(22)}_\lambda(y,z) \text{ see 1.1.22.5}$$

1.22.6
$0 < x$

$$\mathbf{P}_x\Big(\int_0^\tau \frac{ds}{(W_s^{(\mu)})^2} \in dg, \int_0^\tau \frac{ds}{W_s^{(\mu)}} \in dy,\ 0 < \inf_{0\le s\le \tau} W_s^{(\mu)},\ W_\tau^{(\mu)} \in dz\Big)$$

$$= \frac{\lambda\sqrt{(2\lambda+\mu^2)xz}}{8\,\mathrm{sh}(y\sqrt{2\lambda+\mu^2}/2)} \exp\Big(\mu(z-x) - \frac{g}{8} - \frac{(x+z)\sqrt{2\lambda+\mu^2}\,\mathrm{ch}(y\sqrt{2\lambda+\mu^2}/2)}{\mathrm{sh}(y\sqrt{2\lambda+\mu^2}/2)}\Big)$$

$$\times\, \mathrm{i}_{g/8}\Big(\frac{2\sqrt{(2\lambda+\mu^2)xz}}{\mathrm{sh}(y\sqrt{2\lambda+\mu^2}/2)}\Big)dgdydz$$

2 $W_s^{(\mu)} = \mu s + W_s$ $\qquad\qquad \tau \sim \operatorname{Exp}(\lambda)$, independent of $W^{(\mu)}$

1.22.7 (1)	$\mathbf{E}_x\Big\{\exp\Big(-\frac{p^2}{2}\int_0^t \frac{ds}{(W_s^{(\mu)})^2}\Big);\ \int_0^t \frac{ds}{W_s^{(\mu)}} \in dy,\ 0 < \inf_{0\le s\le t} W_s^{(\mu)},\ W_t^{(\mu)} \in dz\Big\}$
$0 < x$	$= \sqrt{xz}\, e^{\mu(z-x)-\mu^2 t/2}\, \mathrm{is}_t(\sqrt{4p^2+1}, y/2, 0, x+z, \sqrt{xz})dydz$
1.22.8	$\mathbf{P}_x\Big(\int_0^t \frac{ds}{(W_s^{(\mu)})^2} \in dg,\ \int_0^t \frac{ds}{W_s^{(\mu)}} \in dy,\ 0 < \inf_{0\le s\le t} W_s^{(\mu)},\ W_t^{(\mu)} \in dz\Big)$
	$= 8^{-1}\sqrt{xz}e^{\mu(z-x)-\mu^2 t/2-g/8}\, \mathrm{ei}_t(g/8, y/2, x+z, 2\sqrt{xz})dgdydz$
1.23.1 $r < b$	$\mathbf{E}_x\Big\{\exp\Big(\int_0^\tau\big(p\mathbb{I}_{(-\infty,r)}(W_s^{(\mu)}) + q\mathbb{I}_{[r,\infty)}(W_s^{(\mu)})\big)ds\Big);\ \sup_{0\le s\le\tau} W_s^{(\mu)} < b\Big\}$
$x \le r$	$= \frac{\lambda}{\lambda+p}$
	$+\frac{\lambda\big((p-q)e^{\mu(r-x)}(\mu+\Upsilon_q\operatorname{cth}((b-r)\Upsilon_q))\operatorname{sh}((b-r)\Upsilon_q) - (\lambda+p)\Upsilon_q e^{\mu(b-x)}\big)e^{(x-r)\Upsilon_p}}{(\lambda+p)(\lambda+q)(\Upsilon_p\operatorname{sh}((b-r)\Upsilon_q) + \Upsilon_q\operatorname{ch}((b-r)\Upsilon_q))}$
$r \le x$ $x \le b$	$= \frac{\lambda}{\lambda+q}\Big(1 - \frac{e^{\mu(b-x)}\operatorname{sh}((x-r)\Upsilon_q)}{\operatorname{sh}((b-r)\Upsilon_q)}\Big)$
	$+\frac{\lambda\big((q-p)e^{\mu(r-x)}(\Upsilon_p-\mu)\operatorname{sh}((b-r)\Upsilon_q) - (\lambda+p)\Upsilon_q e^{\mu(b-x)}\big)\operatorname{sh}((b-x)\Upsilon_q)}{(\lambda+p)(\lambda+q)(\Upsilon_p\operatorname{sh}((b-r)\Upsilon_q) + \Upsilon_q\operatorname{ch}((b-r)\Upsilon_q))\operatorname{sh}((b-r)\Upsilon_q)}$
1.23.3 (1) $\mu < 0$	$\mathbf{E}_x\Big\{\exp\Big(-q\int_0^\infty \mathbb{I}_{[r,\infty)}(W_s^{(\mu)})ds\Big);\ \sup_{0\le s<\infty} W_s^{(\mu)} < b\Big\}$ $\qquad r \vee x \le b$
$x \le r$	$= 1 - e^{2\mu(r-x)} + \frac{2\lvert\mu\rvert e^{2\mu(r-x)}\operatorname{sh}((b-r)\sqrt{\mu^2+2q})}{\sqrt{\mu^2+2q}\operatorname{ch}((b-r)\sqrt{\mu^2+2q}) + \lvert\mu\rvert\operatorname{sh}((b-r)\sqrt{\mu^2+2q})}$
$r \le x$ $x \le b$	$= \frac{2\lvert\mu\rvert e^{\mu(r-x)}\operatorname{sh}((b-x)\sqrt{\mu^2+2q})}{\sqrt{\mu^2+2q}\operatorname{ch}((b-r)\sqrt{\mu^2+2q}) + \lvert\mu\rvert\operatorname{sh}((b-r)\sqrt{\mu^2+2q})}$
1.23.4 (1) $\mu < 0$	$\mathbf{P}_x\Big(\int_0^\infty \mathbb{I}_{[r,\infty)}(W_s^{(\mu)})ds \in dy,\ \sup_{0\le s<\infty} W_s^{(\mu)} < b\Big)$ $\qquad r \vee x \le b$
$x \le r$	$= 2e^{2\mu(r-x)-\mu^2 y/2}\,\mathrm{rc}_y(0, b-r, b-r, 1/\lvert\mu\rvert)dy$

$$\Upsilon_s := \sqrt{2\lambda + \mu^2 + 2s}$$

2 $W_s^{(\mu)} = \mu s + W_s$ $\tau \sim \text{Exp}(\lambda)$, independent of $W^{(\mu)}$

$r \le x$, $x \le b$

$$= 2e^{\mu(r-x)-\mu^2 y/2}\,\text{rc}_y(0, b-r, b-x, 1/|\mu|)dy$$

1.23.5, $r < b$, $z < b$

$$\mathbf{E}_x\Big\{\exp\Big(-\int_0^\tau \big(p\mathbb{I}_{(-\infty,r)}(W_s^{(\mu)}) + q\mathbb{I}_{[r,\infty)}(W_s^{(\mu)})\big)ds\Big); \sup_{0\le s\le\tau} W_s^{(\mu)} < b,\, W_\tau^{(\mu)} \in dz\Big\} = \lambda e^{\mu(z-x)} Q^{(23)}_{\lambda+\mu^2/2}(z)dz$$

for $Q_\lambda^{(23)}(z)$ see 1.1.23.5

1.24.1, $a < r$

$$\mathbf{E}_x\Big\{\exp\Big(-\int_0^\tau \big(p\mathbb{I}_{(-\infty,r)}(W_s^{(\mu)}) + q\mathbb{I}_{[r,\infty)}(W_s^{(\mu)})\big)ds\Big); a < \inf_{0\le s\le\tau} W_s^{(\mu)}\Big\}$$

$a \le x$, $x \le r$

$$= \frac{\lambda}{\lambda+p}\Big(1 - \frac{e^{\mu(a-x)}\,\text{sh}((r-x)\Upsilon_p)}{\text{sh}((r-a)\Upsilon_p)}\Big)$$

$$+\frac{\lambda\big((p-q)e^{\mu(r-x)}(\Upsilon_q+\mu)\,\text{sh}((r-a)\Upsilon_p) - (\lambda+q)\Upsilon_p e^{\mu(a-x)}\big)\,\text{sh}((x-a)\Upsilon_p)}{(\lambda+p)(\lambda+q)(\Upsilon_q\,\text{sh}((r-a)\Upsilon_p) + \Upsilon_p\,\text{ch}((r-a)\Upsilon_p))\,\text{sh}((r-a)\Upsilon_p)}$$

$r \le x$

$$= \frac{\lambda}{\lambda+q}$$

$$+\frac{\lambda\big((q-p)e^{\mu(r-x)}(\Upsilon_p\,\text{cth}((r-a)\Upsilon_p) - \mu)\,\text{sh}((r-a)\Upsilon_p) - (\lambda+q)\Upsilon_p e^{\mu(a-x)}\big)e^{(r-x)\Upsilon_q}}{(\lambda+p)(\lambda+q)(\Upsilon_q\,\text{sh}((r-a)\Upsilon_p) + \Upsilon_p\,\text{ch}((r-a)\Upsilon_p))}$$

1.24.3 (1), $\mu > 0$

$$\mathbf{E}_x\Big\{\exp\Big(-p\int_0^\infty \mathbb{I}_{(-\infty,r)}(W_s^{(\mu)})ds\Big); a < \inf_{0\le s<\infty} W_s^{(\mu)}\Big\} \qquad a \le r \wedge x$$

$a \le x$, $x \le r$

$$= \frac{2\mu e^{\mu(r-x)}\,\text{sh}((x-a)\sqrt{\mu^2+2p})}{\sqrt{\mu^2+2p}\,\text{ch}((r-a)\sqrt{\mu^2+2p}) + \mu\,\text{sh}((r-a)\sqrt{\mu^2+2p})}$$

$r \le x$

$$= 1 - e^{2\mu(r-x)} + \frac{2\mu e^{2\mu(r-x)}\,\text{sh}((r-a)\sqrt{\mu^2+2p})}{\sqrt{\mu^2+2p}\,\text{ch}((r-a)\sqrt{\mu^2+2p}) + \mu\,\text{sh}((r-a)\sqrt{\mu^2+2p})}$$

1.24.4 (1), $\mu > 0$

$$\mathbf{P}_x\Big(\int_0^\infty \mathbb{I}_{(-\infty,r)}(W_s^{(\mu)})ds \in dy, a < \inf_{0\le s<\infty} W_s^{(\mu)}\Big)$$

$a \le x$, $x \le r$

$$= 2e^{\mu(r-x)-\mu^2 y/2}\,\text{rc}_y(0, r-a, x-a, 1/|\mu|)dy$$

$$\Upsilon_s := \sqrt{2\lambda + \mu^2 + 2s}$$

2	$W_s^{(\mu)} = \mu s + W_s$ $\qquad \tau \sim \text{Exp}(\lambda)$, independent of $W^{(\mu)}$
$r \le x$	$= 2e^{2\mu(r-x)-\mu^2 y/2}\,\text{rc}_y(0, r-a, r-a, 1/\lvert\mu\rvert)dy$
1.24.5 $a < r$	$\mathbf{E}_x\Big\{\exp\Big(-\int_0^\tau \big(p\mathbb{I}_{(-\infty,r)}\big(W_s^{(\mu)}\big) + q\mathbb{I}_{[r,\infty)}\big(W_s^{(\mu)}\big)\big)ds\Big);$
$a < z$	$a < \inf_{0\le s\le\tau} W_s^{(\mu)},\ W_\tau^{(\mu)} \in dz\Big\} = \lambda e^{\mu(z-x)} Q^{(24)}_{a,\lambda+\mu^2/2}(z)dz$ for $Q^{(24)}_{a,\lambda}(z)$ see 1.1.24.5
1.25.1 $r < b$	$\mathbf{E}_x\big\{e^{-\gamma\ell(\tau,r)}; a < \inf_{0\le s\le\tau} W_s^{(\mu)}, \sup_{0\le s\le\tau} W_s^{(\mu)} < b\big\}$
$a \le x$ $x \le r$	$= 1 - \dfrac{e^{\mu(a-x)}\,\text{sh}((r-x)\Upsilon_0) + e^{\mu(r-x)}\,\text{sh}((x-a)\Upsilon_0)}{\text{sh}((r-a)\Upsilon_0)}$ $+ \dfrac{(e^{\mu(r-x)}\,\text{sh}((b-a)\Upsilon_0) - e^{\mu(b-x)}\,\text{sh}((r-a)\Upsilon_0) - e^{\mu(a-x)}\,\text{sh}((b-r)\Upsilon_0))\,\text{sh}((x-a)\Upsilon_0)}{\Upsilon_0^{-1}(2\gamma\,\text{sh}((b-r)\Upsilon_0)\,\text{sh}((r-a)\Upsilon_0) + \Upsilon_0\,\text{sh}((b-a)\Upsilon_0))\,\text{sh}((r-a)\Upsilon_0)}$
$r \le x$ $x \le b$	$= 1 - \dfrac{e^{\mu(r-x)}\,\text{sh}((b-x)\Upsilon_0) + e^{\mu(b-x)}\,\text{sh}((x-r)\Upsilon_0)}{\text{sh}((b-r)\Upsilon_0)}$ $+ \dfrac{(e^{\mu(r-x)}\,\text{sh}((b-a)\Upsilon_0) - e^{\mu(b-x)}\,\text{sh}((r-a)\Upsilon_0) - e^{\mu(a-x)}\,\text{sh}((b-r)\Upsilon_0))\,\text{sh}((b-x)\Upsilon_0)}{\Upsilon_0^{-1}(2\gamma\,\text{sh}((b-r)\Upsilon_0)\,\text{sh}((r-a)\Upsilon_0) + \Upsilon_0\,\text{sh}((b-a)\Upsilon_0))\,\text{sh}((b-r)\Upsilon_0)}$
1.25.2 $r < b$	$\mathbf{P}_x\big(\ell(\tau,r) \in dy, a < \inf_{0\le s\le\tau} W_s^{(\mu)}, \sup_{0\le s\le\tau} W_s^{(\mu)} < b\big)$ $= \exp\Big(-\dfrac{y\Upsilon_0\,\text{sh}((b-a)\Upsilon_0)}{2\,\text{sh}((b-r)\Upsilon_0)\,\text{sh}((r-a)\Upsilon_0)}\Big)$ $\times\begin{cases} \dfrac{\Upsilon_0(e^{\mu r}\,\text{sh}((b-a)\Upsilon_0) - e^{\mu b}\,\text{sh}((r-a)\Upsilon_0) - e^{\mu a}\,\text{sh}((b-r)\Upsilon_0))\,\text{sh}((x-a)\Upsilon_0)}{2e^{\mu x}\,\text{sh}((b-r)\Upsilon_0)\,\text{sh}^2((r-a)\Upsilon_0)}dy, & a \le x \le r \\ \dfrac{\Upsilon_0(e^{\mu r}\,\text{sh}((b-a)\Upsilon_0) - e^{\mu b}\,\text{sh}((r-a)\Upsilon_0) - e^{\mu a}\,\text{sh}((b-r)\Upsilon_0))\,\text{sh}((b-x)\Upsilon_0)}{2e^{\mu x}\,\text{sh}^2((b-r)\Upsilon_0)\,\text{sh}((r-a)\Upsilon_0)}dy, & r \le x \le b \end{cases}$
	$\Upsilon_0 := \sqrt{2\lambda + \mu^2}$

2 $W_s^{(\mu)} = \mu s + W_s$ $\quad\tau \sim \text{Exp}(\lambda)$, independent of $W^{(\mu)}$

(1) $\mathbf{P}_x\big(\ell(\tau,r)=0, \sup_{0\le s\le\tau} W_s^{(\mu)} < b\big)$

$$= \begin{cases} 1 - \dfrac{e^{\mu(a-x)}\,\text{sh}((r-x)\Upsilon_0) + e^{\mu(r-x)}\,\text{sh}((x-a)\Upsilon_0)}{\text{sh}((r-a)\Upsilon_0)}, & a \le x \le r \\ 1 - \dfrac{e^{\mu(r-x)}\,\text{sh}((b-x)\Upsilon_0) + e^{\mu(b-x)}\,\text{sh}((x-r)\Upsilon_0)}{\text{sh}((b-r)\Upsilon_0)}, & r \le x \le b \end{cases}$$

1.25.5 $\mathbf{E}_x\{e^{-\gamma\ell(\tau,r)}; a < \inf_{0\le s\le\tau} W_s^{(\mu)}, \sup_{0\le s\le\tau} W_s^{(\mu)} < b, W_\tau^{(\mu)} \in dz\}$

$r < b$

$z < b$

$$= \lambda e^{\mu(z-x)} Q^{(25)}_{a,\lambda+\mu^2/2}(z)dz, \qquad \text{for } Q^{(25)}_{a,\lambda}(z) \text{ see 1.1.25.5}$$

1.25.6 $\mathbf{P}_x\big(\ell(\tau,r) \in dy, a < \inf_{0\le s\le\tau} W_s^{(\mu)}, \sup_{0\le s\le\tau} W_s^{(\mu)} < b, W_\tau^{(\mu)} \in dz\big)$

$r < b$

$z < b$

$$= \lambda e^{\mu(z-x)} B^{(25)}_{a,\lambda+\mu^2/2}(y,z)dydz, \qquad \text{for } B^{(25)}_{a,\lambda}(y,z) \text{ see 1.1.25.6}$$

(1) $\mathbf{P}_x\big(\ell(\tau,r)=0, a < \inf_{0\le s\le\tau} W_s^{(\mu)}, \sup_{0\le s\le\tau} W_s^{(\mu)} < b, W_\tau^{(\mu)} \in dz\big)$

$z \le r$

$x \le r$

$$= \frac{\lambda e^{\mu(z-x)}\big(\text{ch}((r-a-|z-x|)\Upsilon_0) - \text{ch}((r+a-x-z)\Upsilon_0)\big)}{\Upsilon_0\,\text{sh}((r-a)\Upsilon_0)}dz$$

$r \le z$

$r \le x$

$$= \frac{\lambda e^{\mu(z-x)}\big(\text{ch}((b-r-|z-x|)\Upsilon_0) - \text{ch}((b+r-z-x)\Upsilon_0)\big)}{\Upsilon_0\,\text{sh}((b-r)\Upsilon_0)}dz$$

1.25.8 $\mathbf{P}_x\big(\ell(t,r) \in dy, a < \inf_{0\le s\le t} W_s^{(\mu)}, \sup_{0\le s\le t} W_s^{(\mu)} < b, W_t^{(\mu)} \in dz\big)$

$$= e^{\mu(z-x)-\mu^2 t/2} F^{(25)}_{a,x}(t,y,z)dydz, \qquad \text{for } F^{(25)}_{a,x}(t,y,z) \text{ see 1.1.25.8}$$

$$\Upsilon_0 := \sqrt{2\lambda + \mu^2}$$

2 $W_s^{(\mu)} = \mu s + W_s$ $\qquad \tau \sim \mathrm{Exp}(\lambda)$, independent of $W^{(\mu)}$

1.26.1 $\quad r<b, \quad a<r$

$$\mathbf{E}_x\Big\{\exp\Big(-\int_0^\tau \big(p\mathbb{I}_{(-\infty,r)}\big(W_s^{(\mu)}\big)+q\mathbb{I}_{[r,\infty)}\big(W_s^{(\mu)}\big)\big)ds\Big); \quad a<\inf_{0\le s\le\tau} W_s^{(\mu)},\ \sup_{0\le s\le\tau} W_s^{(\mu)}<b\Big\}$$

$a\le x,\ x\le r$

$$=\frac{\lambda}{\lambda+p}\Big(1-\frac{e^{\mu(a-x)}\operatorname{sh}((r-x)\Upsilon_p)}{\operatorname{sh}((r-a)\Upsilon_p)}\Big)$$

$$+\frac{\lambda(p-q)e^{\mu(r-x)}(\Upsilon_q\operatorname{cth}((b-r)\Upsilon_q)+\mu)\operatorname{sh}((x-a)\Upsilon_p)}{(\lambda+p)(\lambda+q)(\Upsilon_p\operatorname{cth}((r-a)\Upsilon_p)+\Upsilon_q\operatorname{cth}((b-r)\Upsilon_q))\operatorname{sh}((r-a)\Upsilon_p)}$$

$$-\frac{\lambda((\lambda+p)\Upsilon_q e^{\mu(b-x)}\operatorname{sh}((r-a)\Upsilon_p)+(\lambda+q)\Upsilon_p e^{\mu(a-x)}\operatorname{sh}((b-r)\Upsilon_q))\operatorname{sh}((x-a)\Upsilon_p)}{(\lambda+p)(\lambda+q)(\Upsilon_p\operatorname{cth}((r-a)\Upsilon_p)+\Upsilon_q\operatorname{cth}((b-r)\Upsilon_q))\operatorname{sh}((b-r)\Upsilon_q)\operatorname{sh}^2((r-a)\Upsilon_p)}$$

$r\le x,\ x\le b$

$$=\frac{\lambda}{\lambda+q}\Big(1-\frac{e^{\mu(b-x)}\operatorname{sh}((x-r)\Upsilon_q)}{\operatorname{sh}((b-r)\Upsilon_q)}\Big)$$

$$+\frac{\lambda(q-p)e^{\mu(r-x)}(\Upsilon_p\operatorname{cth}((r-a)\Upsilon_p)-\mu)\operatorname{sh}((b-x)\Upsilon_q)}{(\lambda+p)(\lambda+q)(\Upsilon_p\operatorname{cth}((r-a)\Upsilon_p)+\Upsilon_q\operatorname{cth}((b-r)\Upsilon_q))\operatorname{sh}((b-r)\Upsilon_q)}$$

$$-\frac{\lambda((\lambda+p)\Upsilon_q e^{\mu(b-x)}\operatorname{sh}((r-a)\Upsilon_p)+(\lambda+q)\Upsilon_p e^{\mu(a-x)}\operatorname{sh}((b-r)\Upsilon_q))\operatorname{sh}((b-x)\Upsilon_q)}{(\lambda+p)(\lambda+q)(\Upsilon_p\operatorname{cth}((r-a)\Upsilon_p)+\Upsilon_q\operatorname{cth}((b-r)\Upsilon_q))\operatorname{sh}^2((b-r)\Upsilon_q)\operatorname{sh}((r-a)\Upsilon_p)}$$

1.26.5 $\quad r<b, \quad a<r$

$$\mathbf{E}_x\Big\{\exp\Big(-\int_0^\tau \big(p\mathbb{I}_{(-\infty,r)}\big(W_s^{(\mu)}\big)+q\mathbb{I}_{[r,\infty)}\big(W_s^{(\mu)}\big)\big)ds\Big); \quad a<\inf_{0\le s\le\tau} W_s^{(\mu)},\ \sup_{0\le s\le\tau} W_s^{(\mu)}<b,\ W_\tau^{(\mu)}\in dz\Big\}$$

$$=\lambda e^{\mu(z-x)}Q^{(26)}_{a,\lambda+\mu^2/2}(z)dz, \qquad \text{for } Q^{(26)}_{a,\lambda}(z) \text{ see 1.1.26.5}$$

1.27.1

$$\mathbf{E}_x\Big\{\exp\Big(-\int_0^\tau \big(p\mathbb{I}_{(-\infty,r)}\big(W_s^{(\mu)}\big)+q\mathbb{I}_{[r,\infty)}\big(W_s^{(\mu)}\big)\big)ds\Big); \tfrac{1}{2}\ell(\tau,r)\in dy\Big\}/dy$$

$x\le r$

$$=\Big(\frac{2\lambda e^{\mu(r-x)}}{\sqrt{2\lambda+\mu^2+2p}+\mu}+\frac{2\lambda e^{\mu(r-x)}}{\sqrt{2\lambda+\mu^2+2q}-\mu}\Big)e^{(x-r-y)\sqrt{2\lambda+\mu^2+2p}-y\sqrt{2\lambda+\mu^2+2q}}$$

$$\Upsilon_s := \sqrt{2\lambda+\mu^2+2s}$$

2 $W_s^{(\mu)} = \mu s + W_s$ $\tau \sim \mathrm{Exp}(\lambda)$, independent of $W^{(\mu)}$

or $$= -\frac{2\lambda e^{\mu(r-x)} e^{(x-r)\sqrt{2\lambda+\mu^2+2p}}}{(\sqrt{2\lambda+\mu^2+2p}+\mu)(\sqrt{2\lambda+\mu^2+2q}-\mu)} \frac{d}{dy} e^{-y\left(\sqrt{2\lambda+\mu^2+2p}+\sqrt{2\lambda+\mu^2+2q}\right)}$$

$r \le x$ $$= \Big(\frac{2\lambda e^{\mu(r-x)}}{\sqrt{2\lambda+\mu^2+2p}+\mu} + \frac{2\lambda e^{\mu(r-x)}}{\sqrt{2\lambda+\mu^2+2q}-\mu}\Big) e^{(r-x-y)\sqrt{2\lambda+\mu^2+2q}-y\sqrt{2\lambda+\mu^2+2p}}$$

or $$= -\frac{2\lambda e^{\mu(r-x)} e^{(r-x)\sqrt{2\lambda+\mu^2+2q}}}{(\sqrt{2\lambda+\mu^2+2p}+\mu)(\sqrt{2\lambda+\mu^2+2q}-\mu)} \frac{d}{dy} e^{-y\left(\sqrt{2\lambda+\mu^2+2p}+\sqrt{2\lambda+\mu^2+2q}\right)}$$

(1) $$\mathbf{E}_x\Big\{\exp\Big(-\int_0^\tau \big(p\mathbb{I}_{(-\infty,r)}(W_s^{(\mu)}) + q\mathbb{I}_{[r,\infty)}(W_s^{(\mu)})\big)ds\Big); \ell(\tau,r)=0\Big\}$$

$x \le r$ $$= \frac{\lambda}{\lambda+p}\Big(1 - e^{(x-r)(\sqrt{2\lambda+\mu^2+2p}-\mu)}\Big)$$

$r \le x$ $$= \frac{\lambda}{\lambda+q}\Big(1 - e^{(r-x)(\sqrt{2\lambda+\mu^2+2q}+\mu)}\Big)$$

1.27.2 $$\mathbf{P}_x\Big(\int_0^\tau \mathbb{I}_{(-\infty,r)}(W_s^{(\mu)})ds \in du,\ \int_0^\tau \mathbb{I}_{[r,\infty)}(W_s^{(\mu)})ds \in dv,\ \frac{1}{2}\ell(\tau,r) \in dy\Big)/dy$$

$$=: \lambda e^{-\lambda(u+v)} B_x^{(27)}(u,v,y)\,du\,dv$$

$x \le r$ $$= -\lambda e^{-\lambda(u+v)} e^{\mu(r-x)} \frac{d}{dy}\Big(\Big(\frac{e^{-y^2/2v-\mu^2 v/2}}{\sqrt{\pi v}} + \frac{\mu}{\sqrt{2}} e^{-\mu y}\,\mathrm{Erfc}\Big(\frac{y}{\sqrt{2v}} - \frac{\mu\sqrt{v}}{\sqrt{2}}\Big)\Big)$$
$$\times\Big(\frac{e^{-(y+r-x)^2/2u-\mu^2 u/2}}{\sqrt{\pi u}} - \frac{\mu}{\sqrt{2}} e^{\mu(y+r-x)}\,\mathrm{Erfc}\Big(\frac{y+r-x}{\sqrt{2u}} + \frac{\mu\sqrt{u}}{\sqrt{2}}\Big)\Big)\Big)du\,dv$$

or $$= \lambda e^{-\lambda(u+v)}\Big(\frac{yu+(y+r-x)v}{\pi v^{3/2}u^{3/2}} e^{\mu(r-x)-(y+r-x)^2/2u-y^2/2v-\mu^2 t/2}$$
$$- \frac{\mu y}{\sqrt{2\pi}v^{3/2}} e^{2\mu(r-x)-(\mu v-y)^2/2v}\,\mathrm{Erfc}\Big(\frac{y+r-x}{\sqrt{2u}} + \frac{\mu\sqrt{u}}{\sqrt{2}}\Big)$$
$$+ \frac{\mu(y+r-x)}{\sqrt{2\pi}u^{3/2}} e^{2\mu(r-x)-(y+r-x+\mu u)^2/2u}\,\mathrm{Erfc}\Big(\frac{y}{\sqrt{2v}} - \frac{\mu\sqrt{v}}{\sqrt{2}}\Big)\Big)du\,dv$$

$r \le x$ $$= -\lambda e^{-\lambda(u+v)} e^{\mu(r-x)} \frac{d}{dy}\Big(\Big(\frac{e^{-y^2/2u-\mu^2 u/2}}{\sqrt{\pi u}} - \frac{\mu}{\sqrt{2}} e^{\mu y}\,\mathrm{Erfc}\Big(\frac{y}{\sqrt{2u}} + \frac{\mu\sqrt{u}}{\sqrt{2}}\Big)\Big)$$

2 $W_s^{(\mu)} = \mu s + W_s$ $\quad\tau \sim \operatorname{Exp}(\lambda)$, independent of $W^{(\mu)}$

$$\times\Big(\frac{e^{-(y+x-r)^2/2v-\mu^2 v/2}}{\sqrt{\pi v}}+\frac{\mu}{\sqrt{2}}e^{-\mu(y+x-r)}\operatorname{Erfc}\Big(\frac{y+x-r}{\sqrt{2v}}-\frac{\mu\sqrt{v}}{\sqrt{2}}\Big)\Big)\Big)dudv$$

or

$$=\lambda e^{-\lambda(u+v)}\Big(\frac{(y+x-r)u+yv}{\pi v^{3/2}u^{3/2}}e^{\mu(r-x)-y^2/2u-(y+x-r)^2/2v-\mu^2 t/2}$$

$$-\frac{\mu(y+x-r)}{\sqrt{2\pi}v^{3/2}}e^{2\mu(r-x)-(y+x-r-\mu v)^2/2v}\operatorname{Erfc}\Big(\frac{y}{\sqrt{2u}}+\frac{\mu\sqrt{u}}{\sqrt{2}}\Big)$$

$$+\frac{\mu y}{\sqrt{2\pi}u^{3/2}}e^{2\mu(r-x)-(y+\mu u)^2/2u}\operatorname{Erfc}\Big(\frac{y+x-r}{\sqrt{2v}}-\frac{\mu\sqrt{v}}{\sqrt{2}}\Big)\Big)dudv$$

(1) $x \le r$

$$\mathbf{P}_x\Big(\int_0^\tau \mathbb{I}_{(-\infty,r)}\big(W_s^{(\mu)}\big)ds\in du,\ \int_0^\tau \mathbb{I}_{[r,\infty)}\big(W_s^{(\mu)}\big)ds=0,\ell(\tau,r)=0\Big)$$

$$=\lambda e^{-\lambda u}\Big(1-\frac{1}{2}\operatorname{Erfc}\Big(\frac{r-x}{\sqrt{2t}}-\frac{\mu\sqrt{u}}{\sqrt{2}}\Big)-\frac{1}{2}e^{2\mu(r-x)}\operatorname{Erfc}\Big(\frac{r-x}{\sqrt{2u}}+\frac{\mu\sqrt{u}}{\sqrt{2}}\Big)\Big)du$$

(2) $x \le r$

$$\mathbf{P}_x\Big(\int_0^\tau \mathbb{I}_{(-\infty,r)}\big(W_s^{(\mu)}\big)ds=0,\ \int_0^\tau \mathbb{I}_{[r,\infty)}\big(W_s^{(\mu)}\big)ds\in dv,\ell(\tau,r)=0\Big)$$

$$=\lambda e^{-\lambda v}\Big(1-\frac{1}{2}\operatorname{Erfc}\Big(\frac{x-r}{\sqrt{2t}}+\frac{\mu\sqrt{v}}{\sqrt{2}}\Big)-\frac{1}{2}e^{2\mu(r-x)}\operatorname{Erfc}\Big(\frac{x-r}{\sqrt{2v}}-\frac{\mu\sqrt{v}}{\sqrt{2}}\Big)\Big)dv$$

1.27.3 (1) $\mu < 0$

$$\mathbf{E}_x\Big\{\exp\Big(-q\int_0^\infty \mathbb{I}_{[r,\infty)}\big(W_s^{(\mu)}\big)ds\Big);\ \frac{1}{2}\ell(\infty,r)\in dy\Big\}$$

$$=\begin{cases} 2|\mu|e^{2\mu(r-x)-y(\sqrt{\mu^2+2q}-\mu)}dy, & x\le r\\ 2|\mu|e^{(r-x)(\sqrt{\mu^2+2q}+\mu)-y(\sqrt{\mu^2+2q}-\mu)}dy, & r\le x\end{cases}$$

(2) $\mu < 0$

$$\mathbf{E}_x\Big\{\exp\Big(-q\int_0^\infty \mathbb{I}_{[r,\infty)}\big(W_s^{(\mu)}\big)ds\Big);\ \ell(\infty,r)=0\Big\}=1-e^{\mu(|r-x|+r-x)}$$

(3) $\mu > 0$

$$\mathbf{E}_x\Big\{\exp\Big(-p\int_0^\infty \mathbb{I}_{(-\infty,r)}\big(W_s^{(\mu)}\big)ds\Big);\ \frac{1}{2}\ell(\infty,r)\in dy\Big\}$$

$$=\begin{cases} 2\mu e^{(x-r)(\sqrt{\mu^2+2p}-\mu)-y(\sqrt{\mu^2+2p}+\mu)}dy, & x\le r\\ 2\mu e^{-2\mu(x-r)-y(\sqrt{\mu^2+2p}+\mu)}dy, & r\le x\end{cases}$$

2 $W_s^{(\mu)} = \mu s + W_s$ $\tau \sim \operatorname{Exp}(\lambda)$, independent of $W^{(\mu)}$

(4) $\mu > 0$

$$\mathbf{E}_x\Big\{\exp\Big(-p\int_0^\infty \mathbb{I}_{(-\infty,r)}(W_s^{(\mu)})ds\Big); \ell(\infty,r)=0\Big\} = 1 - e^{-\mu(|x-r|+x-r)}$$

1.27.4

$$\mathbf{P}_x\Big(\int_0^t (p\mathbb{I}_{(-\infty,r)}(W_s^{(\mu)}) + q\mathbb{I}_{[r,\infty)}(W_s^{(\mu)}))ds \in dv, \frac{1}{2}\ell(t,r) \in dy\Big)$$

$$= \frac{1}{|p-q|}B_x^{(27)}\Big(\frac{|qt-v|}{|p-q|}, \frac{|pt-v|}{|p-q|}, y\Big)\mathbb{I}_{((q\wedge p)t,(q\vee p)t)}(v)\,dvdy$$

for $B_x^{(27)}(u,v,y)$ see 1.27.2

(1) $x \le r$

$$\mathbf{P}_x\Big(\int_0^t (p\mathbb{I}_{(-\infty,r)}(W_s^{(\mu)}) + q\mathbb{I}_{[r,\infty)}(W_s^{(\mu)}))ds = pt, \ell(t,r) = 0\Big)$$

$$= 1 - \frac{1}{2}\operatorname{Erfc}\Big(\frac{r-x}{\sqrt{2t}} - \frac{\mu\sqrt{t}}{\sqrt{2}}\Big) - \frac{1}{2}e^{2\mu(r-x)}\operatorname{Erfc}\Big(\frac{r-x}{\sqrt{2t}} + \frac{\mu\sqrt{t}}{\sqrt{2}}\Big)$$

(2) $r \le x$

$$\mathbf{P}_x\Big(\int_0^t (p\mathbb{I}_{(-\infty,r)}(W_s^{(\mu)}) + q\mathbb{I}_{[r,\infty)}(W_s^{(\mu)}))ds = qt, \ell(t,r) = 0\Big)$$

$$= 1 - \frac{1}{2}\operatorname{Erfc}\Big(\frac{x-r}{\sqrt{2t}} + \frac{\mu\sqrt{t}}{\sqrt{2}}\Big) - \frac{1}{2}e^{2\mu(r-x)}\operatorname{Erfc}\Big(\frac{x-r}{\sqrt{2t}} - \frac{\mu\sqrt{t}}{\sqrt{2}}\Big)$$

(3) $\mu < 0$

$$\mathbf{P}_x\Big(\int_0^\infty \mathbb{I}_{[r,\infty)}(W_s^{(\mu)})ds \in du, \frac{1}{2}\ell(\infty,r) \in dy\Big)$$

$$= \begin{cases} \dfrac{2|\mu|y}{\sqrt{2\pi}u^{3/2}}e^{2\mu(r-x)}e^{-(y-\mu u)^2/2u}dudy, & x \le r \\ \dfrac{2|\mu|(y+x-r)}{\sqrt{2\pi}u^{3/2}}e^{2\mu(r-x)}e^{-(y+x-r-\mu u)^2/2u}dudy, & r \le x \end{cases}$$

(4) $\mu < 0$

$$\mathbf{P}_x\Big(\int_0^\infty \mathbb{I}_{[r,\infty)}(W_s^{(\mu)})ds = 0, \ell(\infty,r) = 0\Big) = 1 - e^{\mu(|r-x|+r-x)}$$

(5) $\mu > 0$

$$\mathbf{P}_x\Big(\int_0^\infty \mathbb{I}_{(-\infty,r)}(W_s^{(\mu)})ds \in du, \frac{1}{2}\ell(\infty,r) \in dy\Big)$$

2 $W_s^{(\mu)} = \mu s + W_s$ $\tau \sim \text{Exp}(\lambda)$, independent of $W^{(\mu)}$

$$= \begin{cases} \frac{2\mu(y+r-x)}{\sqrt{2\pi}u^{3/2}} e^{2\mu(r-x)} e^{-(y+r-x+\mu u)^2/2u} du dy, & x \le r \\ \frac{2\mu y}{\sqrt{2\pi}u^{3/2}} e^{-2\mu(x-r)} e^{-(y+\mu u)^2/2u} du dy, & r \le x \end{cases}$$

(6) $\mu > 0$

$$\mathbf{P}_x\Big(\int_0^\infty \mathbb{I}_{(-\infty,r)}(W_s^{(\mu)})ds = 0, \ell(\infty,r) = 0\Big) = 1 - e^{-\mu(|x-r|+x-r)}$$

1.27.5

$$\mathbf{E}_x\Big\{\exp\Big(-\int_0^\tau \big(p\mathbb{I}_{(-\infty,r)}(W_s^{(\mu)}) + q\mathbb{I}_{[r,\infty)}(W_s^{(\mu)})\big)ds\Big); \tfrac{1}{2}\ell(\tau,r) \in dy, W_\tau^{(\mu)} \in dz\Big\}$$

$x \le r$, $z \le r$

$$= 2\lambda e^{\mu(z-x)} e^{-y\sqrt{2\lambda+\mu^2+2q}} e^{-(y+2r-x-z)\sqrt{2\lambda+\mu^2+2p}} dy dz$$

$x \le r$, $r \le z$

$$= 2\lambda e^{\mu(z-x)} e^{-(y+z-r)\sqrt{2\lambda+\mu^2+2q}} e^{-(y+r-x)\sqrt{2\lambda+\mu^2+2p}} dy dz$$

$r \le x$, $z \le r$

$$= 2\lambda e^{\mu(z-x)} e^{-(y+x-r)\sqrt{2\lambda+\mu^2+2q}} e^{-(y+r-z)\sqrt{2\lambda+\mu^2+2p}} dy dz$$

$r \le x$, $r \le z$

$$= 2\lambda e^{\mu(z-x)} e^{-(y+x+z-2r)\sqrt{2\lambda+\mu^2+2q}} e^{-y\sqrt{2\lambda+\mu^2+2p}} dy dz$$

(1)

$$\mathbf{E}_x\Big\{\exp\Big(-\int_0^\tau \big(p\mathbb{I}_{(-\infty,r)}(W_s^{(\mu)}) + q\mathbb{I}_{[r,\infty)}(W_s^{(\mu)})\big)ds\Big); \ell(\tau,r) = 0, W_\tau^{(\mu)} \in dz\Big\}$$

$x \le r$, $z \le r$

$$= \frac{\lambda}{\sqrt{2\lambda+\mu^2+2p}} e^{\mu(z-x)} \Big(e^{-|z-x|\sqrt{2\lambda+\mu^2+2p}} - e^{(z+x-2r)\sqrt{2\lambda+\mu^2+2p}}\Big) dz$$

$r \le x$, $r \le z$

$$= \frac{\lambda}{\sqrt{2\lambda+\mu^2+2q}} e^{\mu(z-x)} \Big(e^{-|z-x|\sqrt{2\lambda+\mu^2+2q}} - e^{(2r-x-z)\sqrt{2\lambda+\mu^2+2q}}\Big) dz$$

1.27.6

$$\mathbf{P}_x\Big(\int_0^\tau \mathbb{I}_{(-\infty,r)}(W_s^{(\mu)})ds \in du, \int_0^\tau \mathbb{I}_{[r,\infty)}(W_s^{(\mu)})ds \in dv,$$

$$\tfrac{1}{2}\ell(\tau,r) \in dy, W_\tau^{(\mu)} \in dz\Big) = \lambda e^{\mu(z-x)} B^{(27)}_{\lambda+\mu^2/2}(u,v,y,z) du dv dy dz$$

for $B^{(27)}_\lambda(u,v,y,z)$ see 1.1.27.6

2 $W_s^{(\mu)} = \mu s + W_s$ $\tau \sim \operatorname{Exp}(\lambda)$, independent of $W^{(\mu)}$

1.27.8 $$\mathbf{P}_x\Big(\int_0^t \big(p\mathbb{I}_{(-\infty,r)}(W_s^{(\mu)}) + q\mathbb{I}_{[r,\infty)}(W_s^{(\mu)})\big)ds \in dv, \frac{1}{2}\ell(t,r) \in dy, W_t^{(\mu)} \in dz\Big)$$
$$= \frac{1}{|p-q|}\, e^{\mu(z-x)-\mu^2 t/2} B_x^{(27)}\Big(\frac{|qt-v|}{|p-q|}, \frac{|pt-v|}{|p-q|}, y, z\Big)\mathbb{I}_{((q\wedge p)t,(q\vee p)t)}(v)\,dvdydz$$

for $F_x^{(27)}(u,v,y,z)$ see 1.1.27.6

1.28.1 $$\mathbf{E}_x\big\{e^{-\gamma \hat H_\mu(\tau)-\eta \check H_\mu(\tau)}; \inf_{0\le s\le\tau} W_s^{(\mu)} \in da, \sup_{0\le s\le \tau} W_s^{(\mu)} \in db\big\}$$
$a < x$
$x < b$
$$= \frac{\Upsilon_\eta \operatorname{sh}((b-x)\Upsilon_{\gamma+\eta})\big[e^{\mu(b-x)}(\Upsilon_0 \operatorname{ch}((b-a)\Upsilon_0) - \mu \operatorname{sh}((b-a)\Upsilon_0)) - e^{\mu(a-x)}\Upsilon_0\big]}{\operatorname{sh}((b-a)\Upsilon_{\gamma+\eta})\operatorname{sh}((b-a)\Upsilon_\eta)\operatorname{sh}((b-a)\Upsilon_0)} dadb$$
$$+ \frac{\Upsilon_\gamma \operatorname{sh}((x-a)\Upsilon_{\gamma+\eta})\big[e^{\mu(a-x)}(\Upsilon_0 \operatorname{ch}((b-a)\Upsilon_0) + \mu \operatorname{sh}((b-a)\Upsilon_0)) - e^{\mu(b-x)}\Upsilon_0\big]}{\operatorname{sh}((b-a)\Upsilon_{\gamma+\eta})\operatorname{sh}((b-a)\Upsilon_\gamma)\operatorname{sh}((b-a)\Upsilon_0)} dadb$$

1.28.2 $$\mathbf{P}_x\big(\hat H(\tau) \in du, \check H(\tau) \in dv, \inf_{0\le s\le\tau} W_s^{(\mu)} \in da, \sup_{0\le s\le\tau} W_s^{(\mu)} \in db\big)$$

$u < v$ $$= \Big(\frac{\sqrt{2\lambda+\mu^2}\big(e^{\mu b}\operatorname{ch}((b-a)\sqrt{2\lambda+\mu^2}) - e^{\mu a}\big)}{\operatorname{sh}((b-a)\sqrt{2\lambda+\mu^2})} - \mu e^{\mu b}\Big)$$
$$\times e^{-\mu x-(\lambda+\mu^2/2)v}\operatorname{ss}_u(b-x, b-a)\operatorname{s}_{v-u}(b-a)\,dudvdadb$$

$v < u$ $$= \Big(\frac{\sqrt{2\lambda+\mu^2}\big(e^{\mu a}\operatorname{ch}((b-a)\sqrt{2\lambda+\mu^2}) - e^{\mu b}\big)}{\operatorname{sh}((b-a)\sqrt{2\lambda+\mu^2})} + \mu e^{\mu a}\Big)$$
$$\times e^{-\mu x-(\lambda+\mu^2/2)u}\operatorname{ss}_v(x-a, b-a)\operatorname{s}_{u-v}(b-a)\,dudvdadb$$

1.28.5 $$\mathbf{E}_x\big\{e^{-\gamma \hat H_\mu(\tau)-\eta \check H_\mu(\tau)}; \inf_{0\le s\le\tau} W_s^{(\mu)} \in da, \sup_{0\le s\le\tau} W_s^{(\mu)} \in db, W_\tau^{(\mu)} \in dz\big\}$$
$a < x$
$x < b$
$$= \frac{2\lambda e^{\mu(z-x)}\Upsilon_\eta \operatorname{sh}((b-x)\Upsilon_{\gamma+\eta})\operatorname{sh}((z-a)\Upsilon_0)}{\operatorname{sh}((b-a)\Upsilon_{\gamma+\eta})\operatorname{sh}((b-a)\Upsilon_\eta)\operatorname{sh}((b-a)\Upsilon_0)} dadbdz$$
$$+ \frac{2\lambda e^{\mu(z-x)}\Upsilon_\gamma \operatorname{sh}((x-a)\Upsilon_{\gamma+\eta})\operatorname{sh}((b-z)\Upsilon_0)}{\operatorname{sh}((b-a)\Upsilon_{\gamma+\eta})\operatorname{sh}((b-a)\Upsilon_\gamma)\operatorname{sh}((b-a)\Upsilon_0)} dadbdz$$

$$\Upsilon_s := \sqrt{2\lambda+\mu^2+2s}$$

2 $W_s^{(\mu)} = \mu s + W_s$ $\tau \sim \operatorname{Exp}(\lambda)$, independent of $W^{(\mu)}$

1.28.6 $\mathbf{P}_x\Big(\hat{H}(\tau) \in du, \check{H}(\tau) \in dv, \inf_{0\le s\le\tau} W_s^{(\mu)} \in da, \sup_{0\le s\le\tau} W_s^{(\mu)} \in db, W_\tau^{(\mu)} \in dz\Big)$

$u < v$ $$= 2\lambda e^{\mu(z-x)-(\lambda+\mu^2/2)v} \frac{\operatorname{sh}((z-a)\sqrt{2\lambda+\mu^2})}{\operatorname{sh}((b-a)\sqrt{2\lambda+\mu^2})} \times \operatorname{ss}_u(b-x, b-a)\, \operatorname{s}_{v-u}(b-a)\, du dv da db dz$$

$v < u$ $$= e^{\mu(z-x)-(\lambda+\mu^2/2)u} \frac{2\lambda \operatorname{sh}((b-z)\sqrt{2\lambda+\mu^2})}{\operatorname{sh}((b-a)\sqrt{2\lambda+\mu^2})} \times \operatorname{ss}_v(x-a, b-a)\, \operatorname{s}_{u-v}(b-a)\, du dv da db dz$$

1.28.8 $\mathbf{P}_x\Big(\hat{H}_\mu(t) \in du, \check{H}_\mu(t) \in dv, \inf_{0\le s\le t} W_s^{(\mu)} \in da, \sup_{0\le s\le t} W_s^{(\mu)} \in db, W_t^{(\mu)} \in dz\Big)$

for $0 < u \wedge v < u \vee v < t, \qquad a < x \wedge z < x \vee z < b$

$u < v$ $$= 2e^{\mu(z-x)-\mu^2 t/2}\, \operatorname{ss}_u(b-x, b-a)\, \operatorname{s}_{v-u}(b-a)\, \operatorname{ss}_{t-v}(z-a, b-a)\, du dv da db dz$$

$v < u$ $$= 2e^{\mu(z-x)-\mu^2 t/2}\, \operatorname{ss}_v(x-a, b-a)\, \operatorname{s}_{u-v}(b-a)\, \operatorname{ss}_{t-u}(b-z, b-a) du dv da db dz$$

1.29.1 $$\mathbf{E}_x\Big\{\exp\Big(-\int_0^\tau \big(p\mathbb{I}_{(-\infty,r)}(W_s^{(\mu)}) + q\mathbb{I}_{[r,\infty)}(W_s^{(\mu)})\big)ds\Big); a < \inf_{0\le s\le\tau} W_s^{(\mu)}, \sup_{0\le s\le\tau} W_s^{(\mu)} < b, \frac{1}{2}\ell(\tau, r) \in dy\Big\} e^{\mu(x-r)}$$

$$= \lambda\Big(\frac{\Upsilon_q(\operatorname{ch}((b-r)\Upsilon_q) - e^{\mu(b-r)})}{(\lambda+q)\operatorname{sh}((b-r)\Upsilon_q)} + \frac{\Upsilon_p(\operatorname{ch}((r-a)\Upsilon_p) - e^{\mu(a-r)})}{(\lambda+p)\operatorname{sh}((r-a)\Upsilon_p)} + \frac{\mu(p-q)}{(\lambda+p)(\lambda+q)}\Big)$$

$$\times \exp\Big(-\frac{y\Upsilon_q \operatorname{ch}((b-r)\Upsilon_q)}{\operatorname{sh}((b-r)\Upsilon_q)} - \frac{y\Upsilon_p \operatorname{ch}((r-a)\Upsilon_p)}{\operatorname{sh}((r-a)\Upsilon_p)}\Big) \begin{cases} \dfrac{\operatorname{sh}((x-a)\Upsilon_p)}{\operatorname{sh}((r-a)\Upsilon_p)} dy, & a \le x \le r \\ \dfrac{\operatorname{sh}((b-x)\Upsilon_q)}{\operatorname{sh}((b-r)\Upsilon_q)} dy, & r \le x \le b \end{cases}$$

$\Upsilon_s := \sqrt{2\lambda + \mu^2 + 2s}$

2 $W_s^{(\mu)} = \mu s + W_s$ $\qquad \tau \sim \operatorname{Exp}(\lambda)$, independent of $W^{(\mu)}$

(1) $$\mathbf{E}_x\Big\{\exp\Big(-\int_0^\tau \big(p\mathbb{I}_{(-\infty,r)}(W_s^{(\mu)}) + q\mathbb{I}_{[r,\infty)}(W_s^{(\mu)})\big)ds\Big); a < \inf_{0\le s\le\tau} W_s^{(\mu)}, \sup_{0\le s\le\tau} W_s^{(\mu)} < b, \ell(\tau,r)=0\Big\}$$

$$= \begin{cases} \dfrac{\lambda}{\lambda+p}\Big(1 - \dfrac{e^{\mu(a-x)}\operatorname{sh}((r-x)\Upsilon_p) + e^{\mu(r-x)}\operatorname{sh}((x-a)\Upsilon_p)}{\operatorname{sh}((r-a)\Upsilon_p)}\Big), & a\le x\le r \\ \dfrac{\lambda}{\lambda+q}\Big(1 - \dfrac{e^{\mu(r-x)}\operatorname{sh}((b-x)\Upsilon_q) + e^{\mu(b-x)}\operatorname{sh}((x-r)\Upsilon_q)}{\operatorname{sh}((b-r)\Upsilon_q)}\Big), & r\le x\le b \end{cases}$$

1.29.3 (1) $\mu < 0$ $$\mathbf{E}_x\Big\{\exp\Big(-q\int_0^\infty \mathbb{I}_{[r,\infty)}(W_s^{(\mu)})ds\Big); \sup_{0\le s<\infty} W_s^{(\mu)} < b, \frac12\ell(\infty,r)\in dy\Big\}$$

$x\le r$ $$= 2|\mu|\exp\Big(2\mu(r-x) - \frac{y\sqrt{\mu^2+2q}\operatorname{ch}((b-r)\sqrt{\mu^2+2q})}{\operatorname{sh}((b-r)\sqrt{\mu^2+2q})} + y\mu\Big)dy$$

$r\le x$, $x\le b$ $$= \frac{2|\mu|\operatorname{sh}((b-x)\sqrt{\mu^2+2q})}{\operatorname{sh}((b-r)\sqrt{\mu^2+2q})}\exp\Big(\mu(r-x+y) - \frac{y\sqrt{\mu^2+2q}\operatorname{ch}((b-r)\sqrt{\mu^2+2q})}{\operatorname{sh}((b-r)\sqrt{\mu^2+2q})}\Big)dy$$

(2) $\mu > 0$ $$\mathbf{E}_x\Big\{\exp\Big(-p\int_0^\infty \mathbb{I}_{(-\infty,r)}(W_s^{(\mu)})ds\Big); a < \inf_{0\le s<\infty} W_s^{(\mu)}, \frac12\ell(\infty,r)\in dy\Big\}$$

$a\le x$, $x\le r$ $$= \frac{2\mu\operatorname{sh}((x-a)\sqrt{\mu^2+2p})}{\operatorname{sh}((r-a)\sqrt{\mu^2+2p})}\exp\Big(\mu(r-x-y) - \frac{y\sqrt{\mu^2+2p}\operatorname{ch}((r-a)\sqrt{\mu^2+2p})}{\operatorname{sh}((r-a)\sqrt{\mu^2+2p})}\Big)dy$$

$r\le x$ $$= 2\mu\exp\Big(2\mu(r-x) - \frac{y\sqrt{\mu^2+2p}\operatorname{ch}((r-a)\sqrt{\mu^2+2p})}{\operatorname{sh}((r-a)\sqrt{\mu^2+2p})} - y\mu\Big)dy$$

1.29.4 (1) $\mu < 0$ $$\mathbf{P}_x\Big(\int_0^\infty \mathbb{I}_{[r,\infty)}(W_s^{(\mu)})ds\in du, \sup_{0\le s<\infty} W_s^{(\mu)} < b, \frac12\ell(\infty,r)\in dy\Big)$$

$$= \begin{cases} 2|\mu|e^{2\mu(r-x)+\mu y}e^{-\mu^2u/2}\operatorname{es}_u(0,0,b-r,0,y)du\,dy, & x\le r \\ 2|\mu|e^{\mu(r-x)+\mu y}e^{-\mu^2u/2}\operatorname{ess}_u(b-x,b-r,0,y)du\,dy, & r\le x\le b \end{cases}$$

(2) $\mu > 0$ $$\mathbf{P}_x\Big(\int_0^\infty \mathbb{I}_{(-\infty,r)}(W_s^{(\mu)})ds\in du,\ a < \inf_{0\le s<\infty} W_s^{(\mu)}, \frac12\ell(\infty,r)\in dy\Big)$$

2 $W_s^{(\mu)} = \mu s + W_s$ $\quad\tau \sim \text{Exp}(\lambda)$, independent of $W^{(\mu)}$

$$= \begin{cases} 2\mu e^{\mu(r-x)-\mu y} e^{-\mu^2 u/2} \operatorname{ess}_u(x-a, r-a, 0, y)dudy, & a \le x \le r \\ 2\mu e^{2\mu(r-x)-\mu y} e^{-\mu^2 u/2} \operatorname{es}_u(0, 0, r-a, 0, y)dudy, & r \le x \end{cases}$$

1.29.5 $$\mathbf{E}_x\Big\{\exp\Big(-\int_0^\tau \big(p\mathbb{I}_{(-\infty,r)}(W_s^{(\mu)}) + q\mathbb{I}_{[r,\infty)}(W_s^{(\mu)})\big)ds\Big); a < \inf_{0\le s\le\tau} W_s^{(\mu)}, \sup_{0\le s\le\tau} W_s^{(\mu)} < b, \frac{1}{2}\ell(\tau,r) \in dy, W_\tau^{(\mu)} \in dz\Big\} = \lambda e^{\mu(z-x)} Q^{(29)}_{\lambda+\mu^2/2}(y,z)dydz$$

for $Q_\lambda^{(29)}(y,z)$ see 1.1.29.5

(1) $$\mathbf{E}_x\Big\{\exp\Big(-\int_0^\tau \big(p\mathbb{I}_{(-\infty,r)}(W_s^{(\mu)}) + q\mathbb{I}_{[r,\infty)}(W_s^{(\mu)})\big)ds\Big); a < \inf_{0\le s\le\tau} W_s^{(\mu)}, \sup_{0\le s\le\tau} W_s^{(\mu)} < b, \ell(\tau,r) = 0, W_\tau^{(\mu)} \in dz\Big\}$$

$x \le r$, $z \le r$ $$= \frac{\lambda(\operatorname{ch}((r-a-|z-x|)\sqrt{2\lambda+\mu^2+2p}) - \operatorname{ch}((r+a-z-x)\sqrt{2\lambda+\mu^2+2p}))}{e^{\mu(x-z)}\sqrt{2\lambda+\mu^2+2p}\operatorname{sh}((r-a)\sqrt{2\lambda+\mu^2+2p})}dz$$

$r \le x$, $r \le z$ $$= \frac{\lambda(\operatorname{ch}((b-r-|z-x|)\sqrt{2\lambda+\mu^2+2q}) - \operatorname{ch}((b+r-z-x)\sqrt{2\lambda+\mu^2+2q}))}{e^{\mu(x-z)}\sqrt{2\lambda+\mu^2+2q}\operatorname{sh}((b-r)\sqrt{2\lambda+\mu^2+2q})}dz$$

1.29.6 $$\mathbf{P}_x\Big(\int_0^\tau \mathbb{I}_{(-\infty,r)}(W_s^{(\mu)})ds \in du, \int_0^\tau \mathbb{I}_{[r,\infty)}(W_s^{(\mu)})ds \in dv, a < \inf_{0\le s\le\tau} W_s^{(\mu)}, \sup_{0\le s\le\tau} W_s^{(\mu)} < b, \frac{1}{2}\ell(\tau,r) \in dy, W_\tau^{(\mu)} \in dz\Big)$$

$$= \lambda e^{\mu(z-x)-(\lambda+\mu^2/2)(u+v)} B_x^{(29)}(u,v,y,z)dudvdydz$$

for $B_x^{(29)}(u,v,y,z)$ see 1.1.29.6

1.29.8 $$\mathbf{P}_x\Big(\int_0^t \big(p\mathbb{I}_{(-\infty,r)}(W_s^{(\mu)}) + q\mathbb{I}_{[r,\infty)}(W_s^{(\mu)})\big)ds \in dv,\ a < \inf_{0\le s\le t} W_s^{(\mu)}, \sup_{0\le s\le t} W_s^{(\mu)} < b, \frac{1}{2}\ell(t,r) \in dy, W_t^{(\mu)} \in dz\Big)$$

2 $W_s^{(\mu)} = \mu s + W_s$ $\tau \sim \operatorname{Exp}(\lambda)$, independent of $W^{(\mu)}$

$$= \frac{1}{|p-q|} e^{\mu(z-x)-\mu^2 t/2} B_x^{(29)}\Big(\frac{|qt-v|}{|p-q|}, \frac{|pt-v|}{|p-q|}, y, z\Big) \mathbb{I}_{((q\wedge p)t,(q\vee p)t)}(v)\, dvdydz$$

for $B_x^{(29)}(u,v,y,z)$ see 1.1.29.6

1.30.1 (1) $\mathbf{E}_x\Big\{\exp\Big(-\frac{p^2}{2}\int_0^\tau e^{2\beta W_s^{(\mu)}}ds\Big); \int_0^\tau e^{\beta W_s^{(\mu)}}ds \in dy\Big\}$

$\beta \neq 0$

$$= \lambda\Big(\frac{p\operatorname{ch}(yp|\beta|/2)}{|\beta|\operatorname{sh}(yp|\beta|/2)}\Big)^{1/2-\mu/\beta} \frac{\Gamma(\mu/\beta+\sqrt{2\lambda+\mu^2}/|\beta|)}{\Gamma(1+2\sqrt{2\lambda+\mu^2}/|\beta|)}$$

$$\times \exp\Big(-\mu x - \frac{\beta x}{2} - \frac{pe^{\beta x}\operatorname{ch}(yp|\beta|)}{|\beta|\operatorname{sh}(yp|\beta|)}\Big) M_{1/2-\mu/\beta,\sqrt{2\lambda+\mu^2}/|\beta|}\Big(\frac{2pe^{\beta x}}{|\beta|\operatorname{sh}(yp|\beta|)}\Big)dy$$

1.30.3 (1) $\mathbf{E}_x\Big\{\exp\Big(-\frac{p^2}{2}\int_0^t e^{2\beta W_s^{(\mu)}}ds\Big); \int_0^t e^{\beta W_s^{(\mu)}}ds \in dy\Big\} = \Big(\frac{p\operatorname{ch}(yp|\beta|/2)}{|\beta|\operatorname{sh}(yp|\beta|/2)}\Big)^{1/2-\mu/\beta}$

$\beta \neq 0$

$$\times \frac{\beta^2}{2}\exp\Big(-\mu x - \frac{\beta x}{2} - \frac{\mu^2 t}{2} - \frac{pe^{\beta x}\operatorname{ch}(yp|\beta|)}{|\beta|\operatorname{sh}(yp|\beta|)}\Big) \mathrm{m}_{\beta^2 t/2}\Big(\frac{\mu}{\beta}-\frac{1}{2}, \frac{pe^{\beta x}}{|\beta|\operatorname{sh}(yp|\beta|)}\Big)dy$$

(2) $\mathbf{E}_x \exp\Big(-\int_0^\infty \Big(\frac{p^2}{2}e^{2\beta W_s^{(\mu)}} + qe^{\beta W_s^{(\mu)}}\Big)ds\Big)$

$\frac{\mu}{\beta} < 0$

$$= \frac{\Gamma\Big(\frac{|\mu|}{|\beta|}+\frac{q}{p|\beta|}+\frac{1}{2}\Big)}{\Gamma\Big(\frac{2|\mu|}{|\beta|}\Big)}\Big(\frac{2p}{|\beta|}\Big)^{|\mu|/|\beta|-1/2} e^{-(\mu+\beta/2)x} W_{-q/p|\beta|,|\mu|/|\beta|}\Big(\frac{2p}{|\beta|}e^{\beta x}\Big)$$

(3) $\mathbf{E}_x\Big\{\exp\Big(-\frac{p^2}{2}\int_0^\infty e^{2\beta W_s^{(\mu)}}ds\Big); \int_0^\infty e^{\beta W_s^{(\mu)}}ds \in dy\Big\}$

$\frac{\mu}{\beta} < 0$

$$= \frac{\beta^2}{2\Gamma(2|\mu|/|\beta|)}\Big(\frac{p}{|\beta|\operatorname{sh}(yp|\beta|/2)}\Big)^{1+2|\mu|/|\beta|} \exp\Big(-2\mu x - \frac{pe^{\beta x}\operatorname{ch}(yp|\beta|/2)}{|\beta|\operatorname{sh}(yp|\beta|/2)}\Big)dy$$

1.30.4 (1) $\mathbf{P}_x\Big(\int_0^\infty e^{2\beta W_s^{(\mu)}}ds \in dg, \int_0^\infty e^{\beta W_s^{(\mu)}}ds \in dy\Big)$

2 $W_s^{(\mu)} = \mu s + W_s$ $\qquad\tau \sim \operatorname{Exp}(\lambda)$, independent of $W^{(\mu)}$

$\frac{\mu}{\beta} < 0$

$$= \frac{|\beta|^{1-2|\mu|/|\beta|} e^{-2\mu x}}{2\Gamma(2|\mu|/|\beta|)} \operatorname{ee}_g\Big(1 + \frac{2|\mu|}{|\beta|}, \frac{y|\beta|}{2}, \frac{e^{\beta x}}{|\beta|}, 0\Big) dg dy$$

1.30.5

$$\mathbf{E}_x\Big\{\exp\Big(-\int_0^\tau \Big(\frac{p^2}{2} e^{2\beta W_s^{(\mu)}} + q e^{\beta W_s^{(\mu)}}\Big) ds\Big);\ W_\tau^{(\mu)} \in dz\Big\}$$

$$= \lambda e^{\mu(z-x)} Q^{(30)}_{\lambda+\mu^2/2}(z) dz, \qquad \text{for } \ Q^{(30)}_\lambda(z) \ \text{ see 1.1.30.5}$$

(1)

$$\mathbf{E}_x\Big\{\exp\Big(-\frac{p^2}{2}\int_0^\tau e^{2\beta W_s^{(\mu)}} ds\Big);\ \int_0^\tau e^{\beta W_s^{(\mu)}} ds \in dy, W_\tau^{(\mu)} \in dz\Big\}$$

$$= \lambda e^{\mu(z-x)} B^{(30)}_{\lambda+\mu^2/2}(y,z) dy dz, \qquad \text{for } \ B^{(30)}_\lambda(y,z) \ \text{ see 1.1.30.5(1)}$$

1.30.6

$$\mathbf{P}_x\Big(\int_0^\tau e^{2\beta W_s^{(\mu)}} ds \in dg,\ \int_0^\tau e^{\beta W_s^{(\mu)}} ds \in dy,\ W_\tau^{(\mu)} \in dz\Big)$$

$\beta \neq 0$

$$= \lambda e^{\mu(z-x)} \operatorname{is}_g\Big(\frac{2\sqrt{2\lambda+\mu^2}}{|\beta|}, \frac{y|\beta|}{2}, 0, \frac{1}{|\beta|}\big(e^{\beta x} + e^{\beta z}\big), \frac{1}{|\beta|} e^{\beta(x+z)/2}\Big) dg dy dz$$

1.30.7 (1)

$$\mathbf{E}_x\Big\{\exp\Big(-\frac{p^2}{2}\int_0^t e^{2\beta W_s^{(\mu)}} ds\Big);\ \int_0^t e^{\beta W_s^{(\mu)}} ds \in dy, W_t^{(\mu)} \in dz\Big\}$$

$$= e^{\mu(z-x)-\mu^2 t/2} F^{(30)}_{x,t}(y,z) dy dz, \qquad \text{for } \ F^{(30)}_{x,t}(y,z) \ \text{ see 1.1.30.7(1)}$$

1.30.8

$$\mathbf{P}_x\Big(\int_0^t e^{2\beta W_s^{(\mu)}} ds \in dg,\ \int_0^t e^{\beta W_s^{(\mu)}} ds \in dy,\ W_t^{(\mu)} \in dz\Big)$$

$\beta \neq 0$

$$= \frac{\beta^2}{8} e^{\mu(z-x)-\mu^2 t/2} \operatorname{ei}_g\Big(\frac{\beta^2 t}{8}, \frac{y|\beta|}{2}, \frac{1}{|\beta|}\big(e^{\beta x} + e^{\beta z}\big), \frac{2}{|\beta|} e^{\beta(x+z)/2}\Big) dg dy dz$$

1.31.1 $\quad 0 < \alpha$

$$\mathbf{E}_{x/\sqrt{2\lambda+\mu^2}} \exp\Big(-\gamma\sqrt{2\lambda+\mu^2} \sup_{0\le s\le\tau} \Big(\ell\Big(s, \frac{u}{\sqrt{2\lambda+\mu^2}}\Big) - \alpha^{-1}\ell\Big(s, \frac{r}{\sqrt{2\lambda+\mu^2}}\Big)\Big)\Big)$$

$$= 1 - \frac{\gamma(2 - 2e^{-2|u-r|} - \Upsilon)e^{|u-r|-|x-r|} + \gamma\Upsilon e^{-|u-x|}}{2\gamma(1 - e^{-2|u-r|}) + \Upsilon} e^{(u-x)\mu/\sqrt{2\lambda+\mu^2}}$$

2 $W_s^{(\mu)} = \mu s + W_s$ $\qquad \tau \sim \text{Exp}(\lambda)$, independent of $W^{(\mu)}$

1.31.2
$$\mathbf{P}_{x/\sqrt{2\lambda+\mu^2}}\Big(\sup_{0\le s\le\tau}\big(\ell(s,\tfrac{u}{\sqrt{2\lambda+\mu^2}}) - \alpha^{-1}\ell(s,\tfrac{r}{\sqrt{2\lambda+\mu^2}})\big) > \frac{h}{\sqrt{2\lambda+\mu^2}}\Big)$$

$0<\alpha$
$$= \Big[e^{|u-r|-|x-r|} - \frac{\Upsilon}{2(1-e^{-2|u-r|})}\big(e^{|u-r|-|x-r|} - e^{-|u-x|}\big)\Big] \times \exp\Big(\frac{\mu(u-x)}{\sqrt{2\lambda+\mu^2}} - \frac{\Upsilon h}{2(1-e^{-2|u-r|})}\Big)$$

1.31.3 (1)
$$\mathbf{E}_{x/|\mu|}\exp\Big(-\gamma|\mu|\sup_{0\le s<\infty}\big(\ell(s,\tfrac{u}{|\mu|}) - \alpha^{-1}\ell(s,\tfrac{r}{|\mu|})\big)\Big)$$
$$= 1 - \frac{\gamma(2-2e^{-2|u-r|}-\Upsilon)e^{|u-r|-|x-r|} + \gamma\Upsilon e^{-|u-x|}}{2\gamma(1-e^{-2|u-r|})+\Upsilon}e^{(u-x)\mu/|\mu|}$$

1.31.4 (1)
$$\mathbf{P}_{x/|\mu|}\Big(\sup_{0\le s<\infty}\big(\ell(s,\tfrac{u}{|\mu|}) - \alpha^{-1}\ell(s,\tfrac{r}{|\mu|})\big) > \frac{h}{|\mu|}\Big) = \exp\Big(-\frac{\Upsilon h}{2(1-e^{-2|u-r|})}\Big)$$

$0<\alpha$
$$\times e^{(u-x)\mu/|\mu|}\Big[e^{|u-r|-|x-r|} - \frac{\Upsilon}{2(1-e^{-2|u-r|})}\big(e^{|u-r|-|x-r|} - e^{-|u-x|}\big)\Big]$$

1.31.5 $0<\alpha$
$$\mathbf{E}_{x/\sqrt{2\lambda+\mu^2}}\Big\{\exp\Big(-\gamma\sqrt{2\lambda+\mu^2}\sup_{0\le s\le\tau}\big(\ell(s,\tfrac{u}{\sqrt{2\lambda+\mu^2}}) - \alpha^{-1}\ell(s,\tfrac{r}{\sqrt{2\lambda+\mu^2}})\big)\Big);$$
$$W_\tau^{(\mu)} \in \frac{dz}{\sqrt{2\lambda+\mu^2}}\Big\} = e^{(z-x)\mu/\sqrt{2\lambda+\mu^2}}\Big(\frac{\lambda}{2\lambda+\mu^2}e^{-|z-x|}$$
$$- \frac{\lambda\gamma e^{-|z-u|}\big[(2-2e^{-2|u-r|}-\Upsilon)e^{|u-r|-|x-r|} + \Upsilon e^{-|u-x|}\big]}{(2\lambda+\mu^2)(2\gamma(1-e^{-2|u-r|})+\Upsilon)}\Big)dz$$

1.31.6 $0<\alpha$
$$\mathbf{P}_{x/\sqrt{2\lambda+\mu^2}}\Big(\sup_{0\le s\le\tau}\big(\ell(s,\tfrac{u}{\sqrt{2\lambda+\mu^2}}) - \alpha^{-1}\ell(s,\tfrac{r}{\sqrt{2\lambda+\mu^2}})\big) > \frac{h}{\sqrt{2\lambda+\mu^2}},$$
$$W_\tau^{(\mu)} \in \frac{dz}{\sqrt{2\lambda+\mu^2}}\Big) = \frac{\lambda}{2\lambda+\mu^2}\exp\Big(\frac{\mu(z-x)}{\sqrt{2\lambda+\mu^2}} - |z-u| - \frac{\Upsilon h}{2(1-e^{-2|u-r|})}\Big)$$
$$\times\Big[e^{|u-r|-|x-r|} - \frac{\Upsilon}{2(1-e^{-2|u-r|})}\Big(e^{|u-r|-|x-r|} - e^{-|u-x|}\Big)\Big]dz$$

$$\Upsilon := \sqrt{(\alpha-1)^2 + 4\alpha(1-e^{-2|u-r|})} + 1 - \alpha$$

2 $W_s^{(\mu)} = \mu s + W_s$ $H_z = \min\{s : W_s^{(\mu)} = z\}$

2. Stopping at first hitting time

2.0.1 $\mathbf{E}_x e^{-\alpha H_z} = \mathbf{E}_x\{e^{-\alpha H_z};\ H_z < \infty\} = e^{\mu(z-x)-|z-x|\sqrt{2\alpha+\mu^2}}$

2.0.2 $\mathbf{P}_x\big(H_z \in dt\big) = \frac{|z-x|}{\sqrt{2\pi}t^{3/2}} \exp\Big(-\frac{(z-x-\mu t)^2}{2t}\Big)dt$

(1) $\mathbf{P}_x\big(H_z = \infty\big) = 1 - e^{\mu(z-x)-|\mu||z-x|}$

2.1.2, $z < y$: $\mathbf{P}_x\big(\sup_{0\le s\le H_z} W_s^{(\mu)} < y, H_z < \infty\big) = \begin{cases} e^{(\mu-|\mu|)(z-x)}, & x \le z \\ e^{\mu(z-x)} \frac{\mathrm{sh}(|\mu|(y-x))}{\mathrm{sh}(|\mu|(y-z))}, & z \le x \le y \end{cases}$

(1), $z \le x$: $\mathbf{P}_x\big(\sup_{0\le s\le H_z} W_s^{(\mu)} = \infty\big) = 1 - e^{-(\mu+|\mu|)(x-z)}$

2.1.4 $\mathbf{E}_x\big\{e^{-\alpha H_z};\ \sup_{0\le s\le H_z} W_s^{(\mu)} < y\big\} = \begin{cases} e^{(z-x)(\mu-\sqrt{2\alpha+\mu^2})}, & x \le z \le y \\ \frac{\mathrm{sh}((y-x)\sqrt{2\alpha+\mu^2})}{\mathrm{sh}((y-z)\sqrt{2\alpha+\mu^2})} e^{\mu(z-x)}, & z \le x \le y \end{cases}$

(1), $z \le x$, $x \le y$: $\mathbf{P}_x\big(\sup_{0\le s\le H_z} W_s^{(\mu)} < y,\ H_z \in dt\big) = e^{\mu(z-x)-\mu^2 t/2}\,\mathrm{ss}_t(y-x, y-z)dt$

2.2.2 $\mathbf{P}_x\big(\inf_{0\le s\le H_z} W_s^{(\mu)} > y,\ H_z < \infty\big) = \begin{cases} e^{\mu(z-x)} \frac{\mathrm{sh}(|\mu|(x-y))}{\mathrm{sh}(|\mu|(z-y))}, & y \le x \le z \\ e^{-(\mu+|\mu|)(x-z)}, & y \le z \le x \end{cases}$

(1), $x \le z$: $\mathbf{P}_x\big(\inf_{0\le s\le H_z} W_s^{(\mu)} = -\infty\big) = 1 - e^{(\mu-|\mu|)(z-x)}$

2.2.4 $\mathbf{E}_x\big\{e^{-\alpha H_z};\ \inf_{0\le s\le H_z} W_s^{(\mu)} > y\big\} = \begin{cases} \frac{\mathrm{sh}((x-y)\sqrt{2\alpha+\mu^2})}{\mathrm{sh}((z-y)\sqrt{2\alpha+\mu^2})} e^{\mu(z-x)}, & y \le x \le z \\ e^{(z-x)(\mu+\sqrt{2\alpha+\mu^2})}, & y \le z \le x \end{cases}$

(1), $z \le x$, $y \le x$: $\mathbf{P}_x\big(\inf_{0\le s\le H_z} W_s^{(\mu)} > y,\ H_z \in dt\big) = e^{\mu(z-x)-\mu^2 t/2}\,\mathrm{ss}_t(x-y, z-y)dt$

2 $W_s^{(\mu)} = \mu s + W_s$ $H_z = \min\{s : W_s^{(\mu)} = z\}$

2.3.1 $\mathbf{E}_x\{e^{-\gamma\ell(H_z,r)};\ H_z < \infty\}$

$$= e^{\mu(z-x)-|\mu||z-x|}\begin{cases} 1, & r \le z \le x, \quad x \le z \le r \\ \dfrac{|\mu| + \gamma(1 - e^{-2|\mu||x-r|})}{|\mu| + \gamma(1 - e^{-2|\mu||z-r|})}, & z \wedge r \le x \le z \vee r \\ \dfrac{|\mu|}{|\mu| + \gamma(1 - e^{-2|\mu||z-r|})}, & z \le r \le x, \quad x \le r \le z \end{cases}$$

(1) $\mathbf{E}_x\{e^{-\gamma\ell(H_z,r)};\ H_z = \infty\}$ for $\mu(z-r) < 0$

$$= \begin{cases} 0, & r \le z \le x, \quad x \le z \le r \\ \dfrac{|\mu|(1 - e^{-2|\mu||x-z|})}{|\mu| + \gamma(1 - e^{-2|\mu||z-r|})}, & z \wedge r \le x \le z \vee r \\ 1 - e^{-2|\mu||x-r|} + \dfrac{|\mu|(e^{-2|\mu||x-r|} - e^{-2|\mu||x-z|})}{|\mu| + \gamma(1 - e^{-2|\mu||z-r|})}, & z \le r \le x, \quad x \le r \le z \end{cases}$$

2.3.2 $\mathbf{P}_x\big(\ell(H_z,r) \in dy,\ H_z < \infty\big) = \exp\Big(\mu(z-x) - \dfrac{|\mu|y}{1 - e^{-2|\mu||z-r|}}\Big)$

$$\times\begin{cases} 0, & x \le z \le r, \quad r \le z \le x \\ \dfrac{|\mu|\operatorname{sh}(|\mu||z-x|)}{2\operatorname{sh}^2(|\mu||z-r|)}dy, & r \wedge z \le x \le r \vee z \\ \dfrac{|\mu|e^{-|\mu||r-x|}}{2\operatorname{sh}(|\mu||z-r|)}dy, & z \le r \le x, \quad x \le r \le z \end{cases}$$

(1) $\mathbf{P}_x\big(\ell(H_z,r) \in dy,\ H_z = \infty\big)$ for $\mu(z-r) < 0$

$$= \begin{cases} 0, & x \le z \le r, \quad r \le z \le x \\ \dfrac{|\mu|(1 - e^{-2|\mu||x-z|})}{1 - e^{-2|\mu||z-r|}}\exp\Big(-\dfrac{|\mu|y}{1 - e^{-2|\mu||z-r|}}\Big)dy, & r \wedge z \le x \le r \vee z \\ \dfrac{|\mu|(e^{-2|\mu||x-r|} - e^{-2|\mu||x-z|})}{1 - e^{-2|\mu||z-r|}}\exp\Big(-\dfrac{|\mu|y}{1 - e^{-2|\mu||z-r|}}\Big)dy, & \begin{matrix} x \le r \le z \\ z \le r \le x \end{matrix} \end{cases}$$

2.3.3 $\mathbf{E}_x e^{-\alpha H_z - \gamma\ell(H_z,r)} = \exp\big(\mu(z-x) - |z-x|\sqrt{2\alpha+\mu^2}\big)$

$$\times\begin{cases} 1, & x \le z \le r, \quad r \le z \le x \\ \dfrac{\sqrt{2\alpha+\mu^2} + \gamma(1 - e^{-2|x-r|\sqrt{2\alpha+\mu^2}})}{\sqrt{2\alpha+\mu^2} + \gamma(1 - e^{-2|z-r|\sqrt{2\alpha+\mu^2}})}, & r \wedge z \le x \le r \vee z \\ \dfrac{\sqrt{2\alpha+\mu^2}}{\sqrt{2\alpha+\mu^2} + \gamma(1 - e^{-2|z-r|\sqrt{2\alpha+\mu^2}})}, & z \le r \le x, \quad x \le r \le z \end{cases}$$

2 $W_s^{(\mu)} = \mu s + W_s$ $H_z = \min\{s : W_s^{(\mu)} = z\}$

2.3.4 $$\mathbf{E}_x\{e^{-\alpha H_z}; \ell(H_z,r)\in dy\} = \exp\Big(\mu(z-x) - \frac{y\sqrt{2\alpha+\mu^2}}{1-e^{-2|r-z|\sqrt{2\alpha+\mu^2}}}\Big)$$

$$\times\begin{cases} 0, & x\le z\le r, \quad r\le z\le x \\ \dfrac{\sqrt{2\alpha+\mu^2}\,\mathrm{sh}(|x-z|\sqrt{2\alpha+\mu^2})}{2\,\mathrm{sh}^2(|r-z|\sqrt{2\alpha+\mu^2})}dy, & z\wedge r\le x\le r\vee z \\ \dfrac{\sqrt{2\alpha+\mu^2}}{2\,\mathrm{sh}(|r-z|\sqrt{2\alpha+\mu^2})}e^{-|r-x|\sqrt{2\alpha+\mu^2}}dy, & z\le r\le x, \quad x\le r\le z \end{cases}$$

(1) $$\mathbf{E}_x\{e^{-\alpha H_z}; \ell(H_z,r)=0\}$$

$$= e^{\mu(z-x)}\begin{cases} e^{-|z-x|\sqrt{2\alpha+\mu^2}}, & x\le z\le r, \quad r\le z\le x \\ \dfrac{\mathrm{sh}(|x-r|\sqrt{2\alpha+\mu^2})}{\mathrm{sh}(|z-r|\sqrt{2\alpha+\mu^2})}, & r\wedge z\le x\le r\vee z \\ 0, & z\le r\le x, \quad x\le r\le z \end{cases}$$

(2) $$\mathbf{P}_x\Big(H_z\in dt,\ \frac{1}{2}\ell(H_z,r)\in dy\Big)$$

$$= e^{\mu(z-x)-\mu^2t/2}\begin{cases} \frac{1}{2}(\mathrm{es}_t(1,2,|r-z|,y-|x-z|,y) \\ \quad -\mathrm{es}_t(1,2,|r-z|,y+|x-z|,y))dtdy, & z\wedge r\le x\le r\vee z \\ \mathrm{es}_t(1,1,|r-z|,y+|x-r|,y)dtdy, & z\le r\le x,\ x\le r\le z \end{cases}$$

2.4.1 $$\mathbf{E}_x\Big\{\exp\Big(-\gamma\int_0^{H_z}\mathbb{I}_{[r,\infty)}(W_s^{(\mu)})ds\Big);\ H_z<\infty\Big\} = e^{\mu(z-x)}Q_z^{(6)}(\mu^2,\mu^2+2\gamma)$$

for $Q_z^{(6)}(2p,2q)$ see 1.2.6.1

(1) $\mu<0$ $r\le z$ $$\mathbf{E}_x\Big\{\exp\Big(-\gamma\int_0^{H_z}\mathbb{I}_{[r,\infty)}(W_s^{(\mu)})ds\Big);\ H_z=\infty\Big\}$$

$$= \begin{cases} 1-\dfrac{e^{2(x-r)|\mu|}(\sqrt{2\gamma+\mu^2}\,\mathrm{ch}((z-r)\sqrt{2\gamma+\mu^2})-|\mu|\,\mathrm{sh}((z-r)\sqrt{2\gamma+\mu^2}))}{\sqrt{2\gamma+\mu^2}\,\mathrm{ch}((z-r)\sqrt{2\gamma+\mu^2})+|\mu|\,\mathrm{sh}((z-r)\sqrt{2\gamma+\mu^2})}, & x\le r \\ \dfrac{2|\mu|e^{(x-r)|\mu|}\,\mathrm{sh}((z-x)\sqrt{2\gamma+\mu^2})}{\sqrt{2\gamma+\mu^2}\,\mathrm{ch}((z-r)\sqrt{2\gamma+\mu^2})+|\mu|\,\mathrm{sh}((z-r)\sqrt{2\gamma+\mu^2})}, & r\le x\le z \\ 0, & z\le x \end{cases}$$

2 $W_s^{(\mu)} = \mu s + W_s$ $H_z = \min\{s : W_s^{(\mu)} = z\}$

2.4.2 $$\mathbf{P}_x\Big(\int_0^{H_z} \mathbb{I}_{[r,\infty)}\big(W_s^{(\mu)}\big)ds \in dy,\ H_z < \infty\Big)e^{\mu(x-z)}$$

$r < z$
$$= \begin{cases} |\mu|^{-1}e^{(x-r)|\mu|-\mu^2y/2}\widetilde{\mathrm{rc}}_y(1, z-r, 0, 1/|\mu|)dy, & x \le r \\ e^{-\mu^2y/2}\big(\mathrm{ss}_y(x-r, z-r) \\ +|\mu|^{-1}\,\mathrm{ss}_y(z-x, z-r) * \widetilde{\mathrm{rc}}_y(1, z-r, 0, 1/|\mu|)\big)dy, & r \le x \le z \\ e^{-\mu^2y/2}\,\mathrm{h}_y(1, x-z)dy, & z \le x \end{cases}$$

$z < r$
$$= \begin{cases} 0, & x \le z \\ \dfrac{|\mu|\,\mathrm{sh}((x-z)|\mu|)}{\mathrm{sh}^2((r-z)|\mu|)}\Big(\dfrac{e^{-\mu^2y/2}}{\sqrt{2\pi y}} - \dfrac{|\mu|\,\mathrm{ch}((r-z)|\mu|)}{2\,\mathrm{sh}((r-z)|\mu|)} \\ \quad\times\exp\Big(\dfrac{\mu^2y}{2\,\mathrm{sh}^2((r-z)|\mu|)}\Big)\,\mathrm{Erfc}\Big(\dfrac{|\mu|\,\mathrm{ch}((r-z)|\mu|)\sqrt{y}}{\sqrt{2}\,\mathrm{sh}((r-z)|\mu|)}\Big)\Big)dy, & z \le x \le r \\ \dfrac{|\mu|e^{-\mu^2y/2-(x-r)^2/2y}}{\sqrt{2\pi y}\,\mathrm{sh}((r-z)|\mu|)}dy - \dfrac{|\mu|\,\mathrm{ch}((r-z)|\mu|)}{2\,\mathrm{sh}^2((r-z)|\mu|)}\exp\Big(\dfrac{(x-r)|\mu|\,\mathrm{ch}((r-z)|\mu|)}{\mathrm{sh}((r-z)|\mu|)} \\ \quad+\dfrac{\mu^2y}{2\,\mathrm{sh}^2((r-z)|\mu|)}\Big)\,\mathrm{Erfc}\Big(\dfrac{x-r}{\sqrt{2y}} + \dfrac{|\mu|\,\mathrm{ch}((r-z)|\mu|)\sqrt{y}}{\sqrt{2}\,\mathrm{sh}((r-z)|\mu|)}\Big)dy, & r \le x \end{cases}$$

(1) $\mu < 0$ $$\mathbf{P}_x\Big(\int_0^{H_z} \mathbb{I}_{[r,\infty)}\big(W_s^{(\mu)}\big)ds \in dy,\ H_z = \infty\Big)$$

$$= \begin{cases} 2e^{2(x-r)|\mu|-\mu^2y/2}\,\mathrm{rc}_y(0, z-r, z-r, 1/|\mu|)dy, & x \le r \\ 2e^{(x-r)|\mu|-\mu^2y/2}\,\mathrm{rc}_y(0, z-r, z-x, 1/|\mu|)dy, & r \le x \le z \end{cases}$$

2.4.4 $$\mathbf{P}_x\Big(\int_0^{H_z} \mathbb{I}_{[r,\infty)}\big(W_s^{(\mu)}\big)ds \in dv,\ H_z \in dt\Big)$$
$$= e^{\mu(z-x)-\mu^2t/2}\int_0^\infty B_z^{(27)}(t-v, v, y)dy\,dvdt,$$ for $B_z^{(27)}(u, v, y)$ see 1.2.27.2

(1) $$\mathbf{P}_x\Big(\int_0^{H_z} \mathbb{I}_{[r,\infty)}\big(W_s^{(\mu)}\big)ds = 0,\ H_z \in dt\Big)$$

$$= \begin{cases} e^{\mu(z-x)-\mu^2t/2}\,\mathrm{h}_t(1, z-x)dt, & x \le z \\ e^{\mu(z-x)-\mu^2t/2}\,\mathrm{ss}_t(r-x, r-z)dt, & z \le x \le r \end{cases}$$

2.5.1 $$\mathbf{E}_x\Big\{\exp\Big(-\gamma\int_0^{H_z} \mathbb{I}_{(-\infty,r]}\big(W_s^{(\mu)}\big)ds\Big);\ H_z < \infty\Big\} = e^{\mu(z-x)}Q_z^{(6)}(\mu^2+2\gamma, \mu^2)$$

for $Q_z^{(6)}(2p, 2q)$ see 1.2.6.1

2 $W_s^{(\mu)} = \mu s + W_s$ $H_z = \min\{s : W_s^{(\mu)} = z\}$

(1) $\mathbf{E}_x\Big\{\exp\Big(-\gamma\int_0^{H_z} \mathbb{I}_{(-\infty,r]}\big(W_s^{(\mu)}\big)ds\Big);\ H_z=\infty\Big\}$

$z \le r$
$$= \begin{cases} 0, & x \le z \\ \dfrac{2|\mu|e^{(r-x)|\mu|}\,\mathrm{sh}((x-z)\sqrt{2\gamma+\mu^2})}{\sqrt{2\gamma+\mu^2}\,\mathrm{ch}((r-z)\sqrt{2\gamma+\mu^2})+|\mu|\,\mathrm{sh}((r-z)\sqrt{2\gamma+\mu^2})}, & z \le x \le r \\ 1-\dfrac{e^{2(r-x)|\mu|}(\sqrt{2\gamma+\mu^2}\,\mathrm{ch}((r-z)\sqrt{2\gamma+\mu^2})-|\mu|\,\mathrm{sh}((r-z)\sqrt{2\gamma+\mu^2}))}{\sqrt{2\gamma+\mu^2}\,\mathrm{ch}((r-z)\sqrt{2\gamma+\mu^2})+|\mu|\,\mathrm{sh}((r-z)\sqrt{2\gamma+\mu^2})}, & r \le x \end{cases}$$

2.5.2 $\mathbf{P}_x\Big(\int_0^{H_z} \mathbb{I}_{(-\infty,r]}\big(W_s^{(\mu)}\big)ds \in dy,\ H_z<\infty\Big)e^{\mu(x-z)}$

$z < r$
$$= \begin{cases} |\mu|^{-1}e^{(r-x)|\mu|-\mu^2y/2}\widetilde{\mathrm{rc}}_y(1,r-z,0,1/|\mu|)dy, & r \le x \\ e^{-\mu^2y/2}\big(\mathrm{ss}_y(r-x,r-z) & \\ +|\mu|^{-1}\,\mathrm{ss}_y(x-z,r-z)*\widetilde{\mathrm{rc}}_y(1,r-z,0,1/|\mu|)\big)dy, & z \le x \le r \\ e^{-\mu^2y/2}\,\mathrm{h}_y(1,z-x)dy, & x \le z \end{cases}$$

$r < z$
$$= \begin{cases} 0, & z \le x \\ \dfrac{|\mu|\,\mathrm{sh}((z-x)|\mu|)}{\mathrm{sh}^2((z-r)|\mu|)}\Big(\dfrac{e^{-\mu^2y/2}}{\sqrt{2\pi y}}-\dfrac{|\mu|\,\mathrm{ch}((z-r)|\mu|)}{2\,\mathrm{sh}((z-r)|\mu|)} & \\ \qquad\times\exp\Big(\dfrac{\mu^2y}{2\,\mathrm{sh}^2((z-r)|\mu|)}\Big)\,\mathrm{Erfc}\Big(\dfrac{|\mu|\,\mathrm{ch}((z-r)|\mu|)\sqrt{y}}{\sqrt{2}\,\mathrm{sh}((z-r)|\mu|)}\Big)\Big)dy, & r \le x \le z \\ \dfrac{|\mu|e^{-\mu^2y/2-(r-x)^2/2y}}{\sqrt{2\pi y}\,\mathrm{sh}((z-r)|\mu|)}dy-\dfrac{|\mu|\,\mathrm{ch}((z-r)|\mu|)}{2\,\mathrm{sh}^2((z-r)|\mu|)}\exp\Big(\dfrac{(r-x)|\mu|\,\mathrm{ch}((z-r)|\mu|)}{\mathrm{sh}((z-r)|\mu|)} & \\ \qquad+\dfrac{\mu^2y}{2\,\mathrm{sh}^2((z-r)|\mu|)}\Big)\,\mathrm{Erfc}\Big(\dfrac{r-x}{\sqrt{2y}}+\dfrac{|\mu|\,\mathrm{ch}((z-r)|\mu|)\sqrt{y}}{\sqrt{2}\,\mathrm{sh}((z-r)|\mu|)}\Big)dy, & x \le r \end{cases}$$

(1)
$\mu > 0$ $\mathbf{P}_x\Big(\int_0^{H_z} \mathbb{I}_{(-\infty,r]}\big(W_s^{(\mu)}\big)ds \in dy,\ H_z=\infty\Big)$

$$= \begin{cases} 2e^{(r-x)|\mu|-\mu^2y/2}\,\mathrm{rc}_y(0,r-z,x-z,1/|\mu|)dy, & z \le x \le r \\ 2e^{2(r-x)|\mu|-\mu^2y/2}\,\mathrm{rc}_y(0,r-z,r-z,1/|\mu|)dy, & r \le x \end{cases}$$

(2)
$r \le z$
$\mu < 0$
$$\mathbf{P}_x\Big(\int_0^{H_z} \mathbb{I}_{(-\infty,r]}\big(W_s^{(\mu)}\big)ds = 0\Big) = \begin{cases} 0, & x \le r \\ e^{\mu(z-x)}\dfrac{\mathrm{sh}((x-r)|\mu|)}{\mathrm{sh}((z-r)|\mu|)}, & r \le x \le z \\ 1, & z \le x \end{cases}$$

2 $W_s^{(\mu)} = \mu s + W_s$ $H_z = \min\{s : W_s^{(\mu)} = z\}$

2.5.4 $$\mathbf{P}_x\Big(\int_0^{H_z} \mathbb{I}_{(-\infty,r)}\big(W_s^{(\mu)}\big)ds \in dv,\ H_z \in dt\Big)$$

$$= e^{\mu(z-x)-\mu^2 t/2}\int_0^\infty B_z^{(27)}(v, t-v, y)dy\, dvdt, \quad \text{for } B_z^{(27)}(u,v,y) \text{ see 1.2.27.2}$$

(1) $$\mathbf{P}_x\Big(\int_0^{H_z} \mathbb{I}_{(-\infty,r)}\big(W_s^{(\mu)}\big)ds = 0,\ H_z \in dt\Big)$$

$$= \begin{cases} e^{\mu(z-x)-\mu^2 t/2}\, \mathrm{ss}_t(x-r, z-r)dt, & r \le x \le z \\ e^{\mu(z-x)-\mu^2 t/2}\, \mathrm{h}_t(1, x-z)dt, & z \le x \end{cases}$$

2.6.1 $$\mathbf{E}_x \exp\Big(-\int_0^{H_z} \big(p\mathbb{I}_{(-\infty,r)}\big(W_s^{(\mu)}\big) + q\mathbb{I}_{[r,\infty)}\big(W_s^{(\mu)}\big)\big)ds\Big)$$

$$= e^{\mu(z-x)}Q_z^{(6)}(2p+\mu^2, 2q+\mu^2), \quad \text{for } Q_z^{(6)}(2p,2q) \text{ see 1.2.6.1}$$

2.6.2 $$\mathbf{P}_x\Big(\int_0^{H_z} \mathbb{I}_{(-\infty,r)}\big(W_s^{(\mu)}\big)ds \in du,\ \int_0^{H_z} \mathbb{I}_{[r,\infty)}\big(W_s^{(\mu)}\big)ds \in dv\Big)$$

$$= e^{\mu(z-x)-\mu^2(u+v)/2}\int_0^\infty B_z^{(27)}(u, v, y)dy\, dudv, \quad \text{for } B_z^{(27)}(u,v,y) \text{ see 1.2.27.2}$$

(1) $$\mathbf{P}_x\Big(\int_0^{H_z} \mathbb{I}_{(-\infty,r)}\big(W_s^{(\mu)}\big)ds \in du,\ \int_0^{H_z} \mathbb{I}_{[r,\infty)}\big(W_s^{(\mu)}\big)ds = 0\Big)$$

$$= \begin{cases} e^{\mu(z-x)-\mu^2 u/2}\, \mathrm{h}_u(1, z-x)du, & x \le z \\ e^{\mu(z-x)-\mu^2 u/2}\, \mathrm{ss}_u(r-x, r-z)du, & z \le x \le r \end{cases}$$

(2) $$\mathbf{P}_x\Big(\int_0^{H_z} \mathbb{I}_{(-\infty,r)}\big(W_s^{(\mu)}\big)ds = 0,\ \int_0^{H_z} \mathbb{I}_{[r,\infty)}\big(W_s^{(\mu)}\big)ds \in dv\Big)$$

$$= \begin{cases} e^{\mu(z-x)-\mu^2 v/2}\, \mathrm{ss}_v(x-r, z-r)dv, & r \le x \le z \\ e^{\mu(z-x)-\mu^2 v/2}\, \mathrm{h}_v(1, x-z)dv, & z \le x \end{cases}$$

2.7.1
$r < u$ $$\mathbf{E}_x\Big\{\exp\Big(-\gamma\int_0^{H_z} \mathbb{I}_{[r,u]}\big(W_s^{(\mu)}\big)ds\Big);\ H_z < \infty\Big\}e^{\mu(x-z)+|\mu||z-x|}$$

2 $W_s^{(\mu)} = \mu s + W_s$ $H_z = \min\{s : W_s^{(\mu)} = z\}$

$z \le r$

$$= \begin{cases} 1, & x \le z \\ \dfrac{(\mu^2 + \gamma(1 - e^{2(x-r)|\mu|}))\operatorname{sh}((u-r)\Upsilon) + |\mu|\Upsilon\operatorname{ch}((u-r)\Upsilon)}{(\mu^2 + \gamma(1 - e^{2(z-r)|\mu|}))\operatorname{sh}((u-r)\Upsilon) + |\mu|\Upsilon\operatorname{ch}((u-r)\Upsilon)}, & z \le x \le r \\ \dfrac{e^{(x-r)|\mu|}(\mu^2\operatorname{sh}((u-x)\Upsilon) + |\mu|\Upsilon\operatorname{ch}((u-x)\Upsilon))}{(\mu^2 + \gamma(1 - e^{2(z-r)|\mu|}))\operatorname{sh}((u-r)\Upsilon) + |\mu|\Upsilon\operatorname{ch}((u-r)\Upsilon)}, & r \le x \le u \\ \dfrac{|\mu|\Upsilon e^{|\mu|(u-r)}}{(\mu^2 + \gamma(1 - e^{2(z-r)|\mu|}))\operatorname{sh}((u-r)\Upsilon) + |\mu|\Upsilon\operatorname{ch}((u-r)\Upsilon)}, & u \le x \end{cases}$$

$u \le z$

$$= \begin{cases} \dfrac{|\mu|\Upsilon e^{|\mu|(u-r)}}{(\mu^2 + \gamma(1 - e^{2(u-z)|\mu|}))\operatorname{sh}((u-r)\Upsilon) + |\mu|\Upsilon\operatorname{ch}((u-r)\Upsilon)}, & x \le r \\ \dfrac{e^{(u-x)|\mu|}(\mu^2\operatorname{sh}((x-r)\Upsilon) + |\mu|\Upsilon\operatorname{ch}((x-r)\Upsilon))}{(\mu^2 + \gamma(1 - e^{2(u-z)|\mu|}))\operatorname{sh}((u-r)\Upsilon) + |\mu|\Upsilon\operatorname{ch}((u-r)\Upsilon)}, & r \le x \le u \\ \dfrac{(\mu^2 + \gamma(1 - e^{2(u-x)|\mu|}))\operatorname{sh}((u-r)\Upsilon) + |\mu|\Upsilon\operatorname{ch}((u-r)\Upsilon)}{(\mu^2 + \gamma(1 - e^{2(u-z)|\mu|}))\operatorname{sh}((u-r)\Upsilon) + |\mu|\Upsilon\operatorname{ch}((u-r)\Upsilon)}, & u \le x \le z \\ 1, & z \le x \end{cases}$$

(1) $r < u$

$$\mathbf{E}_x\Big\{\exp\Big(-\gamma\int_0^{H_z} \mathbb{I}_{[r,u]}\big(W_s^{(\mu)}\big)ds\Big);\ H_z = \infty\Big\}$$

for $\mu > 0, \quad z \le r$

$z \le x$, $x \le r$

$$= \frac{2|\mu|\Upsilon e^{(u-x)|\mu|}\operatorname{sh}((x-z)|\mu|)}{(\mu^2 e^{(r-z)|\mu|} + 2\gamma\operatorname{sh}((r-z)|\mu|))\operatorname{sh}((u-r)\Upsilon) + |\mu|\Upsilon e^{(r-z)|\mu|}\operatorname{ch}((u-r)\Upsilon)}$$

$r \le x$, $x \le u$

$$= \frac{2e^{(u-x)|\mu|}[\mu^2\operatorname{ch}((r-z)|\mu|)\operatorname{sh}((x-r)\Upsilon) + |\mu|\Upsilon\operatorname{sh}((r-z)|\mu|)\operatorname{ch}((x-r)\Upsilon)]}{(\mu^2 e^{(r-z)|\mu|} + 2\gamma\operatorname{sh}((r-z)|\mu|))\operatorname{sh}((u-r)\Upsilon) + |\mu|\Upsilon e^{(r-z)|\mu|}\operatorname{ch}((u-r)\Upsilon)}$$

$u \le x$

$$= 1 - \frac{e^{2(u-x+z-r)|\mu|}[(\gamma(1 - e^{2(z-r)|\mu|}) - \mu^2)\operatorname{sh}((u-r)\Upsilon) + |\mu|\Upsilon\operatorname{ch}((u-r)\Upsilon)]}{(\mu^2 + \gamma(1 - e^{2(z-r)|\mu|}))\operatorname{sh}((u-r)\Upsilon) + |\mu|\Upsilon\operatorname{ch}((u-r)\Upsilon)}$$

for $\mu < 0, \quad u \le z$

$x \le r$

$$= 1 - \frac{e^{2(u-z+x-r)|\mu|}[(\gamma(1 - e^{2(u-z)|\mu|}) - \mu^2)\operatorname{sh}((u-r)\Upsilon) + |\mu|\Upsilon\operatorname{ch}((u-r)\Upsilon)]}{(\mu^2 + \gamma(1 - e^{2(u-z)|\mu|}))\operatorname{sh}((u-r)\Upsilon) + |\mu|\Upsilon\operatorname{ch}((u-r)\Upsilon)}$$

$r \le x$, $x \le u$

$$= \frac{2e^{(x-r)|\mu|}[\mu^2\operatorname{ch}((z-u)|\mu|)\operatorname{sh}((u-x)\Upsilon) + |\mu|\Upsilon\operatorname{sh}((z-u)|\mu|)\operatorname{ch}((u-x)\Upsilon)]}{(\mu^2 e^{(z-u)|\mu|} + 2\gamma\operatorname{sh}((z-u)|\mu|))\operatorname{sh}((u-r)\Upsilon) + |\mu|\Upsilon e^{(z-u)|\mu|}\operatorname{ch}((u-r)\Upsilon)}$$

$u \le x$, $x \le z$

$$= \frac{2|\mu|\Upsilon e^{(x-r)|\mu|}\operatorname{sh}((z-x)|\mu|)}{(\mu^2 e^{(z-u)|\mu|} + 2\gamma\operatorname{sh}((z-u)|\mu|))\operatorname{sh}((u-r)\Upsilon) + |\mu|\Upsilon e^{(z-u)|\mu|}\operatorname{ch}((u-r)\Upsilon)}$$

$$\Upsilon := \sqrt{\mu^2 + 2\gamma}$$

2 $W_s^{(\mu)} = \mu s + W_s$ $H_z = \min\{s : W_s^{(\mu)} = z\}$

2.8.1 $$\mathbf{E}_x\Big\{\exp\Big(-\gamma\int_0^{H_0}|W_s^{(\mu)}|ds\Big);\ H_0<\infty\Big\} = e^{-\mu x}\frac{\operatorname{Ai}(2^{1/3}\gamma^{-2/3}(\mu^2/2+\gamma|x|))}{\operatorname{Ai}((2\gamma)^{-2/3}\mu^2)}$$

2.9.3 $$\mathbf{E}_x\exp\Big(-\alpha H_z - \frac{\gamma^2}{2}\int_0^{H_z}(W_s^{(\mu)})^2ds\Big) = \begin{cases} \dfrac{e^{\mu(z-x)}D_{-\frac12-\frac{2\alpha+\mu^2}{2\gamma}}(-x\sqrt{2\gamma})}{D_{-\frac12-\frac{2\alpha+\mu^2}{2\gamma}}(-z\sqrt{2\gamma})}, & x\le z\\ \dfrac{e^{\mu(z-x)}D_{-\frac12-\frac{2\alpha+\mu^2}{2\gamma}}(x\sqrt{2\gamma})}{D_{-\frac12-\frac{2\alpha+\mu^2}{2\gamma}}(z\sqrt{2\gamma})}, & z\le x\end{cases}$$

2.9.4 (1) $$\mathbf{E}_x\Big\{\exp\Big(-\frac{\gamma^2}{2}\int_0^{H_0}(W_s^{(\mu)})^2ds\Big);H_0\in dt\Big\} = \frac{|x|\gamma^{3/2}e^{-\mu x-\mu^2t/2}}{\sqrt{2\pi}\operatorname{sh}^{3/2}(t\gamma)}\exp\Big(-\frac{x^2\gamma\operatorname{ch}(t\gamma)}{2\operatorname{sh}(t\gamma)}\Big)dt$$

(2) $$\mathbf{P}_x\Big(\int_0^{H_0}(W_s^{(\mu)})^2ds\in dy, H_0\in dt\Big) = \frac{|x|}{\sqrt{2\pi}}e^{-\mu x-\mu^2t/2}\operatorname{es}_y\Big(\frac32,\frac32,t,0,\frac{x^2}{2}\Big)dydt$$

2.10.3 $$\mathbf{E}_x\exp\Big(-\alpha H_z-\gamma\int_0^{H_z}e^{2\beta W_s^{(\mu)}}ds\Big) = \begin{cases} e^{\mu(z-x)}\dfrac{I_{\sqrt{2\alpha+\mu^2}/|\beta|}\Big(\frac{\sqrt{2\gamma}}{|\beta|}e^{\beta x}\Big)}{I_{\sqrt{2\alpha+\mu^2}/|\beta|}\Big(\frac{\sqrt{2\gamma}}{|\beta|}e^{\beta z}\Big)}, & \dfrac{z}{\beta}\ge\dfrac{x}{\beta}\\ e^{\mu(z-x)}\dfrac{K_{\sqrt{2\alpha+\mu^2}/|\beta|}\Big(\frac{\sqrt{2\gamma}}{|\beta|}e^{\beta x}\Big)}{K_{\sqrt{2\alpha+\mu^2}/|\beta|}\Big(\frac{\sqrt{2\gamma}}{|\beta|}e^{\beta z}\Big)}, & \dfrac{z}{\beta}\le\dfrac{x}{\beta}\end{cases}$$

2.11.2 $$\lambda\int_0^\infty e^{-\lambda z}\mathbf{P}_0\big(\sup_{y\in\mathbb{R}}\ell(H_z,y)<h, H_z<\infty\big)dz$$
$$= 1-\frac{|\mu|e^{-\mu h/2}}{(\lambda-2\mu)\operatorname{sh}(h|\mu|/2)}+\frac{|\mu|e^{|\mu|h/2}}{(\lambda-2\mu)\operatorname{sh}(h|\mu|/2)M\big(\frac12+\frac{\lambda-\mu}{2|\mu|},1,h|\mu|\big)}$$

2.13.1 $z<x$ $$\mathbf{E}_x\big\{e^{-\gamma\check{H}_\mu(H_z)};\ \sup_{0\le s\le H_z}W_s^{(\mu)}\in dy\big\} = \frac{|\mu|e^{\mu(z-x)}\operatorname{sh}((x-z)\sqrt{2\gamma+\mu^2})}{\operatorname{sh}(|\mu|(y-z))\operatorname{sh}((y-z)\sqrt{2\gamma+\mu^2})}dy$$

2.13.2 $z<x$ $x<y$ $$\mathbf{P}_x\big(\check{H}_\mu(H_z)\in du,\ \sup_{0\le s\le H_z}W_s^{(\mu)}\in dy\big) = |\mu|e^{\mu(z-x)-\mu^2u/2}\frac{\operatorname{ss}_u(x-z,y-z)}{\operatorname{sh}(|\mu|(y-z))}dudy$$

2.13.4 $$\mathbf{E}_x\big\{e^{-\alpha H_z-\gamma\check{H}_\mu(H_z)};\ \sup_{0\le s\le H_z}W_s^{(\mu)}\in dy\big\}\qquad z<x<y$$
$$=\frac{\sqrt{2\alpha+\mu^2}e^{\mu(z-x)}\operatorname{sh}((x-z)\sqrt{2\gamma+2\alpha+\mu^2})}{\operatorname{sh}((y-z)\sqrt{2\alpha+\mu^2})\operatorname{sh}((y-z)\sqrt{2\gamma+2\alpha+\mu^2})}dy$$

(1) $u<t$ $$\mathbf{P}_x\big(\check{H}_\mu(H_z)\in du,\ H_z\in dt,\ \sup_{0\le s\le H_z}W_s^{(\mu)}\in dy\big)$$

2 $W_s^{(\mu)} = \mu s + W_s$ $\qquad H_z = \min\{s : W_s^{(\mu)} = z\}$

$$= e^{\mu(z-x)-\mu^2 t/2}\, \mathrm{ss}_u(x-z, y-z)\, \mathrm{s}_{t-u}(y-z)dudtdy$$

2.14.1 ($x < z$)

$$\mathbf{E}_x\{e^{-\gamma \hat{H}_\mu(H_z)}; \inf_{0\le s\le H_z} W_s^{(\mu)} \in dy\} = \frac{|\mu| e^{\mu(z-x)} \operatorname{sh}((z-x)\sqrt{2\gamma+\mu^2})}{\operatorname{sh}(|\mu|(z-y)) \operatorname{sh}((z-y)\sqrt{2\gamma+\mu^2})} dy$$

2.14.2 ($x < z$, $y < x$)

$$\mathbf{P}_x\big(\hat{H}_\mu(H_z) \in du, \inf_{0\le s\le H_z} W_s^{(\mu)} \in dy\big) = |\mu| e^{\mu(z-x)-\mu^2 u/2} \frac{\mathrm{ss}_u(z-x, z-y)}{\operatorname{sh}(|\mu|(z-y))} dudy$$

2.14.4

$$\mathbf{E}_x\{e^{-\alpha H_z - \gamma \hat{H}_\mu(H_z)}; \inf_{0\le s\le H_z} W_s^{(\mu)} \in dy\} \qquad y < x < z$$

$$= \frac{\sqrt{2\alpha+\mu^2} e^{\mu(z-x)} \operatorname{sh}((z-x)\sqrt{2\gamma+2\alpha+\mu^2})}{\operatorname{sh}((z-y)\sqrt{2\alpha+\mu^2}) \operatorname{sh}((z-y)\sqrt{2\gamma+2\alpha+\mu^2})} dy$$

(1) ($u < t$)

$$\mathbf{P}_x\big(\hat{H}_\mu(H_z) \in du,\, H_z \in dt,\, \inf_{0\le s\le H_z} W_s^{(\mu)} \in dy\big)$$

$$= e^{\mu(z-x)-\mu^2 t/2}\, \mathrm{ss}_u(z-x, z-y)\, \mathrm{s}_{t-u}(z-y)dudtdy$$

2.15.2

$$\mathbf{P}_x\big(a < \inf_{0\le s\le H_z} W_s^{(\mu)}, \sup_{0\le s\le H_z} W_s^{(\mu)} < b\big) = e^{\mu(z-x)} \begin{cases} \dfrac{\operatorname{sh}((x-a)|\mu|)}{\operatorname{sh}((z-a)|\mu|)}, & a \le x \le z \\ \dfrac{\operatorname{sh}((b-x)|\mu|)}{\operatorname{sh}((b-z)|\mu|)}, & z \le x \le b \end{cases}$$

2.15.4

$$\mathbf{E}_x\{e^{-\alpha H_z}; a < \inf_{0\le s\le H_z} W_s^{(\mu)}, \sup_{0\le s\le H_z} W_s^{(\mu)} < b\}$$

$$= e^{\mu(z-x)} \begin{cases} \dfrac{\operatorname{sh}((x-a)\sqrt{2\alpha+\mu^2})}{\operatorname{sh}((z-a)\sqrt{2\alpha+\mu^2})}, & a \le x \le z \\ \dfrac{\operatorname{sh}((b-x)\sqrt{2\alpha+\mu^2})}{\operatorname{sh}((b-z)\sqrt{2\alpha+\mu^2})}, & z \le x \le b \end{cases}$$

(1)

$$\mathbf{P}_x\big(H_z \in dt, a < \inf_{0\le s\le t} W_s^{(\mu)}, \sup_{0\le s\le t} W_s^{(\mu)} < b\big)$$

$$= e^{\mu(z-x)-\mu^2 t/2} \begin{cases} \mathrm{ss}_t(x-a, z-a)dt, & a \le x \le z \\ \mathrm{ss}_t(b-x, b-z)dt, & z \le x \le b \end{cases}$$

2.16.1 ($r < b$, $z < r$)

$$\mathbf{E}_x\{e^{-\gamma \ell(H_z, r)}; \sup_{0\le s\le H_z} W_s^{(\mu)} < b\}$$

2 $W_s^{(\mu)} = \mu s + W_s$ $H_z = \min\{s : W_s^{(\mu)} = z\}$

$$= e^{\mu(z-x)} \begin{cases} \dfrac{|\mu| \operatorname{sh}((b-x)|\mu|)}{|\mu| \operatorname{sh}((b-z)|\mu|) + 2\gamma \operatorname{sh}((b-r)|\mu|) \operatorname{sh}((r-z)|\mu|)}, & r \le x \le b \\ \dfrac{|\mu| \operatorname{sh}((b-x)|\mu|) + 2\gamma \operatorname{sh}((b-r)|\mu|) \operatorname{sh}((r-x)|\mu|)}{|\mu| \operatorname{sh}((b-z)|\mu|) + 2\gamma \operatorname{sh}((b-r)|\mu|) \operatorname{sh}((r-z)|\mu|)}, & z \le x \le r \end{cases}$$

2.16.2 $r < b$ $z < r$

$$\mathbf{P}_x\Big(\ell(H_z, r) \in dy, \sup_{0 \le s \le H_z} W_s^{(\mu)} < b\Big) = \exp\Big(-\frac{|\mu| \operatorname{sh}((b-z)|\mu|) y}{2 \operatorname{sh}((b-r)|\mu|) \operatorname{sh}((r-z)|\mu|)}\Big)$$

$$\times e^{\mu(z-x)} \begin{cases} \dfrac{|\mu| \operatorname{sh}((b-x)|\mu|)}{2 \operatorname{sh}((b-r)|\mu|) \operatorname{sh}((r-z)|\mu|)} dy, & r \le x \le b \\ \dfrac{|\mu| \operatorname{sh}((x-z)|\mu|)}{2 \operatorname{sh}^2((r-z)|\mu|)} dy, & z \le x \le r \end{cases}$$

(1) $$\mathbf{P}_x\Big(\ell(H_z, r) = 0, \sup_{0 \le s \le H_z} W_s^{(\mu)} < b\Big) = \begin{cases} 0, & r \le x \le b \\ e^{\mu(z-x)} \dfrac{\operatorname{sh}((r-x)|\mu|)}{\operatorname{sh}((r-z)|\mu|)}, & z \le x \le r \end{cases}$$

2.17.1 $a < r$ $r < z$

$$\mathbf{E}_x\{e^{-\gamma \ell(H_z, r)};\ a < \inf_{0 \le s \le H_z} W_s^{(\mu)}\}$$

$$= e^{\mu(z-x)} \begin{cases} \dfrac{|\mu| \operatorname{sh}((x-a)|\mu|) + 2\gamma \operatorname{sh}((x-r)|\mu|) \operatorname{sh}((r-a)|\mu|)}{|\mu| \operatorname{sh}((z-a)|\mu|) + 2\gamma \operatorname{sh}((z-r)|\mu|) \operatorname{sh}((r-a)|\mu|)}, & r \le x \le z \\ \dfrac{|\mu| \operatorname{sh}((x-a)|\mu|)}{|\mu| \operatorname{sh}((z-a)|\mu|) + 2\gamma \operatorname{sh}((z-r)|\mu|) \operatorname{sh}((r-a)|\mu|)}, & a \le x \le r \end{cases}$$

2.17.2 $a < r$ $r < z$

$$\mathbf{P}_x\Big(\ell(H_z, r) \in dy, a < \inf_{0 \le s \le H_z} W_s^{(\mu)}\Big) = \exp\Big(-\frac{|\mu| \operatorname{sh}((z-a)|\mu|) y}{2 \operatorname{sh}((z-r)|\mu|) \operatorname{sh}((r-a)|\mu|)}\Big)$$

$$\times e^{\mu(z-x)} \begin{cases} \dfrac{|\mu| \operatorname{sh}((z-x)|\mu|)}{2 \operatorname{sh}^2((z-r)|\mu|)} dy, & r \le x \le z \\ \dfrac{|\mu| \operatorname{sh}((x-a)|\mu|)}{2 \operatorname{sh}((z-r)|\mu|) \operatorname{sh}((r-a)|\mu|)} dy, & a \le x \le r \end{cases}$$

(1) $$\mathbf{P}_x\Big(\ell(H_z, r) = 0, a < \inf_{0 \le s \le H_z} W_s^{(\mu)}\Big) = \begin{cases} e^{\mu(z-x)} \dfrac{\operatorname{sh}((x-r)|\mu|)}{\operatorname{sh}((z-r)|\mu|)}, & r \le x \le z \\ 0, & a \le x \le r \end{cases}$$

2 $W_s^{(\mu)} = \mu s + W_s$	$H_z = \min\{s : W_s^{(\mu)} = z\}$
2.18.1 $r < u$	$\mathbf{E}_x\{\exp(-\gamma\ell(H_z,r) - \eta\ell(H_z,u));\ H_z < \infty\}e^{\mu(x-z)+\lvert\mu\rvert\lvert z-x\rvert}$
	for $z < r$
$z \le x$ $x \le r$	$= \dfrac{\mu^2 + \lvert\mu\rvert\gamma(1-e^{2(x-r)\lvert\mu\rvert}) + \lvert\mu\rvert\eta(1-e^{2(x-u)\lvert\mu\rvert}) + \gamma\eta(1-e^{2(r-u)\lvert\mu\rvert})(1-e^{2(x-r)\lvert\mu\rvert})}{\mu^2 + \lvert\mu\rvert\gamma(1-e^{2(z-r)\lvert\mu\rvert}) + \lvert\mu\rvert\eta(1-e^{2(z-u)\lvert\mu\rvert}) + \gamma\eta(1-e^{2(r-u)\lvert\mu\rvert})(1-e^{2(z-r)\lvert\mu\rvert})}$
$r \le x$ $x \le u$	$= \dfrac{\mu^2 + \lvert\mu\rvert\eta(1-e^{2(x-u)\lvert\mu\rvert})}{\mu^2 + \lvert\mu\rvert\gamma(1-e^{2(z-r)\lvert\mu\rvert}) + \lvert\mu\rvert\eta(1-e^{2(z-u)\lvert\mu\rvert}) + \gamma\eta(1-e^{2(r-u)\lvert\mu\rvert})(1-e^{2(z-r)\lvert\mu\rvert})}$
$u \le x$	$= \dfrac{\mu^2}{\mu^2 + \lvert\mu\rvert\gamma(1-e^{2(z-r)\lvert\mu\rvert}) + \lvert\mu\rvert\eta(1-e^{2(z-u)\lvert\mu\rvert}) + \gamma\eta(1-e^{2(r-u)\lvert\mu\rvert})(1-e^{2(z-r)\lvert\mu\rvert})}$
	for $u < z$
$x \le r$	$= \dfrac{\mu^2}{\mu^2 + \lvert\mu\rvert\gamma(1-e^{2(r-z)\lvert\mu\rvert}) + \lvert\mu\rvert\eta(1-e^{2(u-z)\lvert\mu\rvert}) + \gamma\eta(1-e^{2(u-z)\lvert\mu\rvert})(1-e^{2(r-u)\lvert\mu\rvert})}$
$r \le x$ $x \le u$	$= \dfrac{\mu^2 + \lvert\mu\rvert\gamma(1-e^{2(r-x)\lvert\mu\rvert})}{\mu^2 + \lvert\mu\rvert\gamma(1-e^{2(r-z)\lvert\mu\rvert}) + \lvert\mu\rvert\eta(1-e^{2(u-z)\lvert\mu\rvert}) + \gamma\eta(1-e^{2(u-z)\lvert\mu\rvert})(1-e^{2(r-u)\lvert\mu\rvert})}$
$u \le x$ $x \le z$	$= \dfrac{\mu^2 + \lvert\mu\rvert\gamma(1-e^{2(r-x)\lvert\mu\rvert}) + \lvert\mu\rvert\eta(1-e^{2(u-x)\lvert\mu\rvert}) + \gamma\eta(1-e^{2(u-x)\lvert\mu\rvert})(1-e^{2(r-u)\lvert\mu\rvert})}{\mu^2 + \lvert\mu\rvert\gamma(1-e^{2(r-z)\lvert\mu\rvert}) + \lvert\mu\rvert\eta(1-e^{2(u-z)\lvert\mu\rvert}) + \gamma\eta(1-e^{2(u-z)\lvert\mu\rvert})(1-e^{2(r-u)\lvert\mu\rvert})}$
(1) $r < u$	$\mathbf{E}_x\{\exp(-\gamma\ell(H_z,r) - \eta\ell(H_z,u));\ H_z = \infty\}$
	for $\mu > 0$, $z < r$
$z \le x$ $x \le r$	$= \dfrac{\mu^2(1-e^{2(z-x)\lvert\mu\rvert})}{\mu^2 + \lvert\mu\rvert\gamma(1-e^{2(z-r)\lvert\mu\rvert}) + \lvert\mu\rvert\eta(1-e^{2(z-u)\lvert\mu\rvert}) + \gamma\eta(1-e^{2(r-u)\lvert\mu\rvert})(1-e^{2(z-r)\lvert\mu\rvert})}$
$r \le x$ $x \le u$	$= \dfrac{\mu^2(1-e^{2(z-x)\lvert\mu\rvert}) + \lvert\mu\rvert\gamma(1-e^{2(z-r)\lvert\mu\rvert})(1-e^{2(r-x)\lvert\mu\rvert})}{\mu^2 + \lvert\mu\rvert\gamma(1-e^{2(z-r)\lvert\mu\rvert}) + \lvert\mu\rvert\eta(1-e^{2(z-u)\lvert\mu\rvert}) + \gamma\eta(1-e^{2(r-u)\lvert\mu\rvert})(1-e^{2(z-r)\lvert\mu\rvert})}$
$u \le x$	$= 1 - e^{2(z-x)\lvert\mu\rvert}$ $\times \dfrac{\mu^2 + \lvert\mu\rvert\gamma(e^{2(r-z)\lvert\mu\rvert}-1) + \lvert\mu\rvert\eta(e^{2(u-z)\lvert\mu\rvert}-1) + \gamma\eta(e^{2(u-r)\lvert\mu\rvert}-1)(e^{2(r-z)\lvert\mu\rvert}-1)}{\mu^2 + \lvert\mu\rvert\gamma(1-e^{2(z-r)\lvert\mu\rvert}) + \lvert\mu\rvert\eta(1-e^{2(z-u)\lvert\mu\rvert}) + \gamma\eta(1-e^{2(r-u)\lvert\mu\rvert})(1-e^{2(z-r)\lvert\mu\rvert})}$

2 $W_s^{(\mu)} = \mu s + W_s$ $H_z = \min\{s : W_s^{(\mu)} = z\}$

for $\mu < 0, \quad u < z$

$x \le r$

$$= 1 - e^{2(x-z)|\mu|}$$

$$\times \frac{\mu^2 + |\mu|\gamma(e^{2(z-r)|\mu|} - 1) + |\mu|\eta(e^{2(z-u)|\mu|} - 1) + \gamma\eta(e^{2(z-u)|\mu|} - 1)(e^{2(u-r)|\mu|} - 1)}{\mu^2 + |\mu|\gamma(1 - e^{2(r-z)|\mu|}) + |\mu|\eta(1 - e^{2(u-z)|\mu|}) + \gamma\eta(1 - e^{2(u-z)|\mu|})(1 - e^{2(r-u)|\mu|})}$$

$r \le x$, $x \le u$

$$= \frac{\mu^2(1 - e^{2(x-z)|\mu|}) + |\mu|\gamma(1 - e^{2(u-z)|\mu|})(1 - e^{2(x-u)|\mu|})}{\mu^2 + |\mu|\gamma(1 - e^{2(r-z)|\mu|}) + |\mu|\eta(1 - e^{2(u-z)|\mu|}) + \gamma\eta(1 - e^{2(u-z)|\mu|})(1 - e^{2(r-u)|\mu|})}$$

$u \le x$, $x \le z$

$$= \frac{\mu^2(1 - e^{2(x-z)|\mu|})}{\mu^2 + |\mu|\gamma(1 - e^{2(r-z)|\mu|}) + |\mu|\eta(1 - e^{2(u-z)|\mu|}) + \gamma\eta(1 - e^{2(u-z)|\mu|})(1 - e^{2(r-u)|\mu|})}$$

2.19.1 $\mathbf{E}_x\Big\{\exp\Big(-\gamma\int_0^{H_z}\frac{ds}{W_s^{(\mu)}}\Big);\ 0 < \inf_{0\le s\le H_z} W_s^{(\mu)}, H_z < \infty\Big\}$

$$= \begin{cases} \dfrac{e^{-\mu x}M_{-\gamma/|\mu|,1/2}(2|\mu|x)}{e^{-\mu z}M_{-\gamma/|\mu|,1/2}(2|\mu|z)}, & 0 < x \le z \\[2ex] \dfrac{e^{-\mu x}W_{-\gamma/|\mu|,1/2}(2|\mu|x)}{e^{-\mu z}W_{-\gamma/|\mu|,1/2}(2|\mu|z)}, & z \le x \end{cases}$$

2.20.1 $\mathbf{E}_x\Big\{\exp\Big(-\frac{\gamma^2}{2}\int_0^{H_z}\frac{ds}{(W_s^{(\mu)})^2}\Big); 0 < \inf_{0\le s\le H_z} W_s^{(\mu)}, H_z < \infty\Big\}$

$$= \begin{cases} \dfrac{e^{-\mu x}\sqrt{x}I_{\sqrt{\gamma^2+1/4}}(|\mu|x)}{e^{-\mu z}\sqrt{z}I_{\sqrt{\gamma^2+1/4}}(|\mu|z)}, & 0 < x \le z \\[2ex] \dfrac{e^{-\mu x}\sqrt{x}K_{\sqrt{\gamma^2+1/4}}(|\mu|x)}{e^{-\mu z}\sqrt{z}K_{\sqrt{\gamma^2+1/4}}(|\mu|z)}, & z \le x \end{cases}$$

2.21.1 $\mathbf{E}_x\Big\{\exp\Big(-\int_0^{H_z}\Big(\frac{p^2}{2(W_s^{(\mu)})^2} + \frac{q^2(W_s^{(\mu)})^2}{2}\Big)ds\Big);\ 0 < \inf_{0\le s\le H_z} W_s^{(\mu)}, H_z < \infty\Big\}$

$$= \begin{cases} \dfrac{e^{-\mu x}\sqrt{z}M_{-\mu^2/4q,\sqrt{4p^2+1}/4}(qx^2)}{e^{-\mu z}\sqrt{x}M_{-\mu^2/4q,\sqrt{4p^2+1}/4}(qz^2)}, & 0 < x \le z \\[2ex] \dfrac{e^{-\mu x}\sqrt{z}W_{-\mu^2/4q,\sqrt{4p^2+1}/4}(qx^2)}{e^{-\mu z}\sqrt{x}W_{-\mu^2/4q,\sqrt{4p^2+1}/4}(qz^2)}, & z \le x \end{cases}$$

2 $W_s^{(\mu)} = \mu s + W_s$ $H_z = \min\{s : W_s^{(\mu)} = z\}$

2.22.1 $$\mathbf{E}_x\Big\{\exp\Big(-\int_0^{H_z}\Big(\frac{p^2}{2(W_s^{(\mu)})^2}+\frac{q}{W_s^{(\mu)}}\Big)ds\Big);\ 0<\inf_{0\le s\le H_z} W_s^{(\mu)}, H_z<\infty\Big\}$$

$$=\begin{cases}\dfrac{e^{-\mu x}M_{-q/|\mu|,\sqrt{p^2+1/4}}(2|\mu|x)}{e^{-\mu z}M_{-q/|\mu|,\sqrt{p^2+1/4}}(2|\mu|z)}, & 0<x\le z\\[2ex] \dfrac{e^{-\mu x}W_{-q/|\mu|,\sqrt{p^2+1/4}}(2|\mu|x)}{e^{-\mu z}W_{-q/|\mu|,\sqrt{p^2+1/4}}(2|\mu|z)}, & z\le x\end{cases}$$

2.27.1 $$\mathbf{E}_x\Big\{\exp\Big(-\int_0^{H_z}\big(p\mathbb{I}_{(-\infty,r)}\big(W_s^{(\mu)}\big)+q\mathbb{I}_{[r,\infty)}\big(W_s^{(\mu)}\big)\big)ds\Big);\tfrac{1}{2}\ell(H_z,r)\in dy\Big\}$$

$$=e^{\mu(z-x)}Q_z^{(27)}(2p+\mu^2,2q+\mu^2,y)dy, \qquad \text{for } Q_z^{(27)}(2p,2q) \text{ see 1.2.27.1}$$

(1) $$\mathbf{E}_x\Big\{\exp\Big(-\int_0^{H_z}\big(p\mathbb{I}_{(-\infty,r)}\big(W_s^{(\mu)}\big)+q\mathbb{I}_{[r,\infty)}\big(W_s^{(\mu)}\big)\big)ds\Big);\ell(H_z,r)=0\Big\}$$

$$=e^{\mu(z-x)}\begin{cases}\dfrac{\operatorname{sh}((r-x)\sqrt{2p+\mu^2})}{\operatorname{sh}((r-z)\sqrt{2p+\mu^2})}, & z\le x\le r\\[2ex] \dfrac{\operatorname{sh}((x-r)\sqrt{2q+\mu^2})}{\operatorname{sh}((z-r)\sqrt{2q+\mu^2})}, & r\le x\le z\end{cases}$$

2.27.2 $$\mathbf{P}_x\Big(\int_0^{H_z}\mathbb{I}_{(-\infty,r)}\big(W_s^{(\mu)}\big)ds\in du,\int_0^{H_z}\mathbb{I}_{[r,\infty)}\big(W_s^{(\mu)}\big)ds\in dv,\frac{1}{2}\ell(H_z,r)\in dy\Big)$$

$$=e^{\mu(z-x)-\mu^2(u+v)/2}B_z^{(27)}(u,v,y)dudvdy, \quad \text{for } B_z^{(27)}(u,v,y) \text{ see 1.2.27.2}$$

2.30.1 $$\mathbf{E}_x\exp\Big(-\int_0^{H_z}\Big(\frac{p^2}{2}e^{2\beta W_s^{(\mu)}}+qe^{\beta W_s^{(\mu)}}\Big)ds\Big)$$

$$=\begin{cases}\dfrac{e^{-(\mu+\beta/2)x}M_{-q/p|\beta|,|\mu|/|\beta|}\big(\frac{2p}{|\beta|}e^{\beta x}\big)}{e^{-(\mu+\beta/2)z}M_{-q/p|\beta|,|\mu|/|\beta|}\big(\frac{2p}{|\beta|}e^{\beta z}\big)}, & (z-x)\beta\ge 0\\[2ex] \dfrac{e^{-(\mu+\beta/2)x}W_{-q/p|\beta|,|\mu|/|\beta|}\big(\frac{2p}{|\beta|}e^{\beta x}\big)}{e^{-(\mu+\beta/2)z}W_{-q/p|\beta|,|\mu|/|\beta|}\big(\frac{2p}{|\beta|}e^{\beta z}\big)}, & (z-x)\beta\le 0\end{cases}$$

2 $W_s^{(\mu)} = \mu s + W_s$ $H_z = \min\{s : W_s^{(\mu)} = z\}$

2.31.1 $0 < \alpha$

$$\mathbf{E}_{z+\frac{x}{|\mu|}} \Big\{\exp\Big(-\gamma \sup_{0\le s\le H_z} \big(\ell\big(s, z+\frac{u}{|\mu|}\big) - \alpha^{-1}\ell\big(s, z+\frac{r}{|\mu|}\big)\big)\Big); H_z < \infty\Big\} e^{x\mu/|\mu|}$$

for $0 \le x\wedge r\wedge u$ or $x\vee r\vee u \le 0$

$$= e^{-|x|} - \frac{\gamma e^{-|u|}(\Upsilon - e^{-2|r|} + \alpha e^{-2|u|})(e^{-|x-u|} - e^{-|x+u|})}{2\gamma(1-e^{-2(|u|\wedge|r|)})(1-e^{-2|u-r|}) + |\mu|(\Upsilon - e^{-2|r|} + \alpha e^{-2|u|})}$$

$$-\frac{\gamma\big(2(1-e^{-2(|u|\wedge|r|)})(1-e^{-2|u-r|}) - (\Upsilon - e^{-2|r|} + \alpha e^{-2|u|})(1-e^{-2|u|})\big)(e^{-|x-r|} - e^{-|x+r|})}{e^{|u|}\big(2\gamma(1-e^{-2(|u|\wedge|r|)})(1-e^{-2|u-r|}) + |\mu|(\Upsilon - e^{-2|r|} + \alpha e^{-2|u|})\big)(e^{-|u-r|} - e^{-|u+r|})}$$

2.31.1 (1)

$$\mathbf{E}_{z+\frac{x}{|\mu|}} \exp\Big(-\gamma \sup_{0\le s\le H_z} \big(\ell\big(s, z+\frac{u}{|\mu|}\big) - \alpha^{-1}\ell\big(s, z+\frac{r}{|\mu|}\big)\big)\Big) e^{(x-u)\mu/|\mu|}$$

for $0 \le x\wedge r\wedge u$ or $x\vee r\vee u \le 0$

$0 < \alpha$

$$= e^{(x-u)\mu/|\mu|} - \frac{\gamma(\Upsilon - e^{-2|r|} + \alpha e^{-2|u|})(e^{-|x-u|} - e^{-|x+u|})}{2\gamma(1-e^{-2(|u|\wedge|r|)})(1-e^{-2|u-r|}) + |\mu|(\Upsilon - e^{-2|r|} + \alpha e^{-2|u|})}$$

$$-\frac{\gamma\big(2(1-e^{-2(|u|\wedge|r|)})(1-e^{-2|u-r|}) - (\Upsilon - e^{-2|r|} + \alpha e^{-2|u|})(1-e^{-2|u|})\big)(e^{-|x-r|} - e^{-|x+r|})}{\big(2\gamma(1-e^{-2(|u|\wedge|r|)})(1-e^{-2|u-r|}) + |\mu|(\Upsilon - e^{-2|r|} + \alpha e^{-2|u|})\big)(e^{-|u-r|} - e^{-|u+r|})}$$

2.31.2 $0 < \alpha$

$$\mathbf{P}_{z+\frac{x}{|\mu|}} \Big(\sup_{0\le s\le H_z} \big(\ell\big(s, z+\frac{u}{|\mu|}\big) - \alpha^{-1}\ell\big(s, z+\frac{r}{|\mu|}\big)\big) > h; H_z < \infty\Big)$$

for $0 \le x\wedge r\wedge u$ or $x\vee r\vee u \le 0$

$$= \exp\Big(-|u| - \frac{x\mu}{|\mu|} - \frac{|\mu|(\Upsilon - e^{-2|r|} + \alpha e^{-2|u|})h}{2(1-e^{-2(|u|\wedge|r|)})(1-e^{-2|u-r|})}\Big)\Big[\frac{e^{-|x-r|} - e^{-|x+r|}}{e^{-|u-r|} - e^{-|u+r|}}$$

$$-\frac{\Upsilon - e^{-2|r|} + \alpha e^{-2|u|}}{2(1-e^{-2(|u|\wedge|r|)})(1-e^{-2|u-r|})}\Big(\frac{(e^{-|x-r|} - e^{-|x+r|})(1-e^{-2|u|})}{e^{-|u-r|} - e^{-|u+r|}} - e^{-|u-x|} + e^{-|x+u|}\Big)\Big]$$

2.31.2 (1)

$$\mathbf{P}_{z+\frac{x}{|\mu|}} \Big(\sup_{0\le s\le H_z} \big(\ell\big(s, z+\frac{u}{|\mu|}\big) - \alpha^{-1}\ell\big(s, z+\frac{r}{|\mu|}\big)\big) > h\Big)$$

for $0 \le x\wedge r\wedge u$ or $x\vee r\vee u \le 0$

$0 < \alpha$

$$= \exp\Big(\frac{(u-x)\mu}{|\mu|} - \frac{|\mu|(\Upsilon - e^{-2|r|} + \alpha e^{-2|u|})h}{2(1-e^{-2(|u|\wedge|r|)})(1-e^{-2|u-r|})}\Big)\Big[\frac{e^{-|x-r|} - e^{-|x+r|}}{e^{-|u-r|} - e^{-|u+r|}}$$

$$-\frac{\Upsilon - e^{-2|r|} + \alpha e^{-2|u|}}{2(1-e^{-2(|u|\wedge|r|)})(1-e^{-2|u-r|})}\Big(\frac{(e^{-|x-r|} - e^{-|x+r|})(1-e^{-2|u|})}{e^{-|u-r|} - e^{-|u+r|}} - e^{-|u-x|} + e^{-|x+u|}\Big)\Big]$$

$$\Upsilon := \sqrt{\big(1-e^{-2|r|} - \alpha + \alpha e^{-2|u|}\big)^2 + 4\alpha\big(1-e^{-2(|u|\wedge|r|)}\big)\big(1-e^{-2|u-r|}\big)} + 1 - \alpha$$

2 $W_s^{(\mu)} = \mu s + W_s$ $\qquad H = H_{a,b} = \min\{s : W_s^{(\mu)} \notin (a,b)\}$

3. Stopping at first exit time

3.0.1 $$\mathbf{E}_x e^{-\alpha H} = \frac{e^{\mu(a-x)} \operatorname{sh}\big((b-x)\sqrt{2\alpha+\mu^2}\big) + e^{\mu(b-x)} \operatorname{sh}\big((x-a)\sqrt{2\alpha+\mu^2}\big)}{\operatorname{sh}\big((b-a)\sqrt{2\alpha+\mu^2}\big)}$$

3.0.2 $$\mathbf{P}_x(H \in dt) = e^{-\mu^2 t/2}\big\{e^{\mu(a-x)} \operatorname{ss}_t(b-x, b-a)dt + e^{\mu(b-x)} \operatorname{ss}_t(x-a, b-a)dt\big\}$$

3.0.3 $$\mathbf{E}_x e^{i\beta W_H^{(\mu)}} = \frac{\operatorname{sh}((b-x)|\mu|)}{\operatorname{sh}((b-a)|\mu|)} e^{\mu(a-x)+i\beta a} + \frac{\operatorname{sh}((x-a)|\mu|)}{\operatorname{sh}((b-a)|\mu|)} e^{\mu(b-x)+i\beta b}$$

3.0.4 (a) $$\mathbf{P}_x\big(W_H^{(\mu)} = a\big) = e^{\mu(a-x)} \frac{\operatorname{sh}((b-x)|\mu|)}{\operatorname{sh}((b-a)|\mu|)}$$

3.0.4 (b) $$\mathbf{P}_x\big(W_H^{(\mu)} = b\big) = e^{\mu(b-x)} \frac{\operatorname{sh}((x-a)|\mu|)}{\operatorname{sh}((b-a)|\mu|)}$$

3.0.5 (a) $$\mathbf{E}_x\big\{e^{-\alpha H};\ W_H^{(\mu)} = a\big\} = e^{\mu(a-x)} \frac{\operatorname{sh}\big((b-x)\sqrt{2\alpha+\mu^2}\big)}{\operatorname{sh}\big((b-a)\sqrt{2\alpha+\mu^2}\big)}$$

3.0.5 (b) $$\mathbf{E}_x\big\{e^{-\alpha H};\ W_H^{(\mu)} = b\big\} = \frac{e^{\mu(b-x)} \operatorname{sh}\big((x-a)\sqrt{2\alpha+\mu^2}\big)}{\operatorname{sh}\big((b-a)\sqrt{2\alpha+\mu^2}\big)}$$

3.0.6 (a) $$\mathbf{P}_x\big(H \in dt, W_H^{(\mu)} = a\big) = e^{\mu(a-x)-\mu^2 t/2} \operatorname{ss}_t(b-x, b-a)dt$$

3.0.6 (b) $$\mathbf{P}_x\big(H \in dt, W_H^{(\mu)} = b\big) = e^{\mu(b-x)-\mu^2 t/2} \operatorname{ss}_t(x-a, b-a)dt$$

3.1.6 $x \le y$ $$\mathbf{P}_x\Big(\sup_{0\le s\le H} W_s^{(\mu)} \ge y, W_H^{(\mu)} = a\Big) = e^{\mu(a-x)} \frac{\operatorname{sh}((b-y)|\mu|)\operatorname{sh}((x-a)|\mu|)}{\operatorname{sh}((b-a)|\mu|)\operatorname{sh}((y-a)|\mu|)}$$

3.1.8 (1) $$\mathbf{E}_x\big\{e^{-\alpha H};\ \sup_{0\le s\le H} W_s^{(\mu)} \ge y, W_H^{(\mu)} = a\big\}$$

$$= e^{\mu(a-x)} \frac{\operatorname{sh}\big((b-y)\sqrt{2\alpha+\mu^2}\big)\operatorname{sh}\big((x-a)\sqrt{2\alpha+\mu^2}\big)}{\operatorname{sh}\big((b-a)\sqrt{2\alpha+\mu^2}\big)\operatorname{sh}\big((y-a)\sqrt{2\alpha+\mu^2}\big)}, \quad a \le x \le y \le b$$

2 $W_s^{(\mu)} = \mu s + W_s$ $H = H_{a,b} = \min\{s : W_s^{(\mu)} \notin (a,b)\}$

(2) $$\mathbf{P}_x\Big(\sup_{0\le s\le H} W_s^{(\mu)} \ge y, W_H^{(\mu)} = a, H \in dt\Big)$$
$$= e^{\mu(a-x)-\mu^2 t/2}\{\mathrm{ss}_t(b-y,b-a) * \mathrm{ss}_t(x-a,y-a)\}dt, \quad a \le x \le y \le b$$

3.2.6 $$\mathbf{P}_x\Big(\inf_{0\le s\le H} W_s^{(\mu)} \le y, W_H^{(\mu)} = b\Big)$$
$$= e^{\mu(b-x)}\frac{\mathrm{sh}((b-x)|\mu|)\,\mathrm{sh}((y-a)|\mu|)}{\mathrm{sh}((b-y)|\mu|)\,\mathrm{sh}((b-a)|\mu|)}, \quad a \le y \le x \le b$$

3.2.8 (1) $$\mathbf{E}_x\Big\{e^{-\alpha H}; \inf_{0\le s\le H} W_s^{(\mu)} \le y, W_H^{(\mu)} = b\Big\}$$
$$= e^{\mu(b-x)}\frac{\mathrm{sh}\big((b-x)\sqrt{2\alpha+\mu^2}\big)\,\mathrm{sh}\big((y-a)\sqrt{2\alpha+\mu^2}\big)}{\mathrm{sh}\big((b-y)\sqrt{2\alpha+\mu^2}\big)\,\mathrm{sh}\big((b-a)\sqrt{2\alpha+\mu^2}\big)}, \quad a \le y \le x \le b$$

(2) $$\mathbf{P}_x\Big(\inf_{0\le s\le H} W_s^{(\mu)} \le y, W_H^{(\mu)} = b, H \in dt\Big)$$
$$= e^{\mu(b-x)-\mu^2 t/2}\{\mathrm{ss}_t(b-x,b-y) * \mathrm{ss}_t(y-a,b-a)\}dt, \quad a \le y \le x \le b$$

3.3.1 $$\mathbf{E}_r e^{-\gamma\ell(H,r)} = \frac{|\mu|\,\mathrm{sh}((b-a)|\mu|)}{|\mu|\,\mathrm{sh}((b-a)|\mu|) + 2\gamma\,\mathrm{sh}((b-r)|\mu|)\,\mathrm{sh}((r-a)|\mu|)} =: L_\mu$$

$$\mathbf{E}_x e^{-\gamma\ell(H,r)} = \begin{cases} e^{\mu(b-x)}\dfrac{\mathrm{sh}((x-r)|\mu|)}{\mathrm{sh}((b-r)|\mu|)} + e^{\mu(r-x)}\dfrac{\mathrm{sh}((b-x)|\mu|)}{\mathrm{sh}((b-r)|\mu|)}L_\mu, & r \le x \le b \\ e^{\mu(a-x)}\dfrac{\mathrm{sh}((r-x)|\mu|)}{\mathrm{sh}((r-a)|\mu|)} + e^{\mu(r-x)}\dfrac{\mathrm{sh}((x-a)|\mu|)}{\mathrm{sh}((r-a)|\mu|)}L_\mu, & a \le x \le r \end{cases}$$

3.3.2 $$\mathbf{P}_x\big(\ell(H,r) \in dy\big) = \exp\Big(\mu(r-x) - \frac{|\mu|\,\mathrm{sh}((b-a)|\mu|)y}{2\,\mathrm{sh}((b-r)|\mu|)\,\mathrm{sh}((r-a)|\mu|)}\Big)$$
$$\times\begin{cases} \dfrac{|\mu|\,\mathrm{sh}((b-x)|\mu|)\,\mathrm{sh}((b-a)|\mu|)}{2\,\mathrm{sh}^2((b-r)|\mu|)\,\mathrm{sh}((r-a)|\mu|)}dy, & r \le x \le b \\ \dfrac{|\mu|\,\mathrm{sh}((x-a)|\mu|)\,\mathrm{sh}((b-a)|\mu|)}{2\,\mathrm{sh}((b-r)|\mu|)\,\mathrm{sh}^2((r-a)|\mu|)}dy, & a \le x \le r \end{cases}$$

(1) $$\mathbf{P}_x\big(\ell(H,r) = 0\big) = \begin{cases} e^{\mu(b-x)}\dfrac{\mathrm{sh}((x-r)|\mu|)}{\mathrm{sh}((b-r)|\mu|)}, & r \le x \le b \\ e^{\mu(a-x)}\dfrac{\mathrm{sh}((r-x)|\mu|)}{\mathrm{sh}((r-a)|\mu|)}, & a \le x \le r \end{cases}$$

2 $W_s^{(\mu)} = \mu s + W_s$ $\qquad H = H_{a,b} = \min\{s : W_s^{(\mu)} \notin (a,b)\}$

3.3.3 $$\mathbf{E}_r e^{-\alpha H - \gamma \ell(H,r)} = \frac{\Upsilon_\alpha [e^{\mu(a-r)} \operatorname{sh}((b-r)\Upsilon_\alpha) + e^{\mu(b-r)} \operatorname{sh}((r-a)\Upsilon_\alpha)]}{\Upsilon_\alpha \operatorname{sh}((b-a)\Upsilon_\alpha) + 2\gamma \operatorname{sh}((b-r)\Upsilon_\alpha)\operatorname{sh}((r-a)\Upsilon_\alpha)} =: L_\mu$$

$$\mathbf{E}_x e^{-\alpha H - \gamma \ell(H,r)}$$

$$= \begin{cases} e^{\mu(b-x)} \dfrac{\operatorname{sh}((x-r)\Upsilon_\alpha)}{\operatorname{sh}((b-r)\Upsilon_\alpha)} + e^{\mu(r-x)} \dfrac{\operatorname{sh}((b-x)\Upsilon_\alpha)}{\operatorname{sh}((b-r)\Upsilon_\alpha)} L_\mu, & r \le x \le b \\ e^{\mu(a-x)} \dfrac{\operatorname{sh}((r-x)\Upsilon_\alpha)}{\operatorname{sh}((r-a)\Upsilon_\alpha)} + e^{\mu(r-x)} \dfrac{\operatorname{sh}((x-a)\Upsilon_\alpha)}{\operatorname{sh}((r-a)\Upsilon_\alpha)} L_\mu, & a \le x \le r \end{cases}$$

3.3.4 $$\mathbf{E}_x \{e^{-\alpha H}; \ell(H,r) \in dy\} = \exp\Big(-\frac{\Upsilon_\alpha \operatorname{sh}((b-a)\Upsilon_\alpha) y}{2\operatorname{sh}((b-r)\Upsilon_\alpha)\operatorname{sh}((r-a)\Upsilon_\alpha)}\Big)$$

$$\times \begin{cases} \dfrac{\Upsilon_\alpha \operatorname{sh}((b-x)\Upsilon_\alpha)[e^{\mu(a-x)} \operatorname{sh}((b-r)\Upsilon_\alpha) + e^{\mu(b-x)} \operatorname{sh}((r-a)\Upsilon_\alpha)]}{2\operatorname{sh}^2((b-r)\Upsilon_\alpha)\operatorname{sh}((r-a)\Upsilon_\alpha)} dy, & r \le x \le b \\ \dfrac{\Upsilon_\alpha \operatorname{sh}((x-a)\Upsilon_\alpha)[e^{\mu(a-x)} \operatorname{sh}((b-r)\Upsilon_\alpha) + e^{\mu(b-x)} \operatorname{sh}((r-a)\Upsilon_\alpha)]}{2\operatorname{sh}((b-r)\Upsilon_\alpha)\operatorname{sh}^2((r-a)\Upsilon_\alpha)} dy, & a \le x \le r \end{cases}$$

(1) $$\mathbf{E}_x \{e^{-\alpha H}; \ell(H,r) = 0\} = \begin{cases} e^{\mu(b-x)} \dfrac{\operatorname{sh}((x-r)\Upsilon_\alpha)}{\operatorname{sh}((b-r)\Upsilon_\alpha)}, & r \le x \le b \\ e^{\mu(a-x)} \dfrac{\operatorname{sh}((r-x)\Upsilon_\alpha)}{\operatorname{sh}((r-a)\Upsilon_\alpha)}, & a \le x \le r \end{cases}$$

3.3.5 (a) $$\mathbf{E}_x \{e^{-\gamma \ell(H,r)};\ W_H^{(\mu)} = a\} = e^{\mu(a-x)} V_a^{(3)}(\mu^2)$$

for $V_a^{(3)}(2\alpha)$ see 1.3.3.7

3.3.5 (b) $$\mathbf{E}_x \{e^{-\gamma \ell(H,r)};\ W_H^{(\mu)} = b\} = e^{\mu(b-x)} V_b^{(3)}(\mu^2)$$

for $V_b^{(3)}(2\alpha)$ see 1.3.3.7

3.3.6 (a) $$\mathbf{P}_x\big(\ell(H,r) \in dy, W_H^{(\mu)} = a\big) = \exp\Big(\mu(a-x) - \frac{|\mu| \operatorname{sh}((b-a)|\mu|) y}{2\operatorname{sh}((b-r)|\mu|)\operatorname{sh}((r-a)|\mu|)}\Big)$$

$$\times \begin{cases} \dfrac{|\mu| \operatorname{sh}((b-x)|\mu|)}{2\operatorname{sh}((b-r)|\mu|)\operatorname{sh}((r-a)|\mu|)} dy, & r \le x \le b \\ \dfrac{|\mu| \operatorname{sh}((x-a)|\mu|)}{2\operatorname{sh}^2((r-a)|\mu|)} dy, & a \le x \le r \end{cases}$$

$$\Upsilon_\alpha := \sqrt{2\alpha + \mu^2}$$

2 $W_s^{(\mu)} = \mu s + W_s$ $H = H_{a,b} = \min\{s : W_s^{(\mu)} \notin (a,b)\}$

(1) $$\mathbf{P}_x\big(\ell(H,r) = 0, W_H^{(\mu)} = a\big) = \begin{cases} 0, & r \le x \le b \\ e^{\mu(a-x)} \dfrac{\operatorname{sh}((r-x)|\mu|)}{\operatorname{sh}((r-a)|\mu|)}, & a \le x \le r \end{cases}$$

3.3.6 (b) $$\mathbf{P}_x\big(\ell(H,r) \in dy, W_H^{(\mu)} = b\big) = \exp\Big(\mu(b-x) - \frac{|\mu| \operatorname{sh}((b-a)|\mu|)y}{2\operatorname{sh}((b-r)|\mu|)\operatorname{sh}((r-a)|\mu|)}\Big)$$

$$\times \begin{cases} \dfrac{|\mu| \operatorname{sh}((b-x)|\mu|)}{2\operatorname{sh}^2((b-r)|\mu|)} dy, & r \le x \le b \\ \dfrac{|\mu| \operatorname{sh}((x-a)|\mu|)}{2\operatorname{sh}((b-r)|\mu|)\operatorname{sh}((r-a)|\mu|)} dy, & a \le x \le r \end{cases}$$

(1) $$\mathbf{P}_x\big(\ell(H,r) = 0, W_H^{(\mu)} = b\big) = \begin{cases} e^{\mu(b-x)} \dfrac{\operatorname{sh}((x-r)|\mu|)}{\operatorname{sh}((b-r)|\mu|)}, & r \le x \le b \\ 0, & a \le x \le r \end{cases}$$

3.3.7 (a) $$\mathbf{E}_x\{e^{-\alpha H - \gamma\ell(H,r)};\ W_H^{(\mu)} = a\} = e^{\mu(a-x)} V_a^{(3)}(2\alpha + \mu^2)$$

for $V_a^{(3)}(2\alpha)$ see 1.3.3.7

3.3.7 (b) $$\mathbf{E}_x\{e^{-\alpha H - \gamma\ell(H,r)};\ W_H^{(\mu)} = b\} = e^{\mu(b-x)} V_b^{(3)}(2\alpha + \mu^2)$$

for $V_b^{(3)}(2\alpha)$ see 1.3.3.7

3.3.8 (a) $$\mathbf{E}_x\{e^{-\alpha H}; \ell(H,r) \in dy, W_H^{(\mu)} = a\} = \exp\Big(-\frac{\Upsilon_\alpha \operatorname{sh}((b-a)\Upsilon_\alpha)y}{2\operatorname{sh}((b-r)\Upsilon_\alpha)\operatorname{sh}((r-a)\Upsilon_\alpha)}\Big)$$

$$\times e^{\mu(a-x)} \begin{cases} \dfrac{\Upsilon_\alpha \operatorname{sh}((b-x)\Upsilon_\alpha)}{2\operatorname{sh}((b-r)\Upsilon_\alpha)\operatorname{sh}((r-a)\Upsilon_\alpha)} dy, & r \le x \le b \\ \dfrac{\Upsilon_\alpha \operatorname{sh}((x-a)\Upsilon_\alpha)}{2\operatorname{sh}^2((r-a)\Upsilon_\alpha)} dy, & a \le x \le r \end{cases}$$

(1) $$\mathbf{E}_x\{e^{-\alpha H}; \ell(H,r) = 0, W_H^{(\mu)} = a\} = \begin{cases} 0, & r \le x \le b \\ e^{\mu(a-x)} \dfrac{\operatorname{sh}((r-x)\Upsilon_\alpha)}{\operatorname{sh}((r-a)\Upsilon_\alpha)}, & a \le x \le r \end{cases}$$

3.3.8 (b) $$\mathbf{E}_x\{e^{-\alpha H}; \ell(H,r) \in dy, W_H^{(\mu)} = b\} = \exp\Big(-\frac{\Upsilon_\alpha \operatorname{sh}((b-a)\Upsilon_\alpha)y}{2\operatorname{sh}((b-r)\Upsilon_\alpha)\operatorname{sh}((r-a)\Upsilon_\alpha)}\Big)$$

2 $W_s^{(\mu)} = \mu s + W_s$ $H = H_{a,b} = \min\{s : W_s^{(\mu)} \notin (a,b)\}$

$$\times e^{\mu(b-x)} \begin{cases} \dfrac{\Upsilon_\alpha \operatorname{sh}((b-x)\Upsilon_\alpha)}{2\operatorname{sh}^2((b-r)\Upsilon_\alpha)} dy, & r \le x \le b \\ \dfrac{\Upsilon_\alpha \operatorname{sh}((x-a)\Upsilon_\alpha)}{2\operatorname{sh}((b-r)\Upsilon_\alpha)\operatorname{sh}((r-a)\Upsilon_\alpha)} dy, & a \le x \le r \end{cases}$$

(1) $$\mathbf{E}_x\{e^{-\alpha H}; \ell(H,r) = 0, W_H^{(\mu)} = b\} = \begin{cases} e^{\mu(b-x)} \dfrac{\operatorname{sh}((x-r)\Upsilon_\alpha)}{\operatorname{sh}((b-r)\Upsilon_\alpha)}, & r \le x \le b \\ 0, & a \le x \le r \end{cases}$$

3.4.1 $$\mathbf{E}_r \exp\Big(-\gamma \int_0^H \mathbb{I}_{[r,b]}\big(W_s^{(\mu)}\big) ds\Big) =: Q_\mu$$

$$= \frac{|\mu| e^{\mu(a-r)} \operatorname{sh}((b-r)\Upsilon_\gamma) + \Upsilon_\gamma e^{\mu(b-r)} \operatorname{sh}((r-a)|\mu|)}{\Upsilon_\gamma \operatorname{ch}((b-r)\Upsilon_\gamma)\operatorname{sh}((r-a)|\mu|) + |\mu| \operatorname{sh}((b-r)\Upsilon_\gamma)\operatorname{ch}((r-a)|\mu|)}$$

$$\mathbf{E}_x \exp\Big(-\gamma \int_0^H \mathbb{I}_{[r,b]}\big(W_s^{(\mu)}\big) ds\Big)$$

$$= \begin{cases} e^{\mu(b-x)} \dfrac{\operatorname{sh}((x-r)\Upsilon_\gamma)}{\operatorname{sh}((b-r)\Upsilon_\gamma)} + e^{\mu(r-x)} \dfrac{\operatorname{sh}((b-x)\Upsilon_\gamma)}{\operatorname{sh}((b-r)\Upsilon_\gamma)} Q_\mu, & r \le x \le b \\ e^{\mu(a-x)} \dfrac{\operatorname{sh}((r-x)|\mu|)}{\operatorname{sh}((r-a)|\mu|)} + e^{\mu(r-x)} \dfrac{\operatorname{sh}((x-a)|\mu|)}{\operatorname{sh}((r-a)|\mu|)} Q_\mu, & a \le x \le r \end{cases}$$

3.4.5 (a) $$\mathbf{E}_x\Big\{\exp\Big(-\gamma \int_0^H \mathbb{I}_{[r,b]}\big(W_s^{(\mu)}\big) ds\Big);\ W_H^{(\mu)} = a\Big\} = e^{\mu(a-x)} Q_a^{(6)}(\mu^2, \mu^2 + 2\gamma)$$

for $Q_a^{(6)}(2p, 2q)$ see 1.3.6.5

3.4.5 (b) $$\mathbf{E}_x\Big\{\exp\Big(-\gamma \int_0^H \mathbb{I}_{[r,b]}\big(W_s^{(\mu)}\big) ds\Big);\ W_H^{(\mu)} = b\Big\} = e^{\mu(b-x)} Q_b^{(6)}(\mu^2, \mu^2 + 2\gamma)$$

for $Q_b^{(6)}(2p, 2q)$ see 1.3.6.5

3.4.6 (a) $$\mathbf{P}_x\Big(\int_0^H \mathbb{I}_{[r,b]}\big(W_s^{(\mu)}\big) ds \in dy, W_H^{(\mu)} = a\Big)$$

$a \le x$, $x \le r$ $$= \frac{\operatorname{sh}((x-a)|\mu|) e^{\mu(a-x) - \mu^2 y/2}}{\operatorname{sh}((r-a)|\mu|)\operatorname{ch}((r-a)|\mu|)} \operatorname{rc}_y(0, b-r, b-r, \operatorname{th}((r-a)|\mu|)/|\mu|) dy$$

$$\Upsilon_s := \sqrt{2s + \mu^2}$$

2 $W_s^{(\mu)}=\mu s+W_s$ $H=H_{a,b}=\min\{s: W_s^{(\mu)}\notin(a,b)\}$

$r\le x$, $x\le b$
$$=\frac{e^{\mu(a-x)-\mu^2y/2}}{\operatorname{ch}((r-a)|\mu|)}\operatorname{rc}_y(0,b-r,b-x,\operatorname{th}((r-a)|\mu|)/|\mu|)dy$$

(1)
$$\mathbf{P}_x\Big(\int_0^H \mathbb{I}_{[r,b]}\big(W_s^{(\mu)}\big)ds=0, W_H^{(\mu)}=a\Big)=\begin{cases}\dfrac{e^{\mu(a-x)}\operatorname{sh}((r-x)|\mu|)}{\operatorname{sh}((r-a)|\mu|)}, & a\le x\le r\\ 0, & r\le x\le b\end{cases}$$

3.4.6 (b)
$$\mathbf{P}_x\Big(\int_0^H \mathbb{I}_{[r,b]}\big(W_s^{(\mu)}\big)ds\in dy, W_H^{(\mu)}=b\Big)$$

$a\le x$, $x\le r$
$$=\frac{\operatorname{sh}((x-a)|\mu|)}{|\mu|\operatorname{ch}((r-a)|\mu|)}e^{\mu(b-x)-\mu^2y/2}\widetilde{\operatorname{rc}}_y(1,b-r,0,\operatorname{th}((r-a)|\mu|)/|\mu|)dy$$

$r\le x$, $x\le b$
$$=e^{\mu(b-x)-\mu^2y/2}\Big(\operatorname{ss}_y(x-r,b-r)$$
$$+\frac{\operatorname{sh}((r-a)|\mu|)}{|\mu|\operatorname{ch}((r-a)|\mu|)}\operatorname{ss}_y(b-x,b-r)*\widetilde{\operatorname{rc}}_y(1,b-r,0,\operatorname{th}((r-a)|\mu|)/|\mu|)\Big)dy$$

3.5.1
$$\mathbf{E}_r\exp\Big(-\gamma\int_0^H \mathbb{I}_{[a,r]}\big(W_s^{(\mu)}\big)ds\Big)=:Q_\mu$$
$$=\frac{e^{\mu(a-r)}\sqrt{2\gamma+\mu^2}\operatorname{sh}((b-r)|\mu|)+e^{\mu(b-r)}|\mu|\operatorname{sh}((r-a)\sqrt{2\gamma+\mu^2})}{\sqrt{2\gamma+\mu^2}\operatorname{sh}((b-r)|\mu|)\operatorname{ch}((r-a)\sqrt{2\gamma+\mu^2})+|\mu|\operatorname{ch}((b-r)|\mu|)\operatorname{sh}((r-a)\sqrt{2\gamma+\mu^2})}$$

$$\mathbf{E}_x\exp\Big(-\gamma\int_0^H \mathbb{I}_{[a,r]}\big(W_s^{(\mu)}\big)ds\Big)$$
$$=\begin{cases} e^{\mu(b-x)}\dfrac{\operatorname{sh}((x-r)|\mu|)}{\operatorname{sh}((b-r)|\mu|)}+e^{\mu(r-x)}\dfrac{\operatorname{sh}((b-x)|\mu|)}{\operatorname{sh}((b-r)|\mu|)}Q_\mu, & r\le x\le b\\ e^{\mu(a-x)}\dfrac{\operatorname{sh}((r-x)\sqrt{2\gamma+\mu^2})}{\operatorname{sh}((r-a)\sqrt{2\gamma+\mu^2})}+e^{\mu(r-x)}\dfrac{\operatorname{sh}((x-a)\sqrt{2\gamma+\mu^2})}{\operatorname{sh}((r-a)\sqrt{2\gamma+\mu^2})}Q_\mu, & a\le x\le r\end{cases}$$

3.5.5 (a)
$$\mathbf{E}_x\Big\{\exp\Big(-\gamma\int_0^H \mathbb{I}_{[a,r]}\big(W_s^{(\mu)}\big)ds\Big);\ W_H^{(\mu)}=a\Big\}=e^{\mu(a-x)}Q_a^{(6)}(\mu^2+2\gamma,\mu^2)$$

for $Q_a^{(6)}(2p,2q)$ see 1.3.6.5

3.5.5 (b)
$$\mathbf{E}_x\Big\{\exp\Big(-\gamma\int_0^H \mathbb{I}_{[a,r]}\big(W_s^{(\mu)}\big)ds\Big);\ W_H^{(\mu)}=b\Big\}=e^{\mu(b-x)}Q_b^{(6)}(\mu^2+2\gamma,\mu^2)$$

for $Q_b^{(6)}(2p,2q)$ see 1.3.6.5

2 $W_s^{(\mu)} = \mu s + W_s$ $\qquad H = H_{a,b} = \min\{s : W_s^{(\mu)} \notin (a,b)\}$

3.5.6 (a)
$$\mathbf{P}_x\Big(\int_0^H \mathbb{I}_{[a,r]}\big(W_s^{(\mu)}\big)ds \in dy, W_H^{(\mu)} = a\Big)$$

$a \le x$, $x \le r$
$$= e^{\mu(a-x)-\mu^2 y/2}\Big(\mathrm{ss}_y(r-x, r-a)$$
$$+\frac{\mathrm{sh}((b-r)|\mu|)}{|\mu|\,\mathrm{ch}((b-r)|\mu|)}\,\mathrm{ss}_y(x-a, r-a) * \widetilde{\mathrm{rc}}_y(1, r-a, 0, \mathrm{th}((b-r)|\mu|)/|\mu|)dy$$

$r \le x$, $x \le b$
$$= \frac{\mathrm{sh}((b-x)|\mu|)}{|\mu|\,\mathrm{ch}((b-r)|\mu|)} e^{\mu(a-x)-\mu^2 y/2}\widetilde{\mathrm{rc}}_y(1, r-a, 0, \mathrm{th}((b-r)|\mu|)/|\mu|)\Big)dy$$

3.5.6 (b)
$$\mathbf{P}_x\Big(\int_0^H \mathbb{I}_{[a,r]}\big(W_s^{(\mu)}\big)ds \in dy, W_H^{(\mu)} = b\Big)$$

$a \le x$, $x \le r$
$$= \frac{e^{\mu(b-x)-\mu^2 y/2}}{\mathrm{ch}((b-r)|\mu|)}\,\mathrm{rc}_y(0, r-a, x-a, \mathrm{th}((b-r)|\mu|)/|\mu|)dy$$

$r \le x$, $x \le b$
$$= \frac{\mathrm{sh}((b-x)|\mu|)e^{\mu(b-x)-\mu^2 y/2}}{\mathrm{sh}((b-r)|\mu|)\,\mathrm{ch}((b-r)|\mu|)}\,\mathrm{rc}_y(0, r-a, r-a, \mathrm{th}((b-r)|\mu|)/|\mu|)dy$$

(1)
$$\mathbf{P}_x\Big(\int_0^H \mathbb{I}_{[a,r]}\big(W_s^{(\mu)}\big)ds = 0, W_H^{(\mu)} = b\Big) = \begin{cases} 0, & a \le x \le r \\ \dfrac{e^{\mu(b-x)}\,\mathrm{sh}((x-r)|\mu|)}{\mathrm{sh}((b-r)|\mu|)}, & r \le x \le b \end{cases}$$

3.5.7 (a)
$$\mathbf{E}_x\Big\{\exp\Big(-\alpha H - \gamma\int_0^H \mathbb{I}_{[a,r]}\big(W_s^{(\mu)}\big)ds\Big); W_H^{(\mu)} = a\Big\}$$
$$= e^{\mu(a-x)}Q_a^{(6)}(2\alpha+\mu^2+2\gamma, 2\alpha+\mu^2), \qquad \text{for } Q_a^{(6)}(2p,2q) \text{ see 1.3.6.5}$$

3.5.7 (b)
$$\mathbf{E}_x\Big\{\exp\Big(-\alpha H - \gamma\int_0^H \mathbb{I}_{[a,r]}\big(W_s^{(\mu)}\big)ds\Big); W_H^{(\mu)} = b\Big\}$$
$$= e^{\mu(b-x)}Q_b^{(6)}(2\alpha+\mu^2+2\gamma, 2\alpha+\mu^2), \qquad \text{for } Q_b^{(6)}(2p,2q) \text{ see 1.3.6.5}$$

3.6.1
$$\mathbf{E}_r \exp\Big(-\int_0^H \big(p\mathbb{I}_{[a,r]}\big(W_s^{(\mu)}\big) + q\mathbb{I}_{[r,b]}\big(W_s^{(\mu)}\big)\big)ds\Big) =: Q$$
$$= \frac{e^{\mu(a-r)}\Upsilon_p\,\mathrm{sh}((b-r)\Upsilon_q) + e^{\mu(b-r)}\Upsilon_q\,\mathrm{sh}((r-a)\Upsilon_p)}{\Upsilon_q\,\mathrm{ch}((b-r)\Upsilon_q)\,\mathrm{sh}((r-a)\Upsilon_p) + \Upsilon_p\,\mathrm{sh}((b-r)\Upsilon_q)\,\mathrm{ch}((r-a)\Upsilon_p)}$$

$$\mathbf{E}_x\Big\{\exp\Big(-\int_0^H \big(p\mathbb{I}_{[a,r]}\big(W_s^{(\mu)}\big) + q\mathbb{I}_{[r,b]}\big(W_s^{(\mu)}\big)\big)ds\Big)\Big\}$$

2 $W_s^{(\mu)} = \mu s + W_s$ $H = H_{a,b} = \min\{s : W_s^{(\mu)} \notin (a,b)\}$

$$= \begin{cases} e^{\mu(b-x)} \dfrac{\operatorname{sh}((x-r)\Upsilon_q)}{\operatorname{sh}((b-r)\Upsilon_q)} + e^{\mu(r-x)} \dfrac{\operatorname{sh}((b-x)\Upsilon_q)}{\operatorname{sh}((b-r)\Upsilon_q)} Q, & r \le x \le b \\ e^{\mu(a-x)} \dfrac{\operatorname{sh}((r-x)\Upsilon_p)}{\operatorname{sh}((r-a)\Upsilon_p)} + e^{\mu(r-x)} \dfrac{\operatorname{sh}((x-a)\Upsilon_p)}{\operatorname{sh}((r-a)\Upsilon_p)} Q, & a \le x \le r \end{cases}$$

3.6.5 (a) $\mathbf{E}_x\Big\{\exp\Big(-\int_0^H \big(p\mathbb{I}_{[a,r]}\big(W_s^{(\mu)}\big) + q\mathbb{I}_{[r,b]}\big(W_s^{(\mu)}\big)\big)ds\Big); W_H^{(\mu)} = a\Big\}$

$$= e^{\mu(a-x)} Q_a^{(6)}(2p+\mu^2, 2q+\mu^2), \qquad \text{for } Q_a^{(6)}(2p,2q) \text{ see 1.3.6.5}$$

3.6.5 (b) $\mathbf{E}_x\Big\{\exp\Big(-\int_0^H \big(p\mathbb{I}_{[a,r]}\big(W_s^{(\mu)}\big) + q\mathbb{I}_{[r,b]}\big(W_s^{(\mu)}\big)\big)ds\Big); W_H^{(\mu)} = b\Big\}$

$$= e^{\mu(b-x)} Q_b^{(6)}(2p+\mu^2, 2q+\mu^2), \qquad \text{for } Q_b^{(6)}(2p,2q) \text{ see 1.3.6.5}$$

3.6.6 (a) $\mathbf{P}_x\Big(\int_0^H \mathbb{I}_{[a,r)}\big(W_s^{(\mu)}\big)ds \in du,\ \int_0^H \mathbb{I}_{[r,b]}\big(W_s^{(\mu)}\big)ds \in dv,\ W_H^{(\mu)} = a\Big)$

$$= e^{\mu(a-x)-\mu^2 u/2-\mu^2 v/2} \int_0^\infty B_a^{(27)}(u,v,y)dy\,dudv$$

for $B_a^{(27)}(u,v,y)$ see 1.3.27.6(a)

3.6.6 (b) $\mathbf{P}_x\Big(\int_0^H \mathbb{I}_{[a,r)}\big(W_s^{(\mu)}\big)ds \in du,\ \int_0^H \mathbb{I}_{[r,b]}\big(W_s^{(\mu)}\big)ds \in dv,\ W_H^{(\mu)} = b\Big)$

$$= e^{\mu(b-x)-\mu^2 u/2-\mu^2 v/2} \int_0^\infty B_b^{(27)}(u,v,y)dy\,dudv$$

for $B_b^{(27)}(u,v,y)$ see 1.3.27.6(b)

3.8.1 $\mathbf{E}_x \exp\Big(-\gamma\int_0^H W_s^{(\mu)}ds\Big)$

$$= \frac{e^{\mu(a-x)}\Upsilon_{\gamma b}^2\Upsilon_{\gamma x}^2 S_{1/3}(\frac{1}{3\gamma}\Upsilon_{\gamma b}^3, \frac{1}{3\gamma}\Upsilon_{\gamma x}^3) + e^{\mu(b-x)}\Upsilon_{\gamma x}^2\Upsilon_{\gamma a}^2 S_{1/3}(\frac{1}{3\gamma}\Upsilon_{\gamma x}^3, \frac{1}{3\gamma}\Upsilon_{\gamma a}^3)}{\Upsilon_{\gamma b}^2\Upsilon_{\gamma a}^2 S_{1/3}(\frac{1}{3\gamma}\Upsilon_{\gamma b}^3, \frac{1}{3\gamma}\Upsilon_{\gamma a}^3)}$$

3.8.5 (a) $\mathbf{E}_x\Big\{\exp\Big(-\gamma\int_0^H W_s^{(\mu)}ds\Big); W_H^{(\mu)} = a\Big\} = \dfrac{e^{\mu(a-x)}\Upsilon_{\gamma x}^2 S_{1/3}(\frac{1}{3\gamma}\Upsilon_{\gamma b}^3, \frac{1}{3\gamma}\Upsilon_{\gamma x}^3)}{\Upsilon_{\gamma a}^2 S_{1/3}(\frac{1}{3\gamma}\Upsilon_{\gamma b}^3, \frac{1}{3\gamma}\Upsilon_{\gamma a}^3)}$

3.8.5 (b) $\mathbf{E}_x\Big\{\exp\Big(-\gamma\int_0^H W_s^{(\mu)}ds\Big); W_H^{(\mu)} = b\Big\} = \dfrac{e^{\mu(b-x)}\Upsilon_{\gamma x}^2 S_{1/3}(\frac{1}{3\gamma}\Upsilon_{\gamma x}^3, \frac{1}{3\gamma}\Upsilon_{\gamma a}^3)}{\Upsilon_{\gamma b}^2 S_{1/3}(\frac{1}{3\gamma}\Upsilon_{\gamma b}^3, \frac{1}{3\gamma}\Upsilon_{\gamma a}^3)}$

2 $W_s^{(\mu)} = \mu s + W_s$ $\qquad H = H_{a,b} = \min\{s : W_s^{(\mu)} \notin (a,b)\}$

3.9.1 $\mathbf{E}_x \exp\Big(-\frac{\gamma^2}{2}\int_0^H (W_s^{(\mu)})^2 ds\Big)$

$$= \frac{e^{\mu a+(a^2-x^2)\gamma/2} S\big(\frac12+\frac{\mu^2}{2\gamma}, b\sqrt{2\gamma}, x\sqrt{2\gamma}\big) + e^{\mu b+(b^2-x^2)\gamma/2} S\big(\frac12+\frac{\mu^2}{2\gamma}, x\sqrt{2\gamma}, a\sqrt{2\gamma}\big)}{e^{\mu x} S\big(\frac12+\frac{\mu^2}{2\gamma}, b\sqrt{2\gamma}, a\sqrt{2\gamma}\big)}$$

3.9.7 (a) $\mathbf{E}_x\Big\{\exp\Big(-\alpha H - \frac{\gamma^2}{2}\int_0^H (W_s^{(\mu)})^2 ds\Big); W_H^{(\mu)} = a\Big\}$

$$= e^{\mu(a-x)+(a^2-x^2)\gamma/2} \frac{S\big(\frac12+\frac{2\alpha+\mu^2}{2\gamma}, b\sqrt{2\gamma}, x\sqrt{2\gamma}\big)}{S\big(\frac12+\frac{2\alpha+\mu^2}{2\gamma}, b\sqrt{2\gamma}, a\sqrt{2\gamma}\big)}$$

3.9.7 (b) $\mathbf{E}_x\Big\{\exp\Big(-\alpha H - \frac{\gamma^2}{2}\int_0^H (W_s^{(\mu)})^2 ds\Big); W_H^{(\mu)} = b\Big\}$

$$= e^{\mu(b-x)+(b^2-x^2)\gamma/2} \frac{S\big(\frac12+\frac{2\alpha+\mu^2}{2\gamma}, x\sqrt{2\gamma}, a\sqrt{2\gamma}\big)}{S\big(\frac12+\frac{2\alpha+\mu^2}{2\gamma}, b\sqrt{2\gamma}, a\sqrt{2\gamma}\big)}$$

3.10.1 $\mathbf{E}_x \exp\Big(-\gamma\int_0^H e^{2\beta W_s^{(\mu)}} ds\Big)$

$$= \frac{e^{(\mu-|\mu|\,\mathrm{sign}\,\beta)a} S_{|\mu|/|\beta|}\Big(\frac{\sqrt{2\gamma}}{|\beta|}e^{\beta b}, \frac{\sqrt{2\gamma}}{|\beta|}e^{\beta x}\Big) + e^{(\mu-|\mu|\,\mathrm{sign}\,\beta)b} S_{|\mu|/|\beta|}\Big(\frac{\sqrt{2\gamma}}{|\beta|}e^{\beta x}, \frac{\sqrt{2\gamma}}{|\beta|}e^{\beta a}\Big)}{e^{(\mu-|\mu|\,\mathrm{sign}\,\beta)x} S_{|\mu|/|\beta|}\Big(\frac{\sqrt{2\gamma}}{|\beta|}e^{\beta b}, \frac{\sqrt{2\gamma}}{|\beta|}e^{\beta a}\Big)}$$

3.10.7 (a) $\mathbf{E}_x\Big\{\exp\Big(-\alpha H - \gamma\int_0^H e^{2\beta W_s^{(\mu)}} ds\Big); W_H^{(\mu)} = a\Big\} = \frac{e^{(\mu-\Upsilon_\alpha\beta/|\beta|)a} S_{\frac{\Upsilon_\alpha}{|\beta|}}\Big(\frac{\sqrt{2\gamma}}{|\beta|}e^{\beta b}, \frac{\sqrt{2\gamma}}{|\beta|}e^{\beta x}\Big)}{e^{(\mu-\Upsilon_\alpha\beta/|\beta|)x} S_{\frac{\Upsilon_\alpha}{|\beta|}}\Big(\frac{\sqrt{2\gamma}}{|\beta|}e^{\beta b}, \frac{\sqrt{2\gamma}}{|\beta|}e^{\beta a}\Big)}$

3.10.7 (b) $\mathbf{E}_x\Big\{\exp\Big(-\alpha H - \gamma\int_0^H e^{2\beta W_s^{(\mu)}} ds\Big); W_H^{(\mu)} = b\Big\} = \frac{e^{(\mu-\Upsilon_\alpha\beta/|\beta|)b} S_{\frac{\Upsilon_\alpha}{|\beta|}}\Big(\frac{\sqrt{2\gamma}}{|\beta|}e^{\beta x}, \frac{\sqrt{2\gamma}}{|\beta|}e^{\beta a}\Big)}{e^{(\mu-\Upsilon_\alpha\beta/|\beta|)x} S_{\frac{\Upsilon_\alpha}{|\beta|}}\Big(\frac{\sqrt{2\gamma}}{|\beta|}e^{\beta b}, \frac{\sqrt{2\gamma}}{|\beta|}e^{\beta a}\Big)}$

3.12.5 (a) $\mathbf{E}_x\{e^{-\gamma \check{H}_\mu(H)};\ W_H^{(\mu)} = a\} = \int_{x-a}^{b-a} \frac{|\mu| e^{\mu(a-x)}\,\mathrm{sh}((x-a)\sqrt{2\gamma+\mu^2})}{\mathrm{sh}(|\mu|y)\,\mathrm{sh}(y\sqrt{2\gamma+\mu^2})} dy$

$\Upsilon_s := \sqrt{2s+\mu^2}$

2 $W_s^{(\mu)} = \mu s + W_s$ $H = H_{a,b} = \min\{s : W_s^{(\mu)} \notin (a,b)\}$

3.12.5 (b)
$$\mathbf{E}_x\{e^{-\gamma \hat{H}_\mu(H)};\, W_H^{(\mu)} = b\} = \int_{b-x}^{b-a} \frac{|\mu| e^{\mu(b-x)} \operatorname{sh}((b-x)\sqrt{2\gamma+\mu^2})}{\operatorname{sh}(|\mu|y)\operatorname{sh}(y\sqrt{2\gamma+\mu^2})} dy$$

3.12.6 (a)
$$\mathbf{P}_x\big(\check{H}_\mu(H) \in du,\, W_H^{(\mu)} = a\big) = |\mu| e^{\mu(a-x)-\mu^2 u/2} \int_{x-a}^{b-a} \frac{\operatorname{ss}_u(x-a,y)}{\operatorname{sh}(|\mu|y)} dy\, du$$

3.12.6 (b)
$$\mathbf{P}_x\big(\hat{H}_\mu(H) \in du,\, W_H^{(\mu)} = b\big) = |\mu| e^{\mu(b-x)-\mu^2 u/2} \int_{b-x}^{b-a} \frac{\operatorname{ss}_u(b-x,y)}{\operatorname{sh}(|\mu|y)} dy\, du$$

3.12.7 (a)
$$\mathbf{E}_x\{e^{-\alpha H-\gamma \check{H}_\mu(H)};\, W_H^{(\mu)} = a\} = \int_{x-a}^{b-a} \frac{\sqrt{2\alpha+\mu^2}\, e^{\mu a} \operatorname{sh}((x-a)\sqrt{2\gamma+2\alpha+\mu^2})}{e^{\mu x} \operatorname{sh}(y\sqrt{2\alpha+\mu^2}) \operatorname{sh}(y\sqrt{2\gamma+2\alpha+\mu^2})} dy$$

3.12.7 (b)
$$\mathbf{E}_x\{e^{-\alpha H-\gamma \hat{H}_\mu(H)};\, W_H^{(\mu)} = b\} = \int_{b-x}^{b-a} \frac{\sqrt{2\alpha+\mu^2}\, e^{\mu b} \operatorname{sh}((b-x)\sqrt{2\gamma+2\alpha+\mu^2})}{e^{\mu x} \operatorname{sh}(y\sqrt{2\alpha+\mu^2}) \operatorname{sh}(y\sqrt{2\gamma+2\alpha+\mu^2})} dy$$

3.12.8 (a)
$$\mathbf{P}_x\big(\check{H}_\mu(H) \in du,\, H \in dt,\, W_H^{(\mu)} = a\big) \qquad u < t$$
$$= e^{\mu(a-x)-\mu^2 t/2} \int_{x-a}^{b-a} \operatorname{ss}_u(x-a,y)\, \operatorname{s}_{t-u}(y) dy\, du dt$$

3.12.8 (b)
$$\mathbf{P}_x\big(\hat{H}_\mu(H) \in du,\, H \in dt,\, W_H^{(\mu)} = b\big) \qquad u < t$$
$$= e^{\mu(b-x)-\mu^2 t/2} \int_{b-x}^{b-a} \operatorname{ss}_u(b-x,y)\, \operatorname{s}_{t-u}(y) dy\, du dt$$

3.18.5 (a)
$$\mathbf{E}_x\{\exp(-\gamma\ell(H,r) - \eta\ell(H,u));\ W_H^{(\mu)} = a\}$$

$a \le x$, $x \le r$
$$= e^{\mu(a-x)} \frac{\mu^2 s(b,x) + 2|\mu|\gamma s(b,r)s(r,x) + 2|\mu|\eta s(b,u)s(u,x) + 4\gamma\eta s(b,u)s(u,r)s(r,x)}{\mu^2 s(b,a) + 2|\mu|\gamma s(b,r)s(r,a) + 2|\mu|\eta s(b,u)s(u,a) + 4\gamma\eta s(b,u)s(u,r)s(r,a)}$$

$r \le x$, $x \le u$
$$= e^{\mu(a-x)} \frac{\mu^2 s(b,x) + 2|\mu|\eta s(b,u)s(u,x)}{\mu^2 s(b,a) + 2|\mu|\gamma s(b,r)s(r,a) + 2|\mu|\eta s(b,u)s(u,a) + 4\gamma\eta s(b,u)s(u,r)s(r,a)}$$

$u \le x$, $x \le b$
$$= e^{\mu(a-x)} \frac{\mu^2 s(b,x)}{\mu^2 s(b,a) + 2|\mu|\gamma s(b,r)s(r,a) + 2|\mu|\eta s(b,u)s(u,a) + 4\gamma\eta s(b,u)s(u,r)s(r,a)}$$

$$s(y,x) := \operatorname{sh}((y-x)|\mu|)$$

2 $W_s^{(\mu)} = \mu s + W_s$ $\qquad H = H_{a,b} = \min\{s : W_s^{(\mu)} \notin (a,b)\}$

3.18.5 (b)

$$\mathbf{E}_x\{\exp(-\gamma\ell(H,r) - \eta\ell(H,u));\ W_H^{(\mu)} = b\}$$

$a \le x$, $x \le r$

$$= e^{\mu(b-x)}\frac{\mu^2 s(x,a)}{\mu^2 s(b,a) + 2|\mu|\gamma s(b,r)s(r,a) + 2|\mu|\eta s(b,u)s(u,a) + 4\gamma\eta s(b,u)s(u,r)s(r,a)}$$

$r \le x$, $x \le u$

$$= e^{\mu(b-x)}\frac{\mu^2 s(x,a) + 2|\mu|\gamma s(x,r)s(r,a)}{\mu^2 s(b,a) + 2|\mu|\gamma s(b,r)s(r,a) + 2|\mu|\eta s(b,u)s(u,a) + 4\gamma\eta s(b,u)s(u,r)s(r,a)}$$

$u \le x$, $x \le b$

$$= e^{\mu(b-x)}\frac{\mu^2 s(x,a) + 2|\mu|\gamma s(x,r)s(r,a) + 2|\mu|\eta s(x,u)s(u,a) + 4\gamma\eta s(x,u)s(u,r)s(r,a)}{\mu^2 s(b,a) + 2|\mu|\gamma s(b,r)s(r,a) + 2|\mu|\eta s(b,u)s(u,a) + 4\gamma\eta s(b,u)s(u,r)s(r,a)}$$

3.19.5 (a) $0 < a$

$$\mathbf{E}_x\Big\{\exp\Big(-\gamma\int_0^H \frac{ds}{W_s^{(\mu)}}\Big);\ W_H^{(\mu)} = a\Big\} = \frac{xe^{-(\mu+|\mu|)x}S(1+\gamma/|\mu|, 2, 2|\mu|b, 2|\mu|x)}{ae^{-(\mu+|\mu|)a}S(1+\gamma/|\mu|, 2, 2|\mu|b, 2|\mu|a)}$$

3.19.5 (b) $0 < a$

$$\mathbf{E}_x\Big\{\exp\Big(-\gamma\int_0^H \frac{ds}{W_s^{(\mu)}}\Big);\ W_H^{(\mu)} = b\Big\} = \frac{xe^{-(\mu+|\mu|)x}S(1+\gamma/|\mu|, 2, 2|\mu|x, 2|\mu|a)}{be^{-(\mu+|\mu|)b}S(1+\gamma/|\mu|, 2, 2|\mu|b, 2|\mu|a)}$$

3.20.5 (a) $0 < a$

$$\mathbf{E}_x\Big\{\exp\Big(-\frac{\gamma^2}{2}\int_0^H \frac{ds}{(W_s^{(\mu)})^2}\Big);\ W_H^{(\mu)} = a\Big\} = \frac{e^{-\mu x}x^{\sqrt{\gamma^2+1/4}}S_{\sqrt{\gamma^2+1/4}}(|\mu|b, |\mu|x)}{e^{-\mu a}a^{\sqrt{\gamma^2+1/4}}S_{\sqrt{\gamma^2+1/4}}(|\mu|b, |\mu|a)}$$

3.20.5 (b) $0 < a$

$$\mathbf{E}_x\Big\{\exp\Big(-\frac{\gamma^2}{2}\int_0^H \frac{ds}{(W_s^{(\mu)})^2}\Big);\ W_H^{(\mu)} = b\Big\} = \frac{e^{-\mu x}x^{\sqrt{\gamma^2+1/4}}S_{\sqrt{\gamma^2+1/4}}(|\mu|x, |\mu|a)}{e^{-\mu b}b^{\sqrt{\gamma^2+1/4}}S_{\sqrt{\gamma^2+1/4}}(|\mu|b, |\mu|a)}$$

3.21.5 (a)

$$\mathbf{E}_x\Big\{\exp\Big(-\int_0^H\Big(\frac{p^2}{2(W_s^{(\mu)})^2} + \frac{q^2(W_s^{(\mu)})^2}{2}\Big)ds\Big);\ W_H^{(\mu)} = a\Big\}$$

$0 < a$

$$= \frac{e^{-\mu x - qx^2/2}x^{\sqrt{4p^2+1}/4q}S\Big(\frac{\mu^2}{4q} + \frac{\sqrt{4p^2+1}}{4} + \frac{1}{2}, \frac{\sqrt{4p^2+1}}{2} + 1, qb^2, qx^2\Big)}{e^{-\mu a - qa^2/2}a^{\sqrt{4p^2+1}/4q}S\Big(\frac{\mu^2}{4q} + \frac{\sqrt{4p^2+1}}{4} + \frac{1}{2}, \frac{\sqrt{4p^2+1}}{2} + 1, qb^2, qa^2\Big)}, \quad a \le x \le b$$

3.21.5 (b)

$$\mathbf{E}_x\Big\{\exp\Big(-\int_0^H\Big(\frac{p^2}{2(W_s^{(\mu)})^2} + \frac{q^2(W_s^{(\mu)})^2}{2}\Big)ds\Big);\ W_H^{(\mu)} = b\Big\}$$

$$s(y,x) := \operatorname{sh}((y-x)|\mu|)$$

2	$W_s^{(\mu)} = \mu s + W_s$ $\qquad H = H_{a,b} = \min\{s : W_s^{(\mu)} \notin (a,b)\}$
$0 < a$	$= \dfrac{e^{-\mu x - qx^2/2} x^{\sqrt{4p^2+1}/4q} S\Big(\frac{\mu^2}{4q} + \frac{\sqrt{4p^2+1}}{4} + \frac{1}{2}, \frac{\sqrt{4p^2+1}}{2} + 1, qx^2, qa^2\Big)}{e^{-\mu b - qb^2/2} b^{\sqrt{4p^2+1}/4q} S\Big(\frac{\mu^2}{4q} + \frac{\sqrt{4p^2+1}}{4} + \frac{1}{2}, \frac{\sqrt{4p^2+1}}{2} + 1, qb^2, qa^2\Big)}, \quad a \le x \le b$
3.22.5 (a)	$\mathbf{E}_x\Big\{\exp\Big(-\int_0^H \Big(\frac{p^2}{2(W_s^{(\mu)})^2} + \frac{q}{W_s^{(\mu)}}\Big) ds\Big);\ W_H^{(\mu)} = a\Big\}$
$0 < a$	$= \dfrac{e^{-(\mu+\lvert\mu\rvert)x} x^{\sqrt{p^2+1/4}+1/2} S\big(\sqrt{p^2+1/4} + \frac{1}{2} + \frac{q}{\lvert\mu\rvert}, \sqrt{8p+1} + 1, 2\lvert\mu\rvert b, 2\lvert\mu\rvert x\big)}{e^{-(\mu+\lvert\mu\rvert)a} a^{\sqrt{p^2+1/4}+1/2} S\big(\sqrt{p^2+1/4} + \frac{1}{2} + \frac{q}{\lvert\mu\rvert}, \sqrt{8p+1} + 1, 2\lvert\mu\rvert b, 2\lvert\mu\rvert a\big)}$
3.22.5 (b)	$\mathbf{E}_x\Big\{\exp\Big(-\int_0^H \Big(\frac{p^2}{2(W_s^{(\mu)})^2} + \frac{q}{W_s^{(\mu)}}\Big) ds\Big);\ W_H^{(\mu)} = b\Big\}$
$0 < a$	$= \dfrac{e^{-(\mu+\lvert\mu\rvert)x} x^{\sqrt{p^2+1/4}+1/2} S\big(\sqrt{p^2+1/4} + \frac{1}{2} + \frac{q}{\lvert\mu\rvert}, \sqrt{8p+1} + 1, 2\lvert\mu\rvert x, 2\lvert\mu\rvert a\big)}{e^{-(\mu+\lvert\mu\rvert)b} b^{\sqrt{p^2+1/4}+1/2} S\big(\sqrt{p^2+1/4} + \frac{1}{2} + \frac{q}{\lvert\mu\rvert}, \sqrt{8p+1} + 1, 2\lvert\mu\rvert b, 2\lvert\mu\rvert a\big)}$
3.27.5 (a)	$\mathbf{E}_x\Big\{\exp\Big(-\int_0^H \big(p\mathbb{I}_{[a,r)}(W_s^{(\mu)}) + q\mathbb{I}_{[r,b]}(W_s^{(\mu)})\big) ds\Big); \frac{1}{2}\ell(H,r) \in dy, W_H = a\Big\}$
	$= \exp\Big(-y\Big(\frac{\sqrt{2q+\mu^2}\,\mathrm{ch}((b-r)\sqrt{2q+\mu^2})}{\mathrm{sh}((b-r)\sqrt{2q+\mu^2})} + \frac{\sqrt{2p+\mu^2}\,\mathrm{ch}((r-a)\sqrt{2p+\mu^2})}{\mathrm{sh}((r-a)\sqrt{2p+\mu^2})}\Big)\Big)$
	$\times e^{\mu(a-x)} \begin{cases} \dfrac{\sqrt{2p+\mu^2}\,\mathrm{sh}((b-x)\sqrt{2q+\mu^2})}{\mathrm{sh}((b-r)\sqrt{2q+\mu^2})\,\mathrm{sh}((r-a)\sqrt{2p+\mu^2})} dy, & r \le x \le b \\ \dfrac{\sqrt{2p+\mu^2}\,\mathrm{sh}((x-a)\sqrt{2p+\mu^2})}{\mathrm{sh}^2((r-a)\sqrt{2p+\mu^2})} dy, & a \le x \le r \end{cases}$
(1)	$\mathbf{E}_x\Big\{\exp\Big(-\int_0^H \big(p\mathbb{I}_{[a,r)}(W_s^{(\mu)}) + q\mathbb{I}_{[r,b]}(W_s^{(\mu)})\big) ds\Big); \ell(H,r) = 0, W_H = a\Big\}$
	$= e^{\mu(a-x)} \dfrac{\mathrm{sh}((r-x)\sqrt{2p+\mu^2})}{\mathrm{sh}((r-a)\sqrt{2p+\mu^2})}, \qquad a \le x \le r$
3.27.5 (b)	$\mathbf{E}_x\Big\{\exp\Big(-\int_0^H \big(p\mathbb{I}_{[a,r)}(W_s^{(\mu)}) + q\mathbb{I}_{[r,b]}(W_s^{(\mu)})\big) ds\Big); \frac{1}{2}\ell(H,r) \in dy, W_H = b\Big\}$
	$= \exp\Big(-y\Big(\frac{\sqrt{2q+\mu^2}\,\mathrm{ch}((b-r)\sqrt{2q+\mu^2})}{\mathrm{sh}((b-r)\sqrt{2q+\mu^2})} + \frac{\sqrt{2p+\mu^2}\,\mathrm{ch}((r-a)\sqrt{2p+\mu^2})}{\mathrm{sh}((r-a)\sqrt{2p+\mu^2})}\Big)\Big)$

2 $W_s^{(\mu)} = \mu s + W_s$ $\qquad H = H_{a,b} = \min\{s : W_s^{(\mu)} \notin (a,b)\}$

$$\times e^{\mu(b-x)} \begin{cases} \dfrac{\sqrt{2q+\mu^2}\,\mathrm{sh}((b-x)\sqrt{2q+\mu^2})}{\mathrm{sh}^2((b-r)\sqrt{2q+\mu^2})} dy, & r \le x \le b \\ \dfrac{\sqrt{2q+\mu^2}\,\mathrm{sh}((x-a)\sqrt{2p+\mu^2})}{\mathrm{sh}((b-r)\sqrt{2q+\mu^2})\,\mathrm{sh}((r-a)\sqrt{2p+\mu^2})} dy, & a \le x \le r \end{cases}$$

(1) $$\mathbf{E}_x\Big\{\exp\Big(-\int_0^H \big(p\mathbb{I}_{[a,r)}\big(W_s^{(\mu)}\big) + q\mathbb{I}_{[r,b]}\big(W_s^{(\mu)}\big)\big)ds\Big); \ell(H,r) = 0, W_H = b\Big\}$$

$$= e^{\mu(b-x)} \frac{\mathrm{sh}((x-r)\sqrt{2q+\mu^2})}{\mathrm{sh}((b-r)\sqrt{2q+\mu^2})}, \qquad r \le x \le b$$

3.27.6 (a) $$\mathbf{P}_x\Big(\int_0^H \mathbb{I}_{[a,r)}\big(W_s^{(\mu)}\big)ds \in du, \int_0^H \mathbb{I}_{[r,b]}\big(W_s^{(\mu)}\big)ds \in dv, \frac{1}{2}\ell(H,r) \in dy, W_H^{(\mu)} = a\Big)$$

$$= e^{\mu(a-x)-\mu^2 u/2-\mu^2 v/2} B_a^{(27)}(u,v,y)dudvdy$$

for $B_a^{(27)}(u,v,y)$ see 1.3.27.6(*a*)

3.27.6 (b) $$\mathbf{P}_x\Big(\int_0^H \mathbb{I}_{[a,r)}\big(W_s^{(\mu)}\big)ds \in du, \int_0^H \mathbb{I}_{[r,b]}\big(W_s^{(\mu)}\big)ds \in dv, \frac{1}{2}\ell\big(H,r\big) \in dy, W_H^{(\mu)} = b\Big)$$

$$= e^{\mu(b-x)-\mu^2 u/2-\mu^2 v/2} B_b^{(27)}(u,v,y)dudvdy$$

for $B_b^{(27)}(u,v,y)$ see 1.3.27.6(*b*)

3.30.5 (a) $$\mathbf{E}_x\Big\{\exp\Big(-\int_0^H \Big(\frac{p^2}{2}e^{2\beta W_s^{(\mu)}} + qe^{\beta W_s^{(\mu)}}\Big)ds\Big);\ W_H^{(\mu)} = a\Big\}$$

$$= \frac{e^{(|\mu|\operatorname{sign}\beta-\mu)x}\exp\Big(-\frac{p}{|\beta|}e^{\beta x}\Big) S\Big(\frac{|\mu|}{|\beta|} + \frac{q}{p|\beta|} + \frac{1}{2}, \frac{2|\mu|}{|\beta|} + 1, \frac{2p}{|\beta|}e^{\beta b}, \frac{2p}{|\beta|}e^{\beta x}\Big)}{e^{(|\mu|\operatorname{sign}\beta-\mu)a}\exp\Big(-\frac{p}{|\beta|}e^{\beta a}\Big) S\Big(\frac{|\mu|}{|\beta|} + \frac{q}{p|\beta|} + \frac{1}{2}, \frac{2|\mu|}{|\beta|} + 1, \frac{2p}{|\beta|}e^{\beta b}, \frac{2p}{|\beta|}e^{\beta a}\Big)}$$

3.30.5 (b) $$\mathbf{E}_x\Big\{\exp\Big(-\int_0^H \Big(\frac{p^2}{2}e^{2\beta W_s^{(\mu)}} + qe^{\beta W_s^{(\mu)}}\Big)ds\Big);\ W_H^{(\mu)} = b\Big\}$$

$$= \frac{e^{(|\mu|\operatorname{sign}\beta-\mu)x}\exp\Big(-\frac{p}{|\beta|}e^{\beta x}\Big) S\Big(\frac{|\mu|}{|\beta|} + \frac{q}{p|\beta|} + \frac{1}{2}, \frac{2|\mu|}{|\beta|} + 1, \frac{2p}{|\beta|}e^{\beta x}, \frac{2p}{|\beta|}e^{\beta a}\Big)}{e^{(|\mu|\operatorname{sign}\beta-\mu)b}\exp\Big(-\frac{p}{|\beta|}e^{\beta b}\Big) S\Big(\frac{|\mu|}{|\beta|} + \frac{q}{p|\beta|} + \frac{1}{2}, \frac{2|\mu|}{|\beta|} + 1, \frac{2p}{|\beta|}e^{\beta b}, \frac{2p}{|\beta|}e^{\beta a}\Big)}$$

2 $W_s^{(\mu)} = \mu s + W_s$ $H = H_{a,b} = \min\{s : W_s^{(\mu)} \notin (a,b)\}$

3.31.2
$0 < \alpha$

$$\mathbf{P}_{\frac{x+a}{|\mu|}}\Big(\sup_{0\le s\le H_{\frac{a}{|\mu|},\frac{b}{|\mu|}}}\Big(\ell\big(s,\frac{u+a}{|\mu|}\big)-\alpha^{-1}\ell\big(s,\frac{r+a}{|\mu|}\big)\Big)>h\Big) \qquad \Delta=b-a$$

$$=\exp\Big(\frac{(u-x)\mu}{|\mu|}-\frac{|\mu|(\Upsilon+\operatorname{sh} r\operatorname{sh}(\Delta-r)-\alpha\operatorname{sh} u\operatorname{sh}(\Delta-u))h}{4\operatorname{sh}|u-r|\operatorname{sh}(u\wedge r)\operatorname{sh}(\Delta-u\vee r)}\Big)\Big[\frac{\operatorname{sh}(x\wedge r)\operatorname{sh}(\Delta-x\vee r)}{\operatorname{sh}(u\wedge r)\operatorname{sh}(\Delta-u\vee r)}$$

$$-\frac{\Upsilon+\operatorname{sh} r\operatorname{sh}(\Delta-r)-\alpha\operatorname{sh} u\operatorname{sh}(\Delta-u)}{2\operatorname{sh}|u-r|\operatorname{sh}(u\wedge r)\operatorname{sh}(\Delta-u\vee r)}\Big(\frac{\operatorname{sh}(x\wedge r)\operatorname{sh}(\Delta-x\vee r)\operatorname{sh} u\operatorname{sh}(\Delta-u)}{\operatorname{sh}(u\wedge r)\operatorname{sh}(\Delta-u\vee r)\operatorname{sh}\Delta}$$

$$-\frac{\operatorname{sh}(x\wedge u)\operatorname{sh}(\Delta-x\vee u)}{\operatorname{sh}\Delta}\Big)\Big]$$

3.31.6
(a)

$$\mathbf{P}_{\frac{x+a}{|\mu|}}\Big(\sup_{0\le s\le H_{\frac{a}{|\mu|},\frac{b}{|\mu|}}}\Big(\ell\big(s,\frac{u+a}{|\mu|}\big)-\alpha^{-1}\ell\big(s,\frac{r+a}{|\mu|}\big)\Big)>h, W^{(\mu)}_{H_{\frac{a}{|\mu|},\frac{b}{|\mu|}}}=\frac{a}{|\mu|}\Big)e^{\frac{x\mu}{|\mu|}}$$

$0 < \alpha$

$$=\frac{\operatorname{sh}(\Delta-u)}{\operatorname{sh}\Delta}\exp\Big(-\frac{|\mu|(\Upsilon+\operatorname{sh} r\operatorname{sh}(\Delta-r)-\alpha\operatorname{sh} u\operatorname{sh}(\Delta-u))h}{4\operatorname{sh}|u-r|\operatorname{sh}(u\wedge r)\operatorname{sh}(\Delta-u\vee r)}\Big)\Big[\frac{\operatorname{sh}(x\wedge r)\operatorname{sh}(\Delta-x\vee r)}{\operatorname{sh}(u\wedge r)\operatorname{sh}(\Delta-u\vee r)}$$

$$-\frac{\Upsilon+\operatorname{sh} r\operatorname{sh}(\Delta-r)-\alpha\operatorname{sh} u\operatorname{sh}(\Delta-u)}{2\operatorname{sh}|u-r|\operatorname{sh}(u\wedge r)\operatorname{sh}(\Delta-u\vee r)}\Big(\frac{\operatorname{sh}(x\wedge r)\operatorname{sh}(\Delta-x\vee r)\operatorname{sh} u\operatorname{sh}(\Delta-u)}{\operatorname{sh}(u\wedge r)\operatorname{sh}(\Delta-u\vee r)\operatorname{sh}\Delta}$$

$$-\frac{\operatorname{sh}(x\wedge u)\operatorname{sh}(\Delta-x\vee u)}{\operatorname{sh}\Delta}\Big)\Big]$$

3.31.6
(b)

$$\mathbf{P}_{\frac{x+a}{|\mu|}}\Big(\sup_{0\le s\le H_{\frac{a}{|\mu|},\frac{b}{|\mu|}}}\Big(\ell\big(s,\frac{u+a}{|\mu|}\big)-\alpha^{-1}\ell\big(s,\frac{r+a}{|\mu|}\big)\Big)>h, W^{(\mu)}_{H_{\frac{a}{|\mu|},\frac{b}{|\mu|}}}=\frac{b}{|\mu|}\Big)e^{\frac{x\mu}{|\mu|}}$$

$0 < \alpha$

$$=\frac{\operatorname{sh} u}{\operatorname{sh}\Delta}\exp\Big(\frac{\Delta\mu}{|\mu|}-\frac{|\mu|(\Upsilon+\operatorname{sh} r\operatorname{sh}(\Delta-r)-\alpha\operatorname{sh} u\operatorname{sh}(\Delta-u))h}{4\operatorname{sh}|u-r|\operatorname{sh}(u\wedge r)\operatorname{sh}(\Delta-u\vee r)}\Big)\Big[\frac{\operatorname{sh}(x\wedge r)\operatorname{sh}(\Delta-x\vee r)}{\operatorname{sh}(u\wedge r)\operatorname{sh}(\Delta-u\vee r)}$$

$$-\frac{\Upsilon+\operatorname{sh} r\operatorname{sh}(\Delta-r)-\alpha\operatorname{sh} u\operatorname{sh}(\Delta-u)}{2\operatorname{sh}|u-r|\operatorname{sh}(u\wedge r)\operatorname{sh}(\Delta-u\vee r)}\Big(\frac{\operatorname{sh}(x\wedge r)\operatorname{sh}(\Delta-x\vee r)\operatorname{sh} u\operatorname{sh}(\Delta-u)}{\operatorname{sh}(u\wedge r)\operatorname{sh}(\Delta-u\vee r)\operatorname{sh}\Delta}$$

$$-\frac{\operatorname{sh}(x\wedge u)\operatorname{sh}(\Delta-x\vee u)}{\operatorname{sh}\Delta}\Big)\Big]$$

$$\Upsilon := \sqrt{(\operatorname{sh} r\operatorname{sh}(\Delta-r)-\alpha\operatorname{sh} u\operatorname{sh}(\Delta-u))^2+4\alpha\operatorname{sh}\Delta\operatorname{sh}|u-r|\operatorname{sh}(u\wedge r)\operatorname{sh}(\Delta-u\vee r)}$$

2 $W_s^{(\mu)} = \mu s + W_s$ $\varrho = \varrho(v,z) = \min\{s : \ell(s,z) = v\}$

4. Stopping at inverse local time

4.0.1 $\mathbf{E}_x e^{-\alpha\varrho} = \mathbf{E}_x\{e^{-\alpha\varrho};\ \varrho < \infty\} = e^{\mu(z-x)-(|x-z|+v)\sqrt{2\alpha+\mu^2}}$

4.0.2 $\mathbf{P}_x(\varrho \in dt) = \frac{|x-z|+v}{\sqrt{2\pi}t^{3/2}} \exp\Big(\mu(z-x) - \frac{\mu^2 t}{2} - \frac{(|x-z|+v)^2}{2t}\Big) dt$

(1) $\mathbf{P}_x(\varrho = \infty) = 1 - e^{\mu(z-x)-|\mu|(|z-x|+v)}$

4.1.2 $\mathbf{P}_x\big(\sup_{0\le s\le\varrho} W_s^{(\mu)} < y,\ \varrho < \infty\big)$
$z < y$

$$= \begin{cases} \exp\Big((\mu - |\mu|)(z-x) - \dfrac{|\mu|v}{1-e^{-2|\mu|(y-z)}}\Big), & x \le z \\ \dfrac{\operatorname{sh}(|\mu|(y-x))}{\operatorname{sh}(|\mu|(y-z))} \exp\Big(\mu(z-x) - \dfrac{|\mu|v}{1-e^{-2|\mu|(y-z)}}\Big), & z \le x \le y \end{cases}$$

4.1.4 $\mathbf{E}_x\{e^{-\alpha\varrho};\ \sup_{0\le s\le\varrho} W_s^{(\mu)} < y\}$

$$= \begin{cases} \exp\Big(\mu(z-x) - (z-x)\sqrt{2\alpha+\mu^2} - \dfrac{v\sqrt{2\alpha+\mu^2}}{1-e^{-2(y-z)\sqrt{2\alpha+\mu^2}}}\Big), & x \le z \le y \\ \dfrac{\operatorname{sh}((y-x)\sqrt{2\alpha+\mu^2})}{\operatorname{sh}((y-z)\sqrt{2\alpha+\mu^2})} \exp\Big(\mu(z-x) - \dfrac{v\sqrt{2\alpha+\mu^2}}{1-e^{-2(y-z)\sqrt{2\alpha+\mu^2}}}\Big), & z \le x \le y \end{cases}$$

(1) $\mathbf{P}_x\big(\sup_{0\le s\le\varrho} W_s^{(\mu)} < y, \varrho \in dt\big)$

$$= e^{\mu(z-x)-\mu^2 t/2} \begin{cases} \operatorname{es}_t\big(0, 0, y-z, z-x+v/2, v/2\big) dt, & x \le z \le y \\ \operatorname{ess}_t\big(y-x, y-z, v/2, v/2\big) dt, & z \le x \le y \end{cases}$$

4.2.2 $\mathbf{P}_x\big(\inf_{0\le s\le\varrho} W_s^{(\mu)} > y,\ \varrho < \infty\big)$
$y < z$

$$= \begin{cases} \dfrac{\operatorname{sh}(|\mu|(x-y))}{\operatorname{sh}(|\mu|(z-y))} \exp\Big(\mu(z-x) - \dfrac{|\mu|v}{1-e^{-2|\mu|(z-y)}}\Big), & y \le x \le z \\ \exp\Big((\mu + |\mu|)(z-x) - \dfrac{|\mu|v}{1-e^{-2|\mu|(z-y)}}\Big), & z \le x \end{cases}$$

2 $W_s^{(\mu)} = \mu s + W_s$ $\varrho = \varrho(v,z) = \min\{s : \ell(s,z) = v\}$

4.2.4 $\mathbf{E}_x\{e^{-\alpha\varrho}; \inf_{0\le s\le\varrho} W_s^{(\mu)} > y\}$

$$= \begin{cases} \dfrac{\operatorname{sh}((x-y)\sqrt{2\alpha+\mu^2})}{\operatorname{sh}((z-y)\sqrt{2\alpha+\mu^2})} \exp\Big(\mu(z-x) - \dfrac{v\sqrt{2\alpha+\mu^2}}{1-e^{-2(z-y)\sqrt{2\alpha+\mu^2}}}\Big), & y\le x\le z \\ \exp\Big(\mu(z-x) - (x-z)\sqrt{2\alpha+\mu^2} - \dfrac{v\sqrt{2\alpha+\mu^2}}{1-e^{-2(z-y)\sqrt{2\alpha+\mu^2}}}\Big), & y\le z\le x \end{cases}$$

(1) $\mathbf{P}_x\big(\inf_{0\le s\le\varrho} W_s^{(\mu)} > y, \varrho\in dt\big)$

$$= e^{\mu(z-x)-\mu^2t/2}\begin{cases} \operatorname{ess}_t\big(x-y, z-y, v/2, v/2\big)dt, & y\le x\le z \\ \operatorname{es}_t\big(0,0,z-y,x-z+v/2,v/2\big)dt, & y\le z\le x \end{cases}$$

4.3.1 $\mathbf{E}_x\{e^{-\gamma\ell(\varrho,r)};\ \varrho<\infty\} = e^{\mu(z-x)}V_z^{(3)}(\mu^2)$

for $V_z^{(3)}(2\alpha)$ see 1.4.3.3

(1) $\mathbf{E}_x\{e^{-\gamma\ell(\varrho,r)};\ \varrho=\infty\}$ for $\mu(z-r)<0$

$$= \begin{cases} \dfrac{|\mu|}{|\mu|+\gamma}\Big(1-\exp\Big(-\dfrac{|\mu|(|\mu|+\gamma)v}{|\mu|+\gamma(1-e^{-2|\mu||z-r|})}\Big)\Big), & x\le z\le r,\quad r\le z\le x \\ \dfrac{|\mu|}{|\mu|+\gamma} - \dfrac{|\mu|e^{-2|\mu||z-x|}\big(|\mu|+\gamma(1-e^{-2|\mu||x-r|})\big)}{(|\mu|+\gamma)\big(|\mu|+\gamma(1-e^{-2|\mu||z-r|})\big)} & \\ \qquad\times\exp\Big(-\dfrac{|\mu|(|\mu|+\gamma)v}{|\mu|+\gamma(1-e^{-2|\mu||z-r|})}\Big), & r\wedge z\le x\le r\vee z \\ \dfrac{|\mu|+\gamma(1-e^{-2|\mu||x-r|})}{|\mu|+\gamma} - \dfrac{|\mu|e^{-2|\mu||z-x|}}{(|\mu|+\gamma)\big(|\mu|+\gamma(1-e^{-2|\mu||z-r|})\big)} & \\ \qquad\times\exp\Big(-\dfrac{|\mu|(|\mu|+\gamma)v}{|\mu|+\gamma(1-e^{-2|\mu||z-r|})}\Big), & z\le r\le x,\quad x\le r\le z \end{cases}$$

4.3.2 $\mathbf{P}_x\big(\ell(\varrho,r)\in dy,\ \varrho<\infty\big) = e^{\mu(z-x)}F_z^{(3)}(\mu^2,y)dy$

for $F_z^{(3)}(2\alpha,y)$ see 1.4.3.4

(1) $\mathbf{P}_x\big(\ell(\varrho,r)=0,\ \varrho<\infty\big)$

$$= \exp\Big(\mu(z-x) - \frac{|\mu|v}{1-e^{-2|\mu||z-r|}}\Big)\begin{cases} e^{-|\mu||z-x|}, & x\le z\le r,\quad r\le z\le x \\ \dfrac{\operatorname{sh}(|\mu||x-r|)}{\operatorname{sh}(|\mu||z-r|)}, & z\wedge r\le x\le r\vee z \\ 0, & z\le r\le x,\quad x\le r\le z \end{cases}$$

2 $W_s^{(\mu)} = \mu s + W_s$ $\varrho = \varrho(v,z) = \min\{s : \ell(s,z) = v\}$

4.3.3 $\mathbf{E}_x e^{-\alpha\varrho - \gamma\ell(\varrho,r)} = e^{\mu(z-x)} V_z^{(3)}(2\alpha + \mu^2)$

for $V_z^{(3)}(2\alpha)$ see 1.4.3.3

4.3.4 $\mathbf{E}_x\{e^{-\alpha\varrho}; \ell(\varrho,r) \in dy\} = e^{\mu(z-x)} F_z^{(3)}(2\alpha + \mu^2, y)dy$

for $F_z^{(3)}(2\alpha, y)$ see 1.4.3.4

(1) $$\mathbf{E}_x\{e^{-\alpha\varrho}; \ell(\varrho,r) = 0\} = \exp\Big(\mu(z-x) - \frac{v\sqrt{2\alpha+\mu^2}}{1 - e^{-2|z-r|\sqrt{2\alpha+\mu^2}}}\Big)$$
$$\times \begin{cases} e^{-|z-x|\sqrt{2\alpha+\mu^2}}, & x \le z \le r, \quad r \le z \le x \\ \dfrac{\operatorname{sh}(|x-r|\sqrt{2\alpha+\mu^2})}{\operatorname{sh}(|z-r|\sqrt{2\alpha+\mu^2})}, & r \wedge z \le x \le r \vee z \\ 0, & z \le r \le x, \quad x \le r \le z \end{cases}$$

(2) $\mathbf{P}_x\big(\ell(\varrho,r) \in dy,\ \varrho \in dt\big) = e^{\mu(z-x) - \mu^2 t/2} F_z^{(3)}(y,t)dydt$

for $F_z^{(3)}(y,t)$ see 1.4.3.4

4.4.1 $$\mathbf{E}_x\Big\{\exp\Big(-\gamma\int_0^\varrho \mathbb{I}_{[r,\infty)}\big(W_s^{(\mu)}\big)\,ds\Big);\ \varrho < \infty\Big\} = e^{\mu(z-x)} Q_z^{(6)}(\mu^2, \mu^2 + 2\gamma)$$

for $Q_z^{(6)}(2p, 2q)$ see 1.4.6.1

4.4.2
(1) $$\mathbf{P}_x\Big(\int_0^\varrho \mathbb{I}_{[r,\infty)}\big(W_s^{(\mu)}\big)\,ds = 0,\ \varrho < \infty\Big)$$

$z \le r$ $$= \exp\Big(\mu(z-x) - \frac{v|\mu|e^{|\mu|(r-z)}}{\operatorname{sh}((r-z)|\mu|)}\Big) \begin{cases} e^{-(z-x)|\mu|}, & x \le z \\ \dfrac{\operatorname{sh}((r-x)|\mu|)}{\operatorname{sh}((r-z)|\mu|)}, & z \le x \le r \\ 0, & r \le x \end{cases}$$

4.5.1 $$\mathbf{E}_x\Big\{\exp\Big(-\gamma\int_0^\varrho \mathbb{I}_{(-\infty,r]}\big(W_s^{(\mu)}\big)\,ds\Big);\ \varrho < \infty\Big\} = e^{\mu(z-x)} Q_z^{(6)}(\mu^2 + 2\gamma, \mu^2)$$

for $Q_z^{(6)}(2p, 2q)$ see 1.4.6.1

4.5.2
(1) $$\mathbf{P}_x\Big(\int_0^\varrho \mathbb{I}_{(-\infty,r]}\big(W_s^{(\mu)}\big)\,ds = 0,\ \varrho < \infty\Big)$$

2 $W_s^{(\mu)} = \mu s + W_s$ $\varrho = \varrho(v,z) = \min\{s : \ell(s,z) = v\}$

$r \le z$

$$= \exp\Big(\mu(z-x) - \frac{v|\mu|e^{|\mu|(z-r)}}{\operatorname{sh}((z-r)|\mu|)}\Big)\begin{cases} 0, & x \le r \\ \dfrac{\operatorname{sh}((x-r)|\mu|)}{\operatorname{sh}((z-r)|\mu|)}, & r \le x \le z \\ e^{-(x-z)|\mu|}, & z \le x \end{cases}$$

4.6.1 $\mathbf{E}_x \exp\Big(-\int_0^{\varrho} \big(p\mathbb{I}_{(-\infty,r)}(W_s^{(\mu)}) + q\mathbb{I}_{[r,\infty)}(W_s^{(\mu)})\big)\,ds\Big)$

$$= e^{\mu(z-x)} Q_z^{(6)}(2p+\mu^2, 2q+\mu^2), \qquad \text{for } Q_z^{(6)}(2p,2q) \text{ see 1.4.6.1}$$

4.6.2 $\mathbf{P}_x\Big(\int_0^{\varrho} \mathbb{I}_{(-\infty,r)}(W_s^{(\mu)})ds \in du, \int_0^{\varrho} \mathbb{I}_{[r,\infty)}(W_s^{(\mu)})ds \in dg\Big)$

$$= e^{\mu(z-x)-\mu^2 u/2 - \mu^2 g/2} \int_0^\infty B_z^{(27)}(u,g,y)dy\,dudg$$

for $B_z^{(27)}(u,g,y)$ see 1.4.27.2

4.7.1 $\mathbf{E}_x\Big\{\exp\Big(-\gamma\int_0^{\varrho} \mathbb{I}_{[r,u]}(W_s^{(\mu)})\,ds\Big);\ \varrho<\infty\Big\} e^{\mu(x-z)+|\mu||z-x|}$

$r < u$

$z \le r$

$$= \exp\Big(-\frac{v|\mu|[(\mu^2+\gamma)\operatorname{sh}((u-r)\Upsilon_\gamma) + |\mu|\Upsilon_\gamma \operatorname{ch}((u-r)\Upsilon_\gamma)]}{(\mu^2+\gamma(1-e^{2(z-r)|\mu|}))\operatorname{sh}((u-r)\Upsilon_\gamma) + |\mu|\Upsilon_\gamma\operatorname{ch}((u-r)\Upsilon_\gamma)}\Big)$$

$$\times\begin{cases} 1, & x \le z \\ \dfrac{(\mu^2+\gamma(1-e^{2(x-r)|\mu|}))\operatorname{sh}((u-r)\Upsilon_\gamma) + |\mu|\Upsilon_\gamma\operatorname{ch}((u-r)\Upsilon_\gamma)}{(\mu^2+\gamma(1-e^{2(z-r)|\mu|}))\operatorname{sh}((u-r)\Upsilon_\gamma) + |\mu|\Upsilon_\gamma\operatorname{ch}((u-r)\Upsilon_\gamma)}, & z \le x \le r \\ \dfrac{e^{(x-r)|\mu|}(\mu^2\operatorname{sh}((u-x)\Upsilon_\gamma) + |\mu|\Upsilon_\gamma\operatorname{ch}((u-x)\Upsilon_\gamma))}{(\mu^2+\gamma(1-e^{2(z-r)|\mu|}))\operatorname{sh}((u-r)\Upsilon_\gamma) + |\mu|\Upsilon_\gamma\operatorname{ch}((u-r)\Upsilon_\gamma)}, & r \le x \le u \\ \dfrac{|\mu|\Upsilon_\gamma e^{|\mu|(u-r)}}{(\mu^2+\gamma(1-e^{2(z-r)|\mu|}))\operatorname{sh}((u-r)\Upsilon_\gamma) + |\mu|\Upsilon_\gamma\operatorname{ch}((u-r)\Upsilon_\gamma)}, & u \le x \end{cases}$$

$u \le z$

$$= \exp\Big(-\frac{v|\mu|[(\mu^2+\gamma)\operatorname{sh}((u-r)\Upsilon_\gamma) + |\mu|\Upsilon_\gamma \operatorname{ch}((u-r)\Upsilon_\gamma)]}{(\mu^2+\gamma(1-e^{2(u-z)|\mu|}))\operatorname{sh}((u-r)\Upsilon_\gamma) + |\mu|\Upsilon_\gamma\operatorname{ch}((u-r)\Upsilon_\gamma)}\Big)$$

$$\times\begin{cases} \dfrac{|\mu|\Upsilon_\gamma e^{|\mu|(u-r)}}{(\mu^2+\gamma(1-e^{2(u-z)|\mu|}))\operatorname{sh}((u-r)\Upsilon_\gamma) + |\mu|\Upsilon_\gamma\operatorname{ch}((u-r)\Upsilon_\gamma)}, & x \le r \\ \dfrac{e^{(u-x)|\mu|}(\mu^2\operatorname{sh}((x-r)\Upsilon_\gamma) + |\mu|\Upsilon_\gamma\operatorname{ch}((x-r)\Upsilon_\gamma))}{(\mu^2+\gamma(1-e^{2(u-z)|\mu|}))\operatorname{sh}((u-r)\Upsilon_\gamma) + |\mu|\Upsilon_\gamma\operatorname{ch}((u-r)\Upsilon_\gamma)}, & r \le x \le u \\ \dfrac{(\mu^2+\gamma(1-e^{2(u-x)|\mu|}))\operatorname{sh}((u-r)\Upsilon_\gamma) + |\mu|\Upsilon_\gamma\operatorname{ch}((u-r)\Upsilon_\gamma)}{(\mu^2+\gamma(1-e^{2(u-z)|\mu|}))\operatorname{sh}((u-r)\Upsilon_\gamma) + |\mu|\Upsilon_\gamma\operatorname{ch}((u-r)\Upsilon_\gamma)}, & u \le x \le z \\ 1, & z \le x \end{cases}$$

$$\Upsilon_s := \sqrt{2s+\mu^2}$$

2 $W_s^{(\mu)} = \mu s + W_s$ $\varrho = \varrho(v,z) = \min\{s : \ell(s,z) = v\}$

4.8.1
$$\mathbf{E}_x\Big\{\exp\Big(-\gamma\int_0^{\varrho(v,0)}|W_s^{(\mu)}|ds\Big);\ \varrho(v,0)<\infty\Big\}$$
$$=\exp\Big(-\mu x-\frac{(2\gamma)^{1/3}\operatorname{Ai}'((2\gamma)^{-2/3}\mu^2)v}{\operatorname{Ai}((2\gamma)^{-2/3}\mu^2)}\Big)\frac{\operatorname{Ai}(2^{1/3}\gamma^{-2/3}(\mu^2/2+\gamma|x|))}{\operatorname{Ai}((2\gamma)^{-2/3}\mu^2)}$$

4.9.3
$$\mathbf{E}_x\exp\Big(-\alpha\varrho-\frac{\gamma^2}{2}\int_0^{\varrho}(W_s^{(\mu)})^2\,ds\Big)$$

$x \le z$
$$=\frac{D_{-\frac12-\frac{2\alpha+\mu^2}{2\gamma}}(-x\sqrt{2\gamma})}{D_{-\frac12-\frac{2\alpha+\mu^2}{2\gamma}}(-z\sqrt{2\gamma})}\exp\Big(\mu(z-x)-\frac{v\sqrt{\pi\gamma}\Gamma^{-1}(1/2+(2\alpha+\mu^2)/2\gamma)}{D_{-\frac12-\frac{2\alpha+\mu^2}{2\gamma}}(-z\sqrt{2\gamma})D_{-\frac12-\frac{2\alpha+\mu^2}{2\gamma}}(z\sqrt{2\gamma})}\Big)$$

$z \le x$
$$=\frac{D_{-\frac12-\frac{2\alpha+\mu^2}{2\gamma}}(x\sqrt{2\gamma})}{D_{-\frac12-\frac{2\alpha+\mu^2}{2\gamma}}(z\sqrt{2\gamma})}\exp\Big(\mu(z-x)-\frac{v\sqrt{\pi\gamma}\Gamma^{-1}(1/2+(2\alpha+\mu^2)/2\gamma)}{D_{-\frac12-\frac{2\alpha+\mu^2}{2\gamma}}(-z\sqrt{2\gamma})D_{-\frac12-\frac{2\alpha+\mu^2}{2\gamma}}(z\sqrt{2\gamma})}\Big)$$

4.10.3
$$\mathbf{E}_x\exp\Big(-\alpha\varrho-\gamma\int_0^{\varrho}e^{2\beta W_s^{(\mu)}}\,ds\Big)e^{\mu(x-z)}$$
$$=\begin{cases}\dfrac{I_{\Upsilon_\alpha/|\beta|}\big(\frac{\sqrt{2\gamma}}{|\beta|}e^{\beta x}\big)}{I_{\Upsilon_\alpha/|\beta|}\big(\frac{\sqrt{2\gamma}}{|\beta|}e^{\beta z}\big)}\exp\Big(-\dfrac{v|\beta|}{2I_{\Upsilon_\alpha/|\beta|}\big(\frac{\sqrt{2\gamma}}{|\beta|}e^{\beta z}\big)K_{\Upsilon_\alpha/|\beta|}\big(\frac{\sqrt{2\gamma}}{|\beta|}e^{\beta z}\big)}\Big), & \dfrac{z-x}{\beta}\ge 0\\[2ex] \dfrac{K_{\Upsilon_\alpha/|\beta|}\big(\frac{\sqrt{2\gamma}}{|\beta|}e^{\beta x}\big)}{K_{\Upsilon_\alpha/|\beta|}\big(\frac{\sqrt{2\gamma}}{|\beta|}e^{\beta z}\big)}\exp\Big(-\dfrac{v|\beta|}{2I_{\Upsilon_\alpha/|\beta|}\big(\frac{\sqrt{2\gamma}}{|\beta|}e^{\beta z}\big)K_{\Upsilon_\alpha/|\beta|}\big(\frac{\sqrt{2\gamma}}{|\beta|}e^{\beta z}\big)}\Big), & \dfrac{z-x}{\beta}\le 0\end{cases}$$

4.11.2
$$\lambda\int_0^\infty e^{-\lambda z}\mathbf{P}_0\big(\sup_{y\in\mathbb{R}}\ell(\varrho,y)<h,\varrho<\infty\big)dz=\frac{\operatorname{sh}((h-v)|\mu|/2)}{\operatorname{sh}^2(h|\mu|/2)}\Big(\operatorname{sh}((h-v)|\mu|/2)$$
$$-\frac{|\mu|e^{-\mu(h-v)/2}}{(\lambda-2\mu)}+\frac{|\mu|e^{|\mu|(h-v)/2}M\big(\frac12+\frac{\lambda-\mu}{2|\mu|},1,v|\mu|\big)}{(\lambda-2\mu)M\big(\frac12+\frac{\lambda-\mu}{2|\mu|},1,h|\mu|\big)}\Big)\mathbb{I}_{[v,\infty)}(h)$$

(1)
$$\mathbf{P}_0\big(\sup_{y\in\mathbb{R}}\ell(\varrho(v,0),y)<h,\varrho(v,0)<\infty\big)=\frac{\operatorname{sh}^2((h-v)|\mu|/2)}{\operatorname{sh}^2(h|\mu|/2)}\mathbb{I}_{[v,\infty)}(h)$$

4.15.2
$$\mathbf{P}_x\big(a<\inf_{0\le s\le\varrho}W_s^{(\mu)},\ \sup_{0\le s\le\varrho}W_s^{(\mu)}<b\big)$$
$$=\begin{cases}\dfrac{\operatorname{sh}(|\mu|(b-x))}{\operatorname{sh}(|\mu|(b-z))}\exp\Big(\mu(z-x)-\dfrac{|\mu|\operatorname{sh}(|\mu|(b-a))v}{2\operatorname{sh}(|\mu|(b-z))\operatorname{sh}(|\mu|(z-a))}\Big), & z\le x\le b\\[2ex] \dfrac{\operatorname{sh}(|\mu|(x-a))}{\operatorname{sh}(|\mu|(z-a))}\exp\Big(\mu(z-x)-\dfrac{|\mu|\operatorname{sh}(|\mu|(b-a))v}{2\operatorname{sh}(|\mu|(b-z))\operatorname{sh}(|\mu|(z-a))}\Big), & a\le x\le z\end{cases}$$

2 $W_s^{(\mu)} = \mu s + W_s$ $\qquad \varrho = \varrho(v,z) = \min\{s : \ell(s,z) = v\}$

(1) $\mathbf{P}_x\big(\sup_{0\le s\le\varrho} |W_s^{(\mu)}| \le b\big)$

$$= \begin{cases} \dfrac{\operatorname{sh}(|\mu|(b-x))}{\operatorname{sh}(|\mu|(b-z))} \exp\Big(\mu(z-x) - \dfrac{|\mu|\operatorname{sh}(2|\mu|b)v}{\operatorname{ch}(2|\mu|b) - \operatorname{ch}(2|\mu|z)}\Big), & z \le x \le b \\ \dfrac{\operatorname{sh}(|\mu|(b+x))}{\operatorname{sh}(|\mu|(b+z))} \exp\Big(\mu(z-x) - \dfrac{|\mu|\operatorname{sh}(2|\mu|b)v}{\operatorname{ch}(2|\mu|b) - \operatorname{ch}(2|\mu|z)}\Big), & -b \le x \le z \end{cases}$$

4.15.4 $\mathbf{E}_x\big\{e^{-\alpha\varrho}; a < \inf_{0\le s\le\varrho} W_s^{(\mu)}, \sup_{0\le s\le\varrho} W_s^{(\mu)} < b\big\}$

$$= \exp\Big(\mu(z-x) - \frac{v\sqrt{2\alpha+\mu^2}\operatorname{sh}((b-a)\sqrt{2\alpha+\mu^2})}{2\operatorname{sh}((b-z)\sqrt{2\alpha+\mu^2})\operatorname{sh}((z-a)\sqrt{2\alpha+\mu^2})}\Big)$$

$$\times \begin{cases} \dfrac{\operatorname{sh}((x-a)\sqrt{2\alpha+\mu^2})}{\operatorname{sh}((z-a)\sqrt{2\alpha+\mu^2})}, & -b \le x \le z \\ \dfrac{\operatorname{sh}((b-x)\sqrt{2\alpha+\mu^2})}{\operatorname{sh}((b-z)\sqrt{2\alpha+\mu^2})}, & z \le x \le b \end{cases}$$

(1) $\mathbf{E}_x\big\{e^{-\alpha\varrho}; \sup_{0\le s\le\varrho} |W_s^{(\mu)}| \le b\big\} e^{\mu(x-z)}$

$$= \begin{cases} \dfrac{\operatorname{sh}((b+x)\sqrt{2\alpha+\mu^2})}{\operatorname{sh}((b+z)\sqrt{2\alpha+\mu^2})} \exp\Big(-\dfrac{v\sqrt{2\alpha+\mu^2}\operatorname{sh}((2b\sqrt{2\alpha+\mu^2})}{\operatorname{ch}(2b\sqrt{2\alpha+\mu^2}) - \operatorname{ch}(2z\sqrt{2\alpha+\mu^2})}\Big), & -b \le x \le z \\ \dfrac{\operatorname{sh}((b-x)\sqrt{2\alpha+\mu^2})}{\operatorname{sh}((b-z)\sqrt{2\alpha+\mu^2})} \exp\Big(-\dfrac{v\sqrt{2\alpha+\mu^2}\operatorname{sh}(2b\sqrt{2\alpha+\mu^2})}{\operatorname{ch}(2b\sqrt{2\alpha+\mu^2}) - \operatorname{ch}(2z\sqrt{2\alpha+\mu^2})}\Big), & z \le x \le b \end{cases}$$

4.16.1 $\mathbf{E}_x\big\{e^{-\gamma\ell(\varrho,r)}; \sup_{0\le s\le\varrho} W_s^{(\mu)} < b,\ \varrho < \infty\big\}$
$r < b$
$z < r$

$$= \exp\Big(\mu(z-x) - \frac{|\mu|e^{(r-z)|\mu|}(|\mu|e^{(b-r)|\mu|} + 2\gamma\operatorname{sh}((b-r)|\mu|))v}{2(|\mu|\operatorname{sh}((b-z)|\mu|) + 2\gamma\operatorname{sh}((b-r)|\mu|)\operatorname{sh}((r-z)|\mu|))}\Big)$$

$$\times \begin{cases} \dfrac{|\mu|\operatorname{sh}((b-x)|\mu|)}{|\mu|\operatorname{sh}((b-z)|\mu|) + 2\gamma\operatorname{sh}((b-r)|\mu|)\operatorname{sh}((r-z)|\mu|)}, & r \le x \le b \\ \dfrac{|\mu|\operatorname{sh}((b-x)|\mu|) + 2\gamma\operatorname{sh}((b-r)|\mu|)\operatorname{sh}((r-x)|\mu|)}{|\mu|\operatorname{sh}((b-z)|\mu|) + 2\gamma\operatorname{sh}((b-r)|\mu|)\operatorname{sh}((r-z)|\mu|)}, & z \le x \le r \end{cases}$$

4.16.2 $\mathbf{P}_x\big(\ell(\varrho,r) \in dy, \sup_{0\le s\le\varrho} W_s^{(\mu)} < b,\ \varrho < \infty\big)$
$r < b$
$z < r$

2 $W_s^{(\mu)} = \mu s + W_s$ $\qquad\varrho = \varrho(v,z) = \min\{s : \ell(s,z) = v\}$

$$= \exp\Big(\mu(z-x) - \frac{|\mu|(e^{(r-z)|\mu|}\operatorname{sh}((b-r)|\mu|)v + \operatorname{sh}((b-z)|\mu|)y)}{2\operatorname{sh}((b-r)|\mu|)\operatorname{sh}((r-z)|\mu|)}\Big)$$

$$\times \begin{cases} \dfrac{|\mu|\operatorname{sh}((b-x)|\mu|)}{2\operatorname{sh}((b-r)|\mu|)\operatorname{sh}((r-z)|\mu|)} I_0\Big(\dfrac{|\mu|\sqrt{vy}}{\operatorname{sh}((r-z)|\mu|)}\Big)dy, & r \le x \le b \\ \dfrac{|\mu|}{2\operatorname{sh}^2((r-z)|\mu|)}\Big[\operatorname{sh}((r-x)|\mu|)\dfrac{\sqrt{v}}{\sqrt{y}} I_1\Big(\dfrac{|\mu|\sqrt{vy}}{\operatorname{sh}((r-z)|\mu|)}\Big) & \\ \qquad + \operatorname{sh}((x-z)|\mu|) I_0\Big(\dfrac{|\mu|\sqrt{vy}}{\operatorname{sh}((r-z)|\mu|)}\Big)\Big]dy, & z \le x \le r \end{cases}$$

(1) $\mathbf{P}_x\big(\ell(\varrho,r) = 0,\ \sup_{0\le s\le\varrho} W_s^{(\mu)} < b,\ \varrho < \infty\big)$

$$= \begin{cases} 0, & r \le x \le b \\ \dfrac{\operatorname{sh}((r-x)|\mu|)}{\operatorname{sh}((r-z)|\mu|)} \exp\Big(\mu(z-x) - \dfrac{|\mu|v}{2\operatorname{sh}((r-z)|\mu|)}\Big), & z \le x \le r \end{cases}$$

4.17.1 $\mathbf{E}_x\big\{e^{-\gamma\ell(\varrho,r)};\ a < \inf_{0\le s\le\varrho} W_s^{(\mu)},\ \varrho < \infty\big\}$
$a < r$
$r < z$

$$= \exp\Big(\mu(z-x) - \frac{|\mu|e^{(z-r)|\mu|}(|\mu|e^{(r-a)|\mu|} + 2\gamma\operatorname{sh}((r-a)|\mu|))v}{2(|\mu|\operatorname{sh}((z-a)|\mu|) + 2\gamma\operatorname{sh}((z-r)|\mu|)\operatorname{sh}((r-a)|\mu|))}\Big)$$

$$\times \begin{cases} \dfrac{|\mu|\operatorname{sh}((x-a)|\mu|) + 2\gamma\operatorname{sh}((x-r)|\mu|)\operatorname{sh}((r-a)|\mu|)}{|\mu|\operatorname{sh}((z-a)|\mu|) + 2\gamma\operatorname{sh}((z-r)|\mu|)\operatorname{sh}((r-a)|\mu|)}, & r \le x \le z \\ \dfrac{|\mu|\operatorname{sh}((x-a)|\mu|)}{|\mu|\operatorname{sh}((z-a)|\mu|) + 2\gamma\operatorname{sh}((z-r)|\mu|)\operatorname{sh}((r-a)|\mu|)}, & a \le x \le r \end{cases}$$

4.17.2 $\mathbf{P}_x\big(\ell(\varrho,r) \in dy, a < \inf_{0\le s\le\varrho} W_s^{(\mu)},\ \varrho < \infty\big)$
$a < r$
$r < z$

$$= \exp\Big(\mu(z-x) - \frac{|\mu|(e^{(z-r)|\mu|}\operatorname{sh}((r-a)|\mu|)v + \operatorname{sh}((z-a)|\mu|)y)}{2\operatorname{sh}((z-r)|\mu|)\operatorname{sh}((r-a)|\mu|)}\Big)$$

$$\times \begin{cases} \dfrac{|\mu|}{2\operatorname{sh}^2((z-r)|\mu|)}\Big[\operatorname{sh}((x-r)|\mu|)\dfrac{\sqrt{v}}{\sqrt{y}} I_1\Big(\dfrac{|\mu|\sqrt{vy}}{\operatorname{sh}((z-r)|\mu|)}\Big) & \\ \qquad + \operatorname{sh}((z-x)|\mu|) I_0\Big(\dfrac{|\mu|\sqrt{vy}}{\operatorname{sh}((z-r)|\mu|)}\Big)\Big]dy, & r \le x \le z \\ \dfrac{|\mu|\operatorname{sh}((x-a)|\mu|)}{2\operatorname{sh}((z-r)|\mu|)\operatorname{sh}((r-a)|\mu|)} I_0\Big(\dfrac{|\mu|\sqrt{vy}}{\operatorname{sh}((z-r)|\mu|)}\Big)dy, & a \le x \le r \end{cases}$$

2 $W_s^{(\mu)} = \mu s + W_s$ $\varrho = \varrho(v, z) = \min\{s : \ell(s, z) = v\}$

(1) $\mathbf{P}_x\big(\ell(\varrho, r) = 0, a < \inf_{0\le s\le\varrho} W_s^{(\mu)},\ \varrho < \infty\big)$

$$= \begin{cases} \dfrac{\operatorname{sh}((x-r)|\mu|)}{\operatorname{sh}((z-r)|\mu|)} \exp\Big(\mu(z-x) - \dfrac{|\mu|v}{2\operatorname{sh}((z-r)|\mu|)}\Big), & r \le x \le z \\ 0, & a \le x \le r \end{cases}$$

4.18.1 $\mathbf{E}_x\big\{\exp\big(-\gamma\ell(\varrho, r) - \eta\ell(\varrho, u)\big);\ \varrho < \infty\big\} e^{\mu(x-z)+|\mu||z-x|}$

$r < u$

$z < r$
$$=\exp\Big(\frac{-v|\mu|(\mu^2 + |\mu|\gamma + |\mu|\eta + \gamma\eta(1 - e^{2(r-u)|\mu|}))}{\mu^2 + |\mu|\gamma(1 - e^{2(z-r)|\mu|}) + |\mu|\eta(1 - e^{2(z-u)|\mu|}) + \gamma\eta(1 - e^{2(r-u)|\mu|})(1 - e^{2(z-r)|\mu|})}\Big)$$

$$\times \begin{cases} 1, & x \le z \\ \dfrac{\mu^2 + |\mu|\gamma(1 - e^{2(x-r)|\mu|}) + |\mu|\eta(1 - e^{2(x-u)|\mu|}) + \gamma\eta(1 - e^{2(r-u)|\mu|})(1 - e^{2(x-r)|\mu|})}{\mu^2 + |\mu|\gamma(1 - e^{2(z-r)|\mu|}) + |\mu|\eta(1 - e^{2(z-u)|\mu|}) + \gamma\eta(1 - e^{2(r-u)|\mu|})(1 - e^{2(z-r)|\mu|})}, & z \le x \le r \\ \dfrac{\mu^2 + |\mu|\eta(1 - e^{2(x-u)|\mu|})}{\mu^2 + |\mu|\gamma(1 - e^{2(z-r)|\mu|}) + |\mu|\eta(1 - e^{2(z-u)|\mu|}) + \gamma\eta(1 - e^{2(r-u)|\mu|})(1 - e^{2(z-r)|\mu|})}, & r \le x \le u \\ \dfrac{\mu^2}{\mu^2 + |\mu|\gamma(1 - e^{2(z-r)|\mu|}) + |\mu|\eta(1 - e^{2(z-u)|\mu|}) + \gamma\eta(1 - e^{2(r-u)|\mu|})(1 - e^{2(z-r)|\mu|})}, & u \le x \end{cases}$$

$u < z$
$$=\exp\Big(\frac{-v|\mu|(\mu^2 + |\mu|\gamma + |\mu|\eta + \gamma\eta(1 - e^{2(r-u)|\mu|}))}{\mu^2 + |\mu|\gamma(1 - e^{2(r-z)|\mu|}) + |\mu|\eta(1 - e^{2(u-z)|\mu|}) + \gamma\eta(1 - e^{2(u-z)|\mu|})(1 - e^{2(r-u)|\mu|})}\Big)$$

$$\times \begin{cases} \dfrac{\mu^2}{\mu^2 + |\mu|\gamma(1 - e^{2(r-z)|\mu|}) + |\mu|\eta(1 - e^{2(u-z)|\mu|}) + \gamma\eta(1 - e^{2(u-z)|\mu|})(1 - e^{2(r-u)|\mu|})}, & x \le r \\ \dfrac{\mu^2 + |\mu|\gamma(1 - e^{2(r-x)|\mu|})}{\mu^2 + |\mu|\gamma(1 - e^{2(r-z)|\mu|}) + |\mu|\eta(1 - e^{2(u-z)|\mu|}) + \gamma\eta(1 - e^{2(u-z)|\mu|})(1 - e^{2(r-u)|\mu|})}, & r \le x \le u \\ \dfrac{\mu^2 + |\mu|\gamma(1 - e^{2(r-x)|\mu|}) + |\mu|\eta(1 - e^{2(u-x)|\mu|}) + \gamma\eta(1 - e^{2(u-x)|\mu|})(1 - e^{2(r-u)|\mu|})}{\mu^2 + |\mu|\gamma(1 - e^{2(r-z)|\mu|}) + |\mu|\eta(1 - e^{2(u-z)|\mu|}) + \gamma\eta(1 - e^{2(u-z)|\mu|})(1 - e^{2(r-u)|\mu|})}, & u \le x \le z \\ 1, & z \le x \end{cases}$$

4.19.1 $\mathbf{E}_x\Big\{\exp\Big(-\gamma\int_0^{\varrho} \frac{ds}{W_s^{(\mu)}}\Big);\ 0 < \inf_{0\le s\le\varrho} W_s^{(\mu)}, \varrho < \infty\Big\}$

2 $W_s^{(\mu)} = \mu s + W_s$ $\qquad \varrho = \varrho(v,z) = \min\{s : \ell(s,z) = v\}$

$$= \exp\Big(\mu(z-x) - \frac{|\mu|v}{\Gamma(1+\gamma/|\mu|)M_{-\gamma/|\mu|,1/2}(2|\mu|z)W_{-\gamma/|\mu|,1/2}(2|\mu|z)}\Big)$$

$$\times \begin{cases} \dfrac{M_{-\gamma/|\mu|,1/2}(2|\mu|x)}{M_{-\gamma/|\mu|,1/2}(2|\mu|z)}, & 0 < x \le z \\[2ex] \dfrac{W_{-\gamma/|\mu|,1/2}(2|\mu|x)}{W_{-\gamma/|\mu|,1/2}(2|\mu|z)}, & z \le x \end{cases}$$

4.20.1 $\mathbf{E}_x\Big\{\exp\Big(-\frac{\gamma^2}{2}\int_0^\varrho \frac{ds}{(W_s^{(\mu)})^2}\Big);\ 0 < \inf_{0\le s\le\varrho} W_s^{(\mu)}, \varrho < \infty\Big\}$

$$= \begin{cases} \dfrac{e^{-\mu x}\sqrt{x}I_{\sqrt{\gamma+1/4}}(|\mu|x)}{e^{-\mu z}\sqrt{z}I_{\sqrt{\gamma+1/4}}(|\mu|z)} \exp\Big(-\dfrac{v}{2zI_{\sqrt{\gamma+1/4}}(|\mu|z)K_{\sqrt{\gamma+1/4}}(|\mu|z)}\Big), & 0 < x \le z \\[2ex] \dfrac{e^{-\mu x}\sqrt{x}K_{\sqrt{\gamma+1/4}}(|\mu|x)}{e^{-\mu z}\sqrt{z}K_{\sqrt{\gamma+1/4}}(|\mu|z)} \exp\Big(-\dfrac{v}{2zI_{\sqrt{\gamma+1/4}}(|\mu|z)K_{\sqrt{\gamma+1/4}}(|\mu|z)}\Big), & z \le x \end{cases}$$

4.21.1 $\mathbf{E}_x\Big\{\exp\Big(-\int_0^\varrho \Big(\frac{p^2}{2(W_s^{(\mu)})^2} + \frac{q^2(W_s^{(\mu)})^2}{2}\Big)ds\Big);\ 0 < \inf_{0\le s\le\varrho} W_s^{(\mu)}, \varrho < \infty\Big\}$

$0 < x$

$$= \exp\Big(-\frac{qzv\Gamma(1+\sqrt{p^2+1/4})\Gamma^{-1}((1+\sqrt{p^2+1/4}+\mu^2/2q)/2)}{M_{-\mu^2/4q,\sqrt{4p^2+1}/4}(qz^2)W_{-\mu^2/4q,\sqrt{4p^2+1}/4}(qz^2)}\Big)$$

$$\times \begin{cases} \dfrac{e^{-\mu x}\sqrt{z}M_{-\mu^2/4q,\sqrt{4p^2+1}/4}(qx^2)}{e^{-\mu z}\sqrt{x}M_{-\mu^2/4q,\sqrt{4p^2+1}/4}(qz^2)}, & 0 < x \le z \\[2ex] \dfrac{e^{-\mu x}\sqrt{z}W_{-\mu^2/4q,\sqrt{4p^2+1}/4}(qx^2)}{e^{-\mu z}\sqrt{x}W_{-\mu^2/4q,\sqrt{4p^2+1}/4}(qz^2)}, & z \le x \end{cases}$$

4.22.1 $\mathbf{E}_x\Big\{\exp\Big(-\int_0^\varrho \Big(\frac{p^2}{2(W_s^{(\mu)})^2} + \frac{q}{W_s^{(\mu)}}\Big)ds\Big);\ 0 < \inf_{0\le s\le\varrho} W_s^{(\mu)}, \varrho < \infty\Big\}$

$0 < x$

$$= \exp\Big(-\frac{|\mu|v\Gamma(1+\sqrt{4p^2+1})\Gamma^{-1}(1/2+\sqrt{p^2+1/4}+q/|\mu|)}{M_{-q/|\mu|,\sqrt{p^2+1/4}}(2|\mu|z)W_{-q/|\mu|,\sqrt{p^2+1/4}}(2|\mu|z)}\Big)$$

$$\times \begin{cases} \dfrac{e^{-\mu x}M_{-q/|\mu|,\sqrt{p^2+1/4}}(2|\mu|x)}{e^{-\mu z}M_{-q/|\mu|,\sqrt{p^2+1/4}}(2|\mu|z)}, & 0 < x \le z \\[2ex] \dfrac{e^{-\mu x}W_{-q/|\mu|,\sqrt{p^2+1/4}}(2|\mu|x)}{e^{-\mu z}W_{-q/|\mu|,\sqrt{p^2+1/4}}(2|\mu|z)}, & z \le x \end{cases}$$

2 $W_s^{(\mu)} = \mu s + W_s$ $\qquad \varrho = \varrho(v,z) = \min\{s : \ell(s,z) = v\}$

4.27.1 $$\mathbf{E}_x\Big\{\exp\Big(-\int_0^{\varrho}\big(p\mathbb{I}_{(-\infty,r)}\big(W_s^{(\mu)}\big) + q\mathbb{I}_{[r,\infty)}\big(W_s^{(\mu)}\big)\big)ds\Big); \ell(\varrho,r)\in dy\Big\}$$

$$= e^{\mu(z-x)}Q_z^{(27)}(2p+\mu^2, 2q+\mu^2)dy, \qquad \text{for } Q_z^{(27)}(2p,2q) \text{ see 1.4.27.1}$$

(1) $$\mathbf{E}_x\Big\{\exp\Big(-\int_0^{\varrho}\big(p\mathbb{I}_{(-\infty,r)}\big(W_s^{(\mu)}\big) + q\mathbb{I}_{[r,\infty)}\big(W_s^{(\mu)}\big)\big)ds\Big); \ell(\varrho,r) = 0\Big\}$$

$z \le r$ $$= \exp\Big(\mu(z-x) - \frac{v\sqrt{2p+\mu^2}e^{(r-z)\sqrt{2p+\mu^2}}}{2\,\mathrm{sh}((r-z)\sqrt{2p+\mu^2})}\Big)\begin{cases} e^{-(z-x)\sqrt{2p+\mu^2}}, & x \le z \\ \dfrac{\mathrm{sh}((r-x)\sqrt{2p+\mu^2})}{\mathrm{sh}((r-z)\sqrt{2p+\mu^2})}, & z \le x \le r \\ 0, & r \le x \end{cases}$$

$r \le z$ $$= \exp\Big(\mu(z-x) - \frac{v\sqrt{2q+\mu^2}e^{(z-r)\sqrt{2q+\mu^2}}}{2\,\mathrm{sh}((z-r)\sqrt{2q+\mu^2})}\Big)\begin{cases} 0, & x \le r \\ \dfrac{\mathrm{sh}((x-r)\sqrt{2q+\mu^2})}{\mathrm{sh}((z-r)\sqrt{2q+\mu^2})}, & r \le x \le z \\ e^{-(x-z)\sqrt{2q+\mu^2}}, & z \le x \end{cases}$$

4.27.2 $$\mathbf{P}_x\Big(\int_0^{\varrho}\mathbb{I}_{(-\infty,r)}\big(W_s^{(\mu)}\big)ds \in du, \int_0^{\varrho}\mathbb{I}_{[r,\infty)}\big(W_s^{(\mu)}\big)ds \in dg, \ell(\varrho,r)\in dy\Big)$$

$$= e^{\mu(z-x)-\mu^2u/2-\mu^2g/2}B_z^{(27)}(u,g,y)dudgdy$$

for $B_z^{(27)}(u,g,y)$ see 1.4.27.2

4.30.1 $$\mathbf{E}_x\Big\{\exp\Big(-\int_0^{\varrho}\Big(\frac{p^2}{2}e^{2\beta W_s^{(\mu)}} + qe^{\beta W_s^{(\mu)}}\Big)ds\Big); \varrho < \infty\Big\}$$

$$= \exp\Big(-\frac{pv\Gamma(1+2|\mu|/|\beta|)e^{\beta z}}{\Gamma(1/2+|\mu|/|\beta|+q/p|\beta|)M_{-q/p|\beta|,|\mu|/|\beta|}\big(\frac{2p}{|\beta|}e^{\beta z}\big)W_{-q/p|\beta|,|\mu|/|\beta|}\big(\frac{2p}{|\beta|}e^{\beta z}\big)}\Big)$$

$$\times\begin{cases} \dfrac{e^{-(\mu+\beta/2)x}M_{-q/p|\beta|,|\mu|/|\beta|}\big(\frac{2p}{|\beta|}e^{\beta x}\big)}{e^{-(\mu+\beta/2)z}M_{-q/p|\beta|,|\mu|/|\beta|}\big(\frac{2p}{|\beta|}e^{\beta z}\big)}, & (z-x)\beta \ge 0 \\ \dfrac{e^{-(\mu+\beta/2)x}W_{-q/p|\beta|,|\mu|/|\beta|}\big(\frac{2p}{|\beta|}e^{\beta x}\big)}{e^{-(\mu+\beta/2)z}W_{-q/p|\beta|,|\mu|/|\beta|}\big(\frac{2p}{|\beta|}e^{\beta z}\big)}, & (z-x)\beta \le 0 \end{cases}$$

3. REFLECTING BROWNIAN MOTION

1. Exponential stopping

1.0.3 $$\mathbf{E}_x e^{-\beta|W_\tau|} = \frac{\sqrt{2\lambda}}{2\lambda-\beta^2}\big(\sqrt{2\lambda}e^{-\beta x} - \beta e^{-\sqrt{2\lambda}x}\big), \qquad 0 \le x$$

1.0.4 $$\mathbf{P}_x(|W_\tau| \in dz) = \frac{\sqrt{\lambda}}{\sqrt{2}}\big(e^{-|z-x|\sqrt{2\lambda}} + e^{-(z+x)\sqrt{2\lambda}}\big)dz, \quad 0 \le z, \quad 0 \le x$$

1.0.5 $$\mathbf{E}_x e^{-\beta|W_t|} = \frac{1}{2}e^{\beta x+\beta^2 t/2}\,\text{Erfc}\Big(\frac{\beta\sqrt{t}}{\sqrt{2}} + \frac{x}{\sqrt{2t}}\Big) + \frac{1}{2}e^{-\beta x+\beta^2 t/2}\,\text{Erfc}\Big(\frac{\beta\sqrt{t}}{\sqrt{2}} - \frac{x}{\sqrt{2t}}\Big)$$

1.0.6 $$\mathbf{P}_x\big(|W_t| \in dz\big) = \frac{1}{\sqrt{2\pi t}}\big(e^{-(z-x)^2/2t} + e^{-(z+x)^2/2t}\big)dz$$

1.1.2 $$\mathbf{P}_x\big(\sup_{0\le s\le\tau}|W_s| \ge y\big) = \frac{\text{ch}(x\sqrt{2\lambda})}{\text{ch}(y\sqrt{2\lambda})}, \quad 0 \le x \le y$$

1.1.4 $$\mathbf{P}_x\big(\sup_{0\le s\le t}|W_s| \ge y\big) = \tilde{\text{cc}}_t(x,y)$$
$$= 1 - \frac{1}{\sqrt{2\pi t}}\sum_{k=-\infty}^{\infty}\int_{-y}^{y}\big(e^{-(z-x+4ky)^2/2t} - e^{-(z-x+2y+4ky)^2/2t}\big)dz$$

1.1.6 $$\mathbf{P}_x\big(\sup_{0\le s\le\tau}|W_s| < y,\ |W_\tau| \in dz\big)$$
$$= \frac{\sqrt{\lambda}(\text{sh}((y-|z-x|)\sqrt{2\lambda}) + \text{sh}((y-x-z)\sqrt{2\lambda}))}{\sqrt{2}\,\text{ch}(y\sqrt{2\lambda})}dz, \quad x \vee z < y$$

1.1.8 $$\mathbf{P}_x\big(\sup_{0\le s\le t}|W_s| < y,\ |W_t| \in dz\big)$$
$$= \frac{1}{\sqrt{2\pi t}}\sum_{k=-\infty}^{\infty}(-1)^k\big(e^{-(z-x+2ky)^2/2t} + e^{-(z+x+2ky)^2/2t}\big)dz, \quad x \vee z < y$$

1.3.1 $$\mathbf{E}_x e^{-\gamma\ell(\tau,r)} = 1 - \frac{\gamma\big(e^{-|r-x|\sqrt{2\lambda}} + e^{-(x+r)\sqrt{2\lambda}}\big)}{\sqrt{2\lambda} + \gamma\big(1 + e^{-2r\sqrt{2\lambda}}\big)}$$

3 $|W|$ $\qquad\tau \sim \operatorname{Exp}(\lambda)$, independent of W

1.3.2 $$\mathbf{P}_x\big(\ell(\tau,r)\in dy\big)=\frac{\sqrt{2\lambda}\big(e^{-|r-x|\sqrt{2\lambda}}+e^{-(x+r)\sqrt{2\lambda}}\big)}{\big(1+e^{-2r\sqrt{2\lambda}}\big)^2}\exp\Big(-\frac{y\sqrt{\lambda}e^{r\sqrt{2\lambda}}}{\sqrt{2}\operatorname{ch}(r\sqrt{2\lambda})}\Big)dy$$

(1) $$\mathbf{P}_x\big(\ell(\tau,r)=0\big)=1-\frac{e^{-|r-x|\sqrt{2\lambda}}+e^{-(x+r)\sqrt{2\lambda}}}{1+e^{-2r\sqrt{2\lambda}}}$$

$$=\begin{cases}1-\dfrac{\operatorname{ch}(x\sqrt{2\lambda})}{\operatorname{ch}(r\sqrt{2\lambda})}, & 0\le x\le r\\ 1-e^{-(x-r)\sqrt{2\lambda}}, & r\le x\end{cases}$$

1.3.3 $$\mathbf{E}_x e^{-\gamma\ell(t,r)}$$

$$=1+\frac{\sqrt{2}}{\sqrt{\pi}}\sum_{k=0}^{\infty}(-\gamma)^{k+1}k!\sum_{l=0}^{k}\frac{y^{(k+1)/2}}{l!(k-l)!}\Big(e^{-(|r-x|+2rl)^2/4y}D_{-k-2}\Big(\frac{|r-x|+2rl}{\sqrt{y}}\Big)$$

$$+e^{-(x+r+2rl)^2/4y}D_{-k-2}\Big(\frac{x+r+2rl}{\sqrt{y}}\Big)\Big)$$

1.3.4 $$\mathbf{P}_x\Big(\frac{1}{2}\ell(t,r)\in dy\Big)$$

$$=\operatorname{ec}_y(-1,2,r,y+|x-r|-2r,y)dy+\operatorname{ec}_y(-1,2,r,y+x-r,y)dy$$

(1) $$\mathbf{P}_x\big(\ell(t,r)=0\big)=\begin{cases}1-\widetilde{\operatorname{cc}}_t(x,r), & 0\le x\le r\\ 1-\operatorname{Erfc}\Big(\dfrac{x-r}{\sqrt{2t}}\Big), & r\le x\end{cases}$$

1.3.5 $$\mathbf{E}_x\big\{e^{-\gamma\ell(\tau,r)};\ |W_\tau|\in dz\big\}=\frac{\sqrt{\lambda}}{\sqrt{2}}\Big\{e^{-|z-x|\sqrt{2\lambda}}+e^{-(z+x)\sqrt{2\lambda}}$$

$$-\frac{\gamma\big(e^{-|x-r|\sqrt{2\lambda}}+e^{-(x+r)\sqrt{2\lambda}}\big)}{\sqrt{2\lambda}+\gamma\big(1+e^{-2r\sqrt{2\lambda}}\big)}\big(e^{-|z-r|\sqrt{2\lambda}}+e^{-(z+r)\sqrt{2\lambda}}\big)\Big\}dz$$

1.3.6 $$\mathbf{P}_x\big(\ell(\tau,r)\in dy,\ |W_\tau|\in dz\big)=\exp\Big(-\frac{y\sqrt{\lambda}e^{r\sqrt{2\lambda}}}{\sqrt{2}\operatorname{ch}(r\sqrt{2\lambda})}\Big)$$

$$\times\frac{\lambda\big(e^{-|x-r|\sqrt{2\lambda}}+e^{-(x+r)\sqrt{2\lambda}}\big)\big(e^{-|z-r|\sqrt{2\lambda}}+e^{-(z+r)\sqrt{2\lambda}}\big)}{\big(1+e^{-2r\sqrt{2\lambda}}\big)^2}dydz$$

(1) $$\mathbf{P}_x\big(\ell(\tau,r)=0,\ |W_\tau|\in dz\big)=\frac{\sqrt{\lambda}}{\sqrt{2}}\Big\{e^{-|z-x|\sqrt{2\lambda}}+e^{-(z+x)\sqrt{2\lambda}}$$

3 $|W|$ $\qquad$ $\tau\ \sim\ \mathrm{Exp}(\lambda)$, independent of W

$$-\frac{e^{-|x-r|\sqrt{2\lambda}}+e^{-(x+r)\sqrt{2\lambda}}}{1+e^{-2r\sqrt{2\lambda}}}\Big(e^{-|z-r|\sqrt{2\lambda}}+e^{-(z+r)\sqrt{2\lambda}}\Big)\Big\}dz$$

1.3.8 $\mathbf{P}_x\Big(\frac{1}{2}\ell(t,r)\in dy,\ |W_t|\in dz\Big)$

$$=\frac{1}{2}(\mathrm{ec}_y(0,2,r,y+|x-r|+|z-r|-2r,y)+\mathrm{ec}_y(0,2,r,y+|x-r|+z-r,y)$$

$$+\mathrm{ec}_y(0,2,r,y+|z-r|+x-r,y)+\mathrm{ec}_y(0,2,r,y+x+z,y))dydz$$

(1) $\mathbf{P}_x\big(\ell(t,0)\in dy,|W_t|\in dz\big)=\dfrac{x+z+y/2}{t\sqrt{2\pi t}}e^{-(x+z+y/2)^2/2t}dydz$

(2) $\mathbf{P}_x\big(\ell(t,r)=0,|W_t|\in dz\big)$

$$=\begin{cases}\dfrac{1}{\sqrt{2\pi t}}e^{-(z-x)^2/2t}dz-\dfrac{1}{\sqrt{2\pi t}}e^{-(z+x-2r)^2/2t}dz, & r<x\wedge z\\ \mathrm{sc}_t(r-|z-x|,r)dz+\mathrm{sc}_t(r-z-x,r)dz, & x\vee z<r\end{cases}$$

1.4.1 $\mathbf{E}_x\exp\Big(-\gamma\int_0^\tau \mathbb{I}_{[r,\infty)}\big(|W_s|\big)ds\Big)$

$0\le x$, $x\le r$
$$=1-\frac{\gamma\,\mathrm{ch}(x\sqrt{2\lambda})}{\sqrt{\lambda+\gamma}(\sqrt{\lambda}\,\mathrm{sh}(r\sqrt{2\lambda})+\sqrt{\lambda+\gamma}\,\mathrm{ch}(r\sqrt{2\lambda}))}$$

$r\le x$
$$=\frac{\lambda}{\lambda+\gamma}+\frac{\gamma\sqrt{\lambda}\,\mathrm{sh}(r\sqrt{2\lambda})e^{-(x-r)\sqrt{2\lambda+2\gamma}}}{(\lambda+\gamma)(\sqrt{\lambda}\,\mathrm{sh}(r\sqrt{2\lambda})+\sqrt{\lambda+\gamma}\,\mathrm{ch}(r\sqrt{2\lambda}))}$$

1.4.2 $\mathbf{P}_x\Big(\int_0^\tau \mathbb{I}_{[r,\infty)}\big(|W_s|\big)ds\in dy\Big)$

$0\le x$, $x\le r$
$$=e^{-\lambda y}\frac{\mathrm{ch}(x\sqrt{2\lambda})}{\mathrm{ch}(r\sqrt{2\lambda})}\Big\{\frac{\sqrt{\lambda}\,\mathrm{th}(r\sqrt{2\lambda})}{\sqrt{\pi y}}+\frac{\lambda e^{\lambda y\,\mathrm{th}^2(r\sqrt{2\lambda})}}{\mathrm{ch}^2(r\sqrt{2\lambda})}\,\mathrm{Erfc}(\sqrt{\lambda y}\,\mathrm{th}(r\sqrt{2\lambda}))\Big\}dy$$

$r\le x$
$$=\lambda e^{-\lambda y}\Big\{1-\mathrm{Erfc}\Big(\frac{x-r}{\sqrt{2y}}\Big)+\frac{\mathrm{th}(r\sqrt{2\lambda})}{\sqrt{\lambda\pi y}}e^{-(x-r)^2/2y}$$
$$+\frac{1}{\mathrm{ch}^2(r\sqrt{2\lambda})}e^{(x-r)\sqrt{2\lambda}\,\mathrm{th}(r\sqrt{2\lambda})+\lambda y\,\mathrm{th}^2(r\sqrt{2\lambda})}\mathrm{Erfc}\Big(\frac{x-r}{\sqrt{2y}}+\sqrt{\lambda y}\,\mathrm{th}(r\sqrt{2\lambda})\Big)\Big\}dy$$

(1) $\mathbf{P}_x\Big(\int_0^\tau \mathbb{I}_{[r,\infty)}\big(|W_s|\big)ds=0\Big)=\begin{cases}1-\dfrac{\mathrm{ch}(x\sqrt{2\lambda})}{\mathrm{ch}(r\sqrt{2\lambda})}, & 0\le x\le r\\ 0, & r\le x\end{cases}$

3 $|W|$ $\tau \sim \text{Exp}(\lambda)$, independent of W

1.4.4 $$\mathbf{P}_x\Big(\int_0^t \mathbb{I}_{[r,\infty)}\big(|W_s|\big)ds \in dv\Big) = \int_0^\infty B_x^{(27)}(t-v,v,y)dy\,\mathbb{I}_{(0,t)}(v)dv$$

for $B_x^{(27)}(u,v,y)$ see 1.27.2

(1) $x \le r$ $$\mathbf{P}_x\Big(\int_0^t \mathbb{I}_{[r,\infty)}\big(|W_s|\big)ds = 0\Big) = 1 - \widetilde{\text{cc}}_t(x,r)$$

(2) $r \le x$ $$\mathbf{P}_x\Big(\int_0^t \mathbb{I}_{[r,\infty)}\big(|W_s|\big)ds = t\Big) = 1 - \text{Erfc}\Big(\frac{x-r}{\sqrt{2t}}\Big)$$

1.4.5 $$\mathbf{E}_x\Big\{\exp\Big(-\gamma\int_0^\tau \mathbb{I}_{[r,\infty)}\big(|W_s|\big)ds\Big);\ |W_\tau| \in dz\Big\} \qquad 0 \le x \wedge z$$

$z \le r$, $x \le r$ $$= \frac{\sqrt{\lambda}}{\sqrt{2}}\Big(e^{-|z-x|\sqrt{2\lambda}} + e^{-(z+x)\sqrt{2\lambda}} - \frac{2e^{-r\sqrt{2\lambda}}(\sqrt{\lambda+\gamma}-\sqrt{\lambda})\,\text{ch}(x\sqrt{2\lambda})\,\text{ch}(z\sqrt{2\lambda})}{\sqrt{\lambda}\,\text{sh}(r\sqrt{2\lambda}) + \sqrt{\lambda+\gamma}\,\text{ch}(r\sqrt{2\lambda})}\Big)dz$$

$z \le r$, $r \le x$ $$= \frac{\sqrt{2}\lambda\,\text{ch}(z\sqrt{2\lambda})e^{-(x-r)\sqrt{2\lambda+2\gamma}}}{\sqrt{\lambda}\,\text{sh}(r\sqrt{2\lambda}) + \sqrt{\lambda+\gamma}\,\text{ch}(r\sqrt{2\lambda})}dz$$

$r \le z$, $x \le r$ $$= \frac{\sqrt{2}\lambda\,\text{ch}(x\sqrt{2\lambda})e^{-(z-r)\sqrt{2\lambda+2\gamma}}}{\sqrt{\lambda}\,\text{sh}(r\sqrt{2\lambda}) + \sqrt{\lambda+\gamma}\,\text{ch}(r\sqrt{2\lambda})}dz$$

$r \le z$, $r \le x$ $$= \frac{\lambda}{\sqrt{2\lambda+2\gamma}}e^{-|z-x|\sqrt{2\lambda+2\gamma}}dz$$
$$+\frac{\lambda(\sqrt{\lambda+\gamma}\,\text{ch}(r\sqrt{2\lambda}) - \sqrt{\lambda}\,\text{sh}(r\sqrt{2\lambda}))e^{-(z+x-2r)\sqrt{2\lambda+2\gamma}}}{\sqrt{2\lambda+2\gamma}(\sqrt{\lambda}\,\text{sh}(r\sqrt{2\lambda}) + \sqrt{\lambda+\gamma}\,\text{ch}(r\sqrt{2\lambda}))}dz$$

1.4.6 $$\mathbf{P}_x\Big(\int_0^\tau \mathbb{I}_{[r,\infty)}\big(|W_s|\big)ds \in dy,\ |W_\tau| \in dz\Big)$$

$z \le r$, $x \le r$ $$= \frac{\lambda\,\text{ch}(x\sqrt{2\lambda})\,\text{ch}(z\sqrt{2\lambda})}{\text{ch}^2(r\sqrt{2\lambda})}\Big(\frac{\sqrt{2}}{\sqrt{\pi y}}e^{-\lambda y}$$
$$-\sqrt{2\lambda}\,\text{th}(r\sqrt{2\lambda})e^{-\lambda y/\,\text{ch}^2(r\sqrt{2\lambda})}\,\text{Erfc}(\sqrt{\lambda y}\,\text{th}(r\sqrt{2\lambda}))\Big)dydz$$

$z \le r$, $r \le x$ $$= \frac{\lambda\,\text{ch}(z\sqrt{2\lambda})}{\text{ch}(r\sqrt{2\lambda})}\Big(\frac{\sqrt{2}}{\sqrt{\pi y}}e^{-\lambda y-(x-r)^2/2y} - \sqrt{2\lambda}\,\text{th}(r\sqrt{2\lambda})$$
$$\times e^{(x-r)\sqrt{2\lambda}\,\text{th}(r\sqrt{2\lambda})-\lambda y/\,\text{ch}^2(r\sqrt{2\lambda})}\,\text{Erfc}\Big(\frac{x-r}{\sqrt{2y}} + \sqrt{\lambda y}\,\text{th}(r\sqrt{2\lambda})\Big)\Big)dydz$$

3 $\lvert W\rvert$	$\tau \sim \operatorname{Exp}(\lambda)$, independent of W
$r \le z$ $x \le r$	$= \frac{\lambda \operatorname{ch}(x\sqrt{2\lambda})}{\operatorname{ch}(r\sqrt{2\lambda})}\Big(\frac{\sqrt{2}}{\sqrt{\pi y}} e^{-\lambda y-(z-r)^2/2y} - \sqrt{2\lambda}\operatorname{th}(r\sqrt{2\lambda})$ $\times e^{(z-r)\sqrt{2\lambda}\operatorname{th}(r\sqrt{2\lambda})-\lambda y/\operatorname{ch}^2(r\sqrt{2\lambda})}\operatorname{Erfc}\Big(\frac{z-r}{\sqrt{2y}}+\sqrt{\lambda y}\operatorname{th}(r\sqrt{2\lambda})\Big)\Big)dydz$
$r \le z$ $r \le x$	$= \frac{\lambda}{\sqrt{2\pi y}}\big(e^{-\lambda y-(z-x)^2/2y}+e^{-\lambda y-(z+x-2r)^2/2y}\big)dydz - \lambda\sqrt{2\lambda}\operatorname{th}(r\sqrt{2\lambda})$ $\times e^{(z+x-2r)\sqrt{2\lambda}\operatorname{th}(r\sqrt{2\lambda})-\lambda y/\operatorname{ch}^2(r\sqrt{2\lambda})}\operatorname{Erfc}\Big(\frac{z+x-2r}{\sqrt{2y}}+\sqrt{\lambda y}\operatorname{th}(r\sqrt{2\lambda})\Big)dydz$
(1) $z \le r$	$\mathbf{P}_x\Big(\int_0^\tau \mathbb{I}_{[r,\infty)}(\lvert W_s\rvert)ds = 0,\quad \lvert W_\tau\rvert \in dz\Big)$
$x \le r$	$= \frac{\sqrt{\lambda}(\operatorname{sh}((r-\lvert z-x\rvert)\sqrt{2\lambda})+\operatorname{sh}((r-x-z)\sqrt{2\lambda}))}{\sqrt{2}\operatorname{ch}(r\sqrt{2\lambda})}dz$
1.4.8 $v < t$	$\mathbf{P}_x\Big(\int_0^t \mathbb{I}_{[r,\infty)}(\lvert W_s\rvert)ds \in dv, \lvert W_t\rvert \in dz\Big) = \int_0^\infty B_x^{(27)}(t-v,v,y,z)dy\,dvdz$ for $B_x^{(27)}(u,v,y,z)$ see 1.27.6
(1) $x \le r$	$\mathbf{P}_x\Big(\int_0^t \mathbb{I}_{[r,\infty)}(\lvert W_s\rvert)ds = 0, \lvert W_t\rvert \in dz\Big)$
$z \le r$	$= (\operatorname{sc}_t(r-\lvert z-x\rvert, r) + \operatorname{sc}_t(r-z-x, r))dz$
(2) $r \le x$ $r \le z$	$\mathbf{P}_x\Big(\int_0^t \mathbb{I}_{[r,\infty)}(\lvert W_s\rvert)ds = t, \lvert W_t\rvert \in dz\Big) = \Big(\frac{e^{-(z-x)^2/2t}}{\sqrt{2\pi t}} - \frac{e^{-(z+x-2r)^2/2t}}{\sqrt{2\pi t}}\Big)dz$
1.5.1	$\mathbf{E}_x \exp\Big(-\gamma\int_0^\tau \mathbb{I}_{[0,r]}(\lvert W_s\rvert)ds\Big)$
$0 \le x$ $x \le r$	$= \frac{\lambda}{\lambda+\gamma} + \frac{\gamma\sqrt{\lambda}\operatorname{ch}(x\sqrt{2\lambda+2\gamma})}{(\lambda+\gamma)(\sqrt{\lambda}\operatorname{ch}(r\sqrt{2\lambda+2\gamma})+\sqrt{\lambda+\gamma}\operatorname{sh}(r\sqrt{2\lambda+2\gamma}))}$
$r \le x$	$= 1 - \frac{\gamma\operatorname{sh}(r\sqrt{2\lambda+2\gamma})e^{-(x-r)\sqrt{2\lambda}}}{\sqrt{\lambda+\gamma}(\sqrt{\lambda}\operatorname{ch}(r\sqrt{2\lambda+2\gamma})+\sqrt{\lambda+\gamma}\operatorname{sh}(r\sqrt{2\lambda+2\gamma}))}$

3 $|W|$ $\tau \sim \text{Exp}(\lambda)$, independent of W

1.5.2 $$\mathbf{P}_x\Big(\int_0^\tau \mathbb{I}_{[0,r]}\big(|W_s|\big)ds \in dy\Big)$$

$0 \le x$, $x \le r$ $$= \lambda e^{-\lambda y}dy + e^{-\lambda y}(\text{rs}_y(0,r,x,1/\sqrt{2\lambda}) - 2\lambda\,\text{rs}_y(-2,r,x,1/\sqrt{2\lambda}))dy$$

$r \le x$ $$= \lambda e^{-\lambda y-(x-r)}dy + e^{-\lambda y-(x-r)\sqrt{2\lambda}}\big(\text{rs}_y(0,r,r,1/\sqrt{2\lambda}) - 2\lambda\,\text{rs}_y(-2,r,r,1/\sqrt{2\lambda})\big)dy$$

(1) $$\mathbf{P}_x\Big(\int_0^\tau \mathbb{I}_{[0,r]}\big(|W_s|\big)ds = 0\Big) = \begin{cases} 0, & 0 \le x \le r \\ 1 - e^{-(x-r)\sqrt{2\lambda}}, & r \le x \end{cases}$$

1.5.4 $$\mathbf{P}_x\Big(\int_0^t \mathbb{I}_{[0,r]}\big(|W_s|\big)ds \in dv\Big) = \int_0^\infty B_x^{(27)}(v,t-v,y)dy\,\mathbb{I}_{(0,t)}(v)dv$$

for $B_x^{(27)}(u,v,y)$ see 1.27.2

(1) $x \le r$ $$\mathbf{P}_x\Big(\int_0^t \mathbb{I}_{[0,r]}\big(|W_s|\big)ds = t\Big) = 1 - \widetilde{\text{cc}}_t(x,r)$$

(2) $r \le x$ $$\mathbf{P}_x\Big(\int_0^t \mathbb{I}_{[0,r]}\big(|W_s|\big)ds = 0\Big) = 1 - \text{Erfc}\Big(\frac{x-r}{\sqrt{2t}}\Big)$$

1.5.5 $0 \le x$, $0 \le z$ $$\mathbf{E}_x\Big\{\exp\Big(-\gamma\int_0^\tau \mathbb{I}_{[0,r]}\big(|W_s|\big)ds\Big);\ |W_\tau| \in dz\Big\}$$

$x \le r$, $z \le r$ $$= \frac{\lambda}{\sqrt{2\lambda+2\gamma}}\big(e^{-|z-x|\sqrt{2\lambda+2\gamma}} + e^{-(z+x)\sqrt{2\lambda+2\gamma}}\big)dz + \frac{2\lambda(\sqrt{\lambda+\gamma}-\sqrt{\lambda})e^{-r\sqrt{2\lambda+2\gamma}}\,\text{ch}(x\sqrt{2\lambda+2\gamma})\,\text{ch}(z\sqrt{2\lambda+2\gamma})}{\sqrt{2\lambda+2\gamma}(\sqrt{\lambda}\,\text{ch}(r\sqrt{2\lambda+2\gamma}) + \sqrt{\lambda+\gamma}\,\text{sh}(r\sqrt{2\lambda+2\gamma}))}dz$$

$x \le r$, $r \le z$ $$= \frac{\lambda\sqrt{2}\,\text{ch}(x\sqrt{2\lambda+2\gamma})e^{-(z-r)\sqrt{2\lambda}}}{\sqrt{\lambda}\,\text{ch}(r\sqrt{2\lambda+2\gamma}) + \sqrt{\lambda+\gamma}\,\text{sh}(r\sqrt{2\lambda+2\gamma})}dz$$

$r \le x$, $z \le r$ $$= \frac{\lambda\sqrt{2}\,\text{ch}(z\sqrt{2\lambda+2\gamma})e^{-(x-r)\sqrt{2\lambda}}}{\sqrt{\lambda}\,\text{ch}(r\sqrt{2\lambda+2\gamma}) + \sqrt{\lambda+\gamma}\,\text{sh}(r\sqrt{2\lambda+2\gamma})}dz$$

3 $|W|$ $\tau \sim \operatorname{Exp}(\lambda)$, independent of W

$r \le x$ $r \le z$	$= \frac{\sqrt{\lambda}}{\sqrt{2}} e^{-\lvert z-x\rvert\sqrt{2\lambda}} dz$ $+ \frac{\sqrt{\lambda}(\sqrt{\lambda}\operatorname{ch}(r\sqrt{2\lambda+2\gamma}) - \sqrt{\lambda+\gamma}\operatorname{sh}(r\sqrt{2\lambda+2\gamma}))e^{-(z+x-2r)\sqrt{2\lambda}}}{\sqrt{2}(\sqrt{\lambda}\operatorname{ch}(r\sqrt{2\lambda+2\gamma}) + \sqrt{\lambda+\gamma}\operatorname{sh}(r\sqrt{2\lambda+2\gamma}))} dz$
1.5.6	$\mathbf{P}_x\Big(\int_0^\tau \mathbb{I}_{[0,r]}(\lvert W_s\rvert)ds \in dy,\ \lvert W_\tau\rvert \in dz\Big)$
$x \le r$ $r \le z$	$= \sqrt{2\lambda} e^{-\lambda y-(z-r)\sqrt{2\lambda}} \operatorname{rs}_y(0,r,x,1/\sqrt{2\lambda})dydz$
$r \le x$ $z \le r$	$= \sqrt{2\lambda} e^{-\lambda y-(x-r)\sqrt{2\lambda}} \operatorname{rs}_y(0,r,z,1/\sqrt{2\lambda})dydz$
(1) $r \le z$	$\mathbf{P}_x\Big(\int_0^\tau \mathbb{I}_{[0,r]}(\lvert W_s\rvert)ds = 0,\ \lvert W_\tau\rvert \in dz\Big)$
$r \le x$	$= \frac{\sqrt{\lambda}}{\sqrt{2}}\big(e^{-\lvert z-x\rvert\sqrt{2\lambda}} - e^{(2r-x-z)\sqrt{2\lambda}}\big)dz$
1.5.8 $v < t$	$\mathbf{P}_x\Big(\int_0^t \mathbb{I}_{[0,r]}(\lvert W_s\rvert)ds \in dv, \lvert W_t\rvert \in dz\Big) = \int_0^\infty B_x^{(27)}(v,t-v,y,z)dy\,dvdz$ for $B_x^{(27)}(u,v,y,z)$ see 1.27.6
(1) $x \le r$	$\mathbf{P}_x\Big(\int_0^t \mathbb{I}_{[0,r]}(\lvert W_s\rvert)ds = t, \lvert W_t\rvert \in dz\Big)$
$z \le r$	$= (\operatorname{sc}_t(r-\lvert z-x\rvert, r) + \operatorname{sc}_t(r-z-x, r))dz$
(2) $r \le x$ $r \le z$	$\mathbf{P}_x\Big(\int_0^t \mathbb{I}_{[0,r]}(\lvert W_s\rvert)ds = 0, \lvert W_t\rvert \in dz\Big) = \Big(\frac{e^{-(z-x)^2/2t}}{\sqrt{2\pi t}} - \frac{e^{-(z+x-2r)^2/2t}}{\sqrt{2\pi t}}\Big)dz$
1.6.1	$\mathbf{E}_x \exp\Big(-\int_0^\tau (p\mathbb{I}_{[0,r)}(\lvert W_s\rvert) + q\mathbb{I}_{[r,\infty)}(\lvert W_s\rvert))ds\Big)$
$0 \le x$ $x \le r$	$= \frac{\lambda}{\lambda+p} + \frac{\lambda(p-q)\operatorname{ch}(x\sqrt{2\lambda+2p})}{(\lambda+p)\sqrt{\lambda+q}(\sqrt{\lambda+q}\operatorname{ch}(r\sqrt{2\lambda+2p}) + \sqrt{\lambda+p}\operatorname{sh}(r\sqrt{2\lambda+2p}))}$

3 $|W|$ $\tau \sim \mathrm{Exp}(\lambda)$, independent of W

$r \le x$

$$= \frac{\lambda}{\lambda+q} + \frac{\lambda(q-p)\operatorname{sh}(r\sqrt{2\lambda+2p})e^{-(x-r)\sqrt{2\lambda+2q}}}{\sqrt{\lambda+p}(\lambda+q)(\sqrt{\lambda+q}\operatorname{ch}(r\sqrt{2\lambda+2p})+\sqrt{\lambda+p}\operatorname{sh}(r\sqrt{2\lambda+2p}))}$$

1.6.2 $\mathbf{P}_x\Big(\int_0^\tau \mathbb{I}_{[0,r]}(|W_s|)ds \in du, \int_0^\tau \mathbb{I}_{[r,\infty)}(|W_s|)ds \in dv\Big)$

$$= \lambda e^{-\lambda(u+v)} \int_0^\infty B_x^{(27)}(u,v,y)dy\,dudv \qquad \text{for } B_x^{(27)}(u,v,y) \text{ see 1.27.2}$$

(1) $x \le r$ $\mathbf{P}_x\Big(\int_0^\tau \mathbb{I}_{[0,r]}(|W_s|)ds \in du, \int_0^\tau \mathbb{I}_{[r,\infty)}(|W_s|)ds = 0\Big)$

$$= \lambda e^{-\lambda u}(1-\widetilde{\operatorname{cc}}_u(x,r))du$$

(2) $r \le x$ $\mathbf{P}_x\Big(\int_0^\tau \mathbb{I}_{[0,r]}(|W_s|)ds = 0, \int_0^\tau \mathbb{I}_{[r,\infty)}(|W_s|)ds \in dv\Big) = \lambda e^{-\lambda v}\operatorname{Erf}\Big(\frac{x-r}{\sqrt{2v}}\Big)dv$

1.6.4 $\mathbf{P}_x\Big(\int_0^t (p\mathbb{I}_{[0,r)}(|W_s|)+q\mathbb{I}_{[r,\infty)}(|W_s|))ds \in dv\Big)$

$$= \frac{1}{|p-q|}\int_0^\infty B_x^{(27)}\Big(\frac{|qt-v|}{|p-q|}, \frac{|pt-v|}{|p-q|}, y\Big)dy\,\mathbb{I}_{((q\wedge p)t,(q\vee p)t)}(v)\,dv$$

for $B_x^{(27)}(u,v,y)$ see 1.27.2

(1) $x \le r$ $\mathbf{P}_x\Big(\int_0^t (p\mathbb{I}_{[0,r)}(|W_s|)+q\mathbb{I}_{[r,\infty)}(|W_s|))ds = pt\Big) = 1-\widetilde{\operatorname{cc}}_t(x,r)$

(2) $r \le x$ $\mathbf{P}_x\Big(\int_0^t (p\mathbb{I}_{[0,r)}(|W_s|)+q\mathbb{I}_{[r,\infty)}(|W_s|))ds = qt\Big) = 1-\operatorname{Erfc}\Big(\frac{x-r}{\sqrt{2t}}\Big)$

1.6.5 $\mathbf{E}_x\Big\{\exp\Big(-\int_0^\tau (p\mathbb{I}_{[0,r)}(|W_s|)+q\mathbb{I}_{[r,\infty)}(|W_s|))ds\Big);\ |W_\tau| \in dz\Big\}$

$x \le r$, $z \le r$

$$= \frac{\lambda}{\sqrt{2\lambda+2p}}\big(e^{-|z-x|\sqrt{2\lambda+2p}} + e^{-(z+x)\sqrt{2\lambda+2p}}\big)dz$$

3 $|W|$ $\tau \sim \operatorname{Exp}(\lambda)$, independent of W

$$+\frac{2\lambda(\sqrt{\lambda+p}-\sqrt{\lambda+q})e^{-r\sqrt{2\lambda+2p}}\operatorname{ch}(x\sqrt{2\lambda+2p})\operatorname{ch}(z\sqrt{2\lambda+2p})}{\sqrt{2\lambda+2p}(\sqrt{\lambda+p}\operatorname{sh}(r\sqrt{2\lambda+2p})+\sqrt{\lambda+q}\operatorname{ch}(r\sqrt{2\lambda+2p}))}dz$$

$x \le r$, $r \le z$

$$=\frac{\sqrt{2}\lambda\operatorname{ch}(x\sqrt{2\lambda+2p})e^{-(z-r)\sqrt{2\lambda+2q}}}{\sqrt{\lambda+p}\operatorname{sh}(r\sqrt{2\lambda+2p})+\sqrt{\lambda+q}\operatorname{ch}(r\sqrt{2\lambda+2p})}dz$$

$r \le x$, $z \le r$

$$=\frac{\sqrt{2}\lambda\operatorname{ch}(z\sqrt{2\lambda+2p})e^{-(x-r)\sqrt{2\lambda+2q}}}{\sqrt{\lambda+p}\operatorname{sh}(r\sqrt{2\lambda+2p})+\sqrt{\lambda+q}\operatorname{ch}(r\sqrt{2\lambda+2p})}dz$$

$r \le x$, $r \le z$

$$=\frac{\lambda}{\sqrt{2\lambda+2q}}e^{-|z-x|\sqrt{2\lambda+2q}}dz$$

$$+\frac{\lambda(\sqrt{\lambda+q}\operatorname{ch}(r\sqrt{2\lambda+2p})-\sqrt{\lambda+p}\operatorname{sh}(r\sqrt{2\lambda+2p}))e^{-(z+x-2r)\sqrt{2\lambda+2q}}}{\sqrt{2\lambda+2q}(\sqrt{\lambda+p}\operatorname{sh}(r\sqrt{2\lambda+2p})+\sqrt{\lambda+q}\operatorname{ch}(r\sqrt{2\lambda+2p}))}dz$$

1.6.6 $\mathbf{P}_x\Big(\int_0^\tau \mathbb{1}_{[0,r]}(|W_s|)ds \in du, \int_0^\tau \mathbb{1}_{[r,\infty)}(|W_s|)ds \in dv, |W_\tau| \in dz\Big)$

$$=\lambda e^{-\lambda(u+v)}\int_0^\infty B_x^{(27)}(u,v,y,z)dy\,dudvdz$$

for $B_x^{(27)}(u,v,y,z)$ see 1.27.6

(1) $x \le r$ $\mathbf{P}_x\Big(\int_0^\tau \mathbb{1}_{[0,r]}(|W_s|)ds \in du, \int_0^\tau \mathbb{1}_{[r,\infty)}(|W_s|)ds = 0, |W_\tau| \in dz\Big)$

$$=\lambda e^{-\lambda u}(\operatorname{sc}_u(r-|z-x|,r)+\operatorname{sc}_u(r-z-x,r))dudz$$

(2) $r \le x$ $\mathbf{P}_x\Big(\int_0^\tau \mathbb{1}_{[0,r]}(|W_s|)ds = 0, \int_0^\tau \mathbb{1}_{[r,\infty)}(|W_s|)ds \in dv, |W_\tau| \in dz\Big)$

$$=\frac{\lambda}{\sqrt{2\pi v}}e^{-\lambda v}\big(e^{-(z-x)^2/2v}-e^{-(z+x-2r)^2/2v}\big)dvdz$$

1.6.8 $\mathbf{P}_x\Big(\int_0^t (p\mathbb{1}_{[0,r)}(|W_s|)+q\mathbb{1}_{[r,\infty)}(|W_s|))ds \in dv, |W_t| \in dz\Big)$

3 $|W|$ $\tau \sim \operatorname{Exp}(\lambda)$, independent of W

$$= \frac{1}{|p-q|}\int_0^\infty B_x^{(27)}\Big(\frac{|qt-v|}{|p-q|}, \frac{|pt-v|}{|p-q|}, y, z\Big) dy\, \mathbb{I}_{((q\wedge p)t,(q\vee p)t)}(v)\, dvdz$$

for $B_x^{(27)}(u,v,y,z)$ see 1.27.6

(1) $x \le r$

$$\mathbf{P}_x\Big(\int_0^t \big(p\mathbb{I}_{[0,r)}(|W_s|) + q\mathbb{I}_{[r,\infty)}(|W_s|)\big)ds = pt, |W_t| \in dz\Big)$$

$$= (\mathrm{sc}_t(r-|z-x|,r) + \mathrm{sc}_t(r-z-x,r))dz$$

(2) $r \le x$

$$\mathbf{P}_x\Big(\int_0^t \big(p\mathbb{I}_{[0,r)}(|W_s|) + q\mathbb{I}_{[r,\infty)}(|W_s|)\big)ds = qt, |W_t| \in dz\Big)$$

$$= \frac{1}{\sqrt{2\pi t}}\big(e^{-(z-x)^2/2t} - e^{-(z+x-2r)^2/2t}\big)dz$$

1.7.1

$$\mathbf{E}_r \exp\Big(-\gamma\int_0^\tau \mathbb{I}_{[r,u]}(|W_s|)ds\Big) =: S_r$$

$$= 1 - \frac{\gamma\Big(1+\frac{\sqrt{\lambda}}{\sqrt{\lambda+\gamma}}\operatorname{th}((u-r)\sqrt{(\lambda+\gamma)/2})\Big)}{(\lambda+\gamma)\Big(1+\frac{\sqrt{\lambda}}{\sqrt{\lambda+\gamma}}(1+\operatorname{th}(r\sqrt{2\lambda}))\operatorname{cth}((u-r)\sqrt{2\lambda+2\gamma}) + \frac{\lambda}{\lambda+\gamma}\operatorname{th}(r\sqrt{2\lambda})\Big)}$$

$$\mathbf{E}_u \exp\Big(-\gamma\int_0^\tau \mathbb{I}_{[r,u]}(|W_s|)ds\Big) =: S_u$$

$$= 1 - \frac{\gamma\Big(1+\frac{\sqrt{\lambda}}{\sqrt{\lambda+\gamma}}\operatorname{th}(r\sqrt{2\lambda})\operatorname{th}((u-r)\sqrt{(\lambda+\gamma)/2})\Big)}{(\lambda+\gamma)\Big(1+\frac{\sqrt{\lambda}}{\sqrt{\lambda+\gamma}}(1+\operatorname{th}(r\sqrt{2\lambda}))\operatorname{cth}((u-r)\sqrt{2\lambda+2\gamma}) + \frac{\lambda}{\lambda+\gamma}\operatorname{th}(r\sqrt{2\lambda})\Big)}$$

$$\mathbf{E}_x \exp\Big(-\gamma\int_0^\tau \mathbb{I}_{[r,u]}(|W_s|)ds\Big)$$

$0 \le x$, $x \le r$

$$= 1 - (1-S_r)\frac{\operatorname{ch}(x\sqrt{2\lambda})}{\operatorname{ch}(r\sqrt{2\lambda})}$$

$r \le x$, $x \le u$

$$= \frac{\lambda}{\lambda+\gamma} + \Big(S_r - \frac{\lambda}{\lambda+\gamma}\Big)\frac{\operatorname{sh}((u-x)\sqrt{2\lambda+2\gamma})}{\operatorname{sh}((u-r)\sqrt{2\lambda+2\gamma})} + \Big(S_u - \frac{\lambda}{\lambda+\gamma}\Big)\frac{\operatorname{sh}((x-r)\sqrt{2\lambda+2\gamma})}{\operatorname{sh}((u-r)\sqrt{2\lambda+2\gamma})}$$

$u \le x$

$$= 1 - (1-S_u)e^{(u-x)\sqrt{2\lambda}}$$

1.8.5

$$\mathbf{E}_x\Big\{\exp\Big(-\gamma\int_0^\tau |W_s|ds\Big);\ |W_\tau| \in dz\Big\}$$

3	$\lvert W\rvert$ $\qquad\qquad \tau \sim \operatorname{Exp}(\lambda)$, independent of W
$x \le z$	$= \dfrac{(3\gamma)^{1/3}\sqrt{\lambda+\gamma z}\,C_{-2/3}\Big(\frac{2\sqrt2}{3\gamma}(\lambda+\gamma x)^{3/2}, \frac{2\sqrt2}{3\gamma}\lambda^{3/2}\Big)K_{1/3}\Big(\frac{2\sqrt2}{3\gamma}(\lambda+\gamma z)^{3/2}\Big)}{\sqrt{\lambda+\gamma x}\,K_{2/3}\Big(\frac{2\sqrt2}{3\gamma}\lambda^{3/2}\Big)}dz$
$z \le x$	$= \dfrac{(3\gamma)^{1/3}\sqrt{\lambda+\gamma x}\,C_{-2/3}\Big(\frac{2\sqrt2}{3\gamma}(\lambda+\gamma z)^{3/2}, \frac{2\sqrt2}{3\gamma}\lambda^{3/2}\Big)K_{1/3}\Big(\frac{2\sqrt2}{3\gamma}(\lambda+\gamma x)^{3/2}\Big)}{\sqrt{\lambda+\gamma z}\,K_{2/3}\Big(\frac{2\sqrt2}{3\gamma}\lambda^{3/2}\Big)}dz$
(1)	$\mathbf{E}_0\Big\{\exp\Big(-\gamma\int_0^\tau \lvert W_s\rvert ds\Big);\ \lvert W_\tau\rvert \in dz\Big\} = -\dfrac{2\lambda\operatorname{Ai}(2^{1/3}\gamma^{-2/3}(\lambda+\gamma z))}{(2\gamma)^{1/3}\operatorname{Ai}'(2^{1/3}\gamma^{-2/3}\lambda)}dz$
1.8.7	$\mathbf{E}_0\Big\{\exp\Big(-\gamma\int_0^t \lvert W_s\rvert ds\Big);\ \lvert W_t\rvert \in dz\Big\}$
	$= -2\sum_{k=1}^\infty \gamma^{1/3}\exp(2^{-1/3}\gamma^{2/3}t\alpha_k')\dfrac{\operatorname{Ai}(\alpha_k'+(2\gamma)^{1/3}z)}{2^{2/3}\alpha_k'\operatorname{Ai}(\alpha_k')}dz$
1.8.8	$\mathbf{P}_0\Big(\int_0^t \lvert W_s\rvert ds < y \,\Big\vert\, \lvert W_t\rvert = 0\Big) = \sum_{k=1}^\infty \dfrac{(9/y)^{1/3}\sqrt{\pi t}}{2^{1/6}\lvert\alpha_k'\rvert}\exp\Big(\dfrac{(\alpha_k')^3t^3}{27y^2}\Big)\operatorname{Ai}\Big(\dfrac{(\alpha_k')^2t^2}{(3\sqrt2 y)^{4/3}}\Big)$
1.9.3	$\mathbf{E}_x\exp\Big(-\dfrac{\gamma^2}{2}\int_0^t \lvert W_s\rvert^2 ds\Big) = \dfrac{1}{\sqrt{\operatorname{ch}(t\gamma)}}\exp\Big(-\dfrac{x^2\gamma\operatorname{sh}(t\gamma)}{2\operatorname{ch}(t\gamma)}\Big)$
(1)	$\mathbf{E}_x\exp\Big(-\beta\lvert W_t\rvert^2 - \dfrac{\gamma^2}{2}\int_0^t \lvert W_s\rvert^2 ds\Big)$
	$= \dfrac{1}{\sqrt{\operatorname{ch}(t\gamma)+2\beta\gamma^{-1}\operatorname{sh}(t\gamma)}}\exp\Big(-\dfrac{x^2(\gamma\operatorname{sh}(t\gamma)+2\beta\operatorname{ch}(t\gamma))}{2(\operatorname{ch}(t\gamma)+2\beta\gamma^{-1}\operatorname{sh}(t\gamma))}\Big)$
1.9.4	$\mathbf{P}_x\Big(\int_0^t \lvert W_s\rvert^2 ds \in dy\Big) = \operatorname{ec}_y(0,1/2,t,0,x^2/2)dy$
1.9.5	$\mathbf{E}_x\Big\{\exp\Big(-\dfrac{\gamma^2}{2}\int_0^\tau \lvert W_s\rvert^2 ds\Big);\ \lvert W_\tau\rvert \in dz\Big\}$
$x \le z$	$= \dfrac{\lambda\Gamma(\frac12+\frac{\lambda}{\gamma})}{\sqrt{\pi\gamma}}\Big(D_{-\frac12-\frac{\lambda}{\gamma}}(-x\sqrt{2\gamma}) + D_{-\frac12-\frac{\lambda}{\gamma}}(x\sqrt{2\gamma})\Big)D_{-\frac12-\frac{\lambda}{\gamma}}(z\sqrt{2\gamma})dz$
$z \le x$	$= \dfrac{\lambda\Gamma(\frac12+\frac{\lambda}{\gamma})}{\sqrt{\pi\gamma}}\Big(D_{-\frac12-\frac{\lambda}{\gamma}}(-z\sqrt{2\gamma}) + D_{-\frac12-\frac{\lambda}{\gamma}}(z\sqrt{2\gamma})\Big)D_{-\frac12-\frac{\lambda}{\gamma}}(x\sqrt{2\gamma})dz$

3 $|W|$ $\tau \sim \mathrm{Exp}(\lambda)$, independent of W

1.9.7 $$\mathbf{E}_x\Big\{\exp\Big(-\frac{\gamma^2}{2}\int_0^t |W_s|^2 ds\Big);\ |W_t|\in dz\Big\}$$

$$= \frac{\sqrt{\gamma}}{\sqrt{2\pi\,\mathrm{sh}(t\gamma)}}\exp\Big(-\frac{(x^2+z^2)\gamma\,\mathrm{ch}(t\gamma)}{2\,\mathrm{sh}(t\gamma)}\Big)\Big(\exp\Big(\frac{xz\gamma}{\mathrm{sh}(t\gamma)}\Big)+\exp\Big(-\frac{xz\gamma}{\mathrm{sh}(t\gamma)}\Big)\Big)dz$$

1.9.8 $$\mathbf{P}_x\Big(\int_0^t |W_s|^2 ds\in dy,\ |W_t|\in dz\Big)$$

$$= (2\pi)^{-1/2}(\mathrm{ee}_y(1/2,t,(x^2+z^2)/2,-xz)+\mathrm{ee}_y(1/2,t,(x^2+z^2)/2,xz))dydz$$

1.10.5 $$\mathbf{E}_0\Big\{\exp\Big(-\gamma\int_0^\tau e^{2\beta|W_s|}ds\Big);\ |W_\tau|\in dz\Big\}$$

$\beta>0$ $$= -\frac{\sqrt{2}K_{\sqrt{2\lambda}/|\beta|}\Big(\frac{\sqrt{2\gamma}}{|\beta|}e^{\beta z}\Big)}{\sqrt{\gamma}K'_{\sqrt{2\lambda}/|\beta|}\Big(\frac{\sqrt{2\gamma}}{|\beta|}\Big)}dz,\qquad z>0$$

$\beta<0$ $$= \frac{\sqrt{2}I_{\sqrt{2\lambda}/|\beta|}\Big(\frac{\sqrt{2\gamma}}{|\beta|}e^{\beta z}\Big)}{\sqrt{\gamma}I'_{\sqrt{2\lambda}/|\beta|}\Big(\frac{\sqrt{2\gamma}}{|\beta|}\Big)}dz,\qquad z>0$$

1.11.2 $$\mathbf{P}_0\Big(\sup_{y\in[0,\infty)}\ell(\tau,y)\ge h\Big) = \frac{\sqrt{\lambda/2}}{\mathrm{sh}(h\sqrt{\lambda/2})I_0(h\sqrt{\lambda/2})}\int_0^h I_0(v\sqrt{\lambda/2})dv$$

1.11.4 $$\mathbf{P}_0\Big(\sup_{y\in[0,\infty)}\ell(t,y)<h\Big)$$

$$= 2\sum_{k=1}^\infty\Big\{\frac{e^{-2j_{0,k}^2t/h^2}}{j_{0,k}J_1(j_{0,k})\sin j_{0,k}}\int_0^{j_{0.k}}J_0(v)dv-(-1)^k\frac{e^{-2\pi^2k^2t/h^2}}{\pi kJ_0(\pi k)}\int_0^{\pi k}J_0(v)dv\Big\}$$

1.13.1
$x<y$ $$\mathbf{E}_x\{e^{-\gamma\check{H}(\tau)};\ \sup_{0\le s\le\tau}|W_s|\in dy\} = \frac{\sqrt{2\lambda}\,\mathrm{ch}(x\sqrt{2\lambda+2\gamma})\,\mathrm{sh}(y\sqrt{2\lambda})}{\mathrm{ch}(y\sqrt{2\lambda+2\gamma})\,\mathrm{ch}(y\sqrt{2\lambda})}dy$$

1.13.2 $$\mathbf{P}_x\big(\check{H}(\tau)\in dv;\ \sup_{0\le s\le\tau}|W_s|\in dy\big) = \frac{\sqrt{2\lambda}\,\mathrm{sh}(y\sqrt{2\lambda})}{\mathrm{ch}(y\sqrt{2\lambda})}e^{-\lambda v}\,\mathrm{cc}_v(x,y)dvdy$$

1.13.3 $$\mathbf{E}_x\{e^{-\gamma\check{H}(t)};\ \sup_{0\le s\le t}|W_s|\in dy\} = 2\big(e^{-\gamma t}\,\mathrm{cc}_t(x,y)\big)*\mathrm{sc}_t(x,y)dy$$

3 $|W|$ $\qquad \tau \sim \text{Exp}(\lambda)$, independent of W

1.13.4 $$\mathbf{P}_x\big(\check H(t)\in dv;\ \sup_{0\le s\le t}|W_s|\in dy\big)=2\mathbb{I}_{[0,t]}(v)\,\mathrm{cc}_v(x,y)\,\mathrm{sc}_{t-v}(x,y)dvdy$$

1.13.5 $x<y$, $z<y$ $$\mathbf{E}_x\big\{e^{-\gamma\check H(\tau)};\ \sup_{0\le s\le\tau}|W_s|\in dy,|W_\tau|\in dz\big\}=\frac{2\lambda\,\mathrm{ch}(x\sqrt{2\lambda+2\gamma})\,\mathrm{ch}(z\sqrt{2\lambda})}{\mathrm{ch}(y\sqrt{2\lambda+2\gamma})\,\mathrm{ch}(y\sqrt{2\lambda})}dydz$$

1.13.6 $$\mathbf{P}_x\big(\check H(\tau)\in dv;\ \sup_{0\le s\le\tau}|W_s|\in dy,|W_\tau|\in dz\big)=\frac{2\lambda\,\mathrm{cc}_v(x,y)\,\mathrm{ch}(z\sqrt{2\lambda})}{e^{\lambda v}\,\mathrm{ch}(y\sqrt{2\lambda})}dvdydz$$

1.13.7 $$\mathbf{E}_x\big\{e^{-\gamma\check H(t)};\ \sup_{0\le s\le t}|W_s|\in dy,\ |W_t|\in dz\big\}=2\big(e^{-\gamma t}\,\mathrm{cc}_t(x,y)\big)*\mathrm{cc}_t(z,y)dydz$$

1.13.8 $v<t$ $$\mathbf{P}_x\big(\check H(t)\in dv,\ \sup_{0\le s\le t}|W_s|\in dy,|W_t|\in dz\big)=2\,\mathrm{cc}_v(x,y)\,\mathrm{cc}_{t-v}(z,y)dvdydz$$

1.16.1 $r<b$ $$\mathbf{E}_x\big\{e^{-\gamma\ell(\tau,r)};\ \sup_{0\le s\le\tau}|W_s|<b\big\}$$

$0\le x$, $x\le r$ $$=1-\frac{(\gamma\,\mathrm{sh}((b-r)\sqrt{2\lambda})+\sqrt{\lambda/2})\,\mathrm{ch}(x\sqrt{2\lambda})}{\gamma\,\mathrm{sh}((b-r)\sqrt{2\lambda})\,\mathrm{ch}(r\sqrt{2\lambda})+\sqrt{\lambda/2}\,\mathrm{ch}(b\sqrt{2\lambda})}$$

$r\le x$, $x\le b$ $$=1-\frac{\mathrm{sh}((x-r)\sqrt{2\lambda})}{\mathrm{sh}((b-r)\sqrt{2\lambda})}-\frac{(\gamma\,\mathrm{sh}((b-r)\sqrt{2\lambda})+\sqrt{\lambda/2})\,\mathrm{ch}(r\sqrt{2\lambda})\,\mathrm{sh}((b-x)\sqrt{2\lambda})}{(\gamma\,\mathrm{sh}((b-r)\sqrt{2\lambda})\,\mathrm{ch}(r\sqrt{2\lambda})+\sqrt{\lambda/2}\,\mathrm{ch}(b\sqrt{2\lambda}))\,\mathrm{sh}((b-r)\sqrt{2\lambda})}$$

1.16.2 $r<b$ $$\mathbf{P}_x\big(\ell(\tau,r)\in dy,\ \sup_{0\le s\le\tau}|W_s|<b\big)=\exp\Big(-\frac{y\sqrt{\lambda/2}\,\mathrm{ch}(b\sqrt{2\lambda})}{\mathrm{sh}((b-r)\sqrt{2\lambda})\,\mathrm{ch}(r\sqrt{2\lambda})}\Big)$$

$$\times\begin{cases}\dfrac{\sqrt{\lambda/2}(\mathrm{ch}(b\sqrt{2\lambda})-\mathrm{ch}(r\sqrt{2\lambda}))\,\mathrm{ch}(x\sqrt{2\lambda})}{\mathrm{sh}((b-r)\sqrt{2\lambda})\,\mathrm{ch}^2(r\sqrt{2\lambda})}dy, & 0\le x\le r\\[2ex] \dfrac{\sqrt{\lambda/2}(\mathrm{ch}(b\sqrt{2\lambda})-\mathrm{ch}(r\sqrt{2\lambda}))\,\mathrm{sh}((b-x)\sqrt{2\lambda})}{\mathrm{sh}^2((b-r)\sqrt{2\lambda})\,\mathrm{ch}(r\sqrt{2\lambda})}dy, & r\le x\le b\end{cases}$$

(1) $$\mathbf{P}_x\big(\ell(\tau,r)=0,\ \sup_{0\le s\le\tau}|W_s|<b\big)=\begin{cases}1-\dfrac{\mathrm{ch}(x\sqrt{2\lambda})}{\mathrm{ch}(r\sqrt{2\lambda})}, & 0\le x\le r\\[2ex] 1-\dfrac{\mathrm{ch}((b+r-2x)\sqrt{\lambda/2})}{\mathrm{sh}((b-r)\sqrt{\lambda/2})}, & r\le x\le b\end{cases}$$

3 $|W|$ $\tau \sim \operatorname{Exp}(\lambda)$, independent of W

1.16.5 $\mathbf{E}_x\{e^{-\gamma\ell(\tau,r)};\ \sup_{0\le s\le\tau}|W_s|<b,\ |W_\tau|\in dz\}$

$r<b$, $z<b$

$z\le r$, $x\le r$

$$=\frac{\sqrt{\lambda}(\operatorname{sh}((r-|z-x|)\sqrt{2\lambda})+\operatorname{sh}((r-x-z)\sqrt{2\lambda}))}{\sqrt{2}\operatorname{ch}(r\sqrt{2\lambda})}dz$$

$$-\frac{\lambda\operatorname{sh}((b-r)\sqrt{2\lambda})\operatorname{ch}(z\sqrt{2\lambda})\operatorname{ch}(x\sqrt{2\lambda})}{(\gamma\operatorname{sh}((b-r)\sqrt{2\lambda})\operatorname{ch}(r\sqrt{2\lambda})+\sqrt{\lambda/2}\operatorname{ch}(b\sqrt{2\lambda}))\operatorname{ch}(r\sqrt{2\lambda})}dz$$

$z\le r$, $r\le x$

$$=\frac{\lambda\operatorname{sh}((b-x)\sqrt{2\lambda})\operatorname{ch}(z\sqrt{2\lambda})}{\gamma\operatorname{sh}((b-r)\sqrt{2\lambda})\operatorname{ch}(r\sqrt{2\lambda})+\sqrt{\lambda/2}\operatorname{ch}(b\sqrt{2\lambda})}dz$$

$r\le z$, $x\le r$

$$=\frac{\lambda\operatorname{sh}((b-z)\sqrt{2\lambda})\operatorname{ch}(x\sqrt{2\lambda})}{\gamma\operatorname{sh}((b-r)\sqrt{2\lambda})\operatorname{ch}(r\sqrt{2\lambda})+\sqrt{\lambda/2}\operatorname{ch}(b\sqrt{2\lambda})}dz$$

$r\le z$, $r\le x$

$$=\frac{\sqrt{\lambda}(\operatorname{ch}((b-r-|z-x|)\sqrt{2\lambda})-\operatorname{ch}((b+r-z-x)\sqrt{2\lambda}))}{\sqrt{2}\operatorname{sh}((b-r)\sqrt{2\lambda})}dz$$

$$-\frac{\lambda\operatorname{sh}((b-z)\sqrt{2\lambda})\operatorname{ch}(r\sqrt{2\lambda})\operatorname{sh}((b-x)\sqrt{2\lambda})}{(\gamma\operatorname{sh}((b-r)\sqrt{2\lambda})\operatorname{ch}(r\sqrt{2\lambda})+\sqrt{\lambda/2}\operatorname{ch}(b\sqrt{2\lambda}))\operatorname{sh}((b-r)\sqrt{2\lambda})}dz$$

1.16.6 $\mathbf{P}_x\big(\ell(\tau,r)\in dy,\ \sup_{0\le s\le\tau}|W_s|<b,\ |W_\tau|\in dz\big)$

$r<b$, $z<b$

$z\le r$, $x\le r$

$$=\frac{\lambda\operatorname{ch}(z\sqrt{2\lambda})\operatorname{ch}(x\sqrt{2\lambda})}{\operatorname{ch}^2(r\sqrt{2\lambda})}\exp\Big(-\frac{y\sqrt{\lambda/2}\operatorname{ch}(b\sqrt{2\lambda})}{\operatorname{sh}((b-r)\sqrt{2\lambda})\operatorname{ch}(r\sqrt{2\lambda})}\Big)dydz$$

$z\le r$, $r\le x$

$$=\frac{\lambda\operatorname{sh}((b-x)\sqrt{2\lambda})\operatorname{ch}(z\sqrt{2\lambda})}{\operatorname{sh}((b-r)\sqrt{2\lambda})\operatorname{ch}(r\sqrt{2\lambda})}\exp\Big(-\frac{y\sqrt{\lambda/2}\operatorname{ch}(b\sqrt{2\lambda})}{\operatorname{sh}((b-r)\sqrt{2\lambda})\operatorname{ch}(r\sqrt{2\lambda})}\Big)dydz$$

$r\le z$, $x\le r$

$$=\frac{\lambda\operatorname{sh}((b-z)\sqrt{2\lambda})\operatorname{ch}(x\sqrt{2\lambda})}{\operatorname{sh}((b-r)\sqrt{2\lambda})\operatorname{ch}(r\sqrt{2\lambda})}\exp\Big(-\frac{y\sqrt{\lambda/2}\operatorname{ch}(b\sqrt{2\lambda})}{\operatorname{sh}((b-r)\sqrt{2\lambda})\operatorname{ch}(r\sqrt{2\lambda})}\Big)dydz$$

$r\le z$, $r\le x$

$$=\frac{\lambda\operatorname{sh}((b-z)\sqrt{2\lambda})\operatorname{sh}((b-x)\sqrt{2\lambda})}{\operatorname{sh}^2((b-r)\sqrt{2\lambda})}\exp\Big(-\frac{y\sqrt{\lambda/2}\operatorname{ch}(b\sqrt{2\lambda})}{\operatorname{sh}((b-r)\sqrt{2\lambda})\operatorname{ch}(r\sqrt{2\lambda})}\Big)dydz$$

(1) $\mathbf{P}_x\big(\ell(\tau,r)=0,\ \sup_{0\le s\le\tau}|W_s|<b,\ |W_\tau|\in dz\big)$

3 $|W|$ $\tau \sim \text{Exp}(\lambda)$, independent of W

$z \le r$, $x \le r$
$$= \frac{\sqrt{\lambda}\big(\operatorname{sh}((r-|z-x|)\sqrt{2\lambda}) + \operatorname{sh}((r-x-z)\sqrt{2\lambda})\big)}{\sqrt{2}\operatorname{ch}(r\sqrt{2\lambda})}dz$$

$r \le z$, $r \le x$
$$= \frac{\sqrt{\lambda}\big(\operatorname{ch}((b-r-|z-x|)\sqrt{2\lambda}) - \operatorname{ch}((b+r-z-x)\sqrt{2\lambda})\big)}{\sqrt{2}\operatorname{sh}((b-r)\sqrt{2\lambda})}dz$$

1.16.8, $0 < y$
$$\mathbf{P}_x\Big(\ell(t,r) \in dy,\ \sup_{0\le s\le t}|W_s| < b,\ |W_t| \in dz\Big)$$

$z \le r$, $x \le r$
$$= \operatorname{es}_t(0,0,b-r,0,y/2) * \operatorname{eccc}_t(z,x,r,y/2)dydz$$

$z \le r$, $r \le x$
$$= \operatorname{ess}_t(b-x,b-r,0,y/2) * \operatorname{ecc}_t(z,r,0,y/2)dydz$$

$r \le z$, $x \le r$
$$= \operatorname{ess}_t(b-z,b-r,0,y/2) * \operatorname{ecc}_t(x,r,0,y/2)dydz$$

$r \le z$, $r \le x$
$$= \operatorname{esss}_t(b-z,b-x,b-r,y/2) * \operatorname{ec}_t(0,0,r,0,y/2)dydz$$

(1)
$$\mathbf{P}_x\Big(\ell(t,r) = 0,\ \sup_{0\le s\le t}|W_s| < b,\ |W_t| \in dz\Big)$$

$z \le r$, $x \le r$
$$= \operatorname{sc}_t(r-|z-x|,r)dz + \operatorname{sc}_t(r-x-z,r)dz$$

$r \le z$, $r \le x$
$$= \operatorname{cs}_t(b-r-|z-x|,b-r)dz - \operatorname{cs}_t(b+r-z-x,b-r)dz$$

1.18.1
$$\mathbf{E}_{x/\sqrt{2\lambda}}\exp\Big(-\gamma\sqrt{2\lambda}\,\ell(\tau,r/\sqrt{2\lambda}) - \eta\sqrt{2\lambda}\,\ell(\tau,u/\sqrt{2\lambda})\Big)$$

$r < u$
$$= 1 - \frac{\gamma(1+\eta(1+e^{-2u}) - \eta(e^{r-u}+e^{-r-u}))(e^{-|x-r|}+e^{-x-r})}{1+\gamma(1+e^{-2r})+\eta(1+e^{-2u})+\gamma\eta(1+e^{-2r})(1-e^{2(r-u)})}$$
$$- \frac{\eta(1+\gamma(1+e^{-2r}) - \gamma(e^{r-u}+e^{-r-u}))(e^{-|u-x|}+e^{-u-x})}{1+\gamma(1+e^{-2r})+\eta(1+e^{-2u})+\gamma\eta(1+e^{-2r})(1-e^{2(r-u)})}$$

1.23.1, $r < b$
$$\mathbf{E}_x\Big\{\exp\Big(-\int_0^\tau \big(p\mathbb{1}_{[0,r)}(|W_s|) + q\mathbb{1}_{[r,\infty)}(|W_s|)\big)ds\Big);\ \sup_{0\le s\le \tau}|W_s| < b\Big\}$$

$0 \le x$, $x \le r$
$$= \frac{2\lambda}{\Upsilon_p^2} - \frac{2\lambda(\Upsilon_p^2 + 2(q-p)\operatorname{ch}((b-r)\Upsilon_q))\operatorname{ch}(x\Upsilon_p)}{\Upsilon_p^2\Upsilon_q(\Upsilon_p\operatorname{sh}((b-r)\Upsilon_q)\operatorname{sh}(r\Upsilon_p) + \Upsilon_q\operatorname{ch}((b-r)\Upsilon_q)\operatorname{ch}(r\Upsilon_p))}$$

3 $|W|$ $\qquad\tau \sim \mathrm{Exp}(\lambda)$, independent of W

$r \le x$, $x \le b$

$$= \frac{2\lambda}{\Upsilon_q^2}\Big(1 - \frac{\operatorname{sh}((x-r)\Upsilon_q)}{\operatorname{sh}((b-r)\Upsilon_q)}\Big) + \frac{2\lambda(\Upsilon_p\Upsilon_q + 2(q-p)\operatorname{sh}((b-r)\Upsilon_q)\operatorname{th}(r\Upsilon_p))\operatorname{sh}((b-x)\Upsilon_q)}{\Upsilon_p\Upsilon_q^2(\Upsilon_p\operatorname{th}(r\Upsilon_p) + \Upsilon_q\operatorname{cth}((b-r)\Upsilon_q))\operatorname{sh}^2((b-r)\Upsilon_q)}$$

1.23.5, $r < b$, $z < b$

$$\mathbf{E}_x\Big\{\exp\Big(-\int_0^\tau \big(p\mathbb{I}_{[0,r)}(|W_s|) + q\mathbb{I}_{[r,\infty)}(|W_s|)\big)ds\Big);\ \sup_{0\le s\le\tau}|W_s| < b,\ |W_\tau| \in dz\Big\}$$

$z \le r$, $x \le r$

$$= \frac{\lambda}{\Upsilon_p}\big(e^{-|z-x|\Upsilon_p} + e^{-(z+x)\Upsilon_p}\big)dz$$

$$+ \frac{2\lambda(\Upsilon_q - \Upsilon_p\operatorname{th}((b-r)\Upsilon_q))\operatorname{ch}(z\Upsilon_p)\operatorname{ch}(x\Upsilon_p)e^{-r\Upsilon_p}}{\Upsilon_p(\Upsilon_p\operatorname{th}((b-r)\Upsilon_q)\operatorname{sh}(r\Upsilon_p) + \Upsilon_q\operatorname{ch}(r\Upsilon_p))}dz$$

$z \le r$, $r \le x$

$$= \frac{2\lambda\operatorname{ch}(z\Upsilon_p)\operatorname{sh}((b-x)\Upsilon_q)}{\Upsilon_p\operatorname{sh}((b-r)\Upsilon_q)\operatorname{sh}(r\Upsilon_p) + \Upsilon_q\operatorname{ch}((b-r)\Upsilon_q)\operatorname{ch}(r\Upsilon_p)}dz$$

$r \le z$, $x \le r$

$$= \frac{2\lambda\operatorname{sh}((b-z)\Upsilon_q)\operatorname{ch}(x\Upsilon_p)}{\Upsilon_p\operatorname{sh}((b-r)\Upsilon_q)\operatorname{sh}(r\Upsilon_p) + \Upsilon_q\operatorname{ch}((b-r)\Upsilon_q)\operatorname{ch}(r\Upsilon_p)}dz$$

$r \le z$, $r \le x$

$$= \frac{\lambda}{\Upsilon_q}\big(e^{-|z-x|\Upsilon_q} - e^{(x+z-2b)\Upsilon_q}\big)dz$$

$$+ \frac{2\lambda(\Upsilon_p - \Upsilon_p\operatorname{th}(r\Upsilon_p))\operatorname{sh}((b-z)\Upsilon_q)\operatorname{sh}((b-x)\Upsilon_q)e^{(r-b)\Upsilon_q}}{\Upsilon_p(\Upsilon_p\operatorname{sh}((b-r)\Upsilon_q)\operatorname{th}(r\Upsilon_p) + \Upsilon_q\operatorname{ch}((b-r)\Upsilon_q))}dz$$

1.27.1

$$\mathbf{E}_x\Big\{\exp\Big(-\int_0^\tau \big(p\mathbb{I}_{[0,r)}(|W_s|) + q\mathbb{I}_{[r,\infty)}(|W_s|)\big)ds\Big);\ \tfrac{1}{2}\ell(\tau,r) \in dy\Big\}$$

$x \le r$

$$= e^{-y\Upsilon_q}\Big(\frac{2\lambda\operatorname{sh}(r\Upsilon_p)}{\Upsilon_p\operatorname{ch}(r\Upsilon_p)} + \frac{2\lambda}{\Upsilon_q}\Big)\exp\Big(-\frac{y\Upsilon_p\operatorname{sh}(r\Upsilon_p)}{\operatorname{ch}(r\Upsilon_p)}\Big)\frac{\operatorname{ch}(x\Upsilon_p)}{\operatorname{ch}(r\Upsilon_p)}dy$$

$r \le x$

$$= e^{-(y+x-r)\Upsilon_q}\Big(\frac{2\lambda\operatorname{sh}(r\Upsilon_p)}{\Upsilon_p\operatorname{ch}(r\Upsilon_p)} + \frac{2\lambda}{\Upsilon_q}\Big)\exp\Big(-\frac{y\Upsilon_p\operatorname{sh}(r\Upsilon_p)}{\operatorname{ch}(r\Upsilon_p)}\Big)dy$$

(1)

$$\mathbf{E}_x\Big\{\exp\Big(-\int_0^\tau \big(p\mathbb{I}_{[0,r)}(|W_s|) + q\mathbb{I}_{[r,\infty)}(|W_s|)\big)ds\Big);\ \ell(\tau,r) = 0\Big\}$$

$$\Upsilon_s := \sqrt{2\lambda + 2s}$$

3 $|W|$ $\tau \sim \operatorname{Exp}(\lambda)$, independent of W

$$= \begin{cases} \dfrac{\lambda}{\lambda+p}\Big(1-\dfrac{\operatorname{ch}(x\Upsilon_p)}{\operatorname{ch}(r\Upsilon_p)}\Big), & x \le r \\ \dfrac{\lambda}{\lambda+q}\big(1-e^{(r-x)\Upsilon_q}\big), & r \le x \end{cases}$$

1.27.2 $\mathbf{P}_x\Big(\int_0^\tau \mathbb{I}_{[0,r)}(|W_s|)ds \in du, \int_0^\tau \mathbb{I}_{[r,\infty)}(|W_s|)ds \in dv, \frac{1}{2}\ell(\tau,r) \in dy\Big)$

$$=: \lambda e^{-\lambda(u+v)} B_x^{(27)}(u,v,y)dudvdy$$

$x \le r$ $$= \lambda e^{-\lambda(u+v)} 2(\operatorname{ecsc}_u(x,r,r,y)\operatorname{h}_v(1,y) + \operatorname{ecc}_u(x,r,0,y)\operatorname{h}_v(0,y))dudvdy$$

$r \le x$ $$= \lambda e^{-\lambda(u+v)} 2(\operatorname{esc}_u(r,r,0,y)\operatorname{h}_v(1,y+x-r)$$
$$+ \operatorname{ec}_u(0,0,r,0,y)\operatorname{h}_v(0,y+x-r))dudvdy$$

(1) $x \le r$ $\mathbf{P}_x\Big(\int_0^\tau \mathbb{I}_{[0,r]}(|W_s|)ds \in du, \int_0^\tau \mathbb{I}_{[r,\infty)}(|W_s|)ds = 0, \ell(\tau,r) = 0\Big)$

$$= \lambda e^{-\lambda u}(1-\widetilde{\operatorname{cc}}_u(x,r))du$$

(2) $r \le x$ $\mathbf{P}_x\Big(\int_0^\tau \mathbb{I}_{[0,r]}(|W_s|)ds = 0, \int_0^\tau \mathbb{I}_{[r,\infty)}(|W_s|)ds \in dv, \ell(\tau,r) = 0\Big)$

$$= \lambda e^{-\lambda v}\operatorname{Erf}\Big(\frac{x-r}{\sqrt{2v}}\Big)dv$$

1.27.4 $\mathbf{P}_x\Big(\int_0^t (p\mathbb{I}_{[0,r)}(|W_s|) + q\mathbb{I}_{[r,\infty)}(|W_s|))ds \in dv, \frac{1}{2}\ell(t,r) \in dy\Big)$

$$= \frac{1}{|p-q|} B_x^{(27)}\Big(\frac{|qt-v|}{|p-q|}, \frac{|pt-v|}{|p-q|}, y\Big)\mathbb{I}_{((q\wedge p)t,(q\vee p)t)}(v)\,dvdy$$

for $B_x^{(27)}(u,v,y)$ see 1.27.2

(1) $x \le r$ $\mathbf{P}_x\Big(\int_0^t (p\mathbb{I}_{[0,r)}(|W_s|) + q\mathbb{I}_{[r,\infty)}(|W_s|))ds = pt, \ell(t,r) = 0\Big) = 1-\widetilde{\operatorname{cc}}_t(x,r)$

(2) $r \le x$ $\mathbf{P}_x\Big(\int_0^t (p\mathbb{I}_{[0,r)}(|W_s|) + q\mathbb{I}_{[r,\infty)}(|W_s|))ds = qt, \ell(t,r) = 0\Big) = \operatorname{Erf}\Big(\frac{x-r}{\sqrt{2t}}\Big)$

3 $|W|$ $\tau \sim \operatorname{Exp}(\lambda)$, independent of W

1.27.5	$\mathbf{E}_x\Big\{\exp\Big(-\int_0^\tau \big(p\mathbb{I}_{[0,r)}(\lvert W_s\rvert)+q\mathbb{I}_{[r,\infty)}(\lvert W_s\rvert)\big)ds\Big);\frac{1}{2}\ell(\tau,r)\in dy, \lvert W_\tau\rvert\in dz\Big\}$
$x\le r$ $z\le r$	$=2\lambda e^{-y\Upsilon_q}\exp\Big(-\frac{y\Upsilon_p\operatorname{sh}(r\Upsilon_p)}{\operatorname{ch}(r\Upsilon_p)}\Big)\frac{\operatorname{ch}(x\Upsilon_p)\operatorname{ch}(z\Upsilon_p)}{\operatorname{ch}^2(r\Upsilon_p)}dydz$
$x\le r$ $r\le z$	$=2\lambda e^{-(y+z-r)\Upsilon_q}\exp\Big(-\frac{y\Upsilon_p\operatorname{sh}(r\Upsilon_p)}{\operatorname{ch}(r\Upsilon_p)}\Big)\frac{\operatorname{ch}(x\Upsilon_p)}{\operatorname{ch}(r\Upsilon_p)}dydz$
$r\le x$ $z\le r$	$=2\lambda e^{-(y+x-r)\Upsilon_q}\exp\Big(-\frac{y\Upsilon_p\operatorname{sh}(r\Upsilon_p)}{\operatorname{ch}(r\Upsilon_p)}\Big)\frac{\operatorname{ch}(z\Upsilon_p)}{\operatorname{ch}(r\Upsilon_p)}dydz$
$r\le x$ $r\le z$	$=2\lambda e^{-(y+x+z-2r)\Upsilon_q}\exp\Big(-\frac{y\Upsilon_p\operatorname{sh}(r\Upsilon_p)}{\operatorname{ch}(r\Upsilon_p)}\Big)dydz$
(1)	$\mathbf{E}_x\Big\{\exp\Big(-\int_0^\tau \big(p\mathbb{I}_{[0,r)}(\lvert W_s\rvert)+q\mathbb{I}_{[r,\infty)}(\lvert W_s\rvert)\big)ds\Big);\ell(\tau,r)=0, \lvert W_\tau\rvert\in dz\Big\}$
$x\le r$ $z\le r$	$=\frac{\lambda(\operatorname{sh}((r-\lvert z-x\rvert)\Upsilon_p)+\operatorname{sh}((r-z-x)\Upsilon_p))}{\Upsilon_p\operatorname{ch}(r\Upsilon_p)}dz$
$r\le x$ $r\le z$	$=\frac{\lambda}{\Upsilon_q}e^{-\lvert z-x\rvert\Upsilon_q}dz-\frac{\lambda}{\Upsilon_q}e^{(2r-x-z)\Upsilon_q}dz$
1.27.6	$\mathbf{P}_x\Big(\int_0^\tau \mathbb{I}_{[0,r)}(\lvert W_s\rvert)ds\in du, \int_0^\tau \mathbb{I}_{[r,\infty)}(\lvert W_s\rvert)ds\in dv, \frac{1}{2}\ell(\tau,r)\in dy, \lvert W_\tau\rvert\in dz\Big)$ $=:\lambda e^{-\lambda(u+v)}B_x^{(27)}(u,v,y,z)dudvdydz$
$x\le r$ $z\le r$	$=\lambda e^{-\lambda(u+v)}2\operatorname{eccc}_u(x,z,r,y)\operatorname{h}_v(1,y)dudvdydz$
$x\le r$ $r\le z$	$=\lambda e^{-\lambda(u+v)}2\operatorname{ecc}_u(x,r,0,y)\operatorname{h}_v(1,y+z-r)dudvdydz$
$r\le x$ $z\le r$	$=\lambda e^{-\lambda(u+v)}2\operatorname{ecc}_u(z,r,0,y)\operatorname{h}_v(1,y+x-r)dudvdydz$
$r\le x$ $r\le z$	$=\lambda e^{-\lambda(u+v)}2\operatorname{ec}_u(0,0,r,0,y)\operatorname{h}_v(1,y+x+z-2r)dudvdydz$
	$\Upsilon_s:=\sqrt{2\lambda+2s}$

3 $\quad |W|$	$\tau \sim \operatorname{Exp}(\lambda)$, independent of W
(1) $x \le r$	$\mathbf{P}_x\Big(\int_0^\tau \mathbb{I}_{[0,r)}(\vert W_s\vert)ds \in du, \int_0^\tau \mathbb{I}_{[r,\infty)}(\vert W_s\vert)ds = 0, \ell(\tau,r)=0, \vert W_\tau\vert \in dz\Big)$
	$= \lambda e^{-\lambda u}(\operatorname{sc}_u(r-\vert z-x\vert, r) + \operatorname{sc}_u(r-z-x,r))dudz$
(2) $r \le x$	$\mathbf{P}_x\Big(\int_0^\tau \mathbb{I}_{[0,r)}(\vert W_s\vert)ds = 0, \int_0^\tau \mathbb{I}_{[r,\infty)}(\vert W_s\vert)ds \in dv, \ell(\tau,r)=0, \vert W_\tau\vert \in dz\Big)$
	$= \frac{\lambda}{\sqrt{2\pi v}} e^{-\lambda v}\big(e^{-(z-x)^2/2v} - e^{-(z+x-2r)^2/2v}\big)dvdz$
1.27.8	$\mathbf{P}_x\Big(\int_0^t (p\mathbb{I}_{[0,r)}(\vert W_s\vert) + q\mathbb{I}_{[r,\infty)}(\vert W_s\vert))ds \in dv, \frac{1}{2}\ell(t,r) \in dy, \vert W_t\vert \in dz\Big)$
	$= \frac{1}{\vert p-q\vert} B_x^{(27)}\Big(\frac{\vert qt-v\vert}{\vert p-q\vert}, \frac{\vert pt-v\vert}{\vert p-q\vert}, y, z\Big)\mathbb{I}_{((q\wedge p)t,(q\vee p)t)}(v)\, dvdydz$
	for $B_x^{(27)}(u,v,y,z)$ see 1.27.6
(1) $x \le r$	$\mathbf{P}_x\Big(\int_0^t (p\mathbb{I}_{[0,r)}(\vert W_s\vert) + q\mathbb{I}_{[r,\infty)}(\vert W_s\vert))ds = pt, \ell(t,r)=0, \vert W_t\vert \in dz\Big)$
	$= (\operatorname{sc}_t(r-\vert z-x\vert, r) + \operatorname{sc}_t(r-z-x,r))dz$
(2) $r \le x$	$\mathbf{P}_x\Big(\int_0^t (p\mathbb{I}_{[0,r)}(\vert W_s\vert) + q\mathbb{I}_{[r,\infty)}(\vert W_s\vert))ds = qt, \ell(t,r)=0, \vert W_t\vert \in dz\Big)$
	$= \frac{1}{\sqrt{2\pi t}}\big(e^{-(z-x)^2/2t} - e^{-(z+x-2r)^2/2t}\big)dz$
1.29.1	$\mathbf{E}_x\Big\{\exp\Big(-\int_0^\tau (p\mathbb{I}_{[0,r)}(\vert W_s\vert) + q\mathbb{I}_{[r,\infty)}(\vert W_s\vert))ds\Big);$
$r < b$	$\sup_{0\le s\le\tau}\vert W_s\vert < b, \frac{1}{2}\ell(\tau,r)\in dy\Big\}$
	$= 2\lambda\exp\Big(-\frac{y\Upsilon_q \operatorname{ch}((b-r)\Upsilon_q)}{\operatorname{sh}((b-r)\Upsilon_q)} - \frac{y\Upsilon_p\operatorname{sh}(r\Upsilon_p)}{\operatorname{ch}(r\Upsilon_p)}\Big)$
	$\times\Big(\frac{\operatorname{ch}((b-r)\Upsilon_q)-1}{\Upsilon_q\operatorname{sh}((b-r)\Upsilon_q)} + \frac{\operatorname{sh}(r\Upsilon_p)}{\Upsilon_p\operatorname{ch}(r\Upsilon_p)}\Big)\begin{cases}\frac{\operatorname{ch}(x\Upsilon_p)}{\operatorname{ch}(r\Upsilon_p)}dy, & 0\le x\le r\\ \frac{\operatorname{sh}((b-x)\Upsilon_q)}{\operatorname{sh}((b-r)\Upsilon_q)}dy, & r\le x\le b\end{cases}$

3 $|W|$ $\tau \sim \text{Exp}(\lambda)$, independent of W

(1)

$$\mathbf{E}_x\Big\{\exp\Big(-\int_0^\tau \big(p\mathbb{I}_{[0,r)}(|W_s|)+q\mathbb{I}_{[r,\infty)}(|W_s|)\big)ds\Big); \sup_{0\le s\le\tau}|W_s|<b, \ell(\tau,r)=0\Big\}$$

$$=\begin{cases}\dfrac{\lambda}{\lambda+p}\Big(1-\dfrac{\operatorname{ch}(x\Upsilon_p)}{\operatorname{ch}(r\Upsilon_p)}\Big), & 0\le x\le r\\ \dfrac{\lambda}{\lambda+q}\Big(1-\dfrac{\operatorname{sh}((b-x)\Upsilon_q)+\operatorname{sh}((x-r)\Upsilon_q)}{\operatorname{sh}((b-r)\Upsilon_q)}\Big), & r\le x\le b\end{cases}$$

1.29.5

$r<b$

$$\mathbf{E}_x\Big\{\exp\Big(-\int_0^\tau \big(p\mathbb{I}_{[0,r)}(|W_s|)+q\mathbb{I}_{[r,\infty)}(|W_s|)\big)ds\Big); \sup_{0\le s\le\tau}|W_s|<b, \frac12\ell(\tau,r)\in dy, |W_\tau|\in dz\Big\}$$

$$=2\lambda\exp\Big(-\frac{y\Upsilon_q\operatorname{ch}((b-r)\Upsilon_q)}{\operatorname{sh}((b-r)\Upsilon_q)}-\frac{y\Upsilon_p\operatorname{sh}(r\Upsilon_p)}{\operatorname{ch}(r\Upsilon_p)}\Big)$$

$$\times\begin{cases}\dfrac{\operatorname{ch}(x\Upsilon_p)\operatorname{ch}(z\Upsilon_p)}{\operatorname{ch}^2(r\Upsilon_p)}dydz, & 0\le x\le r,\ 0\le z\le r\\ \dfrac{\operatorname{sh}((b-x)\Upsilon_q)\operatorname{ch}(z\Upsilon_p)}{\operatorname{sh}((b-r)\Upsilon_q)\operatorname{ch}(r\Upsilon_p)}dydz, & 0\le z\le r\le x\le b\\ \dfrac{\operatorname{sh}((b-z)\Upsilon_q)\operatorname{ch}(x\Upsilon_p)}{\operatorname{sh}((b-r)\Upsilon_q)\operatorname{ch}(r\Upsilon_p)}dydz, & 0\le x\le r\le z\le b\\ \dfrac{\operatorname{sh}((b-x)\Upsilon_q)\operatorname{sh}((b-z)\Upsilon_q)}{\operatorname{sh}^2((b-r)\Upsilon_q)}dydz, & r\le x\le b,\ r\le z\le b\end{cases}$$

(1)

$$\mathbf{E}_x\Big\{\exp\Big(-\int_0^\tau \big(p\mathbb{I}_{[0,r)}(|W_s|)+q\mathbb{I}_{[r,\infty)}(|W_s|)\big)ds\Big); \sup_{0\le s\le\tau}|W_s|<b, \ell(\tau,r)=0, |W_\tau|\in dz\Big\}$$

$$=\begin{cases}\dfrac{\lambda(\operatorname{sh}((r-|z-x|)\Upsilon_p)+\operatorname{sh}((r-z-x)\Upsilon_p))}{\Upsilon_p\operatorname{ch}(r\Upsilon_p)}dz, & x\vee z\le r\\ \dfrac{\lambda(\operatorname{ch}((b-r-|z-x|)\Upsilon_q)-\operatorname{ch}((b+r-z-x)\Upsilon_q))}{\Upsilon_q\operatorname{sh}((b-r)\Upsilon_q)}dz, & r\le x\wedge z\end{cases}$$

1.29.6

$$\mathbf{P}_x\Big(\int_0^\tau \mathbb{I}_{[0,r)}(|W_s|)ds\in du, \int_0^\tau \mathbb{I}_{[r,\infty)}(|W_s|)ds\in dv, \sup_{0\le s\le\tau}|W_s|<b, \frac12\ell(\tau,r)\in dy, |W_\tau|\in dz\Big)$$

$$\Upsilon_s:=\sqrt{2\lambda+2s}$$

3 $|W|$ $\qquad\tau \sim \mathrm{Exp}(\lambda)$, independent of W

$z \le r$, $x \le r$: $$= 2\lambda e^{-\lambda(u+v)}\,\mathrm{eccc}_u(x,z,r,y)\,\mathrm{es}_v(0,0,b-r,0,y)dudvdydz$$

$z \le r$, $r \le x$: $$= 2\lambda e^{-\lambda(u+v)}\,\mathrm{ecc}_u(z,r,0,y)\,\mathrm{ess}_v(b-x,b-r,0,y)dudvdydz$$

$r \le z$, $x \le r$: $$= 2\lambda e^{-\lambda(u+v)}\,\mathrm{ecc}_u(x,r,0,y)\,\mathrm{ess}_v(b-z,b-r,0,y)dudvdydz$$

$r \le z$, $r \le x$: $$= 2\lambda e^{-\lambda(u+v)}\,\mathrm{ec}_u(0,0,r,0,y)\,\mathrm{esss}_v(b-z,b-x,b-r,y)dudvdydz$$

1.29.8 $$\mathbf{P}_x\Big(\int_0^t \big(p\mathbb{I}_{[0,r)}(|W_s|) + q\mathbb{I}_{[r,\infty)}(|W_s|)\big)ds \in dv,$$

$r < b$ $$\sup_{0\le s\le t}|W_s| < b, \frac{1}{2}\ell(t,r) \in dy, |W_t| \in dz\Big)$$

for $(p\wedge q)t \le v \le (p\vee q)t$

$x \le r$, $z \le r$: $$= \frac{2}{|p-q|}\mathrm{es}_{|pt-v|/|p-q|}(0,0,b-r,0,y)\,\mathrm{eccc}_{|v-qt|/|p-q|}(x,z,r,y)dvdydz$$

$z \le r$, $r \le x$: $$= \frac{2}{|p-q|}\mathrm{ess}_{|pt-v|/|p-q|}(b-x,b-r,0,y)\,\mathrm{ecc}_{|v-qt|/|p-q|}(z,r,0,y)dvdydz$$

$r \le z$, $x \le r$: $$= \frac{2}{|p-q|}\mathrm{ess}_{|pt-v|/|p-q|}(b-z,b-r,0,y)\,\mathrm{ecc}_{|v-qt|/|p-q|}(x,r,0,y)dvdydz$$

$r \le x$, $r \le z$: $$= \frac{2}{|p-q|}\mathrm{esss}_{|pt-v|/|p-q|}(b-x,b-z,b-r,y)\,\mathrm{ec}_{|v-qt|/|p-q|}(0,0,r,0,y)dvdydz$$

(1) $$\mathbf{P}_x\Big(\int_0^t \mathbb{I}_{[0,r)}(|W_s|)ds = t,\ \int_0^t \mathbb{I}_{[r,\infty)}(|W_s|)ds = 0$$
$$\sup_{0\le s\le t}|W_s| < b, \ell(t,r) = 0, |W_t| \in dz\Big)$$
$$= \mathrm{sc}_t(r-|z-x|,r)dz + \mathrm{sc}_t(r-z-x,r)dz, \quad 0 \le x\wedge z \le x\vee z < r$$

(2) $$\mathbf{P}_x\Big(\int_0^t \mathbb{I}_{[0,r)}(|W_s|)ds = 0,\ \int_0^t \mathbb{I}_{[r,\infty)}(|W_s|)ds = t$$
$$\sup_{0\le s\le t}|W_s| < b, \ell(t,r) = 0, |W_t| \in dz\Big)$$
$$= \mathrm{cs}_t(b-r-|z-x|,b-r)dz - \mathrm{cs}_t(b+r-z-x,b-r)dz, \quad r < x\wedge z,\ x\vee z < b$$

3 $|W|$ $\tau \sim \text{Exp}(\lambda)$, independent of W

1.31.1 $0<\alpha$

$$\mathbf{E}_{x/\sqrt{2\lambda}}\exp\Big(-\gamma\sqrt{2\lambda}\sup_{0\le s\le\tau}\big(\ell(s,u/\sqrt{2\lambda})-\alpha^{-1}\ell(s,r/\sqrt{2\lambda})\big)\Big)$$

$$=1-\frac{\gamma(\Upsilon+e^{-2r}-\alpha e^{-2u})\big(e^{-|x-u|}+e^{-x-u}\big)}{2\gamma\big(1+e^{-2(u\wedge r)}\big)\big(1-e^{-2|u-r|}\big)+\Upsilon+e^{-2r}-\alpha e^{-2u}}$$

$$-\frac{\gamma\big(2\big(1+e^{-2(u\wedge r)}\big)\big(1-e^{-2|u-r|}\big)-(\Upsilon+e^{-2r}-\alpha e^{-2u})(1+e^{-2u})\big)\big(e^{-|x-r|}+e^{-x-r}\big)}{\big(2\gamma\big(1+e^{-2(u\wedge r)}\big)\big(1-e^{-2|u-r|}\big)+\Upsilon+e^{-2r}-\alpha e^{-2u}\big)\big(e^{-|u-r|}+e^{-u-r}\big)}$$

1.31.2 $0<\alpha$

$$\mathbf{P}_{x/\sqrt{2\lambda}}\Big(\sup_{0\le s\le\tau}\big(\ell(s,u/\sqrt{2\lambda})-\alpha^{-1}\ell(s,r/\sqrt{2\lambda})\big)>h/\sqrt{2\lambda}\Big)$$

$$=\exp\Big(-\frac{(\Upsilon+e^{-2r}-\alpha e^{-2u})h}{2\big(1+e^{-2(u\wedge r)}\big)\big(1-e^{-2|u-r|}\big)}\Big)\Big[\frac{e^{-|x-r|}+e^{-x-r}}{e^{-|u-r|}+e^{-u-r}}$$

$$-\frac{\Upsilon+e^{-2r}-\alpha e^{-2u}}{2\big(1+e^{-2(u\wedge r)}\big)\big(1-e^{-2|u-r|}\big)}\Big(\frac{(e^{-|x-r|}+e^{-x-r})(1+e^{-2u})}{e^{-|u-r|}+e^{-u-r}}-e^{-|u-x|}-e^{-x-u}\Big)\Big]$$

1.31.5 $0<\alpha$

$$\mathbf{E}_{x/\sqrt{2\lambda}}\Big\{\exp\Big(-\gamma\sqrt{2\lambda}\sup_{0\le s\le\tau}\big(\ell(s,u/\sqrt{2\lambda})-\alpha^{-1}\ell(s,r/\sqrt{2\lambda})\big)\Big);|W_\tau|\in dz/\sqrt{2\lambda}\Big\}$$

$$=\frac{1}{2}\big(e^{-|z-x|}+e^{-z-x}\big)dz$$

$$-\frac{\gamma}{2}\big(e^{-|z-u|}+e^{-z-u}\big)\Big\{\frac{(\Upsilon+e^{-2r}-\alpha e^{-2u})\big(e^{-|x-u|}+e^{-x-u}\big)}{2\gamma\big(1+e^{-2(u\wedge r)}\big)\big(1-e^{-2|u-r|}\big)+\Upsilon+e^{-2r}-\alpha e^{-2u}}$$

$$-\frac{\big(2\big(1+e^{-2(u\wedge r)}\big)\big(1-e^{-2|u-r|}\big)-(\Upsilon+e^{-2r}-\alpha e^{-2u})(1+e^{-2u})\big)\big(e^{-|x-r|}+e^{-x-r}\big)}{\big(2\gamma\big(1+e^{-2(u\wedge r)}\big)\big(1-e^{-2|u-r|}\big)+\Upsilon+e^{-2r}-\alpha e^{-2u}\big)\big(e^{-|u-r|}+e^{-u-r}\big)}\Big\}dz$$

1.31.6 $0<\alpha$

$$\mathbf{P}_{x/\sqrt{2\lambda}}\Big(\sup_{0\le s\le\tau}\big(\ell(s,u/\sqrt{2\lambda})-\alpha^{-1}\ell(s,r/\sqrt{2\lambda})\big)>h/\sqrt{2\lambda},\ |W_\tau|\in dz/\sqrt{2\lambda}\Big)$$

$$=\frac{1}{2}\big(e^{-|z-u|}+e^{-z-u}\big)\exp\Big(-\frac{(\Upsilon+e^{-2r}-\alpha e^{-2u})h}{2\big(1+e^{-2(u\wedge r)}\big)\big(1-e^{-2|u-r|}\big)}\Big)\Big[\frac{e^{-|x-r|}+e^{-x-r}}{e^{-|u-r|}+e^{-u-r}}$$

$$-\frac{\Upsilon+e^{-2r}-\alpha e^{-2u}}{2\big(1+e^{-2(u\wedge r)}\big)\big(1-e^{-2|u-r|}\big)}\Big(\frac{(e^{-|x-r|}+e^{-x-r})(1+e^{-2u})}{e^{-|u-r|}+e^{-u-r}}-e^{-|u-x|}-e^{-x-u}\Big)\Big]dz$$

$$\Upsilon:=\sqrt{\big(1+e^{-2r}-\alpha-\alpha e^{-2u}\big)^2+4\alpha\big(1+e^{-2(u\wedge r)}\big)\big(1-e^{-2|u-r|}\big)}+1-\alpha$$

3 $|W|$ $H_z = \min\{s : |W_s| = z\}$

2. Stopping at first hitting time

2.0.1 $$\mathbf{E}_x e^{-\alpha H_z} = \begin{cases} \dfrac{\operatorname{ch}(x\sqrt{2\alpha})}{\operatorname{ch}(z\sqrt{2\alpha})}, & 0 \le x \le z \\ e^{-(x-z)\sqrt{2\alpha}}, & z \le x \end{cases}$$

2.0.2 $z \le b$ $$\mathbf{P}_x\big(H_z \in dt\big) = \begin{cases} \operatorname{cc}_t(x,z)dt, & 0 \le x \le z \\ \dfrac{x-z}{\sqrt{2\pi}t^{3/2}} \exp\Big(-\dfrac{(x-z)^2}{2t}\Big)dt, & z \le x \end{cases}$$

2.1.2 $z \le y$ $$\mathbf{P}_x\Big(\sup_{0 \le s \le H_z} |W_s| < y\Big) = \begin{cases} 1, & 0 \le x \le z \\ \dfrac{y-x}{y-z}, & z \le x \le y \end{cases}$$

2.1.4 $z \le y$ $$\mathbf{E}_x\Big\{e^{-\alpha H_z}; \sup_{0 \le s \le H_z} |W_s| < y\Big\} = \begin{cases} \dfrac{\operatorname{ch}(x\sqrt{2\alpha})}{\operatorname{ch}(z\sqrt{2\alpha})}, & 0 \le x \le z \\ \dfrac{\operatorname{sh}((y-x)\sqrt{2\alpha})}{\operatorname{sh}((y-z)\sqrt{2\alpha})}, & 0 \le z \le x \le y \end{cases}$$

(1) $$\mathbf{P}_x\Big(\sup_{0 \le s \le H_z} |W_s| < y, H_z \in dt\Big) = \begin{cases} \operatorname{cc}_t(x,z)dt, & 0 \le x \le z \\ \operatorname{ss}_t(y-x, y-z)dt, & z \le x \le y \end{cases}$$

2.3.1 $$\mathbf{E}_x e^{-\gamma\ell(H_z,r)} = \begin{cases} 1, & 0 \le x \le z \le r, \quad r \le z \le x \\ \dfrac{1+2\gamma|x-r|}{1+2\gamma|z-r|}, & z \wedge r \le x \le z \vee r \\ \dfrac{1}{1+2\gamma|z-r|}, & 0 \le x \le r \le z, \quad z \le r \le x \end{cases}$$

2.3.2 $$\mathbf{P}_x\big(\ell(H_z,r) \in dy\big) = \begin{cases} 0, & x \le z \le r, \quad r \le z \le x \\ \dfrac{|z-x|}{2(z-r)^2} \exp\Big(-\dfrac{y}{2|z-r|}\Big)dy, & z \wedge r \le x \le z \vee r \\ \dfrac{1}{2|z-r|} \exp\Big(-\dfrac{y}{2|z-r|}\Big)dy, & x \le r \le z, \quad z \le r \le x \end{cases}$$

(1) $$\mathbf{P}_x\big(\ell(H_z,r) = 0\big) = \begin{cases} 1, & 0 \le x \le z \le r, \quad r \le z \le x \\ \dfrac{|x-r|}{|z-r|}, & z \wedge r \le x \le z \vee r \\ 0, & 0 \le x \le r \le z, \quad z \le r \le x \end{cases}$$

3 $|W|$ $H_z = \min\{s : |W_s| = z\}$

2.3.3 $\mathbf{E}_x e^{-\alpha H_z - \gamma\ell(H_z,r)}$

$z \le r$

$$= \begin{cases} \dfrac{\operatorname{ch}(x\sqrt{2\alpha})}{\operatorname{ch}(z\sqrt{2\alpha})}, & 0 \le x \le z \\ \dfrac{\sqrt{2\alpha}e^{(r-x)\sqrt{2\alpha}} + 2\gamma \operatorname{sh}((r-x)\sqrt{2\alpha})}{\sqrt{2\alpha}e^{(r-z)\sqrt{2\alpha}} + 2\gamma \operatorname{sh}((r-z)\sqrt{2\alpha})}, & z \le x \le r \\ \dfrac{\sqrt{2\alpha}e^{-(x-r)\sqrt{2\alpha}}}{\sqrt{2\alpha}e^{(r-z)\sqrt{2\alpha}} + 2\gamma \operatorname{sh}((r-z)\sqrt{2\alpha})}, & r \le x, \end{cases}$$

$r \le z$

$$= \begin{cases} \dfrac{\sqrt{2\alpha}\operatorname{ch}(x\sqrt{2\alpha})}{\sqrt{2\alpha}\operatorname{ch}(z\sqrt{2\alpha}) + 2\gamma \operatorname{sh}((z-r)\sqrt{2\alpha})\operatorname{ch}(r\sqrt{2\alpha})}, & 0 \le x \le r \\ \dfrac{\sqrt{2\alpha}\operatorname{ch}(x\sqrt{2\alpha}) + 2\gamma \operatorname{sh}((x-r)\sqrt{2\alpha})\operatorname{ch}(r\sqrt{2\alpha})}{\sqrt{2\alpha}\operatorname{ch}(z\sqrt{2\alpha}) + 2\gamma \operatorname{sh}((z-r)\sqrt{2\alpha})\operatorname{ch}(r\sqrt{2\alpha})}, & r \le x \le z \\ e^{-(x-z)\sqrt{2\alpha}}, & z \le x, \end{cases}$$

2.3.4 $\mathbf{E}_x\{e^{-\alpha H_z};\ \ell(H_z,r) \in dy\}$

$z \le r$

$$= \exp\Big(-\frac{y\sqrt{2\alpha}}{1 - e^{-2(r-z)\sqrt{2\alpha}}}\Big) \begin{cases} 0, & 0 \le x \le z \\ \dfrac{\sqrt{2\alpha}\operatorname{sh}((x-z)\sqrt{2\alpha})}{2\operatorname{sh}^2((r-z)\sqrt{2\alpha})}dy, & z \le x \le r \\ \dfrac{\sqrt{2\alpha}e^{-(x-r)\sqrt{2\alpha}}}{2\operatorname{sh}((r-z)\sqrt{2\alpha})}dy, & r \le x \end{cases}$$

$r \le z$

$$= \exp\Big(-\frac{y\sqrt{2\alpha}\operatorname{ch}(z\sqrt{2\alpha})}{2\operatorname{sh}((z-r)\sqrt{2\alpha})\operatorname{ch}(r\sqrt{2\alpha})}\Big) \begin{cases} 0, & z \le x \\ \dfrac{\sqrt{2\alpha}\operatorname{sh}((z-x)\sqrt{2\alpha})}{2\operatorname{sh}^2((z-r)\sqrt{2\alpha})}dy, & r \le x \le z \\ \dfrac{\sqrt{2\alpha}\operatorname{ch}(x\sqrt{2\alpha})}{2\operatorname{sh}((z-r)\sqrt{2\alpha})\operatorname{ch}(r\sqrt{2\alpha})}dy, & 0 \le x \le r \end{cases}$$

(1)
$$\mathbf{E}_x\{e^{-\alpha H_z};\ \ell(H_z,r) = 0\} = \begin{cases} \dfrac{\operatorname{ch}(x\sqrt{2\alpha})}{\operatorname{ch}(z\sqrt{2\alpha})}, & 0 \le x \le z \le r \\ e^{-(x-z)\sqrt{2\alpha}}, & r \le z \le x \\ \dfrac{\operatorname{sh}(|x-r|\sqrt{2\gamma})}{\operatorname{sh}(|z-r|\sqrt{2\gamma})}, & z \wedge r \le x \le z \vee r \\ 0, & 0 \le x \le r \le z,\ z \le r \le x \end{cases}$$

2.4.1 $\mathbf{E}_x \exp\Big(-\gamma \int_0^{H_z} \mathbb{I}_{[r,\infty)}(|W_s|)ds\Big)$

3 $|W|$ $\qquad H_z = \min\{s : |W_s| = z\}$

$z \le r$

$$= \begin{cases} 1, & 0 \le x \le z \\ \dfrac{r-x}{r-z} + \dfrac{x-z}{(r-z)(1+(r-z)\sqrt{2\gamma})}, & z \le x \le r \\ \dfrac{1}{1+(r-z)\sqrt{2\gamma}} e^{-(x-r)\sqrt{2\gamma}}, & r \le x \end{cases}$$

$r \le z$

$$= \begin{cases} \dfrac{1}{\operatorname{ch}((z-r)\sqrt{2\gamma})}, & 0 \le x \le r \\ \dfrac{\operatorname{ch}((x-r)\sqrt{2\gamma})}{\operatorname{ch}((z-r)\sqrt{2\gamma})}, & r \le x \le z \\ e^{(z-x)\sqrt{2\gamma}}, & z \le x \end{cases}$$

2.4.2 $\mathbf{P}_x\Big(\int_0^{H_z} \mathbb{I}_{[r,\infty)}(|W_s|)ds \in dy\Big)$

$z < r$

$$= \begin{cases} 0, & 0 \le x \le z \\ \dfrac{x-z}{(r-z)^2}\Big(\dfrac{1}{\sqrt{2\pi y}} - \dfrac{1}{2(r-z)} e^{y/2(r-z)^2} \operatorname{Erfc}\Big(\dfrac{\sqrt{y}}{(r-z)\sqrt{2}}\Big)\Big)dy, & z \le x \le r \\ \dfrac{1}{(r-z)\sqrt{2\pi y}} e^{-(x-r)^2/2y} dy - \\ -\dfrac{1}{2(r-z)^2} \exp\Big(\dfrac{x-r}{r-z} + \dfrac{y}{2(r-z)^2}\Big) \operatorname{Erfc}\Big(\dfrac{x-r}{\sqrt{2y}} + \dfrac{\sqrt{y}}{(r-z)\sqrt{2}}\Big)dy, & r \le x \end{cases}$$

$r < z$

$$= \begin{cases} \operatorname{cc}_y(0, z-r)dy, & 0 \le x \le r \\ \operatorname{cc}_y(x-r, z-r)dy, & r \le x \le z \\ \dfrac{x-z}{\sqrt{2\pi}y^{3/2}} \exp\Big(-\dfrac{(x-z)^2}{2y}\Big)dy, & z \le x \end{cases}$$

(1)
$z \le r$

$$\mathbf{P}_x\Big(\int_0^{H_z} \mathbb{I}_{[r,\infty)}(|W_s|)ds = 0\Big) = \begin{cases} 1, & 0 \le x \le z \\ \dfrac{r-x}{r-z}, & z \le x \le r \\ 0, & r \le x \end{cases}$$

2.4.4 $\mathbf{P}_x\Big(\int_0^{H_z} \mathbb{I}_{[r,\infty)}(|W_s|)ds \in dv,\ H_z \in dt\Big)$

$$= \int_0^\infty B_z^{(27)}(t-v, v, y)dy\,dvdt, \qquad \text{for } B_z^{(27)}(u,v,y) \text{ see 2.27.2}$$

(1) $\mathbf{P}_x\Big(\int_0^{H_z} \mathbb{I}_{[r,\infty)}(|W_s|)ds = 0,\ H_z \in dt\Big)$

$$= \begin{cases} \operatorname{cc}_t(x,z)dt, & 0 \le x \le z \\ \operatorname{ss}_t(r-x, r-z)dt, & z \le x \le r \end{cases}$$

3 $|W|$ $H_z = \min\{s : |W_s| = z\}$

2.5.1 $\mathbf{E}_x \exp\Big(-\gamma \int_0^{H_z} \mathbb{I}_{[0,r]}\big(|W_s|\big)ds\Big)$

$z \le r$
$$= \begin{cases} \dfrac{\operatorname{ch}(x\sqrt{2\gamma})}{\operatorname{ch}(z\sqrt{2\gamma})}, & 0 \le x \le z \\ \dfrac{\operatorname{ch}((r-x)\sqrt{2\gamma})}{\operatorname{ch}((r-z)\sqrt{2\gamma})}, & z \le x \le r \\ \dfrac{1}{\operatorname{ch}((r-z)\sqrt{2\gamma})}, & r \le x \end{cases}$$

$r \le z$
$$= \begin{cases} \dfrac{\operatorname{ch}(x\sqrt{2\gamma})}{\operatorname{ch}(r\sqrt{2\gamma}) + \sqrt{2\gamma}(z-r)\operatorname{sh}(r\sqrt{2\gamma})}, & 0 \le x \le r \\ \dfrac{\operatorname{ch}(r\sqrt{2\gamma}) + \sqrt{2\gamma}(x-r)\operatorname{sh}(r\sqrt{2\gamma})}{\operatorname{ch}(r\sqrt{2\gamma}) + \sqrt{2\gamma}(z-r)\operatorname{sh}(r\sqrt{2\gamma})}, & r \le x \le z \\ 1, & z \le x \end{cases}$$

2.5.2 $\mathbf{P}_x\Big(\int_0^{H_z} \mathbb{I}_{[0,r]}\big(|W_s|\big)ds \in dy\Big)$

$z < r$
$$= \begin{cases} \operatorname{cc}_y(x,z)dy, & 0 \le x < z \\ \operatorname{cc}_y(r-x, r-z)dy, & z < x \le r \\ \operatorname{cc}_y(0, r-z)dy, & r \le x \end{cases}$$

$r < z$
$$= \begin{cases} \operatorname{rs}_y(0,r,x,z-r)dy, & 0 \le x \le r \\ \dfrac{z-x}{z-r}\operatorname{rs}_y(0,r,r,z-r))dy, & r \le x \le z \\ 0, & z \le x \end{cases}$$

(1) $r \le z$
$$\mathbf{P}_x\Big(\int_0^{H_z} \mathbb{I}_{[0,r]}\big(|W_s|\big)ds = 0\Big) = \begin{cases} 0, & 0 \le x \le r \\ \dfrac{x-r}{z-r}, & r \le x \le z \\ 1, & z \le x \end{cases}$$

2.5.4 $\mathbf{P}_x\Big(\int_0^{H_z} \mathbb{I}_{[0,r]}\big(|W_s|\big)ds \in dv,\ H_z \in dt\Big)$

$$= \int_0^\infty B_z^{(27)}(v, t-v, y)dy\,dvdt,$$ for $B_z^{(27)}(u,v,y)$ see 2.27.2

(1) $\mathbf{P}_x\Big(\int_0^{H_z} \mathbb{I}_{[0,r]}\big(|W_s|\big)ds = 0,\ H_z \in dt\Big)$

3 $|W|$ $H_z = \min\{s : |W_s| = z\}$

$$= \begin{cases} \mathrm{ss}_t(x-r, z-r)dt, & r \le x \le z \\ \mathrm{h}_t(1, x-z)dt, & z \le x \end{cases}$$

2.6.1 $\mathbf{E}_x \exp\Big(-\int_0^{H_z} \big(p\mathbb{I}_{[0,r)}(|W_s|) + q\mathbb{I}_{[r,\infty)}(|W_s|)\big)ds\Big)$

$z \le r$
$$= \begin{cases} \dfrac{\mathrm{ch}(x\sqrt{2p})}{\mathrm{ch}(z\sqrt{2p})}, & 0 \le x \le z \\ \dfrac{\sqrt{p}\,\mathrm{ch}((r-x)\sqrt{2p}) + \sqrt{q}\,\mathrm{sh}((r-x)\sqrt{2p})}{\sqrt{p}\,\mathrm{ch}((r-z)\sqrt{2p}) + \sqrt{q}\,\mathrm{sh}((r-z)\sqrt{2p})}, & z \le x \le r \\ \dfrac{\sqrt{p}e^{-(x-r)\sqrt{2q}}}{\sqrt{p}\,\mathrm{ch}((r-z)\sqrt{2p}) + \sqrt{q}\,\mathrm{sh}((r-z)\sqrt{2p})}, & r \le x \end{cases}$$

$r \le z$
$$= \begin{cases} \dfrac{\sqrt{q}\,\mathrm{ch}(x\sqrt{2p})}{\sqrt{q}\,\mathrm{ch}(r\sqrt{2p})\,\mathrm{ch}((z-r)\sqrt{2q}) + \sqrt{p}\,\mathrm{sh}(r\sqrt{2p})\,\mathrm{sh}((z-r)\sqrt{2q})}, & 0 \le x \le r \\ \dfrac{\sqrt{q}\,\mathrm{ch}(r\sqrt{2p})\,\mathrm{ch}((x-r)\sqrt{2q}) + \sqrt{p}\,\mathrm{sh}(r\sqrt{2p})\,\mathrm{sh}((x-r)\sqrt{2q})}{\sqrt{q}\,\mathrm{ch}(r\sqrt{2p})\,\mathrm{ch}((z-r)\sqrt{2q}) + \sqrt{p}\,\mathrm{sh}(r\sqrt{2p})\,\mathrm{sh}((z-r)\sqrt{2q})}, & r \le x \le z \\ e^{-(x-z)\sqrt{2q}}, & z \le x \end{cases}$$

2.6.2 $\mathbf{P}_x\Big(\int_0^{H_z} \mathbb{I}_{[0,r)}(|W_s|)ds \in du, \int_0^{H_z} \mathbb{I}_{[r,\infty)}(|W_s|)ds \in dv\Big)$

$$= \int_0^\infty B_z^{(27)}(u,v,y)dy\,dudv, \qquad \text{for } B_z^{(27)}(u,v,y) \text{ see 2.27.2}$$

(1) $\mathbf{P}_x\Big(\int_0^{H_z} \mathbb{I}_{[0,r)}(|W_s|)ds \in du, \int_0^{H_z} \mathbb{I}_{[r,\infty)}(|W_s|)ds = 0\Big)$

$$= \begin{cases} \mathrm{cc}_u(x,z)du, & 0 \le x \le z \\ \mathrm{ss}_u(r-x, r-z)du, & z \le x \le r \end{cases}$$

(2) $\mathbf{P}_x\Big(\int_0^{H_z} \mathbb{I}_{[0,r)}(|W_s|)ds = 0, \int_0^{H_z} \mathbb{I}_{[r,\infty)}(|W_s|)ds \in dv\Big)$

$$= \begin{cases} \mathrm{ss}_v(x-r, z-r)dv, & r \le x \le z \\ \mathrm{h}_v(1, x-z)dv, & z \le x \end{cases}$$

3 $|W|$ $H_z = \min\{s : |W_s| = z\}$

2.7.1 $r<u$

$$\mathbf{E}_x \exp\Big(-\gamma \int_0^{H_z} \mathbb{I}_{[r,u]}\big(|W_s|\big)ds\Big)$$

$z \le r$

$$= \begin{cases} 1, & 0 \le x \le z \\ \dfrac{(r-x)\sqrt{2\gamma}\,\mathrm{sh}((u-r)\sqrt{2\gamma}) + \mathrm{ch}((u-r)\sqrt{2\gamma})}{(r-z)\sqrt{2\gamma}\,\mathrm{sh}((u-r)\sqrt{2\gamma}) + \mathrm{ch}((u-r)\sqrt{2\gamma})}, & z \le x \le r \\ \dfrac{\mathrm{ch}((u-x)\sqrt{2\gamma})}{(r-z)\sqrt{2\gamma}\,\mathrm{sh}((u-r)\sqrt{2\gamma}) + \mathrm{ch}((u-r)\sqrt{2\gamma})}, & r \le x \le u \\ \dfrac{1}{(r-z)\sqrt{2\gamma}\,\mathrm{sh}((u-r)\sqrt{2\gamma}) + \mathrm{ch}((u-r)\sqrt{2\gamma})}, & u \le x \end{cases}$$

$u \le z$

$$= \begin{cases} \dfrac{1}{(z-u)\sqrt{2\gamma}\,\mathrm{sh}((u-r)\sqrt{2\gamma}) + \mathrm{ch}((u-r)\sqrt{2\gamma})}, & 0 \le x \le r \\ \dfrac{\mathrm{ch}((x-r)\sqrt{2\gamma})}{(z-u)\sqrt{2\gamma}\,\mathrm{sh}((u-r)\sqrt{2\gamma}) + \mathrm{ch}((u-r)\sqrt{2\gamma})}, & r \le x \le u \\ \dfrac{(x-u)\sqrt{2\gamma}\,\mathrm{sh}((u-r)\sqrt{2\gamma}) + \mathrm{ch}((u-r)\sqrt{2\gamma})}{(z-u)\sqrt{2\gamma}\,\mathrm{sh}((u-r)\sqrt{2\gamma}) + \mathrm{ch}((u-r)\sqrt{2\gamma})}, & u \le x \le z \\ 1, & z \le x \end{cases}$$

2.7.2 $r<u$

$$\mathbf{P}_x\Big(\int_0^{H_z} \mathbb{I}_{[r,u]}\big(|W_s|\big)ds \in dy\Big) = B_z^{(7)}(y)dy, \qquad \text{for } B_z^{(7)}(y) \text{ see 1.2.7.2}$$

2.8.1

$$\mathbf{E}_x \exp\Big(-\gamma \int_0^{H_z} |W_s|ds\Big) = \begin{cases} \dfrac{\sqrt{x}I_{-1/3}(\frac{2}{3}x^{3/2}\sqrt{2\gamma})}{\sqrt{z}I_{-1/3}(\frac{2}{3}z^{3/2}\sqrt{2\gamma})}, & 0 \le x \le z \\ \dfrac{\sqrt{x}K_{1/3}(\frac{2}{3}x^{3/2}\sqrt{2\gamma})}{\sqrt{z}K_{1/3}(\frac{2}{3}z^{3/2}\sqrt{2\gamma})}, & z \le x \end{cases}$$

2.8.2 (1)

$$\mathbf{P}_x\Big(\int_0^{H_0} |W_s|ds \in dy\Big) = \Big(\frac{2}{9y^4}\Big)^{1/3} \frac{|x|}{\Gamma(1/3)} e^{-2|x|^3/9y} dy$$

2.9.1

$$\mathbf{E}_x \exp\Big(-2\gamma \int_0^{H_z} |W_s|^2ds\Big) = \begin{cases} \dfrac{\sqrt{x}I_{-1/4}(x^2\sqrt{\gamma})}{\sqrt{z}I_{-1/4}(z^2\sqrt{\gamma})}, & 0 \le x \le z \\ \dfrac{\sqrt{x}K_{1/4}(x^2\sqrt{\gamma})}{\sqrt{z}K_{1/4}(z^2\sqrt{\gamma})}, & z \le x \end{cases}$$

2.9.4 (1)

$$\mathbf{E}_x\Big\{\exp\Big(-\frac{\gamma^2}{2}\int_0^{H_0} |W_s|^2ds\Big); H_0 \in dt\Big\} = \frac{x\gamma^{3/2}}{\sqrt{2\pi}\,\mathrm{sh}^{3/2}(t\gamma)} \exp\Big(-\frac{x^2\gamma\,\mathrm{ch}(t\gamma)}{2\,\mathrm{sh}(t\gamma)}\Big)dt$$

(2)

$$\mathbf{P}_x\Big(\int_0^{H_0} |W_s|^2ds \in dy, H_0 \in dt\Big) = \frac{x}{\sqrt{2\pi}}\,\mathrm{es}_y\Big(\frac{3}{2}, \frac{3}{2}, t, 0, \frac{x^2}{2}\Big)dydt$$

3 $|W|$ $H_z = \min\{s : |W_s| = z\}$

2.10.3 $$\mathbf{E}_x \exp\Big(-\alpha H_z - \gamma\int_0^{H_z} e^{2\beta|W_s|}ds\Big) = \begin{cases} \dfrac{C_{\sqrt{2\alpha}/|\beta|}\Big(\frac{\sqrt{2\gamma}}{|\beta|}, \frac{\sqrt{2\gamma}}{|\beta|}e^{\beta x}\Big)}{C_{\sqrt{2\alpha}/|\beta|}\Big(\frac{\sqrt{2\gamma}}{|\beta|}, \frac{\sqrt{2\gamma}}{|\beta|}e^{\beta z}\Big)}, & 0 \le x \le z \\ \dfrac{K_{\sqrt{2\alpha}/|\beta|}\Big(\frac{\sqrt{2\gamma}}{|\beta|}e^{\beta x}\Big)}{K_{\sqrt{2\alpha}/|\beta|}\Big(\frac{\sqrt{2\gamma}}{|\beta|}e^{\beta z}\Big)}, & z \le x, \quad \beta > 0 \\ \dfrac{I_{\sqrt{2\alpha}/|\beta|}\Big(\frac{\sqrt{2\gamma}}{|\beta|}e^{\beta x}\Big)}{I_{\sqrt{2\alpha}/|\beta|}\Big(\frac{\sqrt{2\gamma}}{|\beta|}e^{\beta z}\Big)}, & z \le x, \quad \beta < 0 \end{cases}$$

2.11.2 $$\lambda\int_0^\infty e^{-\lambda z}\mathbf{P}_0\big(\sup_{y\in[0,z)} \ell(H_z, y) < h\big)dz = 1 - \frac{1}{I_0(\sqrt{2\lambda h})}$$

(1) $$\mathbf{P}_0\big(\sup_{y\in[0,z)} \ell(H_z, y) < h\big) = 2\sum_{k=1}^\infty \frac{e^{-j_{0,k}^2 z/2h}}{j_{0,k}J_1(j_{0,k})}$$

2.18.1 $r < u$ $$\mathbf{E}_x \exp\big(-\gamma\ell(H_z, r) - \eta\ell(H_z, u)\big)$$

$z < r$ $$= \begin{cases} 1, & 0 \le x \le z \\ \dfrac{1 + 2\gamma(r-x) + 2\eta(u-x) + 4\gamma\eta(u-r)(r-x)}{1 + 2\gamma(r-z) + 2\eta(u-z) + 4\gamma\eta(u-r)(r-z)}, & z \le x \le r \\ \dfrac{1 + 2\eta(u-x)}{1 + 2\gamma(r-z) + 2\eta(u-z) + 4\gamma\eta(u-r)(r-z)}, & r \le x \le u \\ \dfrac{1}{1 + 2\gamma(r-z) + 2\eta(u-z) + 4\gamma\eta(u-r)(r-z)}, & u \le x \end{cases}$$

$u < z$ $$= \begin{cases} \dfrac{1}{1 + 2\gamma(z-r) + 2\eta(z-u) + 4\gamma\eta(z-u)(u-r)}, & 0 \le x \le r \\ \dfrac{1 + 2\gamma(x-r)}{1 + 2\gamma(z-r) + 2\eta(z-u) + 4\gamma\eta(z-u)(u-r)}, & r \le x \le u \\ \dfrac{1 + 2\gamma(x-r) + 2\eta(x-u) + 4\gamma\eta(x-u)(u-r)}{1 + 2\gamma(z-r) + 2\eta(z-u) + 4\gamma\eta(z-u)(u-r)}, & u \le x \le z \\ 1, & z \le x \end{cases}$$

2.18.2 $r < u$ $$\mathbf{P}_x\big(\ell(H_z, r) \in dy,\ \ell(H_z, u) \in dg\big) = B^{(18)}_{x,z}(y, g)dydg$$

for $B^{(18)}_{x,z}(y, g)$ see 1.2.18.2

2.27.1 $$\mathbf{E}_x\Big\{\exp\Big(-\int_0^{H_z}\big(p\mathbb{1}_{[0,r)}(|W_s|) + q\mathbb{1}_{[r,\infty)}(|W_s|)\big)ds\Big); \tfrac{1}{2}\ell(H_z, r) \in dy\Big\}$$

3 $|W|$ $H_z = \min\{s : |W_s| = z\}$

$z \le r$, $x \le r$	$= \dfrac{\sqrt{2p}\,\mathrm{sh}((x-z)\sqrt{2p})}{\mathrm{sh}^2((r-z)\sqrt{2p})} \exp\Bigl(-y\dfrac{\sqrt{2p}\,\mathrm{ch}((r-z)\sqrt{2p})}{\mathrm{sh}((r-z)\sqrt{2q})} - y\sqrt{2q}\Bigr)dy$
$z \le r$, $r \le x$	$= \dfrac{\sqrt{2p}}{\mathrm{sh}((r-z)\sqrt{2p})} \exp\Bigl(-y\dfrac{\sqrt{2p}\,\mathrm{ch}((r-z)\sqrt{2p})}{\mathrm{sh}((r-z)\sqrt{2q})} - (y+x-r)\sqrt{2q}\Bigr)dy$
$r \le z$, $x \le r$	$= \dfrac{\sqrt{2q}\,\mathrm{ch}(x\sqrt{2p})}{\mathrm{ch}(r\sqrt{2p})\,\mathrm{sh}((z-r)\sqrt{2q})} \exp\Bigl(-y\Bigl(\dfrac{\sqrt{2q}\,\mathrm{ch}((z-r)\sqrt{2q})}{\mathrm{sh}((z-r)\sqrt{2q})} + \dfrac{\sqrt{2p}\,\mathrm{sh}(r\sqrt{2p})}{\mathrm{ch}(r\sqrt{2p})}\Bigr)\Bigr)dy$
$r \le z$, $r \le x$	$= \dfrac{\sqrt{2q}\,\mathrm{sh}((z-x)\sqrt{2q})}{\mathrm{sh}^2((z-r)\sqrt{2q})} \exp\Bigl(-y\Bigl(\dfrac{\sqrt{2q}\,\mathrm{ch}((z-r)\sqrt{2q})}{\mathrm{sh}((z-r)\sqrt{2q})} + \dfrac{\sqrt{2p}\,\mathrm{sh}(r\sqrt{2p})}{\mathrm{ch}(r\sqrt{2p})}\Bigr)\Bigr)dy$
(1)	$\mathbf{E}_x\Bigl\{\exp\Bigl(-\int_0^{H_z}\bigl(p\mathbb{I}_{[0,r)}(\lvert W_s\rvert) + q\mathbb{I}_{[r,\infty)}(\lvert W_s\rvert)\bigr)ds\Bigr); \ell(H_z,r)=0\Bigr\}$
	$= \begin{cases} \dfrac{\mathrm{ch}(x\sqrt{2p})}{\mathrm{ch}(z\sqrt{2p})}, & 0 \le x \le z \\ \dfrac{\mathrm{sh}((r-x)\sqrt{2p})}{\mathrm{sh}((r-z)\sqrt{2p})}, & z \le x \le r \\ \dfrac{\mathrm{sh}((x-r)\sqrt{2q})}{\mathrm{sh}((z-r)\sqrt{2q})}, & r \le x \le z \end{cases}$
2.27.2	$\mathbf{P}_x\Bigl(\int_0^{H_z}\mathbb{I}_{[0,r)}(\lvert W_s\rvert)ds \in du,\ \int_0^{H_z}\mathbb{I}_{[r,\infty)}(\lvert W_s\rvert)ds \in dv,\ \frac{1}{2}\ell(H_z,r) \in dy\Bigr)$
	$=: B_z^{(27)}(u,v,y)dudvdy$
$z \le r$, $x \le r$	$= \mathrm{escs}_u(x-z,0,r-z,y)\,\mathrm{h}_v(1,y)dudvdy$
$z \le r$, $r \le x$	$= \mathrm{es}_u(1,1,r-z,0,y)\,\mathrm{h}_v(1,y+x-r)dudvdy$
$r \le z$, $x \le r$	$= \mathrm{ecc}_u(x,r,0,y)\,\mathrm{es}_v(1,1,z-r,0,y)dudvdy$
$r \le z$, $r \le x$	$= \mathrm{ec}_u(0,0,r,0,y)\,\mathrm{escs}_v(z-x,0,z-r,y)dudvdy$
(1)	$\mathbf{P}_x\Bigl(\int_0^{H_z}\mathbb{I}_{[0,r)}(\lvert W_s\rvert)ds \in du,\ \int_0^{H_z}\mathbb{I}_{[r,\infty)}(\lvert W_s\rvert)ds = 0,\ \ell(H_z,r)=0\Bigr)$

3 $|W|$ $H_z = \min\{s : |W_s| = z\}$

$$= \begin{cases} \mathrm{cc}_u(x,z)du, & 0 \le x \le z \\ \mathrm{ss}_u(r-x, r-z)du, & z \le x \le r \end{cases}$$

(2) $$\mathbf{P}_x\Big(\int_0^{H_z} \mathbb{I}_{[0,r)}\big(|W_s|\big)ds = 0,\ \int_0^{H_z} \mathbb{I}_{[r,\infty)}\big(|W_s|\big)ds \in dv,\ \ell\big(H_z, r\big) = 0\Big)$$

$$= \begin{cases} \mathrm{ss}_v(x-r, z-r)dv, & r \le x \le z \\ \mathrm{h}_v(1, x-z)dv, & z \le x \end{cases}$$

2.31.1 $0 < \alpha$ $$\mathbf{E}_x \exp\Big(-\frac{\gamma}{2} \sup_{0\le s\le H_z} \big(\ell(s,u) - \alpha^{-1}\ell(s,r)\big)\Big)$$

for $z \le x \wedge r \wedge u$ or $x \vee r \vee u \le z$

$$= 1 - \frac{\gamma(\Upsilon + |r-z| - \alpha|u-z|)(|u-z| \wedge |x-z|)}{2\gamma(|u-z| \wedge |r-z|)|u-r| + \Upsilon + |r-z| - \alpha|u-z|}$$

$$- \frac{\gamma(2(|u-z| \wedge |r-z|)|u-r| - (\Upsilon + |r-z| - \alpha|u-z|)|u-z|)(|x-z| \wedge |r-z|)}{(2\gamma(|u-z| \wedge |r-z|)|u-r| + \Upsilon + |r-z| - \alpha|u-z|)(|u-z| \wedge |r-z|)}$$

2.31.2 $0 < \alpha$ $$\mathbf{P}_x\Big(\sup_{0\le s\le H_z} \big(\ell(s,u) - \alpha^{-1}\ell(s,r)\big) > h\Big)$$

for $z \le x \wedge r \wedge u$ or $x \vee r \vee u \le z$

$$= \exp\Big(-\frac{(\Upsilon + |r-z| - \alpha|u-z|)h}{4(|u-z| \wedge |r-z|)|u-r|}\Big)\Big[\frac{|x-z| \wedge |r-z|}{|u-z| \wedge |r-z|}$$

$$- \frac{\Upsilon + |r-z| - \alpha|u-z|}{2(|u-z| \wedge |r-z|)|u-r|}\Big(\frac{(|x-z| \wedge |r-z|)|u-z|}{|u-z| \wedge |r-z|} - |u-z| \wedge |x-z|\Big)\Big]$$

$$\Upsilon := \sqrt{(|r-z| - \alpha|u-z|)^2 + 4\alpha(|u-z| \wedge |r-z|)|u-r|}$$

3 $|W|$ $\varrho = \varrho(v,z) = \min\{s : \ell(s,z) = v\}$

4. Stopping at inverse local time

4.0.1 $$\mathbf{E}_x e^{-\alpha\varrho} = \begin{cases} \dfrac{\operatorname{ch}(x\sqrt{2\alpha})}{\operatorname{ch}(z\sqrt{2\alpha})} \exp\Big(-\dfrac{v\sqrt{\alpha}e^{z\sqrt{2\alpha}}}{\sqrt{2}\operatorname{ch}(z\sqrt{2\alpha})}\Big), & 0 \le x \le z \\ e^{-(x-z)\sqrt{2\alpha}} \exp\Big(-\dfrac{v\sqrt{\alpha}e^{z\sqrt{2\alpha}}}{\sqrt{2}\operatorname{ch}(z\sqrt{2\alpha})}\Big), & z \le x \end{cases}$$

4.0.2 $$\mathbf{P}_x(\varrho \in dt) = \begin{cases} \operatorname{ecc}_t(x, z, v/2, v/2)dt, & 0 \le x \le z \\ \operatorname{ec}_t(0, 0, z, x - z + v/2, v/2)dt, & z \le x \end{cases}$$

4.1.1 $x \le z$ $$\mathbf{E}_x \exp\big(-\gamma \sup_{0\le s\le\varrho} |W_s|\big) = e^{-\gamma z}\sqrt{2v\gamma}K_1(\sqrt{2v\gamma}), \qquad 0 \le x \le z$$

4.1.2 $z \le y$ $$\mathbf{P}_x\big(\sup_{0\le s\le\varrho} |W_s| < y\big) = \begin{cases} \exp\Big(-\dfrac{v}{2(y-z)}\Big), & 0 \le x \le z \\ \dfrac{y-x}{y-z}\exp\Big(-\dfrac{v}{2(y-z)}\Big), & 0 \le z \le x \le y \end{cases}$$

4.1.4 $z \le y$ $$\mathbf{E}_x\{e^{-\alpha\varrho}; \sup_{0\le s\le\varrho} |W_s| < y\}$$

$$= \begin{cases} \dfrac{\operatorname{ch}(x\sqrt{2\alpha})}{\operatorname{ch}(z\sqrt{2\alpha})} \exp\Big(-\dfrac{v\sqrt{2\alpha}\operatorname{ch}(y\sqrt{2\alpha})}{2\operatorname{ch}(z\sqrt{2\alpha})\operatorname{sh}((y-z)\sqrt{2\alpha})}\Big), & 0 \le x \le z \le y \\ \dfrac{\operatorname{sh}((y-x)\sqrt{2\alpha})}{\operatorname{sh}((y-z)\sqrt{2\alpha})} \exp\Big(-\dfrac{v\sqrt{2\alpha}\operatorname{ch}(y\sqrt{2\alpha})}{2\operatorname{ch}(z\sqrt{2\alpha})\operatorname{sh}((y-z)\sqrt{2\alpha})}\Big), & 0 \le z \le x \le y \end{cases}$$

4.3.1 $$\mathbf{E}_x e^{-\gamma\ell(\varrho,r)} = \begin{cases} \exp\Big(-\dfrac{\gamma v}{1+2\gamma|z-r|}\Big), & x \le r \le z, \quad r \le z \le x \\ \dfrac{1+2\gamma|r-x|}{1+2\gamma|z-r|}\exp\Big(-\dfrac{\gamma v}{1+2\gamma|z-r|}\Big), & r \wedge z \le x \le r \vee z \\ \dfrac{1}{1+2\gamma|z-r|}\exp\Big(-\dfrac{\gamma v}{1+2\gamma|z-r|}\Big), & z \le r \le x, \quad x \le r \le z \end{cases}$$

4.3.2 $$\mathbf{P}_x(\ell(\varrho, r) \in dy)$$

3 $|W|$ $\varrho = \varrho(v,z) = \min\{s : \ell(s,z) = v\}$

$$= \begin{cases} \dfrac{\sqrt{v}}{2|r-z|\sqrt{y}} \exp\Big(-\dfrac{(v+y)}{2|r-z|}\Big) I_1\Big(\dfrac{\sqrt{vy}}{|r-z|}\Big) dy, & x \le z \le r, \quad r \le z \le x \\ \dfrac{1}{2(r-z)^2} \exp\Big(-\dfrac{(v+y)}{2|r-z|}\Big) \Big[|r-x|\dfrac{\sqrt{v}}{\sqrt{y}} I_1\Big(\dfrac{\sqrt{vy}}{|r-z|}\Big) \\ \qquad\qquad + |x-z| I_0\Big(\dfrac{\sqrt{vy}}{|r-z|}\Big)\Big] dy, & z \wedge r \le x \le r \vee z \\ \dfrac{1}{2|r-z|} \exp\Big(-\dfrac{(v+y)}{2|r-z|}\Big) I_0\Big(\dfrac{\sqrt{vy}}{|r-z|}\Big) dy, & z \le r \le x, \quad x \le r \le z \end{cases}$$

(1) $$\mathbf{P}_x\big(\ell(\varrho, r) = 0\big) = \begin{cases} \exp\Big(-\dfrac{v}{2|r-z|}\Big), & x \le z \le r, \quad r \le z \le x \\ \dfrac{|r-x|}{|r-z|} \exp\Big(-\dfrac{v}{2|r-z|}\Big), & z \wedge r \le x \le r \vee z \\ 0, & z \le r \le x, \quad x \le r \le z \end{cases}$$

4.3.3 $\mathbf{E}_x e^{-\alpha\varrho - \gamma\ell(\varrho,r)}$

$z \le r$
$$= \exp\Big(-\frac{v(\alpha e^{r\sqrt{2\alpha}} + \gamma\sqrt{2\alpha}\,\mathrm{ch}(r\sqrt{2\alpha}))}{\mathrm{ch}(z\sqrt{2\alpha})\big(\sqrt{2\alpha} e^{(r-z)\sqrt{2\alpha}} + 2\gamma\,\mathrm{sh}((r-z)\sqrt{2\alpha})\big)}\Big)$$

$$\times \begin{cases} \dfrac{\mathrm{ch}(x\sqrt{2\alpha})}{\mathrm{ch}(z\sqrt{2\alpha})}, & 0 \le x \le z \\ \dfrac{\sqrt{2\alpha} e^{(r-x)\sqrt{2\alpha}} + 2\gamma\,\mathrm{sh}((r-x)\sqrt{2\alpha})}{\sqrt{2\alpha} e^{(r-z)\sqrt{2\alpha}} + 2\gamma\,\mathrm{sh}((r-z)\sqrt{2\alpha})}, & z \le x \le r \\ \dfrac{\sqrt{2\alpha} e^{-(x-r)\sqrt{2\alpha}}}{\sqrt{2\alpha} e^{(r-z)\sqrt{2\alpha}} + 2\gamma\,\mathrm{sh}((r-z)\sqrt{2\alpha})}, & r \le x, \end{cases}$$

$r \le z$
$$= \exp\Big(-\frac{v e^{(z-r)\sqrt{2\alpha}}(\alpha e^{r\sqrt{2\alpha}} + \gamma\sqrt{2\alpha}\,\mathrm{ch}(r\sqrt{2\alpha}))}{\sqrt{2\alpha}\,\mathrm{ch}(z\sqrt{2\alpha}) + 2\gamma\,\mathrm{sh}((z-r)\sqrt{2\alpha})\,\mathrm{ch}(r\sqrt{2\alpha})}\Big)$$

$$\times \begin{cases} \dfrac{\sqrt{2\alpha}\,\mathrm{ch}(x\sqrt{2\alpha})}{\sqrt{2\alpha}\,\mathrm{ch}(z\sqrt{2\alpha}) + 2\gamma\,\mathrm{sh}((z-r)\sqrt{2\alpha})\,\mathrm{ch}(r\sqrt{2\alpha})}, & 0 \le x \le r \\ \dfrac{\sqrt{2\alpha}\,\mathrm{ch}(x\sqrt{2\alpha}) + 2\gamma\,\mathrm{sh}((x-r)\sqrt{2\alpha})\,\mathrm{ch}(r\sqrt{2\alpha})}{\sqrt{2\alpha}\,\mathrm{ch}(z\sqrt{2\alpha}) + 2\gamma\,\mathrm{sh}((z-r)\sqrt{2\alpha})\,\mathrm{ch}(r\sqrt{2\alpha})}, & r \le x \le z \\ e^{-(x-z)\sqrt{2\alpha}}, & z \le x, \end{cases}$$

4.3.4 $\mathbf{E}_x\{e^{-\alpha\varrho};\ \ell(\varrho, r) \in dy\}$

$z \le r$
$$= \exp\Big(-\frac{\sqrt{\alpha}\,\mathrm{ch}(r\sqrt{2\alpha}) v}{\sqrt{2}\,\mathrm{sh}((r-z)\sqrt{2\alpha})\,\mathrm{ch}(z\sqrt{2\alpha})} - \frac{\sqrt{\alpha} e^{(r-z)\sqrt{2\alpha}} y}{\sqrt{2}\,\mathrm{sh}((r-z)\sqrt{2\alpha})}\Big)$$

3 $|W|$ $\varrho = \varrho(v,z) = \min\{s : \ell(s,z) = v\}$

$$\times\begin{cases} \dfrac{\operatorname{ch}(x\sqrt{2\alpha})\sqrt{\alpha v}}{\sqrt{2}\operatorname{sh}((r-z)\sqrt{2\alpha})\operatorname{ch}(z\sqrt{2\alpha})\sqrt{y}} I_1\Big(\dfrac{\sqrt{2\alpha v y}}{\operatorname{sh}((r-z)\sqrt{2\alpha})}\Big)dy, & 0 \le x \le z \\ \dfrac{\sqrt{\alpha}}{\sqrt{2}\operatorname{sh}^2((r-z)\sqrt{2\alpha})}\Big[\operatorname{sh}((r-x)\sqrt{2\alpha})\dfrac{\sqrt{v}}{\sqrt{y}} I_1\Big(\dfrac{\sqrt{2\alpha v y}}{\operatorname{sh}((r-z)\sqrt{2\alpha})}\Big) & \\ \qquad + \operatorname{sh}((x-z)\sqrt{2\alpha}) I_0\Big(\dfrac{\sqrt{2\alpha v y}}{\operatorname{sh}((r-z)\sqrt{2\alpha})}\Big)\Big]dy, & z \le x \le r \\ \dfrac{\sqrt{\alpha}e^{-(x-r)\sqrt{2\alpha}}}{\sqrt{2}\operatorname{sh}((r-z)\sqrt{2\alpha})} I_0\Big(\dfrac{\sqrt{2\alpha v y}}{\operatorname{sh}((r-z)\sqrt{2\alpha})}\Big)dy, & r \le x \end{cases}$$

$r \le z$

$$= \exp\Big(-\frac{\sqrt{\alpha}e^{(z-r)\sqrt{2\alpha}}v}{\sqrt{2}\operatorname{sh}((z-r)\sqrt{2\alpha})} - \frac{\sqrt{\alpha}\operatorname{ch}(z\sqrt{2\alpha})y}{\sqrt{2}\operatorname{sh}((z-r)\sqrt{2\alpha})\operatorname{ch}(r\sqrt{2\alpha})}\Big)$$

$$\times\begin{cases} \dfrac{e^{-(x-z)\sqrt{2\alpha}}\sqrt{\alpha v}}{\sqrt{2}\operatorname{sh}((z-r)\sqrt{2\alpha})\sqrt{y}} I_1\Big(\dfrac{\sqrt{2\alpha v y}}{\operatorname{sh}((z-r)\sqrt{2\alpha})}\Big)dy, & z \le x \\ \dfrac{\sqrt{\alpha}}{\sqrt{2}\operatorname{sh}^2((z-r)\sqrt{2\alpha})}\Big[\operatorname{sh}((x-r)\sqrt{2\alpha})\sqrt{\dfrac{v}{y}} I_1\Big(\dfrac{\sqrt{2\alpha v y}}{\operatorname{sh}((z-r)\sqrt{2\alpha})}\Big) & \\ \qquad + \operatorname{sh}((z-x)\sqrt{2\alpha}) I_0\Big(\dfrac{\sqrt{2\alpha v y}}{\operatorname{sh}((z-r)\sqrt{2\alpha})}\Big)\Big]dy, & r \le x \le z \\ \dfrac{\sqrt{\alpha}\operatorname{ch}(x\sqrt{2\alpha})}{\sqrt{2}\operatorname{sh}((z-r)\sqrt{2\alpha})\operatorname{ch}(r\sqrt{2\alpha})} I_0\Big(\dfrac{\sqrt{2\alpha v y}}{\operatorname{sh}((z-r)\sqrt{2\alpha})}\Big)dy, & 0 \le x \le r \end{cases}$$

(1) $$\mathbf{E}_x\{e^{-\alpha\varrho};\ \ell(\varrho,r)=0\} = \exp\Big(-\frac{v\sqrt{2\alpha}(e^{(|r-z|+z)\sqrt{2\alpha}} + e^{-r\sqrt{2\alpha}})}{4\operatorname{ch}(z\sqrt{2\alpha})\operatorname{sh}(|z-r|\sqrt{2\alpha})}\Big)$$

$$\times\begin{cases} \dfrac{\operatorname{ch}(x\sqrt{2\alpha})}{\operatorname{ch}(z\sqrt{2\alpha})}, & 0 \le x \le z \le r \\ e^{-(x-z)\sqrt{2\alpha}}, & r \le z \le x \\ \dfrac{\operatorname{sh}(|x-r|\sqrt{2\gamma})}{\operatorname{sh}(|z-r|\sqrt{2\gamma})}, & z \wedge r \le x \le z \vee r \\ 0, & 0 \le x \le r \le z, \quad z \le r \le x \end{cases}$$

4.4.1 $$\mathbf{E}_x \exp\Big(-\gamma\int_0^\varrho \mathbb{I}_{[r,\infty)}(|W_s|)ds\Big)$$

$z \le r$

$$= \begin{cases} \exp\Big(-\dfrac{v\sqrt{2\gamma}}{2(1+(r-z)\sqrt{2\gamma})}\Big), & 0 \le x \le r \\ \dfrac{1+(r-x)\sqrt{2\gamma}}{1+(r-z)\sqrt{2\gamma}} \exp\Big(-\dfrac{v\sqrt{2\gamma}}{2(1+(r-z)\sqrt{2\gamma})}\Big), & z \le x \le r \\ \dfrac{1}{1+(r-z)\sqrt{2\gamma}} \exp\Big(-(x-r)\sqrt{2\gamma} - \dfrac{v\sqrt{2\gamma}}{2(1+(r-z)\sqrt{2\gamma})}\Big), & r \le x \end{cases}$$

3 $|W|$ $\quad\varrho = \varrho(v,z) = \min\{s : \ell(s,z) = v\}$

$r \le z$

$$= \begin{cases} \dfrac{1}{\operatorname{ch}((z-r)\sqrt{2\gamma}}\exp\Big(-\dfrac{v\sqrt{2\gamma}}{\exp(2(r-z)\sqrt{2\gamma})+1}\Big), & 0 \le x \le r \\ \dfrac{\operatorname{ch}((x-r)\sqrt{2\gamma}}{\operatorname{ch}((z-r)\sqrt{2\gamma}}\exp\Big(-\dfrac{v\sqrt{2\gamma}}{\exp(2(r-z)\sqrt{2\gamma})+1}\Big), & r \le x \le z \\ \exp\Big((z-x)\sqrt{2\gamma} - \dfrac{v\sqrt{2\gamma}}{\exp(2(r-z)\sqrt{2\gamma})+1}\Big), & z \le x \end{cases}$$

4.4.4 (1) $\quad \mathbf{E}_x\Big\{e^{-\alpha\varrho};\ \int_0^{\varrho} \mathbb{I}_{[r,\infty)}\big(|W_s|\big)ds = 0\Big\}$

$z \le r$

$$= \exp\Big(-\frac{v\sqrt{2\alpha}\operatorname{ch}(r\sqrt{2\alpha})}{2\operatorname{ch}(z\sqrt{2\alpha})\operatorname{sh}((z-r)\sqrt{2\alpha})}\Big)\begin{cases} \dfrac{\operatorname{ch}(x\sqrt{2\alpha})}{\operatorname{ch}(z\sqrt{2\alpha})}, & 0 \le x \le z \\ \dfrac{\operatorname{sh}((r-x)\sqrt{2\alpha})}{\operatorname{sh}((r-z)\sqrt{2\alpha})}, & z \le x \le r \\ 0, & r \le x \end{cases}$$

4.5.1 $\quad \mathbf{E}_x \exp\Big(-\gamma\int_0^{\varrho} \mathbb{I}_{[0,r]}\big(|W_s|\big)ds\Big)$

$z \le r$

$$= \exp\Big(-\frac{v\sqrt{2\gamma}\operatorname{sh}(r\sqrt{2\gamma})}{2\operatorname{ch}(z\sqrt{2\gamma})\operatorname{ch}((r-z)\sqrt{2\gamma})}\Big)\begin{cases} \dfrac{\operatorname{ch}(x\sqrt{2\gamma})}{\operatorname{ch}(z\sqrt{2\gamma})}, & 0 \le x \le z \\ \dfrac{\operatorname{ch}((r-x)\sqrt{2\gamma})}{\operatorname{ch}((r-z)\sqrt{2\gamma})}, & z \le x \le r \\ \dfrac{1}{\operatorname{ch}((r-z)\sqrt{2\gamma})}, & r \le x \end{cases}$$

$r \le z$

$$= \exp\Big(-\frac{v\sqrt{2\gamma}\operatorname{th}(r\sqrt{2\gamma})}{2(1+\sqrt{2\gamma}(z-r)\operatorname{th}(r\sqrt{2\gamma}))}\Big)$$

$$\times\begin{cases} \dfrac{\operatorname{ch}(x\sqrt{2\gamma})}{\operatorname{ch}(r\sqrt{2\gamma})+\sqrt{2\gamma}(z-r)\operatorname{sh}(r\sqrt{2\gamma})}, & 0 \le x \le r \\ \dfrac{\operatorname{ch}(r\sqrt{2\gamma})+\sqrt{2\gamma}(x-r)\operatorname{sh}(r\sqrt{2\gamma})}{\operatorname{ch}(r\sqrt{2\gamma})+\sqrt{2\gamma}(z-r)\operatorname{sh}(r\sqrt{2\gamma})}, & r \le x \le z \\ 1 & z \le x \end{cases}$$

4.5.4 (1) $\quad \mathbf{E}_x\Big\{e^{-\alpha\varrho};\ \int_0^{\varrho} \mathbb{I}_{[0,r]}\big(|W_s|\big)ds = 0\Big\}$

3 $|W|$ $\qquad\varrho = \varrho(v, z) = \min\{s : \ell(s, z) = v\}$

$r \le z$

$$= \exp\Big(-\frac{v\sqrt{2\alpha}e^{\sqrt{2\alpha}(z-r)}}{2\,\mathrm{sh}((z-r)\sqrt{2\alpha})}\Big)\begin{cases} 0, & x \le r \\ \dfrac{\mathrm{sh}((x-r)\sqrt{2\alpha})}{\mathrm{sh}((z-r)\sqrt{2\alpha})}, & r \le x \le z \\ e^{-(x-z)\sqrt{2\alpha}}, & z \le x \end{cases}$$

4.6.1 $\quad \mathbf{E}_x \exp\Big(-\int_0^{\varrho}\big(p\mathbb{1}_{[0,r)}(|W_s|) + q\mathbb{1}_{[r,\infty)}(|W_s|)\big)ds\Big)$

$z \le r$

$$= \exp\Big(-\frac{v\sqrt{p}(\sqrt{p}\,\mathrm{sh}(r\sqrt{2p}) + \sqrt{q}\,\mathrm{ch}(r\sqrt{2p}))}{\mathrm{ch}(z\sqrt{2p})(\sqrt{2p}\,\mathrm{ch}((r-z)\sqrt{2p}) + \sqrt{2q}\,\mathrm{sh}((r-z)\sqrt{2p}))}\Big)$$

$$\times\begin{cases} \dfrac{\mathrm{ch}(x\sqrt{2p})}{\mathrm{ch}(z\sqrt{2p})}, & 0 \le x \le z \\ \dfrac{\sqrt{p}\,\mathrm{ch}((r-x)\sqrt{2p}) + \sqrt{q}\,\mathrm{sh}((r-x)\sqrt{2p})}{\sqrt{p}\,\mathrm{ch}((r-z)\sqrt{2p}) + \sqrt{q}\,\mathrm{sh}((r-z)\sqrt{2p})}, & z \le x \le r \\ \dfrac{\sqrt{p}e^{-(x-z)\sqrt{2q}}}{\sqrt{p}\,\mathrm{ch}((r-z)\sqrt{2p}) + \sqrt{q}\,\mathrm{sh}((r-z)\sqrt{2p})}, & r \le x \end{cases}$$

$r \le z$

$$= \exp\Big(-\frac{v\sqrt{q}(\sqrt{p}\,\mathrm{sh}(r\sqrt{2p}) + \sqrt{q}\,\mathrm{ch}(r\sqrt{2p}))e^{-(z-r)\sqrt{2q}}}{\sqrt{2q}\,\mathrm{ch}(r\sqrt{2p})\,\mathrm{ch}((z-r)\sqrt{2q}) + \sqrt{2p}\,\mathrm{sh}((z-r)\sqrt{2q})\,\mathrm{sh}(r\sqrt{2p})}\Big)$$

$$\times\begin{cases} \dfrac{\sqrt{q}\,\mathrm{ch}(x\sqrt{2p})}{\sqrt{q}\,\mathrm{ch}(r\sqrt{2p})\,\mathrm{ch}((z-r)\sqrt{2q}) + \sqrt{p}\,\mathrm{sh}(r\sqrt{2p})\,\mathrm{sh}((z-r)\sqrt{2q})}, & x \le r \\ \dfrac{\sqrt{q}\,\mathrm{ch}(r\sqrt{2p})\,\mathrm{ch}((x-r)\sqrt{2q}) + \sqrt{p}\,\mathrm{sh}(r\sqrt{2p})\,\mathrm{sh}((x-r)\sqrt{2q})}{\sqrt{q}\,\mathrm{ch}(r\sqrt{2p})\,\mathrm{ch}((z-r)\sqrt{2q}) + \sqrt{p}\,\mathrm{sh}(r\sqrt{2p})\,\mathrm{sh}((z-r)\sqrt{2q})}, & r \le x \le z \\ e^{-(x-z)\sqrt{2q}}, & z \le x \end{cases}$$

4.7.1 $\quad \mathbf{E}_x \exp\Big(-\gamma\int_0^{\varrho}\mathbb{1}_{[r,u]}(|W_s|)ds\Big)$

$r < u$

$z \le r$

$$= \exp\Big(-\frac{v\sqrt{2\gamma}\,\mathrm{sh}((u-r)\sqrt{2\gamma})}{2((r-x)\sqrt{2\gamma}\,\mathrm{sh}((u-r)\sqrt{2\gamma}) + \mathrm{ch}((u-r)\sqrt{2\gamma}))}\Big)$$

$$\times\begin{cases} 1, & 0 \le x \le z \\ \dfrac{(r-x)\sqrt{2\gamma}\,\mathrm{sh}((u-r)\sqrt{2\gamma}) + \mathrm{ch}((u-r)\sqrt{2\gamma})}{(r-z)\sqrt{2\gamma}\,\mathrm{sh}((u-r)\sqrt{2\gamma}) + \mathrm{ch}((u-r)\sqrt{2\gamma})}, & z \le x \le r \\ \dfrac{\mathrm{ch}((u-x)\sqrt{2\gamma})}{(r-z)\sqrt{2\gamma}\,\mathrm{sh}(u-r)\sqrt{2\gamma}) + \mathrm{ch}((u-r)\sqrt{2\gamma})}, & r \le x \le u \\ \dfrac{1}{(r-z)\sqrt{2\gamma}\,\mathrm{sh}(u-r)\sqrt{2\gamma}) + \mathrm{ch}((u-r)\sqrt{2\gamma})}, & u \le x \end{cases}$$

3 $|W|$ $\varrho = \varrho(v,z) = \min\{s : \ell(s,z) = v\}$

$u \le z$

$$= \exp\Big(-\frac{v\sqrt{2\gamma}\,\mathrm{sh}((u-r)\sqrt{2\gamma})}{2((z-u)\sqrt{2\gamma}\,\mathrm{sh}((u-r)\sqrt{2\gamma}) + \mathrm{ch}((u-r)\sqrt{2\gamma}))}\Big)$$

$$\times\begin{cases} \dfrac{1}{(z-u)\sqrt{2\gamma}\,\mathrm{sh}((u-r)\sqrt{2\gamma}) + \mathrm{ch}((u-r)\sqrt{2\gamma})}, & 0 \le x \le r \\ \dfrac{\mathrm{ch}((x-r)\sqrt{2\gamma})}{(z-u)\sqrt{2\gamma}\,\mathrm{sh}((u-r)\sqrt{2\gamma}) + \mathrm{ch}((u-r)\sqrt{2\gamma})}, & r \le x \le u \\ \dfrac{(x-u)\sqrt{2\gamma}\,\mathrm{sh}((u-r)\sqrt{2\gamma}) + \mathrm{ch}((u-r)\sqrt{2\gamma})}{(z-u)\sqrt{2\gamma}\,\mathrm{sh}((u-r)\sqrt{2\gamma}) + \mathrm{ch}((u-r)\sqrt{2\gamma})}, & u \le x \le z \\ 1, & z \le x \end{cases}$$

4.8.1 $\mathbf{E}_x \exp\Big(-\gamma\int_0^\varrho |W_s|\,ds\Big)$

$$= \exp\Big(-\frac{3z^{-1}v}{4I_{-1/3}(\frac{2}{3}z^{3/2}\sqrt{2\gamma})K_{1/3}(\frac{2}{3}z^{3/2}\sqrt{2\gamma})}\Big)\begin{cases} \dfrac{\sqrt{x}I_{-1/3}(\frac{2}{3}x^{3/2}\sqrt{2\gamma})}{\sqrt{z}I_{-1/3}(\frac{2}{3}z^{3/2}\sqrt{2\gamma})}, & 0 \le x \le z \\ \dfrac{\sqrt{x}K_{1/3}(\frac{2}{3}x^{3/2}\sqrt{2\gamma})}{\sqrt{z}K_{1/3}(\frac{2}{3}z^{3/2}\sqrt{2\gamma})}, & z \le x \end{cases}$$

4.9.1 $\mathbf{E}_x \exp\Big(-2\gamma\int_0^\varrho |W_s|^2\,ds\Big)$

$$= \exp\Big(-\frac{v}{zI_{-1/4}(z^2\sqrt{\gamma})K_{1/4}(z^2\sqrt{\gamma})}\Big)\begin{cases} \dfrac{\sqrt{x}I_{-1/4}(x^2\sqrt{\gamma})}{\sqrt{z}I_{-1/4}(z^2\sqrt{\gamma})}, & 0 \le x \le z \\ \dfrac{\sqrt{x}K_{1/4}(x^2\sqrt{\gamma})}{\sqrt{z}K_{1/4}(z^2\sqrt{\gamma})}, & z \le x \end{cases}$$

4.10.1 $\mathbf{E}_x \exp\Big(-\gamma\int_0^\varrho e^{2\beta|W_s|}\,ds\Big)$

$\beta > 0$

$$= \exp\Big(-\frac{\beta K_1\big(\frac{\sqrt{2\gamma}}{|\beta|}\big)v}{2C_0\big(\frac{\sqrt{2\gamma}}{|\beta|}, \frac{\sqrt{2\gamma}}{|\beta|}e^{\beta z}\big)K_0\big(\frac{\sqrt{2\gamma}}{|\beta|}e^{\beta z}\big)}\Big)\begin{cases} \dfrac{C_0\big(\frac{\sqrt{2\gamma}}{|\beta|}, \frac{\sqrt{2\gamma}}{|\beta|}e^{\beta x}\big)}{C_0\big(\frac{\sqrt{2\gamma}}{|\beta|}, \frac{\sqrt{2\gamma}}{|\beta|}e^{\beta z}\big)}, & 0 \le x \le z \\ \dfrac{K_0\big(\frac{\sqrt{2\gamma}}{|\beta|}e^{\beta x}\big)}{K_0\big(\frac{\sqrt{2\gamma}}{|\beta|}e^{\beta z}\big)}, & z \le x \end{cases}$$

$\beta < 0$

$$= \exp\Big(-\frac{\beta I_1\big(\frac{\sqrt{2\gamma}}{|\beta|}\big)v}{2C_0\big(\frac{\sqrt{2\gamma}}{|\beta|}, \frac{\sqrt{2\gamma}}{|\beta|}e^{\beta z}\big)I_0\big(\frac{\sqrt{2\gamma}}{|\beta|}e^{\beta z}\big)}\Big)\begin{cases} \dfrac{C_0\big(\frac{\sqrt{2\gamma}}{|\beta|}, \frac{\sqrt{2\gamma}}{|\beta|}e^{\beta x}\big)}{C_0\big(\frac{\sqrt{2\gamma}}{|\beta|}, \frac{\sqrt{2\gamma}}{|\beta|}e^{\beta z}\big)}, & 0 \le x \le z \\ \dfrac{I_0\big(\frac{\sqrt{2\gamma}}{|\beta|}e^{\beta x}\big)}{I_0\big(\frac{\sqrt{2\gamma}}{|\beta|}e^{\beta z}\big)}, & z \le x \end{cases}$$

3 $|W|$ $\varrho = \varrho(v,z) = \min\{s : \ell(s,z) = v\}$

4.11.2
$$\lambda \int_0^\infty e^{-\lambda z} \mathbf{P}_0\Big(\sup_{y\in[0,\infty)} \ell(\varrho, y) < h\Big) dz = \Big(1 - \frac{v}{h}\Big)\Big(1 - \frac{I_0(\sqrt{2\lambda v})}{I_0(\sqrt{2\lambda h})}\Big) \mathbb{I}_{[0,h]}(v)$$

(1)
$$\mathbf{P}_0\Big(\sup_{y\in[0,\infty)} \ell(\varrho, y) < h\Big) = 2\Big(1 - \frac{v}{h}\Big) \sum_{k=1}^\infty \frac{e^{-j_{0,k}^2 z/2h} J_0(j_{0,k}\sqrt{v/h})}{j_{0,k} J_1(j_{0,k})} \mathbb{I}_{[0,h]}(v)$$

(2)
$$\mathbf{P}_0\Big(\sup_{y\in[0,\infty)} \ell(\varrho(v,0), y) < h\Big) = \Big(1 - \frac{v}{h}\Big) \mathbb{I}_{[0,h]}(v)$$

4.13.1
$$\mathbf{E}_x\Big\{e^{-\gamma \check{H}(\varrho)}; \sup_{0\le s\le \varrho} |W_s| \in dy\Big\} = \frac{\operatorname{sh}((x-z)^+\sqrt{2\gamma})}{(y-z)\operatorname{sh}((y-z)\sqrt{2\gamma})} \exp\Big(-\frac{v}{2(y-z)}\Big) dy$$
$$+ \frac{\exp\Big(-\frac{v\sqrt{2\gamma}\operatorname{ch}(y\sqrt{2\gamma})}{2\operatorname{sh}((z-y)\sqrt{2\gamma})\operatorname{ch}(z\sqrt{2\gamma})}\Big) - \exp\Big(-\frac{v}{2(y-z)}\Big)}{\operatorname{sh}((y-z)\sqrt{2\gamma})\operatorname{ch}(z\sqrt{2\gamma}) - (y-z)\sqrt{2\gamma}\operatorname{ch}(y\sqrt{2\gamma})}$$
$$\times \begin{cases} \sqrt{2\gamma}\operatorname{ch}(x\sqrt{2\gamma}) dy, & 0 \le x \le z \\ \dfrac{\sqrt{2\gamma}\operatorname{sh}((y-x)\sqrt{2\gamma})\operatorname{ch}(z\sqrt{2\gamma})}{\operatorname{sh}((y-z)\sqrt{2\gamma})} dy, & z \le x \le y \end{cases}$$

4.18.1 $r < u$
$$\mathbf{E}_x \exp\big(-\gamma \ell(\varrho, r) - \eta \ell(\varrho, u)\big)$$

$z \le r$
$$= \exp\Big(-\frac{v(\gamma + \eta + 2\gamma\eta(u-r))}{1 + 2\gamma(r-z) + 2\eta(u-z) + 4\gamma\eta(u-r)(r-z)}\Big)$$
$$\times \begin{cases} 1, & 0 \le x \le z \\ \dfrac{1 + 2\gamma(r-x) + 2\eta(u-x) + 4\gamma\eta(u-r)(r-x)}{1 + 2\gamma(r-z) + 2\eta(u-z) + 4\gamma\eta(u-r)(r-z)}, & z \le x \le r \\ \dfrac{1 + 2\eta(u-x)}{1 + 2\gamma(r-z) + 2\eta(u-z) + 4\gamma\eta(u-r)(r-z)}, & r \le x \le u \\ \dfrac{1}{1 + 2\gamma(r-z) + 2\eta(u-z) + 4\gamma\eta(u-r)(r-z)}, & u \le x \end{cases}$$

$u \le z$
$$= \exp\Big(-\frac{v(\gamma + \eta + 2\gamma\eta(u-r))}{1 + 2\gamma(z-r) + 2\eta(z-u) + 4\gamma\eta(z-u)(u-r)}\Big)$$
$$\times \begin{cases} \dfrac{1}{1 + 2\gamma(z-r) + 2\eta(z-u) + 4\gamma\eta(z-u)(u-r)}, & 0 \le x \le r \\ \dfrac{1 + 2\gamma(x-r))}{1 + 2\gamma(z-r) + 2\eta(z-u) + 4\gamma\eta(z-u)(u-r)}, & r \le x \le u \\ \dfrac{1 + 2\gamma(x-r) + 2\eta(x-u) + 4\gamma\eta(x-u)(u-r)}{1 + 2\gamma(z-r) + 2\eta(z-u) + 4\gamma\eta(z-u)(u-r)}, & u \le x \le z \\ 1, & z \le x \end{cases}$$

3 $|W|$ $\varrho=\varrho(v,z)=\min\{s:\ell(s,z)=v\}$

4.27.1 $\mathbf{E}_x\Big\{\exp\Big(-\int_0^{\varrho}\big(p\mathbb{I}_{[0,r)}\big(|W_s|\big)+q\mathbb{I}_{[r,\infty)}\big(|W_s|\big)\big)ds\Big);\ell\big(\varrho,r\big)\in dy\Big\}$

$z\le r$
$$=\exp\Big(-\frac{v\sqrt{p}\,\mathrm{sh}(z\sqrt{2p})}{\sqrt{2}\,\mathrm{ch}(z\sqrt{2p})}-\frac{(y+v)\sqrt{p}\,\mathrm{ch}((r-z)\sqrt{2p})}{\sqrt{2}\,\mathrm{sh}((r-z)\sqrt{2p})}-\frac{y\sqrt{q}}{\sqrt{2}}\Big)$$
$$\times\begin{cases}\dfrac{\sqrt{pv}\,\mathrm{ch}(x\sqrt{2p})}{\sqrt{2}\,\mathrm{sh}((r-z)\sqrt{2p})\,\mathrm{ch}(z\sqrt{2p})\sqrt{y}}I_1\Big(\dfrac{\sqrt{2pvy}}{\mathrm{sh}((r-z)\sqrt{2p})}\Big)dy, & 0\le x\le z\\ \dfrac{\sqrt{p}}{\sqrt{2}\,\mathrm{sh}^2((r-z)\sqrt{2p})}\Big[\mathrm{sh}((r-x)\sqrt{2p})\dfrac{\sqrt{v}}{\sqrt{y}}I_1\Big(\dfrac{\sqrt{2pvy}}{\mathrm{sh}((r-z)\sqrt{2p})}\Big) & \\ \qquad+\mathrm{sh}((x-z)\sqrt{2p})I_0\Big(\dfrac{\sqrt{2pvy}}{\mathrm{sh}((r-z)\sqrt{2p})}\Big)\Big]dy, & z\le x\le r\\ \dfrac{\sqrt{p}e^{-(x-r)\sqrt{2q}}}{\sqrt{2}\,\mathrm{sh}((r-z)\sqrt{2p})}I_0\Big(\dfrac{\sqrt{2pvy}}{\mathrm{sh}((r-z)\sqrt{2p})}\Big)dy, & r\le x\end{cases}$$

$r\le z$
$$=\exp\Big(-\frac{v\sqrt{q}}{\sqrt{2}}-\frac{(v+y)\sqrt{q}\,\mathrm{ch}((z-r)\sqrt{2q})}{\sqrt{2}\,\mathrm{sh}((z-r)\sqrt{2q})}-\frac{y\sqrt{p}\,\mathrm{sh}(r\sqrt{2p})}{\sqrt{2}\,\mathrm{ch}(r\sqrt{2p})}\Big)$$
$$\times\begin{cases}\dfrac{\sqrt{q}\,\mathrm{ch}(x\sqrt{2p})}{\sqrt{2}\,\mathrm{sh}((z-r)\sqrt{2q})\,\mathrm{ch}(r\sqrt{2p})}I_0\Big(\dfrac{\sqrt{2qvy}}{\mathrm{sh}((z-r)\sqrt{2q})}\Big)dy, & 0\le x\le r\\ \dfrac{\sqrt{q}}{\sqrt{2}\,\mathrm{sh}^2((z-r)\sqrt{2q})}\Big[\mathrm{sh}((x-r)\sqrt{2q})\dfrac{\sqrt{v}}{\sqrt{y}}I_1\Big(\dfrac{\sqrt{2qvy}}{\mathrm{sh}((z-r)\sqrt{2q})}\Big) & \\ \qquad+\mathrm{sh}((z-x)\sqrt{2q})I_0\Big(\dfrac{\sqrt{2qvy}}{\mathrm{sh}((z-r)\sqrt{2q})}\Big)\Big]dy, & r\le x\le z\\ \dfrac{\sqrt{qv}e^{-(x-z)\sqrt{2q}}}{\sqrt{2}\,\mathrm{sh}((z-r)\sqrt{2q})\sqrt{y}}I_1\Big(\dfrac{\sqrt{2qvy}}{\mathrm{sh}((z-r)\sqrt{2q})}\Big)dy, & z\le x\end{cases}$$

(1) $\mathbf{E}_x\Big\{\exp\Big(-\int_0^{\varrho}\big(p\mathbb{I}_{[0,r)}\big(|W_s|\big)+q\mathbb{I}_{[r,\infty)}\big(|W_s|\big)\big)ds\Big);\ell\big(\varrho,r\big)=0\Big\}$

$z\le r$
$$=\exp\Big(-\frac{v\sqrt{p}\,\mathrm{ch}(r\sqrt{2p})}{\sqrt{2}\,\mathrm{sh}((r-z)\sqrt{2p})\,\mathrm{ch}(z\sqrt{2p})}\Big)\begin{cases}\dfrac{\mathrm{ch}(x\sqrt{2p})}{\mathrm{ch}(z\sqrt{2p})}, & 0\le x\le z\\ \dfrac{\mathrm{sh}((r-x)\sqrt{2p})}{\mathrm{sh}((r-z)\sqrt{2p})}, & z\le x\le r\\ 0, & r\le x\end{cases}$$

3	$\lvert W\rvert$ $\qquad\qquad \varrho=\varrho(v,z)=\min\{s:\ell(s,z)=v\}$
$r\le z$	$$=\exp\Big(-\frac{v\sqrt{q}e^{(z-r)\sqrt{2q}}}{\sqrt{2}\,\mathrm{sh}((z-r)\sqrt{2q})}\Big)\begin{cases}0, & 0\le x\le r\\ \dfrac{\mathrm{sh}((x-r)\sqrt{2q})}{\mathrm{sh}((z-r)\sqrt{2q})}, & r\le x\le z\\ e^{-(x-z)\sqrt{2q}}, & z\le x\end{cases}$$
4.27.2	$$\mathbf{P}_x\Big(\int_0^\varrho \mathbb{I}_{[0,r)}(\lvert W_s\rvert)ds\in du,\ \int_0^\varrho \mathbb{I}_{[r,\infty)}(\lvert W_s\rvert)ds\in dg,\ \frac{1}{2}\ell(\varrho,r)\in dy\Big)$$
$x\le z$ $z\le r$	$$=\frac{\sqrt{v}}{\sqrt{y}}(\mathrm{is}_u(1,r-z,0,y+v,\sqrt{vy})*\mathrm{ecc}_u(x,z,0,v))\,\mathrm{h}_g(1,y)dudgdy$$
$z\le r$ $r\le x$	$$=(\mathrm{is}_u(0,r-z,0,y+v,\sqrt{vy})*\mathrm{ec}_u(0,0,z,0,v))\,\mathrm{h}_g(1,x-r+y)dudgdy$$
$x\le r$ $r\le z$	$$=\mathrm{ecc}_u(x,r,0,y)\,\mathrm{is}_g(0,z-r,v,y+v,\sqrt{vy})dudgdy$$
$r\le z$ $z\le x$	$$=\frac{\sqrt{v}}{\sqrt{y}}\mathrm{ec}_u(0,0,r,0,y)\,\mathrm{is}_g(1,z-r,x-z+v,y+v,\sqrt{vy})dudgdy$$
(1)	$$\mathbf{P}_x\Big(\int_0^\varrho \mathbb{I}_{[0,r)}(\lvert W_s\rvert)ds\in du,\ \int_0^\varrho \mathbb{I}_{[r,\infty)}(\lvert W_s\rvert)ds=0,\ \ell(\varrho,r)=0\Big)$$
	$$=\begin{cases}\mathrm{ecc}_u(x,z,0,v/2)*\mathrm{es}_u(0,0,r-z,0,v/2)du, & 0\le x\le z\\ \mathrm{ec}_u(0,0,z,0,v/2)*\mathrm{ess}_u(r-x,r-z,0,v/2)du, & z\le x\le r\end{cases}$$
(2)	$$\mathbf{P}_x\Big(\int_0^\varrho \mathbb{I}_{[0,r)}(\lvert W_s\rvert)ds=0,\ \int_0^\varrho \mathbb{I}_{[r,\infty)}(\lvert W_s\rvert)ds\in dg,\ \ell(\varrho,r)=0\Big)$$
	$$=\begin{cases}\mathrm{ess}_g(x-r,z-r,v/2,v/2)dg, & r\le x\le z\\ \mathrm{es}_g(0,0,z-r,x-z+v/2,v/2)dg, & z\le x\end{cases}$$

4. BESSEL PROCESS OF ORDER ν

1. Exponential stopping

1.0.5 $$\mathbf{P}_x(R_\tau^{(n)} \in dz) = \begin{cases} 2\lambda z^{\nu+1} x^{-\nu} I_\nu(x\sqrt{2\lambda}) K_\nu(z\sqrt{2\lambda}) dz, & 0 \le x \le z \\ 2\lambda z^{\nu+1} x^{-\nu} K_\nu(x\sqrt{2\lambda}) I_\nu(z\sqrt{2\lambda}) dz, & z \le x \end{cases}$$

or $$= 2\lambda z^{\nu+1} x^{-\nu} I_\nu((z + x - |z-x|)\sqrt{\lambda/2}) K_\nu((z + x + |z-x|)\sqrt{\lambda/2}) dz$$

1.0.6 $$\mathbf{P}_x\big(R_t^{(n)} \in dz\big) = z^{\nu+1} x^{-\nu} t^{-1} e^{-(x^2+z^2)/2t} I_\nu(xz/t) dz$$

1.1.2 $$\mathbf{P}_x\Big(\sup_{0\le s\le\tau} R_s^{(n)} \ge y\Big) = \frac{x^{-\nu} I_\nu(x\sqrt{2\lambda})}{y^{-\nu} I_\nu(y\sqrt{2\lambda})}, \qquad 0 \le x \le y$$

1.1.4 $$\mathbf{P}_x\Big(\sup_{0\le s\le t} R_s^{(n)} \ge y\Big) = 1 - \sum_{k=1}^{\infty} \frac{2x^{-\nu} J_\nu(j_{\nu,k} x/y)}{j_{\nu,k} y^{-\nu} J_{\nu+1}(j_{\nu,k})} e^{-j_{\nu,k}^2 t/2y^2}, \qquad 0 \le x \le y$$

1.1.6 $$\mathbf{P}_x\Big(\sup_{0\le s\le\tau} R_s^{(n)} \ge y, R_\tau^{(n)} \in dz\Big)$$
$$= \frac{2\lambda z^{\nu+1} x^{-\nu}}{I_\nu(y\sqrt{2\lambda})} I_\nu(x\sqrt{2\lambda}) I_\nu(z\sqrt{2\lambda}) K_\nu(y\sqrt{2\lambda}) dz, \qquad x \vee z \le y$$

1.2.2 $$\mathbf{P}_x\Big(\inf_{0\le s\le\tau} R_s^{(n)} \le y\Big) = \frac{x^{-\nu} K_\nu(x\sqrt{2\lambda})}{y^{-\nu} K_\nu(y\sqrt{2\lambda})}$$
$y \le x$

1.2.4 $$\mathbf{P}_x\Big(\inf_{0\le s<\infty} R_s^{(n)} \le y\Big) = \frac{x^{-2\nu}}{y^{-2\nu}} \qquad y \le x$$
(1)

1.2.6 $$\mathbf{P}_x\Big(\inf_{0\le s\le\tau} R_s^{(n)} \le y, R_\tau^{(n)} \in dz\Big)$$
$$= \frac{2\lambda z^{\nu+1} x^{-\nu}}{K_\nu(y\sqrt{2\lambda})} K_\nu(x\sqrt{2\lambda}) K_\nu(z\sqrt{2\lambda}) I_\nu(y\sqrt{2\lambda}) dz, \qquad y \le x \wedge z$$

1.3.1 $$\mathbf{E}_x e^{-\gamma\ell(\tau,r)} = \begin{cases} 1 - \dfrac{2\gamma r^{\nu+1} x^{-\nu} I_\nu(x\sqrt{2\lambda}) K_\nu(r\sqrt{2\lambda})}{1 + 2\gamma r I_\nu(r\sqrt{2\lambda}) K_\nu(r\sqrt{2\lambda})}, & 0 \le x \le r \\ 1 - \dfrac{2\gamma r^{\nu+1} x^{-\nu} K_\nu(x\sqrt{2\lambda}) I_\nu(r\sqrt{2\lambda})}{1 + 2\gamma r I_\nu(r\sqrt{2\lambda}) K_\nu(r\sqrt{2\lambda})}, & r \le x \end{cases}$$

4 $R_s^{(n)}$ $\nu \ge 0,\ n = 2\nu + 2 \ge 2,$ $\tau \sim \mathrm{Exp}(\lambda)$, independent of $R^{(n)}$

1.3.2 $$\mathbf{P}_x\big(\ell(\tau,r) \in dy\big) = \exp\Big(-\frac{yI_\nu^{-1}(r\sqrt{2\lambda})}{2rK_\nu(r\sqrt{2\lambda})}\Big)\begin{cases} \dfrac{r^{\nu-1}x^{-\nu}I_\nu(x\sqrt{2\lambda})}{2I_\nu^2(r\sqrt{2\lambda})K_\nu(r\sqrt{2\lambda})}dy, & 0 \le x \le r \\ \dfrac{r^{\nu-1}x^{-\nu}K_\nu(x\sqrt{2\lambda})}{2I_\nu(r\sqrt{2\lambda})K_\nu^2(r\sqrt{2\lambda})}dy, & r \le x \end{cases}$$

(1) $$\mathbf{P}_x\big(\ell(\tau,r) = 0\big) = \begin{cases} 1 - \dfrac{x^{-\nu}I_\nu(x\sqrt{2\lambda})}{r^{-\nu}I_\nu(r\sqrt{2\lambda})}, & 0 \le x \le r \\ 1 - \dfrac{x^{-\nu}K_\nu(x\sqrt{2\lambda})}{r^{-\nu}K_\nu(r\sqrt{2\lambda})}, & r \le x \end{cases}$$

1.3.3 (1) $$\mathbf{E}_x e^{-\gamma\ell(\infty,r)} = \begin{cases} 1 - \dfrac{\gamma r}{\nu + \gamma r}, & 0 \le x \le r \\ 1 - \dfrac{\gamma r^{2\nu+1}x^{-2\nu}}{\nu + \gamma r}, & r \le x \end{cases}$$

1.3.4 (1) $$\mathbf{P}_x\big(\ell(\infty,r) \in dy\big) = \begin{cases} \dfrac{\nu}{r}e^{-\nu y/r}dy, & 0 \le x \le r \\ \dfrac{\nu x^{-2\nu}}{r^{1-2\nu}}e^{-\nu y/r}dy, & r \le x \end{cases}$$

(2) $r \le x$ $$\mathbf{P}_x\big(\ell(\infty,r) = 0\big) = 1 - \frac{x^{-2\nu}}{r^{-2\nu}}$$

1.3.5 $$\mathbf{E}_x\{e^{-\gamma\ell(\tau,r)}, R_\tau^{(n)} \in dz\} = (2\lambda)^{\nu+1}z^{2\nu+1}F_\nu(x\sqrt{2\lambda}, z\sqrt{2\lambda})dz$$
$$-\frac{2\gamma(2\lambda rz)^{2\nu+1}F_\nu(r\sqrt{2\lambda}, z\sqrt{2\lambda})F_\nu(x\sqrt{2\lambda}, r\sqrt{2\lambda})}{1 + 2\gamma r^{2\nu+1}(2\lambda)^\nu F_\nu(r\sqrt{2\lambda}, r\sqrt{2\lambda})}dz$$

1.3.6 $$\mathbf{P}_x\big(\ell(\tau,r) \in dy, R_\tau^{(n)} \in dz\big)$$
$$= \frac{\lambda F_\nu(r\sqrt{2\lambda}, z\sqrt{2\lambda})F_\nu(x\sqrt{2\lambda}, r\sqrt{2\lambda})}{r^{2\nu+1}z^{-2\nu-1}F_\nu^2(r\sqrt{2\lambda}, r\sqrt{2\lambda})}\exp\Big(-\frac{y}{2(2\lambda)^\nu r^{2\nu+1}F_\nu(r\sqrt{2\lambda}, r\sqrt{2\lambda})}\Big)dydz$$

(1) $$\mathbf{P}_x\big(\ell(\tau,r) = 0, R_\tau^{(n)} \in dz\big) = (2\lambda)^{\nu+1}z^{2\nu+1}F_\nu(x\sqrt{2\lambda}, z\sqrt{2\lambda})dz$$
$$-(2\lambda)^{\nu+1}z^{2\nu+1}\frac{F_\nu(r\sqrt{2\lambda}, z\sqrt{2\lambda})F_\nu(x\sqrt{2\lambda}, r\sqrt{2\lambda})}{F_\nu(r\sqrt{2\lambda}, r\sqrt{2\lambda})}dz$$

4	$R_s^{(n)}$ $\qquad \nu \ge 0,\ n = 2\nu + 2 \ge 2, \qquad \tau \sim \operatorname{Exp}(\lambda)$, independent of $R^{(n)}$
1.4.1	$\mathbf{E}_x \exp\Big(-\gamma \int_0^\tau \mathbb{1}_{[r,\infty)}\big(R_s^{(n)}\big)ds\Big)$
$0 \le x$ $x \le r$	$= 1 - \dfrac{\gamma(\lambda+\gamma)^{-1/2} r^{\nu} x^{-\nu} I_\nu(x\sqrt{2\lambda}) K_{\nu+1}(r\sqrt{2\lambda+2\gamma})}{\sqrt{\lambda} I_{\nu+1}(r\sqrt{2\lambda}) K_\nu(r\sqrt{2\lambda+2\gamma}) + \sqrt{\lambda+\gamma} K_{\nu+1}(r\sqrt{2\lambda+2\gamma}) I_\nu(r\sqrt{2\lambda})}$
$r \le x$	$= \dfrac{\lambda}{\lambda+\gamma} + \dfrac{\gamma(\lambda+\gamma)^{-1}\sqrt{\lambda} r^{\nu} x^{-\nu} K_\nu(x\sqrt{2\lambda+2\gamma}) I_{\nu+1}(r\sqrt{2\lambda})}{\sqrt{\lambda} I_{\nu+1}(r\sqrt{2\lambda}) K_\nu(r\sqrt{2\lambda+2\gamma}) + \sqrt{\lambda+\gamma} K_{\nu+1}(r\sqrt{2\lambda+2\gamma}) I_\nu(r\sqrt{2\lambda})}$
1.4.2 (1)	$\mathbf{P}_x\Big(\int_0^\tau \mathbb{1}_{[r,\infty)}\big(R_s^{(n)}\big)ds = 0\Big) = \begin{cases} 1 - \dfrac{x^{-\nu} I_\nu(x\sqrt{2\lambda})}{r^{-\nu} I_\nu(r\sqrt{2\lambda})}, & 0 \le x \le r \\ 0, & r \le x \end{cases}$
(2)	$\mathbf{P}_x\Big(\int_0^\tau \mathbb{1}_{[r,\infty)}\big(R_s^{(n)}\big)ds = \tau\Big) = \begin{cases} 0, & 0 \le x \le r \\ 1 - \dfrac{x^{-\nu} K_\nu(x\sqrt{2\lambda})}{r^{-\nu} K_\nu(r\sqrt{2\lambda})}, & r \le x \end{cases}$
1.4.5	$\mathbf{E}_x\Big\{\exp\Big(-\gamma \int_0^\tau \mathbb{1}_{[r,\infty)}\big(R_s^{(n)}\big)ds\Big), R_\tau^{(n)} \in dz\Big\}$
$z \le r$ $x \le r$	$= (2\lambda)^{\nu+1} z^{2\nu+1} F_\nu(x\sqrt{2\lambda}, z\sqrt{2\lambda})dz + 2\lambda z^{\nu+1} x^{-\nu} I_\nu(x\sqrt{2\lambda}) I_\nu(z\sqrt{2\lambda})$ $\times \dfrac{\sqrt{\lambda} K_{\nu+1}(r\sqrt{2\lambda}) K_\nu(r\sqrt{2\lambda+2\gamma}) - \sqrt{\lambda+\gamma} K_{\nu+1}(r\sqrt{2\lambda+2\gamma}) K_\nu(r\sqrt{2\lambda})}{\sqrt{\lambda} I_{\nu+1}(r\sqrt{2\lambda}) K_\nu(r\sqrt{2\lambda+2\gamma}) + \sqrt{\lambda+\gamma} K_{\nu+1}(r\sqrt{2\lambda+2\gamma}) I_\nu(r\sqrt{2\lambda})} dz$
$z \le r$ $r \le x$	$= \dfrac{\lambda\sqrt{2} r^{-1} z^{\nu+1} x^{-\nu} K_\nu(x\sqrt{2\lambda+2\gamma}) I_\nu(z\sqrt{2\lambda})}{\sqrt{\lambda} I_{\nu+1}(r\sqrt{2\lambda}) K_\nu(r\sqrt{2\lambda+2\gamma}) + \sqrt{\lambda+\gamma} K_{\nu+1}(r\sqrt{2\lambda+2\gamma}) I_\nu(r\sqrt{2\lambda})} dz$
$r \le z$ $x \le r$	$= \dfrac{\lambda\sqrt{2} r^{-1} z^{\nu+1} x^{-\nu} I_\nu(x\sqrt{2\lambda}) K_\nu(z\sqrt{2\lambda+2\gamma})}{\sqrt{\lambda} I_{\nu+1}(r\sqrt{2\lambda}) K_\nu(r\sqrt{2\lambda+2\gamma}) + \sqrt{\lambda+\gamma} K_{\nu+1}(r\sqrt{2\lambda+2\gamma}) I_\nu(r\sqrt{2\lambda})} dz$
$r \le z$ $r \le x$	$= 2\lambda(2\lambda+2\gamma)^{\nu} z^{2\nu+1} F_\nu(x\sqrt{2\lambda+2\gamma}, z\sqrt{2\lambda+2\gamma})dz$ $+2\lambda z^{\nu+1} x^{-\nu} K_\nu(x\sqrt{2\lambda+2\gamma}) K_\nu(z\sqrt{2\lambda+2\gamma})$ $\times \dfrac{\sqrt{\lambda+\gamma} I_{\nu+1}(r\sqrt{2\lambda+2\gamma}) I_\nu(r\sqrt{2\lambda}) - \sqrt{\lambda} I_{\nu+1}(r\sqrt{2\lambda}) I_\nu(r\sqrt{2\lambda+2\gamma})}{\sqrt{\lambda} I_{\nu+1}(r\sqrt{2\lambda}) K_\nu(r\sqrt{2\lambda+2\gamma}) + \sqrt{\lambda+\gamma} K_{\nu+1}(r\sqrt{2\lambda+2\gamma}) I_\nu(r\sqrt{2\lambda})} dz$

4 $R_s^{(n)}$ $\nu\ge 0,\ n=2\nu+2\ge 2,$ $\tau\sim \operatorname{Exp}(\lambda)$, independent of $R^{(n)}$

1.5.1 $\mathbf{E}_x \exp\Big(-\gamma\int_0^\tau \mathbb{I}_{[0,r]}\big(R_s^{(n)}\big)ds\Big)$

$0\le x$, $x\le r$
$$=\frac{\lambda}{\lambda+\gamma}+\frac{\gamma(\lambda+\gamma)^{-1}\sqrt{\lambda}r^\nu x^{-\nu}I_\nu(x\sqrt{2\lambda+2\gamma})K_{\nu+1}(r\sqrt{2\lambda})}{\sqrt{\lambda+\gamma}I_{\nu+1}(r\sqrt{2\lambda+2\gamma})K_\nu(r\sqrt{2\lambda})+\sqrt{\lambda}K_{\nu+1}(r\sqrt{2\lambda})I_\nu(r\sqrt{2\lambda+2\gamma})}$$

$r\le x$
$$=1-\frac{\gamma(\lambda+\gamma)^{-1/2}r^\nu x^{-\nu}K_\nu(x\sqrt{2\lambda})I_{\nu+1}(r\sqrt{2\lambda+2\gamma})}{\sqrt{\lambda+\gamma}I_{\nu+1}(r\sqrt{2\lambda+2\gamma})K_\nu(r\sqrt{2\lambda})+\sqrt{\lambda}K_{\nu+1}(r\sqrt{2\lambda})I_\nu(r\sqrt{2\lambda+2\gamma})}$$

1.5.2 (1)
$$\mathbf{P}_x\Big(\int_0^\tau \mathbb{I}_{[0,r]}\big(R_s^{(n)}\big)ds=0\Big)=\begin{cases}0, & 0\le x\le r\\ 1-\dfrac{x^{-\nu}K_\nu(x\sqrt{2\lambda})}{r^{-\nu}K_\nu(r\sqrt{2\lambda})}, & r\le x\end{cases}$$

1.5.3 (1)
$$\mathbf{E}_x \exp\Big(-\gamma\int_0^\infty \mathbb{I}_{[0,r]}\big(R_s^{(n)}\big)ds\Big)=\begin{cases}\dfrac{2\nu x^{-\nu}I_\nu(x\sqrt{2\gamma})}{\sqrt{2\gamma}r^{1-\nu}I_{\nu-1}(r\sqrt{2\gamma})}, & x\le r\\ 1-\dfrac{x^{-2\nu}I_{\nu+1}(r\sqrt{2\gamma})}{r^{-2\nu}I_{\nu-1}(r\sqrt{2\gamma})}, & r\le x\end{cases}$$

1.5.5 $\mathbf{E}_x\Big\{\exp\Big(-\gamma\int_0^\tau \mathbb{I}_{[0,r]}\big(R_s^{(n)}\big)ds\Big), R_\tau^{(n)}\in dz\Big\}$

$z\le r$, $x\le r$
$$=2\lambda(2\lambda+2\gamma)^{\nu+1}z^{2\nu+1}F_\nu(x\sqrt{2\lambda+2\gamma},z\sqrt{2\lambda+2\gamma})dz$$
$$+2\lambda z^{\nu+1}x^{-\nu}I_\nu(x\sqrt{2\lambda+2\gamma})I_\nu(z\sqrt{2\lambda+2\gamma})$$
$$\times\frac{\sqrt{\lambda+\gamma}K_{\nu+1}(r\sqrt{2\lambda+2\gamma})K_\nu(r\sqrt{2\lambda})-\sqrt{\lambda}K_{\nu+1}(r\sqrt{2\lambda})K_\nu(r\sqrt{2\lambda+2\gamma})}{\sqrt{\lambda+\gamma}I_{\nu+1}(r\sqrt{2\lambda+2\gamma})K_\nu(r\sqrt{2\lambda})+\sqrt{\lambda}K_{\nu+1}(r\sqrt{2\lambda})I_\nu(r\sqrt{2\lambda+2\gamma})}dz$$

$z\le r$, $r\le x$
$$=\frac{\lambda\sqrt{2}r^{-1}z^{\nu+1}x^{-\nu}K_\nu(x\sqrt{2\lambda})I_\nu(z\sqrt{2\lambda+2\gamma})}{\sqrt{\lambda+\gamma}I_{\nu+1}(r\sqrt{2\lambda+2\gamma})K_\nu(r\sqrt{2\lambda})+\sqrt{\lambda}K_{\nu+1}(r\sqrt{2\lambda})I_\nu(r\sqrt{2\lambda+2\gamma})}dz$$

$r\le z$, $x\le r$
$$=\frac{\lambda\sqrt{2}r^{-1}z^{\nu+1}x^{-\nu}I_\nu(x\sqrt{2\lambda+2\gamma})K_\nu(z\sqrt{2\lambda})}{\sqrt{\lambda+\gamma}I_{\nu+1}(r\sqrt{2\lambda+2\gamma})K_\nu(r\sqrt{2\lambda})+\sqrt{\lambda}K_{\nu+1}(r\sqrt{2\lambda})I_\nu(r\sqrt{2\lambda+2\gamma})}dz$$

$r\le z$, $r\le x$
$$=(2\lambda)^{\nu+1}z^{2\nu+1}F_\nu(x\sqrt{2\lambda},z\sqrt{2\lambda})dz+2\lambda z^{\nu+1}x^{-\nu}K_\nu(x\sqrt{2\lambda})K_\nu(z\sqrt{2\lambda})$$
$$\times\frac{\sqrt{\lambda}I_{\nu+1}(r\sqrt{2\lambda})I_\nu(r\sqrt{2\lambda+2\gamma})-\sqrt{\lambda+\gamma}I_{\nu+1}(r\sqrt{2\lambda+2\gamma})I_\nu(r\sqrt{2\lambda})}{\sqrt{\lambda+\gamma}I_{\nu+1}(r\sqrt{2\lambda+2\gamma})K_\nu(r\sqrt{2\lambda})+\sqrt{\lambda}K_{\nu+1}(r\sqrt{2\lambda})I_\nu(r\sqrt{2\lambda+2\gamma})}dz$$

4 $\quad R_s^{(n)}$	$\nu \geq 0, \ n = 2\nu + 2 \geq 2, \qquad \tau \sim \mathrm{Exp}(\lambda)$, independent of $R^{(n)}$
1.6.1	$\mathbf{E}_x \exp\Big(-\int_0^\tau \big(p\mathbb{I}_{[0,r)}\big(R_s^{(n)}\big) + q\mathbb{I}_{[r,\infty)}\big(R_s^{(n)}\big)\big)ds\Big)$
$0 \leq x$ $x \leq r$	$= \frac{\lambda}{\lambda + p} - \frac{4\lambda(q-p)r^\nu x^{-\nu} I_\nu(x\Upsilon_p)K_{\nu+1}(r\Upsilon_q)}{\Upsilon_p^2 \Upsilon_q(\Upsilon_p I_{\nu+1}(r\Upsilon_p)K_\nu(r\Upsilon_q) + \Upsilon_q K_{\nu+1}(r\Upsilon_q)I_\nu(r\Upsilon_p))}$
$r \leq x$	$= \frac{\lambda}{\lambda + q} + \frac{4\lambda(q-p)r^\nu x^{-\nu} K_\nu(x\Upsilon_q)I_{\nu+1}(r\Upsilon_p)}{\Upsilon_p \Upsilon_q^2(\Upsilon_p I_{\nu+1}(r\Upsilon_p)K_\nu(r\Upsilon_q) + \Upsilon_q K_{\nu+1}(r\Upsilon_q)I_\nu(r\Upsilon_p))}$
1.6.5	$\mathbf{E}_x\Big\{\exp\Big(-\int_0^\tau \big(p\mathbb{I}_{[0,r)}\big(R_s^{(n)}\big) + q\mathbb{I}_{[r,\infty)}\big(R_s^{(n)}\big)\big)ds\Big), R_\tau^{(n)} \in dz\Big\}$
$z \leq r$ $x \leq r$	$= 2\lambda\Upsilon_p^{2\nu} z^{2\nu+1} F_\nu(x\Upsilon_p, z\Upsilon_p)dz$ $+ \frac{2\lambda z^{\nu+1} I_\nu(x\Upsilon_p)I_\nu(z\Upsilon_p)(\Upsilon_p K_{\nu+1}(r\Upsilon_p)K_\nu(r\Upsilon_q) - \Upsilon_q K_{\nu+1}(r\Upsilon_q)K_\nu(r\Upsilon_p))}{x^\nu(\Upsilon_p I_{\nu+1}(r\Upsilon_p)K_\nu(r\Upsilon_q) + \Upsilon_q K_{\nu+1}(r\Upsilon_q)I_\nu(r\Upsilon_p))}dz$
$z \leq r$ $r \leq x$	$= \frac{2\lambda r^{-1} z^{\nu+1} x^{-\nu} K_\nu(x\Upsilon_q)I_\nu(z\Upsilon_p)}{\Upsilon_p I_{\nu+1}(r\Upsilon_p)K_\nu(r\Upsilon_q) + \Upsilon_q K_{\nu+1}(r\Upsilon_q)I_\nu(r\Upsilon_p)}dz$
$r \leq z$ $x \leq r$	$= \frac{2\lambda r^{-1} z^{\nu+1} x^{-\nu} I_\nu(x\Upsilon_p)K_\nu(z\Upsilon_q)}{\Upsilon_p I_{\nu+1}(r\Upsilon_p)K_\nu(r\Upsilon_q) + \Upsilon_q K_{\nu+1}(r\Upsilon_q)I_\nu(r\Upsilon_p)}dz$
$r \leq z$ $r \leq x$	$= 2\lambda\Upsilon_q^{2\nu} z^{2\nu+1} F_\nu(x\Upsilon_q, z\Upsilon_q)dz$ $+ \frac{2\lambda z^{\nu+1} K_\nu(x\Upsilon_q)K_\nu(z\Upsilon_q)(\Upsilon_q I_{\nu+1}(r\Upsilon_q)I_\nu(r\Upsilon_p) - \Upsilon_p I_{\nu+1}(r\Upsilon_p)I_\nu(r\Upsilon_q))}{x^\nu(\Upsilon_p I_{\nu+1}(r\Upsilon_p)K_\nu(r\Upsilon_q) + \Upsilon_q K_{\nu+1}(r\Upsilon_q)I_\nu(r\Upsilon_p))}dz$
1.9.3	$\mathbf{E}_x \exp\Big(-\frac{\gamma^2}{2}\int_0^t (R_s^{(n)})^2 ds\Big) = \frac{1}{(\mathrm{ch}(t\gamma))^{\nu+1}} \exp\Big(-\frac{x^2\gamma\,\mathrm{sh}(t\gamma)}{2\,\mathrm{ch}(t\gamma)}\Big)$
(1)	$\mathbf{E}_x \exp\Big(-\beta(R_t^{(n)})^2 - \frac{\gamma^2}{2}\int_0^t (R_s^{(n)})^2 ds\Big)$ $= \frac{1}{(\mathrm{ch}(t\gamma) + 2\beta\gamma^{-1}\mathrm{sh}(t\gamma))^{\nu+1}} \exp\Big(-\frac{x^2(\gamma\,\mathrm{sh}(t\gamma) + 2\beta\,\mathrm{ch}(t\gamma))}{2(\mathrm{ch}(t\gamma) + 2\beta\gamma^{-1}\mathrm{sh}(t\gamma))}\Big)$
1.9.4	$\mathbf{P}_x\Big(\int_0^t \big(R_s^{(n)}\big)^2 ds \in dy\Big) = \mathrm{ec}_y(0, \nu+1, t, 0, x^2/2)dy$
	$\Upsilon_s := \sqrt{2\lambda + 2s}$

4 $R_s^{(n)}$ $\nu \geq 0,\ n = 2\nu + 2 \geq 2,$ $\tau \sim \mathrm{Exp}(\lambda)$, independent of $R^{(n)}$

(1) $$\mathbf{P}_0\Big(\int_0^t \big(R_s^{(n)}\big)^2 ds < y\Big) = 2^{\nu+1}\sum_{k=0}^{\infty} \frac{(-1)^k\Gamma(\nu+k+1)}{\Gamma(\nu+1)\,k!}\,\mathrm{Erfc}\Big(\frac{(\nu+2k+1)t}{\sqrt{2y}}\Big)$$

1.9.5 $$\mathbf{E}_x\Big\{\exp\Big(-\frac{\gamma^2}{2}\int_0^\tau (R_s^{(n)})^2 ds\Big); R_\tau^{(n)} \in dz\Big\}$$

$0 < x$

$x \leq z$ $$= \frac{\lambda\Gamma(1/2+\nu/2+\lambda/2\gamma)}{\gamma\Gamma(1+\nu)x^{\nu+1}z^{-\nu}} M_{-\lambda/2\gamma,\nu/2}(\gamma x^2) W_{-\lambda/2\gamma,\nu/2}(\gamma z^2)dz$$

$z \leq x$ $$= \frac{\lambda\Gamma(1/2+\nu/2+\lambda/2\gamma)}{\gamma\Gamma(1+\nu)x^{\nu+1}z^{-\nu}} W_{-\lambda/2\gamma,\nu/2}(\gamma x^2) M_{-\lambda/2\gamma,\nu/2}(\gamma z^2)dz$$

1.9.7 $$\mathbf{E}_x\Big\{\exp\Big(-\frac{\gamma^2}{2}\int_0^t (R_s^{(n)})^2 ds\Big);\ R_t^{(n)} \in dz\Big\}$$
$$= \frac{\gamma z^{\nu+1}x^{-\nu}}{\mathrm{sh}(t\gamma)}\exp\Big(-\frac{(x^2+z^2)\gamma\,\mathrm{ch}(t\gamma)}{2\,\mathrm{sh}(t\gamma)}\Big) I_\nu\Big(\frac{xz\gamma}{\mathrm{sh}(t\gamma)}\Big)dz$$

1.9.8 $$\mathbf{P}_x\Big(\int_0^t \big(R_s^{(n)}\big)^2 ds \in dy, R_t^{(n)} \in dz\Big) = \frac{z^{\nu+1}}{x^\nu}\,\mathrm{is}_y(\nu, t, 0, (x^2+z^2)/2, xz/2)dydz$$

1.12.1 $$\mathbf{E}_x e^{-\gamma\check{H}(\tau)} = \int_x^\infty \frac{\sqrt{2\lambda}x^{-\nu} I_\nu(x\sqrt{2\lambda+2\gamma}) I_{\nu+1}(y\sqrt{2\lambda})}{y^{-\nu} I_\nu(y\sqrt{2\lambda+2\gamma}) I_\nu(y\sqrt{2\lambda})}dy$$

(1) $$\mathbf{E}_x e^{-\gamma\hat{H}(\tau)} = \int_0^x \frac{\sqrt{2\lambda}x^{-\nu} K_\nu(x\sqrt{2\lambda+2\gamma}) K_{\nu+1}(y\sqrt{2\lambda})}{y^{-\nu} K_\nu(y\sqrt{2\lambda+2\gamma}) K_\nu(y\sqrt{2\lambda})}dy$$

1.12.5 $$\mathbf{E}_x\big\{e^{-\gamma\check{H}(\tau)};\ R_\tau^{(n)} \in dz\big\} = \int_{x\vee z}^\infty \frac{2\lambda z^{\nu+1}x^{-\nu} I_\nu(x\sqrt{2\lambda+2\gamma}) I_\nu(z\sqrt{2\lambda})}{y I_\nu(y\sqrt{2\lambda+2\gamma}) I_\nu(y\sqrt{2\lambda})}dy\ dz$$

(1) $$\mathbf{E}_x\big\{e^{-\gamma\hat{H}(\tau)};\ R_\tau^{(n)} \in dz\big\} = \int_0^{x\wedge z} \frac{2\lambda z^{\nu+1}x^{-\nu} K_\nu(x\sqrt{2\lambda+2\gamma}) K_\nu(z\sqrt{2\lambda})}{y K_\nu(y\sqrt{2\lambda+2\gamma}) K_\nu(y\sqrt{2\lambda})}dy\ dz$$

1.13.1 $$\mathbf{E}_x\big\{e^{-\gamma\check{H}(\tau)};\ \sup_{0\leq s\leq\tau} R_s^{(n)} \in dy\big\} = \frac{\sqrt{2\lambda}x^{-\nu} I_\nu(x\sqrt{2\lambda+2\gamma}) I_{\nu+1}(y\sqrt{2\lambda})}{y^{-\nu} I_\nu(y\sqrt{2\lambda+2\gamma}) I_\nu(y\sqrt{2\lambda})}dy$$

$x < y$

1.13.2 $$\mathbf{P}_x\big(\check{H}(\tau) \in dv,\ \sup_{0\leq s\leq\tau} R_s^{(n)} \in dy\big)$$

$x < y$

$$= \frac{\sqrt{2\lambda}I_{\nu+1}(y\sqrt{2\lambda})}{I_\nu(y\sqrt{2\lambda})}\sum_{k=1}^\infty \frac{x^{-\nu} j_{\nu,k} J_\nu(j_{\nu,k}x/y)}{y^{2-\nu} J_{\nu+1}(j_{\nu,k})} e^{-\lambda v - j_{\nu,k}^2 v/2y^2}dvdy$$

4 $R_s^{(n)}$ $\nu \ge 0,\ n = 2\nu + 2 \ge 2,$ $\tau \sim \mathrm{Exp}(\lambda)$, independent of $R^{(n)}$

1.13.3 $x < y$

$$\mathbf{E}_x\Big\{e^{-\gamma\check{H}(t)};\ \sup_{0\le s\le t} R_s^{(n)} \in dy\Big\}$$

$$= 2\sum_{k=1}^{\infty} \frac{x^{-\nu} j_{\nu,k} J_\nu(j_{\nu,k}x/y)}{y^{3-\nu} J_{\nu+1}(j_{\nu,k})} e^{-\gamma t - j_{\nu,k}^2 t/2y^2} * \sum_{k=1}^{\infty} e^{-j_{\nu,k}^2 t/2y^2} dy$$

1.13.4 $x < y$

$$\mathbf{P}_x\Big(\check{H}(t) \in dv,\ \sup_{0\le s\le t} R_s^{(n)} \in dy\Big)$$

$$= 2\mathbb{1}_{[0,t]}(v) \sum_{k=1}^{\infty} \frac{x^{-\nu} j_{\nu,k} J_\nu(j_{\nu,k}x/y)}{y^{3-\nu} J_{\nu+1}(j_{\nu,k})} e^{-j_{\nu,k}^2 v/2y^2} \sum_{k=1}^{\infty} e^{-j_{\nu,k}^2 (t-v)/2y^2} dvdy$$

1.13.5

$$\mathbf{E}_x\Big\{e^{-\gamma\check{H}(\tau)};\ \sup_{0\le s\le \tau} R_s^{(n)} \in dy,\ R_\tau^{(n)} \in dz\Big\} \qquad x \vee z < y$$

$$= \frac{2\lambda z^{\nu+1} x^{-\nu} I_\nu(x\sqrt{2\lambda+2\gamma}) I_\nu(z\sqrt{2\lambda})}{y I_\nu(y\sqrt{2\lambda+2\gamma}) I_\nu(y\sqrt{2\lambda})} dydz$$

1.13.6

$$\mathbf{P}_x\Big(\check{H}(\tau) \in dv;\ \sup_{0\le s\le \tau} R_s^{(n)} \in dy,\ R_\tau^{(n)} \in dz\Big) \qquad x \vee z < y$$

$$= 2\lambda z^{\nu+1} x^{-\nu} \sum_{k=1}^{\infty} \frac{j_{\nu,k} J_\nu(j_{\nu,k}x/y)}{y^3 J_{\nu+1}(j_{\nu,k})} e^{-\lambda v - j_{\nu,k}^2 v/2y^2} \frac{I_\nu(z\sqrt{2\lambda})}{I_\nu(y\sqrt{2\lambda})} dvdydz$$

1.13.7

$$\mathbf{E}_x\Big\{e^{-\gamma\check{H}(t)};\ \sup_{0\le s\le t} R_s^{(n)} \in dy,\ R_t^{(n)} \in dz\Big\} \qquad x \vee z < y$$

$$= \frac{2z^{\nu+1}}{y^5 x^\nu} \sum_{k=1}^{\infty} \frac{j_{\nu,k} J_\nu(j_{\nu,k}x/y)}{J_{\nu+1}(j_{\nu,k})} e^{-\gamma t - j_{\nu,k}^2 t/2y^2} * \sum_{k=1}^{\infty} \frac{j_{\nu,k} J_\nu(j_{\nu,k}x/y)}{J_{\nu+1}(j_{\nu,k})} e^{-j_{\nu,k}^2 t/2y^2} dydz$$

1.13.8 $v < t$

$$\mathbf{P}_x\Big(\check{H}(t) \in dv,\ \sup_{0\le s\le t} R_s^{(n)} \in dy, R_t^{(n)} \in dz\Big) \qquad x \vee z < y$$

$$= \frac{2z^{\nu+1}}{y^5 x^\nu} \sum_{k=1}^{\infty} \frac{j_{\nu,k} J_\nu(j_{\nu,k}x/y)}{J_{\nu+1}(j_{\nu,k})} e^{-j_{\nu,k}^2 v/2y^2} \sum_{k=1}^{\infty} \frac{j_{\nu,k} J_\nu(j_{\nu,k}z/y)}{J_{\nu+1}(j_{\nu,k})} e^{-j_{\nu,k}^2 (t-v)/2y^2} dvdydz$$

1.14.1 $y < x$

$$\mathbf{E}_x\Big\{e^{-\gamma\hat{H}(\tau)};\ \inf_{0\le s\le \tau} R_s^{(n)} \in dy\Big\} = \frac{\sqrt{2\lambda} x^{-\nu} K_\nu(x\sqrt{2\lambda+2\gamma}) K_{\nu+1}(y\sqrt{2\lambda})}{y^{-\nu} K_\nu(y\sqrt{2\lambda+2\gamma}) K_\nu(y\sqrt{2\lambda})} dy$$

4 $R_s^{(n)}$ $\nu \ge 0,\ n = 2\nu + 2 \ge 2,$ $\tau \sim \operatorname{Exp}(\lambda)$, independent of $R^{(n)}$

1.14.3 (1) $y < x$

$$\mathbf{E}_x\{e^{-\gamma \hat{H}(\infty)};\ \inf_{0\le s<\infty} R_s^{(n)} \in dy\} = \frac{2\nu x^{-\nu} K_\nu(x\sqrt{2\gamma})}{y^{1-\nu} K_\nu(y\sqrt{2\gamma})} dy$$

1.14.5

$$\mathbf{E}_x\{e^{-\gamma \hat{H}(\tau)};\ \inf_{0\le s\le\tau} R_s^{(n)} \in dy,\ R_\tau^{(n)} \in dz\} \qquad 0 < y < x \wedge z$$

$$= \frac{2\lambda z^{\nu+1} x^{-\nu} K_\nu(x\sqrt{2\lambda+2\gamma}) K_\nu(z\sqrt{2\lambda})}{y K_\nu(y\sqrt{2\lambda+2\gamma}) K_\nu(y\sqrt{2\lambda})} dydz$$

1.15.2

$$\mathbf{P}_x\bigl(a < \inf_{0\le s\le\tau} R_s^{(n)},\ \sup_{0\le s\le\tau} R_s^{(n)} < b\bigr) = 1 - \frac{S_\nu(b\sqrt{2\lambda}, x\sqrt{2\lambda}) + S_\nu(x\sqrt{2\lambda}, a\sqrt{2\lambda})}{S_\nu(b\sqrt{2\lambda}, a\sqrt{2\lambda})}$$

1.15.6

$$\mathbf{P}_x\bigl(a < \inf_{0\le s\le\tau} R_s^{(n)},\ \sup_{0\le s\le\tau} R_s^{(n)} < b, R_\tau^{(n)} \in dz\bigr)$$

$$= \frac{(2\lambda)^{\nu+1} z^{2\nu+1} S_\nu(b\sqrt{2\lambda}, (z+x+|z-x|)\sqrt{\lambda/2}) S_\nu((z+x-|z-x|)\sqrt{\lambda/2}, a\sqrt{2\lambda})}{S_\nu(b\sqrt{2\lambda}, a\sqrt{2\lambda})} dz$$

or

$$= \begin{cases} (2\lambda)^{\nu+1} z^{2\nu+1} \dfrac{S_\nu(b\sqrt{2\lambda}, z\sqrt{2\lambda}) S_\nu(x\sqrt{2\lambda}, a\sqrt{2\lambda})}{S_\nu(b\sqrt{2\lambda}, a\sqrt{2\lambda})} dz, & a \le x \le z \\ (2\lambda)^{\nu+1} z^{2\nu+1} \dfrac{S_\nu(b\sqrt{2\lambda}, x\sqrt{2\lambda}) S_\nu(z\sqrt{2\lambda}, a\sqrt{2\lambda})}{S_\nu(b\sqrt{2\lambda}, a\sqrt{2\lambda})} dz, & z \le x \le b \end{cases}$$

1.16.1 $r < b$

$$\mathbf{E}_x\{e^{-\gamma \ell(\tau,r)};\ \sup_{0\le s\le\tau} R_s^{(n)} < b\}$$

$0 \le x$, $x \le r$

$$= 1 - \frac{(2\gamma S_\nu(b\sqrt{2\lambda}, r\sqrt{2\lambda}) + (2\lambda)^{-\nu} r^{-2\nu-1}) x^{-\nu} r^\nu I_\nu(x\sqrt{2\lambda})}{2\gamma S_\nu(b\sqrt{2\lambda}, r\sqrt{2\lambda}) I_\nu(r\sqrt{2\lambda}) + (2\lambda)^{-\nu} r^{-\nu-1} b^{-\nu} I_\nu(b\sqrt{2\lambda})}$$

$r \le x$, $x \le b$

$$= 1 - \frac{S_\nu(x\sqrt{2\lambda}, r\sqrt{2\lambda})}{S_\nu(b\sqrt{2\lambda}, r\sqrt{2\lambda})}$$

$$- \frac{(2\gamma S_\nu(b\sqrt{2\lambda}, r\sqrt{2\lambda}) + (2\lambda)^{-\nu} r^{-2\nu-1}) I_\nu(r\sqrt{2\lambda}) S_\nu(b\sqrt{2\lambda}, x\sqrt{2\lambda})}{(2\gamma S_\nu(b\sqrt{2\lambda}, r\sqrt{2\lambda}) I_\nu(r\sqrt{2\lambda}) + (2\lambda)^{-\nu} r^{-\nu-1} b^{-\nu} I_\nu(b\sqrt{2\lambda})) S_\nu(b\sqrt{2\lambda}, r\sqrt{2\lambda})}$$

1.16.2 $r < b$

$$\mathbf{P}_x\bigl(\ell(\tau,r) \in dy,\ \sup_{0\le s\le\tau} R_s^{(n)} < b\bigr) = \exp\Bigl(-\frac{y(2\lambda)^{-\nu} r^{-\nu-1} b^{-\nu} I_\nu(b\sqrt{2\lambda})}{2S_\nu(b\sqrt{2\lambda}, r\sqrt{2\lambda}) I_\nu(r\sqrt{2\lambda})}\Bigr)$$

$$\times \begin{cases} \dfrac{(b^{-\nu} I_\nu(b\sqrt{2\lambda}) - r^{-\nu} I_\nu(r\sqrt{2\lambda})) x^{-\nu} I_\nu(x\sqrt{2\lambda})}{2(2\lambda)^\nu r S_\nu(b\sqrt{2\lambda}, r\sqrt{2\lambda}) I_\nu^2(r\sqrt{2\lambda})} dy, & 0 \le x \le r \\ \dfrac{(b^{-\nu} I_\nu(b\sqrt{2\lambda}) - r^{-\nu} I_\nu(r\sqrt{2\lambda})) S_\nu(b\sqrt{2\lambda}, x\sqrt{2\lambda})}{2(2\lambda)^\nu r^{\nu+1} S_\nu^2(b\sqrt{2\lambda}, r\sqrt{2\lambda}) I_\nu(r\sqrt{2\lambda})} dy, & r \le x \le b \end{cases}$$

4 $R_s^{(n)}$ $\nu \ge 0,\ n = 2\nu + 2 \ge 2,$ $\tau \sim \text{Exp}(\lambda)$, independent of $R^{(n)}$

(1) $\mathbf{P}_x\big(\ell(\tau,r)=0,\ \sup_{0\le s\le\tau} R_s^{(n)} < b\big)$

$$= \begin{cases} 1 - \dfrac{x^{-\nu} I_\nu(x\sqrt{2\lambda})}{r^{-\nu} I_\nu(r\sqrt{2\lambda})}, & 0 \le x \le r \\ 1 - \dfrac{S_\nu(b\sqrt{2\lambda}, x\sqrt{2\lambda}) + S_\nu(x\sqrt{2\lambda}, r\sqrt{2\lambda})}{S_\nu(b\sqrt{2\lambda}, r\sqrt{2\lambda})}, & r \le x \le b \end{cases}$$

1.16.5 $\mathbf{E}_x\big\{e^{-\gamma\ell(\tau,r)};\ \sup_{0\le s\le\tau} R_s^{(n)} < b,\ R_\tau^{(n)} \in dz\big\}$ $\qquad r \vee z < b$

$z \le r$, $x \le r$:
$$= \frac{(2\lambda)^{\nu+1} z^{2\nu+1} S_\nu(r\sqrt{2\lambda}, (z+x+|z-x|)\sqrt{\lambda/2}) I_\nu((z+x-|z-x|)\sqrt{\lambda/2})}{((z+x-|z-x|)/2)^\nu r^{-\nu} I_\nu(r\sqrt{2\lambda})} dz$$
$$+ \frac{2\lambda z^{\nu+1} r^{-1} x^{-\nu} S_\nu(b\sqrt{2\lambda}, r\sqrt{2\lambda}) I_\nu(z\sqrt{2\lambda}) I_\nu(x\sqrt{2\lambda})}{(2\gamma S_\nu(b\sqrt{2\lambda}, r\sqrt{2\lambda}) I_\nu(r\sqrt{2\lambda}) + (2\lambda)^{-\nu} r^{-\nu-1} b^{-\nu} I_\nu(b\sqrt{2\lambda})) I_\nu(r\sqrt{2\lambda})} dz$$

$z \le r$, $r \le x$:
$$= \frac{2\lambda z^{\nu+1} r^{-\nu-1} S_\nu(b\sqrt{2\lambda}, x\sqrt{2\lambda}) I_\nu(z\sqrt{2\lambda})}{2\gamma S_\nu(b\sqrt{2\lambda}, r\sqrt{2\lambda}) I_\nu(r\sqrt{2\lambda}) + (2\lambda)^{-\nu} r^{-\nu-1} b^{-\nu} I_\nu(b\sqrt{2\lambda})} dz$$

$r \le z$, $x \le r$:
$$= \frac{2\lambda z^{2\nu+1} r^{-\nu-1} x^{-\nu} S_\nu(b\sqrt{2\lambda}, z\sqrt{2\lambda}) I_\nu(x\sqrt{2\lambda})}{2\gamma S_\nu(b\sqrt{2\lambda}, r\sqrt{2\lambda}) I_\nu(r\sqrt{2\lambda}) + (2\lambda)^{-\nu} r^{-\nu-1} b^{-\nu} I_\nu(b\sqrt{2\lambda})} dz$$

$r \le z$, $r \le x$:
$$= \frac{(2\lambda)^{\nu+1} z^{2\nu+1} S_\nu(b\sqrt{2\lambda}, (z+x+|z-x|)\sqrt{\lambda/2}) S_\nu((z+x-|z-x|)\sqrt{\lambda/2}, r\sqrt{2\lambda})}{S_\nu(b\sqrt{2\lambda}, r\sqrt{2\lambda})} dz$$
$$+ \frac{2\lambda z^{2\nu+1} r^{-2\nu-1} S_\nu(b\sqrt{2\lambda}, z\sqrt{2\lambda}) I_\nu(r\sqrt{2\lambda}) S_\nu(b\sqrt{2\lambda}, x\sqrt{2\lambda})}{(2\gamma S_\nu(b\sqrt{2\lambda}, r\sqrt{2\lambda}) I_\nu(r\sqrt{2\lambda}) + (2\lambda)^{-\nu} r^{-\nu-1} b^{-\nu} I_\nu(b\sqrt{2\lambda})) S_\nu(b\sqrt{2\lambda}, r\sqrt{2\lambda})} dz$$

1.16.6 $\mathbf{P}_x\big(\ell(\tau,r) \in dy,\ \sup_{0\le s\le\tau} R_s^{(n)} < b,\ R_\tau^{(n)} \in dz\big)$ $\qquad r \vee z < b$

$z \le r$, $x \le r$:
$$= \frac{\lambda z^{\nu+1} r^{-1} x^{-\nu} I_\nu(z\sqrt{2\lambda}) I_\nu(x\sqrt{2\lambda})}{I_\nu^2(r\sqrt{2\lambda})} \exp\Big(-\frac{y(2\lambda)^{-\nu} r^{-\nu-1} b^{-\nu} I_\nu(b\sqrt{2\lambda})}{2S_\nu(b\sqrt{2\lambda}, r\sqrt{2\lambda}) I_\nu(r\sqrt{2\lambda})}\Big) dydz$$

$z \le r$, $r \le x$:
$$= \frac{\lambda z^{\nu+1} r^{-\nu-1} S_\nu(b\sqrt{2\lambda}, x\sqrt{2\lambda}) I_\nu(z\sqrt{2\lambda})}{S_\nu(b\sqrt{2\lambda}, r\sqrt{2\lambda}) I_\nu(r\sqrt{2\lambda})} \exp\Big(-\frac{y(2\lambda)^{-\nu} r^{-\nu-1} b^{-\nu} I_\nu(b\sqrt{2\lambda})}{2S_\nu(b\sqrt{2\lambda}, r\sqrt{2\lambda}) I_\nu(r\sqrt{2\lambda})}\Big) dydz$$

$r \le z$, $x \le r$:
$$= \frac{\lambda z^{2\nu+1} S_\nu(b\sqrt{2\lambda}, z\sqrt{2\lambda}) I_\nu(x\sqrt{2\lambda})}{r^{\nu+1} x^\nu S_\nu(b\sqrt{2\lambda}, r\sqrt{2\lambda}) I_\nu(r\sqrt{2\lambda})} \exp\Big(-\frac{y(2\lambda)^{-\nu} r^{-\nu-1} b^{-\nu} I_\nu(b\sqrt{2\lambda})}{2S_\nu(b\sqrt{2\lambda}, r\sqrt{2\lambda}) I_\nu(r\sqrt{2\lambda})}\Big) dydz$$

4 $R_s^{(n)}$ $\nu \ge 0,\ n = 2\nu + 2 \ge 2,$ $\tau \sim \text{Exp}(\lambda)$, independent of $R^{(n)}$

$r \le z$, $r \le x$

$$= \frac{\lambda S_\nu(b\sqrt{2\lambda}, z\sqrt{2\lambda}) S_\nu(b\sqrt{2\lambda}, x\sqrt{2\lambda})}{z^{-2\nu-1} r^{2\nu+1} S_\nu^2(b\sqrt{2\lambda}, r\sqrt{2\lambda})} \exp\Big(-\frac{y(2\lambda)^{-\nu} r^{-\nu-1} b^{-\nu} I_\nu(b\sqrt{2\lambda})}{2S_\nu(b\sqrt{2\lambda}, r\sqrt{2\lambda}) I_\nu(r\sqrt{2\lambda})}\Big) dy dz$$

(1) $\mathbf{P}_x\big(\ell(\tau, r) = 0, \sup_{0\le s\le\tau} R_s^{(n)} < b, R_\tau^{(n)} \in dz\big)$

$z \le r$, $x \le r$

$$= \frac{(2\lambda)^{\nu+1} z^{2\nu+1} S_\nu(r\sqrt{2\lambda}, (z + x + |z - x|)\sqrt{\lambda/2}) I_\nu((z + x - |z - x|)\sqrt{\lambda/2})}{((z + x - |z - x|)/2)^\nu r^{-\nu} I_\nu(r\sqrt{2\lambda})} dz$$

$r \le z$, $r \le x$

$$= \frac{(2\lambda)^{\nu+1} z^{2\nu+1} S_\nu(b\sqrt{2\lambda}, (z + x + |z - x|)\sqrt{\lambda/2}) S_\nu((z + x - |z - x|)\sqrt{\lambda/2}, r\sqrt{2\lambda})}{S_\nu(b\sqrt{2\lambda}, r\sqrt{2\lambda})} dz$$

1.17.1 $\mathbf{E}_x\{e^{-\gamma\ell(\tau,r)}; a < \inf_{0\le s\le\tau} R_s^{(n)}\}$

$a < r$

$a \le x$, $x \le r$

$$= 1 - \frac{S_\nu(r\sqrt{2\lambda}, x\sqrt{2\lambda}) + S_\nu(x\sqrt{2\lambda}, a\sqrt{2\lambda})}{S_\nu(r\sqrt{2\lambda}, a\sqrt{2\lambda})}$$

$$+ \frac{(a^{-\nu} K_\nu(a\sqrt{2\lambda}) - r^{-\nu} K_\nu(r\sqrt{2\lambda})) S_\nu(x\sqrt{2\lambda}, a\sqrt{2\lambda})}{(2(2\lambda)^\nu r^{\nu+1} \gamma K_\nu(r\sqrt{2\lambda}) S_\nu(r\sqrt{2\lambda}, a\sqrt{2\lambda}) + a^{-\nu} K_\nu(a\sqrt{2\lambda})) S_\nu(r\sqrt{2\lambda}, a\sqrt{2\lambda})}$$

$r \le x$

$$= 1 - \frac{x^{-\nu} K_\nu(x\sqrt{2\lambda})}{r^{-\nu} K_\nu(r\sqrt{2\lambda})}$$

$$+ \frac{(a^{-\nu} K_\nu(a\sqrt{2\lambda}) - r^{-\nu} K_\nu(r\sqrt{2\lambda})) x^{-\nu} K_\nu(x\sqrt{2\lambda})}{(2(2\lambda)^\nu r^{\nu+1} \gamma K_\nu(r\sqrt{2\lambda}) S_\nu(r\sqrt{2\lambda}, a\sqrt{2\lambda}) + a^{-\nu} K_\nu(a\sqrt{2\lambda})) r^{-\nu} K_\nu(r\sqrt{2\lambda})}$$

1.17.2 $\mathbf{P}_x\big(\ell(\tau, r) \in dy, a < \inf_{0\le s\le\tau} R_s^{(n)}\big) = \exp\Big(-\frac{y(2\lambda)^{-\nu} r^{-\nu-1} a^{-\nu} K_\nu(a\sqrt{2\lambda})}{2K_\nu(r\sqrt{2\lambda}) S_\nu(r\sqrt{2\lambda}, a\sqrt{2\lambda})}\Big)$

$a < r$

$$\times \begin{cases} \dfrac{(a^{-\nu} K_\nu(a\sqrt{2\lambda}) - r^{-\nu} K_\nu(r\sqrt{2\lambda})) S_\nu(x\sqrt{2\lambda}, a\sqrt{2\lambda})}{2(2\lambda)^\nu r^{\nu+1} K_\nu(r\sqrt{2\lambda}) S_\nu^2(r\sqrt{2\lambda}, a\sqrt{2\lambda})} dy, & a \le x \le r \\[2ex] \dfrac{(a^{-\nu} K_\nu(a\sqrt{2\lambda}) - r^{-\nu} K_\nu(r\sqrt{2\lambda})) x^{-\nu} K_\nu(x\sqrt{2\lambda})}{2(2\lambda)^\nu r K_\nu^2(r\sqrt{2\lambda}) S_\nu(r\sqrt{2\lambda}, a\sqrt{2\lambda})} dy, & r \le x \end{cases}$$

(1) $\mathbf{P}_x\big(\ell(\tau, r) = 0, a < \inf_{0\le s\le\tau} R_s^{(n)}\big)$

4 $R_s^{(n)}$ $\nu \ge 0,\ n = 2\nu + 2 \ge 2,$ $\tau \sim \text{Exp}(\lambda)$, independent of $R^{(n)}$

$$= \begin{cases} 1 - \dfrac{S_\nu(r\sqrt{2\lambda}, x\sqrt{2\lambda}) + S_\nu(x\sqrt{2\lambda}, a\sqrt{2\lambda})}{S_\nu(r\sqrt{2\lambda}, a\sqrt{2\lambda})}, & a \le x \le r \\ 1 - \dfrac{x^{-\nu}K_\nu(x\sqrt{2\lambda})}{r^{-\nu}K_\nu(r\sqrt{2\lambda})}, & r \le x \end{cases}$$

1.17.3 (1) $a < r$

$$\mathbf{E}_x\{e^{-\gamma\ell(\infty,r)}; a < \inf_{0\le s<\infty} R_s^{(n)}\}$$

$$= \begin{cases} \dfrac{a^{-2\nu} - x^{-2\nu}}{a^{-2\nu} + \nu^{-1}r\gamma(a^{-2\nu} - r^{-2\nu})}, & a \le x \le r \\ 1 - \dfrac{x^{-2\nu}}{r^{-2\nu}} + \dfrac{(a^{-2\nu} - r^{-2\nu})x^{-2\nu}}{(a^{-2\nu} + \nu^{-1}r\gamma(a^{-2\nu} - r^{-2\nu}))r^{-2\nu}}, & r \le x \le b \end{cases}$$

1.17.4 (1) $a < r$

$$\mathbf{P}_x\big(\ell(\infty,r) \in dy, a < \inf_{0\le s<\infty} R_s^{(n)}\big) =$$

$$= \begin{cases} \dfrac{\nu(a^{-2\nu} - x^{-2\nu})}{r(a^{-2\nu} - r^{-2\nu})} \exp\Big(-\dfrac{\nu a^{-2\nu}y}{r(a^{-2\nu} - r^{-2\nu})}\Big)dy, & a \le x \le r \\ \dfrac{\nu x^{-2\nu}}{r^{1-2\nu}} \exp\Big(-\dfrac{\nu a^{-2\nu}y}{r(a^{-2\nu} - r^{-2\nu})}\Big)dy, & r \le x. \end{cases}$$

1.17.5

$$\mathbf{E}_x\{e^{-\gamma\ell(\tau,r)}; a < \inf_{0\le s\le\tau} R_s^{(n)}, R_\tau^{(n)} \in dz\} \qquad a < r \wedge z$$

$z \le r$, $x \le r$

$$= \frac{(2\lambda)^{\nu+1}z^{2\nu+1}S_\nu(r\sqrt{2\lambda}, (z+x+|z-x|)\sqrt{\lambda/2})S_\nu((z+x-|z-x|)\sqrt{\lambda/2}, a\sqrt{2\lambda})}{S_\nu(r\sqrt{2\lambda}, a\sqrt{2\lambda})}dz$$

$$+ \frac{2\lambda z^{2\nu+1}r^{-\nu}K_\nu(r\sqrt{2\lambda})S_\nu(z\sqrt{2\lambda}, a\sqrt{2\lambda})S_\nu(x\sqrt{2\lambda}, a\sqrt{2\lambda})}{(2\gamma r^{\nu+1}K_\nu(r\sqrt{2\lambda})S_\nu(r\sqrt{2\lambda}, a\sqrt{2\lambda}) + (2\lambda)^{-\nu}a^{-\nu}K_\nu(a\sqrt{2\lambda}))S_\nu(r\sqrt{2\lambda}, a\sqrt{2\lambda})}dz$$

$z \le r$, $r \le x$

$$= \frac{2\lambda z^{2\nu+1}x^{-\nu}K_\nu(x\sqrt{2\lambda})S_\nu(z\sqrt{2\lambda}, a\sqrt{2\lambda})}{2\gamma r^{\nu+1}K_\nu(r\sqrt{2\lambda})S_\nu(r\sqrt{2\lambda}, a\sqrt{2\lambda}) + (2\lambda)^{-\nu}a^{-\nu}K_\nu(a\sqrt{2\lambda})}dz$$

$r \le z$, $x \le r$

$$= \frac{2\lambda z^{\nu+1}K_\nu(z\sqrt{2\lambda})S_\nu(x\sqrt{2\lambda}, a\sqrt{2\lambda})}{2\gamma r^{\nu+1}K_\nu(r\sqrt{2\lambda})S_\nu(r\sqrt{2\lambda}, a\sqrt{2\lambda}) + (2\lambda)^{-\nu}a^{-\nu}K_\nu(a\sqrt{2\lambda})}dz$$

$r \le z$, $r \le x$

$$= \frac{(2\lambda)^{\nu+1}z^{2\nu+1}K_\nu((z+x+|z-x|)\sqrt{\lambda/2})S_\nu((z+x-|z-x|)\sqrt{\lambda/2}, r\sqrt{2\lambda})}{((z+x+|z-x|)/2)^\nu r^{-\nu}K_\nu(r\sqrt{2\lambda})}dz$$

4 $R_s^{(n)}$ $\nu \ge 0,\ n = 2\nu + 2 \ge 2,$ $\tau \sim \operatorname{Exp}(\lambda)$, independent of $R^{(n)}$

$$+\frac{2\lambda z^{\nu+1}x^{-\nu}K_\nu(z\sqrt{2\lambda})S_\nu(r\sqrt{2\lambda},a\sqrt{2\lambda})K_\nu(x\sqrt{2\lambda})}{(2\gamma r^{\nu+1}K_\nu(r\sqrt{2\lambda})S_\nu(r\sqrt{2\lambda},a\sqrt{2\lambda})+(2\lambda)^{-\nu}a^{-\nu}K_\nu(a\sqrt{2\lambda}))r^{-\nu}K_\nu(r\sqrt{2\lambda})}dz$$

1.17.6 $\mathbf{P}_x\big(\ell(\tau,r)\in dy, a<\inf_{0\le s\le\tau} R_s^{(n)}, R_\tau^{(n)}\in dz\big)$ $a < r \wedge z$

$z \le r$, $x \le r$:
$$=\frac{\lambda S_\nu(z\sqrt{2\lambda},a\sqrt{2\lambda})S_\nu(x\sqrt{2\lambda},a\sqrt{2\lambda})}{z^{-2\nu-1}r^{2\nu+1}S_\nu^2(r\sqrt{2\lambda},a\sqrt{2\lambda})}\exp\Big(-\frac{y(2\lambda)^{-\nu}r^{-\nu-1}a^{-\nu}K_\nu(a\sqrt{2\lambda})}{2K_\nu(r\sqrt{2\lambda})S_\nu(r\sqrt{2\lambda},a\sqrt{2\lambda})}\Big)dydz$$

$z \le r$, $r \le x$:
$$=\frac{\lambda z^{2\nu+1}x^{-\nu}K_\nu(x\sqrt{2\lambda})S_\nu(z\sqrt{2\lambda},a\sqrt{2\lambda})}{r^{\nu+1}K_\nu(r\sqrt{2\lambda})S_\nu(r\sqrt{2\lambda},a\sqrt{2\lambda})}\exp\Big(-\frac{y(2\lambda)^{-\nu}r^{-\nu-1}a^{-\nu}K_\nu(a\sqrt{2\lambda})}{2K_\nu(r\sqrt{2\lambda})S_\nu(r\sqrt{2\lambda},a\sqrt{2\lambda})}\Big)dydz$$

$r \le z$, $x \le r$:
$$=\frac{\lambda z^{\nu+1}K_\nu(z\sqrt{2\lambda})S_\nu(x\sqrt{2\lambda},a\sqrt{2\lambda})}{r^{\nu+1}K_\nu(r\sqrt{2\lambda})S_\nu(r\sqrt{2\lambda},a\sqrt{2\lambda})}\exp\Big(-\frac{y(2\lambda)^{-\nu}r^{-\nu-1}a^{-\nu}K_\nu(a\sqrt{2\lambda})}{2K_\nu(r\sqrt{2\lambda})S_\nu(r\sqrt{2\lambda},a\sqrt{2\lambda})}\Big)dydz$$

$r \le z$, $r \le x$:
$$=\frac{\lambda z^{\nu+1}x^{-\nu}K_\nu(z\sqrt{2\lambda})K_\nu(x\sqrt{2\lambda})}{rK_\nu^2(r\sqrt{2\lambda})}\exp\Big(-\frac{y(2\lambda)^{-\nu}r^{-\nu-1}a^{-\nu}K_\nu(a\sqrt{2\lambda})}{2K_\nu(r\sqrt{2\lambda})S_\nu(r\sqrt{2\lambda},a\sqrt{2\lambda})}\Big)dydz$$

(1) $\mathbf{P}_x\big(\ell(\tau,r)=0, a<\inf_{0\le s\le\tau} R_s^{(n)}, R_\tau^{(n)}\in dz\big)$

$z \le r$, $x \le r$:
$$=\frac{(2\lambda)^{\nu+1}z^{2\nu+1}S_\nu(r\sqrt{2\lambda},(z+x+|z-x|)\sqrt{\lambda/2})S_\nu((z+x-|z-x|)\sqrt{\lambda/2},a\sqrt{2\lambda})}{S_\nu(r\sqrt{2\lambda},a\sqrt{2\lambda})}dz$$

$r \le z$, $r \le x$:
$$=\frac{(2\lambda)^{\nu+1}z^{2\nu+1}K_\nu((z+x+|z-x|)\sqrt{\lambda/2})S_\nu((z+x-|z-x|)\sqrt{\lambda/2},r\sqrt{2\lambda})}{((z+x+|z-x|)/2)^\nu r^{-\nu}K_\nu(r\sqrt{2\lambda})}dz$$

1.18.3 (1) $\mathbf{E}_x\exp\big(-\gamma\ell(\infty,r)-\eta\ell(\infty,u)\big)=1-\dfrac{\gamma\nu r^{2\nu+1}\big(r^{-2\nu}+x^{-2\nu}-|r^{-2\nu}-x^{-2\nu}|\big)}{2(\gamma\eta ru(1-(r/u)^{2\nu})+\nu r\gamma+\nu u\eta+\nu^2)}$

$r < u$:
$$-\frac{\eta\big(\nu u^{2\nu+1}+\gamma ur\big(u^{2\nu}-r^{2\nu}\big)\big)\big(x^{-2\nu}+u^{-2\nu}-|u^{-2\nu}-x^{-2\nu}|\big)}{2(\gamma\eta ru(1-(r/u)^{2\nu})+\nu r\gamma+\nu u\eta+\nu^2)}$$

1.18.4 (1) $\mathbf{P}_x\big(\ell(\infty,r)\in dy,\ \ell(\infty,u)\in dg\big)$

4	$R_s^{(n)}$ $\quad \nu \ge 0,\ n = 2\nu+2 \ge 2,\qquad \tau \sim \operatorname{Exp}(\lambda)$, independent of $R^{(n)}$
$x \le r$ $r < u$	$= \frac{\nu^2 r^{-2\nu-1}}{u(r^{-2\nu}-u^{-2\nu})}\exp\Big(-\frac{\nu r^{-2\nu-1}(uy+rg)}{u(r^{-2\nu}-u^{-2\nu})}\Big) I_0\Big(\frac{\nu(ru)^{-\nu-1/2}\sqrt{yg}}{r^{-2\nu}-u^{-2\nu}}\Big)dydg$
$r < x$ $x < u$	$= \frac{\nu^2}{u(r^{-2\nu}-u^{-2\nu})^2}\exp\Big(-\frac{\nu r^{-2\nu-1}(uy+rg)}{u(r^{-2\nu}-u^{-2\nu})}\Big)\Big[\frac{x^{-2\nu}-u^{-2\nu}}{r^{2\nu+1}} I_0\Big(\frac{\nu(ru)^{-\nu-1/2}\sqrt{yg}}{r^{-2\nu}-u^{-2\nu}}\Big)$ $+\frac{r^{-2\nu}-x^{-2\nu}}{(ru)^{\nu+1/2}}\frac{\sqrt{g}}{\sqrt{y}} I_1\Big(\frac{\nu(ru)^{-\nu-1/2}\sqrt{yg}}{r^{-2\nu}-u^{-2\nu}}\Big)\Big]dydg$
$r < u$ $u \le x$	$= \frac{\nu^2 x^{-2\nu} r^{-\nu-1/2}\sqrt{g}}{u^{-\nu+3/2}(r^{-2\nu}-u^{-2\nu})\sqrt{y}}\exp\Big(-\frac{\nu r^{-2\nu-1}(uy+rg)}{u(r^{-2\nu}-u^{-2\nu})}\Big) I_1\Big(\frac{\nu(ru)^{-\nu-1/2}\sqrt{yg}}{r^{-2\nu}-u^{-2\nu}}\Big)dydg$
(2)	$\mathbf{P}_x\big(\ell(\infty,r)=0,\ \ell(\infty,u)\in dg\big)$
	$=\begin{cases} \frac{\nu x^{-2\nu}(r^{-2\nu}-x^{-2\nu})}{u^{1-2\nu}(r^{-2\nu}-u^{-2\nu})}\exp\Big(-\frac{\nu r^{-2\nu}g}{u(r^{-2\nu}-u^{-2\nu})}\Big)dg, & r\le x\le u\\ \frac{\nu x^{-2\nu}}{u^{1-2\nu}}\exp\Big(-\frac{\nu r^{-2\nu}g}{u(r^{-2\nu}-u^{-2\nu})}\Big)dg, & u\le x\end{cases}$
(3)	$\mathbf{P}_x\big(\ell(\infty,r)=0,\ \ell(\infty,u)=0\big) = 1-\frac{x^{-2\nu}}{u^{-2\nu}},\qquad u\le x$
1.19.2 $0 < x$	$\mathbf{P}_x\Big(\int_0^\tau \frac{ds}{R_s^{(n)}}\in dy\Big) = \frac{(2\lambda)^{1/4-\nu/2}(2\nu+1)\operatorname{sh}^{\nu+3/2}(y\sqrt{\lambda/2})}{2x^{\nu+1/2}\operatorname{ch}^{\nu+3/2}(y\sqrt{\lambda/2})}$ $\times\exp\Big(-\frac{x\sqrt{2\lambda}\operatorname{ch}(y\sqrt{2\lambda})}{\operatorname{sh}(y\sqrt{2\lambda})}\Big)M_{-\nu-3/2,\nu}\Big(\frac{2x\sqrt{2\lambda}}{\operatorname{sh}(y\sqrt{2\lambda})}\Big)dy$
1.19.5 $0 < x$	$\mathbf{E}_x\Big\{\exp\Big(-\gamma\int_0^\tau\frac{ds}{R_s^{(n)}}\Big);R_\tau^{(n)}\in dz\Big\}$
$x \le z$	$=\frac{\sqrt{\lambda}\Gamma(\nu+1/2+\gamma/\sqrt{2\lambda})z^{\nu+1/2}}{\sqrt{2}\Gamma(2\nu+1)x^{\nu+1/2}}M_{-\gamma/\sqrt{2\lambda},\nu}(2x\sqrt{2\lambda})W_{-\gamma/\sqrt{2\lambda},\nu}(2z\sqrt{2\lambda})dz$
$z \le x$	$=\frac{\sqrt{\lambda}\Gamma(\nu+1/2+\gamma/\sqrt{2\lambda})z^{\nu+1/2}}{\sqrt{2}\Gamma(2\nu+1)x^{\nu+1/2}}W_{-\gamma/\sqrt{2\lambda},\nu}(2x\sqrt{2\lambda})M_{-\gamma/\sqrt{2\lambda},\nu}(2z\sqrt{2\lambda})dz$

4 $R_s^{(n)}$ $\nu \ge 0, \ n = 2\nu + 2 \ge 2,$ $\tau \sim \mathrm{Exp}(\lambda)$, independent of $R^{(n)}$

1.19.6 $0 < x$

$$\mathbf{P}_x\Big(\int_0^\tau \frac{ds}{R_s^{(n)}} \in dy,\ R_\tau^{(n)} \in dz\Big)$$

$$= \frac{\lambda\sqrt{2\lambda} z^{\nu+1}}{x^\nu \operatorname{sh}(y\sqrt{\lambda/2})} \exp\Big(-\frac{(x+z)\sqrt{2\lambda}\operatorname{ch}(y\sqrt{\lambda/2})}{\operatorname{sh}(y\sqrt{\lambda/2})}\Big) I_{2\nu}\Big(\frac{2\sqrt{2\lambda xz}}{\operatorname{sh}(y\sqrt{\lambda/2})}\Big)dydz$$

1.19.8 $0 < x$

$$\mathbf{P}_x\Big(\int_0^t \frac{ds}{R_s^{(n)}} \in dy,\ R_t^{(n)} \in dz\Big) = \frac{z^{\nu+1}}{x^\nu} \operatorname{is}_t(2\nu, y/2, 0, x+z, \sqrt{xz})dydz$$

1.20.3 $0 < x$

$$\mathbf{E}_x \exp\Big(-\frac{\gamma^2}{2}\int_0^t \frac{ds}{(R_s^{(n)})^2}\Big)$$

$$= \frac{(2t)^{(\nu+1)/2}\Gamma(1+\nu/2+\sqrt{\nu^2+\gamma^2}/2)}{x^{\nu+1}\Gamma(1+\sqrt{\nu^2+\gamma^2})} e^{-x^2/4t} M_{-\nu/2-1/2,\sqrt{\nu^2+\gamma^2}/2}\Big(\frac{x^2}{2t}\Big)$$

1.20.4 $0 < x$

$$\mathbf{P}_x\Big(\int_0^t \frac{ds}{(R_s^{(n)})^2} \in dy\Big) = \frac{2(2t)^{(\nu+1)/2}}{x^{\nu+1}} e^{-\nu^2 y/2 - x^2/4t}\ \operatorname{m}_{2y}\Big(\frac{\nu+1}{2}, \frac{x^2}{4t}\Big)dy$$

1.20.5

$$\mathbf{E}_x\Big\{\exp\Big(-\frac{\gamma^2}{2}\int_0^\tau \frac{ds}{(R_s^{(n)})^2}\Big); R_\tau^{(n)} \in dz\Big\}$$

$$= \begin{cases} 2\lambda x^{-\nu} z^{\nu+1} I_{\sqrt{\nu^2+\gamma^2}}(x\sqrt{2\lambda}) K_{\sqrt{\nu^2+\gamma^2}}(z\sqrt{2\lambda})dz, & 0 < x \le z \\ 2\lambda x^{-\nu} z^{\nu+1} K_{\sqrt{\nu^2+\gamma^2}}(x\sqrt{2\lambda}) I_{\sqrt{\nu^2+\gamma^2}}(z\sqrt{2\lambda})dz, & z \le x \end{cases}$$

1.20.6 $0 < x$

$$\mathbf{P}_x\Big(\int_0^\tau \frac{ds}{(R_s^{(n)})^2} \in dy,\ R_\tau^{(n)} \in dz\Big) = \frac{\lambda z^{\nu+1}}{x^\nu} e^{-\nu^2 y/2} \operatorname{ki}_{y/2}(x\sqrt{2\lambda}, z\sqrt{2\lambda})dydz$$

1.20.7

$$\mathbf{E}_x\Big\{\exp\Big(-\frac{\gamma^2}{2}\int_0^t \frac{ds}{(R_s^{(n)})^2}\Big); R_t^{(n)} \in dz\Big\} = \frac{z^{\nu+1}}{x^\nu t} e^{-(x^2+z^2)/2t} I_{\sqrt{\nu^2+\gamma^2}}\Big(\frac{xz}{t}\Big)dz$$

1.20.8 $0 < x$

$$\mathbf{P}_x\Big(\int_0^t \frac{ds}{(R_s^{(n)})^2} \in dy,\ R_t^{(n)} \in dz\Big) = \frac{z^{\nu+1}}{2tx^\nu} e^{-(x^2+z^2)/2t - \nu^2 y/2}\ \operatorname{i}_{y/2}\Big(\frac{xz}{t}\Big)dydz$$

1.21.3 $0 < x$

$$\mathbf{E}_x \exp\Big(-\int_0^t \Big(\frac{p^2}{2(R_s^{(n)})^2} + \frac{q^2 (R_s^{(n)})^2}{2}\Big)ds\Big) = \frac{(2\operatorname{sh}(tq))^{(\nu+1)/2}}{x^{\nu+1}(q\operatorname{ch}(tq))^{(\nu+1)/2}}$$

$$\times \frac{\Gamma(1+\nu/2+\sqrt{\nu^2+p^2}/2)}{\Gamma(1+\sqrt{\nu^2+p^2})} \exp\Big(-\frac{x^2 q \operatorname{ch}(2tq)}{2\operatorname{sh}(2tq)}\Big) M_{-\frac{\nu+1}{2}, \frac{\sqrt{\nu^2+p^2}}{2}}\Big(\frac{x^2 q}{\operatorname{sh}(2tq)}\Big)$$

4	$R_s^{(n)}$ $\quad \nu \ge 0,\ n = 2\nu + 2 \ge 2,\qquad \tau \sim \mathrm{Exp}(\lambda)$, independent of $R^{(n)}$
(1)	$\mathbf{E}_x\Big\{\exp\Big(-\frac{q^2}{2}\int_0^t \big(R_s^{(n)}\big)^2 ds\Big);\ \int_0^t \frac{ds}{\big(R_s^{(n)}\big)^2} \in dy\Big\}$
$0 < x$	$= \frac{2(2\,\mathrm{sh}(tq))^{(\nu+1)/2}}{x^{\nu+1}(q\,\mathrm{ch}(tq))^{(\nu+1)/2}}\exp\Big(-\frac{\nu^2 y}{2} - \frac{x^2 q\,\mathrm{ch}(2tq)}{2\,\mathrm{sh}(2tq)}\Big)\,\mathrm{m}_{2y}\Big(\frac{\nu+1}{2}, \frac{x^2 q}{2\,\mathrm{sh}(2tq)}\Big)dy$
1.21.5 $0 < x$	$\mathbf{E}_x\Big\{\exp\Big(-\int_0^\tau \Big(\frac{p^2}{2\big(R_s^{(n)}\big)^2} + \frac{q^2\big(R_s^{(n)}\big)^2}{2}\Big)ds\Big);\ R_\tau^{(n)} \in dz\Big\}$
$x \le z$	$= \frac{\lambda\Gamma((1+\sqrt{p^2+\nu^2}+\lambda/q)/2)}{q\Gamma(1+\sqrt{p^2+\nu^2})x^{\nu+1}z^{-\nu}} M_{-\lambda/2q,\sqrt{p^2+\nu^2}/2}(qx^2) W_{-\lambda/2q,\sqrt{p^2+\nu^2}/2}(qz^2)dz$
$z \le x$	$= \frac{\lambda\Gamma((1+\sqrt{p^2+\nu^2}+\lambda/q)/2)}{q\Gamma(1+\sqrt{p^2+\nu^2})x^{\nu+1}z^{-\nu}} W_{-\lambda/2q,\sqrt{p^2+\nu^2}/2}(qx^2) M_{-\lambda/2q,\sqrt{p^2+\nu^2}/2}(qz^2)dz$
1.21.7	$\mathbf{E}_x\Big\{\exp\Big(-\int_0^t \Big(\frac{p^2}{2\big(R_s^{(n)}\big)^2} + \frac{q^2\big(R_s^{(n)}\big)^2}{2}\Big)ds\Big);\ R_t^{(n)} \in dz\Big\}$
	$= \frac{qz^{\nu+1}}{x^\nu\,\mathrm{sh}(tq)}\exp\Big(-\frac{(x^2+z^2)q\,\mathrm{ch}(tq)}{2\,\mathrm{sh}(tq)}\Big) I_{\sqrt{p^2+\nu^2}}\Big(\frac{xzq}{\mathrm{sh}(tq)}\Big)dz$
(1)	$\mathbf{E}_x\Big\{\exp\Big(-\frac{q^2}{2}\int_0^t \big(R_s^{(n)}\big)^2 ds\Big);\ \int_0^t \frac{ds}{\big(R_s^{(n)}\big)^2} \in dy,\ R_t^{(n)} \in dz\Big\}$ $= \frac{qz^{\nu+1}}{2x^\nu\,\mathrm{sh}(tq)}\exp\Big(-\frac{\nu^2 y}{2} - \frac{(x^2+z^2)q\,\mathrm{ch}(tq)}{2\,\mathrm{sh}(tq)}\Big)\,\mathrm{i}_{y/2}\Big(\frac{xzq}{\mathrm{sh}(tq)}\Big)dydz$
(2)	$\mathbf{E}_x\Big\{\exp\Big(-\frac{p^2}{2}\int_0^t \frac{ds}{\big(R_s^{(n)}\big)^2}\Big);\ \int_0^t \big(R_s^{(n)}\big)^2 ds \in dy,\ R_t^{(n)} \in dz\Big\}$ $= \frac{z^{\nu+1}}{x^\nu}\,\mathrm{is}_y(\sqrt{p^2+\nu^2}, t, 0, (x^2+z^2)/2, xz/2)dydz$
1.21.8 $0 < x$	$\mathbf{P}_x\Big(\int_0^t \big(R_s^{(n)}\big)^2 ds \in dg,\ \int_0^t \frac{ds}{\big(R_s^{(n)}\big)^2} \in dy,\ R_t^{(n)} \in dz\Big)$ $= \frac{z^{\nu+1}}{2x^\nu} e^{-\nu^2 y/2}\,\mathrm{ei}_g(y/2, t, (x^2+z^2)/2, xz)dgdydz$
1.22.1 (1)	$\mathbf{E}_x\Big\{\exp\Big(-\frac{p^2}{2}\int_0^\tau \frac{ds}{\big(R_s^{(n)}\big)^2}\Big);\ \int_0^\tau \frac{ds}{R_s^{(n)}} \in dy\Big\} = \frac{(2\lambda)^{1/4-\nu/2}\,\mathrm{sh}^{\nu+3/2}(y\sqrt{\lambda/2})}{2x^{\nu+1/2}\,\mathrm{ch}^{\nu+3/2}(y\sqrt{\lambda/2})}$

4 $R_s^{(n)}$ $\nu\ge 0,\ n=2\nu+2\ge 2,$ $\tau\sim\operatorname{Exp}(\lambda)$, independent of $R^{(n)}$

0 < x

$$\times\frac{\Gamma(2+\nu+\sqrt{\nu^2+p^2})}{\Gamma(1+2\sqrt{\nu^2+p^2})}\exp\Big(-\frac{x\sqrt{2\lambda}\operatorname{ch}(y\sqrt{2\lambda})}{\operatorname{sh}(y\sqrt{2\lambda})}\Big)M_{-\nu-\frac32,\sqrt{\nu^2+p^2}}\Big(\frac{2x\sqrt{2\lambda}}{\operatorname{sh}(y\sqrt{2\lambda})}\Big)dy$$

1.22.2
0 < x

$$\mathbf{P}_x\Big(\int_0^\tau\frac{ds}{(R_s^{(n)})^2}\in dg,\ \int_0^\tau\frac{ds}{R_s^{(n)}}\in dy\Big)=\frac{(2\lambda)^{1/4-\nu/2}\operatorname{sh}^{\nu+3/2}(y\sqrt{\lambda/2})}{4x^{\nu+1/2}\operatorname{ch}^{\nu+3/2}(y\sqrt{\lambda/2})}$$

$$\times\exp\Big(-\frac{\nu^2g}{2}-\frac{x\sqrt{2\lambda}\operatorname{ch}(y\sqrt{2\lambda})}{\operatorname{sh}(y\sqrt{2\lambda})}\Big)\,\mathrm{m}_{g/2}\Big(\nu+\frac32,\frac{x\sqrt{2\lambda}}{\operatorname{sh}(y\sqrt{2\lambda})}\Big)dgdy$$

1.22.5

$$\mathbf{E}_x\Big\{\exp\Big(-\int_0^\tau\Big(\frac{p^2}{2(R_s^{(n)})^2}+\frac{q}{R_s^{(n)}}\Big)ds\Big);R_\tau^{(n)}\in dz\Big\}$$

0 < x

$$=\frac{\sqrt\lambda\,\Gamma(\sqrt{\nu^2+p^2}+1/2+q/\sqrt{2\lambda})z^{\nu+1/2}}{\sqrt2\,\Gamma(2\sqrt{\nu^2+p^2}+1)x^{\nu+1/2}}$$

$$\times\begin{cases}M_{-q/\sqrt{2\lambda},\sqrt{\nu^2+p^2}}(2x\sqrt{2\lambda})W_{-q/\sqrt{2\lambda},\sqrt{\nu^2+p^2}}(2z\sqrt{2\lambda})dz, & x\le z\\ W_{-q/\sqrt{2\lambda},\sqrt{\nu^2+p^2}}(2x\sqrt{2\lambda})M_{-q/\sqrt{2\lambda},\sqrt{\nu^2+p^2}}(2z\sqrt{2\lambda})dz, & z\le x\end{cases}$$

(1)

$$\mathbf{E}_x\Big\{\exp\Big(-\frac{p^2}{2}\int_0^\tau\frac{ds}{(R_s^{(n)})^2}\Big);\ \int_0^\tau\frac{ds}{R_s^{(n)}}\in dy,\ R_\tau^{(n)}\in dz\Big\}$$

$$=\frac{\lambda\sqrt{2\lambda}z^{\nu+1}}{x^\nu\operatorname{sh}(y\sqrt{\lambda/2})}\exp\Big(-\frac{(x+z)\sqrt{2\lambda}\operatorname{ch}(y\sqrt{\lambda/2})}{\operatorname{sh}(y\sqrt{\lambda/2})}\Big)I_{2\sqrt{\nu^2+p^2}}\Big(\frac{2\sqrt{2\lambda xz}}{\operatorname{sh}(y\sqrt{\lambda/2})}\Big)dydz$$

1.22.6
0 < x

$$\mathbf{P}_x\Big(\int_0^\tau\frac{ds}{(R_s^{(n)})^2}\in dg,\ \int_0^\tau\frac{ds}{R_s^{(n)}}\in dy,\ R_\tau^{(n)}\in dz\Big)$$

$$=\frac{\lambda\sqrt{2\lambda}z^{\nu+1}}{8x^\nu\operatorname{sh}(y\sqrt{\lambda/2})}\exp\Big(-\frac{\nu^2g}{2}-\frac{(x+z)\sqrt{2\lambda}\operatorname{ch}(y\sqrt{\lambda/2})}{\operatorname{sh}(y\sqrt{\lambda/2})}\Big)\,\mathrm{i}_{g/8}\Big(\frac{2\sqrt{2\lambda xz}}{\operatorname{sh}(y\sqrt{\lambda/2})}\Big)dgdydz$$

1.22.7
(1)

$$\mathbf{E}_x\Big\{\exp\Big(-\frac{p^2}{2}\int_0^t\frac{ds}{(R_s^{(n)})^2}\Big);\frac{ds}{R_s^{(n)}}\in dy,\ R_t^{(n)}\in dz\Big\}$$

0 < x

$$=\frac{z^{\nu+1}}{x^\nu}\,\mathrm{is}_t(\sqrt{4p^2+4\nu^2},y/2,0,x+z,\sqrt{xz})dydz$$

1.22.8
0 < x

$$\mathbf{P}_x\Big(\int_0^t\frac{ds}{(R_s^{(n)})^2}\in dg,\ \int_0^t\frac{ds}{R_s^{(n)}}\in dy,\ R_t^{(n)}\in dz\Big)$$

$$=\frac{z^{\nu+1}}{8x^\nu}e^{-\nu^2g/2}\,\mathrm{ei}_t(g/8,y/2,x+z,2\sqrt{xz})dgdydz$$

4 $R_s^{(n)}$ $\nu \ge 0,\ n = 2\nu + 2 \ge 2,$ $\tau \sim \text{Exp}(\lambda)$, independent of $R^{(n)}$

1.23.1 $r < b$

$$\mathbf{E}_x\Big\{\exp\Big(-\int_0^\tau \big(p\mathbb{I}_{[0,r)}\big(R_s^{(n)}\big) + q\mathbb{I}_{[r,\infty)}\big(R_s^{(n)}\big)\big)ds\Big);\ \sup_{0\le s\le\tau} R_s^{(n)} < b\Big\}$$

$x \le r$

$$= \frac{2\lambda}{\Upsilon_p^2} + \frac{2\lambda(2(p-q)C_\nu(r\Upsilon_q, b\Upsilon_q) - (r\Upsilon_q)^{-2\nu-1}\Upsilon_p^2)r^\nu x^{-\nu} I_\nu(x\Upsilon_p)}{\Upsilon_p^2\Upsilon_q(\Upsilon_p S_\nu(b\Upsilon_q, r\Upsilon_q)I_{\nu+1}(r\Upsilon_p) + \Upsilon_q C_\nu(r\Upsilon_q, b\Upsilon_q)I_\nu(r\Upsilon_p))}$$

$r \le x$, $x \le b$

$$= \frac{2\lambda}{\Upsilon_q^2}\Big(1 - \frac{S_\nu(x\Upsilon_q, r\Upsilon_q)}{S_\nu(b\Upsilon_q, r\Upsilon_q)}\Big) + \frac{4\lambda\Upsilon_p^{-1}\Upsilon_q^{-2}(q-p)I_{\nu+1}(r\Upsilon_p)S_\nu(b\Upsilon_q, x\Upsilon_q)}{\Upsilon_p S_\nu(b\Upsilon_q, r\Upsilon_q)I_{\nu+1}(r\Upsilon_p) + \Upsilon_q C_\nu(r\Upsilon_q, b\Upsilon_q)I_\nu(r\Upsilon_p)}$$

$$- \frac{2\lambda r^{-2\nu-1}\Upsilon_q^{-2\nu-2}I_\nu(r\Upsilon_p)S_\nu(b\Upsilon_q, x\Upsilon_q)}{(\Upsilon_p S_\nu(b\Upsilon_q, r\Upsilon_q)I_{\nu+1}(r\Upsilon_p) + \Upsilon_q C_\nu(r\Upsilon_q, b\Upsilon_q)I_\nu(r\Upsilon_p))S_\nu(b\Upsilon_q, r\Upsilon_q)}$$

1.23.5 $r < b$, $z < b$

$$\mathbf{E}_x\Big\{\exp\Big(-\int_0^\tau \big(p\mathbb{I}_{[0,r)}\big(R_s^{(n)}\big) + q\mathbb{I}_{[r,\infty)}\big(R_s^{(n)}\big)\big)ds\Big);\ \sup_{0\le s\le\tau} R_s^{(n)} < b,\ R_\tau^{(n)} \in dz\Big\}$$

$z \le r$, $x \le r$

$$= \frac{2\lambda\Upsilon_p^{2\nu}z^{2\nu+1}S_\nu(r\Upsilon_p, (z+x+|z-x|)\Upsilon_p/2)I_\nu((z+x-|z-x|)\Upsilon_p/2)}{((z+x-|z-x|)/2)^\nu r^{-\nu}I_\nu(r\Upsilon_p)}dz$$

$$+ \frac{2\lambda z^{\nu+1}r^{-1}x^{-\nu}S_\nu(b\Upsilon_q, r\Upsilon_q)I_\nu(z\Upsilon_p)I_\nu(x\Upsilon_p)}{(\Upsilon_p S_\nu(b\Upsilon_q, r\Upsilon_q)I_{\nu+1}(r\Upsilon_p) + \Upsilon_q C_\nu(r\Upsilon_q, b\Upsilon_q)I_\nu(r\Upsilon_p))I_\nu(r\Upsilon_p)}dz$$

$z \le r$, $r \le x$

$$= \frac{2\lambda z^{\nu+1}r^{-\nu-1}I_\nu(z\Upsilon_p)S_\nu(b\Upsilon_q, x\Upsilon_q)}{\Upsilon_p S_\nu(b\Upsilon_q, r\Upsilon_q)I_{\nu+1}(r\Upsilon_p) + \Upsilon_q C_\nu(r\Upsilon_q, b\Upsilon_q)I_\nu(r\Upsilon_p)}dz$$

$r \le z$, $x \le r$

$$= \frac{2\lambda z^{2\nu+1}r^{-\nu-1}x^{-\nu}S_\nu(b\Upsilon_q, z\Upsilon_q)I_\nu(x\Upsilon_p)}{\Upsilon_p S_\nu(b\Upsilon_q, r\Upsilon_q)I_{\nu+1}(r\Upsilon_p) + \Upsilon_q C_\nu(r\Upsilon_q, b\Upsilon_q)I_\nu(r\Upsilon_p)}dz$$

$r \le z$, $r \le x$

$$= \frac{2\lambda\Upsilon_q^{2\nu}z^{2\nu+1}S_\nu(b\Upsilon_q, (z+x+|z-x|)\Upsilon_q/2)S_\nu((z+x-|z-x|)\Upsilon_q/2, r\Upsilon_q)}{S_\nu(b\Upsilon_q, r\Upsilon_q)}dz$$

$$+ \frac{2\lambda z^{2\nu+1}r^{-2\nu-1}I_\nu(r\Upsilon_p)S_\nu(b\Upsilon_q, z\Upsilon_q)S_\nu(b\Upsilon_q, x\Upsilon_q)}{(\Upsilon_p S_\nu(b\Upsilon_q, r\Upsilon_q)I_{\nu+1}(r\Upsilon_p) + \Upsilon_q C_\nu(r\Upsilon_q, b\Upsilon_q)I_\nu(r\Upsilon_p))S_\nu(b\Upsilon_q, r\Upsilon_q)}dz$$

1.24.1 $a < r$

$$\mathbf{E}_x\Big\{\exp\Big(-\int_0^\tau \big(p\mathbb{I}_{[0,r)}\big(R_s^{(n)}\big) + q\mathbb{I}_{[r,\infty)}\big(R_s^{(n)}\big)\big)ds\Big);\ a < \inf_{0\le s\le\tau} R_s^{(n)}\Big\}$$

$a \le x$, $x \le r$

$$= \frac{2\lambda}{\Upsilon_p^2}\Big(1 - \frac{S_\nu(r\Upsilon_p, x\Upsilon_p)}{S_\nu(r\Upsilon_p, a\Upsilon_p)}\Big) + \frac{4\lambda\Upsilon_p^{-2}\Upsilon_q^{-1}(p-q)K_{\nu+1}(r\Upsilon_q)S_\nu(x\Upsilon_p, a\Upsilon_p)}{\Upsilon_p K_\nu(r\Upsilon_q)C_\nu(r\Upsilon_p, a\Upsilon_p) + \Upsilon_q K_{\nu+1}(r\Upsilon_q)S_\nu(r\Upsilon_p, a\Upsilon_p)}$$

$$\Upsilon_s := \sqrt{2\lambda + 2s}$$

4 $R_s^{(n)}$ $\nu \ge 0,\ n = 2\nu + 2 \ge 2,$ $\tau \sim \mathrm{Exp}(\lambda)$, independent of $R^{(n)}$

$$-\frac{2\lambda r^{-2\nu-1}\Upsilon_p^{-2\nu-2}K_\nu(r\Upsilon_q)S_\nu(x\Upsilon_p,a\Upsilon_p)}{(\Upsilon_p K_\nu(r\Upsilon_q)C_\nu(r\Upsilon_p,a\Upsilon_p)+\Upsilon_q K_{\nu+1}(r\Upsilon_q)S_\nu(r\Upsilon_p,a\Upsilon_p))S_\nu(r\Upsilon_p,a\Upsilon_p)}$$

$r \le x$

$$=\frac{2\lambda}{\Upsilon_q^2}+\frac{2\lambda(2(q-p)C_\nu(r\Upsilon_p,a\Upsilon_p)-(r\Upsilon_p)^{-2\nu-1}\Upsilon_q^2)r^\nu x^{-\nu}K_\nu(x\Upsilon_q)}{\Upsilon_p\Upsilon_q^2(\Upsilon_p K_\nu(r\Upsilon_q)C_\nu(r\Upsilon_p,a\Upsilon_p)+\Upsilon_q K_{\nu+1}(r\Upsilon_q)S_\nu(r\Upsilon_p,a\Upsilon_p))}$$

1.24.3 (1) $a < r$

$$\mathbf{E}_x\Big\{\exp\Big(-p\int_0^\infty \mathbb{I}_{[0,r)}\big(R_s^{(n)}\big)ds\Big);a<\inf_{0\le s<\infty}R_s^{(n)}\Big\}$$

$$=\begin{cases}\dfrac{2\nu a\sqrt{2p}S_\nu(x\sqrt{2p},a\sqrt{2p})}{C_{\nu-1}(a\sqrt{2p},r\sqrt{2p})}, & a\le x\le r\\[2ex] 1-\dfrac{x^{-2\nu}}{r^{-2\nu}}+\dfrac{2\nu a\sqrt{2p}x^{-2\nu}S_\nu(r\sqrt{2p},a\sqrt{2p})}{r^{-2\nu}C_{\nu-1}(a\sqrt{2p},r\sqrt{2p})}, & r\le x\end{cases}$$

1.24.5 $a < r$ $a < z$

$$\mathbf{E}_x\Big\{\exp\Big(-\int_0^\tau\big(p\mathbb{I}_{[0,r)}\big(R_s^{(n)}\big)+q\mathbb{I}_{[r,\infty)}\big(R_s^{(n)}\big)\big)ds\Big);\quad a<\inf_{0\le s\le\tau}R_s^{(n)},\ R_\tau^{(n)}\in dz\Big\}$$

$z \le r$, $x \le r$

$$=\frac{2\lambda\Upsilon_p^{2\nu}z^{2\nu+1}S_\nu(r\Upsilon_p,(z+x+|z-x|)\Upsilon_p/2)S_\nu((z+x-|z-x|)\Upsilon_p/2,a\Upsilon_p)}{S_\nu(r\Upsilon_p,a\Upsilon_p)}dz$$

$$+\frac{2\lambda z^{2\nu+1}r^{-2\nu-1}K_\nu(r\Upsilon_q)S_\nu(z\Upsilon_p,a\Upsilon_p)S_\nu(x\Upsilon_p,a\Upsilon_p)}{(\Upsilon_p K_\nu(r\Upsilon_q)C_\nu(r\Upsilon_p,a\Upsilon_p)+\Upsilon_q K_{\nu+1}(r\Upsilon_q)S_\nu(r\Upsilon_p,a\Upsilon_p))S_\nu(r\Upsilon_p,a\Upsilon_p)}dz$$

$z \le r$, $r \le x$

$$=\frac{2\lambda z^{2\nu+1}r^{-\nu-1}x^{-\nu}S_\nu(z\Upsilon_p,a\Upsilon_p)K_\nu(x\Upsilon_q)}{\Upsilon_p K_\nu(r\Upsilon_q)C_\nu(r\Upsilon_p,a\Upsilon_p)+\Upsilon_q K_{\nu+1}(r\Upsilon_q)S_\nu(r\Upsilon_p,a\Upsilon_p)}dz$$

$r \le z$, $x \le r$

$$=\frac{2\lambda z^{\nu+1}r^{-\nu-1}K_\nu(z\Upsilon_q)S_\nu(x\Upsilon_p,a\Upsilon_p)}{\Upsilon_p K_\nu(r\Upsilon_q)C_\nu(r\Upsilon_p,a\Upsilon_p)+\Upsilon_q K_{\nu+1}(r\Upsilon_q)S_\nu(r\Upsilon_p,a\Upsilon_p)}dz$$

$r \le z$, $r \le x$

$$=\frac{2\lambda\Upsilon_q^{2\nu}z^{2\nu+1}K_\nu((z+x+|z-x|)\Upsilon_q/2)S_\nu((z+x-|z-x|)\Upsilon_q/2,r\Upsilon_q)}{((z+x+|z-x|)/2)^\nu r^{-\nu}K_\nu(r\Upsilon_q)}dz$$

$$+\frac{2\lambda z^{\nu+1}r^{-1}x^{-\nu}S_\nu(r\Upsilon_p,a\Upsilon_p)K_\nu(z\Upsilon_q)K_\nu(x\Upsilon_q)}{(\Upsilon_p K_\nu(r\Upsilon_q)C_\nu(r\Upsilon_p,a\Upsilon_p)+\Upsilon_q K_{\nu+1}(r\Upsilon_q)S_\nu(r\Upsilon_p,a\Upsilon_p))K_\nu(r\Upsilon_q)}dz$$

$\Upsilon_s := \sqrt{2\lambda+2s}$

4 $R_s^{(n)}$ $\nu \ge 0,\ n = 2\nu + 2 \ge 2,$ $\tau \sim \mathrm{Exp}(\lambda)$, independent of $R^{(n)}$

1.25.1 $\mathbf{E}_x\{e^{-\gamma\ell(\tau,r)}; a < \inf_{0\le s\le\tau} R_s^{(n)}, \sup_{0\le s\le\tau} R_s^{(n)} < b\}$ $a < r < b$

$a \le x$, $x \le r$

$$= 1 - \frac{S_\nu(r\sqrt{2\lambda}, x\sqrt{2\lambda}) + S_\nu(x\sqrt{2\lambda}, a\sqrt{2\lambda})}{S_\nu(r\sqrt{2\lambda}, a\sqrt{2\lambda})}$$

$$+ \frac{(S_\nu(b\sqrt{2\lambda}, a\sqrt{2\lambda}) - S_\nu(b\sqrt{2\lambda}, r\sqrt{2\lambda}) - S_\nu(r\sqrt{2\lambda}, a\sqrt{2\lambda}))S_\nu(x\sqrt{2\lambda}, a\sqrt{2\lambda})}{(2(2\lambda)^\nu r^{2\nu+1}\gamma S_\nu(b\sqrt{2\lambda}, r\sqrt{2\lambda})S_\nu(r\sqrt{2\lambda}, a\sqrt{2\lambda}) + S_\nu(b\sqrt{2\lambda}, a\sqrt{2\lambda}))S_\nu(r\sqrt{2\lambda}, a\sqrt{2\lambda})}$$

$r \le x$, $x \le b$

$$= 1 - \frac{S_\nu(b\sqrt{2\lambda}, x\sqrt{2\lambda}) + S_\nu(x\sqrt{2\lambda}, r\sqrt{2\lambda})}{S_\nu(b\sqrt{2\lambda}, r\sqrt{2\lambda})}$$

$$+ \frac{(S_\nu(b\sqrt{2\lambda}, a\sqrt{2\lambda}) - S_\nu(b\sqrt{2\lambda}, r\sqrt{2\lambda}) - S_\nu(r\sqrt{2\lambda}, a\sqrt{2\lambda}))S_\nu(b\sqrt{2\lambda}, x\sqrt{2\lambda})}{(2(2\lambda)^\nu r^{2\nu+1}\gamma S_\nu(b\sqrt{2\lambda}, r\sqrt{2\lambda})S_\nu(r\sqrt{2\lambda}, a\sqrt{2\lambda}) + S_\nu(b\sqrt{2\lambda}, a\sqrt{2\lambda}))S_\nu(b\sqrt{2\lambda}, r\sqrt{2\lambda})}$$

1.25.2 $\mathbf{P}_x\big(\ell(\tau,r) \in dy, a < \inf_{0\le s\le\tau} R_s^{(n)}, \sup_{0\le s\le\tau} R_s^{(n)} < b\big)$

$$= \exp\Big(-\frac{y(2\lambda)^{-\nu} r^{-2\nu-1} S_\nu(b\sqrt{2\lambda}, a\sqrt{2\lambda})}{2S_\nu(b\sqrt{2\lambda}, r\sqrt{2\lambda})S_\nu(r\sqrt{2\lambda}, a\sqrt{2\lambda})}\Big)$$

$$\times \begin{cases} \dfrac{(S_\nu(b\sqrt{2\lambda}, a\sqrt{2\lambda}) - S_\nu(b\sqrt{2\lambda}, r\sqrt{2\lambda}) - S_\nu(r\sqrt{2\lambda}, a\sqrt{2\lambda}))S_\nu(x\sqrt{2\lambda}, a\sqrt{2\lambda})}{2(2\lambda)^\nu r^{2\nu+1} S_\nu(b\sqrt{2\lambda}, r\sqrt{2\lambda})S_\nu^2(r\sqrt{2\lambda}, a\sqrt{2\lambda})} dy, & a \le x \le r \\ \dfrac{(S_\nu(b\sqrt{2\lambda}, a\sqrt{2\lambda}) - S_\nu(b\sqrt{2\lambda}, r\sqrt{2\lambda}) - S_\nu(r\sqrt{2\lambda}, a\sqrt{2\lambda}))S_\nu(b\sqrt{2\lambda}, x\sqrt{2\lambda})}{2(2\lambda)^\nu r^{2\nu+1} S_\nu^2(b\sqrt{2\lambda}, r\sqrt{2\lambda})S_\nu(r\sqrt{2\lambda}, a\sqrt{2\lambda})} dy, & r \le x \le b \end{cases}$$

(1) $\mathbf{P}_x\big(\ell(\tau,r) = 0, a < \inf_{0\le s\le\tau} R_s^{(n)}, \sup_{0\le s\le\tau} R_s^{(n)} < b\big)$

$$= \begin{cases} 1 - \dfrac{S_\nu(r\sqrt{2\lambda}, x\sqrt{2\lambda}) + S_\nu(x\sqrt{2\lambda}, a\sqrt{2\lambda})}{S_\nu(r\sqrt{2\lambda}, a\sqrt{2\lambda})}, & a \le x \le r \\ 1 - \dfrac{S_\nu(b\sqrt{2\lambda}, x\sqrt{2\lambda}) + S_\nu(x\sqrt{2\lambda}, r\sqrt{2\lambda})}{S_\nu(b\sqrt{2\lambda}, r\sqrt{2\lambda})}, & r \le x \le b \end{cases}$$

1.25.5 $\mathbf{E}_x\{e^{-\gamma\ell(\tau,r)}; a < \inf_{0\le s\le\tau} R_s^{(n)}, \sup_{0\le s\le\tau} R_s^{(n)} < b, R_\tau^{(n)} \in dz\}$ $a < r < b$

4 $R_s^{(n)}$ $\nu \geq 0,\ n = 2\nu+2 \geq 2,$ $\tau \sim \mathrm{Exp}(\lambda)$, independent of $R^{(n)}$

$z \leq r$, $x \leq r$

$$= \frac{(2\lambda)^{\nu+1} z^{2\nu+1} S_\nu(r\sqrt{2\lambda}, (z+x+|z-x|)\sqrt{\lambda/2}) S_\nu((z+x-|z-x|)\sqrt{\lambda/2}, a\sqrt{2\lambda})}{S_\nu(r\sqrt{2\lambda}, a\sqrt{2\lambda})} dz$$

$$+ \frac{(2\lambda)^{\nu+1} z^{2\nu+1} S_\nu(b\sqrt{2\lambda}, r\sqrt{2\lambda}) S_\nu(z\sqrt{2\lambda}, a\sqrt{2\lambda}) S_\nu(x\sqrt{2\lambda}, a\sqrt{2\lambda}) dz}{(2\gamma(2\lambda)^\nu r^{2\nu+1} S_\nu(b\sqrt{2\lambda}, r\sqrt{2\lambda}) S_\nu(r\sqrt{2\lambda}, a\sqrt{2\lambda}) + S_\nu(b\sqrt{2\lambda}, a\sqrt{2\lambda})) S_\nu(r\sqrt{2\lambda}, a\sqrt{2\lambda})}$$

$z \leq r$, $r \leq x$

$$= \frac{(2\lambda)^{\nu+1} z^{2\nu+1} S_\nu(b\sqrt{2\lambda}, x\sqrt{2\lambda}) S_\nu(z\sqrt{2\lambda}, a\sqrt{2\lambda})}{2\gamma(2\lambda)^\nu r^{2\nu+1} S_\nu(b\sqrt{2\lambda}, r\sqrt{2\lambda}) S_\nu(r\sqrt{2\lambda}, a\sqrt{2\lambda}) + S_\nu(b\sqrt{2\lambda}, a\sqrt{2\lambda})} dz$$

$r \leq z$, $x \leq r$

$$= \frac{(2\lambda)^{\nu+1} z^{2\nu+1} S_\nu(b\sqrt{2\lambda}, z\sqrt{2\lambda}) S_\nu(x\sqrt{2\lambda}, a\sqrt{2\lambda})}{2\gamma(2\lambda)^\nu r^{2\nu+1} S_\nu(b\sqrt{2\lambda}, r\sqrt{2\lambda}) S_\nu(r\sqrt{2\lambda}, a\sqrt{2\lambda}) + S_\nu(b\sqrt{2\lambda}, a\sqrt{2\lambda})} dz$$

$r \leq z$, $r \leq x$

$$= \frac{(2\lambda)^{\nu+1} z^{2\nu+1} S_\nu(b\sqrt{2\lambda}, (z+x+|z-x|)\sqrt{\lambda/2}) S_\nu((z+x-|z-x|)\sqrt{\lambda/2}, r\sqrt{2\lambda})}{S_\nu(b\sqrt{2\lambda}, r\sqrt{2\lambda})} dz$$

$$+ \frac{(2\lambda)^{\nu+1} z^{2\nu+1} S_\nu(b\sqrt{2\lambda}, z\sqrt{2\lambda}) S_\nu(r\sqrt{2\lambda}, a\sqrt{2\lambda}) S_\nu(b\sqrt{2\lambda}, x\sqrt{2\lambda}) dz}{(2\gamma(2\lambda)^\nu r^{2\nu+1} S_\nu(b\sqrt{2\lambda}, r\sqrt{2\lambda}) S_\nu(r\sqrt{2\lambda}, a\sqrt{2\lambda}) + S_\nu(b\sqrt{2\lambda}, a\sqrt{2\lambda})) S_\nu(b\sqrt{2\lambda}, r\sqrt{2\lambda})}$$

1.25.6 $\mathbf{P}_x\big(\ell(\tau, r) \in dy, a < \inf_{0 \leq s \leq \tau} R_s^{(n)}, \sup_{0 \leq s \leq \tau} R_s^{(n)} < b, R_\tau^{(n)} \in dz\big)$

$a < r$, $r < b$

$z \leq r$

$$= \exp\Big(-\frac{y(2\lambda)^{-\nu} r^{-2\nu-1} S_\nu(b\sqrt{2\lambda}, a\sqrt{2\lambda})}{2S_\nu(b\sqrt{2\lambda}, r\sqrt{2\lambda}) S_\nu(r\sqrt{2\lambda}, a\sqrt{2\lambda})}\Big)$$

$$\times \begin{cases} \dfrac{\lambda z^{2\nu+1} S_\nu(z\sqrt{2\lambda}, a\sqrt{2\lambda}) S_\nu(x\sqrt{2\lambda}, a\sqrt{2\lambda})}{r^{2\nu+1} S_\nu^2(r\sqrt{2\lambda}, a\sqrt{2\lambda})} dydz, & 0 \leq x \leq r \\ \dfrac{\lambda z^{2\nu+1} S_\nu(b\sqrt{2\lambda}, x\sqrt{2\lambda}) S_\nu(z\sqrt{2\lambda}, a\sqrt{2\lambda})}{r^{2\nu+1} S_\nu(b\sqrt{2\lambda}, r\sqrt{2\lambda}) S_\nu(r\sqrt{2\lambda}, a\sqrt{2\lambda})} dydz, & r \leq x \end{cases}$$

$r \leq z$

$$= \exp\Big(-\frac{y(2\lambda)^{-\nu} r^{-2\nu-1} S_\nu(b\sqrt{2\lambda}, a\sqrt{2\lambda})}{2S_\nu(b\sqrt{2\lambda}, r\sqrt{2\lambda}) S_\nu(r\sqrt{2\lambda}, a\sqrt{2\lambda})}\Big)$$

$$\times \begin{cases} \dfrac{\lambda z^{2\nu+1} S_\nu(b\sqrt{2\lambda}, z\sqrt{2\lambda}) S_\nu(x\sqrt{2\lambda}, a\sqrt{2\lambda})}{r^{2\nu+1} S_\nu(b\sqrt{2\lambda}, r\sqrt{2\lambda}) S_\nu(r\sqrt{2\lambda}, a\sqrt{2\lambda})} dydz, & 0 \leq x \leq r \\ \dfrac{\lambda z^{2\nu+1} S_\nu(b\sqrt{2\lambda}, z\sqrt{2\lambda}) S_\nu(b\sqrt{2\lambda}, x\sqrt{2\lambda})}{r^{2\nu+1} S_\nu^2(b\sqrt{2\lambda}, r\sqrt{2\lambda})} dydz, & r \leq x \end{cases}$$

4 $R_s^{(n)}$ $\nu \ge 0,\ n = 2\nu + 2 \ge 2,$ $\tau \sim \operatorname{Exp}(\lambda)$, independent of $R^{(n)}$

(1) $\mathbf{P}_x\big(\ell(\tau,r)=0, a<\inf_{0\le s\le\tau} R_s^{(n)}, \sup_{0\le s\le\tau} R_s^{(n)}<b, R_\tau^{(n)}\in dz\big)$

$z \le r$, $x \le r$

$$= \frac{(2\lambda)^{\nu+1} z^{2\nu+1} S_\nu(r\sqrt{2\lambda}, (z+x+|z-x|)\sqrt{\lambda/2}) S_\nu((z+x-|z-x|)\sqrt{\lambda/2}, a\sqrt{2\lambda})}{S_\nu(r\sqrt{2\lambda}, a\sqrt{2\lambda})} dz$$

$r \le z$, $r \le x$

$$= \frac{(2\lambda)^{\nu+1} z^{2\nu+1} S_\nu(b\sqrt{2\lambda}, (z+x+|z-x|)\sqrt{\lambda/2}) S_\nu((z+x-|z-x|)\sqrt{\lambda/2}, r\sqrt{2\lambda})}{S_\nu(b\sqrt{2\lambda}, r\sqrt{2\lambda})} dz$$

1.26.1, $a < r$, $r < b$

$$\mathbf{E}_x\Big\{\exp\Big(-\int_0^\tau \big(p\mathbb{I}_{[0,r)}\big(R_s^{(n)}\big) + q\mathbb{I}_{[r,\infty)}\big(R_s^{(n)}\big)\big)ds\Big);\ a<\inf_{0\le s\le\tau} R_s^{(n)}, \sup_{0\le s\le\tau} R_s^{(n)}<b\Big\}$$

$a \le x$, $x \le r$

$$= \frac{2\lambda}{\Upsilon_p^2}\Big(1-\frac{S_\nu(r\Upsilon_p, x\Upsilon_p)}{S_\nu(r\Upsilon_p, a\Upsilon_p)}\Big) - \frac{2\lambda r^{-2\nu-1}\Upsilon_p^{-2\nu-2} S_\nu(b\Upsilon_q, r\Upsilon_q) S_\nu(x\Upsilon_p, a\Upsilon_p) S_\nu^{-1}(r\Upsilon_p, a\Upsilon_p)}{\Upsilon_p S_\nu(b\Upsilon_q, r\Upsilon_q) C_\nu(r\Upsilon_p, a\Upsilon_p) + \Upsilon_q C_\nu(r\Upsilon_q, b\Upsilon_q) S_\nu(r\Upsilon_p, a\Upsilon_p)}$$
$$+ \frac{2\lambda(2(p-q)C_\nu(r\Upsilon_q, b\Upsilon_q) - (r\Upsilon_q)^{-2\nu-1}\Upsilon_p^2) S_\nu(x\Upsilon_p, a\Upsilon_p)}{\Upsilon_p^2\Upsilon_q(\Upsilon_p S_\nu(b\Upsilon_q, r\Upsilon_q) C_\nu(r\Upsilon_p, a\Upsilon_p) + \Upsilon_q C_\nu(r\Upsilon_q, b\Upsilon_q) S_\nu(r\Upsilon_p, a\Upsilon_p))}$$

$r \le x$, $x \le b$

$$= \frac{2\lambda}{\Upsilon_q^2}\Big(1-\frac{S_\nu(x\Upsilon_q, r\Upsilon_q)}{S_\nu(b\Upsilon_q, r\Upsilon_q)}\Big) - \frac{2\lambda r^{-2\nu-1}\Upsilon_q^{-2\nu-2} S_\nu(r\Upsilon_p, a\Upsilon_p) S_\nu(b\Upsilon_q, x\Upsilon_q) S_\nu^{-1}(b\Upsilon_q, r\Upsilon_q)}{\Upsilon_p S_\nu(b\Upsilon_q, r\Upsilon_q) C_\nu(r\Upsilon_p, a\Upsilon_p) + \Upsilon_q C_\nu(r\Upsilon_q, b\Upsilon_q) S_\nu(r\Upsilon_p, a\Upsilon_p)}$$
$$+ \frac{2\lambda(2(q-p)C_\nu(r\Upsilon_p, a\Upsilon_p) - (r\Upsilon_p)^{-2\nu-1}\Upsilon_q^2) S_\nu(b\Upsilon_q, x\Upsilon_q)}{\Upsilon_p\Upsilon_q^2(\Upsilon_p S_\nu(b\Upsilon_q, r\Upsilon_q) C_\nu(r\Upsilon_p, a\Upsilon_p) + \Upsilon_q C_\nu(r\Upsilon_q, b\Upsilon_q) S_\nu(r\Upsilon_p, a\Upsilon_p))}$$

1.26.5, $r < b$, $a < r$

$$\mathbf{E}_x\Big\{\exp\Big(-\int_0^\tau \big(p\mathbb{I}_{[0,r)}\big(R_s^{(n)}\big) + q\mathbb{I}_{[r,\infty)}\big(R_s^{(n)}\big)\big)ds\Big);\ a<\inf_{0\le s\le\tau} R_s^{(n)}, \sup_{0\le s\le\tau} R_s^{(n)}<b,\ R_\tau^{(n)}\in dz\Big\}$$

$z \le r$, $x \le r$

$$= \frac{2\lambda\Upsilon_p^{2\nu} z^{2\nu+1} S_\nu(r\Upsilon_p, (z+x+|z-x|)\Upsilon_p/2) S_\nu((z+x-|z-x|)\Upsilon_p/2, a\Upsilon_p)}{S_\nu(r\Upsilon_p, a\Upsilon_p)} dz$$
$$+ \frac{2\lambda z^{2\nu+1} r^{-2\nu-1} S_\nu(b\Upsilon_q, r\Upsilon_q) S_\nu(z\Upsilon_p, a\Upsilon_p) S_\nu(x\Upsilon_p, a\Upsilon_p)}{(\Upsilon_p S_\nu(b\Upsilon_q, r\Upsilon_q) C_\nu(r\Upsilon_p, a\Upsilon_p) + \Upsilon_q C_\nu(r\Upsilon_q, b\Upsilon_q) S_\nu(r\Upsilon_p, a\Upsilon_p)) S_\nu(r\Upsilon_p, a\Upsilon_p)} dz$$

$\Upsilon_s := \sqrt{2\lambda + 2s}$

4 $R_s^{(n)}$ $\nu \ge 0,\ n = 2\nu + 2 \ge 2,$ $\tau \sim \mathrm{Exp}(\lambda)$, independent of $R^{(n)}$

$z \le r$, $r \le x$
$$= \frac{2\lambda z^{2\nu+1} r^{-2\nu-1} S_\nu(z\Upsilon_p, a\Upsilon_p) S_\nu(b\Upsilon_q, x\Upsilon_q)}{\Upsilon_p S_\nu(b\Upsilon_q, r\Upsilon_q) C_\nu(r\Upsilon_p, a\Upsilon_p) + \Upsilon_q C_\nu(r\Upsilon_q, b\Upsilon_q) S_\nu(r\Upsilon_p, a\Upsilon_p)} dz$$

$r \le z$, $x \le r$
$$= \frac{2\lambda z^{2\nu+1} r^{-2\nu-1} S_\nu(b\Upsilon_q, z\Upsilon_q) S_\nu(x\Upsilon_p, a\Upsilon_p)}{\Upsilon_p S_\nu(b\Upsilon_q, r\Upsilon_q) C_\nu(r\Upsilon_p, a\Upsilon_p) + \Upsilon_q C_\nu(r\Upsilon_q, b\Upsilon_q) S_\nu(r\Upsilon_p, a\Upsilon_p)} dz$$

$r \le z$, $r \le x$
$$= \frac{2\lambda \Upsilon_q^{2\nu} z^{2\nu+1} S_\nu(b\Upsilon_q, (z+x+|z-x|)\Upsilon_q/2) S_\nu((z+x-|z-x|)\Upsilon_q/2, r\Upsilon_q)}{S_\nu(b\Upsilon_q, r\Upsilon_q)} dz$$
$$+ \frac{2\lambda z^{2\nu+1} r^{-2\nu-1} S_\nu(r\Upsilon_p, a\Upsilon_p) S_\nu(b\Upsilon_q, z\Upsilon_q) S_\nu(b\Upsilon_q, x\Upsilon_q)}{(\Upsilon_p S_\nu(b\Upsilon_q, r\Upsilon_q) C_\nu(r\Upsilon_p, a\Upsilon_p) + \Upsilon_q C_\nu(r\Upsilon_q, b\Upsilon_q) S_\nu(r\Upsilon_p, a\Upsilon_p)) S_\nu(b\Upsilon_q, r\Upsilon_q)} dz$$

1.27.1 $$\mathbf{E}_x\Big\{\exp\Big(-\int_0^\tau \big(p\mathbb{I}_{[0,r)}(R_s^{(n)}) + q\mathbb{I}_{[r,\infty)}(R_s^{(n)})\big)ds\Big); \ell(\tau, r) \in dy\Big\}$$
$$= \exp\Big(-\frac{y\Upsilon_p I_{\nu+1}(r\Upsilon_p)}{2I_\nu(r\Upsilon_p)} - \frac{y\Upsilon_q K_{\nu+1}(r\Upsilon_q)}{2K_\nu(r\Upsilon_q)}\Big)$$
$$\times \begin{cases} \dfrac{\lambda r^\nu I_\nu(x\Upsilon_p)}{x^\nu I_\nu(r\Upsilon_p)}\Big(\dfrac{I_{\nu+1}(r\Upsilon_p)}{\Upsilon_p I_\nu(r\Upsilon_p)} + \dfrac{K_{\nu+1}(r\Upsilon_q)}{\Upsilon_q K_\nu(r\Upsilon_q)}\Big)dy, & 0 \le x \le r \\ \dfrac{\lambda r^\nu K_\nu(x\Upsilon_q)}{x^\nu K_\nu(r\Upsilon_q)}\Big(\dfrac{I_{\nu+1}(r\Upsilon_p)}{\Upsilon_p I_\nu(r\Upsilon_p)} + \dfrac{K_{\nu+1}(r\Upsilon_q)}{\Upsilon_q K_\nu(r\Upsilon_q)}\Big)dy, & r \le x \end{cases}$$

(1) $$\mathbf{E}_x\Big\{\exp\Big(-\int_0^\tau \big(p\mathbb{I}_{[0,r)}(R_s^{(n)}) + q\mathbb{I}_{[r,\infty)}(R_s^{(n)})\big)ds\Big); \ell(\tau, r) = 0\Big\}$$
$$= \begin{cases} \dfrac{\lambda}{\lambda+p}\Big(1 - \dfrac{x^{-\nu} I_\nu(x\Upsilon_p)}{r^{-\nu} I_\nu(r\Upsilon_p)}\Big), & 0 \le x \le r \\ \dfrac{\lambda}{\lambda+q}\Big(1 - \dfrac{x^{-\nu} K_\nu(x\Upsilon_q)}{r^{-\nu} K_\nu(r\Upsilon_q)}\Big), & r \le x \end{cases}$$

1.27.3 (1) $$\mathbf{E}_x\Big\{\exp\Big(-p\int_0^\infty \mathbb{I}_{[0,r)}(R_s^{(n)})ds\Big); \ell(\infty, r) \in dy\Big\}$$
$$= \exp\Big(-\frac{y\nu}{r} - \frac{y\sqrt{p} I_{\nu+1}(r\sqrt{2p})}{\sqrt{2} I_\nu(r\sqrt{2p})}\Big) \begin{cases} \dfrac{\nu r^{\nu-1} I_\nu(x\sqrt{2p})}{x^\nu I_\nu(r\sqrt{2p})}dy, & 0 \le x \le r \\ \dfrac{\nu x^{-2\nu}}{r^{1-2\nu}}dy, & r \le x \end{cases}$$

$\Upsilon_s := \sqrt{2\lambda + 2s}$

4 $R_s^{(n)}$ $\nu \ge 0,\ n = 2\nu + 2 \ge 2,$ $\tau \sim \operatorname{Exp}(\lambda)$, independent of $R^{(n)}$

1.27.5

$$\mathbf{E}_x\Big\{\exp\Big(-\int_0^\tau \big(p\mathbb{I}_{[0,r)}\big(R_s^{(n)}\big) + q\mathbb{I}_{[r,\infty)}\big(R_s^{(n)}\big)\big)ds\Big);\ \ell(\tau,r)\in dy,\ R_\tau^{(n)}\in dz\Big\}$$

$z \le r$, $x \le r$

$$= \frac{\lambda z^{\nu+1} I_\nu(x\Upsilon_p) I_\nu(z\Upsilon_p)}{x^\nu r I_\nu^2(r\Upsilon_p)} \exp\Big(-\frac{y\Upsilon_p I_{\nu+1}(r\Upsilon_p)}{2I_\nu(r\Upsilon_p)} - \frac{y\Upsilon_q K_{\nu+1}(r\Upsilon_q)}{2K_\nu(r\Upsilon_q)}\Big)dydz$$

$z \le r$, $r \le x$

$$= \frac{\lambda z^{\nu+1} I_\nu(z\Upsilon_p) K_\nu(x\Upsilon_q)}{x^\nu r I_\nu(r\Upsilon_p) K_\nu(r\Upsilon_q)} \exp\Big(-\frac{y\Upsilon_p I_{\nu+1}(r\Upsilon_p)}{2I_\nu(r\Upsilon_p)} - \frac{y\Upsilon_q K_{\nu+1}(r\Upsilon_q)}{2K_\nu(r\Upsilon_q)}\Big)dydz$$

$r \le z$, $x \le r$

$$= \frac{\lambda z^{\nu+1} I_\nu(x\Upsilon_p) K_\nu(z\Upsilon_q)}{x^\nu r I_\nu(r\Upsilon_p) K_\nu(r\Upsilon_q)} \exp\Big(-\frac{y\Upsilon_p I_{\nu+1}(r\Upsilon_p)}{2I_\nu(r\Upsilon_p)} - \frac{y\Upsilon_q K_{\nu+1}(r\Upsilon_q)}{2K_\nu(r\Upsilon_q)}\Big)dydz$$

$r \le z$, $r \le x$

$$= \frac{\lambda z^{\nu+1} K_\nu(x\Upsilon_q) K_\nu(z\Upsilon_q)}{x^\nu r K_\nu^2(r\Upsilon_q)} \exp\Big(-\frac{y\Upsilon_p I_{\nu+1}(r\Upsilon_p)}{2I_\nu(r\Upsilon_p)} - \frac{y\Upsilon_q K_{\nu+1}(r\Upsilon_q)}{2K_\nu(r\Upsilon_q)}\Big)dydz$$

(1)

$$\mathbf{E}_x\Big\{\exp\Big(-\int_0^\tau \big(p\mathbb{I}_{[0,r)}\big(R_s^{(n)}\big) + q\mathbb{I}_{[r,\infty)}\big(R_s^{(n)}\big)\big)ds\Big);\ \ell(\tau,r)=0,\ R_\tau^{(n)}\in dz\Big\}$$

$z \le r$, $x \le r$

$$= \frac{2\lambda z^{\nu+1}}{x^\nu}\Big\{F_\nu(x\Upsilon_p, z\Upsilon_p) - \frac{I_\nu(x\Upsilon_p) I_\nu(z\Upsilon_p) K_\nu(r\Upsilon_p)}{I_\nu(r\Upsilon_p)}\Big\}dz$$

$r \le z$, $r \le x$

$$= \frac{2\lambda z^{\nu+1}}{x^\nu}\Big\{F_\nu(x\Upsilon_q, z\Upsilon_q) - \frac{K_\nu(x\Upsilon_q) K_\nu(z\Upsilon_q) I_\nu(r\Upsilon_q)}{K_\nu(r\Upsilon_q)}\Big\}dz$$

1.28.1, $a < x$, $x < b$

$$\mathbf{E}_x\big\{e^{-\gamma\hat H(\tau)-\eta\check H(\tau)};\ \inf_{0\le s\le\tau} R_s^{(n)}\in da,\ \sup_{0\le s\le\tau} R_s^{(n)}\in db\big\}/dadb$$

$$= \frac{(2\lambda+2\eta)^{-\nu} S_\nu(b\sqrt{2\lambda+2\gamma+2\eta}, x\sqrt{2\lambda+2\gamma+2\eta})\big(\sqrt{2\lambda}b^{2\nu+1}C_\nu(b\sqrt{2\lambda}, a\sqrt{2\lambda}) - (2\lambda)^{-\nu}\big)}{(ba)^{2\nu+1} S_\nu(b\sqrt{2\lambda+2\gamma+2\eta}, a\sqrt{2\lambda+2\gamma+2\eta}) S_\nu(b\sqrt{2\lambda+2\eta}, a\sqrt{2\lambda+2\eta}) S_\nu(b\sqrt{2\lambda}, a\sqrt{2\lambda})}$$

$$+ \frac{(2\lambda+2\gamma)^{-\nu} S_\nu(x\sqrt{2\lambda+2\gamma+2\eta}, a\sqrt{2\lambda+2\gamma+2\eta})\big(\sqrt{2\lambda}a^{2\nu+1}C_\nu(a\sqrt{2\lambda}, b\sqrt{2\lambda}) - (2\lambda)^{-\nu}\big)}{(ba)^{2\nu+1} S_\nu(b\sqrt{2\lambda+2\gamma+2\eta}, a\sqrt{2\lambda+2\gamma+2\eta}) S_\nu(b\sqrt{2\lambda+2\gamma}, a\sqrt{2\lambda+2\gamma}) S_\nu(b\sqrt{2\lambda}, a\sqrt{2\lambda})}$$

1.28.5, $a < x$, $x < b$

$$\mathbf{E}_x\big\{e^{-\gamma\hat H(\tau)-\eta\check H(\tau)};\ \inf_{0\le s\le\tau} R_s^{(n)}\in da,\ \sup_{0\le s\le\tau} R_s^{(n)}\in db,\ R_\tau^{(n)}\in dz\big\}/dadbdz$$

$$= \frac{2\lambda(2\lambda+2\eta)^{-\nu}(ba)^{-2\nu-1}z^{2\nu+1} S_\nu(b\sqrt{2\lambda+2\gamma+2\eta}, x\sqrt{2\lambda+2\gamma+2\eta}) S_\nu(z\sqrt{2\lambda}, a\sqrt{2\lambda})}{S_\nu(b\sqrt{2\lambda+2\gamma+2\eta}, a\sqrt{2\lambda+2\gamma+2\eta}) S_\nu(b\sqrt{2\lambda+2\eta}, a\sqrt{2\lambda+2\eta}) S_\nu(b\sqrt{2\lambda}, a\sqrt{2\lambda})}$$

$$+ \frac{2\lambda(2\lambda+2\gamma)^{-\nu}(ba)^{-2\nu-1}z^{2\nu+1} S_\nu(x\sqrt{2\lambda+2\gamma+2\eta}, a\sqrt{2\lambda+2\gamma+2\eta}) S_\nu(b\sqrt{2\lambda}, z\sqrt{2\lambda})}{S_\nu(b\sqrt{2\lambda+2\gamma+2\eta}, a\sqrt{2\lambda+2\gamma+2\eta}) S_\nu(b\sqrt{2\lambda+2\gamma}, a\sqrt{2\lambda+2\gamma}) S_\nu(z\sqrt{2\lambda}, a\sqrt{2\lambda})}$$

$$\Upsilon_s := \sqrt{2\lambda + 2s}$$

4 $R_s^{(n)}$ $\nu \ge 0,\ n = 2\nu + 2 \ge 2,$ $\tau \sim \text{Exp}(\lambda)$, independent of $R^{(n)}$

1.29.1

$0 \le a$

$$\mathbf{E}_x\Big\{\exp\Big(-\int_0^\tau \big(p\mathbb{1}_{[0,r)}\big(R_s^{(n)}\big) + q\mathbb{1}_{[r,\infty)}\big(R_s^{(n)}\big)\big)ds\Big); a < \inf_{0\le s\le\tau} R_s^{(n)}, \sup_{0\le s\le\tau} R_s^{(n)} < b, \ell(\tau,r) \in dy\Big\}$$

$$= \lambda\Big(\frac{C_\nu(r\Upsilon_q, b\Upsilon_q) - (r\Upsilon_q)^{-2\nu-1}}{\Upsilon_q S_\nu(b\Upsilon_q, r\Upsilon_q)} + \frac{C_\nu(r\Upsilon_p, a\Upsilon_p) - (r\Upsilon_p)^{-2\nu-1}}{\Upsilon_p S_\nu(r\Upsilon_p, a\Upsilon_p)}\Big)$$

$$\times \exp\Big(-\frac{y\Upsilon_q C_\nu(r\Upsilon_q, b\Upsilon_q)}{2S_\nu(b\Upsilon_q, r\Upsilon_q)} - \frac{y\Upsilon_p C_\nu(r\Upsilon_p, a\Upsilon_p)}{2S_\nu(r\Upsilon_p, a\Upsilon_p)}\Big)\begin{cases} \dfrac{S_\nu(x\Upsilon_p, a\Upsilon_p)}{S_\nu(r\Upsilon_p, a\Upsilon_p)}dy, & a \le x \le r \\ \dfrac{S_\nu(b\Upsilon_q, x\Upsilon_q)}{S_\nu(b\Upsilon_q, r\Upsilon_q)}dy, & r \le x \le b \end{cases}$$

(1)

$$\mathbf{E}_x\Big\{\exp\Big(-\int_0^\tau \big(p\mathbb{1}_{[0,r)}\big(R_s^{(n)}\big) + q\mathbb{1}_{[r,\infty)}\big(R_s^{(n)}\big)\big)ds\Big); a < \inf_{0\le s\le\tau} R_s^{(n)}, \sup_{0\le s\le\tau} R_s^{(n)} < b, \ell(\tau,r) = 0\Big\}$$

$$= \begin{cases} \dfrac{2\lambda}{\Upsilon_p^2}\Big(1 - \dfrac{S_\nu(r\Upsilon_p, x\Upsilon_p) + S_\nu(x\Upsilon_p, a\Upsilon_p)}{S_\nu(r\Upsilon_p, a\Upsilon_p)}\Big), & a \le x \le r \\ \dfrac{2\lambda}{\Upsilon_q^2}\Big(1 - \dfrac{S_\nu(b\Upsilon_q, x\Upsilon_q) + S_\nu(x\Upsilon_q, r\Upsilon_q)}{S_\nu(b\Upsilon_q, r\Upsilon_q)}\Big), & r \le x \le b \end{cases}$$

(2)

$$\mathbf{E}_x\Big\{\exp\Big(-p\int_0^\infty \mathbb{1}_{[0,r)}\big(R_s^{(n)}\big)ds\Big); a < \inf_{0\le s<\infty} R_s^{(n)}, \ell(\infty,r) \in dy\Big\}$$

$$= \exp\Big(-\frac{y\nu}{r} - \frac{y\sqrt{p}C_\nu(r\sqrt{2p}, a\sqrt{2p})}{\sqrt{2}S_\nu(r\sqrt{2p}, a\sqrt{2p})}\Big)\begin{cases} \dfrac{\nu S_\nu(x\sqrt{2p}, a\sqrt{2p})}{rS_\nu(r\sqrt{2p}, a\sqrt{2p})}dy, & a \le x \le r \\ \dfrac{\nu x^{-2\nu}}{r^{1-2\nu}}dy, & r \le x \end{cases}$$

(3)

$$\mathbf{E}_x\Big\{\exp\Big(-p\int_0^\infty \mathbb{1}_{[0,r)}\big(R_s^{(n)}\big)ds\Big); a < \inf_{0\le s<\infty} R_s^{(n)}, \ell(\infty,r) = 0\Big\}$$

$$= \begin{cases} 0, & a \le x \le r \\ 1 - \dfrac{x^{-2\nu}}{r^{-2\nu}}, & r \le x \end{cases}$$

$\Upsilon_s := \sqrt{2\lambda + 2s}$

4 $R_s^{(n)}$ $\nu \ge 0,\ n = 2\nu + 2 \ge 2,$ $\tau \sim \text{Exp}(\lambda)$, independent of $R^{(n)}$

1.29.5

$0 \le a$

$$\mathbf{E}_x\Big\{\exp\Big(-\int_0^\tau \big(p\mathbb{I}_{[0,r)}\big(R_s^{(n)}\big) + q\mathbb{I}_{[r,\infty)}\big(R_s^{(n)}\big)\big)ds\Big); a < \inf_{0\le s\le\tau} R_s^{(n)}, \sup_{0\le s\le\tau} R_s^{(n)} < b, \ell(\tau,r)\in dy, W_\tau \in dz\Big\}$$

$$= \frac{\lambda z^{2\nu+1}}{r^{2\nu+1}} \exp\Big(-\frac{y\Upsilon_q C_\nu(r\Upsilon_q, b\Upsilon_q)}{2S_\nu(b\Upsilon_q, r\Upsilon_q)} - \frac{y\Upsilon_p C_\nu(r\Upsilon_p, a\Upsilon_p)}{2S_\nu(r\Upsilon_p, a\Upsilon_p)}\Big)$$

$$\times \begin{cases} \dfrac{S_\nu(x\Upsilon_p, a\Upsilon_p)S_\nu(z\Upsilon_p, a\Upsilon_p)}{S_\nu^2(r\Upsilon_p, a\Upsilon_p)} dydz, & a \le x \le r,\ a \le z \le r \\ \dfrac{S_\nu(b\Upsilon_q, x\Upsilon_q)S_\nu(z\Upsilon_p, a\Upsilon_p)}{S_\nu(b\Upsilon_q, r\Upsilon_q)S_\nu(r\Upsilon_p, a\Upsilon_p)} dydz, & a \le z \le r \le x \le b \\ \dfrac{S_\nu(b\Upsilon_q, z\Upsilon_q)S_\nu(x\Upsilon_p, a\Upsilon_p)}{S_\nu(b\Upsilon_q, r\Upsilon_q)S_\nu(r\Upsilon_p, a\Upsilon_p)} dydz, & a \le x \le r \le z \le b \\ \dfrac{S_\nu(b\Upsilon_q, x\Upsilon_q)S_\nu(b\Upsilon_q, z\Upsilon_q)}{S_\nu^2(b\Upsilon_q, r\Upsilon_q)} dydz, & r \le x \le b,\ r \le z \le b \end{cases}$$

(1)

$$\mathbf{E}_x\Big\{\exp\Big(-\int_0^\tau \big(p\mathbb{I}_{[0,r)}\big(R_s^{(n)}\big) + q\mathbb{I}_{[r,\infty)}\big(R_s^{(n)}\big)\big)ds\Big); a < \inf_{0\le s\le\tau} R_s^{(n)}, \sup_{0\le s\le\tau} R_s^{(n)} < b, \ell(\tau,r) = 0, W_\tau \in dz\Big\}$$

$x \le r$, $z \le r$:

$$= \frac{2\lambda z^{2\nu+1}\Upsilon_p^{2\nu} S_\nu(r\Upsilon_p, (z+x+|z-x|)\Upsilon_p/2)S_\nu((z+x-|z-x|)\Upsilon_p/2, a\Upsilon_p)}{S_\nu(r\Upsilon_p, a\Upsilon_p)} dz$$

$r \le x$, $r \le z$:

$$= \frac{2\lambda z^{2\nu+1}\Upsilon_q^{2\nu} S_\nu(b\Upsilon_q, z+x+|z-x|)\Upsilon_q/2)S_\nu((z+x-|z-x|)\Upsilon_q/2, r\Upsilon_q)}{S_\nu(b\Upsilon_q, r\Upsilon_q)} dz$$

$$\Upsilon_s := \sqrt{2\lambda + 2s}$$

4 $R_s^{(n)}$ $\nu > 0,\ n = 2\nu + 2 > 2$ $H_z = \min\{s : R_s^{(n)} = z\}$

2. Stopping at first hitting time

2.0.1
$$\mathbf{E}_x e^{-\alpha H_z} = \begin{cases} \dfrac{x^{-\nu} I_\nu(x\sqrt{2\alpha})}{z^{-\nu} I_\nu(z\sqrt{2\alpha})}, & x \le z \\ \dfrac{x^{-\nu} K_\nu(x\sqrt{2\alpha})}{z^{-\nu} K_\nu(z\sqrt{2\alpha})}, & z \le x \end{cases}$$

2.0.2 $x \le z$
$$\mathbf{P}_x\big(H_z \in dt\big) = \sum_{k=1}^{\infty} \frac{j_{\nu,k} x^{-\nu} J_\nu(j_{\nu,k} x/z)}{z^{2-\nu} J_{\nu+1}(j_{\nu,k})} e^{-j_{\nu,k}^2 t/2z^2}\, dt$$

(1)
$$\mathbf{P}_x\big(H_z = \infty\big) = \begin{cases} 0, & x \le z \\ 1 - \dfrac{x^{-2\nu}}{z^{-2\nu}}, & z \le x \end{cases}$$

2.1.2 $z \le y$
$$\mathbf{P}_x\Big(\sup_{0\le s\le H_z} R_s^{(n)} < y\Big) = \begin{cases} 1, & x \le z \\ \dfrac{x^{-2\nu} - y^{-2\nu}}{z^{-2\nu} - y^{-2\nu}}, & z \le x \le y \end{cases}$$

2.1.4 $z \le y$
$$\mathbf{E}_x\Big\{e^{-\alpha H_z};\ \sup_{0\le s\le H_z} R_s^{(n)} < y\Big\} = \begin{cases} \dfrac{x^{-\nu} I_\nu(x\sqrt{2\alpha})}{z^{-\nu} I_\nu(z\sqrt{2\alpha})}, & x \le z \\ \dfrac{S_\nu(y\sqrt{2\alpha}, x\sqrt{2\alpha})}{S_\nu(y\sqrt{2\alpha}, z\sqrt{2\alpha})}, & z \le x \le y \end{cases}$$

2.2.2 $y \le z$
$$\mathbf{P}_x\Big(\inf_{0\le s\le H_z} R_s^{(n)} > y,\ H_z < \infty\Big) = \begin{cases} \dfrac{y^{-2\nu} - x^{-2\nu}}{y^{-2\nu} - z^{-2\nu}}, & y \le x \le z \\ \dfrac{x^{-2\nu}}{z^{-2\nu}}, & z \le x \end{cases}$$

(1) $z \le y$
$$\mathbf{P}_x\Big(\inf_{0\le s\le H_z} R_s^{(n)} > y,\ H_z = \infty\Big) = \begin{cases} 1 - \dfrac{x^{-2\nu}}{y^{-2\nu}}, & z \le y \le x \\ 1 - \dfrac{x^{-2\nu}}{z^{-2\nu}}, & y \le z \le x \end{cases}$$

2.2.4 $y \le z$
$$\mathbf{E}_x\Big\{e^{-\alpha H_z};\ \inf_{0\le s\le H_z} R_s^{(n)} > y\Big\} = \begin{cases} \dfrac{S_\nu(x\sqrt{2\alpha}, y\sqrt{2\alpha})}{S_\nu(z\sqrt{2\alpha}, y\sqrt{2\alpha})}, & y \le x \le z \\ \dfrac{x^{-\nu} K_\nu(x\sqrt{2\alpha})}{z^{-\nu} K_\nu(z\sqrt{2\alpha})}, & z \le x \end{cases}$$

2.3.1
$$\mathbf{E}_x\big\{e^{-\gamma \ell(H_z, r)};\ H_z < \infty\big\}$$

4 $R_s^{(n)}$ $\nu > 0,\ n = 2\nu + 2 > 2$ $H_z = \min\{s : R_s^{(n)} = z\}$

$r \le z$
$$= \begin{cases} \dfrac{x^{-2\nu}}{z^{-2\nu}}, & z \le x \\ \dfrac{\nu + r^{2\nu+1}\gamma(r^{-2\nu} - x^{-2\nu})}{\nu + r^{2\nu+1}\gamma(r^{-2\nu} - z^{-2\nu})}, & r \le x \le z \\ \dfrac{\nu}{\nu + r^{2\nu+1}\gamma(r^{-2\nu} - z^{-2\nu})}, & 0 \le x \le r \end{cases}$$

$z \le r$
$$= \begin{cases} \dfrac{\nu x^{-2\nu}}{\nu z^{-2\nu} + \gamma r(z^{-2\nu} - r^{-2\nu})}, & r \le x \\ \dfrac{\nu x^{-2\nu} + \gamma r(x^{-2\nu} - r^{-2\nu})}{\nu z^{-2\nu} + \gamma r(z^{-2\nu} - r^{-2\nu})}, & z \le x \le r \\ 1, & 0 \le x \le z \end{cases}$$

(1) $z \le r$ $\mathbf{E}_x\{e^{-\gamma \ell(H_z,r)};\ H_z = \infty\}$

$$= \begin{cases} 1 - \dfrac{x^{-2\nu}}{r^{-2\nu}} + \dfrac{\nu x^{-2\nu}(z^{-2\nu} - r^{-2\nu})}{r^{-2\nu}(\nu z^{-2\nu} + \gamma r(z^{-2\nu} - r^{-2\nu}))}, & r \le x \\ \dfrac{\nu(z^{-2\nu} - x^{-2\nu})}{\nu z^{-2\nu} + \gamma r(z^{-2\nu} - r^{-2\nu})}, & z \le x \le r \\ 0, & 0 \le x \le z \end{cases}$$

2.3.2 $\mathbf{P}_x\big(\ell(H_z, r) \in dy, H_z < \infty\big)$

$z \le r$
$$= \begin{cases} 0, & 0 \le x \le z \\ \dfrac{\nu r^{-2\nu-1}(z^{-2\nu} - x^{-2\nu})}{(z^{-2\nu} - r^{-2\nu})^2} \exp\Big(-\dfrac{\nu z^{-2\nu} y}{r(z^{-2\nu} - r^{-2\nu})}\Big)dy, & z \le x \le r \\ \dfrac{\nu x^{-2\nu}}{r(z^{-2\nu} - r^{-2\nu})} \exp\Big(-\dfrac{\nu z^{-2\nu} y}{r(z^{-2\nu} - r^{-2\nu})}\Big)dy, & r \le x \end{cases}$$

$r \le z$
$$= \begin{cases} 0, & z \le x \\ \dfrac{\nu(x^{-2\nu} - z^{-2\nu})}{r^{2\nu+1}(r^{-2\nu} - z^{-2\nu})^2} \exp\Big(-\dfrac{\nu y}{r^{2\nu+1}(r^{-2\nu} - z^{-2\nu})}\Big)dy, & r \le x \le z \\ \dfrac{\nu}{r^{2\nu+1}(r^{-2\nu} - z^{-2\nu})} \exp\Big(-\dfrac{\nu y}{r^{2\nu+1}(r^{-2\nu} - z^{-2\nu})}\Big)dy, & 0 \le x \le r \end{cases}$$

(1) $\mathbf{P}_x\big(\ell(H_z, r) \in dy, H_z = \infty\big)$

$z \le r$
$$= \begin{cases} 0, & 0 \le x \le z \\ \dfrac{\nu(z^{-2\nu} - x^{-2\nu})}{r(z^{-2\nu} - r^{-2\nu})} \exp\Big(-\dfrac{\nu z^{-2\nu} y}{r(z^{-2\nu} - r^{-2\nu})}\Big)dy, & z \le x \le r \\ \dfrac{\nu x^{-2\nu}}{r^{1-2\nu}} \exp\Big(-\dfrac{\nu z^{-2\nu} y}{r(z^{-2\nu} - r^{-2\nu})}\Big)dy, & r \le x \end{cases}$$

4 $R_s^{(n)}$ $\nu > 0,\ n = 2\nu + 2 > 2$ $H_z = \min\{s : R_s^{(n)} = z\}$

(2)
$$\mathbf{P}_x\big(\ell(H_z,r)=0, H_z<\infty\big)=\begin{cases}\dfrac{x^{-2\nu}}{z^{-2\nu}}, & r\le z\le x\\ 1, & 0\le x\le z\le r,\\ \dfrac{|x^{-2\nu}-r^{-2\nu}|}{|z^{-2\nu}-r^{-2\nu}|}, & z\wedge r\le x\le r\vee z\\ 0, & z\le r\le x, 0\le x\le r\le z\end{cases}$$

(3) $z \le r$
$$\mathbf{P}_x\big(\ell(H_z,r)=0, H_z=\infty\big)=1-\frac{x^{-2\nu}}{r^{-2\nu}},\qquad r\le x$$

2.3.3 $\mathbf{E}_x e^{-\alpha H_z-\gamma\ell(H_z,r)}$

$r \le z$
$$=\begin{cases}\dfrac{x^{-\nu}K_\nu(x\sqrt{2\alpha})}{z^{-\nu}K_\nu(z\sqrt{2\alpha})}, & z\le x\\ \dfrac{x^{-\nu}I_\nu(x\sqrt{2\alpha})+2\gamma r^{\nu+1}(2\alpha)^\nu S_\nu(x\sqrt{2\alpha},r\sqrt{2\alpha})I_\nu(r\sqrt{2\alpha})}{z^{-\nu}I_\nu(z\sqrt{2\alpha})+2\gamma r^{\nu+1}(2\alpha)^\nu S_\nu(z\sqrt{2\alpha},r\sqrt{2\alpha})I_\nu(r\sqrt{2\alpha})}, & r\le x\le z\\ \dfrac{x^{-\nu}I_\nu(x\sqrt{2\alpha})}{z^{-\nu}I_\nu(z\sqrt{2\alpha})+2\gamma r^{\nu+1}(2\alpha)^\nu S_\nu(z\sqrt{2\alpha},r\sqrt{2\alpha})I_\nu(r\sqrt{2\alpha})}, & 0\le x\le r\end{cases}$$

$z \le r$
$$=\begin{cases}\dfrac{x^{-\nu}I_\nu(x\sqrt{2\alpha})}{z^{-\nu}I_\nu(z\sqrt{2\alpha})}, & x\le z\\ \dfrac{x^{-\nu}K_\nu(x\sqrt{2\alpha})+2\gamma r^{\nu+1}(2\alpha)^\nu S_\nu(r\sqrt{2\alpha},x\sqrt{2\alpha})K_\nu(r\sqrt{2\alpha})}{z^{-\nu}K_\nu(z\sqrt{2\alpha})+2\gamma r^{\nu+1}(2\alpha)^\nu S_\nu(r\sqrt{2\alpha},z\sqrt{2\alpha})K_\nu(r\sqrt{2\alpha})}, & z\le x\le r\\ \dfrac{x^{-\nu}K_\nu(x\sqrt{2\alpha})}{z^{-\nu}K_\nu(z\sqrt{2\alpha})+2\gamma r^{\nu+1}(2\alpha)^\nu S_\nu(r\sqrt{2\alpha},z\sqrt{2\alpha})K_\nu(r\sqrt{2\alpha})}, & r\le x\end{cases}$$

2.3.4 $\mathbf{E}_x\{e^{-\alpha H_z};\ \ell(H_z,r)\in dy\}$

$r \le z$
$$=\exp\Big(-\frac{(2\alpha z)^{-\nu}r^{-\nu-1}I_\nu(z\sqrt{2\alpha})y}{2S_\nu(z\sqrt{2\alpha},r\sqrt{2\alpha})I_\nu(r\sqrt{2\alpha})}\Big)\begin{cases}\dfrac{(2\alpha x)^{-\nu}r^{-\nu-1}I_\nu(x\sqrt{2\alpha})}{2S_\nu(z\sqrt{2\alpha},r\sqrt{2\alpha})I_\nu(r\sqrt{2\alpha})}dy, & 0\le x\le r\\ \dfrac{(2\alpha)^{-\nu}S_\nu(z\sqrt{2\alpha},x\sqrt{2\alpha})}{2r^{2\nu+1}S_\nu^2(z\sqrt{2\alpha},r\sqrt{2\alpha})}dy, & r\le x\le z\\ 0, & z\le x\end{cases}$$

$z \le r$
$$=\exp\Big(-\frac{(2\alpha z)^{-\nu}r^{-\nu-1}K_\nu(z\sqrt{2\alpha})y}{2S_\nu(r\sqrt{2\alpha},z\sqrt{2\alpha})K_\nu(r\sqrt{2\alpha})}\Big)\begin{cases}\dfrac{(2\alpha x)^{-\nu}r^{-\nu-1}K_\nu(x\sqrt{2\alpha})}{2S_\nu(r\sqrt{2\alpha},z\sqrt{2\alpha})K_\nu(r\sqrt{2\alpha})}dy, & r\le x\\ \dfrac{(2\alpha)^{-\nu}S_\nu(x\sqrt{2\alpha},z\sqrt{2\alpha})}{2r^{2\nu+1}S_\nu^2(r\sqrt{2\alpha},z\sqrt{2\alpha})}dy, & z\le x\le r\\ 0, & 0\le x\le z\end{cases}$$

4 $R_s^{(n)}$ $\nu > 0,\ n = 2\nu + 2 > 2$ $H_z = \min\{s : R_s^{(n)} = z\}$

(1)
$$\mathbf{E}_x\{e^{-\alpha H_z}; \ell(H_z, r) = 0\} = \begin{cases} 0, & 0 \le x \le r \le z \quad z \le r \le x \\ \dfrac{|S_\nu(x\sqrt{2\alpha}, r\sqrt{2\alpha})|}{|S_\nu(z\sqrt{2\alpha}, r\sqrt{2\alpha})|}, & r \wedge z \le x \le r \vee z \\ \dfrac{z^{-\nu} K_\nu(x\sqrt{2\alpha})}{z^{-\nu} K_\nu(z\sqrt{2\alpha})}, & r \le z \le x \\ \dfrac{z^{-\nu} I_\nu(x\sqrt{2\alpha})}{z^{-\nu} I_\nu(z\sqrt{2\alpha})}, & 0 \le x \le z \le r \end{cases}$$

2.4.1
$$\mathbf{E}_x\Big\{\exp\Big(-\gamma \int_0^{H_z} \mathbb{I}_{[r,\infty)}\big(R_s^{(n)}\big) ds\Big);\ H_z < \infty\Big\}$$

$z \le r$
$$= \begin{cases} 1, & 0 \le x \le z \\ \dfrac{x^{-2\nu} K_{\nu+1}(r\sqrt{2\gamma}) - r^{-2\nu} K_{\nu-1}(r\sqrt{2\gamma})}{z^{-2\nu} K_{\nu+1}(r\sqrt{2\gamma}) - r^{-2\nu} K_{\nu-1}(r\sqrt{2\gamma})}, & z \le x \le r \\ \dfrac{2\nu(2\gamma)^{-1/2} r^{-\nu-1} x^{-\nu} K_\nu(x\sqrt{2\gamma})}{z^{-2\nu} K_{\nu+1}(r\sqrt{2\gamma}) - r^{-2\nu} K_{\nu-1}(r\sqrt{2\gamma})}, & r \le x \end{cases}$$

$r \le z$
$$= \begin{cases} \dfrac{(r\sqrt{2\gamma})^{-2\nu-1}}{C_\nu(r\sqrt{2\gamma}, z\sqrt{2\gamma})}, & 0 < x \le r \\ \dfrac{C_\nu(r\sqrt{2\gamma}, x\sqrt{2\gamma})}{C_\nu(r\sqrt{2\gamma}, z\sqrt{2\gamma})}, & r \le x \le z \\ \dfrac{x^{-\nu} K_\nu(x\sqrt{2\gamma})}{z^{-\nu} K_\nu(z\sqrt{2\gamma})}, & z \le x \end{cases}$$

(1)
$$\mathbf{E}_x\Big\{\exp\Big(-\gamma \int_0^{H_z} \mathbb{I}_{[r,\infty)}\big(R_s^{(n)}\big) ds\Big);\ H_z = \infty\Big\} = 0$$

2.4.2
(1)
$z \le r$
$$\mathbf{P}_x\Big(\int_0^{H_z} \mathbb{I}_{[r,\infty)}\big(R_s^{(n)}\big) ds = 0\Big) = \begin{cases} 1, & 0 \le x \le z \\ \dfrac{x^{-2\nu} - r^{-2\nu}}{z^{-2\nu} - r^{-2\nu}}, & z \le x \le r \\ 0, & r \le x \end{cases}$$

2.5.1
$$\mathbf{E}_x\Big\{\exp\Big(-\gamma \int_0^{H_z} \mathbb{I}_{[0,r]}\big(R_s^{(n)}\big) ds\Big);\ H_z < \infty\Big\}$$

$r \le z$
$$= \begin{cases} \dfrac{2\nu(2\gamma)^{-1/2} r^{-\nu-1} x^{-\nu} I_\nu(x\sqrt{2\gamma})}{r^{-2\nu} I_{\nu-1}(r\sqrt{2\gamma}) - z^{-2\nu} I_{\nu+1}(r\sqrt{2\gamma})}, & 0 \le x \le r \\ \dfrac{r^{-2\nu} I_{\nu-1}(r\sqrt{2\gamma}) - x^{-2\nu} I_{\nu+1}(r\sqrt{2\gamma})}{r^{-2\nu} I_{\nu-1}(r\sqrt{2\gamma}) - z^{-2\nu} I_{\nu+1}(r\sqrt{2\gamma})}, & r \le x \le z \\ \dfrac{x^{-2\nu}}{z^{-2\nu}}, & z \le x \end{cases}$$

4 $R_s^{(n)}$ $\nu > 0,\ n = 2\nu + 2 > 2$ $H_z = \min\{s : R_s^{(n)} = z\}$

$z \le r$

$$= \begin{cases} \dfrac{x^{-\nu} I_\nu(x\sqrt{2\gamma})}{z^{-\nu} I_\nu(z\sqrt{2\gamma})}, & 0 \le x \le z \\ \dfrac{x^{-1} C_{\nu-1}(x\sqrt{2\gamma}, r\sqrt{2\gamma})}{z^{-1} C_{\nu-1}(z\sqrt{2\gamma}, r\sqrt{2\gamma})}, & z \le x \le r \\ \dfrac{(2\gamma)^{-\nu+1/2} x^{-2\nu}}{z^{-1} C_{\nu-1}(z\sqrt{2\gamma}, r\sqrt{2\gamma})}, & r \le x \end{cases}$$

(1)
$z \le r$

$$\mathbf{E}_x\Big\{\exp\Big(-\gamma \int_0^{H_z} \mathbb{1}_{[0,r]}\big(R_s^{(n)}\big) ds\Big);\ H_z = \infty\Big\}$$

$$= \begin{cases} 0, & 0 \le x \le z \\ \dfrac{2\nu\sqrt{2\gamma} S_\nu(x\sqrt{2\gamma}, z\sqrt{2\gamma})}{z^{-1} C_{\nu-1}(z\sqrt{2\gamma}, r\sqrt{2\gamma})}, & z \le x \le r \\ 1 - \dfrac{2\gamma x^{-2\nu} r^{2\nu+1} C_\nu(r\sqrt{2\gamma}, z\sqrt{2\gamma})}{z^{-1} C_{\nu-1}(z\sqrt{2\gamma}, r\sqrt{2\gamma})}, & r \le x \end{cases}$$

2.5.2
(2)
$r \le z$

$$\mathbf{P}_x\Big(\int_0^{H_z} \mathbb{1}_{[0,r]}\big(R_s^{(n)}\big) ds = 0\Big) = \begin{cases} 0, & x \le r \\ \dfrac{r^{-2\nu} - x^{-2\nu}}{r^{-2\nu} - z^{-2\nu}}, & r \le x \le z \\ 1, & z \le x \end{cases}$$

2.6.1

$$\mathbf{E}_x\Big\{\exp\Big(-\int_0^{H_z} \big(p\mathbb{1}_{[0,r)}\big(R_s^{(n)}\big) + q\mathbb{1}_{[r,\infty)}\big(R_s^{(n)}\big)\big) ds\Big);\ H_z < \infty\Big\}$$

$r \le z$

$$= \begin{cases} \dfrac{\sqrt{q}(r\sqrt{2q})^{-2\nu-1}(x/r)^{-\nu} I_\nu(x\sqrt{2p})}{\sqrt{p} S_\nu(z\sqrt{2q}, r\sqrt{2q}) I_{\nu+1}(r\sqrt{2p}) + \sqrt{q} C_\nu(r\sqrt{2q}, z\sqrt{2q}) I_\nu(r\sqrt{2p})}, & 0 \le x \le r \\ \dfrac{\sqrt{p} S_\nu(x\sqrt{2q}, r\sqrt{2q}) I_{\nu+1}(r\sqrt{2p}) + \sqrt{q} C_\nu(r\sqrt{2q}, x\sqrt{2q}) I_\nu(r\sqrt{2p})}{\sqrt{p} S_\nu(z\sqrt{2q}, r\sqrt{2q}) I_{\nu+1}(r\sqrt{2p}) + \sqrt{q} C_\nu(r\sqrt{2q}, z\sqrt{2q}) I_\nu(r\sqrt{2p})}, & r \le x \le z \\ \dfrac{x^{-\nu} K_\nu(x\sqrt{2q})}{z^{-\nu} K_\nu(z\sqrt{2q})}, & z \le x \end{cases}$$

$z \le r$

$$= \begin{cases} \dfrac{x^{-\nu} I_\nu(x\sqrt{2p})}{z^{-\nu} I_\nu(z\sqrt{2p})}, & 0 \le x \le z \\ \dfrac{\sqrt{p} K_\nu(r\sqrt{2q}) C_\nu(r\sqrt{2p}, x\sqrt{2p}) + \sqrt{q} K_{\nu+1}(r\sqrt{2q}) S_\nu(r\sqrt{2p}, x\sqrt{2p})}{\sqrt{p} K_\nu(r\sqrt{2q}) C_\nu(r\sqrt{2p}, z\sqrt{2p}) + \sqrt{q} K_{\nu+1}(r\sqrt{2q}) S_\nu(r\sqrt{2p}, z\sqrt{2p})}, & z \le x \le r \\ \dfrac{\sqrt{p}(r\sqrt{2p})^{-2\nu-1}(x/r)^{-\nu} K_\nu(x\sqrt{2q})}{\sqrt{p} K_\nu(r\sqrt{2q}) C_\nu(r\sqrt{2p}, z\sqrt{2p}) + \sqrt{q} K_{\nu+1}(r\sqrt{2q}) S_\nu(r\sqrt{2p}, z\sqrt{2p})}, & r \le x \end{cases}$$

2.8.1

$$\mathbf{E}_x\Big\{\exp\Big(-\gamma \int_0^{H_z} R_s^{(n)} ds\Big); H_z < \infty\Big\} = \begin{cases} \dfrac{x^{-\nu} I_{2\nu/3}(\frac{2}{3}x^{3/2}\sqrt{2\gamma})}{z^{-\nu} I_{2\nu/3}(\frac{2}{3}z^{3/2}\sqrt{2\gamma})}, & 0 \le x \le z \\ \dfrac{x^{-\nu} K_{2\nu/3}(\frac{2}{3}x^{3/2}\sqrt{2\gamma})}{z^{-\nu} K_{2\nu/3}(\frac{2}{3}z^{3/2}\sqrt{2\gamma})}, & z \le x \end{cases}$$

4 $R_s^{(n)}$ $\nu > 0,\ n = 2\nu + 2 > 2$ $H_z = \min\{s : R_s^{(n)} = z\}$

2.9.1

$$\mathbf{E}_x\Big\{\exp\Big(-\frac{\gamma^2}{2}\int_0^{H_z}(R_s^{(n)})^2ds\Big); H_z < \infty\Big\} = \begin{cases} \dfrac{x^{-\nu}I_{\nu/2}(x^2\gamma/2)}{z^{-\nu}I_{\nu/2}(z^2\gamma/2)}, & 0 \le x \le z \\ \dfrac{x^{-\nu}K_{\nu/2}(x^2\gamma/2)}{z^{-\nu}K_{\nu/2}(z^2\gamma/2)}, & z \le x \end{cases}$$

2.9.3

$$\mathbf{E}_x \exp\Big(-\alpha H_z - \frac{\gamma^2}{2}\int_0^{H_z}(R_s^{(n)})^2ds\Big) = \begin{cases} \dfrac{x^{-\nu-1}M_{-\alpha/2\gamma,\nu/2}(\gamma x^2)}{z^{-\nu-1}M_{-\alpha/2\gamma,\nu/2}(\gamma z^2)}, & 0 \le x \le z \\ \dfrac{x^{-\nu-1}W_{-\alpha/2\gamma,\nu/2}(\gamma x^2)}{z^{-\nu-1}W_{-\alpha/2\gamma,\nu/2}(\gamma z^2)}, & z \le x \end{cases}$$

2.13.1
$z < x$
$x < y$

$$\mathbf{E}_x\{e^{-\gamma\check{H}(H_z)}; \sup_{0\le s\le H_z} R_s^{(n)} \in dy\} = \frac{2\nu y^{-1-2\nu}S_\nu(x\sqrt{2\gamma}, z\sqrt{2\gamma})}{(z^{-2\nu}-y^{-2\nu})S_\nu(y\sqrt{2\gamma}, z\sqrt{2\gamma})}dy$$

2.13.4

$$\mathbf{E}_x\{e^{-\gamma\check{H}(H_z)-\eta H_z}; \sup_{0\le s\le H_z} R_s^{(n)} \in dy\} \qquad z < x < y$$

$$= \frac{(2\eta)^{-\nu}y^{-1-2\nu}S_\nu(x\sqrt{2\gamma+2\eta}, z\sqrt{2\gamma+2\eta})}{S_\nu(y\sqrt{2\eta}, z\sqrt{2\eta})S_\nu(y\sqrt{2\gamma+2\eta}, z\sqrt{2\gamma+2\eta})}dy$$

2.14.1
$x < z$
$y < x$

$$\mathbf{E}_x\{e^{-\gamma\hat{H}(H_z)}; \inf_{0\le s\le H_z} R_s^{(n)} \in dy\} = \frac{2\nu y^{-1-2\nu}S_\nu(z\sqrt{2\gamma}, x\sqrt{2\gamma})}{(y^{-2\nu}-z^{-2\nu})S_\nu(z\sqrt{2\gamma}, y\sqrt{2\gamma})}dy$$

2.14.4

$$\mathbf{E}_x\{e^{-\gamma\hat{H}(H_z)-\eta H_z}; \inf_{0\le s\le H_z} R_s^{(n)} \in dy\} \qquad y < x < z$$

$$= \frac{(2\eta)^{-\nu}y^{-1-2\nu}S_\nu(z\sqrt{2\gamma+2\eta}, x\sqrt{2\gamma+2\eta})}{S_\nu(z\sqrt{2\eta}, y\sqrt{2\eta})S_\nu(z\sqrt{2\gamma+2\eta}, y\sqrt{2\gamma+2\eta})}dy$$

2.18.1
$r < u$

$$\mathbf{E}_x\{\exp(-\gamma\ell(H_z, r) - \eta\ell(H_z, u));\ H_z < \infty\}$$

for $z < r$

$z \le x$
$x \le r$

$$= \frac{\nu^2x^{-2\nu} + \nu\gamma r(x^{-2\nu} - r^{-2\nu}) + \nu\eta u(x^{-2\nu} - u^{-2\nu}) + \gamma\eta u(r^{2\nu+1}x^{-2\nu} - r)(r^{-2\nu} - u^{-2\nu})}{\nu^2z^{-2\nu} + \nu\gamma r(z^{-2\nu} - r^{-2\nu}) + \nu\eta u(z^{-2\nu} - u^{-2\nu}) + \gamma\eta u(r^{2\nu+1}z^{-2\nu} - r)(r^{-2\nu} - u^{-2\nu})}$$

$r \le x$
$x \le u$

$$= \frac{\nu^2x^{-2\nu} + \nu\eta u(x^{-2\nu} - u^{-2\nu})}{\nu^2z^{-2\nu} + \nu\gamma r(z^{-2\nu} - r^{-2\nu}) + \nu\eta u(z^{-2\nu} - u^{-2\nu}) + \gamma\eta u(r^{2\nu+1}z^{-2\nu} - r)(r^{-2\nu} - u^{-2\nu})}$$

4 $R_s^{(n)}$ $\nu > 0,\ n = 2\nu + 2 > 2$ $H_z = \min\{s : R_s^{(n)} = z\}$

$u \le x$

$$= \frac{\nu^2 x^{-2\nu}}{\nu^2 z^{-2\nu} + \nu\gamma r(z^{-2\nu} - r^{-2\nu}) + \nu\eta u(z^{-2\nu} - u^{-2\nu}) + \gamma\eta u(r^{2\nu+1} z^{-2\nu} - r)(r^{-2\nu} - u^{-2\nu})}$$

for $u < z$

$0 \le x$, $x \le r$

$$= \frac{\nu^2}{\nu^2 + \nu\gamma(r - r^{2\nu+1} z^{-2\nu}) + \nu\eta(u - u^{2\nu+1} z^{-2\nu}) + \gamma\eta(r - r^{2\nu+1} u^{-2\nu})(u - u^{2\nu+1} z^{-2\nu})}$$

$r \le x$, $x \le u$

$$= \frac{\nu^2 + \nu\gamma r^{2\nu+1}(r^{-2\nu} - x^{-2\nu})}{\nu^2 + \nu\gamma(r - r^{2\nu+1} z^{-2\nu}) + \nu\eta(u - u^{2\nu+1} z^{-2\nu}) + \gamma\eta(r - r^{2\nu+1} u^{-2\nu})(u - u^{2\nu+1} z^{-2\nu})}$$

$u \le x$, $x \le z$

$$= \frac{\nu^2 + \nu\gamma(r - r^{2\nu+1} x^{-2\nu}) + \nu\eta(u - u^{2\nu+1} x^{-2\nu}) + \gamma\eta(r - r^{2\nu+1} u^{-2\nu})(u - u^{2\nu+1} x^{-2\nu})}{\nu^2 + \nu\gamma(r - r^{2\nu+1} z^{-2\nu}) + \nu\eta(u - u^{2\nu+1} z^{-2\nu}) + \gamma\eta(r - r^{2\nu+1} u^{-2\nu})(u - u^{2\nu+1} z^{-2\nu})}$$

$z \le x$

$$= \frac{x^{-2\nu}}{z^{-2\nu}}$$

(1) $\mathbf{E}_x\{\exp(-\gamma\ell(H_z, r) - \eta\ell(H_z, u));\ H_z = \infty\}$

$z < r$, $r < u$

$z \le x$, $x \le r$

$$= \frac{\nu^2(z^{-2\nu} - x^{-2\nu})}{\nu^2 z^{-2\nu} + \nu\gamma r(z^{-2\nu} - r^{-2\nu}) + \nu\eta u(z^{-2\nu} - u^{-2\nu}) + \gamma\eta u(r^{2\nu+1} z^{-2\nu} - r)(r^{-2\nu} - u^{-2\nu})}$$

$r \le x$, $x \le u$

$$= \frac{\nu^2(z^{-2\nu} - x^{-2\nu}) + \nu\gamma r^{2\nu+1}(z^{-2\nu} - r^{-2\nu})(r^{-2\nu} - x^{-2\nu})}{\nu^2 z^{-2\nu} + \nu\gamma r(z^{-2\nu} - r^{-2\nu}) + \nu\eta u(z^{-2\nu} - u^{-2\nu}) + \gamma\eta u(r^{2\nu+1} z^{-2\nu} - r)(r^{-2\nu} - u^{-2\nu})}$$

$u \le x$

$$= \frac{\nu\eta u^{2\nu+1}(z^{-2\nu} - u^{-2\nu})(u^{-2\nu} - x^{-2\nu}) + \nu\gamma r^{2\nu+1}(z^{-2\nu} - r^{-2\nu})(r^{-2\nu} - x^{-2\nu})}{\nu^2 z^{-2\nu} + \nu\gamma r(z^{-2\nu} - r^{-2\nu}) + \nu\eta u(z^{-2\nu} - u^{-2\nu}) + \gamma\eta u(r^{2\nu+1} z^{-2\nu} - r)(r^{-2\nu} - u^{-2\nu})}$$

$$+ \frac{\nu^2(z^{-2\nu} - x^{-2\nu}) + \gamma\eta(ru)^{2\nu+1}(z^{-2\nu} - r^{-2\nu})(r^{-2\nu} - u^{-2\nu})(u^{-2\nu} - x^{-2\nu})}{\nu^2 z^{-2\nu} + \nu\gamma r(z^{-2\nu} - r^{-2\nu}) + \nu\eta u(z^{-2\nu} - u^{-2\nu}) + \gamma\eta u(r^{2\nu+1} z^{-2\nu} - r)(r^{-2\nu} - u^{-2\nu})}$$

2.18.2 $\mathbf{P}_x(r^{-2\nu-1}\ell(H_z, r) \in dy,\ u^{-2\nu-1}\ell(H_z, u) \in dg,\ H_z < \infty)$

$r < u$

$z < r$

$$= \begin{cases} \dfrac{z^{-2\nu} - x^{-2\nu}}{z^{-2\nu} - r^{-2\nu}} p_r dydg, & z \le x \le r \\ \dfrac{x^{-2\nu} - u^{-2\nu}}{r^{-2\nu} - u^{-2\nu}} p_r dydg + \dfrac{r^{-2\nu} - x^{-2\nu}}{r^{-2\nu} - u^{-2\nu}} p_u dydg, & r \le x \le u \\ p_u dydg, & u \le x \end{cases}$$

4 $R_s^{(n)}$ $\nu > 0,\ n = 2\nu + 2 > 2$ $H_z = \min\{s : R_s^{(n)} = z\}$

$u < z$

$$= \begin{cases} q_r dydg, & x \le r \\ \dfrac{x^{-2\nu} - u^{-2\nu}}{r^{-2\nu} - u^{-2\nu}} q_r dydg + \dfrac{r^{-2\nu} - x^{-2\nu}}{r^{-2\nu} - u^{-2\nu}} q_u dydg, & r \le x \le u \\ \dfrac{x^{-2\nu} - z^{-2\nu}}{u^{-2\nu} - z^{-2\nu}} q_u dydg, & u \le x \le z \end{cases}$$

where

$$p_r := \frac{\nu^2 (uzr^2)^{2\nu} \sqrt{y/g}}{(u^{2\nu} - r^{2\nu})(r^{2\nu} - z^{2\nu})} \exp\Bigl(-\frac{\nu(y + g(u/r)^{2\nu})}{r^{-2\nu} - u^{-2\nu}} - \frac{\nu y}{z^{-2\nu} - r^{-2\nu}}\Bigr) I_1\Bigl(\frac{\nu\sqrt{yg}}{r^{-2\nu} - u^{-2\nu}}\Bigr)$$

$$q_r := \frac{\nu^2 (zru^2)^{2\nu}}{(z^{2\nu} - u^{2\nu})(u^{2\nu} - r^{2\nu})} \exp\Bigl(-\frac{\nu(y + g)}{r^{-2\nu} - u^{-2\nu}} - \frac{\nu g}{u^{-2\nu} - z^{-2\nu}}\Bigr) I_0\Bigl(\frac{\nu\sqrt{yg}}{r^{-2\nu} - u^{-2\nu}}\Bigr)$$

$$p_u := \frac{\nu^2 (uzr^2)^{2\nu}}{(u^{2\nu} - r^{2\nu})(r^{2\nu} - z^{2\nu})} \exp\Bigl(-\frac{\nu(y + g(u/r)^{2\nu})}{r^{-2\nu} - u^{-2\nu}} - \frac{\nu y}{z^{-2\nu} - r^{-2\nu}}\Bigr) I_0\Bigl(\frac{\nu\sqrt{yg}}{r^{-2\nu} - u^{-2\nu}}\Bigr)$$

$$q_u := \frac{\nu^2 (zru^2)^{2\nu} \sqrt{g/y}}{(z^{2\nu} - u^{2\nu})(u^{2\nu} - r^{2\nu})} \exp\Bigl(-\frac{\nu(y + g)}{r^{-2\nu} - u^{-2\nu}} - \frac{\nu g}{u^{-2\nu} - z^{-2\nu}}\Bigr) I_1\Bigl(\frac{\nu\sqrt{yg}}{r^{-2\nu} - u^{-2\nu}}\Bigr)$$

(1) $\mathbf{P}_x\bigl(r^{-2\nu-1}\ell(H_z, r) \in dy,\ \ell(H_z, u) = 0,\ H_z < \infty\bigr)$

$$= \begin{cases} \dfrac{\nu(z^{-2\nu} - x^{-2\nu})}{(z^{-2\nu} - r^{-2\nu})^2} \exp\Bigl(-\dfrac{\nu y(z^{-2\nu} - u^{-2\nu})}{(z^{-2\nu} - r^{-2\nu})(r^{-2\nu} - u^{-2\nu})}\Bigr) dy, & z \le x \le r \\ \dfrac{\nu(x^{-2\nu} - u^{-2\nu})}{(z^{-2\nu} - r^{-2\nu})(r^{-2\nu} - u^{-2\nu})} \exp\Bigl(-\dfrac{\nu y(z^{-2\nu} - u^{-2\nu})}{(z^{-2\nu} - r^{-2\nu})(r^{-2\nu} - u^{-2\nu})}\Bigr) dy, & r \le x \le u \end{cases}$$

(2) $\mathbf{P}_x\bigl(\ell(H_z, r) = 0,\ u^{-2\nu-1}\ell(H_z, u) \in dg,\ H_z < \infty\bigr)$

$$= \begin{cases} \dfrac{\nu(r^{-2\nu} - x^{-2\nu})}{(r^{-2\nu} - u^{-2\nu})(u^{-2\nu} - z^{-2\nu})} \exp\Bigl(-\dfrac{\nu g(r^{-2\nu} - z^{-2\nu})}{(r^{-2\nu} - u^{-2\nu})(u^{-2\nu} - z^{-2\nu})}\Bigr) dg, & r \le x \le u \\ \dfrac{\nu(x^{-2\nu} - z^{-2\nu})}{(u^{-2\nu} - z^{-2\nu})^2} \exp\Bigl(-\dfrac{\nu g(r^{-2\nu} - z^{-2\nu})}{(r^{-2\nu} - u^{-2\nu})(u^{-2\nu} - z^{-2\nu})}\Bigr) dg, & u \le x \le z \end{cases}$$

(3) $$\mathbf{P}_x\bigl(\ell(H_z, r) = 0,\ \ell(H_z, u) = 0,\ H_z < \infty\bigr) = \begin{cases} \dfrac{x^{-2\nu} - r^{-2\nu}}{z^{-2\nu} - r^{-2\nu}}, & z \le x \le r \\ \dfrac{u^{-2\nu} - x^{-2\nu}}{u^{-2\nu} - z^{-2\nu}}, & u \le x \le z \end{cases}$$

(4) $\mathbf{P}_x\bigl(r^{-2\nu-1}\ell(H_z, r) \in dy,\ u^{-2\nu-1}\ell(H_z, u) \in dg,\ H_z = \infty\bigr)$

4 $R_s^{(n)}$ $\nu > 0,\ n = 2\nu + 2 > 2$ $H_z = \min\{s : R_s^{(n)} = z\}$

$z < r$

$$= \begin{cases} \dfrac{z^{-2\nu} - x^{-2\nu}}{z^{-2\nu} - r^{-2\nu}} p_r^\infty dydg, & z \le x \le r \\ \dfrac{x^{-2\nu} - u^{-2\nu}}{r^{-2\nu} - u^{-2\nu}} p_r^\infty dydg + \dfrac{r^{-2\nu} - x^{-2\nu}}{r^{-2\nu} - u^{-2\nu}} p_u^\infty dydg, & r \le x \le u \\ \dfrac{u^{2\nu}}{x^{2\nu}} p_u^\infty dydg, & u \le x \end{cases}$$

where

$$p_r^\infty := \frac{\nu^2 u^{2\nu}}{r^{-2\nu} - u^{-2\nu}} \exp\Big(-\frac{\nu(y + g(u/r)^{2\nu})}{r^{-2\nu} - u^{-2\nu}} - \frac{\nu y}{z^{-2\nu} - r^{-2\nu}}\Big) I_0\Big(\frac{\nu\sqrt{yg}}{r^{-2\nu} - u^{-2\nu}}\Big)$$

$$p_u^\infty := \frac{\nu^2 u^{2\nu}\sqrt{g/y}}{r^{-2\nu} - u^{-2\nu}} \exp\Big(-\frac{\nu(y + g(u/r)^{2\nu})}{r^{-2\nu} - u^{-2\nu}} - \frac{\nu y}{z^{-2\nu} - r^{-2\nu}}\Big) I_1\Big(\frac{\nu\sqrt{yg}}{r^{-2\nu} - u^{-2\nu}}\Big)$$

(5) $z < r$ $\mathbf{P}_x\big(\ell(H_z, r) = 0,\ \ell(H_z, u) \in dg,\ H_z = \infty\big)$

$$= \begin{cases} \dfrac{\nu(r^{-2\nu} - x^{-2\nu})}{u(r^{-2\nu} - u^{-2\nu})} \exp\Big(-\dfrac{\nu g r^{-2\nu}}{u(r^{-2\nu} - u^{-2\nu})}\Big) dg, & r \le x \le u \\ \dfrac{\nu x^{-2\nu}}{u^{1-2\nu}} \exp\Big(-\dfrac{\nu g r^{-2\nu}}{u(r^{-2\nu} - u^{-2\nu})}\Big) dg, & u \le x \end{cases}$$

(6) $\mathbf{P}_x\big(\ell(H_z, r) = 0,\ \ell(H_z, u) = 0,\ H_z = \infty\big) = 1 - \dfrac{u^{2\nu}}{x^{2\nu}}, \qquad u \le x$

2.19.1 $$\mathbf{E}_x\Big\{\exp\Big(-\gamma \int_0^{H_z} \frac{ds}{R_s^{(n)}}\Big);\ H_z < \infty\Big\} = \begin{cases} \dfrac{x^{-\nu} I_{2\nu}(2\sqrt{2\gamma x})}{z^{-\nu} I_{2\nu}(2\sqrt{2\gamma z})}, & 0 < x \le z \\ \dfrac{x^{-\nu} K_{2\nu}(2\sqrt{2\gamma x})}{z^{-\nu} K_{2\nu}(2\sqrt{2\gamma z})}, & z \le x \end{cases}$$

2.19.3 $$\mathbf{E}_x \exp\Big(-\alpha H_z - \gamma \int_0^{H_z} \frac{ds}{R_s^{(n)}}\Big) = \begin{cases} \dfrac{x^{-\nu-1/2} M_{-\gamma/\sqrt{2\alpha},\nu}(2x\sqrt{2\alpha})}{z^{-\nu-1/2} M_{-\gamma/\sqrt{2\alpha},\nu}(2z\sqrt{2\alpha})}, & 0 < x \le z \\ \dfrac{x^{-\nu-1/2} W_{-\gamma/\sqrt{2\alpha},\nu}(2x\sqrt{2\alpha})}{z^{-\nu-1/2} W_{-\gamma/\sqrt{2\alpha},\nu}(2z\sqrt{2\alpha})}, & z \le x \end{cases}$$

2.20.1 $$\mathbf{E}_x\Big\{\exp\Big(-\frac{\gamma^2}{2} \int_0^{H_z} \frac{ds}{\big(R_s^{(n)}\big)^2}\Big);\ H_z < \infty\Big\} = \begin{cases} \Big(\dfrac{x}{z}\Big)^{-\nu+\sqrt{\nu^2+\gamma^2}}, & 0 < x \le z \\ \Big(\dfrac{x}{z}\Big)^{-\nu-\sqrt{\nu^2+\gamma^2}}, & z \le x \end{cases}$$

4 $R_s^{(n)}$ $\nu > 0,\ n = 2\nu + 2 > 2$ $H_z = \min\{s : R_s^{(n)} = z\}$

2.20.3

$$\mathbf{E}_x \exp\Big(-\alpha H_z - \frac{\gamma^2}{2}\int_0^{H_z} \frac{ds}{\big(R_s^{(n)}\big)^2}\Big) = \begin{cases} \dfrac{x^{-\nu} I_{\sqrt{\nu^2+\gamma^2}}(x\sqrt{2\alpha})}{z^{-\nu} I_{\sqrt{\nu^2+\gamma^2}}(z\sqrt{2\alpha})}, & 0 < x \le z \\ \dfrac{x^{-\nu} K_{\sqrt{\nu^2+\gamma^2}}(x\sqrt{2\alpha})}{z^{-\nu} K_{\sqrt{\nu^2+\gamma^2}}(z\sqrt{2\alpha})}, & z \le x \end{cases}$$

2.20.4
$0 < y$

$$\mathbf{P}_x\Big(\int_0^{H_z} \frac{ds}{\big(R_s^{(n)}\big)^2} \in dy\Big) = \begin{cases} \dfrac{z^\nu \ln(z/x)}{x^\nu \sqrt{2\pi} y^{3/2}} e^{-\nu^2 y/2} e^{-\ln^2(z/x)/2y} dy, & 0 < x \le z \\ \dfrac{z^\nu \ln(x/z)}{x^\nu \sqrt{2\pi} y^{3/2}} e^{-\nu^2 y/2} e^{-\ln^2(x/z)/2y} dy, & z \le x \end{cases}$$

2.21.1

$$\mathbf{E}_x\Big\{\exp\Big(-\int_0^{H_z}\Big(\frac{p^2}{2\big(R_s^{(n)}\big)^2} + \frac{q^2\big(R_s^{(n)}\big)^2}{2}\Big)ds\Big); H_z < \infty\Big\}$$

$$= \begin{cases} \dfrac{x^{-\nu} I_{\sqrt{\nu^2+p^2}/2}(qx^2/2)}{z^{-\nu} I_{\sqrt{\nu^2+p^2}/2}(qz^2/2)}, & 0 < x \le z \\ \dfrac{x^{-\nu} K_{\sqrt{\nu^2+p^2}/2}(qx^2/2)}{z^{-\nu} K_{\sqrt{\nu^2+p^2}/2}(qz^2/2)}, & z \le x \end{cases}$$

2.21.3

$$\mathbf{E}_x \exp\Big(-\alpha H_z - \int_0^{H_z}\Big(\frac{p^2}{2\big(R_s^{(n)}\big)^2} + \frac{q^2\big(R_s^{(n)}\big)^2}{2}\Big)ds\Big)$$

$$= \begin{cases} \dfrac{x^{-\nu-1} M_{-\alpha/2q,\sqrt{p^2+\nu^2}/2}(qx^2)}{z^{-\nu-1} M_{-\alpha/2q,\sqrt{p^2+\nu^2}/2}(qz^2)}, & 0 < x \le z \\ \dfrac{x^{-\nu-1} W_{-\alpha/2q,\sqrt{p^2+\nu^2}/2}(qx^2)}{z^{-\nu-1} W_{-\alpha/2q,\sqrt{p^2+\nu^2}/2}(qz^2)}, & z \le x \end{cases}$$

2.22.1

$$\mathbf{E}_x\Big\{\exp\Big(-\int_0^{H_z}\Big(\frac{p^2}{2\big(R_s^{(n)}\big)^2} + \frac{q}{R_s^{(n)}}\Big)ds\Big); H_z < \infty\Big\}$$

$$= \begin{cases} \dfrac{x^{-\nu} I_{2\sqrt{\nu^2+p^2}}(2\sqrt{2qx})}{z^{-\nu} I_{2\sqrt{\nu^2+p^2}}(2\sqrt{2qz})}, & 0 < x \le z \\ \dfrac{x^{-\nu} K_{2\sqrt{\nu^2+p^2}}(2\sqrt{2qx})}{z^{-\nu} K_{2\sqrt{\nu^2+p^2}}(2\sqrt{2qz})}, & z \le x \end{cases}$$

4 $R_s^{(n)}$ $\nu > 0,\ n = 2\nu + 2 > 2$ $H_z = \min\{s : R_s^{(n)} = z\}$

2.22.3 $$\mathbf{E}_x \exp\Big(-\alpha H_z - \int_0^{H_z} \Big(\frac{p^2}{2\big(R_s^{(n)}\big)^2} + \frac{q}{R_s^{(n)}}\Big) ds\Big)$$

$$= \begin{cases} \dfrac{x^{-\nu-1/2} M_{-q/\sqrt{2\alpha},\sqrt{\nu^2+p^2}}(2x\sqrt{2\alpha})}{z^{-\nu-1/2} M_{-q/\sqrt{2\alpha},\sqrt{\nu^2+p^2}}(2z\sqrt{2\alpha})}, & 0 < x \le z \\ \dfrac{x^{-\nu-1/2} W_{-q/\sqrt{2\alpha},\sqrt{\nu^2+p^2}}(2x\sqrt{2\alpha})}{z^{-\nu-1/2} W_{-q/\sqrt{2\alpha},\sqrt{\nu^2+p^2}}(2z\sqrt{2\alpha})}, & z \le x \end{cases}$$

2.27.1 $$\mathbf{E}_x\Big\{\exp\Big(-\int_0^{H_z} \big(p\mathbb{I}_{(-\infty,r)}\big(R_s^{(n)}\big) + q\mathbb{I}_{[r,\infty)}\big(R_s^{(n)}\big)\big) ds\Big); \ell(H_z, r) \in dy\Big\}$$

$z \le r$ $$= \exp\Big(-y\Big(\frac{\sqrt{p}C_\nu(r\sqrt{2p}, z\sqrt{2p})}{\sqrt{2}S_\nu(r\sqrt{2p}, z\sqrt{2p})} + \frac{\sqrt{q}K_{\nu+1}(r\sqrt{2q})}{\sqrt{2}K_\nu(r\sqrt{2q})}\Big)\Big)$$

$$\times \begin{cases} \dfrac{(2p)^{-\nu} S_\nu(x\sqrt{2p}, z\sqrt{2p})}{2r^{2\nu+1} S_\nu^2(r\sqrt{2p}, z\sqrt{2p})} dy, & z \le x \le r \\ \dfrac{(2p)^{-\nu} x^{-\nu} K_\nu(x\sqrt{2q})}{2r^{\nu+1} K_\nu(r\sqrt{2q}) S_\nu(r\sqrt{2p}, z\sqrt{2p})} dy, & r \le x \end{cases}$$

$r \le z$ $$= \exp\Big(-y\Big(\frac{\sqrt{q}C_\nu(r\sqrt{2q}, z\sqrt{2q})}{\sqrt{2}S_\nu(z\sqrt{2q}, r\sqrt{2q})} + \frac{\sqrt{p}I_{\nu+1}(r\sqrt{2p})}{\sqrt{2}I_\nu(r\sqrt{2p})}\Big)\Big)$$

$$\times \begin{cases} \dfrac{(2q)^{-\nu} x^{-\nu} I_\nu(x\sqrt{2p})}{2r^{\nu+1} I_\nu(r\sqrt{2p}) S_\nu(z\sqrt{2q}, r\sqrt{2q})} dy, & 0 \le x \le r \\ \dfrac{(2q)^{-\nu} S_\nu(z\sqrt{2q}, x\sqrt{2q})}{2r^{2\nu+1} S_\nu^2(z\sqrt{2q}, r\sqrt{2q})} dy, & r \le x \le z \end{cases}$$

(1) $$\mathbf{E}_x\Big\{\exp\Big(-\int_0^{H_z} \big(p\mathbb{I}_{(-\infty,r)}\big(R_s^{(n)}\big) + q\mathbb{I}_{[r,\infty)}\big(R_s^{(n)}\big)\big) ds\Big); \ell(H_z, r) = 0\Big\}$$

$$= \begin{cases} \dfrac{S_\nu(r\sqrt{2p}, x\sqrt{2p})}{S_\nu(r\sqrt{2p}, z\sqrt{2p})}, & z \le x \le r \\ \dfrac{S_\nu(x\sqrt{2q}, r\sqrt{2q})}{S_\nu(z\sqrt{2q}, r\sqrt{2q})}, & r \le x \le z \end{cases}$$

4 $R_s^{(n)}$ $\nu > 0,\ n = 2\nu + 2 > 2$ $H = H_{a,b} = \min\{s : R_s^{(n)} \notin (a,b)\}$

3. Stopping at first exit time

3.0.1 $$\mathbf{E}_x e^{-\alpha H} = \frac{S_\nu(b\sqrt{2\alpha}, x\sqrt{2\alpha}) + S_\nu(x\sqrt{2\alpha}, a\sqrt{2\alpha})}{S_\nu(b\sqrt{2\alpha}, a\sqrt{2\alpha})}$$

3.0.3 $$\mathbf{E}_x e^{-\beta R_H^{(n)}} = \frac{x^{-2\nu} - b^{-2\nu}}{a^{-2\nu} - b^{-2\nu}} e^{-\beta a} + \frac{a^{-2\nu} - x^{-2\nu}}{a^{-2\nu} - b^{-2\nu}} e^{-\beta b}$$

3.0.4 (a) $$\mathbf{P}_x\big(R_H^{(n)} = a\big) = \frac{x^{-2\nu} - b^{-2\nu}}{a^{-2\nu} - b^{-2\nu}}$$

3.0.4 (b) $$\mathbf{P}_x\big(R_H^{(n)} = b\big) = \frac{a^{-2\nu} - x^{-2\nu}}{a^{-2\nu} - b^{-2\nu}}$$

3.0.5 (a) $$\mathbf{E}_x\big\{e^{-\alpha H}; R_H^{(n)} = a\big\} = \frac{S_\nu(b\sqrt{2\alpha}, x\sqrt{2\alpha})}{S_\nu(b\sqrt{2\alpha}, a\sqrt{2\alpha})}$$

3.0.5 (b) $$\mathbf{E}_x\big\{e^{-\alpha H}; R_H^{(n)} = b\big\} = \frac{S_\nu(x\sqrt{2\alpha}, a\sqrt{2\alpha})}{S_\nu(b\sqrt{2\alpha}, a\sqrt{2\alpha})}$$

3.1.6 $$\mathbf{P}_x\Big(\sup_{0\le s\le H} R_s^{(n)} \ge y, R_H^{(n)} = a\Big) = \frac{(y^{-2\nu} - b^{-2\nu})(a^{-2\nu} - x^{-2\nu})}{(a^{-2\nu} - b^{-2\nu})(a^{-2\nu} - y^{-2\nu})}, \qquad a \le x \le y \le b$$

3.1.8 (1) $$\mathbf{E}_x\Big\{e^{-\alpha H}; \sup_{0\le s\le H} R_s^{(n)} \ge y, R_H^{(n)} = a\Big\}$$
$$= \frac{S_\nu(b\sqrt{2\alpha}, y\sqrt{2\alpha}) S_\nu(x\sqrt{2\alpha}, a\sqrt{2\alpha})}{S_\nu(b\sqrt{2\alpha}, a\sqrt{2\alpha}) S_\nu(y\sqrt{2\alpha}, a\sqrt{2\alpha})}, \qquad a \le x \le y \le b$$

3.2.6 $$\mathbf{P}_x\Big(\inf_{0\le s\le H} R_s^{(n)} \le y, R_H^{(n)} = b\Big) = \frac{(x^{-2\nu} - b^{-2\nu})(a^{-2\nu} - y^{-2\nu})}{(y^{-2\nu} - b^{-2\nu})(a^{-2\nu} - b^{-2\nu})}, \qquad a \le y \le x \le b$$

3.2.8 (1) $$\mathbf{E}_x\Big\{e^{-\alpha H}; \inf_{0\le s\le H} R_s^{(n)} \le y, R_H^{(n)} = b\Big\}$$
$$= \frac{S_\nu(b\sqrt{2\alpha}, x\sqrt{2\alpha}) S_\nu(y\sqrt{2\alpha}, a\sqrt{2\alpha})}{S_\nu(b\sqrt{2\alpha}, y\sqrt{2\alpha}) S_\nu(b\sqrt{2\alpha}, a\sqrt{2\alpha})}, \qquad a \le y \le x \le b$$

4 $R_s^{(n)}$ $\nu>0,\ n=2\nu+2>2$ $H=H_{a,b}=\min\{s: R_s^{(n)}\notin(a,b)\}$

3.3.1 $\mathbf{E}_x e^{-\gamma\ell(H,r)}$

$$=\begin{cases}\dfrac{\nu(a^{-2\nu}-b^{-2\nu})+\gamma r^{2\nu+1}(r^{-2\nu}-x^{-2\nu})(a^{-2\nu}-r^{-2\nu})}{\nu(a^{-2\nu}-b^{-2\nu})+\gamma r^{2\nu+1}(r^{-2\nu}-b^{-2\nu})(a^{-2\nu}-r^{-2\nu})}, & r\le x\le b\\ \dfrac{\nu(a^{-2\nu}-b^{-2\nu})+\gamma r^{2\nu+1}(r^{-2\nu}-b^{-2\nu})(x^{-2\nu}-r^{-2\nu})}{\nu(a^{-2\nu}-b^{-2\nu})+\gamma r^{2\nu+1}(r^{-2\nu}-b^{-2\nu})(a^{-2\nu}-r^{-2\nu})}, & a\le x\le r\end{cases}$$

3.3.2 $\mathbf{P}_x\big(\ell(H,r)\in dy\big)=\exp\Big(-\dfrac{\nu(a^{-2\nu}-b^{-2\nu})y}{r^{2\nu+1}(r^{-2\nu}-b^{-2\nu})(a^{-2\nu}-r^{-2\nu})}\Big)$

$$\times\begin{cases}\dfrac{\nu(x^{-2\nu}-b^{-2\nu})(a^{-2\nu}-b^{-2\nu})}{r^{2\nu+1}(r^{-2\nu}-b^{-2\nu})^2(a^{-2\nu}-r^{-2\nu})}dy, & r\le x\le b\\ \dfrac{\nu(a^{-2\nu}-x^{-2\nu})(a^{-2\nu}-b^{-2\nu})}{r^{2\nu+1}(r^{-2\nu}-b^{-2\nu})(a^{-2\nu}-r^{-2\nu})^2}dy, & a\le x\le r\end{cases}$$

(1) $\mathbf{P}_x\big(\ell(H,r)=0\big)=\begin{cases}\dfrac{r^{-2\nu}-x^{-2\nu}}{r^{-2\nu}-b^{-2\nu}}, & r\le x\le b\\ \dfrac{x^{-2\nu}-r^{-2\nu}}{a^{-2\nu}-r^{-2\nu}}, & a\le x\le r\end{cases}$

3.3.3 $\mathbf{E}_r e^{-\alpha H-\gamma\ell(H,r)}$

$$=\frac{S_\nu(b\sqrt{2\alpha},r\sqrt{2\alpha})+S_\nu(r\sqrt{2\alpha},a\sqrt{2\alpha})}{S_\nu(b\sqrt{2\alpha},a\sqrt{2\alpha})+2\gamma r^{2\nu+1}(2\alpha)^\nu S_\nu(b\sqrt{2\alpha},r\sqrt{2\alpha})S_\nu(r\sqrt{2\alpha},a\sqrt{2\alpha})}=:L$$

$$\mathbf{E}_x e^{-\alpha H-\gamma\ell(H,r)}=\begin{cases}\dfrac{S_\nu(x\sqrt{2\alpha},r\sqrt{2\alpha})}{S_\nu(b\sqrt{2\alpha},r\sqrt{2\alpha})}+\dfrac{S_\nu(b\sqrt{2\alpha},x\sqrt{2\alpha})}{S_\nu(b\sqrt{2\alpha},r\sqrt{2\alpha})}L, & r\le x\le b\\ \dfrac{S_\nu(r\sqrt{2\alpha},x\sqrt{2\alpha})}{S_\nu(r\sqrt{2\alpha},a\sqrt{2\alpha})}+\dfrac{S_\nu(x\sqrt{2\alpha},a\sqrt{2\alpha})}{S_\nu(r\sqrt{2\alpha},a\sqrt{2\alpha})}L, & a\le x\le r\end{cases}$$

3.3.4 $\mathbf{E}_x\{e^{-\alpha H};\ell(H,r)\in dy\}=\exp\Big(-\dfrac{(2\alpha)^{-\nu}S_\nu(b\sqrt{2\alpha},a\sqrt{2\alpha})y}{2r^{2\nu+1}S_\nu(b\sqrt{2\alpha},r\sqrt{2\alpha})S_\nu(r\sqrt{2\alpha},a\sqrt{2\alpha})}\Big)$

$$\times\begin{cases}\dfrac{(2\alpha)^{-\nu}S_\nu(b\sqrt{2\alpha},x\sqrt{2\alpha})\big[S_\nu(b\sqrt{2\alpha},r\sqrt{2\alpha})+S_\nu(r\sqrt{2\alpha},a\sqrt{2\alpha})\big]}{2r^{2\nu+1}S_\nu^2(b\sqrt{2\alpha},r\sqrt{2\alpha})S_\nu(r\sqrt{2\alpha},a\sqrt{2\alpha})}dy, & r\le x\le b\\ \dfrac{(2\alpha)^{-\nu}S_\nu(x\sqrt{2\alpha},a\sqrt{2\alpha})\big[S_\nu(b\sqrt{2\alpha},r\sqrt{2\alpha})+S_\nu(r\sqrt{2\alpha},a\sqrt{2\alpha})\big]}{2r^{2\nu+1}S_\nu(b\sqrt{2\alpha},r\sqrt{2\alpha})S_\nu^2(r\sqrt{2\alpha},a\sqrt{2\alpha})}dy, & a\le x\le r\end{cases}$$

4 $R_s^{(n)}$ $\nu > 0,\ n = 2\nu + 2 > 2$ $H = H_{a,b} = \min\{s : R_s^{(n)} \notin (a,b)\}$

(1) $$\mathbf{E}_x\{e^{-\alpha H}; \ell(H,r) = 0\} = \begin{cases} \dfrac{S_\nu(x\sqrt{2\alpha}, r\sqrt{2\alpha})}{S_\nu(b\sqrt{2\alpha}, r\sqrt{2\alpha})}, & r \le x \le b \\ \dfrac{S_\nu(r\sqrt{2\alpha}, x\sqrt{2\alpha})}{S_\nu(r\sqrt{2\alpha}, a\sqrt{2\alpha})}, & a \le x \le r \end{cases}$$

3.3.5 (a) $\mathbf{E}_x\{e^{-\gamma\ell(H,r)}; R_H^{(n)} = a\}$

$$= \begin{cases} \dfrac{\nu(x^{-2\nu} - b^{-2\nu})}{\nu(a^{-2\nu} - b^{-2\nu}) + \gamma r^{2\nu+1}(r^{-2\nu} - b^{-2\nu})(a^{-2\nu} - r^{-2\nu})}, & r \le x \le b \\ \dfrac{\nu(x^{-2\nu} - b^{-2\nu}) + \gamma r^{2\nu+1}(r^{-2\nu} - b^{-2\nu})(x^{-2\nu} - r^{-2\nu})}{\nu(a^{-2\nu} - b^{-2\nu}) + \gamma r^{2\nu+1}(r^{-2\nu} - b^{-2\nu})(a^{-2\nu} - r^{-2\nu})}, & a \le x \le r \end{cases}$$

3.3.5 (b) $\mathbf{E}_x\{e^{-\gamma\ell(H,r)}; R_H^{(n)} = b\}$

$$= \begin{cases} \dfrac{\nu(a^{-2\nu} - x^{-2\nu}) + \gamma r^{2\nu+1}(r^{-2\nu} - x^{-2\nu})(a^{-2\nu} - r^{-2\nu})}{\nu(a^{-2\nu} - b^{-2\nu}) + \gamma r^{2\nu+1}(r^{-2\nu} - b^{-2\nu})(a^{-2\nu} - r^{-2\nu})}, & r \le x \le b \\ \dfrac{\nu(a^{-2\nu} - x^{-2\nu})}{\nu(a^{-2\nu} - b^{-2\nu}) + \gamma r^{2\nu+1}(r^{-2\nu} - b^{-2\nu})(a^{-2\nu} - r^{-2\nu})}, & a \le x \le r \end{cases}$$

3.3.6 (a) $$\mathbf{P}_x\big(\ell(H,r) \in dy, R_H^{(n)} = a\big) = \exp\Big(-\frac{\nu(a^{-2\nu} - b^{-2\nu})y}{r^{2\nu+1}(r^{-2\nu} - b^{-2\nu})(a^{-2\nu} - r^{-2\nu})}\Big)$$

$$\times \begin{cases} \dfrac{\nu(x^{-2\nu} - b^{-2\nu})}{r^{2\nu+1}(r^{-2\nu} - b^{-2\nu})(a^{-2\nu} - r^{-2\nu})}dy, & r \le x \le b \\ \dfrac{\nu(a^{-2\nu} - x^{-2\nu})}{r^{2\nu+1}(a^{-2\nu} - r^{-2\nu})^2}dy, & a \le x \le r \end{cases}$$

(1) $$\mathbf{P}_x\big(\ell(H,r) = 0, R_H^{(n)} = a\big) = \begin{cases} 0, & r \le x \le b \\ \dfrac{x^{-2\nu} - r^{-2\nu}}{a^{-2\nu} - r^{-2\nu}}, & a \le x \le r \end{cases}$$

3.3.6 (b) $$\mathbf{P}_x\big(\ell(H,r) \in dy, R_H^{(n)} = b\big) = \exp\Big(-\frac{\nu(a^{-2\nu} - b^{-2\nu})y}{r^{2\nu+1}(r^{-2\nu} - b^{-2\nu})(a^{-2\nu} - r^{-2\nu})}\Big)$$

4 $R_s^{(n)}$ $\nu > 0,\ n = 2\nu + 2 > 2$ $H = H_{a,b} = \min\{s : R_s^{(n)} \notin (a,b)\}$

$$\times\begin{cases} \dfrac{\nu(x^{-2\nu} - b^{-2\nu})}{r^{2\nu+1}(r^{-2\nu} - b^{-2\nu})^2}dy, & r \le x \le b \\ \dfrac{\nu(a^{-2\nu} - x^{-2\nu})}{r^{2\nu+1}(r^{-2\nu} - b^{-2\nu})(a^{-2\nu} - r^{-2\nu})}dy, & a \le x \le r \end{cases}$$

(1) $$\mathbf{P}_x\big(\ell(H,r) = 0, R_H^{(n)} = b\big) = \begin{cases} \dfrac{r^{-2\nu} - x^{-2\nu}}{r^{-2\nu} - b^{-2\nu}}, & r \le x \le b \\ 0, & a \le x \le r \end{cases}$$

3.3.7 (a) $\mathbf{E}_x\{e^{-\alpha H - \gamma\ell(H,r)}; R_H^{(n)} = a\}$

$$=\begin{cases} \dfrac{S_\nu(b\sqrt{2\alpha}, x\sqrt{2\alpha})}{S_\nu(b\sqrt{2\alpha}, a\sqrt{2\alpha}) + 2\gamma r^{2\nu+1}(2\alpha)^\nu S_\nu(b\sqrt{2\alpha}, r\sqrt{2\alpha})S_\nu(r\sqrt{2\alpha}, a\sqrt{2\alpha})}, & r \le x \le b \\ \dfrac{S_\nu(b\sqrt{2\alpha}, x\sqrt{2\alpha}) + 2\gamma r^{2\nu+1}(2\alpha)^\nu S_\nu(b\sqrt{2\alpha}, r\sqrt{2\alpha})S_\nu(r\sqrt{2\alpha}, x\sqrt{2\alpha})}{S_\nu(b\sqrt{2\alpha}, a\sqrt{2\alpha}) + 2\gamma r^{2\nu+1}(2\alpha)^\nu S_\nu(b\sqrt{2\alpha}, r\sqrt{2\alpha})S_\nu(r\sqrt{2\alpha}, a\sqrt{2\alpha})}, & a \le x \le r \end{cases}$$

3.3.7 (b) $\mathbf{E}_x\{e^{-\alpha H - \gamma\ell(H,r)}; R_H^{(n)} = b\}$

$$=\begin{cases} \dfrac{S_\nu(x\sqrt{2\alpha}, a\sqrt{2\alpha}) + 2\gamma r^{2\nu+1}(2\alpha)^\nu S_\nu(x\sqrt{2\alpha}, r\sqrt{2\alpha})S_\nu(r\sqrt{2\alpha}, a\sqrt{2\alpha})}{S_\nu(b\sqrt{2\alpha}, a\sqrt{2\alpha}) + 2\gamma r^{2\nu+1}(2\alpha)^\nu S_\nu(b\sqrt{2\alpha}, r\sqrt{2\alpha})S_\nu(r\sqrt{2\alpha}, a\sqrt{2\alpha})}, & r \le x \le b \\ \dfrac{S_\nu(x\sqrt{2\alpha}, a\sqrt{2\alpha})}{S_\nu(b\sqrt{2\alpha}, a\sqrt{2\alpha}) + 2\gamma r^{2\nu+1}(2\alpha)^\nu S_\nu(b\sqrt{2\alpha}, r\sqrt{2\alpha})S_\nu(r\sqrt{2\alpha}, a\sqrt{2\alpha})}, & a \le x \le r \end{cases}$$

3.3.8 (a) $\mathbf{E}_x\{e^{-\alpha H}; \ell(H,r) \in dy, R_H^{(n)} = a\}$

$$= \exp\Big(-\frac{(2\alpha)^{-\nu} S_\nu(b\sqrt{2\alpha}, a\sqrt{2\alpha})y}{2r^{2\nu+1}S_\nu(b\sqrt{2\alpha}, r\sqrt{2\alpha})S_\nu(r\sqrt{2\alpha}, a\sqrt{2\alpha})}\Big)$$

$$\times\begin{cases} \dfrac{(2\alpha)^{-\nu} S_\nu(b\sqrt{2\alpha}, x\sqrt{2\alpha})}{2r^{2\nu+1}S_\nu(b\sqrt{2\alpha}, r\sqrt{2\alpha})S_\nu(r\sqrt{2\alpha}, a\sqrt{2\alpha})}dy, & r \le x \le b \\ \dfrac{(2\alpha)^{-\nu} S_\nu(x\sqrt{2\alpha}, a\sqrt{2\alpha})}{2r^{2\nu+1}S_\nu^2(r\sqrt{2\alpha}, a\sqrt{2\alpha})}dy, & a \le x \le r \end{cases}$$

(1) $$\mathbf{E}_x\{e^{-\alpha H}; \ell(H,r) = 0, R_H^{(n)} = a\} = \begin{cases} 0, & r \le x \le b \\ \dfrac{S_\nu(r\sqrt{2\alpha}, x\sqrt{2\alpha})}{S_\nu(r\sqrt{2\alpha}, a\sqrt{2\alpha})}, & a \le x \le r \end{cases}$$

4 $R_s^{(n)}$ $\nu > 0,\ n = 2\nu + 2 > 2$ $H = H_{a,b} = \min\{s : R_s^{(n)} \notin (a,b)\}$

3.3.8 (b) $\mathbf{E}_x\{e^{-\alpha H}; \ell(H,r) \in dy, R_H^{(n)} = b\}$

$$= \exp\Big(-\frac{(2\alpha)^{-\nu} S_\nu(b\sqrt{2\alpha}, a\sqrt{2\alpha})y}{2r^{2\nu+1} S_\nu(b\sqrt{2\alpha}, r\sqrt{2\alpha}) S_\nu(r\sqrt{2\alpha}, a\sqrt{2\alpha})}\Big)$$

$$\times \begin{cases} \dfrac{(2\alpha)^{-\nu} S_\nu(b\sqrt{2\alpha}, x\sqrt{2\alpha})}{2r^{2\nu+1} S_\nu^2(b\sqrt{2\alpha}, r\sqrt{2\alpha})} dy, & r \le x \le b \\ \dfrac{(2\alpha)^{-\nu} S_\nu(x\sqrt{2\alpha}, a\sqrt{2\alpha})}{2r^{2\nu+1} S_\nu(b\sqrt{2\alpha}, r\sqrt{2\alpha}) S_\nu(r\sqrt{2\alpha}, a\sqrt{2\alpha})} dy, & a \le x \le r \end{cases}$$

(1) $$\mathbf{E}_x\{e^{-\alpha H}; \ell(H,r) = 0, R_H^{(n)} = b\} = \begin{cases} \dfrac{S_\nu(x\sqrt{2\alpha}, r\sqrt{2\alpha})}{S_\nu(b\sqrt{2\alpha}, r\sqrt{2\alpha})}, & r \le x \le b \\ 0, & a \le x \le r \end{cases}$$

3.4.1 $\mathbf{E}_r \exp\Big(-\gamma \int_0^H \mathbb{I}_{[r,b]}\big(R_s^{(n)}\big) ds\Big)$

$$= \frac{2\nu S_\nu(b\sqrt{2\gamma}, r\sqrt{2\gamma}) + (2\gamma)^{-\nu}(a^{-2\nu} - r^{-2\nu})}{2\nu S_\nu(b\sqrt{2\gamma}, r\sqrt{2\gamma}) + \sqrt{2\gamma} r^{2\nu+1}(a^{-2\nu} - r^{-2\nu}) C_\nu(r\sqrt{2\gamma}, b\sqrt{2\gamma})} =: Q$$

$\mathbf{E}_x \exp\Big(-\gamma \int_0^H \mathbb{I}_{[r,b]}\big(R_s^{(n)}\big) ds\Big)$

$$= \begin{cases} \dfrac{x^{-2\nu} - r^{-2\nu}}{a^{-2\nu} - r^{-2\nu}} + \dfrac{a^{-2\nu} - x^{-2\nu}}{a^{-2\nu} - r^{-2\nu}} Q, & a \le x \le r \\ \dfrac{S_\nu(x\sqrt{2\gamma}, r\sqrt{2\gamma})}{S_\nu(b\sqrt{2\gamma}, r\sqrt{2\gamma})} + \dfrac{S_\nu(b\sqrt{2\gamma}, x\sqrt{2\gamma})}{S_\nu(b\sqrt{2\gamma}, r\sqrt{2\gamma})} Q, & r \le x \le b \end{cases}$$

3.4.2 (1) $$\mathbf{P}_x\Big(\int_0^H \mathbb{I}_{[r,b]}\big(R_s^{(n)}\big) ds = 0\Big) = \begin{cases} \dfrac{x^{-2\nu} - r^{-2\nu}}{a^{-2\nu} - r^{-2\nu}}, & a \le x \le r \\ 0, & r \le x \le b \end{cases}$$

3.4.5 (a) $\mathbf{E}_x\Big\{\exp\Big(-\gamma \int_0^H \mathbb{I}_{[r,b]}\big(R_s^{(n)}\big) ds\Big); R_H^{(n)} = a\Big\}$

4 $R_s^{(n)}$ $\nu > 0,\ n = 2\nu + 2 > 2$ $H = H_{a,b} = \min\{s : R_s^{(n)} \notin (a,b)\}$

$$= \begin{cases} \dfrac{2\nu S_\nu(b\sqrt{2\gamma}, r\sqrt{2\gamma}) + \sqrt{2\gamma} r^{2\nu+1}(x^{-2\nu} - r^{-2\nu}) C_\nu(r\sqrt{2\gamma}, b\sqrt{2\gamma})}{2\nu S_\nu(b\sqrt{2\gamma}, r\sqrt{2\gamma}) + \sqrt{2\gamma} r^{2\nu+1}(a^{-2\nu} - r^{-2\nu}) C_\nu(r\sqrt{2\gamma}, b\sqrt{2\gamma})}, & a \le x \le r \\ \dfrac{2\nu S_\nu(b\sqrt{2\gamma}, x\sqrt{2\gamma})}{2\nu S_\nu(b\sqrt{2\gamma}, r\sqrt{2\gamma}) + \sqrt{2\gamma} r^{2\nu+1}(a^{-2\nu} - r^{-2\nu}) C_\nu(r\sqrt{2\gamma}, b\sqrt{2\gamma})}, & r \le x \le b \end{cases}$$

3.4.5 (b) $\mathbf{E}_x\Big\{\exp\Big(-\gamma \int_0^H \mathbb{1}_{[r,b]}(R_s^{(n)})ds\Big); R_H^{(n)} = b\Big\}$

$$= \begin{cases} \dfrac{(2\gamma)^{-\nu}(a^{-2\nu} - x^{-2\nu})}{2\nu S_\nu(b\sqrt{2\gamma}, r\sqrt{2\gamma}) + \sqrt{2\gamma} r^{2\nu+1}(a^{-2\nu} - r^{-2\nu}) C_\nu(r\sqrt{2\gamma}, b\sqrt{2\gamma})}, & a \le x \le r \\ \dfrac{2\nu S_\nu(x\sqrt{2\gamma}, r\sqrt{2\gamma}) + \sqrt{2\gamma} r^{2\nu+1}(a^{-2\nu} - r^{-2\nu}) C_\nu(r\sqrt{2\gamma}, x\sqrt{2\gamma})}{2\nu S_\nu(b\sqrt{2\gamma}, r\sqrt{2\gamma}) + \sqrt{2\gamma} r^{2\nu+1}(a^{-2\nu} - r^{-2\nu}) C_\nu(r\sqrt{2\gamma}, b\sqrt{2\gamma})}, & r \le x \le b \end{cases}$$

3.4.6 (1) $\mathbf{P}_x\Big(\int_0^H \mathbb{1}_{[r,b]}(R_s^{(n)})ds = 0, R_H^{(n)} = a\Big) = \begin{cases} \dfrac{x^{-2\nu} - r^{-2\nu}}{a^{-2\nu} - r^{-2\nu}}, & a \le x \le r \\ 0, & r \le x \le b \end{cases}$

3.5.1 $\mathbf{E}_r \exp\Big(-\gamma \int_0^H \mathbb{1}_{[a,r]}(R_s^{(n)})ds\Big)$

$$= \frac{2\nu S_\nu(r\sqrt{2\gamma}, a\sqrt{2\gamma}) + (2\gamma)^{-\nu}(r^{-2\nu} - b^{-2\nu})}{2\nu S_\nu(r\sqrt{2\gamma}, a\sqrt{2\gamma}) + (r^{-2\nu} - b^{-2\nu})\sqrt{2\gamma} r^{2\nu+1} C_\nu(r\sqrt{2\gamma}, a\sqrt{2\gamma})} =: Q$$

$\mathbf{E}_x \exp\Big(-\gamma \int_0^H \mathbb{1}_{[a,r]}(R_s^{(n)})ds\Big)$

$$= \begin{cases} \dfrac{S_\nu(r\sqrt{2\gamma}, x\sqrt{2\gamma})}{S_\nu(r\sqrt{2\gamma}, a\sqrt{2\gamma})} + \dfrac{S_\nu(x\sqrt{2\gamma}, a\sqrt{2\gamma})}{S_\nu(r\sqrt{2\gamma}, a\sqrt{2\gamma})} Q, & a \le x \le r \\ \dfrac{r^{-2\nu} - x^{-2\nu}}{r^{-2\nu} - b^{-2\nu}} + \dfrac{x^{-2\nu} - b^{-2\nu}}{r^{-2\nu} - b^{-2\nu}} Q, & r \le x \le b \end{cases}$$

3.5.2 (1) $\mathbf{P}_x\Big(\int_0^H \mathbb{1}_{[a,r]}(R_s^{(n)})ds = 0\Big) = \begin{cases} 0, & a \le x \le r \\ \dfrac{r^{-2\nu} - x^{-2\nu}}{r^{-2\nu} - b^{-2\nu}}, & r \le x \le b \end{cases}$

3.5.5 (a) $\mathbf{E}_x\Big\{\exp\Big(-\gamma \int_0^H \mathbb{1}_{[a,r]}(R_s^{(n)})ds\Big); R_H^{(n)} = a\Big\}$

4 $R_s^{(n)}$ $\nu > 0,\ n = 2\nu + 2 > 2$ $H = H_{a,b} = \min\{s : R_s^{(n)} \notin (a,b)\}$

$$= \begin{cases} \dfrac{2\nu S_\nu(r\sqrt{2\gamma}, x\sqrt{2\gamma}) + (r^{-2\nu} - b^{-2\nu})\sqrt{2\gamma}r^{2\nu+1}C_\nu(r\sqrt{2\gamma}, x\sqrt{2\gamma})}{2\nu S_\nu(r\sqrt{2\gamma}, a\sqrt{2\gamma}) + (r^{-2\nu} - b^{-2\nu})\sqrt{2\gamma}r^{2\nu+1}C_\nu(r\sqrt{2\gamma}, a\sqrt{2\gamma})}, & a \le x \le r \\ \dfrac{(x^{-2\nu} - b^{-2\nu})(2\gamma)^{-\nu}}{2\nu S_\nu(r\sqrt{2\gamma}, a\sqrt{2\gamma}) + (r^{-2\nu} - b^{-2\nu})\sqrt{2\gamma}r^{2\nu+1}C_\nu(r\sqrt{2\gamma}, a\sqrt{2\gamma})}, & r \le x \le b \end{cases}$$

3.5.5 (b) $$\mathbf{E}_x\Big\{\exp\Big(-\gamma\int_0^H \mathbb{I}_{[a,r]}\big(R_s^{(n)}\big)ds\Big); R_H^{(n)} = b\Big\}$$

$$= \begin{cases} \dfrac{2\nu S_\nu(x\sqrt{2\gamma}, a\sqrt{2\gamma})}{2\nu S_\nu(r\sqrt{2\gamma}, a\sqrt{2\gamma}) + (r^{-2\nu} - b^{-2\nu})\sqrt{2\gamma}r^{2\nu+1}C_\nu(r\sqrt{2\gamma}, a\sqrt{2\gamma})}, & a \le x \le r \\ \dfrac{2\nu S_\nu(r\sqrt{2\gamma}, a\sqrt{2\gamma}) + (r^{-2\nu} - x^{-2\nu})\sqrt{2\gamma}r^{2\nu+1}C_\nu(r\sqrt{2\gamma}, a\sqrt{2\gamma})}{2\nu S_\nu(r\sqrt{2\gamma}, a\sqrt{2\gamma}) + (r^{-2\nu} - b^{-2\nu})\sqrt{2\gamma}r^{2\nu+1}C_\nu(r\sqrt{2\gamma}, a\sqrt{2\gamma})}, & r \le x \le b \end{cases}$$

3.5.6 (1) $$\mathbf{P}_x\Big(\int_0^H \mathbb{I}_{[a,r]}\big(R_s^{(n)}\big)ds = 0, R_H^{(n)} = b\Big) = \begin{cases} 0, & a \le x \le r \\ \dfrac{r^{-2\nu} - x^{-2\nu}}{r^{-2\nu} - b^{-2\nu}}, & r \le x \le b \end{cases}$$

3.6.1 $$\mathbf{E}_r \exp\Big(-\int_0^H \Big(p\mathbb{I}_{[a,r]}\big(R_s^{(n)}\big) + q\mathbb{I}_{[r,b]}\big(R_s^{(n)}\big)\Big)ds\Big) =: Q$$

$$= \frac{r^{-2\nu-1}2^{-\nu-1/2}(p^{-\nu}S_\nu(b\sqrt{2q}, r\sqrt{2q}) + q^{-\nu}S_\nu(r\sqrt{2p}, a\sqrt{2p}))}{\sqrt{p}S_\nu(b\sqrt{2q}, r\sqrt{2q})C_\nu(r\sqrt{2p}, a\sqrt{2p}) + \sqrt{q}C_\nu(r\sqrt{2q}, b\sqrt{2q})S_\nu(r\sqrt{2p}, a\sqrt{2p})}$$

$$\mathbf{E}_x \exp\Big(-\int_0^H \Big(p\mathbb{I}_{[a,r]}\big(R_s^{(n)}\big) + q\mathbb{I}_{[r,b]}\big(R_s^{(n)}\big)\Big)ds\Big)$$

$$= \begin{cases} \dfrac{S_\nu(x\sqrt{2q}, r\sqrt{2q})}{S_\nu(b\sqrt{2q}, r\sqrt{2q})} + \dfrac{S_\nu(b\sqrt{2q}, x\sqrt{2q})}{S_\nu(b\sqrt{2q}, r\sqrt{2q})}Q, & r \le x \le b \\ \dfrac{S_\nu(r\sqrt{2p}, x\sqrt{2p})}{S_\nu(r\sqrt{2p}, a\sqrt{2p})} + \dfrac{S_\nu(x\sqrt{2p}, a\sqrt{2p})}{S_\nu(r\sqrt{2p}, a\sqrt{2p})}Q, & a \le x \le r \end{cases}$$

3.6.5 (a) $$\mathbf{E}_x\Big\{\exp\Big(-\int_0^H \Big(p\mathbb{I}_{[a,r]}\big(R_s^{(n)}\big) + q\mathbb{I}_{[r,b]}\big(R_s^{(n)}\big)\Big)ds\Big); R_H^{(n)} = a\Big\}$$

$a \le x$, $x \le r$: $$= \frac{\sqrt{p}S_\nu(b\sqrt{2q}, r\sqrt{2q})C_\nu(r\sqrt{2p}, x\sqrt{2p}) + \sqrt{q}C_\nu(r\sqrt{2q}, b\sqrt{2q})S_\nu(r\sqrt{2p}, x\sqrt{2p})}{\sqrt{p}S_\nu(b\sqrt{2q}, r\sqrt{2q})C_\nu(r\sqrt{2p}, a\sqrt{2p}) + \sqrt{q}C_\nu(r\sqrt{2q}, b\sqrt{2q})S_\nu(r\sqrt{2p}, a\sqrt{2p})}$$

$r \le x$, $x \le b$: $$= \frac{r^{-2\nu-1}2^{-\nu-1/2}p^{-\nu}S_\nu(b\sqrt{2q}, x\sqrt{2q})}{\sqrt{p}S_\nu(b\sqrt{2q}, r\sqrt{2q})C_\nu(r\sqrt{2p}, a\sqrt{2p}) + \sqrt{q}C_\nu(r\sqrt{2q}, b\sqrt{2q})S_\nu(r\sqrt{2p}, a\sqrt{2p})}$$

4 $R_s^{(n)}$ $\nu > 0,\ n = 2\nu + 2 > 2$ $H = H_{a,b} = \min\{s : R_s^{(n)} \notin (a,b)\}$

3.6.5 (*b*)	$\mathbf{E}_x\Big\{\exp\Big(-\int_0^H \Big(p\mathbb{I}_{[a,r]}\big(R_s^{(n)}\big) + q\mathbb{I}_{[r,b]}\big(R_s^{(n)}\big)\Big)ds\Big); R_H^{(n)} = b\Big\}$
$a \le x$, $x \le r$	$= \dfrac{r^{-2\nu-1}2^{-\nu-1/2}q^{-\nu}S_\nu(x\sqrt{2p}, a\sqrt{2p})}{\sqrt{p}S_\nu(b\sqrt{2q}, r\sqrt{2q})C_\nu(r\sqrt{2p}, a\sqrt{2p}) + \sqrt{q}C_\nu(r\sqrt{2q}, b\sqrt{2q})S_\nu(r\sqrt{2p}, a\sqrt{2p})}$
$r \le x$, $x \le b$	$= \dfrac{\sqrt{p}S_\nu(x\sqrt{2q}, r\sqrt{2q})C_\nu(r\sqrt{2p}, a\sqrt{2p}) + \sqrt{q}C_\nu(r\sqrt{2q}, x\sqrt{2q})S_\nu(r\sqrt{2p}, a\sqrt{2p})}{\sqrt{p}S_\nu(b\sqrt{2q}, r\sqrt{2q})C_\nu(r\sqrt{2p}, a\sqrt{2p}) + \sqrt{q}C_\nu(r\sqrt{2q}, b\sqrt{2q})S_\nu(r\sqrt{2p}, a\sqrt{2p})}$
3.8.5 (*a*)	$\mathbf{E}_x\Big\{\exp\Big(-\gamma\int_0^H R_s^{(n)}ds\Big);\ R_H^{(n)} = a\Big\} = \dfrac{S_{2\nu/3}(\frac{2}{3}b^{3/2}\sqrt{2\gamma}, \frac{2}{3}x^{3/2}\sqrt{2\gamma})}{S_{2\nu/3}(\frac{2}{3}b^{3/2}\sqrt{2\gamma}, \frac{2}{3}a^{3/2}\sqrt{2\gamma})}$
3.8.5 (*b*)	$\mathbf{E}_x\Big\{\exp\Big(-\gamma\int_0^H R_s^{(n)}ds\Big);\ R_H^{(n)} = b\Big\} = \dfrac{S_{2\nu/3}(\frac{2}{3}x^{3/2}\sqrt{2\gamma}, \frac{2}{3}a^{3/2}\sqrt{2\gamma})}{S_{2\nu/3}(\frac{2}{3}b^{3/2}\sqrt{2\gamma}, \frac{2}{3}a^{3/2}\sqrt{2\gamma})}$
3.9.1	$\mathbf{E}_x \exp\Big(-2\gamma\int_0^H (R_s^{(n)})^2 ds\Big) = \dfrac{S_{\nu/2}(b^2\sqrt{\gamma}, x^2\sqrt{\gamma}) + S_{\nu/2}(x^2\sqrt{\gamma}, a^2\sqrt{\gamma})}{S_{\nu/2}(b^2\sqrt{\gamma}, a^2\sqrt{\gamma})}$
3.9.5 (*a*)	$\mathbf{E}_x\Big\{\exp\Big(-2\gamma\int_0^H (R_s^{(n)})^2 ds\Big); R_H^{(n)} = a\Big\} = \dfrac{S_{\nu/2}(b^2\sqrt{\gamma}, x^2\sqrt{\gamma})}{S_{\nu/2}(b^2\sqrt{\gamma}, a^2\sqrt{\gamma})}$
3.9.5 (*b*)	$\mathbf{E}_x\Big\{\exp\Big(-2\gamma\int_0^H (R_s^{(n)})^2 ds\Big); R_H^{(n)} = b\Big\} = \dfrac{S_{\nu/2}(x^2\sqrt{\gamma}, a^2\sqrt{\gamma})}{S_{\nu/2}(b^2\sqrt{\gamma}, a^2\sqrt{\gamma})}$
3.12.5 (*a*)	$\mathbf{E}_x\{e^{-\gamma\check{H}(H)};\ R_H^{(n)} = a\} = \displaystyle\int_x^b \frac{2\nu y^{-1-2\nu}S_\nu(x\sqrt{2\gamma+2\alpha}, a\sqrt{2\gamma+2\alpha})}{(a^{-2\nu} - y^{-2\nu})S_\nu(y\sqrt{2\gamma+2\alpha}, a\sqrt{2\gamma+2\alpha})}dy$
3.12.5 (*b*)	$\mathbf{E}_x\{e^{-\gamma\hat{H}(H)};\ R_H^{(n)} = b\} = \displaystyle\int_a^x \frac{2\nu y^{-1-2\nu}S_\nu(b\sqrt{2\gamma+2\alpha}, x\sqrt{2\gamma+2\alpha})}{(y^{-2\nu} - b^{-2\nu})S_\nu(b\sqrt{2\gamma+2\alpha}, y\sqrt{2\gamma+2\alpha})}dy$
3.12.7 (*a*)	$\mathbf{E}_x\{e^{-\alpha H-\gamma\check{H}(H)};\ R_H^{(n)} = a\} = \displaystyle\int_x^b \frac{(2\alpha)^{-\nu}y^{-1-2\nu}S_\nu(x\sqrt{2\gamma+2\alpha}, a\sqrt{2\gamma+2\alpha})}{S_\nu(y\sqrt{2\alpha}, a\sqrt{2\alpha})S_\nu(y\sqrt{2\gamma+2\alpha}, a\sqrt{2\gamma+2\alpha})}dy$
3.12.7 (*b*)	$\mathbf{E}_x\{e^{-\alpha H-\gamma\hat{H}(H)};\ R_H^{(n)} = b\} = \displaystyle\int_a^x \frac{(2\alpha)^{-\nu}y^{-1-2\nu}S_\nu(b\sqrt{2\gamma+2\alpha}, x\sqrt{2\gamma+2\alpha})}{S_\nu(b\sqrt{2\alpha}, y\sqrt{2\alpha})S_\nu(b\sqrt{2\gamma+2\alpha}, y\sqrt{2\gamma+2\alpha})}dy$

4	$R_s^{(n)}$ $\quad \nu > 0, \ n = 2\nu + 2 > 2$ $\quad H = H_{a,b} = \min\{s : R_s^{(n)} \notin (a,b)\}$
3.18.1 $r \le u$	$\mathbf{E}_x \exp\big(-\gamma\ell(H,r) - \eta\ell(H,u)\big)$
$a \le x$ $x \le r$	$= \dfrac{\nu^2(b,a) + \nu\gamma r^{2\nu+1}(b,r)(r,x) + \nu\eta u^{2\nu+1}(b,u)(u,x) + \gamma\eta(ru)^{2\nu+1}(b,u)(u,r)(r,x)}{\nu^2(b,a) + \nu\gamma r^{2\nu+1}(b,r)(r,a) + \nu\eta u^{2\nu+1}(b,u)(u,a) + \gamma\eta(ru)^{2\nu+1}(b,u)(u,r)(r,a)}$
$r \le x$ $x \le u$	$= \dfrac{\nu^2(b,a) + \nu\gamma r^{2\nu+1}(x,r)(r,a) + \nu\eta u^{2\nu+1}(b,u)(u,x)}{\nu^2(b,a) + \nu\gamma r^{2\nu+1}(b,r)(r,a) + \nu\eta u^{2\nu+1}(b,u)(u,a) + \gamma\eta(ru)^{2\nu+1}(b,u)(u,r)(r,a)}$
$u \le x$ $x \le b$	$= \dfrac{\nu^2(b,a) + \nu\gamma r^{2\nu+1}(x,r)(r,a) + \nu\eta u^{2\nu+1}(x,u)(u,a) + \gamma\eta(ru)^{2\nu+1}(x,u)(u,r)(r,a)}{\nu^2(b,a) + \nu\gamma r^{2\nu+1}(b,r)(r,a) + \nu\eta u^{2\nu+1}(b,u)(u,a) + \gamma\eta(ru)^{2\nu+1}(b,u)(u,r)(r,a)}$
3.18.5 (a) $r \le u$	$\mathbf{E}_x\big\{\exp\big(-\gamma\ell(H,r) - \eta\ell(H,u)\big);\ W_H = a\big\}$
$a \le x$ $x \le r$	$= \dfrac{\nu^2(b,x) + \nu\gamma r^{2\nu+1}(b,r)(r,x) + \nu\eta u^{2\nu+1}(b,u)(u,x) + \gamma\eta(ru)^{2\nu+1}(b,u)(u,r)(r,x)}{\nu^2(b,a) + \nu\gamma r^{2\nu+1}(b,r)(r,a) + \nu\eta u^{2\nu+1}(b,u)(u,a) + \gamma\eta(ru)^{2\nu+1}(b,u)(u,r)(r,a)}$
$r \le x$ $x \le u$	$= \dfrac{\nu^2(b,x) + \nu\eta u^{2\nu+1}(b,u)(u,x)}{\nu^2(b,a) + \nu\gamma r^{2\nu+1}(b,r)(r,a) + \nu\eta u^{2\nu+1}(b,u)(u,a) + \gamma\eta(ru)^{2\nu+1}(b,u)(u,r)(r,a)}$
$u \le x$ $x \le b$	$= \dfrac{\nu^2(b,x)}{\nu^2(b,a) + \nu\gamma r^{2\nu+1}(b,r)(r,a) + \nu\eta u^{2\nu+1}(b,u)(u,a) + \gamma\eta(ru)^{2\nu+1}(b,u)(u,r)(r,a)}$
3.18.5 (b) $r \le u$	$\mathbf{E}_x\big\{\exp\big(-\gamma\ell(H,r) - \eta\ell(H,u)\big);\ W_H = b\big\}$
$a \le x$ $x \le r$	$= \dfrac{\nu^2(x,a)}{\nu^2(b,a) + \nu\gamma r^{2\nu+1}(b,r)(r,a) + \nu\eta u^{2\nu+1}(b,u)(u,a) + \gamma\eta(ru)^{2\nu+1}(b,u)(u,r)(r,a)}$
$r \le x$ $x \le u$	$= \dfrac{\nu^2(x,a) + \nu\gamma r^{2\nu+1}(x,r)(r,a)}{\nu^2(b,a) + \nu\gamma r^{2\nu+1}(b,r)(r,a) + \nu\eta u^{2\nu+1}(b,u)(u,a) + \gamma\eta(ru)^{2\nu+1}(b,u)(u,r)(r,a)}$
$u \le x$ $x \le b$	$= \dfrac{\nu^2(x,a) + \nu\gamma r^{2\nu+1}(x,r)(r,a) + \nu\eta u^{2\nu+1}(x,u)(u,a) + \gamma\eta(ru)^{2\nu+1}(x,u)(u,r)(r,a)}{\nu^2(b,a) + \nu\gamma r^{2\nu+1}(b,r)(r,a) + \nu\eta u^{2\nu+1}(b,u)(u,a) + \gamma\eta(ru)^{2\nu+1}(b,u)(u,r)(r,a)}$
	$(y,x) := x^{-2\nu} - y^{-2\nu}$

4 $R_s^{(n)}$ $\nu > 0,\ n = 2\nu + 2 > 2$ $H = H_{a,b} = \min\{s : R_s^{(n)} \notin (a,b)\}$

3.19.5 (a)
$$\mathbf{E}_x\Big\{\exp\Big(-\gamma\int_0^H \frac{ds}{R_s^{(n)}}\Big);\ R_H^{(n)} = a\Big\} = \frac{S_{2\nu}(2\sqrt{2\gamma b}, 2\sqrt{2\gamma x})}{S_{2\nu}(2\sqrt{2\gamma b}, 2\sqrt{2\gamma a})}$$

3.19.5 (b)
$$\mathbf{E}_x\Big\{\exp\Big(-\gamma\int_0^H \frac{ds}{R_s^{(n)}}\Big);\ R_H^{(n)} = b\Big\} = \frac{S_{2\nu}(2\sqrt{2\gamma x}, 2\sqrt{2\gamma a})}{S_{2\nu}(2\sqrt{2\gamma b}, 2\sqrt{2\gamma a})}$$

3.20.5 (a)
$$\mathbf{E}_x\Big\{\exp\Big(-\frac{\gamma^2}{2}\int_0^H \frac{ds}{(R_s^{(n)})^2}\Big);\ R_H^{(n)} = a\Big\} = \frac{x^{-\nu}\big((b/x)^{\sqrt{\nu^2+\gamma^2}} - (x/b)^{\sqrt{\nu^2+\gamma^2}}\big)}{a^{-\nu}\big((b/a)^{\sqrt{\nu^2+\gamma^2}} - (a/b)^{\sqrt{\nu^2+\gamma^2}}\big)}$$

3.20.5 (b)
$$\mathbf{E}_x\Big\{\exp\Big(-\frac{\gamma^2}{2}\int_0^H \frac{ds}{(R_s^{(n)})^2}\Big);\ R_H^{(n)} = b\Big\} = \frac{x^{-\nu}\big((x/a)^{\sqrt{\nu^2+\gamma^2}} - (a/x)^{\sqrt{\nu^2+\gamma^2}}\big)}{b^{-\nu}\big((b/a)^{\sqrt{\nu^2+\gamma^2}} - (a/b)^{\sqrt{\nu^2+\gamma^2}}\big)}$$

3.21.5 (a)
$$\mathbf{E}_x\Big\{\exp\Big(-\int_0^H \Big(\frac{p^2}{2(R_s^{(n)})^2} + \frac{q^2(R_s^{(n)})^2}{2}\Big)ds\Big);\ R_H^{(n)} = a\Big\}$$

$$= \frac{x^{\sqrt{\nu^2+p^2}-\nu}\, S_{\sqrt{\nu^2+p^2}/2}(qb^2/2, qx^2/2)}{a^{\sqrt{\nu^2+p^2}-\nu}\, S_{\sqrt{\nu^2+p^2}/2}(qb^2/2, qa^2/2)}$$

3.21.5 (b)
$$\mathbf{E}_x\Big\{\exp\Big(-\int_0^H \Big(\frac{p^2}{2(R_s^{(n)})^2} + \frac{q^2(R_s^{(n)})^2}{2}\Big)ds\Big);\ R_H^{(n)} = b\Big\}$$

$$= \frac{x^{\sqrt{\nu^2+p^2}-\nu}\, S_{\sqrt{\nu^2+p^2}/2}(qx^2/2, qa^2/2)}{b^{\sqrt{\nu^2+p^2}-\nu}\, S_{\sqrt{\nu^2+p^2}/2}(qb^2/2, qa^2/2)}$$

3.22.5 (a)
$$\mathbf{E}_x\Big\{\exp\Big(-\int_0^H \Big(\frac{p^2}{2(R_s^{(n)})^2} + \frac{q}{R_s^{(n)}}\Big)ds\Big);\ R_H^{(n)} = a\Big\}$$

$$= \frac{x^{\sqrt{\nu^2+p^2}-\nu}\, S_{2\sqrt{\nu^2+p^2}}(2\sqrt{2qb}, 2\sqrt{2qx})}{a^{\sqrt{\nu^2+p^2}-\nu}\, S_{2\sqrt{\nu^2+p^2}}(2\sqrt{2qb}, 2\sqrt{2qa})}$$

3.22.5 (b)
$$\mathbf{E}_x\Big\{\exp\Big(-\int_0^H \Big(\frac{p^2}{2(R_s^{(n)})^2} + \frac{q}{R_s^{(n)}}\Big)ds\Big);\ R_H^{(n)} = b\Big\}$$

$$= \frac{x^{\sqrt{\nu^2+p^2}-\nu}\, S_{2\sqrt{\nu^2+p^2}}(2\sqrt{2qx}, 2\sqrt{2qa})}{b^{\sqrt{\nu^2+p^2}-\nu}\, S_{2\sqrt{\nu^2+p^2}}(2\sqrt{2qb}, 2\sqrt{2qa})}$$

4 $R_s^{(n)}$ $\nu > 0,\ n = 2\nu + 2 > 2$ $H = H_{a,b} = \min\{s : R_s^{(n)} \notin (a,b)\}$

3.27.1
$$\mathbf{E}_x\Big\{\exp\Big(-\int_0^H \big(p\mathbb{I}_{[a,r)}\big(R_s^{(n)}\big) + q\mathbb{I}_{[r,b]}\big(R_s^{(n)}\big)\big)ds\Big); \ell(H,r) \in dy\Big\}$$
$$= \exp\Big(-y\Big(\frac{\sqrt{p}C_\nu(r\sqrt{2p}, a\sqrt{2p})}{\sqrt{2}S_\nu(r\sqrt{2p}, a\sqrt{2p})} + \frac{\sqrt{q}C_\nu(r\sqrt{2q}, b\sqrt{2q})}{\sqrt{2}S_\nu(b\sqrt{2q}, r\sqrt{2q})}\Big)\Big)$$
$$\times \begin{cases} \dfrac{S_\nu(b\sqrt{2q}, x\sqrt{2q})\big[q^\nu S_\nu(b\sqrt{2q}, r\sqrt{2q}) + p^\nu S_\nu(r\sqrt{2p}, a\sqrt{2p})\big]}{2(2pq)^\nu r^{2\nu+1} S_\nu^2(b\sqrt{2q}, r\sqrt{2q}) S_\nu(r\sqrt{2p}, a\sqrt{2p})} dy, & r \le x \le b \\ \dfrac{S_\nu(x\sqrt{2p}, a\sqrt{2p})\big[q^\nu S_\nu(b\sqrt{2q}, r\sqrt{2q}) + p^\nu S_\nu(r\sqrt{2p}, a\sqrt{2p})\big]}{2(2pq)^\nu r^{2\nu+1} S_\nu(b\sqrt{2q}, r\sqrt{2q}) S_\nu^2(r\sqrt{2p}, a\sqrt{2p})} dy, & a \le x \le r \end{cases}$$

3.27.5 (a)
$$\mathbf{E}_x\Big\{\exp\Big(-\int_0^H \big(p\mathbb{I}_{[a,r)}\big(R_s^{(n)}\big) + q\mathbb{I}_{[r,b]}\big(R_s^{(n)}\big)\big)ds\Big); \ell(H,r) \in dy, R_H^{(n)} = a\Big\}$$
$$= \exp\Big(-y\Big(\frac{\sqrt{p}C_\nu(r\sqrt{2p}, a\sqrt{2p})}{\sqrt{2}S_\nu(r\sqrt{2p}, a\sqrt{2p})} + \frac{\sqrt{q}C_\nu(r\sqrt{2q}, b\sqrt{2q})}{\sqrt{2}S_\nu(b\sqrt{2q}, r\sqrt{2q})}\Big)\Big)$$
$$\times \begin{cases} \dfrac{(2p)^{-\nu} S_\nu(b\sqrt{2q}, x\sqrt{2q})}{2r^{2\nu+1} S_\nu(b\sqrt{2q}, r\sqrt{2q}) S_\nu(r\sqrt{2p}, a\sqrt{2p})} dy, & r \le x \le b \\ \dfrac{(2p)^{-\nu} S_\nu(x\sqrt{2p}, a\sqrt{2p})}{2r^{2\nu+1} S_\nu^2(r\sqrt{2p}, a\sqrt{2p})} dy, & a \le x \le r \end{cases}$$

(1)
$$\mathbf{E}_x\Big\{\exp\Big(-\int_0^H \big(p\mathbb{I}_{[a,r)}\big(R_s^{(n)}\big) + q\mathbb{I}_{[r,b]}\big(R_s^{(n)}\big)\big)ds\Big); \ell(H_z,r) = 0, R_H^{(n)} = a\Big\}$$
$$= \frac{S_\nu(r\sqrt{2p}, x\sqrt{2p})}{S_\nu(r\sqrt{2p}, a\sqrt{2p})}, \quad a \le x \le r$$

3.27.5 (b)
$$\mathbf{E}_x\Big\{\exp\Big(-\int_0^H \big(p\mathbb{I}_{[a,r)}\big(R_s^{(n)}\big) + q\mathbb{I}_{[r,b]}\big(R_s^{(n)}\big)\big)ds\Big); \ell(H_z,r) \in dy, R_H^{(n)} = b\Big\}$$
$$= \exp\Big(-y\Big(\frac{\sqrt{p}C_\nu(r\sqrt{2p}, a\sqrt{2p})}{\sqrt{2}S_\nu(r\sqrt{2p}, a\sqrt{2p})} + \frac{\sqrt{q}C_\nu(r\sqrt{2q}, b\sqrt{2q})}{\sqrt{2}S_\nu(b\sqrt{2q}, r\sqrt{2q})}\Big)\Big)$$
$$\times \begin{cases} \dfrac{(2q)^{-\nu} S_\nu(b\sqrt{2q}, x\sqrt{2q})}{2r^{2\nu+1} S_\nu^2(b\sqrt{2q}, r\sqrt{2q})} dy, & r \le x \le b \\ \dfrac{(2q)^{-\nu} S_\nu(x\sqrt{2p}, a\sqrt{2p})}{2r^{2\nu+1} S_\nu(b\sqrt{2q}, r\sqrt{2q}) S_\nu(r\sqrt{2p}, a\sqrt{2p})} dy, & a \le x \le r \end{cases}$$

(1)
$$\mathbf{E}_x\Big\{\exp\Big(-\int_0^H \big(p\mathbb{I}_{[a,r)}\big(R_s^{(n)}\big) + q\mathbb{I}_{[r,b]}\big(R_s^{(n)}\big)\big)ds\Big); \ell(H,r) = 0, R_H^{(n)} = b\Big\}$$
$$= \frac{S_\nu(x\sqrt{2q}, r\sqrt{2q})}{S_\nu(b\sqrt{2q}, r\sqrt{2q})}, \quad r \le x \le b$$

4 $R_s^{(n)}$ $\nu > 0,\ n = 2\nu + 2 > 2$ $\varrho = \varrho(v,z) = \min\{s : \ell(s,z) = v\}$

4. Stopping at inverse local time

4.0.1 $$\mathbf{E}_x e^{-\alpha\varrho} = \begin{cases} \dfrac{x^{-\nu} I_\nu(x\sqrt{2\alpha})}{z^{-\nu} I_\nu(z\sqrt{2\alpha})} \exp\Big(-\dfrac{v}{2z I_\nu(z\sqrt{2\alpha}) K_\nu(z\sqrt{2\alpha})}\Big), & 0 \le x \le z \\ \dfrac{x^{-\nu} K_\nu(x\sqrt{2\alpha})}{z^{-\nu} K_\nu(z\sqrt{2\alpha})} \exp\Big(-\dfrac{v}{2z I_\nu(z\sqrt{2\alpha}) K_\nu(z\sqrt{2\alpha})}\Big), & z \le x \end{cases}$$

4.0.2 (1) $$\mathbf{P}_x\big(\varrho = \infty\big) = \begin{cases} 1 - e^{-\nu v/z}, & 0 \le x \le z \\ 1 - \dfrac{x^{-2\nu}}{z^{-2\nu}} e^{-\nu v/z}, & z \le x \end{cases}$$

4.1.2 $z \le y$ $$\mathbf{P}_x\big(\sup_{0\le s\le\varrho} R_s^{(n)} < y\big)$$

$$= \begin{cases} \exp\Big(-\dfrac{\nu v}{z^{2\nu+1}(z^{-2\nu} - y^{-2\nu})}\Big), & 0 \le x \le z \\ \dfrac{x^{-2\nu} - y^{-2\nu}}{z^{-2\nu} - y^{-2\nu}} \exp\Big(-\dfrac{\nu v}{z^{2\nu+1}(z^{-2\nu} - y^{-2\nu})}\Big), & z \le x \le y \end{cases}$$

4.1.4 $z \le y$ $$\mathbf{E}_x\{e^{-\alpha\varrho}; \sup_{0\le s\le\varrho} R_s^{(n)} < y\}$$

$$= \begin{cases} \dfrac{x^{-\nu} I_\nu(x\sqrt{2\alpha})}{z^{-\nu} I_\nu(z\sqrt{2\alpha})} \exp\Big(-\dfrac{y^{-\nu}(2\alpha)^{-\nu} I_\nu(y\sqrt{2\alpha}) v}{2z^{\nu+1} I_\nu(z\sqrt{2\alpha}) S_\nu(y\sqrt{2\alpha}, z\sqrt{2\alpha})}\Big), & 0 \le x \le z \\ \dfrac{S_\nu(y\sqrt{2\alpha}, x\sqrt{2\alpha})}{S_\nu(y\sqrt{2\alpha}, z\sqrt{2\alpha})} \exp\Big(-\dfrac{y^{-\nu}(2\alpha)^{-\nu} I_\nu(y\sqrt{2\alpha}) v}{2z^{\nu+1} I_\nu(z\sqrt{2\alpha}) S_\nu(y\sqrt{2\alpha}, z\sqrt{2\alpha})}\Big), & z \le x \le y \end{cases}$$

4.2.2 $0 \le y$ $$\mathbf{P}_x\big(\inf_{0\le s\le\varrho} R_s^{(n)} > y,\ \varrho < \infty\big)$$

$$= \begin{cases} \dfrac{y^{-2\nu} - x^{-2\nu}}{y^{-2\nu} - z^{-2\nu}} \exp\Big(-\dfrac{\nu y^{-2\nu} v}{z(y^{-2\nu} - z^{-2\nu})}\Big), & y \le x \le z \\ \dfrac{x^{-2\nu}}{z^{-2\nu}} \exp\Big(-\dfrac{\nu y^{-2\nu} v}{z(y^{-2\nu} - z^{-2\nu})}\Big), & z \le x \end{cases}$$

(1) $$\mathbf{P}_x\big(\inf_{0\le s\le\varrho} R_s^{(n)} > y,\ \varrho = \infty\big)$$

4 $R_s^{(n)}$ $\nu > 0,\ n = 2\nu + 2 > 2$ $\varrho = \varrho(v,z) = \min\{s : \ell(s,z) = v\}$

$$= \begin{cases} 1 - \dfrac{x^{-2\nu}}{y^{-2\nu}}, & z \le y \le x \\ \Big(1 - \dfrac{x^{-2\nu}}{y^{-2\nu}}\Big) \exp\Big(-\dfrac{\nu y^{-2\nu} v}{z(y^{-2\nu} - z^{-2\nu})}\Big), & y \le x \le z \\ 1 - \dfrac{x^{-2\nu}}{y^{-2\nu}} - \Big(\dfrac{x^{-2\nu}}{z^{-2\nu}} - \dfrac{x^{-2\nu}}{y^{-2\nu}}\Big) \exp\Big(-\dfrac{\nu y^{-2\nu} v}{z(y^{-2\nu} - z^{-2\nu})}\Big), & y \le z \le x \end{cases}$$

4.2.4 $\mathbf{E}_x\{e^{-\alpha\varrho};\ \inf_{0 \le s \le \varrho} R_s^{(n)} > y\}$

$y \le z$

$$= \begin{cases} \dfrac{S_\nu(x\sqrt{2\alpha}, y\sqrt{2\alpha})}{S_\nu(z\sqrt{2\alpha}, y\sqrt{2\alpha})} \exp\Big(-\dfrac{y^{-\nu}(2\alpha)^{-\nu} K_\nu(y\sqrt{2\alpha}) v}{2z^{\nu+1} K_\nu(z\sqrt{2\alpha}) S_\nu(z\sqrt{2\alpha}, y\sqrt{2\alpha})}\Big), & y \le x \le z \\ \dfrac{x^{-\nu} K_\nu(x\sqrt{2\alpha})}{z^{-\nu} K_\nu(z\sqrt{2\alpha})} \exp\Big(-\dfrac{y^{-\nu}(2\alpha)^{-\nu} K_\nu(y\sqrt{2\alpha}) v}{2z^{\nu+1} K_\nu(z\sqrt{2\alpha}) S_\nu(z\sqrt{2\alpha}, y\sqrt{2\alpha})}\Big), & z \le x \end{cases}$$

4.3.1 $\mathbf{E}_x\{e^{-\gamma\ell(\varrho,r)};\ \varrho < \infty\}$

$z \le r$

$$= \begin{cases} \exp\Big(-\dfrac{\nu z^{-2\nu-1}(\nu + \gamma r) v}{\nu z^{-2\nu} + \gamma r(z^{-2\nu} - r^{-2\nu})}\Big), & 0 \le x \le z \\ \dfrac{\nu x^{-2\nu} + \gamma r(x^{-2\nu} - r^{-2\nu})}{\nu z^{-2\nu} + \gamma r(z^{-2\nu} - r^{-2\nu})} \exp\Big(-\dfrac{\nu z^{-2\nu-1}(\nu + \gamma r) v}{\nu z^{-2\nu} + \gamma r(z^{-2\nu} - r^{-2\nu})}\Big), & z \le x \le r \\ \dfrac{\nu x^{-2\nu}}{\nu z^{-2\nu} + \gamma r(z^{-2\nu} - r^{-2\nu})} \exp\Big(-\dfrac{\nu z^{-2\nu-1}(\nu + \gamma r) v}{\nu z^{-2\nu} + \gamma r(z^{-2\nu} - r^{-2\nu})}\Big), & r \le x \end{cases}$$

$r \le z$

$$= \begin{cases} \dfrac{\nu}{\nu + r^{2\nu+1}\gamma(r^{-2\nu} - z^{-2\nu})} \exp\Big(-\dfrac{\nu(\nu + \gamma r) v}{z(\nu + \gamma r^{2\nu+1}(r^{-2\nu} - z^{-2\nu}))}\Big), & 0 \le x \le r \\ \dfrac{\nu + r^{2\nu+1}\gamma(r^{-2\nu} - x^{-2\nu})}{\nu + r^{2\nu+1}\gamma(r^{-2\nu} - z^{-2\nu})} \exp\Big(-\dfrac{\nu(\nu + \gamma r) v}{z(\nu + \gamma r^{2\nu+1}(r^{-2\nu} - z^{-2\nu}))}\Big), & r \le x \le z \\ \dfrac{x^{-2\nu}}{z^{-2\nu}} \exp\Big(-\dfrac{\nu(\nu + \gamma r) v}{z(\nu + \gamma r^{2\nu+1}(r^{-2\nu} - z^{-2\nu}))}\Big), & z \le x \end{cases}$$

(1) $\mathbf{E}_x\{e^{-\gamma\ell(\varrho,r)};\ \varrho = \infty\}$

$z \le r$

$$= \begin{cases} \dfrac{\nu}{\nu + \gamma r}\Big(1 - \exp\Big(-\dfrac{\nu z^{-2\nu-1}(\nu + \gamma r) v}{\nu z^{-2\nu} + \gamma r(z^{-2\nu} - r^{-2\nu})}\Big)\Big), & 0 \le x \le z \\ \dfrac{\nu}{\nu + \gamma r}\Big(1 - \dfrac{\nu x^{-2\nu} + \gamma r(x^{-2\nu} - r^{-2\nu})}{\nu z^{-2\nu} + \gamma r(z^{-2\nu} - r^{-2\nu})} \exp\Big(-\dfrac{\nu z^{-2\nu-1}(\nu + \gamma r) v}{\nu z^{-2\nu} + \gamma r(z^{-2\nu} - r^{-2\nu})}\Big)\Big), & \begin{matrix} z \le x \\ x \le r \end{matrix} \\ 1 - \dfrac{\gamma r^{2\nu+1} x^{-2\nu}}{\nu + \gamma r} - \dfrac{\nu x^{-2\nu}}{2(\nu + \gamma r)(\nu z^{-2\nu} + \gamma r(z^{-2\nu} - r^{-2\nu}))} & \\ \qquad \times \exp\Big(-\dfrac{\nu z^{-2\nu-1}(\nu + \gamma r) v}{\nu z^{-2\nu} + \gamma r(z^{-2\nu} - r^{-2\nu})}\Big), & r \le x \end{cases}$$

4 $R_s^{(n)}$ $\nu > 0,\ n = 2\nu + 2 > 2$ $\varrho = \varrho(v, z) = \min\{s : \ell(s, z) = v\}$

$r \le z$

$$= \begin{cases} \dfrac{\nu}{\nu+\gamma r}\Big(1 - \exp\Big(-\dfrac{\nu(\nu+\gamma r)v}{z(\nu+\gamma r^{2\nu+1}(r^{-2\nu}-z^{-2\nu}))}\Big)\Big), & 0 \le x \le r \\ \dfrac{\nu+\gamma r^{2\nu+1}(r^{-2\nu}-x^{-2\nu})}{\nu+\gamma r}\Big(1 - \exp\Big(-\dfrac{\nu(\nu+\gamma r)v}{z(\nu+\gamma r^{2\nu+1}(r^{-2\nu}-z^{-2\nu}))}\Big)\Big), & r \le x \le z \\ \dfrac{\nu+\gamma r^{2\nu+1}(r^{-2\nu}-x^{-2\nu})}{\nu+\gamma r} - \dfrac{\nu x^{-2\nu}+\gamma r^{2\nu+1}(r^{-2\nu}-z^{-2\nu})}{\nu+\gamma r} & \\ \qquad \times \exp\Big(-\dfrac{\nu(\nu+\gamma r)v}{z(\nu+\gamma r^{2\nu+1}(r^{-2\nu}-z^{-2\nu}))}\Big), & z \le x \end{cases}$$

4.3.2 $\mathbf{P}_x\big(\ell(\varrho, r) \in dy,\ \varrho < \infty\big)$

$z \le r$

$$= \exp\Big(-\frac{\nu(v/z+y/r)}{z^{2\nu}(z^{-2\nu}-r^{-2\nu})}\Big)$$

$$\times \begin{cases} \dfrac{\nu(rz)^{-\nu-1/2}}{(z^{-2\nu}-r^{-2\nu})}\dfrac{\sqrt{v}}{\sqrt{y}} I_1\Big(\dfrac{2\nu(rz)^{-\nu-1/2}\sqrt{vy}}{z^{-2\nu}-r^{-2\nu}}\Big)dy, & 0 \le x \le z \\ \dfrac{\nu}{(z^{-2\nu}-r^{-2\nu})^2}\Big[\dfrac{x^{-2\nu}-r^{-2\nu}}{(rz)^{\nu+1/2}}\dfrac{\sqrt{v}}{\sqrt{y}} I_1\Big(\dfrac{2\nu(rz)^{-\nu-1/2}\sqrt{vy}}{z^{-2\nu}-r^{-2\nu}}\Big) & \\ \qquad +(z^{-2\nu}-x^{-2\nu})r^{-2\nu-1} I_0\Big(\dfrac{2\nu(rz)^{-\nu-1/2}\sqrt{vy}}{z^{-2\nu}-r^{-2\nu}}\Big)\Big]dy, & z \le x \le r \\ \dfrac{\nu x^{-2\nu}}{r(z^{-2\nu}-r^{-2\nu})} I_0\Big(\dfrac{2\nu(rz)^{-\nu-1/2}\sqrt{vy}}{z^{-2\nu}-r^{-2\nu}}\Big)dy, & r \le x \end{cases}$$

$r \le z$

$$= \exp\Big(-\frac{\nu(v/z+y/r)}{r^{2\nu}(r^{-2\nu}-z^{-2\nu})}\Big)$$

$$\times \begin{cases} \dfrac{\nu x^{-2\nu}(rz)^{-\nu-1/2}}{z^{-2\nu}(r^{-2\nu}-z^{-2\nu})}\dfrac{\sqrt{v}}{\sqrt{y}} I_1\Big(\dfrac{2\nu(rz)^{-\nu-1/2}\sqrt{vy}}{r^{-2\nu}-z^{-2\nu}}\Big)dy, & z \le x \\ \dfrac{\nu}{(r^{-2\nu}-z^{-2\nu})^2}\Big[\dfrac{r^{-2\nu}-x^{-2\nu}}{(rz)^{\nu+1/2}}\dfrac{\sqrt{v}}{\sqrt{y}} I_1\Big(\dfrac{2\nu(rz)^{-\nu-1/2}\sqrt{vy}}{r^{-2\nu}-z^{-2\nu}}\Big) & \\ \qquad +(x^{-2\nu}-z^{-2\nu})r^{-2\nu-1} I_0\Big(\dfrac{2\nu(rz)^{-\nu-1/2}\sqrt{vy}}{r^{-2\nu}-z^{-2\nu}}\Big)\Big]dy, & r \le x \le z \\ \dfrac{\nu}{r^{2\nu+1}(r^{-2\nu}-z^{-2\nu})} I_0\Big(\dfrac{2\nu(rz)^{-\nu-1/2}\sqrt{vy}}{r^{-2\nu}-z^{-2\nu}}\Big)dy, & 0 \le x \le r \end{cases}$$

(2) $\mathbf{P}_x\big(\ell(\varrho, r) = 0,\ \varrho < \infty\big)$

$z \le r$

$$= \begin{cases} \exp\Big(-\dfrac{\nu v}{z^{2\nu+1}(z^{-2\nu}-r^{-2\nu})}\Big), & 0 \le x \le z \\ \dfrac{(x^{-2\nu}-r^{-2\nu})}{(z^{-2\nu}-r^{-2\nu})}\exp\Big(-\dfrac{\nu v}{z^{2\nu+1}(z^{-2\nu}-r^{-2\nu})}\Big), & z \le x \le r \\ 0, & r \le x \end{cases}$$

4 $R_s^{(n)}$ $\nu > 0,\ n = 2\nu + 2 > 2$ $\varrho = \varrho(v, z) = \min\{s : \ell(s, z) = v\}$

$r \le z$

$$= \begin{cases} \dfrac{x^{-2\nu}}{z^{-2\nu}} \exp\Big(-\dfrac{\nu r^{-2\nu} v}{z(r^{-2\nu} - z^{-2\nu})}\Big), & z \le x \\ \dfrac{(r^{-2\nu} - x^{-2\nu})}{(r^{-2\nu} - z^{-2\nu})} \exp\Big(-\dfrac{\nu r^{-2\nu} v}{z(r^{-2\nu} - z^{-2\nu})}\Big), & r \le x \le z \\ 0, & 0 \le x \le r \end{cases}$$

(3) $\mathbf{P}_x\big(\ell(\varrho, r) = 0,\ \varrho = \infty\big)$

$$= \begin{cases} 1 - \dfrac{x^{-2\nu}}{r^{-2\nu}}, & z \le r \le x \\ \Big(1 - \dfrac{x^{-2\nu}}{r^{-2\nu}}\Big) \exp\Big(-\dfrac{\nu r^{-2\nu} v}{z(r^{-2\nu} - z^{-2\nu})}\Big), & r \le x \le z \\ 1 - \dfrac{x^{-2\nu}}{r^{-2\nu}} - \Big(\dfrac{x^{-2\nu}}{z^{-2\nu}} - \dfrac{x^{-2\nu}}{r^{-2\nu}}\Big) \exp\Big(-\dfrac{\nu r^{-2\nu} v}{z(r^{-2\nu} - z^{-2\nu})}\Big), & r \le z \le x \end{cases}$$

4.3.3 $\mathbf{E}_x e^{-\alpha\varrho - \gamma\ell(\varrho, r)}$

$z \le r$

$$= \exp\Big(-\frac{(1 + 2\gamma r I_\nu(r\sqrt{2\alpha}) K_\nu(r\sqrt{2\alpha}))v}{2z I_\nu(z\sqrt{2\alpha})(K_\nu(z\sqrt{2\alpha}) + 2\gamma r^{\nu+1}(2\alpha z)^\nu S_\nu(r\sqrt{2\alpha}, z\sqrt{2\alpha}) K_\nu(r\sqrt{2\alpha}))}\Big)$$

$$\times \begin{cases} \dfrac{x^{-\nu} I_\nu(x\sqrt{2\alpha})}{z^{-\nu} I_\nu(z\sqrt{2\alpha})}, & 0 \le x \le z \\ \dfrac{x^{-\nu} K_\nu(x\sqrt{2\alpha}) + 2\gamma r^{\nu+1}(2\alpha)^\nu S_\nu(r\sqrt{2\alpha}, x\sqrt{2\alpha}) K_\nu(r\sqrt{2\alpha})}{z^{-\nu} K_\nu(z\sqrt{2\alpha}) + 2\gamma r^{\nu+1}(2\alpha)^\nu S_\nu(r\sqrt{2\alpha}, z\sqrt{2\alpha}) K_\nu(r\sqrt{2\alpha})}, & z \le x \le r \\ \dfrac{x^{-\nu} K_\nu(x\sqrt{2\alpha})}{z^{-\nu} K_\nu(z\sqrt{2\alpha}) + 2\gamma r^{\nu+1}(2\alpha)^\nu S_\nu(r\sqrt{2\alpha}, z\sqrt{2\alpha}) K_\nu(r\sqrt{2\alpha})}, & r \le x \end{cases}$$

$r \le z$

$$= \exp\Big(-\frac{(1 + 2\gamma r I_\nu(r\sqrt{2\alpha}) K_\nu(r\sqrt{2\alpha}))v}{2z K_\nu(z\sqrt{2\alpha})\big(I_\nu(z\sqrt{2\alpha}) + 2\gamma r^{\nu+1}(2\alpha z)^\nu S_\nu(z\sqrt{2\alpha}, r\sqrt{2\alpha}) I_\nu(r\sqrt{2\alpha})\big)}\Big)$$

$$\times \begin{cases} \dfrac{x^{-\nu} K_\nu(x\sqrt{2\alpha})}{z^{-\nu} K_\nu(z\sqrt{2\alpha})}, & z \le x \\ \dfrac{x^{-\nu} I_\nu(x\sqrt{2\alpha}) + 2\gamma r^{\nu+1}(2\alpha)^\nu S_\nu(x\sqrt{2\alpha}, r\sqrt{2\alpha}) I_\nu(r\sqrt{2\alpha})}{z^{-\nu} I_\nu(z\sqrt{2\alpha}) + 2\gamma r^{\nu+1}(2\alpha)^\nu S_\nu(z\sqrt{2\alpha}, r\sqrt{2\alpha}) I_\nu(r\sqrt{2\alpha})}, & r \le x \le z \\ \dfrac{x^{-\nu} I_\nu(x\sqrt{2\alpha})}{z^{-\nu} I_\nu(z\sqrt{2\alpha}) + 2\gamma r^{\nu+1}(2\alpha)^\nu S_\nu(z\sqrt{2\alpha}, r\sqrt{2\alpha}) I_\nu(r\sqrt{2\alpha})}, & 0 \le x \le r \end{cases}$$

4.3.4 $\mathbf{E}_x\big\{e^{-\alpha\varrho}; \ell(\varrho, r) \in dy\big\}$

$z \le r$

$$= \exp\Big(-\frac{(2\alpha r)^{-\nu} I_\nu(r\sqrt{2\alpha})v}{2z^{\nu+1} S_\nu(r\sqrt{2\alpha}, z\sqrt{2\alpha}) I_\nu(z\sqrt{2\alpha})} - \frac{(2\alpha z)^{-\nu} K_\nu(z\sqrt{2\alpha})y}{2r^{\nu+1} S_\nu(r\sqrt{2\alpha}, z\sqrt{2\alpha}) K_\nu(r\sqrt{2\alpha})}\Big)$$

4 $R_s^{(n)}$ $\nu > 0,\ n = 2\nu + 2 > 2$ $\varrho = \varrho(v, z) = \min\{s : \ell(s, z) = v\}$

$$\times\begin{cases} \dfrac{(2\alpha xr)^{-\nu} I_\nu(x\sqrt{2\alpha})}{2I_\nu(z\sqrt{2\alpha})S_\nu(r\sqrt{2\alpha}, z\sqrt{2\alpha})}\dfrac{\sqrt{v}}{\sqrt{yrz}}I_1\Big(\dfrac{(2\alpha rz)^{-\nu}\sqrt{vy}}{S_\nu(r\sqrt{2\alpha}, z\sqrt{2\alpha})\sqrt{rz}}\Big)dy, & 0 \le x \le z \\ \dfrac{(2\alpha rz)^{-\nu}}{2S_\nu^2(r\sqrt{2\alpha}, z\sqrt{2\alpha})}\Big[S_\nu(r\sqrt{2\alpha}, x\sqrt{2\alpha})\dfrac{\sqrt{v}}{\sqrt{yrz}}I_1\Big(\dfrac{(2\alpha rz)^{-\nu}\sqrt{vy}}{S_\nu(r\sqrt{2\alpha}, z\sqrt{2\alpha})\sqrt{rz}}\Big) & \\ \qquad +S_\nu(x\sqrt{2\alpha}, z\sqrt{2\alpha})\dfrac{z^\nu}{r^{\nu+1}}I_0\Big(\dfrac{(2\alpha rz)^{-\nu}\sqrt{vy}}{S_\nu(r\sqrt{2\alpha}, z\sqrt{2\alpha})\sqrt{rz}}\Big)\Big]dy, & z \le x \le r \\ \dfrac{(2\alpha x)^{-\nu} K_\nu(x\sqrt{2\alpha})}{2r^{\nu+1}K_\nu(r\sqrt{2\alpha})S_\nu(r\sqrt{2\alpha}, z\sqrt{2\alpha})}I_0\Big(\dfrac{(2\alpha rz)^{-\nu}\sqrt{vy}}{S_\nu(r\sqrt{2\alpha}, z\sqrt{2\alpha})\sqrt{rz}}\Big)dy, & r \le x \end{cases}$$

$r \le z$ $$= \exp\Big(-\frac{(2\alpha r)^{-\nu}K_\nu(r\sqrt{2\alpha})v}{2z^{\nu+1}S_\nu(z\sqrt{2\alpha}, r\sqrt{2\alpha})K_\nu(z\sqrt{2\alpha})} - \frac{(2\alpha z)^{-\nu}I_\nu(z\sqrt{2\alpha})y}{2r^{\nu+1}S_\nu(z\sqrt{2\alpha}, r\sqrt{2\alpha})I_\nu(r\sqrt{2\alpha})}\Big)$$

$$\times\begin{cases} \dfrac{(2\alpha xr)^{-\nu} K_\nu(x\sqrt{2\alpha})}{2K_\nu(z\sqrt{2\alpha})S_\nu(z\sqrt{2\alpha}, r\sqrt{2\alpha})}\dfrac{\sqrt{v}}{\sqrt{yrz}}I_1\Big(\dfrac{(2\alpha rz)^{-\nu}\sqrt{vy}}{S_\nu(z\sqrt{2\alpha}, r\sqrt{2\alpha})\sqrt{rz}}\Big)dy, & z \le x \\ \dfrac{(2\alpha rz)^{-\nu}}{2S_\nu^2(z\sqrt{2\alpha}, r\sqrt{2\alpha})}\Big[S_\nu(x\sqrt{2\alpha}, r\sqrt{2\alpha})\dfrac{\sqrt{v}}{\sqrt{yrz}}I_1\Big(\dfrac{(2\alpha rz)^{-\nu}\sqrt{vy}}{S_\nu(z\sqrt{2\alpha}, r\sqrt{2\alpha})\sqrt{rz}}\Big) & \\ \qquad +S_\nu(z\sqrt{2\alpha}, x\sqrt{2\alpha})\dfrac{z^\nu}{r^{\nu+1}}I_0\Big(\dfrac{(2\alpha rz)^{-\nu}\sqrt{vy}}{S_\nu(z\sqrt{2\alpha}, r\sqrt{2\alpha})\sqrt{rz}}\Big)\Big]dy, & r \le x \le z \\ \dfrac{(2\alpha x)^{-\nu} I_\nu(x\sqrt{2\alpha})}{2r^{\nu+1}I_\nu(r\sqrt{2\alpha})S_\nu(z\sqrt{2\alpha}, r\sqrt{2\alpha})}I_0\Big(\dfrac{(2\alpha rz)^{-\nu}\sqrt{vy}}{S_\nu(z\sqrt{2\alpha}, r\sqrt{2\alpha})\sqrt{rz}}\Big)dy, & 0 \le x \le r \end{cases}$$

(1) $\mathbf{E}_x\{e^{-\alpha\varrho}; \ell(\varrho, r) = 0\}$

$z \le r$ $$= \begin{cases} 0, & r \le x \\ \dfrac{S_\nu(r\sqrt{2\alpha}, x\sqrt{2\alpha})}{S_\nu(r\sqrt{2\alpha}, z\sqrt{2\alpha})}\exp\Big(-\dfrac{(2\alpha r)^{-\nu}I_\nu(r\sqrt{2\alpha})v}{2z^{\nu+1}S_\nu(r\sqrt{2\alpha}, z\sqrt{2\alpha})I_\nu(z\sqrt{2\alpha})}\Big), & z \le x \le r \\ \dfrac{I_\nu(x\sqrt{2\alpha})}{I_\nu(z\sqrt{2\alpha})}\exp\Big(-\dfrac{(2\alpha r)^{-\nu}I_\nu(r\sqrt{2\alpha})v}{2z^{\nu+1}S_\nu(r\sqrt{2\alpha}, z\sqrt{2\alpha})I_\nu(z\sqrt{2\alpha})}\Big), & 0 \le x \le z \end{cases}$$

$r \le z$ $$= \begin{cases} 0, & 0 \le x \le r \\ \dfrac{S_\nu(x\sqrt{2\alpha}, r\sqrt{2\alpha})}{S_\nu(z\sqrt{2\alpha}, r\sqrt{2\alpha})}\exp\Big(-\dfrac{(2\alpha r)^{-\nu}K_\nu(r\sqrt{2\alpha})v}{2z^{\nu+1}K_\nu(z\sqrt{2\alpha})S_\nu(z\sqrt{2\alpha}, r\sqrt{2\alpha})}\Big), & r \le x \le z \\ \dfrac{K_\nu(x\sqrt{2\alpha})}{K_\nu(z\sqrt{2\alpha})}\exp\Big(-\dfrac{(2\alpha r)^{-\nu}K_\nu(r\sqrt{2\alpha})v}{2z^{\nu+1}K_\nu(z\sqrt{2\alpha})S_\nu(z\sqrt{2\alpha}, r\sqrt{2\alpha})}\Big), & z \le x \end{cases}$$

4.4.1 $\mathbf{E}_x\Big\{\exp\Big(-\gamma\int_0^\varrho \mathbb{I}_{[r,\infty)}\big(R_s^{(n)}\big)ds\Big);\ \varrho < \infty\Big\}$

$z \le r$ $$= \exp\Big(-\frac{\nu z^{-2\nu-1}K_{\nu+1}(r\sqrt{2\gamma})v}{z^{-2\nu}K_{\nu+1}(r\sqrt{2\gamma}) - r^{-2\nu}K_{\nu-1}(r\sqrt{2\gamma})}\Big)$$

4 $R_s^{(n)}$ $\nu > 0,\ n = 2\nu + 2 > 2$ $\varrho = \varrho(v, z) = \min\{s : \ell(s, z) = v\}$

$$\times \begin{cases} 1, & 0 \le x \le z \\ \dfrac{x^{-2\nu} K_{\nu+1}(r\sqrt{2\gamma}) - r^{-2\nu} K_{\nu-1}(r\sqrt{2\gamma})}{z^{-2\nu} K_{\nu+1}(r\sqrt{2\gamma}) - r^{-2\nu} K_{\nu-1}(r\sqrt{2\gamma})}, & z \le x \le r \\ \dfrac{2\nu(2\gamma)^{-1/2} r^{-\nu-1} x^{-\nu} K_\nu(x\sqrt{2\gamma})}{z^{-2\nu} K_{\nu+1}(r\sqrt{2\gamma}) - r^{-2\nu} K_{\nu-1}(r\sqrt{2\gamma})}, & r \le x \end{cases}$$

$r \le z$

$$= \exp\Big(-\frac{(2\gamma r)^{-\nu} K_{\nu+1}(r\sqrt{2\gamma}) v}{2z^{\nu+1} K_\nu(z\sqrt{2\gamma}) C_\nu(r\sqrt{2\gamma}, z\sqrt{2\gamma})}\Big) \begin{cases} \dfrac{(r\sqrt{2\gamma})^{-2\nu-1}}{C_\nu(r\sqrt{2\gamma}, z\sqrt{2\gamma})}, & 0 < x \le r \\ \dfrac{C_\nu(r\sqrt{2\gamma}, x\sqrt{2\gamma})}{C_\nu(r\sqrt{2\gamma}, z\sqrt{2\gamma})}, & r \le x \le z \\ \dfrac{x^{-\nu} K_\nu(x\sqrt{2\gamma})}{z^{-\nu} K_\nu(z\sqrt{2\gamma})}, & z \le x \end{cases}$$

4.4.2 (1) $\mathbf{P}_x\Big(\int_0^\varrho \mathbb{I}_{[r,\infty)}\big(R_s^{(n)}\big) ds = 0,\ \varrho < \infty\Big)$

$$= \begin{cases} \exp\Big(-\dfrac{\nu z^{-2\nu-1} v}{z^{-2\nu} - r^{-2\nu}}\Big), & 0 \le x \le z \\ \dfrac{x^{-2\nu} - r^{-2\nu}}{z^{-2\nu} - r^{-2\nu}} \exp\Big(-\dfrac{\nu z^{-2\nu-1} v}{z^{-2\nu} - r^{-2\nu}}\Big), & z \le x \le r \\ 0, & r \le x \end{cases}$$

4.5.1 $\mathbf{E}_x\Big\{\exp\Big(-\gamma \int_0^\varrho \mathbb{I}_{[0,r]}\big(R_s^{(n)}\big) ds\Big);\ \varrho < \infty\Big\}$

$z \le r$

$$= \exp\Big(-\frac{(2\gamma r)^{-\nu} I_{\nu-1}(r\sqrt{2\gamma}) v}{2z^\nu I_\nu(z\sqrt{2\gamma}) C_{\nu-1}(x\sqrt{2\gamma}, r\sqrt{2\gamma})}\Big)$$

$$\times \begin{cases} \dfrac{x^{-\nu} I_\nu(x\sqrt{2\gamma})}{z^{-\nu} I_\nu(z\sqrt{2\gamma})}, & 0 \le x \le z \\ \dfrac{x^{-1} C_{\nu-1}(x\sqrt{2\gamma}, r\sqrt{2\gamma})}{z^{-1} C_{\nu-1}(z\sqrt{2\gamma}, r\sqrt{2\gamma})}, & z \le x \le r \\ \dfrac{(2\gamma)^{-\nu+1/2} x^{-2\nu}}{z^{-1} C_{\nu-1}(z\sqrt{2\gamma}, r\sqrt{2\gamma})}, & r \le x \end{cases}$$

$r \le z$

$$= \exp\Big(-\frac{\nu r^{-2\nu} I_{\nu-1}(r\sqrt{2\gamma}) v}{z(r^{-2\nu} I_{\nu-1}(r\sqrt{2\gamma}) - z^{-2\nu} I_{\nu+1}(r\sqrt{2\gamma}))}\Big)$$

$$\times \begin{cases} \dfrac{2\nu(2\gamma)^{-1/2} r^{-\nu-1} x^{-\nu} I_\nu(x\sqrt{2\gamma})}{r^{-2\nu} I_{\nu-1}(r\sqrt{2\gamma}) - z^{-2\nu} I_{\nu+1}(r\sqrt{2\gamma})}, & 0 \le x \le r \\ \dfrac{r^{-2\nu} I_{\nu-1}(r\sqrt{2\gamma}) - x^{-2\nu} I_{\nu+1}(r\sqrt{2\gamma})}{r^{-2\nu} I_{\nu-1}(r\sqrt{2\gamma}) - z^{-2\nu} I_{\nu+1}(r\sqrt{2\gamma})}, & r \le x \le z \\ \dfrac{x^{-2\nu}}{z^{-2\nu}}, & z \le x \end{cases}$$

4 $R_s^{(n)}$ $\nu>0,\ n=2\nu+2>2$ $\varrho=\varrho(v,z)=\min\{s:\ell(s,z)=v\}$

4.5.2 (1) $r\le z$

$$\mathbf{P}_x\Big(\int_0^\varrho \mathbb{1}_{[0,r]}\big(R_s^{(n)}\big)ds=0\Big)=\begin{cases}0, & 0\le x\le r\\ \dfrac{r^{-2\nu}-x^{-2\nu}}{r^{-2\nu}-z^{-2\nu}}, & r\le x\le z\\ 1, & z\le x\end{cases}$$

4.6.1

$$\mathbf{E}_x\Big\{\exp\Big(-\int_0^\varrho\big(p\mathbb{1}_{[0,r)}\big(R_s^{(n)}\big)+q\mathbb{1}_{[r,\infty)}\big(R_s^{(n)}\big)\big)ds\Big);\ \varrho<\infty\Big\}$$

$r\le z$

$$=\exp\Big(-\frac{(2qrz)^{-\nu}K_\nu^{-1}(z\sqrt{2q})\sqrt{p}I_{\nu+1}(r\sqrt{2p})K_\nu(r\sqrt{2q})v}{2z(\sqrt{p}S_\nu(z\sqrt{2q},r\sqrt{2q})I_{\nu+1}(r\sqrt{2p})+\sqrt{q}C_\nu(r\sqrt{2q},z\sqrt{2q})I_\nu(r\sqrt{2p}))}$$

$$-\frac{(2qrz)^{-\nu}K_\nu^{-1}(z\sqrt{2q})\sqrt{q}I_\nu(r\sqrt{2p})K_{\nu+1}(r\sqrt{2q})v}{2z(\sqrt{p}S_\nu(z\sqrt{2q},r\sqrt{2q})I_{\nu+1}(r\sqrt{2p})+\sqrt{q}C_\nu(r\sqrt{2q},z\sqrt{2q})I_\nu(r\sqrt{2p}))}\Big)$$

$$\times\begin{cases}\dfrac{\sqrt{q}(r\sqrt{2q})^{-2\nu-1}(x/r)^{-\nu}I_\nu(x\sqrt{2p})}{\sqrt{p}S_\nu(z\sqrt{2q},r\sqrt{2q})I_{\nu+1}(r\sqrt{2p})+\sqrt{q}C_\nu(r\sqrt{2q},z\sqrt{2q})I_\nu(r\sqrt{2p})}, & 0\le x\le r\\ \dfrac{\sqrt{p}S_\nu(x\sqrt{2q},r\sqrt{2q})I_{\nu+1}(r\sqrt{2p})+\sqrt{q}C_\nu(r\sqrt{2q},x\sqrt{2q})I_\nu(r\sqrt{2p})}{\sqrt{p}S_\nu(z\sqrt{2q},r\sqrt{2q})I_{\nu+1}(r\sqrt{2p})+\sqrt{q}C_\nu(r\sqrt{2q},z\sqrt{2q})I_\nu(r\sqrt{2p})}, & r\le x\le z\\ \dfrac{x^{-\nu}K_\nu(x\sqrt{2q})}{z^{-\nu}K_\nu(z\sqrt{2q})}, & z\le x\end{cases}$$

$z\le r$

$$=\exp\Big(-\frac{(2prz)^{-\nu}I_\nu^{-1}(z\sqrt{2p})\sqrt{p}I_{\nu+1}(r\sqrt{2p})K_\nu(r\sqrt{2q})v}{2z(\sqrt{p}K_\nu(r\sqrt{2q})C_\nu(r\sqrt{2p},z\sqrt{2p})+\sqrt{q}K_{\nu+1}(r\sqrt{2q})S_\nu(r\sqrt{2p},z\sqrt{2p}))}$$

$$-\frac{(2prz)^{-\nu}I_\nu^{-1}(z\sqrt{2p})\sqrt{q}I_\nu(r\sqrt{2p})K_{\nu+1}(r\sqrt{2q})v}{2z(\sqrt{p}K_\nu(r\sqrt{2q})C_\nu(r\sqrt{2p},z\sqrt{2p})+\sqrt{q}K_{\nu+1}(r\sqrt{2q})S_\nu(r\sqrt{2p},z\sqrt{2p}))}\Big)$$

$$\times\begin{cases}\dfrac{x^{-\nu}I_\nu(x\sqrt{2p})}{z^{-\nu}I_\nu(z\sqrt{2p})}, & 0\le x\le z\\ \dfrac{\sqrt{p}K_\nu(r\sqrt{2q})C_\nu(r\sqrt{2p},x\sqrt{2p})+\sqrt{q}K_{\nu+1}(r\sqrt{2q})S_\nu(r\sqrt{2p},x\sqrt{2p})}{\sqrt{p}K_\nu(r\sqrt{2q})C_\nu(r\sqrt{2p},z\sqrt{2p})+\sqrt{q}K_{\nu+1}(r\sqrt{2q})S_\nu(r\sqrt{2p},z\sqrt{2p})}, & z\le x\le r\\ \dfrac{\sqrt{p}(r\sqrt{2p})^{-2\nu-1}(x/r)^{-\nu}K_\nu(x\sqrt{2q})}{\sqrt{p}K_\nu(r\sqrt{2q})C_\nu(r\sqrt{2p},z\sqrt{2p})+\sqrt{q}K_{\nu+1}(r\sqrt{2q})S_\nu(r\sqrt{2p},z\sqrt{2p})}, & r\le x\end{cases}$$

4.8.1

$$\mathbf{E}_x\Big\{\exp\Big(-\gamma\int_0^\varrho R_s^{(n)}ds\Big);\ \varrho<\infty\Big\}$$

$$=\exp\Big(-\frac{3v}{4zI_{2\nu/3}(\frac23 z^{3/2}\sqrt{2\gamma})K_{2\nu/3}(\frac23 z^{3/2}\sqrt{2\gamma})}\Big)\begin{cases}\dfrac{z^\nu I_{2\nu/3}(\frac23 x^{3/2}\sqrt{2\gamma})}{x^\nu I_{2\nu/3}(\frac23 z^{3/2}\sqrt{2\gamma})}, & 0\le x\le z\\ \dfrac{z^\nu K_{2\nu/3}(\frac23 x^{3/2}\sqrt{2\gamma})}{x^\nu K_{2\nu/3}(\frac23 z^{3/2}\sqrt{2\gamma})}, & z\le x\end{cases}$$

4.9.1

$$\mathbf{E}_x\Big\{\exp\Big(-2\gamma\int_0^\varrho (R_s^{(n)})^2ds\Big);\ \varrho<\infty\Big\}$$

4 $R_s^{(n)}$ $\nu > 0,\ n = 2\nu + 2 > 2$ $\varrho = \varrho(v, z) = \min\{s : \ell(s, z) = v\}$

$$= \begin{cases} \dfrac{x^{-\nu} I_{\nu/2}(x^2\sqrt{\gamma})}{z^{-\nu} I_{\nu/2}(z^2\sqrt{\gamma})} \exp\Big(-\dfrac{v}{zI_{\nu/2}(z^2\sqrt{\gamma})K_{\nu/2}(z^2\sqrt{\gamma})}\Big), & 0 \le x \le z \\ \dfrac{x^{-\nu} K_{\nu/2}(x^2\sqrt{\gamma})}{z^{-\nu} K_{\nu/2}(z^2\sqrt{\gamma})} \exp\Big(-\dfrac{v}{zI_{\nu/2}(z^2\sqrt{\gamma})K_{\nu/2}(z^2\sqrt{\gamma})}\Big), & z \le x \end{cases}$$

4.19.1 $$\mathbf{E}_x\Big\{\exp\Big(-\gamma \int_0^\varrho \frac{ds}{R_s^{(n)}}\Big);\ \varrho < \infty\Big\}$$

$$= \begin{cases} \dfrac{x^{-\nu} I_{2\nu}(2\sqrt{2\gamma x})}{z^{-\nu} I_{2\nu}(2\sqrt{2\gamma z})} \exp\Big(-\dfrac{v}{4zI_{2\nu}(2\sqrt{2\gamma z})K_{2\nu}(2\sqrt{2\gamma z})}\Big), & 0 < x \le z \\ \dfrac{x^{-\nu} K_{2\nu}(2\sqrt{2\gamma x})}{z^{-\nu} K_{2\nu}(2\sqrt{2\gamma z})} \exp\Big(-\dfrac{v}{4zI_{2\nu}(2\sqrt{2\gamma z})K_{2\nu}(2\sqrt{2\gamma z})}\Big), & z \le x \end{cases}$$

4.20.1 $$\mathbf{E}_x\Big\{\exp\Big(-\frac{\gamma^2}{2} \int_0^\varrho \frac{ds}{(R_s^{(n)})^2}\Big);\ \varrho < \infty\Big\}$$

$$= \begin{cases} \Big(\dfrac{x}{z}\Big)^{-\nu+\sqrt{\nu^2+\gamma^2}} \exp\Big(-\dfrac{v\sqrt{\nu^2+\gamma^2}}{z}\Big), & 0 < x \le z \\ \Big(\dfrac{x}{z}\Big)^{-\nu-\sqrt{\nu^2+\gamma^2}} \exp\Big(-\dfrac{v\sqrt{\nu^2+\gamma^2}}{z}\Big), & z \le x \end{cases}$$

4.20.4
$0 < y$ $$\mathbf{P}_x\Big(\int_0^\varrho \frac{ds}{(R_s^{(n)})^2} \in dy\Big) = \begin{cases} \dfrac{z^\nu(\ln(z/x) + v/z)}{x^\nu\sqrt{2\pi}y^{3/2}} e^{-\nu^2 y/2 - (\ln(z/x)+v/z)^2/2y} dy, & x \le z \\ \dfrac{z^\nu(\ln(x/z) + v/z)}{x^\nu\sqrt{2\pi}y^{3/2}} e^{-\nu^2 y/2 - (\ln(x/z)+v/z)^2/2y} dy, & z \le x \end{cases}$$

4.21.1
$0 < x$ $$\mathbf{E}_x\Big\{\exp\Big(-\int_0^\varrho \Big(\frac{p^2}{2(R_s^{(n)})^2} + \frac{q^2(R_s^{(n)})^2}{2}\Big)ds\Big);\ \varrho < \infty\Big\}$$

$$= \begin{cases} \dfrac{x^{-\nu} I_{\sqrt{\nu^2+p^2}/2}(qx^2/2)}{z^{-\nu} I_{\sqrt{\nu^2+p^2}/2}(qz^2/2)} \exp\Big(-\dfrac{v}{zI_{\sqrt{\nu^2+p^2}/2}(qz^2/2)K_{\sqrt{\nu^2+p^2}/2}(qz^2/2)}\Big), & x \le z \\ \dfrac{x^{-\nu} K_{\sqrt{\nu^2+p^2}/2}(qx^2/2)}{z^{-\nu} K_{\sqrt{\nu^2+p^2}/2}(qz^2/2)} \exp\Big(-\dfrac{v}{zI_{\sqrt{\nu^2+p^2}/2}(qz^2/2)K_{\sqrt{\nu^2+p^2}/2}(qz^2/2)}\Big), & z \le x \end{cases}$$

4.22.1
$0 < x$ $$\mathbf{E}_x\Big\{\exp\Big(-\int_0^\varrho \big(\frac{p^2}{2(R_s^{(n)})^2} + \frac{q}{R_s^{(n)}}\big)ds\Big);\ \varrho < \infty\Big\}$$

$$= \begin{cases} \dfrac{x^{-\nu} I_{2\sqrt{\nu^2+p^2}}(2\sqrt{2qx})}{z^{-\nu} I_{2\sqrt{\nu^2+p^2}}(2\sqrt{2qz})} \exp\Big(-\dfrac{v}{4zI_{2\sqrt{\nu^2+p^2}}(2\sqrt{2qz})K_{2\sqrt{\nu^2+p^2}}(2\sqrt{2qz})}\Big), & x \le z \\ \dfrac{x^{-\nu} K_{2\sqrt{\nu^2+p^2}}(2\sqrt{2qx})}{z^{-\nu} K_{2\sqrt{\nu^2+p^2}}(2\sqrt{2qz})} \exp\Big(-\dfrac{v}{4zI_{2\sqrt{\nu^2+p^2}}(2\sqrt{2qz})K_{2\sqrt{\nu^2+p^2}}(2\sqrt{2qz})}\Big), & z \le x \end{cases}$$

4 $R_s^{(n)}$ $\nu>0,\ n=2\nu+2>2$ $\varrho=\varrho(v,z)=\min\{s:\ell(s,z)=v\}$

4.27.1 $$\mathbf{E}_x\Big\{\exp\Big(-\int_0^{\varrho}\big(p\mathbb{1}_{[0,r)}\big(R_s^{(n)}\big)+q\mathbb{1}_{[r,\infty)}\big(R_s^{(n)}\big)\big)ds\Big);\ell(\varrho,r)\in dy\Big\}$$

$z\le r$
$$=\exp\Big(-\frac{(2pr)^{-\nu}I_\nu(r\sqrt{2p})v}{2z^{\nu+1}S_\nu(r\sqrt{2p},z\sqrt{2p})I_\nu(z\sqrt{2p})}-\frac{\sqrt{q}K_{\nu+1}(r\sqrt{2q})y}{\sqrt{2}K_\nu(r\sqrt{2q})}-\frac{\sqrt{p}C_\nu(r\sqrt{2p},z\sqrt{2p})y}{\sqrt{2}S_\nu(r\sqrt{2p},z\sqrt{2p})}\Big)$$
$$\times\begin{cases}\dfrac{(2pxr)^{-\nu}I_\nu(x\sqrt{2p})}{2I_\nu(z\sqrt{2p})S_\nu(r\sqrt{2p},z\sqrt{2p})}\dfrac{\sqrt{v}}{\sqrt{yrz}}I_1\Big(\dfrac{(2prz)^{-\nu}\sqrt{vy}}{S_\nu(r\sqrt{2p},z\sqrt{2p})\sqrt{rz}}\Big)dy, & 0\le x\le z\\[2mm] \dfrac{(2prz)^{-\nu}}{2S_\nu^2(r\sqrt{2p},z\sqrt{2p})}\Big[S_\nu(r\sqrt{2p},x\sqrt{2p})\dfrac{\sqrt{v}}{\sqrt{yrz}}I_1\Big(\dfrac{(2prz)^{-\nu}\sqrt{vy}}{S_\nu(r\sqrt{2p},z\sqrt{2p})\sqrt{rz}}\Big) & \\ \qquad+S_\nu(x\sqrt{2p},z\sqrt{2p})\dfrac{z^\nu}{r^{\nu+1}}I_0\Big(\dfrac{(2prz)^{-\nu}\sqrt{vy}}{S_\nu(r\sqrt{2p},z\sqrt{2p})\sqrt{rz}}\Big)\Big]dy, & z\le x\le r\\[2mm] \dfrac{(2px)^{-\nu}K_\nu(x\sqrt{2q})}{2r^{\nu+1}K_\nu(r\sqrt{2q})S_\nu(r\sqrt{2p},z\sqrt{2p})}I_0\Big(\dfrac{(2prz)^{-\nu}\sqrt{vy}}{S_\nu(r\sqrt{2p},z\sqrt{2p})\sqrt{rz}}\Big)dy, & r\le x\end{cases}$$

$r\le z$
$$=\exp\Big(-\frac{(2qr)^{-\nu}K_\nu(r\sqrt{2q})v}{2z^{\nu+1}S_\nu(z\sqrt{2q},r\sqrt{2q})K_\nu(z\sqrt{2q})}-\frac{\sqrt{q}C_\nu(r\sqrt{2q},z\sqrt{2q})y}{\sqrt{2}S_\nu(z\sqrt{2q},r\sqrt{2q})}-\frac{\sqrt{p}I_{\nu+1}(r\sqrt{2p})y}{\sqrt{2}I_\nu(r\sqrt{2p})}\Big)$$
$$\times\begin{cases}\dfrac{(2qxr)^{-\nu}K_\nu(x\sqrt{2q})}{2K_\nu(z\sqrt{2q})S_\nu(z\sqrt{2q},r\sqrt{2q})}\dfrac{\sqrt{v}}{\sqrt{yrz}}I_1\Big(\dfrac{(2qrz)^{-\nu}\sqrt{vy}}{S_\nu(z\sqrt{2q},r\sqrt{2q})\sqrt{rz}}\Big)dy, & z\le x\\[2mm] \dfrac{(2qrz)^{-\nu}}{2S_\nu^2(z\sqrt{2q},r\sqrt{2q})}\Big[S_\nu(x\sqrt{2q},r\sqrt{2q})\dfrac{\sqrt{v}}{\sqrt{yrz}}I_1\Big(\dfrac{(2qrz)^{-\nu}\sqrt{vy}}{S_\nu(z\sqrt{2q},r\sqrt{2q})\sqrt{rz}}\Big) & \\ \qquad+S_\nu(z\sqrt{2q},x\sqrt{2q})\dfrac{z^\nu}{r^{\nu+1}}I_0\Big(\dfrac{(2qrz)^{-\nu}\sqrt{vy}}{S_\nu(z\sqrt{2q},r\sqrt{2q})\sqrt{rz}}\Big)\Big]dy, & r\le x\le z\\[2mm] \dfrac{(2qx)^{-\nu}I_\nu(x\sqrt{2p})}{2r^{\nu+1}I_\nu(r\sqrt{2p})S_\nu(z\sqrt{2q},r\sqrt{2q})}I_0\Big(\dfrac{(2qrz)^{-\nu}\sqrt{vy}}{S_\nu(z\sqrt{2q},r\sqrt{2q})\sqrt{rz}}\Big)dy, & 0\le x\le r\end{cases}$$

(1) $$\mathbf{E}_x\Big\{\exp\Big(-\int_0^{\varrho}\big(p\mathbb{1}_{[0,r)}\big(R_s^{(n)}\big)+q\mathbb{1}_{[r,\infty)}\big(R_s^{(n)}\big)\big)ds\Big);\ell(\varrho,r)=0\Big\}$$

$z\le r$
$$=\begin{cases}0, & r\le x\\ \dfrac{S_\nu(r\sqrt{2p},x\sqrt{2p})}{S_\nu(r\sqrt{2p},z\sqrt{2p})}\exp\Big(-\dfrac{(2pr)^{-\nu}I_\nu(r\sqrt{2p})v}{2z^{\nu+1}S_\nu(r\sqrt{2p},z\sqrt{2p})I_\nu(z\sqrt{2p})}\Big), & z\le x\le r\\ \dfrac{x^{-\nu}I_\nu(x\sqrt{2p})}{z^{-\nu}I_\nu(z\sqrt{2p})}\exp\Big(-\dfrac{(2pr)^{-\nu}I_\nu(r\sqrt{2p})v}{2z^{\nu+1}S_\nu(r\sqrt{2p},z\sqrt{2p})I_\nu(z\sqrt{2p})}\Big), & 0\le x\le z\end{cases}$$

$r\le z$
$$=\begin{cases}0, & 0\le x\le r\\ \dfrac{S_\nu(x\sqrt{2q},r\sqrt{2q})}{S_\nu(z\sqrt{2q},r\sqrt{2q})}\exp\Big(-\dfrac{(2qr)^{-\nu}K_\nu(r\sqrt{2q})v}{2z^{\nu+1}K_\nu(z\sqrt{2q})S_\nu(z\sqrt{2q},r\sqrt{2q})}\Big), & r\le x\le z\\ \dfrac{x^{-\nu}K_\nu(x\sqrt{2q})}{z^{-\nu}K_\nu(z\sqrt{2q})}\exp\Big(-\dfrac{(2qr)^{-\nu}K_\nu(r\sqrt{2q})v}{2z^{\nu+1}K_\nu(z\sqrt{2q})S_\nu(z\sqrt{2q},r\sqrt{2q})}\Big), & z\le x\end{cases}$$

5. BESSEL PROCESS OF ORDER 1/2

1. Exponential stopping

1.0.3 $$\mathbf{E}_x e^{-\beta R_\tau^{(3)}} = \frac{\lambda}{\lambda-\beta^2/2} e^{-\beta x} + \frac{\lambda\beta}{x(\lambda-\beta^2/2)^2}\left(e^{-x\sqrt{2\lambda}} - e^{-\beta x}\right)$$

1.0.4 $$\mathbf{E}_x e^{-\beta R_t^{(3)}} = e^{-\beta x+\beta^2 t/2}\Big\{1 - \frac{\beta t}{x} + (\beta t + x)e^{2\beta x}\operatorname{Erfc}\Big(\frac{x}{\sqrt{2t}} + \frac{\beta\sqrt{t}}{\sqrt{2}}\Big) + (\beta t - x)\operatorname{Erfc}\Big(\frac{x}{\sqrt{2t}} - \frac{\beta\sqrt{t}}{\sqrt{2}}\Big)\Big\}$$

1.0.5 $$\mathbf{P}_x(R_\tau^{(3)} \in dz) = \frac{z\sqrt{\lambda}}{x\sqrt{2}}\left(e^{-|z-x|\sqrt{2\lambda}} - e^{-(z+x)\sqrt{2\lambda}}\right)dz, \qquad 0 \le z, \quad 0 \le x$$

1.0.6 $$\mathbf{P}_x\big(R_t^{(3)} \in dz\big) = \frac{z}{x\sqrt{2\pi t}}\left(e^{-(z-x)^2/2t} - e^{-(z+x)^2/2t}\right)dz$$

1.1.2 $x \le y$ $$\mathbf{P}_x\Big(\sup_{0\le s\le\tau} R_s^{(3)} \ge y\Big) = \frac{y\operatorname{sh}(x\sqrt{2\lambda})}{x\operatorname{sh}(y\sqrt{2\lambda})}$$

1.1.4 $x \le y$ $$\mathbf{P}_x\Big(\sup_{0\le s\le t} R_s^{(3)} \ge y\Big) = \frac{y}{x}\widetilde{\operatorname{ss}}_t(x,y)$$

1.1.6 $$\mathbf{P}_x\Big(\sup_{0\le s\le\tau} R_s^{(3)} < y, R_\tau^{(3)} \in dz\Big) = \frac{z\sqrt{\lambda}\left(\operatorname{ch}((y-|z-x|)\sqrt{2\lambda}) - \operatorname{ch}((y-x-z)\sqrt{2\lambda})\right)}{x\sqrt{2}\operatorname{sh}(y\sqrt{2\lambda})}dz, \qquad x \vee z < y$$

1.1.8 $$\mathbf{P}_x\Big(\sup_{0\le s\le t} R_s^{(3)} < y, R_t^{(3)} \in dz\Big) = \frac{z}{x\sqrt{2\pi t}}\sum_{k=-\infty}^{\infty}\left(e^{-(z-x+2ky)^2/2t} - e^{-(z+x+2ky)^2/2t}\right)dz, \qquad x \vee z < y$$

1.2.2 $y \le x$ $$\mathbf{P}_x\Big(\inf_{0\le s\le\tau} R_s^{(3)} \le y\Big) = \frac{y}{x}e^{-(x-y)\sqrt{2\lambda}}$$

1.2.4 $y \le x$ $$\mathbf{P}_x\Big(\inf_{0\le s\le t} R_s^{(3)} \le y\Big) = \frac{y}{x}\operatorname{Erfc}\Big(\frac{x-y}{\sqrt{2t}}\Big)$$

5 $R_s^{(3)} = \big(\sum_{k=1}^{3}(W_s^{(k)})^2\big)^{1/2}$ $\qquad \tau \sim \operatorname{Exp}(\lambda)$, independent of $R^{(3)}$

(1) $y \le x$

$$\mathbf{P}_x\Big(\inf_{0\le s<\infty} R_s^{(3)} \le y\Big) = \frac{y}{x}$$

1.2.6

$$\mathbf{P}_x\Big(\inf_{0\le s\le\tau} R_s^{(3)} \le y, R_\tau^{(3)} \in dz\Big) = \frac{z\sqrt{2\lambda}}{x}e^{(y-z-x)\sqrt{2\lambda}}\operatorname{sh}(y\sqrt{2\lambda})dz, \quad y \le x\wedge z$$

1.2.8 $y \le x$

$$\mathbf{P}_x\Big(\inf_{0\le s\le t} R_s^{(3)} \le y, R_t^{(3)} \in dz\Big) = \frac{z}{x\sqrt{2\pi t}}\big(e^{-(z+x-2y)^2/2t} - e^{-(z+x)^2/2t}\big)dz$$

1.3.1

$$\mathbf{E}_x e^{-\gamma\ell(\tau,r)} = 1 - \frac{\gamma r\big(e^{-|x-r|\sqrt{2\lambda}} - e^{-(x+r)\sqrt{2\lambda}}\big)}{x\big(\sqrt{2\lambda}+\gamma(1-e^{-2r\sqrt{2\lambda}})\big)}$$

1.3.2

$$\mathbf{P}_x\big(\ell(\tau,r)\in dy\big) = \exp\Big(-\frac{\sqrt{\lambda}e^{r\sqrt{2\lambda}}y}{\sqrt{2}\operatorname{sh}(r\sqrt{2\lambda})}\Big)\frac{r\sqrt{2\lambda}\big(e^{-|x-r|\sqrt{2\lambda}} - e^{-(x+r)\sqrt{2\lambda}}\big)}{x\big(1-e^{-2r\sqrt{2\lambda}}\big)^2}dy$$

(1)

$$\mathbf{P}_x\big(\ell(\tau,r)=0\big) = 1 - \frac{r\big(e^{-|x-r|\sqrt{2\lambda}} - e^{-(x+r)\sqrt{2\lambda}}\big)}{x\big(1-e^{-2r\sqrt{2\lambda}}\big)}$$

1.3.3

$$\mathbf{E}_x e^{-\gamma\ell(t,r)} = 1 + \frac{r\sqrt{2}}{x\sqrt{\pi}}\sum_{k=0}^{\infty}(-\gamma)^{k+1}k!\sum_{l=0}^{k}\frac{(-1)^l t^{(k+1)/2}}{l!(k-l)!}\Big(e^{-(|r-x|+2rl)^2/4t}D_{-k-2}\Big(\frac{|r-x|+2rl}{\sqrt{t}}\Big) - e^{-(x+r+2rl)^2/4t}D_{-k-2}\Big(\frac{x+r+2rl}{\sqrt{t}}\Big)\Big)$$

(1)

$$\mathbf{E}_x e^{-\gamma\ell(\infty,r)} = 1 - \frac{\gamma r(x+r-|x-r|)}{x(1+2r\gamma)}$$

1.3.4

$$\mathbf{P}_x\Big(\frac{1}{2}\ell(t,r)\in dy\Big) = \frac{r}{x}\big(\operatorname{es}_y(-1,2,r,y+|x-r|-2r,y) - \operatorname{es}_y(-1,2,r,y+x-r,y)\big)dy$$

(1)

$$\mathbf{P}_x\big(\ell(t,r)=0\big) = \begin{cases} 1-\dfrac{r}{x}\widetilde{\operatorname{ss}}_t(x,r), & 0\le x\le r\\[2mm] 1-\dfrac{r}{x}\operatorname{Erfc}\Big(\dfrac{x-r}{\sqrt{2t}}\Big), & r\le x\end{cases}$$

5	$R_s^{(3)} = \left(\sum_{k=1}^3 (W_s^{(k)})^2\right)^{1/2}$ $\qquad\tau \sim \text{Exp}(\lambda)$, independent of $R^{(3)}$
(2)	$\mathbf{P}_x\big(\ell(\infty,r)\in dy\big) = \frac{x+r-\lvert x-r\rvert}{4xr}\, e^{-y/2r}dy$
(3)	$\mathbf{P}_x\big(\ell(\infty,r)=0\big) = 1-\frac{x+r-\lvert x-r\rvert}{2x}$
1.3.5	$\mathbf{E}_x\big\{e^{-\gamma\ell(\tau,r)}; R_\tau^{(3)}\in dz\big\} = \frac{z\sqrt{\lambda}}{x\sqrt{2}}\Big\{e^{-\lvert z-x\rvert\sqrt{2\lambda}} - e^{-(z+x)\sqrt{2\lambda}}$ $-\frac{\gamma(e^{-\lvert x-r\rvert\sqrt{2\lambda}}-e^{-(x+r)\sqrt{2\lambda}})}{\sqrt{2\lambda}+\gamma(1-e^{-2r\sqrt{2\lambda}})}\big(e^{-\lvert z-r\rvert\sqrt{2\lambda}}-e^{-(z+r)\sqrt{2\lambda}}\big)\Big\}dz$
1.3.6 $0<y$	$\mathbf{P}_x\big(\ell(\tau,r)\in dy, R_\tau^{(3)}\in dz\big) = \exp\Big(-\frac{y\sqrt{2\lambda}}{1-e^{-2r\sqrt{2\lambda}}}\Big)$ $\times\frac{z\lambda(e^{-\lvert x-r\rvert\sqrt{2\lambda}}-e^{-(x+r)\sqrt{2\lambda}})(e^{-\lvert z-r\rvert\sqrt{2\lambda}}-e^{-(z+r)\sqrt{2\lambda}})}{x(1-e^{-2r\sqrt{2\lambda}})^2}dydz$
(1)	$\mathbf{P}_x\big(\ell(\tau,r)=0, R_\tau^{(3)}\in dz\big) = \frac{z\sqrt{\lambda}}{x\sqrt{2}}\Big\{e^{-\lvert z-x\rvert\sqrt{2\lambda}} - e^{-(z+x)\sqrt{2\lambda}}$ $-\frac{e^{-\lvert x-r\rvert\sqrt{2\lambda}}-e^{-(x+r)\sqrt{2\lambda}}}{1-e^{-2r\sqrt{2\lambda}}}\big(e^{-\lvert z-r\rvert\sqrt{2\lambda}}-e^{-(z+r)\sqrt{2\lambda}}\big)\Big\}dz$
1.3.8	$\mathbf{P}_x\Big(\frac{1}{2}\ell(t,r)\in dy,\ R_t^{(3)}\in dz\Big)$ $=\frac{z}{2x}(\text{es}_y(0,2,r,y+\lvert x-r\rvert+\lvert z-r\rvert-2r,y)-\text{es}_y(0,2,r,y+\lvert x-r\rvert+z-r,y)$ $-\text{es}_y(0,2,r,y+\lvert z-r\rvert+x-r,y)+\text{es}_y(0,2,r,y+x+z,y))dydz$
1.4.1	$\mathbf{E}_x\exp\Big(-\gamma\int_0^\tau \mathbb{I}_{[r,\infty)}\big(R_s^{(3)}\big)ds\Big)$
$0\le x$ $x\le r$	$=1-\frac{\gamma(1+r\sqrt{2\lambda+2\gamma})\,\text{sh}(x\sqrt{2\lambda})}{x(\lambda+\gamma)(\sqrt{2\lambda}\,\text{ch}(r\sqrt{2\lambda})+\sqrt{2\lambda+2\gamma}\,\text{sh}(r\sqrt{2\lambda}))}$
$r\le x$	$=\frac{\lambda}{\lambda+\gamma}+\frac{\gamma(r\sqrt{2\lambda}\,\text{ch}(r\sqrt{2\lambda})-\text{sh}(r\sqrt{2\lambda}))e^{-(x-r)\sqrt{2\lambda+2\gamma}}}{x(\lambda+\gamma)(\sqrt{2\lambda}\,\text{ch}(r\sqrt{2\lambda})+\sqrt{2\lambda+2\gamma}\,\text{sh}(r\sqrt{2\lambda}))}$

5 $R_s^{(3)} = \left(\sum_{k=1}^3 (W_s^{(k)})^2\right)^{1/2}$ $\tau \sim \operatorname{Exp}(\lambda)$, independent of $R^{(3)}$

1.4.2 $\mathbf{P}_x\Big(\int_0^\tau \mathbb{I}_{[r,\infty)}\big(R_s^{(3)}\big)ds \in dy\Big)$

$0 \le x$, $x \le r$

$$= e^{-\lambda y}\frac{\operatorname{sh}(x\sqrt{2\lambda})}{x\operatorname{sh}(r\sqrt{2\lambda})}\Big\{r\lambda + \frac{r\sqrt{2\lambda}\operatorname{ch}(r\sqrt{2\lambda}) - \operatorname{sh}(r\sqrt{2\lambda})}{\sqrt{2}\operatorname{sh}(r\sqrt{2\lambda})}\Big\{\frac{1}{\sqrt{\pi y}} - \sqrt{\lambda}\operatorname{th}(r\sqrt{2\lambda}) - \frac{\lambda e^{\lambda y \operatorname{cth}^2(r\sqrt{2\lambda})}}{\operatorname{sh}(r\sqrt{2\lambda})\operatorname{ch}(r\sqrt{2\lambda})}\operatorname{Erfc}(\sqrt{\lambda y}\operatorname{th}(r\sqrt{2\lambda}))\Big\}\Big\}dy$$

$r \le x$

$$= e^{-\lambda y}\Big\{\lambda + \frac{r\sqrt{2\lambda}\operatorname{ch}(r\sqrt{2\lambda}) - \operatorname{sh}(r\sqrt{2\lambda})}{x\sqrt{2}\operatorname{sh}(r\sqrt{2\lambda})}\Big\{\frac{e^{-(x-r)^2/2y}}{\sqrt{\pi y}} - \lambda\operatorname{cth}(r\sqrt{2\lambda})\operatorname{Erfc}\Big(\frac{x-r}{\sqrt{2y}}\Big) + \frac{\sqrt{\lambda}e^{(x-r)\sqrt{2\lambda}\operatorname{cth}(r\sqrt{2\lambda}) + \lambda y\operatorname{cth}^2(r\sqrt{2\lambda})}}{\operatorname{sh}(r\sqrt{2\lambda})\operatorname{ch}(r\sqrt{2\lambda})}\operatorname{Erfc}\Big(\frac{x-r}{\sqrt{2y}} + \sqrt{\lambda y}\operatorname{cth}(r\sqrt{2\lambda})\Big)\Big\}\Big\}dy$$

(1) $$\mathbf{P}_x\Big(\int_0^\tau \mathbb{I}_{[r,\infty)}\big(R_s^{(3)}\big)ds = 0\Big) = \begin{cases} 1 - \dfrac{r\operatorname{sh}(x\sqrt{2\lambda})}{x\operatorname{sh}(r\sqrt{2\lambda})}, & 0 \le x \le r \\ 0, & r \le x \end{cases}$$

1.4.4 $$\mathbf{P}_x\Big(\int_0^t \mathbb{I}_{[r,\infty)}\big(R_s^{(3)}\big)ds \in dv\Big) = \int_0^\infty B_x^{(27)}(t-v,v,y)dy\,\mathbb{I}_{(0,t)}(v)dv$$

for $B_x^{(27)}(u,v,y)$ see 1.27.2

(1) $x \le r$ $$\mathbf{P}_x\Big(\int_0^t \mathbb{I}_{[r,\infty)}\big(R_s^{(3)}\big)ds = 0\Big) = 1 - \frac{r}{x}\widetilde{\operatorname{ss}}_t(x,r)$$

(2) $r \le x$ $$\mathbf{P}_x\Big(\int_0^t \mathbb{I}_{[r,\infty)}\big(R_s^{(3)}\big)ds = t\Big) = 1 - \frac{r}{x}\operatorname{Erfc}\Big(\frac{x-r}{\sqrt{2t}}\Big)$$

1.4.5 $\mathbf{E}_x\Big\{\exp\Big(-\gamma\int_0^\tau \mathbb{I}_{[r,\infty)}\big(R_s^{(3)}\big)ds\Big); R_\tau^{(3)} \in dz\Big\}$

$x \le r$, $z \le r$

$$= \frac{z\sqrt{\lambda}}{x\sqrt{2}}\big(e^{-|z-x|\sqrt{2\lambda}} - e^{-(z+x)\sqrt{2\lambda}}\big)dz - \frac{z\sqrt{2\lambda}(\sqrt{\lambda+\gamma} - \sqrt{\lambda})\operatorname{sh}(z\sqrt{2\lambda})\operatorname{sh}(x\sqrt{2\lambda})e^{-r\sqrt{2\lambda}}}{x(\sqrt{\lambda+\gamma}\operatorname{sh}(r\sqrt{2\lambda}) + \sqrt{\lambda}\operatorname{ch}(r\sqrt{2\lambda}))}dz$$

$x \le r$, $r \le z$

$$= \frac{z\lambda\sqrt{2}\operatorname{sh}(x\sqrt{2\lambda})e^{(r-z)\sqrt{2\lambda+2\gamma}}}{x(\sqrt{\lambda+\gamma}\operatorname{sh}(r\sqrt{2\lambda}) + \sqrt{\lambda}\operatorname{ch}(r\sqrt{2\lambda}))}dz$$

$r \le x$, $z \le r$

$$= \frac{z\lambda\sqrt{2}\operatorname{sh}(z\sqrt{2\lambda})e^{(r-x)\sqrt{2\lambda+2\gamma}}}{x(\sqrt{\lambda+\gamma}\operatorname{sh}(r\sqrt{2\lambda}) + \sqrt{\lambda}\operatorname{ch}(r\sqrt{2\lambda}))}dz$$

5	$R_s^{(3)} = \left(\sum_{k=1}^3 (W_s^{(k)})^2\right)^{1/2}$ $\qquad\tau \sim \operatorname{Exp}(\lambda)$, independent of $R^{(3)}$
$r \le x$ $r \le z$	$= \frac{z\lambda}{x\sqrt{2\lambda+2\gamma}} e^{-\lvert z-x\rvert\sqrt{2\lambda+2\gamma}} dz - \frac{z\lambda}{x\sqrt{2\lambda+2\gamma}} e^{(2r-x-z)\sqrt{2\lambda+2\gamma}} dz$ $+ \frac{z\lambda\sqrt{2}\operatorname{sh}(r\sqrt{2\lambda}) e^{(2r-x-z)\sqrt{2\lambda+2\gamma}}}{x(\sqrt{\lambda+\gamma}\operatorname{sh}(r\sqrt{2\lambda}) + \sqrt{\lambda}\operatorname{ch}(r\sqrt{2\lambda}))} dz$
1.4.6	$\mathbf{P}_x\Big(\int_0^\tau \mathbb{I}_{[r,\infty)}\big(R_s^{(3)}\big) ds \in dy, R_\tau^{(3)} \in dz\Big)$
$z \le r$ $x \le r$	$= \frac{z\lambda \operatorname{sh}(x\sqrt{2\lambda})\operatorname{sh}(z\sqrt{2\lambda})}{x\operatorname{sh}^2(r\sqrt{2\lambda})}\Big(\frac{\sqrt{2}}{\sqrt{\pi y}} e^{-\lambda y}$ $-\sqrt{2\lambda}\operatorname{cth}(r\sqrt{2\lambda}) e^{-\lambda y/\operatorname{sh}^2(r\sqrt{2\lambda})} \operatorname{Erfc}(\sqrt{\lambda y}\operatorname{cth}(r\sqrt{2\lambda}))\Big) dydz$
$z \le r$ $r \le x$	$= \frac{z\lambda \operatorname{sh}(z\sqrt{2\lambda})}{x\operatorname{sh}(r\sqrt{2\lambda})}\Big(\frac{\sqrt{2}}{\sqrt{\pi y}} e^{-\lambda y-(x-r)^2/2y} - \sqrt{2\lambda}\operatorname{cth}(r\sqrt{2\lambda})$ $\times e^{(x-r)\sqrt{2\lambda}\operatorname{cth}(r\sqrt{2\lambda})-\lambda y/\operatorname{sh}^2(r\sqrt{2\lambda})} \operatorname{Erfc}\Big(\frac{x-r}{\sqrt{2y}} + \sqrt{\lambda y}\operatorname{cth}(r\sqrt{2\lambda})\Big)\Big) dydz$
$r \le z$ $x \le r$	$= \frac{z\lambda \operatorname{sh}(x\sqrt{2\lambda})}{x\operatorname{sh}(r\sqrt{2\lambda})}\Big(\frac{\sqrt{2}}{\sqrt{\pi y}} e^{-\lambda y-(z-r)^2/2y} - \sqrt{2\lambda}\operatorname{cth}(r\sqrt{2\lambda})$ $\times e^{(z-r)\sqrt{2\lambda}\operatorname{cth}(r\sqrt{2\lambda})-\lambda y/\operatorname{sh}^2(r\sqrt{2\lambda})} \operatorname{Erfc}\Big(\frac{z-r}{\sqrt{2y}} + \sqrt{\lambda y}\operatorname{cth}(r\sqrt{2\lambda})\Big)\Big) dydz$
$r \le z$ $r \le x$	$= \frac{z\lambda}{x\sqrt{2\pi y}}\big(e^{-\lambda y-(z-x)^2/2y} + e^{-\lambda y-(z+x-2r)^2/2y}\big) dydz - \frac{z\lambda\sqrt{2\lambda}}{x}\operatorname{cth}(r\sqrt{2\lambda})$ $\times e^{(z+x-2r)\sqrt{2\lambda}\operatorname{cth}(r\sqrt{2\lambda})-\lambda y/\operatorname{sh}^2(r\sqrt{2\lambda})} \operatorname{Erfc}\Big(\frac{z+x-2r}{\sqrt{2y}} + \sqrt{\lambda y}\operatorname{cth}(r\sqrt{2\lambda})\Big) dydz$
(1) $x \le r$	$\mathbf{P}_x\Big(\int_0^\tau \mathbb{I}_{[r,\infty)}\big(R_s^{(3)}\big) ds = 0,\ R_\tau^{(3)} \in dz\Big)$ $= \frac{z\sqrt{\lambda}\big(\operatorname{ch}((r-\lvert z-x\rvert)\sqrt{2\lambda}) - \operatorname{ch}((r-x-z)\sqrt{2\lambda})\big)}{x\sqrt{2}\operatorname{sh}(r\sqrt{2\lambda})} dz, \quad z \le r$

5 $R_s^{(3)} = \left(\sum_{k=1}^3 (W_s^{(k)})^2\right)^{1/2}$ $\qquad \tau \sim \operatorname{Exp}(\lambda)$, independent of $R^{(3)}$

1.4.8 $v<t$

$$\mathbf{P}_x\Big(\int_0^t \mathbb{I}_{[r,\infty)}\big(R_s^{(3)}\big)ds \in dv, R_t^{(3)} \in dz\Big) = \int_0^\infty B_x^{(27)}(t-v,v,y,z)dy\,dvdz$$

for $B_x^{(27)}(u,v,y,z)$ see 1.27.6

(1) $x\le r$, $z \le r$

$$\mathbf{P}_x\Big(\int_0^t \mathbb{I}_{[r,\infty)}\big(R_s^{(3)}\big)ds = 0, R_t^{(3)} \in dz\Big)$$

$$= \frac{z}{x}(\operatorname{cs}_t(r-|z-x|,r) - \operatorname{cs}_t(r-z-x,r))dz$$

(2) $r\le x$, $r \le z$

$$\mathbf{P}_x\Big(\int_0^t \mathbb{I}_{[r,\infty)}\big(R_s^{(3)}\big)ds = t, R_t^{(3)} \in dz\Big) = \Big(\frac{ze^{-(z-x)^2/2t}}{x\sqrt{2\pi t}} - \frac{ze^{-(z+x-2r)^2/2t}}{x\sqrt{2\pi t}}\Big)dz$$

1.5.1

$$\mathbf{E}_x \exp\Big(-\gamma\int_0^\tau \mathbb{I}_{[0,r]}\big(R_s^{(3)}\big)ds\Big)$$

$0\le x$, $x \le r$

$$= \frac{\lambda}{\lambda+\gamma} + \frac{\gamma(1+r\sqrt{2\lambda})\operatorname{sh}(x\sqrt{2\lambda+2\gamma})}{x(\lambda+\gamma)(\sqrt{2\lambda}\operatorname{sh}(r\sqrt{2\lambda+2\gamma}) + \sqrt{2\lambda+2\gamma}\operatorname{ch}(r\sqrt{2\lambda+2\gamma}))}$$

$r \le x$

$$= 1 - \frac{\gamma r e^{(r-x)\sqrt{2\lambda}}}{x(\lambda+\gamma)} + \frac{\gamma(1+r\sqrt{2\lambda})\operatorname{sh}(r\sqrt{2\lambda+2\gamma})e^{(r-x)\sqrt{2\lambda}}}{x(\lambda+\gamma)(\sqrt{2\lambda}\operatorname{sh}(r\sqrt{2\lambda+2\gamma}) + \sqrt{2\lambda+2\gamma}\operatorname{ch}(r\sqrt{2\lambda+2\gamma}))}$$

1.5.2

$$\mathbf{P}_x\Big(\int_0^\tau \mathbb{I}_{[0,r]}\big(R_s^{(3)}\big)ds \in dy\Big)$$

$0\le x$, $x \le r$

$$= \lambda e^{-\lambda y}dy + \frac{1+r\sqrt{2\lambda}}{x\sqrt{2\lambda}}e^{-\lambda y}(\operatorname{rc}_y(0,r,x,1/\sqrt{2\lambda}) - 2\lambda\operatorname{rc}_y(-2,r,x,1/\sqrt{2\lambda}))dy$$

$r \le x$

$$= \frac{r\lambda}{x}e^{-\lambda y-(x-r)\sqrt{2\lambda}}dy + \frac{1+r\sqrt{2\lambda}}{x\sqrt{2\lambda}}e^{-\lambda y-(x-r)\sqrt{2\lambda}}\big(\operatorname{rc}_y(0,r,r,1/\sqrt{2\lambda})$$

$$-2\lambda\operatorname{rc}_y(-2,r,r,1/\sqrt{2\lambda})\big)dy$$

(1)

$$\mathbf{P}_x\Big(\int_0^\tau \mathbb{I}_{[0,r]}\big(R_s^{(3)}\big)ds = 0\Big) = \begin{cases} 0, & 0\le x\le r \\ 1-\frac{r}{x}e^{(r-x)\sqrt{2\lambda}}, & r\le x \end{cases}$$

5 $R_s^{(3)} = \left(\sum_{k=1}^3 (W_s^{(k)})^2\right)^{1/2}$ $\qquad \tau \sim \operatorname{Exp}(\lambda)$, independent of $R^{(3)}$

1.5.3 (1) $$\mathbf{E}_x \exp\Big(-\gamma \int_0^\infty \mathbb{I}_{[0,r]}\big(R_s^{(3)}\big)ds\Big) = \begin{cases} \dfrac{\operatorname{sh}(x\sqrt{2\gamma})}{x\sqrt{2\gamma}\operatorname{ch}(r\sqrt{2\gamma})}, & 0 \le x \le r \\ 1 - \dfrac{r}{x} + \dfrac{\operatorname{sh}(r\sqrt{2\gamma})}{x\sqrt{2\gamma}\operatorname{ch}(r\sqrt{2\gamma})}, & r \le x \end{cases}$$

1.5.4 $$\mathbf{P}_x\Big(\int_0^t \mathbb{I}_{[0,r]}\big(R_s^{(3)}\big)ds \in dv\Big) = \int_0^\infty B_x^{(27)}(v, t-v, y)dy\, \mathbb{I}_{(0,t)}(v)dv$$

for $B_x^{(27)}(u, v, y)$ see 1.27.2

(1) $x \le r$ $$\mathbf{P}_x\Big(\int_0^t \mathbb{I}_{[0,r]}\big(R_s^{(3)}\big)ds = t\Big) = 1 - \frac{r}{x}\widetilde{\operatorname{ss}}_t(x, r)$$

(2) $r \le x$ $$\mathbf{P}_x\Big(\int_0^t \mathbb{I}_{[0,r]}\big(R_s^{(3)}\big)ds = 0\Big) = 1 - \frac{r}{x}\operatorname{Erfc}\Big(\frac{x-r}{\sqrt{2t}}\Big)$$

(3) $$\mathbf{P}_x\Big(\int_0^\infty \mathbb{I}_{[0,r]}\big(R_s^{(3)}\big)ds \in dy\Big) = \begin{cases} \dfrac{1}{x}\operatorname{sc}_y(x, r)dy, & 0 \le x \le r \\ \dfrac{1}{x}\operatorname{sc}_y(r, r)dy, & r \le x \end{cases}$$

1.5.5 $$\mathbf{E}_x\Big\{\exp\Big(-\gamma \int_0^\tau \mathbb{I}_{[0,r]}\big(R_s^{(3)}\big)ds\Big); R_\tau^{(3)} \in dz\Big\}$$

$x \le r$, $z \le r$ $$= \frac{z\lambda}{x\sqrt{2\lambda+2\gamma}} e^{-|z-x|\sqrt{2\lambda+2\gamma}}dz - \frac{z\lambda}{x\sqrt{2\lambda+2\gamma}} e^{-(z+x)\sqrt{2\lambda+2\gamma}}dz$$
$$+ \frac{z2\lambda(\sqrt{\lambda+\gamma}-\sqrt{\lambda})\operatorname{sh}(z\sqrt{2\lambda+2\gamma})\operatorname{sh}(x\sqrt{2\lambda+2\gamma})e^{-r\sqrt{2\lambda+2\gamma}}}{x\sqrt{2\lambda+2\gamma}(\sqrt{\lambda}\operatorname{sh}(r\sqrt{2\lambda+2\gamma}) + \sqrt{\lambda+\gamma}\operatorname{ch}(r\sqrt{2\lambda+2\gamma}))}dz$$

$x \le r$, $r \le z$ $$= \frac{z\lambda\sqrt{2}\operatorname{sh}(x\sqrt{2\lambda+2\gamma})e^{(r-z)\sqrt{2\lambda}}}{x(\sqrt{\lambda}\operatorname{sh}(r\sqrt{2\lambda+2\gamma}) + \sqrt{\lambda+\gamma}\operatorname{ch}(r\sqrt{2\lambda+2\gamma}))}dz$$

$r \le x$, $z \le r$ $$= \frac{z\lambda\sqrt{2}\operatorname{sh}(z\sqrt{2\lambda+2\gamma})e^{(r-x)\sqrt{2\lambda}}}{x(\sqrt{\lambda}\operatorname{sh}(r\sqrt{2\lambda+2\gamma}) + \sqrt{\lambda+\gamma}\operatorname{ch}(r\sqrt{2\lambda+2\gamma}))}dz$$

5 $R_s^{(3)} = \left(\sum_{k=1}^{3}(W_s^{(k)})^2\right)^{1/2}$ $\tau \sim \mathrm{Exp}(\lambda)$, independent of $R^{(3)}$

$r \le x$ $r \le z$	$= \frac{z\sqrt{\lambda}}{x\sqrt{2}} e^{-\lvert z-x\rvert\sqrt{2\lambda}}dz - \frac{z\sqrt{\lambda}}{x\sqrt{2}} e^{(2r-x-z)\sqrt{2\lambda}}dz + \frac{z\lambda\sqrt{2}\,\mathrm{sh}(r\sqrt{2\lambda+2\gamma})e^{(2r-x-z)\sqrt{2\lambda}}}{x(\sqrt{\lambda}\,\mathrm{sh}(r\sqrt{2\lambda+2\gamma}) + \sqrt{\lambda+\gamma}\,\mathrm{ch}(r\sqrt{2\lambda+2\gamma}))}dz$
1.5.6	$\mathbf{P}_x\Big(\int_0^\tau \mathbb{I}_{[0,r]}\big(R_s^{(3)}\big)ds \in dy,\ R_\tau^{(3)} \in dz\Big)$
$x \le r$ $r \le z$	$= \frac{z\sqrt{2\lambda}}{x} e^{-\lambda y-(z-r)\sqrt{2\lambda}}\,\mathrm{rc}_y(0,r,x,1/\sqrt{2\lambda})dydz$
$r \le x$ $z \le r$	$= \frac{z\sqrt{2\lambda}}{x} e^{-\lambda y-(x-r)\sqrt{2\lambda}}\,\mathrm{rc}_y(0,r,z,1/\sqrt{2\lambda})dydz$
1.5.8 $v < t$	$\mathbf{P}_x\Big(\int_0^t \mathbb{I}_{[0,r]}\big(R_s^{(3)}\big)ds \in dv, R_t^{(3)} \in dz\Big) = \int_0^\infty B_x^{(27)}(v, t-v, y, z)dy\,dvdz$ for $B_x^{(27)}(u,v,y,z)$ see 1.27.6
(1) $x \le r$ $z \le r$	$\mathbf{P}_x\Big(\int_0^t \mathbb{I}_{[0,r]}\big(R_s^{(3)}\big)ds = t, R_t^{(3)} \in dz\Big)$ $= \frac{z}{x}(\mathrm{cs}_t(r-\lvert z-x\rvert, r) - \mathrm{cs}_t(r-z-x, r))dz$
(2) $r \le x$ $r \le z$	$\mathbf{P}_x\Big(\int_0^t \mathbb{I}_{[0,r]}\big(R_s^{(3)}\big)ds = 0, R_t^{(3)} \in dz\Big) = \Big(\frac{ze^{-(z-x)^2/2t}}{x\sqrt{2\pi t}} - \frac{ze^{-(z+x-2r)^2/2t}}{x\sqrt{2\pi t}}\Big)dz$
1.6.1	$\mathbf{E}_x \exp\Big(-\int_0^\tau \big(p\mathbb{I}_{[0,r)}\big(R_s^{(3)}\big) + q\mathbb{I}_{[r,\infty)}\big(R_s^{(3)}\big)\big)ds\Big)$
$0 \le x$ $x \le r$	$= \frac{\lambda}{\lambda+p} + \frac{\lambda(p-q)(1+r\sqrt{2\lambda+2q})\,\mathrm{sh}(x\sqrt{2\lambda+2p})}{x(\lambda+p)(\lambda+q)(\sqrt{2\lambda+2p}\,\mathrm{ch}(r\sqrt{2\lambda+2p}) + \sqrt{2\lambda+2q}\,\mathrm{sh}(r\sqrt{2\lambda+2p}))}$
$r \le x$	$= \frac{\lambda}{\lambda+q} + \frac{\lambda(q-p)(r\sqrt{2\lambda+2p}\,\mathrm{ch}(r\sqrt{2\lambda+2p}) - \mathrm{sh}(r\sqrt{2\lambda+2p}))e^{(r-x)\sqrt{2\lambda+2q}}}{x(\lambda+p)(\lambda+q)(\sqrt{2\lambda+2p}\,\mathrm{ch}(r\sqrt{2\lambda+2p}) + \sqrt{2\lambda+2q}\,\mathrm{sh}(r\sqrt{2\lambda+2p}))}$
1.6.2	$\mathbf{P}_x\Big(\int_0^\tau \mathbb{I}_{[0,r]}\big(R_s^{(3)}\big)ds \in du, \int_0^\tau \mathbb{I}_{[r,\infty)}\big(R_s^{(3)}\big)ds \in dv\Big)$

5	$R_s^{(3)} = \left(\sum_{k=1}^{3}(W_s^{(k)})^2\right)^{1/2}$ $\qquad \tau \sim \text{Exp}(\lambda)$, independent of $R^{(3)}$
	$= \lambda e^{-\lambda(u+v)} \int_0^\infty B_x^{(27)}(u,v,y)dy\,dudv \qquad$ for $B_x^{(27)}(u,v,y)$ see 1.27.2
(1) $x \le r$	$\mathbf{P}_x\Big(\int_0^\tau \mathbb{I}_{[0,r]}(R_s^{(3)})ds \in du, \int_0^\tau \mathbb{I}_{[r,\infty)}(R_s^{(3)})ds = 0\Big)$ $= \lambda e^{-\lambda u}(1 - \frac{r}{x}\widetilde{\text{ss}}_u(x,r))du$
(2) $r \le x$	$\mathbf{P}_x\Big(\int_0^\tau \mathbb{I}_{[0,r]}(R_s^{(3)})ds = 0, \int_0^\tau \mathbb{I}_{[r,\infty)}(R_s^{(3)})ds \in dv\Big)$ $= \lambda e^{-\lambda v}\Big(1 - \frac{r}{x}\,\text{Erfc}\Big(\frac{x-r}{\sqrt{2v}}\Big)\Big)dv$
1.6.4	$\mathbf{P}_x\Big(\int_0^t \big(p\mathbb{I}_{[0,r)}(R_s^{(3)}) + q\mathbb{I}_{[r,\infty)}(R_s^{(3)})\big)ds \in dv\Big)$ $= \frac{1}{\|p-q\|}\int_0^\infty B_x^{(27)}\Big(\frac{\|qt-v\|}{\|p-q\|}, \frac{\|pt-v\|}{\|p-q\|}, y\Big)dy\,\mathbb{I}_{((q\wedge p)t,(q\vee p)t)}(v)\,dv$ for $B_x^{(27)}(u,v,y)$ see 1.27.2
(1) $x \le r$	$\mathbf{P}_x\Big(\int_0^t \big(p\mathbb{I}_{[0,r)}(R_s^{(3)}) + q\mathbb{I}_{[r,\infty)}(R_s^{(3)})\big)ds = pt\Big) = 1 - \frac{r}{x}\widetilde{\text{ss}}_t(x,r)$
(2) $r \le x$	$\mathbf{P}_x\Big(\int_0^t \big(p\mathbb{I}_{[0,r)}(R_s^{(3)}) + q\mathbb{I}_{[r,\infty)}(R_s^{(3)})\big)ds = qt\Big) = 1 - \frac{r}{x}\,\text{Erfc}\Big(\frac{x-r}{\sqrt{2t}}\Big)$
1.6.5	$\mathbf{E}_x\Big\{\exp\Big(-\int_0^\tau \big(p\mathbb{I}_{[0,r)}(R_s^{(3)}) + q\mathbb{I}_{[r,\infty)}(R_s^{(3)})\big)ds\Big); R_\tau^{(3)} \in dz\Big\} = \frac{z\lambda}{x}Q_{0,\lambda}^{(24)}(z)dz$ for $Q_{a,\lambda}^{(24)}(z)$ see 1.1.24.5
1.6.6	$\mathbf{P}_x\Big(\int_0^\tau \mathbb{I}_{[0,r]}(R_s^{(3)})ds \in du, \int_0^\tau \mathbb{I}_{[r,\infty)}(R_s^{(3)})ds \in dv, R_\tau^{(3)} \in dz\Big)$ $= \lambda e^{-\lambda(u+v)} \int_0^\infty B_x^{(27)}(u,v,y,z)dy\,dudvdz$ for $B_x^{(27)}(u,v,y,z)$ see 1.27.6

5 $R_s^{(3)} = \left(\sum_{k=1}^3 (W_s^{(k)})^2\right)^{1/2}$ $\tau \sim \mathrm{Exp}(\lambda)$, independent of $R^{(3)}$

(1) $x \le r$

$$\mathbf{P}_x\Big(\int_0^\tau \mathbb{I}_{[0,r]}\big(R_s^{(3)}\big)ds \in du, \int_0^\tau \mathbb{I}_{[r,\infty)}\big(R_s^{(3)}\big)ds = 0, R_\tau^{(3)} \in dz\Big)$$

$$= \lambda e^{-\lambda u} \frac{z}{x}(\mathrm{cs}_u(r - |z-x|, r) - \mathrm{cs}_u(r - z - x, r))dudz$$

(2) $r \le x$

$$\mathbf{P}_x\Big(\int_0^\tau \mathbb{I}_{[0,r]}\big(R_s^{(3)}\big)ds = 0, \int_0^\tau \mathbb{I}_{[r,\infty)}\big(R_s^{(3)}\big)ds \in dv, R_\tau^{(3)} \in dz\Big)$$

$$= \frac{\lambda z}{x\sqrt{2\pi v}} e^{-\lambda v}\big(e^{-(z-x)^2/2v} - e^{-(z+x-2r)^2/2v}\big)dvdz$$

1.6.8

$$\mathbf{P}_x\Big(\int_0^t \big(p\mathbb{I}_{[0,r)}\big(R_s^{(3)}\big) + q\mathbb{I}_{[r,\infty)}\big(R_s^{(3)}\big)\big)ds \in dv, R_t^{(3)} \in dz\Big)$$

$$= \frac{1}{|p-q|}\int_0^\infty B_x^{(27)}\Big(\frac{|qt-v|}{|p-q|}, \frac{|pt-v|}{|p-q|}, y, z\Big)dy\, \mathbb{I}_{((q\wedge p)t,(q\vee p)t)}(v)\, dvdz$$

for $B_x^{(27)}(u, v, y, z)$ see 1.27.6

(1) $x \le r$

$$\mathbf{P}_x\Big(\int_0^t \big(p\mathbb{I}_{[0,r)}\big(R_s^{(3)}\big) + q\mathbb{I}_{[r,\infty)}\big(R_s^{(3)}\big)\big)ds = pt, R_t^{(3)} \in dz\Big)$$

$$= \frac{z}{x}(\mathrm{cs}_t(r - |z-x|, r) - \mathrm{cs}_t(r - z - x, r))dz$$

(2) $r \le x$

$$\mathbf{P}_x\Big(\int_0^t \big(p\mathbb{I}_{[0,r)}\big(R_s^{(3)}\big) + q\mathbb{I}_{[r,\infty)}\big(R_s^{(3)}\big)\big)ds = qt, R_t^{(3)} \in dz\Big)$$

$$= \frac{z}{x\sqrt{2\pi t}}\big(e^{-(z-x)^2/2t} - e^{-(z+x-2r)^2/2t}\big)dz$$

1.7.1

$$\mathbf{E}_r \exp\Big(-\gamma\int_0^\tau \mathbb{I}_{[r,u]}\big(R_s^{(3)}\big)ds\Big) =: S_r$$

$$= 1 - \frac{2\gamma\big((r\Upsilon_\gamma^2 + \Upsilon_0 + \Upsilon_\gamma(r\Upsilon_0 + 1)\,\mathrm{th}((u-r)\Upsilon_\gamma/2))\,\mathrm{sh}((u-r)\Upsilon_\gamma) - \Upsilon_0\Upsilon_\gamma(u-r)\big)}{r\Upsilon_\gamma^2\big(\Upsilon_\gamma^2 + \Upsilon_0\Upsilon_\gamma(1 + \mathrm{cth}(r\Upsilon_0))\,\mathrm{cth}((u-r)\Upsilon_\gamma) + \Upsilon_0^2\,\mathrm{cth}(r\Upsilon_0)\big)\,\mathrm{sh}((u-r)\Upsilon_\gamma)}$$

$$\mathbf{E}_u \exp\Big(-\gamma\int_0^\tau \mathbb{I}_{[r,u]}\big(R_s^{(3)}\big)ds\Big) =: S_u$$

$$= 1 - \frac{2\gamma(u\Upsilon_\gamma^2 - \Upsilon_0\,\mathrm{cth}(r\Upsilon_0) + \Upsilon_\gamma(u\Upsilon_0\,\mathrm{cth}(r\Upsilon_0) - 1)\,\mathrm{th}((u-r)\Upsilon_\gamma/2))}{u\Upsilon_\gamma^2\big(\Upsilon_\gamma^2 + \Upsilon_0\Upsilon_\gamma(1 + \mathrm{cth}(r\Upsilon_0))\,\mathrm{cth}((u-r)\Upsilon_\gamma) + \Upsilon_0^2\,\mathrm{cth}(r\Upsilon_0)\big)}$$

$$- \frac{2\gamma\Upsilon_0(u-r)\,\mathrm{cth}(r\Upsilon_0)}{u\Upsilon_\gamma\big(\Upsilon_\gamma^2 + \Upsilon_0\Upsilon_\gamma(1 + \mathrm{cth}(r\Upsilon_0))\,\mathrm{cth}((u-r)\Upsilon_\gamma) + \Upsilon_0^2\,\mathrm{cth}(r\Upsilon_0)\big)\,\mathrm{sh}((u-r)\Upsilon_\gamma)}$$

5 $R_s^{(3)} = \left(\sum_{k=1}^3 (W_s^{(k)})^2\right)^{1/2}$ $\tau \sim \mathrm{Exp}(\lambda)$, independent of $R^{(3)}$

$$\mathbf{E}_x \exp\Bigl(-\gamma\int_0^\tau \mathbb{I}_{[r,u]}\bigl(R_s^{(3)}\bigr)ds\Bigr)$$

$0 \le x$, $x \le r$
$$= 1 - \frac{r}{x}(1 - S_r)\frac{\mathrm{sh}(x\Upsilon_0)}{\mathrm{sh}(r\Upsilon_0)}$$

$r \le x$, $x \le u$
$$= \frac{\lambda}{\lambda+\gamma} + \frac{r}{x}\Bigl(S_r - \frac{\lambda}{\lambda+\gamma}\Bigr)\frac{\mathrm{sh}((u-x)\Upsilon_\gamma)}{\mathrm{sh}((u-r)\Upsilon_\gamma)} + \frac{u}{x}\Bigl(S_u - \frac{\lambda}{\lambda+\gamma}\Bigr)\frac{\mathrm{sh}((x-r)\Upsilon_\gamma)}{\mathrm{sh}((u-r)\Upsilon_\gamma)}$$

$u \le x$
$$= 1 - \frac{u}{x}(1 - S_u)e^{(u-x)\Upsilon_0}$$

1.7.3 (1)
$$\mathbf{E}_x \exp\Bigl(-\gamma\int_0^\infty \mathbb{I}_{[r,u]}\bigl(R_s^{(3)}\bigr)ds\Bigr)$$

$0 \le x$, $x \le r$
$$= \frac{1}{r\sqrt{2\gamma}\,\mathrm{sh}((u-r)\sqrt{2\gamma}) + \mathrm{ch}((u-r)\sqrt{2\gamma})}$$

$r \le x$, $x \le u$
$$= \frac{r\,\mathrm{sh}((u-x)\sqrt{2\gamma})}{x(r\sqrt{2\gamma}\,\mathrm{sh}((u-r)\sqrt{2\gamma}) + \mathrm{ch}((u-r)\sqrt{2\gamma}))\,\mathrm{sh}((u-r)\sqrt{2\gamma})}$$
$$+ \frac{(\mathrm{sh}((u-r)\sqrt{2\gamma}) + r\sqrt{2\gamma}\,\mathrm{ch}((u-r)\sqrt{2\gamma}))\,\mathrm{sh}((x-r)\sqrt{2\gamma})}{x\sqrt{2\gamma}(r\sqrt{2\gamma}\,\mathrm{sh}((u-r)\sqrt{2\gamma}) + \mathrm{ch}((u-r)\sqrt{2\gamma}))\,\mathrm{sh}((u-r)\sqrt{2\gamma})}$$

$u \le x$
$$= 1 - \frac{u}{x}\Bigl(1 - \frac{(\mathrm{sh}((u-r)\sqrt{2\gamma}) + r\sqrt{2\gamma}\,\mathrm{ch}((u-r)\sqrt{2\gamma}))}{u\sqrt{2\gamma}(r\sqrt{2\gamma}\,\mathrm{sh}((u-r)\sqrt{2\gamma}) + \mathrm{ch}((u-r)\sqrt{2\gamma}))}\Bigr)$$

1.7.4 (1)
$$\mathbf{P}_x\Bigl(\int_0^\infty \mathbb{I}_{[r,u]}\bigl(R_s^{(3)}\bigr)ds \in dy\Bigr)$$

$0 \le x$, $x \le r$
$$= \widetilde{\mathrm{rs}}_y(0, u-r, 0, r)dy$$

$r \le x$, $x \le u$
$$= \frac{r}{x}\,\mathrm{ss}_y(u-x, u-r) * \widetilde{\mathrm{rs}}_y(0, u-r, 0, r)dy + \frac{1}{2rx}\,\mathrm{ss}_y(x-r, u-r)$$
$$*(1 - 2\,\mathrm{rs}_y(-2, u-r, u-r, r) + 2r^2\,\mathrm{rs}_y(0, u-r, u-r, r))dy$$

$u \le x$
$$= \frac{1}{2rx}(1 - 2\,\mathrm{rs}_y(-2, u-r, u-r, r) + 2r^2\,\mathrm{rs}_y(0, u-r, u-r, r))dy$$

$$\Upsilon_s := \sqrt{2\lambda + 2s}$$

5 $R_s^{(3)} = \big(\sum_{k=1}^3 (W_s^{(k)})^2\big)^{1/2}$ $\tau \sim \operatorname{Exp}(\lambda)$, independent of $R^{(3)}$

1.8.5 $$\mathbf{E}_x\Big\{\exp\Big(-\gamma\int_0^\tau R_s^{(3)}ds\Big);\ R_\tau^{(3)}\in dz\Big\}$$

$x \le z$ $$= \frac{8z\lambda\sqrt{\lambda}(\lambda+\gamma x)\sqrt{\lambda+\gamma z}\,S_{1/3}\Big(\frac{2\sqrt2}{3\gamma}(\lambda+\gamma x)^{3/2}, \frac{2\sqrt2}{3\gamma}\lambda^{3/2}\Big)K_{1/3}\Big(\frac{2\sqrt2}{3\gamma}(\lambda+\gamma z)^{3/2}\Big)}{(3\gamma)^{5/3}xK_{1/3}\Big(\frac{2\sqrt2}{3\gamma}\lambda^{3/2}\Big)}dz$$

$z \le x$ $$= \frac{8z\lambda\sqrt{\lambda}(\lambda+\gamma z)\sqrt{\lambda+\gamma x}\,S_{1/3}\Big(\frac{2\sqrt2}{3\gamma}(\lambda+\gamma z)^{3/2}, \frac{2\sqrt2}{3\gamma}\lambda^{3/2}\Big)K_{1/3}\Big(\frac{2\sqrt2}{3\gamma}(\lambda+\gamma x)^{3/2}\Big)}{(3\gamma)^{5/3}xK_{1/3}\Big(\frac{2\sqrt2}{3\gamma}\lambda^{3/2}\Big)}dz$$

(1) $$\mathbf{E}_0\Big\{\exp\Big(-\gamma\int_0^\tau R_s^{(3)}ds\Big);\ R_\tau^{(3)}\in dz\Big\} = \frac{2\lambda z\operatorname{Ai}(2^{1/3}\gamma^{-2/3}(\lambda+\gamma z))}{2^{2/3}\operatorname{Ai}(2^{1/3}\gamma^{-2/3}\lambda)}dz$$

1.8.7 (1) $$\mathbf{E}_0\Big\{\exp\Big(-\gamma\int_0^t R_s^{(3)}ds\Big); R_t^{(3)}\in dz\Big\} = \sum_{k=1}^\infty \frac{z\operatorname{Ai}(\alpha_k' + (2\gamma)^{1/3}z)e^{(\gamma^2/2)^{1/3}t\alpha_k'}}{(2\gamma)^{-2/3}\operatorname{Ai}'(\alpha_k)}dz$$

1.9.3 $$\mathbf{E}_x\exp\Big(-\frac{\gamma^2}{2}\int_0^t (R_s^{(3)})^2ds\Big) = \frac{1}{(\operatorname{ch}(t\gamma))^{3/2}}\exp\Big(-\frac{x^2\gamma\operatorname{sh}(t\gamma)}{2\operatorname{ch}(t\gamma)}\Big)$$

(1) $$\mathbf{E}_x\exp\Big(-\beta(R_t^{(3)})^2 - \frac{\gamma^2}{2}\int_0^t (R_s^{(3)})^2ds\Big)$$

$$= \frac{1}{(\operatorname{ch}(t\gamma)+2\beta\gamma^{-1}\operatorname{sh}(t\gamma))^{3/2}}\exp\Big(-\frac{x^2(\gamma\operatorname{sh}(t\gamma)+2\beta\operatorname{ch}(t\gamma))}{2(\operatorname{ch}(t\gamma)+2\beta\gamma^{-1}\operatorname{sh}(t\gamma))}\Big)$$

1.9.4 $$\mathbf{P}_x\Big(\int_0^t \big(R_s^{(3)}\big)^2ds\in dy\Big) = \operatorname{ec}_y(0,3/2,t,0,x^2/2)dy$$

(1) $$\mathbf{P}_0\Big(\int_0^t \big(R_s^{(3)}\big)^2ds < y\Big) = \frac{4\sqrt2}{\sqrt\pi}\sum_{k=0}^\infty \frac{(-1)^k\Gamma(k+3/2)}{k!}\operatorname{Erfc}\Big(\frac{(4k+3)t}{2\sqrt{2y}}\Big)$$

1.9.5 $$\mathbf{E}_x\Big\{\exp\Big(-\frac{\gamma^2}{2}\int_0^\tau (R_s^{(3)})^2ds\Big);\ R_\tau^{(3)}\in dz\Big\}$$

$x \le z$ $$= \frac{z\lambda\Gamma(\frac12+\frac\lambda\gamma)}{x\sqrt{\pi\gamma}}\big(D_{-\frac12-\frac\lambda\gamma}(-x\sqrt{2\gamma}) - D_{-\frac12-\frac\lambda\gamma}(x\sqrt{2\gamma})\big)D_{-\frac12-\frac\lambda\gamma}(z\sqrt{2\gamma})dz$$

$z \le x$ $$= \frac{z\lambda\Gamma(\frac12+\frac\lambda\gamma)}{x\sqrt{\pi\gamma}}\big(D_{-\frac12-\frac\lambda\gamma}(-z\sqrt{2\gamma}) - D_{-\frac12-\frac\lambda\gamma}(z\sqrt{2\gamma})\big)D_{-\frac12-\frac\lambda\gamma}(x\sqrt{2\gamma})dz$$

5 $R_s^{(3)} = \left(\sum_{k=1}^3 (W_s^{(k)})^2\right)^{1/2}$ $\tau \sim \text{Exp}(\lambda)$, independent of $R^{(3)}$

1.9.7 $$\mathbf{E}_x\Big\{\exp\Big(-\frac{\gamma^2}{2}\int_0^t (R_s^{(3)})^2 ds\Big);\ R_t^{(3)} \in dz\Big\}$$

$$= \frac{z\sqrt{\gamma}}{x\sqrt{2\pi\,\text{sh}(t\gamma)}}\exp\Big(-\frac{(x^2+z^2)\gamma\,\text{ch}(t\gamma)}{2\,\text{sh}(t\gamma)}\Big)\Big(\exp\Big(\frac{xz\gamma}{\text{sh}(t\gamma)}\Big) - \exp\Big(-\frac{xz\gamma}{\text{sh}(t\gamma)}\Big)\Big)dz$$

1.9.8 $$\mathbf{P}_x\Big(\int_0^t \big(R_s^{(3)}\big)^2 ds \in dy,\ R_t^{(3)} \in dz\Big)$$

$$= \frac{z}{x\sqrt{2\pi}}(\text{ee}_y(1/2, t, (x^2+z^2)/2, -xz) - \text{ee}_y(1/2, t, (x^2+z^2)/2, xz))dydz$$

1.10.3 (1) $$\mathbf{E}_x \exp\Big(-\gamma\int_0^\infty e^{2\beta R_s^{(3)}} ds\Big) = \frac{S_0\Big(\frac{\sqrt{2\gamma}}{|\beta|}, \frac{\sqrt{2\gamma}}{|\beta|}e^{\beta x}\Big)}{x|\beta| I_0\Big(\frac{\sqrt{2\gamma}}{|\beta|}\Big)}, \qquad \beta < 0$$

1.10.5 $\beta \neq 0$ $$\mathbf{E}_x\Big\{\exp\Big(-\gamma\int_0^\tau e^{2\beta R_s^{(3)}} ds\Big);\ R_\tau^{(3)} \in dz\Big\}$$

for $(z-x)\beta \geq 0$

$$= \frac{2z\lambda e^{x\,\text{sign}\,\beta\sqrt{2\lambda}}}{x|\beta|}\Big(\frac{2\gamma}{\beta^2}\Big)^{\sqrt{2\lambda}/|\beta|} S_{\sqrt{2\lambda}/|\beta|}\Big(\frac{\sqrt{2\gamma}}{|\beta|}e^{\beta x}, \frac{\sqrt{2\gamma}}{|\beta|}\Big)\frac{K_{\sqrt{2\lambda}/|\beta|}\Big(\frac{\sqrt{2\gamma}}{|\beta|}e^{\beta z}\Big)}{K_{\sqrt{2\lambda}/|\beta|}\Big(\frac{\sqrt{2\gamma}}{|\beta|}\Big)}dz$$

for $(z-x)\beta \leq 0$

$$= \frac{2z\lambda e^{z\,\text{sign}\,\beta\sqrt{2\lambda}}}{x|\beta|}\Big(\frac{2\gamma}{\beta^2}\Big)^{\sqrt{2\lambda}/|\beta|} S_{\sqrt{2\lambda}/|\beta|}\Big(\frac{\sqrt{2\gamma}}{|\beta|}e^{\beta z}, \frac{\sqrt{2\gamma}}{|\beta|}\Big)\frac{K_{\sqrt{2\lambda}/|\beta|}\Big(\frac{\sqrt{2\gamma}}{|\beta|}e^{\beta x}\Big)}{K_{\sqrt{2\lambda}/|\beta|}\Big(\frac{\sqrt{2\gamma}}{|\beta|}\Big)}dz$$

1.11.2 $$\mathbf{P}_0\big(\sup_{y\in(0,\infty)} \ell(\tau, y) > h\big) = \frac{h\sqrt{\lambda}}{\sqrt{2}\,\text{sh}(h\sqrt{\lambda/2})I_0(h\sqrt{\lambda/2})}$$

$$\times\Big\{\int_0^h I_0(v\sqrt{\lambda/2})K_0(v\sqrt{\lambda/2})dv - \frac{K_0(h\sqrt{\lambda/2})}{I_0(h\sqrt{\lambda/2})}\int_0^h I_0^2(v\sqrt{\lambda/2})dv\Big\}$$

1.12.1 $$\mathbf{E}_x e^{-\gamma\hat{H}(\tau)} = \frac{\sqrt{\lambda}}{\sqrt{\lambda+\gamma}} + \Big(1 - \frac{\sqrt{\lambda}}{\sqrt{\lambda+\gamma}}\Big)\frac{1-e^{-x\sqrt{2\lambda+2\gamma}}}{x\sqrt{2\lambda+2\gamma}}$$

1.12.2 $$\mathbf{P}_x\big(\hat{H}(\tau) \in dv\big) = \frac{e^{-\lambda v}}{x}\Big\{\frac{x\sqrt{\lambda}}{\sqrt{\pi v}} + \frac{1-e^{-x^2/2v}}{\sqrt{2\pi v}} - \frac{\sqrt{\lambda}}{\sqrt{2}}\Big(1 - \text{Erfc}\Big(\frac{x}{\sqrt{2v}}\Big)\Big)\Big\}dv$$

5	$R_s^{(3)} = \big(\sum_{k=1}^3 (W_s^{(k)})^2\big)^{1/2}$ $\tau \sim \mathrm{Exp}(\lambda)$, independent of $R^{(3)}$
1.12.3 (1)	$\mathbf{E}_x e^{-\gamma \hat H(\infty)} = \frac{1-e^{-x\sqrt{2\gamma}}}{x\sqrt{2\gamma}}$
1.12.4	$\mathbf{P}_x\big(\hat H(t)\in dv\big) = \frac{\mathbb{I}_{[0,t]}(v)}{x}\Big\{\frac{x}{\sqrt{\pi(t-v)v}} + \frac{1-e^{-x^2/2v}}{\sqrt{2\pi v}} - \frac{1-\mathrm{Erfc}(x/\sqrt{2v})}{\sqrt{2\pi(t-v)}}\Big\}dv$
(1)	$\mathbf{P}_x\big(\hat H(\infty)\in dv\big) = \frac{1}{x\sqrt{2\pi v}}\Big(1-e^{-x^2/2v}\Big)dv$
1.12.5	$\mathbf{E}_x\big\{e^{-\gamma\hat H(\tau)}; R_\tau^{(3)}\in dz\big\}$ $x\vee z<y$
$x\le z$	$= \frac{\lambda z\sqrt2}{x(\sqrt{\lambda+\gamma}+\sqrt\lambda)}\Big(e^{(x-z)\sqrt{2\lambda}} - e^{-x\sqrt{2\lambda+2\gamma}-z\sqrt{2\lambda}}\Big)dz$
$z\le x$	$= \frac{\lambda z\sqrt2}{x(\sqrt{\lambda+\gamma}+\sqrt\lambda)}\Big(e^{(z-x)\sqrt{2\lambda+2\gamma}} - e^{-x\sqrt{2\lambda+2\gamma}-z\sqrt{2\lambda}}\Big)dz$
1.13.1 $x<y$	$\mathbf{E}_x\big\{e^{-\gamma\check H(\tau)}; \sup_{0\le s\le\tau} R_s^{(3)}\in dy\big\} = \frac{\mathrm{sh}(x\sqrt{2\lambda+2\gamma})(y\sqrt{2\lambda}\,\mathrm{cth}(y\sqrt{2\lambda})-1)}{x\,\mathrm{sh}(y\sqrt{2\lambda+2\gamma})}dy$
1.13.2 $x<y$	$\mathbf{P}_x\big(\check H(\tau)\in dv, \sup_{0\le s\le\tau} R_s^{(3)}\in dy\big) = \frac{e^{-\lambda v}}{x}(y\sqrt{2\lambda}\,\mathrm{cth}(y\sqrt{2\lambda})-1)\,\mathrm{ss}_v(x,y)dvdy$
1.13.3 $x<y$	$\mathbf{E}_x\big\{e^{-\gamma\check H(t)}; \sup_{0\le s\le t} R_s^{(3)}\in dy\big\} = \big(e^{-\gamma t}\,\mathrm{ss}_t(x,y)\big) * (2y\ \mathrm{cs}_t(y,y)-1)dy$
1.13.4 $x<y$	$\mathbf{P}_x\big(\check H(t)\in dv, \sup_{0\le s\le t} R_s^{(3)}\in dy\big) = \frac1x \mathbb{I}_{[0,t]}(v)\,\mathrm{ss}_v(x,y)(2y\ \mathrm{cs}_{t-v}(y,y)-1)dvdy$
1.13.5 $x<y$ $z<y$	$\mathbf{E}_x\big\{e^{-\gamma\check H(\tau)}; \sup_{0\le s\le\tau} R_s^{(3)}\in dy, R_\tau^{(3)}\in dz\big\} = \frac{2\lambda z\,\mathrm{sh}(x\sqrt{2\lambda+2\gamma})\,\mathrm{sh}(z\sqrt{2\lambda})}{x\,\mathrm{sh}(y\sqrt{2\lambda+2\gamma})\,\mathrm{sh}(y\sqrt{2\lambda})}dydz$
1.13.6 $x<y$ $z<y$	$\mathbf{P}_x\big(\check H(\tau)\in dv; \sup_{0\le s\le\tau} R_s^{(3)}\in dy, R_\tau^{(3)}\in dz\big) = \frac{2\lambda z\ \mathrm{ss}_v(x,y)\,\mathrm{sh}(z\sqrt{2\lambda})}{xe^{\lambda v}\,\mathrm{sh}(y\sqrt{2\lambda})}dvdydz$
1.13.7 $x<y$ $z<y$	$\mathbf{E}_x\big\{e^{-\gamma\check H(t)}; \sup_{0\le s\le t} R_s^{(3)}\in dy,\ R_t^{(3)}\in dz\big\} = \frac{2z}{x}\big(e^{-\gamma t}\,\mathrm{ss}_t(x,y)\big) * \mathrm{ss}_t(z,y)dydz$

5	$R_s^{(3)} = \left(\sum_{k=1}^{3}(W_s^{(k)})^2\right)^{1/2}$ $\qquad\qquad \tau \sim \operatorname{Exp}(\lambda)$, independent of $R^{(3)}$
1.13.8 $v<t$	$\mathbf{P}_x\left(\check{H}(t)\in dv,\ \sup_{0\le s\le t} R_s^{(3)}\in dy, R_t^{(3)}\in dz\right) = 2\,\mathrm{ss}_v(x,y)\,\mathrm{ss}_{t-v}(z,y)dvdydz$
1.14.1 $y<x$	$\mathbf{E}_x\left\{e^{-\gamma\hat{H}(\tau)};\ \inf_{0\le s\le\tau} R_s^{(3)}\in dy\right\} = \frac{1}{x}(y\sqrt{2\lambda}+1)e^{-(x-y)\sqrt{2\lambda+2\gamma}}dy$
1.14.2 $y<x$	$\mathbf{P}_x\left(\hat{H}(\tau)\in dv,\ \inf_{0\le s\le\tau} R_s^{(3)}\in dy\right) = \frac{(y\sqrt{2\lambda}+1)(x-y)}{x\sqrt{2\pi}v^{3/2}}\exp\left(-\lambda v-\frac{(x-y)^2}{2v}\right)dvdy$
1.14.3 $y<x$	$\mathbf{E}_x\left\{e^{-\gamma\hat{H}(t)};\ \inf_{0\le s\le t} R_s^{(3)}\in dy\right\} = \frac{(x-y)e^{-\gamma t-(x-y)^2/2t}}{x\sqrt{2\pi}t^{3/2}} * \left(\frac{y\sqrt{2}}{\sqrt{\pi t}}+1\right)dy$
(1)	$\mathbf{E}_x\left\{e^{-\gamma\hat{H}(\infty)};\ \inf_{0\le s<\infty} R_s^{(3)}\in dy\right\} = \frac{1}{x}e^{-(x-y)\sqrt{2\gamma}}dy$
1.14.4 $y<x$	$\mathbf{P}_x\left(\hat{H}(t)\in dv,\ \inf_{0\le s\le t} R_s^{(3)}\in dy\right)$ $= \frac{\mathbb{I}_{[0,t]}(v)(x-y)}{x\sqrt{2\pi}v^{3/2}}\exp\left(-\frac{(x-y)^2}{2v}\right)\left(\frac{y\sqrt{2}}{\sqrt{\pi(t-v)}}+1\right)dvdy$
(1) $y<x$	$\mathbf{P}_x\left(\hat{H}(\infty)\in dv,\ \inf_{0\le s<\infty} R_s^{(3)}\in dy\right) = \frac{x-y}{x\sqrt{2\pi}v^{3/2}}\exp\left(-\frac{(x-y)^2}{2v}\right)dvdy$
1.14.5 $y<x$ $y<z$	$\mathbf{E}_x\left\{e^{-\gamma\hat{H}(\tau)};\ \inf_{0\le s\le\tau} R_s^{(3)}\in dy, R_\tau^{(3)}\in dz\right\} = \frac{2z\lambda}{x}e^{(y-x)\sqrt{2\lambda+2\gamma}}e^{(y-z)\sqrt{2\lambda}}dydz$
1.14.6 $y<x$ $y<z$	$\mathbf{P}_x\left(\hat{H}(\tau)\in dv;\ \inf_{0\le s\le\tau} R_s^{(3)}\in dy, R_\tau^{(3)}\in dz\right)$ $= \frac{2\lambda z}{x}\frac{(x-y)}{\sqrt{2\pi}v^{3/2}}\exp\left(-\lambda v-\frac{(x-y)^2}{2v}\right)e^{-(z-y)\sqrt{2\lambda}}dvdydz$
1.14.7	$\mathbf{E}_x\left\{e^{-\gamma\hat{H}(t)};\ \inf_{0\le s\le t} R_s^{(3)}\in dy,\ R_t^{(3)}\in dz\right\}$ $\qquad y<x\wedge z$ $= \frac{z(x-y)(z-y)}{x\pi}\left(\frac{1}{t^{3/2}}\exp\left(-\gamma t-\frac{(x-y)^2}{2t}\right)\right) * \left(\frac{1}{t^{3/2}}\exp\left(-\frac{(z-y)^2}{2t}\right)\right)dydz$
1.14.8 $v<t$	$\mathbf{P}_x\left(\hat{H}(t)\in dv,\ \inf_{0\le s\le t} R_s^{(3)}\in dy, R_t^{(3)}\in dz\right)$

5 $R_s^{(3)} = \left(\sum_{k=1}^3 (W_s^{(k)})^2\right)^{1/2}$ $\qquad \tau \sim \operatorname{Exp}(\lambda)$, independent of $R^{(3)}$

$$= \frac{z(x-y)(z-y)}{x\pi v^{3/2}(t-v)^{3/2}} \exp\Big(-\frac{(x-y)^2}{2v} - \frac{(z-y)^2}{2(t-v)}\Big) dv dy dz, \qquad y < x \wedge z$$

1.15.2 $$\mathbf{P}_x\big(a < \inf_{0\le s\le\tau} R_s^{(3)}, \sup_{0\le s\le\tau} R_s^{(3)} < b\big) = 1 - \frac{a \operatorname{sh}\big((b-x)\sqrt{2\lambda}\big) + b \operatorname{sh}\big((x-a)\sqrt{2\lambda}\big)}{x \operatorname{sh}\big((b-a)\sqrt{2\lambda}\big)}$$

1.15.4 $$\mathbf{P}_x\big(a < \inf_{0\le s\le t} R_s^{(3)}, \sup_{0\le s\le t} R_s^{(3)} < b\big) = 1 - \frac{a}{x}\widetilde{\operatorname{ss}}_t(b-x, b-a) - \frac{b}{x}\widetilde{\operatorname{ss}}_t(x-a, b-a)$$

1.15.6 $$\mathbf{P}_x\big(a < \inf_{0\le s\le\tau} R_s^{(3)}, \sup_{0\le s\le\tau} R_s^{(3)} < b, R_\tau^{(3)} \in dz\big)$$

$$= \frac{z\sqrt{\lambda}\big(\operatorname{ch}\big((b-a-|z-x|)\sqrt{2\lambda}\big) - \operatorname{ch}\big((a+b-x-z)\sqrt{2\lambda}\big)\big)}{x\sqrt{2}\operatorname{sh}\big((b-a)\sqrt{2\lambda}\big)} dz, \qquad a < x \wedge z,\ x \vee z < b$$

1.15.8 $$\mathbf{P}_x\big(a < \inf_{0\le s\le t} R_s^{(3)}, \sup_{0\le s\le t} R_s^{(3)} < b, R_t^{(3)} \in dz\big)$$

$$= \frac{z}{x\sqrt{2\pi t}} \sum_{k=-\infty}^{\infty} \big(e^{-(z-x+2k(b-a))^2/2t} - e^{-(z+x-2a+2k(b-a))^2/2t}\big) dz, \qquad a < x \wedge z,\ x \vee z < b$$

1.16.1 $r < b$ $$\mathbf{E}_x\big\{e^{-\gamma\ell(\tau,r)}; \sup_{0\le s\le\tau} R_s^{(3)} < b\big\}$$

$0 \le x$, $x \le r$ $$= 1 - \frac{(\gamma r \operatorname{sh}((b-r)\sqrt{2\lambda}) + b\sqrt{\lambda/2}) \operatorname{sh}(x\sqrt{2\lambda})}{x(\gamma \operatorname{sh}((b-r)\sqrt{2\lambda}) \operatorname{sh}(r\sqrt{2\lambda}) + \sqrt{\lambda/2}\operatorname{sh}(b\sqrt{2\lambda}))}$$

$r \le x$, $x \le b$ $$= 1 - \frac{b \operatorname{sh}((x-r)\sqrt{2\lambda})}{x \operatorname{sh}((b-r)\sqrt{2\lambda})} - \frac{(\gamma r \operatorname{sh}((b-r)\sqrt{2\lambda}) + b\sqrt{\lambda/2}) \operatorname{sh}(r\sqrt{2\lambda}) \operatorname{sh}((b-x)\sqrt{2\lambda})}{x(\gamma \operatorname{sh}((b-r)\sqrt{2\lambda}) \operatorname{sh}(r\sqrt{2\lambda}) + \sqrt{\lambda/2}\operatorname{sh}(b\sqrt{2\lambda})) \operatorname{sh}((b-r)\sqrt{2\lambda})}$$

1.16.2 $r < b$ $$\mathbf{P}_x\big(\ell(\tau,r) \in dy, \sup_{0\le s\le\tau} R_s^{(3)} < b\big) = \exp\Big(-\frac{y\sqrt{\lambda/2}\operatorname{sh}(b\sqrt{2\lambda})}{\operatorname{sh}((b-r)\sqrt{2\lambda})\operatorname{sh}(r\sqrt{2\lambda})}\Big)$$

$$\times \begin{cases} \dfrac{\sqrt{\lambda/2}(r \operatorname{sh}(b\sqrt{2\lambda}) - b \operatorname{sh}(r\sqrt{2\lambda})) \operatorname{sh}(x\sqrt{2\lambda})}{x \operatorname{sh}((b-r)\sqrt{2\lambda}) \operatorname{sh}^2(r\sqrt{2\lambda})} dy, & 0 \le x \le r \\ \dfrac{\sqrt{\lambda/2}(r \operatorname{sh}(b\sqrt{2\lambda}) - b \operatorname{sh}(r\sqrt{2\lambda})) \operatorname{sh}((b-x)\sqrt{2\lambda})}{x \operatorname{sh}^2((b-r)\sqrt{2\lambda}) \operatorname{sh}(r\sqrt{2\lambda})} dy, & r \le x \le b \end{cases}$$

5 $R_s^{(3)} = \left(\sum_{k=1}^3 (W_s^{(k)})^2\right)^{1/2}$ $\qquad \tau \sim \text{Exp}(\lambda)$, independent of $R^{(3)}$

(1)	$\mathbf{P}_x\big(\ell(\tau,r)=0, \sup_{0\le s\le\tau} R_s^{(3)} < b\big) = \begin{cases} 1-\dfrac{r\,\mathrm{sh}(x\sqrt{2\lambda})}{x\,\mathrm{sh}(r\sqrt{2\lambda})}, & 0\le x\le r \\ 1-\dfrac{r\,\mathrm{sh}\big((b-x)\sqrt{2\lambda}\big)+b\,\mathrm{sh}\big((x-r)\sqrt{2\lambda}\big)}{x\,\mathrm{sh}\big((b-r)\sqrt{2\lambda}\big)}, & r\le x\le b \end{cases}$
1.16.5 $r<b$	$\mathbf{E}_x\{e^{-\gamma\ell(\tau,r)}; \sup_{0\le s\le\tau} R_s^{(3)} < b, R_\tau^{(3)}\in dz\} = \frac{z\lambda}{x} Q_{0,\lambda}^{(25)}(z)dz$
$z<b$	for $Q_{a,\lambda}^{(25)}(z)$ see 1.1.25.5
1.16.6 $r<b$	$\mathbf{P}_x\big(\ell(\tau,r)\in dy, \sup_{0\le s\le\tau} R_s^{(3)} < b, R_\tau^{(3)}\in dz\big) = \frac{z\lambda}{x} B_{0,\lambda}^{(25)}(y,z)dydz$
$z<b$	for $B_{a,\lambda}^{(25)}(y,z)$ see 1.1.25.6
(1)	$\mathbf{P}_x\big(\ell(\tau,r)=0, \sup_{0\le s\le\tau} R_s^{(3)} < b, R_\tau^{(3)}\in dz\big)$
$z\le r$ $x\le r$	$= \dfrac{z\sqrt{\lambda}\big(\mathrm{ch}((r-\lvert z-x\rvert)\sqrt{2\lambda}) - \mathrm{ch}((r-x-z)\sqrt{2\lambda})\big)}{x\sqrt{2}\,\mathrm{sh}(r\sqrt{2\lambda})}dz$
$r\le z$ $r\le x$	$= \dfrac{z\sqrt{\lambda}\big(\mathrm{ch}\big((b-r-\lvert z-x\rvert)\sqrt{2\lambda}\big) - \mathrm{ch}\big((b+r-z-x)\sqrt{2\lambda}\big)\big)}{x\sqrt{2}\,\mathrm{sh}\big((b-r)\sqrt{2\lambda}\big)}dz$
1.16.8 $r<b$	$\mathbf{P}_x\big(\ell(t,r)\in dy, \sup_{0\le s\le t} R_s^{(3)} < b, R_t^{(3)}\in dz\big) = \frac{z}{x} F_{0,x}^{(25)}(t,y,z)dydz$
$z<b$	for $F_{a,x}^{(25)}(t,y,z)$ see 1.1.25.8
1.17.1 $a<r$	$\mathbf{E}_x\{e^{-\gamma\ell(\tau,r)}; a < \inf_{0\le s\le\tau} R_s^{(3)}\} = 1-\frac{a}{x}e^{-(x-a)\sqrt{2\lambda}}$
$a\le x$	$-\dfrac{\gamma(r-ae^{-(r-a)\sqrt{2\lambda}})}{x\big(\sqrt{2\lambda}+\gamma(1-e^{-2(r-a)\sqrt{2\lambda}})\big)}\big(e^{-\lvert x-r\rvert\sqrt{2\lambda}} - e^{-(x+r-2a)\sqrt{2\lambda}}\big)$
1.17.2 $a<r$ $a\le x$	$\mathbf{P}_x\big(\ell(\tau,r)\in dy, a < \inf_{0\le s\le\tau} R_s^{(3)}\big) = \exp\Big(-\dfrac{y\sqrt{2\lambda}}{1-e^{-2(r-a)\sqrt{2\lambda}}}\Big)$

5 $R_s^{(3)} = \left(\sum_{k=1}^3 (W_s^{(k)})^2\right)^{1/2}$ $\tau \sim \mathrm{Exp}(\lambda)$, independent of $R^{(3)}$

$0<y$	$\times \dfrac{\sqrt{2\lambda}(r-ae^{-(r-a)\sqrt{2\lambda}})(e^{-\lvert x-r\rvert\sqrt{2\lambda}}-e^{-(x+r-2a)\sqrt{2\lambda}})}{x\big(1-e^{-2(r-a)\sqrt{2\lambda}}\big)^2}dy$
(1)	$\mathbf{P}_x\big(\ell(\tau,r)=0, a<\inf_{0\le s\le\tau} R_s^{(3)}\big) = 1-\dfrac{a\,\mathrm{sh}\big((r-x)\sqrt{2\lambda}\big)+r\,\mathrm{sh}\big((x-a)\sqrt{2\lambda}\big)}{x\,\mathrm{sh}\big((r-a)\sqrt{2\lambda}\big)}$
1.17.3 (1) $a\le x$	$\mathbf{E}_x\big\{e^{-\gamma\ell(\infty,r)};\ a<\inf_{0\le s<\infty} R_s^{(3)}\big\} = 1-\dfrac{x+r-\lvert x-r\rvert}{2x}+\dfrac{x+r-2a-\lvert x-r\rvert}{2x(1+2\gamma(r-a))}$
1.17.4 (1)	$\mathbf{P}_x\big(\ell(\infty,r)\in dy,\ a<\inf_{0\le s<\infty} R_s^{(3)}\big) = \dfrac{(x+r-2a-\lvert x-r\rvert)}{4x(r-a)}e^{-y/2(r-a)}dy$
(2) $r\le x$ $a<r$	$\mathbf{P}_x\big(\ell(\infty,r)=0,\ a<\inf_{0\le s<\infty} R_s^{(3)}\big) = 1-\dfrac{r}{x}$
1.17.5 $a<r$ $a<z$	$\mathbf{E}_x\big\{e^{-\gamma\ell(\tau,r)}; a<\inf_{0\le s\le\tau} R_s^{(3)}, R_\tau^{(3)}\in dz\big\} = \dfrac{z\sqrt{\lambda}}{x\sqrt{2}}\Big\{e^{-\lvert z-x\rvert\sqrt{2\lambda}}-e^{-(z+x-2a)\sqrt{2\lambda}}$ $-\dfrac{z\gamma(e^{-\lvert x-r\rvert\sqrt{2\lambda}}-e^{-(x+r-2a)\sqrt{2\lambda}})}{x\big(\sqrt{2\lambda}+\gamma(1-e^{-2(r-a)\sqrt{2\lambda}})\big)}\big(e^{-\lvert z-r\rvert\sqrt{2\lambda}}-e^{-(z+r-2a)\sqrt{2\lambda}}\big)\Big\}dz$
1.17.6 $a<r$ $a<z$	$\mathbf{P}_x\big(\ell(\tau,r)\in dy, a<\inf_{0\le s\le\tau} R_s^{(3)}, R_\tau^{(3)}\in dz\big) = \exp\Big(-\dfrac{y\sqrt{2\lambda}}{1-e^{-2(r-a)\sqrt{2\lambda}}}\Big)$
$0<y$	$\times\dfrac{z\lambda(e^{-\lvert x-r\rvert\sqrt{2\lambda}}-e^{-(x+r-2a)\sqrt{2\lambda}})(e^{-\lvert z-r\rvert\sqrt{2\lambda}}-e^{-(z+r-2a)\sqrt{2\lambda}})}{x\big(1-e^{-2(r-a)\sqrt{2\lambda}}\big)}dydz$
(1)	$\mathbf{P}_x\big(\ell(\tau,r)=0, a<\inf_{0\le s\le\tau} R_s^{(3)}, R_\tau^{(3)}\in dz\big) = \dfrac{z\sqrt{\lambda}}{x\sqrt{2}}\big\{e^{-\lvert z-x\rvert\sqrt{2\lambda}}-e^{-(z+x-2a)\sqrt{2\lambda}}$ $-\dfrac{z\big(e^{-\lvert x-r\rvert\sqrt{2\lambda}}-e^{-(x+r-2a)\sqrt{2\lambda}}\big)}{x\big(1-e^{-2(r-a)\sqrt{2\lambda}}\big)}\big(e^{-\lvert z-r\rvert\sqrt{2\lambda}}-e^{-(z+r-2a)\sqrt{2\lambda}}\big)\big\}dz$
1.17.8 $a<r$	$\mathbf{P}_x\big(\ell(t,r)\in dy, a<\inf_{0\le s\le t} R_s^{(3)}, R_t^{(3)}\in dz\big) = \dfrac{z}{x}F_{a,x}^{(17)}(t,y,z)dydz$
$a<z$	for $F_{a,x}^{(17)}(t,y,z)$ see 1.1.17.8

5 $R_s^{(3)} = \left(\sum_{k=1}^{3}(W_s^{(k)})^2\right)^{1/2}$ $\tau \sim \mathrm{Exp}(\lambda)$, independent of $R^{(3)}$

1.18.1	$\mathbf{E}_{x/\sqrt{2\lambda}}\exp\left(-\gamma\sqrt{2\lambda}\,\ell(\tau,r/\sqrt{2\lambda})-\eta\sqrt{2\lambda}\,\ell(\tau,u/\sqrt{2\lambda})\right)$
$r<u$	$=1-\dfrac{\gamma(r+\eta r(1-e^{-2u})-\eta u(e^{r-u}-e^{-r-u}))(e^{-\lvert x-r\rvert}-e^{-x-r})}{x(1+\gamma(1-e^{-2r})+\eta(1-e^{-2u})+\gamma\eta(1-e^{-2r})(1-e^{2(r-u)}))}$
	$-\dfrac{\eta(u+\gamma u(1-e^{-2r})-\gamma r(e^{r-u}-e^{-r-u}))(e^{-\lvert u-x\rvert}-e^{-u-x})}{x(1+\gamma(1-e^{-2r})+\eta(1-e^{-2u})+\gamma\eta(1-e^{-2r})(1-e^{2(r-u)}))}$
1.18.3 (1)	$\mathbf{E}_x\exp\left(-\gamma\ell(\infty,r)-\eta\ell(\infty,u)\right)$
$r<u$	$=1-\dfrac{r\gamma(x+r-\lvert x-r\rvert)+\eta(u+2\gamma r(u-r))(x+u-\lvert x-u\rvert)}{x(4\gamma\eta r(u-r)+2r\gamma+2u\eta+1)}$
1.18.4 (1)	$\mathbf{P}_x\left(\ell(\infty,r)\in dy,\ \ell(\infty,u)\in dg\right)$
$x\le r$ $r<u$	$=\dfrac{1}{4r(u-r)}\exp\left(-\dfrac{uy+rg}{2r(u-r)}\right)I_0\left(\dfrac{\sqrt{yg}}{u-r}\right)dydg$
$r<x$ $x<u$	$=\dfrac{1}{4x(u-r)^2}\exp\left(-\dfrac{uy+rg}{2r(u-r)}\right)\left[(u-x)I_0\left(\dfrac{\sqrt{yg}}{u-r}\right)+(x-r)\dfrac{\sqrt{g}}{\sqrt{y}}I_1\left(\dfrac{\sqrt{yg}}{u-r}\right)\right]dydg$
$r<u$ $u\le x$	$=\dfrac{\sqrt{g}}{4x(u-r)\sqrt{y}}\exp\left(-\dfrac{uy+rg}{2r(u-r)}\right)I_1\left(\dfrac{\sqrt{yg}}{u-r}\right)dydg$
(2)	$\mathbf{P}_x\left(\ell(\infty,r)=0,\ \ell(\infty,u)\in dg\right)=\begin{cases}\dfrac{x-r}{2x(u-r)}\exp\left(-\dfrac{g}{2(u-r)}\right)dg, & r\le x\le u\\ \dfrac{1}{2x}\exp\left(-\dfrac{g}{2(u-r)}\right)dg, & u\le x\end{cases}$
(3)	$\mathbf{P}_x\left(\ell(\infty,r)=0,\ \ell(\infty,u)=0\right)=1-\dfrac{u}{x},\qquad u\le x$
1.19.2 $0<x$	$\mathbf{P}_x\left(\int_0^\tau\dfrac{ds}{R_s^{(3)}}\in dy\right)$
	$=\dfrac{\operatorname{sh}^2(y\sqrt{\lambda/2})}{2x\operatorname{ch}^2(y\sqrt{\lambda/2})}\exp\left(-\dfrac{x\sqrt{2\lambda}\operatorname{ch}(y\sqrt{2\lambda})}{\operatorname{sh}(y\sqrt{2\lambda})}\right)M_{-2,1/2}\left(\dfrac{2\sqrt{2\lambda}x}{\operatorname{sh}(y\sqrt{2\lambda})}\right)dy$

5 $R_s^{(3)} = \left(\sum_{k=1}^3 (W_s^{(k)})^2\right)^{1/2}$ $\tau \sim \mathrm{Exp}(\lambda)$, independent of $R^{(3)}$

1.19.5, $0 < x$

$$\mathbf{E}_x\Big\{\exp\Big(-\gamma\int_0^\tau \frac{ds}{R_s^{(3)}}\Big); R_\tau^{(3)} \in dz\Big\}$$

$x \le z$

$$= \frac{\sqrt{\lambda}\Gamma(1+\gamma/\sqrt{2\lambda})z}{\sqrt{2}x} M_{-\gamma/\sqrt{2\lambda},1/2}(2x\sqrt{2\lambda}) W_{-\gamma/\sqrt{2\lambda},1/2}(2z\sqrt{2\lambda})dz$$

$z \le x$

$$= \frac{\sqrt{\lambda}\Gamma(1+\gamma/\sqrt{2\lambda})z}{\sqrt{2}x} W_{-\gamma/\sqrt{2\lambda},1/2}(2x\sqrt{2\lambda}) M_{-\gamma/\sqrt{2\lambda},1/2}(2z\sqrt{2\lambda})dz$$

1.19.6, $0 < x$

$$\mathbf{P}_x\Big(\int_0^\tau \frac{ds}{R_s^{(3)}} \in dy,\ R_\tau^{(3)} \in dz\Big)$$

$$= \frac{\lambda\sqrt{2\lambda}z^{3/2}}{x^{1/2}\,\mathrm{sh}(y\sqrt{\lambda/2})} \exp\Big(-\frac{(x+z)\sqrt{2\lambda}\,\mathrm{ch}(y\sqrt{\lambda/2})}{\mathrm{sh}(y\sqrt{\lambda/2})}\Big) I_1\Big(\frac{2\sqrt{2\lambda xz}}{\mathrm{sh}(y\sqrt{\lambda/2})}\Big)dydz$$

1.19.8, $0 < x$

$$\mathbf{P}_x\Big(\int_0^t \frac{ds}{R_s^{(3)}} \in dy,\ R_t^{(3)} \in dz\Big) = \frac{z^{3/2}}{x^{1/2}}\,\mathrm{is}_t(1, y/2, 0, x+z, \sqrt{xz})dydz$$

1.20.3, $0 < x$

$$\mathbf{E}_x \exp\Big(-\frac{\gamma^2}{2}\int_0^t \frac{ds}{(R_s^{(3)})^2}\Big) = \frac{(2t)^{3/4}\Gamma(5/4+\sqrt{1+4\gamma^2}/4)}{x^{3/2}\Gamma(\sqrt{\gamma^2+1/4}+1)} e^{-x^2/4t} M_{-3/4,\sqrt{4\gamma^2+1}/4}\Big(\frac{x^2}{2t}\Big)$$

1.20.4, $0 < x$

$$\mathbf{P}_x\Big(\int_0^t \frac{ds}{(R_s^{(3)})^2} \in dy\Big) = \frac{2(2t)^{3/4}}{x^{3/2}} e^{-y/8-x^2/4t}\ \mathrm{m}_{2y}\Big(\frac{3}{4}, \frac{x^2}{4t}\Big)dy$$

1.20.5

$$\mathbf{E}_x\Big\{\exp\Big(-\frac{\gamma^2}{2}\int_0^\tau \frac{ds}{(R_s^{(3)})^2}\Big);\ R_\tau^{(3)} \in dz\Big\}$$

$$= \begin{cases} \dfrac{2\lambda z\sqrt{z}}{\sqrt{x}} I_{\sqrt{\gamma^2+1/4}}(x\sqrt{2\lambda}) K_{\sqrt{\gamma^2+1/4}}(z\sqrt{2\lambda})dz, & 0 < x \le z \\ \dfrac{2\lambda z\sqrt{z}}{\sqrt{x}} K_{\sqrt{\gamma^2+1/4}}(x\sqrt{2\lambda}) I_{\sqrt{\gamma^2+1/4}}(z\sqrt{2\lambda})dz, & z \le x \end{cases}$$

1.20.6, $0 < x$

$$\mathbf{P}_x\Big(\int_0^\tau \frac{ds}{(R_s^{(3)})^2} \in dy,\ R_\tau^{(3)} \in dz\Big) = \frac{\lambda z^{3/2}}{x^{1/2}} e^{-y/8}\,\mathrm{ki}_{y/2}(x\sqrt{2\lambda}, z\sqrt{2\lambda})dydz$$

1.20.7, $0 < x$

$$\mathbf{E}_x\Big\{\exp\Big(-\frac{\gamma^2}{2}\int_0^t \frac{ds}{(R_s^{(3)})^2}\Big);\ R_t^{(3)} \in dz\Big\} = \frac{z\sqrt{z}}{t\sqrt{x}} e^{-(x^2+z^2)/2t} I_{\sqrt{\gamma^2+1/4}}\Big(\frac{xz}{t}\Big)dz$$

5 $R_s^{(3)} = \left(\sum_{k=1}^3 (W_s^{(k)})^2\right)^{1/2}$ $\qquad \tau \sim \operatorname{Exp}(\lambda)$, independent of $R^{(3)}$

1.20.8
$0<x$

$$\mathbf{P}_x\Big(\int_0^t \frac{ds}{(R_s^{(3)})^2} \in dy,\ R_t^{(3)} \in dz\Big) = \frac{z^{3/2}}{2tx^{1/2}} e^{-(x^2+z^2)/2t - y/8}\ \mathrm{i}_{y/2}\Big(\frac{xz}{t}\Big)dydz$$

1.21.3
$0<x$

$$\mathbf{E}_x \exp\Big(-\int_0^t \Big(\frac{p^2}{2(R_s^{(3)})^2} + \frac{q^2 (R_s^{(3)})^2}{2}\Big)ds\Big)$$

$$= \frac{(2\operatorname{sh}(tq))^{3/4}\Gamma(5/4+\sqrt{4p^2+1}/4)}{x^{3/2}(q\operatorname{ch}(tq))^{3/4}\Gamma(1+\sqrt{p^2+1/4})} \exp\Big(-\frac{x^2 q \operatorname{ch}(2tq)}{2\operatorname{sh}(2tq)}\Big) M_{-\frac{3}{4},\sqrt{4p^2+1}/4}\Big(\frac{x^2 q}{\operatorname{sh}(2tq)}\Big)$$

1.21.3
(1)

$$\mathbf{E}_x\Big\{\exp\Big(-\frac{q^2}{2}\int_0^t (R_s^{(3)})^2 ds\Big);\ \int_0^t \frac{ds}{(R_s^{(3)})^2} \in dy\Big\}$$

$0<x$

$$= \frac{2(2\operatorname{sh}(tq))^{3/4}}{x^{3/2}(q\operatorname{ch}(tq))^{3/4}} \exp\Big(-\frac{y}{8} - \frac{x^2 q \operatorname{ch}(2tq)}{2\operatorname{sh}(2tq)}\Big)\ \mathrm{m}_{2y}\Big(\frac{3}{4}, \frac{x^2 q}{2\operatorname{sh}(2tq)}\Big)dy$$

1.21.5
$0<x$

$$\mathbf{E}_x\Big\{\exp\Big(-\int_0^\tau \Big(\frac{p^2}{2(R_s^{(3)})^2} + \frac{q^2 (R_s^{(3)})^2}{2}\Big)ds\Big);\ R_\tau^{(3)} \in dz\Big\} = \frac{\lambda z}{x} Q_\lambda^{(21)}(z)dz$$

for $Q_\lambda^{(21)}(z)$ see 1.1.21.5

1.21.7
$0<x$

$$\mathbf{E}_x\Big\{\exp\Big(-\int_0^t \Big(\frac{p^2}{2(R_s^{(3)})^2} + \frac{q^2 (R_s^{(3)})^2}{2}\Big)ds\Big);\ R_t^{(3)} \in dz\Big\}$$

$$= \frac{qz^{3/2}}{x^{1/2}\operatorname{sh}(tq)} \exp\Big(-\frac{(x^2+z^2) q \operatorname{ch}(tq)}{2\operatorname{sh}(tq)}\Big) I_{\sqrt{p^2+1/4}}\Big(\frac{xzq}{\operatorname{sh}(tq)}\Big)dz$$

(1)

$$\mathbf{E}_x\Big\{\exp\Big(-\frac{q^2}{2}\int_0^t (R_s^{(3)})^2 ds\Big);\ \int_0^t \frac{ds}{(R_s^{(3)})^2} \in dy,\ R_t^{(3)} \in dz\Big\}$$

$$= \frac{qz^{3/2}}{2x^{1/2}\operatorname{sh}(tq)} \exp\Big(-\frac{y}{8} - \frac{(x^2+z^2) q \operatorname{ch}(tq)}{2\operatorname{sh}(tq)}\Big)\ \mathrm{i}_{y/2}\Big(\frac{xzq}{\operatorname{sh}(tq)}\Big)dydz$$

(2)

$$\mathbf{E}_x\Big\{\exp\Big(-\frac{p^2}{2}\int_0^t \frac{ds}{(R_s^{(3)})^2}\Big);\ \int_0^t (R_s^{(3)})^2 ds \in dy,\ R_t^{(3)} \in dz\Big\}$$

$$= \frac{z^{3/2}}{x^{1/2}}\ \mathrm{is}_y(\sqrt{p^2+1/4}, t, 0, (x^2+z^2)/2, xz/2)dydz$$

5 $R_s^{(3)} = \big(\sum_{k=1}^{3}(W_s^{(k)})^2\big)^{1/2}$ $\tau \sim \mathrm{Exp}(\lambda)$, independent of $R^{(3)}$

1.21.8
$0<x$

$$\mathbf{P}_x\Big(\int_0^t \big(R_s^{(3)}\big)^2 ds \in dg,\ \int_0^t \frac{ds}{\big(R_s^{(3)}\big)^2} \in dy,\ R_t^{(3)} \in dz\Big)$$

$$= \frac{z^{3/2}}{2x^{1/2}} e^{-y/8}\,\mathrm{ei}_g(y/2, t, (x^2+z^2)/2, xz)\,dg\,dy\,dz$$

1.22.1
(1)
$0<x$

$$\mathbf{E}_x\Big\{\exp\Big(-\frac{p^2}{2}\int_0^\tau \frac{ds}{\big(R_s^{(3)}\big)^2}\Big);\ \int_0^\tau \frac{ds}{R_s^{(3)}} \in dy\Big\} = \frac{\mathrm{sh}^2(y\sqrt{\lambda/2})}{2x\,\mathrm{ch}^2(y\sqrt{\lambda/2})}$$

$$\times \frac{\Gamma(5/2+\sqrt{p^2+1/4})}{\Gamma(1+\sqrt{4p^2+1})} \exp\Big(-\frac{x\sqrt{2\lambda}\,\mathrm{ch}(y\sqrt{2\lambda})}{\mathrm{sh}(y\sqrt{2\lambda})}\Big) M_{-2,\sqrt{p^2+1/4}}\Big(\frac{2x\sqrt{2\lambda}}{\mathrm{sh}(y\sqrt{2\lambda})}\Big)dy$$

1.22.2
$0<x$

$$\mathbf{P}_x\Big(\int_0^\tau \frac{ds}{\big(R_s^{(3)}\big)^2} \in dg,\ \int_0^\tau \frac{ds}{R_s^{(3)}} \in dy\Big)$$

$$= \frac{\mathrm{sh}^2(y\sqrt{\lambda/2})}{4x\,\mathrm{ch}^2(y\sqrt{\lambda/2})} \exp\Big(-\frac{g}{8} - \frac{x\sqrt{2\lambda}\,\mathrm{ch}(y\sqrt{2\lambda})}{\mathrm{sh}(y\sqrt{2\lambda})}\Big)\,\mathrm{m}_{g/2}\Big(2, \frac{x\sqrt{2\lambda}}{\mathrm{sh}(y\sqrt{2\lambda})}\Big)dg\,dy$$

1.22.5
$0<x$

$$\mathbf{E}_x\Big\{\exp\Big(-\int_0^\tau \big(\frac{p^2}{2\big(R_s^{(3)}\big)^2} + \frac{q}{R_s^{(3)}}\big)ds\Big); R_\tau^{(3)} \in dz\Big\} = \frac{\lambda z}{x} Q_\lambda^{(22)}(z)dz$$

for $Q_\lambda^{(22)}(z)$ see 1.1.22.5

(1)

$$\mathbf{E}_x\Big\{\exp\Big(-\frac{p^2}{2}\int_0^\tau \frac{ds}{\big(R_s^{(3)}\big)^2}\Big);\ \int_0^\tau \frac{ds}{R_s^{(3)}} \in dy,\ R_\tau^{(3)} \in dz\Big\}$$

$$= \frac{\lambda\sqrt{2\lambda}z^{3/2}}{x^{1/2}\,\mathrm{sh}(y\sqrt{\lambda/2})} \exp\Big(-\frac{(x+z)\sqrt{2\lambda}\,\mathrm{ch}(y\sqrt{\lambda/2})}{\mathrm{sh}(y\sqrt{\lambda/2})}\Big) I_{\sqrt{4p^2+1}}\Big(\frac{2\sqrt{2\lambda xz}}{\mathrm{sh}(y\sqrt{\lambda/2})}\Big)dy\,dz$$

1.22.6
$0<x$

$$\mathbf{P}_x\Big(\int_0^\tau \frac{ds}{\big(R_s^{(3)}\big)^2} \in dg,\ \int_0^\tau \frac{ds}{R_s^{(3)}} \in dy,\ R_\tau^{(3)} \in dz\Big)$$

$$= \frac{\lambda\sqrt{2\lambda}z^{3/2}}{8x^{1/2}\,\mathrm{sh}(y\sqrt{\lambda/2})} \exp\Big(-\frac{g}{8} - \frac{(x+z)\sqrt{2\lambda}\,\mathrm{ch}(y\sqrt{\lambda/2})}{\mathrm{sh}(y\sqrt{\lambda/2})}\Big)\,\mathrm{i}_{g/8}\Big(\frac{2\sqrt{2\lambda xz}}{\mathrm{sh}(y\sqrt{\lambda/2})}\Big)dg\,dy\,dz$$

1.22.7
(1)
$0<x$

$$\mathbf{E}_x\Big\{\exp\Big(-\frac{p^2}{2}\int_0^t \frac{ds}{\big(R_s^{(3)}\big)^2}\Big);\ \int_0^t \frac{ds}{R_s^{(3)}} \in dy,\ R_t^{(3)} \in dz\Big\}$$

$$= \frac{z^{3/2}}{x^{1/2}}\,\mathrm{is}_t(\sqrt{4p^2+1}, y/2, 0, x+z, \sqrt{xz})\,dy\,dz$$

5 $R_s^{(3)} = \left(\sum_{k=1}^{3}(W_s^{(k)})^2\right)^{1/2}$ $\qquad\tau \sim \mathrm{Exp}(\lambda)$, independent of $R^{(3)}$

1.22.8 $0<x$

$$\mathbf{P}_x\Big(\int_0^t \frac{ds}{(R_s^{(3)})^2}\in dg,\ \int_0^t \frac{ds}{R_s^{(3)}}\in dy,\ R_t^{(3)}\in dz\Big)$$

$$=\frac{z^{3/2}}{8x^{1/2}}e^{-g/8}\,\mathrm{ei}_t(g/8,y/2,x+z,2\sqrt{xz})\,dg\,dy\,dz$$

1.23.1 $r<b$

$$\mathbf{E}_x\Big\{\exp\Big(-\int_0^\tau\big(p\mathbb{I}_{[0,r)}(R_s^{(3)})+q\mathbb{I}_{[r,\infty)}(R_s^{(3)})\big)ds\Big);\ \sup_{0\le s\le\tau}R_s^{(3)}<b\Big\}$$

$0\le x$, $x\le r$

$$=\frac{2\lambda}{\Upsilon_p^2}-\frac{4\lambda((q-p)(r\Upsilon_q\operatorname{ch}((b-r)\Upsilon_q)+\operatorname{sh}((b-r)\Upsilon_q))+\lambda b\Upsilon_q)\operatorname{sh}(x\Upsilon_p)}{x\Upsilon_p^2\Upsilon_q^2(\Upsilon_p\operatorname{sh}((b-r)\Upsilon_q)\operatorname{ch}(r\Upsilon_p)+\Upsilon_q\operatorname{ch}((b-r)\Upsilon_q)\operatorname{sh}(r\Upsilon_p))}$$

$r\le x$, $x\le b$

$$=\frac{2\lambda}{\Upsilon_q^2}\Big(1-\frac{b\operatorname{sh}((x-r)\Upsilon_q)}{x\operatorname{sh}((b-r)\Upsilon_q)}\Big)$$

$$+\frac{4\lambda((q-p)(r\Upsilon_p\operatorname{cth}(r\Upsilon_p)-1)\operatorname{sh}((b-r)\Upsilon_q)-\lambda b\Upsilon_p)\operatorname{sh}((b-x)\Upsilon_q)}{x\Upsilon_p^2\Upsilon_q^2(\Upsilon_p\operatorname{cth}(r\Upsilon_p)+\Upsilon_q\operatorname{cth}((b-r)\Upsilon_q))\operatorname{sh}^2((b-r)\Upsilon_q)}$$

1.23.5 $r<b$

$$\mathbf{E}_x\Big\{\exp\Big(-\int_0^\tau\big(p\mathbb{I}_{[0,r)}(R_s^{(3)})+q\mathbb{I}_{[r,\infty)}(R_s^{(3)})\big)ds\Big);\ \sup_{0\le s\le\tau}R_s^{(3)}<b,\ R_\tau^{(3)}\in dz\Big\}$$

$z<b$

$$=\frac{z\lambda}{x}Q_{0,\lambda}^{(26)}(z)dz,\qquad \text{for } Q_{a,\lambda}^{(26)}(z) \text{ see 1.1.26.5}$$

1.24.1 $a<r$

$$\mathbf{E}_x\Big\{\exp\Big(-\int_0^\tau\big(p\mathbb{I}_{[0,r)}(R_s^{(3)})+q\mathbb{I}_{[r,\infty)}(R_s^{(3)})\big)ds\Big);a<\inf_{0\le s\le\tau}R_s^{(3)}\Big\}$$

$a\le x$, $x\le r$

$$=\frac{2\lambda}{\Upsilon_p^2}\Big(1-\frac{a}{x}e^{(a-x)\Upsilon_p}\Big)-\frac{2\lambda(2(q-p)(1+r\Upsilon_q)+a\Upsilon_q^2(\Upsilon_p-\Upsilon_q)e^{(a-r)\Upsilon_p})\operatorname{sh}((x-a)\Upsilon_p)}{x\Upsilon_p\Upsilon_q(\Upsilon_q\operatorname{sh}((r-a)\Upsilon_p)+\Upsilon_p\operatorname{ch}((r-a)\Upsilon_p))}$$

$r\le x$

$$=\frac{2\lambda}{\Upsilon_q^2}-\frac{2\lambda(2(q-p)(r\Upsilon_p\operatorname{ch}((r-a)\Upsilon_p)-\operatorname{sh}((r-a)\Upsilon_p))-a\Upsilon_q^2\Upsilon_p)e^{(r-x)\Upsilon_q}}{x\Upsilon_p\Upsilon_q(\Upsilon_q\operatorname{sh}((r-a)\Upsilon_p)+\Upsilon_p\operatorname{ch}((r-a)\Upsilon_p))}$$

1.24.3 (1)

$$\mathbf{E}_x\Big\{\exp\Big(-p\int_0^\infty\mathbb{I}_{[0,r)}(R_s^{(3)})ds\Big);\ a<\inf_{0\le s<\infty}R_s^{(3)}\Big\}$$

$$=\begin{cases}\dfrac{\operatorname{sh}((x-a)\sqrt{2p})}{x\sqrt{2p}\operatorname{ch}((r-a)\sqrt{2p})}, & a\le x\le r\\[2ex] 1-\dfrac{r}{x}+\dfrac{\operatorname{sh}((r-a)\sqrt{2p})}{x\sqrt{2p}\operatorname{ch}((r-a)\sqrt{2p})}, & r\le x\end{cases}$$

$$\Upsilon_s:=\sqrt{2\lambda+2s}$$

5 $R_s^{(3)} = \big(\sum_{k=1}^3 (W_s^{(k)})^2\big)^{1/2}$ $\tau \sim \operatorname{Exp}(\lambda)$, independent of $R^{(3)}$

1.24.4 (1)

$$\mathbf{P}_x\Big(\int_0^\infty \mathbb{I}_{[0,r)}\big(R_s^{(3)}\big)ds \in dy,\ a < \inf_{0\le s<\infty} R_s^{(3)}\Big)$$

$$= \begin{cases} \dfrac{1}{x}\operatorname{sc}_y(x-a, r-a)dy, & a \le x \le r \\ \dfrac{1}{x}\operatorname{sc}_y(r-a, r-a)dy, & r \le x \end{cases}$$

1.24.5 $a < r$, $a < z$

$$\mathbf{E}_x\Big\{\exp\Big(-\int_0^\tau \big(p\mathbb{I}_{[0,r)}\big(R_s^{(3)}\big) + q\mathbb{I}_{[r,\infty)}\big(R_s^{(3)}\big)\big)ds\Big); a < \inf_{0\le s\le\tau} R_s^{(3)},\ R_\tau^{(3)} \in dz\Big\}$$

$$= \frac{z\lambda}{x} Q_{a,\lambda}^{(24)}(z)dz, \qquad \text{for } Q_{a,\lambda}^{(24)}(z) \text{ see 1.1.24.5}$$

1.25.1 $r < b$

$$\mathbf{E}_x\big\{e^{-\gamma\ell(\tau,r)}; a < \inf_{0\le s\le\tau} R_s^{(3)},\ \sup_{0\le s\le\tau} R_s^{(3)} < b\big\}$$

$a \le x$, $x \le r$

$$= 1 - \frac{a\operatorname{sh}((r-x)\sqrt{2\lambda}) + r\operatorname{sh}((x-a)\sqrt{2\lambda})}{x\operatorname{sh}((r-a)\sqrt{2\lambda})}$$

$$+ \frac{\sqrt{\lambda/2}(r\operatorname{sh}((b-a)\sqrt{2\lambda}) - b\operatorname{sh}((r-a)\sqrt{2\lambda}) - a\operatorname{sh}((b-r)\sqrt{2\lambda}))\operatorname{sh}((x-a)\sqrt{2\lambda})}{x(\gamma\operatorname{sh}((b-r)\sqrt{2\lambda})\operatorname{sh}((r-a)\sqrt{2\lambda}) + \sqrt{\lambda/2}\operatorname{sh}((b-a)\sqrt{2\lambda}))\operatorname{sh}((r-a)\sqrt{2\lambda})}$$

$r \le x$, $x \le b$

$$= 1 - \frac{r\operatorname{sh}((b-x)\sqrt{2\lambda}) + b\operatorname{sh}((x-r)\sqrt{2\lambda})}{x\operatorname{sh}((b-r)\sqrt{2\lambda})}$$

$$+ \frac{\sqrt{\lambda/2}(r\operatorname{sh}((b-a)\sqrt{2\lambda}) - b\operatorname{sh}((r-a)\sqrt{2\lambda}) - a\operatorname{sh}((b-r)\sqrt{2\lambda}))\operatorname{sh}((b-x)\sqrt{2\lambda})}{x(\gamma\operatorname{sh}((b-r)\sqrt{2\lambda})\operatorname{sh}((r-a)\sqrt{2\lambda}) + \sqrt{\lambda/2}\operatorname{sh}((b-a)\sqrt{2\lambda}))\operatorname{sh}((b-r)\sqrt{2\lambda})}$$

1.25.2 $r < b$

$$\mathbf{P}_x\big(\ell(\tau,r) \in dy, a < \inf_{0\le s\le\tau} R_s^{(3)},\ \sup_{0\le s\le\tau} R_s^{(3)} < b\big)$$

$$= \exp\Big(-\frac{y\sqrt{\lambda}\operatorname{sh}((b-a)\sqrt{2\lambda})}{\sqrt{2}\operatorname{sh}((b-r)\sqrt{2\lambda})\operatorname{sh}((r-a)\sqrt{2\lambda})}\Big)$$

$$\times \begin{cases} \dfrac{\sqrt{\lambda}(r\operatorname{sh}((b-a)\sqrt{2\lambda}) - b\operatorname{sh}((r-a)\sqrt{2\lambda}) - a\operatorname{sh}((b-r)\sqrt{2\lambda}))\operatorname{sh}((x-a)\sqrt{2\lambda})}{x\sqrt{2}\operatorname{sh}((b-r)\sqrt{2\lambda})\operatorname{sh}^2((r-a)\sqrt{2\lambda})}dy, \\ \hfill a \le x \le r \\ \dfrac{\sqrt{\lambda}(r\operatorname{sh}((b-a)\sqrt{2\lambda}) - b\operatorname{sh}((r-a)\sqrt{2\lambda}) - a\operatorname{sh}((b-r)\sqrt{2\lambda}))\operatorname{sh}((b-x)\sqrt{2\lambda})}{x\sqrt{2}\operatorname{sh}^2((b-r)\sqrt{2\lambda})\operatorname{sh}((r-a)\sqrt{2\lambda})}dy, \\ \hfill r \le x \le b \end{cases}$$

5 $R_s^{(3)} = \left(\sum_{k=1}^3 (W_s^{(k)})^2\right)^{1/2}$ $\tau \sim \operatorname{Exp}(\lambda)$, independent of $R^{(3)}$

(1) $\mathbf{P}_x\big(\ell(\tau,r)=0, a<\inf_{0\le s\le\tau} R_s^{(3)}, \sup_{0\le s\le\tau} R_s^{(3)}<b\big)$

$$= \begin{cases} 1-\dfrac{a\operatorname{sh}((r-x)\sqrt{2\lambda})+r\operatorname{sh}((x-a)\sqrt{2\lambda})}{x\operatorname{sh}((r-a)\sqrt{2\lambda})}, & a\le x\le r \\ 1-\dfrac{r\operatorname{sh}((b-x)\sqrt{2\lambda})+b\operatorname{sh}((x-r)\sqrt{2\lambda})}{x\operatorname{sh}((b-r)\sqrt{2\lambda})}, & r\le x\le b \end{cases}$$

1.25.5 $\mathbf{E}_x\big\{e^{-\gamma\ell(\tau,r)}; a<\inf_{0\le s\le\tau} R_s^{(3)}, \sup_{0\le s\le\tau} R_s^{(3)}<b, R_\tau^{(3)}\in dz\big\} = \frac{z\lambda}{x} Q_{a,\lambda}^{(25)}(z)dz$

$r<b$

$z<b$ for $Q_{a,\lambda}^{(25)}(z)$ see 1.1.25.5

1.25.6 $\mathbf{P}_x\big(\ell(\tau,r)\in dy, a<\inf_{0\le s\le\tau} R_s^{(3)}, \sup_{0\le s\le\tau} R_s^{(3)}<b, R_\tau^{(3)}\in dz\big)$

$r<b$

$z<b$ $= \frac{z\lambda}{x} B_{a,\lambda}^{(25)}(y,z)dydz,$ for $B_{a,\lambda}^{(25)}(y,z)$ see 1.1.25.6

(1) $\mathbf{P}_x\big(\ell(\tau,r)=0, a<\inf_{0\le s\le\tau} R_s^{(3)}, \sup_{0\le s\le\tau} R_s^{(3)}<b, R_\tau^{(3)}\in dz\big)$

$z\le r$, $x\le r$:
$$= \frac{z\sqrt{\lambda}\big(\operatorname{ch}((r-a-|z-x|)\sqrt{2\lambda})-\operatorname{ch}((r+a-x-z)\sqrt{2\lambda})\big)}{x\sqrt{2}\operatorname{sh}((r-a)\sqrt{2\lambda})}dz$$

$r\le z$, $r\le x$:
$$= \frac{z\sqrt{\lambda}\big(\operatorname{ch}((b-r-|z-x|)\sqrt{2\lambda})-\operatorname{ch}((b+r-z-x)\sqrt{2\lambda})\big)}{x\sqrt{2}\operatorname{sh}((b-r)\sqrt{2\lambda})}dz$$

1.25.8 $\mathbf{P}_x\big(\ell(t,r)\in dy, a<\inf_{0\le s\le t} R_s^{(3)}, \sup_{0\le s\le t} R_s^{(3)}<b, R_t^{(3)}\in dz\big)$

$r<b$

$z<b$ $= \frac{z}{x} F_{a,x}^{(25)}(t,y,z)dydz,$ for $F_{a,x}^{(25)}(t,y,z)$ see 1.1.25.8

1.26.1 $\mathbf{E}_x\Big\{\exp\Big(-\int_0^\tau \big(p\mathbb{I}_{[0,r)}(R_s^{(3)})+q\mathbb{I}_{[r,\infty)}(R_s^{(3)})\big)ds\Big);$

$a<r$

$r<b$ $a<\inf_{0\le s\le\tau} R_s^{(3)}, \sup_{0\le s\le\tau} R_s^{(3)}<b\Big\}$

5 $R_s^{(3)}=\left(\sum_{k=1}^3 (W_s^{(k)})^2\right)^{1/2}$ $\qquad \tau \sim \mathrm{Exp}(\lambda)$, independent of $R^{(3)}$

$a\le x$, $x\le r$

$$=\frac{\lambda}{\lambda+p}\Big(1-\frac{a\,\mathrm{sh}((r-x)\Upsilon_p)}{x\,\mathrm{sh}((r-a)\Upsilon_p)}\Big)$$

$$+\frac{\lambda(p-q)(r\Upsilon_q\,\mathrm{cth}((b-r)\Upsilon_q)+1)\,\mathrm{sh}((x-a)\Upsilon_p)}{x(\lambda+p)(\lambda+q)(\Upsilon_p\,\mathrm{cth}((r-a)\Upsilon_p)+\Upsilon_q\,\mathrm{cth}((b-r)\Upsilon_q))\,\mathrm{sh}((r-a)\Upsilon_p)}$$

$$-\frac{\lambda((\lambda+p)\Upsilon_q b\,\mathrm{sh}((r-a)\Upsilon_p)+(\lambda+q)\Upsilon_p a\,\mathrm{sh}((b-r)\Upsilon_q))\,\mathrm{sh}((x-a)\Upsilon_p)}{x(\lambda+p)(\lambda+q)(\Upsilon_p\,\mathrm{cth}((r-a)\Upsilon_p)+\Upsilon_q\,\mathrm{cth}((b-r)\Upsilon_q))\,\mathrm{sh}((b-r)\Upsilon_q)\,\mathrm{sh}^2((r-a)\Upsilon_p)}$$

$r\le x$, $x\le b$

$$=\frac{\lambda}{\lambda+q}\Big(1-\frac{b\,\mathrm{sh}((x-r)\Upsilon_q)}{x\,\mathrm{sh}((b-r)\Upsilon_q)}\Big)$$

$$+\frac{\lambda(q-p)b(\Upsilon_p\,\mathrm{cth}((r-a)\Upsilon_p)-1)\,\mathrm{sh}((b-x)\Upsilon_q)}{x(\lambda+p)(\lambda+q)(\Upsilon_p\,\mathrm{cth}((r-a)\Upsilon_p)+\Upsilon_q\,\mathrm{cth}((b-r)\Upsilon_q))\,\mathrm{sh}((b-r)\Upsilon_q)}$$

$$-\frac{\lambda((\lambda+p)\Upsilon_q b\,\mathrm{sh}((r-a)\Upsilon_p)+(\lambda+q)\Upsilon_p a\,\mathrm{sh}((b-r)\Upsilon_q))\,\mathrm{sh}((b-x)\Upsilon_q)}{x(\lambda+p)(\lambda+q)(\Upsilon_p\,\mathrm{cth}((r-a)\Upsilon_p)+\Upsilon_q\,\mathrm{cth}((b-r)\Upsilon_q))\,\mathrm{sh}^2((b-r)\Upsilon_q)\,\mathrm{sh}((r-a)\Upsilon_p)}$$

1.26.5 $a<r$, $r<b$

$$\mathbf{E}_x\Big\{\exp\Big(-\int_0^\tau \big(p\mathbb{I}_{[0,r)}(R_s^{(3)})+q\mathbb{I}_{[r,\infty)}(R_s^{(3)})\big)ds\Big);a<\inf_{0\le s\le\tau}R_s^{(3)},\ \sup_{0\le s\le\tau}R_s^{(3)}<b,\ R_\tau^{(3)}\in dz\Big\}=\frac{z\lambda}{x}Q_{a,\lambda}^{(26)}(z)dz$$

for $Q_{a,\lambda}^{(26)}(z)$ see 1.1.26.5

1.27.1

$$\mathbf{E}_x\Big\{\exp\Big(-\int_0^\tau \big(p\mathbb{I}_{[0,r)}(R_s^{(3)})+q\mathbb{I}_{[r,\infty)}(R_s^{(3)})\big)ds\Big);\tfrac12\ell(\tau,r)\in dy\Big\}$$

$x\le r$

$$=\frac{2\lambda}{x}e^{-y\Upsilon_q}\Big(\frac{r\,\mathrm{ch}(r\Upsilon_p)}{\Upsilon_p\,\mathrm{sh}(r\Upsilon_p)}-\frac{1}{\Upsilon_p^2}+\frac{r}{\Upsilon_q}+\frac{1}{\Upsilon_q^2}\Big)\exp\Big(-\frac{y\Upsilon_p\,\mathrm{ch}(r\Upsilon_p)}{\mathrm{sh}(r\Upsilon_p)}\Big)\frac{\mathrm{sh}(x\Upsilon_p)}{\mathrm{sh}(r\Upsilon_p)}dy$$

$r\le x$

$$=\frac{2\lambda}{x}e^{-(y+x-r)\Upsilon_q}\Big(\frac{r\,\mathrm{ch}(r\Upsilon_p)}{\Upsilon_p\,\mathrm{sh}(r\Upsilon_p)}-\frac{1}{\Upsilon_p^2}+\frac{r}{\Upsilon_q}+\frac{1}{\Upsilon_q^2}\Big)\exp\Big(-\frac{y\Upsilon_p\,\mathrm{ch}(r\Upsilon_p)}{\mathrm{sh}(r\Upsilon_p)}\Big)dy$$

(1)

$$\mathbf{E}_x\Big\{\exp\Big(-\int_0^\tau \big(p\mathbb{I}_{[0,r)}(R_s^{(3)})+q\mathbb{I}_{[r,\infty)}(R_s^{(3)})\big)ds\Big);\ell(\tau,r)=0\Big\}$$

$$=\begin{cases}\dfrac{\lambda}{\lambda+p}\Big(1-\dfrac{r\,\mathrm{sh}(x\Upsilon_p)}{x\,\mathrm{sh}(r\Upsilon_p)}\Big), & x\le r\\[2ex] \dfrac{\lambda}{\lambda+q}\Big(1-\dfrac{r}{x}e^{(r-x)\Upsilon_q}\Big), & r\le x\end{cases}$$

$$\Upsilon_s:=\sqrt{2\lambda+2s}$$

5 $R_s^{(3)} = \left(\sum_{k=1}^{3}(W_s^{(k)})^2\right)^{1/2}$ $\tau \sim \mathrm{Exp}(\lambda)$, independent of $R^{(3)}$

1.27.2 $$\mathbf{P}_x\Big(\int_0^\tau \mathbb{I}_{[0,r)}(R_s^{(3)})ds \in du, \int_0^\tau \mathbb{I}_{[r,\infty)}(R_s^{(3)})ds \in dv, \frac{1}{2}\ell(\tau,r) \in dy\Big)$$

$$=: \lambda e^{-\lambda(u+v)} B_x^{(27)}(u,v,y)dudvdy$$

$x \le r$ $$= \lambda e^{-\lambda(u+v)}\frac{2}{x}((r\,\mathrm{escs}_u(x,r,r,y) - \widetilde{\mathrm{ess}}_u(x,r,0,y))\,\mathrm{h}_v(1,y)$$

$$+\,\mathrm{ess}_u(x,r,0,y)(r\,\mathrm{h}_v(0,y) + \mathrm{h}_v(-1,y)))dudvdy$$

$r \le x$ $$= \lambda e^{-\lambda(u+v)}\frac{2}{x}((r\,\mathrm{ecs}_u(r,r,0,y) - \mathrm{es}_u(-2,0,r,0,y))\,\mathrm{h}_v(1,y+x-r)$$

$$+\,\mathrm{es}_u(0,0,r,0,y)(r\,\mathrm{h}_v(0,y+x-r) + \mathrm{h}_v(-1,y+x-r)))dudvdy$$

(1) $x \le r$ $$\mathbf{P}_x\Big(\int_0^\tau \mathbb{I}_{[0,r]}(R_s^{(3)})ds \in du, \int_0^\tau \mathbb{I}_{[r,\infty)}(R_s^{(3)})ds = 0, \ell(\tau,r) = 0\Big)$$

$$= \lambda e^{-\lambda u}\big(1 - \frac{r}{x}\widetilde{\mathrm{ss}}_u(x,r)\big)du$$

(2) $r \le x$ $$\mathbf{P}_x\Big(\int_0^\tau \mathbb{I}_{[0,r]}(R_s^{(3)})ds = 0, \int_0^\tau \mathbb{I}_{[r,\infty)}(R_s^{(3)})ds \in dv, \ell(\tau,r) = 0\Big)$$

$$= \lambda e^{-\lambda v}\Big(1 - \frac{r}{x}\,\mathrm{Erfc}\Big(\frac{x-r}{\sqrt{2v}}\Big)\Big)dv$$

1.27.3 (1) $$\mathbf{E}_x\Big\{\exp\Big(-p\int_0^\infty \mathbb{I}_{[0,r)}(R_s^{(3)})ds\Big);\ \ell(\infty,r) \in dy\Big\}$$

$$= \begin{cases} \dfrac{\mathrm{sh}(x\sqrt{2p})}{2x\,\mathrm{sh}(r\sqrt{2p})}\exp\Big(-\dfrac{y\sqrt{p}\,\mathrm{ch}(r\sqrt{2p})}{\sqrt{2}\,\mathrm{sh}(r\sqrt{2p})}\Big)dy, & 0 \le x \le r \\ \dfrac{1}{2x}\exp\Big(-\dfrac{y\sqrt{p}\,\mathrm{ch}(r\sqrt{2p})}{\sqrt{2}\,\mathrm{sh}(r\sqrt{2p})}\Big)dy, & r \le x \end{cases}$$

1.27.4 $$\mathbf{P}_x\Big(\int_0^t (p\mathbb{I}_{[0,r)}(R_s^{(3)}) + q\mathbb{I}_{[r,\infty)}(R_s^{(3)}))ds \in dv, \frac{1}{2}\ell(t,r) \in dy\Big)$$

$$= \frac{1}{|p-q|}B_x^{(27)}\Big(\frac{|qt-v|}{|p-q|}, \frac{|pt-v|}{|p-q|}, y\Big)\mathbb{I}_{((q\wedge p)t,(q\vee p)t)}(v)\,dvdy$$

for $B_x^{(27)}(u,v,y)$ see 1.27.2

(1) $x \le r$ $$\mathbf{P}_x\Big(\int_0^t (p\mathbb{I}_{[0,r)}(R_s^{(3)}) + q\mathbb{I}_{[r,\infty)}(R_s^{(3)}))ds = pt, \ell(t,r) = 0\Big) = 1 - \frac{r}{x}\widetilde{\mathrm{ss}}_t(x,r)$$

5 $R_s^{(3)} = \big(\sum_{k=1}^3 (W_s^{(k)})^2\big)^{1/2}$ $\tau \sim \mathrm{Exp}(\lambda)$, independent of $R^{(3)}$

(2) $r \le x$

$$\mathbf{P}_x\Big(\int_0^t \big(p\mathbb{I}_{[0,r)}(R_s^{(3)}) + q\mathbb{I}_{[r,\infty)}(R_s^{(3)})\big)ds = qt, \ell(t,r) = 0\Big) = 1 - \frac{r}{x}\operatorname{Erfc}\Big(\frac{x-r}{\sqrt{2t}}\Big)$$

(3)

$$\mathbf{P}_x\Big(\int_0^\infty \mathbb{I}_{[0,r)}(R_s^{(3)})ds \in du,\ \ell(\infty,r) \in dy\Big)$$

$$= \begin{cases} \dfrac{1}{2x}\operatorname{ess}_u(x,r,0,y/2)dudy, & 0 \le x \le r \\ \dfrac{1}{2x}\operatorname{es}_u(0,0,r,0,y/2)dudy, & r \le x \end{cases}$$

1.27.5

$$\mathbf{E}_x\Big\{\exp\Big(-\int_0^\tau \big(p\mathbb{I}_{[0,r)}(R_s^{(3)}) + q\mathbb{I}_{[r,\infty)}(R_s^{(3)})\big)ds\Big); \tfrac{1}{2}\ell(\tau,r) \in dy, R_\tau^{(3)} \in dz\Big\}$$

$x \le r$, $z \le r$

$$= \frac{2z\lambda}{x} e^{-y\Upsilon_q} \exp\Big(-\frac{y\Upsilon_p \operatorname{ch}(r\Upsilon_p)}{\operatorname{sh}(r\Upsilon_p)}\Big)\frac{\operatorname{sh}(x\Upsilon_p)\operatorname{sh}(z\Upsilon_p)}{\operatorname{sh}^2(r\Upsilon_p)}dydz$$

$x \le r$, $r \le z$

$$= \frac{2z\lambda}{x} e^{-(y+z-r)\Upsilon_q} \exp\Big(-\frac{y\Upsilon_p \operatorname{ch}(r\Upsilon_p)}{\operatorname{sh}(r\Upsilon_p)}\Big)\frac{\operatorname{sh}(x\Upsilon_p)}{\operatorname{sh}(r\Upsilon_p)}dydz$$

$r \le x$, $z \le r$

$$= \frac{2z\lambda}{x} e^{-(y+x-r)\Upsilon_q} \exp\Big(-\frac{y\Upsilon_p \operatorname{ch}(r\Upsilon_p)}{\operatorname{sh}(r\Upsilon_p)}\Big)\frac{\operatorname{sh}(z\Upsilon_p)}{\operatorname{sh}(r\Upsilon_p)}dydz$$

$r \le x$, $r \le z$

$$= \frac{2z\lambda}{x} e^{-(y+x+z-2r)\Upsilon_q} \exp\Big(-\frac{y\Upsilon_p \operatorname{ch}(r\Upsilon_p)}{\operatorname{sh}(r\Upsilon_p)}\Big)dydz$$

(1)

$$\mathbf{E}_x\Big\{\exp\Big(-\int_0^\tau \big(p\mathbb{I}_{[0,r)}(R_s^{(3)}) + q\mathbb{I}_{[r,\infty)}(R_s^{(3)})\big)ds\Big); \ell(\tau,r) = 0, R_\tau^{(3)} \in dz\Big\}$$

$x \le r$, $z \le r$

$$= \frac{z\lambda(\operatorname{ch}((r-|z-x|)\Upsilon_p) - \operatorname{ch}((r-z-x)\Upsilon_p))}{x\Upsilon_p \operatorname{sh}(r\Upsilon_p)}dz$$

$r \le x$, $r \le z$

$$= \frac{z\lambda}{x\Upsilon_q} e^{-|z-x|\Upsilon_q}dz - \frac{z\lambda}{x\Upsilon_q} e^{(2r-x-z)\Upsilon_q}dz$$

1.27.6

$$\mathbf{P}_x\Big(\int_0^\tau \mathbb{I}_{[0,r)}(R_s^{(3)})ds \in du, \int_0^\tau \mathbb{I}_{[r,\infty)}(R_s^{(3)})ds \in dv, \tfrac{1}{2}\ell(\tau,r) \in dy, R_\tau^{(3)} \in dz\Big)$$

$$=: \lambda e^{-\lambda(u+v)} B_x^{(27)}(u,v,y,z)dudvdydz$$

$$\Upsilon_s := \sqrt{2\lambda + 2s}$$

5 $R_s^{(3)} = \left(\sum_{k=1}^3 (W_s^{(k)})^2\right)^{1/2}$	$\tau \sim \mathrm{Exp}(\lambda)$, independent of $R^{(3)}$
$x \le r$ $z \le r$	$= \lambda e^{-\lambda(u+v)} \frac{2z}{x} \operatorname{esss}_u(x,z,r,y)\, \mathrm{h}_v(1,y) du dv dy dz$
$x \le r$ $r \le z$	$= \lambda e^{-\lambda(u+v)} \frac{2z}{x} \operatorname{ess}_u(x,r,0,y)\, \mathrm{h}_v(1,y+z-r) du dv dy dz$
$r \le x$ $z \le r$	$= \lambda e^{-\lambda(u+v)} \frac{2z}{x} \operatorname{ess}_u(z,r,0,y)\, \mathrm{h}_v(1,y+x-r) du dv dy dz$
$r \le x$ $r \le z$	$= \lambda e^{-\lambda(u+v)} \frac{2z}{x} \operatorname{es}_u(0,0,r,0,y)\, \mathrm{h}_v(1,y+x+z-2r) du dv dy dz$
(1) $x \le r$	$\mathbf{P}_x\Big(\int_0^\tau \mathbb{I}_{[0,r)}(R_s^{(3)}) ds \in du, \int_0^\tau \mathbb{I}_{[r,\infty)}(R_s^{(3)}) ds = 0, \ell(\tau,r) = 0, R_\tau^{(3)} \in dz\Big)$
	$= \lambda e^{-\lambda u} \frac{z}{x} (\operatorname{cs}_u(r-\lvert z-x\rvert, r) - \operatorname{cs}_u(r-z-x,r)) du dz$
(2) $r \le x$	$\mathbf{P}_x\Big(\int_0^\tau \mathbb{I}_{[0,r)}(R_s^{(3)}) ds = 0, \int_0^\tau \mathbb{I}_{[r,\infty)}(R_s^{(3)}) ds \in dv, \ell(\tau,r) = 0, R_\tau^{(3)} \in dz\Big)$
	$= \lambda e^{-\lambda v} \frac{z}{x\sqrt{2\pi v}} \big(e^{-(z-x)^2/2v} - e^{-(z+x-2r)^2/2v}\big) dv dz$
1.27.8	$\mathbf{P}_x\Big(\int_0^t \big(p\mathbb{I}_{[0,r)}(R_s^{(3)}) + q\mathbb{I}_{[r,\infty)}(R_s^{(3)})\big) ds \in dv, \frac{1}{2}\ell(t,r) \in dy, R_t^{(3)} \in dz\Big)$
	$= \frac{1}{\lvert p-q\rvert} B_x^{(27)}\Big(\frac{\lvert qt-v\rvert}{\lvert p-q\rvert}, \frac{\lvert pt-v\rvert}{\lvert p-q\rvert}, y, z\Big) \mathbb{I}_{((q\wedge p)t,(q\vee p)t)}(v)\, dv dy dz$ for $B_x^{(27)}(u,v,y,z)$ see 1.27.6
(1) $x \le r$	$\mathbf{P}_x\Big(\int_0^t \big(p\mathbb{I}_{[0,r)}(R_s^{(3)}) + q\mathbb{I}_{[r,\infty)}(R_s^{(3)})\big) ds = pt, \ell(t,r) = 0, R_t^{(3)} \in dz\Big)$
$z \le r$	$= \frac{z}{x} (\operatorname{cs}_t(r-\lvert z-x\rvert, r) - \operatorname{cs}_t(r-z-x,r)) dz$
(2) $r \le x$	$\mathbf{P}_x\Big(\int_0^t \big(p\mathbb{I}_{[0,r)}(R_s^{(3)}) + q\mathbb{I}_{[r,\infty)}(R_s^{(3)})\big) ds = qt, \ell(t,r) = 0, R_t^{(3)} \in dz\Big)$

5 $R_s^{(3)} = \big(\sum_{k=1}^3 (W_s^{(k)})^2\big)^{1/2}$ $\tau \sim \mathrm{Exp}(\lambda)$, independent of $R^{(3)}$

$r \le z$

$$= \frac{z}{x\sqrt{2\pi t}} e^{-(z-x)^2/2t} dz - \frac{z}{x\sqrt{2\pi t}} e^{-(x+z-2r)^2/2t} dz$$

1.28.1 $\mathbf{E}_x\big\{e^{-\gamma\hat{H}(\tau)-\eta\check{H}(\tau)};\ \inf_{0\le s\le\tau} R_s^{(3)} \in da,\ \sup_{0\le s\le\tau} R_s^{(3)} \in db\big\}\big/ dadb$

$a < x$

$x < b$

$$= \frac{\sqrt{2\lambda+2\eta}\,\mathrm{sh}((b-x)\sqrt{2\lambda+2\gamma+2\eta})(b\sqrt{2\lambda}\,\mathrm{ch}((b-a)\sqrt{2\lambda}) - \mathrm{sh}((b-a)\sqrt{2\lambda}) - a\sqrt{2\lambda})}{x\,\mathrm{sh}((b-a)\sqrt{2\lambda+2\gamma+2\eta})\,\mathrm{sh}((b-a)\sqrt{2\lambda+2\eta})\,\mathrm{sh}((b-a)\sqrt{2\lambda})}$$

$$+ \frac{\sqrt{2\lambda+2\gamma}\,\mathrm{sh}((x-a)\sqrt{2\lambda+2\gamma+2\eta})(a\sqrt{2\lambda}\,\mathrm{ch}((b-a)\sqrt{2\lambda}) + \mathrm{sh}((b-a)\sqrt{2\lambda}) - b\sqrt{2\lambda})}{x\,\mathrm{sh}((b-a)\sqrt{2\lambda+2\gamma+2\eta})\,\mathrm{sh}((b-a)\sqrt{2\lambda+2\gamma})\,\mathrm{sh}((b-a)\sqrt{2\lambda})}$$

1.28.2 $\mathbf{P}_x\big(\hat{H}(\tau) \in du, \check{H}(\tau) \in dv,\ \inf_{0\le s\le\tau} R_s^{(3)} \in da,\ \sup_{0\le s\le\tau} R_s^{(3)} \in db\big)$

$u < v$

$$= \frac{b\sqrt{2\lambda}\,\mathrm{ch}((b-a)\sqrt{2\lambda}) - \mathrm{sh}((b-a)\sqrt{2\lambda}) - a\sqrt{2\lambda}}{x\,\mathrm{sh}((b-a)\sqrt{2\lambda})}$$

$$\times e^{-\lambda v}\,\mathrm{ss}_u(b-x, b-a)\,\mathrm{s}_{v-u}(b-a)\, dudvdadb$$

$v < u$

$$= \frac{a\sqrt{2\lambda}\,\mathrm{ch}((b-a)\sqrt{2\lambda}) + \mathrm{sh}((b-a)\sqrt{2\lambda}) - b\sqrt{2\lambda}}{x\,\mathrm{sh}((b-a)\sqrt{2\lambda})}$$

$$\times e^{-\lambda u}\,\mathrm{ss}_v(x-a, b-a)\,\mathrm{s}_{u-v}(b-a)\, dudvdadb$$

1.28.4 $\mathbf{P}_x\big(\hat{H}(t) \in du, \check{H}(t) \in dv,\ \inf_{0\le s\le t} R_s^{(3)} \in da,\ \sup_{0\le s\le t} R_s^{(3)} \in db\big)$

for $0 < u\wedge v < u \vee v < t, \qquad a < x < b$

$u < v$

$$= \frac{1}{x}\,\mathrm{ss}_u(b-x, b-a)\,\mathrm{s}_{v-u}(b-a)$$

$$\times\big(2b\ \mathrm{cs}_{t-v}(b-a, b-a) - 1 - 2a\ \mathrm{cs}_{t-v}(0, b-a)\big)\, dudvdadb$$

$v < u$

$$= \frac{1}{x}\,\mathrm{ss}_v(x-a, b-a)\,\mathrm{s}_{u-v}(b-a)$$

$$\times\big(2a\ \mathrm{cs}_{t-u}(b-a, b-a) + 1 - 2b\ \mathrm{cs}_{t-u}(0, b-a)\big)\, dudvdadb$$

5 $R_s^{(3)} = \left(\sum_{k=1}^3 (W_s^{(k)})^2\right)^{1/2}$ $\qquad\tau \sim \operatorname{Exp}(\lambda)$, independent of $R^{(3)}$

1.28.5 $\quad a < x$, $x < b$

$$\mathbf{E}_x\left\{e^{-\gamma\hat{H}(\tau)-\eta\check{H}(\tau)}; \inf_{0\le s\le\tau} R_s^{(3)} \in da,\ \sup_{0\le s\le\tau} R_s^{(3)} \in db,\ R_\tau^{(3)} \in dz\right\}$$

$$= \frac{2\lambda z\sqrt{2\lambda+2\eta}\,\operatorname{sh}((b-x)\sqrt{2\lambda+2\gamma+2\eta})\operatorname{sh}((z-a)\sqrt{2\lambda})}{x\operatorname{sh}((b-a)\sqrt{2\lambda+2\gamma+2\eta})\operatorname{sh}((b-a)\sqrt{2\lambda+2\eta})\operatorname{sh}((b-a)\sqrt{2\lambda})}dadbdz$$
$$+\frac{2\lambda z\sqrt{2\lambda+2\gamma}\,\operatorname{sh}((x-a)\sqrt{2\lambda+2\gamma+2\eta})\operatorname{sh}((b-z)\sqrt{2\lambda})}{x\operatorname{sh}((b-a)\sqrt{2\lambda+2\gamma+2\eta})\operatorname{sh}((b-a)\sqrt{2\lambda+2\gamma})\operatorname{sh}((b-a)\sqrt{2\lambda})}dadbdz$$

1.28.6

$$\mathbf{P}_x\left(\hat{H}(\tau) \in du, \check{H}(\tau) \in dv, \inf_{0\le s\le\tau} R_s^{(3)} \in da,\ \sup_{0\le s\le\tau} R_s^{(3)} \in db, R_\tau^{(3)} \in dz\right)$$

$u < v$
$$= \frac{2\lambda z}{x}e^{-\lambda v}\frac{\operatorname{sh}((z-a)\sqrt{2\lambda})}{\operatorname{sh}((b-a)\sqrt{2\lambda})}\operatorname{ss}_u(b-x,b-a)\operatorname{s}_{v-u}(b-a)\,dudvdadbdz$$

$v < u$
$$= \frac{2\lambda z}{x}e^{-\lambda u}\frac{\operatorname{sh}((b-z)\sqrt{2\lambda})}{\operatorname{sh}((b-a)\sqrt{2\lambda})}\operatorname{ss}_v(x-a,b-a)\operatorname{s}_{u-v}(b-a)\,dudvdadbdz$$

1.28.8

$$\mathbf{P}_x\left(\hat{H}(t) \in du, \check{H}(t) \in dv, \inf_{0\le s\le t} R_s^{(3)} \in da,\ \sup_{0\le s\le t} R_s^{(3)} \in db, R_t^{(3)} \in dz\right)$$

for $\quad 0 < u\wedge v < u\vee v < t, \qquad a < x\wedge z < x\vee z < b$

$u < v$
$$= \frac{2z}{x}\operatorname{ss}_u(b-x,b-a)\operatorname{s}_{v-u}(b-a)\operatorname{ss}_{t-v}(z-a,b-a)dudvdadbdz$$

$v < u$
$$= \frac{2z}{x}\operatorname{ss}_v(x-a,b-a)\operatorname{s}_{u-v}(b-a)\operatorname{ss}_{t-u}(b-z,b-a)dudvdadbdz$$

1.29.1

$$\mathbf{E}_x\left\{\exp\left(-\int_0^\tau \left(p\mathbb{I}_{[0,r)}\left(R_s^{(3)}\right)+q\mathbb{I}_{[r,\infty)}\left(R_s^{(3)}\right)\right)ds\right); a < \inf_{0\le s\le\tau} R_s^{(3)},\ \sup_{0\le s\le\tau} R_s^{(3)} < b, \frac{1}{2}\ell(\tau,r) \in dy\right\}$$

$$= 2\lambda\left(\frac{r\operatorname{ch}((b-r)\Upsilon_q)-b}{x\Upsilon_q\operatorname{sh}((b-r)\Upsilon_q)}+\frac{r\operatorname{ch}((r-a)\Upsilon_p)-a}{x\Upsilon_p\operatorname{sh}((r-a)\Upsilon_p)}+\frac{2(p-q)}{x\Upsilon_p^2\Upsilon_q^2}\right)$$
$$\times\exp\left(-\frac{y\Upsilon_q\operatorname{ch}((b-r)\Upsilon_q)}{\operatorname{sh}((b-r)\Upsilon_q)}-\frac{y\Upsilon_p\operatorname{ch}((r-a)\Upsilon_p)}{\operatorname{sh}((r-a)\Upsilon_p)}\right)\begin{cases}\dfrac{\operatorname{sh}((x-a)\Upsilon_p)}{\operatorname{sh}((r-a)\Upsilon_p)}dy, & a\le x\le r\\ \dfrac{\operatorname{sh}((b-x)\Upsilon_q)}{\operatorname{sh}((b-r)\Upsilon_q)}dy, & r\le x\le b\end{cases}$$

$\Upsilon_s := \sqrt{2\lambda+2s}$

5 $R_s^{(3)} = \left(\sum_{k=1}^3 (W_s^{(k)})^2\right)^{1/2}$ $\tau \sim \mathrm{Exp}(\lambda)$, independent of $R^{(3)}$

(1) $$\mathbf{E}_x\Big\{\exp\Big(-\int_0^\tau \big(p\mathbb{I}_{[0,r)}(R_s^{(3)}) + q\mathbb{I}_{[r,\infty)}(R_s^{(3)})\big)ds\Big); a < \inf_{0\le s\le\tau} R_s^{(3)}, \sup_{0\le s\le\tau} R_s^{(3)} < b, \ell(\tau,r)=0\Big\}$$

$$= \begin{cases} \dfrac{\lambda}{\lambda+p}\Big(1 - \dfrac{a\,\mathrm{sh}((r-x)\Upsilon_p) + r\,\mathrm{sh}((x-a)\Upsilon_p)}{x\,\mathrm{sh}((r-a)\Upsilon_p)}\Big), & a \le x \le r \\ \dfrac{\lambda}{\lambda+q}\Big(1 - \dfrac{r\,\mathrm{sh}((b-x)\Upsilon_q) + b\,\mathrm{sh}((x-r)\Upsilon_q)}{x\,\mathrm{sh}((b-r)\Upsilon_q)}\Big), & r \le x \le b \end{cases}$$

1.29.3 (1) $$\mathbf{E}_x\Big\{\exp\Big(-p\int_0^\infty \mathbb{I}_{[0,r)}(R_s^{(3)})ds\Big);\ a < \inf_{0\le s<\infty} R_s^{(3)},\ \ell(\infty,r)\in dy\Big\}$$

$$= \begin{cases} \dfrac{\mathrm{sh}((x-a)\sqrt{2p})}{2x\,\mathrm{sh}((r-a)\sqrt{2p})}\exp\Big(-\dfrac{y\sqrt{p}\,\mathrm{ch}((r-a)\sqrt{2p})}{\sqrt{2}\,\mathrm{sh}((r-a)\sqrt{2p})}\Big)dy, & a \le x \le r \\ \dfrac{1}{2x}\exp\Big(-\dfrac{y\sqrt{p}\,\mathrm{ch}((r-a)\sqrt{2p})}{\sqrt{2}\,\mathrm{sh}((r-a)\sqrt{2p})}\Big)dy, & r \le x \end{cases}$$

1.29.4 (1) $$\mathbf{P}_x\Big(\int_0^\infty \mathbb{I}_{[0,r)}(R_s^{(3)})ds \in du,\ a < \inf_{0\le s<\infty} R_s^{(3)},\ \ell(\infty,r)\in dy\Big)$$

$$= \begin{cases} \dfrac{1}{2x}\,\mathrm{ess}_u(x-a, r-a, 0, y/2)du dy, & a \le x \le r \\ \dfrac{1}{2x}\,\mathrm{es}_u(0,0,r-a,0,y/2)du dy, & r \le x \end{cases}$$

1.29.5 $$\mathbf{E}_x\Big\{\exp\Big(-\int_0^\tau \big(p\mathbb{I}_{[0,r)}(R_s^{(3)}) + q\mathbb{I}_{[r,\infty)}(R_s^{(3)})\big)ds\Big); a < \inf_{0\le s\le\tau} R_s^{(3)}, \sup_{0\le s\le\tau} R_s^{(3)} < b, \frac{1}{2}\ell(\tau,r)\in dy, R_\tau^{(3)}\in dz\Big\} = \frac{\lambda z}{x} Q_\lambda^{(29)}(y,z)dydz$$

for $Q_\lambda^{(29)}(y,z)$ see 1.1.29.5

(1) $$\mathbf{E}_x\Big\{\exp\Big(-\int_0^\tau \big(p\mathbb{I}_{[0,r)}(R_s^{(3)}) + q\mathbb{I}_{[r,\infty)}(R_s^{(3)})\big)ds\Big); a < \inf_{0\le s\le\tau} R_s^{(3)}, \sup_{0\le s\le\tau} R_s^{(3)} < b, \ell(\tau,r)=0, R_\tau^{(3)}\in dz\Big\}$$

$$\Upsilon_s := \sqrt{2\lambda + 2s}$$

5 $R_s^{(3)} = \left(\sum_{k=1}^3 (W_s^{(k)})^2\right)^{1/2}$ $\qquad \tau \sim \operatorname{Exp}(\lambda)$, independent of $R^{(3)}$

$x \le r$, $z \le r$

$$= \frac{\lambda z(\operatorname{ch}((r-a-|z-x|)\Upsilon_p) - \operatorname{ch}((r+a-z-x)\Upsilon_p))}{x\Upsilon_p \operatorname{sh}((r-a)\Upsilon_p)}$$

$r \le x$, $r \le z$

$$= \frac{\lambda z(\operatorname{ch}((b-r-|z-x|)\Upsilon_q) - \operatorname{ch}((b+r-z-x)\Upsilon_q))}{x\Upsilon_q \operatorname{sh}((b-r)\Upsilon_q)}$$

1.29.6

$$\mathbf{P}_x\Big(\int_0^\tau \mathbb{I}_{[0,r)}\big(R_s^{(3)}\big)ds \in du, \int_0^\tau \mathbb{I}_{[r,\infty)}\big(R_s^{(3)}\big)ds \in dv,\ a < \inf_{0\le s\le\tau} R_s^{(3)},$$

$$\sup_{0\le s\le\tau} R_s^{(3)} < b, \frac{\ell(\tau,r)}{2} \in dy, R_\tau^{(3)} \in dz\Big) = \frac{\lambda z e^{-\lambda(u+v)}}{x} B_x^{(29)}(u,v,y,z)dudvdydz$$

for $B_x^{(29)}(u,v,y,z)$ see 1.1.29.6

1.29.8

$$\mathbf{P}_x\Big(\int_0^t \big(p\mathbb{I}_{[0,r)}\big(R_s^{(3)}\big) + q\mathbb{I}_{[r,\infty)}\big(R_s^{(3)}\big)\big)ds \in dv, a < \inf_{0\le s\le t} R_s^{(3)}, \sup_{0\le s\le t} R_s^{(3)} < b,$$

$$\frac{\ell(t,r)}{2} \in dy, R_t^{(3)} \in dz\Big) = \frac{z\mathbb{I}_{((q\wedge p)t,(q\vee p)t)}(v)}{x|p-q|} B_x^{(29)}\Big(\frac{|qt-v|}{|p-q|}, \frac{|pt-v|}{|p-q|}, y, z\Big)\, dvdydz$$

for $B_x^{(29)}(u,v,y,z)$ see 1.1.29.6

1.30.3, $\beta < 0$

$$\mathbf{E}_x \exp\Big(-\int_0^\infty \Big(\frac{p^2}{2}e^{2\beta R_s^{(3)}} + qe^{\beta R_s^{(3)}}\Big)ds\Big) = \frac{\exp\Big(-\frac{pe^{\beta x}}{|\beta|}\Big)S\Big(\frac12 + \frac{q}{p|\beta|}, 1, \frac{2p}{|\beta|}, \frac{2pe^{\beta x}}{|\beta|}\Big)}{x|\beta|M\Big(\frac12 + \frac{q}{p|\beta|}, 1, \frac{2p}{|\beta|}\Big)}$$

1.30.5

$$\mathbf{E}_x\Big\{\exp\Big(-\int_0^\tau \Big(\frac{p^2}{2}e^{2\beta R_s^{(3)}} + qe^{\beta R_s^{(3)}}\Big)ds\Big);\ R_\tau^{(3)} \in dz\Big\}\Big(\frac{|\beta|}{2p}\Big)^{1+2\sqrt{2\lambda}/|\beta|}$$

for $(z-x)\beta \ge 0$

$$= \frac{\lambda z}{px} e^{x\operatorname{sign}\beta\sqrt{2\lambda} - p/|\beta|} \exp\Big(-\frac{p}{|\beta|}e^{\beta x} - \frac{\beta z}{2}\Big)$$

$$\times S\Big(\frac{q}{p|\beta|} + \frac{\sqrt{2\lambda}}{|\beta|} + \frac12, \frac{2\sqrt{2\lambda}}{|\beta|} + 1, \frac{2p}{|\beta|}e^{\beta x}, \frac{2p}{|\beta|}\Big) \frac{W_{-q/p|\beta|,\sqrt{2\lambda}/|\beta|}\Big(\frac{2p}{|\beta|}e^{\beta z}\Big)}{W_{-q/p|\beta|,\sqrt{2\lambda}/|\beta|}\Big(\frac{2p}{|\beta|}\Big)} dz$$

for $(z-x)\beta \le 0$

$$= \frac{\lambda z}{px} e^{z\operatorname{sign}\beta\sqrt{2\lambda} - p/|\beta|} \exp\Big(-\frac{p}{|\beta|}e^{\beta z} - \frac{\beta x}{2}\Big)$$

$$\times S\Big(\frac{q}{p|\beta|} + \frac{\sqrt{2\lambda}}{|\beta|} + \frac12, \frac{2\sqrt{2\lambda}}{|\beta|} + 1, \frac{2p}{|\beta|}e^{\beta z}, \frac{2p}{|\beta|}\Big) \frac{W_{-q/p|\beta|,\sqrt{2\lambda}/|\beta|}\Big(\frac{2p}{|\beta|}e^{\beta x}\Big)}{W_{-q/p|\beta|,\sqrt{2\lambda}/|\beta|}\Big(\frac{2p}{|\beta|}\Big)} dz$$

1.31.1, $0 < \alpha$

$$\mathbf{E}_{x/\sqrt{2\lambda}} \exp\Big(-\gamma\sqrt{2\lambda} \sup_{0\le s\le\tau} \big(\ell(s,u/\sqrt{2\lambda}) - \alpha^{-1}\ell(s,r/\sqrt{2\lambda})\big)\Big)$$

5 $R_s^{(3)} = \big(\sum_{k=1}^{3}(W_s^{(k)})^2\big)^{1/2}$ $\tau \sim \operatorname{Exp}(\lambda)$, independent of $R^{(3)}$

$$= 1 - \frac{\gamma u(\Upsilon - e^{-2r} + \alpha e^{-2u})(e^{-|x-u|} - e^{-x-u})}{x(2\gamma(1-e^{-2(u\wedge r)})(1-e^{-2|u-r|}) + \Upsilon - e^{-2r} + \alpha e^{-2u})}$$

$$- \frac{\gamma u(2(1-e^{-2(u\wedge r)})(1-e^{-2|u-r|}) - (\Upsilon - e^{-2r} + \alpha e^{-2u})(1-e^{-2u}))(e^{-|x-r|} - e^{-x-r})}{x(2\gamma(1-e^{-2(u\wedge r)})(1-e^{-2|u-r|}) + \Upsilon - e^{-2r} + \alpha e^{-2u})(e^{-|u-r|} - e^{-u-r})}$$

1.31.2 $0<\alpha$

$$\mathbf{P}_{x/\sqrt{2\lambda}}\Big(\sup_{0\le s\le\tau}\big(\ell(s,u/\sqrt{2\lambda}) - \alpha^{-1}\ell(s,r/\sqrt{2\lambda})\big) > h/\sqrt{2\lambda}\Big)$$

$$= \frac{u}{x}\exp\Big(-\frac{(\Upsilon - e^{-2r} + \alpha e^{-2u})h}{2(1-e^{-2(u\wedge r)})(1-e^{-2|u-r|})}\Big)\Big[\frac{e^{-|x-r|} - e^{-x-r}}{e^{-|u-r|} - e^{-u-r}}$$

$$- \frac{\Upsilon - e^{-2r} + \alpha e^{-2u}}{2(1-e^{-2(u\wedge r)})(1-e^{-2|u-r|})}\Big(\frac{(e^{-|x-r|} - e^{-x-r})(1-e^{-2u})}{e^{-|u-r|} - e^{-u-r}} - e^{-|u-x|} + e^{-x-u}\Big)\Big]$$

(1) $0<\alpha$

$$\mathbf{P}_x\Big(\sup_{0\le s<\infty}\big(\ell(s,u) - \frac{1}{\alpha}\ell(s,r)\big) > h\Big) = \exp\Big(-\frac{\sqrt{(r-\alpha u)^2 + 4\alpha(u\wedge r)|u-r|} + r - \alpha u}{4(u\wedge r)|u-r|h^{-1}}\Big)$$

$$\times\frac{u}{x}\Big[\frac{x\wedge r}{u\wedge r} - \frac{\sqrt{(r-\alpha u)^2 + 4\alpha(u\wedge r)|u-r|} + r - \alpha u}{2(u\wedge r)|u-r|}\Big(\frac{(x\wedge r)|u|}{u\wedge r} - u\wedge x\Big)\Big]$$

1.31.5 $0<\alpha$

$$\mathbf{E}_{x/\sqrt{2\lambda}}\Big\{\exp\big(-\gamma\sqrt{2\lambda}\sup_{0\le s\le\tau}\big(\ell(s,u/\sqrt{2\lambda}) - \alpha^{-1}\ell(s,r/\sqrt{2\lambda})\big)\big); R_\tau^{(3)} \in dz/\sqrt{2\lambda}\Big\}$$

$$= \frac{z}{2x}\big(e^{-|z-x|} - e^{-z-x}\big)dz$$

$$-\frac{z\gamma}{2x}\big(e^{-|z-u|} - e^{-z-u}\big)\Big\{\frac{(\Upsilon - e^{-2r} + \alpha e^{-2u})(e^{-|x-u|} - e^{-x-u})}{2\gamma(1-e^{-2(u\wedge r)})(1-e^{-2|u-r|}) + \Upsilon - e^{-2r} + \alpha e^{-2u}}$$

$$- \frac{(2(1-e^{-2(u\wedge r)})(1-e^{-2|u-r|}) - (\Upsilon - e^{-2r} + \alpha e^{-2u})(1-e^{-2u}))(e^{-|x-r|} - e^{-x-r})}{(2\gamma(1-e^{-2(u\wedge r)})(1-e^{-2|u-r|}) + \Upsilon - e^{-2r} + \alpha e^{-2u})(e^{-|u-r|} - e^{-u-r})}\Big\}dz$$

1.31.6 $0<\alpha$

$$\mathbf{P}_{x/\sqrt{2\lambda}}\Big(\sup_{0\le s\le\tau}\big(\ell(s,u/\sqrt{2\lambda}) - \alpha^{-1}\ell(s,r/\sqrt{2\lambda})\big) > h/\sqrt{2\lambda},\ R_\tau^{(3)} \in dz/\sqrt{2\lambda}\Big)$$

$$= \frac{z}{2x}\big(e^{-|z-u|} - e^{-z-u}\big)\exp\Big(-\frac{(\Upsilon - e^{-2r} + \alpha e^{-2u})h}{2(1-e^{-2(u\wedge r)})(1-e^{-2|u-r|})}\Big)\Big[\frac{e^{-|x-r|} - e^{-x-r}}{e^{-|u-r|} - e^{-u-r}}$$

$$- \frac{\Upsilon - e^{-2r} + \alpha e^{-2u}}{2(1-e^{-2(u\wedge r)})(1-e^{-2|u-r|})}\Big(\frac{(e^{-|x-r|} - e^{-x-r})(1-e^{-2u})}{e^{-|u-r|} - e^{-u-r}} - e^{-|u-x|} + e^{-x-u}\Big)\Big]dz$$

$$\Upsilon := \sqrt{\big(1-e^{-2r} - \alpha + \alpha e^{-2u}\big)^2 + 4\alpha\big(1-e^{-2(u\wedge r)}\big)\big(1-e^{-2|u-r|}\big)} + 1 - \alpha$$

5 $R_s^{(3)} = \left(\sum_{k=1}^3 (W_s^{(k)})^2\right)^{1/2}$ $H_z = \min\{s : R_s^{(3)} = z\}$

2. Stopping at first hitting time

2.0.1 $$\mathbf{E}_x\Big\{e^{-\alpha H_z};\ H_z < \infty\Big\} = \begin{cases} \dfrac{z\sqrt{2\alpha}}{\operatorname{sh}(z\sqrt{2\alpha})}, & x = 0 < z \\ \dfrac{z\operatorname{sh}(x\sqrt{2\alpha})}{x\operatorname{sh}(z\sqrt{2\alpha})}, & 0 < x \le z \\ \dfrac{z}{x}e^{-(x-z)\sqrt{2\alpha}}, & z \le x \end{cases}$$

2.0.2 $$\mathbf{P}_x\big(H_z \in dt\big) = \begin{cases} z\,\mathrm{s}_t(z), & x = 0 < z \\ \dfrac{z}{x}\,\mathrm{ss}_t(x,z)dt, & 0 < x \le z \\ \dfrac{z(x-z)}{x\sqrt{2\pi}t^{3/2}}\exp\Big(-\dfrac{(x-z)^2}{2t}\Big)dt, & z \le x \end{cases}$$

(1) $$\mathbf{P}_x\big(H_z = \infty\big) = \begin{cases} 0, & 0 \le x \le z \\ 1 - \dfrac{z}{x}, & z \le x \end{cases}$$

2.1.2 $z \le y$ $$\mathbf{P}_x\Big(\sup_{0\le s\le H_z} R_s^{(3)} < y,\ H_z < \infty\Big) = \begin{cases} 1, & 0 \le x \le z \\ \dfrac{z(y-x)}{x(y-z)}, & z \le x \le y \end{cases}$$

2.1.4 $z \le y$ $$\mathbf{E}_x\{e^{-\alpha H_z}; \sup_{0\le s\le H_z} R_s^{(3)} < y\} = \begin{cases} \dfrac{z\operatorname{sh}(x\sqrt{2\alpha})}{x\operatorname{sh}(z\sqrt{2\alpha})}, & 0 \le x \le z \\ \dfrac{z\operatorname{sh}((y-x)\sqrt{2\alpha})}{x\operatorname{sh}((y-z)\sqrt{2\alpha})}, & z \le x \le y \end{cases}$$

(1) $z \le y$ $$\mathbf{P}_x\Big(\sup_{0\le s\le H_z} R_s^{(3)} < y, H_z \in dt\Big) = \begin{cases} z\,\mathrm{s}_t(z), & x = 0 < z \\ \dfrac{z}{x}\,\mathrm{ss}_t(x,z)dt, & 0 < x \le z \\ \dfrac{z}{x}\,\mathrm{ss}_t(y-x, y-z)dt, & z \le x \le y \end{cases}$$

2.2.2 $0 \le y$ $$\mathbf{P}_x\Big(\inf_{0\le s\le H_z} R_s^{(3)} > y,\ H_z < \infty\Big) = \begin{cases} \dfrac{z(x-y)}{x(z-y)}, & y \le x \le z \\ \dfrac{z}{x}, & y \le z \le x \end{cases}$$

(1) $$\mathbf{P}_x\Big(\inf_{0\le s\le H_z} R_s^{(3)} > y,\ H_z = \infty\Big) = \begin{cases} 1 - \dfrac{y}{x}, & z \le y \le x \\ 1 - \dfrac{z}{x}, & y \le z \le x \end{cases}$$

5 $R_s^{(3)}=\left(\sum_{k=1}^3 (W_s^{(k)})^2\right)^{1/2}$ $H_z=\min\{s: R_s^{(3)}=z\}$

2.2.4 $0\le y$

$$\mathbf{E}_x\Big\{e^{-\alpha H_z};\inf_{0\le s\le H_z} R_s^{(3)}>y\Big\}=\begin{cases}\dfrac{z\,\mathrm{sh}((x-y)\sqrt{2\alpha})}{x\,\mathrm{sh}((z-y)\sqrt{2\alpha})}, & y\le x\le z\\ \dfrac{z}{x}e^{-(x-z)\sqrt{2\alpha}}, & y\le z\le x\end{cases}$$

(1)

$$\mathbf{P}_x\Big(\inf_{0\le s\le H_z} R_s^{(3)}>y, H_z\in dt\Big)=\begin{cases}\dfrac{z}{x}\,\mathrm{ss}_t(x-y,z-y)dt, & y\le x\le z\\ \dfrac{z(x-z)}{x\sqrt{2\pi}t^{3/2}}\exp\Big(-\dfrac{(x-z)^2}{2t}\Big)dt, & y\le z\le x\end{cases}$$

2.3.1 $\mathbf{E}_x\big\{e^{-\gamma\ell(H_z,r)};\ H_z<\infty\big\}$

$z\le r$

$$=\begin{cases}1, & 0\le x\le z\\ \dfrac{z(1+2\gamma(r-x))}{x(1+2\gamma(r-z))}, & z\le x\le r\\ \dfrac{z}{x(1+2\gamma(r-z))}, & r\le x\end{cases}$$

$r\le z$

$$=\begin{cases}\dfrac{z}{z+2\gamma(z-r)r}, & 0\le x\le r\\ \dfrac{z(x+2\gamma(x-r)r)}{x(z+2\gamma(z-r)r)}, & r\le x\le z\\ \dfrac{z}{x}, & z\le x\end{cases}$$

(1) $z\le r$

$$\mathbf{E}_x\big\{e^{-\gamma\ell(H_z,r)};\ H_z=\infty\big\}=\begin{cases}1-\dfrac{r}{x}+\dfrac{r-z}{x(1+2\gamma(r-z))}, & r\le x\\ \dfrac{x-z}{x(1+2\gamma(r-z))}, & z\le x\le r\\ 0, & 0\le x\le z\end{cases}$$

2.3.2 $\mathbf{P}_x\big(\ell(H_z,r)\in dy,\ H_z<\infty\big)$

$z\le r$

$$=\begin{cases}0, & 0\le x\le z\\ \dfrac{z(x-z)}{2x(r-z)^2}\exp\Big(-\dfrac{y}{2(r-z)}\Big)dy, & z\le x\le r\\ \dfrac{z}{2x(r-z)}\exp\Big(-\dfrac{y}{2(r-z)}\Big)dy, & r\le x\end{cases}$$

$r\le z$

$$=\begin{cases}\dfrac{z}{2r(z-r)}\exp\Big(-\dfrac{zy}{2r(z-r)}\Big)dy, & 0\le x\le r\\ \dfrac{z(z-x)}{2x(z-r)^2}\exp\Big(-\dfrac{zy}{2r(z-r)}\Big)dy, & r\le x\le z\\ 0, & z\le x\end{cases}$$

5 $R_s^{(3)} = \left(\sum_{k=1}^3 (W_s^{(k)})^2\right)^{1/2}$ $H_z = \min\{s : R_s^{(3)} = z\}$

(1) $z \le r$

$$\mathbf{P}_x\big(\ell(H_z,r) \in dy,\ H_z = \infty\big) = \begin{cases} 0, & 0 \le x \le z \\ \dfrac{x-z}{2x(r-z)} \exp\Big(-\dfrac{y}{2(r-z)}\Big)dy, & z \le x \le r \\ \dfrac{1}{2x} \exp\Big(-\dfrac{y}{2(r-z)}\Big)dy, & r \le x \end{cases}$$

(2)

$$\mathbf{P}_x\big(\ell(H_z,r) = 0,\ H_z < \infty\big) = \begin{cases} \dfrac{z}{x}, & r \le z \le x \\ 1, & 0 \le x \le z \le r, \\ \dfrac{z|r-x|}{x|r-z|}, & z \wedge r \le x \le r \vee z \\ 0, & z \le r \le x,\ 0 \le x \le r \le z \end{cases}$$

(3) $z \le r$

$$\mathbf{P}_x\big(\ell(H_z,r) = 0,\ H_z = \infty\big) = 1 - \frac{r}{x}, \quad r \le x$$

2.3.3 $\mathbf{E}_x\big\{e^{-\alpha H_z - \gamma\ell(H_z,r)};\ H_z < \infty\big\}$

$z \le r$

$$= \begin{cases} \dfrac{z\,\mathrm{sh}(x\sqrt{2\alpha})}{x\,\mathrm{sh}(z\sqrt{2\alpha})}, & 0 \le x \le z \\ \dfrac{z\big(\sqrt{2\alpha}e^{(r-x)\sqrt{2\alpha}} + 2\gamma\,\mathrm{sh}((r-x)\sqrt{2\alpha})\big)}{x\big(\sqrt{2\alpha}e^{(r-z)\sqrt{2\alpha}} + 2\gamma\,\mathrm{sh}((r-z)\sqrt{2\alpha})\big)}, & z \le x \le r \\ \dfrac{z\sqrt{2\alpha}e^{-(x-r)\sqrt{2\alpha}}}{x\big(\sqrt{2\alpha}e^{(r-z)\sqrt{2\alpha}} + 2\gamma\,\mathrm{sh}((r-z)\sqrt{2\alpha})\big)}, & r \le x \end{cases}$$

$r \le z$

$$= \begin{cases} \dfrac{z\sqrt{2\alpha}\,\mathrm{sh}(x\sqrt{2\alpha})}{x\big(\sqrt{2\alpha}\,\mathrm{sh}(z\sqrt{2\alpha}) + 2\gamma\,\mathrm{sh}((z-r)\sqrt{2\alpha})\,\mathrm{sh}(r\sqrt{2\alpha})\big)}, & 0 \le x \le r \\ \dfrac{z\big(\sqrt{2\alpha}\,\mathrm{sh}(x\sqrt{2\alpha}) + 2\gamma\,\mathrm{sh}((x-r)\sqrt{2\alpha})\,\mathrm{sh}(r\sqrt{2\alpha})\big)}{x\big(\sqrt{2\alpha}\,\mathrm{sh}(z\sqrt{2\alpha}) + 2\gamma\,\mathrm{sh}((z-r)\sqrt{2\alpha})\,\mathrm{sh}(r\sqrt{2\alpha})\big)}, & r \le x \le z \\ \dfrac{z}{x}e^{-(x-z)\sqrt{2\alpha}}, & z \le x \end{cases}$$

2.3.4 $\mathbf{E}_x\big\{e^{-\alpha H_z}; \ell(H_z,r) \in dy,\ H_z < \infty\big\}$

$z \le r$

$$= \exp\Big(-\frac{y\sqrt{2\alpha}}{1 - e^{-2(r-z)\sqrt{2\alpha}}}\Big) \begin{cases} 0, & 0 \le x \le z \\ \dfrac{z\sqrt{2\alpha}\,\mathrm{sh}((x-z)\sqrt{2\alpha})}{2x\,\mathrm{sh}^2((r-z)\sqrt{2\alpha})}dy, & z \le x \le r \\ \dfrac{z\sqrt{2\alpha}e^{-(x-r)\sqrt{2\alpha}}}{2x\,\mathrm{sh}((r-z)\sqrt{2\alpha})}dy, & r \le x \end{cases}$$

5 $R_s^{(3)} = \left(\sum_{k=1}^3 (W_s^{(k)})^2\right)^{1/2}$ $H_z = \min\{s : R_s^{(3)} = z\}$

$r \le z$

$$= \exp\Big(-\frac{y\sqrt{2\alpha}\operatorname{sh}(z\sqrt{2\alpha})}{2\operatorname{sh}((z-r)\sqrt{2\alpha})\operatorname{sh}(r\sqrt{2\alpha})}\Big)\begin{cases} 0, & z \le x \\ \dfrac{z\sqrt{2\alpha}\operatorname{sh}((z-x)\sqrt{2\alpha})}{2x\operatorname{sh}^2((z-r)\sqrt{2\alpha})}dy, & r \le x \le z \\ \dfrac{z\sqrt{2\alpha}\operatorname{sh}(x\sqrt{2\alpha})}{2x\operatorname{sh}((z-r)\sqrt{2\alpha})\operatorname{sh}(r\sqrt{2\alpha})}dy, & 0 \le x \le r \end{cases}$$

(1) $\mathbf{E}_x\{e^{-\alpha H_z}; \ell(H_z, r) = 0\}$

$r \le z$

$$= \begin{cases} 0, & 0 \le x \le r \\ \dfrac{z\operatorname{sh}((x-r)\sqrt{2\alpha})}{x\operatorname{sh}((z-r)\sqrt{2\alpha})}, & r \le x \le z \\ \dfrac{z}{x}e^{-(x-z)\sqrt{2\alpha}}, & z \le x \end{cases}$$

$z \le r$

$$= \begin{cases} \dfrac{z\operatorname{sh}(x\sqrt{2\alpha})}{x\operatorname{sh}(z\sqrt{2\alpha})}, & 0 \le x \le z \\ \dfrac{z\operatorname{sh}((r-x)\sqrt{2\alpha})}{x\operatorname{sh}((r-z)\sqrt{2\alpha})}, & z \le x \le r \\ 0, & r \le x \end{cases}$$

2.4.1 $\mathbf{E}_x\Big\{\exp\Big(-\gamma\int_0^{H_z} \mathbb{I}_{[r,\infty)}\big(R_s^{(3)}\big)ds\Big);\ H_z < \infty\Big\}$

$z \le r$

$$= \begin{cases} 1, & 0 \le x \le z \\ \dfrac{z(1+(r-x)\sqrt{2\gamma})}{x(1+(r-z)\sqrt{2\gamma})}, & z \le x \le r \\ \dfrac{ze^{-(x-r)\sqrt{2\gamma}}}{x(1+(r-z)\sqrt{2\gamma})}, & r \le x \end{cases}$$

$r \le z$

$$= \begin{cases} \dfrac{z\sqrt{2\gamma}}{r\sqrt{2\gamma}\operatorname{ch}((z-r)\sqrt{2\gamma}) + \operatorname{sh}((z-r)\sqrt{2\gamma})}, & 0 \le x \le r \\ \dfrac{z\big(r\sqrt{2\gamma}\operatorname{ch}((x-r)\sqrt{2\gamma}) + \operatorname{sh}((x-r)\sqrt{2\gamma})\big)}{x\big(r\sqrt{2\gamma}\operatorname{ch}((z-r)\sqrt{2\gamma}) + \operatorname{sh}((z-r)\sqrt{2\gamma})\big)}, & r \le x \le z \\ \dfrac{z}{x}e^{-(x-z)\sqrt{2\gamma}}, & z \le x \end{cases}$$

(1) $\mathbf{E}_x\Big\{\exp\Big(-\gamma\int_0^{H_z} \mathbb{I}_{[r,\infty)}\big(R_s^{(3)}\big)ds\Big);\ H_z = \infty\Big\} = 0$

2.4.2 $\mathbf{P}_x\Big(\int_0^{H_z} \mathbb{I}_{[r,\infty)}\big(R_s^{(3)}\big)ds \in dy,\ H_z < \infty\Big)$

5 $R_s^{(3)} = \left(\sum_{k=1}^3 (W_s^{(k)})^2\right)^{1/2}$ $\qquad H_z = \min\{s : R_s^{(3)} = z\}$

$z < r$

$$= \begin{cases} 0, & 0 \le x \le z \\ \frac{z(x-z)}{x(r-z)^2}\Big(\frac{1}{\sqrt{2\pi y}} - \frac{1}{2(r-z)}e^{y/2(r-z)^2}\operatorname{Erfc}\Big(\frac{\sqrt{y}}{(r-z)\sqrt{2}}\Big)\Big)dy, & z \le x \le r \\ \frac{z}{x(r-z)\sqrt{2\pi y}}e^{-(x-r)^2/2y}dy- & \\ -\frac{z}{2x(r-z)^2}\exp\Big(\frac{x-r}{r-z}+\frac{y}{2(r-z)^2}\Big)\operatorname{Erfc}\Big(\frac{x-r}{\sqrt{2y}}+\frac{\sqrt{y}}{(r-z)\sqrt{2}}\Big)dy, & r \le x \end{cases}$$

$r < z$

$$= \begin{cases} z\widetilde{\mathrm{rc}}_y(1, z-r, 0, r)dy, & 0 \le x \le r \\ \frac{z}{x}(\mathrm{ss}_y(x-r, b-r) + r\,\mathrm{ss}_y(z-x, z-r) * \widetilde{\mathrm{rc}}_y(1, z-r, 0, r))dy, & r \le x \le z \\ \frac{z}{x}\,\mathrm{h}_y(1, x-z)dy, & z \le x \end{cases}$$

(1) $z \le r$

$$\mathbf{P}_x\Big(\int_0^{H_z} \mathbb{I}_{[r,\infty)}\big(R_s^{(3)}\big)ds = 0\Big) = \begin{cases} 1, & x \le z \\ \frac{z(r-x)}{x(r-z)}, & z \le x \le r \\ 0, & r \le x \end{cases}$$

2.4.4

$$\mathbf{P}_x\Big(\int_0^{H_z} \mathbb{I}_{[r,\infty)}\big(R_s^{(3)}\big)ds \in dv,\ H_z \in dt\Big)$$

$$= \int_0^\infty B_z^{(27)}(t-v, v, y)dy\,dvdt, \qquad \text{for } B_z^{(27)}(u,v,y) \text{ see 2.27.2}$$

(1)

$$\mathbf{P}_x\Big(\int_0^{H_z} \mathbb{I}_{[r,\infty)}\big(R_s^{(3)}\big)ds = 0,\ H_z \in dt\Big)$$

$$= \begin{cases} \frac{z}{x}\,\mathrm{ss}_t(x,z)dt, & 0 \le x \le z \\ \frac{z}{x}\,\mathrm{ss}_t(r-x, r-z)dt, & z \le x \le r \end{cases}$$

2.5.1

$$\mathbf{E}_x\Big\{\exp\Big(-\gamma\int_0^{H_z} \mathbb{I}_{[0,r]}\big(R_s^{(3)}\big)ds\Big);\ H_z < \infty\Big\}$$

$z \le r$

$$= \begin{cases} \frac{z\,\mathrm{sh}(x\sqrt{2\gamma})}{x\,\mathrm{sh}(z\sqrt{2\gamma})}, & 0 \le x \le z \\ \frac{z\,\mathrm{ch}((r-x)\sqrt{2\gamma})}{x\,\mathrm{ch}((r-z)\sqrt{2\gamma})}, & z \le x \le r \\ \frac{z}{x\,\mathrm{ch}((r-z)\sqrt{2\gamma})}, & r \le x \end{cases}$$

5 $R_s^{(3)} = \big(\sum_{k=1}^3 (W_s^{(k)})^2\big)^{1/2}$ $H_z = \min\{s : R_s^{(3)} = z\}$

$r \le z$

$$= \begin{cases} \dfrac{z\,\mathrm{sh}(x\sqrt{2\gamma})}{x\big(\mathrm{sh}(r\sqrt{2\gamma}) + \sqrt{2\gamma}(z-r)\,\mathrm{ch}(r\sqrt{2\gamma})\big)}, & 0 \le x \le r \\ \dfrac{z\big(\mathrm{sh}(r\sqrt{2\gamma}) + \sqrt{2\gamma}(x-r)\,\mathrm{ch}(r\sqrt{2\gamma})\big)}{x\big(\mathrm{sh}(r\sqrt{2\gamma}) + \sqrt{2\gamma}(z-r)\,\mathrm{ch}(r\sqrt{2\gamma})\big)}, & r \le x \le z \\ \dfrac{z}{x}, & z \le x \end{cases}$$

(1) $z \le r$

$$\mathbf{E}_x\Big\{\exp\Big(-\gamma\int_0^{H_z} \mathbb{I}_{[0,r]}\big(R_s^{(3)}\big)ds\Big);\ H_z = \infty\Big\}$$

$$= \begin{cases} 0, & 0 \le x \le z \\ \dfrac{\mathrm{sh}((x-z)\sqrt{2\gamma})}{x\sqrt{2\gamma}\,\mathrm{ch}((r-z)\sqrt{2\gamma})}, & z \le x \le r \\ 1 - \dfrac{r}{x} + \dfrac{\mathrm{sh}((r-z)\sqrt{2\gamma})}{x\sqrt{2\gamma}\,\mathrm{ch}((r-z)\sqrt{2\gamma})}, & r \le x \end{cases}$$

2.5.2

$$\mathbf{P}_x\Big(\int_0^{H_z} \mathbb{I}_{[0,r]}\big(R_s^{(3)}\big)ds \in dy,\ H_z < \infty\Big)$$

$r < z$

$$= \begin{cases} \dfrac{z}{x}\,\mathrm{rc}_y(0,r,x,z-r)dy, & 0 \le x \le r \\ \dfrac{z(z-x)}{x(z-r)}\,\mathrm{rc}_y(0,r,r,z-r)dy, & r \le x \le z \end{cases}$$

$z < r$

$$= \begin{cases} zx^{-1}\,\mathrm{ss}_y(x,z)dy, & 0 \le x < z \\ zx^{-1}\,\mathrm{cc}_y(r-x,r-z)dy, & z < x \le r \\ zx^{-1}\,\mathrm{cc}_y(0,r-z)dy, & r \le x \end{cases}$$

(1) $z \le r$

$$\mathbf{P}_x\Big(\int_0^{H_z} \mathbb{I}_{[0,r]}\big(R_s^{(3)}\big)ds \in dy,\ H_z = \infty\Big) = \begin{cases} 0, & 0 \le x \le z \\ \dfrac{1}{x}\,\mathrm{sc}_y(x-z,r-z)dy, & z \le x \le r \\ \dfrac{1}{x}\,\mathrm{sc}_y(r-z,r-z)dy, & r \le x \end{cases}$$

(2) $r \le z$

$$\mathbf{P}_x\Big(\int_0^{H_z} \mathbb{I}_{[0,r]}\big(R_s^{(3)}\big)ds = 0\Big) = \begin{cases} 0, & x \le r \\ \dfrac{z(x-r)}{x(z-r)}, & r \le x \le z \\ zx^{-1}, & z \le x \end{cases}$$

2.5.4

$$\mathbf{P}_x\Big(\int_0^{H_z} \mathbb{I}_{[0,r]}\big(R_s^{(3)}\big)ds \in dv,\ H_z \in dt\Big)$$

$$= \int_0^\infty B_z^{(27)}(v,t-v,y)dy\,dvdt,$$ for $B_z^{(27)}(u,v,y)$ see 2.27.2

5 $R_s^{(3)} = \left(\sum_{k=1}^{3}(W_s^{(k)})^2\right)^{1/2}$ $H_z = \min\{s : R_s^{(3)} = z\}$

(1) $$\mathbf{P}_x\Big(\int_0^{H_z} \mathbb{I}_{[0,r]}\big(R_s^{(3)}\big)ds = 0,\ H_z \in dt\Big)$$

$$= \begin{cases} \dfrac{z}{x}\,\mathrm{ss}_t(x-r, z-r)dt, & r \le x \le z \\ \dfrac{z}{x}\,\mathrm{h}_t(1, x-z)dt, & z \le x \end{cases}$$

2.6.1 $$\mathbf{E}_x\Big\{\exp\Big(-\int_0^{H_z} \big(p\mathbb{I}_{[0,r)}\big(R_s^{(3)}\big) + q\mathbb{I}_{[r,\infty)}\big(R_s^{(3)}\big)\big)ds\Big);\ H_z < \infty\Big\}$$

$z \le r$ $$= \begin{cases} \dfrac{z\,\mathrm{sh}(x\sqrt{2p})}{x\,\mathrm{sh}(z\sqrt{2p})}, & 0 \le x \le z \\ \dfrac{z\big(\sqrt{p}\,\mathrm{ch}((r-x)\sqrt{2p}) + \sqrt{q}\,\mathrm{sh}((r-x)\sqrt{2p})\big)}{x\big(\sqrt{p}\,\mathrm{ch}((r-z)\sqrt{2p}) + \sqrt{q}\,\mathrm{sh}((r-z)\sqrt{2p})\big)}, & z \le x \le r \\ \dfrac{z\sqrt{p}e^{-(x-r)\sqrt{2q}}}{x\big(\sqrt{p}\,\mathrm{ch}((r-z)\sqrt{2p}) + \sqrt{q}\,\mathrm{sh}((r-z)\sqrt{2p})\big)}, & r \le x \end{cases}$$

$r \le z$ $$= \begin{cases} \dfrac{z\sqrt{q}\,\mathrm{sh}(x\sqrt{2p})}{x\big(\sqrt{q}\,\mathrm{sh}(r\sqrt{2p})\,\mathrm{ch}((z-r)\sqrt{2q}) + \sqrt{p}\,\mathrm{ch}(r\sqrt{2p})\,\mathrm{sh}((z-r)\sqrt{2q})\big)}, & 0 \le x \le r \\ \dfrac{z\big(\sqrt{q}\,\mathrm{sh}(r\sqrt{2p})\,\mathrm{ch}((x-r)\sqrt{2q}) + \sqrt{p}\,\mathrm{ch}(r\sqrt{2p})\,\mathrm{sh}((x-r)\sqrt{2q})\big)}{x\big(\sqrt{q}\,\mathrm{sh}(r\sqrt{2p})\,\mathrm{ch}((z-r)\sqrt{2q}) + \sqrt{p}\,\mathrm{ch}(r\sqrt{2p})\,\mathrm{sh}((z-r)\sqrt{2q})\big)}, & r \le x \le z \\ zx^{-1}e^{-(x-z)\sqrt{2q}}, & z \le x \end{cases}$$

2.6.2 $$\mathbf{P}_x\Big(\int_0^{H_z} \mathbb{I}_{[0,r)}\big(R_s^{(3)}\big)ds \in du,\ \int_0^{H_z} \mathbb{I}_{[r,\infty)}\big(R_s^{(3)}\big)ds \in dv\Big)$$

$$= \int_0^{\infty} B_z^{(27)}(u,v,y)dy\,dudv,$$ for $B_z^{(27)}(u,v,y)$ see 2.27.2

(1) $$\mathbf{P}_x\Big(\int_0^{H_z} \mathbb{I}_{[0,r)}\big(R_s^{(3)}\big)ds \in du,\ \int_0^{H_z} \mathbb{I}_{[r,\infty)}\big(R_s^{(3)}\big)ds = 0\Big)$$

$$= \begin{cases} \dfrac{z}{x}\,\mathrm{ss}_u(x, z)du, & 0 \le x \le z \\ \dfrac{z}{x}\,\mathrm{ss}_u(r-x, r-z)du, & z \le x \le r \end{cases}$$

(2) $$\mathbf{P}_x\Big(\int_0^{H_z} \mathbb{I}_{[0,r)}\big(R_s^{(3)}\big)ds = 0,\ \int_0^{H_z} \mathbb{I}_{[r,\infty)}\big(R_s^{(3)}\big)ds \in dv\Big)$$

$$= \begin{cases} \dfrac{z}{x}\,\mathrm{ss}_v(x-r, z-r)dv, & r \le x \le z \\ \dfrac{z}{x}\,\mathrm{h}_v(1, x-z)dv, & z \le x \end{cases}$$

5 $R_s^{(3)} = \left(\sum_{k=1}^3 (W_s^{(k)})^2\right)^{1/2}$ $H_z = \min\{s : R_s^{(3)} = z\}$

2.7.1 $r < u$

$$\mathbf{E}_x\Big\{\exp\Big(-\gamma\int_0^{H_z} \mathbb{1}_{[r,u]}\big(R_s^{(3)}\big)ds\Big),\ H_z < \infty\Big\}$$

$z \le r$

$$=\begin{cases} 1, & 0 \le x \le z\\ \dfrac{z((r-x)\sqrt{2\gamma}\,\mathrm{sh}((u-r)\sqrt{2\gamma}) + \mathrm{ch}((u-r)\sqrt{2\gamma}))}{x((r-z)\sqrt{2\gamma}\,\mathrm{sh}((u-r)\sqrt{2\gamma}) + \mathrm{ch}((u-r)\sqrt{2\gamma}))}, & z \le x \le r\\ \dfrac{z\,\mathrm{ch}((u-x)\sqrt{2\gamma})}{x((r-z)\sqrt{2\gamma}\,\mathrm{sh}((u-r)\sqrt{2\gamma}) + \mathrm{ch}((u-r)\sqrt{2\gamma}))}, & r \le x \le u\\ \dfrac{z}{x((r-z)\sqrt{2\gamma}\,\mathrm{sh}((u-r)\sqrt{2\gamma}) + \mathrm{ch}((u-r)\sqrt{2\gamma}))}, & u \le x \end{cases}$$

$u \le z$

$$=\begin{cases} \dfrac{z\sqrt{2\gamma}}{(1+(z-u)r2\gamma)\,\mathrm{sh}((u-r)\sqrt{2\gamma}) + (r+z-u)\sqrt{2\gamma}\,\mathrm{ch}((u-r)\sqrt{2\gamma})}, & 0 \le x \le r\\ \dfrac{z(\mathrm{sh}((x-r)\sqrt{2\gamma}) + r\sqrt{2\gamma}\,\mathrm{ch}((x-r)\sqrt{2\gamma}))}{x((1+(z-u)r2\gamma)\,\mathrm{sh}((u-r)\sqrt{2\gamma}) + (r+z-u)\sqrt{2\gamma}\,\mathrm{ch}((u-r)\sqrt{2\gamma}))}, & r \le x \le u\\ \dfrac{z((1+(x-u)r2\gamma)\,\mathrm{sh}((u-r)\sqrt{2\gamma}) + (r+x-u)\sqrt{2\gamma}\,\mathrm{ch}((u-r)\sqrt{2\gamma}))}{x((1+(z-u)r2\gamma)\,\mathrm{sh}((u-r)\sqrt{2\gamma}) + (r+z-u)\sqrt{2\gamma}\,\mathrm{ch}((u-r)\sqrt{2\gamma}))}, & u \le x \le z\\ \dfrac{z}{x}, & z \le x \end{cases}$$

2.8.3

$$\mathbf{E}_x \exp\Big(-\alpha H_z - \gamma\int_0^{H_z} R_s^{(3)}ds\Big) = \begin{cases} \dfrac{z\sqrt{\alpha+\gamma x}\,I_{\frac13}\Big(\frac{2\sqrt2}{3\gamma}(\alpha+\gamma x)^{3/2}\Big)}{x\sqrt{\alpha+\gamma z}\,I_{\frac13}\Big(\frac{2\sqrt2}{3\gamma}(\alpha+\gamma z)^{3/2}\Big)}, & 0 \le x \le z\\ \dfrac{z\sqrt{\alpha+\gamma x}\,K_{\frac13}\Big(\frac{2\sqrt2}{3\gamma}(\alpha+\gamma x)^{3/2}\Big)}{x\sqrt{\alpha+\gamma z}\,K_{\frac13}\Big(\frac{2\sqrt2}{3\gamma}(\alpha+\gamma z)^{3/2}\Big)}, & z \le x \end{cases}$$

2.9.1

$$\mathbf{E}_x\Big\{\exp\Big(-2\gamma\int_0^{H_z} (R_s^{(3)})^2 ds\Big);\ H_z < \infty\Big\} = \begin{cases} \dfrac{\sqrt{z}\,I_{1/4}(x^2\sqrt{\gamma})}{\sqrt{x}\,I_{1/4}(z^2\sqrt{\gamma})}, & 0 \le x \le z\\ \dfrac{\sqrt{z}\,K_{1/4}(x^2\sqrt{\gamma})}{\sqrt{x}\,K_{1/4}(z^2\sqrt{\gamma})}, & z \le x \end{cases}$$

2.9.3

$$\mathbf{E}_x \exp\Big(-\alpha H_z - 2\gamma\int_0^{H_z} (R_s^{(3)})^2 ds\Big)$$

$$=\begin{cases} \dfrac{z\big(D_{-\frac12-\frac{\alpha}{\gamma}}(-x\sqrt{2\gamma}) - D_{-\frac12-\frac{\alpha}{\gamma}}(x\sqrt{2\gamma})\big)}{x\big(D_{-\frac12-\frac{\alpha}{\gamma}}(-z\sqrt{2\gamma}) - D_{-\frac12-\frac{\alpha}{\gamma}}(z\sqrt{2\gamma})\big)}, & 0 \le x \le z\\ \dfrac{zD_{-\frac12-\frac{\alpha}{\gamma}}(x\sqrt{2\gamma})}{xD_{-\frac12-\frac{\alpha}{\gamma}}(z\sqrt{2\gamma})}, & z \le x \end{cases}$$

5 $R_s^{(3)} = \left(\sum_{k=1}^3 (W_s^{(k)})^2\right)^{1/2}$ $\qquad H_z = \min\{s : R_s^{(3)} = z\}$

2.10.3

$$\mathbf{E}_x \exp\Bigl(-\alpha H_z - \gamma\int_0^{H_z} e^{2\beta R_s^{(3)}} ds\Bigr) = \begin{cases} \dfrac{zS_{\sqrt{2\alpha}/|\beta|}\Bigl(\frac{\sqrt{2\gamma}}{|\beta|}e^{\beta x}, \frac{\sqrt{2\gamma}}{|\beta|}\Bigr)}{xS_{\sqrt{2\alpha}/|\beta|}\Bigl(\frac{\sqrt{2\gamma}}{|\beta|}e^{\beta z}, \frac{\sqrt{2\gamma}}{|\beta|}\Bigr)}, & 0 \le x \le z \\ \dfrac{zK_{\sqrt{2\alpha}/|\beta|}\Bigl(\frac{\sqrt{2\gamma}}{|\beta|}e^{\beta x}\Bigr)}{xK_{\sqrt{2\alpha}/|\beta|}\Bigl(\frac{\sqrt{2\gamma}}{|\beta|}e^{\beta z}\Bigr)}, & z \le x, \quad \beta > 0 \\ \dfrac{zI_{\sqrt{2\alpha}/|\beta|}\Bigl(\frac{\sqrt{2\gamma}}{|\beta|}e^{\beta x}\Bigr)}{xI_{\sqrt{2\alpha}/|\beta|}\Bigl(\frac{\sqrt{2\gamma}}{|\beta|}e^{\beta z}\Bigr)}, & z \le x, \quad \beta < 0 \end{cases}$$

2.11.2 $z < x$

$$\mathbf{P}_x\Bigl(\sup_{y\in[z,\infty)} \ell(H_z, y) < h,\ H_z < \infty\Bigr) = \frac{8z}{x}\sum_{k=1}^{\infty} \frac{\exp(-zj_{0,k}^2/2h)}{j_{0,k}^3 J_1(j_{0,k})}$$

2.13.1 $z < x$

$$\mathbf{E}_x\bigl\{e^{-\gamma \check{H}(H_z)};\ \sup_{0\le s\le H_z} R_s^{(3)} \in dy\bigr\} = \frac{z\,\mathrm{sh}((x-z)\sqrt{2\gamma})}{x(y-z)\,\mathrm{sh}((y-z)\sqrt{2\gamma})}dy, \qquad x < y$$

2.13.2 $z < x$, $x < y$

$$\mathbf{P}_x\bigl(\check{H}(H_z) \in du,\ \sup_{0\le s\le H_z} R_s^{(3)} \in dy\bigr) = \frac{z}{x(y-z)}\,\mathrm{ss}_u(x-z, y-z)dudy$$

2.13.4 $z < x$, $x < y$

$$\mathbf{E}_x\bigl\{e^{-\alpha H_z - \gamma \check{H}(H_z)};\ \sup_{0\le s\le H_z} R_s^{(3)} \in dy\bigr\} = \frac{z\sqrt{2\alpha}\,\mathrm{sh}((x-z)\sqrt{2\gamma+2\alpha})}{x\,\mathrm{sh}((y-z)\sqrt{2\alpha})\,\mathrm{sh}((y-z)\sqrt{2\gamma+2\alpha})}dy$$

(1) $u < t$

$$\mathbf{P}_x\bigl(\check{H}(H_z) \in du,\ H_z \in dt,\ \sup_{0\le s\le H_z} R_s^{(3)} \in dy\bigr)$$
$$= \frac{z}{x}\,\mathrm{ss}_u(x-z, y-z)\,\mathrm{s}_{t-u}(y-z)dudtdy$$

2.14.1 $x < z$

$$\mathbf{E}_x\bigl\{e^{-\gamma \hat{H}(H_z)};\ \inf_{0\le s\le H_z} R_s^{(3)} \in dy\bigr\} = \frac{z\,\mathrm{sh}((z-x)\sqrt{2\gamma})}{x(z-y)\,\mathrm{sh}((z-y)\sqrt{2\gamma})}dy, \qquad y < x$$

2.14.2 $y < x$, $x < z$

$$\mathbf{P}_x\bigl(\hat{H}(H_z) \in du,\ \inf_{0\le s\le H_z} R_s^{(3)} \in dy\bigr) = \frac{z}{x(z-y)}\,\mathrm{ss}_u(z-x, z-y)dudy$$

2.14.4 $y < x$, $x < z$

$$\mathbf{E}_x\bigl\{e^{-\alpha H_z - \gamma \hat{H}(H_z)};\ \inf_{0\le s\le H_z} R_s^{(3)} \in dy\bigr\} = \frac{z\sqrt{2\alpha}\,\mathrm{sh}((z-x)\sqrt{2\gamma+2\alpha})}{x\,\mathrm{sh}((z-y)\sqrt{2\alpha})\,\mathrm{sh}((z-y)\sqrt{2\gamma+2\alpha})}dy$$

5 $R_s^{(3)} = \left(\sum_{k=1}^{3}(W_s^{(k)})^2\right)^{1/2}$ $H_z = \min\{s : R_s^{(3)} = z\}$

(1)
$u < t$

$$\mathbf{P}_x\big(\hat{H}(H_z) \in du,\, H_z \in dt,\, \inf_{0\le s\le H_z} R_s^{(3)} \in dy\big)$$

$$= \frac{z}{x}\,\mathrm{ss}_u(z-x, z-y)\,\mathrm{s}_{t-u}(z-y)dudtdy$$

2.15.2

$$\mathbf{P}_x\big(a < \inf_{0\le s\le H_z} R_s^{(3)},\, \sup_{0\le s\le H_z} R_s^{(3)} < b\big) = \begin{cases} \dfrac{z(x-a)}{x(z-a)}, & a \le x \le z \\ \dfrac{z(b-x)}{x(b-z)}, & z \le x \le b \end{cases}$$

2.15.4

$$\mathbf{E}_x\big\{e^{-\alpha H_z}, a < \inf_{0\le s\le H_z} R_s^{(3)},\, \sup_{0\le s\le H_z} R_s^{(3)} < b\big\}$$

$$= \begin{cases} \dfrac{z\,\mathrm{sh}((x-a)\sqrt{2\alpha})}{x\,\mathrm{sh}((z-a)\sqrt{2\alpha})}, & -b \le x \le z \\ \dfrac{z\,\mathrm{sh}((b-x)\sqrt{2\alpha})}{x\,\mathrm{sh}((b-z)\sqrt{2\alpha})}, & z \le x \le b \end{cases}$$

(1)

$$\mathbf{P}_x\big(H_z \in dt, a < \inf_{0\le s\le t} R_s^{(3)},\, \sup_{0\le s\le t} R_s^{(3)} < b\big)$$

$$= \begin{cases} \dfrac{z}{x}\,\mathrm{ss}_t(x-a, z-a)dt, & a \le x \le z \\ \dfrac{z}{x}\,\mathrm{ss}_t(b-x, b-z)dt, & z \le x \le b \end{cases}$$

2.18.1
$r < u$

$$\mathbf{E}_x\big\{\exp(-\gamma\ell(H_z,r) - \eta\ell(H_z,u));\ H_z < \infty\big\}$$

$z < r$

$$= \begin{cases} 1, & 0 \le x \le z \\ \dfrac{z(1+2\gamma(r-x)+2\eta(u-x)+4\gamma\eta(u-r)(r-x))}{x(1+2\gamma(r-z)+2\eta(u-z)+4\gamma\eta(u-r)(r-z))}, & z \le x \le r \\ \dfrac{z(1+2\eta(u-x))}{x(1+2\gamma(r-z)+2\eta(u-z)+4\gamma\eta(u-r)(r-z))}, & r \le x \le u \\ \dfrac{z}{x(1+2\gamma(r-z)+2\eta(u-z)+4\gamma\eta(u-r)(r-z))}, & u \le x \end{cases}$$

$u < z$

$$= \begin{cases} \dfrac{z}{z+2\gamma(z-r)r+2\eta(z-u)u+4\gamma\eta(z-u)(u-r)r}, & 0 \le x \le r \\ \dfrac{z(x+2\gamma(x-r)r)}{x(z+2\gamma(z-r)r+2\eta(z-u)u+4\gamma\eta(z-u)(u-r)r)}, & r \le x \le u \\ \dfrac{z(x+2\gamma(x-r)r+2\eta(x-u)u+4\gamma\eta(x-u)(u-r)r)}{x(z+2\gamma(z-r)r+2\eta(z-u)u+4\gamma\eta(z-u)(u-r)r)}, & u \le x \le z \\ \dfrac{z}{x}, & z \le x \end{cases}$$

5 $R_s^{(3)} = \left(\sum_{k=1}^{3}(W_s^{(k)})^2\right)^{1/2}$ $\qquad H_z = \min\{s : R_s^{(3)} = z\}$

(1) $r < u$

$$\mathbf{E}_x\{\exp(-\gamma\ell(H_z,r) - \eta\ell(H_z,u));\ H_z = \infty\}$$

$z < r$

$$= \begin{cases} 0, & 0 \le x \le z \\ \dfrac{x-z}{x(1+2\gamma(r-z)+2\eta(u-z)+4\gamma\eta(u-r)(r-z))}, & z \le x \le r \\ \dfrac{x-z+2\gamma(x-r)(r-z)}{x(1+2\gamma(r-z)+2\eta(u-z)+4\gamma\eta(u-r)(r-z))}, & r \le x \le u \\ 1 - \dfrac{z+2\gamma r(r-z)+2\eta u(u-z)+4\gamma\eta u(u-r)(r-z)}{x(1+2\gamma(r-z)+2\eta(u-z)+4\gamma\eta(u-r)(r-z))}, & u \le x \end{cases}$$

2.18.2 $r < u$

$$\mathbf{P}_x(\ell(H_z,r) \in dy,\ \ell(H_z,u) \in dg,\ H_z < \infty)$$

$z < r$

$$= \tfrac{z}{x} B_{x,z}^{(18)}(y,g)dydg, \qquad \text{for } B_{x,z}^{(18)}(y,g) \text{ see 1.2.18.2}$$

$u < z$

$$= \begin{cases} q_r dydg, & 0 \le x \le r \\ \dfrac{r(u-x)}{x(u-r)} q_r dydg + \dfrac{u(x-r)}{x(u-r)} q_u dydg, & r \le x \le u \\ \dfrac{u(z-x)}{x(z-u)} q_u dydg, & u \le x \le z \end{cases}$$

where

$$q_r := \frac{z}{4r(z-u)(u-r)} \exp\Big(-\frac{uy}{2r(u-r)} - \frac{(z-r)g}{2(z-u)(u-r)}\Big) I_0\Big(\frac{\sqrt{yg}}{u-r}\Big)$$

$$q_u := \frac{z\sqrt{g}}{4u(z-u)(u-r)\sqrt{y}} \exp\Big(-\frac{uy}{2r(u-r)} - \frac{(z-r)g}{2(z-u)(u-r)}\Big) I_1\Big(\frac{\sqrt{yg}}{u-r}\Big)$$

(1)

$$\mathbf{P}_x(\ell(H_z,r) \in dy,\ \ell(H_z,u) = 0,\ H_z < \infty)$$

$$= \begin{cases} \dfrac{z(x-z)}{2x(r-z)^2} \exp\Big(-\dfrac{y(u-z)}{2(u-r)(r-z)}\Big)dy, & z \le x \le r \\ \dfrac{z(u-x)}{2x(u-r)(r-z)} \exp\Big(-\dfrac{y(u-z)}{2(u-r)(r-z)}\Big)dy, & r \le x \le u \end{cases}$$

(2)

$$\mathbf{P}_x(\ell(H_z,r) = 0,\ \ell(H_z,u) \in dg,\ H_z < \infty)$$

$$= \begin{cases} \dfrac{z(x-r)}{2x(z-u)(u-r)} \exp\Big(-\dfrac{g(z-r)}{2(z-u)(u-r)}\Big)dg, & r \le x \le u \\ \dfrac{z(z-x)}{2x(z-u)^2} \exp\Big(-\dfrac{g(z-r)}{2(z-u)(u-r)}\Big)dg, & u \le x \le z \end{cases}$$

5 $R_s^{(3)} = \left(\sum_{k=1}^3 (W_s^{(k)})^2\right)^{1/2}$ $H_z = \min\{s : R_s^{(3)} = z\}$

(3) $$\mathbf{P}_x\big(\ell(H_z,r)=0,\ \ell(H_z,u)=0,\ H_z<\infty\big) = \begin{cases} \dfrac{z(r-x)}{x(r-z)}, & z\le x\le r \\ \dfrac{z(x-u)}{x(z-u)}, & u\le x\le z \end{cases}$$

(4) $$\mathbf{P}_x\big(\ell(H_z,r)\in dy,\ \ell(H_z,u)\in dg,\ H_z=\infty\big)$$

$z<r$ $$= \begin{cases} \dfrac{r(x-z)}{x(r-z)}p_r^\infty dydg, & z\le x\le r \\ \dfrac{r(u-x)}{x(u-r)}p_r^\infty dydg + \dfrac{u(x-r)}{x(u-r)}p_u^\infty dydg, & r\le x\le u \\ \dfrac{u}{x}p_u^\infty dydg, & u\le x \end{cases}$$

where

$$p_r^\infty := \frac{1}{4r(u-r)}\exp\Big(-\frac{y+g}{2(u-r)}-\frac{y}{2(r-z)}\Big)I_0\Big(\frac{\sqrt{yg}}{u-r}\Big)$$

$$p_u^\infty := \frac{\sqrt{g}}{4u(u-r)\sqrt{y}}\exp\Big(-\frac{y+g}{2(u-r)}-\frac{y}{2(r-z)}\Big)I_1\Big(\frac{\sqrt{yg}}{u-r}\Big)$$

(5) $$\mathbf{P}_x\big(\ell(H_z,r)=0,\ \ell(H_z,u)\in dg,\ H_z=\infty\big)$$

$$= \begin{cases} \dfrac{x-r}{2x(u-r)}\exp\Big(-\dfrac{g}{2(u-r)}\Big)dg, & r\le x\le u \\ \dfrac{1}{2x}\exp\Big(-\dfrac{g}{2(u-r)}\Big)dg, & u\le x \end{cases}$$

(6) $$\mathbf{P}_x\big(\ell(H_z,r)=0,\ \ell(H_z,u)=0,\ H_z=\infty\big) = 1-\frac{u}{x}, \qquad u\le x$$

2.19.1 $$\mathbf{E}_x\Big\{\exp\Big(-\gamma\int_0^{H_z}\frac{ds}{R_s^{(3)}}\Big);\ H_z<\infty\Big\} = \begin{cases} \dfrac{x^{-1/2}I_1(2\sqrt{2\gamma x})}{z^{-1/2}I_1(2\sqrt{2\gamma z})}, & 0<x\le z \\ \dfrac{x^{-1/2}K_1(2\sqrt{2\gamma x})}{z^{-1/2}K_1(2\sqrt{2\gamma z})}, & z\le x \end{cases}$$

2.19.3 $$\mathbf{E}_x\exp\Big(-\alpha H_z-\gamma\int_0^{H_z}\frac{ds}{R_s^{(3)}}\Big) = \begin{cases} \dfrac{zM_{-\gamma/\sqrt{2\alpha},1/2}(2x\sqrt{2\alpha})}{xM_{-\gamma/\sqrt{2\alpha},1/2}(2z\sqrt{2\alpha})}, & 0<x\le z \\ \dfrac{zW_{-\gamma/\sqrt{2\alpha},1/2}(2x\sqrt{2\alpha})}{xW_{-\gamma/\sqrt{2\alpha},1/2}(2z\sqrt{2\alpha})}, & z\le x \end{cases}$$

5 $R_s^{(3)} = \left(\sum_{k=1}^{3}(W_s^{(k)})^2\right)^{1/2}$ $\qquad H_z = \min\{s : R_s^{(3)} = z\}$

2.20.1

$$\mathbf{E}_x\Big\{\exp\Big(-\frac{\gamma^2}{2}\int_0^{H_z}\frac{ds}{(R_s^{(3)})^2}\Big); H_z < \infty\Big\} = \begin{cases} \Big(\dfrac{x}{z}\Big)^{-1/2+\sqrt{\gamma^2+1/4}}, & 0 < x \le z \\ \Big(\dfrac{x}{z}\Big)^{-1/2-\sqrt{\gamma^2+1/4}}, & z \le x \end{cases}$$

2.20.3

$$\mathbf{E}_x \exp\Big(-\alpha H_z - \frac{\gamma^2}{2}\int_0^{H_z}\frac{ds}{(R_s^{(3)})^2}\Big) = \begin{cases} \dfrac{\sqrt{z}I_{\sqrt{\gamma^2+1/4}}(x\sqrt{2\alpha})}{\sqrt{x}I_{\sqrt{\gamma^2+1/4}}(z\sqrt{2\alpha})}, & 0 < x \le z \\ \dfrac{\sqrt{z}K_{\sqrt{\gamma^2+1/4}}(x\sqrt{2\alpha})}{\sqrt{x}K_{\sqrt{\gamma^2+1/4}}(z\sqrt{2\alpha})}, & z \le x \end{cases}$$

2.20.4
$0 < y$

$$\mathbf{P}_x\Big(\int_0^{H_z}\frac{ds}{(R_s^{(3)})^2} \in dy\Big) = \begin{cases} \dfrac{z^{1/2}\ln(z/x)}{x^{1/2}\sqrt{2\pi}y^{3/2}}e^{-y/8}e^{-\ln^2(z/x)/2y}dy, & 0 < x \le z \\ \dfrac{z^{1/2}\ln(x/z)}{x^{1/2}\sqrt{2\pi}y^{3/2}}e^{-y/8}e^{-\ln^2(x/z)/2y}dy, & z \le x \end{cases}$$

2.21.1

$$\mathbf{E}_x\Big\{\exp\Big(-\int_0^{H_z}\Big(\frac{p^2}{2(R_s^{(3)})^2} + \frac{q^2(R_s^{(3)})^2}{2}\Big)ds\Big); H_z < \infty\Big\}$$

$$= \begin{cases} \dfrac{x^{-1/2}I_{\sqrt{4p^2+1}/4}(qx^2/2)}{z^{-1/2}I_{\sqrt{4p^2+1}/4}(qz^2/2)}, & 0 < x \le z \\ \dfrac{x^{-1/2}K_{\sqrt{4p^2+1}/4}(qx^2/2)}{z^{-1/2}K_{\sqrt{4p^2+1}/4}(qz^2/2)}, & z \le x \end{cases}$$

2.21.3

$$\mathbf{E}_x \exp\Big(-\alpha H_z - \int_0^{H_z}\Big(\frac{p^2}{2(R_s^{(3)})^2} + \frac{q^2(R_s^{(3)})^2}{2}\Big)ds\Big)$$

$$= \begin{cases} \dfrac{x^{-3/2}M_{-\alpha/2q,\sqrt{4p^2+1}/4}(qx^2)}{z^{-3/2}M_{-\alpha/2q,\sqrt{4p^2+1}/4}(qz^2)}, & 0 < x \le z \\ \dfrac{x^{-3/2}W_{-\alpha/2q,\sqrt{4p^2+1}/4}(qx^2)}{z^{-3/2}W_{-\alpha/2q,\sqrt{4p^2+1}/4}(qx^2)}, & z \le x \end{cases}$$

2.22.1

$$\mathbf{E}_x\Big\{\exp\Big(-\int_0^{H_z}\Big(\frac{p^2}{2(R_s^{(3)})^2} + \frac{q}{R_s^{(3)}}\Big)ds\Big); H_z < \infty\Big\}$$

5 $R_s^{(3)} = \left(\sum_{k=1}^3 (W_s^{(k)})^2\right)^{1/2}$ $H_z = \min\{s : R_s^{(3)} = z\}$

$$= \begin{cases} \dfrac{x^{-1/2} I_{\sqrt{4p^2+1}}(2\sqrt{2qx})}{z^{-1/2} I_{\sqrt{4p^2+1}}(2\sqrt{2qz})}, & 0 < x \le z \\ \dfrac{x^{-1/2} K_{\sqrt{4p^2+1}}(2\sqrt{2qx})}{z^{-1/2} K_{\sqrt{4p^2+1}}(2\sqrt{2qz})}, & z \le x \end{cases}$$

2.22.3 $$\mathbf{E}_x \exp\Bigl(-\alpha H_z - \int_0^{H_z} \Bigl(\frac{p^2}{2(R_s^{(3)})^2} + \frac{q}{R_s^{(3)}}\Bigr) ds\Bigr)$$

$$= \begin{cases} \dfrac{x^{-1} M_{-q/\sqrt{2\alpha}, \sqrt{p^2+1/4}}(2x\sqrt{2\alpha})}{z^{-1} M_{-q/\sqrt{2\alpha}, \sqrt{p^2+1/4}}(2z\sqrt{2\alpha})}, & 0 < x \le z \\ \dfrac{x^{-1} W_{-q/\sqrt{2\alpha}, \sqrt{p^2+1/4}}(2x\sqrt{2\alpha})}{z^{-1} W_{-q/\sqrt{2\alpha}, \sqrt{p^2+1/4}}(2z\sqrt{2\alpha})}, & z \le x \end{cases}$$

2.27.1 $$\mathbf{E}_x\Bigl\{\exp\Bigl(-\int_0^{H_z} \bigl(p\mathbb{I}_{[0,r)}(R_s^{(3)}) + q\mathbb{I}_{[r,\infty)}(R_s^{(3)})\bigr) ds\Bigr); \tfrac{1}{2}\ell(H_z, r) \in dy\Bigr\}$$

$z \le r$ $$= \exp\Bigl(-y\sqrt{2p}\operatorname{cth}((r-z)\sqrt{2p}) - y\sqrt{2q}\Bigr)$$

$$\times \begin{cases} \dfrac{z\sqrt{2p}\operatorname{sh}((x-z)\sqrt{2p})}{x\operatorname{sh}^2((r-z)\sqrt{2p})} dy, & z \le x \le r \\ \dfrac{z\sqrt{2p}}{x\operatorname{sh}((r-z)\sqrt{2p})} e^{-(x-r)\sqrt{2q}} dy, & r \le x \end{cases}$$

$r \le z$ $$= \exp\bigl(-y(\sqrt{2q}\operatorname{cth}((z-r)\sqrt{2q}) + \sqrt{2p}\operatorname{cth}(r\sqrt{2p}))\bigr)$$

$$\times \begin{cases} \dfrac{z\sqrt{2q}\operatorname{sh}(x\sqrt{2p})}{x\operatorname{sh}((z-r)\sqrt{2q})\operatorname{sh}(r\sqrt{2p})} dy, & 0 \le x \le r \\ \dfrac{z\sqrt{2q}\operatorname{sh}((z-x)\sqrt{2q})}{x\operatorname{sh}^2((z-r)\sqrt{2q})} dy, & r \le x \le z \end{cases}$$

(1) $$\mathbf{E}_x\Bigl\{\exp\Bigl(-\int_0^{H_z} \bigl(p\mathbb{I}_{[0,r)}(R_s^{(3)}) + q\mathbb{I}_{[r,\infty)}(R_s^{(3)})\bigr) ds\Bigr); \ell(H_z, r) = 0\Bigr\}$$

$$= \begin{cases} \dfrac{z\operatorname{sh}(x\sqrt{2p})}{x\operatorname{sh}(z\sqrt{2p})}, & 0 \le x \le z \\ \dfrac{z\operatorname{sh}((r-x)\sqrt{2p})}{x\operatorname{sh}((r-z)\sqrt{2p})}, & z \le x \le r \\ \dfrac{z\operatorname{sh}((x-r)\sqrt{2q})}{x\operatorname{sh}((z-r)\sqrt{2q})}, & r \le x \le z \\ \dfrac{z}{x} e^{-(x-z)\sqrt{2q}}, & z \le x \end{cases}$$

5 $R_s^{(3)} = \left(\sum_{k=1}^3 (W_s^{(k)})^2\right)^{1/2}$ $\qquad H_z = \min\{s : R_s^{(3)} = z\}$

2.27.2 $$\mathbf{P}_x\Big(\int_0^{H_z} \mathbb{I}_{[0,r)}\big(R_s^{(3)}\big)ds \in du,\ \int_0^{H_z} \mathbb{I}_{[r,\infty)}\big(R_s^{(3)}\big)ds \in dv,\ \frac{1}{2}\ell\big(H_z, r\big) \in dy\Big)$$
$$=: B_z^{(27)}(u,v,y)dudvdy$$

$z \le r$, $x \le r$: $$= \frac{z}{x}\,\mathrm{escs}_u(x-z, 0, r-z, y)\,\mathrm{h}_v(1,y)dudvdy$$

$z \le r$, $r \le x$: $$= \frac{z}{x}\,\mathrm{es}_u(1,1,r-z,0,y)\,\mathrm{h}_v(1,y+x-r)dudvdy$$

$r \le z$, $x \le r$: $$= \frac{z}{x}\,\mathrm{ess}_u(x,r,0,y)\,\mathrm{es}_v(1,1,z-r,0,y)dudvdy$$

$r \le z$, $r \le x$: $$= \frac{z}{x}\,\mathrm{es}_u(0,0,r,0,y)\,\mathrm{escs}_v(z-x,0,z-r,y)dudvdy$$

(1) $$\mathbf{P}_x\Big(\int_0^{H_z} \mathbb{I}_{[0,r)}\big(R_s^{(3)}\big)ds \in du,\ \int_0^{H_z} \mathbb{I}_{[r,\infty)}\big(R_s^{(3)}\big)ds = 0,\ \ell\big(H_z, r\big) = 0\Big)$$
$$= \begin{cases} \dfrac{z}{x}\,\mathrm{ss}_u(x,z)du, & 0 \le x \le z \\ \dfrac{z}{x}\,\mathrm{ss}_u(r-x,r-z)du, & z \le x \le r \end{cases}$$

(2) $$\mathbf{P}_x\Big(\int_0^{H_z} \mathbb{I}_{[0,r)}\big(R_s^{(3)}\big)ds = 0,\ \int_0^{H_z} \mathbb{I}_{[r,\infty)}\big(R_s^{(3)}\big)ds \in dv,\ \ell\big(H_z, r\big) = 0\Big)$$
$$= \begin{cases} \dfrac{z}{x}\,\mathrm{ss}_v(x-r,z-r)dv, & r \le x \le z \\ \dfrac{z}{x}\,\mathrm{h}_v(1,x-z)dv, & z \le x \end{cases}$$

2.30.1 $$\mathbf{E}_x\Big\{\exp\Big(-\int_0^{H_z}\Big(\frac{p^2}{2}e^{2\beta R_s^{(3)}} + qe^{\beta R_s^{(3)}}\Big)ds\Big); H_z < \infty\Big\}$$
$$= \begin{cases} \dfrac{z\exp\Big(-\frac{p}{|\beta|}e^{\beta x}\Big)S\Big(\frac{1}{2}+\frac{q}{p|\beta|}, 1, \frac{2p}{|\beta|}e^{\beta x}, \frac{2p}{|\beta|}\Big)}{x\exp\Big(-\frac{p}{|\beta|}e^{\beta z}\Big)S\Big(\frac{1}{2}+\frac{q}{p|\beta|}, 1, \frac{2p}{|\beta|}e^{\beta z}, \frac{2p}{|\beta|}\Big)}, & 0 \le x \le z \\ \dfrac{ze^{-\beta x/2}M_{-q/p|\beta|,0}\Big(\frac{2p}{|\beta|}e^{\beta x}\Big)}{xe^{-\beta z/2}M_{-q/p|\beta|,0}\Big(\frac{2p}{|\beta|}e^{\beta z}\Big)}, & z \le x, \quad \beta < 0 \\ \dfrac{ze^{-\beta x/2}W_{-q/p|\beta|,0}\Big(\frac{2p}{|\beta|}e^{\beta x}\Big)}{xe^{-\beta z/2}W_{-q/p|\beta|,0}\Big(\frac{2p}{|\beta|}e^{\beta z}\Big)}, & z \le x, \quad \beta > 0 \end{cases}$$

5 $R_s^{(3)} = \left(\sum_{k=1}^3 (W_s^{(k)})^2\right)^{1/2}$ $H_z = \min\{s : R_s^{(3)} = z\}$

2.31.1 (1) $\mathbf{E}_x\Big\{\exp\Big(-\frac{\gamma}{2}\sup_{0\le s\le H_z}\big(\ell(s,u)-\alpha^{-1}\ell(s,r)\big)\Big);\ H_z<\infty\Big\}$ $z \le x\wedge r\wedge u$

$0<\alpha$

$$= \frac{z}{x} - \frac{\gamma z(\Upsilon_1 + r - z - \alpha u + \alpha z)(u\wedge x - z)}{x(2\gamma(u\wedge r - z)|u-r| + \Upsilon_1 + r - z - \alpha u + \alpha z)}$$

$$- \frac{\gamma z(2(u\wedge r - z)|u-r| - (\Upsilon_1 + r - z - \alpha u + \alpha z)(u - z))(x\wedge r - z)}{x(2\gamma(u\wedge r - z)|u-r| + \Upsilon_1 + r - z - \alpha u + \alpha z)(u\wedge r - z)}$$

2.31.1 (2) $\mathbf{E}_x\Big\{\exp\Big(-\frac{\gamma}{2}\sup_{0\le s\le H_z}\big(\ell(s,u)-\alpha^{-1}\ell(s,r)\big)\Big);\ H_z=\infty\Big\}$ $z \le x\wedge r\wedge u$

$0<\alpha$

$$= 1 - \frac{z}{x} - \frac{\gamma(u-z)(\Upsilon_1 + r - z - \alpha u + \alpha z)(u\wedge x - z)}{x(2\gamma(u\wedge r - z)|u-r| + \Upsilon_1 + r - z - \alpha u + \alpha z)}$$

$$- \frac{\gamma(u-z)(2(u\wedge r - z)|u-r| - (\Upsilon_1 + r - z - \alpha u + \alpha z)(u - z))(x\wedge r - z)}{x(2\gamma(u\wedge r - z)|u-r| + \Upsilon_1 + r - z - \alpha u + \alpha z)(u\wedge r - z)}$$

2.31.1 (3) $\mathbf{E}_{xz}\exp\Big(-\frac{\gamma}{2z}\sup_{0\le s\le H_z}\big(\ell(s,uz)-\alpha^{-1}\ell(s,rz)\big)\Big)$ $x\vee r\vee u \le z$

$0<\alpha$

$$= 1 - \frac{\gamma u(\Upsilon_2 + r - r^2 - \alpha u + \alpha u^2)(u\wedge x - ux)}{x(2\gamma(u\wedge r - ur)|u-r| + \Upsilon_2 + r - r^2 - \alpha u + \alpha u^2)}$$

$$- \frac{\gamma u(2(u\wedge r - ur)|u-r| - (\Upsilon_2 + r - r^2 - \alpha u + \alpha u^2)(u - u^2))(x\wedge r - xr)}{x(2\gamma(u\wedge r - ur)|u-r| + \Upsilon_2 + r - r^2 - \alpha u + \alpha u^2)(u\wedge r - ur)}$$

2.31.2 (1) $\mathbf{P}_x\Big(\sup_{0\le s\le H_z}\big(\ell(s,u)-\alpha^{-1}\ell(s,r)\big) > h,\ H_z<\infty\Big)$ $z \le x\wedge r\wedge u$

$0<\alpha$

$$= \frac{z}{x}\exp\Big(-\frac{(\Upsilon_1 + r - z - \alpha u + \alpha z)h}{4(u\wedge r - z)|u-r|}\Big)\Big[\frac{x\wedge r - z}{u\wedge r - z}$$

$$- \frac{\Upsilon_1 + r - z - \alpha u + \alpha z}{2(u\wedge r - z)|u-r|}\Big(\frac{(x\wedge r - z)|u-z|}{u\wedge r - z} - u\wedge x + z\Big)\Big]$$

$$\Upsilon_1 := \sqrt{(r - z - \alpha u + \alpha z)^2 + 4\alpha(u\wedge r - z)|u-r|}$$

$$\Upsilon_2 := \sqrt{(r - r^2 - \alpha u + \alpha u^2)^2 + 4\alpha(u\wedge r - ur)|u-r|}$$

5 $R_s^{(3)} = \left(\sum_{k=1}^{3}(W_s^{(k)})^2\right)^{1/2}$ $H_z = \min\{s : R_s^{(3)} = z\}$

2.31.2 (2) $0 < \alpha$ $z \le x \wedge r \wedge u$

$$\mathbf{P}_x\Big(\sup_{0\le s\le H_z}\big(\ell(s,u) - \alpha^{-1}\ell(s,r)\big) > h,\ H_z = \infty\Big)$$

$$= \frac{(u-z)}{x}\exp\Big(-\frac{(\Upsilon_1 + r - z - \alpha u + \alpha z)h}{4(u\wedge r - z)|u-r|}\Big)\Big[\frac{x\wedge r - z}{u \wedge r - z}$$

$$-\frac{\Upsilon_1 + r - z - \alpha u + \alpha z}{2(u\wedge r - z)|u-r|}\Big(\frac{(x\wedge r - z)|u-z|}{u\wedge r - z} - u\wedge x + z\Big)\Big]$$

2.31.2 (3) $0 < \alpha$ $x \vee r \vee u \le z$

$$\mathbf{P}_{xz}\Big(\sup_{0\le s\le H_z}\big(\ell(s,uz) - \alpha^{-1}\ell(s,rz)\big) > hz\Big)$$

$$= \frac{u}{x}\exp\Big(-\frac{(\Upsilon_2 + r - r^2 - \alpha u + \alpha u^2)h}{4(u\wedge r - ur)|u-r|}\Big)\Big[\frac{x\wedge r - xr}{u\wedge r - ur}$$

$$-\frac{\Upsilon_2 + r - r^2 - \alpha u + \alpha u^2}{2(u\wedge r - ur)|u-r|}\Big(\frac{(x\wedge r - xr)(u-u^2)}{u\wedge r - ur} - u\wedge x + ux\Big)\Big]$$

$$\Upsilon_1 := \sqrt{(r - z - \alpha u + \alpha z)^2 + 4\alpha(u\wedge r - z)|u-r|}$$

$$\Upsilon_2 := \sqrt{(r - r^2 - \alpha u + \alpha u^2)^2 + 4\alpha(u\wedge r - ur)|u-r|}$$

5 $R_s^{(3)} = \left(\sum_{k=1}^{3}(W_s^{(k)})^2\right)^{1/2}$ $H = H_{a,b} = \min\{s : R_s^{(3)} \notin (a,b)\}$

3. Stopping at first exit time

3.0.1 $$\mathbf{E}_x e^{-\alpha H} = \frac{a\,\mathrm{sh}\big((b-x)\sqrt{2\alpha}\big) + b\,\mathrm{sh}\big((x-a)\sqrt{2\alpha}\big)}{x\,\mathrm{sh}\big((b-a)\sqrt{2\alpha}\big)}, \quad 0 \le a \le x \le b$$

3.0.2 $$\mathbf{P}_x(H \in dt) = \frac{a}{x}\,\mathrm{ss}_t(b-x, b-a)dt + \frac{b}{x}\,\mathrm{ss}_t(x-a, b-a)dt$$

3.0.3 $$\mathbf{E}_x e^{-\beta R_H^{(3)}} = \frac{a(b-x)}{x(b-a)}e^{-\beta a} + \frac{b(x-a)}{x(b-a)}e^{-\beta b}$$

3.0.4 (a) $$\mathbf{P}_x\big(R_H^{(3)} = a\big) = \frac{a(b-x)}{x(b-a)}$$

3.0.4 (b) $$\mathbf{P}_x\big(R_H^{(3)} = b\big) = \frac{b(x-a)}{x(b-a)}$$

3.0.5 (a) $$\mathbf{E}_x\big\{e^{-\alpha H};\ R_H^{(3)} = a\big\} = \frac{a\,\mathrm{sh}\big((b-x)\sqrt{2\alpha}\big)}{x\,\mathrm{sh}\big((b-a)\sqrt{2\alpha}\big)}$$

3.0.5 (b) $$\mathbf{E}_x\big\{e^{-\alpha H};\ R_H^{(3)} = b\big\} = \frac{b\,\mathrm{sh}\big((x-a)\sqrt{2\alpha}\big)}{x\,\mathrm{sh}\big((b-a)\sqrt{2\alpha}\big)}$$

3.0.6 (a) $$\mathbf{P}_x\big(H \in dt, R_H^{(3)} = a\big) = \frac{a}{x}\,\mathrm{ss}_t(b-x, b-a)dt$$

3.0.6 (b) $$\mathbf{P}_x\big(H \in dt, R_H^{(3)} = b\big) = \frac{b}{x}\,\mathrm{ss}_t(x-a, b-a)dt$$

3.1.6 $$\mathbf{P}_x\Big(\sup_{0\le s\le H} R_s^{(3)} \ge y, R_H^{(3)} = a\Big) = \frac{a(b-y)(x-a)}{x(b-a)(y-a)}, \quad a \le x \le y \le b$$

3.1.8 (1) $$\mathbf{E}_x\Big\{e^{-\alpha H};\ \sup_{0\le s\le H} R_s^{(3)} \ge y, R_H^{(3)} = a\Big\}$$
$$= \frac{a\,\mathrm{sh}\big((b-y)\sqrt{2\alpha}\big)\,\mathrm{sh}\big((x-a)\sqrt{2\alpha}\big)}{x\,\mathrm{sh}\big((b-a)\sqrt{2\alpha}\big)\,\mathrm{sh}\big((y-a)\sqrt{2\alpha}\big)}, \quad a \le x \le y \le b$$

3.1.8 (2) $$\mathbf{P}_x\Big(\sup_{0\le s\le H} R_s^{(3)} \ge y, R_H^{(3)} = a, H \in dt\Big)$$

5 $R_s^{(3)} = \big(\sum_{k=1}^3 (W_s^{(k)})^2\big)^{1/2}$ $\qquad H = H_{a,b} = \min\{s : R_s^{(3)} \notin (a,b)\}$

$$= \frac{a}{x}\,\mathrm{ss}_t(b-y, b-a) * \mathrm{ss}_t(x-a, y-a)dt, \quad a \le x \le y \le b$$

3.2.6 $$\mathbf{P}_x\Big(\inf_{0\le s\le H} R_s^{(3)} \le y, R_H^{(3)} = b\Big) = \frac{b(b-x)(y-a)}{x(b-y)(b-a)}, \quad a \le y \le x \le b$$

3.2.8 (1) $$\mathbf{E}_x\Big\{e^{-\alpha H};\ \inf_{0\le s\le H} R_s^{(3)} \le y, R_H^{(3)} = b\Big\}$$

$$= \frac{b\,\mathrm{sh}\big((b-x)\sqrt{2\alpha}\big)\,\mathrm{sh}\big((y-a)\sqrt{2\alpha}\big)}{x\,\mathrm{sh}\big((b-y)\sqrt{2\alpha}\big)\,\mathrm{sh}\big((b-a)\sqrt{2\alpha}\big)}, \quad a \le y \le x \le b$$

3.2.8 (2) $$\mathbf{P}_x\Big(\inf_{0\le s\le H} R_s^{(3)} \le y, R_H^{(3)} = b, H \in dt\Big)$$

$$= \frac{b}{x}\,\mathrm{ss}_t(b-x, b-y) * \mathrm{ss}_t(y-a, b-a)dt, \quad a \le y \le x \le b$$

3.3.1 $$\mathbf{E}_x e^{-\gamma\ell(H,r)} = \begin{cases} \dfrac{b(x-r)}{x(b-r)} + \dfrac{r(b-a)(b-x)}{x(b-r)(b-a+2\gamma(b-r)(r-a))}, & r \le x \le b \\[2ex] \dfrac{a(r-x)}{x(r-a)} + \dfrac{r(x-a)(b-a)}{x(r-a)(b-a+2\gamma(b-r)(r-a))}, & a \le x \le r \end{cases}$$

3.3.2 $$\mathbf{P}_x\big(\ell(H,r) \in dy\big) = \begin{cases} \dfrac{r(b-x)(b-a)}{2x(b-r)^2(r-a)} \exp\Big(-\dfrac{(b-a)y}{2(b-r)(r-a)}\Big)dy, & r \le x \le b \\[2ex] \dfrac{r(x-a)(b-a)}{2x(b-r)(r-a)^2} \exp\Big(-\dfrac{(b-a)y}{2(b-r)(r-a)}\Big)dy, & a \le x \le r \end{cases}$$

(1) $$\mathbf{P}_x\big(\ell(H,r) = 0\big) = \begin{cases} \dfrac{b(x-r)}{x(b-r)}, & r \le x \le b \\[2ex] \dfrac{a(r-x)}{x(r-a)}, & a \le x \le r \end{cases}$$

3.3.3 $$\mathbf{E}_r e^{-\alpha H - \gamma\ell(H,r)}$$

$$= \frac{\sqrt{2\alpha}\big(a\,\mathrm{sh}((b-r)\sqrt{2\alpha}) + b\,\mathrm{sh}((r-a)\sqrt{2\alpha})\big)}{r\big(\sqrt{2\alpha}\,\mathrm{sh}((b-a)\sqrt{2\alpha}) + 2\gamma\,\mathrm{sh}((b-r)\sqrt{2\alpha})\,\mathrm{sh}((r-a)\sqrt{2\alpha})\big)} =: L$$

$$\mathbf{E}_x e^{-\alpha H - \gamma\ell(H,r)} = \begin{cases} \dfrac{b\,\mathrm{sh}((x-r)\sqrt{2\alpha})}{x\,\mathrm{sh}((b-r)\sqrt{2\alpha})} + \dfrac{r\,\mathrm{sh}((b-x)\sqrt{2\alpha})}{x\,\mathrm{sh}((b-r)\sqrt{2\alpha})}L, & r \le x \le b \\[2ex] \dfrac{a\,\mathrm{sh}((r-x)\sqrt{2\alpha})}{x\,\mathrm{sh}((r-a)\sqrt{2\alpha})} + \dfrac{r\,\mathrm{sh}((x-a)\sqrt{2\alpha})}{x\,\mathrm{sh}((r-a)\sqrt{2\alpha})}L, & a \le x \le r \end{cases}$$

5 $R_s^{(3)}=\left(\sum_{k=1}^3(W_s^{(k)})^2\right)^{1/2}$ $H=H_{a,b}=\min\{s: R_s^{(3)}\notin(a,b)\}$

3.3.4 $$\mathbf{E}_x\{e^{-\alpha H};\ \ell(H,r)\in dy\}=\exp\Big(-\frac{\sqrt{2\alpha}\,\mathrm{sh}((b-a)\sqrt{2\alpha})y}{2\,\mathrm{sh}((b-r)\sqrt{2\alpha})\,\mathrm{sh}((r-a)\sqrt{2\alpha})}\Big)$$

$$\times\begin{cases}\dfrac{\sqrt{2\alpha}\,\mathrm{sh}((b-x)\sqrt{2\alpha})[a\,\mathrm{sh}((b-r)\sqrt{2\alpha})+b\,\mathrm{sh}((r-a)\sqrt{2\alpha})]}{2x\,\mathrm{sh}^2((b-r)\sqrt{2\alpha})\,\mathrm{sh}((r-a)\sqrt{2\alpha})}dy, & r\le x\le b\\ \dfrac{\sqrt{2\alpha}\,\mathrm{sh}((x-a)\sqrt{2\alpha})[a\,\mathrm{sh}((b-r)\sqrt{2\alpha})+b\,\mathrm{sh}((r-a)\sqrt{2\alpha})]}{2x\,\mathrm{sh}((b-r)\sqrt{2\alpha})\,\mathrm{sh}^2((r-a)\sqrt{2\alpha})}dy, & a\le x\le r\end{cases}$$

(1) $$\mathbf{E}_x\{e^{-\alpha H};\ \ell(H,r)=0\}=\begin{cases}\dfrac{b\,\mathrm{sh}((x-r)\sqrt{2\alpha})}{x\,\mathrm{sh}((b-r)\sqrt{2\alpha})}, & r\le x\le b\\ \dfrac{a\,\mathrm{sh}((r-x)\sqrt{2\alpha})}{x\,\mathrm{sh}((r-a)\sqrt{2\alpha})}, & a\le x\le r\end{cases}$$

3.3.5 (a) $\mathbf{E}_x\{e^{-\gamma\ell(H,r)};\ R_H^{(3)}=a\}=\frac{a}{x}Q_a^{(3)}(2\gamma)$, for $Q_a^{(3)}(2\gamma)$ see 1.3.3.5

3.3.5 (b) $\mathbf{E}_x\{e^{-\gamma\ell(H,r)};\ R_H^{(3)}=b\}=\frac{b}{x}Q_b^{(3)}(2\gamma)$, for $Q_b^{(3)}(2\gamma)$ see 1.3.3.5

3.3.6 (a) $\mathbf{P}_x\big(\ell(H,r)\in dy, R_H^{(3)}=a\big)=\frac{a}{x}B_a^{(3)}(y)dy$, for $B_a^{(3)}(y)$ see 1.3.3.6

(1) $$\mathbf{P}_x\big(\ell(H,r)=0, R_H^{(3)}=a\big)=\begin{cases}0, & r\le x\le b\\ \dfrac{a(r-x)}{x(r-a)}, & a\le x\le r\end{cases}$$

3.3.6 (b) $\mathbf{P}_x\big(\ell(H,r)\in dy, R_H^{(3)}=b\big)=\frac{b}{x}B_b^{(3)}(y)dy$, for $B_b^{(3)}(y)$ see 1.3.3.6

(1) $$\mathbf{P}_x\big(\ell(H,r)=0, R_H^{(3)}=b\big)=\begin{cases}\dfrac{b(x-r)}{x(b-r)}, & r\le x\le b\\ 0, & a\le x\le r\end{cases}$$

3.3.7 (a) $\mathbf{E}_x\{e^{-\alpha H-\gamma\ell(H,r)};\ R_H^{(3)}=a\}=\frac{a}{x}V_a^{(3)}(2\alpha)$, for $V_a^{(3)}(2\alpha)$ see 1.3.3.7

3.3.7 (b) $\mathbf{E}_x\{e^{-\alpha H-\gamma\ell(H,r)};\ R_H^{(3)}=b\}=\frac{b}{x}V_b^{(3)}(2\alpha)$, for $V_b^{(3)}(2\alpha)$ see 1.3.3.7

5 $R_s^{(3)} = \big(\sum_{k=1}^3 (W_s^{(k)})^2\big)^{1/2}$ $\qquad H = H_{a,b} = \min\{s : R_s^{(3)} \notin (a,b)\}$

3.3.8 (a)

$$\mathbf{E}_x\{e^{-\alpha H};\ \ell(H,r) \in dy, R_H^{(3)} = a\} = \exp\Big(-\frac{\sqrt{2\alpha}\,\mathrm{sh}((b-a)\sqrt{2\alpha})y}{2\,\mathrm{sh}((b-r)\sqrt{2\alpha})\,\mathrm{sh}((r-a)\sqrt{2\alpha})}\Big)$$

$$\times\begin{cases} \dfrac{a\sqrt{2\alpha}\,\mathrm{sh}((b-x)\sqrt{2\alpha})}{2x\,\mathrm{sh}((b-r)\sqrt{2\alpha})\,\mathrm{sh}((r-a)\sqrt{2\alpha})}dy, & r \le x \le b \\ \dfrac{a\sqrt{2\alpha}\,\mathrm{sh}((x-a)\sqrt{2\alpha})}{2x\,\mathrm{sh}^2((r-a)\sqrt{2\alpha})}dy, & a \le x \le r \end{cases}$$

(1)

$$\mathbf{E}_x\{e^{-\alpha H};\ \ell(H,r) = 0, R_H^{(3)} = a\} = \begin{cases} 0, & r \le x \le b \\ \dfrac{a\,\mathrm{sh}((r-x)\sqrt{2\alpha})}{x\,\mathrm{sh}((r-a)\sqrt{2\alpha})}, & a \le x \le r \end{cases}$$

3.3.8 (b)

$$\mathbf{E}_x\{e^{-\alpha H};\ \ell(H,r) \in dy, R_H^{(3)} = b\} = \exp\Big(-\frac{\sqrt{2\alpha}\,\mathrm{sh}((b-a)\sqrt{2\alpha})y}{2\,\mathrm{sh}((b-r)\sqrt{2\alpha})\,\mathrm{sh}((r-a)\sqrt{2\alpha})}\Big)$$

$$\times\begin{cases} \dfrac{b\sqrt{2\alpha}\,\mathrm{sh}((b-x)\sqrt{2\alpha})}{2x\,\mathrm{sh}^2((b-r)\sqrt{2\alpha})}dy, & r \le x \le b \\ \dfrac{b\sqrt{2\alpha}\,\mathrm{sh}((x-a)\sqrt{2\alpha})}{2x\,\mathrm{sh}((b-r)\sqrt{2\alpha})\,\mathrm{sh}((r-a)\sqrt{2\alpha})}dy, & a \le x \le r \end{cases}$$

(1)

$$\mathbf{E}_x\{e^{-\alpha H}; \ell(H,r) = 0, R_H^{(3)} = b\} = \begin{cases} \dfrac{b\,\mathrm{sh}((x-r)\sqrt{2\alpha})}{x\,\mathrm{sh}((b-r)\sqrt{2\alpha})}, & r \le x \le b \\ 0, & a \le x \le r \end{cases}$$

3.4.1

$$\mathbf{E}_x \exp\Big(-\gamma\int_0^H \mathbb{1}_{[r,b]}\big(R_s^{(3)}\big)ds\Big)$$

$$=\begin{cases} \dfrac{a(r-x)}{x(r-a)} + \dfrac{(x-a)(a\,\mathrm{sh}((b-r)\sqrt{2\gamma}) + b\sqrt{2\gamma}(r-a))}{x(r-a)(\mathrm{sh}((b-r)\sqrt{2\gamma}) + \sqrt{2\gamma}(r-a)\,\mathrm{ch}((b-r)\sqrt{2\gamma}))}, & a \le x \le r \\ \dfrac{b\,\mathrm{sh}((x-r)\sqrt{2\gamma})}{x\,\mathrm{sh}((b-r)\sqrt{2\gamma})} & \\ \qquad + \dfrac{\mathrm{sh}((b-x)\sqrt{2\gamma})(a\,\mathrm{sh}((b-r)\sqrt{2\gamma}) + b\sqrt{2\gamma}(r-a))}{x\,\mathrm{sh}((b-r)\sqrt{2\gamma})(\mathrm{sh}((b-r)\sqrt{2\gamma}) + \sqrt{2\gamma}(r-a)\,\mathrm{ch}((b-r)\sqrt{2\gamma}))}, & r \le x \le b \end{cases}$$

3.4.2

$$\mathbf{P}_x\Big(\int_0^H \mathbb{1}_{[r,b]}\big(R_s^{(3)}\big)ds \in dy\Big)$$

$$=\begin{cases} \dfrac{x-a}{x}\Big(\dfrac{a}{r-a}\,\mathrm{rc}_y(0,b-r,b-r,r-a) + b\widetilde{\mathrm{rc}}_y(1,b-r,0,r-a)\Big)dy, & x \le r \\ \dfrac{b}{x}\,\mathrm{ss}_y(x-r,b-r)dy + \dfrac{1}{x}\big(\mathrm{ss}_y(b-x,b-r) & \\ \quad *\big(a\,\mathrm{rc}_y(0,b-r,b-r,r-a) + b(r-a)\widetilde{\mathrm{rc}}_y(1,b-r,0,r-a)\big)dy, & r \le x \end{cases}$$

5 $R_s^{(3)} = \left(\sum_{k=1}^3 (W_s^{(k)})^2\right)^{1/2}$ $H = H_{a,b} = \min\{s : R_s^{(3)} \notin (a,b)\}$

(1) $$\mathbf{P}_x\Big(\int_0^H \mathbb{I}_{[r,b]}\big(R_s^{(3)}\big)ds = 0\Big) = \begin{cases} \dfrac{a(r-x)}{x(r-a)}, & a \le x \le r \\ 0, & r \le x \le b \end{cases}$$

3.4.5 (a) $$\mathbf{E}_x\Big\{\exp\Big(-\gamma\int_0^H \mathbb{I}_{[r,b]}\big(R_s^{(3)}\big)ds\Big);\ R_H^{(3)} = a\Big\} = \frac{a}{x} Q_a^{(4)}(2\gamma)$$

for $Q_a^{(4)}(2\gamma)$ see 1.3.4.5

3.4.5 (b) $$\mathbf{E}_x\Big\{\exp\Big(-\gamma\int_0^H \mathbb{I}_{[r,b]}\big(R_s^{(3)}\big)ds\Big);\ R_H^{(3)} = b\Big\} = \frac{b}{x} Q_b^{(4)}(2\gamma)$$

for $Q_b^{(4)}(2\gamma)$ see 1.3.4.5

3.4.6 (a) $$\mathbf{P}_x\Big(\int_0^H \mathbb{I}_{[r,b]}\big(R_s^{(3)}\big)ds \in dy, R_H^{(3)} = a\Big) = \frac{a}{x} B_a^{(4)}(y)dy$$

for $B_a^{(4)}(y)$ see 1.3.4.6

(1) $$\mathbf{P}_x\Big(\int_0^H \mathbb{I}_{[r,b]}\big(R_s^{(3)}\big)ds = 0, R_H^{(3)} = a\Big) = \begin{cases} \dfrac{a(r-x)}{x(r-a)}, & a \le x \le r \\ 0, & r \le x \le b \end{cases}$$

3.4.6 (b) $$\mathbf{P}_x\Big(\int_0^H \mathbb{I}_{[r,b]}\big(R_s^{(3)}\big)ds \in dy, R_H^{(3)} = b\Big) = \frac{b}{x} B_b^{(4)}(y)dy$$

for $B_b^{(4)}(y)$ see 1.3.4.6

3.5.1 $$\mathbf{E}_x \exp\Big(-\gamma\int_0^H \mathbb{I}_{[a,r]}\big(R_s^{(3)}\big)ds\Big)$$

$$= \begin{cases} \dfrac{a\,\mathrm{sh}((r-x)\sqrt{2\gamma})}{x\,\mathrm{sh}((r-a)\sqrt{2\gamma})} \\ \quad + \dfrac{\mathrm{sh}((x-a)\sqrt{2\gamma})(b\,\mathrm{sh}((r-a)\sqrt{2\gamma}) + a\sqrt{2\gamma}(b-r))}{x\,\mathrm{sh}((r-a)\sqrt{2\gamma})(\mathrm{sh}((r-a)\sqrt{2\gamma}) + \sqrt{2\gamma}(b-r)\,\mathrm{ch}((r-a)\sqrt{2\gamma}))}, & a \le x \le r \\ \dfrac{b(x-r)}{x(b-r)} + \dfrac{(b-x)(b\,\mathrm{sh}((r-a)\sqrt{2\gamma}) + a\sqrt{2\gamma}(b-r))}{x(b-r)(\mathrm{sh}((r-a)\sqrt{2\gamma}) + \sqrt{2\gamma}(b-r)\,\mathrm{ch}((r-a)\sqrt{2\gamma}))}, & r \le x \le b \end{cases}$$

5 $R_s^{(3)} = \left(\sum_{k=1}^3 (W_s^{(k)})^2\right)^{1/2}$ $\qquad H = H_{a,b} = \min\{s : R_s^{(3)} \notin (a,b)\}$

3.5.2 $$\mathbf{P}_x\Big(\int_0^H \mathbb{I}_{[a,r]}\big(R_s^{(3)}\big)ds \in dy\Big)$$

$$= \begin{cases} \dfrac{a}{x}\operatorname{ss}_y(r-x, r-a)dy + \dfrac{1}{x}\big(\operatorname{ss}_y(x-a, r-a) \\ \quad *\big(b\operatorname{rc}_y(0, r-a, r-a, b-r) + a(b-r)\widetilde{\operatorname{rc}}_y(1, r-a, 0, b-r)\big)dy, & x \le r \\ \dfrac{b-x}{x}\Big(\dfrac{b}{b-r}\operatorname{rc}_y(0, r-a, r-a, b-r) + a\widetilde{\operatorname{rc}}_y(1, r-a, 0, b-r)\Big)dy, & r \le x \end{cases}$$

(1) $$\mathbf{P}_x\Big(\int_0^H \mathbb{I}_{[a,r]}\big(R_s^{(3)}\big)ds = 0\Big) = \begin{cases} 0, & a \le x \le r \\ \dfrac{b(x-r)}{x(b-r)}, & r \le x \le b \end{cases}$$

3.5.5 (a) $$\mathbf{E}_x\Big\{\exp\Big(-\gamma\int_0^H \mathbb{I}_{[a,r]}\big(R_s^{(3)}\big)ds\Big);\ R_H^{(3)} = a\Big\} = \frac{a}{x}Q_a^{(5)}(2\gamma)$$

for $Q_a^{(5)}(2\gamma)$ see 1.3.5.5

3.5.5 (b) $$\mathbf{E}_x\Big\{\exp\Big(-\gamma\int_0^H \mathbb{I}_{[a,r]}\big(R_s^{(3)}\big)ds\Big);\ R_H^{(3)} = b\Big\} = \frac{b}{x}Q_b^{(5)}(2\gamma)$$

for $Q_b^{(5)}(2\gamma)$ see 1.3.5.5

3.5.6 (a) $$\mathbf{P}_x\Big(\int_0^H \mathbb{I}_{[a,r]}\big(R_s^{(3)}\big)ds \in dy, R_H^{(3)} = a\Big) = \frac{a}{x}B_a^{(5)}(y)dy$$

for $B_a^{(5)}(y)$ see 1.3.5.6

(1) $$\mathbf{P}_x\Big(\int_0^H \mathbb{I}_{[a,r]}\big(R_s^{(3)}\big)ds = 0, R_H^{(3)} = a\Big) = 0$$

3.5.6 (b) $$\mathbf{P}_x\Big(\int_0^H \mathbb{I}_{[a,r]}\big(R_s^{(3)}\big)ds \in dy, R_H^{(3)} = b\Big) = \frac{b}{x}B_b^{(5)}(y)dy$$

for $B_b^{(5)}(y)$ see 1.3.5.6

(1) $$\mathbf{P}_x\Big(\int_0^H \mathbb{I}_{[a,r]}\big(R_s^{(3)}\big)ds = 0, R_H^{(3)} = b\Big) = \begin{cases} 0, & a \le x \le r \\ \dfrac{b(x-r)}{x(b-r)}, & r \le x \le b \end{cases}$$

5 $R_s^{(3)} = \left(\sum_{k=1}^{3}(W_s^{(k)})^2\right)^{1/2}$ $H = H_{a,b} = \min\{s : R_s^{(3)} \notin (a,b)\}$

3.6.1 $$\mathbf{E}_r \exp\Big(-\int_0^H \Big(p\mathbb{1}_{[a,r]}\big(R_s^{(3)}\big) + q\mathbb{1}_{[r,b]}\big(R_s^{(3)}\big)\Big)ds\Big) =: Q$$

$$= \frac{a\sqrt{p}\,\mathrm{sh}((b-r)\sqrt{2q}) + b\sqrt{q}\,\mathrm{sh}((r-a)\sqrt{2p})}{r(\sqrt{q}\,\mathrm{ch}((b-r)\sqrt{2q})\,\mathrm{sh}((r-a)\sqrt{2p}) + \sqrt{p}\,\mathrm{sh}((b-r)\sqrt{2q})\,\mathrm{ch}((r-a)\sqrt{2p}))}$$

$$\mathbf{E}_x \exp\Big(-\int_0^H \Big(p\mathbb{1}_{[a,r]}\big(R_s^{(3)}\big) + q\mathbb{1}_{[r,b]}\big(R_s^{(3)}\big)\Big)ds\Big)$$

$$= \begin{cases} \dfrac{b\,\mathrm{sh}((x-r)\sqrt{2q})}{x\,\mathrm{sh}((b-r)\sqrt{2q})} + \dfrac{r\,\mathrm{sh}((b-x)\sqrt{2q})}{x\,\mathrm{sh}((b-r)\sqrt{2q})}Q, & r \le x \le b \\ \dfrac{a\,\mathrm{sh}((r-x)\sqrt{2p})}{x\,\mathrm{sh}((r-a)\sqrt{2p})} + \dfrac{r\,\mathrm{sh}((x-a)\sqrt{2p})}{x\,\mathrm{sh}((r-a)\sqrt{2p})}Q, & a \le x \le r \end{cases}$$

3.6.5 (a) $$\mathbf{E}_x\Big\{\exp\Big(-\int_0^H\Big(p\mathbb{1}_{[a,r]}\big(R_s^{(3)}\big) + q\mathbb{1}_{[r,b]}\big(R_s^{(3)}\big)\Big)ds\Big);\ R_H^{(3)} = a\Big\} = \frac{a}{x}Q_a^{(6)}(2p,2q)$$

for $Q_a^{(6)}(2p,2q)$ see 1.3.6.5

3.6.5 (b) $$\mathbf{E}_x\Big\{\exp\Big(-\int_0^H\Big(p\mathbb{1}_{[a,r]}\big(R_s^{(3)}\big) + q\mathbb{1}_{[r,b]}\big(R_s^{(3)}\big)\Big)ds\Big);\ R_H^{(3)} = b\Big\} = \frac{b}{x}Q_b^{(6)}(2p,2q)$$

for $Q_b^{(6)}(2p,2q)$ see 1.3.6.5

3.7.1 $a \le r$ $u \le b$ $$\mathbf{E}_r \exp\Big(-\gamma\int_0^H \mathbb{1}_{[r,u]}\big(R_s^{(3)}\big)ds\Big)$$

$$= \frac{a\,\mathrm{sh}((u-r)\sqrt{2\gamma}) + a(b-u)\sqrt{2\gamma}\,\mathrm{ch}((u-r)\sqrt{2\gamma}) + b(r-a)\sqrt{2\gamma}}{r((1+(b-u)(r-a)2\gamma)\,\mathrm{sh}((u-r)\sqrt{2\gamma}) + (b-u+r-a)\sqrt{2\gamma}\,\mathrm{ch}((u-r)\sqrt{2\gamma}))} =: Q^r$$

$$\mathbf{E}_u \exp\Big(-\gamma\int_0^H \mathbb{1}_{[r,u]}\big(R_s^{(3)}\big)ds\Big)$$

$$= \frac{b\,\mathrm{sh}((u-r)\sqrt{2\gamma}) + a(b-u)\sqrt{2\gamma} + b(r-a)\sqrt{2\gamma}\,\mathrm{ch}((u-r)\sqrt{2\gamma})}{u((1+(b-u)(r-a)2\gamma)\,\mathrm{sh}((u-r)\sqrt{2\gamma}) + (b-u+r-a)\sqrt{2\gamma}\,\mathrm{ch}((u-r)\sqrt{2\gamma}))} =: Q^u$$

$$\mathbf{E}_x \exp\Big(-\gamma\int_0^H \mathbb{1}_{[r,u]}\big(R_s^{(3)}\big)ds\Big)$$

5 $R_s^{(3)} = \left(\sum_{k=1}^3 (W_s^{(k)})^2\right)^{1/2}$ $\qquad H = H_{a,b} = \min\{s : R_s^{(3)} \notin (a,b)\}$

$$= \begin{cases} \dfrac{a(r-x)}{x(r-a)} + \dfrac{r(x-a)}{x(r-a)} Q^r, & a \le x \le r \\ \dfrac{r \operatorname{sh}((u-x)\sqrt{2\gamma})}{x \operatorname{sh}((u-r)\sqrt{2\gamma})} Q^r + \dfrac{u \operatorname{sh}((x-r)\sqrt{2\gamma})}{x \operatorname{sh}((u-r)\sqrt{2\gamma})} Q^u, & r \le x \le u \\ \dfrac{b(x-u)}{x(b-u)} + \dfrac{u(b-x)}{x(b-u)} Q^u, & u \le x \le b \end{cases}$$

3.7.5 (a) $\mathbf{E}_x\left\{\exp\left(-\gamma \int_0^H \mathbb{I}_{[r,u]}(R_s^{(3)})ds\right);\ R_H^{(3)} = a\right\} = \frac{a}{x} Q_a^{(7)}(2\gamma)$

for $Q_a^{(7)}(2\gamma)$ see 1.3.7.5

3.7.5 (b) $\mathbf{E}_x\left\{\exp\left(-\gamma \int_0^H \mathbb{I}_{[r,u]}(R_s^{(3)})ds\right);\ R_H^{(3)} = b\right\} = \frac{b}{x} Q_b^{(7)}(2\gamma)$

for $Q_b^{(7)}(2\gamma)$ see 1.3.7.5

3.8.1 $\mathbf{E}_x \exp\left(-\gamma \int_0^H R_s^{(3)} ds\right) = \dfrac{S_{\frac{1}{3}}(\frac{2}{3}b^{3/2}\sqrt{2\gamma}, \frac{2}{3}x^{3/2}\sqrt{2\gamma}) + S_{\frac{1}{3}}(\frac{2}{3}x^{3/2}\sqrt{2\gamma}, \frac{2}{3}a^{3/2}\sqrt{2\gamma})}{S_{1/3}(\frac{2}{3}b^{3/2}\sqrt{2\gamma}, \frac{2}{3}a^{3/2}\sqrt{2\gamma})}$

3.8.5 (a) $\mathbf{E}_x\left\{\exp\left(-\gamma \int_0^H R_s^{(3)} ds\right);\ R_H^{(3)} = a\right\} = \dfrac{S_{1/3}(\frac{2}{3}b^{3/2}\sqrt{2\gamma}, \frac{2}{3}x^{3/2}\sqrt{2\gamma})}{S_{1/3}(\frac{2}{3}b^{3/2}\sqrt{2\gamma}, \frac{2}{3}a^{3/2}\sqrt{2\gamma})}$

3.8.5 (b) $\mathbf{E}_x\left\{\exp\left(-\gamma \int_0^H R_s^{(3)} ds\right);\ R_H^{(3)} = b\right\} = \dfrac{S_{1/3}(\frac{2}{3}x^{3/2}\sqrt{2\gamma}, \frac{2}{3}a^{3/2}\sqrt{2\gamma})}{S_{1/3}(\frac{2}{3}b^{3/2}\sqrt{2\gamma}, \frac{2}{3}a^{3/2}\sqrt{2\gamma})}$

3.9.1 $\mathbf{E}_x \exp\left(-2\gamma \int_0^H (R_s^{(3)})^2 ds\right) = \dfrac{S_{1/4}(b^2\sqrt{\gamma}, x^2\sqrt{\gamma}) + S_{1/4}(x^2\sqrt{\gamma}, a^2\sqrt{\gamma})}{S_{1/4}(b^2\sqrt{\gamma}, a^2\sqrt{\gamma})}$

3.9.5 (a) $\mathbf{E}_x\left\{\exp\left(-2\gamma \int_0^H (R_s^{(3)})^2 ds\right);\ R_H^{(3)} = a\right\} = \dfrac{S_{1/4}(b^2\sqrt{\gamma}, x^2\sqrt{\gamma})}{S_{1/4}(b^2\sqrt{\gamma}, a^2\sqrt{\gamma})}$

3.9.5 (b) $\mathbf{E}_x\left\{\exp\left(-2\gamma \int_0^H (R_s^{(3)})^2 ds\right);\ R_H^{(3)} = b\right\} = \dfrac{S_{1/4}(x^2\sqrt{\gamma}, a^2\sqrt{\gamma})}{S_{1/4}(b^2\sqrt{\gamma}, a^2\sqrt{\gamma})}$

5 $R_s^{(3)}=\left(\sum_{k=1}^3 (W_s^{(k)})^2\right)^{1/2}$ $H=H_{a,b}=\min\{s: R_s^{(3)}\notin(a,b)\}$

3.10.1
$$\mathbf{E}_x\exp\Big(-\gamma\int_0^H e^{2\beta R_s^{(3)}}ds\Big)=\frac{aS_0\Big(\frac{\sqrt{2\gamma}}{|\beta|}e^{\beta b},\frac{\sqrt{2\gamma}}{|\beta|}e^{\beta x}\Big)+bS_0\Big(\frac{\sqrt{2\gamma}}{|\beta|}e^{\beta x},\frac{\sqrt{2\gamma}}{|\beta|}e^{\beta a}\Big)}{xS_0\Big(\frac{\sqrt{2\gamma}}{|\beta|}e^{\beta b},\frac{\sqrt{2\gamma}}{|\beta|}e^{\beta a}\Big)}$$

3.10.7 (a)
$$\mathbf{E}_x\Big\{\exp\Big(-\alpha H-\gamma\int_0^H e^{2\beta R_s^{(3)}}ds\Big);\ R_H^{(3)}=a\Big\}=\frac{aS_{\sqrt{2\alpha}/|\beta|}\Big(\frac{\sqrt{2\gamma}}{|\beta|}e^{\beta b},\frac{\sqrt{2\gamma}}{|\beta|}e^{\beta x}\Big)}{xS_{\sqrt{2\alpha}/|\beta|}\Big(\frac{\sqrt{2\gamma}}{|\beta|}e^{\beta b},\frac{\sqrt{2\gamma}}{|\beta|}e^{\beta a}\Big)}$$

3.10.7 (b)
$$\mathbf{E}_x\Big\{\exp\Big(-\alpha H-\gamma\int_0^H e^{2\beta R_s^{(3)}}ds\Big);\ R_H^{(3)}=b\Big\}=\frac{bS_{\sqrt{2\alpha}/|\beta|}\Big(\frac{\sqrt{2\gamma}}{|\beta|}e^{\beta x},\frac{\sqrt{2\gamma}}{|\beta|}e^{\beta a}\Big)}{xS_{\sqrt{2\alpha}/|\beta|}\Big(\frac{\sqrt{2\gamma}}{|\beta|}e^{\beta b},\frac{\sqrt{2\gamma}}{|\beta|}e^{\beta a}\Big)}$$

3.12.5 (a)
$$\mathbf{E}_x\{e^{-\gamma\check{H}(H)};\ R_H^{(3)}=a\}=\int_x^b\frac{a\,\mathrm{sh}((x-a)\sqrt{2\gamma})}{xy\,\mathrm{sh}(y\sqrt{2\gamma})}dy$$

3.12.5 (b)
$$\mathbf{E}_x\{e^{-\gamma\hat{H}(H)};\ R_H^{(3)}=b\}=\int_a^x\frac{b\,\mathrm{sh}((b-x)\sqrt{2\gamma})}{xy\,\mathrm{sh}(y\sqrt{2\gamma})}dy$$

3.12.6 (a)
$$\mathbf{P}_x\big(\check{H}(H)\in du, R_H^{(3)}=a\big)=\int_x^b\frac{a}{xy}\,\mathrm{ss}_u(x-a,y)dy\,du$$

3.12.6 (b)
$$\mathbf{P}_x\big(\hat{H}(H)\in du, R_H^{(3)}=b\big)=\int_a^x\frac{b}{xy}\,\mathrm{ss}_u(b-x,y)dy\,du$$

3.12.7 (a)
$$\mathbf{E}_x\{e^{-\alpha H-\gamma\check{H}(H)};\ R_H^{(3)}=a\}=\int_x^b\frac{a\sqrt{2\alpha}\,\mathrm{sh}((x-a)\sqrt{2\gamma+2\alpha})}{x\,\mathrm{sh}(y\sqrt{2\alpha})\,\mathrm{sh}(y\sqrt{2\gamma+2\alpha})}dy$$

3.12.7 (b)
$$\mathbf{E}_x\{e^{-\alpha H-\gamma\hat{H}(H)};\ R_H^{(3)}=b\}=\int_a^x\frac{b\sqrt{2\alpha}\,\mathrm{sh}((b-x)\sqrt{2\gamma+2\alpha})}{x\,\mathrm{sh}(y\sqrt{2\alpha})\,\mathrm{sh}(y\sqrt{2\gamma+2\alpha})}dy$$

3.12.8 (a) $u<t$
$$\mathbf{P}_x\big(\check{H}(H)\in du,\ H\in dt,\ R_H^{(3)}=a\big)=\frac{a}{x}\int_x^b\mathrm{ss}_u(x-a,y)\,\mathrm{s}_{t-u}(y)dy\,dudt$$

3.12.8 (b) $u<t$
$$\mathbf{P}_x\big(\hat{H}(H)\in du,\ H\in dt,\ R_H^{(3)}=b\big)=\frac{b}{x}\int_a^x\mathrm{ss}_u(b-x,y)\,\mathrm{s}_{t-u}(y)dy\,dudt$$

3.18.1 $r\le u$
$$\mathbf{E}_x\exp\big(-\gamma\ell(H,r)-\eta\ell(H,u)\big)$$

$a\le x$, $x\le r$
$$=\frac{a}{x}-\frac{(x-a)(a-b+2a\gamma(b-r)+2a\eta(b-u)+4a\gamma\eta(b-u)(u-r))}{x((b-a)+2\gamma(b-r)(r-a)+2\eta(b-u)(u-a)+4\gamma\eta(b-u)(u-r)(r-a))}$$

5 $R_s^{(3)} = \left(\sum_{k=1}^{3}(W_s^{(k)})^2\right)^{1/2}$ $H = H_{a,b} = \min\{s : R_s^{(3)} \notin (a,b)\}$

$r \le x$, $x \le u$: $$= \frac{x(b-a) + 2\gamma b(x-r)(r-a) + 2\eta a(b-u)(u-x)}{x((b-a) + 2\gamma(b-r)(r-a) + 2\eta(b-u)(u-a) + 4\gamma\eta(b-u)(u-r)(r-a))}$$

$u \le x$, $x \le b$: $$= \frac{b}{x} - \frac{(b-x)(b-a+2b\eta(u-a)+2b\gamma(r-a)+4b\gamma\eta(u-r)(r-a))}{x((b-a) + 2\gamma(b-r)(r-a) + 2\eta(b-u)(u-a) + 4\gamma\eta(b-u)(u-r)(r-a))}$$

3.18.5 (a) $$\mathbf{E}_x\{\exp(-\gamma\ell(H,r) - \eta\ell(H,u));\ R_H^{(3)} = a\} = \frac{a}{x}Q_a^{(18)}(2\gamma, 2\eta)$$

for $Q_a^{(18)}(2\gamma, 2\eta)$ see 1.3.18.5

3.18.5 (b) $$\mathbf{E}_x\{\exp(-\gamma\ell(H,r) - \eta\ell(H,u));\ R_H^{(3)} = b\} = \frac{b}{x}Q_b^{(18)}(2\gamma, 2\eta)$$

for $Q_b^{(18)}(2\gamma, 2\eta)$ see 1.3.18.5

3.18.6 (a) $$\mathbf{P}_x(\ell(H,r) \in dy,\ \ell(H,u) \in dv,\ R_H^{(3)} = a) = \frac{a}{x}B_{x,a}^{(18)}(y,v)dydv$$

for $B_{x,a}^{(18)}(y,v)$ see 1.1.18.6

3.18.6 (b) $$\mathbf{P}_x(\ell(H,r) \in dy,\ \ell(H,u) \in dv,\ R_H^{(3)} = b) = \frac{b}{x}B_{x,b}^{(18)}(y,v)dydv$$

for $B_{x,b}^{(18)}(y,v)$ see 1.1.18.6

3.19.5 (a) $$\mathbf{E}_x\Big\{\exp\Big(-\gamma\int_0^H \frac{ds}{R_s^{(3)}}\Big);\ R_H^{(3)} = a\Big\} = \frac{S_1(2\sqrt{2\gamma b}, 2\sqrt{2\gamma x})}{S_1(2\sqrt{2\gamma b}, 2\sqrt{2\gamma a})}$$

3.19.5 (b) $$\mathbf{E}_x\Big\{\exp\Big(-\gamma\int_0^H \frac{ds}{R_s^{(3)}}\Big);\ R_H^{(3)} = b\Big\} = \frac{S_1(2\sqrt{2\gamma x}, 2\sqrt{2\gamma a})}{S_1(2\sqrt{2\gamma b}, 2\sqrt{2\gamma a})}$$

3.20.5 (a) $$\mathbf{E}_x\Big\{\exp\Big(-\frac{\gamma^2}{2}\int_0^H \frac{ds}{(R_s^{(3)})^2}\Big);\ R_H^{(3)} = a\Big\} = \frac{\sqrt{a}((b/x)^{\sqrt{\gamma^2+1/4}} - (x/b)^{\sqrt{\gamma^2+1/4}})}{\sqrt{x}((b/a)^{\sqrt{\gamma^2+1/4}} - (a/b)^{\sqrt{\gamma^2+1/4}})}$$

5 $R_s^{(3)} = \left(\sum_{k=1}^3 (W_s^{(k)})^2\right)^{1/2}$ $H = H_{a,b} = \min\{s : R_s^{(3)} \notin (a,b)\}$

3.20.5 (b)

$$\mathbf{E}_x\Big\{\exp\Big(-\frac{\gamma^2}{2}\int_0^H \frac{ds}{(R_s^{(3)})^2}\Big);\ R_H^{(3)} = b\Big\} = \frac{\sqrt{b}((x/a)^{\sqrt{\gamma^2+1/4}} - (a/x)^{\sqrt{\gamma^2+1/4}})}{\sqrt{x}((b/a)^{\sqrt{\gamma^2+1/4}} - (a/b)^{\sqrt{\gamma^2+1/4}})}$$

3.21.5 (a)

$$\mathbf{E}_x\Big\{\exp\Big(-\int_0^H \Big(\frac{p^2}{2(R_s^{(3)})^2} + \frac{q^2(R_s^{(3)})^2}{2}\Big)ds\Big);\ R_H^{(3)} = a\Big\}$$

$$= \frac{x^{\sqrt{p^2+1/4}-1/2} S_{\sqrt{4p^2+1}/4}(qb^2/2, qx^2/2)}{a^{\sqrt{p^2+1/4}-1/2} S_{\sqrt{4p^2+1}/4}(qb^2/2, qa^2/2)}$$

3.21.5 (b)

$$\mathbf{E}_x\Big\{\exp\Big(-\int_0^H \Big(\frac{p^2}{2(R_s^{(3)})^2} + \frac{q^2(R_s^{(3)})^2}{2}\Big)ds\Big);\ R_H^{(3)} = b\Big\}$$

$$= \frac{x^{\sqrt{p^2+1/4}-1/2} S_{\sqrt{4p^2+1}/4}(qx^2/2, qa^2/2)}{b^{\sqrt{p^2+1/4}-1/2} S_{\sqrt{4p^2+1}/4}(qb^2/2, qa^2/2)}$$

3.22.5 (a)

$$\mathbf{E}_x\Big\{\exp\Big(-\int_0^H \Big(\frac{p^2}{2(R_s^{(3)})^2} + \frac{q}{R_s^{(3)}}\Big)ds\Big);\ R_H^{(3)} = a\Big\}$$

$$= \frac{x^{\sqrt{p^2+1/4}-1/2} S_{\sqrt{4p^2+1}}(2\sqrt{2qb}, 2\sqrt{2qx})}{a^{\sqrt{p^2+1/4}-1/2} S_{\sqrt{4p^2+1}}(2\sqrt{2qb}, 2\sqrt{2qa})}$$

3.22.5 (b)

$$\mathbf{E}_x\Big\{\exp\Big(-\int_0^H \Big(\frac{p^2}{2(R_s^{(3)})^2} + \frac{q}{R_s^{(3)}}\Big)ds\Big);\ R_H^{(3)} = b\Big\}$$

$$= \frac{x^{\sqrt{p^2+1/4}-1/2} S_{\sqrt{4p^2+1}}(2\sqrt{2qx}, 2\sqrt{2qa})}{b^{\sqrt{p^2+1/4}-1/2} S_{\sqrt{4p^2+1}}(2\sqrt{2qb}, 2\sqrt{2qa})}$$

3.27.1

$$\mathbf{E}_x\Big\{\exp\Big(-\int_0^H \big(p\mathbb{1}_{[a,r)}(R_s^{(3)}) + q\mathbb{1}_{[r,b]}(R_s^{(3)})\big)ds\Big);\ \tfrac{1}{2}\ell(H,r) \in dy\Big\}$$

$$= \exp\big(-y(\sqrt{2q}\operatorname{cth}((b-r)\sqrt{2q}) + \sqrt{2p}\operatorname{cth}((r-a)\sqrt{2p}))\big)$$

$$\times \begin{cases} \dfrac{\sqrt{2}\operatorname{sh}((b-x)\sqrt{2q})[a\sqrt{p}\operatorname{sh}((b-r)\sqrt{2q}) + b\sqrt{q}\operatorname{sh}((r-a)\sqrt{2p})]}{x\operatorname{sh}^2((b-r)\sqrt{2q})\operatorname{sh}((r-a)\sqrt{2p})}dy, & r \le x \le b \\ \dfrac{\sqrt{2}\operatorname{sh}((x-a)\sqrt{2p})[a\sqrt{p}\operatorname{sh}((b-r)\sqrt{2q}) + b\sqrt{q}\operatorname{sh}((r-a)\sqrt{2p})]}{x\operatorname{sh}((b-r)\sqrt{2q})\operatorname{sh}^2((r-a)\sqrt{2p})}dy, & a \le x \le r \end{cases}$$

5 $R_s^{(3)} = \left(\sum_{k=1}^3 (W_s^{(k)})^2\right)^{1/2}$ $\qquad H = H_{a,b} = \min\{s : R_s^{(3)} \notin (a,b)\}$

3.27.5 (a)

$$\mathbf{E}_x\Big\{\exp\Big(-\int_0^H \big(p\mathbb{I}_{[a,r)}(R_s^{(3)}) + q\mathbb{I}_{[r,b]}(R_s^{(3)})\big)ds\Big); \frac{1}{2}\ell(H,r) \in dy, R_H^{(3)} = a\Big\}$$

$$= \exp\big(-y(\sqrt{2q}\operatorname{cth}((b-r)\sqrt{2q}) + \sqrt{2p}\operatorname{cth}((r-a)\sqrt{2p}))\big)$$

$$\times \begin{cases} \dfrac{a\sqrt{2p}\operatorname{sh}((b-x)\sqrt{2q})}{x\operatorname{sh}((b-r)\sqrt{2q})\operatorname{sh}((r-a)\sqrt{2p})}dy, & r \le x \le b \\ \dfrac{a\sqrt{2p}\operatorname{sh}((x-a)\sqrt{2p})}{x\operatorname{sh}^2((r-a)\sqrt{2p})}dy, & a \le x \le r \end{cases}$$

(1)

$$\mathbf{E}_x\Big\{\exp\Big(-\int_0^H \big(p\mathbb{I}_{[a,r)}(R_s^{(3)}) + q\mathbb{I}_{[r,b]}(R_s^{(3)})\big)ds\Big); \ell(H_z,r) = 0, R_H^{(3)} = a\Big\}$$

$$= \frac{a\operatorname{sh}((r-x)\sqrt{2p})}{x\operatorname{sh}((r-a)\sqrt{2p})}, \qquad a \le x \le r$$

3.27.5 (b)

$$\mathbf{E}_x\Big\{\exp\Big(-\int_0^H \big(p\mathbb{I}_{[a,r)}(R_s^{(3)}) + q\mathbb{I}_{[r,b]}(R_s^{(3)})\big)ds\Big); \frac{1}{2}\ell(H_z,r) \in dy, R_H^{(3)} = b\Big\}$$

$$= \exp\big(-y(\sqrt{2q}\operatorname{cth}((b-r)\sqrt{2q}) + \sqrt{2p}\operatorname{cth}((r-a)\sqrt{2p}))\big)$$

$$\times \begin{cases} \dfrac{b\sqrt{2q}\operatorname{sh}((b-x)\sqrt{2q})}{x\operatorname{sh}^2((b-r)\sqrt{2q})}dy, & r \le x \le b \\ \dfrac{b\sqrt{2q}\operatorname{sh}((x-a)\sqrt{2p})}{x\operatorname{sh}((b-r)\sqrt{2q})\operatorname{sh}((r-a)\sqrt{2p})}dy, & a \le x \le r \end{cases}$$

(1)

$$\mathbf{E}_x\Big\{\exp\Big(-\int_0^H \big(p\mathbb{I}_{[a,r)}(R_s^{(3)}) + q\mathbb{I}_{[r,b]}(R_s^{(3)})\big)ds\Big); \ell(H,r) = 0, R_H^{(3)} = b\Big\}$$

$$= \frac{b\operatorname{sh}((x-r)\sqrt{2q})}{x\operatorname{sh}((b-r)\sqrt{2q})}, \qquad r \le x \le b$$

3.27.6 (a)

$$\mathbf{P}_x\Big(\int_0^H \mathbb{I}_{[a,r)}(R_s^{(3)})ds \in du, \int_0^H \mathbb{I}_{[r,b]}(R_s^{(3)})ds \in dv, \frac{1}{2}\ell(H,r) \in dy, R_H^{(3)} = a\Big)$$

$$= \frac{a}{x}B_a^{(27)}(u,v,y)du\,dv\,dy, \qquad \text{for } B_a^{(27)}(u,v,y) \text{ see 1.3.27.6}$$

3.27.6 (b)

$$\mathbf{P}_x\Big(\int_0^H \mathbb{I}_{[a,r)}(R_s^{(3)})ds \in du, \int_0^H \mathbb{I}_{[r,b]}(R_s^{(3)})ds \in dv, \frac{1}{2}\ell(H,r) \in dy, R_H^{(3)} = b\Big)$$

$$= \frac{b}{x}B_b^{(27)}(u,v,y)du\,dv\,dy, \qquad \text{for } B_b^{(27)}(u,v,y) \text{ see 1.3.27.6}$$

5 $R_s^{(3)} = \left(\sum_{k=1}^3 (W_s^{(k)})^2\right)^{1/2}$ $H = H_{a,b} = \min\{s : R_s^{(3)} \notin (a,b)\}$

3.31.5 (a)

$$\mathbf{E}_{x(b-a)+a}\Big\{ \exp\Big(-\frac{\gamma}{2(b-a)} \sup_{0\le s\le H}\big(\ell(s,u(b-a)+a)-\alpha^{-1}\ell(s,r(b-a)+a)\big)\Big);$$

$0<\alpha$

$$R_H^{(3)}=a\Big\}=\frac{a(1-x)}{x(b-a)+a}-\frac{\gamma a(1-u)(\Upsilon+r-r^2-\alpha u+\alpha u^2)(u\wedge x-ux)}{(x(b-a)+a)(2\gamma(u\wedge r-ur)|u-r|+\Upsilon+r-r^2-\alpha u+\alpha u^2)}$$

$$-\frac{\gamma a(1-u)(2(u\wedge r-ur)|u-r|-(\Upsilon+r-r^2-\alpha u+\alpha u^2)(u-u^2))(x\wedge r-xr)}{(x(b-a)+a)(2\gamma(u\wedge r-ur)|u-r|+\Upsilon+r-r^2-\alpha u+\alpha u^2)(u\wedge r-ur)}$$

3.31.5 (b)

$$\mathbf{E}_{x(b-a)+a}\Big\{ \exp\Big(-\frac{\gamma}{2(b-a)} \sup_{0\le s\le H}\big(\ell(s,u(b-a)+a)-\alpha^{-1}\ell(s,r(b-a)+a)\big)\Big);$$

$0<\alpha$

$$R_H^{(3)}=b\Big\}=\frac{bx}{x(b-a)+a}-\frac{\gamma bu(\Upsilon+r-r^2-\alpha u+\alpha u^2)(u\wedge x-ux)}{(x(b-a)+a)(2\gamma(u\wedge r-ur)|u-r|+\Upsilon+r-r^2-\alpha u+\alpha u^2)}$$

$$-\frac{\gamma bu(2(u\wedge r-ur)|u-r|-(\Upsilon+r-r^2-\alpha u+\alpha u^2)(u-u^2))(x\wedge r-xr)}{(x(b-a)+a)(2\gamma(u\wedge r-ur)|u-r|+\Upsilon+r-r^2-\alpha u+\alpha u^2)(u\wedge r-ur)}$$

3.31.6 (a)

$$\mathbf{P}_{x(b-a)+a}\Big(\sup_{0\le s\le H}\big(\ell(s,u(b-a)+a)-\alpha^{-1}\ell(s,r(b-a)+a)\big)>(b-a)h,$$

$0<\alpha$

$$R_H^{(3)}=a\Big)=\frac{a(1-u)}{x(b-a)+a}\exp\Big(-\frac{(\Upsilon+r-r^2-\alpha u+\alpha u^2)h}{4(u\wedge r-ur)|u-r|}\Big)\Big[\frac{x\wedge r-xr}{u\wedge r-ur}$$

$$-\frac{\Upsilon+r-r^2-\alpha u+\alpha u^2}{2(u\wedge r-ur)|u-r|}\Big(\frac{(x\wedge r-xr)(u-u^2)}{u\wedge r-ur}-u\wedge x+ux\Big)\Big]$$

3.31.6 (b)

$$\mathbf{P}_{x(b-a)+a}\Big(\sup_{0\le s\le H}\big(\ell(s,u(b-a)+a)-\alpha^{-1}\ell(s,r(b-a)+a)\big)>(b-a)h,$$

$0<\alpha$

$$R_H^{(3)}=b\Big)=\frac{bu}{x(b-a)+a}\exp\Big(-\frac{(\Upsilon+r-r^2-\alpha u+\alpha u^2)h}{4(u\wedge r-ur)|u-r|}\Big)\Big[\frac{x\wedge r-xr}{u\wedge r-ur}$$

$$-\frac{\Upsilon+r-r^2-\alpha u+\alpha u^2}{2(u\wedge r-ur)|u-r|}\Big(\frac{(x\wedge r-xr)(u-u^2)}{u\wedge r-ur}-u\wedge x+ux\Big)\Big]$$

$$\Upsilon:=\sqrt{(r-r^2-\alpha u+\alpha u^2)^2+4\alpha(u\wedge r-ur)|u-r|}$$

5 $R_s^{(3)} = \big(\sum_{k=1}^3 (W_s^{(k)})^2\big)^{1/2}$ $\qquad \varrho = \varrho(v,z) = \min\{s : \ell(s,z) = v\}$

4. Stopping at inverse local time

4.0.1 $$\mathbf{E}_x\{e^{-\alpha\varrho};\ \varrho<\infty\} = \begin{cases} \dfrac{z\,\mathrm{sh}(x\sqrt{2\alpha})}{x\,\mathrm{sh}(z\sqrt{2\alpha})}\exp\Big(-\dfrac{v\sqrt{2\alpha}}{1-e^{-2z\sqrt{2\alpha}}}\Big), & 0\le x\le z\\ \dfrac{z}{x}\exp\Big(-(x-z)\sqrt{2\alpha}-\dfrac{v\sqrt{2\alpha}}{1-e^{-2z\sqrt{2\alpha}}}\Big), & z\le x \end{cases}$$

4.0.2 $$\mathbf{P}_x(\varrho\in dt) = \begin{cases} \dfrac{z}{x}\,\mathrm{ess}_t(x,z,v/2,v/2)dt, & 0\le x\le z\\ \dfrac{z}{x}\,\mathrm{es}_t(0,0,z,x-z+v/2,v/2)dt, & z\le x \end{cases}$$

(1) $$\mathbf{P}_x(\varrho=\infty) = \begin{cases} 1-\exp\Big(-\dfrac{v}{2z}\Big), & 0\le x\le z\\ 1-\dfrac{z}{x}\exp\Big(-\dfrac{v}{2z}\Big), & z\le x \end{cases}$$

4.1.1 $x\le z$ $$\mathbf{E}_x\Big\{\exp\Big(-\gamma\sup_{0\le s\le\varrho}R_s^{(3)}\Big);\ \varrho<\infty\Big\} = e^{-\gamma z}\frac{\sqrt{2v\gamma}}{\sqrt{z}}K_1\Big(\frac{\sqrt{2v\gamma}}{\sqrt{z}}\Big)$$

4.1.2 $z\le y$ $$\mathbf{P}_x\Big(\sup_{0\le s\le\varrho}R_s^{(3)}<y,\ \varrho<\infty\Big) = \begin{cases} \exp\Big(-\dfrac{vy}{2z(y-z)}\Big), & 0\le x\le z\\ \dfrac{z(y-x)}{x(y-z)}\exp\Big(-\dfrac{vy}{2z(y-z)}\Big), & z\le x\le y \end{cases}$$

4.1.4 $z\le y$ $$\mathbf{E}_x\Big\{e^{-\alpha\varrho};\ \sup_{0\le s\le\varrho}R_s^{(3)}<y,\ \varrho<\infty\Big\}$$

$$= \begin{cases} \dfrac{z\,\mathrm{sh}(x\sqrt{2\alpha})}{x\,\mathrm{sh}(z\sqrt{2\alpha})}\exp\Big(-\dfrac{v\sqrt{2\alpha}\,\mathrm{sh}(y\sqrt{2\alpha})}{2\,\mathrm{sh}(z\sqrt{2\alpha})\,\mathrm{sh}((y-z)\sqrt{2\alpha})}\Big), & 0\le x\le z\\ \dfrac{z\,\mathrm{sh}((y-x)\sqrt{2\alpha})}{x\,\mathrm{sh}((y-z)\sqrt{2\alpha})}\exp\Big(-\dfrac{v\sqrt{2\alpha}\,\mathrm{sh}(y\sqrt{2\alpha})}{2\,\mathrm{sh}(z\sqrt{2\alpha})\,\mathrm{sh}((y-z)\sqrt{2\alpha})}\Big), & z\le x\le y \end{cases}$$

4.2.2 $0\le y$ $$\mathbf{P}_x\Big(\inf_{0\le s\le\varrho}R_s^{(3)}>y,\ \varrho<\infty\Big) = \begin{cases} \dfrac{z(x-y)}{x(z-y)}\exp\Big(-\dfrac{v}{2(z-y)}\Big), & y\le x\le z\\ \dfrac{z}{x}\exp\Big(-\dfrac{v}{2(z-y)}\Big), & y\le z\le x \end{cases}$$

5 $R_s^{(3)} = \left(\sum_{k=1}^{3}(W_s^{(k)})^2\right)^{1/2}$ $\varrho = \varrho(v,z) = \min\{s : \ell(s,z) = v\}$

(1) $\mathbf{P}_x\left(\inf_{0\le s\le\varrho} R_s^{(3)} > y,\ \varrho = \infty\right)$

$$= \begin{cases} 1 - \dfrac{y}{x}, & z \le y \le x \\ \left(1-\dfrac{y}{x}\right)\left(1-\exp\left(-\dfrac{v}{2(z-y)}\right)\right), & y \le x \le z \\ 1 - \dfrac{y}{x} - \dfrac{z-y}{x}\exp\left(-\dfrac{v}{2(z-y)}\right), & y \le z \le x \end{cases}$$

4.2.4 $\mathbf{E}_x\left\{e^{-\alpha\varrho};\ \inf_{0\le s\le\varrho} R_s^{(3)} > y\right\}$

$0 \le y$

$$= \begin{cases} \dfrac{z\,\mathrm{sh}((x-y)\sqrt{2\alpha})}{x\,\mathrm{sh}((z-y)\sqrt{2\alpha})}\exp\left(-\dfrac{v\sqrt{2\alpha}}{1-e^{-2(z-y)\sqrt{2\alpha}}}\right), & y \le x \le z \\ \dfrac{z}{x}\exp\left(-(x-z)\sqrt{2\alpha} - \dfrac{v\sqrt{2\alpha}}{1-e^{-2(z-y)\sqrt{2\alpha}}}\right), & y \le z \le x \end{cases}$$

(1) $0 \le x$

$$\mathbf{P}_x\left(\inf_{0\le s\le\varrho} R_s^{(3)} > y, \varrho \in dt\right) = \begin{cases} \dfrac{z}{x}\,\mathrm{ess}_t(x-y, z-y, v/2, v/2)dt, & x \le z \\ \dfrac{z}{x}\,\mathrm{es}_t(0,0,z-y,x-z+v/2,v/2)dt, & z \le x \end{cases}$$

4.3.1 $\mathbf{E}_x\left\{e^{-\gamma\ell(\varrho,r)};\ \varrho < \infty\right\}$

$z \le r$

$$= \begin{cases} \exp\left(-\dfrac{v(1+2r\gamma)}{2z(1+2\gamma(r-z))}\right), & 0 \le x \le z \\ \dfrac{z(1+2\gamma(r-x))}{x(1+2\gamma(r-z))}\exp\left(-\dfrac{v(1+2r\gamma)}{2z(1+2\gamma(r-z))}\right), & z \le x \le r \\ \dfrac{z}{x(1+2\gamma(r-z))}\exp\left(-\dfrac{v(1+2r\gamma)}{2z(1+2\gamma(r-z))}\right), & r \le x \end{cases}$$

$r \le z$

$$= \begin{cases} \dfrac{z}{z+2\gamma(z-r)r}\exp\left(-\dfrac{v(1+2r\gamma)}{2(z+2\gamma(z-r)r)}\right), & 0 \le x \le r \\ \dfrac{z(x+2\gamma(x-r)r)}{x(z+2\gamma(z-r)r)}\exp\left(-\dfrac{v(1+2r\gamma)}{2(z+2\gamma(z-r)r)}\right), & r \le x \le z \\ \dfrac{z}{x}\exp\left(-\dfrac{v(1+2r\gamma)}{2(z+2\gamma(z-r)r)}\right), & z \le x \end{cases}$$

(1) $\mathbf{E}_x\left\{e^{-\gamma\ell(\varrho,r)};\ \varrho = \infty\right\}$

5 $R_s^{(3)} = \left(\sum_{k=1}^3 (W_s^{(k)})^2\right)^{1/2}$ $\qquad\varrho = \varrho(v,z) = \min\{s : \ell(s,z) = v\}$

$z \le r$

$$= \begin{cases} \dfrac{1}{1+2\gamma r}\Big(1 - \exp\Big(-\dfrac{v(1+2r\gamma)}{2z(1+2\gamma(r-z))}\Big)\Big), & 0 \le x \le z \\ \dfrac{1}{1+2\gamma r}\Big(1 - \dfrac{z(1+2\gamma(r-x))}{x(1+2\gamma(r-z))}\exp\Big(-\dfrac{v(1+2r\gamma)}{2z(1+2\gamma(r-z))}\Big)\Big), & z \le x \le r \\ \dfrac{x+2\gamma r(x-r)}{1+2\gamma r} & \\ \quad -\dfrac{z}{x(1+2\gamma r)(1+2\gamma(r-z))}\exp\Big(-\dfrac{v(1+2r\gamma)}{2z(1+2\gamma(r-z))}\Big), & r \le x \end{cases}$$

$r \le z$

$$= \begin{cases} \dfrac{1}{1+2\gamma r}\Big(1 - \exp\Big(-\dfrac{v(1+2r\gamma)}{2(z+2\gamma(z-r)r)}\Big)\Big), & 0 \le x \le r \\ \dfrac{x+2\gamma(x-r)r}{x(1+2\gamma r)}\Big(1 - \exp\Big(-\dfrac{v(1+2r\gamma)}{2(z+2\gamma(z-r)r)}\Big)\Big), & r \le x \le z \\ \dfrac{x+2\gamma(x-r)r}{x(1+2\gamma r)} - \dfrac{z+2\gamma(z-r)r}{x(1+2\gamma r)}\exp\Big(-\dfrac{v(1+2r\gamma)}{2(z+2\gamma(z-r)r)}\Big), & z \le x \end{cases}$$

4.3.2 $\mathbf{P}_x\big(\ell(\varrho,r) \in dy,\ \varrho < \infty\big)$

$z \le r$

$$= \begin{cases} \dfrac{1}{2(r-z)}\dfrac{\sqrt{v}}{\sqrt{y}}\exp\Big(-\dfrac{(v+y)}{2(r-z)}\Big)I_1\Big(\dfrac{\sqrt{vy}}{(r-z)}\Big)dy, & 0 \le x \le z, \\ \dfrac{z}{2x(r-z)^2}\exp\Big(-\dfrac{(v+y)}{2(r-z)}\Big)\Big[(r-x)\dfrac{\sqrt{v}}{\sqrt{y}}I_1\Big(\dfrac{\sqrt{vy}}{(r-z)}\Big) & \\ \qquad +(x-z)I_0\Big(\dfrac{\sqrt{vy}}{(r-z)}\Big)\Big]dy, & z \le x \le r \\ \dfrac{z}{2x(r-z)}\exp\Big(-\dfrac{(v+y)}{2(r-z)}\Big)I_0\Big(\dfrac{\sqrt{vy}}{(r-z)}\Big)dy, & r \le x \end{cases}$$

$r \le z$

$$= \begin{cases} \dfrac{z}{2x(z-r)}\dfrac{\sqrt{v}}{\sqrt{y}}\exp\Big(-\dfrac{(vr+yz)}{2(z-r)r}\Big)I_1\Big(\dfrac{\sqrt{vy}}{(z-r)}\Big)dy, & z \le x \\ \dfrac{z}{2x(z-r)^2}\exp\Big(-\dfrac{(vr+yz)}{2(z-r)r}\Big)\Big[(x-r)\dfrac{\sqrt{v}}{\sqrt{y}}I_1\Big(\dfrac{\sqrt{vy}}{(z-r)}\Big) & \\ \qquad +(z-x)I_0\Big(\dfrac{\sqrt{vy}}{(z-r)}\Big)\Big]dy, & r \le x \le z \\ \dfrac{z}{2r(z-r)}\exp\Big(-\dfrac{(vr+yz)}{2(z-r)r}\Big)I_0\Big(\dfrac{\sqrt{vy}}{(z-r)}\Big)dy, & 0 \le x \le r \end{cases}$$

(2) $\mathbf{P}_x\big(\ell(\varrho,r) = 0,\ \varrho < \infty\big)$

$z \le r$

$$= \begin{cases} \exp\Big(-\dfrac{vr}{2z(r-z)}\Big), & 0 \le x \le z \\ \dfrac{z(r-x)}{x(r-z)}\exp\Big(-\dfrac{vr}{2z(r-z)}\Big), & z \le x \le r \\ 0, & r \le x \end{cases}$$

5 $R_s^{(3)} = \left(\sum_{k=1}^3 (W_s^{(k)})^2\right)^{1/2}$ $\qquad \varrho = \varrho(v,z) = \min\{s : \ell(s,z) = v\}$

$r \le z$
$$= \begin{cases} 0, & 0 \le x \le r \\ \dfrac{z(x-r)}{x(z-r)} \exp\Big(-\dfrac{v}{2(z-r)}\Big), & r \le x \le z \\ \dfrac{z}{x} \exp\Big(-\dfrac{v}{2(z-r)}\Big), & z \le x \end{cases}$$

(3)
$$\mathbf{P}_x\big(\ell(\varrho,r) = 0,\ \varrho = \infty\big) = \begin{cases} 1 - \dfrac{r}{x}, & z \le r \le x \\ \Big(1 - \dfrac{r}{x}\Big)\Big(1 - \exp\Big(-\dfrac{v}{2(z-r)}\Big)\Big), & r \le x \le z \\ 1 - \dfrac{r}{x} - \dfrac{z-r}{x} \exp\Big(-\dfrac{v}{2(z-r)}\Big), & r \le z \le x \end{cases}$$

4.3.3 $\mathbf{E}_x e^{-\alpha\varrho - \gamma\ell(\varrho,r)}$

$z \le r$
$$= \exp\Big(-\frac{v\big(\alpha e^{r\sqrt{2\alpha}} + \gamma\sqrt{2\alpha}\,\mathrm{sh}(r\sqrt{2\alpha})\big)}{\mathrm{sh}(z\sqrt{2\alpha})\big(\sqrt{2\alpha}e^{(r-z)\sqrt{2\alpha}} + 2\gamma\,\mathrm{sh}((r-z)\sqrt{2\alpha})\big)}\Big)$$

$$\times \begin{cases} \dfrac{z\,\mathrm{sh}(x\sqrt{2\alpha})}{x\,\mathrm{sh}(z\sqrt{2\alpha})}, & 0 \le x \le z \\ \dfrac{z\big(\sqrt{2\alpha}e^{(r-x)\sqrt{2\alpha}} + 2\gamma\,\mathrm{sh}((r-x)\sqrt{2\alpha})\big)}{x\big(\sqrt{2\alpha}e^{(r-z)\sqrt{2\alpha}} + 2\gamma\,\mathrm{sh}((r-z)\sqrt{2\alpha})\big)}, & z \le x \le r \\ \dfrac{z\sqrt{2\alpha}e^{-(x-r)\sqrt{2\alpha}}}{x\big(\sqrt{2\alpha}e^{(r-z)\sqrt{2\alpha}} + 2\gamma\,\mathrm{sh}((r-z)\sqrt{2\alpha})\big)}, & r \le x \end{cases}$$

$r \le z$
$$= \exp\Big(-\frac{ve^{(z-r)\sqrt{2\alpha}}\big(\alpha e^{r\sqrt{2\alpha}} + \gamma\sqrt{2\alpha}\,\mathrm{sh}(r\sqrt{2\alpha})\big)}{\sqrt{2\alpha}\,\mathrm{sh}(z\sqrt{2\alpha}) + 2\gamma\,\mathrm{sh}((z-r)\sqrt{2\alpha})\,\mathrm{sh}(r\sqrt{2\alpha})}\Big)$$

$$\times \begin{cases} \dfrac{z\sqrt{2\alpha}\,\mathrm{sh}(x\sqrt{2\alpha})}{x\big(\sqrt{2\alpha}\,\mathrm{sh}(z\sqrt{2\alpha}) + 2\gamma\,\mathrm{sh}((z-r)\sqrt{2\alpha})\,\mathrm{sh}(r\sqrt{2\alpha})\big)}, & 0 \le x \le r \\ \dfrac{z\big(\sqrt{2\alpha}\,\mathrm{sh}(x\sqrt{2\alpha}) + 2\gamma\,\mathrm{sh}((x-r)\sqrt{2\alpha})\,\mathrm{sh}(r\sqrt{2\alpha})\big)}{x\big(\sqrt{2\alpha}\,\mathrm{sh}(z\sqrt{2\alpha}) + 2\gamma\,\mathrm{sh}((z-r)\sqrt{2\alpha})\,\mathrm{sh}(r\sqrt{2\alpha})\big)}, & r \le x \le z \\ \dfrac{z}{x}e^{-(x-z)\sqrt{2\alpha}}, & z \le x \end{cases}$$

4.3.4 $\mathbf{E}_x\{e^{-\alpha\varrho};\ \ell(\varrho,r) \in dy\}$

$z \le r$
$$= \exp\Big(-\frac{\sqrt{2\alpha}\,\mathrm{sh}(r\sqrt{2\alpha})v}{2\,\mathrm{sh}((r-z)\sqrt{2\alpha})\,\mathrm{sh}(z\sqrt{2\alpha})} - \frac{\sqrt{2\alpha}e^{(r-z)\sqrt{2\alpha}}y}{2\,\mathrm{sh}((r-z)\sqrt{2\alpha})}\Big)$$

5 $R_s^{(3)}=\left(\sum_{k=1}^{3}(W_s^{(k)})^2\right)^{1/2}$ $\qquad\qquad$ $\varrho=\varrho(v,z)=\min\{s:\ell(s,z)=v\}$

$$\times\begin{cases}\dfrac{z\,\mathrm{sh}(x\sqrt{2\alpha})\sqrt{2\alpha v}}{2x\,\mathrm{sh}(z\sqrt{2\alpha})\,\mathrm{sh}((r-z)\sqrt{2\alpha})\sqrt{y}}I_1\Bigl(\dfrac{\sqrt{2\alpha vy}}{\mathrm{sh}((r-z)\sqrt{2\alpha})}\Bigr)dy, & 0\le x\le z\\[2ex] \dfrac{z\sqrt{2\alpha}}{2x\,\mathrm{sh}^2((r-z)\sqrt{2\alpha})}\Bigl[\mathrm{sh}((r-x)\sqrt{2\alpha})\dfrac{\sqrt{v}}{\sqrt{y}}I_1\Bigl(\dfrac{\sqrt{2\alpha vy}}{\mathrm{sh}((r-z)\sqrt{2\alpha})}\Bigr) & \\ \qquad+\,\mathrm{sh}((x-z)\sqrt{2\alpha})I_0\Bigl(\dfrac{\sqrt{2\alpha vy}}{\mathrm{sh}((r-z)\sqrt{2\alpha})}\Bigr)\Bigr]dy, & z\le x\le r\\[2ex] \dfrac{z\sqrt{2\alpha}e^{-(x-r)\sqrt{2\alpha}}}{2x\,\mathrm{sh}((r-z)\sqrt{2\alpha})}I_0\Bigl(\dfrac{\sqrt{2\alpha vy}}{\mathrm{sh}((r-z)\sqrt{2\alpha})}\Bigr)dy, & r\le x\end{cases}$$

$r\le z$

$$=\exp\Bigl(-\frac{\sqrt{2\alpha}e^{(z-r)\sqrt{2\alpha}}v}{2\,\mathrm{sh}((z-r)\sqrt{2\alpha})}-\frac{\sqrt{2\alpha}\,\mathrm{sh}(z\sqrt{2\alpha})y}{2\,\mathrm{sh}((z-r)\sqrt{2\alpha})\,\mathrm{sh}(r\sqrt{2\alpha})}\Bigr)$$

$$\times\begin{cases}\dfrac{z\sqrt{2\alpha v}e^{-(x-z)\sqrt{2\alpha}}}{2x\,\mathrm{sh}((z-r)\sqrt{2\alpha})\sqrt{y}}I_1\Bigl(\dfrac{\sqrt{2\alpha vy}}{\mathrm{sh}((z-r)\sqrt{2\alpha})}\Bigr)dy, & z\le x\\[2ex] \dfrac{z\sqrt{2\alpha}}{2x\,\mathrm{sh}^2((z-r)\sqrt{2\alpha})}\Bigl[\mathrm{sh}((x-r)\sqrt{2\alpha})\sqrt{\dfrac{v}{y}}I_1\Bigl(\dfrac{\sqrt{2\alpha vy}}{\mathrm{sh}((z-r)\sqrt{2\alpha})}\Bigr) & \\ \qquad+\,\mathrm{sh}((z-x)\sqrt{2\alpha})I_0\Bigl(\dfrac{\sqrt{2\alpha vy}}{\mathrm{sh}((z-r)\sqrt{2\alpha})}\Bigr)\Bigr]dy, & r\le x\le z\\[2ex] \dfrac{z\sqrt{2\alpha}\,\mathrm{sh}(x\sqrt{2\alpha})}{2x\,\mathrm{sh}((z-r)\sqrt{2\alpha})\,\mathrm{sh}(r\sqrt{2\alpha})}I_0\Bigl(\dfrac{\sqrt{2\alpha vy}}{\mathrm{sh}((z-r)\sqrt{2\alpha})}\Bigr)dy, & 0\le x\le r\end{cases}$$

(1) $\mathbf{E}_x\{e^{-\alpha\varrho};\ \ell(\varrho,r)=0\}$

$z\le r$

$$=\exp\Bigl(-\frac{v\sqrt{2\alpha}\,\mathrm{sh}(r\sqrt{2\alpha})}{2\,\mathrm{sh}(z\sqrt{2\alpha})\,\mathrm{sh}((r-z)\sqrt{2\alpha})}\Bigr)\begin{cases}\dfrac{z\,\mathrm{sh}(x\sqrt{2\alpha})}{x\,\mathrm{sh}(z\sqrt{2\alpha})}, & 0\le x\le z\\[2ex] \dfrac{z\,\mathrm{sh}((r-x)\sqrt{2\alpha})}{x\,\mathrm{sh}((r-z)\sqrt{2\alpha})}, & z\le x\le r\\[2ex] 0, & r\le x\end{cases}$$

$r\le z$

$$=\exp\Bigl(-\frac{v\sqrt{2\alpha}e^{(z-r)\sqrt{2\alpha}}}{2\,\mathrm{sh}((z-r)\sqrt{2\alpha})}\Bigr)\begin{cases}\dfrac{z}{x}e^{-(x-z)\sqrt{2\alpha}}, & z\le x\\[2ex] \dfrac{z\,\mathrm{sh}((x-r)\sqrt{2\alpha})}{x\,\mathrm{sh}((z-r)\sqrt{2\alpha})}, & r\le x\le z\\[2ex] 0, & 0\le x\le r\end{cases}$$

4.4.1 $\mathbf{E}_x\Bigl\{\exp\Bigl(-\gamma\displaystyle\int_0^{\varrho}\mathbb{I}_{[r,\infty)}\bigl(R_s^{(3)}\bigr)ds\Bigr);\varrho<\infty\Bigr\}$

5 $R_s^{(3)} = \left(\sum_{k=1}^3 (W_s^{(k)})^2\right)^{1/2}$ $\varrho = \varrho(v,z) = \min\{s : \ell(s,z) = v\}$

$z \le r$

$$= \exp\Big(-\frac{v(1+\sqrt{2\gamma}r)}{2z(1+(r-z)\sqrt{2\gamma})}\Big)\begin{cases} 1, & 0 \le x \le z \\ \dfrac{z(1+(r-x)\sqrt{2\gamma})}{x(1+(r-z)\sqrt{2\gamma})}, & z \le x \le r \\ \dfrac{ze^{-(x-r)\sqrt{2\gamma}}}{x(1+(r-z)\sqrt{2\gamma})}, & r \le x \end{cases}$$

$r \le z$

$$= \exp\Big(-\frac{v(\sqrt{2\gamma}+2r\gamma)e^{(z-r)\sqrt{2\gamma}}}{2(r\sqrt{2\gamma}\operatorname{ch}((z-r)\sqrt{2\gamma})+\operatorname{sh}((z-r)\sqrt{2\gamma}))}\Big)$$

$$\times\begin{cases} \dfrac{z\sqrt{2\gamma}}{r\sqrt{2\gamma}\operatorname{ch}((z-r)\sqrt{2\gamma})+\operatorname{sh}((z-r)\sqrt{2\gamma})}, & 0 \le x \le r \\ \dfrac{z(r\sqrt{2\gamma}\operatorname{ch}((x-r)\sqrt{2\gamma})+\operatorname{sh}((x-r)\sqrt{2\gamma}))}{x(r\sqrt{2\gamma}\operatorname{ch}((z-r)\sqrt{2\gamma})+\operatorname{sh}((z-r)\sqrt{2\gamma}))}, & r \le x \le z \\ \dfrac{z}{x}e^{-(x-z)\sqrt{2\gamma}}, & z \le x \end{cases}$$

4.4.2
(1)
$z \le r$

$$\mathbf{P}_x\Big(\int_0^\varrho \mathrm{1\!I}_{[r,\infty)}(R_s^{(3)})ds = 0\Big) = \begin{cases} \exp\Big(-\dfrac{v}{2(r-z)}\Big), & 0 \le x \le z \\ \dfrac{r-x}{r-z}\exp\Big(-\dfrac{v}{2(r-z)}\Big), & z \le x \le r \\ 0, & r \le x \end{cases}$$

4.4.4
(1)

$$\mathbf{E}_x\Big\{e^{-\alpha\varrho};\ \int_0^\varrho \mathrm{1\!I}_{[r,\infty)}(R_s^{(3)})ds = 0\Big\}$$

$z \le r$

$$= \exp\Big(-\frac{v\sqrt{2\alpha}\operatorname{sh}(r\sqrt{2\alpha})}{\operatorname{sh}(z\sqrt{2\alpha})\operatorname{sh}((r-z)\sqrt{2\alpha})}\Big)\begin{cases} \dfrac{z\operatorname{sh}(x\sqrt{2\alpha})}{x\operatorname{sh}(z\sqrt{2\alpha})}, & 0 \le x \le z \\ \dfrac{z\operatorname{sh}((r-x)\sqrt{2\alpha})}{x\operatorname{sh}((r-z)\sqrt{2\alpha})}, & z \le x \le r \\ 0, & r \le x \end{cases}$$

4.5.1

$$\mathbf{E}_x\Big\{\exp\Big(-\gamma\int_0^\varrho \mathrm{1\!I}_{[0,r]}(R_s^{(3)})ds\Big);\ \varrho < \infty\Big\}$$

$z \le r$

$$= \exp\Big(-\frac{v\sqrt{2\gamma}\operatorname{ch}(r\sqrt{2\gamma})}{2\operatorname{sh}(z\sqrt{2\gamma})\operatorname{ch}((r-z)\sqrt{2\gamma})}\Big)\begin{cases} \dfrac{z\operatorname{sh}(x\sqrt{2\gamma})}{x\operatorname{sh}(z\sqrt{2\gamma})}, & 0 \le x \le z \\ \dfrac{z\operatorname{ch}((r-x)\sqrt{2\gamma})}{x\operatorname{ch}((r-z)\sqrt{2\gamma})}, & z \le x \le r \\ \dfrac{z}{x\operatorname{ch}((r-z)\sqrt{2\gamma})}, & r \le x \end{cases}$$

5 $R_s^{(3)} = \left(\sum_{k=1}^{3}(W_s^{(k)})^2\right)^{1/2}$ $\quad\quad$ $\varrho = \varrho(v,z) = \min\{s : \ell(s,z) = v\}$

$r \le z$

$$= \exp\Bigl(-\frac{v\sqrt{2\gamma}\,\mathrm{ch}(r\sqrt{2\gamma})}{2(\mathrm{sh}(r\sqrt{2\gamma}) + \sqrt{2\gamma}(z-r)\,\mathrm{ch}(r\sqrt{2\gamma})}\Bigr)$$

$$\times \begin{cases} \dfrac{z\,\mathrm{sh}(x\sqrt{2\gamma})}{x(\mathrm{sh}(r\sqrt{2\gamma}) + \sqrt{2\gamma}(z-r)\,\mathrm{ch}(r\sqrt{2\gamma}))}, & 0 \le x \le r \\ \dfrac{z(\mathrm{sh}(r\sqrt{2\gamma}) + \sqrt{2\gamma}(x-r)\,\mathrm{ch}(r\sqrt{2\gamma}))}{x(\mathrm{sh}(r\sqrt{2\gamma}) + \sqrt{2\gamma}(z-r)\,\mathrm{ch}(r\sqrt{2\gamma}))}, & r \le x \le z \\ \dfrac{z}{x}, & z \le x \end{cases}$$

(1) $\mathbf{E}_x\Bigl\{\exp\Bigl(-\gamma\int_0^{\varrho}\mathbb{I}_{[0,r]}(R_s^{(3)})\,ds\Bigr);\ \varrho = \infty\Bigr\}$

$z \le r$

$$= \begin{cases} \dfrac{\mathrm{sh}(x\sqrt{2\gamma})}{x\sqrt{2\gamma}\,\mathrm{ch}(r\sqrt{2\gamma})}\Bigl(1 - \exp\Bigl(-\dfrac{v\sqrt{2\gamma}\,\mathrm{ch}(r\sqrt{2\gamma})}{2\,\mathrm{sh}(z\sqrt{2\gamma})\,\mathrm{ch}((r-z)\sqrt{2\gamma})}\Bigr)\Bigr), & 0 \le x \le z \\ \dfrac{\mathrm{sh}(x\sqrt{2\gamma})}{x\sqrt{2\gamma}\,\mathrm{ch}(r\sqrt{2\gamma})} - \dfrac{\mathrm{sh}(z\sqrt{2\gamma})\,\mathrm{ch}((r-x)\sqrt{2\gamma})}{x\sqrt{2\gamma}\,\mathrm{ch}(r\sqrt{2\gamma})\,\mathrm{ch}((r-z)\sqrt{2\gamma})} \\ \qquad \times \exp\Bigl(-\dfrac{v\sqrt{2\gamma}\,\mathrm{ch}(r\sqrt{2\gamma})}{2\,\mathrm{sh}(z\sqrt{2\gamma})\,\mathrm{ch}((r-z)\sqrt{2\gamma})}\Bigr), & z \le x \le r \\ \dfrac{\mathrm{sh}(r\sqrt{2\gamma}) + \sqrt{2\gamma}(x-r)\,\mathrm{ch}(r\sqrt{2\gamma})}{x\sqrt{2\gamma}\,\mathrm{ch}(r\sqrt{2\gamma})} - \dfrac{\mathrm{sh}(z\sqrt{2\gamma})}{x\sqrt{2\gamma}\,\mathrm{ch}(r\sqrt{2\gamma})\,\mathrm{ch}((r-z)\sqrt{2\gamma})} \\ \qquad \times \exp\Bigl(-\dfrac{v\sqrt{2\gamma}\,\mathrm{ch}(r\sqrt{2\gamma})}{2\,\mathrm{sh}(z\sqrt{2\gamma})\,\mathrm{ch}((r-z)\sqrt{2\gamma})}\Bigr), & r \le x \end{cases}$$

$r \le z$

$$= \begin{cases} \dfrac{\mathrm{sh}(x\sqrt{2\gamma})}{x\sqrt{2\gamma}\,\mathrm{ch}(r\sqrt{2\gamma})}\Bigl(1 - \exp\Bigl(-\dfrac{v\sqrt{2\gamma}\,\mathrm{ch}(r\sqrt{2\gamma})}{2(\mathrm{sh}(r\sqrt{2\gamma}) + \sqrt{2\gamma}(z-r)\,\mathrm{ch}(r\sqrt{2\gamma})}\Bigr)\Bigr), & 0 \le x \le r \\ \dfrac{\mathrm{sh}(r\sqrt{2\gamma}) + \sqrt{2\gamma}(x-r)\,\mathrm{ch}(r\sqrt{2\gamma})}{x\sqrt{2\gamma}\,\mathrm{ch}(r\sqrt{2\gamma})} \\ \qquad \times \Bigl(1 - \exp\Bigl(-\dfrac{v\sqrt{2\gamma}\,\mathrm{ch}(r\sqrt{2\gamma})}{2(\mathrm{sh}(r\sqrt{2\gamma}) + \sqrt{2\gamma}(z-r)\,\mathrm{ch}(r\sqrt{2\gamma})}\Bigr)\Bigr), & r \le x \le z \\ \dfrac{\mathrm{sh}(r\sqrt{2\gamma}) + \sqrt{2\gamma}(x-r)\,\mathrm{ch}(r\sqrt{2\gamma})}{x\sqrt{2\gamma}\,\mathrm{ch}(r\sqrt{2\gamma})} - \dfrac{\mathrm{sh}(r\sqrt{2\gamma}) + \sqrt{2\gamma}(z-r)\,\mathrm{ch}(r\sqrt{2\gamma})}{x\sqrt{2\gamma}\,\mathrm{ch}(r\sqrt{2\gamma})} \\ \qquad \times \exp\Bigl(-\dfrac{v\sqrt{2\gamma}\,\mathrm{ch}(r\sqrt{2\gamma})}{2(\mathrm{sh}(r\sqrt{2\gamma}) + \sqrt{2\gamma}(z-r)\,\mathrm{ch}(r\sqrt{2\gamma})}\Bigr), & z \le x \end{cases}$$

4.5.2
(1)
$r \le z$

$$\mathbf{P}_x\Bigl(\int_0^{\varrho}\mathbb{I}_{[0,r]}(R_s^{(3)})\,ds = 0\Bigr) = \begin{cases} 0, & 0 \le x \le r \\ \dfrac{z(x-r)}{x(z-r)}\exp\Bigl(-\dfrac{v}{2(z-r)}\Bigr), & r \le x \le z \\ \dfrac{z}{x}\exp\Bigl(-\dfrac{v}{2(z-r)}\Bigr), & z \le x \end{cases}$$

5 $R_s^{(3)} = \left(\sum_{k=1}^3 (W_s^{(k)})^2\right)^{1/2}$ $\varrho = \varrho(v,z) = \min\{s : \ell(s,z) = v\}$

4.5.4 (1) $\mathbf{E}_x\Big\{e^{-\alpha\varrho};\ \int_0^{\varrho} \mathbb{I}_{[0,r]}\big(R_s^{(3)}\big)ds = 0\Big\}$

$r \le z$
$$= \exp\Big(-\frac{v\sqrt{\alpha}e^{\sqrt{2\alpha}(z-r)}}{\sqrt{2}\,\mathrm{sh}((z-r)\sqrt{2\alpha})}\Big)\begin{cases} 0, & 0 \le x \le r \\ \dfrac{z\,\mathrm{sh}((x-r)\sqrt{2\alpha})}{x\,\mathrm{sh}((z-r)\sqrt{2\alpha})}, & r \le x \le z \\ \dfrac{z}{x}e^{-(x-z)\sqrt{2\alpha}}, & z \le x \end{cases}$$

4.6.1 $\mathbf{E}_x \exp\Big(-\int_0^{\varrho}\big(p\mathbb{I}_{[0,r)}\big(R_s^{(3)}\big) + q\mathbb{I}_{[r,\infty)}\big(R_s^{(3)}\big)\big)ds\Big)$

$z \le r$
$$= \exp\Big(-\frac{v\sqrt{p}\big(\sqrt{p}\,\mathrm{ch}(r\sqrt{2p}) + \sqrt{q}\,\mathrm{sh}(r\sqrt{2p})\big)}{\mathrm{sh}(z\sqrt{2p})\big(\sqrt{2p}\,\mathrm{ch}((r-z)\sqrt{2p}) + \sqrt{2q}\,\mathrm{sh}((r-z)\sqrt{2p})\big)}\Big)$$

$$\times\begin{cases} \dfrac{z\,\mathrm{sh}(x\sqrt{2p})}{x\,\mathrm{sh}(z\sqrt{2p})}, & 0 \le x \le z \\ \dfrac{z\big(\sqrt{p}\,\mathrm{ch}((r-x)\sqrt{2p}) + \sqrt{q}\,\mathrm{sh}((r-x)\sqrt{2p})\big)}{x\big(\sqrt{p}\,\mathrm{ch}((r-z)\sqrt{2p}) + \sqrt{q}\,\mathrm{sh}((r-z)\sqrt{2p})\big)}, & z \le x \le r \\ \dfrac{z\sqrt{p}e^{-(x-r)\sqrt{2q}}}{x\big(\sqrt{p}\,\mathrm{ch}((r-z)\sqrt{2p}) + \sqrt{q}\,\mathrm{sh}((r-z)\sqrt{2p})\big)}, & r \le x \end{cases}$$

$r \le z$
$$= \exp\Big(-\frac{v\sqrt{q}\big(\sqrt{p}\,\mathrm{ch}(r\sqrt{2p}) + \sqrt{q}\,\mathrm{sh}(r\sqrt{2p})\big)e^{(z-r)\sqrt{2q}}}{\sqrt{2q}\,\mathrm{sh}(r\sqrt{2p})\,\mathrm{ch}((z-r)\sqrt{2q}) + \sqrt{2p}\,\mathrm{ch}(r\sqrt{2p})\,\mathrm{sh}((z-r)\sqrt{2q})}\Big)$$

$$\times\begin{cases} \dfrac{z\sqrt{q}\,\mathrm{sh}(x\sqrt{2p})}{x\big(\sqrt{q}\,\mathrm{sh}(r\sqrt{2p})\,\mathrm{ch}((z-r)\sqrt{2q}) + \sqrt{p}\,\mathrm{ch}(r\sqrt{2p})\,\mathrm{sh}((z-r)\sqrt{2q})\big)}, & 0 \le x \le r \\ \dfrac{z\big(\sqrt{q}\,\mathrm{sh}(r\sqrt{2p})\,\mathrm{ch}((x-r)\sqrt{2q}) + \sqrt{p}\,\mathrm{ch}(r\sqrt{2p})\,\mathrm{sh}((x-r)\sqrt{2q})\big)}{x\big(\sqrt{q}\,\mathrm{sh}(r\sqrt{2p})\,\mathrm{ch}((z-r)\sqrt{2q}) + \sqrt{p}\,\mathrm{ch}(r\sqrt{2p})\,\mathrm{sh}((z-r)\sqrt{2q})\big)}, & r \le x \le z \\ \dfrac{z}{x}e^{-(x-z)\sqrt{2q}}, & z \le x \end{cases}$$

4.7.1 $r < u$ $\mathbf{E}_x\Big\{\exp\Big(-\gamma\int_0^{\varrho} \mathbb{I}_{[r,u]}\big(R_s^{(3)}\big)ds\Big);\ \varrho < \infty\Big\}$

$z \le r$
$$= \exp\Big(-\frac{v\big(r\sqrt{2\gamma}\,\mathrm{sh}((u-r)\sqrt{2\gamma}) + \mathrm{ch}((u-r)\sqrt{2\gamma})\big)}{2z\big((r-z)\sqrt{2\gamma}\,\mathrm{sh}((u-r)\sqrt{2\gamma}) + \mathrm{ch}((u-r)\sqrt{2\gamma})\big)}\Big)$$

5 $R_s^{(3)} = \left(\sum_{k=1}^3 (W_s^{(k)})^2\right)^{1/2}$ $\qquad \varrho = \varrho(v,z) = \min\{s : \ell(s,z) = v\}$

$$\times \begin{cases} 1, & 0 \le x \le z \\ \frac{z((r-x)\sqrt{2\gamma}\,\mathrm{sh}((u-r)\sqrt{2\gamma}) + \mathrm{ch}((u-r)\sqrt{2\gamma}))}{x((r-z)\sqrt{2\gamma}\,\mathrm{sh}((u-r)\sqrt{2\gamma}) + \mathrm{ch}((u-r)\sqrt{2\gamma}))}, & z \le x \le r \\ \frac{z\,\mathrm{ch}((u-x)\sqrt{2\gamma})}{x((r-z)\sqrt{2\gamma}\,\mathrm{sh}((u-r)\sqrt{2\gamma}) + \mathrm{ch}((u-r)\sqrt{2\gamma}))}, & r \le x \le u \\ \frac{z}{x((r-z)\sqrt{2\gamma}\,\mathrm{sh}((u-r)\sqrt{2\gamma}) + \mathrm{ch}((u-r)\sqrt{2\gamma}))}, & u \le x \end{cases}$$

$u \le z$

$$= \exp\Big(-\frac{v\big(r2\gamma\,\mathrm{sh}((u-r)\sqrt{2\gamma}) + \sqrt{2\gamma}\,\mathrm{ch}((u-r)\sqrt{2\gamma})\big)}{2((1+(z-u)r2\gamma)\,\mathrm{sh}((u-r)\sqrt{2\gamma}) + (r+z-u)\sqrt{2\gamma}\,\mathrm{ch}((u-r)\sqrt{2\gamma}))}\Big)$$

$$\times \begin{cases} \frac{z\sqrt{2\gamma}}{(1+(z-u)r2\gamma)\,\mathrm{sh}((u-r)\sqrt{2\gamma}) + (r+z-u)\sqrt{2\gamma}\,\mathrm{ch}((u-r)\sqrt{2\gamma})}, & 0 \le x \le r \\ \frac{z(\mathrm{sh}((x-r)\sqrt{2\gamma}) + r\sqrt{2\gamma}\,\mathrm{ch}((x-r)\sqrt{2\gamma}))}{x((1+(z-u)r2\gamma)\,\mathrm{sh}((u-r)\sqrt{2\gamma}) + (r+z-u)\sqrt{2\gamma}\,\mathrm{ch}((u-r)\sqrt{2\gamma}))}, & r \le x \le u \\ \frac{z((1+(x-u)r2\gamma)\,\mathrm{sh}((u-r)\sqrt{2\gamma}) + (r+x-u)\sqrt{2\gamma}\,\mathrm{ch}((u-r)\sqrt{2\gamma}))}{x((1+(z-u)r2\gamma)\,\mathrm{sh}((u-r)\sqrt{2\gamma}) + (r+z-u)\sqrt{2\gamma}\,\mathrm{ch}((u-r)\sqrt{2\gamma}))}, & u \le x \le z \\ \frac{z}{x}, & z \le x \end{cases}$$

4.8.1 $\mathbf{E}_x\Big\{\exp\Big(-\gamma\int_0^\varrho R_s^{(3)}ds\Big);\ \varrho < \infty\Big\}$

$$= \exp\Big(-\frac{3z^{-1}v}{4I_{1/3}(\frac{2}{3}z^{3/2}\sqrt{2\gamma})K_{1/3}(\frac{2}{3}z^{3/2}\sqrt{2\gamma})}\Big)\begin{cases} \frac{\sqrt{z}I_{1/3}(\frac{2}{3}x^{3/2}\sqrt{2\gamma})}{\sqrt{x}I_{1/3}(\frac{2}{3}z^{3/2}\sqrt{2\gamma})}, & 0 \le x \le z \\ \frac{\sqrt{z}K_{1/3}(\frac{2}{3}x^{3/2}\sqrt{2\gamma})}{\sqrt{x}K_{1/3}(\frac{2}{3}z^{3/2}\sqrt{2\gamma})}, & z \le x \end{cases}$$

4.9.1 $\mathbf{E}_x\Big\{\exp\Big(-2\gamma\int_0^\varrho (R_s^{(3)})^2ds\Big);\ \varrho < \infty\Big\}$

$$= \exp\Big(-\frac{v}{zI_{1/4}(z^2\sqrt{\gamma})K_{1/4}(z^2\sqrt{\gamma})}\Big)\begin{cases} \frac{\sqrt{z}I_{1/4}(x^2\sqrt{\gamma})}{\sqrt{x}I_{1/4}(z^2\sqrt{\gamma})}, & 0 \le x \le z \\ \frac{\sqrt{z}K_{1/4}(x^2\sqrt{\gamma})}{\sqrt{x}K_{1/4}(z^2\sqrt{\gamma})}, & z \le x \end{cases}$$

4.10.3 $\mathbf{E}_x \exp\Big(-\alpha\varrho - \gamma\int_0^\varrho e^{2\beta R_s^{(3)}}ds\Big) = \exp\Bigg(-\frac{v\beta 2^{-1}K_{\sqrt{2\alpha}/\beta}\Big(\frac{\sqrt{2\gamma}}{|\beta|}\Big)}{S_{\sqrt{2\alpha}/\beta}\Big(\frac{\sqrt{2\gamma}}{|\beta|}e^{\beta z}, \frac{\sqrt{2\gamma}}{|\beta|}\Big)K_{\sqrt{2\alpha}/\beta}\Big(\frac{\sqrt{2\gamma}}{|\beta|}e^{\beta z}\Big)}\Bigg)$

$\beta > 0$

$$\times \begin{cases} \frac{zS_{\sqrt{2\alpha}/\beta}\Big(\frac{\sqrt{2\gamma}}{|\beta|}e^{\beta x}, \frac{\sqrt{2\gamma}}{|\beta|}\Big)}{xS_{\sqrt{2\alpha}/\beta}\Big(\frac{\sqrt{2\gamma}}{|\beta|}e^{\beta z}, \frac{\sqrt{2\gamma}}{|\beta|}\Big)}, & 0 \le x \le z \\ \frac{zK_{\sqrt{2\alpha}/\beta}\Big(\frac{\sqrt{2\gamma}}{|\beta|}e^{\beta x}\Big)}{xK_{\sqrt{2\alpha}/\beta}\Big(\frac{\sqrt{2\gamma}}{|\beta|}e^{\beta z}\Big)}, & z \le x \end{cases}$$

5 $R_s^{(3)} = \left(\sum_{k=1}^3 (W_s^{(k)})^2\right)^{1/2}$ $\varrho = \varrho(v,z) = \min\{s : \ell(s,z) = v\}$

4.13.1
$$\mathbf{E}_x\{e^{-\gamma\check{H}(\varrho)}; \sup_{0\le s\le\varrho} R_s^{(3)} \in dy\} = \frac{z\,\mathrm{sh}((x-z)^+\sqrt{2\gamma})}{x(y-z)\,\mathrm{sh}((y-z)\sqrt{2\gamma})}\exp\Big(-\frac{yv}{2z(y-z)}\Big)dy$$

$$+\frac{z^2\sqrt{2\gamma}\Big(\exp\Big(-\frac{v\sqrt{2\gamma}\,\mathrm{sh}(y\sqrt{2\gamma})}{2\,\mathrm{sh}((y-z)\sqrt{2\gamma})\,\mathrm{sh}(z\sqrt{2\gamma})}\Big)-\exp\Big(-\frac{yv}{2z(y-z)}\Big)\Big)}{x(y\,\mathrm{sh}((y-z)\sqrt{2\gamma})\,\mathrm{sh}(z\sqrt{2\gamma})-z(z-y)\sqrt{2\gamma}\,\mathrm{sh}(y\sqrt{2\gamma}))}$$

$$\times\begin{cases}\mathrm{sh}(x\sqrt{2\gamma})dy, & 0\le x\le z\\ \dfrac{\mathrm{sh}((y-x)\sqrt{2\gamma})\,\mathrm{sh}(z\sqrt{2\gamma})}{\mathrm{sh}((y-z)\sqrt{2\gamma})}dy, & z\le x\le y\end{cases}$$

4.14.1
$$\mathbf{E}_x\{e^{-\gamma\hat{H}(\varrho)}; \inf_{0\le s\le\varrho} R_s^{(3)} \in dy, \varrho<\infty\}$$

$$=\frac{z\,\mathrm{sh}((z-x)^+\sqrt{2\gamma})}{x(z-y)\,\mathrm{sh}((z-y)\sqrt{2\gamma})}\exp\Big(-\frac{v}{2(z-y)}\Big)dy$$

$$+\frac{2\sqrt{2\gamma}e^{-(z-y)\sqrt{2\gamma}}}{1-2(z-y)\sqrt{2\gamma}-e^{2(y-z)\sqrt{2\gamma}}}\Big(\exp\Big(-\frac{v\sqrt{2\gamma}}{1-e^{2(y-z)\sqrt{2\gamma}}}\Big)-\exp\Big(-\frac{v}{2(z-y)}\Big)\Big)$$

$$\times\begin{cases}\dfrac{z\,\mathrm{sh}((x-y)\sqrt{2\gamma})}{x\,\mathrm{sh}((z-y)\sqrt{2\gamma})}dy, & y\le x\le z\\ \dfrac{z}{x}e^{-(x-z)\sqrt{2\gamma}}dy, & z\le x\end{cases}$$

(1)
$$\mathbf{E}_x\{e^{-\gamma\hat{H}(\varrho)}; \inf_{0\le s\le\varrho} R_s^{(3)} \in dy, \varrho=\infty\}$$

$$=\frac{\mathrm{sh}((z-x)^+\sqrt{2\gamma})}{x\,\mathrm{sh}((z-y)\sqrt{2\gamma})}\Big(1-\exp\Big(-\frac{v}{2(z-y)}\Big)\Big)dy$$

$$+\Big[1-\exp\Big(-\frac{v}{2(z-y)}\Big)-\frac{2(z-y)\sqrt{2\gamma}\Big(\exp\Big(-\frac{v\sqrt{2\gamma}}{1-e^{2(y-z)\sqrt{2\gamma}}}\Big)-\exp\Big(-\frac{v}{2(z-y)}\Big)\Big)}{1-2(z-y)\sqrt{2\gamma}-e^{2(y-z)\sqrt{2\gamma}}}\Big]$$

$$\times\begin{cases}\dfrac{e^{-(z-y)\sqrt{2\gamma}}\,\mathrm{sh}((x-y)\sqrt{2\gamma})}{x\,\mathrm{sh}((z-y)\sqrt{2\gamma})}dy, & y\le x\le z\\ \dfrac{1}{x}e^{-(x-y)\sqrt{2\gamma}}dy, & z\le x\end{cases}$$

5 $R_s^{(3)} = \left(\sum_{k=1}^3 (W_s^{(k)})^2\right)^{1/2}$ $\qquad \varrho = \varrho(v,z) = \min\{s : \ell(s,z) = v\}$

4.15.2 $\mathbf{P}_x\big(a < \inf_{0\le s\le\varrho} R_s^{(3)},\ \sup_{0\le s\le\varrho} R_s^{(3)} < b\big)$

$$= \begin{cases} \dfrac{z(x-a)}{x(z-a)} \exp\Big(-\dfrac{(b-a)v}{2(z-a)(b-z)}\Big), & a \le x \le z \\ \dfrac{z(b-x)}{x(b-z)} \exp\Big(-\dfrac{(b-a)v}{2(z-a)(b-z)}\Big), & z \le x \le b \end{cases}$$

4.15.4 $\mathbf{E}_x\big\{e^{-\alpha\varrho}; a < \inf_{0\le s\le\varrho} R_s^{(3)},\ \sup_{0\le s\le\varrho} R_s^{(3)} < b\big\}$

$$= \begin{cases} \dfrac{z\,\mathrm{sh}((x-a)\sqrt{2\alpha})}{x\,\mathrm{sh}((z-a)\sqrt{2\alpha})} \exp\Big(-\dfrac{v\sqrt{2\alpha}\,\mathrm{sh}((b-a)\sqrt{2\alpha})}{2\,\mathrm{sh}((b-z)\sqrt{2\alpha})\,\mathrm{sh}((z-a)\sqrt{2\alpha})}\Big), & -b \le x \le z \\ \dfrac{z\,\mathrm{sh}((b-x)\sqrt{2\alpha})}{x\,\mathrm{sh}((b-z)\sqrt{2\alpha})} \exp\Big(-\dfrac{v\sqrt{2\alpha}\,\mathrm{sh}((b-a)\sqrt{2\alpha})}{2\,\mathrm{sh}((b-z)\sqrt{2\alpha})\,\mathrm{sh}((z-a)\sqrt{2\alpha})}\Big), & z \le x \le b \end{cases}$$

4.18.1 $\mathbf{E}_x\big\{\exp(-\gamma\ell(\varrho,r) - \eta\ell(\varrho,u));\ \varrho < \infty\big\}$

$r < u$

$z < r$
$$= \exp\Big(-\frac{v(1+2\gamma r+2\eta u+4\gamma\eta(u-r)r)}{2z(1+2\gamma(r-z)+2\eta(u-z)+4\gamma\eta(u-r)(r-z))}\Big)$$

$$\times \begin{cases} 1, & 0 \le x \le z \\ \dfrac{z(1+2\gamma(r-x)+2\eta(u-x)+4\gamma\eta(u-r)(r-x))}{x(1+2\gamma(r-z)+2\eta(u-z)+4\gamma\eta(u-r)(r-z))}, & z \le x \le r \\ \dfrac{z(1+2\eta(u-x))}{x(1+2\gamma(r-z)+2\eta(u-z)+4\gamma\eta(u-r)(r-z))}, & r \le x \le u \\ \dfrac{z}{x(1+2\gamma(r-z)+2\eta(u-z)+4\gamma\eta(u-r)(r-z))}, & u \le x \end{cases}$$

$u < z$
$$= \exp\Big(-\frac{v(1+2\gamma r+2\eta u+4\gamma\eta(u-r)r)}{2(z+2\gamma(z-r)r+2\eta(z-u)u+4\gamma\eta(z-u)(u-r)r)}\Big)$$

$u < z$
$$\times \begin{cases} \dfrac{z}{x(z+2\gamma(z-r)r+2\eta(z-u)u+4\gamma\eta(z-u)(u-r)r)}, & 0 \le x \le r \\ \dfrac{z(x+2\gamma(x-r)r)}{x(1+2\gamma(z-r)r+2\eta(z-u)u+4\gamma\eta(z-u)(u-r)r)}, & r \le x \le u \\ \dfrac{z(x+2\gamma(x-r)r+2\eta(x-u)u+4\gamma\eta(x-u)(u-r)r)}{x(z+2\gamma(z-r)r+2\eta(z-u)u+4\gamma\eta(z-u)(u-r)r)}, & u \le x \le z \\ \dfrac{z}{x}, & z \le x \end{cases}$$

5 $R_s^{(3)} = \big(\sum_{k=1}^3 (W_s^{(k)})^2\big)^{1/2}$ $\qquad \varrho = \varrho(v,z) = \min\{s : \ell(s,z) = v\}$

4.19.1 $\mathbf{E}_x\Big\{\exp\Big(-\gamma\int_0^\varrho \frac{ds}{R_s^{(3)}}\Big);\ \varrho<\infty\Big\}$

$$= \begin{cases} \dfrac{x^{-1/2}I_1(2\sqrt{2\gamma x})}{z^{-1/2}I_1(2\sqrt{2\gamma z})}\exp\Big(-\dfrac{v}{4zI_1(2\sqrt{2\gamma z})K_1(2\sqrt{2\gamma z})}\Big), & 0<x\le z\\ \dfrac{x^{-1/2}K_1(2\sqrt{2\gamma x})}{z^{-1/2}K_1(2\sqrt{2\gamma z})}\exp\Big(-\dfrac{v}{4zI_1(2\sqrt{2\gamma z})K_1(2\sqrt{2\gamma z})}\Big), & z\le x \end{cases}$$

4.20.1 $\mathbf{E}_x\Big\{\exp\Big(-\frac{\gamma^2}{2}\int_0^\varrho \frac{ds}{(R_s^{(3)})^2}\Big);\ \varrho<\infty\Big\}$

$$= \begin{cases} \Big(\dfrac{x}{z}\Big)^{-1/2+\sqrt{\gamma^2+1/4}}\exp\Big(-\dfrac{v\sqrt{\gamma^2+1/4}}{z}\Big), & 0<x\le z\\ \Big(\dfrac{x}{z}\Big)^{-1/2-\sqrt{\gamma^2+1/4}}\exp\Big(-\dfrac{v\sqrt{\gamma^2+1/4}}{z}\Big), & z\le x \end{cases}$$

4.20.4 $\mathbf{P}_x\Big(\int_0^\varrho \frac{ds}{(R_s^{(3)})^2}\in dy\Big)$
$0<y$

$$= \begin{cases} \dfrac{z^{1/2}(\ln(z/x)+v/z)}{x^{1/2}\sqrt{2\pi}y^{3/2}}e^{-y/8}e^{-(\ln(z/x)+v/z)^2/2y}dy, & 0<x\le z\\ \dfrac{z^{1/2}(\ln(x/z)+v/z)}{x^{1/2}\sqrt{2\pi}y^{3/2}}e^{-y/8}e^{-(\ln(x/z)+v/z)^2/2y}dy, & z\le x \end{cases}$$

4.21.1 $\mathbf{E}_x\Big\{\exp\Big(-\int_0^\varrho\Big(\frac{p^2}{2(R_s^{(3)})^2}+\frac{q^2(R_s^{(3)})^2}{2}\Big)ds\Big);\ \varrho<\infty\Big\}$
$0<x$

$$= \begin{cases} \dfrac{\sqrt{z}I_{\sqrt{4p^2+1}/4}(qx^2/2)}{\sqrt{x}I_{\sqrt{4p^2+1}/4}(qz^2/2)}\exp\Big(-\dfrac{v}{zI_{\sqrt{4p^2+1}/4}(qz^2/2)K_{\sqrt{4p^2+1}/4}(qz^2/2)}\Big), & x\le z\\ \dfrac{\sqrt{z}K_{\sqrt{4p^2+1}/4}(qx^2/2)}{\sqrt{x}K_{\sqrt{4p^2+1}/4}(qz^2/2)}\exp\Big(-\dfrac{v}{zI_{\sqrt{4p^2+1}/4}(qz^2/2)K_{\sqrt{4p^2+1}/4}(qz^2/2)}\Big), & z\le x \end{cases}$$

4.22.1 $\mathbf{E}_x\Big\{\exp\Big(-\int_0^\varrho\Big(\frac{p^2}{2(R_s^{(3)})^2}+\frac{q}{R_s^{(3)}}\Big)ds\Big);\ \varrho<\infty\Big\}$
$0<x$

$$= \begin{cases} \dfrac{\sqrt{z}I_{\sqrt{4p^2+1}}(2\sqrt{2qx})}{\sqrt{x}I_{\sqrt{4p^2+1}}(2\sqrt{2qz})}\exp\Big(-\dfrac{v}{4zI_{\sqrt{4p^2+1}}(2\sqrt{2qz})K_{\sqrt{4p^2+1}}(2\sqrt{2qz})}\Big), & x\le z\\ \dfrac{\sqrt{z}K_{\sqrt{4p^2+1}}(2\sqrt{2qx})}{\sqrt{x}K_{\sqrt{4p^2+1}}(2\sqrt{2qz})}\exp\Big(-\dfrac{v}{4zI_{\sqrt{4p^2+1}}(2\sqrt{2qz})K_{\sqrt{4p^2+1}}(2\sqrt{2qz})}\Big), & z\le x \end{cases}$$

5 $R_s^{(3)} = \left(\sum_{k=1}^3 (W_s^{(k)})^2\right)^{1/2}$ $\varrho = \varrho(v,z) = \min\{s : \ell(s,z) = v\}$

4.27.1 $$\mathbf{E}_x\Big\{\exp\Big(-\int_0^\varrho \big(p\mathbb{1}_{[0,r)}(R_s^{(3)}) + q\mathbb{1}_{[r,\infty)}(R_s^{(3)})\big)ds\Big);\ \ell(\varrho,r)\in dy\Big\}$$

$z \le r$ $$= \exp\Big(-\frac{v\sqrt{p}\,\mathrm{ch}(z\sqrt{2p})}{\mathrm{sh}(z\sqrt{2p})} - \frac{(y+v)\sqrt{p}\,\mathrm{ch}((r-z)\sqrt{2p})}{\sqrt{2}\,\mathrm{sh}((r-z)\sqrt{2p})} - \frac{y\sqrt{q}}{\sqrt{2}}\Big)$$
$$\times \begin{cases} \dfrac{z\,\mathrm{sh}(x\sqrt{2p})\sqrt{2pv}}{2x\,\mathrm{sh}(z\sqrt{2p})\,\mathrm{sh}((r-z)\sqrt{2p})\sqrt{y}} I_1\Big(\dfrac{\sqrt{2pvy}}{\mathrm{sh}((r-z)\sqrt{2p})}\Big)dy, & 0\le x\le z \\ \dfrac{z\sqrt{2p}}{2x\,\mathrm{sh}^2((r-z)\sqrt{2p})}\Big[\mathrm{sh}((r-x)\sqrt{2p})\dfrac{\sqrt{v}}{\sqrt{y}} I_1\Big(\dfrac{\sqrt{2pvy}}{\mathrm{sh}((r-z)\sqrt{2p})}\Big) & \\ \qquad + \mathrm{sh}((x-z)\sqrt{2p}) I_0\Big(\dfrac{\sqrt{2pvy}}{\mathrm{sh}((r-z)\sqrt{2p})}\Big)\Big]dy, & z\le x\le r \\ \dfrac{z\sqrt{2p}e^{-(x-r)\sqrt{2q}}}{2x\,\mathrm{sh}((r-z)\sqrt{2p})} I_0\Big(\dfrac{\sqrt{2pvy}}{\mathrm{sh}((r-z)\sqrt{2p})}\Big)dy, & r\le x \end{cases}$$

$r \le z$ $$= \exp\Big(-\frac{v\sqrt{q}}{\sqrt{2}} - \frac{(y+v)\sqrt{q}\,\mathrm{ch}((z-r)\sqrt{2q})}{\sqrt{2}\,\mathrm{sh}((z-r)\sqrt{2q})} - \frac{y\sqrt{p}\,\mathrm{ch}(r\sqrt{2p})}{\sqrt{2}\,\mathrm{sh}((r\sqrt{2p})}\Big)$$
$$\times \begin{cases} \dfrac{z\sqrt{2q}\,\mathrm{sh}(x\sqrt{2p})}{2x\,\mathrm{sh}((z-r)\sqrt{2q})\,\mathrm{sh}(r\sqrt{2p})} I_0\Big(\dfrac{\sqrt{2qvy}}{\mathrm{sh}((z-r)\sqrt{2q})}\Big)dy, & 0\le x\le r \\ \dfrac{z\sqrt{2q}}{2x\,\mathrm{sh}^2((z-r)\sqrt{2q})}\Big[\mathrm{sh}((x-r)\sqrt{2q})\sqrt{\dfrac{v}{y}} I_1\Big(\dfrac{\sqrt{2qvy}}{\mathrm{sh}((z-r)\sqrt{2q})}\Big) & \\ \qquad + \mathrm{sh}((z-x)\sqrt{2q}) I_0\Big(\dfrac{\sqrt{2qvy}}{\mathrm{sh}((z-r)\sqrt{2q})}\Big)\Big]dy, & r\le x\le z \\ \dfrac{z\sqrt{2qv}e^{-(x-z)\sqrt{2q}}}{2x\,\mathrm{sh}((z-r)\sqrt{2q})\sqrt{y}} I_1\Big(\dfrac{\sqrt{2qvy}}{\mathrm{sh}((z-r)\sqrt{2q})}\Big)dy, & z\le x \end{cases}$$

(1) $$\mathbf{E}_x\Big\{\exp\Big(-\int_0^\varrho \big(p\mathbb{1}_{[0,r)}(R_s^{(3)}) + q\mathbb{1}_{[r,\infty)}(R_s^{(3)})\big)ds\Big);\ \ell(\varrho,r)=0\Big\}$$

$z \le r$ $$= \exp\Big(-\frac{v\sqrt{2p}\,\mathrm{sh}(r\sqrt{2p})}{2\,\mathrm{sh}(z\sqrt{2p})\,\mathrm{sh}((r-z)\sqrt{2p})}\Big)\begin{cases} \dfrac{z\,\mathrm{sh}(x\sqrt{2p})}{x\,\mathrm{sh}(z\sqrt{2p})}, & 0\le x\le z \\ \dfrac{z\,\mathrm{sh}((r-x)\sqrt{2p})}{x\,\mathrm{sh}((r-z)\sqrt{2p})}, & z\le x\le r \\ 0, & r\le x \end{cases}$$

$r \le z$ $$= \exp\Big(-\frac{v\sqrt{2q}e^{(z-r)\sqrt{2q}}}{2\,\mathrm{sh}((z-r)\sqrt{2q})}\Big)\begin{cases} 0, & 0\le x\le r \\ \dfrac{z\,\mathrm{sh}((x-r)\sqrt{2q})}{x\,\mathrm{sh}((z-r)\sqrt{2q})}, & r\le x\le z \\ \dfrac{z}{x}e^{-(x-z)\sqrt{2q}}, & z\le x \end{cases}$$

5 $R_s^{(3)}=\left(\sum_{k=1}^3 (W_s^{(k)})^2\right)^{1/2}$ $\qquad \varrho=\varrho(v,z)=\min\{s:\ell(s,z)=v\}$

4.27.2 $\mathbf{P}_x\Big(\int_0^\varrho \mathbb{I}_{[0,r)}\big(R_s^{(3)}\big)ds\in du,\ \int_0^\varrho \mathbb{I}_{[r,\infty)}\big(R_s^{(3)}\big)ds\in dg,\ \frac12\ell(\varrho,r)\in dy\Big)$

$x\le z$, $z\le r$: $\quad =\frac{z\sqrt v}{x\sqrt y}(\mathrm{is}_u(1,r-z,0,y+v,\sqrt{vy})*\mathrm{ess}_u(x,z,0,v))\,\mathrm{h}_g(1,y)dudgdy$

$z\le r$, $r\le x$: $\quad =\frac{z}{x}(\mathrm{is}_u(0,r-z,0,y+v,\sqrt{vy})*\mathrm{es}_u(0,0,z,0,v))\,\mathrm{h}_g(1,x-r+y)dudgdy$

$x\le r$, $r\le z$: $\quad =\frac{z}{x}\,\mathrm{ess}_u(x,r,0,y)\,\mathrm{is}_g(0,z-r,v,y+v,\sqrt{vy})dudgdy$

$r\le z$, $z\le x$: $\quad =\frac{z\sqrt v}{x\sqrt y}\,\mathrm{es}_u(0,0,r,0,y)\,\mathrm{is}_g(1,z-r,x-z+v,y+v,\sqrt{vy})dudgdy$

(1) $\mathbf{P}_x\Big(\int_0^\varrho \mathbb{I}_{[0,r)}\big(R_s^{(3)}\big)ds\in du,\ \int_0^\varrho \mathbb{I}_{[r,\infty)}\big(R_s^{(3)}\big)ds=0,\ \ell(\varrho,r)=0\Big)$

$$=\begin{cases}\frac{z}{x}\,\mathrm{ess}_u(x,z,0,v/2)*\mathrm{es}_u(0,0,r-z,0,v/2)du, & 0\le x\le z\\ \frac{z}{x}\,\mathrm{es}_u(0,0,z,0,v/2)*\mathrm{ess}_u(r-x,r-z,0,v/2)du, & z\le x\le r\end{cases}$$

(2) $\mathbf{P}_x\Big(\int_0^\varrho \mathbb{I}_{[0,r)}\big(R_s^{(3)}\big)ds=0,\ \int_0^\varrho \mathbb{I}_{[r,\infty)}\big(R_s^{(3)}\big)ds\in dg,\ \ell(\varrho,r)=0\Big)$

$$=\begin{cases}\frac{z}{x}\,\mathrm{ess}_g(x-r,z-r,v/2,v/2)dg, & r\le x\le z\\ \frac{z}{x}\,\mathrm{es}_g(0,0,z-r,x-z+v/2,v/2)dg, & z\le x\end{cases}$$

4.30.1 $\mathbf{E}_x\Big\{\exp\Big(-\int_0^\varrho\Big(\frac{p^2}{2}e^{2\beta R_s^{(3)}}+qe^{\beta R_s^{(3)}}\Big)ds\Big);\ \varrho<\infty\Big\}$

$\beta>0$

$$=\exp\Bigg(-\frac{\beta vU\big(\frac12+\frac{q}{p\beta},1,\frac{2p}{\beta}\big)}{2\exp\big(-\frac{p}{\beta}e^{\beta z}\big)U\big(\frac12+\frac{q}{p\beta},1,\frac{2p}{\beta}e^{\beta z}\big)S\big(\frac12+\frac{q}{p\beta},1,\frac{2p}{\beta}e^{\beta z},\frac{2p}{\beta}\big)}\Bigg)$$

$$\times\begin{cases}\dfrac{z\exp\big(-\frac{p}{\beta}e^{\beta x}\big)S\big(\frac12+\frac{q}{p\beta},1,\frac{2p}{\beta}e^{\beta x},\frac{2p}{\beta}\big)}{x\exp\big(-\frac{p}{\beta}e^{\beta z}\big)S\big(\frac12+\frac{q}{p\beta},1,\frac{2p}{\beta}e^{\beta z},\frac{2p}{\beta}\big)}, & 0\le x\le z\\[2ex] \dfrac{ze^{-\beta x/2}W_{-q/p\beta,0}\big(\frac{2p}{\beta}e^{\beta x}\big)}{xe^{-\beta z/2}W_{-q/p\beta,0}\big(\frac{2p}{\beta}e^{\beta z}\big)}, & z\le x\end{cases}$$

6. BESSEL PROCESS OF ORDER ZERO

2. Stopping at first hitting time

2.0.1 $$\mathbf{E}_x e^{-\alpha H_z} = \begin{cases} \dfrac{I_0(x\sqrt{2\alpha})}{I_0(z\sqrt{2\alpha})}, & x \le z \\ \dfrac{K_0(x\sqrt{2\alpha})}{K_0(z\sqrt{2\alpha})}, & z \le x \end{cases}$$

2.1.2 $z \le y$ $$\mathbf{P}_x\Big(\sup_{0\le s\le H_z} R_s^{(2)} < y\Big) = \begin{cases} 1, & x \le z \\ \dfrac{\ln(y/x)}{\ln(y/z)}, & z \le x \le y \end{cases}$$

2.2.2 $y \le z$ $$\mathbf{P}_x\Big(\inf_{0\le s\le H_z} R_s^{(2)} > y\Big) = \begin{cases} \dfrac{\ln(x/y)}{\ln(z/y)}, & y \le x \le z \\ 1, & z \le x \end{cases}$$

2.3.1 $$\mathbf{E}_x e^{-\gamma\ell(H_z,r)} = \begin{cases} 1, & x \le z \le r, \quad r \le z \le x \\ \dfrac{1 + 2\gamma r|\ln(r/x)|}{1 + 2\gamma r|\ln(r/z)|}, & r\wedge z \le x \le r \vee z \\ \dfrac{1}{1 + 2\gamma r|\ln(r/z)|}, & z \le r \le x, \quad x \le r \le z \end{cases}$$

2.3.2 $\mathbf{P}_x\big(\ell(H_z,r) \in dy\big)$

$$= \begin{cases} 0, & x \le z \le r, \quad r \le z \le x \\ \dfrac{|\ln(x/z)|}{2r\ln^2(z/r)} \exp\Big(-\dfrac{y}{2r|\ln(z/r)|}\Big)dy, & z\wedge r \le x \le r\vee z \\ \dfrac{1}{2r|\ln(z/r)|} \exp\Big(-\dfrac{y}{2r|\ln(z/r)|}\Big)dy, & z \le r \le x, \quad x \le r \le z \end{cases}$$

(1) $$\mathbf{P}_x\big(\ell(H_z,r) = 0\big) = \begin{cases} 1, & x \le z \le r, \quad r \le z \le x \\ \dfrac{|\ln(r/x)|}{|\ln(r/z)|}, & z\wedge r \le x \le r\vee z \\ 0, & z \le r \le x, \quad x \le r \le z \end{cases}$$

2.4.1 $\mathbf{E}_x \exp\Big(-\gamma \int_0^{H_z} \mathbb{I}_{[r,\infty)}\big(R_s^{(2)}\big)ds\Big)$

$r \le z$ $$= \begin{cases} \dfrac{K_0(x\sqrt{2\gamma})}{K_0(z\sqrt{2\gamma})}, & z \le x \\ \dfrac{C_0(r\sqrt{2\gamma}, x\sqrt{2\gamma})}{C_0(r\sqrt{2\gamma}, z\sqrt{2\gamma})}, & r \le x \le z \\ \dfrac{1}{r\sqrt{2\gamma}C_0(r\sqrt{2\gamma}, z\sqrt{2\gamma})}, & 0 \le x \le r \end{cases}$$

6 $R_s^{(2)} = \big((W_s^{(1)})^2 + (W_s^{(2)})^2\big)^{1/2}$ $\qquad H_z = \min\{s : R_s^{(2)} = z\}$

$z \le r$

$$= \begin{cases} \dfrac{K_0(x\sqrt{2\gamma})}{K_0(r\sqrt{2\gamma}) + \ln(r/z) r\sqrt{2\gamma} K_1(r\sqrt{2\gamma})}, & r \le x \\ \dfrac{K_0(r\sqrt{2\gamma}) + \ln(r/x) r\sqrt{2\gamma} K_1(r\sqrt{2\gamma})}{K_0(r\sqrt{2\gamma}) + \ln(r/z) r\sqrt{2\gamma} K_1(r\sqrt{2\gamma})}, & z \le x \le r \\ 1, & 0 \le x \le z \end{cases}$$

2.4.2 $z \le r$

$$\mathbf{P}_x\Big(\int_0^{H_z} \mathbb{1}_{[r,\infty)}\big(R_s^{(2)}\big) ds = 0\Big) = \begin{cases} 1, & x \le z \\ \dfrac{\ln(r/x)}{\ln(r/z)}, & z \le x \le r \\ 0, & r \le x \end{cases}$$

2.5.1

$$\mathbf{E}_x \exp\Big(-\gamma \int_0^{H_z} \mathbb{1}_{[0,r]}\big(R_s^{(2)}\big) ds\Big)$$

$r \le z$

$$= \begin{cases} \dfrac{I_0(x\sqrt{2\gamma})}{I_0(r\sqrt{2\gamma}) + \ln(z/r) r\sqrt{2\gamma} I_1(r\sqrt{2\gamma})}, & 0 \le x \le r \\ \dfrac{I_0(r\sqrt{2\gamma}) + \ln(x/r) r\sqrt{2\gamma} I_1(r\sqrt{2\gamma})}{I_0(r\sqrt{2\gamma}) + \ln(z/r) r\sqrt{2\gamma} I_1(r\sqrt{2\gamma})}, & r \le x \le z \\ 1, & z \le x \end{cases}$$

$z \le r$

$$= \begin{cases} \dfrac{I_0(x\sqrt{2\gamma})}{I_0(z\sqrt{2\gamma})}, & 0 \le x \le z \\ \dfrac{C_0(r\sqrt{2\gamma}, x\sqrt{2\gamma})}{C_0(r\sqrt{2\gamma}, z\sqrt{2\gamma})}, & z \le x \le r \\ \dfrac{1}{r\sqrt{2\gamma} C_0(r\sqrt{2\gamma}, z\sqrt{2\gamma})}, & r \le x \end{cases}$$

2.5.2 $r \le z$

$$\mathbf{P}_x\Big(\int_0^{H_z} \mathbb{1}_{[0,r]}\big(R_s^{(2)}\big) ds = 0\Big) = \begin{cases} 0, & x \le r \\ \dfrac{\ln(x/r)}{\ln(z/r)}, & r \le x \le z \\ 1, & z \le x \end{cases}$$

2.6.1

$$\mathbf{E}_x \exp\Big(-\int_0^{H_z} \big(p\mathbb{1}_{[0,r)}\big(R_s^{(2)}\big) + q\mathbb{1}_{[r,\infty)}\big(R_s^{(2)}\big)\big) ds\Big)$$

$r \le z$

$$= \begin{cases} \dfrac{(r\sqrt{2})^{-1} I_0(x\sqrt{2p})}{\sqrt{p} S_0(z\sqrt{2q}, r\sqrt{2q}) I_1(r\sqrt{2p}) + \sqrt{q} C_0(r\sqrt{2q}, z\sqrt{2q}) I_0(r\sqrt{2p})}, & 0 \le x \le r \\ \dfrac{\sqrt{p} S_0(x\sqrt{2q}, r\sqrt{2q}) I_1(r\sqrt{2p}) + \sqrt{q} C_0(r\sqrt{2q}, x\sqrt{2q}) I_0(r\sqrt{2p})}{\sqrt{p} S_0(z\sqrt{2q}, r\sqrt{2q}) I_1(r\sqrt{2p}) + \sqrt{q} C_0(r\sqrt{2q}, z\sqrt{2q}) I_0(r\sqrt{2p})}, & r \le x \le z \\ \dfrac{K_0(x\sqrt{2q})}{K_0(z\sqrt{2q})}, & z \le x \end{cases}$$

6 $\quad R_s^{(2)} = \big((W_s^{(1)})^2 + (W_s^{(2)})^2\big)^{1/2} \qquad H_z = \min\{s : R_s^{(2)} = z\}$

$z \le r$

$$= \begin{cases} \dfrac{I_0(x\sqrt{2p})}{I_0(z\sqrt{2p})}, & 0 \le x \le z \\ \dfrac{\sqrt{p}K_0(r\sqrt{2q})C_0(r\sqrt{2p}, x\sqrt{2p}) + \sqrt{q}K_1(r\sqrt{2q})S_0(r\sqrt{2p}, x\sqrt{2p})}{\sqrt{p}K_0(r\sqrt{2q})C_0(r\sqrt{2p}, z\sqrt{2p}) + \sqrt{q}K_1(r\sqrt{2q})S_0(r\sqrt{2p}, z\sqrt{2p})}, & z \le x \le r \\ \dfrac{(r\sqrt{2})^{-1}K_0(x\sqrt{2q})}{\sqrt{p}K_0(r\sqrt{2q})C_0(r\sqrt{2p}, z\sqrt{2p}) + \sqrt{q}K_1(r\sqrt{2q})S_0(r\sqrt{2p}, z\sqrt{2p})}, & r \le x \end{cases}$$

2.8.1

$$\mathbf{E}_x \exp\Big(-\gamma \int_0^{H_z} R_s^{(2)} ds\Big) = \begin{cases} \dfrac{I_0(\frac{2}{3}x^{3/2}\sqrt{2\gamma})}{I_0(\frac{2}{3}z^{3/2}\sqrt{2\gamma})}, & 0 \le x \le z \\ \dfrac{K_0(\frac{2}{3}x^{3/2}\sqrt{2\gamma})}{K_0(\frac{2}{3}z^{3/2}\sqrt{2\gamma})}, & z \le x \end{cases}$$

2.9.1

$$\mathbf{E}_x \exp\Big(-\frac{\gamma^2}{2} \int_0^{H_z} (R_s^{(2)})^2 ds\Big) = \begin{cases} \dfrac{I_0(x^2\gamma/2)}{I_0(z^2\gamma/2)}, & x \le z \\ \dfrac{K_0(x^2\gamma/2)}{K_0(z^2\gamma/2)}, & z \le x \end{cases}$$

2.13.1
$z < x$
$x < y$

$$\mathbf{E}_x\big\{e^{-\gamma \check{H}(H_z)}; \sup_{0 \le s \le H_z} R_s^{(2)} \in dy\big\} = \frac{S_0(x\sqrt{2\gamma}, z\sqrt{2\gamma})}{y\ln(y/z)S_0(y\sqrt{2\gamma}, z\sqrt{2\gamma})}dy$$

2.14.1
$x < z$
$y < x$

$$\mathbf{E}_x\big\{e^{-\gamma \hat{H}(H_z)}; \inf_{0 \le s \le H_z} R_s^{(2)} \in dy\big\} = \frac{S_0(z\sqrt{2\gamma}, x\sqrt{2\gamma})}{y\ln(z/y)S_0(z\sqrt{2\gamma}, y\sqrt{2\gamma})}dy$$

2.18.1
$r < u$

$$\mathbf{E}_x \exp\big(-\gamma\ell(H_z, r) - \eta\ell(H_z, u)\big)$$

$z < r$

$$= \begin{cases} 1, & x \le z \\ \dfrac{1 + 2\gamma r\ln(r/x) + 2\eta u\ln(u/x) + 4\gamma r\eta u\ln(u/r)\ln(r/x)}{1 + 2\gamma r\ln(r/z) + 2\eta u\ln(u/z) + 4\gamma r\eta u\ln(u/r)\ln(r/z)}, & z \le x \le r \\ \dfrac{1 + 2\eta u\ln(u/x)}{1 + 2\gamma r\ln(r/z) + 2\eta u\ln(u/z) + 4\gamma r\eta u\ln(u/r)\ln(r/z)}, & r \le x \le u \\ \dfrac{1}{1 + 2\gamma r\ln(r/z) + 2\eta u\ln(u/z) + 4\gamma r\eta u\ln(u/r)\ln(r/z)}, & u \le x \end{cases}$$

$u < z$

$$= \begin{cases} \dfrac{1}{1 + 2\gamma r\ln(z/r) + 2\eta u\ln(z/u) + 4\gamma r\eta u\ln(z/u)\ln(u/r)}, & x \le r \\ \dfrac{1 + 2\gamma r\ln(x/r)}{1 + 2\gamma r\ln(z/r) + 2\eta u\ln(z/u) + 4\gamma r\eta u\ln(z/u)\ln(u/r)}, & r \le x \le u \\ \dfrac{1 + 2\gamma r\ln(x/r) + 2\eta u\ln(x/u) + 4\gamma r\eta u\ln(x/u)\ln(u/r)}{1 + 2\gamma r\ln(z/r) + 2\eta u\ln(z/u) + 4\gamma r\eta u\ln(z/u)\ln(u/r)}, & u \le x \le z \\ 1, & z \le x \end{cases}$$

6 $R_s^{(2)} = \big((W_s^{(1)})^2 + (W_s^{(2)})^2\big)^{1/2}$ $H_z = \min\{s : R_s^{(2)} = z\}$

2.18.2 $r < u$

$$\mathbf{P}_x\Big(\frac{1}{r}\ell(H_z,r) \in dy, \frac{1}{u}\ell(H_z,u) \in dg\Big)$$

$z < r$

$$= \begin{cases} \frac{\ln(x/z)}{\ln(r/z)} p_r dydg, & z \le x \le r \\ \frac{\ln(u/x)}{\ln(u/r)} p_r dydg + \frac{\ln(x/r)}{\ln(u/r)} p_u dydg, & r \le x \le u \\ p_u dydg, & u \le x \end{cases}$$

$u < z$

$$= \begin{cases} q_r dydg, & 0 \le x \le r \\ \frac{\ln(u/x)}{\ln(u/r)} q_r dydg + \frac{\ln(x/r)}{\ln(u/r)} q_u dydg, & r \le x \le u \\ \frac{\ln(z/x)}{\ln(z/u)} q_u dydg, & u \le x \le z \end{cases}$$

where

$$p_r := \frac{\sqrt{y}}{4\ln(u/r)\ln(r/z)\sqrt{g}} \exp\Big(-\frac{y+g}{2\ln(u/r)} - \frac{y}{2\ln(r/z)}\Big) I_1\Big(\frac{\sqrt{yg}}{\ln(u/r)}\Big)$$

$$q_r := \frac{1}{4\ln(z/u)\ln(u/r)} \exp\Big(-\frac{y+g}{2\ln(u/r)} - \frac{g}{2\ln(z/u)}\Big) I_0\Big(\frac{\sqrt{yg}}{\ln(u/r)}\Big)$$

$$p_u := \frac{1}{4\ln(u/r)\ln(r/z)} \exp\Big(-\frac{y+g}{2\ln(u/r)} - \frac{y}{2\ln(r/z)}\Big) I_0\Big(\frac{\sqrt{yg}}{\ln(u/r)}\Big)$$

$$q_u := \frac{\sqrt{g}}{4\ln(z/u)\ln(u/r)\sqrt{y}} \exp\Big(-\frac{y+g}{2\ln(u/r)} - \frac{g}{2\ln(z/u)}\Big) I_1\Big(\frac{\sqrt{yg}}{\ln(u/r)}\Big)$$

(1) $$\mathbf{P}_x\Big(\frac{1}{r}\ell(H_z,r) \in dy, \ell(H_z,u) = 0\Big)$$

$$= \begin{cases} \frac{\ln(x/z)}{2\ln^2(r/z)} \exp\Big(-\frac{y\ln(u/z)}{2\ln(u/r)\ln(r/z)}\Big)dy, & z \le x \le r \\ \frac{\ln(u/x)}{2\ln(u/r)\ln(r/z)} \exp\Big(-\frac{y\ln(u/z)}{2\ln(u/r)\ln(r/z)}\Big)dy, & r \le x \le u \end{cases}$$

(2) $$\mathbf{P}_x\Big(\ell(H_z,r) = 0, \frac{1}{u}\ell(H_z,u) \in dg\Big)$$

$$= \begin{cases} \frac{\ln(x/r)}{2\ln(z/u)\ln(u/r)} \exp\Big(-\frac{g\ln(z/r)}{2\ln(z/u)\ln(u/r)}\Big)dg, & r \le x \le u \\ \frac{\ln(z/x)}{2\ln^2(z/u)} \exp\Big(-\frac{g\ln(z/r)}{2\ln(z/u)\ln(u/r)}\Big)dg, & u \le x \le z \end{cases}$$

6 $R_s^{(2)} = \big((W_s^{(1)})^2 + (W_s^{(2)})^2\big)^{1/2}$ $\qquad H_z = \min\{s : R_s^{(2)} = z\}$

(3) $$\mathbf{P}_x\big(\ell(H_z, r) = 0,\ \ell(H_z, u) = 0\big) = \begin{cases} \dfrac{\ln(r/x)}{\ln(r/z)}, & z \le x \le r \\ \dfrac{\ln(x/u)}{\ln(z/u)}, & u \le x \le z \end{cases}$$

2.19.1 $$\mathbf{E}_x \exp\Big(-\gamma \int_0^{H_z} \frac{ds}{R_s^{(2)}}\Big) = \begin{cases} \dfrac{I_0(2\sqrt{2\gamma x})}{I_0(2\sqrt{2\gamma z})}, & 0 < x \le z \\ \dfrac{K_0(2\sqrt{2\gamma x})}{K_0(2\sqrt{2\gamma z})}, & z \le x \end{cases}$$

2.20.1 $$\mathbf{E}_x \exp\Big(-\frac{\gamma^2}{2} \int_0^{H_z} \frac{ds}{(R_s^{(2)})^2}\Big) = \begin{cases} \Big(\dfrac{x}{z}\Big)^{|\gamma|}, & 0 < x \le z \\ \Big(\dfrac{z}{x}\Big)^{|\gamma|}, & z \le x \end{cases}$$

2.20.4 $0 < y$ $$\mathbf{P}_x\Big(\int_0^{H_z} \frac{ds}{(R_s^{(2)})^2} \in dy\Big) = \begin{cases} \dfrac{\ln(z/x)}{\sqrt{2\pi} y^{3/2}} e^{-\ln^2(z/x)/2y} dy, & 0 < x \le z \\ \dfrac{\ln(x/z)}{\sqrt{2\pi} y^{3/2}} e^{-\ln^2(x/z)/2y} dy, & z \le x \end{cases}$$

2.21.1 $$\mathbf{E}_x \exp\Big(-\int_0^{H_z} \Big(\frac{p^2}{2(R_s^{(2)})^2} + \frac{q^2}{2}(R_s^{(2)})^2\Big) ds\Big) = \begin{cases} \dfrac{I_{|p|/2}(qx^2/2)}{I_{|p|/2}(qz^2/2)}, & 0 < x \le z \\ \dfrac{K_{|p|/2}(qx^2/2)}{K_{|p|/2}(qz^2/2)}, & z \le x \end{cases}$$

2.21.3 $$\mathbf{E}_x \exp\Big(-\alpha H_z - \int_0^{H_z} \Big(\frac{p^2}{2(R_s^{(2)})^2} + \frac{q^2}{2}(R_s^{(2)})^2\Big) ds\Big)$$

$$= \begin{cases} \dfrac{z M_{-\alpha/2q, |p|/2}(qx^2)}{x M_{-\alpha/2q, |p|/2}(qz^2)}, & 0 < x \le z \\ \dfrac{z W_{-\alpha/2q, |p|/2}(qx^2)}{x W_{-\alpha/2q, |p|/2}(qz^2)}, & z \le x \end{cases}$$

2.22.1 $$\mathbf{E}_x \exp\Big(-\int_0^{H_z} \Big(\frac{p^2}{2(R_s^{(2)})^2} + \frac{q}{R_s^{(2)}}\Big) ds\Big) = \begin{cases} \dfrac{I_{2|p|}(2\sqrt{2qx})}{I_{2|p|}(2\sqrt{2qz})}, & 0 < x \le z \\ \dfrac{K_{2|p|}(2\sqrt{2qx})}{K_{2|p|}(2\sqrt{2qz})}, & z \le x \end{cases}$$

2.22.3 $$\mathbf{E}_x \exp\Big(-\alpha H_z - \int_0^{H_z} \Big(\frac{p^2}{2(R_s^{(2)})^2} + \frac{q}{R_s^{(2)}}\Big) ds\Big) = \begin{cases} \dfrac{\sqrt{z} M_{-q/\sqrt{2\alpha}, |p|}(2x\sqrt{2\alpha})}{\sqrt{x} M_{-q/\sqrt{2\alpha}, |p|}(2z\sqrt{2\alpha})}, & 0 < x \le z \\ \dfrac{\sqrt{z} W_{-q/\sqrt{2\alpha}, |p|}(2x\sqrt{2\alpha})}{\sqrt{x} W_{-q/\sqrt{2\alpha}, |p|}(2z\sqrt{2\alpha})}, & z \le x \end{cases}$$

6 $R_s^{(2)} = \big((W_s^{(1)})^2 + (W_s^{(2)})^2\big)^{1/2}$ $\qquad H = H_{a,b} = \min\{s : R_s^{(2)} \notin (a,b)\}$

3. Stopping at first exit time

3.0.1 $$\mathbf{E}_x e^{-\alpha H} = \frac{S_0(b\sqrt{2\alpha}, x\sqrt{2\alpha}) + S_0(x\sqrt{2\alpha}, a\sqrt{2\alpha})}{S_0(b\sqrt{2\alpha}, a\sqrt{2\alpha})}$$

3.0.3 $$\mathbf{E}_x e^{-\beta R_H^{(2)}} = \frac{\ln(b/x)}{\ln(b/a)} e^{-\beta a} + \frac{\ln(x/a)}{\ln(b/a)} e^{-\beta b}$$

3.0.4 (a) $$\mathbf{P}_x\big(R_H^{(2)} = a\big) = \frac{\ln(b/x)}{\ln(b/a)}$$

3.0.4 (b) $$\mathbf{P}_x\big(R_H^{(2)} = b\big) = \frac{\ln(x/a)}{\ln(b/a)}$$

3.1.6 $$\mathbf{P}_x\Big(\sup_{0\le s\le H} R_s^{(2)} \ge y, R_H^{(2)} = a\Big) = \frac{\ln(b/y)\ln(x/a)}{\ln(b/a)\ln(y/a)}, \qquad a \le x \le y \le b$$

3.2.6 $$\mathbf{P}_x\Big(\inf_{0\le s\le H} R_s^{(2)} \le y, R_H^{(2)} = b\Big) = \frac{\ln(b/x)\ln(y/a)}{\ln(b/y)\ln(b/a)}, \qquad a \le y \le x \le b$$

3.3.1 $$\mathbf{E}_x e^{-\gamma\ell(H,r)} = \begin{cases} \dfrac{\ln(b/a) + 2\gamma r\ln(x/r)\ln(r/a)}{\ln(b/a) + 2\gamma r\ln(b/r)\ln(r/a)}, & r \le x \le b \\[2ex] \dfrac{\ln(b/a) + 2\gamma r\ln(b/r)\ln(r/x)}{\ln(b/a) + 2\gamma r\ln(b/r)\ln(r/a)}, & a \le x \le r \end{cases}$$

3.3.2 $$\mathbf{P}_x\big(\ell(H,r) \in dy\big) = \exp\Big(-\frac{\ln(b/a)y}{2r\ln(b/r)\ln(r/a)}\Big)\begin{cases} \dfrac{\ln(b/x)\ln(b/a)}{2r\ln^2(b/r)\ln(r/a)}dy, & r \le x \le b \\[2ex] \dfrac{\ln(x/a)\ln(b/a)}{2r\ln(b/r)\ln^2(r/a)}dy, & a \le x \le r \end{cases}$$

(1) $$\mathbf{P}_x\big(\ell(H,r) = 0\big) = \begin{cases} \dfrac{\ln(x/r)}{\ln(b/r)}, & r \le x \le b \\[2ex] \dfrac{\ln(r/x)}{\ln(r/a)}, & a \le x \le r \end{cases}$$

3.3.5 (a) $$\mathbf{E}_x\big\{e^{-\gamma\ell(H,r)};\ R_H^{(2)} = a\big\} = \begin{cases} \dfrac{\ln(b/x)}{\ln(b/a) + 2\gamma r\ln(b/r)\ln(r/a)}, & r \le x \le b \\[2ex] \dfrac{\ln(b/x) + 2\gamma r\ln(b/r)\ln(r/x)}{\ln(b/a) + 2\gamma r\ln(b/r)\ln(r/a)}, & a \le x \le r \end{cases}$$

6 $R_s^{(2)} = \big((W_s^{(1)})^2 + (W_s^{(2)})^2\big)^{1/2}$ $H = H_{a,b} = \min\{s : R_s^{(2)} \notin (a,b)\}$

3.3.5 (b)

$$\mathbf{E}_x\big\{e^{-\gamma\ell(H,r)};\ R_H^{(2)} = b\big\} = \begin{cases} \dfrac{\ln(x/a) + 2\gamma r \ln(x/r)\ln(r/a)}{\ln(b/a) + 2\gamma r\ln(b/r)\ln(r/a)}, & r \le x \le b \\ \dfrac{\ln(x/a)}{\ln(b/a) + 2\gamma r\ln(b/r)\ln(r/a)}, & a \le x \le r \end{cases}$$

3.3.6 (a)

$$\mathbf{P}_x\big(\ell(H,r) \in dy, R_H^{(2)} = a\big)$$

$$= \exp\Big(-\frac{\ln(b/a)y}{2r\ln(b/r)\ln(r/a)}\Big)\begin{cases} \dfrac{\ln(b/x)}{2r\ln(b/r)\ln(r/a)}dy, & r \le x \le b \\ \dfrac{\ln(x/a)}{2r\ln^2(r/a)}dy, & a \le x \le r \end{cases}$$

(1)

$$\mathbf{P}_x\big(\ell(H,r) = 0, R_H^{(2)} = a\big) = \begin{cases} 0, & r \le x \le b \\ \dfrac{\ln(r/x)}{\ln(r/a)}, & a \le x \le r \end{cases}$$

3.3.6 (b)

$$\mathbf{P}_x\big(\ell(H,r) \in dy, R_H^{(2)} = b\big)$$

$$= \exp\Big(-\frac{\ln(b/a)y}{2r\ln(b/r)\ln(r/a)}\Big)\begin{cases} \dfrac{\ln(b/x)}{2r\ln^2(b/r)}dy, & r \le x \le b \\ \dfrac{\ln(x/a)}{2r\ln(b/r)\ln(r/a)}dy, & a \le x \le r \end{cases}$$

(1)

$$\mathbf{P}_x\big(\ell(H,r) = 0, R_H^{(2)} = b\big) = \begin{cases} \dfrac{\ln(x/r)}{\ln(b/r)}, & r \le x \le b \\ 0, & a \le x \le r \end{cases}$$

3.4.1

$$\mathbf{E}_r \exp\Big(-\gamma\int_0^H \mathbb{I}_{[r,b]}\big(R_s^{(2)}\big)ds\Big)$$

$$= \frac{S_0(b\sqrt{2\gamma}, r\sqrt{2\gamma}) + \ln(r/a)}{S_0(b\sqrt{2\gamma}, r\sqrt{2\gamma}) + \ln(r/a)r\sqrt{2\gamma}C_0(r\sqrt{2\gamma}, b\sqrt{2\gamma})} =: Q$$

$$\mathbf{E}_x \exp\Big(-\gamma\int_0^H \mathbb{I}_{[r,b]}\big(R_s^{(2)}\big)ds\Big)$$

$$= \begin{cases} \dfrac{\ln(r/x)}{\ln(r/a)} + \dfrac{\ln(x/a)}{\ln(r/a)}Q, & a \le x \le r \\ \dfrac{S_0(x\sqrt{2\gamma}, r\sqrt{2\gamma})}{S_0(b\sqrt{2\gamma}, r\sqrt{2\gamma})} + \dfrac{S_0(b\sqrt{2\gamma}, x\sqrt{2\gamma})}{S_0(b\sqrt{2\gamma}, r\sqrt{2\gamma})}Q, & r \le x \le b \end{cases}$$

6 $R_s^{(2)} = \big((W_s^{(1)})^2 + (W_s^{(2)})^2\big)^{1/2}$ $H = H_{a,b} = \min\{s : R_s^{(2)} \notin (a,b)\}$

3.4.2 (1)

$$\mathbf{P}_x\Big(\int_0^H \mathbb{I}_{[r,b]}\big(R_s^{(2)}\big)ds = 0\Big) = \begin{cases} \dfrac{\ln(r/x)}{\ln(r/a)}, & a \le x \le r \\ 0, & r \le x \le b \end{cases}$$

3.4.5 (a)

$$\mathbf{E}_x\Big\{\exp\Big(-\gamma\int_0^H \mathbb{I}_{[r,b]}\big(R_s^{(2)}\big)ds\Big);\ R_H^{(2)} = a\Big\}$$

$$= \begin{cases} \dfrac{S_0(b\sqrt{2\gamma}, r\sqrt{2\gamma}) + \ln(r/x)r\sqrt{2\gamma}C_0(r\sqrt{2\gamma}, b\sqrt{2\gamma})}{S_0(b\sqrt{2\gamma}, r\sqrt{2\gamma}) + \ln(r/a)r\sqrt{2\gamma}C_0(r\sqrt{2\gamma}, b\sqrt{2\gamma})}, & a \le x \le r \\ \dfrac{S_0(b\sqrt{2\gamma}, x\sqrt{2\gamma})}{S_0(b\sqrt{2\gamma}, r\sqrt{2\gamma}) + \ln(r/a)r\sqrt{2\gamma}C_0(r\sqrt{2\gamma}, b\sqrt{2\gamma})}, & r \le x \le b \end{cases}$$

3.4.5 (b)

$$\mathbf{E}_x\Big\{\exp\Big(-\gamma\int_0^H \mathbb{I}_{[r,b]}\big(R_s^{(2)}\big)ds\Big);\ R_H^{(2)} = b\Big\}$$

$$= \begin{cases} \dfrac{\ln(x/a)}{S_0(b\sqrt{2\gamma}, r\sqrt{2\gamma}) + \ln(r/a)r\sqrt{2\gamma}C_0(r\sqrt{2\gamma}, b\sqrt{2\gamma})}, & a \le x \le r \\ \dfrac{S_0(x\sqrt{2\gamma}, r\sqrt{2\gamma}) + \ln(r/a)r\sqrt{2\gamma}C_0(r\sqrt{2\gamma}, x\sqrt{2\gamma})}{S_0(b\sqrt{2\gamma}, r\sqrt{2\gamma}) + \ln(r/a)r\sqrt{2\gamma}C_0(r\sqrt{2\gamma}, b\sqrt{2\gamma})}, & r \le x \le b \end{cases}$$

3.4.6 (1)

$$\mathbf{P}_x\Big(\int_0^H \mathbb{I}_{[r,b]}\big(R_s^{(2)}\big)ds = 0, R_H^{(2)} = a\Big) = \begin{cases} \dfrac{\ln(r/x)}{\ln(r/a)}, & a \le x \le r \\ 0, & r \le x \le b \end{cases}$$

3.5.1

$$\mathbf{E}_r \exp\Big(-\gamma\int_0^H \mathbb{I}_{[a,r]}\big(R_s^{(2)}\big)ds\Big)$$

$$= \frac{S_0(r\sqrt{2\gamma}, a\sqrt{2\gamma}) + \ln(b/r)}{S_0(r\sqrt{2\gamma}, a\sqrt{2\gamma}) + \ln(b/r)r\sqrt{2\gamma}C_0(r\sqrt{2\gamma}, a\sqrt{2\gamma})} =: Q$$

$$\mathbf{E}_x \exp\Big(-\gamma\int_0^H \mathbb{I}_{[a,r]}\big(R_s^{(2)}\big)ds\Big)$$

$$= \begin{cases} \dfrac{S_0(r\sqrt{2\gamma}, x\sqrt{2\gamma})}{S_0(r\sqrt{2\gamma}, a\sqrt{2\gamma})} + \dfrac{S_0(x\sqrt{2\gamma}, a\sqrt{2\gamma})}{S_0(r\sqrt{2\gamma}, a\sqrt{2\gamma})}Q, & a \le x \le r \\ \dfrac{\ln(x/r)}{\ln(b/r)} + \dfrac{\ln(b/x)}{\ln(b/r)}Q, & r \le x \le b \end{cases}$$

6 $R_s^{(2)} = \big((W_s^{(1)})^2 + (W_s^{(2)})^2\big)^{1/2}$ $\qquad H = H_{a,b} = \min\{s : R_s^{(2)} \notin (a,b)\}$

3.5.2 (1)
$$\mathbf{P}_x\Big(\int_0^H \mathbb{I}_{[a,r]}\big(R_s^{(2)}\big)ds = 0\Big) = \begin{cases} 0, & a \le x \le r \\ \dfrac{\ln(x/r)}{\ln(b/r)}, & r \le x \le b \end{cases}$$

3.5.5 (a)
$$\mathbf{E}_x\Big\{\exp\Big(-\gamma\int_0^H \mathbb{I}_{[a,r]}\big(R_s^{(2)}\big)ds\Big);\ R_H^{(2)} = a\Big\}$$

$$= \begin{cases} \dfrac{S_0(r\sqrt{2\gamma}, x\sqrt{2\gamma}) + \ln(b/r)r\sqrt{2\gamma}C_0(r\sqrt{2\gamma}, x\sqrt{2\gamma})}{S_0(r\sqrt{2\gamma}, a\sqrt{2\gamma}) + \ln(b/r)r\sqrt{2\gamma}C_0(r\sqrt{2\gamma}, a\sqrt{2\gamma})}, & a \le x \le r \\ \dfrac{\ln(b/x)}{S_0(r\sqrt{2\gamma}, a\sqrt{2\gamma}) + \ln(b/r)r\sqrt{2\gamma}C_0(r\sqrt{2\gamma}, a\sqrt{2\gamma})}, & r \le x \le b \end{cases}$$

3.5.5 (b)
$$\mathbf{E}_x\Big\{\exp\Big(-\gamma\int_0^H \mathbb{I}_{[a,r]}\big(R_s^{(2)}\big)ds\Big);\ R_H^{(2)} = b\Big\}$$

$$= \begin{cases} \dfrac{S_0(x\sqrt{2\gamma}, a\sqrt{2\gamma})}{S_0(r\sqrt{2\gamma}, a\sqrt{2\gamma}) + \ln(b/r)r\sqrt{2\gamma}C_0(r\sqrt{2\gamma}, a\sqrt{2\gamma})}, & a \le x \le r \\ \dfrac{S_0(r\sqrt{2\gamma}, a\sqrt{2\gamma}) + \ln(x/r)r\sqrt{2\gamma}C_0(r\sqrt{2\gamma}, a\sqrt{2\gamma})}{S_0(r\sqrt{2\gamma}, a\sqrt{2\gamma}) + \ln(b/r)r\sqrt{2\gamma}C_0(r\sqrt{2\gamma}, a\sqrt{2\gamma})}, & r \le x \le b \end{cases}$$

3.5.6 (1)
$$\mathbf{P}_x\Big(\int_0^H \mathbb{I}_{[a,r]}\big(R_s^{(2)}\big)ds = 0, R_H^{(2)} = b\Big) = \begin{cases} 0, & a \le x \le r \\ \dfrac{\ln(x/r)}{\ln(b/r)}, & r \le x \le b \end{cases}$$

3.8.5 (a)
$$\mathbf{E}_x\Big\{\exp\Big(-\gamma\int_0^H R_s^{(2)}ds\Big);\ R_H^{(2)} = a\Big\} = \frac{S_0(\frac{2}{3}b^{3/2}\sqrt{2\gamma}, \frac{2}{3}x^{3/2}\sqrt{2\gamma})}{S_0(\frac{2}{3}b^{3/2}\sqrt{2\gamma}, \frac{2}{3}a^{3/2}\sqrt{2\gamma})}$$

3.8.5 (b)
$$\mathbf{E}_x\Big\{\exp\Big(-\gamma\int_0^H R_s^{(2)}ds\Big);\ R_H^{(2)} = b\Big\} = \frac{S_0(\frac{2}{3}x^{3/2}\sqrt{2\gamma}, \frac{2}{3}a^{3/2}\sqrt{2\gamma})}{S_0(\frac{2}{3}b^{3/2}\sqrt{2\gamma}, \frac{2}{3}a^{3/2}\sqrt{2\gamma})}$$

3.9.5 (a)
$$\mathbf{E}_x\Big\{\exp\Big(-\frac{\gamma^2}{2}\int_0^H (R_s^{(2)})^2 ds\Big);\ R_H^{(2)} = a\Big\} = \frac{S_0(b^2\gamma/2, x^2\gamma/2)}{S_0(b^2\gamma/2, a^2\gamma/2)}$$

3.9.5 (b)
$$\mathbf{E}_x\Big\{\exp\Big(-\frac{\gamma^2}{2}\int_0^H (R_s^{(2)})^2 ds\Big);\ R_H^{(2)} = b\Big\} = \frac{S_0(x^2\gamma/2, a^2\gamma/2)}{S_0(b^2\gamma/2, a^2\gamma/2)}$$

3.12.5 (a)
$$\mathbf{E}_x\{e^{-\gamma\check{H}(H)};\ R_H^{(2)} = a\} = \int_x^b \frac{S_0(x\sqrt{2\gamma}, a\sqrt{2\gamma})}{y\ln(y/a)S_0(y\sqrt{2\gamma}, a\sqrt{2\gamma})}dy$$

6 $R_s^{(2)} = \left((W_s^{(1)})^2 + (W_s^{(2)})^2\right)^{1/2}$ $H = H_{a,b} = \min\{s : R_s^{(2)} \notin (a,b)\}$

3.12.5 (b)

$$\mathbf{E}_x\{e^{-\gamma \hat{H}(H)};\ R_H^{(2)} = b\} = \int_a^x \frac{S_0(b\sqrt{2\gamma}, x\sqrt{2\gamma})}{y \ln(b/y) S_0(b\sqrt{2\gamma}, y\sqrt{2\gamma})} dy$$

3.18.1 $r \le u$

$$\mathbf{E}_x \exp\big(-\gamma\ell(H,r) - \eta\ell(H,u)\big)$$

$a \le x$, $x \le r$

$$= \frac{\ln(b/a) + 2\gamma r \ln(b/r)\ln(r/x) + 2\eta u \ln(b/u)\ln(u/x) + 4\gamma r\eta u \ln(b/u)\ln(u/r)\ln(r/x)}{\ln(b/a) + 2\gamma r \ln(b/r)\ln(r/a) + 2\eta u \ln(b/u)\ln(u/a) + 4\gamma r\eta u \ln(b/u)\ln(u/r)\ln(r/a)}$$

$r \le x$, $x \le u$

$$= \frac{\ln(b/a) + 2\gamma r \ln(x/r)\ln(r/a) + 2\eta u \ln(b/u)\ln(u/x)}{\ln(b/a) + 2\gamma r \ln(b/r)\ln(r/a) + 2\eta u \ln(b/u)\ln(u/a) + 4\gamma r\eta u \ln(b/u)\ln(u/r)\ln(r/a)}$$

$u \le x$, $x \le b$

$$= \frac{\ln(b/a) + 2\gamma r \ln(x/r)\ln(r/a) + 2\eta u \ln(x/u)\ln(u/a) + 4\gamma r\eta u \ln(x/u)\ln(u/r)\ln(r/a)}{\ln(b/a) + 2\gamma r \ln(b/r)\ln(r/a) + 2\eta u \ln(b/u)\ln(u/a) + 4\gamma r\eta u \ln(b/u)\ln(u/r)\ln(r/a)}$$

3.18.5 (a)

$$\mathbf{E}_x\big\{\exp\big(-\gamma\ell(H,r) - \eta\ell(H,u)\big);\ R_H^{(2)} = a\big\}$$

$a \le x$, $x \le r$

$$= \frac{\ln(b/x) + 2\gamma r \ln(b/r)\ln(r/x) + 2\eta u \ln(b/u)\ln(u/x) + 4\gamma r\eta u \ln(b/u)\ln(u/r)\ln(r/x)}{\ln(b/a) + 2\gamma r \ln(b/r)\ln(r/a) + 2\eta u \ln(b/u)\ln(u/a) + 4\gamma r\eta u \ln(b/u)\ln(u/r)\ln(r/a)}$$

$r \le x$, $x \le u$

$$= \frac{\ln(b/x) + 2\eta u \ln(b/u)\ln(u/x)}{\ln(b/a) + 2\gamma r \ln(b/r)\ln(r/a) + 2\eta u \ln(b/u)\ln(u/a) + 4\gamma r\eta u \ln(b/u)\ln(u/r)\ln(r/a)}$$

$u \le x$, $x \le b$

$$= \frac{\ln(b/x)}{\ln(b/a) + 2\gamma r \ln(b/r)\ln(r/a) + 2\eta u \ln(b/u)\ln(u/a) + 4\gamma r\eta u \ln(b/u)\ln(u/r)\ln(r/a)}$$

3.18.5 (b)

$$\mathbf{E}_x\big\{\exp\big(-\gamma\ell(H,r) - \eta\ell(H,u)\big);\ R_H^{(2)} = b\big\}$$

$a \le x$, $x \le r$

$$= \frac{\ln(x/a)}{\ln(b/a) + 2\gamma r \ln(b/r)\ln(r/a) + 2\eta u \ln(b/u)\ln(u/a) + 4\gamma r\eta u \ln(b/u)\ln(u/r)\ln(r/a)}$$

$r \le x$, $x \le u$

$$= \frac{\ln(x/a) + 2\gamma r \ln(x/r)\ln(r/a)}{\ln(b/a) + 2\gamma r \ln(b/r)\ln(r/a) + 2\eta u \ln(b/u)\ln(u/a) + 4\gamma r\eta u \ln(b/u)\ln(u/r)\ln(r/a)}$$

$u \le x$, $x \le b$

$$= \frac{\ln(x/a) + 2\gamma r \ln(x/r)\ln(r/a) + 2\eta u \ln(x/u)\ln(u/a) + 4\gamma r\eta u \ln(x/u)\ln(u/r)\ln(r/a)}{\ln(b/a) + 2\gamma r \ln(b/r)\ln(r/a) + 2\eta u \ln(b/u)\ln(u/a) + 4\gamma r\eta u \ln(b/u)\ln(u/r)\ln(r/a)}$$

6 $R_s^{(2)} = \big((W_s^{(1)})^2 + (W_s^{(2)})^2\big)^{1/2}$ $\qquad H = H_{a,b} = \min\{s : R_s^{(2)} \notin (a,b)\}$

3.18.6 (a) $\mathbf{P}_x\Big(\frac{1}{r}\ell(H,r) \in dy, \frac{1}{u}\ell(H,u) \in dg, R_H^{(2)} = a\Big)$

$$= \begin{cases} \dfrac{\ln(x/a)}{\ln(r/a)} p_r\,dydg, & a \le x \le r \\ \dfrac{\ln(u/x)}{\ln(u/r)} p_r\,dydg + \dfrac{\ln(x/r)}{\ln(u/r)} p_u\,dydg, & r \le x \le u \\ \dfrac{\ln(b/x)}{\ln(b/u)} p_u\,dydg, & u \le x \le b \end{cases}$$

where

$$p_r := \frac{\sqrt{y}}{4\ln(u/r)\ln(r/a)\sqrt{g}} \exp\Big(-\frac{y+g}{2\ln(u/r)} - \frac{y}{2\ln(r/a)} - \frac{g}{2\ln(b/u)}\Big) I_1\Big(\frac{\sqrt{yg}}{\ln(u/r)}\Big)$$

$$p_u := \frac{1}{4\ln(u/r)\ln(r/a)} \exp\Big(-\frac{y+g}{2\ln(u/r)} - \frac{y}{2\ln(r/a)} - \frac{g}{2\ln(b/u)}\Big) I_0\Big(\frac{\sqrt{yg}}{\ln(u/r)}\Big)$$

(1) $\mathbf{P}_x\Big(\frac{1}{r}\ell(H,r) \in dy, \ell(H,u) = 0, R_H^{(2)} = a\Big)$

$$= \begin{cases} \dfrac{\ln(x/a)}{2\ln^2(r/a)} \exp\Big(-\dfrac{y\ln(u/a)}{2\ln(u/r)\ln(r/a)}\Big)dy, & a \le x \le r \\ \dfrac{\ln(u/x)}{2\ln(u/r)\ln(r/a)} \exp\Big(-\dfrac{y\ln(u/a)}{2\ln(u/r)\ln(r/a)}\Big)dy, & r \le x \le u \end{cases}$$

(2) $\mathbf{P}_x\big(\ell(H,r) = 0,\ \ell(H,u) = 0, R_H^{(2)} = a\big) = \dfrac{\ln(r/x)}{\ln(r/a)}, \qquad a \le x \le r$

3.18.6 (b) $\mathbf{P}_x\Big(\frac{1}{r}\ell(H,r) \in dy, \frac{1}{u}\ell(H,u) \in dg, R_H^{(2)} = b\Big)$

$$= \begin{cases} \dfrac{\ln(x/a)}{\ln(r/a)} q_r\,dydg, & a \le x \le r \\ \dfrac{\ln(u/x)}{\ln(u/r)} q_r\,dydg + \dfrac{\ln(x/r)}{\ln(u/r)} q_u\,dydg, & r \le x \le u \\ \dfrac{\ln(b/x)}{\ln(b/u)} q_u\,dydg, & u \le x \le b \end{cases}$$

where

$$q_r := \frac{1}{4\ln(b/u)\ln(u/r)} \exp\Big(-\frac{y+g}{2\ln(u/r)} - \frac{y}{2\ln(r/a)} - \frac{g}{2\ln(b/u)}\Big) I_0\Big(\frac{\sqrt{yg}}{\ln(u/r)}\Big)$$

6 $R_s^{(2)} = \big((W_s^{(1)})^2 + (W_s^{(2)})^2\big)^{1/2}$ $\qquad H = H_{a,b} = \min\{s : R_s^{(2)} \notin (a,b)\}$

$$q_u := \frac{\sqrt{g}}{4\ln(b/u)\ln(u/r)\sqrt{y}}\exp\Big(-\frac{y+g}{2\ln(u/r)} - \frac{y}{2\ln(r/a)} - \frac{g}{2\ln(b/u)}\Big) I_1\Big(\frac{\sqrt{yg}}{\ln(u/r)}\Big)$$

(1) $\mathbf{P}_x\Big(\ell(H,r)=0, \frac{1}{u}\ell(H,u)\in dg, R_H^{(2)}=b\Big)$

$$= \begin{cases} \dfrac{\ln(x/r)}{2\ln(b/u)\ln(u/r)}\exp\Big(-\dfrac{g\ln(b/r)}{2\ln(b/u)\ln(u/r)}\Big)dg, & r\le x\le u \\ \dfrac{\ln(b/x)}{2\ln^2(b/u)}\exp\Big(-\dfrac{g\ln(b/r)}{2\ln(b/u)\ln(u/r)}\Big)dg, & u\le x\le b \end{cases}$$

(2) $\mathbf{P}_x\big(\ell(H,r)=0,\ \ell(H,u)=0, R_H^{(2)}=b\big) = \dfrac{\ln(x/u)}{\ln(b/u)}, \qquad u\le x\le b$

3.19.5 (a) $\mathbf{E}_x\Big\{\exp\Big(-\gamma\int_0^H \frac{ds}{R_s^{(2)}}\Big);\ R_H^{(2)}=a\Big\} = \dfrac{S_0(2\sqrt{2\gamma b}, 2\sqrt{2\gamma x})}{S_0(2\sqrt{2\gamma b}, 2\sqrt{2\gamma a})}$

3.19.5 (b) $\mathbf{E}_x\Big\{\exp\Big(-\gamma\int_0^H \frac{ds}{R_s^{(2)}}\Big);\ R_H^{(2)}=b\Big\} = \dfrac{S_0(2\sqrt{2\gamma x}, 2\sqrt{2\gamma a})}{S_0(2\sqrt{2\gamma b}, 2\sqrt{2\gamma a})}$

3.20.5 (a) $\mathbf{E}_x\Big\{\exp\Big(-\frac{\gamma^2}{2}\int_0^H \frac{ds}{(R_s^{(2)})^2}\Big);\ R_H^{(2)}=a\Big\} = \dfrac{(b/x)^{|\gamma|}-(x/b)^{|\gamma|}}{(b/a)^{|\gamma|}-(a/b)^{|\gamma|}}, \qquad a\le x\le b$

3.20.5 (b) $\mathbf{E}_x\Big\{\exp\Big(-\frac{\gamma^2}{2}\int_0^H \frac{ds}{(R_s^{(2)})^2}\Big);\ R_H^{(2)}=b\Big\} = \dfrac{(x/a)^{|\gamma|}-(a/x)^{|\gamma|}}{(b/a)^{|\gamma|}-(a/b)^{|\gamma|}}, \qquad a\le x\le b$

3.21.5 (a) $\mathbf{E}_x\Big\{\exp\Big(-\int_0^H\Big(\frac{p^2}{2(R_s^{(2)})^2}+\frac{(qR_s^{(2)})^2}{2}\Big)ds\Big);\ R_H^{(2)}=a\Big\} = \dfrac{x^{|p|}S_{|p|/2}(qb^2/2, qx^2/2)}{a^{|p|}S_{|p|/2}(qb^2/2, qa^2/2)}$

3.21.5 (b) $\mathbf{E}_x\Big\{\exp\Big(-\int_0^H\Big(\frac{p^2}{2(R_s^{(2)})^2}+\frac{(qR_s^{(2)})^2}{2}\Big)ds\Big);\ R_H^{(2)}=b\Big\} = \dfrac{x^{|p|}S_{|p|/2}(qx^2/2, qa^2/2)}{b^{|p|}S_{|p|/2}(qb^2/2, qa^2/2)}$

3.22.5 (a) $\mathbf{E}_x\Big\{\exp\Big(-\int_0^H\Big(\frac{p^2}{2(R_s^{(2)})^2}+\frac{q}{R_s^{(2)}}\Big)ds\Big);\ R_H^{(2)}=a\Big\} = \dfrac{x^{|p|}S_{2|p|}(2\sqrt{2qb}, 2\sqrt{2qx})}{a^{|p|}S_{2|p|}(2\sqrt{2qb}, 2\sqrt{2qa})}$

3.22.5 (b) $\mathbf{E}_x\Big\{\exp\Big(-\int_0^H\Big(\frac{p^2}{2(R_s^{(2)})^2}+\frac{q}{R_s^{(2)}}\Big)ds\Big);\ R_H^{(2)}=b\Big\} = \dfrac{x^{|p|}S_{2|p|}(2\sqrt{2qx}, 2\sqrt{2qa})}{b^{|p|}S_{2|p|}(2\sqrt{2qb}, 2\sqrt{2qa})}$

6 $R_s^{(2)} = \big((W_s^{(1)})^2 + (W_s^{(2)})^2\big)^{1/2}$ $\qquad \varrho = \varrho(v,z) = \min\{s : \ell(s,z) = v\}$

4. Stopping at inverse local time

4.0.1
$$\mathbf{E}_x e^{-\alpha\varrho} = \begin{cases} \dfrac{I_0(x\sqrt{2\alpha})}{I_0(z\sqrt{2\alpha})}\exp\Big(-\dfrac{v}{2zI_0(z\sqrt{2\alpha})K_0(z\sqrt{2\alpha})}\Big), & x \le z \\ \dfrac{K_0(x\sqrt{2\alpha})}{K_0(z\sqrt{2\alpha})}\exp\Big(-\dfrac{v}{2zI_0(z\sqrt{2\alpha})K_0(z\sqrt{2\alpha})}\Big), & z \le x \end{cases}$$

4.1.2 $z \le y$
$$\mathbf{P}_x\Big(\sup_{0\le s\le\varrho} R_s^{(2)} < y\Big) = \begin{cases} \exp\Big(-\dfrac{v}{2z\ln(y/z)}\Big), & x \le z \\ \dfrac{\ln(y/x)}{\ln(y/z)}\exp\Big(-\dfrac{v}{2z\ln(y/z)}\Big), & z \le x \le y \end{cases}$$

4.2.2 $y \le z$
$$\mathbf{P}_x\Big(\inf_{0\le s\le\varrho} R_s^{(2)} > y\Big) = \begin{cases} \dfrac{\ln(x/y)}{\ln(z/y)}\exp\Big(-\dfrac{v}{2z\ln(z/y)}\Big), & y \le x \le z \\ \exp\Big(-\dfrac{v}{2z\ln(z/y)}\Big), & z \le x \end{cases}$$

4.3.1 $\mathbf{E}_x e^{-\gamma\ell(\varrho,r)}$

$$= \begin{cases} \exp\Big(-\dfrac{\gamma rv}{z(1+2\gamma r|\ln(r/z)|)}\Big), & x \le z \le r, \quad r \le z \le x \\ \dfrac{1+2\gamma r|\ln(r/x)|}{1+2\gamma r|\ln(r/z)|}\exp\Big(-\dfrac{\gamma rv}{z(1+2\gamma r|\ln(r/z)|)}\Big), & r\wedge z \le x \le r\vee z \\ \dfrac{1}{1+2\gamma r|\ln(r/z)|}\exp\Big(-\dfrac{\gamma rv}{z(1+2\gamma r|\ln(r/z)|)}\Big), & z \le r \le x, \quad x \le r \le z \end{cases}$$

4.3.2 $\mathbf{P}\big(\ell(\varrho,r) \in dy\big)$

$$= \begin{cases} \dfrac{1}{2|\ln(r/z)|}\dfrac{\sqrt{v}}{\sqrt{yrz}}\exp\Big(-\dfrac{v/z+y/r}{2|\ln(r/z)|}\Big)I_1\Big(\dfrac{\sqrt{vy}}{|\ln(r/z)|\sqrt{rz}}\Big)dy, & \begin{matrix} r \le z \le x \\ x \le z \le r \end{matrix} \\ \dfrac{1}{2\ln^2(r/z)}\exp\Big(-\dfrac{v/z+y/r}{2|\ln(r/z)|}\Big)\Big[|\ln(r/x)|\dfrac{\sqrt{v}}{\sqrt{yrz}}I_1\Big(\dfrac{\sqrt{vy}}{|\ln(r/z)|\sqrt{rz}}\Big) & \\ \qquad +|\ln(x/z)|\dfrac{1}{r}I_0\Big(\dfrac{\sqrt{vy}}{|\ln(r/z)|\sqrt{rz}}\Big)\Big]dy, & z\wedge r \le x \le r\vee z \\ \dfrac{1}{2r|\ln(r/z)|}\exp\Big(-\dfrac{v/z+y/r}{2|\ln(r/z)|}\Big)I_0\Big(\dfrac{\sqrt{vy}}{|\ln(r/z)|\sqrt{rz}}\Big)dy, & z \le r \le x, x \le r \le z \end{cases}$$

(1)
$$\mathbf{P}_x\big(\ell(\varrho,r) = 0\big) = \begin{cases} \exp\Big(-\dfrac{v}{2z|\ln(r/z)|}\Big), & x \le z \le r, \quad r \le z \le x \\ \dfrac{|\ln(r/x)|}{|\ln(r/z)|}\exp\Big(-\dfrac{v}{2z|\ln(r/z)|}\Big), & z\wedge r \le x \le r\vee z \\ 0, & z \le r \le x, \quad x \le r \le z \end{cases}$$

4.4.1 $\mathbf{E}_x \exp\Big(-\gamma\displaystyle\int_0^\varrho \mathbb{I}_{[r,\infty)}\big(R_s^{(2)}\big)ds\Big)$

6 $R_s^{(2)}=\big((W_s^{(1)})^2+(W_s^{(2)})^2\big)^{1/2}$ $\qquad \varrho=\varrho(v,z)=\min\{s:\ell(s,z)=v\}$

$r\le z$

$$=\begin{cases}\dfrac{K_0(x\sqrt{2\gamma})}{K_0(z\sqrt{2\gamma})}\exp\Big(-\dfrac{K_1(r\sqrt{2\gamma})v}{2zC_0(r\sqrt{2\gamma},z\sqrt{2\gamma})K_0(z\sqrt{2\gamma})}\Big), & z\le x\\ \dfrac{C_0(r\sqrt{2\gamma},x\sqrt{2\gamma})}{C_0(r\sqrt{2\gamma},z\sqrt{2\gamma})}\exp\Big(-\dfrac{K_1(r\sqrt{2\gamma})v}{2zC_0(r\sqrt{2\gamma},z\sqrt{2\gamma})K_0(z\sqrt{2\gamma})}\Big), & r\le x\le z\\ \dfrac{1}{r\sqrt{2\gamma}C_0(r\sqrt{2\gamma},z\sqrt{2\gamma})}\exp\Big(-\dfrac{K_1(r\sqrt{2\gamma})v}{2zC_0(r\sqrt{2\gamma},z\sqrt{2\gamma})K_0(z\sqrt{2\gamma})}\Big), & 0\le x\le r\end{cases}$$

$z\le r$

$$=\exp\Big(-\frac{r\sqrt{2\gamma}K_1(r\sqrt{2\gamma})v}{2z(K_0(r\sqrt{2\gamma})+\ln(r/z)r\sqrt{2\gamma}K_1(r\sqrt{2\gamma}))}\Big)$$

$$\times\begin{cases}\dfrac{K_0(x\sqrt{2\gamma})}{K_0(r\sqrt{2\gamma})+\ln(r/z)r\sqrt{2\gamma}K_1(r\sqrt{2\gamma})}, & r\le x\\ \dfrac{K_0(r\sqrt{2\gamma})+\ln(r/x)r\sqrt{2\gamma}K_1(r\sqrt{2\gamma})}{K_0(r\sqrt{2\gamma})+\ln(r/z)r\sqrt{2\gamma}K_1(r\sqrt{2\gamma})}, & z\le x\le r\\ 1, & 0\le x\le z\end{cases}$$

4.4.2 $z\le r$

$$\mathbf{P}_x\Big(\int_0^{\varrho}\mathbb{I}_{[r,\infty)}\big(R_s^{(2)}\big)ds=0\Big)=\begin{cases}\exp\Big(-\dfrac{v}{2z\ln(r/z)}\Big), & x\le z\\ \dfrac{\ln(r/x)}{\ln(r/z)}\exp\Big(-\dfrac{v}{2z\ln(r/z)}\Big), & z\le x\le r\\ 0, & r\le x\end{cases}$$

4.5.1

$$\mathbf{E}_x\exp\Big(-\gamma\int_0^{\varrho}\mathbb{I}_{[0,r]}\big(R_s^{(2)}\big)ds\Big)$$

$r\le z$

$$=\exp\Big(-\frac{r\sqrt{2\gamma}I_1(r\sqrt{2\gamma})v}{2z(I_0(r\sqrt{2\gamma})+\ln(z/r)r\sqrt{2\gamma}I_1(r\sqrt{2\gamma}))}\Big)$$

$$\times\begin{cases}\dfrac{I_0(x\sqrt{2\gamma})}{I_0(r\sqrt{2\gamma})+\ln(z/r)r\sqrt{2\gamma}I_1(r\sqrt{2\gamma})}, & 0\le x\le r\\ \dfrac{I_0(r\sqrt{2\gamma})+\ln(x/r)r\sqrt{2\gamma}I_1(r\sqrt{2\gamma})}{I_0(r\sqrt{2\gamma})+\ln(z/r)r\sqrt{2\gamma}I_1(r\sqrt{2\gamma})}, & r\le x\le z\\ 1, & z\le x\end{cases}$$

$z\le r$

$$=\begin{cases}\dfrac{I_0(x\sqrt{2\gamma})}{I_0(z\sqrt{2\gamma})}\exp\Big(-\dfrac{I_1(r\sqrt{2\gamma})v}{2zC_0(r\sqrt{2\gamma},z\sqrt{2\gamma})I_0(z\sqrt{2\gamma})}\Big), & 0\le x\le z\\ \dfrac{C_0(r\sqrt{2\gamma},x\sqrt{2\gamma})}{C_0(r\sqrt{2\gamma},z\sqrt{2\gamma})}\exp\Big(-\dfrac{I_1(r\sqrt{2\gamma})v}{2zC_0(r\sqrt{2\gamma},z\sqrt{2\gamma})I_0(z\sqrt{2\gamma})}\Big), & z\le x\le r\\ \dfrac{1}{r\sqrt{2\gamma}C_0(r\sqrt{2\gamma},z\sqrt{2\gamma})}\exp\Big(-\dfrac{I_1(r\sqrt{2\gamma})v}{2zC_0(r\sqrt{2\gamma},z\sqrt{2\gamma})I_0(z\sqrt{2\gamma})}\Big), & r\le x\end{cases}$$

4.5.2 $r\le z$

$$\mathbf{P}_x\Big(\int_0^{\varrho}\mathbb{I}_{[0,r]}\big(R_s^{(2)}\big)ds=0\Big)=\begin{cases}0, & x\le r\\ \dfrac{\ln(x/r)}{\ln(z/r)}\exp\Big(-\dfrac{v}{2z\ln(z/r)}\Big), & r\le x\le z\\ \exp\Big(-\dfrac{v}{2z\ln(z/r)}\Big), & z\le x\end{cases}$$

6 $R_s^{(2)} = \big((W_s^{(1)})^2 + (W_s^{(2)})^2\big)^{1/2}$ $\qquad \varrho = \varrho(v,z) = \min\{s : \ell(s,z) = v\}$

4.18.1 $\mathbf{E}_x \exp\big(-\gamma\ell(\varrho,r) - \eta\ell(\varrho,u)\big)$ $\qquad r < u$

$z < r$

$$= \exp\Big(-\frac{(\gamma r + \eta u + 2\gamma r\eta u\ln(u/r))v}{z(1 + 2\gamma r\ln(r/z) + 2\eta u\ln(u/z) + 4\gamma r\eta u\ln(u/r)\ln(r/z))}\Big)$$

$$\times \begin{cases} 1, & x \le z \\ \dfrac{1 + 2\gamma r\ln(r/x) + 2\eta u\ln(u/x) + 4\gamma r\eta u\ln(u/r)\ln(r/x)}{1 + 2\gamma r\ln(r/z) + 2\eta u\ln(u/z) + 4\gamma r\eta u\ln(u/r)\ln(r/z)}, & z \le x \le r \\ \dfrac{1 + 2\eta u\ln(u/x)}{1 + 2\gamma r\ln(r/z) + 2\eta u\ln(u/z) + 4\gamma r\eta u\ln(u/r)\ln(r/z)}, & r \le x \le u \\ \dfrac{1}{1 + 2\gamma r\ln(r/z) + 2\eta u\ln(u/z) + 4\gamma r\eta u\ln(u/r)\ln(r/z)}, & u \le x \end{cases}$$

$u < z$

$$= \exp\Big(-\frac{(\gamma r + \eta u + 2\gamma r\eta u\ln(u/r))v}{z(1 + 2\gamma r\ln(z/r) + 2\eta u\ln(z/u) + 4\gamma r\eta u\ln(z/u)\ln(u/r))}\Big)$$

$$\times \begin{cases} \dfrac{1}{1 + 2\gamma r\ln(z/r) + 2\eta u\ln(z/u) + 4\gamma r\eta u\ln(z/u)\ln(u/r)}, & x \le r \\ \dfrac{1 + 2\gamma r\ln(x/r)}{1 + 2\gamma r\ln(z/r) + 2\eta u\ln(z/u) + 4\gamma r\eta u\ln(z/u)\ln(u/r)}, & r \le x \le u \\ \dfrac{1 + 2\gamma r\ln(x/r) + 2\eta u\ln(x/u) + 4\gamma r\eta u\ln(x/u)\ln(u/r)}{1 + 2\gamma r\ln(z/r) + 2\eta u\ln(z/u) + 4\gamma r\eta u\ln(z/u)\ln(u/r)}, & u \le x \le z \\ 1, & z \le x \end{cases}$$

4.20.1 $\mathbf{E}_x \exp\Big(-\dfrac{\gamma^2}{2}\displaystyle\int_0^\varrho \frac{ds}{(R_s^{(2)})^2}\Big) = \begin{cases} x^{|\gamma|}z^{-|\gamma|}e^{-v|\gamma|/z}, & 0 < x \le z \\ z^{|\gamma|}x^{-|\gamma|}e^{-v|\gamma|/z}, & z \le x \end{cases}$

4.20.4
$0 < y$

$$\mathbf{P}_x\Big(\int_0^\varrho \frac{ds}{(R_s^{(2)})^2} \in dy\Big) = \begin{cases} \dfrac{(\ln(z/x) + v/z)}{\sqrt{2\pi}y^{3/2}}e^{-(\ln(z/x)+v/z)^2/2y}dy, & 0 < x \le z \\ \dfrac{(\ln(x/z) + v/z)}{\sqrt{2\pi}y^{3/2}}e^{-(\ln(x/z)+v/z)^2/2y}dy, & z \le x \end{cases}$$

4.21.1
$0 < x$

$$\mathbf{E}_x \exp\Big(-\int_0^\varrho \Big(\frac{p^2}{2(R_s^{(2)})^2} + \frac{q^2(R_s^{(2)})^2}{2}\Big)ds\Big)$$

$$= \begin{cases} \dfrac{I_{|p|/2}(qx^2/2)}{I_{|p|/2}(qz^2/2)}\exp\Big(-\dfrac{v}{zI_{|p|/2}(qz^2/2)K_{|p|/2}(qz^2/2)}\Big), & x \le z \\ \dfrac{K_{|p|/2}(qx^2/2)}{K_{|p|/2}(qz^2/2)}\exp\Big(-\dfrac{v}{zI_{|p|/2}(qz^2/2)K_{|p|/2}(qz^2/2)}\Big), & z \le x \end{cases}$$

4.22.1
$0 < x$

$$\mathbf{E}_x \exp\Big(-\int_0^\varrho \Big(\frac{p^2}{2(R_s^{(2)})^2} + \frac{q}{R_s^{(2)}}\Big)ds\Big)$$

$$= \begin{cases} \dfrac{I_{2|p|}(2\sqrt{2qx})}{I_{2|p|}(2\sqrt{2qz})}\exp\Big(-\dfrac{v}{4zI_{2|p|}(2\sqrt{2qz})K_{2|p|}(2\sqrt{2qz})}\Big), & x \le z \\ \dfrac{K_{2|p|}(2\sqrt{2qx})}{K_{2|p|}(2\sqrt{2qz})}\exp\Big(-\dfrac{v}{4zI_{2|p|}(2\sqrt{2qz})K_{2|p|}(2\sqrt{2qz})}\Big), & z \le x \end{cases}$$

7. ORNSTEIN–UHLENBECK PROCESS

1. Exponential stopping

1.0.5 $$\mathbf{P}_x(U_\tau \in dz) = \begin{cases} \dfrac{\lambda\Gamma(\lambda/\theta)}{\theta\sigma\sqrt{2\pi}} e^{(x^2-z^2)/4\sigma^2} D_{-\lambda/\theta}(-\frac{x}{\sigma}) D_{-\lambda/\theta}(\frac{z}{\sigma})dz, & x \le z \\ \dfrac{\lambda\Gamma(\lambda/\theta)}{\theta\sigma\sqrt{2\pi}} e^{(x^2-z^2)/4\sigma^2} D_{-\lambda/\theta}(\frac{x}{\sigma}) D_{-\lambda/\theta}(-\frac{z}{\sigma})dz, & z \le x \end{cases}$$

or $$= \frac{\lambda\Gamma(\lambda/\theta)}{\theta\sigma\sqrt{2\pi}} e^{(x^2-z^2)/4\sigma^2} D_{-\lambda/\theta}\big(-\tfrac{z+x-|z-x|}{2\sigma}\big) D_{-\lambda/\theta}\big(\tfrac{z+x+|z-x|}{2\sigma}\big)dz$$

1.0.6 $$\mathbf{P}_x\big(U_t \in dz\big) = \frac{1}{(2\pi\sigma^2(1-e^{-2\theta t}))^{1/2}} \exp\Big(-\frac{(z-xe^{-\theta t})^2}{2\sigma^2(1-e^{-2\theta t})}\Big)dz$$

1.1.2 $x \le y$ $$\mathbf{P}_x\big(\sup_{0\le s\le\tau} U_s \ge y\big) = \frac{e^{x^2/4\sigma^2} D_{-\lambda/\theta}(-\frac{x}{\sigma})}{e^{y^2/4\sigma^2} D_{-\lambda/\theta}(-\frac{y}{\sigma})}$$

1.1.4 (1) $$\mathbf{P}_x\big(\sup_{0\le s\le t} U_s < 0\big) = \text{Erf}\Big(\frac{|x|}{\sigma\sqrt{2(e^{2\theta t}-1)}}\Big)$$ $x \le 0$

1.1.6 $$\mathbf{P}_x\big(\sup_{0\le s\le\tau} U_s \ge y, U_\tau \in dz\big)$$ $x \vee z \le y$

$$= \frac{\lambda\Gamma(\lambda/\theta)}{\theta\sigma\sqrt{2\pi} D_{-\lambda/\theta}(-\frac{y}{\sigma})} e^{(x^2-z^2)/4\sigma^2} D_{-\lambda/\theta}(-\tfrac{x}{\sigma}) D_{-\lambda/\theta}(-\tfrac{z}{\sigma}) D_{-\lambda/\theta}(\tfrac{y}{\sigma})dz$$

1.1.8 $x \le 0$ $$\mathbf{P}_x\big(\sup_{0\le s\le t} U_s \ge 0, U_t \in dz\big) = \frac{1}{(2\pi\sigma^2(1-e^{-2\theta t}))^{1/2}} \exp\Big(-\frac{(|z|-xe^{-\theta t})^2}{2\sigma^2(1-e^{-2\theta t})}\Big)dz$$

1.2.2 $y \le x$ $$\mathbf{P}_x\big(\inf_{0\le s\le\tau} U_s \le y\big) = \frac{e^{x^2/4\sigma^2} D_{-\lambda/\theta}(\frac{x}{\sigma})}{e^{y^2/4\sigma^2} D_{-\lambda/\theta}(\frac{y}{\sigma})}$$

1.2.4 (1) $$\mathbf{P}_x\big(0 < \inf_{0\le s\le t} U_s\big) = \text{Erf}\Big(\frac{x}{\sigma\sqrt{2(e^{2\theta t}-1)}}\Big)$$ $0 \le x$

1.2.6 $$\mathbf{P}_x\big(\inf_{0\le s\le\tau} U_s \le y, U_\tau \in dz\big)$$ $y \le x \wedge z$

$$= \frac{\lambda\Gamma(\lambda/\theta)}{\theta\sigma\sqrt{2\pi} D_{-\lambda/\theta}(\frac{y}{\sigma})} e^{(x^2-z^2)/4\sigma^2} D_{-\lambda/\theta}(\tfrac{x}{\sigma}) D_{-\lambda/\theta}(\tfrac{z}{\sigma}) D_{-\lambda/\theta}(-\tfrac{y}{\sigma})dz$$

1.2.8 $0 \le x$ $$\mathbf{P}_x\big(\inf_{0\le s\le t} U_s \le 0, U_t \in dz\big) = \frac{1}{(2\pi\sigma^2(1-e^{-2\theta t}))^{1/2}} \exp\Big(-\frac{(|z|+xe^{-\theta t})^2}{2\sigma^2(1-e^{-2\theta t})}\Big)dz$$

7 $U_s = \sigma e^{-\theta s} W_{e^{2\theta s}-1}$ $\sigma, \theta > 0$ $\tau \sim \text{Exp}(\lambda)$, independent of U

1.3.1 $$\mathbf{E}_x e^{-\gamma \ell(\tau,r)} = \begin{cases} 1 - \dfrac{\gamma \Gamma(\lambda/\theta) e^{(x^2-r^2)/4\sigma^2} D_{-\lambda/\theta}(-\frac{x}{\sigma}) D_{-\lambda/\theta}(\frac{r}{\sigma})}{\theta\sigma\sqrt{2\pi} + \gamma\Gamma(\lambda/\theta) D_{-\lambda/\theta}(-\frac{r}{\sigma}) D_{-\lambda/\theta}(\frac{r}{\sigma})}, & x \le r \\ 1 - \dfrac{\gamma \Gamma(\lambda/\theta) e^{(x^2-r^2)/4\sigma^2} D_{-\lambda/\theta}(\frac{x}{\sigma}) D_{-\lambda/\theta}(-\frac{r}{\sigma})}{\theta\sigma\sqrt{2\pi} + \gamma\Gamma(\lambda/\theta) D_{-\lambda/\theta}(-\frac{r}{\sigma}) D_{-\lambda/\theta}(\frac{r}{\sigma})}, & r \le x \end{cases}$$

1.3.2 $$\mathbf{P}_x\big(\ell(\tau,r) \in dy\big) = \exp\Big(-\frac{\theta\sigma\sqrt{2\pi} y}{\Gamma(\lambda/\theta) D_{-\lambda/\theta}(-\frac{r}{\sigma}) D_{-\lambda/\theta}(\frac{r}{\sigma})}\Big)$$

$$\times \begin{cases} \dfrac{\theta\sigma\sqrt{2\pi} e^{(x^2-r^2)/4\sigma^2} D_{-\lambda/\theta}(-\frac{x}{\sigma})}{\Gamma(\lambda/\theta) D^2_{-\lambda/\theta}(-\frac{r}{\sigma}) D_{-\lambda/\theta}(\frac{r}{\sigma})} dy, & x \le r \\ \dfrac{\theta\sigma\sqrt{2\pi} e^{(x^2-r^2)/4\sigma^2} D_{-\lambda/\theta}(\frac{x}{\sigma})}{\Gamma(\lambda/\theta) D_{-\lambda/\theta}(-\frac{r}{\sigma}) D^2_{-\lambda/\theta}(\frac{r}{\sigma})} dy, & r \le x \end{cases}$$

(1) $$\mathbf{P}_x\big(\ell(\tau,r) = 0\big) = \begin{cases} 1 - \dfrac{e^{x^2/4\sigma^2} D_{-\lambda/\theta}(-\frac{x}{\sigma})}{e^{r^2/4\sigma^2} D_{-\lambda/\theta}(-\frac{r}{\sigma})}, & x \le r \\ 1 - \dfrac{e^{x^2/4\sigma^2} D_{-\lambda/\theta}(\frac{x}{\sigma})}{e^{r^2/4\sigma^2} D_{-\lambda/\theta}(\frac{r}{\sigma})}, & r \le x \end{cases}$$

1.3.5 $$\mathbf{E}_x\big\{e^{-\gamma\ell(\tau,r)}; U_\tau \in dz\big\}$$

$$= \frac{\lambda\sqrt{\pi}}{\theta\sigma\sqrt{2}} e^{-z^2/2\sigma^2} F\big(\tfrac{\lambda}{\theta}, \tfrac{x}{\sigma}, \tfrac{z}{\sigma}\big) dz - \frac{\gamma\lambda\pi e^{-(z^2+r^2)/2\sigma^2} F(\frac{\lambda}{\theta}, \frac{r}{\sigma}, \frac{z}{\sigma}) F(\frac{\lambda}{\theta}, \frac{x}{\sigma}, \frac{r}{\sigma})}{2\theta^2\sigma^2 + \theta\sigma\sqrt{2\pi}\gamma e^{-r^2/2\sigma^2} F(\frac{\lambda}{\theta}, \frac{r}{\sigma}, \frac{r}{\sigma})} dz$$

1.3.6 $$\mathbf{P}_x\big(\ell(\tau,r) \in dy, U_\tau \in dz\big) = \frac{\lambda F(\frac{\lambda}{\theta}, \frac{r}{\sigma}, \frac{z}{\sigma}) F(\frac{\lambda}{\theta}, \frac{x}{\sigma}, \frac{r}{\sigma})}{e^{(z^2-r^2)/2\sigma^2} F^2(\frac{\lambda}{\theta}, \frac{r}{\sigma}, \frac{r}{\sigma})} \exp\Big(-\frac{\theta\sigma\sqrt{2} e^{r^2/2\sigma^2} y}{\sqrt{\pi} F(\frac{\lambda}{\theta}, \frac{r}{\sigma}, \frac{r}{\sigma})}\Big) dydz$$

(1) $$\mathbf{P}_x\big(\ell(\tau,r) = 0, U_\tau \in dz\big)$$

$$= \frac{\lambda\sqrt{\pi}}{\theta\sigma\sqrt{2}} e^{-z^2/2\sigma^2} F\big(\tfrac{\lambda}{\theta}, \tfrac{x}{\sigma}, \tfrac{z}{\sigma}\big) dz - \frac{\lambda\sqrt{\pi} e^{-z^2/2\sigma^2} F(\frac{\lambda}{\theta}, \frac{r}{\sigma}, \frac{z}{\sigma}) F(\frac{\lambda}{\theta}, \frac{x}{\sigma}, \frac{r}{\sigma})}{\theta\sigma\sqrt{2} F(\frac{\lambda}{\theta}, \frac{r}{\sigma}, \frac{r}{\sigma})} dz$$

1.6.1 $$\mathbf{E}_x \exp\Big(-\int_0^\tau \big(p\mathbb{I}_{(-\infty,r)}(U_s) + q\mathbb{I}_{[r,\infty)}(U_s)\big) ds\Big)$$

$x \le r$ $$= \frac{\lambda}{\lambda+p} - \frac{\lambda(\lambda+p)^{-1}(q-p) e^{(x^2-r^2)/4\sigma^2} D_{-\frac{\lambda+p}{\theta}}(-\frac{x}{\sigma}) D_{-\frac{\theta+\lambda+q}{\theta}}(\frac{r}{\sigma})}{(\lambda+p) D_{-\frac{\theta+\lambda+p}{\theta}}(-\frac{r}{\sigma}) D_{-\frac{\lambda+q}{\theta}}(\frac{r}{\sigma}) + (\lambda+q) D_{-\frac{\theta+\lambda+q}{\theta}}(\frac{r}{\sigma}) D_{-\frac{\lambda+p}{\theta}}(-\frac{r}{\sigma})}$$

7 $U_s = \sigma e^{-\theta s} W_{e^{2\theta s}-1}$ $\sigma, \theta > 0$ $\tau \sim \mathrm{Exp}(\lambda)$, independent of U

$r \le x$

$$= \frac{\lambda}{\lambda+q} + \frac{\lambda(\lambda+q)^{-1}(q-p)e^{(x^2-r^2)/4\sigma^2} D_{-\frac{\lambda+q}{\theta}}(\frac{x}{\sigma}) D_{-\frac{\theta+\lambda+p}{\theta}}(-\frac{r}{\sigma})}{(\lambda+p)D_{-\frac{\theta+\lambda+p}{\theta}}(-\frac{r}{\sigma}) D_{-\frac{\lambda+q}{\theta}}(\frac{r}{\sigma}) + (\lambda+q) D_{-\frac{\theta+\lambda+q}{\theta}}(\frac{r}{\sigma}) D_{-\frac{\lambda+p}{\theta}}(-\frac{r}{\sigma})}$$

1.6.2 (1) $x \le r$

$$\mathbf{P}_x\Big(\int_0^\tau \big(p\mathbb{1}_{(-\infty,r)}(U_s) + q\mathbb{1}_{[r,\infty)}(U_s)\big)ds = p\tau\Big) = 1 - \frac{e^{x^2/4\sigma^2} D_{-\lambda/\theta}(-\frac{x}{\sigma})}{e^{r^2/4\sigma^2} D_{-\lambda/\theta}(-\frac{r}{\sigma})}$$

(2) $r \le x$

$$\mathbf{P}_x\Big(\int_0^\tau \big(p\mathbb{1}_{(-\infty,r)}(U_s) + q\mathbb{1}_{[r,\infty)}(U_s)\big)ds = q\tau\Big) = 1 - \frac{e^{x^2/4\sigma^2} D_{-\lambda/\theta}(\frac{x}{\sigma})}{e^{r^2/4\sigma^2} D_{-\lambda/\theta}(\frac{r}{\sigma})}$$

1.6.5

$$\mathbf{E}_x\Big\{\exp\Big(-\int_0^\tau \big(p\mathbb{1}_{(-\infty,r)}(U_s) + q\mathbb{1}_{[r,\infty)}(U_s)\big)ds\Big); U_\tau \in dz\Big\}$$

$z \le r$, $x \le r$

$$= \frac{\lambda\sqrt{\pi}}{\theta\sigma\sqrt{2}} e^{-z^2/2\sigma^2} F\big(\tfrac{\lambda+p}{\theta}, \tfrac{x}{\sigma}, \tfrac{z}{\sigma}\big)dz + \frac{\lambda\Gamma((\lambda+p)/\theta)}{\theta\sigma\sqrt{2\pi}} e^{(x^2-z^2)/4\sigma^2} D_{-\frac{\lambda+p}{\theta}}(-\tfrac{x}{\sigma}) D_{-\frac{\lambda+p}{\theta}}(-\tfrac{z}{\sigma})$$
$$\times \frac{(\lambda+p)D_{-\frac{\theta+\lambda+p}{\theta}}(\frac{r}{\sigma}) D_{-\frac{\lambda+q}{\theta}}(\frac{r}{\sigma}) - (\lambda+q) D_{-\frac{\theta+\lambda+q}{\theta}}(\frac{r}{\sigma}) D_{-\frac{\lambda+p}{\theta}}(\frac{r}{\sigma})}{(\lambda+p)D_{-\frac{\theta+\lambda+p}{\theta}}(-\frac{r}{\sigma}) D_{-\frac{\lambda+q}{\theta}}(\frac{r}{\sigma}) + (\lambda+q) D_{-\frac{\theta+\lambda+q}{\theta}}(\frac{r}{\sigma}) D_{-\frac{\lambda+p}{\theta}}(-\frac{r}{\sigma})} dz$$

$z \le r$, $r \le x$

$$= \frac{\lambda\sigma^{-1} e^{(x^2-z^2)/4\sigma^2} D_{-\frac{\lambda+q}{\theta}}(\frac{x}{\sigma}) D_{-\frac{\lambda+p}{\theta}}(-\frac{z}{\sigma})}{(\lambda+p)D_{-\frac{\theta+\lambda+p}{\theta}}(-\frac{r}{\sigma}) D_{-\frac{\lambda+q}{\theta}}(\frac{r}{\sigma}) + (\lambda+q) D_{-\frac{\theta+\lambda+q}{\theta}}(\frac{r}{\sigma}) D_{-\frac{\lambda+p}{\theta}}(-\frac{r}{\sigma})} dz$$

$r \le z$, $x \le r$

$$= \frac{\lambda\sigma^{-1} e^{(x^2-z^2)/4\sigma^2} D_{-\frac{\lambda+p}{\theta}}(-\frac{x}{\sigma}) D_{-\frac{\lambda+q}{\theta}}(\frac{z}{\sigma})}{(\lambda+p)D_{-\frac{\theta+\lambda+p}{\theta}}(-\frac{r}{\sigma}) D_{-\frac{\lambda+q}{\theta}}(\frac{r}{\sigma}) + (\lambda+q) D_{-\frac{\theta+\lambda+q}{\theta}}(\frac{r}{\sigma}) D_{-\frac{\lambda+p}{\theta}}(-\frac{r}{\sigma})} dz$$

$r \le z$, $r \le x$

$$= \frac{\lambda\sqrt{\pi}}{\theta\sigma\sqrt{2}} e^{-z^2/2\sigma^2} F\big(\tfrac{\lambda+q}{\theta}, \tfrac{x}{\sigma}, \tfrac{z}{\sigma}\big)dz + \frac{\lambda\Gamma((\lambda+q)/\theta)}{\theta\sigma\sqrt{2\pi}} e^{(x^2-z^2)/4\sigma^2} D_{-\frac{\lambda+q}{\theta}}(\tfrac{x}{\sigma}) D_{-\frac{\lambda+q}{\theta}}(\tfrac{z}{\sigma})$$
$$\times \frac{(\lambda+q)D_{-\frac{\theta+\lambda+q}{\theta}}(-\frac{r}{\sigma}) D_{-\frac{\lambda+p}{\theta}}(-\frac{r}{\sigma}) - (\lambda+p) D_{-\frac{\theta+\lambda+p}{\theta}}(-\frac{r}{\sigma}) D_{-\frac{\lambda+q}{\theta}}(-\frac{r}{\sigma})}{(\lambda+p)D_{-\frac{\theta+\lambda+p}{\theta}}(-\frac{r}{\sigma}) D_{-\frac{\lambda+q}{\theta}}(\frac{r}{\sigma}) + (\lambda+q) D_{-\frac{\theta+\lambda+q}{\theta}}(\frac{r}{\sigma}) D_{-\frac{\lambda+p}{\theta}}(-\frac{r}{\sigma})} dz$$

7 $U_s = \sigma e^{-\theta s} W_{e^{2\theta s}-1}$ $\quad \sigma, \theta > 0$ $\quad \tau \sim \mathrm{Exp}(\lambda)$, independent of U

1.8.3
$$\mathbf{E}_x \exp\Big(-\gamma \int_0^t U_s ds\Big) = \exp\Big(-\frac{\gamma x}{\theta}(1-e^{-\theta t}) + \frac{\gamma^2\sigma^2}{2\theta^2}\big(2\theta t + 1 - (2-e^{-\theta t})^2\big)\Big)$$

1.8.4
$$\mathbf{P}_x\Big(\int_0^t U_s ds \in dy\Big)$$
$$= \frac{\theta}{\sigma\big(2\pi\big(2\theta t + 1 - (2-e^{-\theta t})^2\big)\big)^{1/2}} \exp\Big(-\frac{\big(\theta y - x(1-e^{-\theta t})\big)^2}{2\sigma^2\big(2\theta t + 1 - (2-e^{-\theta t})^2\big)}\Big) dy$$

1.8.5
$$\mathbf{E}_x\Big\{\exp\Big(-\frac{\gamma\theta}{\sigma}\int_0^\tau U_s ds\Big);\ U_\tau \in dz\Big\}$$

$x \le z$
$$= \frac{\lambda\Gamma(\frac{\lambda}{\theta} - \gamma^2)}{\theta\sigma\sqrt{2\pi}} e^{(x^2-z^2)/4\sigma^2} D_{\gamma^2-\lambda/\theta}\big(-\tfrac{x}{\sigma} - 2\gamma\big) D_{\gamma^2-\lambda/\theta}\big(\tfrac{z}{\sigma} + 2\gamma\big) dz$$

$z \le x$
$$= \frac{\lambda\Gamma(\frac{\lambda}{\theta} - \gamma^2)}{\theta\sigma\sqrt{2\pi}} e^{(x^2-z^2)/4\sigma^2} D_{\gamma^2-\lambda/\theta}\big(\tfrac{x}{\sigma} + 2\gamma\big) D_{\gamma^2-\lambda/\theta}\big(-\tfrac{z}{\sigma} - 2\gamma\big) dz$$

1.8.7
$$\mathbf{E}_x\Big\{\exp\Big(-\frac{\gamma\theta}{\sigma}\int_0^t U_s ds\Big);\ U_t \in dz\Big\}$$
$$= \frac{e^{\gamma^2\theta t + \gamma(z-x)/\sigma}}{(2\pi\sigma^2(1-e^{-2\theta t}))^{1/2}} \exp\Big(-\frac{(z - xe^{-\theta t} + 2\gamma\sigma(1-e^{-\theta t}))^2}{2\sigma^2(1-e^{-2\theta t})}\Big) dz$$

(1)
$$\mathbf{E}_x\Big\{\exp\Big(-\gamma\int_0^t U_s ds\Big)\ \Big|\ U_t = z\Big\}$$
$$= \exp\Big(-\frac{\gamma(x+z)}{\theta}\operatorname{th}(\theta t/2) + \frac{\gamma^2\sigma^2}{\theta^2}\big(\theta t - 2\operatorname{th}(\theta t/2)\big)\Big)$$

1.8.8
$$\mathbf{P}_x\Big(\int_0^t U_s ds \in dy\ \Big|\ U_t = z\Big)$$
$$= \frac{\theta}{2\sigma\big(\pi(\theta t - 2\operatorname{th}(\theta t/2))\big)^{1/2}} \exp\Big(-\frac{\big(\theta y - (x+z)\operatorname{th}(\theta t/2)\big)^2}{4\sigma^2(\theta t - 2\operatorname{th}(\theta t/2))}\Big) dy$$

1.9.3
$1 \le \gamma$
$$\mathbf{E}_x \exp\Big(-\frac{(\gamma^2-1)\theta}{4\sigma^2}\int_0^t U_s^2 ds\Big) = \frac{\sqrt{\gamma}e^{\theta t/2}}{\sqrt{\operatorname{sh}(t\gamma\theta) + \gamma\operatorname{ch}(t\gamma\theta)}} \exp\Big(-\frac{x^2(\gamma^2-1)\operatorname{sh}(t\gamma\theta)}{4\sigma^2(\operatorname{sh}(t\gamma\theta) + \gamma\operatorname{ch}(t\gamma\theta))}\Big)$$

7 $U_s = \sigma e^{-\theta s} W_{e^{2\theta s}-1}$ $\sigma, \theta > 0$ $\tau \sim \text{Exp}(\lambda)$, independent of U

(1) $0 < x$

$$\mathbf{E}_x\Big\{\exp\Big(-\frac{(\gamma^2-1)\theta}{4\sigma^2}\int_0^t U_s^2 ds\Big);\ 0 < \inf_{0\le s\le t} U_s\Big\} = \frac{\sqrt{\gamma}e^{\theta t/2}}{\sqrt{\operatorname{sh}(t\gamma\theta)+\gamma\operatorname{ch}(t\gamma\theta)}}$$

$$\times\exp\Big(-\frac{x^2(\gamma^2-1)\operatorname{sh}(t\gamma\theta)}{4\sigma^2(\operatorname{sh}(t\gamma\theta)+\gamma\operatorname{ch}(t\gamma\theta))}\Big)\operatorname{Erf}\Big(\frac{x\gamma}{2\sigma\sqrt{\operatorname{sh}(t\gamma\theta)(\operatorname{sh}(t\gamma\theta)+\gamma\operatorname{ch}(t\gamma\theta))}}\Big)$$

1.9.5 $1 \le \gamma$

$$\mathbf{E}_x\Big\{\exp\Big(-\frac{(\gamma^2-1)\theta}{4\sigma^2}\int_0^\tau U_s^2 ds\Big); U_\tau \in dz\Big\}$$

$x \le z$

$$= \frac{\lambda\Gamma(\frac12-\frac{1}{2\gamma}+\frac{\lambda}{\theta\gamma})}{\theta\sigma\sqrt{2\pi\gamma}}e^{(x^2-z^2)/4\sigma^2}D_{-\frac12+\frac{1}{2\gamma}-\frac{\lambda}{\theta\gamma}}\big(-\tfrac{x\sqrt{\gamma}}{\sigma}\big)D_{-\frac12+\frac{1}{2\gamma}-\frac{\lambda}{\theta\gamma}}\big(\tfrac{z\sqrt{\gamma}}{\sigma}\big)dz$$

$z \le x$

$$= \frac{\lambda\Gamma(\frac12-\frac{1}{2\gamma}+\frac{\lambda}{\theta\gamma})}{\theta\sigma\sqrt{2\pi\gamma}}e^{(x^2-z^2)/4\sigma^2}D_{-\frac12+\frac{1}{2\gamma}-\frac{\lambda}{\theta\gamma}}\big(\tfrac{x\sqrt{\gamma}}{\sigma}\big)D_{-\frac12+\frac{1}{2\gamma}-\frac{\lambda}{\theta\gamma}}\big(-\tfrac{z\sqrt{\gamma}}{\sigma}\big)dz$$

1.9.7

$$\mathbf{E}_x\Big\{\exp\Big(-\frac{(\gamma^2-1)\theta}{4\sigma^2}\int_0^t U_s^2 ds\Big);\ U_t \in dz\Big\}$$

$$= \frac{\sqrt{\gamma}e^{t\theta/2+(x^2-z^2)/4\sigma^2}}{2\sigma\sqrt{\pi\operatorname{sh}(t\gamma\theta)}}\exp\Big(-\frac{(x^2+z^2)\gamma\operatorname{ch}(t\gamma\theta)-2xz\gamma}{4\sigma^2\operatorname{sh}(t\gamma\theta)}\Big)dz$$

1.9.8

$$\mathbf{P}_x\Big(\int_0^t U_s^2 ds \in dy,\ U_t \in dz\Big)$$

$$= \frac{1}{4\sqrt{\pi}\theta^{3/2}\sigma^3}e^{t\theta/2+(x^2-z^2-\theta y)/4\sigma^2}\operatorname{ee}_{y/2\theta\sigma^2}\Big(\frac12, t, \frac{x^2+z^2}{4\theta\sigma^2}, -\frac{xz}{2\theta\sigma^2}\Big)dydz$$

(1)

$$\mathbf{P}_0\Big(\int_0^t U_s^2 ds \in dy\Big| U_t = 0\Big)e^{-\theta t/2+\theta y/4\sigma^2}$$

$$= \frac{\sqrt{\sigma}\sqrt{1-e^{-2\theta t}}}{(2\theta)^{1/4}\pi y^{5/4}}\sum_{k=0}^\infty \frac{\Gamma(k+1/2)}{k!}e^{-(4k+1)^2t^2\sigma^2\theta/8y}D_{3/2}\Big(\frac{(4k+1)t\sigma\sqrt{\theta}}{\sqrt{2y}}\Big)dy$$

1.13.1 $x < y$

$$\mathbf{E}_x\big\{e^{-\gamma\check{H}(\tau)};\ \sup_{0\le s\le\tau} U_s \in dy\big\} = \frac{\lambda e^{(x^2-y^2)/4\sigma^2}D_{-(\lambda+\gamma)/\theta}(-\frac{x}{\sigma})D_{-(\theta+\lambda)/\theta}(-\frac{y}{\sigma})}{\theta\sigma D_{-(\lambda+\gamma)/\theta}(-\frac{y}{\sigma})D_{-\lambda/\theta}(-\frac{y}{\sigma})}dy$$

1.13.4 $x < 0$

$$\frac{\partial}{\partial y}\mathbf{P}_x\big(\check{H}(t) \in dv,\ \sup_{0\le s\le t} U_s < y\big)\big|_{y=0}$$

7 $U_s = \sigma e^{-\theta s} W_{e^{2\theta s}-1}$ $\sigma, \theta > 0$ $\tau \sim \mathrm{Exp}(\lambda)$, independent of U

$$= \frac{\theta|x|e^{v\theta - t\theta/2 + x^2/4\sigma^2}\mathbb{I}_{[0,t]}(v)}{2\pi\sigma^2 \operatorname{sh}^{3/2}(v\theta)\operatorname{sh}^{1/2}((t-v)\theta)} \exp\Big(-\frac{x^2 \operatorname{ch}(v\theta)}{4\sigma^2 \operatorname{sh}(v\theta)}\Big)dv$$

1.13.5 $\mathbf{E}_x\big\{e^{-\gamma \check{H}(\tau)};\ \sup_{0\le s\le\tau} U_s \in dy,\ U_\tau \in dz\big\}$ $x \vee z < y$

$$= \frac{\lambda e^{(x^2-z^2)/4\sigma^2} D_{-(\lambda+\gamma)/\theta}(-\frac{x}{\sigma}) D_{-\lambda/\theta}(-\frac{z}{\sigma})}{\theta\sigma^2 D_{-(\lambda+\gamma)/\theta}(-\frac{y}{\sigma}) D_{-\lambda/\theta}(-\frac{y}{\sigma})} dydz$$

1.13.8 (1) $\frac{\partial}{\partial y}\mathbf{P}_x\big(\check{H}(t) \in dv,\ \sup_{0\le s\le t} U_s < y, U_t \in dz\big)\big|_{y=0}$ $x \vee z \le 0$

$$= \frac{\theta x z e^{t\theta/2}\mathbb{I}_{[0,t]}(v)}{2\pi\sigma^4 \operatorname{sh}^{3/2}(v\theta)\operatorname{sh}^{3/2}((t-v)\theta)} \exp\Big(\frac{x^2 - z^2}{4\sigma^2} - \frac{x^2 \operatorname{ch}(v\theta)}{4\sigma^2 \operatorname{sh}(v\theta)} - \frac{z^2 \operatorname{ch}((t-v)\theta)}{4\sigma^2 \operatorname{sh}((t-v)\theta)}\Big)dvdz$$

1.14.1 $y < x$ $\mathbf{E}_x\big\{e^{-\gamma \hat{H}(\tau)};\ \inf_{0\le s\le\tau} U_s \in dy\big\} = \frac{\lambda e^{(x^2-y^2)/4\sigma^2} D_{-(\lambda+\gamma)/\theta}(\frac{x}{\sigma}) D_{-(\theta+\lambda)/\theta}(\frac{y}{\sigma})}{\theta\sigma D_{-(\lambda+\gamma)/\theta}(\frac{y}{\sigma}) D_{-\lambda/\theta}(\frac{y}{\sigma})} dy$

1.14.4 $0 < x$ $\frac{\partial}{\partial y}\mathbf{P}_x\big(\hat{H}(t) \in dv,\ \inf_{0\le s\le t} U_s < y\big)\big|_{y=0}$

$$= \frac{\theta x e^{v\theta - t\theta/2 + x^2/4\sigma^2}\mathbb{I}_{[0,t]}(v)}{2\pi\sigma^2 \operatorname{sh}^{3/2}(v\theta)\operatorname{sh}^{1/2}((t-v)\theta)} \exp\Big(-\frac{x^2 \operatorname{ch}(v\theta)}{4\sigma^2 \operatorname{sh}(v\theta)}\Big)dv$$

1.14.5 $\mathbf{E}_x\big\{e^{-\gamma \hat{H}(\tau)};\ \inf_{0\le s\le\tau} U_s \in dy,\ U_\tau \in dz\big\}$ $y < x \wedge z$

$$= \frac{\lambda e^{(x^2-z^2)/4\sigma^2} D_{-(\lambda+\gamma)/\theta}(\frac{x}{\sigma}) D_{-\lambda/\theta}(\frac{z}{\sigma})}{\theta\sigma^2 D_{-(\lambda+\gamma)/\theta}(\frac{y}{\sigma}) D_{-\lambda/\theta}(\frac{y}{\sigma})} dydz$$

1.14.8 (1) $\frac{\partial}{\partial y}\mathbf{P}_x\big(\check{H}(t) \in dv,\ \inf_{0\le s\le t} U_s < y, U_t \in dz\big)\big|_{y=0}$ $0 \le x \wedge z$

$$= \frac{\theta x z e^{t\theta/2}\mathbb{I}_{[0,t]}(v)}{2\pi\sigma^4 \operatorname{sh}^{3/2}(v\theta)\operatorname{sh}^{3/2}((t-v)\theta)} \exp\Big(\frac{x^2 - z^2}{4\sigma^2} - \frac{x^2 \operatorname{ch}(v\theta)}{4\sigma^2 \operatorname{sh}(v\theta)} - \frac{z^2 \operatorname{ch}((t-v)\theta)}{4\sigma^2 \operatorname{sh}((t-v)\theta)}\Big)dvdz$$

1.15.2 $\mathbf{P}_x\big(a < \inf_{0\le s\le\tau} U_s,\ \sup_{0\le s\le\tau} U_s < b\big) = 1 - \frac{S(\frac{\lambda}{\theta}, \frac{b}{\sigma}, \frac{x}{\sigma}) + S(\frac{\lambda}{\theta}, \frac{x}{\sigma}, \frac{a}{\sigma})}{S(\frac{\lambda}{\theta}, \frac{b}{\sigma}, \frac{a}{\sigma})}$

7 $U_s = \sigma e^{-\theta s} W_{e^{2\theta s}-1}$ $\sigma, \theta > 0$ $\tau \sim \mathrm{Exp}(\lambda)$, independent of U

1.15.6 $\mathbf{P}_x\big(a < \inf_{0\le s\le\tau} U_s,\ \sup_{0\le s\le\tau} U_s < b, U_\tau \in dz\big)$

$$= \frac{\lambda\sqrt{\pi}e^{-z^2/2\sigma^2} S(\frac{\lambda}{\theta}, \frac{b}{\sigma}, \frac{z+x+|z-x|}{2\sigma}) S(\frac{\lambda}{\theta}, \frac{z+x-|z-x|}{2\sigma}, \frac{a}{\sigma})}{\theta\sigma\sqrt{2}S(\frac{\lambda}{\theta}, \frac{b}{\sigma}, \frac{a}{\sigma})} dz$$

1.16.1 $\mathbf{E}_x\{e^{-\gamma\ell(\tau,r)};\ \sup_{0\le s\le\tau} U_s < b\}$

$r < b$

$x \le r$
$$= 1 - \frac{(\gamma\sqrt{\pi}e^{-r^2/2\sigma^2} S(\frac{\lambda}{\theta}, \frac{b}{\sigma}, \frac{r}{\sigma}) + \theta\sigma\sqrt{2}) e^{x^2/4\sigma^2} D_{-\lambda/\theta}(-\frac{x}{\sigma})}{\gamma\sqrt{\pi}e^{-r^2/4\sigma^2} S(\frac{\lambda}{\theta}, \frac{b}{\sigma}, \frac{r}{\sigma}) D_{-\lambda/\theta}(-\frac{r}{\sigma}) + \theta\sigma\sqrt{2}e^{b^2/4\sigma^2} D_{-\lambda/\theta}(-\frac{b}{\sigma})}$$

$r \le x$
$x \le b$
$$= 1 - \frac{S(\frac{\lambda}{\theta}, \frac{b}{\sigma}, \frac{x}{\sigma}) + S(\frac{\lambda}{\theta}, \frac{x}{\sigma}, \frac{r}{\sigma})}{S(\frac{\lambda}{\theta}, \frac{b}{\sigma}, \frac{r}{\sigma})}$$

$$+ \frac{\theta\sigma\sqrt{2}(e^{b^2/4\sigma^2} D_{-\lambda/\theta}(-\frac{b}{\sigma}) - e^{r^2/4\sigma^2} D_{-\lambda/\theta}(-\frac{r}{\sigma})) S(\frac{\lambda}{\theta}, \frac{b}{\sigma}, \frac{x}{\sigma})}{(\gamma\sqrt{\pi}e^{-r^2/4\sigma^2} S(\frac{\lambda}{\theta}, \frac{b}{\sigma}, \frac{r}{\sigma}) D_{-\lambda/\theta}(-\frac{r}{\sigma}) + \theta\sigma\sqrt{2}e^{b^2/4\sigma^2} D_{-\lambda/\theta}(-\frac{b}{\sigma})) S(\frac{\lambda}{\theta}, \frac{b}{\sigma}, \frac{r}{\sigma})}$$

1.16.2 $\mathbf{P}_x\big(\ell(\tau,r) \in dy,\ \sup_{0\le s\le\tau} U_s < b\big) = \exp\Big(-\frac{y\theta\sigma\sqrt{2}e^{(b^2+r^2)/4\sigma^2} D_{-\lambda/\theta}(-\frac{b}{\sigma})}{\sqrt{\pi}S(\frac{\lambda}{\theta}, \frac{b}{\sigma}, \frac{r}{\sigma}) D_{-\lambda/\theta}(-\frac{r}{\sigma})}\Big)$

$r < b$

$$\times \begin{cases} \dfrac{\theta\sigma\sqrt{2}(e^{b^2/4\sigma^2} D_{-\lambda/\theta}(-\frac{b}{\sigma}) - e^{r^2/4\sigma^2} D_{-\lambda/\theta}(-\frac{r}{\sigma})) e^{x^2/4\sigma^2} D_{-\lambda/\theta}(-\frac{x}{\sigma})}{\sqrt{\pi}S(\frac{\lambda}{\theta}, \frac{b}{\sigma}, \frac{r}{\sigma}) D^2_{-\lambda/\theta}(-\frac{r}{\sigma})} dy, & x \le r \\ \dfrac{\theta\sigma\sqrt{2}(e^{b^2/4\sigma^2} D_{-\lambda/\theta}(-\frac{b}{\sigma}) - e^{r^2/4\sigma^2} D_{-\lambda/\theta}(-\frac{r}{\sigma})) S(\frac{\lambda}{\theta}, \frac{b}{\sigma}, \frac{x}{\sigma})}{\sqrt{\pi}S^2(\frac{\lambda}{\theta}, \frac{b}{\sigma}, \frac{r}{\sigma}) e^{-r^2/4\sigma^2} D_{-\lambda/\theta}(-\frac{r}{\sigma})} dy, & r \le x \le b \end{cases}$$

(1)
$$\mathbf{P}_x\big(\ell(\tau,r) = 0,\ \sup_{0\le s\le\tau} U_s < b\big) = \begin{cases} 1 - \dfrac{e^{x^2/4\sigma^2} D_{-\lambda/\theta}(-\frac{x}{\sigma})}{e^{r^2/4\sigma^2} D_{-\lambda/\theta}(-\frac{r}{\sigma})}, & x \le r \\ 1 - \dfrac{S(\frac{\lambda}{\theta}, \frac{b}{\sigma}, \frac{x}{\sigma}) + S(\frac{\lambda}{\theta}, \frac{x}{\sigma}, \frac{r}{\sigma})}{S(\frac{\lambda}{\theta}, \frac{b}{\sigma}, \frac{r}{\sigma})}, & r \le x \le b \end{cases}$$

1.16.5 $\mathbf{E}_x\{e^{-\gamma\ell(\tau,r)};\ \sup_{0\le s\le\tau} U_s < b, U_\tau \in dz\}$ $r \vee z < b$

$z \le r$
$x \le r$
$$= \frac{\lambda\sqrt{\pi}e^{-z^2/2\sigma^2} S(\frac{\lambda}{\theta}, \frac{r}{\sigma}, \frac{z+x+|z-x|}{2\sigma}) e^{(z+x-|z-x|)^2/4\sigma^2} D_{-\lambda/\theta}(-\frac{z+x-|z-x|}{2\sigma})}{\theta\sigma\sqrt{2}e^{r^2/4\sigma^2} D_{-\lambda/\theta}(-\frac{r}{\sigma})} dz$$

$$+ \frac{\lambda\sqrt{\pi}e^{(x^2-z^2-r^2)/4\sigma^2} S(\frac{\lambda}{\theta}, \frac{b}{\sigma}, \frac{r}{\sigma}) D_{-\lambda/\theta}(-\frac{z}{\sigma}) D_{-\lambda/\theta}(-\frac{x}{\sigma})}{(\gamma\sqrt{\pi}e^{-r^2/4\sigma^2} S(\frac{\lambda}{\theta}, \frac{b}{\sigma}, \frac{r}{\sigma}) D_{-\lambda/\theta}(-\frac{r}{\sigma}) + \theta\sigma\sqrt{2}e^{b^2/4\sigma^2} D_{-\lambda/\theta}(-\frac{b}{\sigma})) D_{-\lambda/\theta}(-\frac{r}{\sigma})} dz$$

7 $U_s = \sigma e^{-\theta s} W_{e^{2\theta s}-1}$ $\sigma, \theta > 0$ $\tau \sim \operatorname{Exp}(\lambda)$, independent of U

$z \le r$, $r \le x$
$$= \frac{\lambda\sqrt{\pi}e^{-z^2/4\sigma^2} S(\frac{\lambda}{\theta},\frac{b}{\sigma},\frac{x}{\sigma}) D_{-\lambda/\theta}(-\frac{z}{\sigma})}{\gamma\sqrt{\pi}e^{-r^2/4\sigma^2} S(\frac{\lambda}{\theta},\frac{b}{\sigma},\frac{r}{\sigma}) D_{-\lambda/\theta}(-\frac{r}{\sigma}) + \theta\sigma\sqrt{2}e^{b^2/4\sigma^2} D_{-\lambda/\theta}(-\frac{b}{\sigma})} dz$$

$r \le z$, $x \le r$
$$= \frac{\lambda\sqrt{\pi}e^{-z^2/2\sigma^2} S(\frac{\lambda}{\theta},\frac{b}{\sigma},\frac{z}{\sigma}) e^{x^2/4\sigma^2} D_{-\lambda/\theta}(-\frac{x}{\sigma})}{\gamma\sqrt{\pi}e^{-r^2/4\sigma^2} S(\frac{\lambda}{\theta},\frac{b}{\sigma},\frac{r}{\sigma}) D_{-\lambda/\theta}(-\frac{r}{\sigma}) + \theta\sigma\sqrt{2}e^{b^2/4\sigma^2} D_{-\lambda/\theta}(-\frac{b}{\sigma})} dz$$

$r \le z$, $r \le x$
$$= \frac{\lambda\sqrt{\pi}e^{-z^2/2\sigma^2} S(\frac{\lambda}{\theta},\frac{b}{\sigma},\frac{z+x+|z-x|}{2\sigma}) S(\frac{\lambda}{\theta},\frac{z+x-|z-x|}{2\sigma},\frac{r}{2\sigma})}{\theta\sigma\sqrt{2}S(\frac{\lambda}{\theta},\frac{b}{\sigma},\frac{r}{\sigma})} dz$$
$$+ \frac{\lambda\sqrt{\pi}e^{-z^2/2\sigma^2} S(\frac{\lambda}{\theta},\frac{b}{\sigma},\frac{z}{\sigma}) e^{r^2/4\sigma^2} D_{-\lambda/\theta}(-\frac{r}{\sigma}) S(\frac{\lambda}{\theta},\frac{b}{\sigma},\frac{x}{\sigma})}{(\gamma\sqrt{\pi}e^{-r^2/4\sigma^2} S(\frac{\lambda}{\theta},\frac{b}{\sigma},\frac{r}{\sigma}) D_{-\lambda/\theta}(-\frac{r}{\sigma}) + \theta\sigma\sqrt{2}e^{b^2/4\sigma^2} D_{-\lambda/\theta}(-\frac{b}{\sigma})) S(\frac{\lambda}{\theta},\frac{b}{\sigma},\frac{r}{\sigma})} dz$$

1.16.6 $\mathbf{P}_x\big(\ell(\tau,r) \in dy, \sup_{0\le s\le\tau} U_s < b, U_\tau \in dz\big)$ $r \vee z < b$

$z \le r$, $x \le r$
$$= \frac{\lambda D_{-\lambda/\theta}(-\frac{z}{\sigma}) D_{-\lambda/\theta}(-\frac{x}{\sigma})}{e^{(z^2-x^2)/4\sigma^2} D^2_{-\lambda/\theta}(-\frac{r}{\sigma})} \exp\Big(-\frac{y\theta\sigma\sqrt{2}e^{(b^2+r^2)/4\sigma^2} D_{-\lambda/\theta}(-\frac{b}{\sigma})}{\sqrt{\pi}S(\frac{\lambda}{\theta},\frac{b}{\sigma},\frac{r}{\sigma}) D_{-\lambda/\theta}(-\frac{r}{\sigma})}\Big) dydz$$

$z \le r$, $r \le x$
$$= \frac{\lambda e^{(r^2-z^2)/4\sigma^2} S(\frac{\lambda}{\theta},\frac{b}{\sigma},\frac{x}{\sigma}) D_{-\lambda/\theta}(-\frac{z}{\sigma})}{S(\frac{\lambda}{\theta},\frac{b}{\sigma},\frac{r}{\sigma}) D_{-\lambda/\theta}(-\frac{r}{\sigma})} \exp\Big(-\frac{y\theta\sigma\sqrt{2}e^{(b^2+r^2)/4\sigma^2} D_{-\lambda/\theta}(-\frac{b}{\sigma})}{\sqrt{\pi}S(\frac{\lambda}{\theta},\frac{b}{\sigma},\frac{r}{\sigma}) D_{-\lambda/\theta}(-\frac{r}{\sigma})}\Big) dydz$$

$r \le z$, $x \le r$
$$= \frac{\lambda e^{(r^2+x^2)/4\sigma^2} S(\frac{\lambda}{\theta},\frac{b}{\sigma},\frac{z}{\sigma}) D_{-\lambda/\theta}(-\frac{x}{\sigma})}{e^{z^2/2\sigma^2} S(\frac{\lambda}{\theta},\frac{b}{\sigma},\frac{r}{\sigma}) D_{-\lambda/\theta}(-\frac{r}{\sigma})} \exp\Big(-\frac{y\theta\sigma\sqrt{2}e^{(b^2+r^2)/4\sigma^2} D_{-\lambda/\theta}(-\frac{b}{\sigma})}{\sqrt{\pi}S(\frac{\lambda}{\theta},\frac{b}{\sigma},\frac{r}{\sigma}) D_{-\lambda/\theta}(-\frac{r}{\sigma})}\Big) dydz$$

$r \le z$, $r \le x$
$$= \frac{\lambda e^{(r^2-z^2)/2\sigma^2} S(\frac{\lambda}{\theta},\frac{b}{\sigma},\frac{z}{\sigma}) S(\frac{\lambda}{\theta},\frac{b}{\sigma},\frac{x}{\sigma})}{S^2(\frac{\lambda}{\sigma},\frac{b}{\sigma},\frac{r}{\sigma})} \exp\Big(-\frac{y\theta\sigma\sqrt{2}e^{(b^2+r^2)/4\sigma^2} D_{-\lambda/\theta}(-\frac{b}{\sigma})}{\sqrt{\pi}S(\frac{\lambda}{\theta},\frac{b}{\sigma},\frac{r}{\sigma}) D_{-\lambda/\theta}(-\frac{r}{\sigma})}\Big) dydz$$

(1) $\mathbf{P}_x\big(\ell(\tau,r) = 0, \sup_{0\le s\le\tau} U_s < b, U_\tau \in dz\big)$

$z \le r$, $x \le r$
$$= \frac{\lambda\sqrt{\pi}e^{-z^2/2\sigma^2} S(\frac{\lambda}{\theta},\frac{r}{\sigma},\frac{z+x+|z-x|}{2\sigma}) e^{(z+x-|z-x|)^2/4\sigma^2} D_{-\lambda/\theta}(-\frac{z+x-|z-x|}{2\sigma})}{\theta\sigma\sqrt{2}e^{r^2/4\sigma^2} D_{-\lambda/\theta}(-\frac{r}{\sigma})} dz$$

$r \le z$, $r \le x$
$$= \frac{2\lambda\sqrt{\pi}e^{-z^2/2\sigma^2} S(\frac{\lambda}{\theta},\frac{b}{\sigma},\frac{z+x+|z-x|}{2\sigma}) S(\frac{\lambda}{\theta},\frac{z+x-|z-x|}{2\sigma},\frac{r}{\sigma})}{\theta\sigma\sqrt{2}S(\frac{\lambda}{\theta},\frac{b}{\sigma},\frac{r}{\sigma})} dz$$

7 $U_s = \sigma e^{-\theta s} W_{e^{2\theta s}-1}$ $\sigma, \theta > 0$ $\tau \sim \text{Exp}(\lambda)$, independent of U

1.17.1 $\mathbf{E}_x\big\{e^{-\gamma \ell(\tau,r)}; a < \inf_{0\le s\le\tau} U_s\big\}$

$a < r$

$a \le x$, $x \le r$

$$= 1 - \frac{S(\frac{\lambda}{\theta},\frac{r}{\sigma},\frac{x}{\sigma}) + S(\frac{\lambda}{\theta},\frac{x}{\sigma},\frac{a}{\sigma})}{S(\frac{\lambda}{\theta},\frac{r}{\sigma},\frac{a}{\sigma})}$$
$$+ \frac{\theta\sigma\sqrt{2}\big(e^{a^2/4\sigma^2} D_{-\lambda/\theta}(\frac{a}{\sigma}) - e^{r^2/4\sigma^2} D_{-\lambda/\theta}(\frac{r}{\sigma})\big) S(\frac{\lambda}{\theta},\frac{x}{\sigma},\frac{a}{\sigma})}{\big(\gamma\sqrt{\pi} e^{-r^2/4\sigma^2} D_{-\lambda/\theta}(\frac{r}{\sigma}) S(\frac{\lambda}{\theta},\frac{r}{\sigma},\frac{a}{\sigma}) + \theta\sigma\sqrt{2} e^{a^2/4\sigma^2} D_{-\lambda/\theta}(\frac{a}{\sigma})\big) S(\frac{\lambda}{\theta},\frac{r}{\sigma},\frac{a}{\sigma})}$$

$r \le x$

$$= 1 - \frac{\big(\gamma\sqrt{\pi} e^{-r^2/2\sigma^2} S(\frac{\lambda}{\theta},\frac{r}{\sigma},\frac{a}{\sigma}) + \theta\sigma\sqrt{2}\big) e^{x^2/4\sigma^2} D_{-\lambda/\theta}(\frac{x}{\sigma})}{\gamma\sqrt{\pi} e^{-r^2/4\sigma^2} D_{-\lambda/\theta}(\frac{r}{\sigma}) S(\frac{\lambda}{\theta},\frac{r}{\sigma},\frac{a}{\sigma}) + \theta\sigma\sqrt{2} e^{a^2/4\sigma^2} D_{-\lambda/\theta}(\frac{a}{\sigma})}$$

1.17.2 $\mathbf{P}_x\big(\ell(\tau,r) \in dy, a < \inf_{0\le s\le\tau} U_s\big) = \exp\Big(-\frac{y\theta\sigma\sqrt{2} e^{(a^2+r^2)/4\sigma^2} D_{-\lambda/\theta}(\frac{a}{\sigma})}{\sqrt{\pi} D_{-\lambda/\theta}(\frac{r}{\sigma}) S(\frac{\lambda}{\theta},\frac{r}{\sigma},\frac{a}{\sigma})}\Big)$

$a < r$

$$\times \begin{cases} \dfrac{\theta\sigma\sqrt{2}\big(e^{a^2/4\sigma^2} D_{-\lambda/\theta}(\frac{a}{\sigma}) - e^{r^2/4\sigma^2} D_{-\lambda/\theta}(\frac{r}{\sigma})\big) S(\frac{\lambda}{\theta},\frac{x}{\sigma},\frac{a}{\sigma})}{\sqrt{\pi} e^{-r^2/4\sigma^2} D_{-\lambda/\theta}(\frac{r}{\sigma}) S^2(\frac{\lambda}{\theta},\frac{r}{\sigma},\frac{a}{\sigma})} dy, & a \le x \le r \\ \dfrac{\theta\sigma\sqrt{2}\big(e^{a^2/4\sigma^2} D_{-\lambda/\theta}(\frac{a}{\sigma}) - e^{r^2/4\sigma^2} D_{-\lambda/\theta}(\frac{r}{\sigma})\big) e^{x^2/4\sigma^2} D_{-\lambda/\theta}(\frac{x}{\sigma})}{\sqrt{\pi} D^2_{-\lambda/\theta}(\frac{r}{\sigma}) S(\frac{\lambda}{\theta},\frac{r}{\sigma},\frac{a}{\sigma})} dy, & r \le x \end{cases}$$

(1)
$$\mathbf{P}_x\big(\ell(\tau,r) = 0, a < \inf_{0\le s\le\tau} U_s\big) = \begin{cases} 1 - \dfrac{S(\frac{\lambda}{\theta},\frac{r}{\sigma},\frac{x}{\sigma}) + S(\frac{\lambda}{\theta},\frac{x}{\sigma},\frac{a}{\sigma})}{S(\frac{\lambda}{\theta},\frac{r}{\sigma},\frac{a}{\sigma})}, & a \le x \le r \\ 1 - \dfrac{e^{x^2/4\sigma^2} D_{-\lambda/\theta}(\frac{x}{\sigma})}{e^{r^2/4\sigma^2} D_{-\lambda/\theta}(\frac{r}{\sigma})}, & r \le x \end{cases}$$

1.17.5 $\mathbf{E}_x\big\{e^{-\gamma \ell(\tau,r)}; a < \inf_{0\le s\le\tau} U_s, U_\tau \in dz\big\}$ $a < r \wedge z$

$z \le r$, $x \le r$

$$= \frac{\lambda\sqrt{\pi} e^{-z^2/2\sigma^2} S(\frac{\lambda}{\theta},\frac{r}{\sigma},\frac{z+x+|z-x|}{2\sigma}) S(\frac{\lambda}{\theta},\frac{z+x-|z-x|}{2\sigma},\frac{a}{\sigma})}{\theta\sigma\sqrt{2} S(\frac{\lambda}{\theta},\frac{r}{\sigma},\frac{a}{\sigma})} dz$$
$$+ \frac{\lambda\sqrt{\pi} e^{-z^2/2\sigma^2} e^{r^2/4\sigma^2} D_{-\lambda/\theta}(\frac{r}{\sigma}) S(\frac{\lambda}{\theta},\frac{z}{\sigma},\frac{a}{\sigma}) S(\frac{\lambda}{\theta},\frac{x}{\sigma},\frac{a}{\sigma})}{\big(\gamma\sqrt{\pi} e^{-r^2/4\sigma^2} D_{-\lambda/\theta}(\frac{r}{\sigma}) S(\frac{\lambda}{\theta},\frac{r}{\sigma},\frac{a}{\sigma}) + \theta\sigma\sqrt{2} e^{a^2/4\sigma^2} D_{-\lambda/\theta}(\frac{a}{\sigma})\big) S(\frac{\lambda}{\theta},\frac{r}{\sigma},\frac{a}{\sigma})} dz$$

$z \le r$, $r \le x$

$$= \frac{\lambda\sqrt{\pi} e^{-z^2/2\sigma^2} e^{x^2/4\sigma^2} D_{-\lambda/\theta}(\frac{x}{\sigma}) S(\frac{\lambda}{\theta},\frac{z}{\sigma},\frac{a}{\sigma})}{\gamma\sqrt{\pi} e^{-r^2/4\sigma^2} D_{-\lambda/\theta}(\frac{r}{\sigma}) S(\frac{\lambda}{\theta},\frac{r}{\sigma},\frac{a}{\sigma}) + \theta\sigma\sqrt{2} e^{a^2/4\sigma^2} D_{-\lambda/\theta}(\frac{a}{\sigma})} dz$$

$r \le z$, $x \le r$

$$= \frac{\lambda\sqrt{\pi} e^{-z^2/4\sigma^2} D_{-\lambda/\theta}(\frac{z}{\sigma}) S(\frac{\lambda}{\theta},\frac{x}{\sigma},\frac{a}{\sigma})}{\gamma\sqrt{\pi} e^{-r^2/4\sigma^2} D_{-\lambda/\theta}(\frac{r}{\sigma}) S(\frac{\lambda}{\theta},\frac{r}{\sigma},\frac{a}{\sigma}) + \theta\sigma\sqrt{2} e^{a^2/4\sigma^2} D_{-\lambda/\theta}(\frac{a}{\sigma})} dz$$

7 $U_s = \sigma e^{-\theta s} W_{e^{2\theta s}-1}$ $\sigma, \theta > 0$ $\tau \sim \mathrm{Exp}(\lambda)$, independent of U

$r \le z$, $r \le x$

$$= \frac{\lambda\sqrt{\pi}e^{-z^2/2\sigma^2}e^{(z+x+|z-x|)^2/4\sigma^2}D_{-\lambda/\theta}(\frac{z+x+|z-x|}{2\sigma})S(\frac{\lambda}{\theta},\frac{z+x-|z-x|}{2\sigma},\frac{r}{\sigma})}{\theta\sigma\sqrt{2}e^{r^2/4\sigma^2}D_{-\lambda/\theta}(\frac{r}{\sigma})}dz$$

$$+\frac{\lambda\sqrt{\pi}e^{(x^2-z^2-r^2)/4\sigma^2}D_{-\lambda/\theta}(\frac{z}{\sigma})S(\frac{\lambda}{\theta},\frac{r}{\sigma},\frac{a}{\sigma})D_{-\lambda/\theta}(\frac{x}{\sigma})}{(\gamma\sqrt{\pi}e^{-r^2/4\sigma^2}D_{-\lambda/\theta}(\frac{r}{\sigma})S(\frac{\lambda}{\theta},\frac{r}{\sigma},\frac{a}{\sigma})+\theta\sigma\sqrt{2}e^{a^2/4\sigma^2}D_{-\lambda/\theta}(\frac{a}{\sigma}))D_{-\lambda/\theta}(\frac{r}{\sigma})}dz$$

1.17.6 $\mathbf{P}_x\big(\ell(\tau,r)\in dy, a<\inf_{0\le s\le\tau}U_s, U_\tau\in dz\big)$ $a < r\wedge z$

$z \le r$, $x \le r$

$$= \frac{\lambda e^{(r^2-z^2)/4\sigma^2}S(\frac{\lambda}{\theta},\frac{z}{\sigma},\frac{a}{\sigma})S(\frac{\lambda}{\theta},\frac{x}{\sigma},\frac{a}{\sigma})}{S^2(\frac{\lambda}{\sigma},\frac{r}{\sigma},\frac{a}{\sigma})}\exp\Big(-\frac{y\theta\sigma\sqrt{2}e^{(a^2+r^2)/4\sigma^2}D_{-\lambda/\theta}(\frac{a}{\sigma})}{\sqrt{\pi}D_{-\lambda/\theta}(\frac{r}{\sigma})S(\frac{\lambda}{\theta},\frac{r}{\sigma},\frac{a}{\sigma})}\Big)dydz$$

$z \le r$, $r \le x$

$$= \frac{\lambda e^{(x^2+r^2)/4\sigma^2}D_{-\lambda/\theta}(\frac{x}{\sigma})S(\frac{\lambda}{\theta},\frac{z}{\sigma},\frac{a}{\sigma})}{e^{z^2/2\sigma^2}D_{-\lambda/\theta}(\frac{r}{\sigma})S(\frac{\lambda}{\theta},\frac{r}{\sigma},\frac{a}{\sigma})}\exp\Big(-\frac{y\theta\sigma\sqrt{2}e^{(a^2+r^2)/4\sigma^2}D_{-\lambda/\theta}(\frac{a}{\sigma})}{\sqrt{\pi}D_{-\lambda/\theta}(\frac{r}{\sigma})S(\frac{\lambda}{\theta},\frac{r}{\sigma},\frac{a}{\sigma})}\Big)dydz$$

$r \le z$, $x \le r$

$$= \frac{\lambda e^{(r^2-z^2)/4\sigma^2}D_{-\lambda/\theta}(\frac{z}{\sigma})S(\frac{\lambda}{\theta},\frac{x}{\sigma},\frac{a}{\sigma})}{D_{-\lambda/\theta}(\frac{r}{\sigma})S(\frac{\lambda}{\theta},\frac{r}{\sigma},\frac{a}{\sigma})}\exp\Big(-\frac{y\theta\sigma\sqrt{2}e^{(a^2+r^2)/4\sigma^2}D_{-\lambda/\theta}(\frac{a}{\sigma})}{\sqrt{\pi}D_{-\lambda/\theta}(\frac{r}{\sigma})S(\frac{\lambda}{\theta},\frac{r}{\sigma},\frac{a}{\sigma})}\Big)dydz$$

$r \le z$, $r \le x$

$$= \frac{\lambda e^{(x^2-z^2)/4\sigma^2}D_{-\lambda/\theta}(\frac{z}{\sigma})D_{-\lambda/\theta}(\frac{x}{\sigma})}{D^2_{-\lambda/\theta}(\frac{r}{\sigma})}\exp\Big(-\frac{y\theta\sigma\sqrt{2}e^{(a^2+r^2)/4\sigma^2}D_{-\lambda/\theta}(\frac{a}{\sigma})}{\sqrt{\pi}D_{-\lambda/\theta}(\frac{r}{\sigma})S(\frac{\lambda}{\theta},\frac{r}{\sigma},\frac{a}{\sigma})}\Big)dydz$$

(1) $\mathbf{P}_x\big(\ell(\tau,r)=0, a<\inf_{0\le s\le\tau}U_s, U_\tau\in dz\big)$

$z \le r$, $x \le r$

$$= \frac{\lambda\sqrt{\pi}e^{-z^2/2\sigma^2}S(\frac{\lambda}{\theta},\frac{r}{\sigma},\frac{z+x+|z-x|}{2\sigma})S(\frac{\lambda}{\theta},\frac{z+x-|z-x|}{2\sigma},\frac{a}{\sigma})}{\theta\sigma\sqrt{2}S(\frac{\lambda}{\theta},\frac{r}{\sigma},\frac{a}{\sigma})}dz$$

$r \le z$, $r \le x$

$$= \frac{\lambda\sqrt{\pi}e^{-z^2/2\sigma^2}e^{(z+x+|z-x|)^2/4\sigma^2}D_{-\lambda/\theta}(\frac{z+x+|z-x|}{2\sigma})S(\frac{\lambda}{\theta},\frac{z+x-|z-x|}{2\sigma},\frac{r}{\sigma})}{\theta\sigma\sqrt{2}e^{r^2/4\sigma^2}D_{-\lambda/\theta}(\frac{r}{\sigma})}dz$$

1.20.3 $\mathbf{E}_x\Big\{\exp\Big(-\int_0^t\frac{\gamma^2\sigma^2\theta}{U_s^2}ds\Big);\ 0<\inf_{0\le s\le t}U_s\Big\}$

$0 < x$

$$= \frac{(4\sigma^2\,\mathrm{sh}(t\theta))^{1/4}\Gamma\Big(\frac34+\sqrt{\frac{1}{16}+\frac{\gamma^2}{4}}\Big)}{\sqrt{x}e^{-t\theta/4}\Gamma\Big(\sqrt{\frac14+\gamma^2}+1\Big)}\exp\Big(-\frac{x^2e^{-t\theta}}{8\sigma^2\,\mathrm{sh}(t\theta)}\Big)M_{-\frac14,\sqrt{\frac{1}{16}+\frac{\gamma^2}{4}}}\Big(\frac{x^2e^{-t\theta}}{4\sigma^2\,\mathrm{sh}(t\theta)}\Big)$$

1.20.4 $\mathbf{P}_x\Big(\int_0^t\frac{ds}{U_s^2}\in dy,\ 0<\inf_{0\le s\le t}U_s\Big)$

$0 < x$

7 $U_s = \sigma e^{-\theta s} W_{e^{2\theta s}-1}$ $\sigma, \theta > 0$ $\tau \sim \operatorname{Exp}(\lambda)$, independent of U

$$= \frac{\theta(4\sigma^2)^{5/4}\operatorname{sh}^{1/4}(t\theta)}{x^{1/2}} \exp\Big(\frac{(t-\sigma^2 y)\theta}{4} - \frac{x^2 e^{-t\theta}}{8\sigma^2\operatorname{sh}(t\theta)}\Big) \operatorname{m}_{4\sigma^2\theta y}\Big(\frac{1}{4}, \frac{x^2 e^{-t\theta}}{8\sigma^2\operatorname{sh}(t\theta)}\Big) dy$$

1.20.5 $0 < x$

$$\mathbf{E}_x\Big\{\exp\Big(-\int_0^\tau \frac{\gamma^2\sigma^2\theta}{U_s^2} ds\Big);\ 0 < \inf_{0\le s\le\tau} U_s,\ U_\tau \in dz\Big\} = \frac{\lambda\Gamma\Big(\sqrt{\frac{1}{16}+\frac{\gamma^2}{4}}+\frac14+\frac{\lambda}{2\theta}\Big)}{\theta\Gamma\Big(\sqrt{\frac14+\gamma^2}+1\Big)\sqrt{xz}}$$

$$\times e^{(x^2-z^2)/4\sigma^2} \begin{cases} M_{\frac{\theta-2\lambda}{4\theta},\sqrt{\frac{1}{16}+\frac{\gamma^2}{4}}}\Big(\frac{x^2}{2\sigma^2}\Big) W_{\frac{\theta-2\lambda}{4\theta},\sqrt{\frac{1}{16}+\frac{\gamma^2}{4}}}\Big(\frac{z^2}{2\sigma^2}\Big) dz, & 0 \le x \le z \\ W_{\frac{\theta-2\lambda}{4\theta},\sqrt{\frac{1}{16}+\frac{\gamma^2}{4}}}\Big(\frac{x^2}{2\sigma^2}\Big) M_{\frac{\theta-2\lambda}{4\theta},\sqrt{\frac{1}{16}+\frac{\gamma^2}{4}}}\Big(\frac{z^2}{2\sigma^2}\Big) dz, & z \le x \end{cases}$$

1.20.7 $0 < x$

$$\mathbf{E}_x\Big\{\exp\Big(-\int_0^t \frac{\gamma^2\sigma^2\theta}{U_s^2} ds\Big);\ 0 < \inf_{0\le s\le t} U_s,\ U_t \in dz\Big\}$$

$$= \frac{\sqrt{xz}}{2\sigma^2\operatorname{sh}(t\theta)} \exp\Big(\frac{t\theta}{2} + \frac{x^2-z^2}{4\sigma^2} - \frac{(x^2+z^2)\operatorname{ch}(t\theta)}{4\sigma^2\operatorname{sh}(t\theta)}\Big) I_{\sqrt{\frac14+\gamma^2}}\Big(\frac{xz}{2\sigma^2\operatorname{sh}(t\theta)}\Big) dz$$

1.20.8 $0 < x$

$$\mathbf{P}_x\Big(\int_0^t \frac{ds}{U_s^2} \in dy,\ 0 < \inf_{0\le s\le t} U_s, U_t \in dz\Big)$$

$$= \frac{\theta\sqrt{xz}}{2\operatorname{sh}(t\theta)} \exp\Big(\frac{t\theta}{2} - \frac{\sigma^2\theta y}{4} + \frac{x^2-z^2}{4\sigma^2} - \frac{(x^2+z^2)\operatorname{ch}(t\theta)}{4\sigma^2\operatorname{sh}(t\theta)}\Big) \operatorname{i}_{\sigma^2\theta y}\Big(\frac{xz}{2\sigma^2\operatorname{sh}(t\theta)}\Big) dydz$$

1.21.3 $0 < x$

$$\mathbf{E}_x\Big\{\exp\Big(-\int_0^t \Big(\frac{p^2\sigma^2\theta}{U_s^2} + \frac{\theta(q^2-1)}{4\sigma^2} U_s^2\Big) ds\Big);\ 0 < \inf_{0\le s\le t} U_s\Big\}$$

$$= \frac{(4\sigma^2\operatorname{sh}(t\theta q))^{1/4} e^{t\theta/2}}{(\operatorname{sh}(t\theta q) + q\operatorname{ch}(t\theta q))^{1/4}} \exp\Big(-\frac{x^2(2(q^2-1)\operatorname{sh}^2(t\theta q)+q^2)}{8\sigma^2\operatorname{sh}(t\theta q)(\operatorname{sh}(t\theta q)+q\operatorname{ch}(t\theta q))}\Big)$$

$$\times \frac{\Gamma\Big(\frac34 + \sqrt{\frac{1}{16}+\frac{p^2}{4}}\Big)}{\sqrt{x}\Gamma\Big(\sqrt{\frac14+p^2}+1\Big)} M_{-\frac14,\sqrt{\frac{1}{16}+\frac{p^2}{4}}}\Big(\frac{x^2q^2}{4\sigma^2\operatorname{sh}(t\theta q)(\operatorname{sh}(t\theta q)+q\operatorname{ch}(t\theta q))}\Big)$$

(1)

$$\mathbf{E}_x\Big\{\exp\Big(-\frac{\theta(q^2-1)}{4\sigma^2}\int_0^t U_s^2 ds\Big);\ \int_0^t \frac{ds}{U_s^2} \in dy,\ 0 < \inf_{0\le s\le t} U_s\Big\}$$

$$= \frac{\theta(4\sigma^2)^{5/4}\operatorname{sh}^{1/4}(t\theta q) e^{t\theta/2}}{x^{1/2}(\operatorname{sh}(t\theta q)+q\operatorname{ch}(t\theta q))^{1/4}} \operatorname{m}_{4\sigma^2\theta y}\Big(\frac14, \frac{x^2q^2}{8\sigma^2\operatorname{sh}(t\theta q)(\operatorname{sh}(t\theta q)+q\operatorname{ch}(t\theta q))}\Big)$$

$$\times \exp\Big(-\frac{\sigma^2\theta y}{4} - \frac{x^2(2(q^2-1)\operatorname{sh}^2(t\theta q)+q^2)}{8\sigma^2\operatorname{sh}(t\theta q)(\operatorname{sh}(t\theta q)+q\operatorname{ch}(t\theta q))}\Big) dy$$

7 $U_s = \sigma e^{-\theta s} W_{e^{2\theta s}-1}$ $\qquad \sigma, \theta > 0$ $\qquad \tau \sim \mathrm{Exp}(\lambda)$, independent of U

1.21.5 $\quad 0 < x \quad q > 1$

$$\mathbf{E}_x\Big\{\exp\Big(-\int_0^\tau \Big(\frac{p^2\sigma^2\theta}{U_s^2} + \frac{\theta(q^2-1)U_s^2}{4\sigma^2}\Big)ds\Big);\ 0 < \inf_{0\le s\le\tau} U_s, U_\tau \in dz\Big\}$$

$$= \frac{\lambda\Gamma\Big(\sqrt{\frac{1}{16}+\frac{p^2}{4}}+\frac{1}{2}+\frac{2\lambda-\theta}{4\theta q}\Big)}{\theta q\Gamma\Big(\sqrt{\frac{1}{4}+p^2}+1\Big)\sqrt{xz}} e^{(x^2-z^2)/4\sigma^2}$$

$$\times \begin{cases} M_{\frac{\theta-2\lambda}{4\theta q},\sqrt{\frac{1}{16}+\frac{p^2}{4}}}\Big(\frac{qx^2}{2\sigma^2}\Big) W_{\frac{\theta-2\lambda}{4\theta q},\sqrt{\frac{1}{16}+\frac{p^2}{4}}}\Big(\frac{qz^2}{2\sigma^2}\Big)dz, & 0 \le x \le z \\ W_{\frac{\theta-2\lambda}{4\theta q},\sqrt{\frac{1}{16}+\frac{p^2}{4}}}\Big(\frac{qx^2}{2\sigma^2}\Big) M_{\frac{\theta-2\lambda}{4\theta q},\sqrt{\frac{1}{16}+\frac{p^2}{4}}}\Big(\frac{qz^2}{2\sigma^2}\Big)dz, & z \le x \end{cases}$$

1.21.7 $\quad 0 < x$

$$\mathbf{E}_x\Big\{\exp\Big(-\int_0^t \Big(\frac{p^2\sigma^2\theta}{U_s^2} + \frac{\theta(q^2-1)}{4\sigma^2}U_s^2\Big)ds\Big);\ 0 < \inf_{0\le s\le t} U_s, U_t \in dz\Big\}$$

$$= \frac{q\sqrt{xz}}{2\sigma^2\,\mathrm{sh}(t\theta q)} \exp\Big(\frac{t\theta}{2} + \frac{x^2-z^2}{4\sigma^2} - \frac{(x^2+z^2)q\,\mathrm{ch}(t\theta q)}{4\sigma^2\,\mathrm{sh}(t\theta q)}\Big) I_{\sqrt{\frac{1}{4}+p^2}}\Big(\frac{xzq}{2\sigma^2\,\mathrm{sh}(t\theta q)}\Big)dz$$

(1)

$$\mathbf{E}_x\Big\{\exp\Big(-\frac{\theta(q^2-1)}{4\sigma^2}\int_0^t U_s^2 ds\Big);\ \int_0^t \frac{ds}{U_s^2} \in dy,\ 0 < \inf_{0\le s\le t} U_s,\ U_t \in dz\Big\}$$

$$= \frac{q\theta\sqrt{xz}}{2\,\mathrm{sh}(t\theta q)} \exp\Big(\frac{t\theta}{2} - \frac{\sigma^2\theta y}{4} + \frac{x^2-z^2}{4\sigma^2} - \frac{(x^2+z^2)q\,\mathrm{ch}(t\theta q)}{4\sigma^2\,\mathrm{sh}(t\theta q)}\Big)\mathrm{i}_{\sigma^2\theta y}\Big(\frac{xzq}{2\sigma^2\,\mathrm{sh}(t\theta q)}\Big)dydz$$

(2)

$$\mathbf{E}_x\Big\{\exp\Big(-\int_0^t \frac{p^2\sigma^2\theta}{U_s^2}ds\Big);\ \int_0^t U_s^2 ds \in dy,\ 0 < \inf_{0\le s\le t} U_s,\ U_t \in dz\Big\}$$

$$= \frac{\sqrt{xz}}{4\sigma^4\theta^2} \exp\Big(\frac{t\theta}{2} + \frac{x^2-z^2-y\theta}{4\sigma^2}\Big)\,\mathrm{is}_{y/2\sigma^2\theta}\Big(\sqrt{\tfrac{1}{4}+p^2}, t, 0, \frac{x^2+z^2}{4\sigma^2\theta}, \frac{xz}{4\sigma^2\theta}\Big)dydz$$

1.21.8 $\quad 0 < x$

$$\mathbf{P}_x\Big(\int_0^t U_s^2 ds \in dg,\ \int_0^t \frac{ds}{U_s^2} \in dy,\ 0 < \inf_{0\le s\le t} U_s,\ U_t \in dz\Big)$$

$$= \frac{\sqrt{xz}}{4\sigma^2\theta} \exp\Big(\frac{t\theta}{2} - \frac{\sigma^2\theta y}{4} + \frac{x^2-z^2-g\theta}{4\sigma^2}\Big)\,\mathrm{ei}_{\frac{g}{2\sigma^2\theta}}\Big(\sigma^2\theta y, t, \frac{x^2+z^2}{4\sigma^2\theta}, \frac{xz}{2\sigma^2\theta}\Big)dgdydz$$

1.23.1 $\quad r < b$

$$\mathbf{E}_x\Big\{\exp\Big(-\int_0^\tau \big(p\mathbb{I}_{(-\infty,r)}(U_s) + q\mathbb{I}_{[r,\infty)}(U_s)\big)ds\Big);\ \sup_{0\le s\le\tau} U_s < b\Big\}$$

$x \le r$

$$= \frac{\lambda}{\lambda+p} + \frac{\lambda(\lambda+p)^{-1}\big((p-q)C(\frac{\lambda+q}{\theta},\frac{r}{\sigma},\frac{b}{\sigma}) - (\lambda+p)\sqrt{2/\pi}e^{r^2/2\sigma^2}\big)D_{-\frac{\lambda+p}{\theta}}(-\frac{x}{\sigma})e^{x^2/4\sigma^2}}{(\lambda+q)\big(S(\frac{\lambda+q}{\theta},\frac{b}{\sigma},\frac{r}{\sigma})\frac{\lambda+p}{\theta}D_{-\frac{\theta+\lambda+p}{\theta}}(-\frac{r}{\sigma}) + C(\frac{\lambda+q}{\theta},\frac{r}{\sigma},\frac{b}{\sigma})D_{-\frac{\lambda+p}{\theta}}(-\frac{r}{\sigma})\big)e^{r^2/4\sigma^2}}$$

7 $U_s = \sigma e^{-\theta s} W_{e^{2\theta s}-1}$ $\sigma, \theta > 0$ $\tau \sim \text{Exp}(\lambda)$, independent of U

$r \le x$, $x \le b$

$$= \frac{\lambda}{\lambda+q}\Big(1 - \frac{S(\frac{\lambda+q}{\theta}, \frac{x}{\sigma}, \frac{r}{\sigma})}{S(\frac{\lambda+q}{\theta}, \frac{b}{\sigma}, \frac{r}{\sigma})}\Big)$$

$$- \frac{\lambda(\lambda+q)^{-1}\sqrt{2/\pi}e^{r^2/2\sigma^2} D_{-\frac{\lambda+p}{\theta}}(-\frac{r}{\sigma}) S(\frac{\lambda+q}{\theta}, \frac{b}{\sigma}, \frac{x}{\sigma})}{(S(\frac{\lambda+q}{\theta}, \frac{b}{\sigma}, \frac{r}{\sigma})\frac{\lambda+p}{\theta} D_{-\frac{\theta+\lambda+p}{\theta}}(-\frac{r}{\sigma}) + C(\frac{\lambda+q}{\theta}, \frac{r}{\sigma}, \frac{b}{\sigma}) D_{-\frac{\lambda+p}{\theta}}(-\frac{r}{\sigma})) S(\frac{\lambda+q}{\theta}, \frac{b}{\sigma}, \frac{r}{\sigma})}$$

$$+ \frac{\lambda(q-p) D_{-\frac{\theta+\lambda+p}{\theta}}(-\frac{r}{\sigma}) S(\frac{\lambda+q}{\theta}, \frac{b}{\sigma}, \frac{x}{\sigma})}{(\lambda+q)(S(\frac{\lambda+q}{\theta}, \frac{b}{\sigma}, \frac{r}{\sigma})\frac{\lambda+p}{\theta} D_{-\frac{\theta+\lambda+p}{\theta}}(-\frac{r}{\sigma}) + C(\frac{\lambda+q}{\theta}, \frac{r}{\sigma}, \frac{b}{\sigma}) D_{-\frac{\lambda+p}{\theta}}(-\frac{r}{\sigma}))}$$

1.23.5 $r < b$, $z < b$

$$\mathbf{E}_x\Big\{\exp\Big(-\int_0^\tau \big(p\mathbb{I}_{(-\infty,r)}(U_s) + q\mathbb{I}_{[r,\infty)}(U_s)\big)ds\Big);\ \sup_{0\le s\le\tau} U_s < b,\ U_\tau \in dz\Big\}$$

$z \le r$, $x \le r$

$$= \frac{\lambda\sqrt{\pi}e^{-z^2/2\sigma^2} S(\frac{\lambda+p}{\theta}, \frac{r}{\sigma}, \frac{z+x+|z-x|}{2\sigma}) e^{(z+x-|z-x|)^2/4\sigma^2} D_{-\frac{\lambda+p}{\theta}}(-\frac{z+x-|z-x|}{2\sigma})}{\theta\sigma\sqrt{2}e^{r^2/4\sigma^2} D_{-\frac{\lambda+p}{\theta}}(-\frac{r}{\sigma})} dz$$

$$+ \frac{\lambda e^{(x^2-z^2)/4\sigma^2} S(\frac{\lambda+q}{\theta}, \frac{b}{\sigma}, \frac{r}{\sigma}) D_{-\frac{\lambda+p}{\theta}}(-\frac{z}{\sigma}) D_{-\frac{\lambda+p}{\theta}}(-\frac{x}{\sigma})}{\theta\sigma(S(\frac{\lambda+q}{\theta}, \frac{b}{\sigma}, \frac{r}{\sigma})\frac{\lambda+p}{\theta} D_{-\frac{\theta+\lambda+p}{\theta}}(-\frac{r}{\sigma}) + C(\frac{\lambda+q}{\theta}, \frac{r}{\sigma}, \frac{b}{\sigma}) D_{-\frac{\lambda+p}{\theta}}(-\frac{r}{\sigma})) D_{-\frac{\lambda+p}{\theta}}(-\frac{r}{\sigma})} dz$$

$z \le r$, $r \le x$

$$= \frac{\lambda e^{(r^2-z^2)/4\sigma^2} D_{-\frac{\lambda+p}{\theta}}(-\frac{z}{\sigma}) S(\frac{\lambda+q}{\theta}, \frac{b}{\sigma}, \frac{x}{\sigma})}{\theta\sigma(S(\frac{\lambda+q}{\theta}, \frac{b}{\sigma}, \frac{r}{\sigma})\frac{\lambda+p}{\theta} D_{-\frac{\theta+\lambda+p}{\theta}}(-\frac{r}{\sigma}) + C(\frac{\lambda+q}{\theta}, \frac{r}{\sigma}, \frac{b}{\sigma}) D_{-\frac{\lambda+p}{\theta}}(-\frac{r}{\sigma}))} dz$$

$r \le z$, $x \le r$

$$= \frac{\lambda e^{-z^2/2\sigma^2} e^{(x^2+r^2)/4\sigma^2} S(\frac{\lambda+q}{\theta}, \frac{b}{\sigma}, \frac{z}{\sigma}) D_{-\frac{\lambda+p}{\theta}}(-\frac{x}{\sigma})}{\theta\sigma(S(\frac{\lambda+q}{\theta}, \frac{b}{\sigma}, \frac{r}{\sigma})\frac{\lambda+p}{\theta} D_{-\frac{\theta+\lambda+p}{\theta}}(-\frac{r}{\sigma}) + C(\frac{\lambda+q}{\theta}, \frac{r}{\sigma}, \frac{b}{\sigma}) D_{-\frac{\lambda+p}{\theta}}(-\frac{r}{\sigma}))} dz$$

$r \le z$, $r \le x$

$$= \frac{\lambda\sqrt{\pi}e^{-z^2/2\sigma^2} S(\frac{\lambda+q}{\theta}, \frac{b}{\sigma}, \frac{z+x+|z-x|}{2\sigma}) S(\frac{\lambda+q}{\theta}, \frac{z+x-|z-x|}{2\sigma}, \frac{r}{\sigma})}{\theta\sigma\sqrt{2}S(\frac{\lambda+q}{\theta}, \frac{b}{\sigma}, \frac{r}{\sigma})} dz$$

$$+ \frac{\lambda e^{(r^2-z^2)/2\sigma^2} D_{-\frac{\lambda+q}{\theta}}(-\frac{r}{\sigma}) S(\frac{\lambda+q}{\theta}, \frac{b}{\sigma}, \frac{z}{\sigma}) S(\frac{\lambda+q}{\theta}, \frac{b}{\sigma}, \frac{x}{\sigma})}{\theta\sigma(S(\frac{\lambda+q}{\theta}, \frac{b}{\sigma}, \frac{r}{\sigma})\frac{\lambda+p}{\theta} D_{-\frac{\theta+\lambda+p}{\theta}}(-\frac{r}{\sigma}) + C(\frac{\lambda+q}{\theta}, \frac{r}{\sigma}, \frac{b}{\sigma}) D_{-\frac{\lambda+p}{\theta}}(-\frac{r}{\sigma})) S(\frac{\lambda+q}{\theta}, \frac{b}{\sigma}, \frac{r}{\sigma})} dz$$

1.24.1 $a < r$

$$\mathbf{E}_x\Big\{\exp\Big(-\int_0^\tau \big(p\mathbb{I}_{(-\infty,r)}(U_s) + q\mathbb{I}_{[r,\infty)}(U_s)\big)ds\Big);\ a < \inf_{0\le s\le\tau} U_s\Big\}$$

$a \le x$, $x \le r$

$$= \frac{\lambda}{\lambda+p}\Big(1 - \frac{S(\frac{\lambda+p}{\theta}, \frac{r}{\sigma}, \frac{x}{\sigma})}{S(\frac{\lambda+p}{\theta}, \frac{r}{\sigma}, \frac{a}{\sigma})}\Big)$$

7 $U_s = \sigma e^{-\theta s} W_{e^{2\theta s}-1}$ $\quad \sigma, \theta > 0 \quad$ $\tau \sim \text{Exp}(\lambda)$, independent of U

$$+\frac{\lambda(p-q)D_{-\frac{\theta+\lambda+p}{\theta}}(\frac{r}{\sigma})S(\frac{\lambda+q}{\theta},\frac{x}{\sigma},\frac{a}{\sigma})}{(\lambda+p)(D_{-\frac{\lambda+q}{\theta}}(\frac{r}{\sigma})C(\frac{\lambda+q}{\theta},\frac{r}{\sigma},\frac{a}{\sigma})+\frac{\lambda+q}{\theta}D_{-\frac{\theta+\lambda+q}{\theta}}(\frac{r}{\sigma})S(\frac{\lambda+p}{\theta},\frac{r}{\sigma},\frac{a}{\sigma}))}$$

$$-\frac{\lambda(\lambda+p)^{-1}\sqrt{2/\pi}e^{r^2/2\sigma^2}D_{-\frac{\lambda+q}{\theta}}(\frac{r}{\sigma})S(\frac{\lambda+p}{\theta},\frac{x}{\sigma},\frac{a}{\sigma})}{(D_{-\frac{\lambda+q}{\theta}}(\frac{r}{\sigma})C(\frac{\lambda+p}{\theta},\frac{r}{\sigma},\frac{a}{\sigma})+\frac{\lambda+q}{\theta}D_{-\frac{\theta+\lambda+q}{\theta}}(\frac{r}{\sigma})S(\frac{\lambda+p}{\theta},\frac{r}{\sigma},\frac{a}{\sigma}))S(\frac{\lambda+p}{\theta},\frac{r}{\sigma},\frac{a}{\sigma})}$$

$r \le x$

$$=\frac{\lambda}{\lambda+q}+\frac{\lambda(\lambda+p)^{-1}((q-p)C(\frac{\lambda+p}{\theta},\frac{r}{\sigma},\frac{a}{\sigma})-(\lambda+q)\sqrt{2/\pi}e^{r^2/2\sigma^2})D_{-\frac{\lambda+q}{\theta}}(\frac{x}{\sigma})e^{x^2/4\sigma^2}}{(\lambda+q)(D_{-\frac{\lambda+q}{\theta}}(\frac{r}{\sigma})C(\frac{\lambda+p}{\theta},\frac{r}{\sigma},\frac{a}{\sigma})+\frac{\lambda+q}{\theta}D_{-\frac{\theta+\lambda+q}{\theta}}(\frac{r}{\sigma})S(\frac{\lambda+p}{\theta},\frac{r}{\sigma},\frac{a}{\sigma}))e^{r^2/4\sigma^2}}$$

1.24.5 $\quad \mathbf{E}_x\Big\{\exp\Big(-\int_0^\tau \big(p\mathbb{I}_{(-\infty,r)}(U_s)+q\mathbb{I}_{[r,\infty)}(U_s)\big)ds\Big); a < \inf_{0\le s\le\tau} U_s,\ U_\tau \in dz\Big\}$

$a < r$

$a < z$

$z \le r$

$x \le r$

$$=\frac{\lambda\sqrt{\pi}e^{-z^2/2\sigma^2}S(\frac{\lambda+p}{\theta},\frac{r}{\sigma},\frac{z+x+|z-x|}{2\sigma})S(\frac{\lambda+p}{\theta},\frac{z+x-|z-x|}{2\sigma},\frac{a}{\sigma})}{\theta\sigma\sqrt{2}S(\frac{\lambda+p}{\theta},\frac{r}{\sigma},\frac{a}{\sigma})}dz$$

$$+\frac{\lambda e^{(r^2-z^2)/2\sigma^2}D_{-\frac{\lambda+q}{\theta}}(\frac{r}{\sigma})S(\frac{\lambda+p}{\theta},\frac{z}{\sigma},\frac{a}{\sigma})S(\frac{\lambda+p}{\theta},\frac{x}{\sigma},\frac{a}{\sigma})}{\theta\sigma(D_{-\frac{\lambda+q}{\theta}}(\frac{r}{\sigma})C(\frac{\lambda+p}{\theta},\frac{r}{\sigma},\frac{a}{\sigma})+\frac{\lambda+q}{\theta}D_{-\frac{\theta+\lambda+q}{\theta}}(\frac{r}{\sigma})S(\frac{\lambda+p}{\theta},\frac{r}{\sigma},\frac{a}{\sigma}))S(\frac{\lambda+p}{\theta},\frac{r}{\sigma},\frac{a}{\sigma})}dz$$

$z \le r$

$r \le x$

$$=\frac{\lambda e^{-z^2/2\sigma^2}e^{(x^2+r^2)/4\sigma^2}S(\frac{\lambda+p}{\theta},\frac{z}{\sigma},\frac{a}{\sigma})D_{-\frac{\lambda+q}{\theta}}(\frac{x}{\sigma})}{\theta\sigma(D_{-\frac{\lambda+q}{\theta}}(\frac{r}{\sigma})C(\frac{\lambda+p}{\theta},\frac{r}{\sigma},\frac{a}{\sigma})+\frac{\lambda+q}{\theta}D_{-\frac{\theta+\lambda+q}{\theta}}(\frac{r}{\sigma})S(\frac{\lambda+p}{\theta},\frac{r}{\sigma},\frac{a}{\sigma}))}dz$$

$r \le z$

$x \le r$

$$=\frac{\lambda e^{(r^2-z^2)/4\sigma^2}D_{-\frac{\lambda+q}{\theta}}(\frac{z}{\sigma})S(\frac{\lambda+p}{\theta},\frac{x}{\sigma},\frac{a}{\sigma})}{\theta\sigma(D_{-\frac{\lambda+q}{\theta}}(\frac{r}{\sigma})C(\frac{\lambda+p}{\theta},\frac{r}{\sigma},\frac{a}{\sigma})+\frac{\lambda+q}{\theta}D_{-\frac{\theta+\lambda+q}{\theta}}(\frac{r}{\sigma})S(\frac{\lambda+p}{\theta},\frac{r}{\sigma},\frac{a}{\sigma}))}dz$$

$r \le z$

$r \le x$

$$=\frac{\lambda\sqrt{\pi}e^{-z^2/2\sigma^2}e^{(z+x+|z-x|)^2/4\sigma^2}D_{-\frac{\lambda+q}{\theta}}(\frac{z+x+|z-x|}{2\sigma})S(\frac{\lambda+q}{\theta},\frac{z+x-|z-x|}{2\sigma},\frac{r}{\sigma})}{\theta\sigma\sqrt{2}e^{r^2/4\sigma^2}D_{-\frac{\lambda+q}{\theta}}(\frac{r}{\sigma})}dz$$

$$+\frac{\lambda e^{(x^2-z^2)/4\sigma^2}S(\frac{\lambda+p}{\theta},\frac{r}{\sigma},\frac{a}{\sigma})D_{-\frac{\lambda+q}{\theta}}(\frac{z}{\sigma})D_{-\frac{\lambda+q}{\theta}}(\frac{x}{\sigma})}{\theta\sigma(D_{-\frac{\lambda+q}{\theta}}(\frac{r}{\sigma})C(\frac{\lambda+p}{\theta},\frac{r}{\sigma},\frac{a}{\sigma})+\frac{\lambda+q}{\theta}D_{-\frac{\theta+\lambda+q}{\theta}}(\frac{r}{\sigma})S(\frac{\lambda+p}{\theta},\frac{r}{\sigma},\frac{a}{\sigma}))D_{-\frac{\lambda+q}{\theta}}(\frac{r}{\sigma})}dz$$

1.25.1 $\quad \mathbf{E}_x\{e^{-\gamma\ell(\tau,r)}; a < \inf_{0\le s\le\tau} U_s,\ \sup_{0\le s\le\tau} U_s < b\}$ $\quad a < r < b$

$a \le x$

$x \le r$

$$=1-\frac{S(\frac{\lambda}{\theta},\frac{r}{\sigma},\frac{x}{\sigma})+S(\frac{\lambda}{\theta},\frac{x}{\sigma},\frac{a}{\sigma})}{S(\frac{\lambda}{\theta},\frac{r}{\sigma},\frac{a}{\sigma})}$$

7 $U_s = \sigma e^{-\theta s} W_{e^{2\theta s}-1}$ $\sigma, \theta > 0$ $\tau \sim \text{Exp}(\lambda)$, independent of U

$$+\frac{\theta\sigma\sqrt{2}\big(S(\frac{\lambda}{\theta},\frac{b}{\sigma},\frac{a}{\sigma})-S(\frac{\lambda}{\theta},\frac{b}{\sigma},\frac{r}{\sigma})-S(\frac{\lambda}{\theta},\frac{r}{\sigma},\frac{a}{\sigma})\big)S(\frac{\lambda}{\theta},\frac{x}{\sigma},\frac{a}{\sigma})}{\big(\gamma\sqrt{\pi}e^{-r^2/2\sigma^2}S(\frac{\lambda}{\theta},\frac{b}{\sigma},\frac{r}{\sigma})S(\frac{\lambda}{\theta},\frac{r}{\sigma},\frac{a}{\sigma})+\theta\sigma\sqrt{2}S(\frac{\lambda}{\theta},\frac{b}{\sigma},\frac{a}{\sigma})\big)S(\frac{\lambda}{\theta},\frac{r}{\sigma},\frac{a}{\sigma})}$$

$r \le x$, $x \le b$

$$=1-\frac{S(\frac{\lambda}{\theta},\frac{b}{\sigma},\frac{x}{\sigma})+S(\frac{\lambda}{\theta},\frac{x}{\sigma},\frac{r}{\sigma})}{S(\frac{\lambda}{\theta},\frac{b}{\sigma},\frac{r}{\sigma})}$$

$$+\frac{\theta\sigma\sqrt{2}\big(S(\frac{\lambda}{\theta},\frac{b}{\sigma},\frac{a}{\sigma})-S(\frac{\lambda}{\theta},\frac{b}{\sigma},\frac{r}{\sigma})-S(\frac{\lambda}{\theta},\frac{r}{\sigma},\frac{a}{\sigma})\big)S(\frac{\lambda}{\theta},\frac{b}{\sigma},\frac{x}{\sigma})}{\big(\gamma\sqrt{\pi}e^{-r^2/2\sigma^2}S(\frac{\lambda}{\theta},\frac{b}{\sigma},\frac{r}{\sigma})S(\frac{\lambda}{\theta},\frac{r}{\sigma},\frac{a}{\sigma})+\theta\sigma\sqrt{2}S(\frac{\lambda}{\theta},\frac{b}{\sigma},\frac{a}{\sigma})\big)S(\frac{\lambda}{\theta},\frac{b}{\sigma},\frac{r}{\sigma})}$$

1.25.2 $\mathbf{P}_x\big(\ell(\tau,r)\in dy, a<\inf_{0\le s\le\tau}U_s, \sup_{0\le s\le\tau}U_s<b\big)$

$$=\exp\Big(-\frac{y\theta\sigma\sqrt{2}e^{r^2/2\sigma^2}S(\frac{\lambda}{\theta},\frac{b}{\sigma},\frac{a}{\sigma})}{\sqrt{\pi}S(\frac{\lambda}{\theta},\frac{b}{\sigma},\frac{r}{\sigma})S(\frac{\lambda}{\theta},\frac{r}{\sigma},\frac{a}{\sigma})}\Big)$$

$$\times\begin{cases}\dfrac{\theta\sigma\sqrt{2}\big(S(\frac{\lambda}{\theta},\frac{b}{\sigma},\frac{a}{\sigma})-S(\frac{\lambda}{\theta},\frac{b}{\sigma},\frac{r}{\sigma})-S(\frac{\lambda}{\theta},\frac{r}{\sigma},\frac{a}{\sigma})\big)S(\frac{\lambda}{\theta},\frac{x}{\sigma},\frac{a}{\sigma})}{\sqrt{\pi}e^{-r^2/2\sigma^2}S(\frac{\lambda}{\theta},\frac{b}{\sigma},\frac{r}{\sigma})S^2(\frac{\lambda}{\theta},\frac{r}{\sigma},\frac{a}{\sigma})}dy, & a\le x\le r\\ \dfrac{\theta\sigma\sqrt{2}\big(S(\frac{\lambda}{\theta},\frac{b}{\sigma},\frac{a}{\sigma})-S(\frac{\lambda}{\theta},\frac{b}{\sigma},\frac{r}{\sigma})-S(\frac{\lambda}{\theta},\frac{r}{\sigma},\frac{a}{\sigma})\big)S(\frac{\lambda}{\theta},\frac{b}{\sigma},\frac{x}{\sigma})}{\sqrt{\pi}e^{-r^2/2\sigma^2}S^2(\frac{\lambda}{\theta},\frac{b}{\sigma},\frac{r}{\sigma})S(\frac{\lambda}{\theta},\frac{r}{\sigma},\frac{a}{\sigma})}dy, & r\le x\le b\end{cases}$$

(1) $\mathbf{P}_x\big(\ell(\tau,r)=0, a<\inf_{0\le s\le\tau}U_s, \sup_{0\le s\le\tau}U_s<b\big)$

$$=\begin{cases}1-\dfrac{S(\frac{\lambda}{\theta},\frac{r}{\sigma},\frac{x}{\sigma})+S(\frac{\lambda}{\theta},\frac{x}{\sigma},\frac{a}{\sigma})}{S(\frac{\lambda}{\theta},\frac{r}{\sigma},\frac{a}{\sigma})}, & a\le x\le r\\ 1-\dfrac{S(\frac{\lambda}{\theta},\frac{b}{\sigma},\frac{x}{\sigma})+S(\frac{\lambda}{\theta},\frac{x}{\sigma},\frac{r}{\sigma})}{S(\frac{\lambda}{\theta},\frac{b}{\sigma},\frac{r}{\sigma})}, & r\le x\le b\end{cases}$$

1.25.5 $\mathbf{E}_x\big\{e^{-\gamma\ell(\tau,r)}; a<\inf_{0\le s\le\tau}U_s, \sup_{0\le s\le\tau}U_s<b, U_\tau\in dz\big\}$ $a<r<b$

$z \le r$, $x \le r$

$$=\frac{\lambda\sqrt{\pi}e^{-z^2/2\sigma^2}S(\frac{\lambda}{\theta},\frac{r}{\sigma},\frac{z+x+|z-x|}{2\sigma})S(\frac{\lambda}{\theta},\frac{z+x-|z-x|}{2\sigma},\frac{a}{\sigma})}{\theta\sigma\sqrt{2}S(\frac{\lambda}{\theta},\frac{r}{\sigma},\frac{a}{\sigma})}dz$$

$$+\frac{\lambda\sqrt{\pi}e^{-z^2/2\sigma^2}S(\frac{\lambda}{\theta},\frac{b}{\sigma},\frac{r}{\sigma})S(\frac{\lambda}{\theta},\frac{z}{\sigma},\frac{a}{\sigma})S(\frac{\lambda}{\theta},\frac{x}{\sigma},\frac{a}{\sigma})}{\big(\gamma\sqrt{\pi}e^{-r^2/2\sigma^2}S(\frac{\lambda}{\theta},\frac{b}{\sigma},\frac{r}{\sigma})S(\frac{\lambda}{\theta},\frac{r}{\sigma},\frac{a}{\sigma})+\theta\sigma\sqrt{2}S(\frac{\lambda}{\theta},\frac{b}{\sigma},\frac{a}{\sigma})\big)S(\frac{\lambda}{\theta},\frac{r}{\sigma},\frac{a}{\sigma})}dz$$

$z \le r$, $r \le x$

$$=\frac{\lambda\sqrt{\pi}e^{-z^2/2\sigma^2}S(\frac{\lambda}{\theta},\frac{b}{\sigma},\frac{x}{\sigma})S(\frac{\lambda}{\theta},\frac{z}{\sigma},\frac{a}{\sigma})}{\gamma\sqrt{\pi}e^{-r^2/2\sigma^2}S(\frac{\lambda}{\theta},\frac{b}{\sigma},\frac{r}{\sigma})S(\frac{\lambda}{\theta},\frac{r}{\sigma},\frac{a}{\sigma})+\theta\sigma\sqrt{2}S(\frac{\lambda}{\theta},\frac{b}{\sigma},\frac{a}{\sigma})}dz$$

7 $U_s = \sigma e^{-\theta s} W_{e^{2\theta s}-1}$ $\quad\sigma,\theta > 0$ $\quad\tau \sim \mathrm{Exp}(\lambda)$, independent of U

$r \le z$, $x \le r$

$$= \frac{\lambda\sqrt{\pi}e^{-z^2/2\sigma^2} S(\frac{\lambda}{\theta},\frac{b}{\sigma},\frac{z}{\sigma}) S(\frac{\lambda}{\theta},\frac{x}{\sigma},\frac{a}{\sigma})}{\gamma\sqrt{\pi}e^{-r^2/2\sigma^2} S(\frac{\lambda}{\theta},\frac{b}{\sigma},\frac{r}{\sigma}) S(\frac{\lambda}{\theta},\frac{r}{\sigma},\frac{a}{\sigma}) + \theta\sigma\sqrt{2} S(\frac{\lambda}{\theta},\frac{b}{\sigma},\frac{a}{\sigma})} dz$$

$r \le z$, $r \le x$

$$= \frac{\lambda\sqrt{\pi}e^{-z^2/2\sigma^2} S(\frac{\lambda}{\theta},\frac{b}{\sigma},\frac{z+x+|z-x|}{2\sigma}) S(\frac{\lambda}{\theta},\frac{z+x-|z-x|}{2\sigma},\frac{r}{\sigma})}{\theta\sigma\sqrt{2} S(\frac{\lambda}{\theta},\frac{b}{\sigma},\frac{r}{\sigma})} dz$$

$$+ \frac{\lambda\sqrt{\pi}e^{-z^2/2\sigma^2} S(\frac{\lambda}{\theta},\frac{b}{\sigma},\frac{z}{\sigma}) S(\frac{\lambda}{\theta},\frac{r}{\sigma},\frac{a}{\sigma}) S(\frac{\lambda}{\theta},\frac{b}{\sigma},\frac{x}{\sigma})}{(\gamma\sqrt{\pi}e^{-r^2/2\sigma^2} S(\frac{\lambda}{\theta},\frac{b}{\sigma},\frac{r}{\sigma}) S(\frac{\lambda}{\theta},\frac{r}{\sigma},\frac{a}{\sigma}) + \theta\sigma\sqrt{2} S(\frac{\lambda}{\theta},\frac{b}{\sigma},\frac{a}{\sigma})) S(\frac{\lambda}{\theta},\frac{b}{\sigma},\frac{r}{\sigma})} dz$$

1.25.6, $a < r$, $r < b$

$$\mathbf{P}_x\big(\ell(\tau,r) \in dy, a < \inf_{0\le s\le\tau} U_s, \sup_{0\le s\le\tau} U_s < b, U_\tau \in dz\big)$$

$z \le r$

$$= \begin{cases} \dfrac{\lambda e^{-z^2/2\sigma^2} S(\frac{\lambda}{\theta},\frac{z}{\sigma},\frac{a}{\sigma}) S(\frac{\lambda}{\theta},\frac{x}{\sigma},\frac{a}{\sigma})}{e^{-r^2/2\sigma^2} S^2(\frac{\lambda}{\theta},\frac{r}{\sigma},\frac{a}{\sigma})} \exp\Big(-\dfrac{y\theta\sigma\sqrt{2}e^{r^2/2\sigma^2} S(\frac{\lambda}{\theta},\frac{b}{\sigma},\frac{a}{\sigma})}{\sqrt{\pi} S(\frac{\lambda}{\theta},\frac{b}{\sigma},\frac{r}{\sigma}) S(\frac{\lambda}{\theta},\frac{r}{\sigma},\frac{a}{\sigma})}\Big) dy dz, & x \le r \\ \dfrac{\lambda e^{-z^2/2\sigma^2} S(\frac{\lambda}{\theta},\frac{b}{\sigma},\frac{x}{\sigma}) S(\frac{\lambda}{\theta},\frac{z}{\sigma},\frac{a}{\sigma})}{e^{-r^2/2\sigma^2} S(\frac{\lambda}{\theta},\frac{b}{\sigma},\frac{r}{\sigma}) S(\frac{\lambda}{\theta},\frac{r}{\sigma},\frac{a}{\sigma})} \exp\Big(-\dfrac{y\theta\sigma\sqrt{2}e^{r^2/2\sigma^2} S(\frac{\lambda}{\theta},\frac{b}{\sigma},\frac{a}{\sigma})}{\sqrt{\pi} S(\frac{\lambda}{\theta},\frac{b}{\sigma},\frac{r}{\sigma}) S(\frac{\lambda}{\theta},\frac{r}{\sigma},\frac{a}{\sigma})}\Big) dy dz, & r \le x \end{cases}$$

$r \le z$

$$= \begin{cases} \dfrac{\lambda e^{-z^2/2\sigma^2} S(\frac{\lambda}{\theta},\frac{b}{\sigma},\frac{z}{\sigma}) S(\frac{\lambda}{\theta},\frac{x}{\sigma},\frac{a}{\sigma})}{e^{-r^2/2\sigma^2} S(\frac{\lambda}{\theta},\frac{b}{\sigma},\frac{r}{\sigma}) S(\frac{\lambda}{\theta},\frac{r}{\sigma},\frac{a}{\sigma})} \exp\Big(-\dfrac{y\theta\sigma\sqrt{2}e^{r^2/2\sigma^2} S(\frac{\lambda}{\theta},\frac{b}{\sigma},\frac{a}{\sigma})}{\sqrt{\pi} S(\frac{\lambda}{\theta},\frac{b}{\sigma},\frac{r}{\sigma}) S(\frac{\lambda}{\theta},\frac{r}{\sigma},\frac{a}{\sigma})}\Big) dy dz, & x \le r \\ \dfrac{\lambda e^{-z^2/2\sigma^2} S(\frac{\lambda}{\theta},\frac{b}{\sigma},\frac{z}{\sigma}) S(\frac{\lambda}{\theta},\frac{b}{\sigma},\frac{x}{\sigma})}{e^{-r^2/2\sigma^2} S^2(\frac{\lambda}{\theta},\frac{b}{\sigma},\frac{r}{\sigma})} \exp\Big(-\dfrac{y\theta\sigma\sqrt{2}e^{r^2/2\sigma^2} S(\frac{\lambda}{\theta},\frac{b}{\sigma},\frac{a}{\sigma})}{\sqrt{\pi} S(\frac{\lambda}{\theta},\frac{b}{\sigma},\frac{r}{\sigma}) S(\frac{\lambda}{\theta},\frac{r}{\sigma},\frac{a}{\sigma})}\Big) dy dz, & r \le x \end{cases}$$

(1)

$$\mathbf{P}_x\big(\ell(\tau,r) = 0, a < \inf_{0\le s\le\tau} U_s, \sup_{0\le s\le\tau} U_s < b, U_\tau \in dz\big)$$

$z \le r$, $x \le r$

$$= \frac{\lambda\sqrt{\pi}e^{-z^2/2\sigma^2} S(\frac{\lambda}{\theta},\frac{r}{\sigma},\frac{z+x+|z-x|}{2\sigma}) S(\frac{\lambda}{\theta},\frac{z+x-|z-x|}{2\sigma},\frac{a}{\sigma})}{\theta\sigma\sqrt{2} S(\frac{\lambda}{\theta},\frac{r}{\sigma},\frac{a}{\sigma})} dz$$

$r \le z$, $r \le x$

$$= \frac{\lambda\sqrt{\pi}e^{-z^2/2\sigma^2} S(\frac{\lambda}{\theta},\frac{b}{\sigma},\frac{z+x+|z-x|}{2\sigma}) S(\frac{\lambda}{\theta},\frac{z+x-|z-x|}{2\sigma},\frac{r}{\sigma})}{\theta\sigma\sqrt{2} S(\frac{\lambda}{\theta},\frac{b}{\sigma},\frac{r}{\sigma})} dz$$

1.26.1, $a < r$, $r < b$

$$\mathbf{E}_x\Big\{\exp\Big(-\int_0^\tau \big(p\mathbb{I}_{(-\infty,r)}(U_s) + q\mathbb{I}_{[r,\infty)}(U_s)\big) ds\Big); a < \inf_{0\le s\le\tau} U_s, \sup_{0\le s\le\tau} U_s < b\Big\}$$

7 $U_s = \sigma e^{-\theta s} W_{e^{2\theta s}-1}$ $\sigma, \theta > 0$ $\tau \sim \text{Exp}(\lambda)$, independent of U

$a \le x$
$x \le r$

$$= \frac{\lambda}{\lambda+p}\Big(1 - \frac{S(\tfrac{\lambda+p}{\theta},\tfrac{r}{\sigma},\tfrac{x}{\sigma})}{S(\tfrac{\lambda+p}{\theta},\tfrac{r}{\sigma},\tfrac{a}{\sigma})}\Big)$$

$$-\frac{\lambda(\lambda+p)^{-1}\sqrt{2/\pi}e^{r^2/2\sigma^2}S(\tfrac{\lambda+q}{\theta},\tfrac{b}{\sigma},\tfrac{r}{\sigma})S(\tfrac{\lambda+p}{\theta},\tfrac{x}{\sigma},\tfrac{a}{\sigma})}{(S(\tfrac{\lambda+q}{\theta},\tfrac{b}{\sigma},\tfrac{r}{\sigma})C(\tfrac{\lambda+p}{\theta},\tfrac{r}{\sigma},\tfrac{a}{\sigma})+C(\tfrac{\lambda+q}{\theta},\tfrac{r}{\sigma},\tfrac{b}{\sigma})S(\tfrac{\lambda+p}{\theta},\tfrac{r}{\sigma},\tfrac{a}{\sigma}))S(\tfrac{\lambda+p}{\theta},\tfrac{r}{\sigma},\tfrac{a}{\sigma})}$$

$$+\frac{\lambda((p-q)C(\tfrac{\lambda+q}{\theta},\tfrac{r}{\sigma},\tfrac{b}{\sigma})-(\lambda+p)\sqrt{2/\pi}e^{r^2/2\sigma^2})S(\tfrac{\lambda+p}{\theta},\tfrac{x}{\sigma},\tfrac{a}{\sigma})}{(\lambda+p)(\lambda+q)(S(\tfrac{\lambda+q}{\theta},\tfrac{b}{\sigma},\tfrac{r}{\sigma})C(\tfrac{\lambda+p}{\theta},\tfrac{r}{\sigma},\tfrac{a}{\sigma})+C(\tfrac{\lambda+q}{\theta},\tfrac{r}{\sigma},\tfrac{b}{\sigma})S(\tfrac{\lambda+p}{\theta},\tfrac{r}{\sigma},\tfrac{a}{\sigma}))}$$

$r \le x$
$x \le b$

$$= \frac{\lambda}{\lambda+q}\Big(1 - \frac{S(\tfrac{\lambda+q}{\theta},\tfrac{x}{\sigma},\tfrac{r}{\sigma})}{S(\tfrac{\lambda+q}{\theta},\tfrac{b}{\sigma},\tfrac{r}{\sigma})}\Big)$$

$$-\frac{\lambda(\lambda+q)^{-1}\sqrt{2/\pi}e^{r^2/2\sigma^2}S(\tfrac{\lambda+p}{\theta},\tfrac{r}{\sigma},\tfrac{a}{\sigma})S(\tfrac{\lambda+q}{\theta},\tfrac{b}{\sigma},\tfrac{x}{\sigma})}{(S(\tfrac{\lambda+q}{\theta},\tfrac{b}{\sigma},\tfrac{r}{\sigma})C(\tfrac{\lambda+p}{\theta},\tfrac{r}{\sigma},\tfrac{a}{\sigma})+C(\tfrac{\lambda+q}{\theta},\tfrac{r}{\sigma},\tfrac{b}{\sigma})S(\tfrac{\lambda+p}{\theta},\tfrac{r}{\sigma},\tfrac{a}{\sigma}))S(\tfrac{\lambda+q}{\theta},\tfrac{b}{\sigma},\tfrac{r}{\sigma})}$$

$$+\frac{\lambda((q-p)C(\tfrac{\lambda+p}{\theta},\tfrac{r}{\sigma},\tfrac{a}{\sigma})-(\lambda+q)\sqrt{2/\pi}e^{r^2/2\sigma^2})S(\tfrac{\lambda+q}{\theta},\tfrac{b}{\sigma},\tfrac{x}{\sigma})}{(\lambda+p)(\lambda+q)(S(\tfrac{\lambda+q}{\theta},\tfrac{b}{\sigma},\tfrac{r}{\sigma})C(\tfrac{\lambda+p}{\theta},\tfrac{r}{\sigma},\tfrac{a}{\sigma})+C(\tfrac{\lambda+q}{\theta},\tfrac{r}{\sigma},\tfrac{b}{\sigma})S(\tfrac{\lambda+p}{\theta},\tfrac{r}{\sigma},\tfrac{a}{\sigma}))}$$

1.26.5
$r < b$
$a < r$

$$\mathbf{E}_x\Big\{\exp\Big(-\int_0^\tau \big(p\mathbb{I}_{(-\infty,r)}(U_s)+q\mathbb{I}_{[r,\infty)}(U_s)\big)ds\Big); \quad a < \inf_{0\le s\le\tau} U_s,\ \sup_{0\le s\le\tau} U_s < b,\ U_\tau \in dz\Big\}$$

$z \le r$
$x \le r$

$$= \frac{\lambda\sqrt{\pi}e^{-z^2/2\sigma^2}S(\tfrac{\lambda+p}{\theta},\tfrac{r}{\sigma},\tfrac{z+x+|z-x|}{2\sigma})S(\tfrac{\lambda+p}{\theta},\tfrac{z+x-|z-x|}{2\sigma},\tfrac{a}{\sigma})}{\theta\sigma\sqrt{2}S(\tfrac{\lambda+p}{\theta},\tfrac{r}{\sigma},\tfrac{a}{\sigma})}dz$$

$$+\frac{\lambda e^{(r^2-z^2)/2\sigma^2}S(\tfrac{\lambda+q}{\theta},\tfrac{b}{\sigma},\tfrac{r}{\sigma})S(\tfrac{\lambda+p}{\theta},\tfrac{z}{\sigma},\tfrac{a}{\sigma})S(\tfrac{\lambda+p}{\theta},\tfrac{x}{\sigma},\tfrac{a}{\sigma})}{\theta\sigma(S(\tfrac{\lambda+q}{\theta},\tfrac{b}{\sigma},\tfrac{r}{\sigma})C(\tfrac{\lambda+p}{\theta},\tfrac{r}{\sigma},\tfrac{a}{\sigma})+C(\tfrac{\lambda+q}{\theta},\tfrac{r}{\sigma},\tfrac{b}{\sigma})S(\tfrac{\lambda+p}{\theta},\tfrac{r}{\sigma},\tfrac{a}{\sigma}))S(\tfrac{\lambda+p}{\theta},\tfrac{r}{\sigma},\tfrac{a}{\sigma})}dz$$

$z \le r$
$r \le x$

$$= \frac{\lambda e^{(r^2-z^2)/2\sigma^2}S(\tfrac{\lambda+p}{\theta},\tfrac{z}{\sigma},\tfrac{a}{\sigma})S(\tfrac{\lambda+q}{\theta},\tfrac{b}{\sigma},\tfrac{x}{\sigma})}{\theta\sigma(S(\tfrac{\lambda+q}{\theta},\tfrac{b}{\sigma},\tfrac{r}{\sigma})C(\tfrac{\lambda+p}{\theta},\tfrac{r}{\sigma},\tfrac{a}{\sigma})+C(\tfrac{\lambda+q}{\theta},\tfrac{r}{\sigma},\tfrac{b}{\sigma})S(\tfrac{\lambda+p}{\theta},\tfrac{r}{\sigma},\tfrac{a}{\sigma}))}dz$$

$r \le z$
$x \le r$

$$= \frac{\lambda e^{(r^2-z^2)/2\sigma^2}S(\tfrac{\lambda+q}{\theta},\tfrac{b}{\sigma},\tfrac{z}{\sigma})S(\tfrac{\lambda+p}{\theta},\tfrac{x}{\sigma},\tfrac{a}{\sigma})}{\theta\sigma(S(\tfrac{\lambda+q}{\theta},\tfrac{b}{\sigma},\tfrac{r}{\sigma})C(\tfrac{\lambda+p}{\theta},\tfrac{r}{\sigma},\tfrac{a}{\sigma})+C(\tfrac{\lambda+q}{\theta},\tfrac{r}{\sigma},\tfrac{b}{\sigma})S(\tfrac{\lambda+p}{\theta},\tfrac{r}{\sigma},\tfrac{a}{\sigma}))}dz$$

$r \le z$
$r \le x$

$$= \frac{\lambda\sqrt{\pi}e^{-z^2/2\sigma^2}S(\tfrac{\lambda+q}{\theta},\tfrac{b}{\sigma},\tfrac{z+x+|z-x|}{2\sigma})S(\tfrac{\lambda+q}{\theta},\tfrac{z+x-|z-x|}{2\sigma},\tfrac{r}{\sigma})}{\theta\sigma\sqrt{2}S(\tfrac{\lambda+q}{\theta},\tfrac{b}{\sigma},\tfrac{r}{\sigma})}dz$$

$$+\frac{\lambda e^{(r^2-z^2)/2\sigma^2}S(\tfrac{\lambda+p}{\theta},\tfrac{r}{\sigma},\tfrac{a}{\sigma})S(\tfrac{\lambda+q}{\theta},\tfrac{b}{\sigma},\tfrac{z}{\sigma})S(\tfrac{\lambda+q}{\theta},\tfrac{b}{\sigma},\tfrac{x}{\sigma})}{\theta\sigma(S(\tfrac{\lambda+q}{\theta},\tfrac{b}{\sigma},\tfrac{r}{\sigma})C(\tfrac{\lambda+p}{\theta},\tfrac{r}{\sigma},\tfrac{a}{\sigma})+C(\tfrac{\lambda+q}{\theta},\tfrac{r}{\sigma},\tfrac{b}{\sigma})S(\tfrac{\lambda+p}{\theta},\tfrac{r}{\sigma},\tfrac{a}{\sigma}))S(\tfrac{\lambda+p}{\theta},\tfrac{b}{\sigma},\tfrac{r}{\sigma})}dz$$

7 $\quad U_s = \sigma e^{-\theta s} W_{e^{2\theta s}-1} \qquad \sigma, \theta > 0 \qquad \tau \sim \mathrm{Exp}(\lambda)$, independent of U

1.27.1

$$\mathbf{E}_x\Big\{\exp\Big(-\int_0^\tau \big(p\mathbb{1}_{(-\infty,r)}(U_s) + q\mathbb{1}_{[r,\infty)}(U_s)\big)ds\Big); \ell(\tau,r) \in dy\Big\}$$

$$= \exp\Bigg(-\frac{y\sigma(\lambda+p)D_{-\frac{\theta+\lambda+p}{\theta}}(-\frac{r}{\sigma})}{D_{-\frac{\lambda+p}{\theta}}(-\frac{r}{\sigma})} - \frac{y\sigma(\lambda+q)D_{-\frac{\theta+\lambda+q}{\theta}}(\frac{r}{\sigma})}{D_{-\frac{\lambda+q}{\theta}}(\frac{r}{\sigma})}\Bigg)$$

$$\times \begin{cases} \dfrac{\lambda\sigma e^{x^2/4\sigma^2} D_{-\frac{\lambda+p}{\theta}}(-\frac{x}{\sigma})}{e^{r^2/4\sigma^2} D_{-\frac{\lambda+p}{\theta}}(-\frac{r}{\sigma})}\Bigg(\dfrac{D_{-\frac{\theta+\lambda+p}{\theta}}(-\frac{r}{\sigma})}{D_{-\frac{\lambda+p}{\theta}}(-\frac{r}{\sigma})} + \dfrac{D_{-\frac{\theta+\lambda+q}{\theta}}(\frac{r}{\sigma})}{D_{-\frac{\lambda+q}{\theta}}(\frac{r}{\sigma})}\Bigg)dy, & x \le r \\[2ex] \dfrac{\lambda\sigma e^{x^2/4\sigma^2} D_{-\frac{\lambda+q}{\theta}}(\frac{x}{\sigma})}{e^{r^2/4\sigma^2} D_{-\frac{\lambda+q}{\theta}}(\frac{r}{\sigma})}\Bigg(\dfrac{D_{-\frac{\theta+\lambda+p}{\theta}}(-\frac{r}{\sigma})}{D_{-\frac{\lambda+p}{\theta}}(-\frac{r}{\sigma})} + \dfrac{D_{-\frac{\theta+\lambda+q}{\theta}}(\frac{r}{\sigma})}{D_{-\frac{\lambda+q}{\theta}}(\frac{r}{\sigma})}\Bigg)dy, & r \le x \end{cases}$$

(1)

$$\mathbf{E}_x\Big\{\exp\Big(-\int_0^\tau \big(p\mathbb{1}_{(-\infty,r)}(U_s) + q\mathbb{1}_{[r,\infty)}(U_s)\big)ds\Big); \ell(\tau,r) = 0\Big\}$$

$$= \begin{cases} \dfrac{\lambda}{\lambda+p}\Bigg(1 - \dfrac{e^{x^2/4\sigma^2} D_{-\frac{\lambda+p}{\theta}}(-\frac{x}{\sigma})}{e^{r^2/4\sigma^2} D_{-\frac{\lambda+p}{\theta}}(-\frac{r}{\sigma})}\Bigg), & x \le r \\[2ex] \dfrac{\lambda}{\lambda+q}\Bigg(1 - \dfrac{e^{x^2/4\sigma^2} D_{-\frac{\lambda+q}{\theta}}(\frac{x}{\sigma})}{e^{r^2/4\sigma^2} D_{-\frac{\lambda+q}{\theta}}(\frac{r}{\sigma})}\Bigg), & r \le x \end{cases}$$

1.27.5

$$\mathbf{E}_x\Big\{\exp\Big(-\int_0^\tau \big(p\mathbb{1}_{(-\infty,r)}(U_s) + q\mathbb{1}_{[r,\infty)}(U_s)\big)ds\Big); \ell(\tau,r) \in dy, U_\tau \in dz\Big\}$$

$z \le r$

$$= \exp\Bigg(-\frac{y\sigma(\lambda+p)D_{-\frac{\theta+\lambda+p}{\theta}}(-\frac{r}{\sigma})}{D_{-\frac{\lambda+p}{\theta}}(-\frac{r}{\sigma})} - \frac{y\sigma(\lambda+q)D_{-\frac{\theta+\lambda+q}{\theta}}(\frac{r}{\sigma})}{D_{-\frac{\lambda+q}{\theta}}(\frac{r}{\sigma})}\Bigg)$$

$$\times \begin{cases} \dfrac{\lambda e^{x^2/4\sigma^2} D_{-\frac{\lambda+p}{\theta}}(-\frac{x}{\sigma}) D_{-\frac{\lambda+p}{\theta}}(-\frac{z}{\sigma})}{e^{z^2/4\sigma^2} D^2_{-\frac{\lambda+p}{\theta}}(-\frac{r}{\sigma})} dydz, & x \le r \\[2ex] \dfrac{\lambda e^{x^2/4\sigma^2} D_{-\frac{\lambda+p}{\theta}}(-\frac{z}{\sigma}) D_{-\frac{\lambda+q}{\theta}}(\frac{x}{\sigma})}{e^{z^2/4\sigma^2} D_{-\frac{\lambda+p}{\theta}}(-\frac{r}{\sigma}) D_{-\frac{\lambda+q}{\theta}}(\frac{r}{\sigma})} dydz, & r \le x \end{cases}$$

$r \le z$

$$= \exp\Bigg(-\frac{y\sigma(\lambda+p)D_{-\frac{\theta+\lambda+p}{\theta}}(-\frac{r}{\sigma})}{D_{-\frac{\lambda+p}{\theta}}(-\frac{r}{\sigma})} - \frac{y\sigma(\lambda+q)D_{-\frac{\theta+\lambda+q}{\theta}}(\frac{r}{\sigma})}{D_{-\frac{\lambda+q}{\theta}}(\frac{r}{\sigma})}\Bigg)$$

$$\times \begin{cases} \dfrac{\lambda e^{x^2/4\sigma^2} D_{-\frac{\lambda+p}{\theta}}(-\frac{x}{\sigma}) D_{-\frac{\lambda+q}{\theta}}(\frac{z}{\sigma})}{e^{z^2/4\sigma^2} D_{-\frac{\lambda+p}{\theta}}(-\frac{r}{\sigma}) D_{-\frac{\lambda+q}{\theta}}(\frac{r}{\sigma})} dydz, & x \le r \\[2ex] \dfrac{\lambda e^{x^2/4\sigma^2} D_{-\frac{\lambda+q}{\theta}}(\frac{x}{\sigma}) D_{-\frac{\lambda+q}{\theta}}(\frac{z}{\sigma})}{e^{z^2/4\sigma^2} D^2_{-\frac{\lambda+q}{\theta}}(\frac{r}{\sigma})} dydz, & r \le x \end{cases}$$

7 $U_s = \sigma e^{-\theta s} W_{e^{2\theta s}-1}$ $\sigma, \theta > 0$ $\tau \sim \mathrm{Exp}(\lambda)$, independent of U

(1) $\mathbf{E}_x\Big\{\exp\Big(-\int_0^\tau \big(p\mathbb{I}_{(-\infty,r)}(U_s) + q\mathbb{I}_{[r,\infty)}(U_s)\big)ds\Big); \ell(\tau,r) = 0, U_\tau \in dz\Big\}$

$z \le r$, $x \le r$

$$= \frac{\lambda\sqrt{\pi}e^{-z^2/2\sigma^2}}{\theta\sigma\sqrt{2}}\Big\{F\big(\tfrac{\lambda+p}{\theta}, \tfrac{x}{\sigma}, \tfrac{z}{\sigma}\big) - \frac{\Gamma(\frac{\lambda+p}{\theta})D_{-\frac{\lambda+p}{\theta}}(-\frac{x}{\sigma})D_{-\frac{\lambda+p}{\theta}}(-\frac{z}{\sigma})D_{-\frac{\lambda+p}{\theta}}(\frac{r}{\sigma})}{e^{(-x^2-z^2)/4\sigma^2}D_{-\frac{\lambda+p}{\theta}}(-\frac{r}{\sigma})}\Big\}dz$$

$r \le z$, $r \le x$

$$= \frac{\lambda\sqrt{\pi}e^{-z^2/2\sigma^2}}{\theta\sigma\sqrt{2}}\Big\{F\big(\tfrac{\lambda+q}{\theta}, \tfrac{x}{\sigma}, \tfrac{z}{\sigma}\big) - \frac{\Gamma(\frac{\lambda+q}{\theta})D_{-\frac{\lambda+q}{\theta}}(\frac{x}{\sigma})D_{-\frac{\lambda+q}{\theta}}(\frac{z}{\sigma})D_{-\frac{\lambda+q}{\theta}}(-\frac{r}{\sigma})}{e^{(-x^2-z^2)/4\sigma^2}D_{-\frac{\lambda+q}{\theta}}(\frac{r}{\sigma})}\Big\}dz$$

1.28.1 $\mathbf{E}_x\big\{e^{-\gamma\check{H}(\tau)-\eta\hat{H}(\tau)}; \inf_{0\le s\le\tau} U_s \in da, \sup_{0\le s\le\tau} U_s \in db\big\}$

$a < x$, $x < b$

$$= \frac{\sqrt{2}e^{(b^2+a^2)/2\sigma^2}S(\frac{\lambda+\gamma+\eta}{\theta}, \frac{b}{\sigma}, \frac{x}{\sigma})(e^{-b^2/2\sigma^2}C(\frac{\lambda}{\theta}, \frac{b}{\sigma}, \frac{a}{\sigma}) - \sqrt{2}/\sqrt{\pi})}{\sigma^2\sqrt{\pi}S(\frac{\lambda+\gamma+\eta}{\theta}, \frac{b}{\sigma}, \frac{a}{\sigma})S(\frac{\lambda+\eta}{\theta}, \frac{b}{\sigma}, \frac{a}{\sigma})S(\frac{\lambda}{\theta}, \frac{b}{\sigma}, \frac{a}{\sigma})}dadb$$

$$+\frac{\sqrt{2}e^{(b^2+a^2)/2\sigma^2}S(\frac{\lambda+\gamma+\eta}{\theta}, \frac{x}{\sigma}, \frac{a}{\sigma})(e^{-a^2/2\sigma^2}C(\frac{\lambda}{\theta}, \frac{a}{\sigma}, \frac{b}{\sigma}) - \sqrt{2}/\sqrt{\pi})}{\sigma^2\sqrt{\pi}S(\frac{\lambda+\gamma+\eta}{\theta}, \frac{b}{\sigma}, \frac{a}{\sigma})S(\frac{\lambda+\gamma}{\theta}, \frac{b}{\sigma}, \frac{a}{\sigma})S(\frac{\lambda}{\theta}, \frac{b}{\sigma}, \frac{a}{\sigma})}dadb$$

1.28.5 $\mathbf{E}_x\big\{e^{-\gamma\check{H}(\tau)-\eta\hat{H}(\tau)}; \inf_{0\le s\le\tau} U_s \in da, \sup_{0\le s\le\tau} U_s \in db, U_\tau \in dz\big\}$

$a < x$, $x < b$

$$= \frac{\lambda\sqrt{2}e^{(b^2+a^2-z^2)/2\sigma^2}S(\frac{\lambda+\gamma+\eta}{\theta}, \frac{b}{\sigma}, \frac{x}{\sigma})S(\frac{\lambda}{\theta}, \frac{z}{\sigma}, \frac{a}{\sigma})}{\theta\sigma^3\sqrt{\pi}S(\frac{\lambda+\gamma+\eta}{\theta}, \frac{b}{\sigma}, \frac{a}{\sigma})S(\frac{\lambda+\eta}{\theta}, \frac{b}{\sigma}, \frac{a}{\sigma})S(\frac{\lambda}{\theta}, \frac{b}{\sigma}, \frac{a}{\sigma})}dadbdz$$

$$+\frac{\lambda\sqrt{2}e^{(b^2+a^2-z^2)/2\sigma^2}S(\frac{\lambda+\gamma+\eta}{\theta}, \frac{x}{\sigma}, \frac{a}{\sigma})S(\frac{\lambda}{\theta}, \frac{b}{\sigma}, \frac{z}{\sigma})}{\theta\sigma^3\sqrt{\pi}S(\frac{\lambda+\gamma+\eta}{\theta}, \frac{b}{\sigma}, \frac{a}{\sigma})S(\frac{\lambda+\gamma}{\theta}, \frac{b}{\sigma}, \frac{a}{\sigma})S(\frac{\lambda}{\theta}, \frac{b}{\sigma}, \frac{a}{\sigma})}dadbdz$$

1.29.1 $\mathbf{E}_x\Big\{\exp\Big(-\int_0^\tau \big(p\mathbb{I}_{(-\infty,r)}(U_s) + q\mathbb{I}_{[r,\infty)}(U_s)\big)ds\Big); a < \inf_{0\le s\le\tau} U_s,$

$\sup_{0\le s\le\tau} U_s < b, \ \ell(\tau,r) \in dy\Big\}$

$$= \lambda\theta\sigma\Big(\frac{C(\frac{\lambda+q}{\theta}, \frac{r}{\sigma}, \frac{b}{\sigma}) - \sqrt{2/\pi}e^{r^2/2\sigma^2}}{(\lambda+q)S(\frac{\lambda+q}{\theta}, \frac{b}{\sigma}, \frac{r}{\sigma})} + \frac{C(\frac{\lambda+p}{\theta}, \frac{r}{\sigma}, \frac{a}{\sigma}) - \sqrt{2/\pi}e^{r^2/2\sigma^2}}{(\lambda+p)S(\frac{\lambda+p}{\theta}, \frac{r}{\sigma}, \frac{a}{\sigma})}\Big)$$

$$\times\exp\Big(-\frac{y\theta\sigma C(\frac{\lambda+q}{\theta}, \frac{r}{\sigma}, \frac{b}{\sigma})}{S(\frac{\lambda+q}{\theta}, \frac{b}{\sigma}, \frac{r}{\sigma})} - \frac{y\theta\sigma C(\frac{\lambda+p}{\theta}, \frac{r}{\sigma}, \frac{a}{\sigma})}{S(\frac{\lambda+p}{\theta}, \frac{r}{\sigma}, \frac{a}{\sigma})}\Big)\begin{cases}\dfrac{S(\frac{\lambda+p}{\theta}, \frac{x}{\sigma}, \frac{a}{\sigma})}{S(\frac{\lambda+p}{\theta}, \frac{r}{\sigma}, \frac{a}{\sigma})}dy, & a \le x \le r\\[2ex] \dfrac{S(\frac{\lambda+q}{\theta}, \frac{b}{\sigma}, \frac{x}{\sigma})}{S(\frac{\lambda+q}{\theta}, \frac{b}{\sigma}, \frac{r}{\sigma})}dy, & r \le x \le b\end{cases}$$

7 $U_s=\sigma e^{-\theta s}W_{e^{2\theta s}-1}$ $\sigma,\theta>0$ $\tau\sim\operatorname{Exp}(\lambda)$, independent of U

(1) $$\mathbf{E}_x\Big\{\exp\Big(-\int_0^\tau\big(p\mathbb{I}_{(-\infty,r)}(U_s)+q\mathbb{I}_{[r,\infty)}(U_s)\big)ds\Big);a<\inf_{0\le s\le\tau}U_s,\ \sup_{0\le s\le\tau}U_s<b,\ \ell(\tau,r)=0\Big\}$$

$$=\begin{cases}\dfrac{\lambda}{\lambda+p}\Big(1-\dfrac{S(\frac{\lambda+p}{\theta},\frac{r}{\sigma},\frac{x}{\sigma})+S(\frac{\lambda+p}{\theta},\frac{x}{\sigma},\frac{a}{\sigma})}{S(\frac{\lambda+p}{\theta},\frac{r}{\sigma},\frac{a}{\sigma})}\Big), & a\le x\le r\\[2ex] \dfrac{\lambda}{\lambda+q}\Big(1-\dfrac{S(\frac{\lambda+q}{\theta},\frac{b}{\sigma},\frac{x}{\sigma})+S(\frac{\lambda+q}{\theta},\frac{x}{\sigma},\frac{r}{\sigma})}{S(\frac{\lambda+q}{\theta},\frac{b}{\sigma},\frac{r}{\sigma})}\Big), & r\le x\le b\end{cases}$$

1.29.5 $$\mathbf{E}_x\Big\{\exp\Big(-\int_0^\tau\big(p\mathbb{I}_{(-\infty,r)}(U_s)+q\mathbb{I}_{[r,\infty)}(U_s)\big)ds\Big);a<\inf_{0\le s\le\tau}U_s,\ \sup_{0\le s\le\tau}U_s<b,\ \ell(\tau,r)\in dy,\ U_\tau\in dz\Big\}$$

$$=\lambda\exp\Big(-\frac{y\theta\sigma C(\frac{\lambda+q}{\theta},\frac{r}{\sigma},\frac{b}{\sigma})}{S(\frac{\lambda+q}{\theta},\frac{b}{\sigma},\frac{r}{\sigma})}-\frac{y\theta\sigma C(\frac{\lambda+p}{\theta},\frac{r}{\sigma},\frac{a}{\sigma})}{S(\frac{\lambda+p}{\theta},\frac{r}{\sigma},\frac{a}{\sigma})}\Big)$$

$$\times\begin{cases}\dfrac{e^{-z^2/2\sigma^2}S(\frac{\lambda+p}{\theta},\frac{z}{\sigma},\frac{a}{\sigma})S(\frac{\lambda+p}{\theta},\frac{x}{\sigma},\frac{a}{\sigma})}{e^{-r^2/2\sigma^2}S^2(\frac{\lambda+p}{\theta},\frac{r}{\sigma},\frac{a}{\sigma})}dydz, & a\le x\le r,\ a\le z\le r\\[2ex] \dfrac{e^{-z^2/2\sigma^2}S(\frac{\lambda+q}{\theta},\frac{b}{\sigma},\frac{x}{\sigma})S(\frac{\lambda+p}{\theta},\frac{z}{\sigma},\frac{a}{\sigma})}{e^{-r^2/2\sigma^2}S(\frac{\lambda+q}{\theta},\frac{b}{\sigma},\frac{r}{\sigma})S(\frac{\lambda+p}{\theta},\frac{r}{\sigma},\frac{a}{\sigma})}dydz, & a\le z\le r\le x\le b\\[2ex] \dfrac{e^{-z^2/2\sigma^2}S(\frac{\lambda+q}{\theta},\frac{b}{\sigma},\frac{z}{\sigma})S(\frac{\lambda+p}{\theta},\frac{x}{\sigma},\frac{a}{\sigma})}{e^{-r^2/2\sigma^2}S(\frac{\lambda+q}{\theta},\frac{b}{\sigma},\frac{r}{\sigma})S(\frac{\lambda+p}{\theta},\frac{r}{\sigma},\frac{a}{\sigma})}dydz, & a\le x\le r\le z\le b\\[2ex] \dfrac{e^{-z^2/2\sigma^2}S(\frac{\lambda+q}{\theta},\frac{b}{\sigma},\frac{z}{\sigma})S(\frac{\lambda+q}{\theta},\frac{b}{\sigma},\frac{x}{\sigma})}{e^{-r^2/2\sigma^2}S^2(\frac{\lambda+q}{\theta},\frac{b}{\sigma},\frac{r}{\sigma})}dydz, & r\le x\le b,\ r\le z\le b\end{cases}$$

(1) $$\mathbf{E}_x\Big\{\exp\Big(-\int_0^\tau\big(p\mathbb{I}_{(-\infty,r)}(U_s)+q\mathbb{I}_{[r,\infty)}(U_s)\big)ds\Big);a<\inf_{0\le s\le\tau}U_s,\ \sup_{0\le s\le\tau}U_s<b,\ \ell(\tau,r)=0,\ U_\tau\in dz\Big\}$$

$z\le r$, $x\le r$: $$=\frac{\lambda\sqrt{\pi}e^{-z^2/2\sigma^2}S(\frac{\lambda+p}{\theta},\frac{r}{\sigma},\frac{z+x+|z-x|}{2\sigma})S(\frac{\lambda+p}{\theta},\frac{z+x-|z-x|}{2\sigma},\frac{a}{\sigma})}{\theta\sigma\sqrt{2}S(\frac{\lambda+p}{\theta},\frac{r}{\sigma},\frac{a}{\sigma})}dz$$

$r\le z$, $r\le x$: $$=\frac{\lambda\sqrt{\pi}e^{-z^2/2\sigma^2}S(\frac{\lambda+q}{\theta},\frac{b}{\sigma},\frac{z+x+|z-x|}{2\sigma})S(\frac{\lambda+q}{\theta},\frac{z+x-|z-x|}{2\sigma},\frac{r}{\sigma})}{\theta\sigma\sqrt{2}S(\frac{\lambda+q}{\theta},\frac{b}{\sigma},\frac{r}{\sigma})}dz$$

7 $U_s = \sigma e^{-\theta s} W_{e^{2\theta s}-1} \qquad \sigma, \theta > 0 \qquad\qquad H_z = \min\{s : U_s = z\}$

2. Stopping at first hitting time

2.0.1 $$\mathbf{E}_x e^{-\alpha H_z} = \begin{cases} \dfrac{e^{x^2/4\sigma^2} D_{-\alpha/\theta}(-\frac{x}{\sigma})}{e^{z^2/4\sigma^2} D_{-\alpha/\theta}(-\frac{z}{\sigma})}, & x \le z \\ \dfrac{e^{x^2/4\sigma^2} D_{-\alpha/\theta}(\frac{x}{\sigma})}{e^{z^2/4\sigma^2} D_{-\alpha/\theta}(\frac{z}{\sigma})}, & z \le x \end{cases}$$

2.0.2 (1) $$\mathbf{P}_x\big(H_0 > t\big) = \operatorname{Erf}\Big(\frac{|x|}{\sigma\sqrt{2(e^{2\theta t}-1)}}\Big)$$

2.1.2 $z \le y$ $$\mathbf{P}_x\Big(\sup_{0\le s\le H_z} U_s \ge y\Big) = \frac{\operatorname{Erfid}(\frac{x}{\sigma}, \frac{z}{\sigma})}{\operatorname{Erfid}(\frac{y}{\sigma}, \frac{z}{\sigma})}, \quad z \le x \le y$$

2.1.4 $z \le y$ $$\mathbf{E}_x\Big\{e^{-\alpha H_z}; \sup_{0\le s\le H_z} U_s \le y, \Big\} = \begin{cases} \dfrac{e^{x^2/4\sigma^2} D_{-\alpha/\theta}(-\frac{x}{\sigma})}{e^{z^2/4\sigma^2} D_{-\alpha/\theta}(-\frac{z}{\sigma})}, & x \le z \\ \dfrac{S(\frac{\alpha}{\theta}, \frac{y}{\sigma}, \frac{x}{\sigma})}{S(\frac{\alpha}{\theta}, \frac{y}{\sigma}, \frac{z}{\sigma})}, & z \le x \le y \end{cases}$$

2.2.2 $y \le z$ $$\mathbf{P}_x\Big(\inf_{0\le s\le H_z} U_s \le y\Big) = \frac{\operatorname{Erfid}(\frac{z}{\sigma}, \frac{x}{\sigma})}{\operatorname{Erfid}(\frac{z}{\sigma}, \frac{y}{\sigma})}, \quad y \le x \le z$$

2.2.4 $y \le z$ $$\mathbf{E}_x\Big\{e^{-\alpha H_z}; \inf_{0\le s\le H_z} U_s \ge y\Big\} = \begin{cases} \dfrac{S(\frac{\alpha}{\theta}, \frac{x}{\sigma}, \frac{y}{\sigma})}{S(\frac{\alpha}{\theta}, \frac{z}{\sigma}, \frac{y}{\sigma})}, & y \le x \le z \\ \dfrac{e^{x^2/4\sigma^2} D_{-\alpha/\theta}(\frac{x}{\sigma})}{e^{z^2/4\sigma^2} D_{-\alpha/\theta}(\frac{z}{\sigma})}, & z \le x \end{cases}$$

2.3.1 $$\mathbf{E}_x e^{-\gamma \ell(H_z, r)} = \begin{cases} 1, & x \le z \le r, \quad r \le z \le x \\ \dfrac{1 + \frac{\gamma\sqrt{\pi}}{\theta\sigma\sqrt{2}} e^{-r^2/2\sigma^2} |\operatorname{Erfid}(\frac{x}{\sigma}, \frac{r}{\sigma})|}{1 + \frac{\gamma\sqrt{\pi}}{\theta\sigma\sqrt{2}} e^{-r^2/2\sigma^2} |\operatorname{Erfid}(\frac{z}{\sigma}, \frac{r}{\sigma})|}, & r \wedge z \le x \le r \vee z \\ \dfrac{1}{1 + \frac{\gamma\sqrt{\pi}}{\theta\sigma\sqrt{2}} e^{-r^2/2\sigma^2} |\operatorname{Erfid}(\frac{z}{\sigma}, \frac{r}{\sigma})|}, & z \le r \le x, \quad x \le r \le z \end{cases}$$

2.3.2 $$\mathbf{P}_x\big(\ell(H_z, r) \in dy\big)$$

$$= \exp\Big(-\frac{\theta\sigma\sqrt{2/\pi}\, e^{r^2/2\sigma^2} y}{|\operatorname{Erfid}(\frac{z}{\sigma}, \frac{r}{\sigma})|}\Big) \begin{cases} 0, & x \le z \le r, \quad r \le z \le x \\ \dfrac{\theta\sigma\sqrt{2/\pi}|\operatorname{Erfid}(\frac{z}{\sigma}, \frac{x}{\sigma})|}{e^{-r^2/2\sigma^2} \operatorname{Erfid}^2(\frac{z}{\sigma}, \frac{r}{\sigma})} dy, & r \wedge z \le x \le r \vee z \\ \dfrac{\theta\sigma\sqrt{2} e^{r^2/2\sigma^2}}{\sqrt{\pi}|\operatorname{Erfid}(\frac{z}{\sigma}, \frac{r}{\sigma})|} dy, & z \le r \le x, \quad x \le r \le z \end{cases}$$

7 $U_s = \sigma e^{-\theta s} W_{e^{2\theta s}-1}$ $\quad \sigma, \theta > 0$ $\quad H_z = \min\{s : U_s = z\}$

(1) $$\mathbf{P}_x\big(\ell(H_z, r) = 0\big) = \begin{cases} 1, & x \le z \le r, \quad r \le z \le x \\ \dfrac{|\operatorname{Erfid}(\frac{x}{\sigma}, \frac{r}{\sigma})|}{|\operatorname{Erfid}(\frac{z}{\sigma}, \frac{r}{\sigma})|}, & r \wedge z \le x \le r \vee z \\ 0, & z \le r \le x, \quad x \le r \le z \end{cases}$$

2.3.3 $\mathbf{E}_x e^{-\alpha H_z - \gamma \ell(H_z, r)}$

$z \le r$
$$= \begin{cases} \dfrac{e^{x^2/4\sigma^2} D_{-\alpha/\theta}(\frac{x}{\sigma})}{e^{z^2/4\sigma^2} D_{-\alpha/\theta}(\frac{z}{\sigma}) + \frac{\gamma\sqrt{\pi}}{\theta\sigma\sqrt{2}} e^{-r^2/4\sigma^2} D_{-\alpha/\theta}(\frac{r}{\sigma}) S(\frac{\alpha}{\theta}, \frac{r}{\sigma}, \frac{z}{\sigma})}, & r \le x \\ \dfrac{e^{x^2/4\sigma^2} D_{-\alpha/\theta}(\frac{x}{\sigma}) + \frac{\gamma\sqrt{\pi}}{\theta\sigma\sqrt{2}} e^{-r^2/4\sigma^2} D_{-\alpha/\theta}(\frac{r}{\sigma}) S(\frac{\alpha}{\theta}, \frac{r}{\sigma}, \frac{x}{\sigma})}{e^{z^2/4\sigma^2} D_{-\alpha/\theta}(\frac{z}{\sigma}) + \frac{\gamma\sqrt{\pi}}{\theta\sigma\sqrt{2}} e^{-r^2/4\sigma^2} D_{-\alpha/\theta}(\frac{r}{\sigma}) S(\frac{\alpha}{\theta}, \frac{r}{\sigma}, \frac{z}{\sigma})}, & z \le x \le r \\ \dfrac{e^{x^2/4\sigma^2} D_{-\alpha/\theta}(-\frac{x}{\sigma})}{e^{z^2/4\sigma^2} D_{-\alpha/\theta}(-\frac{z}{\sigma})}, & x \le z \end{cases}$$

$r \le z$
$$= \begin{cases} \dfrac{e^{x^2/4\sigma^2} D_{-\alpha/\theta}(\frac{x}{\sigma})}{e^{z^2/4\sigma^2} D_{-\alpha/\theta}(\frac{z}{\sigma})}, & z \le x \\ \dfrac{e^{x^2/4\sigma^2} D_{-\alpha/\theta}(-\frac{x}{\sigma}) + \frac{\gamma\sqrt{\pi}}{\theta\sigma\sqrt{2}} e^{-r^2/4\sigma^2} D_{-\alpha/\theta}(-\frac{r}{\sigma}) S(\frac{\alpha}{\theta}, \frac{x}{\sigma}, \frac{r}{\sigma})}{e^{z^2/4\sigma^2} D_{-\alpha/\theta}(-\frac{z}{\sigma}) + \frac{\gamma\sqrt{\pi}}{\theta\sigma\sqrt{2}} e^{-r^2/4\sigma^2} D_{-\alpha/\theta}(-\frac{r}{\sigma}) S(\frac{\alpha}{\theta}, \frac{z}{\sigma}, \frac{r}{\sigma})}, & r \le x \le z \\ \dfrac{e^{x^2/4\sigma^2} D_{-\alpha/\theta}(-\frac{x}{\sigma})}{e^{z^2/4\sigma^2} D_{-\alpha/\theta}(-\frac{z}{\sigma}) + \frac{\gamma\sqrt{\pi}}{\theta\sigma\sqrt{2}} e^{-r^2/4\sigma^2} D_{-\alpha/\theta}(-\frac{r}{\sigma}) S(\frac{\alpha}{\theta}, \frac{r}{\sigma}, \frac{z}{\sigma})}, & x \le r \end{cases}$$

2.3.4 $\mathbf{E}_x\{e^{-\alpha H_z}; \ell(H_z, r) \in dy\}$

$z \le r$
$$= \exp\Big(-\frac{\theta\sigma\sqrt{2} e^{(z^2+r^2)/4\sigma^2} D_{-\alpha/\theta}(\frac{z}{\sigma}) y}{\sqrt{\pi} D_{-\alpha/\theta}(\frac{r}{\sigma}) S(\frac{\alpha}{\theta}, \frac{r}{\sigma}, \frac{z}{\sigma})}\Big) \begin{cases} \dfrac{\theta\sigma\sqrt{2} e^{(x^2+r^2)/4\sigma^2} D_{-\alpha/\theta}(\frac{x}{\sigma})}{\sqrt{\pi} S(\frac{\alpha}{\theta}, \frac{r}{\sigma}, \frac{z}{\sigma}) D_{-\alpha/\theta}(\frac{r}{\sigma})} dy, & r \le x \\ \dfrac{\theta\sigma\sqrt{2} e^{r^2/2\sigma^2} S(\frac{\alpha}{\theta}, \frac{x}{\sigma}, \frac{z}{\sigma})}{\sqrt{\pi} S^2(\frac{\alpha}{\theta}, \frac{r}{\sigma}, \frac{z}{\sigma})} dy, & z \le x \le r \\ 0, & x \le z \end{cases}$$

$r \le z$
$$= \exp\Big(-\frac{\theta\sigma e^{(z^2+r^2)/4\sigma^2} D_{-\alpha/\theta}(-\frac{z}{\sigma}) y}{\sqrt{\pi/2} D_{-\alpha/\theta}(-\frac{r}{\sigma}) S(\frac{\alpha}{\theta}, \frac{z}{\sigma}, \frac{r}{\sigma})}\Big) \begin{cases} 0, & z \le x \\ \dfrac{\theta\sigma\sqrt{2} e^{r^2/2\sigma^2} S(\frac{\alpha}{\theta}, \frac{z}{\sigma}, \frac{x}{\sigma})}{\sqrt{\pi} S^2(\frac{\alpha}{\theta}, \frac{z}{\sigma}, \frac{r}{\sigma})} dy, & r \le x \le z \\ \dfrac{\theta\sigma\sqrt{2} e^{(x^2+r^2)/4\sigma^2} D_{-\alpha/\theta}(-\frac{x}{\sigma})}{\sqrt{\pi} S(\frac{\alpha}{\theta}, \frac{z}{\sigma}, \frac{r}{\sigma}) D_{-\alpha/\theta}(-\frac{r}{\sigma})} dy, & x \le r \end{cases}$$

7 $U_s = \sigma e^{-\theta s} W_{e^{2\theta s}-1}$ $\sigma, \theta > 0$ $H_z = \min\{s : U_s = z\}$

(1)
$z \le r$

$$\mathbf{E}_x\{e^{-\alpha H_z}; \ell(H_z, r) = 0\} = \begin{cases} 0 & r \le x \\ \dfrac{S(\frac{\alpha}{\theta}, \frac{r}{\sigma}, \frac{x}{\sigma})}{S(\frac{\alpha}{\theta}, \frac{r}{\sigma}, \frac{z}{\sigma})}, & z \le x \le r \\ \dfrac{e^{x^2/4\sigma^2} D_{-\alpha/\theta}(-\frac{x}{\sigma})}{e^{z^2/4\sigma^2} D_{-\alpha/\theta}(-\frac{z}{\sigma})}, & x \le z \end{cases}$$

$r \le z$

$$\mathbf{E}_x\{e^{-\alpha H_z}; \ell(H_z, r) = 0\} = \begin{cases} 0 & x \le r \\ \dfrac{S(\frac{\alpha}{\theta}, \frac{x}{\sigma}, \frac{r}{\sigma})}{S(\frac{\alpha}{\theta}, \frac{z}{\sigma}, \frac{r}{\sigma})}, & r \le x \le z \\ \dfrac{e^{x^2/4\sigma^2} D_{-\alpha/\theta}(\frac{x}{\sigma})}{e^{z^2/4\sigma^2} D_{-\alpha/\theta}(\frac{z}{\sigma})}, & z \le x \end{cases}$$

2.4.1

$$\mathbf{E}_x \exp\Big(-\gamma \int_0^{H_z} \mathbb{I}_{[r,\infty)}(U_s)ds\Big)$$

$z \le r$

$$= \begin{cases} 1, & x \le z \\ \dfrac{e^{r^2/2\sigma^2} D_{-\gamma/\theta}(\frac{r}{\sigma}) + \frac{\gamma\sqrt{\pi}}{\theta\sqrt{2}} \operatorname{Erfid}(\frac{r}{\sigma}, \frac{x}{\sigma}) D_{-(\theta+\gamma)/\theta}(\frac{r}{\sigma})}{e^{r^2/2\sigma^2} D_{-\gamma/\theta}(\frac{r}{\sigma}) + \frac{\gamma\sqrt{\pi}}{\theta\sqrt{2}} \operatorname{Erfid}(\frac{r}{\sigma}, \frac{z}{\sigma}) D_{-(\theta+\gamma)/\theta}(\frac{r}{\sigma})}, & z \le x \le r \\ \dfrac{e^{(x^2+r^2)/4\sigma^2} D_{-\gamma/\theta}(\frac{x}{\sigma})}{e^{r^2/2\sigma^2} D_{-\gamma/\theta}(\frac{r}{\sigma}) + \frac{\gamma\sqrt{\pi}}{\theta\sqrt{2}} \operatorname{Erfid}(\frac{r}{\sigma}, \frac{z}{\sigma}) D_{-(\theta+\gamma)/\theta}(\frac{r}{\sigma})}, & r \le x \end{cases}$$

$r \le z$

$$= \begin{cases} \dfrac{\sqrt{2/\pi} e^{r^2/2\sigma^2}}{C(\frac{\gamma}{\theta}, \frac{r}{\sigma}, \frac{z}{\sigma})} & x \le r \\ \dfrac{C(\frac{\gamma}{\theta}, \frac{r}{\sigma}, \frac{x}{\sigma})}{C(\frac{\gamma}{\theta}, \frac{r}{\sigma}, \frac{z}{\sigma})}, & r \le x \le z \\ \dfrac{e^{x^2/4\sigma^2} D_{-\gamma/\theta}(\frac{x}{\sigma})}{e^{z^2/4\sigma^2} D_{-\gamma/\theta}(\frac{z}{\sigma})}, & z \le x \end{cases}$$

2.4.2
(1)
$z \le r$

$$\mathbf{P}_x\Big(\int_0^{H_z} \mathbb{I}_{[r,\infty)}(U_s)ds = 0\Big) = \begin{cases} 1, & x \le z \\ \dfrac{\operatorname{Erfid}(\frac{r}{\sigma}, \frac{x}{\sigma})}{\operatorname{Erfid}(\frac{r}{\sigma}, \frac{z}{\sigma})}, & z \le x \le r \\ 0, & r \le x \end{cases}$$

2.5.1

$$\mathbf{E}_x \exp\Big(-\gamma \int_0^{H_z} \mathbb{I}_{(-\infty,r]}(U_s)ds\Big)$$

$z \le r$

$$= \begin{cases} \dfrac{e^{x^2/4\sigma^2} D_{-\gamma/\theta}(-\frac{x}{\sigma})}{e^{z^2/4\sigma^2} D_{-\gamma/\theta}(-\frac{z}{\sigma})}, & x \le z \\ \dfrac{C(\frac{\gamma}{\theta}, \frac{r}{\sigma}, \frac{x}{\sigma})}{C(\frac{\gamma}{\theta}, \frac{r}{\sigma}, \frac{z}{\sigma})}, & z \le x \le r \\ \dfrac{\sqrt{2/\pi} e^{r^2/2\sigma^2}}{C(\frac{\gamma}{\theta}, \frac{r}{\sigma}, \frac{z}{\sigma})} & r \le x \end{cases}$$

7 $\quad U_s = \sigma e^{-\theta s} W_{e^{2\theta s}-1} \qquad \sigma, \theta > 0 \qquad H_z = \min\{s : U_s = z\}$

$r \le z$

$$= \begin{cases} \dfrac{e^{(x^2+r^2)/4\sigma^2} D_{-\gamma/\theta}(-\frac{x}{\sigma})}{e^{r^2/2\sigma^2} D_{-\gamma/\theta}(-\frac{r}{\sigma}) + \frac{\gamma\sqrt{\pi}}{\theta\sqrt{2}} \operatorname{Erfid}(\frac{z}{\sigma}, \frac{r}{\sigma}) D_{-(\theta+\gamma)/\theta}(-\frac{r}{\sigma})}, & x \le r \\ \dfrac{e^{r^2/2\sigma^2} D_{-\gamma/\theta}(-\frac{r}{\sigma}) + \frac{\gamma\sqrt{\pi}}{\theta\sqrt{2}} \operatorname{Erfid}(\frac{x}{\sigma}, \frac{r}{\sigma}) D_{-(\theta+\gamma)/\theta}(-\frac{r}{\sigma})}{e^{r^2/2\sigma^2} D_{-\gamma/\theta}(-\frac{r}{\sigma}) + \frac{\gamma\sqrt{\pi}}{\theta\sqrt{2}} \operatorname{Erfid}(\frac{z}{\sigma}, \frac{r}{\sigma}) D_{-(\theta+\gamma)/\theta}(-\frac{r}{\sigma})}, & r \le x \le z \\ 1, & z \le x \end{cases}$$

2.5.2 (1) $r \le z$

$$\mathbf{P}_x\Big(\int_0^{H_z} \mathbb{I}_{(-\infty,r]}(U_s)ds = 0\Big) = \begin{cases} 0, & x \le r \\ \dfrac{\operatorname{Erfid}(\frac{x}{\sigma}, \frac{r}{\sigma})}{\operatorname{Erfid}(\frac{z}{\sigma}, \frac{r}{\sigma})}, & r \le x \le z \\ 1, & z \le x \end{cases}$$

2.6.1

$$\mathbf{E}_x \exp\Big(-\int_0^{H_z} \big(p\mathbb{I}_{(-\infty,r]}(U_s) + q\mathbb{I}_{[r,\infty)}(U_s)\big)ds\Big)$$

$z \le r$

$$= \begin{cases} \dfrac{e^{x^2/4\sigma^2} D_{-p/\theta}(-\frac{x}{\sigma})}{e^{z^2/4\sigma^2} D_{-p/\theta}(-\frac{z}{\sigma})}, & x \le z \\ \dfrac{D_{-q/\theta}(\frac{r}{\sigma}) C(\frac{p}{\theta}, \frac{r}{\sigma}, \frac{x}{\sigma}) + \frac{q}{\theta} D_{-(\theta+q)/\theta}(\frac{r}{\sigma}) S(\frac{p}{\theta}, \frac{r}{\sigma}, \frac{x}{\sigma})}{D_{-q/\theta}(\frac{r}{\sigma}) C(\frac{p}{\theta}, \frac{r}{\sigma}, \frac{z}{\sigma}) + \frac{q}{\theta} D_{-(\theta+q)/\theta}(\frac{r}{\sigma}) S(\frac{p}{\theta}, \frac{r}{\sigma}, \frac{z}{\sigma})}, & z \le x \le r \\ \dfrac{\sqrt{2/\pi} e^{(x^2+r^2)/4\sigma^2} D_{-q/\theta}(\frac{x}{\sigma})}{D_{-q/\theta}(\frac{r}{\sigma}) C(\frac{p}{\theta}, \frac{r}{\sigma}, \frac{z}{\sigma}) + \frac{q}{\theta} D_{-(\theta+q)/\theta}(\frac{r}{\sigma}) S(\frac{p}{\theta}, \frac{r}{\sigma}, \frac{z}{\sigma})}, & r \le x \end{cases}$$

$r \le z$

$$= \begin{cases} \dfrac{\sqrt{2/\pi} e^{(x^2+r^2)/4\sigma^2} D_{-p/\theta}(-\frac{x}{\sigma})}{C(\frac{q}{\theta}, \frac{r}{\sigma}, \frac{z}{\sigma}) D_{-p/\theta}(-\frac{r}{\sigma}) + \frac{p}{\theta} S(\frac{q}{\theta}, \frac{z}{\sigma}, \frac{r}{\sigma}) D_{-(\theta+p)/\theta}(-\frac{r}{\sigma})}, & x \le r \\ \dfrac{C(\frac{q}{\theta}, \frac{r}{\sigma}, \frac{x}{\sigma}) D_{-p/\theta}(-\frac{r}{\sigma}) + \frac{p}{\theta} S(\frac{q}{\theta}, \frac{x}{\sigma}, \frac{r}{\sigma}) D_{-(\theta+p)/\theta}(-\frac{r}{\sigma})}{C(\frac{q}{\theta}, \frac{r}{\sigma}, \frac{z}{\sigma}) D_{-p/\theta}(-\frac{r}{\sigma}) + \frac{p}{\theta} S(\frac{q}{\theta}, \frac{z}{\sigma}, \frac{r}{\sigma}) D_{-(\theta+p)/\theta}(-\frac{r}{\sigma})}, & r \le x \le z \\ \dfrac{e^{x^2/4\sigma^2} D_{-q/\theta}(\frac{x}{\sigma})}{e^{z^2/4\sigma^2} D_{-q/\theta}(\frac{z}{\sigma})}, & z \le x \end{cases}$$

2.9.3 $1 \le \gamma$

$$\mathbf{E}_x \exp\Big(-\alpha H_z - \frac{(\gamma^2-1)\theta}{4\sigma^2} \int_0^{H_z} U_s^2 ds\Big) = \begin{cases} \dfrac{e^{x^2/4\sigma^2} D_{-\frac{1}{2}+\frac{1}{2\gamma}-\frac{\alpha}{\theta\gamma}}(-\frac{x\sqrt{\gamma}}{\sigma})}{e^{z^2/4\sigma^2} D_{-\frac{1}{2}+\frac{1}{2\gamma}-\frac{\alpha}{\theta\gamma}}(-\frac{z\sqrt{\gamma}}{\sigma})}, & x \le z \\ \dfrac{e^{x^2/4\sigma^2} D_{-\frac{1}{2}+\frac{1}{2\gamma}-\frac{\alpha}{\theta\gamma}}(\frac{x\sqrt{\gamma}}{\sigma})}{e^{z^2/4\sigma^2} D_{-\frac{1}{2}+\frac{1}{2\gamma}-\frac{\alpha}{\theta\gamma}}(\frac{z\sqrt{\gamma}}{\sigma})}, & z \le x \end{cases}$$

2.9.4 (1)

$$\mathbf{E}_x\Big\{\exp\Big(-\frac{(\gamma^2-1)\theta}{4\sigma^2} \int_0^{H_0} U_s^2 ds\Big); H_0 \in dt\Big\}$$

$$= \frac{|x|\theta\gamma^{3/2}}{2\sqrt{\pi}\sigma \operatorname{sh}^{3/2}(t\theta\gamma)} \exp\Big(\frac{t\theta}{2} + \frac{x^2}{4\sigma^2} - \frac{x^2\gamma \operatorname{ch}(t\theta\gamma)}{4\sigma^2 \operatorname{sh}(t\theta\gamma)}\Big)dt$$

7 $U_s = \sigma e^{-\theta s} W_{e^{2\theta s}-1}$ $\sigma, \theta > 0$ $H_z = \min\{s : U_s = z\}$

(2) $$\mathbf{P}_x\Big(\int_0^{H_0} U_s^2 ds \in dy, H_0 \in dt\Big)$$

$$= \frac{|x|}{\sqrt{2\pi}(2\sigma^2\theta)^{3/2}} \exp\Big(\frac{t\theta}{2} + \frac{x^2 - y\theta}{4\sigma^2}\Big) \operatorname{es}_{y/2\sigma^2\theta}\Big(\frac{3}{2}, \frac{3}{2}, t, 0, \frac{x^2}{4\sigma^2\theta}\Big) dy dt$$

2.13.1 $z < x$ $x < y$ $$\mathbf{E}_x\big\{e^{-\gamma \check{H}(H_z)}; \sup_{0 \le s \le H_z} U_s \in dy\big\} = \frac{\sqrt{2}\, e^{y^2/2\sigma^2} S(\frac{\gamma}{\theta}, \frac{x}{\sigma}, \frac{z}{\sigma})}{\sigma\sqrt{\pi}\operatorname{Erfid}(\frac{y}{\sigma}, \frac{z}{\sigma}) S(\frac{\gamma}{\theta}, \frac{y}{\sigma}, \frac{z}{\sigma})} dy$$

2.13.4 $z < x$ $x < y$ $$\mathbf{E}_x\big\{e^{-\alpha H_z - \gamma \check{H}(H_z)}; \sup_{0 \le s \le H_z} U_s \in dy\big\} = \frac{\sqrt{2}\, e^{y^2/2\sigma^2} S(\frac{\gamma+\alpha}{\theta}, \frac{x}{\sigma}, \frac{z}{\sigma})}{\sigma\sqrt{\pi} S(\frac{\alpha}{\theta}, \frac{y}{\sigma}, \frac{z}{\sigma}) S(\frac{\gamma+\alpha}{\theta}, \frac{y}{\sigma}, \frac{z}{\sigma})} dy$$

2.14.1 $x < z$ $y < x$ $$\mathbf{E}_x\big\{e^{-\gamma \hat{H}(H_z)}; \inf_{0 \le s \le H_z} U_s \in dy\big\} = \frac{\sqrt{2}\, e^{y^2/2\sigma^2} S(\frac{\gamma}{\theta}, \frac{z}{\sigma}, \frac{x}{\sigma})}{\sigma\sqrt{\pi}\operatorname{Erfid}(\frac{z}{\sigma}, \frac{y}{\sigma}) S(\frac{\gamma}{\theta}, \frac{z}{\sigma}, \frac{y}{\sigma})} dy$$

2.14.4 $x < z$ $y < x$ $$\mathbf{E}_x\big\{e^{-\alpha H_z - \gamma \hat{H}(H_z)}; \inf_{0 \le s \le H_z} U_s \in dy\big\} = \frac{\sqrt{2}\, e^{y^2/2\sigma^2} S(\frac{\gamma+\alpha}{\theta}, \frac{z}{\sigma}, \frac{x}{\sigma})}{\sigma\sqrt{\pi} S(\frac{\alpha}{\theta}, \frac{z}{\sigma}, \frac{y}{\sigma}) S(\frac{\gamma+\alpha}{\theta}, \frac{z}{\sigma}, \frac{y}{\sigma})} dy$$

2.20.3 $$\mathbf{E}_x\Big\{\exp\Big(-\alpha H_z - p^2\sigma^2\theta \int_0^{H_z} \frac{ds}{U_s^2}\Big); \ 0 < \inf_{0 \le s \le H_z} U_s\Big\}$$

$$= \begin{cases} \dfrac{x^{-1/2} e^{x^2/4\sigma^2} M_{\frac{\theta-2\alpha}{4\theta}, \frac{1}{4}\sqrt{1+4p^2}}\big(\frac{x^2}{2\sigma^2}\big)}{z^{-1/2} e^{z^2/4\sigma^2} M_{\frac{\theta-2\alpha}{4\theta}, \frac{1}{4}\sqrt{1+4p^2}}\big(\frac{z^2}{2\sigma^2}\big)}, & 0 < x \le z \\[2ex] \dfrac{x^{-1/2} e^{x^2/4\sigma^2} W_{\frac{\theta-2\alpha}{4\theta}, \frac{1}{4}\sqrt{1+4p^2}}\big(\frac{x^2}{2\sigma^2}\big)}{z^{-1/2} e^{z^2/4\sigma^2} W_{\frac{\theta-2\alpha}{4\theta}, \frac{1}{4}\sqrt{1+4p^2}}\big(\frac{z^2}{2\sigma^2}\big)}, & z \le x \end{cases}$$

2.21.3 $$\mathbf{E}_x\Big\{\exp\Big(-\alpha H_z - \int_0^{H_z} \Big(\frac{p^2\sigma^2\theta}{U_s^2} + \frac{\theta(q^2-1)}{4\sigma^2} U_s^2\Big) ds\Big); \ 0 < \inf_{0 \le s \le H_z} U_s\Big\}$$

$$= \begin{cases} \dfrac{x^{-1/2} e^{x^2/4\sigma^2} M_{\frac{\theta-2\alpha}{4\theta q}, \frac{1}{4}\sqrt{1+4p^2}}\big(\frac{qx^2}{2\sigma^2}\big)}{z^{-1/2} e^{z^2/4\sigma^2} M_{\frac{\theta-2\alpha}{4\theta q}, \frac{1}{4}\sqrt{1+4p^2}}\big(\frac{qz^2}{2\sigma^2}\big)}, & 0 < x \le z \\[2ex] \dfrac{x^{-1/2} e^{x^2/4\sigma^2} W_{\frac{\theta-2\alpha}{4\theta q}, \frac{1}{4}\sqrt{1+4p^2}}\big(\frac{qx^2}{2\sigma^2}\big)}{z^{-1/2} e^{z^2/4\sigma^2} W_{\frac{\theta-2\alpha}{4\theta q}, \frac{1}{4}\sqrt{1+4p^2}}\big(\frac{qz^2}{2\sigma^2}\big)}, & z \le x \end{cases}$$

7 $U_s = \sigma e^{-\theta s} W_{e^{2\theta s}-1} \qquad \sigma, \theta > 0 \qquad H_z = \min\{s : U_s = z\}$

2.27.1 $$\mathbf{E}_x\Big\{\exp\Big(-\int_0^{H_z} \big(p\mathbb{1}_{(-\infty,r)}(U_s) + q\mathbb{1}_{[r,\infty)}(U_s)\big)ds\Big); \ell(H_z,r) \in dy\Big\}$$

$z \le r$ $$= \exp\Big(-y\theta\sigma\Big(\frac{C(\frac{p}{\theta},\frac{r}{\sigma},\frac{z}{\sigma})}{S(\frac{p}{\theta},\frac{r}{\sigma},\frac{z}{\sigma})} + \frac{qD_{-(\theta+q)/\theta}(\frac{r}{\sigma})}{\theta D_{-q/\theta}(\frac{r}{\sigma})}\Big)\Big)$$

$$\times \begin{cases} \dfrac{\theta\sigma\sqrt{2}e^{r^2/2\sigma^2} S(\frac{p}{\theta},\frac{x}{\sigma},\frac{z}{\sigma})}{\sqrt{\pi} S^2(\frac{p}{\theta},\frac{r}{\sigma},\frac{z}{\sigma})} dy, & z \le x \le r \\ \dfrac{\theta\sigma\sqrt{2}e^{(x^2+r^2)/4\sigma^2} D_{-q/\theta}(\frac{x}{\sigma})}{\sqrt{\pi} D_{-q/\theta}(\frac{r}{\sigma}) S(\frac{p}{\theta},\frac{r}{\sigma},\frac{z}{\sigma})} dy, & r \le x \end{cases}$$

$r \le z$ $$= \exp\Big(-y\theta\sigma\Big(\frac{C(\frac{q}{\theta},\frac{r}{\sigma},\frac{z}{\sigma})}{S(\frac{q}{\theta},\frac{z}{\sigma},\frac{r}{\sigma})} + \frac{pD_{-(\theta+p)/\theta}(-\frac{r}{\sigma})}{\theta D_{-p/\theta}(-\frac{r}{\sigma})}\Big)\Big)$$

$$\times \begin{cases} \dfrac{\theta\sigma\sqrt{2}e^{(x^2+r^2)/4\sigma^2} D_{-p/\theta}(-\frac{x}{\sigma})}{\sqrt{\pi} D_{-p/\theta}(-\frac{r}{\sigma}) S(\frac{q}{\theta},\frac{z}{\sigma},\frac{r}{\sigma})} dy, & x \le r \\ \dfrac{\theta\sigma\sqrt{2}e^{r^2/2\sigma^2} S(\frac{q}{\theta},\frac{z}{\sigma},\frac{x}{\sigma})}{\sqrt{\pi} S^2(\frac{q}{\theta},\frac{z}{\sigma},\frac{r}{\sigma})} dy, & r \le x \le z \end{cases}$$

(1) $$\mathbf{E}_x\Big\{\exp\Big(-\int_0^{H_z} \big(p\mathbb{1}_{(-\infty,r)}(U_s) + q\mathbb{1}_{[r,\infty)}(U_s)\big)ds\Big); \ell(H_z,r) = 0\Big\}$$

$$= \begin{cases} \dfrac{e^{x^2/4\sigma^2} D_{-p/\theta}(-\frac{x}{\sigma})}{e^{z^2/4\sigma^2} D_{-p/\theta}(-\frac{z}{\sigma})}, & x \le z \le r \\ \dfrac{S(\frac{p}{\theta},\frac{r}{\sigma},\frac{x}{\sigma})}{S(\frac{p}{\theta},\frac{r}{\sigma},\frac{z}{\sigma})}, & z \le x \le r \\ \dfrac{S(\frac{q}{\theta},\frac{x}{\sigma},\frac{r}{\sigma})}{S(\frac{q}{\theta},\frac{z}{\sigma},\frac{r}{\sigma})}, & r \le x \le z \\ \dfrac{e^{x^2/4\sigma^2} D_{-q/\theta}(\frac{x}{\sigma})}{e^{z^2/4\sigma^2} D_{-q/\theta}(\frac{z}{\sigma})}, & r \le z \le x \end{cases}$$

7 $U_s = \sigma e^{-\theta s} W_{e^{2\theta s}-1}$ $\sigma, \theta > 0$ $H = H_{a,b} = \min\{s : U_s \notin (a,b)\}$

3. Stopping at first exit time

3.0.1 $$\mathbf{E}_x e^{-\alpha H} = \frac{S(\frac{\alpha}{\theta}, \frac{b}{\sigma}, \frac{x}{\sigma}) + S(\frac{\alpha}{\theta}, \frac{x}{\sigma}, \frac{a}{\sigma})}{S(\frac{\alpha}{\theta}, \frac{b}{\sigma}, \frac{a}{\sigma})}, \quad a \le x \le b$$

3.0.3 $$\mathbf{E}_x e^{-\mu U_H} = \frac{\operatorname{Erfid}(\frac{b}{\sigma}, \frac{x}{\sigma})}{\operatorname{Erfid}(\frac{b}{\sigma}, \frac{a}{\sigma})} e^{-\mu a} + \frac{\operatorname{Erfid}(\frac{x}{\sigma}, \frac{a}{\sigma})}{\operatorname{Erfid}(\frac{b}{\sigma}, \frac{a}{\sigma})} e^{-\mu b}, \quad a \le x \le b$$

3.0.4 (a) $$\mathbf{P}_x\big(U_H = a\big) = \frac{\operatorname{Erfid}(\frac{b}{\sigma}, \frac{x}{\sigma})}{\operatorname{Erfid}(\frac{b}{\sigma}, \frac{a}{\sigma})}, \quad a \le x \le b$$

3.0.4 (b) $$\mathbf{P}_x\big(U_H = b\big) = \frac{\operatorname{Erfid}(\frac{x}{\sigma}, \frac{a}{\sigma})}{\operatorname{Erfid}(\frac{b}{\sigma}, \frac{a}{\sigma})}, \quad a \le x \le b$$

3.0.5 (a) $$\mathbf{E}_x\big\{e^{-\alpha H}; U_H = a\big\} = \frac{S(\frac{\alpha}{\theta}, \frac{b}{\sigma}, \frac{x}{\sigma})}{S(\frac{\alpha}{\theta}, \frac{b}{\sigma}, \frac{a}{\sigma})}, \quad a \le x \le b$$

3.0.5 (b) $$\mathbf{E}_x\big\{e^{-\alpha H}; U_H = b\big\} = \frac{S(\frac{\alpha}{\theta}, \frac{x}{\sigma}, \frac{a}{\sigma})}{S(\frac{\alpha}{\theta}, \frac{b}{\sigma}, \frac{a}{\sigma})}, \quad a \le x \le b$$

3.1.6 $$\mathbf{P}_x\Big(\sup_{0 \le s \le H} U_s \ge y, U_H = a\Big) = \frac{\operatorname{Erfid}(\frac{b}{\sigma}, \frac{y}{\sigma}) \operatorname{Erfid}(\frac{x}{\sigma}, \frac{a}{\sigma})}{\operatorname{Erfid}(\frac{b}{\sigma}, \frac{a}{\sigma}) \operatorname{Erfid}(\frac{y}{\sigma}, \frac{a}{\sigma})}, \quad a \le x \le y \le b$$

3.1.8 (1) $$\mathbf{E}_x\Big\{e^{-\alpha H}; \sup_{0 \le s \le H} U_s \ge y, U_H = a\Big\}$$
$$= \frac{S(\frac{\alpha}{\theta}, \frac{b}{\sigma}, \frac{y}{\sigma}) S(\frac{\alpha}{\theta}, \frac{x}{\sigma}, \frac{a}{\sigma})}{S(\frac{\alpha}{\theta}, \frac{b}{\sigma}, \frac{a}{\sigma}) S(\frac{\alpha}{\theta}, \frac{y}{\sigma}, \frac{a}{\sigma})}, \quad a \le x \le y \le b$$

3.2.6 $$\mathbf{P}_x\Big(\inf_{0 \le s \le H} U_s \le y, U_H = b\Big) = \frac{\operatorname{Erfid}(\frac{b}{\sigma}, \frac{x}{\sigma}) \operatorname{Erfid}(\frac{y}{\sigma}, \frac{a}{\sigma})}{\operatorname{Erfid}(\frac{b}{\sigma}, \frac{y}{\sigma}) \operatorname{Erfid}(\frac{b}{\sigma}, \frac{a}{\sigma})}, \quad a \le y \le x \le b$$

3.2.8 (1) $$\mathbf{E}_x\Big\{e^{-\alpha H}; \inf_{0 \le s \le H} U_s \le y, U_H = b\Big\}$$
$$= \frac{S(\frac{\alpha}{\theta}, \frac{b}{\sigma}, \frac{x}{\sigma}) S(\frac{\alpha}{\theta}, \frac{y}{\sigma}, \frac{a}{\sigma})}{S(\frac{\alpha}{\theta}, \frac{b}{\sigma}, \frac{y}{\sigma}) S(\frac{\alpha}{\theta}, \frac{b}{\sigma}, \frac{a}{\sigma})}, \quad a \le y \le x \le b$$

7 $\quad U_s = \sigma e^{-\theta s} W_{e^{2\theta s}-1} \qquad \sigma, \theta > 0 \qquad H = H_{a,b} = \min\{s : U_s \notin (a,b)\}$

3.3.1
$$\mathbf{E}_x e^{-\gamma \ell(H,r)} = \begin{cases} \dfrac{\operatorname{Erfid}\big(\frac{b}{\sigma}, \frac{a}{\sigma}\big) + \frac{\gamma\sqrt{\pi}}{\theta\sigma\sqrt{2}} e^{-r^2/2\sigma^2} \operatorname{Erfid}\big(\frac{x}{\sigma}, \frac{r}{\sigma}\big) \operatorname{Erfid}\big(\frac{r}{\sigma}, \frac{a}{\sigma}\big)}{\operatorname{Erfid}\big(\frac{b}{\sigma}, \frac{a}{\sigma}\big) + \frac{\gamma\sqrt{\pi}}{\theta\sigma\sqrt{2}} e^{-r^2/2\sigma^2} \operatorname{Erfid}\big(\frac{b}{\sigma}, \frac{r}{\sigma}\big) \operatorname{Erfid}\big(\frac{r}{\sigma}, \frac{a}{\sigma}\big)}, & r \le x \le b \\[2ex] \dfrac{\operatorname{Erfid}\big(\frac{b}{\sigma}, \frac{a}{\sigma}\big) + \frac{\gamma\sqrt{\pi}}{\theta\sigma\sqrt{2}} e^{-r^2/2\sigma^2} \operatorname{Erfid}\big(\frac{b}{\sigma}, \frac{r}{\sigma}\big) \operatorname{Erfid}\big(\frac{r}{\sigma}, \frac{x}{\sigma}\big)}{\operatorname{Erfid}\big(\frac{b}{\sigma}, \frac{a}{\sigma}\big) + \frac{\gamma\sqrt{\pi}}{\theta\sigma\sqrt{2}} e^{-r^2/2\sigma^2} \operatorname{Erfid}\big(\frac{b}{\sigma}, \frac{r}{\sigma}\big) \operatorname{Erfid}\big(\frac{r}{\sigma}, \frac{a}{\sigma}\big)}, & a \le x \le r \end{cases}$$

3.3.2
$$\mathbf{P}_x\big(\ell(H,r) \in dy\big) = \exp\Big(-\frac{\theta\sigma\sqrt{2} e^{r^2/2\sigma^2} \operatorname{Erfid}\big(\frac{b}{\sigma}, \frac{a}{\sigma}\big) y}{\sqrt{\pi} \operatorname{Erfid}\big(\frac{b}{\sigma}, \frac{r}{\sigma}\big) \operatorname{Erfid}\big(\frac{r}{\sigma}, \frac{a}{\sigma}\big)}\Big)$$

$$\times \begin{cases} \dfrac{\theta\sigma\sqrt{2} e^{r^2/2\sigma^2} \operatorname{Erfid}\big(\frac{b}{\sigma}, \frac{x}{\sigma}\big) \operatorname{Erfid}\big(\frac{b}{\sigma}, \frac{a}{\sigma}\big)}{\sqrt{\pi} \operatorname{Erfid}^2\big(\frac{b}{\sigma}, \frac{r}{\sigma}\big) \operatorname{Erfid}\big(\frac{r}{\sigma}, \frac{a}{\sigma}\big)} dy, & r \le x \le b \\[2ex] \dfrac{\theta\sigma\sqrt{2} e^{r^2/2\sigma^2} \operatorname{Erfid}\big(\frac{x}{\sigma}, \frac{a}{\sigma}\big) \operatorname{Erfid}\big(\frac{b}{\sigma}, \frac{a}{\sigma}\big)}{\sqrt{\pi} \operatorname{Erfid}\big(\frac{b}{\sigma}, \frac{r}{\sigma}\big) \operatorname{Erfid}^2\big(\frac{r}{\sigma}, \frac{a}{\sigma}\big)} dy, & a \le x \le r \end{cases}$$

(1)
$$\mathbf{P}_x\big(\ell(H,r) = 0\big) = \begin{cases} \dfrac{\operatorname{Erfid}\big(\frac{x}{\sigma}, \frac{r}{\sigma}\big)}{\operatorname{Erfid}\big(\frac{b}{\sigma}, \frac{r}{\sigma}\big)}, & r \le x \le b \\[2ex] \dfrac{\operatorname{Erfid}\big(\frac{r}{\sigma}, \frac{x}{\sigma}\big)}{\operatorname{Erfid}\big(\frac{r}{\sigma}, \frac{a}{\sigma}\big)}, & a \le x \le r \end{cases}$$

3.3.3
$$\mathbf{E}_r e^{-\alpha\,\mathrm{m} - \gamma \ell(H,r)} = \frac{S\big(\frac{\alpha}{\theta}, \frac{b}{\sigma}, \frac{r}{\sigma}\big) + S\big(\frac{\alpha}{\theta}, \frac{r}{\sigma}, \frac{a}{\sigma}\big)}{S\big(\frac{\alpha}{\theta}, \frac{b}{\sigma}, \frac{a}{\sigma}\big) + \frac{\gamma\sqrt{\pi}}{\theta\sigma\sqrt{2}} e^{-r^2/2\sigma^2} S\big(\frac{\alpha}{\theta}, \frac{b}{\sigma}, \frac{r}{\sigma}\big) S\big(\frac{\alpha}{\theta}, \frac{r}{\sigma}, \frac{a}{\sigma}\big)} =: L$$

$$\mathbf{E}_x e^{-\alpha\,\mathrm{m} - \gamma \ell(H,r)} = \begin{cases} \dfrac{S\big(\frac{\alpha}{\theta}, \frac{x}{\sigma}, \frac{r}{\sigma}\big)}{S\big(\frac{\alpha}{\theta}, \frac{b}{\sigma}, \frac{r}{\theta}\big)} + \dfrac{S\big(\frac{\alpha}{\theta}, \frac{b}{\sigma}, \frac{x}{\sigma}\big)}{S\big(\frac{\alpha}{\theta}, \frac{b}{\sigma}, \frac{r}{\sigma}\big)} L, & r \le x \le b \\[2ex] \dfrac{S\big(\frac{\alpha}{\theta}, \frac{r}{\sigma}, \frac{x}{\sigma}\big)}{S\big(\frac{\alpha}{\theta}, \frac{r}{\sigma}, \frac{a}{\sigma}\big)} + \dfrac{S\big(\frac{\alpha}{\theta}, \frac{x}{\sigma}, \frac{a}{\sigma}\big)}{S\big(\frac{\alpha}{\theta}, \frac{r}{\sigma}, \frac{a}{\sigma}\big)} L, & a \le x \le r \end{cases}$$

3.3.4
$$\mathbf{E}_x\big\{e^{-\alpha H}; \ell(H,r) \in dy\big\} = \exp\Big(-\frac{\theta\sigma\sqrt{2} e^{r^2/2\sigma^2} S\big(\frac{\alpha}{\theta}, \frac{b}{\sigma}, \frac{a}{\sigma}\big) y}{\sqrt{\pi} S\big(\frac{\alpha}{\theta}, \frac{b}{\sigma}, \frac{r}{\sigma}\big) S\big(\frac{\alpha}{\theta}, \frac{r}{\sigma}, \frac{a}{\sigma}\big)}\Big)$$

$$\times \begin{cases} \dfrac{\theta\sigma\sqrt{2} e^{r^2/2\sigma^2} S\big(\frac{\alpha}{\theta}, \frac{b}{\sigma}, \frac{x}{\sigma}\big) \Big[S\big(\frac{\alpha}{\theta}, \frac{b}{\sigma}, \frac{r}{\sigma}\big) + S\big(\frac{\alpha}{\theta}, \frac{r}{\sigma}, \frac{a}{\sigma}\big)\Big]}{\sqrt{\pi} S^2\big(\frac{\alpha}{\theta}, \frac{b}{\sigma}, \frac{r}{\sigma}\big) S\big(\frac{\alpha}{\theta}, \frac{r}{\sigma}, \frac{a}{\sigma}\big)} dy, & r \le x \le b \\[2ex] \dfrac{\theta\sigma\sqrt{2} e^{r^2/2\sigma^2} S\big(\frac{\alpha}{\theta}, \frac{x}{\sigma}, \frac{a}{\sigma}\big) \Big[S\big(\frac{\alpha}{\theta}, \frac{b}{\sigma}, \frac{r}{\sigma}\big) + S\big(\frac{\alpha}{\theta}, \frac{r}{\sigma}, \frac{a}{\sigma}\big)\Big]}{\sqrt{\pi} S\big(\frac{\alpha}{\theta}, \frac{b}{\sigma}, \frac{r}{\sigma}\big) S^2\big(\frac{\alpha}{\theta}, \frac{r}{\sigma}, \frac{a}{\sigma}\big)} dy, & a \le x \le r \end{cases}$$

(1)
$$\mathbf{E}_x\big\{e^{-\alpha H}; \ell(H,r) = 0\big\} = \begin{cases} \dfrac{S\big(\frac{\alpha}{\theta}, \frac{x}{\sigma}, \frac{r}{\sigma}\big)}{S\big(\frac{\alpha}{\theta}, \frac{b}{\sigma}, \frac{r}{\sigma}\big)}, & r \le x \le b \\[2ex] \dfrac{S\big(\frac{\alpha}{\theta}, \frac{r}{\sigma}, \frac{x}{\sigma}\big)}{S\big(\frac{\alpha}{\theta}, \frac{r}{\sigma}, \frac{a}{\sigma}\big)}, & a \le x \le r \end{cases}$$

7 $U_s = \sigma e^{-\theta s} W_{e^{2\theta s}-1}$ $\sigma, \theta > 0$ $H = H_{a,b} = \min\{s : U_s \notin (a,b)\}$

3.3.5 (a) $\mathbf{E}_x\{e^{-\gamma\ell(H,r)}; U_H = a\}$

$$= \begin{cases} \dfrac{\operatorname{Erfid}(\frac{b}{\sigma},\frac{x}{\sigma})}{\operatorname{Erfid}(\frac{b}{\sigma},\frac{a}{\sigma}) + \frac{\gamma\sqrt{\pi}}{\theta\sigma\sqrt{2}}e^{-r^2/2\sigma^2}\operatorname{Erfid}(\frac{b}{\sigma},\frac{r}{\sigma})\operatorname{Erfid}(\frac{r}{\sigma},\frac{a}{\sigma})}, & r \le x \le b \\ \dfrac{\operatorname{Erfid}(\frac{b}{\sigma},\frac{x}{\sigma}) + \frac{\gamma\sqrt{\pi}}{\theta\sigma\sqrt{2}}e^{-r^2/2\sigma^2}\operatorname{Erfid}(\frac{b}{\sigma},\frac{r}{\sigma})\operatorname{Erfid}(\frac{r}{\sigma},\frac{x}{\sigma})}{\operatorname{Erfid}(\frac{b}{\sigma},\frac{a}{\sigma}) + \frac{\gamma\sqrt{\pi}}{\theta\sigma\sqrt{2}}e^{-r^2/2\sigma^2}\operatorname{Erfid}(\frac{b}{\sigma},\frac{r}{\sigma})\operatorname{Erfid}(\frac{r}{\sigma},\frac{a}{\sigma})}, & a \le x \le r \end{cases}$$

3.3.5 (b) $\mathbf{E}_x\{e^{-\gamma\ell(H,r)}; U_H = b\}$

$$= \begin{cases} \dfrac{\operatorname{Erfid}(\frac{x}{\sigma},\frac{a}{\sigma}) + \frac{\gamma\sqrt{\pi}}{\theta\sigma\sqrt{2}}e^{-r^2/2\sigma^2}\operatorname{Erfid}(\frac{x}{\sigma},\frac{r}{\sigma})\operatorname{Erfid}(\frac{r}{\sigma},\frac{a}{\sigma})}{\operatorname{Erfid}(\frac{b}{\sigma},\frac{a}{\sigma}) + \frac{\gamma\sqrt{\pi}}{\theta\sigma\sqrt{2}}e^{-r^2/2\sigma^2}\operatorname{Erfid}(\frac{b}{\sigma},\frac{r}{\sigma})\operatorname{Erfid}(\frac{r}{\sigma},\frac{a}{\sigma})}, & r \le x \le b \\ \dfrac{\operatorname{Erfid}(\frac{x}{\sigma},\frac{a}{\sigma})}{\operatorname{Erfid}(\frac{b}{\sigma},\frac{a}{\sigma}) + \frac{\gamma\sqrt{\pi}}{\theta\sigma\sqrt{2}}e^{-r^2/2\sigma^2}\operatorname{Erfid}(\frac{b}{\sigma},\frac{r}{\sigma})\operatorname{Erfid}(\frac{r}{\sigma},\frac{a}{\sigma})}, & a \le x \le r \end{cases}$$

3.3.6 (a) $\mathbf{P}_x\big(\ell(H,r) \in dy, U_H = a\big) = \exp\Big(-\dfrac{\theta\sigma\sqrt{2}e^{r^2/2\sigma^2}\operatorname{Erfid}(\frac{b}{\sigma},\frac{a}{\sigma})y}{\sqrt{\pi}\operatorname{Erfid}(\frac{b}{\sigma},\frac{r}{\sigma})\operatorname{Erfid}(\frac{r}{\sigma},\frac{a}{\sigma})}\Big)$

$$\times \begin{cases} \dfrac{\theta\sigma\sqrt{2}e^{r^2/2\sigma^2}\operatorname{Erfid}(\frac{b}{\sigma},\frac{x}{\sigma})}{\sqrt{\pi}\operatorname{Erfid}(\frac{b}{\sigma},\frac{r}{\sigma})\operatorname{Erfid}(\frac{r}{\sigma},\frac{a}{\sigma})}dy, & r \le x \le b \\ \dfrac{\theta\sigma\sqrt{2}e^{r^2/2\sigma^2}\operatorname{Erfid}(\frac{x}{\sigma},\frac{a}{\sigma})}{\sqrt{\pi}\operatorname{Erfid}^2(\frac{r}{\sigma},\frac{a}{\sigma})}dy, & a \le x \le r \end{cases}$$

(1) $\mathbf{P}_x\big(\ell(H,r) = 0, U_H = a\big) = \begin{cases} 0, & r \le x \le b \\ \dfrac{\operatorname{Erfid}(\frac{r}{\sigma},\frac{x}{\sigma})}{\operatorname{Erfid}(\frac{r}{\sigma},\frac{a}{\sigma})}, & a \le x \le r \end{cases}$

3.3.6 (b) $\mathbf{P}_x\big(\ell(H,r) \in dy, U_H = b\big) = \exp\Big(-\dfrac{\theta\sigma\sqrt{2}e^{r^2/2\sigma^2}\operatorname{Erfid}(\frac{b}{\sigma},\frac{a}{\sigma})y}{\sqrt{\pi}\operatorname{Erfid}(\frac{b}{\sigma},\frac{r}{\sigma})\operatorname{Erfid}(\frac{r}{\sigma},\frac{a}{\sigma})}\Big)$

$$\times \begin{cases} \dfrac{\theta\sigma\sqrt{2}e^{r^2/2\sigma^2}\operatorname{Erfid}(\frac{b}{\sigma},\frac{x}{\sigma})}{\sqrt{\pi}\operatorname{Erfid}^2(\frac{b}{\sigma},\frac{r}{\sigma})}dy, & r \le x \le b \\ \dfrac{\theta\sigma\sqrt{2}e^{r^2/2\sigma^2}\operatorname{Erfid}(\frac{x}{\sigma},\frac{a}{\sigma})}{\sqrt{\pi}\operatorname{Erfid}(\frac{b}{\sigma},\frac{r}{\sigma})\operatorname{Erfid}(\frac{r}{\sigma},\frac{a}{\sigma})}dy, & a \le x \le r \end{cases}$$

(1) $\mathbf{P}_x\big(\ell(H,r) = 0, U_H = b\big) = \begin{cases} \dfrac{\operatorname{Erfid}(\frac{x}{\sigma},\frac{r}{\sigma})}{\operatorname{Erfid}(\frac{b}{\sigma},\frac{r}{\sigma})}, & r \le x \le b \\ 0, & a \le x \le r \end{cases}$

7 $U_s = \sigma e^{-\theta s} W_{e^{2\theta s}-1}$ $\sigma, \theta > 0$ $H = H_{a,b} = \min\{s : U_s \notin (a,b)\}$

3.3.7 (a) $\mathbf{E}_x\{e^{-\alpha \mathrm{m} - \gamma\ell(H,r)}; U_H = a\}$

$$= \begin{cases} \dfrac{S(\frac{\alpha}{\theta}, \frac{b}{\sigma}, \frac{x}{\sigma})}{S(\frac{\alpha}{\theta}, \frac{b}{\sigma}, \frac{a}{\sigma}) + \frac{\gamma\sqrt{\pi}}{\theta\sigma\sqrt{2}} e^{-r^2/2\sigma^2} S(\frac{\alpha}{\theta}, \frac{b}{\sigma}, \frac{r}{\sigma}) S(\frac{\alpha}{\theta}, \frac{r}{\sigma}, \frac{a}{\sigma})}, & r \le x \le b \\ \dfrac{S(\frac{\alpha}{\theta}, \frac{b}{\sigma}, \frac{x}{\sigma}) + \frac{\gamma\sqrt{\pi}}{\theta\sigma\sqrt{2}} e^{-r^2/2\sigma^2} S(\frac{\alpha}{\theta}, \frac{b}{\sigma}, \frac{r}{\sigma}) S(\frac{\alpha}{\theta}, \frac{r}{\sigma}, \frac{x}{\sigma})}{S(\frac{\alpha}{\theta}, \frac{b}{\sigma}, \frac{a}{\sigma}) + \frac{\gamma\sqrt{\pi}}{\theta\sigma\sqrt{2}} e^{-r^2/2\sigma^2} S(\frac{\alpha}{\theta}, \frac{b}{\sigma}, \frac{r}{\sigma}) S(\frac{\alpha}{\theta}, \frac{r}{\sigma}, \frac{a}{\sigma})}, & a \le x \le r \end{cases}$$

3.3.7 (b) $\mathbf{E}_x\{e^{-\alpha \mathrm{m} - \gamma\ell(H,r)}; U_H = b\}$

$$= \begin{cases} \dfrac{S(\frac{\alpha}{\theta}, \frac{x}{\sigma}, \frac{a}{\sigma}) + \frac{\gamma\sqrt{\pi}}{\theta\sigma\sqrt{2}} e^{-r^2/2\sigma^2} S(\frac{\alpha}{\theta}, \frac{x}{\sigma}, \frac{r}{\sigma}) S(\frac{\alpha}{\theta}, \frac{r}{\sigma}, \frac{a}{\sigma})}{S(\frac{\alpha}{\theta}, \frac{b}{\sigma}, \frac{a}{\sigma}) + \frac{\gamma\sqrt{\pi}}{\theta\sigma\sqrt{2}} e^{-r^2/2\sigma^2} S(\frac{\alpha}{\theta}, \frac{b}{\sigma}, \frac{r}{\sigma}) S(\frac{\alpha}{\theta}, \frac{r}{\sigma}, \frac{a}{\sigma})}, & r \le x \le b \\ \dfrac{S(\frac{\alpha}{\theta}, \frac{x}{\sigma}, \frac{a}{\sigma})}{S(\frac{\alpha}{\theta}, \frac{b}{\sigma}, \frac{a}{\sigma}) + \frac{\gamma\sqrt{\pi}}{\theta\sigma\sqrt{2}} e^{-r^2/2\sigma^2} S(\frac{\alpha}{\theta}, \frac{b}{\sigma}, \frac{r}{\sigma}) S(\frac{\alpha}{\theta}, \frac{r}{\sigma}, \frac{a}{\sigma})}, & a \le x \le r \end{cases}$$

3.3.8 (a) $\mathbf{E}_x\{e^{-\alpha H}; \ell(H,r) \in dy, U_H = a\}$

$$= \exp\Big(-\frac{\theta\sigma\sqrt{2}e^{r^2/2\sigma^2} S(\frac{\alpha}{\theta}, \frac{b}{\sigma}, \frac{a}{\sigma}) y}{\sqrt{\pi} S(\frac{\alpha}{\theta}, \frac{b}{\sigma}, \frac{r}{\sigma}) S(\frac{\alpha}{\theta}, \frac{r}{\sigma}, \frac{a}{\sigma})}\Big) \begin{cases} \dfrac{\theta\sigma\sqrt{2}e^{r^2/2\sigma^2} S(\frac{\alpha}{\theta}, \frac{b}{\sigma}, \frac{x}{\sigma})}{\sqrt{\pi} S(\frac{\alpha}{\theta}, \frac{b}{\sigma}, \frac{r}{\sigma}) S(\frac{\alpha}{\theta}, \frac{r}{\sigma}, \frac{a}{\sigma})} dy, & r \le x \le b \\ \dfrac{\theta\sigma\sqrt{2}e^{r^2/2\sigma^2} S(\frac{\alpha}{\theta}, \frac{x}{\sigma}, \frac{a}{\sigma})}{\sqrt{\pi} S^2(\frac{\alpha}{\theta}, \frac{r}{\sigma}, \frac{a}{\sigma})} dy, & a \le x \le r \end{cases}$$

(1) $$\mathbf{E}_x\{e^{-\alpha H}; \ell(H,r) = 0, U_H = a\} = \begin{cases} 0, & r \le x \le b \\ \dfrac{S(\frac{\alpha}{\theta}, \frac{r}{\sigma}, \frac{x}{\sigma})}{S(\frac{\alpha}{\theta}, \frac{r}{\sigma}, \frac{a}{\sigma})}, & a \le x \le r \end{cases}$$

3.3.8 (b) $\mathbf{E}_x\{e^{-\alpha H}; \ell(H,r) \in dy, U_H = b\}$

$$= \exp\Big(-\frac{\theta\sigma\sqrt{2}e^{r^2/2\sigma^2} S(\frac{\alpha}{\theta}, \frac{b}{\sigma}, \frac{a}{\sigma}) y}{\sqrt{\pi} S(\frac{\alpha}{\theta}, \frac{b}{\sigma}, \frac{r}{\sigma}) S(\frac{\alpha}{\theta}, \frac{r}{\sigma}, \frac{a}{\sigma})}\Big) \begin{cases} \dfrac{\theta\sigma\sqrt{2}e^{r^2/2\sigma^2} S(\frac{\alpha}{\theta}, \frac{b}{\sigma}, \frac{x}{\sigma})}{\sqrt{\pi} S^2(\frac{\alpha}{\theta}, \frac{b}{\sigma}, \frac{r}{\sigma})} dy, & r \le x \le b \\ \dfrac{\theta\sigma\sqrt{2}e^{r^2/2\sigma^2} S(\frac{\alpha}{\theta}, \frac{x}{\sigma}, \frac{a}{\sigma})}{\sqrt{\pi} S(\frac{\alpha}{\theta}, \frac{b}{\sigma}, \frac{r}{\sigma}) S(\frac{\alpha}{\theta}, \frac{r}{\sigma}, \frac{a}{\sigma})} dy, & a \le x \le r \end{cases}$$

(1) $$\mathbf{E}_x\{e^{-\alpha H}; \ell(H,r) = 0, U_H = b\} = \begin{cases} \dfrac{S(\frac{\alpha}{\theta}, \frac{x}{\sigma}, \frac{r}{\sigma})}{S(\frac{\alpha}{\theta}, \frac{b}{\sigma}, \frac{r}{\sigma})}, & r \le x \le b \\ 0, & a \le x \le r \end{cases}$$

7 $U_s = \sigma e^{-\theta s} W_{e^{2\theta s}-1}$ $\sigma, \theta > 0$ $H = H_{a,b} = \min\{s : U_s \notin (a,b)\}$

3.4.1 $$\mathbf{E}_r \exp\Big(-\gamma \int_0^H \mathbb{I}_{[r,b]}(U_s)ds\Big)$$

$$= \frac{S(\frac{\gamma}{\theta}, \frac{b}{\sigma}, \frac{r}{\sigma}) + \operatorname{Erfid}(\frac{r}{\sigma}, \frac{a}{\sigma})}{S(\frac{\gamma}{\theta}, \frac{b}{\sigma}, \frac{r}{\sigma}) + \frac{\sqrt{\pi}}{\sqrt{2}} e^{-r^2/2\sigma^2} \operatorname{Erfid}(\frac{r}{\sigma}, \frac{a}{\sigma}) C(\frac{\gamma}{\theta}, \frac{r}{\sigma}, \frac{b}{\sigma})} =: U$$

$$\mathbf{E}_x \exp\Big(-\gamma \int_0^H \mathbb{I}_{[r,b]}(U_s)ds\Big) = \begin{cases} \dfrac{\operatorname{Erfid}(\frac{r}{\sigma}, \frac{x}{\sigma})}{\operatorname{Erfid}(\frac{r}{\sigma}, \frac{a}{\sigma})} + \dfrac{\operatorname{Erfid}(\frac{x}{\sigma}, \frac{a}{\sigma})}{\operatorname{Erfid}(\frac{r}{\sigma}, \frac{a}{\sigma})} U, & a \le x \le r \\ \dfrac{S(\frac{\gamma}{\theta}, \frac{x}{\sigma}, \frac{r}{\sigma})}{S(\frac{\gamma}{\theta}, \frac{b}{\sigma}, \frac{r}{\sigma})} + \dfrac{S(\frac{\gamma}{\theta}, \frac{b}{\sigma}, \frac{x}{\sigma})}{S(\frac{\gamma}{\theta}, \frac{b}{\sigma}, \frac{r}{\sigma})} U, & r \le x \le b \end{cases}$$

3.4.2 (1) $$\mathbf{P}_x\Big(\int_0^H \mathbb{I}_{[r,b]}(U_s)ds = 0\Big) = \begin{cases} \dfrac{\operatorname{Erfid}(\frac{r}{\sigma}, \frac{x}{\sigma})}{\operatorname{Erfid}(\frac{r}{\sigma}, \frac{a}{\sigma})}, & a \le x \le r \\ 0, & r \le x \le b \end{cases}$$

3.4.5 (a) $$\mathbf{E}_x\Big\{\exp\Big(-\gamma \int_0^H \mathbb{I}_{[r,b]}(U_s)ds\Big); U_H = a\Big\}$$

$$= \begin{cases} \dfrac{S(\frac{\gamma}{\theta}, \frac{b}{\sigma}, \frac{r}{\sigma}) + \frac{\sqrt{\pi}}{\sqrt{2}} e^{-r^2/2\sigma^2} \operatorname{Erfid}(\frac{r}{\sigma}, \frac{x}{\sigma}) C(\frac{\gamma}{\theta}, \frac{r}{\sigma}, \frac{b}{\sigma})}{S(\frac{\gamma}{\theta}, \frac{b}{\sigma}, \frac{r}{\sigma}) + \frac{\sqrt{\pi}}{\sqrt{2}} e^{-r^2/2\sigma^2} \operatorname{Erfid}(\frac{r}{\sigma}, \frac{a}{\sigma}) C(\frac{\gamma}{\theta}, \frac{r}{\sigma}, \frac{b}{\sigma})}, & a \le x \le r \\ \dfrac{S(\frac{\gamma}{\theta}, \frac{b}{\sigma}, \frac{x}{\sigma})}{S(\frac{\gamma}{\theta}, \frac{b}{\sigma}, \frac{r}{\sigma}) + \frac{\sqrt{\pi}}{\sqrt{2}} e^{-r^2/2\sigma^2} \operatorname{Erfid}(\frac{r}{\sigma}, \frac{a}{\sigma}) C(\frac{\gamma}{\theta}, \frac{r}{\sigma}, \frac{b}{\sigma})}, & r \le x \le b \end{cases}$$

3.4.5 (b) $$\mathbf{E}_x\Big\{\exp\Big(-\gamma \int_0^H \mathbb{I}_{[r,b]}(U_s)ds\Big); U_H = b\Big\}$$

$$= \begin{cases} \dfrac{\operatorname{Erfid}(\frac{x}{\sigma}, \frac{a}{\sigma})}{S(\frac{\gamma}{\theta}, \frac{b}{\sigma}, \frac{r}{\sigma}) + \frac{\sqrt{\pi}}{\sqrt{2}} e^{-r^2/2\sigma^2} \operatorname{Erfid}(\frac{r}{\sigma}, \frac{a}{\sigma}) C(\frac{\gamma}{\theta}, \frac{r}{\sigma}, \frac{b}{\sigma})}, & a \le x \le r \\ \dfrac{S(\frac{\gamma}{\theta}, \frac{x}{\sigma}, \frac{r}{\sigma}) + \frac{\sqrt{\pi}}{\sqrt{2}} e^{-r^2/2\sigma^2} \operatorname{Erfid}(\frac{r}{\sigma}, \frac{a}{\sigma}) C(\frac{\gamma}{\theta}, \frac{r}{\sigma}, \frac{x}{\sigma})}{S(\frac{\gamma}{\theta}, \frac{b}{\sigma}, \frac{r}{\sigma}) + \frac{\sqrt{\pi}}{\sqrt{2}} e^{-r^2/2\sigma^2} \operatorname{Erfid}(\frac{r}{\sigma}, \frac{a}{\sigma}) C(\frac{\gamma}{\theta}, \frac{r}{\sigma}, \frac{b}{\sigma})}, & r \le x \le b \end{cases}$$

3.4.6 (1) $$\mathbf{P}_x\Big(\int_0^H \mathbb{I}_{[r,b]}(U_s)ds = 0, U_H = a\Big) = \begin{cases} \dfrac{\operatorname{Erfid}(\frac{r}{\sigma}, \frac{x}{\sigma})}{\operatorname{Erfid}(\frac{r}{\sigma}, \frac{a}{\sigma})}, & a \le x \le r \\ 0, & r \le x \le b \end{cases}$$

3.5.1 $$\mathbf{E}_r \exp\Big(-\gamma \int_0^H \mathbb{I}_{[a,r]}(U_s)ds\Big)$$

7 $U_s = \sigma e^{-\theta s} W_{e^{2\theta s}-1}$ $\quad \sigma, \theta > 0$ $\quad H = H_{a,b} = \min\{s : U_s \notin (a,b)\}$

$$= \frac{S(\frac{\gamma}{\theta}, \frac{r}{\sigma}, \frac{a}{\sigma}) + \operatorname{Erfid}(\frac{b}{\sigma}, \frac{r}{\sigma})}{S(\frac{\gamma}{\theta}, \frac{r}{\sigma}, \frac{a}{\sigma}) + \frac{\sqrt{\pi}}{\sqrt{2}} e^{-r^2/2\sigma^2} \operatorname{Erfid}(\frac{b}{\sigma}, \frac{r}{\sigma}) C(\frac{\gamma}{\theta}, \frac{r}{\sigma}, \frac{a}{\sigma})} =: U$$

$$\mathbf{E}_x \exp\Big(-\gamma \int_0^H \mathbb{I}_{[a,r]}(U_s)ds\Big) = \begin{cases} \dfrac{S(\frac{\gamma}{\theta}, \frac{r}{\sigma}, \frac{x}{\sigma})}{S(\frac{\gamma}{\theta}, \frac{r}{\sigma}, \frac{a}{\sigma})} + \dfrac{S(\frac{\gamma}{\theta}, \frac{x}{\sigma}, \frac{a}{\sigma})}{S(\frac{\gamma}{\theta}, \frac{r}{\sigma}, \frac{a}{\sigma})} U, & a \le x \le r \\ \dfrac{\operatorname{Erfid}(\frac{x}{\sigma}, \frac{r}{\sigma})}{\operatorname{Erfid}(\frac{b}{\sigma}, \frac{r}{\sigma})} + \dfrac{\operatorname{Erfid}(\frac{b}{\sigma}, \frac{x}{\sigma})}{\operatorname{Erfid}(\frac{b}{\sigma}, \frac{r}{\sigma})} U, & r \le x \le b \end{cases}$$

3.5.2 (1)

$$\mathbf{P}_x\Big(\int_0^H \mathbb{I}_{[a,r]}(U_s)ds = 0\Big) = \begin{cases} 0, & a \le x \le r \\ \dfrac{\operatorname{Erfid}(\frac{x}{\sigma}, \frac{r}{\sigma})}{\operatorname{Erfid}(\frac{b}{\sigma}, \frac{r}{\sigma})}, & r \le x \le b \end{cases}$$

3.5.5 (a)

$$\mathbf{E}_x\Big\{\exp\Big(-\gamma \int_0^H \mathbb{I}_{[a,r]}(U_s)ds\Big); U_H = a\Big\}$$

$$= \begin{cases} \dfrac{S(\frac{\gamma}{\theta}, \frac{r}{\sigma}, \frac{x}{\sigma}) + \frac{\sqrt{\pi}}{\sqrt{2}} e^{-r^2/2\sigma^2} \operatorname{Erfid}(\frac{b}{\sigma}, \frac{r}{\sigma}) C(\frac{\gamma}{\theta}, \frac{r}{\sigma}, \frac{x}{\sigma})}{S(\frac{\gamma}{\theta}, \frac{r}{\sigma}, \frac{a}{\sigma}) + \frac{\sqrt{\pi}}{\sqrt{2}} e^{-r^2/2\sigma^2} \operatorname{Erfid}(\frac{b}{\sigma}, \frac{r}{\sigma}) C(\frac{\gamma}{\theta}, \frac{r}{\sigma}, \frac{a}{\sigma})}, & a \le x \le r \\ \dfrac{\operatorname{Erfid}(\frac{b}{\sigma}, \frac{x}{\sigma})}{S(\frac{\gamma}{\theta}, \frac{r}{\sigma}, \frac{a}{\sigma}) + \frac{\sqrt{\pi}}{\sqrt{2}} e^{-r^2/2\sigma^2} \operatorname{Erfid}(\frac{b}{\sigma}, \frac{r}{\sigma}) C(\frac{\gamma}{\theta}, \frac{r}{\sigma}, \frac{a}{\sigma})}, & r \le x \le b \end{cases}$$

3.5.5 (b)

$$\mathbf{E}_x\Big\{\exp\Big(-\gamma \int_0^H \mathbb{I}_{[a,r]}(U_s)ds\Big); U_H = b\Big\}$$

$$= \begin{cases} \dfrac{S(\frac{\gamma}{\theta}, \frac{x}{\sigma}, \frac{a}{\sigma})}{S(\frac{\gamma}{\theta}, \frac{r}{\sigma}, \frac{a}{\sigma}) + \frac{\sqrt{\pi}}{\sqrt{2}} e^{-r^2/2\sigma^2} \operatorname{Erfid}(\frac{b}{\sigma}, \frac{r}{\sigma}) C(\frac{\gamma}{\theta}, \frac{r}{\sigma}, \frac{a}{\sigma})}, & a \le x \le r \\ \dfrac{S(\frac{\gamma}{\theta}, \frac{r}{\sigma}, \frac{a}{\sigma}) + \frac{\sqrt{\pi}}{\sqrt{2}} e^{-r^2/2\sigma^2} \operatorname{Erfid}(\frac{x}{\sigma}, \frac{r}{\sigma}) C(\frac{\gamma}{\theta}, \frac{r}{\sigma}, \frac{a}{\sigma})}{S(\frac{\gamma}{\theta}, \frac{r}{\sigma}, \frac{a}{\sigma}) + \frac{\sqrt{\pi}}{\sqrt{2}} e^{-r^2/2\sigma^2} \operatorname{Erfid}(\frac{b}{\sigma}, \frac{r}{\sigma}) C(\frac{\gamma}{\theta}, \frac{r}{\sigma}, \frac{a}{\sigma})}, & r \le x \le b \end{cases}$$

3.5.6 (1)

$$\mathbf{P}_x\Big(\int_0^H \mathbb{I}_{[a,r]}(U_s)ds = 0, U_H = b\Big) = \begin{cases} 0, & a \le x \le r \\ \dfrac{\operatorname{Erfid}(\frac{x}{\sigma}, \frac{r}{\sigma})}{\operatorname{Erfid}(\frac{b}{\sigma}, \frac{r}{\sigma})}, & r \le x \le b \end{cases}$$

3.6.1

$$\mathbf{E}_r \exp\Big(-\int_0^H \big(p\mathbb{I}_{[a,r]}(U_s) + q\mathbb{I}_{[r,b]}(U_s)\big)ds\Big) =: U$$

$$= \frac{\sqrt{2} e^{r^2/2\sigma^2} \big(S(\frac{q}{\theta}, \frac{b}{\sigma}, \frac{r}{\sigma}) + S(\frac{p}{\theta}, \frac{r}{\sigma}, \frac{a}{\sigma})\big)}{\sqrt{\pi}\big(S(\frac{q}{\theta}, \frac{b}{\sigma}, \frac{r}{\sigma}) C(\frac{p}{\theta}, \frac{r}{\sigma}, \frac{a}{\sigma}) + C(\frac{q}{\theta}, \frac{r}{\sigma}, \frac{b}{\sigma}) S(\frac{p}{\theta}, \frac{r}{\sigma}, \frac{a}{\sigma})\big)}$$

7 $U_s = \sigma e^{-\theta s} W_{e^{2\theta s}-1} \qquad \sigma, \theta > 0 \qquad H = H_{a,b} = \min\{s : U_s \notin (a,b)\}$

$$\mathbf{E}_x \exp\Big(-\int_0^H \Big(p\mathbb{I}_{[a,r]}(U_s) + q\mathbb{I}_{[r,b]}(U_s)\Big)ds\Big)$$
$$= \begin{cases} \dfrac{S(\frac{q}{\theta},\frac{x}{\sigma},\frac{r}{\sigma})}{S(\frac{q}{\theta},\frac{b}{\sigma},\frac{r}{\sigma})} + \dfrac{S(\frac{q}{\theta},\frac{b}{\sigma},\frac{x}{\sigma})}{S(\frac{q}{\theta},\frac{b}{\sigma},\frac{r}{\sigma})}U, & r \le x \le b \\ \dfrac{S(\frac{p}{\theta},\frac{r}{\sigma},\frac{x}{\sigma})}{S(\frac{p}{\theta},\frac{r}{\sigma},\frac{a}{\sigma})} + \dfrac{S(\frac{p}{\theta},\frac{x}{\sigma},\frac{a}{\sigma})}{S(\frac{p}{\theta},\frac{r}{\sigma},\frac{a}{\sigma})}U, & a \le x \le r \end{cases}$$

3.6.5 (a)

$$\mathbf{E}_x\Big\{\exp\Big(-\int_0^H \Big(p\mathbb{I}_{[a,r]}(U_s) + q\mathbb{I}_{[r,b]}(U_s)\Big)ds\Big); U_H = a\Big\}$$
$$= \begin{cases} \dfrac{\sqrt{2}e^{r^2/2\sigma^2}S(\frac{q}{\theta},\frac{b}{\sigma},\frac{x}{\sigma})}{\sqrt{\pi}(S(\frac{q}{\theta},\frac{b}{\sigma},\frac{r}{\sigma})C(\frac{p}{\theta},\frac{r}{\sigma},\frac{a}{\sigma}) + C(\frac{q}{\theta},\frac{r}{\sigma},\frac{b}{\sigma})S(\frac{p}{\theta},\frac{r}{\sigma},\frac{a}{\sigma}))}, & r \le x \le b \\ \dfrac{S(\frac{q}{\theta},\frac{b}{\sigma},\frac{r}{\sigma})C(\frac{p}{\theta},\frac{r}{\sigma},\frac{x}{\sigma}) + C(\frac{q}{\theta},\frac{r}{\sigma},\frac{b}{\sigma})S(\frac{p}{\theta},\frac{r}{\sigma},\frac{x}{\sigma})}{S(\frac{q}{\theta},\frac{b}{\sigma},\frac{r}{\sigma})C(\frac{p}{\theta},\frac{r}{\sigma},\frac{a}{\sigma}) + C(\frac{q}{\theta},\frac{r}{\sigma},\frac{b}{\sigma})S(\frac{p}{\theta},\frac{r}{\sigma},\frac{a}{\sigma})} & a \le x \le r \end{cases}$$

3.6.5 (b)

$$\mathbf{E}_x\Big\{\exp\Big(-\int_0^H \Big(p\mathbb{I}_{[a,r]}(U_s) + q\mathbb{I}_{[r,b]}(U_s)\Big)ds\Big); U_H = b\Big\}$$
$$= \begin{cases} \dfrac{S(\frac{q}{\theta},\frac{x}{\sigma},\frac{r}{\sigma})C(\frac{p}{\theta},\frac{r}{\sigma},\frac{a}{\sigma}) + C(\frac{q}{\theta},\frac{r}{\sigma},\frac{x}{\sigma})S(\frac{p}{\theta},\frac{r}{\sigma},\frac{a}{\sigma})}{S(\frac{q}{\theta},\frac{b}{\sigma},\frac{r}{\sigma})C(\frac{p}{\theta},\frac{r}{\sigma},\frac{a}{\sigma}) + C(\frac{q}{\theta},\frac{r}{\sigma},\frac{b}{\sigma})S(\frac{p}{\theta},\frac{r}{\sigma},\frac{a}{\sigma})}, & r \le x \le b \\ \dfrac{\sqrt{2}e^{r^2/2\sigma^2}S(\frac{p}{\theta},\frac{x}{\sigma},\frac{a}{\sigma})}{\sqrt{\pi}(S(\frac{q}{\theta},\frac{b}{\sigma},\frac{r}{\sigma})C(\frac{p}{\theta},\frac{r}{\sigma},\frac{a}{\sigma}) + C(\frac{q}{\theta},\frac{r}{\sigma},\frac{b}{\sigma})S(\frac{p}{\theta},\frac{r}{\sigma},\frac{a}{\sigma}))}, & a \le x \le r \end{cases}$$

3.9.1 $1 \le \gamma$

$$\mathbf{E}_x \exp\Big(-\frac{(\gamma^2-1)\theta}{4\sigma^2}\int_0^H U_s^2 ds\Big)$$
$$= \frac{e^{a^2(\gamma-1)/4\sigma^2}S(\frac{1}{2}-\frac{1}{2\gamma},\frac{b\sqrt{\gamma}}{\sigma},\frac{x\sqrt{\gamma}}{\sigma}) + e^{b^2(\gamma-1)/4\sigma^2}S(\frac{1}{2}-\frac{1}{2\gamma},\frac{x\sqrt{\gamma}}{\sigma},\frac{a\sqrt{\gamma}}{\sigma})}{e^{x^2(\gamma-1)/4\sigma^2}S(\frac{1}{2}-\frac{1}{2\gamma},\frac{b\sqrt{\gamma}}{\sigma},\frac{a\sqrt{\gamma}}{\sigma})}$$

3.9.7 (a)

$$\mathbf{E}_x\Big\{\exp\Big(-\alpha H - \frac{(\gamma^2-1)\theta}{4\sigma^2}\int_0^H U_s^2 ds\Big); U_H = a\Big\} \qquad \gamma \ge 1$$
$$= \frac{e^{a^2(\gamma-1)/4\sigma^2}S(\frac{1}{2}-\frac{1}{2\gamma}+\frac{\alpha}{\theta\gamma},\frac{b\sqrt{\gamma}}{\sigma},\frac{x\sqrt{\gamma}}{\sigma})}{e^{x^2(\gamma-1)/4\sigma^2}S(\frac{1}{2}-\frac{1}{2\gamma}+\frac{\alpha}{\theta\gamma},\frac{b\sqrt{\gamma}}{\sigma},\frac{a\sqrt{\gamma}}{\sigma})}$$

3.9.7 (b)

$$\mathbf{E}_x\Big\{\exp\Big(-\alpha H - \frac{(\gamma^2-1)\theta}{4\sigma^2}\int_0^H U_s^2 ds\Big); U_H = b\Big\} \qquad \gamma \ge 1$$
$$= \frac{e^{b^2(\gamma-1)/4\sigma^2}S(\frac{1}{2}-\frac{1}{2\gamma}+\frac{\alpha}{\theta\gamma},\frac{x\sqrt{\gamma}}{\sigma},\frac{a\sqrt{\gamma}}{\sigma})}{e^{x^2(\gamma-1)/4\sigma^2}S(\frac{1}{2}-\frac{1}{2\gamma}+\frac{\alpha}{\theta\gamma},\frac{b\sqrt{\gamma}}{\sigma},\frac{a\sqrt{\gamma}}{\sigma})}$$

7 $U_s = \sigma e^{-\theta s} W_{e^{2\theta s}-1}$ $\sigma, \theta > 0$ $H = H_{a,b} = \min\{s : U_s \notin (a,b)\}$

3.12.5 (a) $$\mathbf{E}_x\{e^{-\gamma \check{H}(H)}; U_H = a\} = \int_x^b \frac{\sqrt{2}\, e^{y^2/2\sigma^2} S(\frac{\gamma}{\theta}, \frac{x}{\sigma}, \frac{a}{\sigma})}{\sigma\sqrt{\pi}\,\mathrm{Erfid}(\frac{y}{\sigma}, \frac{a}{\sigma}) S(\frac{\gamma}{\theta}, \frac{y}{\sigma}, \frac{a}{\sigma})} dy$$

3.12.5 (b) $$\mathbf{E}_x\{e^{-\gamma \hat{H}(H)}; U_H = b\} = \int_a^x \frac{\sqrt{2}\, e^{y^2/2\sigma^2} S(\frac{\gamma}{\theta}, \frac{b}{\sigma}, \frac{x}{\sigma})}{\sigma\sqrt{\pi}\,\mathrm{Erfid}(\frac{b}{\sigma}, \frac{y}{\sigma}) S(\frac{\gamma}{\theta}, \frac{b}{\sigma}, \frac{y}{\sigma})} dy$$

3.12.7 (a) $$\mathbf{E}_x\{e^{-\alpha H - \gamma \check{H}(H)}; U_H = a\} = \int_x^b \frac{\sqrt{2}\, e^{y^2/2\sigma^2} S(\frac{\gamma+\alpha}{\theta}, \frac{x}{\sigma}, \frac{a}{\sigma})}{\sigma\sqrt{\pi} S(\frac{\alpha}{\theta}, \frac{y}{\sigma}, \frac{a}{\sigma}) S(\frac{\gamma+\alpha}{\theta}, \frac{y}{\sigma}, \frac{a}{\sigma})} dy$$

3.12.7 (b) $$\mathbf{E}_x\{e^{-\alpha H - \gamma \hat{H}(H)}; U_H = b\} = \int_a^x \frac{\sqrt{2}\, e^{y^2/2\sigma^2} S(\frac{\gamma+\alpha}{\theta}, \frac{b}{\sigma}, \frac{x}{\sigma})}{\sigma\sqrt{\pi} S(\frac{\alpha}{\theta}, \frac{b}{\sigma}, \frac{y}{\sigma}) S(\frac{\gamma+\alpha}{\theta}, \frac{b}{\sigma}, \frac{y}{\sigma})} dy$$

3.20.5 (a) $$\mathbf{E}_x\Big\{\exp\Big(-\gamma^2\sigma^2\theta \int_0^H \frac{ds}{U_s^2}\Big); U_H = a\Big\}$$

$$= \frac{x^{\sqrt{\gamma^2+1/4}+1/2} S(\frac14\sqrt{4\gamma^2+1} + \frac14, \frac12\sqrt{4\gamma^2+1}+1, \frac{b^2}{4\sigma^2}, \frac{x^2}{4\sigma^2})}{a^{\sqrt{\gamma^2+1/4}+1/2} S(\frac14\sqrt{4\gamma^2+1} + \frac14, \frac12\sqrt{4\gamma^2+1}+1, \frac{b^2}{4\sigma^2}, \frac{a^2}{4\sigma^2})}, \qquad 0 < a \le x \le b$$

3.20.5 (b) $$\mathbf{E}_x\Big\{\exp\Big(-\gamma^2\sigma^2\theta \int_0^H \frac{ds}{U_s^2}\Big); U_H = b\Big\}$$

$$= \frac{x^{\sqrt{\gamma^2+1/4}+1/2} S(\frac14\sqrt{4\gamma^2+1} + \frac14, \frac12\sqrt{4\gamma^2+1}+1, \frac{x^2}{4\sigma^2}, \frac{a^2}{4\sigma^2})}{b^{\sqrt{\gamma^2+1/4}+1/2} S(\frac14\sqrt{4\gamma^2+1} + \frac14, \frac12\sqrt{4\gamma^2+1}+1, \frac{b^2}{4\sigma^2}, \frac{a^2}{4\sigma^2})}, \qquad 0 < a \le x \le b$$

3.21.5 (a) $$\mathbf{E}_x\Big\{\exp\Big(-\int_0^H \Big(\frac{p^2\sigma^2\theta}{U_s^2} + \frac{\theta(q^2-1)}{4\sigma^2} U_s^2\Big) ds\Big); U_H = a\Big\}$$

$0 < a$ $$= \frac{x^{\sqrt{p^2+1/4}+1/2} e^{x^2(1-q)/4\sigma^2} S(\frac14\sqrt{4p^2+1} + \frac{2q-1}{4q}, \frac12\sqrt{4p^2+1}+1, \frac{qb^2}{4\sigma^2}, \frac{qx^2}{4\sigma^2})}{a^{\sqrt{p^2+1/4}+1/2} e^{a^2(1-q)/4\sigma^2} S(\frac14\sqrt{4p^2+1} + \frac{2q-1}{4q}, \frac12\sqrt{4p^2+1}+1, \frac{qb^2}{4\sigma^2}, \frac{qa^2}{4\sigma^2})}$$

3.21.5 (b) $$\mathbf{E}_x\Big\{\exp\Big(-\int_0^H \Big(\frac{p^2\sigma^2\theta}{U_s^2} + \frac{\theta(q^2-1)}{4\sigma^2} U_s^2\Big) ds\Big); U_H = b\Big\}$$

$0 < a$ $$= \frac{x^{\sqrt{p^2+1/4}+1/2} e^{x^2(1-q)/4\sigma^2} S(\frac14\sqrt{4p^2+1} + \frac{2q-1}{4q}, \frac12\sqrt{4p^2+1}+1, \frac{qx^2}{4\sigma^2}, \frac{qa^2}{4\sigma^2})}{b^{\sqrt{p^2+1/4}+1/2} e^{b^2(1-q)/4\sigma^2} S(\frac14\sqrt{4p^2+1} + \frac{2q-1}{4q}, \frac12\sqrt{4p^2+1}+1, \frac{qb^2}{4\sigma^2}, \frac{qa^2}{4\sigma^2})}$$

7 $U_s = \sigma e^{-\theta s} W_{e^{2\theta s}-1}$ $\sigma, \theta > 0$ $H = H_{a,b} = \min\{s : U_s \notin (a,b)\}$

3.27.1 $$\mathbf{E}_x\Big\{\exp\Big(-\int_0^H \big(p\mathbb{1}_{[a,r)}(U_s) + q\mathbb{1}_{[r,b]}(U_s)\big)ds\Big); \ell(H,r) \in dy\Big\}$$

$$= \exp\Big(-y\theta\sigma\Big(\frac{C(\frac{p}{\theta},\frac{r}{\sigma},\frac{a}{\sigma})}{S(\frac{p}{\theta},\frac{r}{\sigma},\frac{a}{\sigma})} + \frac{C(\frac{q}{\theta},\frac{r}{\sigma},\frac{b}{\sigma})}{S(\frac{q}{\theta},\frac{b}{\sigma},\frac{r}{\sigma})}\Big)\Big)$$

$$\times \begin{cases} \dfrac{\theta\sigma\sqrt{2}e^{r^2/\sigma^2} S(\frac{q}{\theta},\frac{b}{\sigma},\frac{x}{\sigma})\Big[S(\frac{q}{\theta},\frac{b}{\sigma},\frac{r}{\sigma}) + S(\frac{p}{\theta},\frac{r}{\sigma},\frac{a}{\sigma})\Big]}{\sqrt{\pi}S^2(\frac{q}{\theta},\frac{b}{\sigma},\frac{r}{\sigma})S(\frac{p}{\theta},\frac{r}{\sigma},\frac{a}{\sigma})}dy, & r \le x \le b \\ \dfrac{\theta\sigma\sqrt{2}e^{r^2/\sigma^2} S(\frac{p}{\theta},\frac{x}{\sigma},\frac{a}{\sigma})\Big[S(\frac{q}{\theta},\frac{b}{\sigma},\frac{r}{\sigma}) + S(\frac{p}{\theta},\frac{r}{\sigma},\frac{a}{\sigma})\Big]}{\sqrt{\pi}S(\frac{q}{\theta},\frac{b}{\sigma},\frac{r}{\sigma})S^2(\frac{p}{\theta},\frac{r}{\sigma},\frac{a}{\sigma})}dy, & a \le x \le r \end{cases}$$

3.27.5 (a) $$\mathbf{E}_x\Big\{\exp\Big(-\int_0^H \big(p\mathbb{1}_{[a,r)}(U_s) + q\mathbb{1}_{[r,b]}(U_s)\big)ds\Big); \ell(H,r) \in dy, U_H = a\Big\}$$

$$= \exp\Big(-y\theta\sigma\Big(\frac{C(\frac{p}{\theta},\frac{r}{\sigma},\frac{a}{\sigma})}{S(\frac{p}{\theta},\frac{r}{\sigma},\frac{a}{\sigma})} + \frac{C(\frac{q}{\theta},\frac{r}{\sigma},\frac{b}{\sigma})}{S(\frac{q}{\theta},\frac{b}{\sigma},\frac{r}{\sigma})}\Big)\Big) \begin{cases} \dfrac{\theta\sigma\sqrt{2}e^{r^2/\sigma^2} S(\frac{q}{\theta},\frac{b}{\sigma},\frac{x}{\sigma})}{\sqrt{\pi}S(\frac{q}{\theta},\frac{b}{\sigma},\frac{r}{\sigma})S(\frac{p}{\theta},\frac{r}{\sigma},\frac{a}{\sigma})}dy, & r \le x \\ & x \le b \\ \dfrac{\theta\sigma\sqrt{2}e^{r^2/\sigma^2} S(\frac{p}{\theta},\frac{x}{\sigma},\frac{a}{\sigma})}{\sqrt{\pi}S^2(\frac{p}{\theta},\frac{r}{\sigma},\frac{a}{\sigma})}dy, & a \le x \\ & x \le r \end{cases}$$

(1) $$\mathbf{E}_x\Big\{\exp\Big(-\int_0^H \big(p\mathbb{1}_{[a,r)}(U_s) + q\mathbb{1}_{[r,b]}(U_s)\big)ds\Big); \ell(H,r) = 0, U_H = a\Big\}$$

$$= \frac{S(\frac{p}{\theta},\frac{r}{\sigma},\frac{x}{\sigma})}{S(\frac{p}{\theta},\frac{r}{\sigma},\frac{a}{\sigma})}, \quad a \le x \le r$$

3.27.5 (b) $$\mathbf{E}_x\Big\{\exp\Big(-\int_0^H \big(p\mathbb{1}_{[a,r)}(U_s) + q\mathbb{1}_{[r,b]}(U_s)\big)ds\Big); \ell(H,r) \in dy, U_H = b\Big\}$$

$$= \exp\Big(-y\theta\sigma\Big(\frac{C(\frac{p}{\theta},\frac{r}{\sigma},\frac{a}{\sigma})}{S(\frac{p}{\theta},\frac{r}{\sigma},\frac{a}{\sigma})} + \frac{C(\frac{q}{\theta},\frac{r}{\sigma},\frac{b}{\sigma})}{S(\frac{q}{\theta},\frac{b}{\sigma},\frac{r}{\sigma})}\Big)\Big) \begin{cases} \dfrac{\theta\sigma\sqrt{2}e^{r^2/\sigma^2} S(\frac{q}{\theta},\frac{b}{\sigma},\frac{x}{\sigma})}{\sqrt{\pi}S^2(\frac{q}{\theta},\frac{b}{\sigma},\frac{r}{\sigma})}dy, & r \le x \\ & x \le b \\ \dfrac{\theta\sigma\sqrt{2}e^{r^2/\sigma^2} S(\frac{p}{\theta},\frac{x}{\sigma},\frac{a}{\sigma})}{\sqrt{\pi}S(\frac{q}{\theta},\frac{b}{\sigma},\frac{r}{\sigma})S(\frac{p}{\theta},\frac{r}{\sigma},\frac{a}{\sigma})}dy, & a \le x \\ & x \le r \end{cases}$$

(1) $$\mathbf{E}_x\Big\{\exp\Big(-\int_0^H \big(p\mathbb{1}_{[a,r)}(U_s) + q\mathbb{1}_{[r,b]}(U_s)\big)ds\Big); \ell(H,r) = 0, U_H = b\Big\}$$

$$= \frac{S(\frac{q}{\theta},\frac{x}{\sigma},\frac{r}{\sigma})}{S(\frac{q}{\theta},\frac{b}{\sigma},\frac{r}{\sigma})}, \quad r \le x \le b$$

7 $U_s = \sigma e^{-\theta s} W_{e^{2\theta s}-1} \qquad \sigma, \theta > 0 \qquad \varrho = \varrho(v,z) = \min\{s : \ell(s,z) = v\}$

4. Stopping at inverse local time

4.0.1
$$\mathbf{E}_x e^{-\alpha\varrho} = \begin{cases} \dfrac{e^{x^2/4\sigma^2} D_{-\alpha/\theta}(-\frac{x}{\sigma})}{e^{z^2/4\sigma^2} D_{-\alpha/\theta}(-\frac{z}{\sigma})} \exp\Big(-\dfrac{\theta\sigma\sqrt{2\pi}v}{\Gamma(\alpha/\theta) D_{-\alpha/\theta}(-\frac{z}{\sigma}) D_{-\alpha/\theta}(\frac{z}{\sigma})}\Big), & x \le z \\ \dfrac{e^{x^2/4\sigma^2} D_{-\alpha/\theta}(\frac{x}{\sigma})}{e^{z^2/4\sigma^2} D_{-\alpha/\theta}(\frac{z}{\sigma})} \exp\Big(-\dfrac{\theta\sigma\sqrt{2\pi}v}{\Gamma(\alpha/\theta) D_{-\alpha/\theta}(-\frac{z}{\sigma}) D_{-\alpha/\theta}(\frac{z}{\sigma})}\Big), & z \le x \end{cases}$$

4.1.2
$z \le y$
$$\mathbf{P}_x\Big(\sup_{0\le s\le\varrho} U_s < y\Big) = \begin{cases} \exp\Big(-\dfrac{\theta\sigma\sqrt{2}e^{z^2/2\sigma^2}v}{\sqrt{\pi}\,\mathrm{Erfid}(\frac{y}{\sigma},\frac{z}{\sigma})}\Big), & x \le z \\ \dfrac{\mathrm{Erfid}(\frac{y}{\sigma},\frac{x}{\sigma})}{\mathrm{Erfid}(\frac{y}{\sigma},\frac{z}{\sigma})} \exp\Big(-\dfrac{\theta\sigma\sqrt{2}e^{z^2/2\sigma^2}v}{\sqrt{\pi}\,\mathrm{Erfid}(\frac{y}{\sigma},\frac{z}{\sigma})}\Big), & z \le x \le y \end{cases}$$

4.1.4
$z \le y$
$$\mathbf{E}_x\{e^{-\alpha\varrho}; \sup_{0\le s\le\varrho} U_s \le y\}$$

$$= \begin{cases} \dfrac{e^{x^2/4\sigma^2} D_{-\alpha/\theta}(-\frac{x}{\sigma})}{e^{z^2/4\sigma^2} D_{-\alpha/\theta}(-\frac{z}{\sigma})} \exp\Big(-\dfrac{\theta\sigma\sqrt{2}e^{(z^2+y^2)/4\sigma^2} D_{-\alpha/\theta}(-\frac{y}{\sigma})v}{\sqrt{\pi} D_{-\alpha/\theta}(-\frac{z}{\sigma}) S(\frac{\alpha}{\theta},\frac{y}{\sigma},\frac{z}{\sigma})}\Big), & x \le z \\ \dfrac{S(\frac{\alpha}{\theta},\frac{y}{\sigma},\frac{x}{\sigma})}{S(\frac{\alpha}{\theta},\frac{y}{\sigma},\frac{z}{\sigma})} \exp\Big(-\dfrac{\theta\sigma\sqrt{2}e^{(z^2+y^2)/4\sigma^2} D_{-\alpha/\theta}(-\frac{y}{\sigma})v}{\sqrt{\pi} D_{-\alpha/\theta}(-\frac{z}{\sigma}) S(\frac{\alpha}{\theta},\frac{y}{\sigma},\frac{z}{\sigma})}\Big), & z \le x \le y \end{cases}$$

4.2.2
$y \le z$
$$\mathbf{P}_x\Big(\inf_{0\le s\le\varrho} U_s \le y\Big) = \begin{cases} \dfrac{\mathrm{Erfid}(\frac{x}{\sigma},\frac{y}{\sigma})}{\mathrm{Erfid}(\frac{z}{\sigma},\frac{y}{\sigma})} \exp\Big(-\dfrac{\theta\sigma\sqrt{2}e^{z^2/2\sigma^2}v}{\sqrt{\pi}\,\mathrm{Erfid}(\frac{z}{\sigma},\frac{y}{\sigma})}\Big), & y \le x \le z \\ \exp\Big(-\dfrac{\theta\sigma\sqrt{2}e^{z^2/2\sigma^2}v}{\sqrt{\pi}\,\mathrm{Erfid}(\frac{z}{\sigma},\frac{y}{\sigma})}\Big), & z \le x \end{cases}$$

4.2.4
$y \le z$
$$\mathbf{E}_x\{e^{-\alpha\varrho}; \inf_{0\le s\le\varrho} U_s \ge y\}$$

$$= \begin{cases} \dfrac{S(\frac{\alpha}{\theta},\frac{x}{\sigma},\frac{y}{\sigma})}{S(\frac{\alpha}{\theta},\frac{z}{\sigma},\frac{y}{\sigma})} \exp\Big(-\dfrac{\theta\sigma\sqrt{2}e^{(z^2+y^2)/4\sigma^2} D_{-\alpha/\theta}(\frac{y}{\sigma})v}{\sqrt{\pi} S(\frac{\alpha}{\theta},\frac{z}{\sigma},\frac{y}{\sigma}) D_{-\alpha/\theta}(\frac{z}{\sigma})}\Big), & y \le x \le z \\ \dfrac{e^{x^2/4\sigma^2} D_{-\alpha/\theta}(\frac{x}{\sigma})}{e^{z^2/4\sigma^2} D_{-\alpha/\theta}(\frac{z}{\sigma})} \exp\Big(-\dfrac{\theta\sigma\sqrt{2}e^{(z^2+y^2)/4\sigma^2} D_{-\alpha/\theta}(\frac{y}{\sigma})v}{\sqrt{\pi} S(\frac{\alpha}{\theta},\frac{z}{\sigma},\frac{y}{\sigma}) D_{-\alpha/\theta}(\frac{z}{\sigma})}\Big), & z \le x \end{cases}$$

4.3.1
$$\mathbf{E}_x e^{-\gamma\ell(\varrho,r)} = \exp\Big(-\dfrac{\gamma e^{z^2/2\sigma^2}v}{e^{r^2/2\sigma^2} + \frac{\gamma\sqrt{\pi}}{\theta\sigma\sqrt{2}}|\,\mathrm{Erfid}(\frac{r}{\sigma},\frac{z}{\sigma})|}\Big)$$

7 $U_s = \sigma e^{-\theta s} W_{e^{2\theta s}-1}$ $\sigma, \theta > 0$ $\varrho = \varrho(v,z) = \min\{s : \ell(s,z) = v\}$

$$\times \begin{cases} 1, & x \le z \le r, \quad r \le z \le x \\ \dfrac{1 + \frac{\gamma\sqrt{\pi}}{\theta\sigma\sqrt{2}} e^{-r^2/2\sigma^2} |\operatorname{Erfid}(\frac{x}{\sigma}, \frac{r}{\sigma})|}{1 + \frac{\gamma\sqrt{\pi}}{\theta\sigma\sqrt{2}} e^{-r^2/2\sigma^2} |\operatorname{Erfid}(\frac{z}{\sigma}, \frac{r}{\sigma})|}, & r \wedge z \le x \le r \vee z \\ \dfrac{1}{1 + \frac{\gamma\sqrt{\pi}}{\theta\sigma\sqrt{2}} e^{-r^2/2\sigma^2} |\operatorname{Erfid}(\frac{z}{\sigma}, \frac{r}{\sigma})|}, & z \le r \le x, \quad x \le r \le z \end{cases}$$

4.3.2 $\mathbf{P}_x\big(\ell(\varrho, r) \in dy\big) = \exp\Big(-\dfrac{\theta\sigma\sqrt{2}(e^{z^2/2\sigma^2} v + e^{r^2/2\sigma^2} y)}{\sqrt{\pi}|\operatorname{Erfid}(\frac{r}{\sigma}, \frac{z}{\sigma})|}\Big)$

$$\times \begin{cases} \dfrac{\sigma\theta\sqrt{2} e^{(z^2+r^2)/4\sigma^2}}{\sqrt{\pi}|\operatorname{Erfid}(\frac{r}{\sigma}, \frac{z}{\sigma})|} \dfrac{\sqrt{v}}{\sqrt{y}} I_1\Big(\dfrac{2\sqrt{2}\theta\sigma e^{(z^2+r^2)/4\sigma^2}\sqrt{vy}}{\sqrt{\pi}|\operatorname{Erfid}(\frac{r}{\sigma}, \frac{z}{\sigma})|}\Big) dy, & \begin{matrix} r \le z \le x \\ x \le z \le r \end{matrix} \\ \dfrac{\theta\sigma\sqrt{2} e^{r^2/4\sigma^2}}{\sqrt{\pi}\operatorname{Erfid}^2(\frac{r}{\sigma}, \frac{z}{\sigma})} \Big[|\operatorname{Erfid}(\frac{r}{\sigma}, \frac{x}{\sigma})| e^{z^2/4\sigma^2} \dfrac{\sqrt{v}}{\sqrt{y}} I_1\Big(\dfrac{2\sqrt{2}\theta\sigma e^{(z^2+r^2)/4\sigma^2}\sqrt{vy}}{\sqrt{\pi}|\operatorname{Erfid}(\frac{r}{\sigma}, \frac{z}{\sigma})|}\Big) & \\ + |\operatorname{Erfid}(\frac{x}{\sigma}, \frac{z}{\sigma})| e^{r^2/4\sigma^2} I_0\Big(\dfrac{2\sqrt{2}\theta\sigma e^{(z^2+r^2)/4\sigma^2}\sqrt{vy}}{\sqrt{\pi}|\operatorname{Erfid}(\frac{r}{\sigma}, \frac{z}{\sigma})|}\Big)\Big] dy, & z \wedge r \le x \le r \vee z \\ \dfrac{\theta\sigma\sqrt{2} e^{r^2/2\sigma^2}}{\sqrt{\pi}|\operatorname{Erfid}(\frac{r}{\sigma}, \frac{z}{\sigma})|} I_0\Big(\dfrac{2\sqrt{2}\theta\sigma e^{(z^2+r^2)/4\sigma^2}\sqrt{vy}}{\sqrt{\pi}|\operatorname{Erfid}(\frac{r}{\sigma}, \frac{z}{\sigma})|}\Big) dy, & z \le r \le x, \ x \le r \le z \end{cases}$$

(1) $\mathbf{P}_x\big(\ell(\varrho, r) = 0\big)$

$$= \begin{cases} \exp\Big(-\dfrac{\theta\sigma\sqrt{2} e^{z^2/2\sigma^2} v}{\sqrt{\pi}|\operatorname{Erfid}(\frac{r}{\sigma}, \frac{z}{\sigma})|}\Big), & x \le z \le r, \quad r \le z \le x \\ \dfrac{|\operatorname{Erfid}(\frac{x}{\sigma}, \frac{r}{\sigma})|}{|\operatorname{Erfid}(\frac{z}{\sigma}, \frac{r}{\sigma})|} \exp\Big(-\dfrac{\theta\sigma\sqrt{2} e^{z^2/2\sigma^2} v}{\sqrt{\pi}|\operatorname{Erfid}(\frac{r}{\sigma}, \frac{z}{\sigma})|}\Big), & r \wedge z \le x \le r \vee z \\ 0, & z \le r \le x, \quad x \le r \le z \end{cases}$$

4.3.3 $\mathbf{E}_x e^{-\alpha\varrho - \gamma\ell(\varrho, r)}$

$z \le r$
$$= \exp\Big(-\frac{\big(\frac{\theta\sigma\sqrt{2\pi}}{\Gamma(\alpha/\theta)} + \gamma D_{-\alpha/\theta}(-\frac{r}{\sigma}) D_{-\alpha/\theta}(\frac{r}{\sigma})\big)v}{D_{-\alpha/\theta}(-\frac{z}{\sigma})\big(D_{-\alpha/\theta}(\frac{z}{\sigma}) + \frac{\gamma\sqrt{\pi}}{\theta\sigma\sqrt{2}} e^{-(z^2+r^2)/4\sigma^2} D_{-\alpha/\theta}(\frac{r}{\sigma}) S(\frac{\alpha}{\theta}, \frac{r}{\sigma}, \frac{z}{\sigma})\big)}\Big)$$

$$\times \begin{cases} \dfrac{e^{x^2/4\sigma^2} D_{-\alpha/\theta}(\frac{x}{\sigma})}{e^{z^2/4\sigma^2} D_{-\alpha/\theta}(\frac{z}{\sigma}) + \frac{\gamma\sqrt{\pi}}{\theta\sigma\sqrt{2}} e^{-r^2/4\sigma^2} D_{-\alpha/\theta}(\frac{r}{\sigma}) S(\frac{\alpha}{\theta}, \frac{r}{\sigma}, \frac{z}{\sigma})}, & r \le x \\ \dfrac{e^{x^2/4\sigma^2} D_{-\alpha/\theta}(\frac{x}{\sigma}) + \frac{\gamma\sqrt{\pi}}{\theta\sigma\sqrt{2}} e^{-r^2/4\sigma^2} D_{-\alpha/\theta}(\frac{r}{\sigma}) S(\frac{\alpha}{\theta}, \frac{r}{\sigma}, \frac{x}{\sigma})}{e^{z^2/4\sigma^2} D_{-\alpha/\theta}(\frac{z}{\sigma}) + \frac{\gamma\sqrt{\pi}}{\theta\sigma\sqrt{2}} e^{-r^2/4\sigma^2} D_{-\alpha/\theta}(\frac{r}{\sigma}) S(\frac{\alpha}{\theta}, \frac{r}{\sigma}, \frac{z}{\sigma})}, & z \le x \le r \\ \dfrac{e^{x^2/4\sigma^2} D_{-\alpha/\theta}(-\frac{x}{\sigma})}{e^{z^2/4\sigma^2} D_{-\alpha/\theta}(-\frac{z}{\sigma})}, & 0 \le x \le z \end{cases}$$

7 $U_s = \sigma e^{-\theta s} W_{e^{2\theta s}-1} \qquad \sigma, \theta > 0 \qquad \varrho = \varrho(v,z) = \min\{s : \ell(s,z) = v\}$

$r \le z$

$$= \exp\Big(-\frac{\big(\frac{\theta\sigma\sqrt{2\pi}}{\Gamma(\alpha/\theta)} + \gamma D_{-\alpha/\theta}(-\frac{r}{\sigma}) D_{-\alpha/\theta}(\frac{r}{\sigma})\big)v}{D_{-\alpha/\theta}(\frac{z}{\sigma})\big(D_{-\alpha/\theta}(-\frac{z}{\sigma}) + \frac{\gamma\sqrt{\pi}}{\theta\sigma\sqrt{2}} e^{-(z^2+r^2)/4\sigma^2} S(\frac{\alpha}{\theta},\frac{z}{\sigma},\frac{r}{\sigma}) D_{-\alpha/\theta}(-\frac{r}{\sigma})\big)}\Big)$$

$$\times \begin{cases} \dfrac{e^{x^2/4\sigma^2} D_{-\alpha/\theta}(\frac{x}{\sigma})}{e^{z^2/4\sigma^2} D_{-\alpha/\theta}(\frac{z}{\sigma})}, & z \le x \\[2ex] \dfrac{e^{x^2/4\sigma^2} D_{-\alpha/\theta}(-\frac{x}{\sigma}) + \frac{\gamma\sqrt{\pi}}{\theta\sigma\sqrt{2}} e^{-r^2/4\sigma^2} D_{-\alpha/\theta}(-\frac{r}{\sigma}) S(\frac{\alpha}{\theta},\frac{x}{\sigma},\frac{r}{\sigma})}{e^{z^2/4\sigma^2} D_{-\alpha/\theta}(-\frac{z}{\sigma}) + \frac{\gamma\sqrt{\pi}}{\theta\sigma\sqrt{2}} e^{-r^2/4\sigma^2} D_{-\alpha/\theta}(-\frac{r}{\sigma}) S(\frac{\alpha}{\theta},\frac{z}{\sigma},\frac{r}{\sigma})}, & r \le x \le z \\[2ex] \dfrac{e^{x^2/4\sigma^2} D_{-\alpha/\theta}(-\frac{x}{\sigma})}{e^{z^2/4\sigma^2} D_{-\alpha/\theta}(-\frac{z}{\sigma}) + \frac{\gamma\sqrt{\pi}}{\theta\sigma\sqrt{2}} e^{-r^2/4\sigma^2} D_{-\alpha/\theta}(-\frac{r}{\sigma}) S(\frac{\alpha}{\theta},\frac{z}{\sigma},\frac{r}{\sigma})}, & x \le r \end{cases}$$

4.3.4 $\mathbf{E}_x\{e^{-\alpha\varrho}; \ell(\varrho, r) \in dy\}$

$z \le r$

$$= \exp\Big(-\frac{\theta\sigma\sqrt{2}e^{(z^2+r^2)/4\sigma^2} D_{-\alpha/\theta}(-\frac{r}{\sigma})v}{\sqrt{\pi}S(\frac{\alpha}{\theta},\frac{r}{\sigma},\frac{z}{\sigma}) D_{-\alpha/\theta}(-\frac{z}{\sigma})} - \frac{\theta\sigma\sqrt{2}e^{(z^2+r^2)/4\sigma^2} D_{-\alpha/\theta}(\frac{z}{\sigma})y}{\sqrt{\pi}D_{-\alpha/\theta}(\frac{r}{\sigma}) S(\frac{\alpha}{\theta},\frac{r}{\sigma},\frac{z}{\sigma})}\Big)$$

$$\times \begin{cases} \dfrac{\theta\sigma\sqrt{2}e^{(x^2+r^2)/4\sigma^2} D_{-\alpha/\theta}(-\frac{x}{\sigma})}{\sqrt{\pi}D_{-\alpha/\theta}(-\frac{z}{\sigma}) S(\frac{\alpha}{\theta},\frac{r}{\sigma},\frac{z}{\sigma})} \dfrac{\sqrt{v}}{\sqrt{y}} I_1\Big(\dfrac{2\theta\sigma\sqrt{2}e^{(z^2+r^2)/4\sigma^2}\sqrt{vy}}{\sqrt{\pi}S(\frac{\alpha}{\theta},\frac{r}{\sigma},\frac{z}{\sigma})}\Big)dy, & 0 \le x \le z \\[2ex] \dfrac{\theta\sigma\sqrt{2}e^{(z^2+r^2)/4\sigma^2}}{\sqrt{\pi}S^2(\frac{\alpha}{\theta},\frac{r}{\sigma},\frac{z}{\sigma})}\Big[S(\tfrac{\alpha}{\theta},\tfrac{r}{\sigma},\tfrac{x}{\sigma}) \dfrac{\sqrt{v}}{\sqrt{y}} I_1\Big(\dfrac{2\theta\sigma\sqrt{2}e^{(z^2+r^2)/4\sigma^2}\sqrt{vy}}{\sqrt{\pi}S(\frac{\alpha}{\theta},\frac{r}{\sigma},\frac{z}{\sigma})}\Big) & \\ \qquad + S(\tfrac{\alpha}{\theta},\tfrac{x}{\sigma},\tfrac{z}{\sigma}) \dfrac{e^{r^2/4\sigma^2}}{e^{z^2/4\sigma^2}} I_0\Big(\dfrac{2\theta\sigma\sqrt{2}e^{(z^2+r^2)/4\sigma^2}\sqrt{vy}}{\sqrt{\pi}S(\frac{\alpha}{\theta},\frac{r}{\sigma},\frac{z}{\sigma})}\Big)\Big]dy, & z \le x \le r \\[2ex] \dfrac{\theta\sigma\sqrt{2}e^{(x^2+r^2)/4\sigma^2} D_{-\alpha/\theta}(\frac{x}{\sigma})}{\sqrt{\pi}D_{-\alpha/\theta}(\frac{r}{\sigma}) S(\frac{\alpha}{\theta},\frac{r}{\sigma},\frac{z}{\sigma})} I_0\Big(\dfrac{2\theta\sigma\sqrt{2}e^{(z^2+r^2)/4\sigma^2}\sqrt{vy}}{\sqrt{\pi}S(\frac{\alpha}{\theta},\frac{r}{\sigma},\frac{z}{\sigma})}\Big)dy, & r \le x \end{cases}$$

$r \le z$

$$= \exp\Big(-\frac{\theta\sigma\sqrt{2}e^{(z^2+r^2)/4\sigma^2} D_{-\alpha/\theta}(\frac{r}{\sigma})v}{\sqrt{\pi}S(\frac{\alpha}{\theta},\frac{z}{\sigma},\frac{r}{\sigma}) D_{-\alpha/\theta}(\frac{z}{\sigma})} - \frac{\theta\sigma\sqrt{2}e^{(z^2+r^2)/4\sigma^2} D_{-\alpha/\theta}(-\frac{z}{\sigma})y}{\sqrt{\pi}D_{-\alpha/\theta}(-\frac{r}{\sigma}) S(\frac{\alpha}{\theta},\frac{z}{\sigma},\frac{r}{\sigma})}\Big)$$

$$\times \begin{cases} \dfrac{\theta\sigma\sqrt{2}e^{(x^2+r^2)/4\sigma^2} D_{-\alpha/\theta}(\frac{x}{\sigma})}{\sqrt{\pi}D_{-\alpha/\theta}(\frac{z}{\sigma}) S(\frac{\alpha}{\theta},\frac{z}{\sigma},\frac{r}{\sigma})} \dfrac{\sqrt{v}}{\sqrt{y}} I_1\Big(\dfrac{2\theta\sigma\sqrt{2}e^{(z^2+r^2)/4\sigma^2}\sqrt{vy}}{\sqrt{\pi}S(\frac{\alpha}{\theta},\frac{z}{\sigma},\frac{r}{\sigma})}\Big)dy, & z \le x \\[2ex] \dfrac{\theta\sigma\sqrt{2}e^{(z^2+r^2)/4\sigma^2}}{\sqrt{\pi}S^2(\frac{\alpha}{\theta},\frac{z}{\sigma},\frac{r}{\sigma})}\Big[S(\tfrac{\alpha}{\theta},\tfrac{x}{\sigma},\tfrac{r}{\sigma}) \dfrac{\sqrt{v}}{\sqrt{y}} I_1\Big(\dfrac{2\theta\sigma\sqrt{2}e^{(z^2+r^2)/4\sigma^2}\sqrt{vy}}{\sqrt{\pi}S(\frac{\alpha}{\theta},\frac{z}{\sigma},\frac{r}{\sigma})}\Big) & \\ \qquad + S(\tfrac{\alpha}{\theta},\tfrac{z}{\sigma},\tfrac{x}{\sigma}) \dfrac{e^{r^2/4\sigma^2}}{e^{z^2/4\sigma^2}} I_0\Big(\dfrac{2\theta\sigma\sqrt{2}e^{(z^2+r^2)/4\sigma^2}\sqrt{vy}}{\sqrt{\pi}S(\frac{\alpha}{\theta},\frac{z}{\sigma},\frac{r}{\sigma})}\Big)\Big]dy, & r \le x \le z \\[2ex] \dfrac{\theta\sigma\sqrt{2}e^{(x^2+r^2)/4\sigma^2} D_{-\alpha/\theta}(-\frac{x}{\sigma})}{\sqrt{\pi}D_{-\alpha/\theta}(-\frac{r}{\sigma}) S(\frac{\alpha}{\theta},\frac{z}{\sigma},\frac{r}{\sigma})} I_0\Big(\dfrac{2\theta\sigma\sqrt{2}e^{(z^2+r^2)/4\sigma^2}\sqrt{vy}}{\sqrt{\pi}S(\frac{\alpha}{\theta},\frac{z}{\sigma},\frac{r}{\sigma})}\Big)dy, & 0 \le x \le r \end{cases}$$

7 $U_s = \sigma e^{-\theta s} W_{e^{2\theta s}-1}$ $\quad \sigma, \theta > 0$ $\quad \varrho = \varrho(v,z) = \min\{s : \ell(s,z) = v\}$

(1) $\mathbf{E}_x\{e^{-\alpha\varrho}; \ell(\varrho, r) = 0\}$

$z \le r$

$$= \begin{cases} 0, & r \le x \\ \dfrac{S(\frac{\alpha}{\theta}, \frac{r}{\sigma}, \frac{x}{\sigma})}{S(\frac{\alpha}{\theta}, \frac{r}{\sigma}, \frac{z}{\sigma})} \exp\Big(-\dfrac{\theta\sigma\sqrt{2}e^{(z^2+r^2)/4\sigma^2} D_{-\alpha/\theta}(-\frac{r}{\sigma})v}{\sqrt{\pi}D_{-\alpha/\theta}(-\frac{z}{\sigma})S(\frac{\alpha}{\theta}, \frac{r}{\sigma}, \frac{z}{\sigma})}\Big), & z \le x \le r \\ \dfrac{e^{x^2/4\sigma^2} D_{-\alpha/\theta}(-\frac{x}{\sigma})}{e^{z^2/4\sigma^2} D_{-\alpha/\theta}(-\frac{z}{\sigma})} \exp\Big(-\dfrac{\theta\sigma\sqrt{2}e^{(z^2+r^2)/4\sigma^2} D_{-\alpha/\theta}(-\frac{r}{\sigma})v}{\sqrt{\pi}D_{-\alpha/\theta}(-\frac{z}{\sigma})S(\frac{\alpha}{\theta}, \frac{r}{\sigma}, \frac{z}{\sigma})}\Big), & x \le z \end{cases}$$

$r \le z$

$$= \begin{cases} 0, & x \le r \\ \dfrac{S(\frac{\alpha}{\theta}, \frac{x}{\sigma}, \frac{r}{\sigma})}{S(\frac{\alpha}{\theta}, \frac{z}{\sigma}, \frac{r}{\sigma})} \exp\Big(-\dfrac{\theta\sigma\sqrt{2}e^{(z^2+r^2)/4\sigma^2} D_{-\alpha/\theta}(\frac{r}{\sigma})v}{\sqrt{\pi}S(\frac{\alpha}{\theta}, \frac{z}{\sigma}, \frac{r}{\sigma})D_{-\alpha/\theta}(\frac{z}{\sigma})}\Big), & r \le x \le z \\ \dfrac{e^{x^2/4\sigma^2} D_{-\alpha/\theta}(\frac{x}{\sigma})}{e^{z^2/4\sigma^2} D_{-\alpha/\theta}(\frac{z}{\sigma})} \exp\Big(-\dfrac{\theta\sigma\sqrt{2}e^{(z^2+r^2)/4\sigma^2} D_{-\alpha/\theta}(\frac{r}{\sigma})v}{\sqrt{\pi}S(\frac{\alpha}{\theta}, \frac{z}{\sigma}, \frac{r}{\sigma})D_{-\alpha/\theta}(\frac{z}{\sigma})}\Big), & z \le x \end{cases}$$

4.4.1 $\mathbf{E}_x \exp\Big(-\gamma \int_0^{\varrho} \mathbb{1}_{[r,\infty)}(U_s)ds\Big)$

$z \le r$

$$= \exp\Big(-\frac{\gamma\sigma e^{z^2/2\sigma^2} D_{-(\theta+\gamma)/\theta}(\frac{r}{\sigma})v}{e^{r^2/2\sigma^2} D_{-\gamma/\theta}(\frac{r}{\sigma}) + \frac{\gamma\sqrt{\pi}}{\theta\sqrt{2}} \operatorname{Erfid}(\frac{r}{\sigma}, \frac{z}{\sigma}) D_{-(\theta+\gamma)/\theta}(\frac{r}{\sigma})}\Big)$$

$$\times \begin{cases} 1, & x \le z \\ \dfrac{e^{r^2/2\sigma^2} D_{-\gamma/\theta}(\frac{r}{\sigma}) + \frac{\gamma\sqrt{\pi}}{\theta\sqrt{2}} \operatorname{Erfid}(\frac{r}{\sigma}, \frac{x}{\sigma}) D_{-(\theta+\gamma)/\theta}(\frac{r}{\sigma})}{e^{r^2/2\sigma^2} D_{-\gamma/\theta}(\frac{r}{\sigma}) + \frac{\gamma\sqrt{\pi}}{\theta\sqrt{2}} \operatorname{Erfid}(\frac{r}{\sigma}, \frac{z}{\sigma}) D_{-(\theta+\gamma)/\theta}(\frac{r}{\sigma})}, & z \le x \le r \\ \dfrac{e^{(x^2+r^2)/4\sigma^2} D_{-\gamma/\theta}(\frac{x}{\sigma})}{e^{r^2/2\sigma^2} D_{-\gamma/\theta}(\frac{r}{\sigma}) + \frac{\gamma\sqrt{\pi}}{\theta\sqrt{2}} \operatorname{Erfid}(\frac{r}{\sigma}, \frac{z}{\sigma}) D_{-(\theta+\gamma)/\theta}(\frac{r}{\sigma})}, & r \le x \end{cases}$$

$r \le z$

$$= \begin{cases} \dfrac{\sqrt{2/\pi}e^{r^2/2\sigma^2}}{C(\frac{\gamma}{\theta}, \frac{r}{\sigma}, \frac{z}{\sigma})} \exp\Big(-\dfrac{\gamma\sigma\sqrt{2}e^{(z^2+r^2)/4\sigma^2} D_{-(\theta+\gamma)/\theta}(\frac{r}{\sigma})v}{\sqrt{\pi}C(\frac{\gamma}{\theta}, \frac{r}{\sigma}, \frac{z}{\sigma})D_{-\gamma/\theta}(\frac{z}{\sigma})}\Big), & x \le r \\ \dfrac{C(\frac{\gamma}{\theta}, \frac{r}{\sigma}, \frac{x}{\sigma})}{C(\frac{\gamma}{\theta}, \frac{r}{\sigma}, \frac{z}{\sigma})} \exp\Big(-\dfrac{\gamma\sigma\sqrt{2}e^{(z^2+r^2)/4\sigma^2} D_{-(\theta+\gamma)/\theta}(\frac{r}{\sigma})v}{\sqrt{\pi}C(\frac{\gamma}{\theta}, \frac{r}{\sigma}, \frac{z}{\sigma})D_{-\gamma/\theta}(\frac{z}{\sigma})}\Big), & r \le x \le z \\ \dfrac{e^{x^2/4\sigma^2} D_{-\gamma/\theta}(\frac{x}{\sigma})}{e^{z^2/4\sigma^2} D_{-\gamma/\theta}(\frac{z}{\sigma})} \exp\Big(-\dfrac{\gamma\sigma\sqrt{2}e^{(z^2+r^2)/4\sigma^2} D_{-(\theta+\gamma)/\theta}(\frac{r}{\sigma})v}{\sqrt{\pi}C(\frac{\gamma}{\theta}, \frac{r}{\sigma}, \frac{z}{\sigma})D_{-\gamma/\theta}(\frac{z}{\sigma})}\Big), & z \le x \end{cases}$$

7 $U_s = \sigma e^{-\theta s} W_{e^{2\theta s}-1}$ $\sigma, \theta > 0$ $\varrho = \varrho(v,z) = \min\{s : \ell(s,z) = v\}$

4.4.2 (1) $\mathbf{P}_x\Big(\int_0^\varrho \mathbb{I}_{[r,\infty)}(U_s)ds = 0\Big)$

$z \le r$

$$= \begin{cases} \exp\Big(-\frac{\theta\sigma\sqrt{2}e^{z^2/2\sigma^2}v}{\sqrt{\pi}\,\mathrm{Erfid}(\frac{r}{\sigma},\frac{z}{\sigma})}\Big), & x \le z \\ \frac{\mathrm{Erfid}(\frac{r}{\sigma},\frac{x}{\sigma})}{\mathrm{Erfid}(\frac{r}{\sigma},\frac{z}{\sigma})}\exp\Big(-\frac{\theta\sigma\sqrt{2}e^{z^2/2\sigma^2}v}{\sqrt{\pi}\,\mathrm{Erfid}(\frac{r}{\sigma},\frac{z}{\sigma})}\Big), & z \le x \le r \\ 0, & r \le x \end{cases}$$

4.5.1 $\mathbf{E}_x \exp\Big(-\gamma\int_0^\varrho \mathbb{I}_{(-\infty,r]}(U_s)ds\Big)$

$z \le r$

$$= \begin{cases} \frac{e^{x^2/4\sigma^2}D_{-\gamma/\theta}(-\frac{x}{\sigma})}{e^{z^2/4\sigma^2}D_{-\gamma/\theta}(-\frac{z}{\sigma})}\exp\Big(-\frac{\gamma\sigma\sqrt{2}e^{(z^2+r^2)/4\sigma^2}D_{-(\theta+\gamma)/\theta}(-\frac{r}{\sigma})v}{\sqrt{\pi}C(\frac{\gamma}{\theta},\frac{r}{\sigma},\frac{z}{\sigma})D_{-\gamma/\theta}(-\frac{z}{\sigma})}\Big), & x \le z \\ \frac{C(\frac{\gamma}{\theta},\frac{r}{\sigma},\frac{x}{\sigma})}{C(\frac{\gamma}{\theta},\frac{r}{\sigma},\frac{z}{\sigma})}\exp\Big(-\frac{\gamma\sigma\sqrt{2}e^{(z^2+r^2)/4\sigma^2}D_{-(\theta+\gamma)/\theta}(-\frac{r}{\sigma})v}{\sqrt{\pi}C(\frac{\gamma}{\theta},\frac{r}{\sigma},\frac{z}{\sigma})D_{-\gamma/\theta}(-\frac{z}{\sigma})}\Big), & z \le x \le r \\ \frac{\sqrt{2/\pi}e^{r^2/2\sigma^2}}{C(\frac{\gamma}{\theta},\frac{r}{\sigma},\frac{z}{\sigma})}\exp\Big(-\frac{\gamma\sigma\sqrt{2}e^{(z^2+r^2)/4\sigma^2}D_{-(\theta+\gamma)/\theta}(-\frac{r}{\sigma})v}{\sqrt{\pi}C(\frac{\gamma}{\theta},\frac{r}{\sigma},\frac{z}{\sigma})D_{-\gamma/\theta}(-\frac{z}{\sigma})}\Big), & r \le x \end{cases}$$

$r \le z$

$$= \exp\Big(-\frac{\gamma\sigma e^{z^2/2\sigma^2}D_{-(\theta+\gamma)/\theta}(-\frac{r}{\sigma})v}{e^{r^2/4\sigma^2}D_{-\gamma/\theta}(-\frac{r}{\sigma}) + \frac{\gamma\sqrt{\pi}}{\theta\sqrt{2}}\mathrm{Erfid}(\frac{z}{\sigma},\frac{r}{\sigma})D_{-(\theta+\gamma)/\theta}(-\frac{r}{\sigma})}\Big)$$

$$\times \begin{cases} \frac{e^{(x^2+r^2)/4\sigma^2}D_{-\gamma/\theta}(-\frac{x}{\sigma})}{e^{r^2/2\sigma^2}D_{-\gamma/\theta}(-\frac{r}{\sigma}) + \frac{\gamma\sqrt{\pi}}{\theta\sqrt{2}}\mathrm{Erfid}(\frac{z}{\sigma},\frac{r}{\sigma})D_{-(\theta+\gamma)/\theta}(-\frac{r}{\sigma})}, & x \le r \\ \frac{e^{r^2/2\sigma^2}D_{-\gamma/\theta}(-\frac{r}{\sigma}) + \frac{\gamma\sqrt{\pi}}{\theta\sqrt{2}}\mathrm{Erfid}(\frac{x}{\sigma},\frac{r}{\sigma})D_{-(\theta+\gamma)/\theta}(-\frac{r}{\sigma})}{e^{r^2/2\sigma^2}D_{-\gamma/\theta}(-\frac{r}{\sigma}) + \frac{\gamma\sqrt{\pi}}{\theta\sqrt{2}}\mathrm{Erfid}(\frac{z}{\sigma},\frac{r}{\sigma})D_{-(\theta+\gamma)/\theta}(-\frac{r}{\sigma})}, & r \le x \le z \\ 1, & z \le x \end{cases}$$

4.5.2 (1) $\mathbf{P}_x\Big(\int_0^\varrho \mathbb{I}_{(-\infty,r]}(U_s)ds = 0\Big)$

$r \le z$

$$= \begin{cases} 0, & x \le r \\ \frac{\mathrm{Erfid}(\frac{x}{\sigma},\frac{r}{\sigma})}{\mathrm{Erfid}(\frac{z}{\sigma},\frac{r}{\sigma})}\exp\Big(-\frac{\theta\sigma\sqrt{2}e^{z^2/2\sigma^2}v}{\sqrt{\pi}\,\mathrm{Erfid}(\frac{z}{\sigma},\frac{r}{\sigma})}\Big), & r \le x \le z \\ \exp\Big(-\frac{\theta\sigma\sqrt{2}e^{z^2/2\sigma^2}v}{\sqrt{\pi}\,\mathrm{Erfid}(\frac{z}{\sigma},\frac{r}{\sigma})}\Big), & z \le x \end{cases}$$

7 $U_s = \sigma e^{-\theta s} W_{e^{2\theta s}-1}$ $\sigma, \theta > 0$ $\varrho = \varrho(v, z) = \min\{s : \ell(s, z) = v\}$

4.6.1 $\mathbf{E}_x \exp\Big(-\int_0^{\varrho} \big(p\mathbb{I}_{(-\infty,r]}(U_s) + q\mathbb{I}_{[r,\infty)}(U_s)\big)ds\Big)$

$z \le r$

$$= \exp\Big(-\frac{\sigma\sqrt{2}e^{(r^2+z^2)/4\sigma^2}(pD_{-(\theta+p)/\theta}(-\frac{r}{\sigma})D_{-q/\theta}(\frac{r}{\sigma}) + qD_{-p/\theta}(-\frac{r}{\sigma})D_{-(\theta+q)/\theta}(\frac{r}{\sigma}))v}{\theta^{-2}\sqrt{\pi}D_{-p/\theta}(-\frac{z}{\sigma})(\theta D_{-q/\theta}(\frac{r}{\sigma})C(\frac{p}{\theta},\frac{r}{\sigma},\frac{z}{\sigma}) + qD_{-(\theta+q)/\theta}(\frac{r}{\sigma})S(\frac{p}{\theta},\frac{r}{\sigma},\frac{z}{\sigma}))}\Big)$$

$$\times \begin{cases} \dfrac{e^{x^2/4\sigma^2}D_{-p/\theta}(-\frac{x}{\sigma})}{e^{z^2/4\sigma^2}D_{-p/\theta}(-\frac{z}{\sigma})}, & x \le z \\ \dfrac{D_{-q/\theta}(\frac{r}{\sigma})C(\frac{p}{\theta},\frac{r}{\sigma},\frac{x}{\sigma}) + \frac{q}{\theta}D_{-(\theta+q)/\theta}(\frac{r}{\sigma})S(\frac{p}{\theta},\frac{r}{\sigma},\frac{x}{\sigma})}{D_{-q/\theta}(\frac{r}{\sigma})C(\frac{p}{\theta},\frac{r}{\sigma},\frac{z}{\sigma}) + \frac{q}{\theta}D_{-(\theta+q)/\theta}(\frac{r}{\sigma})S(\frac{p}{\theta},\frac{r}{\sigma},\frac{z}{\sigma})}, & z \le x \le r \\ \dfrac{\sqrt{2/\pi}e^{(x^2+r^2)/4\sigma^2}D_{-q/\theta}(\frac{x}{\sigma})}{D_{-q/\theta}(\frac{r}{\sigma})C(\frac{p}{\theta},\frac{r}{\sigma},\frac{z}{\sigma}) + \frac{q}{\theta}D_{-(\theta+q)/\theta}(\frac{r}{\sigma})S(\frac{p}{\theta},\frac{r}{\sigma},\frac{z}{\sigma})}, & r \le x \end{cases}$$

$r \le z$

$$= \exp\Big(-\frac{\sigma\sqrt{2}e^{(r^2+z^2)/4\sigma^2}(pD_{-(\theta+p)/\theta}(-\frac{r}{\sigma})D_{-q/\theta}(\frac{r}{\sigma}) + qD_{-p/\theta}(-\frac{r}{\sigma})D_{-(\theta+q)/\theta}(\frac{r}{\sigma}))v}{\sqrt{\pi}\theta^{-2}D_{-q/\theta}(\frac{z}{\sigma})(pS(\frac{q}{\theta},\frac{z}{\sigma},\frac{r}{\sigma})D_{-(\theta+p)/\theta}(-\frac{r}{\sigma}) + \theta C(\frac{q}{\theta},\frac{r}{\sigma},\frac{z}{\sigma})D_{-p/\theta}(-\frac{r}{\sigma}))}\Big)$$

$$\times \begin{cases} \dfrac{\sqrt{2/\pi}e^{(x^2+r^2)/4\sigma^2}D_{-p/\theta}(-\frac{x}{\sigma})}{C(\frac{q}{\theta},\frac{r}{\sigma},\frac{z}{\sigma})D_{-p/\theta}(-\frac{r}{\sigma}) + \frac{p}{\theta}S(\frac{q}{\theta},\frac{z}{\sigma},\frac{r}{\sigma})D_{-(\theta+p)/\theta}(-\frac{r}{\sigma})}, & x \le r \\ \dfrac{C(\frac{q}{\theta},\frac{r}{\sigma},\frac{x}{\sigma})D_{-p/\theta}(-\frac{r}{\sigma}) + \frac{p}{\theta}S(\frac{q}{\theta},\frac{x}{\sigma},\frac{r}{\sigma})D_{-(\theta+p)/\theta}(-\frac{r}{\sigma})}{C(\frac{q}{\theta},\frac{r}{\sigma},\frac{z}{\sigma})D_{-p/\theta}(-\frac{r}{\sigma}) + \frac{p}{\theta}S(\frac{q}{\theta},\frac{z}{\sigma},\frac{r}{\sigma})D_{-(\theta+p)/\theta}(-\frac{r}{\sigma})}, & r \le x \le z \\ \dfrac{e^{x^2/4\sigma^2}D_{-q/\theta}(\frac{x}{\sigma})}{e^{z^2/4\sigma^2}D_{-q/\theta}(\frac{z}{\sigma})}, & z \le x \end{cases}$$

4.9.3
$1 \le \gamma$ $\mathbf{E}_x \exp\Big(-\alpha\varrho - \frac{(\gamma^2-1)\theta}{4\sigma^2}\int_0^{\varrho} U_s^2 ds\Big)$

$x \le z$

$$= \frac{e^{x^2/4\sigma^2}D_{-\frac12+\frac{1}{2\gamma}-\frac{\alpha}{\theta\gamma}}(-\frac{x\sqrt{\gamma}}{\sigma})}{e^{z^2/4\sigma^2}D_{-\frac12+\frac{1}{2\gamma}-\frac{\alpha}{\theta\gamma}}(-\frac{z\sqrt{\gamma}}{\sigma})}\exp\Big(-\frac{\theta\sigma\sqrt{2\pi\gamma}\,\Gamma^{-1}(\frac12-\frac{1}{2\gamma}+\frac{\alpha}{\theta\gamma})v}{D_{-\frac12+\frac{1}{2\gamma}-\frac{\alpha}{\theta\gamma}}(\frac{z\sqrt{\gamma}}{\sigma})D_{-\frac12+\frac{1}{2\gamma}-\frac{\alpha}{\theta\gamma}}(-\frac{z\sqrt{\gamma}}{\sigma})}\Big)$$

$z \le x$

$$= \frac{e^{x^2/4\sigma^2}D_{-\frac12+\frac{1}{2\gamma}-\frac{\alpha}{\theta\gamma}}(\frac{x\sqrt{\gamma}}{\sigma})}{e^{z^2/4\sigma^2}D_{-\frac12+\frac{1}{2\gamma}-\frac{\alpha}{\theta\gamma}}(\frac{z\sqrt{\gamma}}{\sigma})}\exp\Big(-\frac{\theta\sigma\sqrt{2\pi\gamma}\,\Gamma^{-1}(\frac12-\frac{1}{2\gamma}+\frac{\alpha}{\theta\gamma})v}{D_{-\frac12+\frac{1}{2\gamma}-\frac{\alpha}{\theta\gamma}}(\frac{z\sqrt{\gamma}}{\sigma})D_{-\frac12+\frac{1}{2\gamma}-\frac{\alpha}{\theta\gamma}}(-\frac{z\sqrt{\gamma}}{\sigma})}\Big)$$

4.20.1 $\mathbf{E}_x\Big\{\exp\Big(-\gamma\sigma^2\theta\int_0^{\varrho}\frac{ds}{U_s^2}\Big);\ 0 < \inf_{0\le s\le\varrho} U_s\Big\}$

$$= \exp\Big(-\frac{\theta z v\Gamma(\frac12\sqrt{1+4\gamma}+1)}{\Gamma(\frac14\sqrt{1+4\gamma}+\frac14)M_{\frac14,\frac14\sqrt{1+4\gamma}}(\frac{z^2}{2\sigma^2})W_{\frac14,\frac14\sqrt{1+4\gamma}}(\frac{z^2}{2\sigma^2})}\Big)$$

7 $U_s = \sigma e^{-\theta s} W_{e^{2\theta s}-1} \qquad \sigma, \theta > 0 \qquad \varrho = \varrho(v,z) = \min\{s : \ell(s,z) = v\}$

$$\times \begin{cases} \dfrac{x^{-1/2} e^{x^2/4\sigma^2} M_{\frac14,\frac14\sqrt{1+4\gamma}}\big(\frac{x^2}{2\sigma^2}\big)}{z^{-1/2} e^{z^2/4\sigma^2} M_{\frac14,\frac14\sqrt{1+4\gamma}}\big(\frac{z^2}{2\sigma^2}\big)}, & 0 < x \le z \\ \dfrac{x^{-1/2} e^{x^2/4\sigma^2} W_{\frac14,\frac14\sqrt{1+4\gamma}}\big(\frac{x^2}{2\sigma^2}\big)}{z^{-1/2} e^{z^2/4\sigma^2} W_{\frac14,\frac14\sqrt{1+4\gamma}}\big(\frac{z^2}{2\sigma^2}\big)}, & z \le x \end{cases}$$

4.21.1 $$\mathbf{E}_x\Big\{\exp\Big(-\int_0^{\varrho}\Big(\frac{p\sigma^2\theta}{U_s^2} + \frac{\theta(q^2-1)}{4\sigma^2}U_s^2\Big)ds\Big);\ 0 < \inf_{0\le s\le\varrho} U_s\Big\}$$

$$= \exp\Big(-\frac{\theta q z v \Gamma\big(\frac12\sqrt{1+4p}+1\big)}{\Gamma\big(\frac14\sqrt{1+4p}+\frac{2q-1}{4q}\big) M_{\frac{1}{4q},\frac14\sqrt{1+4p}}\big(\frac{qz^2}{2\sigma^2}\big) W_{\frac{1}{4q},\frac14\sqrt{1+4p}}\big(\frac{qz^2}{2\sigma^2}\big)}\Big)$$

$$\times \begin{cases} \dfrac{x^{-1/2} e^{x^2/4\sigma^2} M_{\frac{1}{4q},\frac14\sqrt{1+4p}}\big(\frac{qx^2}{2\sigma^2}\big)}{z^{-1/2} e^{z^2/4\sigma^2} M_{\frac{1}{4q},\frac14\sqrt{1+4p}}\big(\frac{qz^2}{2\sigma^2}\big)}, & 0 < x \le z \\ \dfrac{x^{-1/2} e^{x^2/4\sigma^2} W_{\frac{1}{4q},\frac14\sqrt{1+4p}}\big(\frac{qx^2}{2\sigma^2}\big)}{z^{-1/2} e^{z^2/4\sigma^2} W_{\frac{1}{4q},\frac14\sqrt{1+4p}}\big(\frac{qz^2}{2\sigma^2}\big)}, & z \le x \end{cases}$$

4.27.1 $$\mathbf{E}_x\Big\{\exp\Big(-\int_0^{\varrho}\big(p\mathbb{1}_{(-\infty,r]}(U_s) + q\mathbb{1}_{[r,\infty)}(U_s)\big)ds\Big); \ell(\varrho,r) \in dy\Big\}$$

$z \le r$ $$= \exp\Big(-\frac{\theta\sigma\sqrt{2}e^{(z^2+r^2)/4\sigma^2} D_{-p/\theta}(-\frac{r}{\sigma})v}{\sqrt{\pi}S(\frac{p}{\theta},\frac{r}{\sigma},\frac{z}{\sigma})D_{-p/\theta}(-\frac{z}{\sigma})} - \frac{\sigma\theta y C(\frac{p}{\theta},\frac{r}{\sigma},\frac{z}{\sigma})}{S(\frac{p}{\theta},\frac{r}{\sigma},\frac{z}{\sigma})} - \frac{\sigma q y D_{-(\theta+q)/\theta}(\frac{r}{\sigma})}{D_{-q/\theta}(\frac{r}{\sigma})}\Big)$$

$$\times \begin{cases} \dfrac{\theta\sigma\sqrt{2}e^{(x^2+r^2)/4\sigma^2} D_{-p/\theta}(-\frac{x}{\sigma})}{\sqrt{\pi}D_{-p/\theta}(-\frac{z}{\sigma})S(\frac{p}{\theta},\frac{r}{\sigma},\frac{z}{\sigma})} \dfrac{\sqrt{v}}{\sqrt{y}} I_1\Big(\dfrac{2\theta\sigma\sqrt{2}e^{(z^2+r^2)/4\sigma^2}\sqrt{vy}}{\sqrt{\pi}S(\frac{p}{\theta},\frac{r}{\sigma},\frac{z}{\sigma})}\Big)dy, & 0 \le x \le z \\ \dfrac{\theta\sigma\sqrt{2}e^{(z^2+r^2)/4\sigma^2}}{\sqrt{\pi}S^2(\frac{p}{\theta},\frac{r}{\sigma},\frac{z}{\sigma})}\Big[S(\frac{p}{\theta},\frac{r}{\sigma},\frac{x}{\sigma})\dfrac{\sqrt{v}}{\sqrt{y}} I_1\Big(\dfrac{2\theta\sigma\sqrt{2}e^{(z^2+r^2)/4\sigma^2}\sqrt{vy}}{\sqrt{\pi}S(\frac{p}{\theta},\frac{r}{\sigma},\frac{z}{\sigma})}\Big) & \\ \quad + S(\frac{p}{\theta},\frac{x}{\sigma},\frac{z}{\sigma})\dfrac{e^{r^2/4\sigma^2}}{e^{z^2/4\sigma^2}} I_0\Big(\dfrac{2\theta\sigma\sqrt{2}e^{(z^2+r^2)/4\sigma^2}\sqrt{vy}}{\sqrt{\pi}S(\frac{p}{\theta},\frac{r}{\sigma},\frac{z}{\sigma})}\Big)\Big]dy, & z \le x \le r \\ \dfrac{\theta\sigma\sqrt{2}e^{(x^2+r^2)/4\sigma^2} D_{-q/\theta}(\frac{x}{\sigma})}{\sqrt{\pi}D_{-q/\theta}(\frac{r}{\sigma})S(\frac{p}{\theta},\frac{r}{\sigma},\frac{z}{\sigma})} I_0\Big(\dfrac{2\theta\sigma\sqrt{2}e^{(z^2+r^2)/4\sigma^2}\sqrt{vy}}{\sqrt{\pi}S(\frac{p}{\theta},\frac{r}{\sigma},\frac{z}{\sigma})}\Big)dy, & r \le x \end{cases}$$

7 $U_s = \sigma e^{-\theta s} W_{e^{2\theta s}-1}$ $\sigma, \theta > 0$ $\varrho = \varrho(v,z) = \min\{s : \ell(s,z) = v\}$

$r \le z$

$$= \exp\Big(-\frac{\theta\sigma\sqrt{2}e^{(z^2+r^2)/4\sigma^2}D_{-q/\theta}(\frac{r}{\sigma})v}{\sqrt{\pi}S(\frac{q}{\theta},\frac{z}{\sigma},\frac{r}{\sigma})D_{-q/\theta}(\frac{z}{\sigma})} - \frac{\sigma\theta y C(\frac{q}{\theta},\frac{r}{\sigma},\frac{z}{\sigma})}{S(\frac{q}{\theta},\frac{z}{\sigma},\frac{r}{\sigma})} - \frac{\sigma p y D_{-(\theta+p)/\theta}(-\frac{r}{\sigma})}{D_{-p/\theta}(-\frac{r}{\sigma})}\Big)$$

$$\times\begin{cases} \dfrac{\theta\sigma\sqrt{2}e^{(x^2+r^2)/4\sigma^2}D_{-q/\theta}(\frac{x}{\sigma})}{\sqrt{\pi}D_{-q/\theta}(\frac{z}{\sigma})S(\frac{q}{\theta},\frac{z}{\sigma},\frac{r}{\sigma})}\dfrac{\sqrt{v}}{\sqrt{y}}I_1\Big(\dfrac{2\theta\sigma\sqrt{2}e^{(z^2+r^2)/4\sigma^2}\sqrt{vy}}{\sqrt{\pi}S(\frac{q}{\theta},\frac{z}{\sigma},\frac{r}{\sigma})}\Big)dy, & z \le x \\ \dfrac{\theta\sigma\sqrt{2}e^{(z^2+r^2)/4\sigma^2}}{\sqrt{\pi}S^2(\frac{q}{\theta},\frac{z}{\sigma},\frac{r}{\sigma})}\Big[S(\tfrac{q}{\theta},\tfrac{x}{\sigma},\tfrac{r}{\sigma})\dfrac{\sqrt{v}}{\sqrt{y}}I_1\Big(\dfrac{2\theta\sigma\sqrt{2}e^{(z^2+r^2)/4\sigma^2}\sqrt{vy}}{\sqrt{\pi}S(\frac{q}{\theta},\frac{z}{\sigma},\frac{r}{\sigma})}\Big) & \\ \quad +S(\tfrac{q}{\theta},\tfrac{z}{\sigma},\tfrac{x}{\sigma})\dfrac{e^{r^2/4\sigma^2}}{e^{z^2/4\sigma^2}}I_0\Big(\dfrac{2\theta\sigma\sqrt{2}e^{(z^2+r^2)/4\sigma^2}\sqrt{vy}}{\sqrt{\pi}S(\frac{q}{\theta},\frac{z}{\sigma},\frac{r}{\sigma})}\Big)\Big]dy, & r \le x \le z \\ \dfrac{\theta\sigma\sqrt{2}e^{(x^2+r^2)/4\sigma^2}D_{-p/\theta}(-\frac{x}{\sigma})}{\sqrt{\pi}D_{-p/\theta}(-\frac{r}{\sigma})S(\frac{q}{\theta},\frac{z}{\sigma},\frac{r}{\sigma})}I_0\Big(\dfrac{2\theta\sigma\sqrt{2}e^{(z^2+r^2)/4\sigma^2}\sqrt{vy}}{\sqrt{\pi}S(\frac{q}{\theta},\frac{z}{\sigma},\frac{r}{\sigma})}\Big)dy, & 0 \le x \le r \end{cases}$$

(1) $$\mathbf{E}_x\Big\{\exp\Big(-\int_0^{\varrho}\big(p\mathbb{1}_{(-\infty,r]}(U_s) + q\mathbb{1}_{[r,\infty)}(U_s)\big)ds\Big); \ell(\varrho, r) = 0\Big\}$$

$z \le r$

$$= \begin{cases} 0, & r \le x \\ \dfrac{S(\frac{p}{\theta},\frac{r}{\sigma},\frac{x}{\sigma})}{S(\frac{p}{\theta},\frac{r}{\sigma},\frac{z}{\sigma})}\exp\Big(-\dfrac{\theta\sigma\sqrt{2}e^{(z^2+r^2)/4\sigma^2}D_{-p/\theta}(-\frac{r}{\sigma})v}{\sqrt{\pi}D_{-p/\theta}(-\frac{z}{\sigma})S(\frac{p}{\theta},\frac{r}{\sigma},\frac{z}{\sigma})}\Big), & z \le x \le r \\ \dfrac{e^{x^2/4\sigma^2}D_{-p/\theta}(-\frac{x}{\sigma})}{e^{z^2/4\sigma^2}D_{-p/\theta}(-\frac{z}{\sigma})}\exp\Big(-\dfrac{\theta\sigma\sqrt{2}e^{(z^2+r^2)/4\sigma^2}D_{-p/\theta}(-\frac{r}{\sigma})v}{\sqrt{\pi}D_{-p/\theta}(-\frac{z}{\sigma})S(\frac{p}{\theta},\frac{r}{\sigma},\frac{z}{\sigma})}\Big), & x \le z \end{cases}$$

$r \le z$

$$= \begin{cases} 0, & x \le r \\ \dfrac{S(\frac{q}{\theta},\frac{x}{\sigma},\frac{r}{\sigma})}{S(\frac{q}{\theta},\frac{z}{\sigma},\frac{r}{\sigma})}\exp\Big(-\dfrac{\theta\sigma\sqrt{2}e^{(z^2+r^2)/4\sigma^2}D_{-q/\theta}(\frac{r}{\sigma})v}{\sqrt{\pi}S(\frac{q}{\theta},\frac{z}{\sigma},\frac{r}{\sigma})D_{-q/\theta}(\frac{z}{\sigma})}\Big), & r \le x \le z \\ \dfrac{e^{x^2/4\sigma^2}D_{-q/\theta}(\frac{x}{\sigma})}{e^{z^2/4\sigma^2}D_{-q/\theta}(\frac{z}{\sigma})}\exp\Big(-\dfrac{\theta\sigma\sqrt{2}e^{(z^2+r^2)/4\sigma^2}D_{-q/\theta}(\frac{r}{\sigma})v}{\sqrt{\pi}S(\frac{q}{\theta},\frac{z}{\sigma},\frac{r}{\sigma})D_{-q/\theta}(\frac{z}{\sigma})}\Big), & z \le x \end{cases}$$

8. RADIAL ORNSTEIN–UHLENBECK PROCESS

1. Exponential stopping

1.0.2 $\mathbf{P}(\tau > t) = e^{-\alpha t}$

1.0.5 $\mathbf{P}_x(Q_\tau^{(n)} \in dz)$

$$= \begin{cases} \dfrac{\lambda\Gamma(\frac{\lambda}{2\theta}) z^{n-1} e^{-z^2/2\sigma^2}}{\theta 2^{n/2}\Gamma(\frac{n}{2})\sigma^n} M\big(\frac{\lambda}{2\theta}, \frac{n}{2}, \frac{x^2}{2\sigma^2}\big) U\big(\frac{\lambda}{2\theta}, \frac{n}{2}, \frac{z^2}{2\sigma^2}\big) dz, & 0 \leq x \leq z \\ \dfrac{\lambda\Gamma(\frac{\lambda}{2\theta}) z^{n-1} e^{-z^2/2\sigma^2}}{\theta 2^{n/2}\Gamma(\frac{n}{2})\sigma^n} U\big(\frac{\lambda}{2\theta}, \frac{n}{2}, \frac{x^2}{2\sigma^2}\big) M\big(\frac{\lambda}{2\theta}, \frac{n}{2}, \frac{z^2}{2\sigma^2}\big) dz, & z \leq x \end{cases}$$

1.0.6 $\mathbf{P}_x\big(Q_t^{(n)} \in dz\big)$

$$= \frac{z^{n/2} e^{\theta n t/2}}{2\sigma^2 x^{(n-2)/2}\,\mathrm{sh}(t\theta)} \exp\Big(\frac{x^2 - z^2}{4\sigma^2} - \frac{(x^2+z^2)\,\mathrm{ch}(t\theta)}{4\sigma^2\,\mathrm{sh}(t\theta)}\Big) I_{\frac{n-2}{2}}\Big(\frac{xz}{2\sigma^2\,\mathrm{sh}(t\theta)}\Big) dz$$

1.1.2 $\mathbf{P}_x\big(\sup_{0\leq s\leq\tau} Q_s^{(n)} \geq y\big) = \dfrac{M\big(\frac{\lambda}{2\theta}, \frac{n}{2}, \frac{x^2}{2\sigma^2}\big)}{M\big(\frac{\lambda}{2\theta}, \frac{n}{2}, \frac{y^2}{2\sigma^2}\big)}, \qquad 0 \leq x \leq y$

1.1.6 $\mathbf{P}_x\big(\sup_{0\leq s\leq\tau} Q_s^{(n)} \geq y, Q_\tau^{(n)} \in dz\big)$ $\qquad x \vee z \leq y$

$$= \frac{\lambda\Gamma(\frac{\lambda}{2\theta}) z^{n-1} e^{-z^2/2\sigma^2} M\big(\frac{\lambda}{2\theta}, \frac{n}{2}, \frac{x^2}{2\sigma^2}\big)}{\theta 2^{n/2}\Gamma(\frac{n}{2})\sigma^n M\big(\frac{\lambda}{2\theta}, \frac{n}{2}, \frac{y^2}{2\sigma^2}\big)} M\big(\tfrac{\lambda}{2\theta}, \tfrac{n}{2}, \tfrac{z^2}{2\sigma^2}\big) U\big(\tfrac{\lambda}{2\theta}, \tfrac{n}{2}, \tfrac{y^2}{2\sigma^2}\big) dz$$

1.2.2 $\mathbf{P}_x\big(\inf_{0\leq s\leq\tau} Q_s^{(n)} \leq y\big) = \dfrac{U\big(\frac{\lambda}{2\theta}, \frac{n}{2}, \frac{x^2}{2\sigma^2}\big)}{U\big(\frac{\lambda}{2\theta}, \frac{n}{2}, \frac{y^2}{2\sigma^2}\big)}, \qquad 0 \leq y \leq x$

1.2.6 $\mathbf{P}_x\big(\inf_{0\leq s\leq\tau} Q_s^{(n)} \leq y, Q_\tau^{(n)} \in dz\big)$ $\qquad y \leq x \wedge z$

$$= \frac{\lambda\Gamma(\frac{\lambda}{2\theta}) z^{n-1} e^{-z^2/2\sigma^2} U\big(\frac{\lambda}{2\theta}, \frac{n}{2}, \frac{x^2}{2\sigma^2}\big)}{\theta 2^{n/2}\Gamma(\frac{n}{2})\sigma^n U\big(\frac{\lambda}{2\theta}, \frac{n}{2}, \frac{y^2}{2\sigma^2}\big)} U\big(\tfrac{\lambda}{2\theta}, \tfrac{n}{2}, \tfrac{z^2}{2\sigma^2}\big) M\big(\tfrac{\lambda}{2\theta}, \tfrac{n}{2}, \tfrac{y^2}{2\sigma^2}\big) dz$$

1.3.1 $\mathbf{E}_x e^{-\gamma \ell(\tau, r)}$

8 $Q_s^{(n)} = \sigma e^{-\theta s} R^{(n)}_{e^{2\theta s}-1}$ $n \ge 2,\ \sigma, \theta > 0$ $\tau \sim \mathrm{Exp}(\lambda)$, independent of $Q^{(n)}$

$x \le r$
$$= 1 - \frac{\gamma r^{n-1} M\big(\frac{\lambda}{2\theta}, \frac{n}{2}, \frac{x^2}{2\sigma^2}\big) U\big(\frac{\lambda}{2\theta}, \frac{n}{2}, \frac{r^2}{2\sigma^2}\big)}{\theta 2^{n/2}\Gamma\big(\frac{n}{2}\big)\sigma^n e^{r^2/2\sigma^2} + \gamma r^{n-1}\Gamma\big(\frac{\lambda}{2\theta}\big) M\big(\frac{\lambda}{2\theta}, \frac{n}{2}, \frac{r^2}{2\sigma^2}\big) U\big(\frac{\lambda}{2\theta}, \frac{n}{2}, \frac{r^2}{2\sigma^2}\big)}$$

$r \le x$
$$= 1 - \frac{\gamma r^{n-1} U\big(\frac{\lambda}{2\theta}, \frac{n}{2}, \frac{x^2}{2\sigma^2}\big) M\big(\frac{\lambda}{2\theta}, \frac{n}{2}, \frac{r^2}{2\sigma^2}\big)}{\theta 2^{n/2}\Gamma\big(\frac{n}{2}\big)\sigma^n e^{r^2/2\sigma^2} + \gamma r^{n-1}\Gamma\big(\frac{\lambda}{2\theta}\big) M\big(\frac{\lambda}{2\theta}, \frac{n}{2}, \frac{r^2}{2\sigma^2}\big) U\big(\frac{\lambda}{2\theta}, \frac{n}{2}, \frac{r^2}{2\sigma^2}\big)}$$

1.3.2
$$\mathbf{P}_x\big(\ell(\tau,r) \in dy\big) = \exp\Big(-\frac{\theta 2^{n/2}\Gamma\big(\frac{n}{2}\big)\sigma^n e^{r^2/2\sigma^2} y}{r^{n-1}\Gamma\big(\frac{\lambda}{2\theta}\big) M\big(\frac{\lambda}{2\theta}, \frac{n}{2}, \frac{r^2}{2\sigma^2}\big) U\big(\frac{\lambda}{2\theta}, \frac{n}{2}, \frac{r^2}{2\sigma^2}\big)}\Big)$$
$$\times \begin{cases} \dfrac{\theta 2^{n/2}\Gamma\big(\frac{n}{2}\big)\sigma^n e^{r^2/2\sigma^2} M\big(\frac{\lambda}{2\theta}, \frac{n}{2}, \frac{x^2}{2\sigma^2}\big)}{r^{n-1}\Gamma\big(\frac{\lambda}{2\theta}\big) M^2\big(\frac{\lambda}{2\theta}, \frac{n}{2}, \frac{r^2}{2\sigma^2}\big) U\big(\frac{\lambda}{2\theta}, \frac{n}{2}, \frac{r^2}{2\sigma^2}\big)} dy, & 0 \le x \le r \\ \dfrac{\theta 2^{n/2}\Gamma\big(\frac{n}{2}\big)\sigma^n e^{r^2/2\sigma^2} U\big(\frac{\lambda}{2\theta}, \frac{n}{2}, \frac{x^2}{2\sigma^2}\big)}{r^{n-1}\Gamma\big(\frac{\lambda}{2\theta}\big) M\big(\frac{\lambda}{2\theta}, \frac{n}{2}, \frac{r^2}{2\sigma^2}\big) U^2\big(\frac{\lambda}{2\theta}, \frac{n}{2}, \frac{r^2}{2\sigma^2}\big)} dy, & r \le x \end{cases}$$

(1)
$$\mathbf{P}_x\big(\ell(\tau,r) = 0\big) = \begin{cases} 1 - \dfrac{M\big(\frac{\lambda}{2\theta}, \frac{n}{2}, \frac{x^2}{2\sigma^2}\big)}{M\big(\frac{\lambda}{2\theta}, \frac{n}{2}, \frac{r^2}{2\sigma^2}\big)}, & 0 \le x \le r \\ 1 - \dfrac{U\big(\frac{\lambda}{2\theta}, \frac{n}{2}, \frac{x^2}{2\sigma^2}\big)}{U\big(\frac{\lambda}{2\theta}, \frac{n}{2}, \frac{r^2}{2\sigma^2}\big)}, & r \le x \end{cases}$$

1.3.5
$$\mathbf{E}_x\big\{e^{-\gamma\ell(\tau,r)}, Q_\tau^{(n)} \in dz\big\} = \frac{\lambda z^{n-1} e^{-z^2/2\sigma^2}}{\theta 2^{n/2}\sigma^n} F\big(\tfrac{\lambda}{2\theta}, \tfrac{n}{2}, \tfrac{x^2}{2\sigma^2}, \tfrac{z^2}{2\sigma^2}\big) dz$$
$$- \frac{\gamma z^{n-1} e^{-z^2/2\sigma^2} F\big(\frac{\lambda}{2\theta}, \frac{n}{2}, \frac{r^2}{2\sigma^2}, \frac{z^2}{2\sigma^2}\big) F\big(\frac{\lambda}{2\theta}, \frac{n}{2}, \frac{x^2}{2\sigma^2}, \frac{r^2}{2\sigma^2}\big)}{\theta 2^{n/2}\sigma^n \Big(\theta 2^{n/2} r^{1-n}\sigma^n e^{r^2/2\sigma^2} + \gamma F\big(\frac{\lambda}{2\theta}, \frac{n}{2}, \frac{r^2}{2\sigma^2}, \frac{r^2}{2\sigma^2}\big)\Big)} dz$$

1.3.6
$$\mathbf{P}_x\big(\ell(\tau,r) \in dy,\ Q_\tau^{(n)} \in dz\big) = \exp\Big(-\frac{\theta 2^{n/2} r^{1-n}\sigma^n e^{r^2/2\sigma^2} y}{F\big(\frac{\lambda}{2\theta}, \frac{n}{2}, \frac{r^2}{2\sigma^2}, \frac{r^2}{2\sigma^2}\big)}\Big)$$
$$\times \frac{\lambda z^{n-1} e^{-z^2/2\sigma^2} F\big(\frac{\lambda}{2\theta}, \frac{n}{2}, \frac{r^2}{2\sigma^2}, \frac{z^2}{2\sigma^2}\big) F\big(\frac{\lambda}{2\theta}, \frac{n}{2}, \frac{x^2}{2\sigma^2}, \frac{r^2}{2\sigma^2}\big)}{r^{n-1} e^{-r^2/2\sigma^2} F^2\big(\frac{\lambda}{2\theta}, \frac{n}{2}, \frac{r^2}{2\sigma^2}, \frac{r^2}{2\sigma^2}\big)} dydz$$

(1)
$$\mathbf{P}_x\big(\ell(\tau,r) = 0, Q_\tau^{(n)} \in dz\big) = \frac{\lambda z^{n-1} e^{-z^2/2\sigma^2}}{\theta 2^{n/2}\sigma^n} F\big(\tfrac{\lambda}{2\theta}, \tfrac{n}{2}, \tfrac{x^2}{2\sigma^2}, \tfrac{z^2}{2\sigma^2}\big) dz$$
$$- \frac{\lambda z^{n-1} e^{-z^2/2\sigma^2} F\big(\frac{\lambda}{2\theta}, \frac{n}{2}, \frac{r^2}{2\sigma^2}, \frac{z^2}{2\sigma^2}\big) F\big(\frac{\lambda}{2\theta}, \frac{n}{2}, \frac{x^2}{2\sigma^2}, \frac{r^2}{2\sigma^2}\big)}{\theta 2^{n/2}\sigma^n F\big(\frac{\lambda}{2\theta}, \frac{n}{2}, \frac{r^2}{2\sigma^2}, \frac{r^2}{2\sigma^2}\big)} dz$$

8 $\quad Q_s^{(n)} = \sigma e^{-\theta s} R^{(n)}_{e^{2\theta s}-1} \quad n \ge 2,\ \sigma, \theta > 0 \qquad \tau \sim \operatorname{Exp}(\lambda)$, independent of $Q^{(n)}$

1.6.1 $\quad \mathbf{E}_x \exp\Big(-\int_0^\tau \big(p\mathbb{I}_{[0,r)}\big(Q_s^{(n)}\big) + q\mathbb{I}_{[r,\infty)}\big(Q_s^{(n)}\big)\big)ds\Big)$

$0 \le x$, $x \le r$

$$= \frac{\lambda}{\lambda+p}$$

$$+ \frac{\lambda(p-q)n2^{-1}(\lambda+p)^{-2}M\big(\frac{\lambda+p}{2\theta},\frac{n}{2},\frac{x^2}{2\sigma^2}\big)U\big(\frac{\lambda+q+2\theta}{2\theta},\frac{n+2}{2},\frac{r^2}{2\sigma^2}\big)}{M\big(\frac{\lambda+p+2\theta}{2\theta},\frac{n+2}{2},\frac{r^2}{2\sigma^2}\big)U\big(\frac{\lambda+q}{2\theta},\frac{n}{2},\frac{r^2}{2\sigma^2}\big)+\frac{n(\lambda+q)}{2(\lambda+p)}U\big(\frac{\lambda+q+2\theta}{2\theta},\frac{n+2}{2},\frac{r^2}{2\sigma^2}\big)M\big(\frac{\lambda+p}{2\theta},\frac{n}{2},\frac{r^2}{2\sigma^2}\big)}$$

$r \le x$

$$= \frac{\lambda}{\lambda+q}$$

$$+ \frac{\lambda(q-p)(\lambda+q)^{-1}(\lambda+p)^{-1}U\big(\frac{\lambda+q}{2\theta},\frac{n}{2},\frac{x^2}{2\sigma^2}\big)M\big(\frac{\lambda+p+2\theta}{2\theta},\frac{n+2}{2},\frac{r^2}{2\sigma^2}\big)}{M\big(\frac{\lambda+p+2\theta}{2\theta},\frac{n+2}{2},\frac{r^2}{2\sigma^2}\big)U\big(\frac{\lambda+q}{2\theta},\frac{n}{2},\frac{r^2}{2\sigma^2}\big)+\frac{n(\lambda+q)}{2(\lambda+p)}U\big(\frac{\lambda+q+2\theta}{2\theta},\frac{n+2}{2},\frac{r^2}{2\sigma^2}\big)M\big(\frac{\lambda+p}{2\theta},\frac{n}{2},\frac{r^2}{2\sigma^2}\big)}$$

1.6.5 $\quad \mathbf{E}_x\Big\{\exp\Big(-\int_0^\tau \big(p\mathbb{I}_{[0,r)}\big(Q_s^{(n)}\big) + q\mathbb{I}_{[r,\infty)}\big(Q_s^{(n)}\big)\big)ds\Big), Q_\tau^{(n)} \in dz\Big\}/dz$

$z \le r$, $x \le r$

$$= \frac{\lambda z^{n-1}e^{-z^2/2\sigma^2}S\big(\frac{\lambda+p}{2\theta},\frac{n}{2},\frac{r^2}{2\sigma^2},\frac{(z+x+|z-x|)^2}{4\sigma^2}\big)M\big(\frac{\lambda+p}{2\theta},\frac{n}{2},\frac{(z+x-|z-x|)^2}{4\sigma^2}\big)}{\theta 2^{n/2}\sigma^n M\big(\frac{\lambda+p}{2\theta},\frac{n}{2},\frac{r^2}{2\sigma^2}\big)}$$

$$+ \frac{M\big(\frac{\lambda+p}{2\theta},\frac{n}{2},\frac{x^2}{2\sigma^2}\big)}{M\big(\frac{\lambda+p}{2\theta},\frac{n}{2},\frac{r^2}{2\sigma^2}\big)}$$

$$\times \frac{\lambda n(\lambda+p)^{-1}z^{n-1}r^{-n}e^{(r^2-z^2)/2\sigma^2}U\big(\frac{\lambda+q}{2\theta},\frac{n+2}{2},\frac{r^2}{2\sigma^2}\big)M\big(\frac{\lambda+q}{2\theta},\frac{n}{2},\frac{z^2}{2\sigma^2}\big)}{M\big(\frac{\lambda+p+2\theta}{2\theta},\frac{n+2}{2},\frac{r^2}{2\sigma^2}\big)U\big(\frac{\lambda+q}{2\theta},\frac{n}{2},\frac{r^2}{2\sigma^2}\big)+\frac{n(\lambda+q)}{2(\lambda+p)}U\big(\frac{\lambda+q+2\theta}{2\theta},\frac{n+2}{2},\frac{r^2}{2\sigma^2}\big)M\big(\frac{\lambda+p}{2\theta},\frac{n}{2},\frac{r^2}{2\sigma^2}\big)}$$

$z \le r$, $r \le x$

$$= \frac{\lambda n(\lambda+p)^{-1}z^{n-1}r^{-n}e^{(r^2-z^2)/2\sigma^2}U\big(\frac{\lambda+q}{2\theta},\frac{n}{2},\frac{x^2}{2\sigma^2}\big)M\big(\frac{\lambda+p}{2\theta},\frac{n}{2},\frac{z^2}{2\sigma^2}\big)}{M\big(\frac{\lambda+p+2\theta}{2\theta},\frac{n+2}{2},\frac{r^2}{2\sigma^2}\big)U\big(\frac{\lambda+q}{2\theta},\frac{n}{2},\frac{r^2}{2\sigma^2}\big)+\frac{n(\lambda+q)}{2(\lambda+p)}U\big(\frac{\lambda+q+2\theta}{2\theta},\frac{n+2}{2},\frac{r^2}{2\sigma^2}\big)M\big(\frac{\lambda+p}{2\theta},\frac{n}{2},\frac{r^2}{2\sigma^2}\big)}$$

$r \le z$, $x \le r$

$$= \frac{\lambda n(\lambda+p)^{-1}z^{n-1}r^{-n}e^{(r^2-z^2)/2\sigma^2}U\big(\frac{\lambda+q}{2\theta},\frac{n}{2},\frac{z^2}{2\sigma^2}\big)M\big(\frac{\lambda+p}{2\theta},\frac{n}{2},\frac{x^2}{2\sigma^2}\big)}{M\big(\frac{\lambda+p+2\theta}{2\theta},\frac{n+2}{2},\frac{r^2}{2\sigma^2}\big)U\big(\frac{\lambda+q}{2\theta},\frac{n}{2},\frac{r^2}{2\sigma^2}\big)+\frac{n(\lambda+q)}{2(\lambda+p)}U\big(\frac{\lambda+q+2\theta}{2\theta},\frac{n+2}{2},\frac{r^2}{2\sigma^2}\big)M\big(\frac{\lambda+p}{2\theta},\frac{n}{2},\frac{r^2}{2\sigma^2}\big)}$$

$r \le z$, $r \le x$

$$= \frac{\lambda z^{n-1}e^{-z^2/2\sigma^2}U\big(\frac{\lambda+q}{2\theta},\frac{n}{2},\frac{(z+x+|z-x|)^2}{4\sigma^2}\big)S\big(\frac{\lambda+q}{2\theta},\frac{n}{2},\frac{(z+x-|z-x|)^2}{4\sigma^2},\frac{r^2}{2\sigma^2}\big)}{\theta 2^{n/2}\sigma^n U\big(\frac{\lambda+q}{2\theta},\frac{n}{2},\frac{r^2}{2\sigma^2}\big)}$$

$$+ \frac{U\big(\frac{\lambda+q}{2\theta},\frac{n}{2},\frac{x^2}{2\sigma^2}\big)}{U\big(\frac{\lambda+q}{2\theta},\frac{n}{2},\frac{r^2}{2\sigma^2}\big)}$$

$$\times \frac{\lambda n(\lambda+p)^{-1}z^{n-1}r^{-n}e^{(r^2-z^2)/2\sigma^2}M\big(\frac{\lambda+p}{2\theta},\frac{n+2}{2},\frac{r^2}{2\sigma^2}\big)U\big(\frac{\lambda+p}{2\theta},\frac{n}{2},\frac{z^2}{2\sigma^2}\big)}{M\big(\frac{\lambda+p+2\theta}{2\theta},\frac{n+2}{2},\frac{r^2}{2\sigma^2}\big)U\big(\frac{\lambda+q}{2\theta},\frac{n}{2},\frac{r^2}{2\sigma^2}\big)+\frac{n(\lambda+q)}{2(\lambda+p)}U\big(\frac{\lambda+q+2\theta}{2\theta},\frac{n+2}{2},\frac{r^2}{2\sigma^2}\big)M\big(\frac{\lambda+p}{2\theta},\frac{n}{2},\frac{r^2}{2\sigma^2}\big)}$$

8 $Q_s^{(n)} = \sigma e^{-\theta s} R^{(n)}_{e^{2\theta s}-1}$ $\quad n \ge 2,\ \sigma, \theta > 0 \quad \tau \sim \operatorname{Exp}(\lambda)$, independent of $Q^{(n)}$

1.9.3 $1 \le \gamma$	$\mathbf{E}_x \exp\Big(-\frac{(\gamma^2-1)\theta}{4\sigma^2}\int_0^t (Q_s^{(n)})^2 ds\Big)$
	$= \frac{\gamma^{n/2} e^{t\theta n/2}}{(\operatorname{sh}(t\theta\gamma)+\gamma\operatorname{ch}(t\theta\gamma))^{n/2}} \exp\Big(-\frac{x^2(\gamma^2-1)\operatorname{sh}(t\theta\gamma)}{4\sigma^2(\operatorname{sh}(t\theta\gamma)+\gamma\operatorname{ch}(t\theta\gamma))}\Big)$
1.9.5 $0 < x$	$\mathbf{E}_x\Big\{\exp\Big(-\frac{(\gamma^2-1)\theta}{4\sigma^2}\int_0^\tau (Q_s^{(n)})^2 ds\Big);\ Q_\tau^{(n)} \in dz\Big\}$
$x \le z$	$= \frac{\lambda\Gamma(\frac{n}{4}-\frac{n}{4\gamma}+\frac{\lambda}{2\theta\gamma})}{\theta\gamma\Gamma(\frac{n}{2})x^{n/2}z^{1-n/2}} e^{(x^2-z^2)/4\sigma^2} M_{\frac{\theta n-2\lambda}{4\theta\gamma},\frac{n-2}{4}}\Big(\frac{\gamma x^2}{2\sigma^2}\Big) W_{\frac{\theta n-2\lambda}{4\theta\gamma},\frac{n-2}{4}}\Big(\frac{\gamma z^2}{2\sigma^2}\Big)dz$
$z \le x$	$= \frac{\lambda\Gamma(\frac{n}{4}-\frac{n}{4\gamma}+\frac{\lambda}{2\theta\gamma})}{\theta\gamma\Gamma(\frac{n}{2})x^{n/2}z^{1-n/2}} e^{(x^2-z^2)/4\sigma^2} W_{\frac{\theta n-2\lambda}{4\theta\gamma},\frac{n-2}{4}}\Big(\frac{\gamma x^2}{2\sigma^2}\Big) M_{\frac{\theta n-2\lambda}{4\theta\gamma},\frac{n-2}{4}}\Big(\frac{\gamma z^2}{2\sigma^2}\Big)dz$
1.9.7 $1 \le \gamma$	$\mathbf{E}_x\Big\{\exp\Big(-\frac{(\gamma^2-1)\theta}{4\sigma^2}\int_0^t (Q_s^{(n)})^2 ds\Big);\ Q_t^{(n)} \in dz\Big\}$
	$= \frac{\gamma x^{1-n/2} z^{n/2}}{2\sigma^2\operatorname{sh}(t\theta\gamma)} \exp\Big(\frac{t\theta n}{2}+\frac{x^2-z^2}{4\sigma^2}-\frac{(x^2+z^2)\gamma\operatorname{ch}(t\theta\gamma)}{4\sigma^2\operatorname{sh}(t\theta\gamma)}\Big) I_{\frac{n-2}{2}}\Big(\frac{xz\gamma}{2\sigma^2\operatorname{sh}(t\theta\gamma)}\Big)dz$
1.9.8	$\mathbf{P}_x\Big(\int_0^t (Q_s^{(n)})^2 ds \in dy,\ Q_t^{(n)} \in dz\Big)$
	$= \frac{x(z/x)^{n/2}}{4\sigma^4\theta^2} \exp\Big(\frac{t\theta n}{2}+\frac{x^2-z^2-y\theta}{4\sigma^2}\Big) \operatorname{is}_{y/2\sigma^2\theta}\Big(\frac{n-2}{2}, t, 0, \frac{x^2+z^2}{4\sigma^2\theta}, \frac{xz}{4\sigma^2\theta}\Big)dydz$
1.13.1 $x < y$	$\mathbf{E}_x\{e^{-\gamma\check{H}(\tau)};\ \sup_{0\le s\le\tau} Q_s^{(n)} \in dy\} = \frac{\lambda y M(\frac{\lambda+\gamma}{2\theta},\frac{n}{2},\frac{x^2}{2\sigma^2}) M(\frac{\lambda+2\theta}{2\theta},\frac{n+2}{2},\frac{y^2}{2\sigma^2})}{\theta\sigma^2 n M(\frac{\lambda+\gamma}{2\theta},\frac{n}{2},\frac{y^2}{2\sigma^2}) M(\frac{\lambda}{2\theta},\frac{n}{2},\frac{y^2}{2\sigma^2})}dy$
1.13.5	$\mathbf{E}_x\{e^{-\gamma\check{H}(\tau)};\ \sup_{0\le s\le\tau} Q_s^{(n)} \in dy,\ Q_\tau^{(n)} \in dz\}$ $\qquad x \vee z < y$
	$= \frac{\lambda y^{1-n} z^{n-1} e^{(y^2-z^2)/2\sigma^2} M(\frac{\lambda+\gamma}{2\theta},\frac{n}{2},\frac{x^2}{2\sigma^2}) M(\frac{\lambda}{2\theta},\frac{n}{2},\frac{z^2}{2\sigma^2})}{\theta\sigma^2 M(\frac{\lambda+\gamma}{2\theta},\frac{n}{2},\frac{y^2}{2\sigma^2}) M(\frac{\lambda}{2\theta},\frac{n}{2},\frac{y^2}{2\sigma^2})}dydz$

8 $Q_s^{(n)} = \sigma e^{-\theta s} R^{(n)}_{e^{2\theta s}-1}$ $n \ge 2,\ \sigma, \theta > 0$ $\tau \sim \operatorname{Exp}(\lambda)$, independent of $Q^{(n)}$

1.14.1 $y < x$

$$\mathbf{E}_x\{e^{-\gamma \hat H(\tau)};\ \inf_{0\le s\le\tau} Q_s^{(n)} \in dy\} = \frac{\lambda y U(\frac{\lambda+\gamma}{2\theta}, \frac{n}{2}, \frac{x^2}{2\sigma^2}) U(\frac{\lambda+2\theta}{2\theta}, \frac{n+2}{2}, \frac{y^2}{2\sigma^2})}{2\theta\sigma^2 U(\frac{\lambda+\gamma}{2\theta}, \frac{n}{2}, \frac{y^2}{2\sigma^2}) U(\frac{\lambda}{2\theta}, \frac{n}{2}, \frac{y^2}{2\sigma^2})} dy$$

1.14.5 $0 < y < x \wedge z$

$$\mathbf{E}_x\{e^{-\gamma \hat H(\tau)};\ \inf_{0\le s\le\tau} Q_s^{(n)} \in dy,\ Q_\tau^{(n)} \in dz\}$$

$$= \frac{\lambda y^{1-n} z^{n-1} e^{(y^2-z^2)/2\sigma^2} U(\frac{\lambda+\gamma}{2\theta}, \frac{n}{2}, \frac{x^2}{2\sigma^2}) U(\frac{\lambda}{2\theta}, \frac{n}{2}, \frac{z^2}{2\sigma^2})}{\theta\sigma^2 U(\frac{\lambda+\gamma}{2\theta}, \frac{n}{2}, \frac{y^2}{2\sigma^2}) U(\frac{\lambda}{2\theta}, \frac{n}{2}, \frac{y^2}{2\sigma^2})} dydz$$

1.15.2

$$\mathbf{P}_x\Big(a < \inf_{0\le s\le\tau} Q_s^{(n)},\ \sup_{0\le s\le\tau} Q_s^{(n)} < b\Big) = 1 - \frac{S(\frac{\lambda}{2\theta}, \frac{n}{2}, \frac{b^2}{2\sigma^2}, \frac{x^2}{2\sigma^2}) + S(\frac{\lambda}{2\theta}, \frac{n}{2}, \frac{x^2}{2\sigma^2}, \frac{a^2}{2\sigma^2})}{S(\frac{\lambda}{2\theta}, \frac{n}{2}, \frac{b^2}{2\sigma^2}, \frac{a^2}{2\sigma^2})}$$

1.15.6

$$\mathbf{P}_x\Big(a < \inf_{0\le s\le\tau} Q_s^{(n)},\ \sup_{0\le s\le\tau} Q_s^{(n)} < b, Q_\tau^{(n)} \in dz\Big)$$

$$= \begin{cases} \dfrac{\lambda z^{n-1} e^{-z^2/2\sigma^2} S(\frac{\lambda}{2\theta}, \frac{n}{2}, \frac{b^2}{2\sigma^2}, \frac{z^2}{2\sigma^2}) S(\frac{\lambda}{2\theta}, \frac{n}{2}, \frac{x^2}{2\sigma^2}, \frac{a^2}{2\sigma^2})}{\theta 2^{n/2}\sigma^n S(\frac{\lambda}{2\theta}, \frac{n}{2}, \frac{b^2}{2\sigma^2}, \frac{a^2}{2\sigma^2})} dz, & a \le x \le z \\ \dfrac{\lambda z^{n-1} e^{-z^2/2\sigma^2} S(\frac{\lambda}{2\theta}, \frac{n}{2}, \frac{b^2}{2\sigma^2}, \frac{x^2}{2\sigma^2}) S(\frac{\lambda}{2\theta}, \frac{n}{2}, \frac{z^2}{2\sigma^2}, \frac{a^2}{2\sigma^2})}{\theta 2^{n/2}\sigma^n S(\frac{\lambda}{2\theta}, \frac{n}{2}, \frac{b^2}{2\sigma^2}, \frac{a^2}{2\sigma^2})} dz, & z \le x \le b \end{cases}$$

1.16.2 $r < b$

$$\mathbf{P}_x\Big(\ell(\tau, r) \in dy,\ \sup_{0\le s\le\tau} Q_s^{(n)} < b\Big) = \exp\Big(-\frac{y\theta r^{1-n} 2^{n/2}\sigma^n e^{r^2/2\sigma^2} M(\frac{\lambda}{2\theta}, \frac{n}{2}, \frac{b^2}{2\sigma^2})}{S(\frac{\lambda}{2\theta}, \frac{n}{2}, \frac{b^2}{2\sigma^2}, \frac{r^2}{2\sigma^2}) M(\frac{\lambda}{2\theta}, \frac{n}{2}, \frac{r^2}{2\sigma^2})}\Big)$$

$$\times \begin{cases} \dfrac{\theta 2^{n/2}\sigma^n \big(M(\frac{\lambda}{2\theta}, \frac{n}{2}, \frac{b^2}{2\sigma^2}) - M(\frac{\lambda}{2\theta}, \frac{n}{2}, \frac{r^2}{2\sigma^2})\big) M(\frac{\lambda}{2\theta}, \frac{n}{2}, \frac{x^2}{2\sigma^2})}{r^{n-1} e^{-r^2/2\sigma^2} S(\frac{\lambda}{2\theta}, \frac{n}{2}, \frac{b^2}{2\sigma^2}, \frac{r^2}{2\sigma^2}) M^2(\frac{\lambda}{2\theta}, \frac{n}{2}, \frac{r^2}{2\sigma^2})} dy, & 0 \le x \le r \\ \dfrac{\theta 2^{n/2}\sigma^n \big(M(\frac{\lambda}{2\theta}, \frac{n}{2}, \frac{b^2}{2\sigma^2}) - M(\frac{\lambda}{2\theta}, \frac{n}{2}, \frac{r^2}{2\sigma^2})\big) S(\frac{\lambda}{2\theta}, \frac{n}{2}, \frac{b^2}{2\sigma^2}, \frac{x^2}{2\sigma^2})}{r^{n-1} e^{-r^2/2\sigma^2} S^2(\frac{\lambda}{2\theta}, \frac{n}{2}, \frac{b^2}{2\sigma^2}, \frac{r^2}{2\sigma^2}) M(\frac{\lambda}{2\theta}, \frac{n}{2}, \frac{r^2}{2\sigma^2})} dy, & r \le x \le b \end{cases}$$

(1)

$$\mathbf{P}_x\Big(\ell(\tau, r) = 0,\ \sup_{0\le s\le\tau} Q_s^{(n)} < b\Big)$$

$$= \begin{cases} 1 - \dfrac{M(\frac{\lambda}{2\theta}, \frac{n}{2}, \frac{x^2}{2\sigma^2})}{M(\frac{\lambda}{2\theta}, \frac{n}{2}, \frac{r^2}{2\sigma^2})}, & 0 \le x \le r \\ 1 - \dfrac{S(\frac{\lambda}{2\theta}, \frac{n}{2}, \frac{b^2}{2\sigma^2}, \frac{x^2}{2\sigma^2}) + S(\frac{\lambda}{2\theta}, \frac{n}{2}, \frac{x^2}{2\sigma^2}, \frac{r^2}{2\sigma^2})}{S(\frac{\lambda}{2\theta}, \frac{n}{2}, \frac{b^2}{2\sigma^2}, \frac{r^2}{2\sigma^2})}, & r \le x \le b \end{cases}$$

8 $Q_s^{(n)} = \sigma e^{-\theta s} R^{(n)}_{e^{2\theta s}-1}$ $n \ge 2,\ \sigma,\theta > 0$ $\tau \sim \operatorname{Exp}(\lambda)$, independent of $Q^{(n)}$

1.16.6 $\mathbf{P}_x\big(\ell(\tau,r) \in dy,\ \sup_{0\le s\le\tau} Q_s^{(n)} < b, Q_\tau^{(n)} \in dz\big)$ $r \vee z < b$

$$= \frac{\lambda z^{n-1} e^{-z^2/2\sigma^2}}{r^{n-1}e^{-r^2/2\sigma^2}} \exp\Big(-\frac{y\theta r^{1-n}2^{n/2}\sigma^n e^{r^2/2\sigma^2} M\big(\frac{\lambda}{2\theta},\frac{n}{2},\frac{b^2}{2\sigma^2}\big)}{S\big(\frac{\lambda}{2\theta},\frac{n}{2},\frac{b^2}{2\sigma^2},\frac{r^2}{2\sigma^2}\big) M\big(\frac{\lambda}{2\theta},\frac{n}{2},\frac{r^2}{2\sigma^2}\big)}\Big)$$

$$\times \begin{cases} \dfrac{M\big(\frac{\lambda}{2\theta},\frac{n}{2},\frac{z^2}{2\sigma^2}\big) M\big(\frac{\lambda}{2\theta},\frac{n}{2},\frac{x^2}{2\sigma^2}\big)}{M^2\big(\frac{\lambda}{2\theta},\frac{n}{2},\frac{r^2}{2\sigma^2}\big)} dydz, & 0 \le x \le r, \quad z \le r \\ \dfrac{S\big(\frac{\lambda}{2\theta},\frac{n}{2},\frac{b^2}{2\sigma^2},\frac{x^2}{2\sigma^2}\big) M\big(\frac{\lambda}{2\theta},\frac{n}{2},\frac{z^2}{2\sigma^2}\big)}{S\big(\frac{\lambda}{2\theta},\frac{n}{2},\frac{b^2}{2\sigma^2},\frac{r^2}{2\sigma^2}\big) M\big(\frac{\lambda}{2\theta},\frac{n}{2},\frac{r^2}{2\sigma^2}\big)} dydz, & z \le r \le x \\ \dfrac{S\big(\frac{\lambda}{2\theta},\frac{n}{2},\frac{b^2}{2\sigma^2},\frac{z^2}{2\sigma^2}\big) M\big(\frac{\lambda}{2\theta},\frac{n}{2},\frac{x^2}{2\sigma^2}\big)}{S\big(\frac{\lambda}{2\theta},\frac{n}{2},\frac{b^2}{2\sigma^2},\frac{r^2}{2\sigma^2}\big) M\big(\frac{\lambda}{2\theta},\frac{n}{2},\frac{r^2}{2\sigma^2}\big)} dydz, & 0 \le x \le r \le z \\ \dfrac{S\big(\frac{\lambda}{2\theta},\frac{n}{2},\frac{b^2}{2\sigma^2},\frac{z^2}{2\sigma^2}\big) S\big(\frac{\lambda}{2\theta},\frac{n}{2},\frac{b^2}{2\sigma^2},\frac{x^2}{2\sigma^2}\big)}{S^2\big(\frac{\lambda}{2\theta},\frac{n}{2},\frac{b^2}{2\sigma^2},\frac{r^2}{2\sigma^2}\big)} dydz, & r \le x, \quad r \le z \end{cases}$$

(1) $\mathbf{P}_x\big(\ell(\tau,r) = 0,\ \sup_{0\le s\le\tau} Q_s^{(n)} < b, Q_\tau^{(n)} \in dz\big)$

$z \le r$, $x \le r$

$$= \frac{\lambda z^{n-1} e^{-z^2/2\sigma^2} S\big(\frac{\lambda}{2\theta},\frac{n}{2},\frac{r^2}{2\sigma^2},\frac{(z+x+|z-x|)^2}{4\sigma^2}\big) M\big(\frac{\lambda}{2\theta},\frac{n}{2},\frac{(z+x-|z-x|)^2}{4\sigma^2}\big)}{\theta 2^{n/2}\sigma^n M\big(\frac{\lambda}{2\theta},\frac{n}{2},\frac{r^2}{2\sigma^2}\big)} dz$$

$r \le z$, $r \le x$

$$= \frac{\lambda z^{n-1} e^{-z^2/2\sigma^2} S\big(\frac{\lambda}{2\theta},\frac{n}{2},\frac{b^2}{2\sigma^2},\frac{(z+x+|z-x|)^2}{4\sigma^2}\big) S\big(\frac{\lambda}{2\theta},\frac{n}{2},\frac{(z+x-|z-x|)^2}{4\sigma^2},\frac{r^2}{2\sigma^2}\big)}{\theta 2^{n/2}\sigma^n S\big(\frac{\lambda}{2\theta},\frac{n}{2},\frac{b^2}{2\sigma^2},\frac{r^2}{2\sigma^2}\big)} dz$$

1.17.2 ($a < r$) $\mathbf{P}_x\big(\ell(\tau,r) \in dy, a < \inf_{0\le s\le\tau} Q_s^{(n)}\big) = \exp\Big(-\dfrac{y\theta r^{1-n}2^{n/2}\sigma^n e^{r^2/2\sigma^2} U\big(\frac{\lambda}{2\theta},\frac{n}{2},\frac{a^2}{2\sigma^2}\big)}{U\big(\frac{\lambda}{2\theta},\frac{n}{2},\frac{r^2}{2\sigma^2}\big) S\big(\frac{\lambda}{2\theta},\frac{n}{2},\frac{r^2}{2\sigma^2},\frac{a^2}{2\sigma^2}\big)}\Big)$

$$\times \begin{cases} \dfrac{\theta 2^{n/2}\sigma^n \big(U\big(\frac{\lambda}{2\theta},\frac{n}{2},\frac{a^2}{2\sigma^2}\big) - U\big(\frac{\lambda}{2\theta},\frac{n}{2},\frac{r^2}{2\sigma^2}\big)\big) S\big(\frac{\lambda}{2\theta},\frac{n}{2},\frac{x^2}{2\sigma^2},\frac{a^2}{2\sigma^2}\big)}{r^{n-1}e^{-r^2/2\sigma^2} U\big(\frac{\lambda}{2\theta},\frac{n}{2},\frac{r^2}{2\sigma^2}\big) S^2\big(\frac{\lambda}{2\theta},\frac{n}{2},\frac{r^2}{2\sigma^2},\frac{a^2}{2\sigma^2}\big)} dy, & a \le x \le r \\ \dfrac{\theta 2^{n/2}\sigma^n \big(U\big(\frac{\lambda}{2\theta},\frac{n}{2},\frac{a^2}{2\sigma^2}\big) - U\big(\frac{\lambda}{2\theta},\frac{n}{2},\frac{r^2}{2\sigma^2}\big)\big) U\big(\frac{\lambda}{2\theta},\frac{n}{2},\frac{x^2}{2\sigma^2}\big)}{r^{n-1}e^{-r^2/2\sigma^2} U^2\big(\frac{\lambda}{2\theta},\frac{n}{2},\frac{r^2}{2\sigma^2}\big) S\big(\frac{\lambda}{2\theta},\frac{n}{2},\frac{r^2}{2\sigma^2},\frac{a^2}{2\sigma^2}\big)} dy, & r \le x \end{cases}$$

(1) $\mathbf{P}_x\big(\ell(\tau,r) = 0, a < \inf_{0\le s\le\tau} Q_s^{(n)}\big)$

8 $Q_s^{(n)} = \sigma e^{-\theta s} R^{(n)}_{e^{2\theta s}-1}$ $\quad n \ge 2,\ \sigma, \theta > 0 \quad \tau \sim \operatorname{Exp}(\lambda)$, independent of $Q^{(n)}$

$$= \begin{cases} 1 - \dfrac{S\big(\frac{\lambda}{2\theta}, \frac{n}{2}, \frac{r^2}{2\sigma^2}, \frac{x^2}{2\sigma^2}\big) + S\big(\frac{\lambda}{2\theta}, \frac{n}{2}, \frac{x^2}{2\sigma^2}, \frac{a^2}{2\sigma^2}\big)}{S\big(\frac{\lambda}{2\theta}, \frac{n}{2}, \frac{r^2}{2\sigma^2}, \frac{a^2}{2\sigma^2}\big)}, & a \le x \le r \\ 1 - \dfrac{U\big(\frac{\lambda}{2\theta}, \frac{n}{2}, \frac{x^2}{2\sigma^2}\big)}{U\big(\frac{\lambda}{2\theta}, \frac{n}{2}, \frac{r^2}{2\sigma^2}\big)}, & r \le x \end{cases}$$

1.17.6 $\mathbf{P}_x\big(\ell(\tau, r) \in dy, a < \inf_{0 \le s \le \tau} Q_s^{(n)}, Q_\tau^{(n)} \in dz\big)$ $\qquad a < r \wedge z$

$$= \frac{\lambda z^{n-1} e^{-z^2/2\sigma^2}}{r^{n-1} e^{-r^2/2\sigma^2}} \exp\Big(-\frac{y\theta r^{1-n} 2^{n/2} \sigma^n e^{r^2/2\sigma^2} U\big(\frac{\lambda}{2\theta}, \frac{n}{2}, \frac{a^2}{2\sigma^2}\big)}{U\big(\frac{\lambda}{2\theta}, \frac{n}{2}, \frac{r^2}{2\sigma^2}\big) S\big(\frac{\lambda}{2\theta}, \frac{n}{2}, \frac{r^2}{2\sigma^2}, \frac{a^2}{2\sigma^2}\big)}\Big)$$

$$\times \begin{cases} \dfrac{S\big(\frac{\lambda}{2\theta}, \frac{n}{2}, \frac{z^2}{2\sigma^2}, \frac{a^2}{2\sigma^2}\big) S\big(\frac{\lambda}{2\theta}, \frac{n}{2}, \frac{x^2}{2\sigma^2}, \frac{a^2}{2\sigma^2}\big)}{S^2\big(\frac{\lambda}{2\theta}, \frac{n}{2}, \frac{r^2}{2\sigma^2}, \frac{a^2}{2\sigma^2}\big)} dydz, & 0 \le x \le r,\ z \le r \\ \dfrac{U\big(\frac{\lambda}{2\theta}, \frac{n}{2}, \frac{x^2}{2\sigma^2}\big) S\big(\frac{\lambda}{2\theta}, \frac{n}{2}, \frac{z^2}{2\sigma^2}, \frac{a^2}{2\sigma^2}\big)}{U\big(\frac{\lambda}{2\theta}, \frac{n}{2}, \frac{r^2}{2\sigma^2}\big) S\big(\frac{\lambda}{2\theta}, \frac{n}{2}, \frac{r^2}{2\sigma^2}, \frac{a^2}{2\sigma^2}\big)} dydz, & z \le r \le x \\ \dfrac{U\big(\frac{\lambda}{2\theta}, \frac{n}{2}, \frac{z^2}{2\sigma^2}\big) S\big(\frac{\lambda}{2\theta}, \frac{n}{2}, \frac{x^2}{2\sigma^2}, \frac{a^2}{2\sigma^2}\big)}{U\big(\frac{\lambda}{2\theta}, \frac{n}{2}, \frac{r^2}{2\sigma^2}\big) S\big(\frac{\lambda}{2\theta}, \frac{n}{2}, \frac{r^2}{2\sigma^2}, \frac{a^2}{2\sigma^2}\big)} dydz, & 0 \le x \le r \le z \\ \dfrac{U\big(\frac{\lambda}{2\theta}, \frac{n}{2}, \frac{z^2}{2\sigma^2}\big) U\big(\frac{\lambda}{2\theta}, \frac{n}{2}, \frac{x^2}{2\sigma^2}\big)}{U^2\big(\frac{\lambda}{2\theta}, \frac{n}{2}, \frac{r^2}{2\sigma^2}\big)} dydz, & r \le x,\ r \le z \end{cases}$$

(1) $\mathbf{P}_x\big(\ell(\tau, r) = 0, a < \inf_{0 \le s \le \tau} Q_s^{(n)}, Q_\tau^{(n)} \in dz\big)$

$z \le r$, $x \le r$:

$$= \frac{\lambda z^{n-1} e^{-z^2/2\sigma^2} S\big(\frac{\lambda}{2\theta}, \frac{n}{2}, \frac{r^2}{2\sigma^2}, \frac{(z+x+|z-x|)^2}{4\sigma^2}\big) S\big(\frac{\lambda}{2\theta}, \frac{n}{2}, \frac{(z+x-|z-x|)^2}{4\sigma^2}, \frac{a^2}{2\sigma^2}\big)}{\theta 2^{n/2} \sigma^n S\big(\frac{\lambda}{2\theta}, \frac{n}{2}, \frac{r^2}{2\sigma^2}, \frac{a^2}{2\sigma^2}\big)} dz$$

$r \le z$, $r \le x$:

$$= \frac{\lambda z^{n-1} e^{-z^2/2\sigma^2} U\big(\frac{\lambda}{2\theta}, \frac{n}{2}, \frac{(z+x+|z-x|)^2}{4\sigma^2}\big) S\big(\frac{\lambda}{2\theta}, \frac{n}{2}, \frac{(z+x-|z-x|)^2}{4\sigma^2}, \frac{r^2}{2\sigma^2}\big)}{\theta 2^{n/2} \sigma^n U\big(\frac{\lambda}{2\theta}, \frac{n}{2}, \frac{r^2}{2\sigma^2}\big)} dz$$

1.20.3 ($0 < x$) $\displaystyle \mathbf{E}_x \exp\Big(-\int_0^t \frac{\gamma^2 \sigma^2 \theta}{(Q_s^{(n)})^2} ds\Big) = (4\sigma^2 e^{t\theta} \operatorname{sh}(t\theta))^{n/4} \exp\Big(-\frac{x^2 e^{-t\theta}}{8\sigma^2 \operatorname{sh}(t\theta)}\Big)$

$$\times \frac{\Gamma\Big(\frac{n}{4} + \sqrt{\frac{(n-2)^2}{16} + \frac{\gamma^2}{4}} + \frac{1}{2}\Big)}{x^{n/2} \Gamma\Big(\sqrt{\frac{(n-2)^2}{4} + \gamma^2} + 1\Big)} M_{-\frac{n}{4}, \sqrt{\frac{(n-2)^2}{16} + \frac{\gamma^2}{4}}}\Big(\frac{x^2 e^{-t\theta}}{4\sigma^2 \operatorname{sh}(t\theta)}\Big)$$

8 $Q_s^{(n)} = \sigma e^{-\theta s} R^{(n)}_{e^{2\theta s}-1}$ $n \ge 2,\ \sigma, \theta > 0$ $\tau \sim \operatorname{Exp}(\lambda)$, independent of $Q^{(n)}$

1.20.4 $0<x$	$\mathbf{P}_x\Big(\int_0^t \frac{ds}{(Q_s^{(n)})^2} \in dy\Big)$
	$= \frac{\theta \operatorname{sh}^{n/4}(t\theta) e^{t\theta n/4}}{x^{n/2}(4\sigma^2)^{-1-n/4}} \exp\Big(-\frac{(n-2)^2\sigma^2\theta y}{4} - \frac{x^2 e^{-t\theta}}{8\sigma^2 \operatorname{sh}(t\theta)}\Big) \operatorname{m}_{4\sigma^2\theta y}\Big(\frac{n}{4}, \frac{x^2 e^{-t\theta}}{8\sigma^2 \operatorname{sh}(t\theta)}\Big) dy$
1.20.5	$\mathbf{E}_x\Big\{\exp\Big(-\int_0^\tau \frac{\gamma^2\sigma^2\theta}{(Q_s^{(n)})^2} ds\Big); Q_\tau^{(n)} \in dz\Big\}$
	$= \frac{\lambda \Gamma\Big(\sqrt{\frac{(n-2)^2}{16} + \frac{\gamma^2}{4}} + \frac{1}{2} + \frac{2\lambda - n\theta}{4\theta}\Big)}{\theta \Gamma\Big(\sqrt{\frac{(n-2)^2}{4} + \gamma^2} + 1\Big) x^{n/2} z^{1-n/2}} e^{(x^2-z^2)^2/4\sigma^2}$
	$\times \begin{cases} M_{\frac{n\theta-2\lambda}{4\theta}, \sqrt{\frac{(n-2)^2}{16}+\frac{\gamma^2}{4}}}\Big(\frac{x^2}{2\sigma^2}\Big) W_{\frac{n\theta-2\lambda}{4\theta}, \sqrt{\frac{(n-2)^2}{16}+\frac{\gamma^2}{4}}}\Big(\frac{z^2}{2\sigma^2}\Big) dz, & 0 \le x \le z \\ W_{\frac{n\theta-2\lambda}{4\theta}, \sqrt{\frac{(n-2)^2}{16}+\frac{\gamma^2}{4}}}\Big(\frac{x^2}{2\sigma^2}\Big) M_{\frac{n\theta-2\lambda}{4\theta}, \sqrt{\frac{(n-2)^2}{16}+\frac{\gamma^2}{4}}}\Big(\frac{z^2}{2\sigma^2}\Big) dz, & z \le x \end{cases}$
1.20.7	$\mathbf{E}_x\Big\{\exp\Big(-\int_0^t \frac{\gamma^2\sigma^2\theta}{(Q_s^{(n)})^2} ds\Big); Q_t^{(n)} \in dz\Big\}$
	$= \frac{x(z/x)^{n/2}}{2\sigma^2 \operatorname{sh}(t\theta)} \exp\Big(\frac{t\theta n}{2} + \frac{x^2-z^2}{4\sigma^2} - \frac{(x^2+z^2)\operatorname{ch}(t\theta)}{4\sigma^2 \operatorname{sh}(t\theta)}\Big) I_{\sqrt{\frac{(n-2)^2}{4}+\gamma^2}}\Big(\frac{xz}{2\sigma^2 \operatorname{sh}(t\theta)}\Big) dz$
1.20.8 $0<x$	$\mathbf{P}_x\Big(\int_0^t \frac{ds}{(Q_s^{(n)})^2} \in dy,\ Q_t^{(n)} \in dz\Big) = \frac{\theta x (z/x)^{n/2}}{2\operatorname{sh}(t\theta)}$
	$\times \exp\Big(\frac{t\theta n}{2} - \frac{(n-2)^2\sigma^2\theta y}{4} + \frac{x^2-z^2}{4\sigma^2} - \frac{(x^2+z^2)\operatorname{ch}(t\theta)}{4\sigma^2 \operatorname{sh}(t\theta)}\Big) \operatorname{i}_{\sigma^2\theta y}\Big(\frac{xz}{2\sigma^2 \operatorname{sh}(t\theta)}\Big) dy dz$
1.21.3 $0<x$	$\mathbf{E}_x \exp\Big(-\int_0^t \Big(\frac{p^2\sigma^2\theta}{(Q_s^{(n)})^2} + \frac{\theta(q^2-1)}{4\sigma^2}(Q_s^{(n)})^2\Big) ds\Big)$
	$= \frac{(4\sigma^2 \operatorname{sh}(t\theta q))^{n/4} e^{t\theta n/2}}{(\operatorname{sh}(t\theta q) + q\operatorname{ch}(t\theta q))^{n/4}} \exp\Big(-\frac{x^2(2(q^2-1)\operatorname{sh}^2(t\theta q) + q^2)}{8\sigma^2 \operatorname{sh}(t\theta q)(\operatorname{sh}(t\theta q) + q\operatorname{ch}(t\theta q))}\Big)$
	$\times \frac{\Gamma\Big(\frac{n}{4} + \sqrt{\frac{(n-2)^2}{16} + \frac{p^2}{4}} + \frac{1}{2}\Big)}{x^{n/2}\Gamma\Big(\sqrt{\frac{(n-2)^2}{4} + p^2} + 1\Big)} M_{-\frac{n}{4}, \sqrt{\frac{(n-2)^2}{16}+\frac{p^2}{4}}}\Big(\frac{x^2 q^2}{4\sigma^2 \operatorname{sh}(t\theta q)(\operatorname{sh}(t\theta q) + q\operatorname{ch}(t\theta q))}\Big)$

8 $Q_s^{(n)} = \sigma e^{-\theta s} R^{(n)}_{e^{2\theta s}-1}$ $n \ge 2,\ \sigma, \theta > 0$ $\tau \sim \operatorname{Exp}(\lambda)$, independent of $Q^{(n)}$

(1) $0 < x$

$$\mathbf{E}_x\Big\{\exp\Big(-\frac{\theta(q^2-1)}{4\sigma^2}\int_0^t \big(Q_s^{(n)}\big)^2 ds\Big);\ \int_0^t \frac{ds}{\big(Q_s^{(n)}\big)^2} \in dy\Big\}$$

$$= \frac{\theta(4\sigma^2)^{1+n/4}\operatorname{sh}^{n/4}(t\theta q)e^{t\theta n/2}}{x^{n/2}(\operatorname{sh}(t\theta q)+q\operatorname{ch}(t\theta q))^{n/4}}\operatorname{m}_{4\sigma^2\theta y}\Big(\frac{n}{4}, \frac{x^2q^2}{8\sigma^2\operatorname{sh}(t\theta q)(\operatorname{sh}(t\theta q)+q\operatorname{ch}(t\theta q))}\Big)$$

$$\times\exp\Big(-\frac{(n-2)^2\sigma^2\theta y}{4} - \frac{x^2(2(q^2-1)\operatorname{sh}^2(t\theta q)+q^2)}{8\sigma^2\operatorname{sh}(t\theta q)(\operatorname{sh}(t\theta q)+q\operatorname{ch}(t\theta q))}\Big)dy$$

1.21.5 $0 < x$

$$\mathbf{E}_x\Big\{\exp\Big(-\int_0^\tau\Big(\frac{p^2\sigma^2\theta}{\big(Q_s^{(n)}\big)^2} + \frac{\theta(q^2-1)}{4\sigma^2}(Q_s^{(n)})^2\Big)ds\Big);\ Q_\tau^{(n)} \in dz\Big\}$$

$$= \frac{\lambda\Gamma\Big(\sqrt{\frac{(n-2)^2}{16}+\frac{p^2}{4}}+\frac12+\frac{2\lambda-n\theta}{4\theta q}\Big)}{\theta q\Gamma\Big(\sqrt{\frac{(n-2)^2}{4}+p^2}+1\Big)x^{n/2}z^{1-n/2}}e^{(x^2-z^2)/4\sigma^2}$$

$$\times\begin{cases} M_{\frac{n\theta-2\lambda}{4\theta q},\sqrt{\frac{(n-2)^2}{16}+\frac{p^2}{4}}}\Big(\frac{qx^2}{2\sigma^2}\Big)W_{\frac{n\theta-2\lambda}{4\theta q},\sqrt{\frac{(n-2)^2}{16}+\frac{p^2}{4}}}\Big(\frac{qz^2}{2\sigma^2}\Big)dz, & 0 \le x \le z\\ W_{\frac{n\theta-2\lambda}{4\theta q},\sqrt{\frac{(n-2)^2}{16}+\frac{p^2}{4}}}\Big(\frac{qx^2}{2\sigma^2}\Big)M_{\frac{n\theta-2\lambda}{4\theta q},\sqrt{\frac{(n-2)^2}{16}+\frac{p^2}{4}}}\Big(\frac{qz^2}{2\sigma^2}\Big)dz, & z \le x\end{cases}$$

1.21.7

$$\mathbf{E}_x\Big\{\exp\Big(-\int_0^t\Big(\frac{p^2\sigma^2\theta}{\big(Q_s^{(n)}\big)^2} + \frac{\theta(q^2-1)}{4\sigma^2}(Q_s^{(n)})^2\Big)ds\Big);\ Q_t^{(n)} \in dz\Big\}$$

$$= \frac{qx(z/x)^{n/2}}{2\sigma^2\operatorname{sh}(t\theta q)}\exp\Big(\frac{t\theta n}{2}+\frac{x^2-z^2}{4\sigma^2}-\frac{(x^2+z^2)q\operatorname{ch}(t\theta q)}{4\sigma^2\operatorname{sh}(t\theta q)}\Big)I_{\sqrt{\frac{(n-2)^2}{4}+p^2}}\Big(\frac{xzq}{2\sigma^2\operatorname{sh}(t\theta q)}\Big)dz$$

(1)

$$\mathbf{E}_x\Big\{\exp\Big(-\frac{\theta(q^2-1)}{4\sigma^2}\int_0^t\big(Q_s^{(n)}\big)^2ds\Big);\ \int_0^t\frac{ds}{\big(Q_s^{(n)}\big)^2} \in dy, Q_t^{(n)} \in dz\Big\} = \frac{q\theta x(z/x)^{n/2}}{2\operatorname{sh}(t\theta q)}$$

$$\times\exp\Big(\frac{t\theta n}{2}-\frac{(n-2)^2\sigma^2\theta y}{4}+\frac{x^2-z^2}{4\sigma^2}-\frac{(x^2+z^2)q\operatorname{ch}(t\theta q)}{4\sigma^2\operatorname{sh}(t\theta q)}\Big)\operatorname{i}_{\sigma^2\theta y}\Big(\frac{xzq}{2\sigma^2\operatorname{sh}(t\theta q)}\Big)dydz$$

(2)

$$\mathbf{E}_x\Big\{\exp\Big(-\int_0^t\frac{p^2\sigma^2\theta}{\big(Q_s^{(n)}\big)^2}ds\Big);\ \int_0^t\big(Q_s^{(n)}\big)^2ds \in dy,\ Q_t^{(n)} \in dz\Big\} = \frac{x(z/x)^{n/2}}{4\sigma^4\theta^2}$$

$$\times\exp\Big(\frac{t\theta n}{2}+\frac{x^2-z^2-y\theta}{4\sigma^2}\Big)\operatorname{is}_{y/2\sigma^2\theta}\Big(\sqrt{\frac{(n-2)^2}{4}+p^2}, t, 0, \frac{x^2+z^2}{4\sigma^2\theta}, \frac{xz}{4\sigma^2\theta}\Big)dydz$$

8 $Q_s^{(n)} = \sigma e^{-\theta s} R^{(n)}_{e^{2\theta s}-1} \quad n \geq 2,\ \sigma, \theta > 0 \qquad \tau \sim \operatorname{Exp}(\lambda)$, independent of $Q^{(n)}$

1.21.8 $0 < x$

$$\mathbf{P}_x\Big(\int_0^t \big(Q_s^{(n)}\big)^2 ds \in dg,\ \int_0^t \frac{ds}{\big(Q_s^{(n)}\big)^2} \in dy,\ Q_t^{(n)} \in dz\Big) = \frac{z^{n/2}}{4\sigma^2\theta x^{(n-2)/2}}$$

$$\times \exp\Big(\frac{t\theta n}{2} - \frac{(n-2)^2\sigma^2\theta y}{4} + \frac{x^2 - z^2 - g\theta}{4\sigma^2}\Big) \operatorname{ei}_{\frac{g}{2\sigma^2\theta}}\Big(\sigma^2\theta y, t, \frac{x^2+z^2}{4\sigma^2\theta}, \frac{xz}{2\sigma^2\theta}\Big) dg dy dz$$

1.25.2 $\mathbf{P}_x\big(\ell(\tau,r) \in dy, a < \inf_{0\le s\le\tau} Q_s^{(n)}, \sup_{0\le s\le\tau} Q_s^{(n)} < b\big)$

$$= \theta r^{1-n} 2^{n/2} \sigma^n e^{r^2/2\sigma^2} \exp\Big(-\frac{y\theta r^{1-n} 2^{n/2}\sigma^n e^{r^2/2\sigma^2} S\big(\frac{\lambda}{2\theta}, \frac{n}{2}, \frac{b^2}{2\sigma^2}, \frac{a^2}{2\sigma^2}\big)}{S\big(\frac{\lambda}{2\theta}, \frac{n}{2}, \frac{b^2}{2\sigma^2}, \frac{r^2}{2\sigma^2}\big) S\big(\frac{\lambda}{2\theta}, \frac{n}{2}, \frac{r^2}{2\sigma^2}, \frac{a^2}{2\sigma^2}\big)}\Big)$$

$$\times \frac{S\big(\frac{\lambda}{2\theta}, \frac{n}{2}, \frac{b^2}{2\sigma^2}, \frac{a^2}{2\sigma^2}\big) - S\big(\frac{\lambda}{2\theta}, \frac{n}{2}, \frac{b^2}{2\sigma^2}, \frac{r^2}{2\sigma^2}\big) - S\big(\frac{\lambda}{2\theta}, \frac{n}{2}, \frac{r^2}{2\sigma^2}, \frac{a^2}{2\sigma^2}\big)}{S\big(\frac{\lambda}{2\theta}, \frac{n}{2}, \frac{b^2}{2\sigma^2}, \frac{r^2}{2\sigma^2}\big) S\big(\frac{\lambda}{2\theta}, \frac{n}{2}, \frac{r^2}{2\sigma^2}, \frac{a^2}{2\sigma^2}\big)}$$

$$\times \begin{cases} \dfrac{S\big(\frac{\lambda}{2\theta}, \frac{n}{2}, \frac{x^2}{2\sigma^2}, \frac{a^2}{2\sigma^2}\big)}{S\big(\frac{\lambda}{2\theta}, \frac{n}{2}, \frac{r^2}{2\sigma^2}, \frac{a^2}{2\sigma^2}\big)} dy, & a \le x \le r \\[2ex] \dfrac{S\big(\frac{\lambda}{2\theta}, \frac{n}{2}, \frac{b^2}{2\sigma^2}, \frac{x^2}{2\sigma^2}\big)}{S\big(\frac{\lambda}{2\theta}, \frac{n}{2}, \frac{b^2}{2\sigma^2}, \frac{r^2}{2\sigma^2}\big)} dy, & r \le x \le b \end{cases}$$

(1) $\mathbf{P}_x\big(\ell(\tau,r) = 0, a < \inf_{0\le s\le\tau} Q_s^{(n)}, \sup_{0\le s\le\tau} Q_s^{(n)} < b\big)$

$$= \begin{cases} 1 - \dfrac{S\big(\frac{\lambda}{2\theta}, \frac{n}{2}, \frac{r^2}{2\sigma^2}, \frac{x^2}{2\sigma^2}\big) + S\big(\frac{\lambda}{2\theta}, \frac{n}{2}, \frac{x^2}{2\sigma^2}, \frac{a^2}{2\sigma^2}\big)}{S\big(\frac{\lambda}{2\theta}, \frac{n}{2}, \frac{r^2}{2\sigma^2}, \frac{a^2}{2\sigma^2}\big)}, & a \le x \le r \\[2ex] 1 - \dfrac{S\big(\frac{\lambda}{2\theta}, \frac{n}{2}, \frac{b^2}{2\sigma^2}, \frac{x^2}{2\sigma^2}\big) + S\big(\frac{\lambda}{2\theta}, \frac{n}{2}, \frac{x^2}{2\sigma^2}, \frac{r^2}{2\sigma^2}\big)}{S\big(\frac{\lambda}{2\theta}, \frac{n}{2}, \frac{b^2}{2\sigma^2}, \frac{r^2}{2\sigma^2}\big)}, & r \le x \le b \end{cases}$$

1.25.5 $\mathbf{E}_x\{e^{-\gamma\ell(\tau,r)}; a < \inf_{0\le s\le\tau} Q_s^{(n)}, \sup_{0\le s\le\tau} Q_s^{(n)} < b, Q_\tau^{(n)} \in dz\}$ $\qquad a < r < b$

$z \le r$, $x \le r$

$$= \frac{\lambda z^{n-1} e^{-z^2/2\sigma^2} S\big(\frac{\lambda}{2\theta}, \frac{n}{2}, \frac{r^2}{2\sigma^2}, \frac{(z+x+|z-x|)^2}{4\sigma^2}\big) S\big(\frac{\lambda}{2\theta}, \frac{n}{2}, \frac{(z+x-|z-x|)^2}{4\sigma^2}, \frac{a^2}{2\sigma^2}\big)}{\theta 2^{n/2}\sigma^n S\big(\frac{\lambda}{2\theta}, \frac{n}{2}, \frac{r^2}{2\sigma^2}, \frac{a^2}{2\sigma^2}\big)} dz$$

8 $\quad Q_s^{(n)} = \sigma e^{-\theta s} R^{(n)}_{e^{2\theta s}-1} \quad n \ge 2,\ \sigma,\theta > 0 \quad \tau \sim \mathrm{Exp}(\lambda)$, independent of $Q^{(n)}$

$$+\frac{S\big(\frac{\lambda}{2\theta},\frac{n}{2},\frac{x^2}{2\sigma^2},\frac{a^2}{2\sigma^2}\big)}{S\big(\frac{\lambda}{2\theta},\frac{n}{2},\frac{r^2}{2\sigma^2},\frac{a^2}{2\sigma^2}\big)}$$

$$\times\frac{\lambda z^{n-1}e^{-z^2/2\sigma^2}S\big(\frac{\lambda}{2\theta},\frac{n}{2},\frac{b^2}{2\sigma^2},\frac{r^2}{2\sigma^2}\big)S\big(\frac{\lambda}{2\theta},\frac{n}{2},\frac{z^2}{2\sigma^2},\frac{a^2}{2\sigma^2}\big)}{\gamma r^{n-1}e^{-r^2/2\sigma^2}S\big(\frac{\lambda}{2\theta},\frac{n}{2},\frac{b^2}{2\sigma^2},\frac{r^2}{2\sigma^2}\big)S\big(\frac{\lambda}{2\theta},\frac{n}{2},\frac{r^2}{2\sigma^2},\frac{a^2}{2\sigma^2}\big)+\theta 2^{n/2}\sigma^n S\big(\frac{\lambda}{2\theta},\frac{n}{2},\frac{b^2}{2\sigma^2},\frac{a^2}{2\sigma^2}\big)}dz$$

$z \le r$, $r \le x$:

$$=\frac{\lambda z^{n-1}e^{-z^2/2\sigma^2}S\big(\frac{\lambda}{2\theta},\frac{n}{2},\frac{b^2}{2\sigma^2},\frac{x^2}{2\sigma^2}\big)S\big(\frac{\lambda}{2\theta},\frac{n}{2},\frac{z^2}{2\sigma^2},\frac{a^2}{2\sigma^2}\big)}{\gamma r^{n-1}e^{-r^2/2\sigma^2}S\big(\frac{\lambda}{2\theta},\frac{n}{2},\frac{b^2}{2\sigma^2},\frac{r^2}{2\sigma^2}\big)S\big(\frac{\lambda}{2\theta},\frac{n}{2},\frac{r^2}{2\sigma^2},\frac{a^2}{2\sigma^2}\big)+\theta 2^{n/2}\sigma^n S\big(\frac{\lambda}{2\theta},\frac{n}{2},\frac{b^2}{2\sigma^2},\frac{a^2}{2\sigma^2}\big)}dz$$

$r \le z$, $x \le r$:

$$=\frac{\lambda z^{n-1}e^{-z^2/2\sigma^2}S\big(\frac{\lambda}{2\theta},\frac{n}{2},\frac{b^2}{2\sigma^2},\frac{z^2}{2\sigma^2}\big)S\big(\frac{\lambda}{2\theta},\frac{n}{2},\frac{x^2}{2\sigma^2},\frac{a^2}{2\sigma^2}\big)}{\gamma r^{n-1}e^{-r^2/2\sigma^2}S\big(\frac{\lambda}{2\theta},\frac{n}{2},\frac{b^2}{2\sigma^2},\frac{r^2}{2\sigma^2}\big)S\big(\frac{\lambda}{2\theta},\frac{n}{2},\frac{r^2}{2\sigma^2},\frac{a^2}{2\sigma^2}\big)+\theta 2^{n/2}\sigma^n S\big(\frac{\lambda}{2\theta},\frac{n}{2},\frac{b^2}{2\sigma^2},\frac{a^2}{2\sigma^2}\big)}dz$$

$r \le z$, $r \le x$:

$$=\frac{\lambda z^{n-1}e^{-z^2/2\sigma^2}S\big(\frac{\lambda}{2\theta},\frac{n}{2},\frac{b^2}{2\sigma^2},\frac{(z+x+|z-x|)^2}{4\sigma^2}\big)S\big(\frac{\lambda}{2\theta},\frac{n}{2},\frac{(z+x-|z-x|)^2}{4\sigma^2},\frac{r^2}{2\sigma^2}\big)}{\theta 2^{n/2}\sigma^n S\big(\frac{\lambda}{2\theta},\frac{n}{2},\frac{b^2}{2\sigma^2},\frac{r^2}{2\sigma^2}\big)}dz$$

$$+\frac{S\big(\frac{\lambda}{2\theta},\frac{n}{2},\frac{b^2}{2\sigma^2},\frac{x^2}{2\sigma^2}\big)}{S\big(\frac{\lambda}{2\theta},\frac{n}{2},\frac{b^2}{2\sigma^2},\frac{r^2}{2\sigma^2}\big)}$$

$$\times\frac{\lambda z^{n-1}e^{-z^2/2\sigma^2}S\big(\frac{\lambda}{2\theta},\frac{n}{2},\frac{b^2}{2\sigma^2},\frac{z^2}{2\sigma^2}\big)S\big(\frac{\lambda}{2\theta},\frac{n}{2},\frac{r^2}{2\sigma^2},\frac{a^2}{2\sigma^2}\big)}{\gamma r^{n-1}e^{-r^2/2\sigma^2}S\big(\frac{\lambda}{2\theta},\frac{n}{2},\frac{b^2}{2\sigma^2},\frac{r^2}{2\sigma^2}\big)S\big(\frac{\lambda}{2\theta},\frac{n}{2},\frac{r^2}{2\sigma^2},\frac{a^2}{2\sigma^2}\big)+\theta 2^{n/2}\sigma^n S\big(\frac{\lambda}{2\theta},\frac{n}{2},\frac{b^2}{2\sigma^2},\frac{a^2}{2\sigma^2}\big)}dz$$

1.25.6 $\quad \mathbf{P}_x\big(\ell(\tau,r)\in dy,\ a<\inf_{0\le s\le\tau}Q_s^{(n)},\ \sup_{0\le s\le\tau}Q_s^{(n)}<b,\ Q_\tau^{(n)}\in dz\big)$

$a<r$, $r<b$

$$=\frac{\lambda z^{n-1}e^{-z^2/2\sigma^2}}{r^{n-1}e^{-r^2/2\sigma^2}}\exp\Big(-\frac{y\theta r^{1-n}2^{n/2}\sigma^n e^{r^2/2\sigma^2}S\big(\frac{\lambda}{2\theta},\frac{n}{2},\frac{b^2}{2\sigma^2},\frac{a^2}{2\sigma^2}\big)}{S\big(\frac{\lambda}{2\theta},\frac{n}{2},\frac{b^2}{2\sigma^2},\frac{r^2}{2\sigma^2}\big)S\big(\frac{\lambda}{2\theta},\frac{n}{2},\frac{r^2}{2\sigma^2},\frac{a^2}{2\sigma^2}\big)}\Big)$$

$$\times\begin{cases}\dfrac{S\big(\frac{\lambda}{2\theta},\frac{n}{2},\frac{z^2}{2\sigma^2},\frac{a^2}{2\sigma^2}\big)S\big(\frac{\lambda}{2\theta},\frac{n}{2},\frac{x^2}{2\sigma^2},\frac{a^2}{2\sigma^2}\big)}{S^2\big(\frac{\lambda}{2\theta},\frac{n}{2},\frac{r^2}{2\sigma^2},\frac{a^2}{2\sigma^2}\big)}dydz, & 0\le x\le r,\ z\le r\\[2ex] \dfrac{S\big(\frac{\lambda}{2\theta},\frac{n}{2},\frac{b^2}{2\sigma^2},\frac{x^2}{2\sigma^2}\big)S\big(\frac{\lambda}{2\theta},\frac{n}{2},\frac{z^2}{2\sigma^2},\frac{a^2}{2\sigma^2}\big)}{S\big(\frac{\lambda}{2\theta},\frac{n}{2},\frac{b^2}{2\sigma^2},\frac{r^2}{2\sigma^2}\big)S\big(\frac{\lambda}{2\theta},\frac{n}{2},\frac{r^2}{2\sigma^2},\frac{a^2}{2\sigma^2}\big)}dydz, & z\le r\le x\\[2ex] \dfrac{S\big(\frac{\lambda}{2\theta},\frac{n}{2},\frac{b^2}{2\sigma^2},\frac{z^2}{2\sigma^2}\big)S\big(\frac{\lambda}{2\theta},\frac{n}{2},\frac{x^2}{2\sigma^2},\frac{a^2}{2\sigma^2}\big)}{S\big(\frac{\lambda}{2\theta},\frac{n}{2},\frac{b^2}{2\sigma^2},\frac{r^2}{2\sigma^2}\big)S\big(\frac{\lambda}{2\theta},\frac{n}{2},\frac{r^2}{2\sigma^2},\frac{a^2}{2\sigma^2}\big)}dydz, & 0\le x\le r\le z\\[2ex] \dfrac{S\big(\frac{\lambda}{2\theta},\frac{n}{2},\frac{b^2}{2\sigma^2},\frac{z^2}{2\sigma^2}\big)S\big(\frac{\lambda}{2\theta},\frac{n}{2},\frac{b^2}{2\sigma^2},\frac{x^2}{2\sigma^2}\big)}{S^2\big(\frac{\lambda}{2\theta},\frac{n}{2},\frac{b^2}{2\sigma^2},\frac{r^2}{2\sigma^2}\big)}dydz, & r\le x,\ r\le z\end{cases}$$

8 $Q_s^{(n)} = \sigma e^{-\theta s} R^{(n)}_{e^{2\theta s}-1} \quad n \geq 2,\ \sigma, \theta > 0 \qquad \tau \sim \text{Exp}(\lambda)$, independent of $Q^{(n)}$

(1) $\mathbf{P}_x\big(\ell(\tau, r) = 0, a < \inf_{0\le s\le\tau} Q_s^{(n)}, \sup_{0\le s\le\tau} Q_s^{(n)} < b, Q_\tau^{(n)} \in dz\big)$

$z \le r$, $x \le r$

$$= \frac{\lambda z^{n-1} e^{-z^2/2\sigma^2} S\big(\frac{\lambda}{2\theta}, \frac{n}{2}, \frac{r^2}{2\sigma^2}, \frac{(z+x+|z-x|)^2}{4\sigma^2}\big) S\big(\frac{\lambda}{2\theta}, \frac{n}{2}, \frac{(z+x-|z-x|)^2}{4\sigma^2}, \frac{a^2}{2\sigma^2}\big)}{\theta 2^{n/2} \sigma^n S\big(\frac{\lambda}{2\theta}, \frac{n}{2}, \frac{r^2}{2\sigma^2}, \frac{a^2}{2\sigma^2}\big)} dz$$

$r \le z$, $r \le x$

$$= \frac{\lambda z^{n-1} e^{-z^2/2\sigma^2} S\big(\frac{\lambda}{2\theta}, \frac{n}{2}, \frac{b^2}{2\sigma^2}, \frac{(z+x+|z-x|)^2}{4\sigma^2}\big) S\big(\frac{\lambda}{2\theta}, \frac{n}{2}, \frac{(z+x-|z-x|)^2}{4\sigma^2}, \frac{r^2}{2\sigma^2}\big)}{\theta 2^{n/2} \sigma^n S\big(\frac{\lambda}{2\theta}, \frac{n}{2}, \frac{b^2}{2\sigma^2}, \frac{r^2}{2\sigma^2}\big)} dz$$

1.26.1 $\mathbf{E}_x\Big\{\exp\Big(-\int_0^\tau \big(p\mathbb{I}_{[0,r)}\big(Q_s^{(n)}\big) + q\mathbb{I}_{[r,\infty)}\big(Q_s^{(n)}\big)\big)ds\Big);$

$a < r$, $r < b$

$$a < \inf_{0\le s\le\tau} Q_s^{(n)}, \sup_{0\le s\le\tau} Q_s^{(n)} < b\Big\}$$

$a \le x$, $x \le r$

$$= \frac{\lambda}{\lambda+p}\Big(1 - \frac{S\big(\frac{\lambda+p}{2\theta}, \frac{n}{2}, \frac{r^2}{2\sigma^2}, \frac{x^2}{2\sigma^2}\big)}{S\big(\frac{\lambda+p}{2\theta}, \frac{n}{2}, \frac{r^2}{2\sigma^2}, \frac{a^2}{2\sigma^2}\big)}\Big) - \frac{S\big(\frac{\lambda+p}{2\theta}, \frac{n}{2}, \frac{x^2}{2\sigma^2}, \frac{a^2}{2\sigma^2}\big)}{S\big(\frac{\lambda+p}{2\theta}, \frac{n}{2}, \frac{r^2}{2\sigma^2}, \frac{a^2}{2\sigma^2}\big)}$$

$$\times \frac{\lambda(\lambda+p)^{-1} 2^{n/2} (\sigma/r)^n S\big(\frac{\lambda+q}{2\theta}, \frac{n}{2}, \frac{b^2}{2\sigma^2}, \frac{r^2}{2\sigma^2}\big)}{S\big(\frac{\lambda+q}{2\theta}, \frac{n}{2}, \frac{b^2}{2\sigma^2}, \frac{r^2}{2\sigma^2}\big) C\big(\frac{\lambda+p}{2\theta}, \frac{n}{2}, \frac{r^2}{2\sigma^2}, \frac{a^2}{2\sigma^2}\big) + C\big(\frac{\lambda+q}{2\theta}, \frac{n}{2}, \frac{r^2}{2\sigma^2}, \frac{b^2}{2\sigma^2}\big) S\big(\frac{\lambda+p}{2\theta}, \frac{n}{2}, \frac{r^2}{2\sigma^2}, \frac{a^2}{2\sigma^2}\big)}$$

$$+ \frac{\lambda(\lambda+q)^{-1}\Big((\lambda+p)^{-1}(p-q) C\big(\frac{\lambda+q}{2\theta}, \frac{n}{2}, \frac{r^2}{2\sigma^2}, \frac{b^2}{2\sigma^2}\big) - \big(\frac{\sigma\sqrt{2}}{r}\big)^n e^{r^2/2\sigma^2}\Big) S\big(\frac{\lambda+p}{2\theta}, \frac{n}{2}, \frac{x^2}{2\sigma^2}, \frac{a^2}{2\sigma^2}\big)}{S\big(\frac{\lambda+q}{2\theta}, \frac{n}{2}, \frac{b^2}{2\sigma^2}, \frac{r^2}{2\sigma^2}\big) C\big(\frac{\lambda+p}{2\theta}, \frac{n}{2}, \frac{r^2}{2\sigma^2}, \frac{a^2}{2\sigma^2}\big) + C\big(\frac{\lambda+q}{2\theta}, \frac{n}{2}, \frac{r^2}{2\sigma^2}, \frac{b^2}{2\sigma^2}\big) S\big(\frac{\lambda+p}{2\theta}, \frac{n}{2}, \frac{r^2}{2\sigma^2}, \frac{a^2}{2\sigma^2}\big)}$$

$r \le x$, $x \le b$

$$= \frac{\lambda}{\lambda+q}\Big(1 - \frac{S\big(\frac{\lambda+q}{2\theta}, \frac{n}{2}, \frac{x^2}{2\sigma^2}, \frac{r^2}{2\sigma^2}\big)}{S\big(\frac{\lambda+q}{2\theta}, \frac{n}{2}, \frac{b^2}{2\sigma^2}, \frac{r^2}{2\sigma^2}\big)}\Big) - \frac{S\big(\frac{\lambda+q}{2\theta}, \frac{n}{2}, \frac{b^2}{2\sigma^2}, \frac{x^2}{2\sigma^2}\big)}{S\big(\frac{\lambda+q}{2\theta}, \frac{n}{2}, \frac{b^2}{2\sigma^2}, \frac{r^2}{2\sigma^2}\big)}$$

$$\times \frac{\lambda(\lambda+q)^{-1} 2^{n/2} (\sigma/r)^n S\big(\frac{\lambda+p}{2\theta}, \frac{n}{2}, \frac{r^2}{2\sigma^2}, \frac{a^2}{2\sigma^2}\big)}{S\big(\frac{\lambda+q}{2\theta}, \frac{n}{2}, \frac{b^2}{2\sigma^2}, \frac{r^2}{2\sigma^2}\big) C\big(\frac{\lambda+p}{2\theta}, \frac{n}{2}, \frac{r^2}{2\sigma^2}, \frac{a^2}{2\sigma^2}\big) + C\big(\frac{\lambda+q}{2\theta}, \frac{n}{2}, \frac{r^2}{2\sigma^2}, \frac{b^2}{2\sigma^2}\big) S\big(\frac{\lambda+p}{2\theta}, \frac{n}{2}, \frac{r^2}{2\sigma^2}, \frac{a^2}{2\sigma^2}\big)}$$

$$+ \frac{\lambda(\lambda+p)^{-1}\Big((\lambda+q)^{-1}(q-p) C\big(\frac{\lambda}{2\theta}, \frac{n}{2}, \frac{r^2}{2\sigma^2}, \frac{a^2}{2\sigma^2}\big) - \big(\frac{\sigma\sqrt{2}}{r}\big)^n e^{r^2/2\sigma^2}\Big) S\big(\frac{\lambda+q}{2\theta}, \frac{n}{2}, \frac{b^2}{2\sigma^2}, \frac{x^2}{2\sigma^2}\big)}{S\big(\frac{\lambda+q}{2\theta}, \frac{n}{2}, \frac{b^2}{2\sigma^2}, \frac{r^2}{2\sigma^2}\big) C\big(\frac{\lambda+p}{2\theta}, \frac{n}{2}, \frac{r^2}{2\sigma^2}, \frac{a^2}{2\sigma^2}\big) + C\big(\frac{\lambda+q}{2\theta}, \frac{n}{2}, \frac{r^2}{2\sigma^2}, \frac{b^2}{2\sigma^2}\big) S\big(\frac{\lambda+p}{2\theta}, \frac{n}{2}, \frac{r^2}{2\sigma^2}, \frac{a^2}{2\sigma^2}\big)}$$

1.26.5 $\mathbf{E}_x\Big\{\exp\Big(-\int_0^\tau \big(p\mathbb{I}_{[0,r)}\big(Q_s^{(n)}\big) + q\mathbb{I}_{[r,\infty)}\big(Q_s^{(n)}\big)\big)ds\Big);$

$r < b$, $a < r$

$$a < \inf_{0\le s\le\tau} Q_s^{(n)}, \sup_{0\le s\le\tau} Q_s^{(n)} < b, Q_\tau^{(n)} \in dz\Big\}$$

8 $Q_s^{(n)} = \sigma e^{-\theta s} R^{(n)}_{e^{2\theta s}-1}$ $\quad n \ge 2,\ \sigma, \theta > 0 \quad \tau \sim \operatorname{Exp}(\lambda)$, independent of $Q^{(n)}$

$z \le r$, $x \le r$

$$= \frac{\lambda z^{n-1} e^{-z^2/2\sigma^2} S\big(\frac{\lambda+p}{2\theta},\frac{n}{2},\frac{r^2}{2\sigma^2},\frac{(z+x+|z-x|)^2}{4\sigma^2}\big) S\big(\frac{\lambda+p}{2\theta},\frac{n}{2},\frac{(z+x-|z-x|)^2}{4\sigma^2},\frac{a^2}{2\sigma^2}\big)}{\theta 2^{n/2}\sigma^n S\big(\frac{\lambda+p}{2\theta},\frac{n}{2},\frac{r^2}{2\sigma^2},\frac{a^2}{2\sigma^2}\big)} dz$$

$$+\frac{S\big(\frac{\lambda+p}{2\theta},\frac{n}{2},\frac{x^2}{2\sigma^2},\frac{a^2}{2\sigma^2}\big)}{S\big(\frac{\lambda+p}{2\theta},\frac{n}{2},\frac{r^2}{2\sigma^2},\frac{a^2}{2\sigma^2}\big)}$$

$$\times \frac{\lambda(\theta r)^{-1}(z/r)^{n-1} e^{(r^2-z^2)/2\sigma^2} S\big(\frac{\lambda+q}{2\theta},\frac{n}{2},\frac{b^2}{2\sigma^2},\frac{r^2}{2\sigma^2}\big) S\big(\frac{\lambda+p}{2\theta},\frac{n}{2},\frac{z^2}{2\sigma^2},\frac{a^2}{2\sigma^2}\big)}{S\big(\frac{\lambda+q}{2\theta},\frac{n}{2},\frac{b^2}{2\sigma^2},\frac{r^2}{2\sigma^2}\big)C\big(\frac{\lambda+p}{2\theta},\frac{n}{2},\frac{r^2}{2\sigma^2},\frac{a^2}{2\sigma^2}\big)+C\big(\frac{\lambda+q}{2\theta},\frac{n}{2},\frac{r^2}{2\sigma^2},\frac{b^2}{2\sigma^2}\big)S\big(\frac{\lambda+p}{2\theta},\frac{n}{2},\frac{r^2}{2\sigma^2},\frac{a^2}{2\sigma^2}\big)} dz$$

$z \le r$, $r \le x$

$$= \frac{\lambda(\theta r)^{-1}(z/r)^{n-1} e^{(r^2-z^2)/2\sigma^2} S\big(\frac{\lambda+p}{2\theta},\frac{n}{2},\frac{z^2}{2\sigma^2},\frac{a^2}{2\sigma^2}\big) S\big(\frac{\lambda+q}{2\theta},\frac{n}{2},\frac{b^2}{2\sigma^2},\frac{x^2}{2\sigma^2}\big)}{S\big(\frac{\lambda+q}{2\theta},\frac{n}{2},\frac{b^2}{2\sigma^2},\frac{r^2}{2\sigma^2}\big)C\big(\frac{\lambda+p}{2\theta},\frac{n}{2},\frac{r^2}{2\sigma^2},\frac{a^2}{2\sigma^2}\big)+C\big(\frac{\lambda+q}{2\theta},\frac{n}{2},\frac{r^2}{2\sigma^2},\frac{b^2}{2\sigma^2}\big)S\big(\frac{\lambda+p}{2\theta},\frac{n}{2},\frac{r^2}{2\sigma^2},\frac{a^2}{2\sigma^2}\big)} dz$$

$r \le z$, $x \le r$

$$= \frac{\lambda(\theta r)^{-1}(z/r)^{n-1} e^{(r^2-z^2)/2\sigma^2} S\big(\frac{\lambda+q}{2\theta},\frac{n}{2},\frac{b^2}{2\sigma^2},\frac{z^2}{2\sigma^2}\big) S\big(\frac{\lambda+p}{2\theta},\frac{n}{2},\frac{x^2}{2\sigma^2},\frac{a^2}{2\sigma^2}\big)}{S\big(\frac{\lambda+q}{2\theta},\frac{n}{2},\frac{b^2}{2\sigma^2},\frac{r^2}{2\sigma^2}\big)C\big(\frac{\lambda+p}{2\theta},\frac{n}{2},\frac{r^2}{2\sigma^2},\frac{a^2}{2\sigma^2}\big)+C\big(\frac{\lambda+q}{2\theta},\frac{n}{2},\frac{r^2}{2\sigma^2},\frac{b^2}{2\sigma^2}\big)S\big(\frac{\lambda+p}{2\theta},\frac{n}{2},\frac{r^2}{2\sigma^2},\frac{a^2}{2\sigma^2}\big)} dz$$

$r \le z$, $r \le x$

$$= \frac{\lambda z^{n-1} e^{-z^2/2\sigma^2} S\big(\frac{\lambda+q}{2\theta},\frac{n}{2},\frac{b^2}{2\sigma^2},\frac{(z+x+|z-x|)^2}{4\sigma^2}\big) S\big(\frac{\lambda+q}{2\theta},\frac{n}{2},\frac{(z+x-|z-x|)^2}{4\sigma^2},\frac{r^2}{2\sigma^2}\big)}{\theta 2^{n/2}\sigma^n S\big(\frac{\lambda+q}{2\theta},\frac{n}{2},\frac{b^2}{2\sigma^2},\frac{r^2}{2\sigma^2}\big)} dz$$

$$+\frac{S\big(\frac{\lambda+q}{2\theta},\frac{n}{2},\frac{b^2}{2\sigma^2},\frac{x^2}{2\sigma^2}\big)}{S\big(\frac{\lambda+q}{2\theta},\frac{n}{2},\frac{b^2}{2\sigma^2},\frac{r^2}{2\sigma^2}\big)}$$

$$\times \frac{\lambda(\theta r)^{-1}(z/r)^{n-1} e^{(r^2-z^2)/2\sigma^2} S\big(\frac{\lambda+p}{2\theta},\frac{n}{2},\frac{r^2}{2\sigma^2},\frac{a^2}{2\sigma^2}\big) S\big(\frac{\lambda+q}{2\theta},\frac{n}{2},\frac{b^2}{2\sigma^2},\frac{z^2}{2\sigma^2}\big)}{S\big(\frac{\lambda+q}{2\theta},\frac{n}{2},\frac{b^2}{2\sigma^2},\frac{r^2}{2\sigma^2}\big)C\big(\frac{\lambda+p}{2\theta},\frac{n}{2},\frac{r^2}{2\sigma^2},\frac{a^2}{2\sigma^2}\big)+C\big(\frac{\lambda+q}{2\theta},\frac{n}{2},\frac{r^2}{2\sigma^2},\frac{b^2}{2\sigma^2}\big)S\big(\frac{\lambda+p}{2\theta},\frac{n}{2},\frac{r^2}{2\sigma^2},\frac{a^2}{2\sigma^2}\big)} dz$$

1.27.1 $\mathbf{E}_x\Big\{\exp\Big(-\int_0^\tau \big(p\mathbb{I}_{[0,r)}\big(Q_s^{(n)}\big) + q\mathbb{I}_{[r,\infty)}\big(Q_s^{(n)}\big)\big)ds\Big); \ell(\tau,r)\in dy\Big\}$

$0 \le x$

$$= \exp\Big(-\frac{yr(\lambda+p)M\big(\frac{\lambda+p+2\theta}{2\theta},\frac{n+2}{2},\frac{r^2}{2\sigma^2}\big)}{nM\big(\frac{\lambda+p}{2\theta},\frac{n}{2},\frac{r^2}{2\sigma^2}\big)} - \frac{yr(\lambda+q)U\big(\frac{\lambda+q+2\theta}{2\theta},\frac{n+2}{2},\frac{r^2}{2\sigma^2}\big)}{2U\big(\frac{\lambda+q}{2\theta},\frac{n}{2},\frac{r^2}{2\sigma^2}\big)}\Big)$$

$$\times\begin{cases} \dfrac{\lambda r M\big(\frac{\lambda+p}{2\theta},\frac{n}{2},\frac{x^2}{2\sigma^2}\big)}{M\big(\frac{\lambda}{2\theta},\frac{n}{2},\frac{r^2}{2\sigma^2}\big)}\Big(\dfrac{M\big(\frac{\lambda+p+2\theta}{2\theta},\frac{n+2}{2},\frac{r^2}{2\sigma^2}\big)}{nM\big(\frac{\lambda+p}{2\theta},\frac{n}{2},\frac{r^2}{2\sigma^2}\big)} + \dfrac{U\big(\frac{\lambda+q+2\theta}{2\theta},\frac{n+2}{2},\frac{r^2}{2\sigma^2}\big)}{2U\big(\frac{\lambda+q}{2\theta},\frac{n}{2},\frac{r^2}{2\sigma^2}\big)}\Big)dy, & x \le r \\ \dfrac{\lambda r U\big(\frac{\lambda+q}{2\theta},\frac{n}{2},\frac{x^2}{2\sigma^2}\big)}{U\big(\frac{\lambda+q}{2\theta},\frac{n}{2},\frac{r^2}{2\sigma^2}\big)}\Big(\dfrac{M\big(\frac{\lambda+p+2\theta}{2\theta},\frac{n+2}{2},\frac{r^2}{2\sigma^2}\big)}{nM\big(\frac{\lambda+p}{2\theta},\frac{n}{2},\frac{r^2}{2\sigma^2}\big)} + \dfrac{U\big(\frac{\lambda+q+2\theta}{2\theta},\frac{n+2}{2},\frac{r^2}{2\sigma^2}\big)}{2U\big(\frac{\lambda+q}{2\theta},\frac{n}{2},\frac{r^2}{2\sigma^2}\big)}\Big)dy, & r \le x \end{cases}$$

8 $Q_s^{(n)} = \sigma e^{-\theta s} R^{(n)}_{e^{2\theta s}-1}$ $n \ge 2,\ \sigma, \theta > 0$ $\tau \sim \mathrm{Exp}(\lambda)$, independent of $Q^{(n)}$

(1) $$\mathbf{E}_x\Big\{\exp\Big(-\int_0^\tau \big(p\mathbb{I}_{[0,r)}(Q_s^{(n)}) + q\mathbb{I}_{[r,\infty)}(Q_s^{(n)})\big)ds\Big); \ell(\tau,r)=0\Big\}$$

$$= \begin{cases} \dfrac{\lambda}{\lambda+p}\Big(1 - \dfrac{M(\frac{\lambda+p}{2\theta},\frac{n}{2},\frac{x^2}{2\sigma^2})}{M(\frac{\lambda+p}{2\theta},\frac{n}{2},\frac{r^2}{2\sigma^2})}\Big), & 0 \le x \le r \\[2ex] \dfrac{\lambda}{\lambda+q}\Big(1 - \dfrac{U(\frac{\lambda+q}{2\theta},\frac{n}{2},\frac{x^2}{2\sigma^2})}{U(\frac{\lambda+q}{2\theta},\frac{n}{2},\frac{r^2}{2\sigma^2})}\Big), & r \le x \end{cases}$$

1.27.5 $$\mathbf{E}_x\Big\{\exp\Big(-\int_0^\tau \big(p\mathbb{I}_{[0,r)}(Q_s^{(n)}) + q\mathbb{I}_{[r,\infty)}(Q_s^{(n)})\big)ds\Big); \ell(\tau,r)\in dy, Q_\tau^{(n)} \in dz\Big\}$$

$$= \exp\Big(-\frac{yr(\lambda+p)M(\frac{\lambda+p+2\theta}{2\theta},\frac{n+2}{2},\frac{r^2}{2\sigma^2})}{nM(\frac{\lambda+p}{2\theta},\frac{n}{2},\frac{r^2}{2\sigma^2})} - \frac{yr(\lambda+q)U(\frac{\lambda+q+2\theta}{2\theta},\frac{n+2}{2},\frac{r^2}{2\sigma^2})}{2U(\frac{\lambda+q}{2\theta},\frac{n}{2},\frac{r^2}{2\sigma^2})}\Big)$$

$$\times \begin{cases} \dfrac{\lambda z^{n-1}e^{-z^2/2\sigma^2}M(\frac{\lambda+p}{2\theta},\frac{n}{2},\frac{x^2}{2\sigma^2})M(\frac{\lambda+p}{2\theta},\frac{n}{2},\frac{z^2}{2\sigma^2})}{r^{n-1}e^{-r^2/2\sigma^2}M^2(\frac{\lambda+p}{2\theta},\frac{n}{2},\frac{r^2}{2\sigma^2})}dydz & z \vee x \le r \\[2ex] \dfrac{\lambda z^{n-1}e^{-z^2/2\sigma^2}M(\frac{\lambda+p}{2\theta},\frac{n}{2},\frac{z^2}{2\sigma^2})U(\frac{\lambda+q}{2\theta},\frac{n}{2},\frac{x^2}{2\sigma^2})}{r^{n-1}e^{-r^2/2\sigma^2}M(\frac{\lambda+p}{2\theta},\frac{n}{2},\frac{r^2}{2\sigma^2})U(\frac{\lambda+q}{2\theta},\frac{n}{2},\frac{r^2}{2\sigma^2})}dydz & z \le r \le x \\[2ex] \dfrac{\lambda z^{n-1}e^{-z^2/2\sigma^2}M(\frac{\lambda+p}{2\theta},\frac{n}{2},\frac{x^2}{2\sigma^2})U(\frac{\lambda+q}{2\theta},\frac{n}{2},\frac{z^2}{2\sigma^2})}{r^{n-1}e^{-r^2/2\sigma^2}M(\frac{\lambda+p}{2\theta},\frac{n}{2},\frac{r^2}{2\sigma^2})U(\frac{\lambda+q}{2\theta},\frac{n}{2},\frac{r^2}{2\sigma^2})}dydz & x \le r \le z \\[2ex] \dfrac{\lambda z^{n-1}e^{-z^2/2\sigma^2}U(\frac{\lambda+q}{2\theta},\frac{n}{2},\frac{x^2}{2\sigma^2})U(\frac{\lambda+q}{2\theta},\frac{n}{2},\frac{z^2}{2\sigma^2})}{r^{n-1}e^{-r^2/2\sigma^2}U^2(\frac{\lambda+q}{2\theta},\frac{n}{2},\frac{r^2}{2\sigma^2})} & r \le z,\ r \le x \end{cases}$$

(1) $$\mathbf{E}_x\Big\{\exp\Big(-\int_0^\tau \big(p\mathbb{I}_{[0,r)}(Q_s^{(n)}) + q\mathbb{I}_{[r,\infty)}(Q_s^{(n)})\big)ds\Big); \ell(\tau,r)=0, Q_\tau^{(n)} \in dz\Big\}$$

$x \le r$, $z \le r$
$$= \frac{\lambda z^{n-1}e^{-z^2/2\sigma^2}S(\frac{\lambda+p}{2\theta},\frac{n}{2},\frac{r^2}{2\sigma^2},\frac{(z+x+|z-x|)^2}{4\sigma^2})M(\frac{\lambda+p}{2\theta},\frac{n}{2},\frac{(z+x-|z-x|)^2}{4\sigma^2})}{\theta 2^{n/2}\sigma^n M(\frac{\lambda+p}{2\theta},\frac{n}{2},\frac{r^2}{2\sigma^2})}$$

$r \le x$, $r \le z$
$$= \frac{\lambda z^{n-1}e^{-z^2/2\sigma^2}U(\frac{\lambda+q}{2\theta},\frac{n}{2},\frac{(z+x+|z-x|)^2}{4\sigma^2})S(\frac{\lambda+q}{2\theta},\frac{n}{2},\frac{(z+x-|z-x|)^2}{4\sigma^2},\frac{r^2}{2\sigma^2})}{\theta 2^{n/2}\sigma^n U(\frac{\lambda+q}{2\theta},\frac{n}{2},\frac{r^2}{2\sigma^2})}$$

1.28.1 $$\mathbf{E}_x\big\{e^{-\gamma\hat H(\tau)-\eta\check H(\tau)}; \inf_{0\le s\le\tau} Q_s^{(n)} \in da, \sup_{0\le s\le\tau} Q_s^{(n)} \in db\big\}/dadb$$

$a < x$, $x < b$

$$= \frac{2^{n/2}\sigma^{n-4}be^{a^2/2\sigma^2}S(\frac{\lambda+\gamma+\eta}{2\theta},\frac{n}{2},\frac{b^2}{2\sigma^2},\frac{x^2}{2\sigma^2})\big(C(\frac{\lambda}{2\theta},\frac{n}{2},\frac{b^2}{2\sigma^2},\frac{a^2}{2\sigma^2}) - 2^{n/2}\sigma^n b^{-n}e^{b^2/2\sigma^2}\big)}{a^{n-1}S(\frac{\lambda+\gamma+\eta}{2\theta},\frac{n}{2},\frac{b^2}{2\sigma^2},\frac{a^2}{2\sigma^2})S(\frac{\lambda+\eta}{2\theta},\frac{n}{2},\frac{b^2}{2\sigma^2},\frac{a^2}{2\sigma^2})S(\frac{\lambda}{2\theta},\frac{n}{2},\frac{b^2}{2\sigma^2},\frac{a^2}{2\sigma^2})}$$

8 $Q_s^{(n)} = \sigma e^{-\theta s} R^{(n)}_{e^{2\theta s}-1} \quad n \ge 2,\ \sigma, \theta > 0 \qquad \tau \sim \operatorname{Exp}(\lambda)$, independent of $Q^{(n)}$

$$+\frac{2^{n/2}\sigma^{n-4} a e^{b^2/2\sigma^2} S(\frac{\lambda+\gamma+\eta}{2\theta},\frac{n}{2},\frac{x^2}{2\sigma^2},\frac{a^2}{2\sigma^2})(C(\frac{\lambda}{2\theta},\frac{n}{2},\frac{a^2}{2\sigma^2},\frac{b^2}{2\sigma^2}) - 2^{n/2}\sigma^n a^{-n} e^{a^2/2\sigma^2})}{b^{n-1} S(\frac{\lambda+\gamma+\eta}{2\theta},\frac{n}{2},\frac{b^2}{2\sigma^2},\frac{a^2}{2\sigma^2}) S(\frac{\lambda+\gamma}{2\theta},\frac{n}{2},\frac{b^2}{2\sigma^2},\frac{a^2}{2\sigma^2}) S(\frac{\lambda}{2\theta},\frac{n}{2},\frac{b^2}{2\sigma^2},\frac{a^2}{2\sigma^2})}$$

1.28.5 $a < x$ $x < b$

$$\mathbf{E}_x\big\{e^{-\gamma\hat H(\tau)-\eta\check H(\tau)};\ \inf_{0\le s\le\tau} Q_s^{(n)} \in da,\ \sup_{0\le s\le\tau} Q_s^{(n)} \in db,\ Q_\tau^{(n)} \in dz\big\}/dadbdz$$

$$=\frac{\lambda 2^{n/2}\sigma^{n-4}(ab)^{1-n} z^{n-1} e^{(b^2+a^2-z^2)/2\sigma^2} S(\frac{\lambda+\gamma+\eta}{2\theta},\frac{n}{2},\frac{b^2}{2\sigma^2},\frac{x^2}{2\sigma^2}) S(\frac{\lambda}{2\theta},\frac{n}{2},\frac{z^2}{2\sigma^2},\frac{a^2}{2\sigma^2})}{S(\frac{\lambda+\gamma+eta}{2\theta},\frac{n}{2},\frac{b^2}{2\sigma^2},\frac{a^2}{2\sigma^2}) S(\frac{\lambda+\eta}{2\theta},\frac{n}{2},\frac{b^2}{2\sigma^2},\frac{a^2}{2\sigma^2}) S(\frac{\lambda}{2\theta},\frac{n}{2},\frac{b^2}{2\sigma^2},\frac{a^2}{2\sigma^2})}$$

$$+\frac{\lambda 2^{n/2}\sigma^{n-4}(ab)^{1-n} z^{n-1} e^{(b^2+a^2-z^2)/2\sigma^2} S(\frac{\lambda+\gamma+\eta}{2\theta},\frac{n}{2},\frac{x^2}{2\sigma^2},\frac{a^2}{2\sigma^2}) S(\frac{\lambda}{2\theta},\frac{n}{2},\frac{b^2}{2\sigma^2},\frac{z^2}{2\sigma^2})}{S(\frac{\lambda+\gamma+\eta}{2\theta},\frac{n}{2},\frac{b^2}{2\sigma^2},\frac{a^2}{2\sigma^2}) S(\frac{\lambda+\gamma}{2\theta},\frac{n}{2},\frac{b^2}{2\sigma^2},\frac{a^2}{2\sigma^2}) S(\frac{\lambda}{2\theta},\frac{n}{2},\frac{b^2}{2\sigma^2},\frac{a^2}{2\sigma^2})}$$

1.29.1 $0 \le a$

$$\mathbf{E}_x\Big\{\exp\Big(-\int_0^\tau \big(p\mathbb{I}_{[0,r)}(Q_s^{(n)}) + q\mathbb{I}_{[r,\infty)}(Q_s^{(n)})\big)ds\Big); a < \inf_{0\le s\le\tau} Q_s^{(n)}, \sup_{0\le s\le\tau} Q_s^{(n)} < b, \ell(\tau,r) \in dy\Big\}$$

$$=\Big(\frac{C(\frac{\lambda+q}{2\theta},\frac{n}{2},\frac{r^2}{2\sigma^2},\frac{b^2}{2\sigma^2}) - (\frac{\sigma\sqrt2}{r})^n e^{r^2/2\sigma^2}}{(\lambda+q) S(\frac{\lambda+q}{2\theta},\frac{n}{2},\frac{b^2}{2\sigma^2},\frac{r^2}{2\sigma^2})} + \frac{C(\frac{\lambda+p}{2\theta},\frac{n}{2},\frac{r^2}{2\sigma^2},\frac{a^2}{2\sigma^2}) - (\frac{\sigma\sqrt2}{r})^n e^{r^2/2\sigma^2}}{(\lambda+p) S(\frac{\lambda+p}{2\theta},\frac{n}{2},\frac{r^2}{2\sigma^2},\frac{a^2}{2\sigma^2})}\Big)$$

$$\times\lambda\theta r\exp\Big(-\frac{y\theta r C(\frac{\lambda+q}{2\theta},\frac{n}{2},\frac{r^2}{2\sigma^2},\frac{b^2}{2\sigma^2})}{S(\frac{\lambda+q}{2\theta},\frac{n}{2},\frac{b^2}{2\sigma^2},\frac{r^2}{2\sigma^2})} - \frac{y\theta r C(\frac{\lambda+p}{2\theta},\frac{n}{2},\frac{r^2}{2\sigma^2},\frac{a^2}{2\sigma^2})}{S(\frac{\lambda+p}{2\theta},\frac{n}{2},\frac{r^2}{2\sigma^2},\frac{a^2}{2\sigma^2})}\Big)$$

$$\times\begin{cases}\dfrac{S(\frac{\lambda+p}{2\theta},\frac{n}{2},\frac{x^2}{2\sigma^2},\frac{a^2}{2\sigma^2})}{S(\frac{\lambda+p}{2\theta},\frac{n}{2},\frac{r^2}{2\sigma^2},\frac{a^2}{2\sigma^2})}dy, & a \le x \le r\\[2ex] \dfrac{S(\frac{\lambda+q}{2\theta},\frac{n}{2},\frac{b^2}{2\sigma^2},\frac{x^2}{2\sigma^2})}{S(\frac{\lambda+q}{2\theta},\frac{n}{2},\frac{b^2}{2\sigma^2},\frac{r^2}{2\sigma^2})}dy, & r \le x \le b\end{cases}$$

(1)

$$\mathbf{E}_x\Big\{\exp\Big(-\int_0^\tau \big(p\mathbb{I}_{[0,r)}(Q_s^{(n)}) + q\mathbb{I}_{[r,\infty)}(Q_s^{(n)})\big)ds\Big); a < \inf_{0\le s\le\tau} Q_s^{(n)}, \sup_{0\le s\le\tau} Q_s^{(n)} < b, \ell(\tau,r) = 0\Big\}$$

$$=\begin{cases}\dfrac{\lambda}{\lambda+p}\Big(1 - \dfrac{S(\frac{\lambda+p}{2\theta},\frac{n}{2},\frac{r^2}{2\sigma^2},\frac{x^2}{2\sigma^2}) + S(\frac{\lambda+p}{2\theta},\frac{n}{2},\frac{x^2}{2\sigma^2},\frac{a^2}{2\sigma^2})}{S(\frac{\lambda+p}{2\theta},\frac{n}{2},\frac{r^2}{2\sigma^2},\frac{a^2}{2\sigma^2})}\Big), & a \le x \le r\\[2ex] \dfrac{\lambda}{\lambda+q}\Big(1 - \dfrac{S(\frac{\lambda+q}{2\theta},\frac{n}{2},\frac{b^2}{2\sigma^2},\frac{x^2}{2\sigma^2}) + S(\frac{\lambda+q}{2\theta},\frac{n}{2},\frac{x^2}{2\sigma^2},\frac{r^2}{2\sigma^2})}{S(\frac{\lambda+q}{2\theta},\frac{n}{2},\frac{b^2}{2\sigma^2},\frac{r^2}{2\sigma^2})}\Big), & r \le x \le b\end{cases}$$

8 $Q_s^{(n)} = \sigma e^{-\theta s} R^{(n)}_{e^{2\theta s}-1} \quad n \geq 2,\ \sigma,\theta > 0 \quad \tau \sim \mathrm{Exp}(\lambda)$, independent of $Q^{(n)}$

1.29.5 $\quad 0 \leq a$

$$\mathbf{E}_x\Big\{\exp\Big(-\int_0^\tau \big(p\mathbb{1}_{[0,r)}\big(Q_s^{(n)}\big) + q\mathbb{1}_{[r,\infty)}\big(Q_s^{(n)}\big)\big)ds\Big); a < \inf_{0\leq s\leq\tau} Q_s^{(n)}, \sup_{0\leq s\leq\tau} Q_s^{(n)} < b, \ell(\tau,r) \in dy, Q_\tau^{(n)} \in dz\Big\}$$

$$= \frac{\lambda z^{n-1}e^{-z^2/2\sigma^2}}{r^{n-1}e^{-r^2/2\sigma^2}} \exp\Big(-\frac{y\theta r C\big(\frac{\lambda+q}{2\theta},\frac{n}{2},\frac{r^2}{2\sigma^2},\frac{b^2}{2\sigma^2}\big)}{S\big(\frac{\lambda+q}{2\theta},\frac{n}{2},\frac{b^2}{2\sigma^2},\frac{r^2}{2\sigma^2}\big)} - \frac{y\theta r C\big(\frac{\lambda+p}{2\theta},\frac{n}{2},\frac{r^2}{2\sigma^2},\frac{a^2}{2\sigma^2}\big)}{S\big(\frac{\lambda+p}{2\theta},\frac{n}{2},\frac{r^2}{2\sigma^2},\frac{a^2}{2\sigma^2}\big)}\Big)$$

$$\times \begin{cases} \dfrac{S\big(\frac{\lambda+p}{2\theta},\frac{n}{2},\frac{x^2}{2\sigma^2},\frac{a^2}{2\sigma^2}\big)S\big(\frac{\lambda+p}{2\theta},\frac{n}{2},\frac{z^2}{2\sigma^2},\frac{a^2}{2\sigma^2}\big)}{S^2\big(\frac{\lambda+p}{2\theta},\frac{n}{2},\frac{r^2}{2\sigma^2},\frac{a^2}{2\sigma^2}\big)}dydz, & a \leq x \leq r,\ a \leq z \leq r \\ \dfrac{S\big(\frac{\lambda+q}{2\theta},\frac{n}{2},\frac{b^2}{2\sigma^2},\frac{x^2}{2\sigma^2}\big)S\big(\frac{\lambda+p}{2\theta},\frac{n}{2},\frac{z^2}{2\sigma^2},\frac{a^2}{2\sigma^2}\big)}{S\big(\frac{\lambda+q}{2\theta},\frac{n}{2},\frac{b^2}{2\sigma^2},\frac{r^2}{2\sigma^2}\big)S\big(\frac{\lambda+p}{2\theta},\frac{n}{2},\frac{r^2}{2\sigma^2},\frac{a^2}{2\sigma^2}\big)}dydz, & a \leq z \leq r \leq x \leq b \\ \dfrac{S\big(\frac{\lambda+q}{2\theta},\frac{n}{2},\frac{b^2}{2\sigma^2},\frac{z^2}{2\sigma^2}\big)S\big(\frac{\lambda+p}{2\theta},\frac{n}{2},\frac{x^2}{2\sigma^2},\frac{a^2}{2\sigma^2}\big)}{S\big(\frac{\lambda+q}{2\theta},\frac{n}{2},\frac{b^2}{2\sigma^2},\frac{r^2}{2\sigma^2}\big)S\big(\frac{\lambda+p}{2\theta},\frac{n}{2},\frac{r^2}{2\sigma^2},\frac{a^2}{2\sigma^2}\big)}dydz, & a \leq x \leq r \leq z \leq b \\ \dfrac{S\big(\frac{\lambda+q}{2\theta},\frac{n}{2},\frac{b^2}{2\sigma^2},\frac{x^2}{2\sigma^2}\big)S\big(\frac{\lambda+q}{2\theta},\frac{n}{2},\frac{b^2}{2\sigma^2},\frac{z^2}{2\sigma^2}\big)}{S^2\big(\frac{\lambda+q}{2\theta},\frac{n}{2},\frac{b^2}{2\sigma^2},\frac{r^2}{2\sigma^2}\big)}dydz, & r \leq x \leq b,\ r \leq z \leq b \end{cases}$$

(1)

$$\mathbf{E}_x\Big\{\exp\Big(-\int_0^\tau \big(p\mathbb{1}_{[0,r)}\big(Q_s^{(n)}\big) + q\mathbb{1}_{[r,\infty)}\big(Q_s^{(n)}\big)\big)ds\Big); a < \inf_{0\leq s\leq\tau} Q_s^{(n)}, \sup_{0\leq s\leq\tau} Q_s^{(n)} < b, \ell(\tau,r) = 0, Q_\tau^{(n)} \in dz\Big\}$$

$x \leq r$, $z \leq r$

$$= \frac{\lambda z^{n-1}e^{-z^2/2\sigma^2} S\big(\frac{\lambda+p}{2\theta},\frac{n}{2},\frac{r^2}{2\sigma^2},\frac{(z+x+|z-x|)^2}{4\sigma^2}\big)S\big(\frac{\lambda+p}{2\theta},\frac{n}{2},\frac{(z+x-|z-x|)^2}{4\sigma^2},\frac{a^2}{2\sigma^2}\big)}{\theta 2^{n/2}\sigma^n S\big(\frac{\lambda+p}{2\theta},\frac{n}{2},\frac{r^2}{2\sigma^2},\frac{a^2}{2\sigma^2}\big)}dz$$

$r \leq x$, $r \leq z$

$$= \frac{\lambda z^{n-1}e^{-z^2/2\sigma^2} S\big(\frac{\lambda+q}{2\theta},\frac{n}{2},\frac{b^2}{2\sigma^2},\frac{(z+x+|z-x|)^2}{4\sigma^2}\big)S\big(\frac{\lambda+q}{2\theta},\frac{n}{2},\frac{(z+x-|z-x|)^2}{4\sigma^2},\frac{r^2}{2\sigma^2}\big)}{\theta 2^{n/2}\sigma^n S\big(\frac{\lambda+q}{2\theta},\frac{n}{2},\frac{b^2}{2\sigma^2},\frac{r^2}{2\sigma^2}\big)}dz$$

8 $Q_s^{(n)} = \sigma e^{-\theta s} R^{(n)}_{e^{2\theta s}-1} \quad n \ge 2, \quad \sigma, \theta > 0 \qquad H_z = \min\{s : Q_s^{(n)} = z\}$

2. Stopping at first hitting time

2.0.1 $$\mathbf{E}_x e^{-\alpha H_z} = \begin{cases} \dfrac{M\left(\frac{\alpha}{2\theta}, \frac{n}{2}, \frac{x^2}{2\sigma^2}\right)}{M\left(\frac{\alpha}{2\theta}, \frac{n}{2}, \frac{z^2}{2\sigma^2}\right)}, & x \le z \\ \dfrac{U\left(\frac{\alpha}{2\theta}, \frac{n}{2}, \frac{x^2}{2\sigma^2}\right)}{U\left(\frac{\alpha}{2\theta}, \frac{n}{2}, \frac{z^2}{2\sigma^2}\right)}, & z \le x \end{cases}$$

2.1.2 $$\mathbf{P}_x\Big(\sup_{0 \le s \le H_z} Q_s^{(n)} < y\Big) = \frac{\operatorname{Ed}\left(\frac{n}{2}, \frac{y}{\sigma}, \frac{x}{\sigma}\right)}{\operatorname{Ed}\left(\frac{n}{2}, \frac{y}{\sigma}, \frac{z}{\sigma}\right)}, \qquad z \le x \le y$$

2.1.4 $z \le y$ $$\mathbf{E}_x\Big\{e^{-\alpha H_z}; \sup_{0 \le s \le H_z} Q_s^{(n)} < y\Big\} = \begin{cases} \dfrac{M\left(\frac{\alpha}{2\theta}, \frac{n}{2}, \frac{x^2}{2\sigma^2}\right)}{M\left(\frac{\alpha}{2\theta}, \frac{n}{2}, \frac{z^2}{2\sigma^2}\right)}, & x \le z \\ \dfrac{S\left(\frac{\alpha}{2\theta}, \frac{n}{2}, \frac{y^2}{2\sigma^2}, \frac{x^2}{2\sigma^2}\right)}{S\left(\frac{\alpha}{2\theta}, \frac{n}{2}, \frac{y^2}{2\sigma^2}, \frac{z^2}{2\sigma^2}\right)}, & z \le x \le y \end{cases}$$

2.2.2 $$\mathbf{P}_x\Big(\inf_{0 \le s \le H_z} Q_s^{(n)} > y\Big) = \frac{\operatorname{Ed}\left(\frac{n}{2}, \frac{x}{\sigma}, \frac{y}{\sigma}\right)}{\operatorname{Ed}\left(\frac{n}{2}, \frac{z}{\sigma}, \frac{y}{\sigma}\right)}, \qquad y \le x \le z$$

2.2.4 $y \le z$ $$\mathbf{E}_x\Big\{e^{-\alpha H_z}; \inf_{0 \le s \le H_z} Q_s^{(n)} > y\Big\} = \begin{cases} \dfrac{S\left(\frac{\alpha}{2\theta}, \frac{n}{2}, \frac{x^2}{2\sigma^2}, \frac{y^2}{2\sigma^2}\right)}{S\left(\frac{\alpha}{2\theta}, \frac{n}{2}, \frac{z^2}{2\sigma^2}, \frac{y^2}{2\sigma^2}\right)}, & y \le x \le z \\ \dfrac{U\left(\frac{\alpha}{2\theta}, \frac{n}{2}, \frac{x^2}{2\sigma^2}\right)}{U\left(\frac{\alpha}{2\theta}, \frac{n}{2}, \frac{z^2}{2\sigma^2}\right)}, & z \le x \end{cases}$$

2.3.1 $\mathbf{E}_x e^{-\gamma \ell(H_z, r)}$

$$= \begin{cases} 1, & x \le z \le r, \quad r \le z \le x \\ \dfrac{\theta 2^{n/2}\sigma^n e^{r^2/2\sigma^2} + \gamma r^{n-1} \left|\operatorname{Ed}\left(\frac{n}{2}, \frac{x}{\sigma}, \frac{r}{\sigma}\right)\right|}{\theta 2^{n/2}\sigma^n e^{r^2/2\sigma^2} + \gamma r^{n-1} \left|\operatorname{Ed}\left(\frac{n}{2}, \frac{z}{\sigma}, \frac{r}{\sigma}\right)\right|}, & r \wedge z \le x \le r \vee z \\ \dfrac{\theta 2^{n/2}\sigma^n e^{r^2/2\sigma^2}}{\theta 2^{n/2}\sigma^n e^{r^2/2\sigma^2} + \gamma r^{n-1} \left|\operatorname{Ed}\left(\frac{n}{2}, \frac{z}{\sigma}, \frac{r}{\sigma}\right)\right|}, & 0 \le x \le r \le z, \quad z \le r \le x \end{cases}$$

2.3.2 $$\mathbf{P}_x\big(\ell(H_z, r) \in dy\big) = \exp\left(-\frac{\theta 2^{n/2}\sigma^n e^{r^2/2\sigma^2} y}{r^{n-1}\left|\operatorname{Ed}\left(\frac{n}{2}, \frac{z}{\sigma}, \frac{r}{\sigma}\right)\right|}\right)$$

8 $Q_s^{(n)} = \sigma e^{-\theta s} R^{(n)}_{e^{2\theta s}-1} \quad n \ge 2, \quad \sigma, \theta > 0$ $\qquad H_z = \min\{s : Q_s^{(n)} = z\}$

$$\times \begin{cases} 0, & 0 \le x \le z \le r, \quad r \le z \le x \\ \dfrac{\theta 2^{n/2}\sigma^n e^{r^2/2\sigma^2} |\operatorname{Ed}(\frac{n}{2}, \frac{x}{\sigma}, \frac{z}{\sigma})|}{r^{n-1}\operatorname{Ed}^2(\frac{n}{2}, \frac{z}{\sigma}, \frac{r}{\sigma})} dy, & r \wedge z \le x \le r \vee z \\ \dfrac{\theta 2^{n/2}\sigma^n e^{r^2/2\sigma^2}}{r^{n-1}|\operatorname{Ed}(\frac{n}{2}, \frac{z}{\sigma}, \frac{r}{\sigma})|} dy, & 0 \le x \le r \le z, \quad z \le r \le x \end{cases}$$

(2) $$\mathbf{P}_x\big(\ell(H_z, r) = 0\big) = \begin{cases} 1, & 0 \le x \le z \le r, \quad r \le z \le x \\ \dfrac{|\operatorname{Ed}(\frac{n}{2}, \frac{r}{\sigma}, \frac{x}{\sigma})|}{|\operatorname{Ed}(\frac{n}{2}, \frac{r}{\sigma}, \frac{z}{\sigma})|}, & z \wedge r \le x \le r \vee z \\ 0, & 0 \le x \le r \le z, \quad z \le r \le x \end{cases}$$

2.3.3 $\mathbf{E}_x e^{-\alpha H_z - \gamma \ell(H_z, r)}$

$0 \le x$

$r \le z$
$$= \begin{cases} \dfrac{U(\frac{\alpha}{2\theta}, \frac{n}{2}, \frac{x^2}{2\sigma^2})}{U(\frac{\alpha}{2\theta}, \frac{n}{2}, \frac{z^2}{2\sigma^2})}, & z \le x \\ \dfrac{\theta 2^{n/2}\sigma^n e^{r^2/2\sigma^2} M(\frac{\alpha}{2\theta}, \frac{n}{2}, \frac{x^2}{2\sigma^2}) + \gamma r^{n-1} S(\frac{\alpha}{2\theta}, \frac{n}{2}, \frac{x^2}{2\sigma^2}, \frac{r^2}{2\sigma^2}) M(\frac{\alpha}{2\theta}, \frac{n}{2}, \frac{r^2}{2\sigma^2})}{\theta 2^{n/2}\sigma^n e^{r^2/2\sigma^2} M(\frac{\alpha}{2\theta}, \frac{n}{2}, \frac{z^2}{2\sigma^2}) + \gamma r^{n-1} S(\frac{\alpha}{2\theta}, \frac{n}{2}, \frac{z^2}{2\sigma^2}, \frac{r^2}{2\sigma^2}) M(\frac{\alpha}{2\theta}, \frac{n}{2}, \frac{r^2}{2\sigma^2})}, & r \le x,\ x \le z \\ \dfrac{\theta 2^{n/2}\sigma^n e^{r^2/2\sigma^2} M(\frac{\alpha}{2\theta}, \frac{n}{2}, \frac{x^2}{2\sigma^2})}{\theta 2^{n/2}\sigma^n e^{r^2/2\sigma^2} M(\frac{\alpha}{2\theta}, \frac{n}{2}, \frac{z^2}{2\sigma^2}) + \gamma r^{n-1} S(\frac{\alpha}{2\theta}, \frac{n}{2}, \frac{z^2}{2\sigma^2}, \frac{r^2}{2\sigma^2}) M(\frac{\alpha}{2\theta}, \frac{n}{2}, \frac{r^2}{2\sigma^2})}, & x \le r \end{cases}$$

$z \le r$
$$= \begin{cases} \dfrac{M(\frac{\alpha}{2\theta}, \frac{n}{2}, \frac{x^2}{2\sigma^2})}{M(\frac{\alpha}{2\theta}, \frac{n}{2}, \frac{z^2}{2\sigma^2})}, & x \le z \\ \dfrac{\theta 2^{n/2}\sigma^n e^{r^2/2\sigma^2} U(\frac{\alpha}{2\theta}, \frac{n}{2}, \frac{x^2}{2\sigma^2}) + \gamma r^{n-1} S(\frac{\alpha}{2\theta}, \frac{n}{2}, \frac{r^2}{2\sigma^2}, \frac{x^2}{2\sigma^2}) U(\frac{\alpha}{2\theta}, \frac{n}{2}, \frac{r^2}{2\sigma^2})}{\theta 2^{n/2}\sigma^n e^{r^2/2\sigma^2} U(\frac{\alpha}{2\theta}, \frac{n}{2}, \frac{z^2}{2\sigma^2}) + \gamma r^{n-1} S(\frac{\alpha}{2\theta}, \frac{n}{2}, \frac{r^2}{2\sigma^2}, \frac{z^2}{2\sigma^2}) U(\frac{\alpha}{2\theta}, \frac{n}{2}, \frac{r^2}{2\sigma^2})}, & z \le x,\ x \le r \\ \dfrac{\theta 2^{n/2}\sigma^n e^{r^2/2\sigma^2} U(\frac{\alpha}{2\theta}, \frac{n}{2}, \frac{x^2}{2\sigma^2})}{\theta 2^{n/2}\sigma^n e^{r^2/2\sigma^2} U(\frac{\alpha}{2\theta}, \frac{n}{2}, \frac{z^2}{2\sigma^2}) + \gamma r^{n-1} S(\frac{\alpha}{2\theta}, \frac{n}{2}, \frac{r^2}{2\sigma^2}, \frac{z^2}{2\sigma^2}) U(\frac{\alpha}{2\theta}, \frac{n}{2}, \frac{r^2}{2\sigma^2})}, & r \le x \end{cases}$$

2.3.4 $\mathbf{E}_x\{e^{-\alpha H_z};\ \ell(H_z, r) \in dy\}$

$r \le z$
$$= \exp\Big(-\frac{\theta 2^{n/2}\sigma^n e^{r^2/2\sigma^2} M(\frac{\alpha}{2\theta}, \frac{n}{2}, \frac{z^2}{2\sigma^2}) y}{r^{n-1} S(\frac{\alpha}{2\theta}, \frac{n}{2}, \frac{z^2}{2\sigma^2}, \frac{r^2}{2\sigma^2}) M(\frac{\alpha}{2\theta}, \frac{n}{2}, \frac{r^2}{2\sigma^2})}\Big)$$
$$\times \begin{cases} \dfrac{\theta 2^{n/2}\sigma^n e^{r^2/2\sigma^2} M(\frac{\alpha}{2\theta}, \frac{n}{2}, \frac{x^2}{2\sigma^2})}{r^{n-1} S(\frac{\alpha}{2\theta}, \frac{n}{2}, \frac{z^2}{2\sigma^2}, \frac{r^2}{2\sigma^2}) M(\frac{\alpha}{2\theta}, \frac{n}{2}, \frac{r^2}{2\sigma^2})} dy, & 0 \le x \le r \\ \dfrac{\theta 2^{n/2}\sigma^n e^{r^2/2\sigma^2} S(\frac{\alpha}{2\theta}, \frac{n}{2}, \frac{z^2}{2\sigma^2}, \frac{x^2}{2\sigma^2})}{r^{n-1} S^2(\frac{\alpha}{2\theta}, \frac{n}{2}, \frac{z^2}{2\sigma^2}, \frac{r^2}{2\sigma^2})} dy, & r \le x \le z \\ 0, & z \le x \end{cases}$$

8 $Q_s^{(n)} = \sigma e^{-\theta s} R^{(n)}_{e^{2\theta s}-1} \quad n \ge 2, \quad \sigma, \theta > 0 \qquad H_z = \min\{s : Q_s^{(n)} = z\}$

$z \le r$

$$= \exp\Big(-\frac{\theta 2^{n/2}\sigma^n e^{r^2/2\sigma^2} U\big(\frac{\alpha}{2\theta}, \frac{n}{2}, \frac{z^2}{2\sigma^2}\big) y}{r^{n-1} S\big(\frac{\alpha}{2\theta}, \frac{n}{2}, \frac{r^2}{2\sigma^2}, \frac{z^2}{2\sigma^2}\big) U\big(\frac{\alpha}{2\theta}, \frac{n}{2}, \frac{r^2}{2\sigma^2}\big)}\Big) \times \begin{cases} \dfrac{\theta 2^{n/2}\sigma^n e^{r^2/2\sigma^2} U\big(\frac{\alpha}{2\theta}, \frac{n}{2}, \frac{x^2}{2\sigma^2}\big)}{r^{n-1} S\big(\frac{\alpha}{2\theta}, \frac{n}{2}, \frac{r^2}{2\sigma^2}, \frac{z^2}{2\sigma^2}\big) U\big(\frac{\alpha}{2\theta}, \frac{n}{2}, \frac{r^2}{2\sigma^2}\big)} dy, & r \le x \\ \dfrac{\theta 2^{n/2}\sigma^n e^{r^2/2\sigma^2} S\big(\frac{\alpha}{2\theta}, \frac{n}{2}, \frac{x^2}{2\sigma^2}, \frac{z^2}{2\sigma^2}\big)}{r^{n-1} S^2\big(\frac{\alpha}{2\theta}, \frac{n}{2}, \frac{r^2}{2\sigma^2}, \frac{z^2}{2\sigma^2}\big)} dy, & z \le x \le r \\ 0, & 0 \le x \le z \end{cases}$$

(1)

$$\mathbf{E}_x\{e^{-\alpha H_z};\ \ell(H_z, r) = 0\} = \begin{cases} 0, & 0 \le x \le r \le z \quad z \le r \le x \\ \dfrac{|S\big(\frac{\alpha}{2\theta}, \frac{n}{2}, \frac{x^2}{2\sigma^2}, \frac{r^2}{2\sigma^2}\big)|}{|S\big(\frac{\alpha}{2\theta}, \frac{n}{2}, \frac{z^2}{2\sigma^2}, \frac{r^2}{2\sigma^2}\big)|}, & r \wedge z \le x \le r \vee z \\ \dfrac{U\big(\frac{\alpha}{2\theta}, \frac{n}{2}, \frac{x^2}{2\sigma^2}\big)}{U\big(\frac{\alpha}{2\theta}, \frac{n}{2}, \frac{z^2}{2\sigma^2}\big)}, & r \le z \le x \\ \dfrac{M\big(\frac{\alpha}{2\theta}, \frac{n}{2}, \frac{x^2}{2\sigma^2}\big)}{M\big(\frac{\alpha}{2\theta}, \frac{n}{2}, \frac{z^2}{2\sigma^2}\big)}, & 0 \le x \le z \le r \end{cases}$$

2.4.1
$0 \le x$

$$\mathbf{E}_x \exp\Big(-\gamma \int_0^{H_z} \mathbb{I}_{[r,\infty)}\big(Q_s^{(n)}\big) ds\Big)$$

$z \le r$

$$= \begin{cases} 1, & x \le z \\ \dfrac{\theta 2^{1+n/2}\sigma^n e^{r^2/2\sigma^2} U\big(\frac{\gamma}{2\theta}, \frac{n}{2}, \frac{r^2}{2\sigma^2}\big) + \gamma r^n \operatorname{Ed}\big(\frac{n}{2}, \frac{r}{\sigma}, \frac{x}{\sigma}\big) U\big(\frac{\gamma+2\theta}{2\theta}, \frac{n+2}{2}, \frac{r^2}{2\sigma^2}\big)}{\theta 2^{1+n/2}\sigma^n e^{r^2/2\sigma^2} U\big(\frac{\gamma}{2\theta}, \frac{n}{2}, \frac{r^2}{2\sigma^2}\big) + \gamma r^n \operatorname{Ed}\big(\frac{n}{2}, \frac{r}{\sigma}, \frac{z}{\sigma}\big) U\big(\frac{\gamma+2\theta}{2\theta}, \frac{n+2}{2}, \frac{r^2}{2\sigma^2}\big)}, & z \le x \\ & x \le r \\ \dfrac{\theta 2^{1+n/2}\sigma^n e^{r^2/2\sigma^2} U\big(\frac{\gamma}{2\theta}, \frac{n}{2}, \frac{x^2}{2\sigma^2}\big)}{\theta 2^{1+n/2}\sigma^n e^{r^2/2\sigma^2} U\big(\frac{\gamma}{2\theta}, \frac{n}{2}, \frac{r^2}{2\sigma^2}\big) + \gamma r^n \operatorname{Ed}\big(\frac{n}{2}, \frac{r}{\sigma}, \frac{z}{\sigma}\big) U\big(\frac{\gamma+2\theta}{2\theta}, \frac{n+2}{2}, \frac{r^2}{2\sigma^2}\big)}, & r \le x \end{cases}$$

$r \le z$

$$= \begin{cases} \dfrac{2^{n/2}\sigma^n r^{-n} e^{r^2/2\sigma^2}}{C\big(\frac{\gamma}{2\theta}, \frac{n}{2}, \frac{r^2}{2\sigma^2}, \frac{z^2}{2\sigma^2}\big)}, & 0 < x \le r \\ \dfrac{C\big(\frac{\gamma}{2\theta}, \frac{n}{2}, \frac{r^2}{2\sigma^2}, \frac{x^2}{2\sigma^2}\big)}{C\big(\frac{\gamma}{2\theta}, \frac{n}{2}, \frac{r^2}{2\sigma^2}, \frac{z^2}{2\sigma^2}\big)}, & r \le x \le z \\ \dfrac{U\big(\frac{\gamma}{2\theta}, \frac{n}{2}, \frac{x^2}{2\sigma^2}\big)}{U\big(\frac{\gamma}{2\theta}, \frac{n}{2}, \frac{z^2}{2\sigma^2}\big)}, & z \le x \end{cases}$$

2.4.2
(1)
$z \le r$

$$\mathbf{P}_x\Big(\int_0^{H_z} \mathbb{I}_{[r,\infty)}\big(Q_s^{(n)}\big) ds = 0\Big) = \begin{cases} 1, & 0 \le x \le z \\ \dfrac{\operatorname{Ed}\big(\frac{n}{2}, \frac{r}{\sigma}, \frac{x}{\sigma}\big)}{\operatorname{Ed}\big(\frac{n}{2}, \frac{r}{\sigma}, \frac{z}{\sigma}\big)}, & z \le x \le r \\ 0, & r \le x \end{cases}$$

8 $\quad Q_s^{(n)} = \sigma e^{-\theta s} R^{(n)}_{e^{2\theta s}-1} \quad n \ge 2, \quad \sigma, \theta > 0 \qquad\qquad H_z = \min\{s : Q_s^{(n)} = z\}$

2.5.1 $0 \le x$

$$\mathbf{E}_x \exp\Big(-\gamma \int_0^{H_z} \mathbb{1}_{[0,r]}\big(Q_s^{(n)}\big)ds\Big)$$

$r \le z$

$$= \begin{cases} \dfrac{\theta n 2^{n/2}\sigma^n e^{r^2/2\sigma^2} M\big(\frac{\gamma}{2\theta},\frac{n}{2},\frac{x^2}{2\sigma^2}\big)}{\theta n 2^{n/2}\sigma^n e^{r^2/2\sigma^2} M\big(\frac{\gamma}{2\theta},\frac{n}{2},\frac{r^2}{2\sigma^2}\big) + \gamma r^n \operatorname{Ed}\big(\frac{n}{2},\frac{z}{\sigma},\frac{r}{\sigma}\big) M\big(\frac{\gamma+2\theta}{2\theta},\frac{n+2}{2},\frac{r^2}{2\sigma^2}\big)}, & x \le r \\[2ex] \dfrac{\theta n 2^{n/2}\sigma^n e^{r^2/2\sigma^2} M\big(\frac{\gamma}{2\theta},\frac{n}{2},\frac{r^2}{2\sigma^2}\big) + \gamma r^n \operatorname{Ed}\big(\frac{n}{2},\frac{x}{\sigma},\frac{r}{\sigma}\big) M\big(\frac{\gamma+2\theta}{2\theta},\frac{n+2}{2},\frac{r^2}{2\sigma^2}\big)}{\theta n 2^{n/2}\sigma^n e^{r^2/2\sigma^2} M\big(\frac{\gamma}{2\theta},\frac{n}{2},\frac{r^2}{2\sigma^2}\big) + \gamma r^n \operatorname{Ed}\big(\frac{n}{2},\frac{z}{\sigma},\frac{r}{\sigma}\big) M\big(\frac{\gamma+2\theta}{2\theta},\frac{n+2}{2},\frac{r^2}{2\sigma^2}\big)}, & r \le x \\ & x \le z \\[1ex] 1, & z \le x \end{cases}$$

$z \le r$

$$= \begin{cases} \dfrac{M\big(\frac{\gamma}{2\theta},\frac{n}{2},\frac{x^2}{2\sigma^2}\big)}{M\big(\frac{\gamma}{2\theta},\frac{n}{2},\frac{z^2}{2\sigma^2}\big)}, & 0 \le x \le z \\[2ex] \dfrac{C\big(\frac{\gamma}{2\theta},\frac{n}{2},\frac{r^2}{2\sigma^2},\frac{x^2}{2\sigma^2}\big)}{C\big(\frac{\gamma}{2\theta},\frac{n}{2},\frac{r^2}{2\sigma^2},\frac{z^2}{2\sigma^2}\big)}, & z \le x \le r \\[2ex] \dfrac{2^{n/2}\sigma^n r^{-n} e^{r^2/2\sigma^2}}{C\big(\frac{\gamma}{2\theta},\frac{n}{2},\frac{r^2}{2\sigma^2},\frac{z^2}{2\sigma^2}\big)}, & r \le x \end{cases}$$

2.5.2 (1) $r \le z$

$$\mathbf{P}_x\Big(\int_0^{H_z} \mathbb{1}_{[0,r]}\big(Q_s^{(n)}\big)ds = 0\Big) = \begin{cases} 0, & x \le r \\ \dfrac{\operatorname{Ed}\big(\frac{n}{2},\frac{x}{\sigma},\frac{r}{\sigma}\big)}{\operatorname{Ed}\big(\frac{n}{2},\frac{z}{\sigma},\frac{r}{\sigma}\big)}, & r \le x \le z \\ 1, & z \le x \end{cases}$$

2.6.1 $0 \le x$

$$\mathbf{E}_x \exp\Big(-\int_0^{H_z} \big(p\mathbb{1}_{[0,r)}\big(Q_s^{(n)}\big) + q\mathbb{1}_{[r,\infty)}\big(Q_s^{(n)}\big)\big)ds\Big)$$

$r \le z$

$$= \begin{cases} \dfrac{2^{n/2}\sigma^n r^{-n} e^{r^2/2\sigma^2} M\big(\frac{p}{2\theta},\frac{n}{2},\frac{x^2}{2\sigma^2}\big)}{\frac{p}{\theta n} S\big(\frac{q}{2\theta},\frac{n}{2},\frac{z^2}{2\sigma^2},\frac{r^2}{2\sigma^2}\big) M\big(\frac{p+2\theta}{2\theta},\frac{n+2}{2},\frac{r^2}{2\sigma^2}\big) + C\big(\frac{q}{2\theta},\frac{n}{2},\frac{r^2}{2\sigma^2},\frac{z^2}{2\sigma^2}\big) M\big(\frac{p}{2\theta},\frac{n}{2},\frac{r^2}{2\sigma^2}\big)}, & x \le r \\[2ex] \dfrac{\frac{p}{\theta n} S\big(\frac{q}{2\theta},\frac{n}{2},\frac{x^2}{2\sigma^2},\frac{r^2}{2\sigma^2}\big) M\big(\frac{p+2\theta}{2\theta},\frac{n+2}{2},\frac{r^2}{2\sigma^2}\big) + C\big(\frac{q}{2\theta},\frac{n}{2},\frac{r^2}{2\sigma^2},\frac{x^2}{2\sigma^2}\big) M\big(\frac{p}{2\theta},\frac{n}{2},\frac{r^2}{2\sigma^2}\big)}{\frac{p}{\theta n} S\big(\frac{q}{2\theta},\frac{n}{2},\frac{z^2}{2\sigma^2},\frac{r^2}{2\sigma^2}\big) M\big(\frac{p+2\theta}{2\theta},\frac{n+2}{2},\frac{r^2}{2\sigma^2}\big) + C\big(\frac{q}{2\theta},\frac{n}{2},\frac{r^2}{2\sigma^2},\frac{z^2}{2\sigma^2}\big) M\big(\frac{p}{2\theta},\frac{n}{2},\frac{r^2}{2\sigma^2}\big)}, & r \le x \\ & x \le z \\[1ex] \dfrac{U\big(\frac{q}{2\theta},\frac{n}{2},\frac{x^2}{2\sigma^2}\big)}{U\big(\frac{q}{2\theta},\frac{n}{2},\frac{z^2}{2\sigma^2}\big)}, & z \le x \end{cases}$$

$z \le r$

$$= \begin{cases} \dfrac{M\big(\frac{p}{2\theta},\frac{n}{2},\frac{x^2}{2\sigma^2}\big)}{M\big(\frac{p}{2\theta},\frac{n}{2},\frac{z^2}{2\sigma^2}\big)}, & x \le z \\[2ex] \dfrac{2\theta U\big(\frac{q}{2\theta},\frac{n}{2},\frac{r^2}{2\sigma^2}\big) C\big(\frac{p}{2\theta},\frac{n}{2},\frac{r^2}{2\sigma^2},\frac{x^2}{2\sigma^2}\big) + qU\big(\frac{q+2\theta}{2\theta},\frac{n}{2},\frac{r^2}{2\sigma^2}\big) S\big(\frac{p}{2\theta},\frac{n}{2},\frac{r^2}{2\sigma^2},\frac{x^2}{2\sigma^2}\big)}{2\theta U\big(\frac{q}{2\theta},\frac{n}{2},\frac{r^2}{2\sigma^2}\big) C\big(\frac{p}{2\theta},\frac{n}{2},\frac{r^2}{2\sigma^2},\frac{z^2}{2\sigma^2}\big) + qU\big(\frac{q+2\theta}{2\theta},\frac{n}{2},\frac{r^2}{2\sigma^2}\big) S\big(\frac{p}{2\theta},\frac{n}{2},\frac{r^2}{2\sigma^2},\frac{z^2}{2\sigma^2}\big)}, & z \le x \\ & x \le r \\[2ex] \dfrac{\theta 2^{1+n/2}\sigma^n r^{-n} e^{r^2/2\sigma^2} U\big(\frac{q}{2\theta},\frac{n}{2},\frac{x^2}{2\sigma^2}\big)}{2\theta U\big(\frac{q}{2\theta},\frac{n}{2},\frac{r^2}{2\sigma^2}\big) C\big(\frac{p}{2\theta},\frac{n}{2},\frac{r^2}{2\sigma^2},\frac{z^2}{2\sigma^2}\big) + qU\big(\frac{q+2\theta}{2\theta},\frac{n}{2},\frac{r^2}{2\sigma^2}\big) S\big(\frac{p}{2\theta},\frac{n}{2},\frac{r^2}{2\sigma^2},\frac{z^2}{2\sigma^2}\big)}, & r \le x \end{cases}$$

8 $Q_s^{(n)} = \sigma e^{-\theta s} R_{e^{2\theta s}-1}^{(n)} \quad n \ge 2, \quad \sigma, \theta > 0$ $\qquad H_z = \min\{s : Q_s^{(n)} = z\}$

2.9.1
$0 < x$

$$\mathbf{E}_x \exp\Big(-\frac{\theta(\gamma^2-1)}{4\sigma^2}\int_0^{H_z}(Q_s^{(n)})^2 ds\Big) = \begin{cases} \dfrac{z^{n/2}e^{x^2/4\sigma^2} M_{\frac{n}{4\gamma},\frac{n-2}{4}}\Big(\dfrac{\gamma x^2}{2\sigma^2}\Big)}{x^{n/2}e^{z^2/4\sigma^2} M_{\frac{n}{4\gamma},\frac{n-2}{4}}\Big(\dfrac{\gamma z^2}{2\sigma^2}\Big)}, & x \le z \\ \dfrac{z^{n/2}e^{x^2/4\sigma^2} W_{\frac{n}{4\gamma},\frac{n-2}{4}}\Big(\dfrac{\gamma x^2}{2\sigma^2}\Big)}{x^{n/2}e^{z^2/4\sigma^2} W_{\frac{n}{4\gamma},\frac{n-2}{4}}\Big(\dfrac{\gamma z^2}{2\sigma^2}\Big)}, & z \le x \end{cases}$$

2.13.1
$z < x$
$x < y$

$$\mathbf{E}_x\big\{e^{-\gamma \check{H}(H_z)}; \sup_{0\le s\le H_z} Q_s^{(n)} \in dy\big\} = \frac{2^{n/2}\sigma^{n-2}y^{1-n}e^{y^2/2\sigma^2} S\big(\frac{\gamma}{2\theta},\frac{n}{2},\frac{x^2}{2\sigma^2},\frac{z^2}{2\sigma^2}\big)}{\mathrm{Ed}\big(\frac{n}{2},\frac{y}{\sigma},\frac{z}{\sigma}\big) S\big(\frac{\gamma}{2\theta},\frac{n}{2},\frac{y^2}{2\sigma^2},\frac{z^2}{2\sigma^2}\big)}dy$$

2.13.4

$$\mathbf{E}_x\big\{e^{-\alpha H_z-\gamma \check{H}(H_z)}; \sup_{0\le s\le H_z} Q_s^{(n)} \in dy\big\} \qquad z < x < y$$

$$= \frac{2^{n/2}\sigma^{n-2}y^{1-n}e^{y^2/2\sigma^2} S\big(\frac{\alpha+\gamma}{2\theta},\frac{n}{2},\frac{x^2}{2\sigma^2},\frac{z^2}{2\sigma^2}\big)}{S\big(\frac{\alpha}{2\theta},\frac{n}{2},\frac{y^2}{2\sigma^2},\frac{z^2}{2\sigma^2}\big) S\big(\frac{\alpha+\gamma}{2\theta},\frac{n}{2},\frac{y^2}{2\sigma^2},\frac{z^2}{2\sigma^2}\big)}dy$$

2.14.1
$x < z$
$y < x$

$$\mathbf{E}_x\big\{e^{-\gamma \hat{H}(H_z)}; \inf_{0\le s\le H_z} Q_s^{(n)} \in dy\big\} = \frac{2^{n/2}\sigma^{n-2}y^{1-n}e^{y^2/2\sigma^2} S\big(\frac{\gamma}{2\theta},\frac{n}{2},\frac{z^2}{2\sigma^2},\frac{x^2}{2\sigma^2}\big)}{\mathrm{Ed}\big(\frac{n}{2},\frac{z}{\sigma},\frac{y}{\sigma}\big) S\big(\frac{\gamma}{2\theta},\frac{n}{2},\frac{z^2}{2\sigma^2},\frac{y^2}{2\sigma^2}\big)}dy$$

2.14.4

$$\mathbf{E}_x\big\{e^{-\alpha H_z-\gamma \hat{H}(H_z)}; \inf_{0\le s\le H_z} Q_s^{(n)} \in dy\big\} \qquad y < x < z$$

$$= \frac{2^{n/2}\sigma^{n-2}y^{1-n}e^{y^2/2\sigma^2} S\big(\frac{\alpha+\gamma}{2\theta},\frac{n}{2},\frac{z^2}{2\sigma^2},\frac{x^2}{2\sigma^2}\big)}{S\big(\frac{\alpha}{2\theta},\frac{n}{2},\frac{z^2}{2\sigma^2},\frac{y^2}{2\sigma^2}\big) S\big(\frac{\alpha+\gamma}{2\theta},\frac{n}{2},\frac{z^2}{2\sigma^2},\frac{y^2}{2\sigma^2}\big)}dy$$

2.20.1
$0 < x$

$$\mathbf{E}_x \exp\Big(-\gamma^2\sigma^2\theta\int_0^{H_z}\frac{ds}{(Q_s^{(n)})^2}\Big) = \begin{cases} \dfrac{z^{n/2}e^{x^2/4\sigma^2} M_{\frac{n}{4},\sqrt{\frac{(n-2)^2}{16}+\frac{\gamma^2}{4}}}\Big(\dfrac{x^2}{2\sigma^2}\Big)}{x^{n/2}e^{z^2/4\sigma^2} M_{\frac{n}{4},\sqrt{\frac{(n-2)^2}{16}+\frac{\gamma^2}{4}}}\Big(\dfrac{z^2}{2\sigma^2}\Big)}, & x \le z \\ \dfrac{z^{n/2}e^{x^2/4\sigma^2} W_{\frac{n}{4},\sqrt{\frac{(n-2)^2}{16}+\frac{\gamma^2}{4}}}\Big(\dfrac{x^2}{2\sigma^2}\Big)}{x^{n/2}e^{z^2/4\sigma^2} W_{\frac{n}{4},\sqrt{\frac{(n-2)^2}{16}+\frac{\gamma^2}{4}}}\Big(\dfrac{z^2}{2\sigma^2}\Big)}, & z \le x \end{cases}$$

8 $Q_s^{(n)} = \sigma e^{-\theta s} R^{(n)}_{e^{2\theta s}-1} \quad n \ge 2, \quad \sigma, \theta > 0$ $\qquad H_z = \min\{s : Q_s^{(n)} = z\}$

2.21.3 $$\mathbf{E}_x \exp\Big(-\alpha H_z - \int_0^{H_z} \Big(\frac{p^2\sigma^2\theta}{(Q_s^{(n)})^2} + \frac{\theta(q^2-1)}{4\sigma^2}(Q_s^{(n)})^2\Big) ds\Big)$$

$$= \begin{cases} \dfrac{x^{-n/2} e^{x^2/4\sigma^2} M_{\frac{n\theta-2\alpha}{4\theta q}, \sqrt{\frac{(n-2)^2}{16}+\frac{p^2}{4}}}\Big(\frac{qx^2}{2\sigma^2}\Big)}{z^{-n/2} e^{z^2/4\sigma^2} M_{\frac{n\theta-2\alpha}{4\theta q}, \sqrt{\frac{(n-2)^2}{16}+\frac{p^2}{4}}}\Big(\frac{qz^2}{2\sigma^2}\Big)}, & 0 < x \le z \\[2ex] \dfrac{x^{-n/2} e^{x^2/4\sigma^2} W_{\frac{n\theta-2\alpha}{4\theta q}, \sqrt{\frac{(n-2)^2}{16}+\frac{p^2}{4}}}\Big(\frac{qx^2}{2\sigma^2}\Big)}{z^{-n/2} e^{z^2/4\sigma^2} W_{\frac{n\theta-2\alpha}{4\theta q}, \sqrt{\frac{(n-2)^2}{16}+\frac{p^2}{4}}}\Big(\frac{qz^2}{2\sigma^2}\Big)}, & z \le x \end{cases}$$

2.27.1 $$\mathbf{E}_x\Big\{\exp\Big(-\int_0^{H_z} \big(p\mathbb{1}_{[0,r)}(Q_s^{(n)}) + q\mathbb{1}_{[r,\infty)}(Q_s^{(n)})\big) ds\Big); \ell(H_z, r) \in dy\Big\}$$

$z \le r$ $$= \exp\Big(-y\theta r\Big(\frac{C(\frac{p}{2\theta}, \frac{n}{2}, \frac{r^2}{2\sigma^2}, \frac{z^2}{2\sigma^2})}{S(\frac{p}{2\theta}, \frac{n}{2}, \frac{r^2}{2\sigma^2}, \frac{z^2}{2\sigma^2})} + \frac{qU(\frac{q+2\theta}{2\theta}, \frac{n+2}{2}, \frac{r^2}{2\sigma^2})}{2\theta U(\frac{q}{2\theta}, \frac{n}{2}, \frac{r^2}{2\sigma^2})}\Big)\Big)$$

$$\times \begin{cases} \dfrac{\theta 2^{n/2}\sigma^n e^{r^2/2\sigma^2} S(\frac{p}{2\theta}, \frac{n}{2}, \frac{x^2}{2\sigma^2}, \frac{z^2}{2\sigma^2})}{r^{n-1} S^2(\frac{p}{2\theta}, \frac{n}{2}, \frac{r^2}{2\sigma^2}, \frac{z^2}{2\sigma^2})} dy, & z \le x \le r \\[2ex] \dfrac{\theta 2^{n/2}\sigma^n e^{r^2/2\sigma^2} U(\frac{q}{2\theta}, \frac{n}{2}, \frac{x^2}{2\sigma^2})}{r^{n-1} U(\frac{q}{2\theta}, \frac{n}{2}, \frac{r^2}{2\sigma^2}) S(\frac{p}{2\theta}, \frac{n}{2}, \frac{r^2}{2\sigma^2}, \frac{z^2}{2\sigma^2})} dy, & r \le x \end{cases}$$

$r \le z$ $$= \exp\Big(-y\theta r\Big(\frac{C(\frac{q}{2\theta}, \frac{n}{2}, \frac{r^2}{2\sigma^2}, \frac{z^2}{2\sigma^2})}{S(\frac{q}{2\theta}, \frac{n}{2}, \frac{z^2}{2\sigma^2}, \frac{r^2}{2\sigma^2})} + \frac{pM(\frac{p+2\theta}{2\theta}, \frac{n+2}{2}, \frac{r^2}{2\sigma^2})}{\theta n M(\frac{p}{2\theta}, \frac{n}{2}, \frac{r^2}{2\sigma^2})}\Big)\Big)$$

$$\times \begin{cases} \dfrac{\theta 2^{n/2}\sigma^n e^{r^2/2\sigma^2} M(\frac{p}{2\theta}, \frac{n}{2}, \frac{x^2}{2\sigma^2})}{r^{n-1} M(\frac{p}{2\theta}, \frac{n}{2}, \frac{r^2}{2\sigma^2}) S(\frac{q}{2\theta}, \frac{n}{2}, \frac{z^2}{2\sigma^2}, \frac{r^2}{2\sigma^2})} dy, & 0 \le x \le r \\[2ex] \dfrac{\theta 2^{n/2}\sigma^n e^{r^2/2\sigma^2} S(\frac{q}{2\theta}, \frac{n}{2}, \frac{z^2}{2\sigma^2}, \frac{x^2}{2\sigma^2})}{r^{n-1} S^2(\frac{q}{2\theta}, \frac{n}{2}, \frac{z^2}{2\sigma^2}, \frac{r^2}{2\sigma^2})} dy, & r \le x \le z \end{cases}$$

(1) $$\mathbf{E}_x\Big\{\exp\Big(-\int_0^{H_z} \big(p\mathbb{1}_{(-\infty,r)}(Q_s^{(n)}) + q\mathbb{1}_{[r,\infty)}(Q_s^{(n)})\big) ds\Big); \ell(H_z, r) = 0\Big\}$$

$$= \begin{cases} \dfrac{S(\frac{p}{2\theta}, \frac{n}{2}, \frac{r^2}{2\sigma^2}, \frac{x^2}{2\sigma^2})}{S(\frac{p}{2\theta}, \frac{n}{2}, \frac{r^2}{2\sigma^2}, \frac{z^2}{2\sigma^2})}, & z \le x \le r \\[2ex] \dfrac{S(\frac{q}{2\theta}, \frac{n}{2}, \frac{x^2}{2\sigma^2}, \frac{r^2}{2\sigma^2})}{S(\frac{q}{2\theta}, \frac{n}{2}, \frac{z^2}{2\sigma^2}, \frac{r^2}{2\sigma^2})}, & r \le x \le z \end{cases}$$

8 $Q_s^{(n)} = \sigma e^{-\theta s} R^{(n)}_{e^{2\theta s}-1} \quad n \ge 2, \quad \sigma, \theta > 0 \quad H = H_{a,b} = \min\{s : Q_s^{(n)} \notin (a,b)\}$

3. Stopping at first exit time

3.0.1 $$\mathbf{E}_x e^{-\alpha H} = \frac{S\big(\frac{\alpha}{2\theta}, \frac{n}{2}, \frac{b^2}{2\sigma^2}, \frac{x^2}{2\sigma^2}\big) + S\big(\frac{\alpha}{2\theta}, \frac{n}{2}, \frac{x^2}{2\sigma^2}, \frac{a^2}{2\sigma^2}\big)}{S\big(\frac{\alpha}{2\theta}, \frac{n}{2}, \frac{b^2}{2\sigma^2}, \frac{a^2}{2\sigma^2}\big)}$$

3.0.3 $$\mathbf{E}_x e^{-\beta Q_H^{(n)}} = \frac{\operatorname{Ed}\big(\frac{n}{2}, \frac{b}{\sigma}, \frac{x}{\sigma}\big)}{\operatorname{Ed}\big(\frac{n}{2}, \frac{b}{\sigma}, \frac{a}{\sigma}\big)} e^{-\beta a} + \frac{\operatorname{Ed}\big(\frac{n}{2}, \frac{x}{\sigma}, \frac{a}{\sigma}\big)}{\operatorname{Ed}\big(\frac{n}{2}, \frac{b}{\sigma}, \frac{a}{\sigma}\big)} e^{-\beta b}$$

3.0.4 (a) $$\mathbf{P}_x\big(Q_H^{(n)} = a\big) = \frac{\operatorname{Ed}\big(\frac{n}{2}, \frac{b}{\sigma}, \frac{x}{\sigma}\big)}{\operatorname{Ed}\big(\frac{n}{2}, \frac{b}{\sigma}, \frac{a}{\sigma}\big)}$$

3.0.4 (b) $$\mathbf{P}_x\big(Q_H^{(n)} = b\big) = \frac{\operatorname{Ed}\big(\frac{n}{2}, \frac{x}{\sigma}, \frac{a}{\sigma}\big)}{\operatorname{Ed}\big(\frac{n}{2}, \frac{b}{\sigma}, \frac{a}{\sigma}\big)}$$

3.0.5 (a) $$\mathbf{E}_x\big\{e^{-\alpha H}; Q_H^{(n)} = a\big\} = \frac{S\big(\frac{\alpha}{2\theta}, \frac{n}{2}, \frac{b^2}{2\sigma^2}, \frac{x^2}{2\sigma^2}\big)}{S\big(\frac{\alpha}{2\theta}, \frac{n}{2}, \frac{b^2}{2\sigma^2}, \frac{a^2}{2\sigma^2}\big)}$$

3.0.5 (b) $$\mathbf{E}_x\big\{e^{-\alpha H}; Q_H^{(n)} = b\big\} = \frac{S\big(\frac{\alpha}{2\theta}, \frac{n}{2}, \frac{x^2}{2\sigma^2}, \frac{a^2}{2\sigma^2}\big)}{S\big(\frac{\alpha}{2\theta}, \frac{n}{2}, \frac{b^2}{2\sigma^2}, \frac{a^2}{2\sigma^2}\big)}$$

3.1.6 $$\mathbf{P}_x\Big(\sup_{0\le s\le H} Q_s^{(n)} \ge y, Q_H^{(n)} = a\Big) = \frac{\operatorname{Ed}\big(\frac{n}{2}, \frac{b}{\sigma}, \frac{y}{\sigma}\big)\operatorname{Ed}\big(\frac{n}{2}, \frac{x}{\sigma}, \frac{a}{\sigma}\big)}{\operatorname{Ed}\big(\frac{n}{2}, \frac{b}{\sigma}, \frac{a}{\sigma}\big)\operatorname{Ed}\big(\frac{n}{2}, \frac{y}{\sigma}, \frac{a}{\sigma}\big)}, \quad a \le x \le y \le b$$

3.1.8 (1) $$\mathbf{E}_x\Big\{e^{-\alpha H}; \sup_{0\le s\le H} Q_s^{(n)} \ge y, Q_H^{(n)} = a\Big\}$$
$$= \frac{S\big(\frac{\alpha}{2\theta}, \frac{n}{2}, \frac{b^2}{2\sigma^2}, \frac{y^2}{2\sigma^2}\big) S\big(\frac{\alpha}{2\theta}, \frac{n}{2}, \frac{x^2}{2\sigma^2}, \frac{a^2}{2\sigma^2}\big)}{S\big(\frac{\alpha}{2\theta}, \frac{n}{2}, \frac{b^2}{2\sigma^2}, \frac{a^2}{2\sigma^2}\big) S\big(\frac{\alpha}{2\theta}, \frac{n}{2}, \frac{y^2}{2\sigma^2}, \frac{a^2}{2\sigma^2}\big)}, \quad a \le x \le y \le b$$

3.2.6 $$\mathbf{P}_x\Big(\inf_{0\le s\le H} Q_s^{(n)} \le y, Q_H^{(n)} = b\Big) = \frac{\operatorname{Ed}\big(\frac{n}{2}, \frac{b}{\sigma}, \frac{x}{\sigma}\big)\operatorname{Ed}\big(\frac{n}{2}, \frac{y}{\sigma}, \frac{a}{\sigma}\big)}{\operatorname{Ed}\big(\frac{n}{2}, \frac{b}{\sigma}, \frac{y}{\sigma}\big)\operatorname{Ed}\big(\frac{n}{2}, \frac{b}{\sigma}, \frac{a}{\sigma}\big)}, \quad a \le y \le x \le b$$

3.2.8 (1) $$\mathbf{E}_x\Big\{e^{-\alpha H}; \inf_{0\le s\le H} Q_s^{(n)} \le y, Q_H^{(n)} = b\Big\}$$
$$= \frac{S\big(\frac{\alpha}{2\theta}, \frac{n}{2}, \frac{b^2}{2\sigma^2}, \frac{x^2}{2\sigma^2}\big) S\big(\frac{\alpha}{2\theta}, \frac{n}{2}, \frac{y^2}{2\sigma^2}, \frac{a^2}{2\sigma^2}\big)}{S\big(\frac{\alpha}{2\theta}, \frac{n}{2}, \frac{b^2}{2\sigma^2}, \frac{y^2}{2\sigma^2}\big) S\big(\frac{\alpha}{2\theta}, \frac{n}{2}, \frac{b^2}{2\sigma^2}, \frac{a^2}{2\sigma^2}\big)}, \quad a \le y \le x \le b$$

8 $Q_s^{(n)}=\sigma e^{-\theta s}R^{(n)}_{e^{2\theta s}-1}\quad n\ge 2,\quad \sigma,\theta>0\quad H=H_{a,b}=\min\{s:Q_s^{(n)}\notin(a,b)\}$

3.3.1 $\mathbf{E}_x e^{-\gamma\ell(H,r)}$

$$=\begin{cases}\dfrac{\theta 2^{n/2}\sigma^n e^{r^2/2\sigma^2}\operatorname{Ed}\left(\frac{n}{2},\frac{b}{\sigma},\frac{a}{\sigma}\right)+\gamma r^{n-1}\operatorname{Ed}\left(\frac{n}{2},\frac{x}{\sigma},\frac{r}{\sigma}\right)\operatorname{Ed}\left(\frac{n}{2},\frac{r}{\sigma},\frac{a}{\sigma}\right)}{\theta 2^{n/2}\sigma^n e^{r^2/2\sigma^2}\operatorname{Ed}\left(\frac{n}{2},\frac{b}{\sigma},\frac{a}{\sigma}\right)+\gamma r^{n-1}\operatorname{Ed}\left(\frac{n}{2},\frac{b}{\sigma},\frac{r}{\sigma}\right)\operatorname{Ed}\left(\frac{n}{2},\frac{r}{\sigma},\frac{a}{\sigma}\right)}, & r\le x\le b\\[2ex] \dfrac{\theta 2^{n/2}\sigma^n e^{r^2/2\sigma^2}\operatorname{Ed}\left(\frac{n}{2},\frac{b}{\sigma},\frac{a}{\sigma}\right)+\gamma r^{n-1}\operatorname{Ed}\left(\frac{n}{2},\frac{b}{\sigma},\frac{r}{\sigma}\right)\operatorname{Ed}\left(\frac{n}{2},\frac{r}{\sigma},\frac{x}{\sigma}\right)}{\theta 2^{n/2}\sigma^n e^{r^2/2\sigma^2}\operatorname{Ed}\left(\frac{n}{2},\frac{b}{\sigma},\frac{a}{\sigma}\right)+\gamma r^{n-1}\operatorname{Ed}\left(\frac{n}{2},\frac{b}{\sigma},\frac{r}{\sigma}\right)\operatorname{Ed}\left(\frac{n}{2},\frac{r}{\sigma},\frac{a}{\sigma}\right)}, & a\le x\le r\end{cases}$$

3.3.2 $$\mathbf{P}_x\big(\ell(H,r)\in dy\big)=\exp\Big(-\frac{\theta 2^{n/2}\sigma^n e^{r^2/2\sigma^2}\operatorname{Ed}\left(\frac{n}{2},\frac{b}{\sigma},\frac{a}{\sigma}\right)y}{r^{n-1}\operatorname{Ed}\left(\frac{n}{2},\frac{b}{\sigma},\frac{r}{\sigma}\right)\operatorname{Ed}\left(\frac{n}{2},\frac{r}{\sigma},\frac{a}{\sigma}\right)}\Big)$$

$$\times\begin{cases}\dfrac{\theta 2^{n/2}\sigma^n e^{r^2/2\sigma^2}\operatorname{Ed}\left(\frac{n}{2},\frac{b}{\sigma},\frac{x}{\sigma}\right)\operatorname{Ed}\left(\frac{n}{2},\frac{b}{\sigma},\frac{a}{\sigma}\right)}{r^{n-1}\operatorname{Ed}^2\left(\frac{n}{2},\frac{b}{\sigma},\frac{r}{\sigma}\right)\operatorname{Ed}\left(\frac{n}{2},\frac{r}{\sigma},\frac{a}{\sigma}\right)}dy, & r\le x\le b\\[2ex] \dfrac{\theta 2^{n/2}\sigma^n e^{r^2/2\sigma^2}\operatorname{Ed}\left(\frac{n}{2},\frac{x}{\sigma},\frac{a}{\sigma}\right)\operatorname{Ed}\left(\frac{n}{2},\frac{b}{\sigma},\frac{a}{\sigma}\right)}{r^{n-1}\operatorname{Ed}\left(\frac{n}{2},\frac{b}{\sigma},\frac{r}{\sigma}\right)\operatorname{Ed}^2\left(\frac{n}{2},\frac{r}{\sigma},\frac{a}{\sigma}\right)}dy, & a\le x\le r\end{cases}$$

(1) $$\mathbf{P}_x\big(\ell(H,r)=0\big)=\begin{cases}\dfrac{\operatorname{Ed}\left(\frac{n}{2},\frac{x}{\sigma},\frac{r}{\sigma}\right)}{\operatorname{Ed}\left(\frac{n}{2},\frac{b}{\sigma},\frac{r}{\sigma}\right)}, & r\le x\le b\\[2ex] \dfrac{\operatorname{Ed}\left(\frac{n}{2},\frac{r}{\sigma},\frac{x}{\sigma}\right)}{\operatorname{Ed}\left(\frac{n}{2},\frac{r}{\sigma},\frac{a}{\sigma}\right)}, & a\le x\le r\end{cases}$$

3.3.3 $\mathbf{E}_r e^{-\alpha H-\gamma\ell(H,r)}=:L$

$$=\frac{\theta 2^{n/2}\sigma^n e^{r^2/2\sigma^2}\Big(S\left(\frac{\alpha}{2\theta},\frac{n}{2},\frac{b^2}{2\sigma^2},\frac{r^2}{2\sigma^2}\right)+S\left(\frac{\alpha}{2\theta},\frac{n}{2},\frac{r^2}{2\sigma^2},\frac{a^2}{2\sigma^2}\right)\Big)}{\theta 2^{n/2}\sigma^n e^{r^2/2\sigma^2}S\left(\frac{\alpha}{2\theta},\frac{n}{2},\frac{b^2}{2\sigma^2},\frac{a^2}{2\sigma^2}\right)+\gamma r^{n-1}S\left(\frac{\alpha}{2\theta},\frac{n}{2},\frac{b^2}{2\sigma^2},\frac{r^2}{2\sigma^2}\right)S\left(\frac{\alpha}{2\theta},\frac{n}{2},\frac{r^2}{2\sigma^2},\frac{a^2}{2\sigma^2}\right)}$$

$$\mathbf{E}_x e^{-\alpha H-\gamma\ell(H,r)}=\begin{cases}\dfrac{S\left(\frac{\alpha}{2\theta},\frac{n}{2},\frac{x^2}{2\sigma^2},\frac{r^2}{2\sigma^2}\right)}{S\left(\frac{\alpha}{2\theta},\frac{n}{2},\frac{b^2}{2\sigma^2},\frac{r^2}{2\sigma^2}\right)}+\dfrac{S\left(\frac{\alpha}{2\theta},\frac{n}{2},\frac{b^2}{2\sigma^2},\frac{x^2}{2\sigma^2}\right)}{S\left(\frac{\alpha}{2\theta},\frac{n}{2},\frac{b^2}{2\sigma^2},\frac{r^2}{2\sigma^2}\right)}L, & r\le x\le b\\[2ex] \dfrac{S\left(\frac{\alpha}{2\theta},\frac{n}{2},\frac{r^2}{2\sigma^2},\frac{x^2}{2\sigma^2}\right)}{S\left(\frac{\alpha}{2\theta},\frac{n}{2},\frac{r^2}{2\sigma^2},\frac{a^2}{2\sigma^2}\right)}+\dfrac{S\left(\frac{\alpha}{2\theta},\frac{n}{2},\frac{x^2}{2\sigma^2},\frac{a^2}{2\sigma^2}\right)}{S\left(\frac{\alpha}{2\theta},\frac{n}{2},\frac{r^2}{2\sigma^2},\frac{a^2}{2\sigma^2}\right)}L, & a\le x\le r\end{cases}$$

3.3.4 $$\mathbf{E}_x\{e^{-\alpha H};\ell(H,r)\in dy\}=\exp\Big(-\frac{\theta 2^{n/2}\sigma^n e^{r^2/2\sigma^2}S\left(\frac{\alpha}{2\theta},\frac{n}{2},\frac{b^2}{2\sigma^2},\frac{a^2}{2\sigma^2}\right)y}{r^{n-1}S\left(\frac{\alpha}{2\theta},\frac{n}{2},\frac{b^2}{2\sigma^2},\frac{r^2}{2\sigma^2}\right)S\left(\frac{\alpha}{2\theta},\frac{n}{2},\frac{r^2}{2\sigma^2},\frac{a^2}{2\sigma^2}\right)}\Big)$$

$$\times\begin{cases}\dfrac{\theta S\left(\frac{\alpha}{2\theta},\frac{n}{2},\frac{b^2}{2\sigma^2},\frac{x^2}{2\sigma^2}\right)\Big[S\left(\frac{\alpha}{2\theta},\frac{n}{2},\frac{b^2}{2\sigma^2},\frac{r^2}{2\sigma^2}\right)+S\left(\frac{\alpha}{2\theta},\frac{n}{2},\frac{r^2}{2\sigma^2},\frac{a^2}{2\sigma^2}\right)\Big]}{2^{-n/2}\sigma^{-n}e^{-r^2/2\sigma^2}r^{n-1}S^2\left(\frac{\alpha}{2\theta},\frac{n}{2},\frac{b^2}{2\sigma^2},\frac{r^2}{2\sigma^2}\right)S\left(\frac{\alpha}{2\theta},\frac{n}{2},\frac{r^2}{2\sigma^2},\frac{a^2}{2\sigma^2}\right)}dy, & r\le x\le b\\[2ex] \dfrac{\theta S\left(\frac{\alpha}{2\theta},\frac{n}{2},\frac{x^2}{2\sigma^2},\frac{a^2}{2\sigma^2}\right)\Big[S\left(\frac{\alpha}{2\theta},\frac{n}{2},\frac{b^2}{2\sigma^2},\frac{r^2}{2\sigma^2}\right)+S\left(\frac{\alpha}{2\theta},\frac{n}{2},\frac{r^2}{2\sigma^2},\frac{a^2}{2\sigma^2}\right)\Big]}{2^{-n/2}\sigma^{-n}e^{-r^2/2\sigma^2}r^{n-1}S\left(\frac{\alpha}{2\theta},\frac{n}{2},\frac{b^2}{2\sigma^2},\frac{r^2}{2\sigma^2}\right)S^2\left(\frac{\alpha}{2\theta},\frac{n}{2},\frac{r^2}{2\sigma^2},\frac{a^2}{2\sigma^2}\right)}dy, & a\le x\le r\end{cases}$$

8 $Q_s^{(n)} = \sigma e^{-\theta s} R^{(n)}_{e^{2\theta s}-1} \quad n \ge 2, \quad \sigma, \theta > 0 \quad H = H_{a,b} = \min\{s : Q_s^{(n)} \notin (a,b)\}$

(1) $$\mathbf{E}_x\{e^{-\alpha H}; \ell(H,r) = 0\} = \begin{cases} \dfrac{S(\frac{\alpha}{2\theta}, \frac{n}{2}, \frac{x^2}{2\sigma^2}, \frac{r^2}{2\sigma^2})}{S(\frac{\alpha}{2\theta}, \frac{n}{2}, \frac{b^2}{2\sigma^2}, \frac{r^2}{2\sigma^2})}, & r \le x \le b \\ \dfrac{S(\frac{\alpha}{2\theta}, \frac{n}{2}, \frac{r^2}{2\sigma^2}, \frac{x^2}{2\sigma^2})}{S(\frac{\alpha}{2\theta}, \frac{n}{2}, \frac{r^2}{2\sigma^2}, \frac{a^2}{2\sigma^2})}, & a \le x \le r \end{cases}$$

3.3.5 (a) $$\mathbf{E}_x\{e^{-\gamma \ell(H,r)}; Q_H^{(n)} = a\}$$

$$= \begin{cases} \dfrac{\theta 2^{n/2}\sigma^n e^{r^2/2\sigma^2} \operatorname{Ed}(\frac{n}{2}, \frac{b}{\sigma}, \frac{x}{\sigma})}{\theta 2^{n/2}\sigma^n e^{r^2/2\sigma^2} \operatorname{Ed}(\frac{n}{2}, \frac{b}{\sigma}, \frac{a}{\sigma}) + \gamma r^{n-1} \operatorname{Ed}(\frac{n}{2}, \frac{b}{\sigma}, \frac{r}{\sigma}) \operatorname{Ed}(\frac{n}{2}, \frac{r}{\sigma}, \frac{a}{\sigma})}, & r \le x \le b \\ \dfrac{\theta 2^{n/2}\sigma^n e^{r^2/2\sigma^2} \operatorname{Ed}(\frac{n}{2}, \frac{b}{\sigma}, \frac{x}{\sigma}) + \gamma r^{n-1} \operatorname{Ed}(\frac{n}{2}, \frac{b}{\sigma}, \frac{r}{\sigma}) \operatorname{Ed}(\frac{n}{2}, \frac{r}{\sigma}, \frac{x}{\sigma})}{\theta 2^{n/2}\sigma^n e^{r^2/2\sigma^2} \operatorname{Ed}(\frac{n}{2}, \frac{b}{\sigma}, \frac{a}{\sigma}) + \gamma r^{n-1} \operatorname{Ed}(\frac{n}{2}, \frac{b}{\sigma}, \frac{r}{\sigma}) \operatorname{Ed}(\frac{n}{2}, \frac{r}{\sigma}, \frac{a}{\sigma})}, & a \le x \le r \end{cases}$$

3.3.5 (b) $$\mathbf{E}_x\{e^{-\gamma \ell(H,r)}; Q_H^{(n)} = b\}$$

$$= \begin{cases} \dfrac{\theta 2^{n/2}\sigma^n e^{r^2/2\sigma^2} \operatorname{Ed}(\frac{n}{2}, \frac{x}{\sigma}, \frac{a}{\sigma}) + \gamma r^{n-1} \operatorname{Ed}(\frac{n}{2}, \frac{x}{\sigma}, \frac{r}{\sigma}) \operatorname{Ed}(\frac{n}{2}, \frac{r}{\sigma}, \frac{a}{\sigma})}{\theta 2^{n/2}\sigma^n e^{r^2/2\sigma^2} \operatorname{Ed}(\frac{n}{2}, \frac{b}{\sigma}, \frac{a}{\sigma}) + \gamma r^{n-1} \operatorname{Ed}(\frac{n}{2}, \frac{b}{\sigma}, \frac{r}{\sigma}) \operatorname{Ed}(\frac{n}{2}, \frac{r}{\sigma}, \frac{a}{\sigma})}, & r \le x \le b \\ \dfrac{\theta 2^{n/2}\sigma^n e^{r^2/2\sigma^2} \operatorname{Ed}(\frac{n}{2}, \frac{x}{\sigma}, \frac{a}{\sigma})}{\theta 2^{n/2}\sigma^n e^{r^2/2\sigma^2} \operatorname{Ed}(\frac{n}{2}, \frac{b}{\sigma}, \frac{a}{\sigma}) + \gamma r^{n-1} \operatorname{Ed}(\frac{n}{2}, \frac{b}{\sigma}, \frac{r}{\sigma}) \operatorname{Ed}(\frac{n}{2}, \frac{r}{\sigma}, \frac{a}{\sigma})}, & a \le x \le r \end{cases}$$

3.3.6 (a) $$\mathbf{P}_x\big(\ell(H,r) \in dy, Q_H^{(n)} = a\big) = \exp\Big(-\frac{\theta 2^{n/2}\sigma^n e^{r^2/2\sigma^2} \operatorname{Ed}(\frac{n}{2}, \frac{b}{\sigma}, \frac{a}{\sigma}) y}{r^{n-1} \operatorname{Ed}(\frac{n}{2}, \frac{b}{\sigma}, \frac{r}{\sigma}) \operatorname{Ed}(\frac{n}{2}, \frac{r}{\sigma}, \frac{a}{\sigma})}\Big)$$

$$\times \begin{cases} \dfrac{\theta 2^{n/2}\sigma^n e^{r^2/2\sigma^2} \operatorname{Ed}(\frac{n}{2}, \frac{b}{\sigma}, \frac{x}{\sigma})}{r^{n-1} \operatorname{Ed}(\frac{n}{2}, \frac{b}{\sigma}, \frac{r}{\sigma}) \operatorname{Ed}(\frac{n}{2}, \frac{r}{\sigma}, \frac{a}{\sigma})} dy, & r \le x \le b \\ \dfrac{\theta 2^{n/2}\sigma^n e^{r^2/2\sigma^2} \operatorname{Ed}(\frac{n}{2}, \frac{x}{\sigma}, \frac{a}{\sigma})}{r^{n-1} \operatorname{Ed}(\frac{n}{2}, \frac{r}{\sigma}, \frac{a}{\sigma})^2} dy, & a \le x \le r \end{cases}$$

(1) $$\mathbf{P}_x\big(\ell(H,r) = 0, Q_H^{(n)} = a\big) = \begin{cases} 0, & r \le x \le b \\ \dfrac{\operatorname{Ed}(\frac{n}{2}, \frac{r}{\sigma}, \frac{x}{\sigma})}{\operatorname{Ed}(\frac{n}{2}, \frac{r}{\sigma}, \frac{a}{\sigma})}, & a \le x \le r \end{cases}$$

3.3.6 (b) $$\mathbf{P}_x\big(\ell(H,r) \in dy, Q_H^{(n)} = b\big) = \exp\Big(-\frac{\theta 2^{n/2}\sigma^n e^{r^2/2\sigma^2} \operatorname{Ed}(\frac{n}{2}, \frac{b}{\sigma}, \frac{a}{\sigma}) y}{r^{n-1} \operatorname{Ed}(\frac{n}{2}, \frac{b}{\sigma}, \frac{r}{\sigma}) \operatorname{Ed}(\frac{n}{2}, \frac{r}{\sigma}, \frac{a}{\sigma})}\Big)$$

8 $Q_s^{(n)} = \sigma e^{-\theta s} R^{(n)}_{e^{2\theta s}-1} \quad n \ge 2, \quad \sigma, \theta > 0 \quad H = H_{a,b} = \min\{s : Q_s^{(n)} \notin (a,b)\}$

$$\times \begin{cases} \dfrac{\theta 2^{n/2}\sigma^n e^{r^2/2\sigma^2}\operatorname{Ed}\big(\frac{n}{2},\frac{b}{\sigma},\frac{x}{\sigma}\big)}{r^{n-1}\operatorname{Ed}\big(\frac{n}{2},\frac{b}{\sigma},\frac{r}{\sigma}\big)^2}dy, & r \le x \le b \\ \dfrac{\theta 2^{n/2}\sigma^n e^{r^2/2\sigma^2}\operatorname{Ed}\big(\frac{n}{2},\frac{x}{\sigma},\frac{a}{\sigma}\big)}{r^{n-1}\operatorname{Ed}\big(\frac{n}{2},\frac{b}{\sigma},\frac{r}{\sigma}\big)\operatorname{Ed}\big(\frac{n}{2},\frac{r}{\sigma},\frac{a}{\sigma}\big)}dy, & a \le x \le r \end{cases}$$

(1) $$\mathbf{P}_x\big(\ell(H,r)=0, Q_H^{(n)}=b\big) = \begin{cases} \dfrac{\operatorname{Ed}\big(\frac{n}{2},\frac{x}{\sigma},\frac{r}{\sigma}\big)}{\operatorname{Ed}\big(\frac{n}{2},\frac{b}{\sigma},\frac{r}{\sigma}\big)}, & r \le x \le b \\ 0, & a \le x \le r \end{cases}$$

3.3.7 (a) $\mathbf{E}_x\{e^{-\alpha H-\gamma\ell(H,r)}; Q_H^{(n)}=a\}$

$a \le x$, $x \le r$
$$= \frac{\theta 2^{n/2}\sigma^n e^{r^2/2\sigma^2}S\big(\frac{\alpha}{2\theta},\frac{n}{2},\frac{b^2}{2\sigma^2},\frac{x^2}{2\sigma^2}\big)+\gamma r^{n-1}S\big(\frac{\alpha}{2\theta},\frac{n}{2},\frac{b^2}{2\sigma^2},\frac{r^2}{2\sigma^2}\big)S\big(\frac{\alpha}{2\theta},\frac{n}{2},\frac{r^2}{2\sigma^2},\frac{x^2}{2\sigma^2}\big)}{\theta 2^{n/2}\sigma^n e^{r^2/2\sigma^2}S\big(\frac{\alpha}{2\theta},\frac{n}{2},\frac{b^2}{2\sigma^2},\frac{a^2}{2\sigma^2}\big)+\gamma r^{n-1}S\big(\frac{\alpha}{2\theta},\frac{n}{2},\frac{b^2}{2\sigma^2},\frac{r^2}{2\sigma^2}\big)S\big(\frac{\alpha}{2\theta},\frac{n}{2},\frac{r^2}{2\sigma^2},\frac{a^2}{2\sigma^2}\big)}$$

$r \le x$, $x \le b$
$$= \frac{\theta 2^{n/2}\sigma^n e^{r^2/2\sigma^2}S\big(\frac{\alpha}{2\theta},\frac{n}{2},\frac{b^2}{2\sigma^2},\frac{x^2}{2\sigma^2}\big)}{\theta 2^{n/2}\sigma^n e^{r^2/2\sigma^2}S\big(\frac{\alpha}{2\theta},\frac{n}{2},\frac{b^2}{2\sigma^2},\frac{a^2}{2\sigma^2}\big)+\gamma r^{n-1}S\big(\frac{\alpha}{2\theta},\frac{n}{2},\frac{b^2}{2\sigma^2},\frac{r^2}{2\sigma^2}\big)S\big(\frac{\alpha}{2\theta},\frac{n}{2},\frac{r^2}{2\sigma^2},\frac{a^2}{2\sigma^2}\big)}$$

3.3.7 (b) $\mathbf{E}_x\{e^{-\alpha H-\gamma\ell(H,r)}; Q_H^{(n)}=b\}$

$a \le x$, $x \le r$
$$= \frac{\theta 2^{n/2}\sigma^n e^{r^2/2\sigma^2}S\big(\frac{\alpha}{2\theta},\frac{n}{2},\frac{x^2}{2\sigma^2},\frac{a^2}{2\sigma^2}\big)}{\theta 2^{n/2}\sigma^n e^{r^2/2\sigma^2}S\big(\frac{\alpha}{2\theta},\frac{n}{2},\frac{b^2}{2\sigma^2},\frac{a^2}{2\sigma^2}\big)+\gamma r^{n-1}S\big(\frac{\alpha}{2\theta},\frac{n}{2},\frac{b^2}{2\sigma^2},\frac{r^2}{2\sigma^2}\big)S\big(\frac{\alpha}{2\theta},\frac{n}{2},\frac{r^2}{2\sigma^2},\frac{a^2}{2\sigma^2}\big)}$$

$r \le x$, $x \le b$
$$= \frac{\theta 2^{n/2}\sigma^n e^{r^2/2\sigma^2}S\big(\frac{\alpha}{2\theta},\frac{n}{2},\frac{x^2}{2\sigma^2},\frac{a^2}{2\sigma^2}\big)+\gamma r^{n-1}S\big(\frac{\alpha}{2\theta},\frac{n}{2},\frac{x^2}{2\sigma^2},\frac{r^2}{2\sigma^2}\big)S\big(\frac{\alpha}{2\theta},\frac{n}{2},\frac{r^2}{2\sigma^2},\frac{a^2}{2\sigma^2}\big)}{\theta 2^{n/2}\sigma^n e^{r^2/2\sigma^2}S\big(\frac{\alpha}{2\theta},\frac{n}{2},\frac{b^2}{2\sigma^2},\frac{a^2}{2\sigma^2}\big)+\gamma r^{n-1}S\big(\frac{\alpha}{2\theta},\frac{n}{2},\frac{b^2}{2\sigma^2},\frac{r^2}{2\sigma^2}\big)S\big(\frac{\alpha}{2\theta},\frac{n}{2},\frac{r^2}{2\sigma^2},\frac{a^2}{2\sigma^2}\big)}$$

3.3.8 (a) $\mathbf{E}_x\{e^{-\alpha H}; \ell(H,r)\in dy, Q_H^{(n)}=a\}$

$$= \exp\Big(-\frac{\theta 2^{n/2}\sigma^n e^{r^2/2\sigma^2}S\big(\frac{\alpha}{2\theta},\frac{n}{2},\frac{b^2}{2\sigma^2},\frac{a^2}{2\sigma^2}\big)y}{r^{n-1}S\big(\frac{\alpha}{2\theta},\frac{n}{2},\frac{b^2}{2\sigma^2},\frac{r^2}{2\sigma^2}\big)S\big(\frac{\alpha}{2\theta},\frac{n}{2},\frac{r^2}{2\sigma^2},\frac{a^2}{2\sigma^2}\big)}\Big)$$

$$\times \begin{cases} \dfrac{\theta 2^{n/2}\sigma^n e^{r^2/2\sigma^2}S\big(\frac{\alpha}{2\theta},\frac{n}{2},\frac{b^2}{2\sigma^2},\frac{x^2}{2\sigma^2}\big)}{r^{n-1}S\big(\frac{\alpha}{2\theta},\frac{n}{2},\frac{b^2}{2\sigma^2},\frac{r^2}{2\sigma^2}\big)S\big(\frac{\alpha}{2\theta},\frac{n}{2},\frac{r^2}{2\sigma^2},\frac{a^2}{2\sigma^2}\big)}dy, & r \le x \le b \\ \dfrac{\theta 2^{n/2}\sigma^n e^{r^2/2\sigma^2}S\big(\frac{\alpha}{2\theta},\frac{n}{2},\frac{x^2}{2\sigma^2},\frac{a^2}{2\sigma^2}\big)}{r^{n-1}S^2\big(\frac{\alpha}{2\theta},\frac{n}{2},\frac{r^2}{2\sigma^2},\frac{a^2}{2\sigma^2}\big)}dy, & a \le x \le r \end{cases}$$

8 $\quad Q_s^{(n)} = \sigma e^{-\theta s} R^{(n)}_{e^{2\theta s}-1} \quad n \ge 2, \quad \sigma, \theta > 0 \quad H = H_{a,b} = \min\{s : Q_s^{(n)} \notin (a,b)\}$

(1) $$\mathbf{E}_x\{e^{-\alpha H}; \ell(H,r) = 0, Q_H^{(n)} = a\} = \begin{cases} 0, & r \le x \le b \\ \dfrac{S(\frac{\alpha}{2\theta}, \frac{n}{2}, \frac{r^2}{2\sigma^2}, \frac{x^2}{2\sigma^2})}{S(\frac{\alpha}{2\theta}, \frac{n}{2}, \frac{r^2}{2\sigma^2}, \frac{a^2}{2\sigma^2})}, & a \le x \le r \end{cases}$$

3.3.8 (b) $$\mathbf{E}_x\{e^{-\alpha H}; \ell(H,r) \in dy, Q_H^{(n)} = b\}$$

$$= \exp\Big(-\frac{\theta 2^{n/2}\sigma^n e^{r^2/2\sigma^2} S(\frac{\alpha}{2\theta}, \frac{n}{2}, \frac{b^2}{2\sigma^2}, \frac{a^2}{2\sigma^2}) y}{r^{n-1} S(\frac{\alpha}{2\theta}, \frac{n}{2}, \frac{b^2}{2\sigma^2}, \frac{r^2}{2\sigma^2}) S(\frac{\alpha}{2\theta}, \frac{n}{2}, \frac{r^2}{2\sigma^2}, \frac{a^2}{2\sigma^2})}\Big)$$

$$\times \begin{cases} \dfrac{\theta 2^{n/2}\sigma^n e^{r^2/2\sigma^2} S(\frac{\alpha}{2\theta}, \frac{n}{2}, \frac{b^2}{2\sigma^2}, \frac{x^2}{2\sigma^2})}{r^{n-1} S^2(\frac{\alpha}{2\theta}, \frac{n}{2}, \frac{b^2}{2\sigma^2}, \frac{r^2}{2\sigma^2})} dy, & r \le x \le b \\ \dfrac{\theta 2^{n/2}\sigma^n e^{r^2/2\sigma^2} S(\frac{\alpha}{2\theta}, \frac{n}{2}, \frac{x^2}{2\sigma^2}, \frac{a^2}{2\sigma^2})}{r^{n-1} S(\frac{\alpha}{2\theta}, \frac{n}{2}, \frac{b^2}{2\sigma^2}, \frac{r^2}{2\sigma^2}) S(\frac{\alpha}{2\theta}, \frac{n}{2}, \frac{r^2}{2\sigma^2}, \frac{a^2}{2\sigma^2})} dy, & a \le x \le r \end{cases}$$

(1) $$\mathbf{E}_x\{e^{-\alpha H}; \ell(H,r) = 0, Q_H^{(n)} = b\} = \begin{cases} \dfrac{S(\frac{\alpha}{2\theta}, \frac{n}{2}, \frac{x^2}{2\sigma^2}, \frac{r^2}{2\sigma^2})}{S(\frac{\alpha}{2\theta}, \frac{n}{2}, \frac{b^2}{2\sigma^2}, \frac{r^2}{2\sigma^2})}, & r \le x \le b \\ 0, & a \le x \le r \end{cases}$$

3.4.1 $$\mathbf{E}_r \exp\Big(-\gamma \int_0^H \mathbb{I}_{[r,b]}(Q_s^{(n)}) ds\Big)$$

$$= \frac{2^{n/2}\sigma^n e^{r^2/2\sigma^2}\Big(S(\frac{\gamma}{2\theta}, \frac{n}{2}, \frac{b^2}{2\sigma^2}, \frac{r^2}{2\sigma^2}) + \mathrm{Ed}(\frac{n}{2}, \frac{r}{\sigma}, \frac{a}{\sigma})\Big)}{2^{n/2}\sigma^n e^{r^2/2\sigma^2} S(\frac{\gamma}{2\theta}, \frac{n}{2}, \frac{b^2}{2\sigma^2}, \frac{r^2}{2\sigma^2}) + r^n \,\mathrm{Ed}(\frac{n}{2}, \frac{r}{\sigma}, \frac{a}{\sigma}) C(\frac{\gamma}{2\theta}, \frac{n}{2}, \frac{r^2}{2\sigma^2}, \frac{b^2}{2\sigma^2})} =: Q$$

$$\mathbf{E}_x \exp\Big(-\gamma \int_0^H \mathbb{I}_{[r,b]}(Q_s^{(n)}) ds\Big)$$

$$= \begin{cases} \dfrac{\mathrm{Ed}(\frac{n}{2}, \frac{r}{\sigma}, \frac{x}{\sigma})}{\mathrm{Ed}(\frac{n}{2}, \frac{r}{\sigma}, \frac{a}{\sigma})} + \dfrac{\mathrm{Ed}(\frac{n}{2}, \frac{x}{\sigma}, \frac{a}{\sigma})}{\mathrm{Ed}(\frac{n}{2}, \frac{r}{\sigma}, \frac{a}{\sigma})} Q, & a \le x \le r \\ \dfrac{S(\frac{\gamma}{2\theta}, \frac{n}{2}, \frac{x^2}{2\sigma^2}, \frac{r^2}{2\sigma^2})}{S(\frac{\gamma}{2\theta}, \frac{n}{2}, \frac{b^2}{2\sigma^2}, \frac{r^2}{2\sigma^2})} + \dfrac{S(\frac{\gamma}{2\theta}, \frac{n}{2}, \frac{b^2}{2\sigma^2}, \frac{x^2}{2\sigma^2})}{S(\frac{\gamma}{2\theta}, \frac{n}{2}, \frac{b^2}{2\sigma^2}, \frac{r^2}{2\sigma^2})} Q, & r \le x \le b \end{cases}$$

3.4.2 (1) $$\mathbf{P}_x\Big(\int_0^H \mathbb{I}_{[r,b]}(Q_s^{(n)}) ds = 0\Big) = \begin{cases} \dfrac{\mathrm{Ed}(\frac{n}{2}, \frac{r}{\sigma}, \frac{x}{\sigma})}{\mathrm{Ed}(\frac{n}{2}, \frac{r}{\sigma}, \frac{a}{\sigma})}, & a \le x \le r \\ 0, & r \le x \le b \end{cases}$$

8 $Q_s^{(n)} = \sigma e^{-\theta s} R^{(n)}_{e^{2\theta s}-1} \quad n \ge 2, \quad \sigma, \theta > 0 \quad H = H_{a,b} = \min\{s : Q_s^{(n)} \notin (a,b)\}$

3.4.5 (a)

$$\mathbf{E}_x\Big\{\exp\Big(-\gamma\int_0^H \mathbb{I}_{[r,b]}\big(Q_s^{(n)}\big)ds\Big); Q_H^{(n)} = a\Big\}$$

$a \le x$, $x \le r$

$$= \frac{2^{n/2}\sigma^n e^{r^2/2\sigma^2} S\big(\frac{\gamma}{2\theta}, \frac{n}{2}, \frac{b^2}{2\sigma^2}, \frac{r^2}{2\sigma^2}\big) + r^n \operatorname{Ed}\big(\frac{n}{2}, \frac{r}{\sigma}, \frac{x}{\sigma}\big) C\big(\frac{\gamma}{2\theta}, \frac{n}{2}, \frac{r^2}{2\sigma^2}, \frac{b^2}{2\sigma^2}\big)}{2^{n/2}\sigma^n e^{r^2/2\sigma^2} S\big(\frac{\gamma}{2\theta}, \frac{n}{2}, \frac{b^2}{2\sigma^2}, \frac{r^2}{2\sigma^2}\big) + r^n \operatorname{Ed}\big(\frac{n}{2}, \frac{r}{\sigma}, \frac{a}{\sigma}\big) C\big(\frac{\gamma}{2\theta}, \frac{n}{2}, \frac{r^2}{2\sigma^2}, \frac{b^2}{2\sigma^2}\big)}$$

$r \le x$, $x \le b$

$$= \frac{2^{n/2}\sigma^n e^{r^2/2\sigma^2} S\big(\frac{\gamma}{2\theta}, \frac{n}{2}, \frac{b^2}{2\sigma^2}, \frac{x^2}{2\sigma^2}\big)}{2^{n/2}\sigma^n e^{r^2/2\sigma^2} S\big(\frac{\gamma}{2\theta}, \frac{n}{2}, \frac{b^2}{2\sigma^2}, \frac{r^2}{2\sigma^2}\big) + r^n \operatorname{Ed}\big(\frac{n}{2}, \frac{r}{\sigma}, \frac{a}{\sigma}\big) C\big(\frac{\gamma}{2\theta}, \frac{n}{2}, \frac{r^2}{2\sigma^2}, \frac{b^2}{2\sigma^2}\big)}$$

3.4.5 (b)

$$\mathbf{E}_x\Big\{\exp\Big(-\gamma\int_0^H \mathbb{I}_{[r,b]}\big(Q_s^{(n)}\big)ds\Big); Q_H^{(n)} = b\Big\}$$

$a \le x$, $x \le r$

$$= \frac{2^{n/2}\sigma^n e^{r^2/2\sigma^2} \operatorname{Ed}\big(\frac{n}{2}, \frac{x}{\sigma}, \frac{a}{\sigma}\big)}{2^{n/2}\sigma^n e^{r^2/2\sigma^2} S\big(\frac{\gamma}{2\theta}, \frac{n}{2}, \frac{b^2}{2\sigma^2}, \frac{r^2}{2\sigma^2}\big) + r^n \operatorname{Ed}\big(\frac{n}{2}, \frac{r}{\sigma}, \frac{a}{\sigma}\big) C\big(\frac{\gamma}{2\theta}, \frac{n}{2}, \frac{r^2}{2\sigma^2}, \frac{b^2}{2\sigma^2}\big)}$$

$r \le x$, $x \le b$

$$= \frac{2^{n/2}\sigma^n e^{r^2/2\sigma^2} S\big(\frac{\gamma}{2\theta}, \frac{n}{2}, \frac{x^2}{2\sigma^2}, \frac{r^2}{2\sigma^2}\big) + r^n \operatorname{Ed}\big(\frac{n}{2}, \frac{r}{\sigma}, \frac{a}{\sigma}\big) C\big(\frac{\gamma}{2\theta}, \frac{n}{2}, \frac{r^2}{2\sigma^2}, \frac{x^2}{2\sigma^2}\big)}{2^{n/2}\sigma^n e^{r^2/2\sigma^2} S\big(\frac{\gamma}{2\theta}, \frac{n}{2}, \frac{b^2}{2\sigma^2}, \frac{r^2}{2\sigma^2}\big) + r^n \operatorname{Ed}\big(\frac{n}{2}, \frac{r}{\sigma}, \frac{a}{\sigma}\big) C\big(\frac{\gamma}{2\theta}, \frac{n}{2}, \frac{r^2}{2\sigma^2}, \frac{b^2}{2\sigma^2}\big)}$$

3.4.6 (1)

$$\mathbf{P}_x\Big(\int_0^H \mathbb{I}_{[r,b]}\big(Q_s^{(n)}\big)ds = 0, Q_H^{(n)} = a\Big) = \begin{cases} \dfrac{\operatorname{Ed}\big(\frac{n}{2}, \frac{r}{\sigma}, \frac{x}{\sigma}\big)}{\operatorname{Ed}\big(\frac{n}{2}, \frac{r}{\sigma}, \frac{a}{\sigma}\big)}, & a \le x \le r \\ 0, & r \le x \le b \end{cases}$$

3.5.1

$$\mathbf{E}_r \exp\Big(-\gamma\int_0^H \mathbb{I}_{[a,r]}\big(Q_s^{(n)}\big)ds\Big)$$

$$= \frac{2^{n/2}\sigma^n e^{r^2/2\sigma^2}\Big(S\big(\frac{\gamma}{2\theta}, \frac{n}{2}, \frac{r^2}{2\sigma^2}, \frac{a^2}{2\sigma^2}\big) + \operatorname{Ed}\big(\frac{n}{2}, \frac{b}{\sigma}, \frac{r}{\sigma}\big)\Big)}{2^{n/2}\sigma^n e^{r^2/2\sigma^2} S\big(\frac{\gamma}{2\theta}, \frac{n}{2}, \frac{r^2}{2\sigma^2}, \frac{a^2}{2\sigma^2}\big) + r^n \operatorname{Ed}\big(\frac{n}{2}, \frac{b}{\sigma}, \frac{r}{\sigma}\big) C\big(\frac{\gamma}{2\theta}, \frac{n}{2}, \frac{r^2}{2\sigma^2}, \frac{a^2}{2\sigma^2}\big)} =: Q$$

$$\mathbf{E}_x \exp\Big(-\gamma\int_0^H \mathbb{I}_{[a,r]}\big(Q_s^{(n)}\big)ds\Big)$$

$$= \begin{cases} \dfrac{S\big(\frac{\gamma}{2\theta}, \frac{n}{2}, \frac{r^2}{2\sigma^2}, \frac{x^2}{2\sigma^2}\big)}{S\big(\frac{\gamma}{2\theta}, \frac{n}{2}, \frac{r^2}{2\sigma^2}, \frac{a^2}{2\sigma^2}\big)} + \dfrac{S\big(\frac{\gamma}{2\theta}, \frac{n}{2}, \frac{x^2}{2\sigma^2}, \frac{a^2}{2\sigma^2}\big)}{S\big(\frac{\gamma}{2\theta}, \frac{n}{2}, \frac{r^2}{2\sigma^2}, \frac{a^2}{2\sigma^2}\big)} Q, & a \le x \le r \\ \dfrac{\operatorname{Ed}\big(\frac{n}{2}, \frac{x}{\sigma}, \frac{r}{\sigma}\big)}{\operatorname{Ed}\big(\frac{n}{2}, \frac{b}{\sigma}, \frac{r}{\sigma}\big)} + \dfrac{\operatorname{Ed}\big(\frac{n}{2}, \frac{b}{\sigma}, \frac{x}{\sigma}\big)}{\operatorname{Ed}\big(\frac{n}{2}, \frac{b}{\sigma}, \frac{r}{\sigma}\big)} Q, & r \le x \le b \end{cases}$$

3.5.2 (1)

$$\mathbf{P}_x\Big(\int_0^H \mathbb{I}_{[a,r]}\big(Q_s^{(n)}\big)ds = 0\Big) = \begin{cases} 0, & a \le x \le r \\ \dfrac{\operatorname{Ed}\big(\frac{n}{2}, \frac{x}{\sigma}, \frac{r}{\sigma}\big)}{\operatorname{Ed}\big(\frac{n}{2}, \frac{b}{\sigma}, \frac{r}{\sigma}\big)}, & r \le x \le b \end{cases}$$

8 $\quad Q_s^{(n)} = \sigma e^{-\theta s} R_{e^{2\theta s}-1}^{(n)} \quad n \ge 2, \quad \sigma, \theta > 0 \quad H = H_{a,b} = \min\{s : Q_s^{(n)} \notin (a,b)\}$

3.5.5 (a)

$$\mathbf{E}_x\Big\{\exp\Big(-\gamma \int_0^H \mathbb{I}_{[a,r]}\big(Q_s^{(n)}\big)ds\Big); Q_H^{(n)} = a\Big\}$$

$a \le x$, $x \le r$

$$= \frac{2^{n/2}\sigma^n e^{r^2/2\sigma^2} S\big(\frac{\gamma}{2\theta}, \frac{n}{2}, \frac{r^2}{2\sigma^2}, \frac{x^2}{2\sigma^2}\big) + r^n \operatorname{Ed}\big(\frac{n}{2}, \frac{b}{\sigma}, \frac{r}{\sigma}\big) C\big(\frac{\gamma}{2\theta}, \frac{n}{2}, \frac{r^2}{2\sigma^2}, \frac{x^2}{2\sigma^2}\big)}{2^{n/2}\sigma^n e^{r^2/2\sigma^2} S\big(\frac{\gamma}{2\theta}, \frac{n}{2}, \frac{r^2}{2\sigma^2}, \frac{a^2}{2\sigma^2}\big) + r^n \operatorname{Ed}\big(\frac{n}{2}, \frac{b}{\sigma}, \frac{r}{\sigma}\big) C\big(\frac{\gamma}{2\theta}, \frac{n}{2}, \frac{r^2}{2\sigma^2}, \frac{a^2}{2\sigma^2}\big)}$$

$r \le x$, $x \le b$

$$= \frac{2^{n/2}\sigma^n e^{r^2/2\sigma^2} \operatorname{Ed}\big(\frac{n}{2}, \frac{b}{\sigma}, \frac{x}{\sigma}\big)}{2^{n/2}\sigma^n e^{r^2/2\sigma^2} S\big(\frac{\gamma}{2\theta}, \frac{n}{2}, \frac{r^2}{2\sigma^2}, \frac{a^2}{2\sigma^2}\big) + r^n \operatorname{Ed}\big(\frac{n}{2}, \frac{b}{\sigma}, \frac{r}{\sigma}\big) C\big(\frac{\gamma}{2\theta}, \frac{n}{2}, \frac{r^2}{2\sigma^2}, \frac{a^2}{2\sigma^2}\big)}$$

3.5.5 (b)

$$\mathbf{E}_x\Big\{\exp\Big(-\gamma \int_0^H \mathbb{I}_{[a,r]}\big(Q_s^{(n)}\big)ds\Big); Q_H^{(n)} = b\Big\}$$

$a \le x$, $x \le r$

$$= \frac{2^{n/2}\sigma^n e^{r^2/2\sigma^2} S\big(\frac{\gamma}{2\theta}, \frac{n}{2}, \frac{x^2}{2\sigma^2}, \frac{a^2}{2\sigma^2}\big)}{2^{n/2}\sigma^n e^{r^2/2\sigma^2} S\big(\frac{\gamma}{2\theta}, \frac{n}{2}, \frac{r^2}{2\sigma^2}, \frac{a^2}{2\sigma^2}\big) + r^n \operatorname{Ed}\big(\frac{n}{2}, \frac{b}{\sigma}, \frac{r}{\sigma}\big) C\big(\frac{\gamma}{2\theta}, \frac{n}{2}, \frac{r^2}{2\sigma^2}, \frac{a^2}{2\sigma^2}\big)}$$

$r \le x$, $x \le b$

$$= \frac{2^{n/2}\sigma^n e^{r^2/2\sigma^2} S\big(\frac{\gamma}{2\theta}, \frac{n}{2}, \frac{r^2}{2\sigma^2}, \frac{a^2}{2\sigma^2}\big) + r^n \operatorname{Ed}\big(\frac{n}{2}, \frac{x}{\sigma}, \frac{r}{\sigma}\big) C\big(\frac{\gamma}{2\theta}, \frac{n}{2}, \frac{r^2}{2\sigma^2}, \frac{a^2}{2\sigma^2}\big)}{2^{n/2}\sigma^n e^{r^2/2\sigma^2} S\big(\frac{\gamma}{2\theta}, \frac{n}{2}, \frac{r^2}{2\sigma^2}, \frac{a^2}{2\sigma^2}\big) + r^n \operatorname{Ed}\big(\frac{n}{2}, \frac{b}{\sigma}, \frac{r}{\sigma}\big) C\big(\frac{\gamma}{2\theta}, \frac{n}{2}, \frac{r^2}{2\sigma^2}, \frac{a^2}{2\sigma^2}\big)}$$

3.5.6 (1)

$$\mathbf{P}_x\Big(\int_0^H \mathbb{I}_{[a,r]}\big(Q_s^{(n)}\big)ds = 0, Q_H^{(n)} = b\Big) = \begin{cases} 0, & a \le x \le r \\ \dfrac{\operatorname{Ed}\big(\frac{n}{2}, \frac{x}{\sigma}, \frac{r}{\sigma}\big)}{\operatorname{Ed}\big(\frac{n}{2}, \frac{b}{\sigma}, \frac{r}{\sigma}\big)}, & r \le x \le b \end{cases}$$

3.6.1

$$\mathbf{E}_r \exp\Big(-\int_0^H \Big(p\mathbb{I}_{[a,r]}\big(Q_s^{(n)}\big) + q\mathbb{I}_{[r,b]}\big(Q_s^{(n)}\big)\Big)ds\Big) =: Q$$

$$= \frac{2^{n/2}\sigma^n r^{-n} e^{r^2/2\sigma^2}\Big(S\big(\frac{q}{2\theta}, \frac{n}{2}, \frac{b^2}{2\sigma^2}, \frac{r^2}{2\sigma^2}\big) + S\big(\frac{p}{2\theta}, \frac{n}{2}, \frac{r^2}{2\sigma^2}, \frac{a^2}{2\sigma^2}\big)\Big)}{S\big(\frac{q}{2\theta}, \frac{n}{2}, \frac{b^2}{2\sigma^2}, \frac{r^2}{2\sigma^2}\big) C\big(\frac{p}{2\theta}, \frac{n}{2}, \frac{r^2}{2\sigma^2}, \frac{a^2}{2\sigma^2}\big) + C\big(\frac{q}{2\theta}, \frac{n}{2}, \frac{r^2}{2\sigma^2}, \frac{b^2}{2\sigma^2}\big) S\big(\frac{p}{2\theta}, \frac{n}{2}, \frac{r^2}{2\sigma^2}, \frac{a^2}{2\sigma^2}\big)}$$

$$\mathbf{E}_x \exp\Big(-\int_0^H \Big(p\mathbb{I}_{[a,r]}\big(Q_s^{(n)}\big) + q\mathbb{I}_{[r,b]}\big(Q_s^{(n)}\big)\Big)ds\Big)$$

$$= \begin{cases} \dfrac{S\big(\frac{q}{2\theta}, \frac{n}{2}, \frac{x^2}{2\sigma^2}, \frac{r^2}{2\sigma^2}\big)}{S\big(\frac{q}{2\theta}, \frac{n}{2}, \frac{b^2}{2\sigma^2}, \frac{r^2}{2\sigma^2}\big)} + \dfrac{S\big(\frac{q}{2\theta}, \frac{n}{2}, \frac{b^2}{2\sigma^2}, \frac{x^2}{2\sigma^2}\big)}{S\big(\frac{q}{2\theta}, \frac{n}{2}, \frac{b^2}{2\sigma^2}, \frac{r^2}{2\sigma^2}\big)} Q, & r \le x \le b \\ \dfrac{S\big(\frac{p}{2\theta}, \frac{n}{2}, \frac{r^2}{2\sigma^2}, \frac{x^2}{2\sigma^2}\big)}{S\big(\frac{p}{2\theta}, \frac{n}{2}, \frac{r^2}{2\sigma^2}, \frac{a^2}{2\sigma^2}\big)} + \dfrac{S\big(\frac{p}{2\theta}, \frac{n}{2}, \frac{x^2}{2\sigma^2}, \frac{a^2}{2\sigma^2}\big)}{S\big(\frac{p}{2\theta}, \frac{n}{2}, \frac{r^2}{2\sigma^2}, \frac{a^2}{2\sigma^2}\big)} Q, & a \le x \le r \end{cases}$$

8 $Q_s^{(n)}=\sigma e^{-\theta s}R^{(n)}_{e^{2\theta s}-1}\quad n\ge 2,\quad \sigma,\theta>0\quad H=H_{a,b}=\min\{s: Q_s^{(n)}\notin(a,b)\}$

3.6.5 (a) $\mathbf{E}_x\Big\{\exp\Big(-\int_0^H\Big(p\mathbb{1}_{[a,r]}\big(Q_s^{(n)}\big)+q\mathbb{1}_{[r,b]}\big(Q_s^{(n)}\big)\Big)ds\Big);Q_H^{(n)}=a\Big\}$

$a\le x$, $x\le r$

$$=\frac{S(\frac{q}{2\theta},\frac n2,\frac{b^2}{2\sigma^2},\frac{r^2}{2\sigma^2})C(\frac{p}{2\theta},\frac n2,\frac{r^2}{2\sigma^2},\frac{x^2}{2\sigma^2})+C(\frac{q}{2\theta},\frac n2,\frac{r^2}{2\sigma^2},\frac{b^2}{2\sigma^2})S(\frac{p}{2\theta},\frac n2,\frac{r^2}{2\sigma^2},\frac{x^2}{2\sigma^2})}{S(\frac{q}{2\theta},\frac n2,\frac{b^2}{2\sigma^2},\frac{r^2}{2\sigma^2})C(\frac{p}{2\theta},\frac n2,\frac{r^2}{2\sigma^2},\frac{a^2}{2\sigma^2})+C(\frac{q}{2\theta},\frac n2,\frac{r^2}{2\sigma^2},\frac{b^2}{2\sigma^2})S(\frac{p}{2\theta},\frac n2,\frac{r^2}{2\sigma^2},\frac{a^2}{2\sigma^2})}$$

$r\le x$, $x\le b$

$$=\frac{2^{n/2}\sigma^n r^{-n}e^{r^2/2\sigma^2}S(\frac{q}{2\theta},\frac n2,\frac{b^2}{2\sigma^2},\frac{x^2}{2\sigma^2})}{S(\frac{q}{2\theta},\frac n2,\frac{b^2}{2\sigma^2},\frac{r^2}{2\sigma^2})C(\frac{p}{2\theta},\frac n2,\frac{r^2}{2\sigma^2},\frac{a^2}{2\sigma^2})+C(\frac{q}{2\theta},\frac n2,\frac{r^2}{2\sigma^2},\frac{b^2}{2\sigma^2})S(\frac{p}{2\theta},\frac n2,\frac{r^2}{2\sigma^2},\frac{a^2}{2\sigma^2})}$$

3.6.5 (b) $\mathbf{E}_x\Big\{\exp\Big(-\int_0^H\Big(p\mathbb{1}_{[a,r]}\big(Q_s^{(n)}\big)+q\mathbb{1}_{[r,b]}\big(Q_s^{(n)}\big)\Big)ds\Big);Q_H^{(n)}=b\Big\}$

$a\le x$, $x\le r$

$$=\frac{2^{n/2}\sigma^n r^{-n}e^{r^2/2\sigma^2}S(\frac{p}{2\theta},\frac n2,\frac{x^2}{2\sigma^2},\frac{a^2}{2\sigma^2})}{S(\frac{q}{2\theta},\frac n2,\frac{b^2}{2\sigma^2},\frac{r^2}{2\sigma^2})C(\frac{p}{2\theta},\frac n2,\frac{r^2}{2\sigma^2},\frac{a^2}{2\sigma^2})+C(\frac{q}{2\theta},\frac n2,\frac{r^2}{2\sigma^2},\frac{b^2}{2\sigma^2})S(\frac{p}{2\theta},\frac n2,\frac{r^2}{2\sigma^2},\frac{a^2}{2\sigma^2})}$$

$r\le x$, $x\le b$

$$=\frac{S(\frac{q}{2\theta},\frac n2,\frac{x^2}{2\sigma^2},\frac{r^2}{2\sigma^2})C(\frac{q}{2\theta},\frac n2,\frac{r^2}{2\sigma^2},\frac{a^2}{2\sigma^2})+C(\frac{q}{2\theta},\frac n2,\frac{r^2}{2\sigma^2},\frac{x^2}{2\sigma^2})S(\frac{q}{2\theta},\frac n2,\frac{r^2}{2\sigma^2},\frac{a^2}{2\sigma^2})}{S(\frac{q}{2\theta},\frac n2,\frac{b^2}{2\sigma^2},\frac{r^2}{2\sigma^2})C(\frac{q}{2\theta},\frac n2,\frac{r^2}{2\sigma^2},\frac{a^2}{2\sigma^2})+C(\frac{q}{2\theta},\frac n2,\frac{r^2}{2\sigma^2},\frac{b^2}{2\sigma^2})S(\frac{q}{2\theta},\frac n2,\frac{r^2}{2\sigma^2},\frac{a^2}{2\sigma^2})}$$

3.9.5 (a) $\mathbf{E}_x\Big\{\exp\Big(-\frac{\theta(\gamma^2-1)}{4\sigma^2}\int_0^H(Q_s^{(n)})^2ds\Big);\ Q_H^{(n)}=a\Big\}$

$$=\frac{e^{x^2(1-\gamma)/4\sigma^2}S(\frac n4-\frac{n}{4q},\frac n2,\frac{b^2\gamma}{2\sigma^2},\frac{x^2\gamma}{2\sigma^2})}{e^{a^2(1-\gamma)/4\sigma^2}S(\frac n4-\frac{n}{4q},\frac n2,\frac{b^2\gamma}{2\sigma^2},\frac{a^2\gamma}{2\sigma^2})}$$

3.9.5 (b) $\mathbf{E}_x\Big\{\exp\Big(-\frac{\theta(\gamma^2-1)}{4\sigma^2}\int_0^H(Q_s^{(n)})^2ds\Big);Q_H^{(n)}=b\Big\}$

$$=\frac{e^{x^2(1-\gamma)/4\sigma^2}S(\frac n4-\frac{n}{4q},\frac n2,\frac{x^2\gamma}{2\sigma^2},\frac{a^2\gamma}{2\sigma^2})}{e^{b^2(1-\gamma)/4\sigma^2}S(\frac n4-\frac{n}{4q},\frac n2,\frac{b^2\gamma}{2\sigma^2},\frac{a^2\gamma}{2\sigma^2})}$$

3.12.7 (a) $\mathbf{E}_x\{e^{-\alpha H-\gamma\check H(H)};Q_H^{(n)}=a\}=\displaystyle\int_x^b\frac{2^{n/2}\sigma^{n-2}y^{1-n}e^{y^2/2\sigma^2}S(\frac{\alpha+\gamma}{2\theta},\frac n2,\frac{x^2}{2\sigma^2},\frac{a^2}{2\sigma^2})}{S(\frac{\alpha}{2\theta},\frac n2,\frac{y^2}{2\sigma^2},\frac{a^2}{2\sigma^2})S(\frac{\alpha+\gamma}{2\theta},\frac n2,\frac{y^2}{2\sigma^2},\frac{a^2}{2\sigma^2})}dy$

3.12.7 (b) $\mathbf{E}_x\{e^{-\alpha H-\gamma\hat H(H)};Q_H^{(n)}=b\}=\displaystyle\int_a^x\frac{2^{n/2}\sigma^{n-2}y^{1-n}e^{y^2/2\sigma^2}S(\frac{\alpha+\gamma}{2\theta},\frac n2,\frac{b^2}{2\sigma^2},\frac{x^2}{2\sigma^2})}{S(\frac{\alpha}{2\theta},\frac n2,\frac{b^2}{2\sigma^2},\frac{y^2}{2\sigma^2})S(\frac{\alpha+\gamma}{2\theta},\frac n2,\frac{b^2}{2\sigma^2},\frac{y^2}{2\sigma^2})}dy$

8 $Q_s^{(n)} = \sigma e^{-\theta s} R^{(n)}_{e^{2\theta s}-1} \quad n \ge 2, \quad \sigma, \theta > 0 \quad H = H_{a,b} = \min\{s : Q_s^{(n)} \notin (a,b)\}$

3.20.5 (a)

$$\mathbf{E}_x\Big\{\exp\Big(-p^2\sigma^2\theta\int_0^H \frac{ds}{(Q_s^{(n)})^2}\Big);\ Q_H^{(n)} = a\Big\}$$

$$= \frac{x^{1-n/2+\sqrt{(n-2)^2+4p^2}/2} S\Big(\sqrt{\frac{(n-2)^2}{16}+\frac{p^2}{4}}+\frac{2-n}{4}, \sqrt{\frac{(n-2)^2}{4}+p^2}+1, \frac{b^2}{2\sigma^2}, \frac{x^2}{2\sigma^2}\Big)}{a^{1-n/2+\sqrt{(n-2)^2+4p^2}/2} S\Big(\sqrt{\frac{(n-2)^2}{16}+\frac{p^2}{4}}+\frac{2-n}{4}, \sqrt{\frac{(n-2)^2}{4}+p^2}+1, \frac{b^2}{2\sigma^2}, \frac{a^2}{2\sigma^2}\Big)}$$

3.20.5 (b)

$$\mathbf{E}_x\Big\{\exp\Big(-p^2\sigma^2\theta\int_0^H \frac{ds}{(Q_s^{(n)})^2}\Big);\ Q_H^{(n)} = b\Big\}$$

$$= \frac{x^{1-n/2+\sqrt{(n-2)^2+4p^2}/2} S\Big(\sqrt{\frac{(n-2)^2}{16}+\frac{p^2}{4}}+\frac{2-n}{4}, \sqrt{\frac{(n-2)^2}{4}+p^2}+1, \frac{x^2}{2\sigma^2}, \frac{a^2}{2\sigma^2}\Big)}{b^{1-n/2+\sqrt{(n-2)^2+4p^2}/2} S\Big(\sqrt{\frac{(n-2)^2}{16}+\frac{p^2}{4}}+\frac{2-n}{4}, \sqrt{\frac{(n-2)^2}{4}+p^2}+1, \frac{b^2}{2\sigma^2}, \frac{a^2}{2\sigma^2}\Big)}$$

3.21.5 (a)

$$\mathbf{E}_x\Big\{\exp\Big(-\int_0^H\Big(\frac{p^2\sigma^2\theta}{(Q_s^{(n)})^2}+\frac{\theta(q^2-1)}{4\sigma^2}(Q_s^{(n)})^2\Big)ds\Big);\ Q_H^{(n)} = a\Big\}\frac{x^{n/2-1}}{a^{n/2-1}}$$

$$= \frac{x^{\sqrt{(n-2)^2+4p^2}/2} e^{x^2(1-q)/4\sigma^2} S\Big(\sqrt{\frac{(n-2)^2}{16}+\frac{p^2}{4}}+\frac{2q-n}{4q}, \sqrt{\frac{(n-2)^2}{4}+p^2}+1, \frac{b^2q}{2\sigma^2}, \frac{x^2q}{2\sigma^2}\Big)}{a^{\sqrt{(n-2)^2+4p^2}/2} e^{a^2(1-q)/4\sigma^2} S\Big(\sqrt{\frac{(n-2)^2}{16}+\frac{p^2}{4}}+\frac{2q-n}{4q}, \sqrt{\frac{(n-2)^2}{4}+p^2}+1, \frac{b^2q}{2\sigma^2}, \frac{a^2q}{2\sigma^2}\Big)}$$

3.21.5 (b)

$$\mathbf{E}_x\Big\{\exp\Big(-\int_0^H\Big(\frac{p^2\sigma^2\theta}{(Q_s^{(n)})^2}+\frac{\theta(q^2-1)}{4\sigma^2}(Q_s^{(n)})^2\Big)ds\Big);\ Q_H^{(n)} = b\Big\}\frac{x^{n/2-1}}{b^{n/2-1}}$$

$$= \frac{x^{\sqrt{(n-2)^2+4p^2}/2} e^{x^2(1-q)/4\sigma^2} S\Big(\sqrt{\frac{(n-2)^2}{16}+\frac{p^2}{4}}+\frac{2q-n}{4q}, \sqrt{\frac{(n-2)^2}{4}+p^2}+1, \frac{x^2q}{2\sigma^2}, \frac{a^2q}{2\sigma^2}\Big)}{b^{\sqrt{(n-2)^2+4p^2}/2} e^{b^2(1-q)/4\sigma^2} S\Big(\sqrt{\frac{(n-2)^2}{16}+\frac{p^2}{4}}+\frac{2q-n}{4q}, \sqrt{\frac{(n-2)^2}{4}+p^2}+1, \frac{b^2q}{2\sigma^2}, \frac{a^2q}{2\sigma^2}\Big)}$$

3.27.1

$$\mathbf{E}_x\Big\{\exp\Big(-\int_0^H\big(p\mathbb{I}_{[a,r)}(Q_s^{(n)})+q\mathbb{I}_{[r,b]}(Q_s^{(n)})\big)ds\Big);\ell(H,r)\in dy\Big\}$$

$$= \theta 2^{n/2}\sigma^n e^{r^2/2\sigma^2}\exp\Big(-y\theta r\Big(\frac{C(\frac{p}{2\theta},\frac{n}{2},\frac{r^2}{2\sigma^2},\frac{a^2}{2\sigma^2})}{S(\frac{p}{2\theta},\frac{n}{2},\frac{r^2}{2\sigma^2},\frac{a^2}{2\sigma^2})}+\frac{C(\frac{q}{2\theta},\frac{n}{2},\frac{r^2}{2\sigma^2},\frac{b^2}{2\sigma^2})}{S(\frac{q}{2\theta},\frac{n}{2},\frac{b^2}{2\sigma^2},\frac{r^2}{2\sigma^2})}\Big)\Big)$$

$$\times\begin{cases}\dfrac{S(\frac{q}{2\theta},\frac{n}{2},\frac{b^2}{2\sigma^2},\frac{x^2}{2\sigma^2})\Big[S(\frac{q}{2\theta},\frac{n}{2},\frac{b^2}{2\sigma^2},\frac{r^2}{2\sigma^2})+S(\frac{p}{2\theta},\frac{n}{2},\frac{r^2}{2\sigma^2},\frac{a^2}{2\sigma^2})\Big]}{r^{n-1}S^2(\frac{q}{2\theta},\frac{n}{2},\frac{b^2}{2\sigma^2},\frac{r^2}{2\sigma^2})S(\frac{p}{2\theta},\frac{n}{2},\frac{r^2}{2\sigma^2},\frac{a^2}{2\sigma^2})}dy, & r\le x\le b\\[2ex] \dfrac{S(\frac{p}{2\theta},\frac{n}{2},\frac{x^2}{2\sigma^2},\frac{a^2}{2\sigma^2})\Big[S(\frac{q}{2\theta},\frac{n}{2},\frac{b^2}{2\sigma^2},\frac{r^2}{2\sigma^2})+S(\frac{p}{2\theta},\frac{n}{2},\frac{r^2}{2\sigma^2},\frac{a^2}{2\sigma^2})\Big]}{r^{n-1}S(\frac{q}{2\theta},\frac{n}{2},\frac{b^2}{2\sigma^2},\frac{r^2}{2\sigma^2})S^2(\frac{p}{2\theta},\frac{n}{2},\frac{r^2}{2\sigma^2},\frac{a^2}{2\sigma^2})}dy, & a\le x\le r\end{cases}$$

8 $Q_s^{(n)} = \sigma e^{-\theta s} R^{(n)}_{e^{2\theta s}-1} \quad n \ge 2, \quad \sigma, \theta > 0 \quad H = H_{a,b} = \min\{s : Q_s^{(n)} \notin (a,b)\}$

3.27.5 (a)

$$\mathbf{E}_x\Big\{\exp\Big(-\int_0^H \big(p\mathbb{1}_{[a,r)}\big(Q_s^{(n)}\big) + q\mathbb{1}_{[r,b]}\big(Q_s^{(n)}\big)\big)ds\Big); \ell(H,r) \in dy, Q_H^{(n)} = a\Big\}$$

$$= \exp\Big(-y\theta r\Big(\frac{C\big(\frac{p}{2\theta}, \frac{n}{2}, \frac{r^2}{2\sigma^2}, \frac{a^2}{2\sigma^2}\big)}{S\big(\frac{p}{2\theta}, \frac{n}{2}, \frac{r^2}{2\sigma^2}, \frac{a^2}{2\sigma^2}\big)} + \frac{C\big(\frac{q}{2\theta}, \frac{n}{2}, \frac{r^2}{2\sigma^2}, \frac{b^2}{2\sigma^2}\big)}{S\big(\frac{q}{2\theta}, \frac{n}{2}, \frac{b^2}{2\sigma^2}, \frac{r^2}{2\sigma^2}\big)}\Big)\Big)$$

$$\times \begin{cases} \dfrac{\theta 2^{n/2}\sigma^n e^{r^2/2\sigma^2} S\big(\frac{q}{2\theta}, \frac{n}{2}, \frac{b^2}{2\sigma^2}, \frac{x^2}{2\sigma^2}\big)}{r^{n-1} S\big(\frac{q}{2\theta}, \frac{n}{2}, \frac{b^2}{2\sigma^2}, \frac{r^2}{2\sigma^2}\big) S\big(\frac{p}{2\theta}, \frac{n}{2}, \frac{r^2}{2\sigma^2}, \frac{a^2}{2\sigma^2}\big)} dy, & r \le x \le b \\ \dfrac{\theta 2^{n/2}\sigma^n e^{r^2/2\sigma^2} S\big(\frac{p}{2\theta}, \frac{n}{2}, \frac{x^2}{2\sigma^2}, \frac{a^2}{2\sigma^2}\big)}{r^{n-1} S^2\big(\frac{p}{2\theta}, \frac{n}{2}, \frac{r^2}{2\sigma^2}, \frac{a^2}{2\sigma^2}\big)} dy, & a \le x \le r \end{cases}$$

(1)

$$\mathbf{E}_x\Big\{\exp\Big(-\int_0^H \big(p\mathbb{1}_{[a,r)}\big(Q_s^{(n)}\big) + q\mathbb{1}_{[r,b]}\big(Q_s^{(n)}\big)\big)ds\Big); \ell(H_z,r) = 0, Q_H^{(n)} = a\Big\}$$

$$= \frac{S\big(\frac{p}{2\theta}, \frac{n}{2}, \frac{r^2}{2\sigma^2}, \frac{x^2}{2\sigma^2}\big)}{S\big(\frac{p}{2\theta}, \frac{n}{2}, \frac{r^2}{2\sigma^2}, \frac{a^2}{2\sigma^2}\big)}, \quad a \le x \le r$$

3.27.5 (b)

$$\mathbf{E}_x\Big\{\exp\Big(-\int_0^H \big(p\mathbb{1}_{[a,r)}\big(Q_s^{(n)}\big) + q\mathbb{1}_{[r,b]}\big(Q_s^{(n)}\big)\big)ds\Big); \ell(H_z,r) \in dy, Q_H^{(n)} = b\Big\}$$

$$= \exp\Big(-y\theta r\Big(\frac{C\big(\frac{p}{2\theta}, \frac{n}{2}, \frac{r^2}{2\sigma^2}, \frac{a^2}{2\sigma^2}\big)}{S\big(\frac{p}{2\theta}, \frac{n}{2}, \frac{r^2}{2\sigma^2}, \frac{a^2}{2\sigma^2}\big)} + \frac{C\big(\frac{q}{2\theta}, \frac{n}{2}, \frac{r^2}{2\sigma^2}, \frac{b^2}{2\sigma^2}\big)}{S\big(\frac{q}{2\theta}, \frac{n}{2}, \frac{b^2}{2\sigma^2}, \frac{r^2}{2\sigma^2}\big)}\Big)\Big)$$

$$\times \begin{cases} \dfrac{\theta 2^{n/2}\sigma^n e^{r^2/2\sigma^2} S\big(\frac{q}{2\theta}, \frac{n}{2}, \frac{b^2}{2\sigma^2}, \frac{x^2}{2\sigma^2}\big)}{r^{n-1} S^2\big(\frac{q}{2\theta}, \frac{n}{2}, \frac{b^2}{2\sigma^2}, \frac{r^2}{2\sigma^2}\big)} dy, & r \le x \le b \\ \dfrac{\theta 2^{n/2}\sigma^n e^{r^2/2\sigma^2} S\big(\frac{p}{2\theta}, \frac{n}{2}, \frac{x^2}{2\sigma^2}, \frac{a^2}{2\sigma^2}\big)}{r^{n-1} S\big(\frac{q}{2\theta}, \frac{n}{2}, \frac{b^2}{2\sigma^2}, \frac{r^2}{2\sigma^2}\big) S\big(\frac{p}{2\theta}, \frac{n}{2}, \frac{r^2}{2\sigma^2}, \frac{a^2}{2\sigma^2}\big)} dy, & a \le x \le r \end{cases}$$

(1)

$$\mathbf{E}_x\Big\{\exp\Big(-\int_0^H \big(p\mathbb{1}_{[a,r)}\big(Q_s^{(n)}\big) + q\mathbb{1}_{[r,b]}\big(Q_s^{(n)}\big)\big)ds\Big); \ell(H,r) = 0, Q_H^{(n)} = b\Big\}$$

$$= \frac{S\big(\frac{q}{2\theta}, \frac{n}{2}, \frac{x^2}{2\sigma^2}, \frac{r^2}{2\sigma^2}\big)}{S\big(\frac{q}{2\theta}, \frac{n}{2}, \frac{b^2}{2\sigma^2}, \frac{r^2}{2\sigma^2}\big)}, \quad r \le x \le b$$

8 $Q_s^{(n)} = \sigma e^{-\theta s} R^{(n)}_{e^{2\theta s}-1}$ $n \ge 2$, $\sigma, \theta > 0$ $\varrho = \varrho(v,z) = \min\{s : \ell(s,z) = v\}$

4. Stopping at inverse local time

4.0.1 $0 \le x$

$$\mathbf{E}_x e^{-\alpha\varrho} = \begin{cases} \dfrac{M(\frac{\alpha}{2\theta}, \frac{n}{2}, \frac{x^2}{2\sigma^2})}{M(\frac{\alpha}{2\theta}, \frac{n}{2}, \frac{z^2}{2\sigma^2})} \exp\Big(-\dfrac{\theta 2^{n/2}\Gamma(\frac{n}{2})\sigma^n z^{1-n} e^{z^2/2\sigma^2} v}{\Gamma(\frac{\alpha}{2\theta}) M(\frac{\alpha}{2\theta}, \frac{n}{2}, \frac{z^2}{2\sigma^2}) U(\frac{\alpha}{2\theta}, \frac{n}{2}, \frac{z^2}{2\sigma^2})}\Big), & x \le z \\ \dfrac{U(\frac{\alpha}{2\theta}, \frac{n}{2}, \frac{x^2}{2\sigma^2})}{U(\frac{\alpha}{2\theta}, \frac{n}{2}, \frac{z^2}{2\sigma^2})} \exp\Big(-\dfrac{\theta 2^{n/2}\Gamma(\frac{n}{2})\sigma^n z^{1-n} e^{z^2/2\sigma^2} v}{\Gamma(\frac{\alpha}{2\theta}) M(\frac{\alpha}{2\theta}, \frac{n}{2}, \frac{z^2}{2\sigma^2}) U(\frac{\alpha}{2\theta}, \frac{n}{2}, \frac{z^2}{2\sigma^2})}\Big), & z \le x \end{cases}$$

4.1.2 $z \le y$

$$\mathbf{P}_x\Big(\sup_{0\le s\le \varrho} Q_s^{(n)} < y\Big) = \begin{cases} \exp\Big(-\dfrac{\theta 2^{n/2}\sigma^n e^{z^2/2\sigma^2} v}{z^{n-1}\operatorname{Ed}(\frac{n}{2}, \frac{y}{\sigma}, \frac{z}{\sigma})}\Big), & 0 \le x \le z \\ \dfrac{\operatorname{Ed}(\frac{n}{2}, \frac{y}{\sigma}, \frac{x}{\sigma})}{\operatorname{Ed}(\frac{n}{2}, \frac{y}{\sigma}, \frac{z}{\sigma})} \exp\Big(-\dfrac{\theta 2^{n/2}\sigma^n e^{z^2/2\sigma^2} v}{z^{n-1}\operatorname{Ed}(\frac{n}{2}, \frac{y}{\sigma}, \frac{z}{\sigma})}\Big), & z \le x \le y \end{cases}$$

4.1.4 $z \le y$

$$\mathbf{E}_x\{e^{-\alpha\varrho}; \sup_{0\le s\le\varrho} Q_s^{(n)} < y\}$$

$$= \begin{cases} \dfrac{M(\frac{\alpha}{2\theta}, \frac{n}{2}, \frac{x^2}{2\sigma^2})}{M(\frac{\alpha}{2\theta}, \frac{n}{2}, \frac{z^2}{2\sigma^2})} \exp\Big(-\dfrac{\theta 2^{n/2}\sigma^n z^{1-n} e^{z^2/2\sigma^2} M(\frac{\alpha}{2\theta}, \frac{n}{2}, \frac{y^2}{2\sigma^2}) v}{M(\frac{\alpha}{2\theta}, \frac{n}{2}, \frac{z^2}{2\sigma^2}) S(\frac{\alpha}{2\theta}, \frac{n}{2}, \frac{y^2}{2\sigma^2}, \frac{z^2}{2\sigma^2})}\Big), & 0 \le x \le z \\ \dfrac{S(\frac{\alpha}{2\theta}, \frac{n}{2}, \frac{y^2}{2\sigma^2}, \frac{x^2}{2\sigma^2})}{S(\frac{\alpha}{2\theta}, \frac{n}{2}, \frac{y^2}{2\sigma^2}, \frac{z^2}{2\sigma^2})} \exp\Big(-\dfrac{\theta 2^{n/2}\sigma^n z^{1-n} e^{z^2/2\sigma^2} M(\frac{\alpha}{2\theta}, \frac{n}{2}, \frac{y^2}{2\sigma^2}) v}{M(\frac{\alpha}{2\theta}, \frac{n}{2}, \frac{z^2}{2\sigma^2}) S(\frac{\alpha}{2\theta}, \frac{n}{2}, \frac{y^2}{2\sigma^2}, \frac{z^2}{2\sigma^2})}\Big), & z \le x \le y \end{cases}$$

4.2.2 $0 \le y$

$$\mathbf{P}_x\Big(\inf_{0\le s\le\varrho} Q_s^{(n)} > y\Big) = \begin{cases} \dfrac{\operatorname{Ed}(\frac{n}{2}, \frac{x}{\sigma}, \frac{y}{\sigma})}{\operatorname{Ed}(\frac{n}{2}, \frac{z}{\sigma}, \frac{y}{\sigma})} \exp\Big(-\dfrac{\theta 2^{n/2}\sigma^n e^{z^2/2\sigma^2} v}{z^{n-1}\operatorname{Ed}(\frac{n}{2}, \frac{z}{\sigma}, \frac{y}{\sigma})}\Big), & y \le x \le z \\ \exp\Big(-\dfrac{\theta 2^{n/2}\sigma^n e^{z^2/2\sigma^2} v}{z^{n-1}\operatorname{Ed}(\frac{n}{2}, \frac{z}{\sigma}, \frac{y}{\sigma})}\Big), & z \le x \end{cases}$$

4.2.4 $y \le z$

$$\mathbf{E}_x\{e^{-\alpha\varrho}; \inf_{0\le s\le\varrho} Q_s^{(n)} > y\}$$

$$= \begin{cases} \dfrac{S(\frac{\alpha}{2\theta}, \frac{n}{2}, \frac{x^2}{2\sigma^2}, \frac{y^2}{2\sigma^2})}{S(\frac{\alpha}{2\theta}, \frac{n}{2}, \frac{z^2}{2\sigma^2}, \frac{y^2}{2\sigma^2})} \exp\Big(-\dfrac{\theta 2^{n/2}\sigma^n z^{1-n} e^{z^2/2\sigma^2} U(\frac{\alpha}{2\theta}, \frac{n}{2}, \frac{y^2}{2\sigma^2}) v}{U(\frac{\alpha}{2\theta}, \frac{n}{2}, \frac{z^2}{2\sigma^2}) S(\frac{\alpha}{2\theta}, \frac{n}{2}, \frac{z^2}{2\sigma^2}, \frac{y^2}{2\sigma^2})}\Big), & y \le x \le z \\ \dfrac{U(\frac{\alpha}{2\theta}, \frac{n}{2}, \frac{x^2}{2\sigma^2})}{U(\frac{\alpha}{2\theta}, \frac{n}{2}, \frac{z^2}{2\sigma^2})} \exp\Big(-\dfrac{\theta 2^{n/2}\sigma^n z^{1-n} e^{z^2/2\sigma^2} U(\frac{\alpha}{2\theta}, \frac{n}{2}, \frac{y^2}{2\sigma^2}) v}{U(\frac{\alpha}{2\theta}, \frac{n}{2}, \frac{z^2}{2\sigma^2}) S(\frac{\alpha}{2\theta}, \frac{n}{2}, \frac{z^2}{2\sigma^2}, \frac{y^2}{2\sigma^2})}\Big), & z \le x \end{cases}$$

8 $Q_s^{(n)} = \sigma e^{-\theta s} R^{(n)}_{e^{2\theta s}-1} \quad n \ge 2, \quad \sigma, \theta > 0 \quad \varrho = \varrho(v,z) = \min\{s : \ell(s,z) = v\}$

4.3.1 $$\mathbf{E}_x e^{-\gamma \ell(\varrho,r)} = \exp\Big(-\frac{\gamma\theta 2^{n/2}\sigma^n r^{n-1} z^{1-n} e^{z^2/2\sigma^2} v}{\theta 2^{n/2}\sigma^n e^{z^2/2\sigma^2} + \gamma r^{n-1} |\operatorname{Ed}(\frac{n}{2},\frac{z}{\sigma},\frac{r}{\sigma})|}\Big)$$

$$\times \begin{cases} 1, & x \le z \le r, \quad r \le z \le x \\ \dfrac{\theta 2^{n/2}\sigma^n e^{r^2/2\sigma^2} + \gamma r^{n-1}|\operatorname{Ed}(\frac{n}{2},\frac{x}{\sigma},\frac{r}{\sigma})|}{\theta 2^{n/2}\sigma^n e^{r^2/2\sigma^2} + \gamma r^{n-1}|\operatorname{Ed}(\frac{n}{2},\frac{z}{\sigma},\frac{r}{\sigma})|}, & r \wedge z \le x \le r \vee z \\ \dfrac{\theta 2^{n/2}\sigma^n e^{r^2/2\sigma^2}}{\theta 2^{n/2}\sigma^n e^{r^2/2\sigma^2} + \gamma r^{n-1}|\operatorname{Ed}(\frac{n}{2},\frac{z}{\sigma},\frac{r}{\sigma})|}, & 0 \le x \le r \le z, \quad z \le r \le x \end{cases}$$

4.3.2 $0 \le x$ $$\mathbf{P}_x\big(\ell(\varrho,r) \in dy\big) = \exp\Big(-\frac{\theta 2^{n/2}\sigma^n\big(z^{1-n}e^{z^2/2\sigma^2} v + r^{1-n} e^{r^2/2\sigma^2} y\big)}{|\operatorname{Ed}(\frac{n}{2},\frac{r}{\sigma},\frac{z}{\sigma})|}\Big)$$

$$\times \begin{cases} \dfrac{\theta 2^{n/2}\sigma^n e^{(r^2+z^2)/4\sigma^2}}{(rz)^{(n-1)/2}|\operatorname{Ed}(\frac{n}{2},\frac{r}{\sigma},\frac{z}{\sigma})|}\dfrac{\sqrt{v}}{\sqrt{y}} I_1\Big(\dfrac{\theta 2^{1+n/2}\sigma^n e^{(r^2+z^2)/4\sigma^2}\sqrt{vy}}{(rz)^{(n-1)/2}|\operatorname{Ed}(\frac{n}{2},\frac{r}{\sigma},\frac{z}{\sigma})|}\Big)dy, \\ \hfill x \le z \le r,\ r \le z \le x \\ \dfrac{\theta 2^{n/2}\sigma^n e^{(r^2+z^2)/4\sigma^2}}{(rz)^{(n-1)/2}\operatorname{Ed}^2(\frac{n}{2},\frac{r}{\sigma},\frac{z}{\sigma})}\Big[\operatorname{Ed}\big(\frac{n}{2},\frac{r}{\sigma},\frac{x}{\sigma}\big)\dfrac{\sqrt{v}}{\sqrt{y}} I_1\Big(\dfrac{\theta 2^{1+n/2}\sigma^n e^{(r^2+z^2)/4\sigma^2}\sqrt{vy}}{(rz)^{(n-1)/2}|\operatorname{Ed}(\frac{n}{2},\frac{r}{\sigma},\frac{z}{\sigma})|}\Big) \\ \qquad + \operatorname{Ed}\big(\frac{n}{2},\frac{x}{\sigma},\frac{z}{\sigma}\big)\dfrac{r^{(1-n)/2}e^{r^2/4\sigma^2}}{z^{(1-n)/2}e^{z^2/4\sigma^2}} I_0\Big(\dfrac{\theta 2^{1+n/2}\sigma^n e^{(r^2+z^2)/4\sigma^2}\sqrt{vy}}{(rz)^{(n-1)/2}|\operatorname{Ed}(\frac{n}{2},\frac{r}{\sigma},\frac{z}{\sigma})|}\Big)\Big]dy, \\ \hfill r \wedge z \le x \le r \vee z \\ \dfrac{\theta 2^{n/2}\sigma^n e^{r^2/2\sigma^2}}{r^{n-1}|\operatorname{Ed}(\frac{n}{2},\frac{r}{\sigma},\frac{z}{\sigma})|} I_0\Big(\dfrac{\theta 2^{1+n/2}\sigma^n e^{(r^2+z^2)/4\sigma^2}\sqrt{vy}}{(rz)^{(n-1)/2}|\operatorname{Ed}(\frac{n}{2},\frac{r}{\sigma},\frac{z}{\sigma})|}\Big)dy, \\ \hfill z \le r \le x,\ x \le r \le z \end{cases}$$

(1) $\mathbf{P}_x\big(\ell(\varrho,r) = 0\big)$

$z \le r$ $$= \begin{cases} \exp\Big(-\dfrac{\theta 2^{n/2}\sigma^n e^{z^2/2\sigma^2} v}{z^{n-1}|\operatorname{Ed}(\frac{n}{2},\frac{r}{\sigma},\frac{z}{\sigma})|}\Big), & 0 \le x \le z \le r, \quad r \le z \le x \\ \dfrac{|\operatorname{Ed}(\frac{n}{2},\frac{r}{\sigma},\frac{x}{\sigma})|}{|\operatorname{Ed}(\frac{n}{2},\frac{r}{\sigma},\frac{z}{\sigma})|}\exp\Big(-\dfrac{\theta 2^{n/2}\sigma^n e^{z^2/2\sigma^2} v}{z^{n-1}|\operatorname{Ed}(\frac{n}{2},\frac{r}{\sigma},\frac{z}{\sigma})|}\Big), & r \wedge z \le x \le r \vee z \\ 0, & 0 \le x \le r \le z, \quad z \le r \le x \end{cases}$$

4.3.3 $\mathbf{E}_x e^{-\alpha\varrho - \gamma\ell(\varrho,r)}$

8 $\quad Q_s^{(n)} = \sigma e^{-\theta s} R^{(n)}_{e^{2\theta s}-1} \quad n \geq 2, \quad \sigma, \theta > 0 \quad \varrho = \varrho(v,z) = \min\{s : \ell(s,z) = v\}$

$r \leq z$

$$= \exp\Big(-\frac{\theta^2 2^n \sigma^{2n} z^{1-n} e^{(r^2+z^2)/2\sigma^2}\Gamma(n/2)\Gamma^{-1}(\alpha/2) M^{-1}\big(\frac{\alpha}{2\theta},\frac{n}{2},\frac{z^2}{2\sigma^2}\big)v}{\theta 2^{n/2}\sigma^n e^{r^2/2\sigma^2} U\big(\frac{\alpha}{2\theta},\frac{n}{2},\frac{z^2}{2\sigma^2}\big) + \gamma r^{n-1} S\big(\frac{\alpha}{2\theta},\frac{n}{2},\frac{r^2}{2\sigma^2},\frac{z^2}{2\sigma^2}\big) U\big(\frac{\alpha}{2\theta},\frac{n}{2},\frac{r^2}{2\sigma^2}\big)}\Big)$$

$$\times \exp\Big(-\frac{\theta\gamma 2^{n/2}\sigma^n z^{1-n} r^{n-1} e^{z^2/2\sigma^2} M\big(\frac{\alpha}{2\theta},\frac{n}{2},\frac{r^2}{2\sigma^2}\big) U\big(\frac{\alpha}{2\theta},\frac{n}{2},\frac{r^2}{2\sigma^2}\big) M^{-1}\big(\frac{\alpha}{2\theta},\frac{n}{2},\frac{z^2}{2\sigma^2}\big)v}{\theta 2^{n/2}\sigma^n e^{r^2/2\sigma^2} U\big(\frac{\alpha}{2\theta},\frac{n}{2},\frac{z^2}{2\sigma^2}\big) + \gamma r^{n-1} S\big(\frac{\alpha}{2\theta},\frac{n}{2},\frac{r^2}{2\sigma^2},\frac{z^2}{2\sigma^2}\big) U\big(\frac{\alpha}{2\theta},\frac{n}{2},\frac{r^2}{2\sigma^2}\big)}\Big)$$

$$\times \begin{cases} \dfrac{U\big(\frac{\alpha}{2\theta},\frac{n}{2},\frac{x^2}{2\sigma^2}\big)}{U\big(\frac{\alpha}{2\theta},\frac{n}{2},\frac{z^2}{2\sigma^2}\big)}, & z \leq x \\ \dfrac{\theta 2^{n/2}\sigma^n e^{r^2/2\sigma^2} M\big(\frac{\alpha}{2\theta},\frac{n}{2},\frac{x^2}{2\sigma^2}\big) + \gamma r^{n-1} S\big(\frac{\alpha}{2\theta},\frac{n}{2},\frac{x^2}{2\sigma^2},\frac{r^2}{2\sigma^2}\big) M\big(\frac{\alpha}{2\theta},\frac{n}{2},\frac{r^2}{2\sigma^2}\big)}{\theta 2^{n/2}\sigma^n e^{r^2/2\sigma^2} M\big(\frac{\alpha}{2\theta},\frac{n}{2},\frac{z^2}{2\sigma^2}\big) + \gamma r^{n-1} S\big(\frac{\alpha}{2\theta},\frac{n}{2},\frac{z^2}{2\sigma^2},\frac{r^2}{2\sigma^2}\big) M\big(\frac{\alpha}{2\theta},\frac{n}{2},\frac{r^2}{2\sigma^2}\big)}, & \begin{array}{l} r \leq x \\ x \leq z \end{array} \\ \dfrac{\theta 2^{n/2}\sigma^n e^{r^2/2\sigma^2} M\big(\frac{\alpha}{2\theta},\frac{n}{2},\frac{x^2}{2\sigma^2}\big)}{\theta 2^{n/2}\sigma^n e^{r^2/2\sigma^2} M\big(\frac{\alpha}{2\theta},\frac{n}{2},\frac{z^2}{2\sigma^2}\big) + \gamma r^{n-1} S\big(\frac{\alpha}{2\theta},\frac{n}{2},\frac{z^2}{2\sigma^2},\frac{r^2}{2\sigma^2}\big) M\big(\frac{\alpha}{2\theta},\frac{n}{2},\frac{r^2}{2\sigma^2}\big)}, & x \leq r \end{cases}$$

$z \leq r$

$$= \exp\Big(-\frac{\theta^2 2^n \sigma^{2n} z^{1-n} e^{(r^2+z^2)/2\sigma^2}\Gamma(n/2)\Gamma^{-1}(\alpha/2) U^{-1}\big(\frac{\alpha}{2\theta},\frac{n}{2},\frac{z^2}{2\sigma^2}\big)v}{\theta 2^{n/2}\sigma^n e^{r^2/2\sigma^2} M\big(\frac{\alpha}{2\theta},\frac{n}{2},\frac{z^2}{2\sigma^2}\big) + \gamma r^{n-1} S\big(\frac{\alpha}{2\theta},\frac{n}{2},\frac{z^2}{2\sigma^2},\frac{r^2}{2\sigma^2}\big) M\big(\frac{\alpha}{2\theta},\frac{n}{2},\frac{r^2}{2\sigma^2}\big)}\Big)$$

$$\times \exp\Big(-\frac{\theta\gamma 2^{n/2}\sigma^n z^{1-n} r^{n-1} e^{z^2/2\sigma^2} M\big(\frac{\alpha}{2\theta},\frac{n}{2},\frac{r^2}{2\sigma^2}\big) U\big(\frac{\alpha}{2\theta},\frac{n}{2},\frac{r^2}{2\sigma^2}\big) U^{-1}\big(\frac{\alpha}{2\theta},\frac{n}{2},\frac{z^2}{2\sigma^2}\big)v}{\theta 2^{n/2}\sigma^n e^{r^2/2\sigma^2} M\big(\frac{\alpha}{2\theta},\frac{n}{2},\frac{z^2}{2\sigma^2}\big) + \gamma r^{n-1} S\big(\frac{\alpha}{2\theta},\frac{n}{2},\frac{z^2}{2\sigma^2},\frac{r^2}{2\sigma^2}\big) M\big(\frac{\alpha}{2\theta},\frac{n}{2},\frac{r^2}{2\sigma^2}\big)}\Big)$$

$$\times \begin{cases} \dfrac{M\big(\frac{\alpha}{2\theta},\frac{n}{2},\frac{x^2}{2\sigma^2}\big)}{M\big(\frac{\alpha}{2\theta},\frac{n}{2},\frac{z^2}{2\sigma^2}\big)}, & x \leq z \\ \dfrac{\theta 2^{n/2}\sigma^n e^{r^2/2\sigma^2} U\big(\frac{\alpha}{2\theta},\frac{n}{2},\frac{x^2}{2\sigma^2}\big) + \gamma r^{n-1} S\big(\frac{\alpha}{2\theta},\frac{n}{2},\frac{r^2}{2\sigma^2},\frac{x^2}{2\sigma^2}\big) U\big(\frac{\alpha}{2\theta},\frac{n}{2},\frac{r^2}{2\sigma^2}\big)}{\theta 2^{n/2}\sigma^n e^{r^2/2\sigma^2} U\big(\frac{\alpha}{2\theta},\frac{n}{2},\frac{z^2}{2\sigma^2}\big) + \gamma r^{n-1} S\big(\frac{\alpha}{2\theta},\frac{n}{2},\frac{r^2}{2\sigma^2},\frac{z^2}{2\sigma^2}\big) U\big(\frac{\alpha}{2\theta},\frac{n}{2},\frac{r^2}{2\sigma^2}\big)}, & \begin{array}{l} z \leq x \\ x \leq r \end{array} \\ \dfrac{\theta 2^{n/2}\sigma^n e^{r^2/2\sigma^2} U\big(\frac{\alpha}{2\theta},\frac{n}{2},\frac{x^2}{2\sigma^2}\big)}{\theta 2^{n/2}\sigma^n e^{r^2/2\sigma^2} U\big(\frac{\alpha}{2\theta},\frac{n}{2},\frac{z^2}{2\sigma^2}\big) + \gamma r^{n-1} S\big(\frac{\alpha}{2\theta},\frac{n}{2},\frac{r^2}{2\sigma^2},\frac{z^2}{2\sigma^2}\big) U\big(\frac{\alpha}{2\theta},\frac{n}{2},\frac{r^2}{2\sigma^2}\big)}, & r \leq x \end{cases}$$

4.3.4 $\quad \mathbf{E}_x\{e^{-\alpha\varrho}; \ell(\varrho, r) \in dy\}$

$0 \leq x$

$r \leq z$

$$= \exp\Big(-\frac{\theta 2^{n/2}\sigma^n z^{1-n} e^{z^2/2\sigma^2} M\big(\frac{\alpha}{2\theta},\frac{n}{2},\frac{r^2}{2\sigma^2}\big)v}{S\big(\frac{\alpha}{2\theta},\frac{n}{2},\frac{r^2}{2\sigma^2},\frac{z^2}{2\sigma^2}\big) M\big(\frac{\alpha}{2\theta},\frac{n}{2},\frac{z^2}{2\sigma^2}\big)} - \frac{\theta 2^{n/2}\sigma^n r^{1-n} e^{r^2/2\sigma^2} U\big(\frac{\alpha}{2\theta},\frac{n}{2},\frac{z^2}{2\sigma^2}\big)y}{S\big(\frac{\alpha}{2\theta},\frac{n}{2},\frac{r^2}{2\sigma^2},\frac{z^2}{2\sigma^2}\big) U\big(\frac{\alpha}{2\theta},\frac{n}{2},\frac{r^2}{2\sigma^2}\big)}\Big)$$

$$\times \frac{\theta 2^{n/2}\sigma^n e^{(r^2+z^2)/4\sigma^2}}{(rz)^{(n-1)/2} S\big(\frac{\alpha}{2\theta},\frac{n}{2},\frac{r^2}{2\sigma^2},\frac{z^2}{2\sigma^2}\big)}$$

8 $\quad Q_s^{(n)} = \sigma e^{-\theta s} R^{(n)}_{e^{2\theta s}-1} \quad n \ge 2, \quad \sigma, \theta > 0 \quad \varrho = \varrho(v,z) = \min\{s : \ell(s,z) = v\}$

$$\times \begin{cases} \dfrac{M(\frac{\alpha}{2\theta},\frac{n}{2},\frac{x^2}{2\sigma^2})}{M(\frac{\alpha}{2\theta},\frac{n}{2},\frac{z^2}{2\sigma^2})}\dfrac{\sqrt{v}}{\sqrt{y}} I_1\Big(\dfrac{\theta 2^{1+n/2}\sigma^n e^{(r^2+z^2)/4\sigma^2}\sqrt{vy}}{S(\frac{\alpha}{2\theta},\frac{n}{2},\frac{r^2}{2\sigma^2},\frac{z^2}{2\sigma^2})(rz)^{(n-1)/2}}\Big)dy, & x \le z \\[2ex] \dfrac{S(\frac{\alpha}{2\theta},\frac{n}{2},\frac{r^2}{2\sigma^2},\frac{x^2}{2\sigma^2})}{S(\frac{\alpha}{2\theta},\frac{n}{2},\frac{r^2}{2\sigma^2},\frac{z^2}{2\sigma^2})}\dfrac{\sqrt{v}}{\sqrt{y}} I_1\Big(\dfrac{\theta 2^{1+n/2}\sigma^n e^{(r^2+z^2)/4\sigma^2}\sqrt{vy}}{S(\frac{\alpha}{2\theta},\frac{n}{2},\frac{r^2}{2\sigma^2},\frac{z^2}{2\sigma^2})(rz)^{(n-1)/2}}\Big)dy \\ \quad + \dfrac{S(\frac{\alpha}{2\theta},\frac{n}{2},\frac{x^2}{2\sigma^2},\frac{z^2}{2\sigma^2})}{S(\frac{\alpha}{2\theta},\frac{n}{2},\frac{r^2}{2\sigma^2},\frac{z^2}{2\sigma^2})}\dfrac{r^{(1-n)/2}e^{r^2/4\sigma^2}}{z^{(1-n)/2}e^{z^2/4\sigma^2}} I_0\Big(\dfrac{\theta 2^{1+n/2}\sigma^n e^{(r^2+z^2)/4\sigma^2}\sqrt{vy}}{S(\frac{\alpha}{2\theta},\frac{n}{2},\frac{r^2}{2\sigma^2},\frac{z^2}{2\sigma^2})(rz)^{(n-1)/2}}\Big)dy, & z \le x \le r \\[2ex] \dfrac{r^{(1-n)/2}e^{r^2/4\sigma^2}U(\frac{\alpha}{2\theta},\frac{n}{2},\frac{x^2}{2\sigma^2})}{z^{(1-n)/2}e^{z^2/4\sigma^2}U(\frac{\alpha}{2\theta},\frac{n}{2},\frac{r^2}{2\sigma^2})} I_0\Big(\dfrac{\theta 2^{1+n/2}\sigma^n e^{(r^2+z^2)/4\sigma^2}\sqrt{vy}}{S(\frac{\alpha}{2\theta},\frac{n}{2},\frac{r^2}{2\sigma^2},\frac{z^2}{2\sigma^2})(rz)^{(n-1)/2}}\Big)dy, & r \le x \end{cases}$$

$z \le r$

$$= \exp\Big(-\frac{\theta 2^{n/2}\sigma^n z^{1-n}e^{z^2/2\sigma^2}U(\frac{\alpha}{2\theta},\frac{n}{2},\frac{r^2}{2\sigma^2})v}{S(\frac{\alpha}{2\theta},\frac{n}{2},\frac{z^2}{2\sigma^2},\frac{r^2}{2\sigma^2})U(\frac{\alpha}{2\theta},\frac{n}{2},\frac{z^2}{2\sigma^2})} - \frac{\theta 2^{n/2}\sigma^n r^{1-n}e^{r^2/2\sigma^2}M(\frac{\alpha}{2\theta},\frac{n}{2},\frac{z^2}{2\sigma^2})y}{S(\frac{\alpha}{2\theta},\frac{n}{2},\frac{z^2}{2\sigma^2},\frac{r^2}{2\sigma^2})M(\frac{\alpha}{2\theta},\frac{n}{2},\frac{r^2}{2\sigma^2})}\Big)$$

$$\times \frac{\theta 2^{n/2}\sigma^n e^{(r^2+z^2)/4\sigma^2}}{(rz)^{(n-1)/2}S(\frac{\alpha}{2\theta},\frac{n}{2},\frac{z^2}{2\sigma^2},\frac{r^2}{2\sigma^2})}$$

$$\times \begin{cases} \dfrac{U(\frac{\alpha}{2\theta},\frac{n}{2},\frac{x^2}{2\sigma^2})}{U(\frac{\alpha}{2\theta},\frac{n}{2},\frac{z^2}{2\sigma^2})}\dfrac{\sqrt{v}}{\sqrt{y}} I_1\Big(\dfrac{\theta 2^{1+n/2}\sigma^n e^{(r^2+z^2)/4\sigma^2}\sqrt{vy}}{S(\frac{\alpha}{2\theta},\frac{n}{2},\frac{z^2}{2\sigma^2},\frac{r^2}{2\sigma^2})(rz)^{(n-1)/2}}\Big)dy, & z \le x \\[2ex] \dfrac{S(\frac{\alpha}{2\theta},\frac{n}{2},\frac{x^2}{2\sigma^2},\frac{r^2}{2\sigma^2})}{S(\frac{\alpha}{2\theta},\frac{n}{2},\frac{z^2}{2\sigma^2},\frac{r^2}{2\sigma^2})}\dfrac{\sqrt{v}}{\sqrt{y}} I_1\Big(\dfrac{\theta 2^{1+n/2}\sigma^n e^{(r^2+z^2)/4\sigma^2}\sqrt{vy}}{S(\frac{\alpha}{2\theta},\frac{n}{2},\frac{z^2}{2\sigma^2},\frac{r^2}{2\sigma^2})(rz)^{(n-1)/2}}\Big)dy \\ \quad + \dfrac{S(\frac{\alpha}{2\theta},\frac{n}{2},\frac{z^2}{2\sigma^2},\frac{x^2}{2\sigma^2})}{S(\frac{\alpha}{2\theta},\frac{n}{2},\frac{z^2}{2\sigma^2},\frac{r^2}{2\sigma^2})}\dfrac{r^{(1-n)/2}e^{r^2/4\sigma^2}}{z^{(1-n)/2}e^{z^2/4\sigma^2}} I_0\Big(\dfrac{\theta 2^{1+n/2}\sigma^n e^{(r^2+z^2)/4\sigma^2}\sqrt{vy}}{S(\frac{\alpha}{2\theta},\frac{n}{2},\frac{z^2}{2\sigma^2},\frac{r^2}{2\sigma^2})(rz)^{(n-1)/2}}\Big)dy, & r \le x \le z \\[2ex] \dfrac{r^{(1-n)/2}e^{r^2/4\sigma^2}M(\frac{\alpha}{2\theta},\frac{n}{2},\frac{x^2}{2\sigma^2})}{z^{(1-n)/2}e^{z^2/4\sigma^2}M(\frac{\alpha}{2\theta},\frac{n}{2},\frac{r^2}{2\sigma^2})} I_0\Big(\dfrac{\theta 2^{1+n/2}\sigma^n e^{(r^2+z^2)/4\sigma^2}\sqrt{vy}}{S(\frac{\alpha}{2\theta},\frac{n}{2},\frac{z^2}{2\sigma^2},\frac{r^2}{2\sigma^2})(rz)^{(n-1)/2}}\Big)dy, & x \le r \end{cases}$$

(1) $\mathbf{E}_x\{e^{-\alpha\varrho}; \ell(\varrho, r) = 0\}$

$z \le r$

$$= \begin{cases} 0, & r \le x \\[1ex] \dfrac{S(\frac{\alpha}{2\theta},\frac{n}{2},\frac{r^2}{2\sigma^2},\frac{x^2}{2\sigma^2})}{S(\frac{\alpha}{2\theta},\frac{n}{2},\frac{r^2}{2\sigma^2},\frac{z^2}{2\sigma^2})}\exp\Big(-\dfrac{\theta 2^{n/2}\sigma^n z^{1-n}e^{z^2/2\sigma^2}M(\frac{\alpha}{2\theta},\frac{n}{2},\frac{r^2}{2\sigma^2})v}{S(\frac{\alpha}{2\theta},\frac{n}{2},\frac{r^2}{2\sigma^2},\frac{z^2}{2\sigma^2})M(\frac{\alpha}{2\theta},\frac{n}{2},\frac{z^2}{2\sigma^2})}\Big), & z \le x \le r \\[2ex] \dfrac{M(\frac{\alpha}{2\theta},\frac{n}{2},\frac{x^2}{2\sigma^2})}{M(\frac{\alpha}{2\theta},\frac{n}{2},\frac{z^2}{2\sigma^2})}\exp\Big(-\dfrac{\theta 2^{n/2}\sigma^n z^{1-n}e^{z^2/2\sigma^2}M(\frac{\alpha}{2\theta},\frac{n}{2},\frac{r^2}{2\sigma^2})v}{S(\frac{\alpha}{2\theta},\frac{n}{2},\frac{r^2}{2\sigma^2},\frac{z^2}{2\sigma^2})M(\frac{\alpha}{2\theta},\frac{n}{2},\frac{z^2}{2\sigma^2})}\Big), & 0 \le x \le z \end{cases}$$

8 $\quad Q_s^{(n)} = \sigma e^{-\theta s} R^{(n)}_{e^{2\theta s}-1} \quad n \ge 2, \quad \sigma, \theta > 0 \quad \varrho = \varrho(v,z) = \min\{s : \ell(s,z) = v\}$

$r \le z$

$$= \begin{cases} 0, & 0 \le x \le r \\ \dfrac{S\left(\frac{\alpha}{2\theta}, \frac{n}{2}, \frac{x^2}{2\sigma^2}, \frac{r^2}{2\sigma^2}\right)}{S\left(\frac{\alpha}{2\theta}, \frac{n}{2}, \frac{z^2}{2\sigma^2}, \frac{r^2}{2\sigma^2}\right)} \exp\Bigl(-\dfrac{\theta 2^{n/2}\sigma^n z^{1-n} e^{z^2/2\sigma^2} U\left(\frac{\alpha}{2\theta}, \frac{n}{2}, \frac{r^2}{2\sigma^2}\right) v}{U\left(\frac{\alpha}{2\theta}, \frac{n}{2}, \frac{z^2}{2\sigma^2}\right) S\left(\frac{\alpha}{2\theta}, \frac{n}{2}, \frac{z^2}{2\sigma^2}, \frac{r^2}{2\sigma^2}\right)}\Bigr), & r \le x \le z \\ \dfrac{U\left(\frac{\alpha}{2\theta}, \frac{n}{2}, \frac{x^2}{2\sigma^2}\right)}{U\left(\frac{\alpha}{2\theta}, \frac{n}{2}, \frac{z^2}{2\sigma^2}\right)} \exp\Bigl(-\dfrac{\theta 2^{n/2}\sigma^n z^{1-n} e^{z^2/2\sigma^2} U\left(\frac{\alpha}{2\theta}, \frac{n}{2}, \frac{r^2}{2\sigma^2}\right) v}{U\left(\frac{\alpha}{2\theta}, \frac{n}{2}, \frac{z^2}{2\sigma^2}\right) S\left(\frac{\alpha}{2\theta}, \frac{n}{2}, \frac{z^2}{2\sigma^2}, \frac{r^2}{2\sigma^2}\right)}\Bigr), & z \le x \end{cases}$$

4.4.1 $\quad 0 \le x$

$$\mathbf{E}_x \exp\Bigl(-\gamma \int_0^{\varrho} \mathbb{I}_{[r,\infty)}\bigl(Q_s^{(n)}\bigr) ds\Bigr)$$

$z \le r$

$$= \exp\Bigl(-\frac{\theta\gamma 2^{n/2}\sigma^n z^{1-n} r^n e^{z^2/2\sigma^2} U\left(\frac{\gamma+2\theta}{2\theta}, \frac{n+2}{2}, \frac{r^2}{2\sigma^2}\right) v}{\theta 2^{1+n/2}\sigma^n e^{r^2/2\sigma^2} U\left(\frac{\gamma}{2\theta}, \frac{n}{2}, \frac{r^2}{2\sigma^2}\right) + \gamma r^n \operatorname{Ed}\left(\frac{n}{2}, \frac{r}{\sigma}, \frac{z}{\sigma}\right) U\left(\frac{\gamma+2\theta}{2\theta}, \frac{n+2}{2}, \frac{r^2}{2\sigma^2}\right)}\Bigr)$$

$$\times \begin{cases} 1, & x \le z \\ \dfrac{\theta 2^{1+n/2}\sigma^n e^{r^2/2\sigma^2} U\left(\frac{\gamma}{2\theta}, \frac{n}{2}, \frac{r^2}{2\sigma^2}\right) + \gamma r^n \operatorname{Ed}\left(\frac{n}{2}, \frac{r}{\sigma}, \frac{x}{\sigma}\right) U\left(\frac{\gamma+2\theta}{2\theta}, \frac{n+2}{2}, \frac{r^2}{2\sigma^2}\right)}{\theta 2^{1+n/2}\sigma^n e^{r^2/2\sigma^2} U\left(\frac{\gamma}{2\theta}, \frac{n}{2}, \frac{r^2}{2\sigma^2}\right) + \gamma r^n \operatorname{Ed}\left(\frac{n}{2}, \frac{r}{\sigma}, \frac{z}{\sigma}\right) U\left(\frac{\gamma+2\theta}{2\theta}, \frac{n+2}{2}, \frac{r^2}{2\sigma^2}\right)}, & \begin{matrix} z \le x \\ x \le r \end{matrix} \\ \dfrac{\theta 2^{1+n/2}\sigma^n e^{r^2/2\sigma^2} U\left(\frac{\gamma}{2\theta}, \frac{n}{2}, \frac{x^2}{2\sigma^2}\right)}{\theta 2^{1+n/2}\sigma^n e^{r^2/2\sigma^2} U\left(\frac{\gamma}{2\theta}, \frac{n}{2}, \frac{r^2}{2\sigma^2}\right) + \gamma r^n \operatorname{Ed}\left(\frac{n}{2}, \frac{r}{\sigma}, \frac{z}{\sigma}\right) U\left(\frac{\gamma+2\theta}{2\theta}, \frac{n+2}{2}, \frac{r^2}{2\sigma^2}\right)}, & r \le x \end{cases}$$

$r \le z$

$$= \exp\Bigl(-\frac{\gamma 2^{n/2}\sigma^n z^{1-n} e^{z^2/2\sigma^2} U\left(\frac{\gamma+2\theta}{2\theta}, \frac{n+2}{2}, \frac{r^2}{2\sigma^2}\right) v}{2U\left(\frac{\gamma}{2\theta}, \frac{n}{2}, \frac{z^2}{2\sigma^2}\right) C\left(\frac{\gamma}{2\theta}, \frac{n}{2}, \frac{r^2}{2\sigma^2}, \frac{z^2}{2\sigma^2}\right)}\Bigr) \begin{cases} \dfrac{2^{n/2}\sigma^n r^{-n} e^{r^2/2\sigma^2}}{C\left(\frac{\gamma}{2\theta}, \frac{n}{2}, \frac{r^2}{2\sigma^2}, \frac{z^2}{2\sigma^2}\right)}, & x \le r \\ \dfrac{C\left(\frac{\gamma}{2\theta}, \frac{n}{2}, \frac{r^2}{2\sigma^2}, \frac{x^2}{2\sigma^2}\right)}{C\left(\frac{\gamma}{2\theta}, \frac{n}{2}, \frac{r^2}{2\sigma^2}, \frac{z^2}{2\sigma^2}\right)}, & \begin{matrix} r \le x \\ x \le z \end{matrix} \\ \dfrac{U\left(\frac{\gamma}{2\theta}, \frac{n}{2}, \frac{x^2}{2\sigma^2}\right)}{U\left(\frac{\gamma}{2\theta}, \frac{n}{2}, \frac{z^2}{2\sigma^2}\right)}, & z \le x \end{cases}$$

4.4.2 $\quad$ (1) $\quad 0 \le x$

$$\mathbf{P}_x\Bigl(\int_0^{\varrho} \mathbb{I}_{[r,\infty)}\bigl(Q_s^{(n)}\bigr) ds = 0\Bigr) = \begin{cases} \exp\Bigl(-\dfrac{\theta 2^{n/2}\sigma^n e^{z^2/2\sigma^2} v}{z^{n-1} \operatorname{Ed}\left(\frac{n}{2}, \frac{r}{\sigma}, \frac{z}{\sigma}\right)}\Bigr), & x \le z \\ \dfrac{\operatorname{Ed}\left(\frac{n}{2}, \frac{r}{\sigma}, \frac{x}{\sigma}\right)}{\operatorname{Ed}\left(\frac{n}{2}, \frac{r}{\sigma}, \frac{z}{\sigma}\right)} \exp\Bigl(-\dfrac{\theta 2^{n/2}\sigma^n e^{z^2/2\sigma^2} v}{z^{n-1} \operatorname{Ed}\left(\frac{n}{2}, \frac{r}{\sigma}, \frac{z}{\sigma}\right)}\Bigr), & \begin{matrix} z \le x \\ x \le r \end{matrix} \\ 0, & r \le x \end{cases}$$

4.5.1

$$\mathbf{E}_x \exp\Bigl(-\gamma \int_0^{\varrho} \mathbb{I}_{[0,r]}\bigl(Q_s^{(n)}\bigr) ds\Bigr)$$

$z \le r$

$$= \exp\Bigl(-\frac{\gamma 2^{n/2}\sigma^n z^{1-n} e^{z^2/2\sigma^2} M\left(\frac{\gamma+2\theta}{2\theta}, \frac{n+2}{2}, \frac{r^2}{2\sigma^2}\right) v}{n M\left(\frac{\gamma}{2\theta}, \frac{n}{2}, \frac{z^2}{2\sigma^2}\right) C\left(\frac{\gamma}{2\theta}, \frac{n}{2}, \frac{r^2}{2\sigma^2}, \frac{z^2}{2\sigma^2}\right)}\Bigr)$$

8 $\quad Q_s^{(n)} = \sigma e^{-\theta s} R^{(n)}_{e^{2\theta s}-1} \quad n \ge 2, \quad \sigma, \theta > 0 \quad \varrho = \varrho(v,z) = \min\{s : \ell(s,z) = v\}$

$$\times \begin{cases} \dfrac{M\big(\frac{\gamma}{2\theta}, \frac{n}{2}, \frac{x^2}{2\sigma^2}\big)}{M\big(\frac{\gamma}{2\theta}, \frac{n}{2}, \frac{z^2}{2\sigma^2}\big)}, & 0 \le x \le z \\[2ex] \dfrac{C\big(\frac{\gamma}{2\theta}, \frac{n}{2}, \frac{r^2}{2\sigma^2}, \frac{x^2}{2\sigma^2}\big)}{C\big(\frac{\gamma}{2\theta}, \frac{n}{2}, \frac{r^2}{2\sigma^2}, \frac{z^2}{2\sigma^2}\big)}, & z \le x \le r \\[2ex] \dfrac{2^{n/2}\sigma^n r^{-n} e^{r^2/2\sigma^2}}{C\big(\frac{\gamma}{2\theta}, \frac{n}{2}, \frac{r^2}{2\sigma^2}, \frac{z^2}{2\sigma^2}\big)}, & r \le x \end{cases}$$

$r \le z$

$$= \exp\Big(-\frac{\theta\gamma 2^{n/2}\sigma^n z^{1-n} r^n e^{z^2/2\sigma^2} M\big(\frac{\gamma+2\theta}{2\theta}, \frac{n+2}{2}, \frac{r^2}{2\sigma^2}\big) v}{\theta n 2^{n/2}\sigma^n e^{r^2/2\sigma^2} M\big(\frac{\gamma}{2\theta}, \frac{n}{2}, \frac{r^2}{2\sigma^2}\big) + \gamma r^n \operatorname{Ed}\big(\frac{n}{2}, \frac{z}{\sigma}, \frac{r}{\sigma}\big) M\big(\frac{\gamma+2\theta}{2\theta}, \frac{n+2}{2}, \frac{r^2}{2\sigma^2}\big)}\Big)$$

$$\times \begin{cases} \dfrac{\theta n 2^{n/2}\sigma^n e^{r^2/2\sigma^2} M\big(\frac{\gamma}{2\theta}, \frac{n}{2}, \frac{x^2}{2\sigma^2}\big)}{\theta n 2^{n/2}\sigma^n e^{r^2/2\sigma^2} M\big(\frac{\gamma}{2\theta}, \frac{n}{2}, \frac{r^2}{2\sigma^2}\big) + \gamma r^n \operatorname{Ed}\big(\frac{n}{2}, \frac{z}{\sigma}, \frac{r}{\sigma}\big) M\big(\frac{\gamma+2\theta}{2\theta}, \frac{n+2}{2}, \frac{r^2}{2\sigma^2}\big)}, & x \le r \\[2ex] \dfrac{\theta n 2^{n/2}\sigma^n e^{r^2/2\sigma^2} M\big(\frac{\gamma}{2\theta}, \frac{n}{2}, \frac{r^2}{2\sigma^2}\big) + \gamma r^n \operatorname{Ed}\big(\frac{n}{2}, \frac{x}{\sigma}, \frac{r}{\sigma}\big) M\big(\frac{\gamma+2\theta}{2\theta}, \frac{n+2}{2}, \frac{r^2}{2\sigma^2}\big)}{\theta n 2^{n/2}\sigma^n e^{r^2/2\sigma^2} M\big(\frac{\gamma}{2\theta}, \frac{n}{2}, \frac{r^2}{2\sigma^2}\big) + \gamma r^n \operatorname{Ed}\big(\frac{n}{2}, \frac{z}{\sigma}, \frac{r}{\sigma}\big) M\big(\frac{\gamma+2\theta}{2\theta}, \frac{n+2}{2}, \frac{r^2}{2\sigma^2}\big)}, & \begin{array}{l} r \le x \\ x \le z \end{array} \\[2ex] 1, & z \le x \end{cases}$$

4.5.2
(1)
$0 \le x$

$$\mathbf{P}_x\Big(\int_0^\varrho \mathbb{1}_{[0,r]}\big(Q_s^{(n)}\big)ds = 0\Big) = \begin{cases} 0, & x \le r \\[1ex] \dfrac{\operatorname{Ed}\big(\frac{n}{2}, \frac{x}{\sigma}, \frac{r}{\sigma}\big)}{\operatorname{Ed}\big(\frac{n}{2}, \frac{z}{\sigma}, \frac{r}{\sigma}\big)} \exp\Big(-\dfrac{\theta 2^{n/2}\sigma^n e^{z^2/2\sigma^2} v}{z^{n-1}\operatorname{Ed}\big(\frac{n}{2}, \frac{z}{\sigma}, \frac{r}{\sigma}\big)}\Big), & \begin{array}{l} r \le x \\ x \le z \end{array} \\[2ex] \exp\Big(-\dfrac{\theta 2^{n/2}\sigma^n e^{z^2/2\sigma^2} v}{z^{n-1}\operatorname{Ed}\big(\frac{n}{2}, \frac{z}{\sigma}, \frac{r}{\sigma}\big)}\Big), & z \le x \end{cases}$$

4.6.1

$$\mathbf{E}_x \exp\Big(-\int_0^\varrho \big(p\mathbb{1}_{[0,r)}\big(Q_s^{(n)}\big) + q\mathbb{1}_{[r,\infty)}\big(Q_s^{(n)}\big)\big)ds\Big)$$

$r \le z$

$$= \exp\Big(-\frac{\theta p 2^{n/2}\sigma^n z^{1-n} e^{z^2/2\sigma^2} M\big(\frac{p+2\theta}{2\theta}, \frac{n+2}{2}, \frac{r^2}{2\sigma^2}\big) U\big(\frac{q}{2\theta}, \frac{n}{2}, \frac{r^2}{2\sigma^2}\big) U^{-1}\big(\frac{q}{2\theta}, \frac{n}{2}, \frac{z^2}{2\sigma^2}\big) v}{pS\big(\frac{q}{2\theta}, \frac{n}{2}, \frac{z^2}{2\sigma^2}, \frac{r^2}{2\sigma^2}\big) M\big(\frac{p+2\theta}{2\theta}, \frac{n+2}{2}, \frac{r^2}{2\sigma^2}\big) + \theta n C\big(\frac{q}{2\theta}, \frac{n}{2}, \frac{r^2}{2\sigma^2}, \frac{z^2}{2\sigma^2}\big) M\big(\frac{p}{2\theta}, \frac{n}{2}, \frac{r^2}{2\sigma^2}\big)}$$

$$-\frac{\theta q n 2^{n/2-1}\sigma^n z^{1-n} e^{z^2/2\sigma^2} U\big(\frac{q+2\theta}{2\theta}, \frac{n+2}{2}, \frac{r^2}{2\sigma^2}\big) M\big(\frac{p}{2\theta}, \frac{n}{2}, \frac{r^2}{2\sigma^2}\big) U^{-1}\big(\frac{q}{2\theta}, \frac{n}{2}, \frac{z^2}{2\sigma^2}\big) v}{pS\big(\frac{q}{2\theta}, \frac{n}{2}, \frac{z^2}{2\sigma^2}, \frac{r^2}{2\sigma^2}\big) M\big(\frac{p+2\theta}{2\theta}, \frac{n+2}{2}, \frac{r^2}{2\sigma^2}\big) + \theta n C\big(\frac{q}{2\theta}, \frac{n}{2}, \frac{r^2}{2\sigma^2}, \frac{z^2}{2\sigma^2}\big) M\big(\frac{p}{2\theta}, \frac{n}{2}, \frac{r^2}{2\sigma^2}\big)}\Big)$$

$$\times \begin{cases} \dfrac{2^{n/2}\sigma^n r^{-n} e^{r^2/2\sigma^2} M\big(\frac{p}{2\theta}, \frac{n}{2}, \frac{x^2}{2\sigma^2}\big)}{\frac{p}{\theta n} S\big(\frac{q}{2\theta}, \frac{n}{2}, \frac{z^2}{2\sigma^2}, \frac{r^2}{2\sigma^2}\big) M\big(\frac{p+2\theta}{2\theta}, \frac{n+2}{2}, \frac{r^2}{2\sigma^2}\big) + C\big(\frac{q}{2\theta}, \frac{n}{2}, \frac{r^2}{2\sigma^2}, \frac{z^2}{2\sigma^2}\big) M\big(\frac{p}{2\theta}, \frac{n}{2}, \frac{r^2}{2\sigma^2}\big)}, & x \le r \\[2ex] \dfrac{\frac{p}{\theta n} S\big(\frac{q}{2\theta}, \frac{n}{2}, \frac{x^2}{2\sigma^2}, \frac{r^2}{2\sigma^2}\big) M\big(\frac{p+2\theta}{2\theta}, \frac{n+2}{2}, \frac{r^2}{2\sigma^2}\big) + C\big(\frac{q}{2\theta}, \frac{n}{2}, \frac{r^2}{2\sigma^2}, \frac{x^2}{2\sigma^2}\big) M\big(\frac{p}{2\theta}, \frac{n}{2}, \frac{r^2}{2\sigma^2}\big)}{\frac{p}{\theta n} S\big(\frac{q}{2\theta}, \frac{n}{2}, \frac{z^2}{2\sigma^2}, \frac{r^2}{2\sigma^2}\big) M\big(\frac{p+2\theta}{2\theta}, \frac{n+2}{2}, \frac{r^2}{2\sigma^2}\big) + C\big(\frac{q}{2\theta}, \frac{n}{2}, \frac{r^2}{2\sigma^2}, \frac{z^2}{2\sigma^2}\big) M\big(\frac{p}{2\theta}, \frac{n}{2}, \frac{r^2}{2\sigma^2}\big)}, & \begin{array}{l} r \le x \\ x \le z \end{array} \\[2ex] \dfrac{U\big(\frac{q}{2\theta}, \frac{n}{2}, \frac{x^2}{2\sigma^2}\big)}{U\big(\frac{q}{2\theta}, \frac{n}{2}, \frac{z^2}{2\sigma^2}\big)}, & z \le x \end{cases}$$

8 $Q_s^{(n)} = \sigma e^{-\theta s} R^{(n)}_{e^{2\theta s}-1} \quad n \ge 2, \quad \sigma, \theta > 0 \quad \varrho = \varrho(v,z) = \min\{s : \ell(s,z) = v\}$

$z \le r$

$$= \exp\Big(-\frac{\theta p n^{-1} 2^{1+n/2}\sigma^n z^{1-n} e^{z^2/2\sigma^2} M\big(\frac{p+2\theta}{2\theta},\frac{n+2}{2},\frac{r^2}{2\sigma^2}\big) U\big(\frac{q}{2\theta},\frac{n}{2},\frac{r^2}{2\sigma^2}\big) M^{-1}\big(\frac{p}{2\theta},\frac{n}{2},\frac{z^2}{2\sigma^2}\big) v}{2\theta U\big(\frac{q}{2\theta},\frac{n}{2},\frac{r^2}{2\sigma^2}\big) C\big(\frac{p}{2\theta},\frac{n}{2},\frac{r^2}{2\sigma^2},\frac{z^2}{2\sigma^2}\big) + q U\big(\frac{q+2\theta}{2\theta},\frac{n}{2},\frac{r^2}{2\sigma^2}\big) S\big(\frac{p}{2\theta},\frac{n}{2},\frac{r^2}{2\sigma^2},\frac{z^2}{2\sigma^2}\big)}$$

$$- \frac{\theta q 2^{n/2}\sigma^n z^{1-n} e^{z^2/2\sigma^2} U\big(\frac{q+2\theta}{2\theta},\frac{n+2}{2},\frac{r^2}{2\sigma^2}\big) M\big(\frac{p}{2\theta},\frac{n}{2},\frac{r^2}{2\sigma^2}\big) M^{-1}\big(\frac{p}{2\theta},\frac{n}{2},\frac{z^2}{2\sigma^2}\big) v}{2\theta U\big(\frac{q}{2\theta},\frac{n}{2},\frac{r^2}{2\sigma^2}\big) C\big(\frac{p}{2\theta},\frac{n}{2},\frac{r^2}{2\sigma^2},\frac{z^2}{2\sigma^2}\big) + q U\big(\frac{q+2\theta}{2\theta},\frac{n}{2},\frac{r^2}{2\sigma^2}\big) S\big(\frac{p}{2\theta},\frac{n}{2},\frac{r^2}{2\sigma^2},\frac{z^2}{2\sigma^2}\big)}\Big)$$

$$\times \begin{cases} \dfrac{M\big(\frac{p}{2\theta},\frac{n}{2},\frac{x^2}{2\sigma^2}\big)}{M\big(\frac{p}{2\theta},\frac{n}{2},\frac{z^2}{2\sigma^2}\big)}, & x \le z \\[2ex] \dfrac{2\theta U\big(\frac{q}{2\theta},\frac{n}{2},\frac{r^2}{2\sigma^2}\big) C\big(\frac{p}{2\theta},\frac{n}{2},\frac{r^2}{2\sigma^2},\frac{x^2}{2\sigma^2}\big) + q U\big(\frac{q+2\theta}{2\theta},\frac{n}{2},\frac{r^2}{2\sigma^2}\big) S\big(\frac{p}{2\theta},\frac{n}{2},\frac{r^2}{2\sigma^2},\frac{x^2}{2\sigma^2}\big)}{2\theta U\big(\frac{q}{2\theta},\frac{n}{2},\frac{r^2}{2\sigma^2}\big) C\big(\frac{p}{2\theta},\frac{n}{2},\frac{r^2}{2\sigma^2},\frac{z^2}{2\sigma^2}\big) + q U\big(\frac{q+2\theta}{2\theta},\frac{n}{2},\frac{r^2}{2\sigma^2}\big) S\big(\frac{p}{2\theta},\frac{n}{2},\frac{r^2}{2\sigma^2},\frac{z^2}{2\sigma^2}\big)}, & \begin{matrix} z \le x \\ x \le r \end{matrix} \\[2ex] \dfrac{\theta 2^{1+n/2}\sigma^n r^{-n} e^{r^2/2\sigma^2} U\big(\frac{q}{2\theta},\frac{n}{2},\frac{x^2}{2\sigma^2}\big)}{2\theta U\big(\frac{q}{2\theta},\frac{n}{2},\frac{r^2}{2\sigma^2}\big) C\big(\frac{p}{2\theta},\frac{n}{2},\frac{r^2}{2\sigma^2},\frac{z^2}{2\sigma^2}\big) + q U\big(\frac{q+2\theta}{2\theta},\frac{n}{2},\frac{r^2}{2\sigma^2}\big) S\big(\frac{p}{2\theta},\frac{n}{2},\frac{r^2}{2\sigma^2},\frac{z^2}{2\sigma^2}\big)}, & r \le x \end{cases}$$

4.9.1
$1 \le \gamma$

$$\mathbf{E}_x \exp\Big(-\frac{\theta(\gamma^2-1)}{4\sigma^2}\int_0^\varrho (Q_s^{(n)})^2 ds\Big) = \exp\Bigg(-\frac{\theta\gamma z v \Gamma\big(\frac{n}{2}\big)\Gamma^{-1}\big(\frac{n}{4}-\frac{n}{4\gamma}\big)}{M_{\frac{n}{4\gamma},\frac{n-2}{4}}\Big(\frac{\gamma z^2}{2\sigma^2}\Big) W_{\frac{n}{4\gamma},\frac{n-2}{4}}\Big(\frac{\gamma z^2}{2\sigma^2}\Big)}\Bigg)$$

$$\times \begin{cases} \dfrac{x^{-n/2} e^{x^2/4\sigma^2} M_{\frac{n}{4\gamma},\frac{n-2}{4}}\Big(\frac{\gamma x^2}{2\sigma^2}\Big)}{z^{-n/2} e^{z^2/4\sigma^2} M_{\frac{n}{4\gamma},\frac{n-2}{4}}\Big(\frac{\gamma z^2}{2\sigma^2}\Big)}, & 0 < x \le z \\[3ex] \dfrac{x^{-n/2} e^{x^2/4\sigma^2} W_{\frac{n}{4\gamma},\frac{n-2}{4}}\Big(\frac{\gamma x^2}{2\sigma^2}\Big)}{z^{-n/2} e^{z^2/4\sigma^2} W_{\frac{n}{4\gamma},\frac{n-2}{4}}\Big(\frac{\gamma z^2}{2\sigma^2}\Big)}, & z \le x \end{cases}$$

4.20.1

$$\mathbf{E}_x \exp\Big(-\gamma^2\sigma^2\theta\int_0^\varrho \frac{ds}{\big(Q_s^{(n)}\big)^2}\Big)$$

$$= \exp\Bigg(-\frac{\theta z v \Gamma\Big(\sqrt{\frac{(n-2)^2}{4}+\gamma^2}+1\Big)}{\Gamma\Big(\sqrt{\frac{(n-2)^2}{16}+\frac{\gamma^2}{4}}-\frac{n}{4}+\frac{1}{2}\Big) M_{\frac{n}{4},\sqrt{\frac{(n-2)^2}{16}+\frac{\gamma^2}{4}}}\Big(\frac{z^2}{2\sigma^2}\Big) W_{\frac{n}{4},\sqrt{\frac{(n-2)^2}{16}+\frac{\gamma^2}{4}}}\Big(\frac{z^2}{2\sigma^2}\Big)}\Bigg)$$

$$\times \begin{cases} \dfrac{x^{-n/2} e^{x^2/4\sigma^2} M_{\frac{n}{4},\sqrt{\frac{(n-2)^2}{16}+\frac{\gamma^2}{4}}}\Big(\frac{x^2}{2\sigma^2}\Big)}{z^{-n/2} e^{z^2/4\sigma^2} M_{\frac{n}{4},\sqrt{\frac{(n-2)^2}{16}+\frac{\gamma^2}{4}}}\Big(\frac{z^2}{2\sigma^2}\Big)}, & 0 < x \le z \\[3ex] \dfrac{x^{-n/2} e^{x^2/4\sigma^2} W_{\frac{n}{4},\sqrt{\frac{(n-2)^2}{16}+\frac{\gamma^2}{4}}}\Big(\frac{x^2}{2\sigma^2}\Big)}{z^{-n/2} e^{z^2/4\sigma^2} W_{\frac{n}{4},\sqrt{\frac{(n-2)^2}{16}+\frac{\gamma^2}{4}}}\Big(\frac{z^2}{2\sigma^2}\Big)}, & z \le x \end{cases}$$

8 $Q_s^{(n)}=\sigma e^{-\theta s}R^{(n)}_{e^{2\theta s}-1}\quad n\ge 2,\quad \sigma,\theta>0\quad \varrho=\varrho(v,z)=\min\{s:\ell(s,z)=v\}$

4.21.1
$0<x$

$$\mathbf{E}_x\exp\Bigl(-\int_0^{\varrho}\Bigl(\frac{p^2\sigma^2\theta}{(Q_s^{(n)})^2}+\frac{\theta(q^2-1)}{4\sigma^2}(Q_s^{(n)})^2\Bigr)ds\Bigr)$$

$$=\exp\Biggl(-\frac{\theta qzv\Gamma\Bigl(\sqrt{\frac{(n-2)^2}{4}+p^2}+1\Bigr)}{\Gamma\Bigl(\sqrt{\frac{(n-2)^2}{16}+\frac{p^2}{4}}+\frac{2q-n}{4q}\Bigr)M_{\frac{n}{4q},\sqrt{\frac{(n-2)^2}{16}+\frac{p^2}{4}}}\Bigl(\frac{qz^2}{2\sigma^2}\Bigr)W_{\frac{n}{4q},\sqrt{\frac{(n-2)^2}{16}+\frac{p^2}{4}}}\Bigl(\frac{qz^2}{2\sigma^2}\Bigr)}\Biggr)$$

$$\times\begin{cases}\dfrac{x^{-n/2}e^{x^2/4\sigma^2}M_{\frac{n}{4q},\sqrt{\frac{(n-2)^2}{16}+\frac{p^2}{4}}}\Bigl(\frac{qx^2}{2\sigma^2}\Bigr)}{z^{-n/2}e^{z^2/4\sigma^2}M_{\frac{n}{4q},\sqrt{\frac{(n-2)^2}{16}+\frac{p^2}{4}}}\Bigl(\frac{qz^2}{2\sigma^2}\Bigr)}, & 0<x\le z\\[2ex] \dfrac{x^{-n/2}e^{x^2/4\sigma^2}W_{\frac{n}{4q},\sqrt{\frac{(n-2)^2}{16}+\frac{p^2}{4}}}\Bigl(\frac{qx^2}{2\sigma^2}\Bigr)}{z^{-n/2}e^{z^2/4\sigma^2}W_{\frac{n}{4q},\sqrt{\frac{(n-2)^2}{16}+\frac{p^2}{4}}}\Bigl(\frac{qz^2}{2\sigma^2}\Bigr)}, & z\le x\end{cases}$$

4.27.1
$0\le x$

$$\mathbf{E}_x\Bigl\{\exp\Bigl(-\int_0^{\varrho}\bigl(p\mathbb{1}_{[0,r)}(Q_s^{(n)})+q\mathbb{1}_{[r,\infty)}(Q_s^{(n)})\bigr)ds\Bigr);\ell(\varrho,r)\in dy\Bigr\}$$

$z\le r$

$$=\exp\Bigl(-\frac{\theta 2^{n/2}\sigma^n z^{1-n}e^{z^2/2\sigma^2}M(\frac{p}{2\theta},\frac{n}{2},\frac{r^2}{2\sigma^2})v}{S(\frac{p}{2\theta},\frac{n}{2},\frac{r^2}{2\sigma^2},\frac{z^2}{2\sigma^2})M(\frac{p}{2\theta},\frac{n}{2},\frac{z^2}{2\sigma^2})}$$

$$-\frac{rqU(\frac{q+2\theta}{2\theta},\frac{n+2}{2},\frac{r^2}{2\sigma^2})y}{2U(\frac{q}{2\theta},\frac{n}{2},\frac{r^2}{2\sigma^2})}-\frac{\theta rC(\frac{p}{2\theta},\frac{n}{2},\frac{r^2}{2\sigma^2},\frac{z^2}{2\sigma^2})y}{S(\frac{p}{2\theta},\frac{n}{2},\frac{r^2}{2\sigma^2},\frac{z^2}{2\sigma^2})}\Bigr)\frac{\theta 2^{n/2}\sigma^n e^{(r^2+z^2)/4\sigma^2}}{(rz)^{(n-1)/2}S(\frac{p}{2\theta},\frac{n}{2},\frac{r^2}{2\sigma^2},\frac{z^2}{2\sigma^2})}$$

$$\times\begin{cases}\dfrac{M(\frac{p}{2\theta},\frac{n}{2},\frac{x^2}{2\sigma^2})}{M(\frac{p}{2\theta},\frac{n}{2},\frac{z^2}{2\sigma^2})}\dfrac{\sqrt{v}}{\sqrt{y}}I_1\Bigl(\dfrac{\theta 2^{1+n/2}\sigma^n e^{(r^2+z^2)/4\sigma^2}\sqrt{vy}}{S(\frac{p}{2\theta},\frac{n}{2},\frac{r^2}{2\sigma^2},\frac{z^2}{2\sigma^2})(rz)^{(n-1)/2}}\Bigr)dy, & x\le z\\[2ex] \dfrac{S(\frac{p}{2\theta},\frac{n}{2},\frac{r^2}{2\sigma^2},\frac{x^2}{2\sigma^2})}{S(\frac{p}{2\theta},\frac{n}{2},\frac{r^2}{2\sigma^2},\frac{z^2}{2\sigma^2})}\dfrac{\sqrt{v}}{\sqrt{y}}I_1\Bigl(\dfrac{\theta 2^{1+n/2}\sigma^n e^{(r^2+z^2)/4\sigma^2}\sqrt{vy}}{S(\frac{p}{2\theta},\frac{n}{2},\frac{r^2}{2\sigma^2},\frac{z^2}{2\sigma^2})(rz)^{(n-1)/2}}\Bigr)dy & \\ \quad+\dfrac{S(\frac{p}{2\theta},\frac{n}{2},\frac{x^2}{2\sigma^2},\frac{z^2}{2\sigma^2})}{S(\frac{p}{2\theta},\frac{n}{2},\frac{r^2}{2\sigma^2},\frac{z^2}{2\sigma^2})}\dfrac{r^{(1-n)/2}e^{r^2/4\sigma^2}}{z^{(1-n)/2}e^{z^2/4\sigma^2}}I_0\Bigl(\dfrac{\theta 2^{1+n/2}\sigma^n e^{(r^2+z^2)/4\sigma^2}\sqrt{vy}}{S(\frac{p}{2\theta},\frac{n}{2},\frac{r^2}{2\sigma^2},\frac{z^2}{2\sigma^2})(rz)^{(n-1)/2}}\Bigr)dy, & z\le x\le r\\[2ex] \dfrac{r^{(1-n)/2}e^{r^2/4\sigma^2}U(\frac{q}{2\theta},\frac{n}{2},\frac{x^2}{2\sigma^2})}{z^{(1-n)/2}e^{z^2/4\sigma^2}U(\frac{q}{2\theta},\frac{n}{2},\frac{r^2}{2\sigma^2})}I_0\Bigl(\dfrac{\theta 2^{1+n/2}\sigma^n e^{(r^2+z^2)/4\sigma^2}\sqrt{vy}}{S(\frac{p}{2\theta},\frac{n}{2},\frac{r^2}{2\sigma^2},\frac{z^2}{2\sigma^2})(rz)^{(n-1)/2}}\Bigr)dy, & r\le x\end{cases}$$

8 $Q_s^{(n)} = \sigma e^{-\theta s} R_{e^{2\theta s}-1}^{(n)} \quad n \ge 2, \quad \sigma,\theta > 0 \quad \varrho = \varrho(v,z) = \min\{s : \ell(s,z) = v\}$

$r \le z$

$$= \exp\Big(-\frac{\theta 2^{n/2}\sigma^n z^{1-n} e^{z^2/2\sigma^2} U(\frac{q}{2\theta},\frac{n}{2},\frac{r^2}{2\sigma^2})v}{S(\frac{q}{2\theta},\frac{n}{2},\frac{z^2}{2\sigma^2},\frac{r^2}{2\sigma^2})U(\frac{q}{2\theta},\frac{n}{2},\frac{z^2}{2\sigma^2})}$$

$$-\frac{\theta r C(\frac{q}{2\theta},\frac{n}{2},\frac{r^2}{2\sigma^2},\frac{z^2}{2\sigma^2})y}{S(\frac{q}{2\theta},\frac{n}{2},\frac{z^2}{2\sigma^2},\frac{r^2}{2\sigma^2})} - \frac{rpM(\frac{p+2\theta}{2\theta},\frac{n+2}{2},\frac{r^2}{2\sigma^2})y}{nM(\frac{p}{2\theta},\frac{n}{2},\frac{r^2}{2\sigma^2})}\Big)\frac{\theta 2^{n/2}\sigma^n e^{(r^2+z^2)/4\sigma^2}}{(rz)^{(n-1)/2}S(\frac{q}{2\theta},\frac{n}{2},\frac{z^2}{2\sigma^2},\frac{r^2}{2\sigma^2})}$$

$$\times\begin{cases} \frac{U(\frac{q}{2\theta},\frac{n}{2},\frac{x^2}{2\sigma^2})}{U(\frac{q}{2\theta},\frac{n}{2},\frac{z^2}{2\sigma^2})}\frac{\sqrt{v}}{\sqrt{y}}I_1\Big(\frac{\theta 2^{1+n/2}\sigma^n e^{(r^2+z^2)/4\sigma^2}\sqrt{vy}}{S(\frac{q}{2\theta},\frac{n}{2},\frac{z^2}{2\sigma^2},\frac{r^2}{2\sigma^2})(rz)^{(n-1)/2}}\Big)dy, & z \le x \\ \frac{S(\frac{q}{2\theta},\frac{n}{2},\frac{x^2}{2\sigma^2},\frac{r^2}{2\sigma^2})}{S(\frac{q}{2\theta},\frac{n}{2},\frac{z^2}{2\sigma^2},\frac{r^2}{2\sigma^2})}\frac{\sqrt{v}}{\sqrt{y}}I_1\Big(\frac{\theta 2^{1+n/2}\sigma^n e^{(r^2+z^2)/4\sigma^2}\sqrt{vy}}{S(\frac{q}{2\theta},\frac{n}{2},\frac{z^2}{2\sigma^2},\frac{r^2}{2\sigma^2})(rz)^{(n-1)/2}}\Big)dy & \\ \quad +\frac{S(\frac{q}{2\theta},\frac{n}{2},\frac{z^2}{2\sigma^2},\frac{x^2}{2\sigma^2})}{S(\frac{q}{2\theta},\frac{n}{2},\frac{z^2}{2\sigma^2},\frac{r^2}{2\sigma^2})}\frac{r^{(1-n)/2}e^{r^2/4\sigma^2}}{z^{(1-n)/2}e^{z^2/4\sigma^2}}I_0\Big(\frac{\theta 2^{1+n/2}\sigma^n e^{(r^2+z^2)/4\sigma^2}\sqrt{vy}}{S(\frac{q}{2\theta},\frac{n}{2},\frac{z^2}{2\sigma^2},\frac{r^2}{2\sigma^2})(rz)^{(n-1)/2}}\Big)dy, & r \le x \le z \\ \frac{r^{(1-n)/2}e^{r^2/4\sigma^2}M(\frac{p}{2\theta},\frac{n}{2},\frac{x^2}{2\sigma^2})}{z^{(1-n)/2}e^{z^2/4\sigma^2}M(\frac{p}{2\theta},\frac{n}{2},\frac{r^2}{2\sigma^2})}I_0\Big(\frac{\theta 2^{1+n/2}\sigma^n e^{(r^2+z^2)/4\sigma^2}\sqrt{vy}}{S(\frac{q}{2\theta},\frac{n}{2},\frac{z^2}{2\sigma^2},\frac{r^2}{2\sigma^2})(rz)^{(n-1)/2}}\Big)dy, & x \le r \end{cases}$$

(1) $\mathbf{E}_x\Big\{\exp\Big(-\int_0^{\varrho}\big(p\mathbb{I}_{[0,r)}(Q_s^{(n)}) + q\mathbb{I}_{[r,\infty)}(Q_s^{(n)})\big)ds\Big); \ell(\varrho,r) = 0\Big\}$

$z \le r$

$$= \begin{cases} 0, & r \le x \\ \frac{S(\frac{p}{2\theta},\frac{n}{2},\frac{r^2}{2\sigma^2},\frac{x^2}{2\sigma^2})}{S(\frac{p}{2\theta},\frac{n}{2},\frac{r^2}{2\sigma^2},\frac{z^2}{2\sigma^2})}\exp\Big(-\frac{\theta 2^{n/2}\sigma^n z^{1-n}e^{z^2/2\sigma^2}M(\frac{p}{2\theta},\frac{n}{2},\frac{r^2}{2\sigma^2})v}{S(\frac{p}{2\theta},\frac{n}{2},\frac{r^2}{2\sigma^2},\frac{z^2}{2\sigma^2})M(\frac{p}{2\theta},\frac{n}{2},\frac{z^2}{2\sigma^2})}\Big), & z \le x \le r \\ \frac{M(\frac{p}{2\theta},\frac{n}{2},\frac{x^2}{2\sigma^2})}{M(\frac{p}{2\theta},\frac{n}{2},\frac{z^2}{2\sigma^2})}\exp\Big(-\frac{\theta 2^{n/2}\sigma^n z^{1-n}e^{z^2/2\sigma^2}M(\frac{p}{2\theta},\frac{n}{2},\frac{r^2}{2\sigma^2})v}{S(\frac{p}{2\theta},\frac{n}{2},\frac{r^2}{2\sigma^2},\frac{z^2}{2\sigma^2})M(\frac{p}{2\theta},\frac{n}{2},\frac{z^2}{2\sigma^2})}\Big), & 0 \le x \le z \end{cases}$$

$r \le z$

$$= \begin{cases} 0, & 0 \le x \le r \\ \frac{S(\frac{q}{2\theta},\frac{n}{2},\frac{x^2}{2\sigma^2},\frac{r^2}{2\sigma^2})}{S(\frac{q}{2\theta},\frac{n}{2},\frac{z^2}{2\sigma^2},\frac{r^2}{2\sigma^2})}\exp\Big(-\frac{\theta 2^{n/2}\sigma^n z^{1-n}e^{z^2/2\sigma^2}U(\frac{q}{2\theta},\frac{n}{2},\frac{r^2}{2\sigma^2})v}{U(\frac{q}{2\theta},\frac{n}{2},\frac{z^2}{2\sigma^2})S(\frac{q}{2\theta},\frac{n}{2},\frac{z^2}{2\sigma^2},\frac{r^2}{2\sigma^2})}\Big), & r \le x \le z \\ \frac{U(\frac{q}{2\theta},\frac{n}{2},\frac{x^2}{2\sigma^2})}{U(\frac{q}{2\theta},\frac{n}{2},\frac{z^2}{2\sigma^2})}\exp\Big(-\frac{\theta 2^{n/2}\sigma^n z^{1-n}e^{z^2/2\sigma^2}U(\frac{q}{2\theta},\frac{n}{2},\frac{r^2}{2\sigma^2})v}{U(\frac{q}{2\theta},\frac{n}{2},\frac{z^2}{2\sigma^2})S(\frac{q}{2\theta},\frac{n}{2},\frac{z^2}{2\sigma^2},\frac{r^2}{2\sigma^2})}\Big), & z \le x \end{cases}$$

9. GEOMETRIC BROWNIAN MOTION

1. Exponential stopping

1.0.5 $$\mathbf{P}_x(V_\tau \in dz) = \begin{cases} \dfrac{\lambda}{z\sigma^2\sqrt{\nu^2+2\lambda/\sigma^2}}\Big(\dfrac{x}{z}\Big)^{\sqrt{\nu^2+2\lambda/\sigma^2}-\nu} dz, & x \le z \\ \dfrac{\lambda}{z\sigma^2\sqrt{\nu^2+2\lambda/\sigma^2}}\Big(\dfrac{z}{x}\Big)^{\sqrt{\nu^2+2\lambda/\sigma^2}+\nu} dz, & z \le x \end{cases}$$

1.0.6 $$\mathbf{P}_x\big(V_t \in dz\big) = \frac{1}{z\sigma\sqrt{2\pi t}} e^{-(\ln(z/x)-\nu\sigma^2 t)^2/2\sigma^2 t} dz$$

1.1.2 $x \le y$ $$\mathbf{P}_x\big(\sup_{0\le s\le\tau} V_s \ge y\big) = \Big(\frac{x}{y}\Big)^{\sqrt{\nu^2+2\lambda/\sigma^2}-\nu}$$

1.1.4 $x \le y$ $$\mathbf{P}_x\big(\sup_{0\le s\le t} V_s \ge y\big) = \frac{1}{2}\,\mathrm{Erfc}\Big(\frac{\ln(y/x)}{\sigma\sqrt{2t}} - \frac{\nu\sigma\sqrt{t}}{\sqrt{2}}\Big) + \frac{1}{2}\Big(\frac{y}{x}\Big)^{2\nu}\mathrm{Erfc}\Big(\frac{\ln(y/x)}{\sigma\sqrt{2t}} + \frac{\nu\sigma\sqrt{t}}{\sqrt{2}}\Big)$$

(1) $x \le y$ $$\mathbf{P}_x\big(\sup_{0\le s<\infty} V_s \ge y\big) = \Big(\frac{x}{y}\Big)^{|\nu|-\nu}$$

1.1.6 $x \le y$ $z \le y$ $$\mathbf{P}_x\big(\sup_{0\le s\le\tau} V_s \ge y, V_\tau \in dz\big) = \frac{\lambda}{z\sigma^2\sqrt{\nu^2+2\lambda/\sigma^2}}\Big(\frac{z}{x}\Big)^{\nu}\Big(\frac{zx}{y^2}\Big)^{\sqrt{\nu^2+2\lambda/\sigma^2}} dz$$

1.1.8 $x \le y$ $$\mathbf{P}_x\big(\sup_{0\le s\le t} V_s \ge y, V_t \in dz\big) = \begin{cases} \dfrac{z^{\nu-1}}{x^\nu\sigma\sqrt{2\pi t}} e^{-\nu^2\sigma^2 t/2} e^{-\ln^2(z/x)/2\sigma^2 t} dz, & y \le z \\ \dfrac{z^{\nu-1}}{x^\nu\sigma\sqrt{2\pi t}} e^{-\nu^2\sigma^2 t/2} e^{-\ln^2(y^2/zx)/2\sigma^2 t} dz, & z \le y \end{cases}$$

1.2.2 $y \le x$ $$\mathbf{P}_x\big(\inf_{0\le s\le\tau} V_s \le y\big) = \Big(\frac{y}{x}\Big)^{\sqrt{\nu^2+2\lambda/\sigma^2}+\nu}$$

1.2.4 $y \le x$ $$\mathbf{P}_x\big(\inf_{0\le s\le t} V_s \le y\big) = \frac{1}{2}\,\mathrm{Erfc}\Big(\frac{\ln(x/y)}{\sigma\sqrt{2t}} + \frac{\nu\sigma\sqrt{t}}{\sqrt{2}}\Big) + \frac{1}{2}\Big(\frac{y}{x}\Big)^{2\nu}\mathrm{Erfc}\Big(\frac{\ln(x/y)}{\sigma\sqrt{2t}} - \frac{\nu\sigma\sqrt{t}}{\sqrt{2}}\Big)$$

9	$V_s = e^{\sigma^2\nu s+\sigma W_s}$ $\quad \sigma > 0$ $\quad \tau \sim \operatorname{Exp}(\lambda)$, independent of V
(1) $y \le x$	$\mathbf{P}_x\Big(\inf_{0\le s<\infty} V_s \le y\Big) = \Big(\frac{y}{x}\Big)^{\lvert\nu\rvert+\nu}$
1.2.6 $y \le x$ $y \le z$	$\mathbf{P}_x\Big(\inf_{0\le s\le\tau} V_s \le y, V_\tau \in dz\Big) = \frac{\lambda}{z\sigma^2\sqrt{\nu^2+2\lambda/\sigma^2}}\Big(\frac{z}{x}\Big)^{\nu}\Big(\frac{y^2}{xz}\Big)^{\sqrt{\nu^2+2\lambda/\sigma^2}} dz$
1.2.8 $y \le x$	$\mathbf{P}_x\Big(\inf_{0\le s\le t} V_s \le y, V_t \in dz\Big) = \begin{cases} \frac{z^{\nu-1}}{x^\nu\sigma\sqrt{2\pi t}} e^{-\nu^2\sigma^2 t/2} e^{-\ln^2(zx/y^2)/2\sigma^2 t} dz, & y \le z \\ \frac{z^{\nu-1}}{x^\nu\sigma\sqrt{2\pi t}} e^{-\nu^2\sigma^2 t/2} e^{-\ln^2(x/z)/2\sigma^2 t} dz, & z \le y \end{cases}$
1.3.1	$\mathbf{E}_x e^{-\gamma\ell(\tau,r)} = \begin{cases} 1 - \frac{\gamma}{\gamma+\sigma r\sqrt{\sigma^2\nu^2+2\lambda}}\Big(\frac{x}{r}\Big)^{\sqrt{\nu^2+2\lambda/\sigma^2}-\nu}, & x \le r \\ 1 - \frac{\gamma}{\gamma+\sigma r\sqrt{\sigma^2\nu^2+2\lambda}}\Big(\frac{r}{x}\Big)^{\sqrt{\nu^2+2\lambda/\sigma^2}+\nu}, & r \le x \end{cases}$
1.3.2 $0 < y$	$\mathbf{P}_x\big(\ell(\tau,r) \in dy\big) = \begin{cases} \frac{\sigma r\sqrt{\sigma^2\nu^2+2\lambda}}{e^{y\sigma r\sqrt{\nu^2\sigma^2+2\lambda}}}\Big(\frac{x}{r}\Big)^{\sqrt{\nu^2+2\lambda/\sigma^2}-\nu} dy, & x \le r \\ \frac{\sigma r\sqrt{\sigma^2\nu^2+2\lambda}}{e^{y\sigma r\sqrt{\nu^2\sigma^2+2\lambda}}}\Big(\frac{r}{x}\Big)^{\sqrt{\nu^2+2\lambda/\sigma^2}+\nu} dy, & r \le x \end{cases}$
(1)	$\mathbf{P}_x\big(\ell(\tau,r) = 0\big) = \begin{cases} 1 - \Big(\frac{x}{r}\Big)^{\sqrt{\nu^2+2\lambda/\sigma^2}-\nu}, & x \le r \\ 1 - \Big(\frac{r}{x}\Big)^{\sqrt{\nu^2+2\lambda/\sigma^2}+\nu}, & r \le x \end{cases}$
1.3.3	$\mathbf{E}_x e^{-\gamma\ell(t,r)} = 1 - \frac{\gamma}{2}\Big(\frac{r}{x}\Big)^{\nu}\Big[\frac{e^{\lvert\nu\rvert\lvert\ln(r/x)\rvert}}{\gamma-\lvert\nu\rvert\sigma^2 r}\operatorname{Erfc}\Big(\frac{\lvert\ln(r/x)\rvert}{\sigma\sqrt{2t}} + \frac{\lvert\nu\rvert\sigma\sqrt{t}}{\sqrt{2}}\Big)$ $+\frac{e^{-\lvert\nu\rvert\lvert\ln(r/x)\rvert}}{\gamma+\lvert\nu\rvert\sigma^2 r}\operatorname{Erfc}\Big(\frac{\lvert\ln(r/x)\rvert}{\sigma\sqrt{2t}} - \frac{\lvert\nu\rvert\sigma\sqrt{t}}{\sqrt{2}}\Big)$ $+\frac{2\gamma e^{\lvert\ln(r/x)\rvert\gamma/\sigma^2 r}}{\gamma^2-\nu^2\sigma^4 r^2} e^{-(\nu^2\sigma^2-\gamma^2/\sigma^2 r^2)t/2}\operatorname{Erfc}\Big(\frac{\lvert\ln(r/x)\rvert}{\sigma\sqrt{2t}} + \frac{\gamma\sqrt{t}}{\sigma r\sqrt{2}}\Big)\Big]$
(1)	$\mathbf{E}_x e^{-\gamma\ell(\infty,r)} = \begin{cases} 1 - \frac{\gamma}{\gamma+\lvert\nu\rvert\sigma r}\Big(\frac{x}{r}\Big)^{\lvert\nu\rvert-\nu}, & x \le r \\ 1 - \frac{\gamma}{\gamma+\lvert\nu\rvert\sigma r}\Big(\frac{r}{x}\Big)^{\lvert\nu\rvert+\nu}, & r \le x \end{cases}$

9 $V_s = e^{\sigma^2 \nu s + \sigma W_s}$ $\sigma > 0$ $\tau \sim \text{Exp}(\lambda)$, independent of V

1.3.4
$0 < y$

$$\mathbf{P}_x\big(\ell(t,r) \in dy\big) = \frac{\sigma r\sqrt{2}}{\sqrt{\pi t}}\Big(\frac{r}{x}\Big)^{\nu} e^{-\nu^2\sigma^2 t/2 - (\sigma^2 ry + |\ln(r/x)|)^2/2\sigma^2 t} dy$$

$$+\frac{|\nu|\sigma^2 r}{2}\Big(\frac{r}{x}\Big)^{\nu}\Big[e^{-|\nu|(\sigma^2 ry + |\ln(r/x)|)}\,\text{Erfc}\Big(\frac{\sigma^2 ry + |\ln(r/x)|}{\sigma\sqrt{2t}} - \frac{|\nu|\sigma\sqrt{t}}{\sqrt{2}}\Big)$$

$$-e^{|\nu|(\sigma^2 ry + |\ln(r/x)|)}\,\text{Erfc}\Big(\frac{\sigma^2 ry + |\ln(r/x)|}{\sigma\sqrt{2t}} + \frac{|\nu|\sigma\sqrt{t}}{\sqrt{2}}\Big)\Big] dy$$

(1)

$$\mathbf{P}_x\big(\ell(t,r) = 0\big) = 1 - \frac{1}{2}\Big(\frac{r}{x}\Big)^{\nu}\Big[e^{|\nu||\ln(r/x)|}\,\text{Erfc}\Big(\frac{|\ln(r/x)|}{\sigma\sqrt{2t}} + \frac{|\nu|\sigma\sqrt{t}}{\sqrt{2}}\Big)$$

$$+e^{-|\nu||\ln(r/x)|}\,\text{Erfc}\Big(\frac{|\ln(r/x)|}{\sigma\sqrt{2t}} - \frac{|\nu|\sigma\sqrt{t}}{\sqrt{2}}\Big)\Big]$$

(2)
$0 < y$

$$\mathbf{P}_x\big(\ell(\infty,r) \in dy\big) = \begin{cases} |\nu|\sigma^2 r\Big(\dfrac{x}{r}\Big)^{|\nu|-\nu} e^{-|\nu|\sigma^2 ry} dy, & x \le r \\ |\nu|\sigma^2 r\Big(\dfrac{r}{x}\Big)^{|\nu|+\nu} e^{-|\nu|\sigma^2 ry} dy, & r \le x \end{cases}$$

(3)

$$\mathbf{P}_x\big(\ell(\infty,r) = 0\big) = \begin{cases} 1 - \Big(\dfrac{x}{r}\Big)^{|\nu|-\nu}, & x \le r \\ 1 - \Big(\dfrac{r}{x}\Big)^{|\nu|+\nu}, & r \le x \end{cases}$$

1.3.8
$0 < y$

$$\mathbf{P}_x\big(\ell(t,r) \in dy, V_t \in dz\big) = \frac{r}{z\sigma t\sqrt{2\pi t}}\big(\sigma^2 ry + |\ln(z/r)| + |\ln(r/x)|\big)$$

$$\times\Big(\frac{z}{x}\Big)^{\nu} e^{-\nu^2\sigma^2 t/2 - (|\ln(z/r)| + |\ln(r/x)| + \sigma^2 ry)^2/2\sigma^2 t} dy dz$$

(1)

$$\mathbf{P}_x\big(\ell(t,r) = 0, V_t \in dz\big) = \frac{1}{z\sigma\sqrt{2\pi t}} e^{-(\ln(z/x) - \nu\sigma^2 t)^2/2\sigma^2 t} dz$$

$$-\frac{1}{z\sigma\sqrt{2\pi t}}\Big(\frac{z}{x}\Big)^{\nu} e^{-\nu^2\sigma^2 t/2 - (|\ln(z/r)| + |\ln(r/x)|)^2/2\sigma^2 t} dz$$

1.4.3
(1)
$\nu < 0$

$$\mathbf{E}_x \exp\Big(-\gamma\int_0^\infty \mathbb{I}_{[r,\infty)}(V_s) ds\Big) = \begin{cases} 1 - \dfrac{\sqrt{\nu^2 + 2\gamma/\sigma^2} + \nu}{\sqrt{\nu^2 + 2\gamma/\sigma^2} - \nu}\Big(\dfrac{x}{r}\Big)^{2|\nu|}, & x \le r \\ \dfrac{2|\nu|}{\sqrt{\nu^2 + 2\gamma/\sigma^2} - \nu}\Big(\dfrac{r}{x}\Big)^{\sqrt{\nu^2 + 2\gamma/\sigma^2} + \nu}, & r \le x \end{cases}$$

9 $V_s = e^{\sigma^2 \nu s + \sigma W_s}$ $\quad \sigma > 0$ $\quad \tau \sim \operatorname{Exp}(\lambda)$, independent of V

1.4.4 $$\mathbf{P}_x\Big(\int_0^t \mathbb{I}_{[r,\infty)}(V_s)ds \in dy\Big) = B_x^{(6)}(t-y,y)\mathbb{I}_{(0,t)}(y)dy$$

for $B_x^{(6)}(u,v)$ see 1.6.2

(1) $x \le r$ $$\mathbf{P}_x\Big(\int_0^t \mathbb{I}_{[r,\infty)}(V_s)ds = 0\Big)$$

$$= 1 - \frac{1}{2}\operatorname{Erfc}\Big(\frac{\ln(r/x)}{\sigma\sqrt{2t}} - \frac{\nu\sigma\sqrt{t}}{\sqrt{2}}\Big) - \frac{1}{2}\Big(\frac{r}{x}\Big)^{2\nu}\operatorname{Erfc}\Big(\frac{\ln(r/x)}{\sigma\sqrt{2t}} + \frac{\nu\sigma\sqrt{t}}{\sqrt{2}}\Big)$$

(2) $r \le x$ $$\mathbf{P}_x\Big(\int_0^t \mathbb{I}_{[r,\infty)}(V_s)ds = t\Big)$$

$$= 1 - \frac{1}{2}\operatorname{Erfc}\Big(\frac{\ln(x/r)}{\sqrt{2t}} + \frac{\nu\sigma\sqrt{t}}{\sqrt{2}}\Big) - \frac{1}{2}\Big(\frac{r}{x}\Big)^{2\nu}\operatorname{Erfc}\Big(\frac{\ln(x/r)}{\sqrt{2t}} - \frac{\nu\sigma\sqrt{t}}{\sqrt{2}}\Big)$$

(3) $\nu < 0$ $$\mathbf{P}_x\Big(\int_0^\infty \mathbb{I}_{[r,\infty)}(V_s)ds \in dy\Big)/dy$$

$x \le r$ $$= \frac{|\nu|\sigma\sqrt{2}}{\sqrt{\pi t}}\Big(\frac{x}{r}\Big)^{2|\nu|} e^{-\nu^2\sigma^2 y/2} - \nu^2\sigma^2\Big(\frac{x}{r}\Big)^{2|\nu|}\operatorname{Erfc}\Big(\frac{|\nu|\sigma\sqrt{y}}{\sqrt{2}}\Big)$$

$r \le x$ $$= \frac{|\nu|\sigma\sqrt{2}}{\sqrt{\pi t}}\Big(\frac{x}{r}\Big)^{|\nu|} e^{-\nu^2\sigma^2 y/2 - \ln^2(x/r)/2\sigma^2 y} - \nu^2\sigma^2\Big(\frac{x}{r}\Big)^{2|\nu|}\operatorname{Erfc}\Big(\frac{\ln(x/r)}{\sigma\sqrt{2y}} - \frac{\nu\sigma\sqrt{y}}{\sqrt{2}}\Big)$$

(4) $x \le r$ $$\mathbf{P}_x\Big(\int_0^\infty \mathbb{I}_{[r,\infty)}(V_s)ds = 0\Big) = 1 - \Big(\frac{x}{r}\Big)^{|\nu|-\nu}$$

1.5.3 (1) $\nu > 0$ $$\mathbf{E}_x \exp\Big(-\gamma\int_0^\infty \mathbb{I}_{(0,r)}(V_s)ds\Big) = \begin{cases} \dfrac{2\nu}{\sqrt{\nu^2+2\gamma/\sigma^2}+\nu}\Big(\dfrac{x}{r}\Big)^{\sqrt{\nu^2+2\gamma/\sigma^2}-\nu}, & x \le r \\ 1 - \dfrac{\sqrt{\nu^2+2\gamma/\sigma^2}-\nu}{\sqrt{\nu^2+2\gamma/\sigma^2}+\nu}\Big(\dfrac{r}{x}\Big)^{2\nu}, & r \le x \end{cases}$$

1.5.4 $$\mathbf{P}_x\Big(\int_0^t \mathbb{I}_{(0,r)}(V_s)ds \in dy\Big) = B_x^{(6)}(y,t-y)\mathbb{I}_{(0,t)}(y)dy$$

for $B_x^{(6)}(u,v)$ see 1.6.2

(1) $r \le x$ $$\mathbf{P}_x\Big(\int_0^t \mathbb{I}_{(0,r)}(V_s)ds = 0\Big)$$

9 $V_s = e^{\sigma^2 \nu s + \sigma W_s}$ $\sigma > 0$ $\tau \sim \text{Exp}(\lambda)$, independent of V

$$= 1 - \frac{1}{2}\operatorname{Erfc}\Big(\frac{\ln(x/r)}{\sigma\sqrt{2t}} + \frac{\nu\sigma\sqrt{t}}{\sqrt{2}}\Big) - \frac{1}{2}\Big(\frac{r}{x}\Big)^{2\nu}\operatorname{Erfc}\Big(\frac{\ln(x/r)}{\sigma\sqrt{2t}} - \frac{\nu\sigma\sqrt{t}}{\sqrt{2}}\Big)$$

(2) $x \le r$

$$\mathbf{P}_x\Big(\int_0^t \mathbb{I}_{(0,r)}(V_s)ds = t\Big)$$

$$= 1 - \frac{1}{2}\operatorname{Erfc}\Big(\frac{\ln(r/x)}{\sqrt{2t}} - \frac{\nu\sigma\sqrt{t}}{\sqrt{2}}\Big) - \frac{1}{2}\Big(\frac{r}{x}\Big)^{2\nu}\operatorname{Erfc}\Big(\frac{\ln(r/x)}{\sqrt{2t}} + \frac{\nu\sigma\sqrt{t}}{\sqrt{2}}\Big)$$

(3) $\nu > 0$

$$\mathbf{P}_x\Big(\int_0^\infty \mathbb{I}_{(0,r)}(V_s)ds \in dy\Big)/dy$$

$x \le r$
$$= \frac{\nu\sigma\sqrt{2}}{\sqrt{\pi t}}\Big(\frac{r}{x}\Big)^{\nu} e^{-\nu^2\sigma^2 y/2 - \ln^2(r/x)/2\sigma^2 y} - \nu^2\sigma^2\Big(\frac{r}{x}\Big)^{2\nu}\operatorname{Erfc}\Big(\frac{\ln(r/x)}{\sigma\sqrt{2y}} + \frac{\nu\sigma\sqrt{y}}{\sqrt{2}}\Big)$$

$r \le x$
$$= \frac{\nu\sigma\sqrt{2}}{\sqrt{\pi t}}\Big(\frac{r}{x}\Big)^{2\nu} e^{-\nu^2\sigma^2 y/2} - \nu^2\sigma^2\Big(\frac{r}{x}\Big)^{2\nu}\operatorname{Erfc}\Big(\frac{\nu\sigma\sqrt{y}}{\sqrt{2}}\Big)$$

1.6.1
$$\mathbf{E}_x \exp\Big(-\int_0^\tau \big(p\mathbb{I}_{(0,r)}(V_s) + q\mathbb{I}_{[r,\infty)}(V_s)\big)ds\Big)$$

$x \le r$
$$= \frac{\lambda}{\lambda+p} - \frac{\lambda(q-p)\big(\sqrt{\nu^2+2(\lambda+q)/\sigma^2}+\nu\big)(x/r)^{\sqrt{\nu^2+2(\lambda+p)/\sigma^2}-\nu}}{(\lambda+p)(\lambda+q)\big(\sqrt{\nu^2+2(\lambda+q)/\sigma^2}+\sqrt{\nu^2+2(\lambda+p)/\sigma^2}\big)}$$

$r \le x$
$$= \frac{\lambda}{\lambda+q} + \frac{\lambda(q-p)\big(\sqrt{\nu^2+2(\lambda+p)/\sigma^2}-\nu\big)(r/x)^{\sqrt{\nu^2+2(\lambda+q)/\sigma^2}+\nu}}{(\lambda+p)(\lambda+q)\big(\sqrt{\nu^2+2(\lambda+q)/\sigma^2}+\sqrt{\nu^2+2(\lambda+p)/\sigma^2}\big)}$$

1.6.2
$$\mathbf{P}_x\Big(\int_0^\tau \mathbb{I}_{(0,r)}(V_s)ds \in du, \int_0^\tau \mathbb{I}_{[r,\infty)}(V_s)ds \in dv\Big)$$

$$=: \lambda e^{-\lambda(u+v)} B_x^{(6)}(u,v)dudv$$

$x \le r$
$$= \lambda e^{-\lambda(u+v)}\Big(\frac{r}{x}\Big)^{\nu}\Big(\frac{1}{\sqrt{\pi v}}\exp\Big(-\frac{\nu^2\sigma^2 v}{2}\Big) + \frac{\nu\sigma}{\sqrt{2}}\operatorname{Erfc}\Big(-\frac{\nu\sigma\sqrt{v}}{\sqrt{2}}\Big)\Big)$$

$$\times\Big(\frac{1}{\sqrt{\pi u}}\exp\Big(-\frac{\ln^2(r/x)}{2\sigma^2 u} - \frac{\nu^2\sigma^2 u}{2}\Big) - \frac{\nu\sigma}{\sqrt{2}}\Big(\frac{r}{x}\Big)^{\nu}\operatorname{Erfc}\Big(\frac{\ln(r/x)}{\sigma\sqrt{2u}} + \frac{\nu\sigma\sqrt{u}}{\sqrt{2}}\Big)\Big)dudv$$

$r \le x$
$$= \lambda e^{-\lambda(u+v)}\Big(\frac{r}{x}\Big)^{\nu}\Big(\frac{1}{\sqrt{\pi u}}\exp\Big(-\frac{\nu^2\sigma^2 u}{2}\Big) - \frac{\nu\sigma}{\sqrt{2}}\operatorname{Erfc}\Big(\frac{\nu\sigma\sqrt{u}}{\sqrt{2}}\Big)\Big)$$

9 $V_s = e^{\sigma^2\nu s+\sigma W_s}$ $\sigma>0$	$\tau \sim \text{Exp}(\lambda)$, independent of V
	$\times\Big(\frac{1}{\sqrt{\pi v}}\exp\Big(-\frac{\ln^2(x/r)}{2\sigma^2 v}-\frac{\nu^2\sigma^2 v}{2}\Big)+\frac{\nu\sigma}{\sqrt{2}}\Big(\frac{r}{x}\Big)^{\nu}\text{Erfc}\Big(\frac{\ln(x/r)}{\sigma\sqrt{2v}}-\frac{\nu\sigma\sqrt{v}}{\sqrt{2}}\Big)\Big)dudv$
(1) $x\le r$	$\mathbf{P}_x\Big(\int_0^\tau \mathbb{I}_{(0,r)}(V_s)ds\in du, \int_0^\tau \mathbb{I}_{[r,\infty)}(V_s)ds=0\Big)$
	$=\lambda e^{-\lambda u}\Big(1-\frac12\text{Erfc}\Big(\frac{\ln(r/x)}{\sigma\sqrt{2t}}-\frac{\nu\sigma\sqrt{t}}{\sqrt2}\Big)-\frac12\Big(\frac{r}{x}\Big)^{2\nu}\text{Erfc}\Big(\frac{\ln(r/x)}{\sigma\sqrt{2t}}+\frac{\nu\sigma\sqrt{t}}{\sqrt2}\Big)\Big)du$
(2) $r\le x$	$\mathbf{P}_x\Big(\int_0^\tau \mathbb{I}_{(0,r)}(V_s)ds=0, \int_0^\tau \mathbb{I}_{[r,\infty)}(V_s)ds\in dv\Big)$
	$=\lambda e^{-\lambda v}\Big(1-\frac12\text{Erfc}\Big(\frac{\ln(x/r)}{\sigma\sqrt{2t}}+\frac{\nu\sigma\sqrt{t}}{\sqrt2}\Big)-\frac12\Big(\frac{r}{x}\Big)^{2\nu}\text{Erfc}\Big(\frac{\ln(x/r)}{\sigma\sqrt{2t}}-\frac{\nu\sigma\sqrt{t}}{\sqrt2}\Big)\Big)dv$
1.6.4	$\mathbf{P}_x\Big(\int_0^t (p\mathbb{I}_{(0,r)}(V_s)+q\mathbb{I}_{[r,\infty)}(V_s))ds\in dv,\Big)$
	$=\frac{1}{\lvert p-q\rvert}B_x^{(6)}\Big(\frac{\lvert qt-v\rvert}{\lvert p-q\rvert},\frac{\lvert pt-v\rvert}{\lvert p-q\rvert}\Big)\mathbb{I}_{((q\wedge p)t,(q\vee p)t)}(v)\,dv$ for $B_x^{(6)}(u,v)$ see 1.6.2
(1) $x\le r$	$\mathbf{P}_x\Big(\int_0^t (p\mathbb{I}_{(0,r)}(V_s)+q\mathbb{I}_{[r,\infty)}(V_s))ds=pt\Big)$
	$=1-\frac12\text{Erfc}\Big(\frac{\ln(r/x)}{\sigma\sqrt{2t}}-\frac{\nu\sigma\sqrt{t}}{\sqrt2}\Big)-\frac12\Big(\frac{r}{x}\Big)^{2\nu}\text{Erfc}\Big(\frac{\ln(r/x)}{\sigma\sqrt{2t}}+\frac{\nu\sigma\sqrt{t}}{\sqrt2}\Big)$
(2) $r\le x$	$\mathbf{P}_x\Big(\int_0^t (p\mathbb{I}_{(0,r)}(V_s)+q\mathbb{I}_{[r,\infty)}(V_s))ds=qt\Big)$
	$=1-\frac12\text{Erfc}\Big(\frac{\ln(x/r)}{\sigma\sqrt{2t}}+\frac{\nu\sigma\sqrt{t}}{\sqrt2}\Big)-\frac12\Big(\frac{r}{x}\Big)^{2\nu}\text{Erfc}\Big(\frac{\ln(x/r)}{\sigma\sqrt{2t}}-\frac{\nu\sigma\sqrt{t}}{\sqrt2}\Big)$
1.6.5	$\mathbf{E}_x\Big\{\exp\Big(-\int_0^\tau (p\mathbb{I}_{(0,r)}(V_s)+q\mathbb{I}_{[r,\infty)}(V_s))ds\Big);V_\tau\in dz\Big\}$
$x\le r$ $z\le r$	$=\frac{\lambda}{z\sigma^2\sqrt{\nu^2+2(\lambda+p)/\sigma^2}}\Big(\frac{z}{x}\Big)^{\nu}e^{-\lvert\ln(z/x)\rvert\sqrt{\nu^2+2(\lambda+p)/\sigma^2}}dz$

9 $V_s = e^{\sigma^2 \nu s + \sigma W_s}$ $\sigma > 0$ $\tau \sim \mathrm{Exp}(\lambda)$, independent of V

$$+\Big(\frac{2\lambda z^{-1}\sigma^{-1}(z/x)^{\nu}}{\sqrt{\nu^2\sigma^2+2\lambda+2p}+\sqrt{\nu^2\sigma^2+2\lambda+2q}}-\frac{\lambda z^{-1}\sigma^{-1}(z/x)^{\nu}}{\sqrt{\nu^2\sigma^2+2\lambda+2p}}\Big)\Big(\frac{zx}{r^2}\Big)^{\sqrt{\nu^2+2(\lambda+p)/\sigma^2}}dz$$

$x \le r$, $r \le z$

$$= \frac{2\lambda z^{-1}\sigma^{-2}(z/x)^{\nu}}{\sqrt{\nu^2+2(\lambda+p)/\sigma^2}+\sqrt{\nu^2+2(\lambda+q)/\sigma^2}}\Big(\frac{x}{r}\Big)^{\sqrt{\nu^2+2(\lambda+p)/\sigma^2}}\Big(\frac{r}{z}\Big)^{\sqrt{\nu^2+2(\lambda+q)/\sigma^2}}dz$$

$r \le x$, $z \le r$

$$= \frac{2\lambda z^{-1}\sigma^{-2}(z/x)^{\nu}}{\sqrt{\nu^2+2(\lambda+p)/\sigma^2}+\sqrt{\nu^2+2(\lambda+q)/\sigma^2}}\Big(\frac{r}{x}\Big)^{\sqrt{\nu^2+2(\lambda+q)/\sigma^2}}\Big(\frac{z}{r}\Big)^{\sqrt{\nu^2+2(\lambda+p)/\sigma^2}}dz$$

$r \le x$, $r \le z$

$$= \frac{\lambda}{\sqrt{\nu^2+2(\lambda+q)/\sigma^2}}\Big(\frac{z}{x}\Big)^{\nu}e^{-|\ln(z/x)|\sqrt{\nu^2+2(\lambda+q)/\sigma^2}}dz$$

$$+\Big(\frac{2\lambda z^{-1}\sigma^{-1}(z/x)^{\nu}}{\sqrt{\nu^2\sigma^2+2\lambda+2p}+\sqrt{\nu^2\sigma^2+2\lambda+2q}}-\frac{\lambda z^{-1}\sigma^{-2}(z/x)^{\nu}}{\sqrt{\nu^2\sigma^2+2\lambda+2q}}\Big)\Big(\frac{r^2}{xz}\Big)^{\sqrt{\nu^2+2(\lambda+q)/\sigma^2}}dz$$

1.8.2

$$\mathbf{P}_x\Big(\int_0^{\tau} V_s ds \in dy\Big) = \frac{\lambda\sigma^{2\nu-1}y^{\nu-1/2}\Gamma(\sqrt{\nu^2+2\lambda/\sigma^2}+\nu)}{x^{\nu+1/2}2^{\nu-1/2}\Gamma(2\sqrt{\nu^2+2\lambda/\sigma^2}+1)}$$

$$\times e^{-x/\sigma^2 y}M_{1/2-\nu,\sqrt{\nu^2+2\lambda/\sigma^2}}\Big(\frac{2x}{\sigma^2 y}\Big)dy$$

1.8.3, $\nu < 0$

$$\mathbf{E}_x \exp\Big(-\gamma\int_0^{\infty} V_s ds\Big) = \frac{2^{|\nu|+1}\gamma^{|\nu|}x^{|\nu|}}{\sigma^{2|\nu|}\Gamma(2|\nu|)}K_{2|\nu|}\Big(\frac{2\sqrt{2\gamma x}}{\sigma}\Big)$$

1.8.4, $-1 < \nu$

$$\mathbf{P}_x\Big(\int_0^{t} V_s ds \in dy\Big) = \frac{\sigma^{2\nu+1}y^{\nu-1/2}}{2^{\nu+1/2}x^{\nu+1/2}}e^{-\nu^2\sigma^2 t/2 - x/\sigma^2 y}\,\mathrm{m}_{\sigma^2 t/2}\Big(\nu-\frac{1}{2},\frac{x}{\sigma^2 y}\Big)dy$$

(1), $\nu < 0$

$$\mathbf{P}_x\Big(\int_0^{\infty} V_s ds \in dy\Big) = \frac{2^{2|\nu|}x^{2|\nu|}}{\sigma^{4|\nu|}y^{2|\nu|+1}\Gamma(2|\nu|)}e^{-2x/\sigma^2 y}dy$$

1.8.5

$$\mathbf{E}_x\Big\{\exp\Big(-\gamma\int_0^{\tau} V_s ds\Big); V_\tau \in dz\Big\}$$

$$= \begin{cases} \dfrac{4\lambda z^{\nu-1}}{x^{\nu}\sigma^2}I_{2\sqrt{\nu^2+2\lambda/\sigma^2}}\Big(\dfrac{2\sqrt{2\gamma x}}{\sigma}\Big)K_{2\sqrt{\nu^2+2\lambda/\sigma^2}}\Big(\dfrac{2\sqrt{2\gamma z}}{\sigma}\Big)dz, & x \le z \\ \dfrac{4\lambda z^{\nu-1}}{x^{\nu}\sigma^2}K_{2\sqrt{\nu^2+2\lambda/\sigma^2}}\Big(\dfrac{2\sqrt{2\gamma x}}{\sigma}\Big)I_{2\sqrt{\nu^2+2\lambda/\sigma^2}}\Big(\dfrac{2\sqrt{2\gamma z}}{\sigma}\Big)dz, & z \le x \end{cases}$$

9	$V_s = e^{\sigma^2\nu s+\sigma W_s}$ $\quad \sigma>0$ $\qquad \tau \sim \operatorname{Exp}(\lambda)$, independent of V
1.8.6	$\mathbf{P}_x\Big(\int_0^\tau V_s ds \in dy, V_\tau \in dz\Big) = \frac{2\lambda z^{\nu-1}}{\sigma^2 y x^\nu} e^{-2(x+z)/\sigma^2 y} I_{2\sqrt{\nu^2+2\lambda/\sigma^2}}\Big(\frac{4\sqrt{xz}}{\sigma^2 y}\Big) dydz$
1.8.7	$\mathbf{E}_x\Big\{\exp\Big(-\gamma\int_0^t V_s ds\Big);\ V_t \in dz\Big\} = \frac{z^{\nu-1}}{2x^\nu} e^{-\nu^2\sigma^2 t/2}\, \mathrm{ki}_{\sigma^2 t/8}\Big(\frac{2\sqrt{2\gamma x}}{\sigma}, \frac{2\sqrt{2\gamma z}}{\sigma}\Big) dz$
1.8.8	$\mathbf{E}_x\Big(\int_0^t V_s ds \in dy,\ V_t \in dz\Big) = \frac{z^{\nu-1}}{4yx^\nu} e^{-\nu^2\sigma^2 t/2-2(x+z)/\sigma^2 y}\, \mathrm{i}_{\sigma^2 t/8}\Big(\frac{4\sqrt{xz}}{\sigma^2 y}\Big) dydz$
1.11.2	$\mathbf{P}_x\Big(\sup_{0<y<\infty}(y\ell(\tau,y)) \ge h\Big) = \frac{h^2\sigma^2(2\lambda+\nu^2\sigma^2)}{8\,\mathrm{sh}^2\big(h\sigma\sqrt{2\lambda+\nu^2\sigma^2}/2\big)}$
	$\times\left(\frac{M\big(\frac32+\frac{\nu\sigma}{2\sqrt{2\lambda+\nu^2\sigma^2}}, 3, h\sigma\sqrt{2\lambda+\nu^2\sigma^2}\big)}{M\big(\frac12+\frac{\nu\sigma}{2\sqrt{2\lambda+\nu^2\sigma^2}}, 1, h\sigma\sqrt{2\lambda+\nu^2\sigma^2}\big)} + \frac{M\big(\frac32-\frac{\nu\sigma}{2\sqrt{2\lambda+\nu^2\sigma^2}}, 3, h\sigma\sqrt{2\lambda+\nu^2\sigma^2}\big)}{M\big(\frac12-\frac{\nu\sigma}{2\sqrt{2\lambda+\nu^2\sigma^2}}, 1, h\sigma\sqrt{2\lambda+\nu^2\sigma^2}\big)}\right)$
1.12.1	$\mathbf{E}_x e^{-\gamma \check{H}(\tau)} = \frac{\sqrt{\nu^2+2\lambda/\sigma^2}-\nu}{\sqrt{\nu^2+2(\lambda+\gamma)/\sigma^2}-\nu}$
(1)	$\mathbf{E}_x e^{-\gamma \hat{H}(\tau)} = \frac{\sqrt{\nu^2+2\lambda/\sigma^2}+\nu}{\sqrt{\nu^2+2(\lambda+\gamma)/\sigma^2}+\nu}$
1.12.2	$\mathbf{P}_x\big(\check{H}(\tau)\in dv\big) = \frac{\sigma(\sqrt{\nu^2+2\lambda/\sigma^2}-\nu)}{e^{\lambda v}}\Big(\frac{e^{-\nu^2\sigma^2 v/2}}{\sqrt{2\pi v}} + \frac{\nu\sigma}{2}\operatorname{Erfc}\Big(-\frac{\nu\sigma\sqrt{v}}{\sqrt2}\Big)\Big)dv$
(1)	$\mathbf{P}_x\big(\hat{H}(\tau)\in dv\big) = \frac{\sigma(\sqrt{\nu^2+2\lambda/\sigma^2}+\nu)}{e^{\lambda v}}\Big(\frac{e^{-\nu^2\sigma^2 v/2}}{\sqrt{2\pi v}} - \frac{\nu\sigma}{2}\operatorname{Erfc}\Big(\frac{\nu\sigma\sqrt{v}}{\sqrt2}\Big)\Big)dv$
1.12.3 (1)	$\mathbf{E}_x e^{-\gamma \check{H}(\infty)} = \frac{\lvert\nu\rvert-\nu}{\sqrt{\nu^2+2\gamma/\sigma^2}-\nu}$
(2)	$\mathbf{E}_x e^{-\gamma \hat{H}(\infty)} = \frac{\lvert\nu\rvert+\nu}{\sqrt{\nu^2+2\gamma/\sigma^2}+\nu}$
1.12.4	$\mathbf{P}_x\big(\check{H}(t)\in dv\big)$ $\qquad 0\le v\le t$
	$= \Big(\frac{e^{-\nu^2\sigma^2 v/2}}{\sqrt{\pi v}} + \frac{\nu\sigma}{\sqrt2}\operatorname{Erfc}\Big(-\frac{\nu\sigma\sqrt{v}}{\sqrt2}\Big)\Big)\Big(\frac{e^{-\nu^2\sigma^2(t-v)/2}}{\sqrt{\pi(t-v)}} - \frac{\nu\sigma}{\sqrt2}\operatorname{Erfc}\Big(\frac{\nu\sigma\sqrt{t-v}}{\sqrt2}\Big)\Big)dv$
(1)	$\mathbf{P}_x\big(\hat{H}(t)\in dv\big)$ $\qquad 0\le v\le t$

9 $V_s = e^{\sigma^2 \nu s + \sigma W_s}$ $\sigma > 0$ $\tau \sim \text{Exp}(\lambda)$, independent of V

$$= \Big(\frac{e^{-\nu^2\sigma^2 v/2}}{\sqrt{\pi v}} - \frac{\nu\sigma}{\sqrt{2}}\operatorname{Erfc}\Big(\frac{\nu\sigma\sqrt{v}}{\sqrt{2}}\Big)\Big)\Big(\frac{e^{-\nu^2\sigma^2(t-v)/2}}{\sqrt{\pi(t-v)}} + \frac{\nu\sigma}{\sqrt{2}}\operatorname{Erfc}\Big(-\frac{\nu\sigma\sqrt{t-v}}{\sqrt{2}}\Big)\Big)dv$$

(2) $$\mathbf{P}_x\big(\check{H}(\infty)\in dv\big) = \sigma(|\nu|-\nu)\Big(\frac{e^{-\nu^2\sigma^2 v/2}}{\sqrt{2\pi v}} + \frac{\nu\sigma}{2}\operatorname{Erfc}\Big(-\frac{\nu\sigma\sqrt{v}}{\sqrt{2}}\Big)\Big)dv$$

(3) $$\mathbf{P}_x\big(\hat{H}(\infty)\in dv\big) = \sigma(|\nu|+\nu)\Big(\frac{e^{-\nu^2\sigma^2 v/2}}{\sqrt{2\pi v}} - \frac{\nu\sigma}{2}\operatorname{Erfc}\Big(\frac{\nu\sigma\sqrt{v}}{\sqrt{2}}\Big)\Big)dv$$

1.12.5 $$\mathbf{E}_x\{e^{-\gamma\check{H}(\tau)}; V_\tau\in dz\} = \begin{cases} \dfrac{2\lambda(x/z)^{\sqrt{\nu^2+2(\lambda+\gamma)/\sigma^2}-\nu}}{z\sigma(\sqrt{\nu^2\sigma^2+2\lambda+2\gamma}+\sqrt{\nu^2\sigma^2+2\lambda})}dz & x\le z\\[2ex] \dfrac{2\lambda(z/x)^{\sqrt{\nu^2+2\lambda/\sigma^2}+\nu}}{z\sigma(\sqrt{\nu^2\sigma^2+2\lambda+2\gamma}+\sqrt{\nu^2\sigma^2+2\lambda})}dz & z\le x \end{cases}$$

(1) $$\mathbf{E}_x\{e^{-\gamma\hat{H}(\tau)}; V_\tau\in dz\} = \begin{cases} \dfrac{2\lambda(x/z)^{\sqrt{\nu^2+2\lambda/\sigma^2}-\nu}}{z\sigma(\sqrt{\nu^2\sigma^2+2\lambda+2\gamma}+\sqrt{\nu^2\sigma^2+2\lambda})}dz & x\le z\\[2ex] \dfrac{2\lambda(z/x)^{\sqrt{\nu^2+2(\lambda+\gamma)/\sigma^2}+\nu}}{z\sigma(\sqrt{\nu^2\sigma^2+2\lambda+2\gamma}+\sqrt{\nu^2\sigma^2+2\lambda})}dz & z\le x \end{cases}$$

1.13.1
$x<y$ $$\mathbf{E}_x\{e^{-\gamma\check{H}(\tau)};\ \sup_{0\le s\le\tau} V_s\in dy\} = \frac{\sqrt{\nu^2+2\lambda/\sigma^2}-\nu}{y}\Big(\frac{x}{y}\Big)^{\sqrt{\nu^2+2(\lambda+\gamma)/\sigma^2}-\nu}dy$$

1.13.2
$x<y$ $$\mathbf{P}_x\big(\check{H}(\tau)\in dv,\ \sup_{0\le s\le\tau} V_s\in dy\big)$$

$$= \frac{2\lambda y^{\nu-1}\ln(y/x)}{x^\nu\sigma^2\sqrt{2\pi}(\sqrt{\nu^2\sigma^2+2\lambda}+\nu\sigma)v^{3/2}}\exp\Big(-\lambda v - \frac{\nu^2\sigma^2 v}{2} - \frac{\ln^2(y/x)}{2\sigma^2 v}\Big)dvdy$$

1.13.3
(1)
$x<y$ $$\mathbf{E}_x\{e^{-\gamma\check{H}(\infty)};\ \sup_{0\le s<\infty} V_s\in dy\} = \frac{|\nu|-\nu}{y}\Big(\frac{x}{y}\Big)^{\sqrt{\nu^2+2\gamma/\sigma^2}-\nu}dy$$

1.13.4
$x<y$ $$\mathbf{P}_x\big(\check{H}(t)\in dv,\ \sup_{0\le s\le t} V_s\in dy\big) \qquad 0\le v\le t$$

$$= \frac{\ln(y/x)}{y\sigma^2\sqrt{\pi}v^{3/2}}\exp\Big(-\frac{(\ln(y/x)-\nu\sigma^2 v)^2}{2\sigma^2 v}\Big)\Big(\frac{e^{-\nu^2\sigma^2(t-v)/2}}{\sqrt{\pi(t-v)}} - \frac{\nu\sigma}{\sqrt{2}}\operatorname{Erfc}\Big(\frac{\nu\sigma\sqrt{t-v}}{\sqrt{2}}\Big)\Big)dvdy$$

(1)
$x<y$ $$\mathbf{P}_x\big(\check{H}(\infty)\in dv,\ \sup_{0\le s<\infty} V_s\in dy\big) = \frac{(|\nu|-\nu)\ln(y/x)}{y\sigma\sqrt{2\pi}v^{3/2}}\exp\Big(-\frac{(\ln(y/x)-\nu\sigma^2 v)^2}{2\sigma^2 v}\Big)dvdy$$

9 $V_s = e^{\sigma^2\nu s+\sigma W_s}$ $\sigma > 0$ $\tau \sim \text{Exp}(\lambda)$, independent of V

1.13.5 $\mathbf{E}_x\{e^{-\gamma\check{H}(\tau)};\ \sup_{0\le s\le\tau} V_s \in dy, V_\tau \in dz\}$ $x \vee z < y$

$$= \frac{2\lambda}{yz\sigma^2}\Big(\frac{z}{x}\Big)^\nu\Big(\frac{x}{y}\Big)^{\sqrt{\nu^2+2(\lambda+\gamma)/\sigma^2}}\Big(\frac{z}{y}\Big)^{\sqrt{\nu^2+2\lambda/\sigma^2}} dydz$$

1.13.6 $\mathbf{P}_x\big(\check{H}(\tau) \in dv,\ \sup_{0\le s\le\tau} V_s \in dy, V_\tau \in dz\big)$ $x \vee z < y$

$$= \frac{2\lambda\ln(y/x)}{yz\sigma^3\sqrt{2\pi}v^{3/2}}\Big(\frac{z}{x}\Big)^\nu\Big(\frac{z}{y}\Big)^{\sqrt{\nu^2+2\lambda/\sigma^2}}\exp\Big(-\frac{\ln^2(y/x)}{2\sigma^2 v} - \lambda v - \frac{\nu^2\sigma^2 v}{2}\Big)dvdydz$$

1.13.8 $v < t$ $\mathbf{P}_x\big(\check{H}(t) \in dv,\ \sup_{0\le s\le t} V_s \in dy, V_t \in dz\big)$ $x \vee z < y$

$$= \frac{\ln(y/x)\ln(y/z)}{yz\sigma^4\pi(v(t-v))^{3/2}}\exp\Big(-\frac{(\ln(y/x)-\nu\sigma^2 v)^2}{2\sigma^2 v} - \frac{(\ln(y/z)+\nu\sigma^2(t-v))^2}{2\sigma^2(t-v)}\Big)dvdydz$$

1.14.1 $y < x$ $\mathbf{E}_x\{e^{-\gamma\hat{H}(\tau)};\ \inf_{0\le s\le\tau} V_s \in dy\} = \frac{\sqrt{\nu^2+2\lambda/\sigma^2}+\nu}{y}\Big(\frac{y}{x}\Big)^{\sqrt{\nu^2+2(\lambda+\gamma)/\sigma^2}+\nu} dy$

1.14.2 $y < x$ $\mathbf{P}_x\big(\hat{H}(\tau) \in dv,\ \inf_{0\le s\le\tau} V_s \in dy\big)$

$$= \frac{2\lambda y^{\nu-1}\ln(x/y)}{x^\nu\sigma^2\sqrt{2\pi}(\sqrt{\nu^2\sigma^2+2\lambda}-\nu\sigma)v^{3/2}}\exp\Big(-\lambda v - \frac{\nu^2\sigma^2 v}{2} - \frac{\ln^2(x/y)}{2\sigma^2 v}\Big)dvdy$$

1.14.3 (1) $y < x$ $\mathbf{E}_x\{e^{-\gamma\hat{H}(\infty)};\ \inf_{0\le s<\infty} V_s \in dy\} = \frac{|\nu|+\nu}{y}\Big(\frac{y}{x}\Big)^{\sqrt{\nu^2+2\gamma/\sigma^2}+\nu} dy$

1.14.4 $y < x$ $\mathbf{P}_x\big(\hat{H}(t) \in dv,\ \inf_{0\le s\le t} V_s \in dy\big)$ $0 \le v \le t$

$$= \frac{\ln(x/y)}{y\sigma^2\sqrt{\pi}v^{3/2}}\exp\Big(-\frac{(\ln(x/y)+\nu\sigma^2 v)^2}{2\sigma^2 v}\Big)\Big(\frac{e^{-\nu^2\sigma^2(t-v)/2}}{\sqrt{\pi(t-v)}} + \frac{\nu\sigma}{\sqrt{2}}\text{Erfc}\Big(-\frac{\nu\sigma\sqrt{t-v}}{\sqrt{2}}\Big)\Big)dvdy$$

(1) $y < x$ $\mathbf{P}_x\big(\hat{H}(\infty) \in dv,\ \inf_{0\le s<\infty} V_s \in dy\big) = \frac{(|\nu|+\nu)\ln(x/y)}{y\sigma\sqrt{2\pi}v^{3/2}}\exp\Big(-\frac{(\ln(x/y)+\nu\sigma^2 v)^2}{2\sigma^2 v}\Big)dvdy$

9 $V_s = e^{\sigma^2 \nu s + \sigma W_s}$ $\sigma > 0$ $\tau \sim \text{Exp}(\lambda)$, independent of V

1.14.5 $\mathbf{E}_x\{e^{-\gamma \hat{H}(\tau)}; \inf_{0\le s\le\tau} V_s \in dy, V_\tau \in dz\}$ $y < x \wedge z$

$$= \frac{2\lambda}{yz\sigma^2}\Big(\frac{z}{x}\Big)^\nu \Big(\frac{y}{x}\Big)^{\sqrt{\nu^2+2(\lambda+\gamma)/\sigma^2}} \Big(\frac{y}{z}\Big)^{\sqrt{\nu^2+2\lambda/\sigma^2}} dydz$$

1.14.6 $\mathbf{P}_x\big(\hat{H}(\tau) \in dv, \inf_{0\le s\le\tau} V_s \in dy, V_\tau \in dz\big)$ $y < x \wedge z$

$$= \frac{2\lambda \ln(x/y)}{yz\sigma^3\sqrt{2\pi}v^{3/2}}\Big(\frac{z}{x}\Big)^\nu \Big(\frac{y}{z}\Big)^{\sqrt{\nu^2+2\lambda/\sigma^2}} \exp\Big(-\frac{\ln^2(x/y)}{2\sigma^2 v} - \lambda v - \frac{\nu^2\sigma^2 v}{2}\Big) dvdydz$$

1.14.8 $\mathbf{P}_x\big(\hat{H}(t) \in dv, \inf_{0\le s\le t} V_s \in dy, V_t \in dz\big)$ $y < x \wedge z$

$v < t$

$$= \frac{\ln(x/y)\ln(z/y)}{yz\sigma^4\pi(v(t-v))^{3/2}} \exp\Big(-\frac{(\ln(x/y)+\nu\sigma^2 v)^2}{2\sigma^2 v} - \frac{(\ln(z/y)-\nu\sigma^2(t-v))^2}{2\sigma^2(t-v)}\Big) dvdydz$$

1.15.2 $\mathbf{P}_x\big(a < \inf_{0\le s\le\tau} V_s; \sup_{0\le s\le\tau} V_s < b\big) = 1$

$$-\frac{\big(\frac{a}{x}\big)^\nu\Big(\big(\frac{b}{x}\big)^{\sqrt{\nu^2+2\lambda/\sigma^2}} - \big(\frac{x}{b}\big)^{\sqrt{\nu^2+2\lambda/\sigma^2}}\Big) + \big(\frac{b}{x}\big)^\nu\Big(\big(\frac{x}{a}\big)^{\sqrt{\nu^2+2\lambda/\sigma^2}} - \big(\frac{a}{x}\big)^{\sqrt{\nu^2+2\lambda/\sigma^2}}\Big)}{\big(\frac{b}{a}\big)^{\sqrt{\nu^2+2\lambda/\sigma^2}} - \big(\frac{a}{b}\big)^{\sqrt{\nu^2+2\lambda/\sigma^2}}}$$

1.15.4 $\mathbf{P}_x\big(a < \inf_{0\le s\le t} V_s; \sup_{0\le s\le t} V_s < b\big) = \dfrac{e^{-\nu^2\sigma^2 t/2}}{\sigma\sqrt{2\pi t}}$

$$\times \sum_{k=-\infty}^{\infty} \int_a^b \frac{z^{\nu-1}}{x^\nu}\big(e^{-(\ln(z/x)+2k\ln(b/a))^2/2\sigma^2 t} - e^{-(\ln(zx/a^2)+2k\ln(b/a))^2/2\sigma^2 t}\big) dz$$

1.15.6 $\mathbf{P}_x\big(a < \inf_{0\le s\le\tau} V_s; \sup_{0\le s\le\tau} V_s < b, V_\tau \in dz\big)$ $a < x \wedge z < x \vee z < b$

$$= \frac{2\lambda z^{\nu-1}\big(\text{ch}\big((\ln(b/a) - |\ln(z/x)|)\sqrt{\nu^2+2\lambda/\sigma^2}\big) - \text{ch}\big(\ln(ba/zx)\sqrt{\nu^2+2\lambda/\sigma^2}\big)\big)}{x^\nu\sigma\sqrt{\nu^2\sigma^2+2\lambda}\big((b/a)^{\sqrt{\nu^2+2\lambda/\sigma^2}} - (a/b)^{\sqrt{\nu^2+2\lambda/\sigma^2}}\big)} dz$$

1.15.8 $\mathbf{P}_x\big(a < \inf_{0\le s\le t} V_s; \sup_{0\le s\le t} V_s < b\,, V_t \in dz\big)/dz$ $a < x \wedge z \le x \vee z < b$

$$= \frac{z^{\nu-1}e^{-\nu^2\sigma^2 t/2}}{x^\nu\sigma\sqrt{2\pi t}} \sum_{k=-\infty}^{\infty}\big(e^{-(\ln(z/x)+2k\ln(b/a))^2/2\sigma^2 t} - e^{-(\ln(zx/a^2)+2k\ln(b/a))^2/2\sigma^2 t}\big)$$

9	$V_s = e^{\sigma^2\nu s+\sigma W_s}$ $\qquad \sigma > 0$ $\qquad \tau \sim \mathrm{Exp}(\lambda)$, independent of V
1.19.2 $0 < x$	$\mathbf{P}_x\Big(\int_0^\tau \frac{ds}{V_s} \in dy\Big)$ $= \frac{2\lambda(\sigma^2 y/2)^{-\nu}\Gamma(\sqrt{\nu^2+2\lambda/\sigma^2}-\nu)}{\sigma x^{\nu-1/2}\sqrt{2y}\,\Gamma(2\sqrt{\nu^2+2\lambda/\sigma^2}+1)} e^{-1/\sigma^2 xy} M_{1/2+\nu,\sqrt{\nu^2+2\lambda/\sigma^2}}\Big(\frac{2}{\sigma^2 xy}\Big)dy$
1.19.3 $0 < \nu$	$\mathbf{E}_x \exp\Big(-\gamma\int_0^\infty \frac{ds}{V_s}\Big) = \frac{2^{\nu+1}\gamma^\nu x^{-\nu}}{\sigma^{2\nu}\Gamma(2\nu)} K_{2\nu}\Big(\frac{2\sqrt{2\gamma}}{\sigma\sqrt{x}}\Big)$
1.19.4 $\nu < 1$	$\mathbf{P}_x\Big(\int_0^t \frac{ds}{V_s} \in dy\Big) = \frac{\sigma(\sigma^2 y/2)^{-\nu}}{\sqrt{2y}x^{\nu-1/2}} e^{-\nu^2\sigma^2 t/2-1/\sigma^2 xy}\, \mathrm{m}_{\sigma^2 t/2}\Big(-\nu-\frac{1}{2},\frac{1}{\sigma^2 xy}\Big)dy$
(1) $0 < \nu$	$\mathbf{P}_x\Big(\int_0^\infty \frac{ds}{V_s} \in dy\Big) = \frac{2^{2\nu}x^{-2\nu}}{\sigma^{4\nu}y^{2\nu+1}\Gamma(2\nu)} e^{-2/\sigma^2 xy}dy$
1.19.5 $0 < x$	$\mathbf{E}_x\Big\{\exp\Big(-\gamma\int_0^\tau \frac{ds}{V_s}\Big); V_\tau \in dz\Big\}$ $= \begin{cases} \frac{4\lambda z^{\nu-1}}{x^\nu\sigma^2} K_{2\sqrt{\nu^2+2\lambda/\sigma^2}}\Big(\frac{2\sqrt{2\gamma}}{\sigma\sqrt{x}}\Big) I_{2\sqrt{\nu^2+2\lambda/\sigma^2}}\Big(\frac{2\sqrt{2\gamma}}{\sigma\sqrt{z}}\Big)dz, & x \le z \\ \frac{4\lambda z^{\nu-1}}{x^\nu\sigma^2} I_{2\sqrt{\nu^2+2\lambda/\sigma^2}}\Big(\frac{2\sqrt{2\gamma}}{\sigma\sqrt{x}}\Big) K_{2\sqrt{\nu^2+2\lambda/\sigma^2}}\Big(\frac{2\sqrt{2\gamma}}{\sigma\sqrt{z}}\Big)dz, & z \le x \end{cases}$
1.19.6 $0 < x$	$\mathbf{P}_x\Big(\int_0^\tau \frac{ds}{V_s} \in dy, V_\tau \in dz\Big) = \frac{2\lambda z^{\nu-1}}{\sigma^2 y x^\nu} e^{-2(x+z)/\sigma^2 xzy} I_{2\sqrt{\nu^2+2\lambda/\sigma^2}}\Big(\frac{4}{\sigma^2 y\sqrt{xz}}\Big)dydz$
1.19.7 $0 < x$	$\mathbf{E}_x\Big\{\exp\Big(-\gamma\int_0^t \frac{ds}{V_s}\Big);\ V_t \in dz\Big\} = \frac{z^{\nu-1}}{2x^\nu} e^{-\nu^2\sigma^2 t/2}\, \mathrm{ki}_{\sigma^2 t/8}\Big(\frac{2\sqrt{2\gamma}}{\sigma\sqrt{x}},\frac{2\sqrt{2\gamma}}{\sigma\sqrt{z}}\Big)dz$
1.19.8 $0 < x$	$\mathbf{P}_x\Big(\int_0^t \frac{ds}{V_s} \in dy, V_t \in dz\Big) = \frac{z^{\nu-1}}{4yx^\nu} e^{-\nu^2\sigma^2 t/2-2(x+z)/\sigma^2 xzy}\, \mathrm{i}_{\sigma^2 t/8}\Big(\frac{4}{\sigma^2 y\sqrt{xz}}\Big)dydz$
1.20.2	$\mathbf{P}_x\Big(\int_0^\tau V_s^{2\beta}ds \in dy\Big) = \frac{\lambda(2\sigma^2\beta^2 y)^{\nu/2\beta-1/2}\Gamma(\sqrt{\nu^2+2\lambda/\sigma^2}/2\lvert\beta\rvert+\nu/2\beta)}{x^{\nu+\beta}\Gamma(\sqrt{\nu^2+2\lambda/\sigma^2}/\lvert\beta\rvert+1)}$ $\times e^{-x^{2\beta}/4\sigma^2\beta^2 y} M_{1/2-\nu/2\beta,\sqrt{\nu^2+2\lambda/\sigma^2}/2\lvert\beta\rvert}\Big(\frac{x^{2\beta}}{2\sigma^2\beta^2 y}\Big)dy$
1.20.3 $\frac{\nu}{\beta} < 0$	$\mathbf{E}_x \exp\Big(-\frac{\gamma^2}{2}\int_0^\infty V_s^{2\beta}ds\Big) = \frac{2(\gamma/2\sigma\lvert\beta\rvert)^{\lvert\nu\rvert/\lvert\beta\rvert}x^{-\nu}}{\Gamma(\lvert\nu\rvert/\lvert\beta\rvert)} K_{\lvert\nu\rvert/\lvert\beta\rvert}\Big(\frac{\gamma x^\beta}{\sigma\lvert\beta\rvert}\Big)$

9 $V_s = e^{\sigma^2 \nu s + \sigma W_s}$ $\sigma > 0$ $\tau \sim \mathrm{Exp}(\lambda)$, independent of V

1.20.4 $\mathbf{P}_x\Big(\int_0^t V_s^{2\beta} ds \in dy\Big)$

$\frac{\nu}{\beta} > -2$

$$= \frac{(2\sigma^2\beta^2)^{\nu/2\beta+1/2}}{x^{\nu+\beta} y^{1/2-\nu/2\beta}} e^{-\nu^2\sigma^2 t/2 - x^{2\beta}/4\sigma^2\beta^2 y}\, \mathrm{m}_{2\sigma^2\beta^2 t}\Big(\frac{\nu}{2\beta} - \frac{1}{2}, \frac{x^{2\beta}}{4\sigma^2\beta^2 y}\Big) dy$$

(1) $\frac{\nu}{\beta} < 0$

$$\mathbf{P}_x\Big(\int_0^\infty V_s^{2\beta} ds \in dy\Big) = \frac{(2\sigma^2\beta^2 y)^{-|\nu|/|\beta|}}{y\Gamma(|\nu|/|\beta|)} x^{-2\nu} e^{-x^{2\beta}/2\sigma^2\beta^2 y} dy$$

1.20.5 $\mathbf{E}_x\Big\{\exp\Big(-\frac{\gamma^2}{2}\int_0^\tau V_s^{2\beta} ds\Big);\ V_\tau \in dz\Big\}$

$$= \begin{cases} \frac{2\lambda z^{\nu-1}}{x^\nu \sigma^2 |\beta|} I_{\sqrt{\nu^2+2\lambda/\sigma^2}/|\beta|}\Big(\frac{\gamma x^\beta}{\sigma|\beta|}\Big) K_{\sqrt{\nu^2+2\lambda/\sigma^2}/|\beta|}\Big(\frac{\gamma z^\beta}{\sigma|\beta|}\Big) dz, & (z-x)\beta \ge 0 \\ \frac{2\lambda z^{\nu-1}}{x^\nu \sigma^2 |\beta|} K_{\sqrt{\nu^2+2\lambda/\sigma^2}/|\beta|}\Big(\frac{\gamma x^\beta}{\sigma|\beta|}\Big) I_{\sqrt{\nu^2+2\lambda/\sigma^2}/|\beta|}\Big(\frac{\gamma z^\beta}{\sigma|\beta|}\Big) dz, & (z-x)\beta \le 0 \end{cases}$$

1.20.6 $\mathbf{P}_x\Big(\int_0^\tau V_s^{2\beta} ds \in dy, V_\tau \in dz\Big)$

$$= \frac{\lambda z^{\nu-1}}{\sigma^2 y x^\nu |\beta|} e^{-(x^{2\beta}+z^{2\beta})/2\sigma^2\beta^2 y} I_{\sqrt{\nu^2+2\lambda/\sigma^2}/|\beta|}\Big(\frac{x^\beta z^\beta}{\sigma^2\beta^2 y}\Big) dy dz$$

1.20.7 $$\mathbf{E}_x\Big\{\exp\Big(-\frac{\gamma^2}{2}\int_0^t V_s^{2\beta} ds\Big); V_t \in dz\Big\} = \frac{z^{\nu-1}|\beta|}{x^\nu} e^{-\nu^2\sigma^2 t/2}\, \mathrm{ki}_{\sigma^2\beta^2 t/2}\Big(\frac{\gamma x^\beta}{\sigma|\beta|}, \frac{\gamma z^\beta}{\sigma|\beta|}\Big) dz$$

1.20.8 $\mathbf{P}_x\Big(\int_0^t V_s^{2\beta} ds \in dy,\ V_t \in dz\Big)$

$$= \frac{z^{\nu-1}|\beta|}{2 y x^\nu} e^{-\nu^2\sigma^2 t/2 - (x^{2\beta}+z^{2\beta})/2\sigma^2\beta^2 y}\, \mathrm{i}_{\sigma^2\beta^2 t/2}\Big(\frac{x^\beta z^\beta}{\sigma^2\beta^2 y}\Big) dy dz$$

1.21.1 (1) $\mathbf{E}_x\Big\{\exp\Big(-\frac{p^2}{2}\int_0^\tau V_s^{2\beta} ds\Big); \int_0^\tau V_s^\beta ds \in dy\Big\}$

$$= \frac{\lambda\Gamma(\sqrt{\nu^2+2\lambda/\sigma^2}/|\beta| + \nu/\beta)}{x^{\nu+\beta/2}\Gamma(2\sqrt{\nu^2+2\lambda/\sigma^2}/|\beta| + 1)} \Big(\frac{p\,\mathrm{ch}(yp\sigma|\beta|/2)}{\sigma|\beta|\,\mathrm{sh}(yp\sigma|\beta|/2)}\Big)^{1/2-\nu/\beta}$$

$$\times \exp\Big(-\frac{p x^\beta\,\mathrm{ch}(yp\sigma|\beta|)}{\sigma|\beta|\,\mathrm{sh}(yp\sigma|\beta|)}\Big) M_{\frac{1}{2}-\frac{\nu}{\beta}, \frac{\sqrt{\nu^2+2\lambda/\sigma^2}}{|\beta|}}\Big(\frac{2p x^\beta}{\sigma|\beta|\,\mathrm{sh}(yp\sigma|\beta|)}\Big) dy$$

9 $V_s = e^{\sigma^2 \nu s + \sigma W_s}$ $\sigma > 0$ $\tau \sim \operatorname{Exp}(\lambda)$, independent of V

1.21.3 (1)

$$\mathbf{E}_x\Big\{\exp\Big(-\frac{p^2}{2}\int_0^t V_s^{2\beta}ds\Big);\ \int_0^t V_s^{\beta}ds \in dy\Big\} = \frac{\sigma^2\beta^2}{2}\Big(\frac{p\operatorname{ch}(yp\sigma|\beta|/2)}{\sigma|\beta|\operatorname{sh}(yp\sigma|\beta|/2)}\Big)^{1/2-\nu/\beta}$$

$\frac{\nu}{\beta} > -2$

$$\times x^{-(\nu+\beta/2)}\exp\Big(-\frac{\nu^2\sigma^2 t}{2} - \frac{px^\beta\operatorname{ch}(yp\sigma|\beta|)}{\sigma|\beta|\operatorname{sh}(yp\sigma|\beta|)}\Big)\operatorname{m}_{\sigma^2\beta^2 t/2}\Big(\frac{\nu}{\beta}-\frac{1}{2}, \frac{px^\beta}{\sigma|\beta|\operatorname{sh}(yp\sigma|\beta|)}\Big)dy$$

(2)

$$\mathbf{E}_x\exp\Big(-\int_0^\infty\Big(\frac{p^2}{2}V_s^{2\beta} + qV_s^{\beta}\Big)ds\Big)$$

$\frac{\nu}{\beta} < 0$

$$= \frac{\Gamma(\frac{|\nu|}{|\beta|} + \frac{q}{p\sigma|\beta|} + \frac{1}{2})}{\Gamma(2|\nu|/|\beta|)}\Big(\frac{2p}{\sigma|\beta|}\Big)^{|\nu|/|\beta|-1/2} x^{-(\nu+\beta/2)}W_{-q/p\sigma|\beta|,|\nu|/|\beta|}\Big(\frac{2px^\beta}{\sigma|\beta|}\Big)$$

(3)

$$\mathbf{E}_x\Big\{\exp\Big(-\frac{p^2}{2}\int_0^\infty V_s^{2\beta}ds\Big);\ \int_0^\infty V_s^{\beta}ds \in dy\Big\}$$

$\frac{\nu}{\beta} < 0$

$$= \frac{\sigma^2\beta^2 x^{-2\nu}}{2\Gamma(2|\nu|/|\beta|)}\Big(\frac{p}{\sigma|\beta|\operatorname{sh}(yp\sigma|\beta|/2)}\Big)^{1+2|\nu|/|\beta|}\exp\Big(-\frac{px^\beta\operatorname{ch}(yp\sigma|\beta|/2)}{\sigma|\beta|\operatorname{sh}(yp\sigma|\beta|/2)}\Big)dy$$

1.21.4 (1)

$$\mathbf{P}_x\Big(\int_0^\infty V_s^{2\beta}ds \in dg,\ \int_0^\infty V_s^{\beta}ds \in dy\Big)$$

$\frac{\nu}{\beta} < 0$

$$= \frac{(\sigma|\beta|)^{1-2|\nu|/|\beta|}x^{-2\nu}}{2\Gamma(2|\nu|/|\beta|)}\operatorname{ee}_g\Big(1+\frac{2|\nu|}{|\beta|}, \frac{y\sigma|\beta|}{2}, \frac{x^\beta}{\sigma|\beta|}, 0\Big)dgdy$$

1.21.5

$$\mathbf{E}_x\Big\{\exp\Big(-\int_0^\tau\Big(\frac{p^2}{2}V_s^{2\beta} + qV_s^{\beta}\Big)ds\Big);\ V_\tau \in dz\Big\} = \frac{\lambda z^{-\beta/2}\Gamma\Big(\frac{\sqrt{\nu^2+2\lambda/\sigma^2}}{|\beta|} + \frac{1}{2} + \frac{q}{p\sigma|\beta|}\Big)}{\sigma p z x^{\beta/2}\Gamma(\frac{2}{|\beta|}\sqrt{\nu^2+2\lambda/\sigma^2}+1)}$$

$$\times\begin{cases}\frac{z^\nu}{x^\nu}M_{-\frac{q}{p\sigma|\beta|},\frac{\sqrt{\nu^2+2\lambda/\sigma^2}}{|\beta|}}\Big(\frac{2px^\beta}{\sigma|\beta|}\Big)W_{-\frac{q}{p\sigma|\beta|},\frac{\sqrt{\nu^2+2\lambda/\sigma^2}}{|\beta|}}\Big(\frac{2pz^\beta}{\sigma|\beta|}\Big)dz, & (z-x)\beta \ge 0\\ \frac{z^\nu}{x^\nu}W_{-\frac{q}{p\sigma|\beta|},\frac{\sqrt{\nu^2+2\lambda/\sigma^2}}{|\beta|}}\Big(\frac{2px^\beta}{\sigma|\beta|}\Big)M_{-\frac{q}{p\sigma|\beta|},\frac{\sqrt{\nu^2+2\lambda/\sigma^2}}{|\beta|}}\Big(\frac{2pz^\beta}{\sigma|\beta|}\Big)dz, & (z-x)\beta \le 0\end{cases}$$

(1)

$$\mathbf{E}_x\Big\{\exp\Big(-\frac{p^2}{2}\int_0^\tau V_s^{2\beta}ds\Big);\ \int_0^\tau V_s^{\beta}ds \in dy, V_\tau \in dz\Big\} = \frac{\lambda p z^{\nu-1}}{x^\nu\sigma\operatorname{sh}(yp\sigma|\beta|/2)}$$

$$\times\exp\Big(-\frac{p(x^\beta+z^\beta)\operatorname{ch}(yp\sigma|\beta|/2)}{\sigma|\beta|\operatorname{sh}(yp\sigma|\beta|/2)}\Big)I_{2\sqrt{\nu^2+2\lambda/\sigma^2}/|\beta|}\Big(\frac{2p(xz)^{\beta/2}}{\sigma|\beta|\operatorname{sh}(yp\sigma|\beta|/2)}\Big)dydz$$

9 $V_s = e^{\sigma^2\nu s+\sigma W_s}$ $\sigma > 0$ $\tau \sim \operatorname{Exp}(\lambda)$, independent of V

1.21.6 $$\mathbf{P}_x\Big(\int_0^\tau V_s^{2\beta}ds \in dg,\ \int_0^\tau V_s^{\beta}ds \in dy,\ V_\tau \in dz\Big)$$

$$= \frac{\lambda}{z\sigma}\Big(\frac{z}{x}\Big)^{\nu} \operatorname{is}_g\Big(\frac{2\sqrt{\nu^2+2\lambda/\sigma^2}}{|\beta|}, \frac{y\sigma|\beta|}{2}, 0, \frac{x^\beta+z^\beta}{\sigma|\beta|}, \frac{(xz)^{\beta/2}}{\sigma|\beta|}\Big)dgdydz$$

1.21.7 (1) $$\mathbf{E}_x\Big\{\exp\Big(-\frac{p^2}{2}\int_0^t V_s^{2\beta}ds\Big);\ \int_0^t V_s^{\beta}ds \in dy, V_t \in dz\Big\} = \frac{\sigma\beta^2 p z^{\nu-1}}{8x^\nu \operatorname{sh}(yp\sigma|\beta|/2)}$$

$$\times e^{-\nu^2\sigma^2 t/2}\exp\Big(-\frac{p(x^\beta+z^\beta)\operatorname{ch}(yp\sigma|\beta|/2)}{\sigma|\beta|\operatorname{sh}(yp\sigma|\beta|/2)}\Big)\operatorname{i}_{\sigma^2\beta^2 t/8}\Big(\frac{2p(xz)^{\beta/2}}{\sigma|\beta|\operatorname{sh}(yp\sigma|\beta|/2)}\Big)dydz$$

1.21.8 $$\mathbf{P}_x\Big(\int_0^t V_s^{2\beta}ds \in dg,\ \int_0^t V_s^{\beta}ds \in dy,\ V_t \in dz\Big)$$

$$= \frac{\sigma\beta^2 z^{\nu-1}}{8x^\nu}e^{-\nu^2\sigma^2 t/2}\operatorname{ei}_g\Big(\frac{\sigma^2\beta^2 t}{8}, \frac{y\sigma|\beta|}{2}, \frac{x^\beta+z^\beta}{\sigma|\beta|}, \frac{2(xz)^{\beta/2}}{\sigma|\beta|}\Big)dgdydz$$

1.27.5 $$\mathbf{E}_x\Big\{\exp\Big(-\int_0^\tau\big(p\mathbb{I}_{(0,r)}(V_s)+q\mathbb{I}_{[r,\infty)}(V_s)\big)ds\Big); \tfrac{1}{2}\ell(\tau,r) \in dy, V_\tau \in dz\Big\}$$

$x \le r$, $z \le r$: $$= \frac{2\lambda r}{z}\Big(\frac{z}{x}\Big)^{\nu} e^{-y\sigma r(\sqrt{\nu^2\sigma^2+2\lambda+2p}+\sqrt{\nu^2\sigma^2+2\lambda+2q})}\Big(\frac{zx}{r^2}\Big)^{\sqrt{\nu^2+2(\lambda+p)/\sigma^2}}dydz$$

$x \le r$, $r \le z$: $$= \frac{2\lambda r}{z}\Big(\frac{z}{x}\Big)^{\nu} e^{-y\sigma r(\sqrt{\nu^2\sigma^2+2\lambda+2p}+\sqrt{\nu^2\sigma^2+2\lambda+2q})}\frac{(r/z)^{\sqrt{\nu^2+2(\lambda+q)/\sigma^2}}}{(x/r)^{\sqrt{\nu^2+2(\lambda+p)/\sigma^2}}}dydz$$

$r \le x$, $z \le r$: $$= \frac{2\lambda r}{z}\Big(\frac{z}{x}\Big)^{\nu} e^{-y\sigma r(\sqrt{\nu^2\sigma^2+2\lambda+2p}+\sqrt{\nu^2\sigma^2+2\lambda+2q})}\frac{(r/x)^{\sqrt{\nu^2+2(\lambda+q)/\sigma^2}}}{(z/r)^{\sqrt{\nu^2+2(\lambda+p)/\sigma^2}}}dydz$$

$r \le x$, $r \le z$: $$= \frac{2\lambda r}{z}\Big(\frac{z}{x}\Big)^{\nu} e^{-y\sigma r(\sqrt{\nu^2\sigma^2+2\lambda+2p}+\sqrt{\nu^2\sigma^2+2\lambda+2q})}\Big(\frac{r^2}{xz}\Big)^{\sqrt{\nu^2+2(\lambda+q)/\sigma^2}}dydz$$

(1) $$\mathbf{E}_x\Big\{\exp\Big(-\int_0^\tau\big(p\mathbb{I}_{(0,r)}(V_s)+q\mathbb{I}_{[r,\infty)}(V_s)\big)ds\Big); \ell(\tau,r)=0, V_\tau \in dz\Big\}$$

$x \le r$, $z \le r$: $$= \frac{\lambda}{z\sigma\sqrt{\nu^2\sigma^2+2\lambda+2p}}\Big(\frac{z}{x}\Big)^{\nu}\Big(e^{-|\ln(z/x)|\sqrt{\nu^2+2(\lambda+p)/\sigma^2}} - \Big(\frac{zx}{r^2}\Big)^{\sqrt{\nu^2+2(\lambda+p)/\sigma^2}}\Big)dz$$

$r \le x$, $r \le z$: $$= \frac{\lambda}{z\sigma\sqrt{\nu^2\sigma^2+2\lambda+2q}}\Big(\frac{z}{x}\Big)^{\nu}\Big(e^{-|\ln(z/x)|\sqrt{\nu^2+2(\lambda+q)/\sigma^2}} - \Big(\frac{r^2}{zx}\Big)^{\sqrt{\nu^2+2(\lambda+q)/\sigma^2}}\Big)dz$$

9 $V_s = e^{\sigma^2\nu s+\sigma W_s}$ $\sigma > 0$ $\tau \sim \text{Exp}(\lambda)$, independent of V

1.31.2 $0<\alpha$

$$\mathbf{P}_x\Big(\sup_{0\le s\le\tau}(\ell(s,u)-\alpha^{-1}\ell(s,r))>h\Big)$$

$r \le u$

$$=\Big(\frac{u}{x}\Big)^{\sqrt{\nu^2+2\lambda/\sigma^2}+\nu}\exp\Big(-\frac{\Upsilon_\lambda h\sigma^2 r\sqrt{\nu^2+2\lambda/\sigma^2}}{2-2(r/u)^{2\sqrt{\nu^2+2\lambda/\sigma^2}}}\Big)$$

$$\times\begin{cases}\Big(\frac{x}{r}\Big)^{2\sqrt{\nu^2+2\lambda/\sigma^2}}-\frac{\Upsilon_\lambda\big((x/r)^{2\sqrt{\nu^2+2\lambda/\sigma^2}}-(x/u)^{2\sqrt{\nu^2+2\lambda/\sigma^2}}\big)}{2-2(r/u)^{2\sqrt{\nu^2+2\lambda/\sigma^2}}}, & x\le r\\ 1-\frac{\Upsilon_\lambda\big(1-(x/u)^{2\sqrt{\nu^2+2\lambda/\sigma^2}}\big)}{2-2(r/u)^{2\sqrt{\nu^2+2\lambda/\sigma^2}}}, & r\le x\le u\\ 1, & u\le x\end{cases}$$

$u \le r$

$$=\Big(\frac{x}{u}\Big)^{\sqrt{\nu^2+2\lambda/\sigma^2}-\nu}\exp\Big(-\frac{\Upsilon_\lambda h\sigma^2 r\sqrt{\nu^2+2\lambda/\sigma^2}}{2-2(u/r)^{2\sqrt{\nu^2+2\lambda/\sigma^2}}}\Big)$$

$$\times\begin{cases}1, & x\le u\\ 1-\frac{\Upsilon_\lambda\big(1-(u/x)^{2\sqrt{\nu^2+2\lambda/\sigma^2}}\big)}{2-2(u/r)^{2\sqrt{\nu^2+2\lambda/\sigma^2}}}, & u\le x\le r\\ \Big(\frac{r}{x}\Big)^{2\sqrt{\nu^2+2\lambda/\sigma^2}}-\frac{\Upsilon_\lambda\big((r/x)^{2\sqrt{\nu^2+2\lambda/\sigma^2}}-(u/x)^{2\sqrt{\nu^2+2\lambda/\sigma^2}}\big)}{2-2(u/r)^{2\sqrt{\nu^2+2\lambda/\sigma^2}}}, & r\le x\end{cases}$$

(2)

$$\mathbf{P}_x\Big(\sup_{0\le s<\infty}(\ell(s,u)-\alpha^{-1}\ell(s,r))>h\Big)\Big(\frac{x}{u}\Big)^{\nu+|\nu|\,\text{sign}(u-r)}$$

$r \le u$

$$=\exp\Big(-\frac{\Upsilon_0 h\sigma^2 r|\nu|}{2-2(r/u)^{2|\nu|}}\Big)\begin{cases}\Big(\frac{x}{r}\Big)^{2|\nu|}-\frac{\Upsilon_0\big((x/r)^{2|\nu|}-(x/u)^{2|\nu|}\big)}{2-2(r/u)^{2|\nu|}}, & x\le r\\ 1-\frac{\Upsilon_0\big(1-(x/u)^{2|\nu|}\big)}{2-2(r/u)^{2|\nu|}}, & r\le x\le u\\ 1, & u\le x\end{cases}$$

$u \le r$

$$=\exp\Big(-\frac{\Upsilon_0 h\sigma^2 r|\nu|}{2-2(u/r)^{2|\nu|}}\Big)\begin{cases}1, & x\le u\\ 1-\frac{\Upsilon_0\big(1-(u/x)^{2|\nu|}\big)}{2-2(u/r)^{2|\nu|}}, & u\le x\le r\\ \Big(\frac{r}{x}\Big)^{2|\nu|}-\frac{\Upsilon_0\big((r/x)^{2|\nu|}-(u/x)^{2|\nu|}\big)}{2-2(u/r)^{2|\nu|}}, & r\le x\end{cases}$$

$$\Upsilon_\lambda := \sqrt{(1-\alpha r u^{-1})^2+4\alpha r u^{-1}\big(1-((r\wedge u)/(r\vee u))^{2\sqrt{\nu^2+2\lambda/\sigma^2}}\big)}+1-\alpha$$

9 $V_s = e^{\sigma^2 \nu s + \sigma W_s}$ $\sigma > 0$ $H_z = \min\{s : V_s = z\}$

2. Stopping at first hitting time

2.0.1 $$\mathbf{E}_x e^{-\alpha H_z} = \mathbf{E}_x\{e^{-\alpha H_z};\ H_z < \infty\} = \begin{cases} \left(\frac{x}{z}\right)^{\sqrt{\nu^2+2\alpha/\sigma^2}-\nu}, & x \le z \\ \left(\frac{z}{x}\right)^{\sqrt{\nu^2+2\alpha/\sigma^2}+\nu}, & z \le x \end{cases}$$

2.0.2 $$\mathbf{P}_x\big(H_z \in dt\big) = \frac{|\ln(z/x)|}{\sigma\sqrt{2\pi}t^{3/2}}\left(\frac{z}{x}\right)^{\nu} \exp\Big(-\frac{\nu^2\sigma^2 t}{2} - \frac{\ln^2(z/x)}{2\sigma^2 t}\Big)dt$$

(1) $$\mathbf{P}_x\big(H_z = \infty\big) = \begin{cases} 1 - \left(\frac{x}{z}\right)^{|\nu|-\nu}, & x \le z \\ 1 - \left(\frac{z}{x}\right)^{|\nu|+\nu}, & z \le x \end{cases}$$

2.1.2
$z < y$ $$\mathbf{P}_x\Big(\sup_{0\le s\le H_z} V_s < y, H_z < \infty\Big) = \begin{cases} \left(\frac{x}{z}\right)^{|\nu|-\nu}, & x \le z \\ \left(\frac{z}{x}\right)^{\nu} \frac{(y/x)^{|\nu|} - (x/y)^{|\nu|}}{(y/z)^{|\nu|} - (z/y)^{|\nu|}}, & z \le x \le y \end{cases}$$

(1)
$z \le x$ $$\mathbf{P}_x\Big(\sup_{0\le s\le H_z} V_s = \infty\Big) = 1 - \left(\frac{z}{x}\right)^{\nu+|\nu|}$$

2.1.4 $$\mathbf{E}_x\Big\{e^{-\alpha H_z};\ \sup_{0\le s\le H_z} V_s < y\Big\} = \begin{cases} \left(\frac{x}{z}\right)^{\sqrt{\nu^2+2\alpha/\sigma^2}-\nu}, & x \le z,\ z \le y \\ \left(\frac{z}{x}\right)^{\nu} \frac{\left(\frac{y}{x}\right)^{\sqrt{\nu^2+2\alpha/\sigma^2}} - \left(\frac{x}{y}\right)^{\sqrt{\nu^2+2\alpha/\sigma^2}}}{\left(\frac{y}{z}\right)^{\sqrt{\nu^2+2\alpha/\sigma^2}} - \left(\frac{z}{y}\right)^{\sqrt{\nu^2+2\alpha/\sigma^2}}}, & z \le x,\ x \le y \end{cases}$$

(1)
$z \le x$
$x \le y$ $$\mathbf{P}_x\Big(\sup_{0\le s\le H_z} V_s < y,\ H_z \in dt\Big) = (z/x)^{\nu} e^{-\nu^2\sigma^2 t/2}\,\mathrm{ss}_t\Big(\frac{\ln(y/x)}{\sigma}, \frac{\ln(y/z)}{\sigma}\Big)dt$$

2.2.2 $$\mathbf{P}_x\Big(\inf_{0\le s\le H_z} V_s > y,\ H_z < \infty\Big) = \begin{cases} \left(\frac{z}{x}\right)^{\nu} \frac{(x/y)^{|\nu|} - (y/x)^{|\nu|}}{(z/y)^{|\nu|} - (y/z)^{|\nu|}}, & y \le x \le z \\ \left(\frac{z}{x}\right)^{|\nu|+\nu}, & y \le z \le x \end{cases}$$

(1)
$x \le z$ $$\mathbf{P}_x\Big(\inf_{0\le s\le H_z} V_s = 0\Big) = 1 - \left(\frac{x}{z}\right)^{|\nu|-\nu}$$

9 $V_s = e^{\sigma^2 \nu s + \sigma W_s}$ $\sigma > 0$ $H_z = \min\{s : V_s = z\}$

2.2.4
$$\mathbf{E}_x\{e^{-\alpha H_z};\ \inf_{0\le s\le H_z} V_s > y\} = \begin{cases} \left(\frac{z}{x}\right)^{\nu} \frac{\left(\frac{x}{y}\right)^{\sqrt{\nu^2+2\alpha/\sigma^2}} - \left(\frac{y}{x}\right)^{\sqrt{\nu^2+2\alpha/\sigma^2}}}{\left(\frac{z}{y}\right)^{\sqrt{\nu^2+2\alpha/\sigma^2}} - \left(\frac{y}{z}\right)^{\sqrt{\nu^2+2\alpha/\sigma^2}}}, & y \le x,\ x \le z \\ \left(\frac{z}{x}\right)^{\sqrt{\nu^2+2\alpha/\sigma^2}+\nu}, & y \le z,\ z \le x \end{cases}$$

(1) $z \le x$, $y \le x$
$$\mathbf{P}_x\Big(\inf_{0\le s\le H_z} V_s > y,\ H_z \in dt\Big) = (z/x)^{\nu} e^{-\nu^2\sigma^2 t/2}\,\mathrm{ss}_t\Big(\frac{\ln(x/y)}{\sigma}, \frac{\ln(z/y)}{\sigma}\Big)dt$$

2.3.1
$$\mathbf{E}_x\{e^{-\gamma\ell(H_z,r)};\ H_z < \infty\}$$

$$= \left(\frac{z}{x}\right)^{\nu} e^{-|\nu||\ln(z/x)|} \begin{cases} 1, & r \le z \le x, \quad x \le z \le r \\ \frac{|\nu|r\sigma^2 + \gamma(1 - e^{-2|\nu||\ln(x/r)|})}{|\nu|r\sigma^2 + \gamma(1 - e^{-2|\nu||\ln(z/r)|})}, & z \wedge r \le x \le z \vee r \\ \frac{|\nu|r\sigma^2}{|\nu|r\sigma^2 + \gamma(1 - e^{-2|\nu||\ln(z/r)|})}, & z \le r \le x, \quad x \le r \le z \end{cases}$$

(1)
$$\mathbf{E}_x\{e^{-\gamma\ell(H_z,r)};\ H_z = \infty\} \qquad \text{for} \quad \nu\ln(z/r) < 0$$

$$= \begin{cases} 0, & r \le z \le x, \quad x \le z \le r \\ \frac{|\nu|r\sigma^2(1 - e^{-2|\nu||\ln(x/z)|})}{|\nu|r\sigma^2 + \gamma(1 - e^{-2|\nu||\ln(z/r)|})}, & z \wedge r \le x \le z \vee r \\ 1 - e^{-2|\nu||\ln(x/r)|} \\ \quad + \frac{|\nu|r\sigma^2(e^{-2|\nu||\ln(x/r)|} - e^{-2|\nu||\ln(x/z)|})}{|\nu|r\sigma^2 + \gamma(1 - e^{-2|\nu||\ln(z/r)|})}, & z \le r \le x, \quad x \le r \le z \end{cases}$$

2.3.2
$$\mathbf{P}_x\big(\ell(H_z,r) \in dy,\ H_z < \infty\big) = \left(\frac{z}{x}\right)^{\nu} \exp\Big(-\frac{|\nu|yr\sigma^2}{1 - e^{-2|\nu||\ln(z/r)|}}\Big)$$

$$\times \begin{cases} 0, & x \le z \le r, \quad r \le z \le x \\ \frac{|\nu|r\sigma^2 \operatorname{sh}(|\nu||\ln(z/x)|)}{2\operatorname{sh}^2(|\nu||\ln(z/r)|)}dy, & r \wedge z \le x \le r \vee z \\ \frac{|\nu|r\sigma^2 e^{-|\nu||\ln(r/x)|}}{2\operatorname{sh}(|\nu||\ln(z/r)|)}dy, & z \le r \le x, \quad x \le r \le z \end{cases}$$

(1)
$$\mathbf{P}_x\big(\ell(H_z,r) \in dy,\ H_z = \infty\big) \qquad \text{for} \quad \nu\ln(z/r) < 0$$

9 $V_s = e^{\sigma^2\nu s+\sigma W_s}$ $\sigma > 0$ $H_z = \min\{s : V_s = z\}$

$$=\begin{cases}0, & x \le z \le r,\quad r \le z \le x\\ \dfrac{|\nu|r\sigma^2\big(1-e^{-2|\nu||\ln(x/z)|}\big)}{1-e^{-2|\nu||\ln(z/r)|}}\exp\Big(-\dfrac{|\nu|yr\sigma^2}{1-e^{-2|\nu||\ln(z/r)|}}\Big)dy, & r\wedge z \le x \le r\vee z\\ \dfrac{|\nu|r\sigma^2\big(e^{-2|\nu||\ln(x/r)|}-e^{-2|\nu||\ln(x/z)|}\big)}{1-e^{-2|\nu||\ln(z/r)|}}\exp\Big(-\dfrac{|\nu|yr\sigma^2}{1-e^{-2|\nu||\ln(z/r)|}}\Big)dy, & x\le r\le z,\ z\le r\le x\end{cases}$$

2.3.3 $$\mathbf{E}_x e^{-\alpha H_z-\gamma\ell(H_z,r)} = \Big(\frac{z}{x}\Big)^{\nu} e^{-|\ln(z/x)|\sqrt{\nu^2+2\alpha/\sigma^2}}$$

$$\times\begin{cases}1, & x\le z\le r,\ r\le z\le x\\ \dfrac{\sigma r\sqrt{\nu^2\sigma^2+2\alpha}+\gamma\big(1-e^{-2|\ln(x/r)|\sqrt{\nu^2+2\alpha/\sigma^2}}\big)}{\sigma r\sqrt{\nu^2\sigma^2+2\alpha}+\gamma\big(1-e^{-2|\ln(z/r)|\sqrt{\nu^2+2\alpha/\sigma^2}}\big)}, & r\wedge z\le x\le r\vee z\\ \dfrac{\sigma r\sqrt{\nu^2\sigma^2+2\alpha}}{\sigma r\sqrt{\nu^2\sigma^2+2\alpha}+\gamma\big(1-e^{-2|\ln(z/r)|\sqrt{\nu^2+2\alpha/\sigma^2}}\big)}, & z\le r\le x,\ x\le r\le z\end{cases}$$

2.6.1 $$\mathbf{E}_x\exp\Big(-\int_0^{H_z}\big(p\mathbb{I}_{(0,r)}(V_s)+q\mathbb{I}_{[r,\infty)}(V_s)\big)ds\Big)$$

$x\le z$, $z\le r$: $$=\Big(\frac{x}{z}\Big)^{\Upsilon_p-\nu}$$

$z\le x$, $x\le r$: $$=\Big(\frac{z}{x}\Big)^{\nu}\frac{\Upsilon_p\big((r/x)^{\Upsilon_p}+(x/r)^{\Upsilon_p}\big)+\Upsilon_q\big((r/x)^{\Upsilon_p}-(x/r)^{\Upsilon_p}\big)}{\Upsilon_p\big((r/z)^{\Upsilon_p}+(z/r)^{\Upsilon_p}\big)+\Upsilon_q\big((r/z)^{\Upsilon_p}-(z/r)^{\Upsilon_p}\big)}$$

$z\le r$, $r\le x$: $$=\Big(\frac{z}{x}\Big)^{\nu}\frac{2\Upsilon_p(r/x)^{\Upsilon_q}}{\Upsilon_p\big((r/z)^{\Upsilon_p}+(z/r)^{\Upsilon_p}\big)+\Upsilon_q\big((r/z)^{\Upsilon_p}-(z/r)^{\Upsilon_p}\big)}$$

$x\le r$, $r\le z$: $$=\Big(\frac{z}{x}\Big)^{\nu}\frac{2\Upsilon_q(x/r)^{\Upsilon_p}}{\Upsilon_q\big((z/r)^{\Upsilon_q}+(r/z)^{\Upsilon_q}\big)+\Upsilon_p\big((z/r)^{\Upsilon_q}-(r/z)^{\Upsilon_q}\big)}$$

$r\le x$, $x\le z$: $$=\Big(\frac{z}{x}\Big)^{\nu}\frac{\Upsilon_q\big((x/r)^{\Upsilon_q}+(r/x)^{\Upsilon_q}\big)+\Upsilon_p\big((x/r)^{\Upsilon_q}-(r/x)^{\Upsilon_q}\big)}{\Upsilon_q\big((z/r)^{\Upsilon_q}+(r/z)^{\Upsilon_q}\big)+\Upsilon_p\big((z/r)^{\Upsilon_q}-(r/z)^{\Upsilon_q}\big)}$$

$r\le z$, $z\le x$: $$=\Big(\frac{z}{x}\Big)^{\Upsilon_q+\nu}$$

2.8.3 $$\mathbf{E}_x\exp\Big(-\alpha H_z-\gamma\int_0^{H_z}V_s ds\Big)=\begin{cases}\Big(\dfrac{z}{x}\Big)^{\nu}\dfrac{I_{2\sqrt{\nu^2+2\alpha/\sigma^2}}(2\sqrt{2\gamma x}/\sigma)}{I_{2\sqrt{\nu^2+2\alpha/\sigma^2}}(2\sqrt{2\gamma z}/\sigma)}, & x\le z\\ \Big(\dfrac{z}{x}\Big)^{\nu}\dfrac{K_{2\sqrt{\nu^2+2\alpha/\sigma^2}}(2\sqrt{2\gamma x}/\sigma)}{K_{2\sqrt{\nu^2+2\alpha/\sigma^2}}(2\sqrt{2\gamma z}/\sigma)}, & z\le x\end{cases}$$

$$\Upsilon_s := \sqrt{\nu^2+2s/\sigma^2}$$

9 $V_s = e^{\sigma^2\nu s+\sigma W_s}$ $\sigma > 0$ $H_z = \min\{s : V_s = z\}$

2.13.1 $\mathbf{E}_x\{e^{-\gamma\check{H}(H_z)}; \sup_{0\le s\le H_z} V_s \in dy\}$ $z < x < y$

$$= \frac{|\nu|(z/x)^\nu\big((x/z)^{\sqrt{\nu^2+2\gamma/\sigma^2}} - (z/x)^{\sqrt{\nu^2+2\gamma/\sigma^2}}\big)}{y\big((y/z)^{|\nu|} - (z/y)^{|\nu|}\big)\big((y/z)^{\sqrt{\nu^2+2\gamma/\sigma^2}} - (z/y)^{\sqrt{\nu^2+2\gamma/\sigma^2}}\big)}dy$$

2.13.2 $z < x$ $x < y$

$$\mathbf{P}_x\big(\check{H}(H_z) \in du, \sup_{0\le s\le H_z} V_s \in dy\big) = \Big(\frac{z}{x}\Big)^\nu e^{-\nu^2\sigma^2u/2}\frac{|\nu|\,\mathrm{ss}_u\big(\frac{\ln(x/z)}{\sigma}, \frac{\ln(y/z)}{\sigma}\big)}{y\big((y/z)^{|\nu|} - (z/y)^{|\nu|}\big)}dudy$$

2.13.4 $\mathbf{E}_x\{e^{-\alpha H_z-\gamma\check{H}(H_z)}; \sup_{0\le s\le H_z} V_s \in dy\}$ $z < x < y$

$$= \frac{y^{-1}\sqrt{\nu^2+2\alpha/\sigma^2}(z/x)^\nu\big((x/z)^{\sqrt{\nu^2+2(\alpha+\gamma)/\sigma^2}} - (z/x)^{\sqrt{\nu^2+2(\alpha+\gamma)/\sigma^2}}\big)}{\big((y/z)^{\sqrt{\nu^2+2\alpha/\sigma^2}} - (z/y)^{\sqrt{\nu^2+2\alpha/\sigma^2}}\big)\big((y/z)^{\sqrt{\nu^2+2(\alpha+\gamma)/\sigma^2}} - (z/y)^{\sqrt{\nu^2+2(\alpha+\gamma)/\sigma^2}}\big)}dy$$

(1) $u < t$ $\mathbf{P}_x\big(\check{H}(H_z) \in du, H_z \in dt, \sup_{0\le s\le H_z} V_s \in dy\big)$

$$= \Big(\frac{z}{x}\Big)^\nu \frac{1}{y\sigma}e^{-\nu^2\sigma^2t/2}\,\mathrm{ss}_u\Big(\frac{\ln(x/z)}{\sigma}, \frac{\ln(y/z)}{\sigma}\Big)\,\mathrm{s}_{t-u}\Big(\frac{\ln(y/z)}{\sigma}\Big)dudtdy$$

2.14.1 $x < z$ $\mathbf{E}_x\{e^{-\gamma\hat{H}(H_z)}; \inf_{0\le s\le H_z} V_s \in dy\}$ $y < x < z$

$$= \frac{|\nu|(z/x)^\nu\big((z/x)^{\sqrt{\nu^2+2\gamma/\sigma^2}} - (x/z)^{\sqrt{\nu^2+2\gamma/\sigma^2}}\big)}{y\big((z/y)^{|\nu|} - (y/z)^{|\nu|}\big)\big((z/y)^{\sqrt{\nu^2+2\gamma/\sigma^2}} - (y/z)^{\sqrt{\nu^2+2\gamma/\sigma^2}}\big)}dy$$

2.14.2 $x < z$ $y < x$

$$\mathbf{P}_x\big(\hat{H}(H_z) \in du, \inf_{0\le s\le H_z} V_s \in dy\big) = \Big(\frac{z}{x}\Big)^\nu e^{-\nu^2\sigma^2u/2}\frac{|\nu|\,\mathrm{ss}_u\big(\frac{\ln(z/x)}{\sigma}, \frac{\ln(z/y)}{\sigma}\big)}{y\big((z/y)^{|\nu|} - (y/z)^{|\nu|}\big)}dudy$$

2.14.4 $\mathbf{E}_x\{e^{-\alpha H_z-\gamma\hat{H}(H_z)}; \inf_{0\le s\le H_z} V_s \in dy\}$ $y < x < z$

$$= \frac{y^{-1}\sqrt{\nu^2+2\alpha/\sigma^2}(z/x)^\nu\big((z/x)^{\sqrt{\nu^2+2(\alpha+\gamma)/\sigma^2}} - (x/z)^{\sqrt{\nu^2+2(\alpha+\gamma)/\sigma^2}}\big)}{\big((z/y)^{\sqrt{\nu^2+2\alpha/\sigma^2}} - (y/z)^{\sqrt{\nu^2+2\alpha/\sigma^2}}\big)\big((z/y)^{\sqrt{\nu^2+2(\alpha+\gamma)/\sigma^2}} - (y/z)^{\sqrt{\nu^2+2(\alpha+\gamma)/\sigma^2}}\big)}dy$$

9 $V_s = e^{\sigma^2 \nu s + \sigma W_s}$ $\sigma > 0$ $H_z = \min\{s : V_s = z\}$

(1) $u < t$

$$\mathbf{P}_x\big(\hat{H}(H_z) \in du,\, H_z \in dt,\, \inf_{0 \le s \le H_z} V_s \in dy\big)$$

$$= \Big(\frac{z}{x}\Big)^{\nu} \frac{1}{y\sigma} e^{-\nu^2\sigma^2 t/2}\, \mathrm{ss}_u\Big(\frac{\ln(z/x)}{\sigma}, \frac{\ln(z/y)}{\sigma}\Big)\, \mathrm{s}_{t-u}\Big(\frac{\ln(z/y)}{\sigma}\Big) du dt dy$$

2.15.2

$$\mathbf{P}_x\big(a < \inf_{0 \le s \le H_z} V_s,\, \sup_{0 \le s \le H_z} V_s < b\big) = \begin{cases} \Big(\frac{z}{x}\Big)^{\nu} \frac{((x/a)^{|\nu|} - (a/x)^{|\nu|})}{((z/a)^{|\nu|} - (a/z)^{|\nu|})}, & a \le x \le z \\ \Big(\frac{z}{x}\Big)^{\nu} \frac{((b/x)^{|\nu|} - (x/b)^{|\nu|})}{((b/z)^{|\nu|} - (z/b)^{|\nu|})}, & z \le x \le b \end{cases}$$

2.19.3

$$\mathbf{E}_x \exp\Big(-\alpha H_z - \gamma \int_0^{H_z} \frac{ds}{V_s}\Big) = \begin{cases} \Big(\frac{z}{x}\Big)^{\nu} \frac{K_{2\sqrt{\nu^2+2\alpha/\sigma^2}}(2\sqrt{2\gamma}/\sigma\sqrt{x})}{K_{2\sqrt{\nu^2+2\alpha/\sigma^2}}(2\sqrt{2\gamma}/\sigma\sqrt{z})}, & x \le z \\ \Big(\frac{z}{x}\Big)^{\nu} \frac{I_{2\sqrt{\nu^2+2\alpha/\sigma^2}}(2\sqrt{2\gamma}/\sigma\sqrt{x})}{I_{2\sqrt{\nu^2+2\alpha/\sigma^2}}(2\sqrt{2\gamma}/\sigma\sqrt{z})}, & z \le x \end{cases}$$

2.20.3

$$\mathbf{E}_x \exp\Big(-\alpha H_z - \frac{\gamma^2}{2} \int_0^{H_z} V_s^{2\beta} ds\Big)$$

$$= \begin{cases} \Big(\frac{z}{x}\Big)^{\nu} \frac{I_{\sqrt{\nu^2+2\alpha/\sigma^2}/|\beta|}(\gamma x^{\beta}/\sigma|\beta|)}{I_{\sqrt{\nu^2+2\alpha/\sigma^2}/|\beta|}(\gamma z^{\beta}/\sigma|\beta|)}, & (z-x)\beta \ge 0 \\ \Big(\frac{z}{x}\Big)^{\nu} \frac{K_{\sqrt{\nu^2+2\alpha/\sigma^2}/|\beta|}(\gamma x^{\beta}/\sigma|\beta|)}{K_{\sqrt{\nu^2+2\alpha/\sigma^2}/|\beta|}(\gamma z^{\beta}/\sigma|\beta|)}, & (z-x)\beta \le 0 \end{cases}$$

2.21.3

$$\mathbf{E}_x \exp\Big(-\alpha H_z - \int_0^{H_z} \big(\frac{p^2}{2} V_s^{2\beta} + q V_s^{\beta}\big) ds\Big)$$

$$= \begin{cases} \Big(\frac{z}{x}\Big)^{\nu} \frac{z^{\beta/2} M_{-q/p\sigma|\beta|,\sqrt{\nu^2+2\alpha/\sigma^2}/|\beta|}(2px^{\beta}/\sigma|\beta|)}{x^{\beta/2} M_{-q/p\sigma|\beta|,\sqrt{\nu^2+2\alpha/\sigma^2}/|\beta|}(2pz^{\beta}/\sigma|\beta|)}, & (z-x)\beta \ge 0 \\ \Big(\frac{z}{x}\Big)^{\nu} \frac{z^{\beta/2} W_{-q/p\sigma|\beta|,\sqrt{\nu^2+2\alpha/\sigma^2}/|\beta|}(2px^{\beta}/\sigma|\beta|)}{x^{\beta/2} W_{-q/p\sigma|\beta|,\sqrt{\nu^2+2\alpha/\sigma^2}/|\beta|}(2pz^{\beta}/\sigma|\beta|)}, & (z-x)\beta \le 0 \end{cases}$$

9 $V_s = e^{\sigma^2\nu s+\sigma W_s}$ $\sigma > 0$ $H = H_{a,b} = \min\{s : V_s \notin (a,b)\}$

3. Stopping at first exit time

3.0.4 (a)
$$\mathbf{P}_x\big(V_H = a\big) = \Big(\frac{a}{x}\Big)^{\nu} \frac{(b/x)^{|\nu|} - (x/b)^{|\nu|}}{(b/a)^{|\nu|} - (a/b)^{|\nu|}}$$

3.0.4 (b)
$$\mathbf{P}_x\big(V_H = b\big) = \Big(\frac{b}{x}\Big)^{\nu} \frac{(x/a)^{|\nu|} - (x/y)^{|\nu|}}{(b/a)^{|\nu|} - (a/b)^{|\nu|}}$$

3.0.5 (a)
$$\mathbf{E}_x\big\{e^{-\alpha H};\ V_H = a\big\} = \Big(\frac{a}{x}\Big)^{\nu} \frac{(b/x)^{\sqrt{\nu^2+2\alpha/\sigma^2}} - (x/b)^{\sqrt{\nu^2+2\alpha/\sigma^2}}}{(b/a)^{\sqrt{\nu^2+2\alpha/\sigma^2}} - (a/b)^{\sqrt{\nu^2+2\alpha/\sigma^2}}}$$

3.0.5 (b)
$$\mathbf{E}_x\big\{e^{-\alpha H};\ V_H = b\big\} = \Big(\frac{b}{x}\Big)^{\nu} \frac{(x/a)^{\sqrt{\nu^2+2\alpha/\sigma^2}} - (a/x)^{\sqrt{\nu^2+2\alpha/\sigma^2}}}{(b/a)^{\sqrt{\nu^2+2\alpha/\sigma^2}} - (a/b)^{\sqrt{\nu^2+2\alpha/\sigma^2}}}$$

3.0.6 (a)
$$\mathbf{P}_x\big(H \in dt, V_H = a\big) = (a/x)^{\nu} e^{-\nu^2\sigma^2 t/2}\, \mathrm{ss}_t\Big(\frac{\ln(b/x)}{\sigma}, \frac{\ln(b/a)}{\sigma}\Big)dt$$

3.0.6 (b)
$$\mathbf{P}_x\big(H \in dt, V_H = b\big) = (b/x)^{\nu} e^{-\nu^2\sigma^2 t/2}\, \mathrm{ss}_t\Big(\frac{\ln(x/a)}{\sigma}, \frac{\ln(b/a)}{\sigma}\Big)dt$$

3.1.6
$$\mathbf{P}_x\Big(\sup_{0\le s\le H} V_s \ge y, V_H = a\Big)$$

$$= \Big(\frac{a}{x}\Big)^{\nu} \frac{\big((b/y)^{|\nu|} - (y/b)^{|\nu|}\big)\big((x/a)^{|\nu|} - (x/y)^{|\nu|}\big)}{\big((b/a)^{|\nu|} - (a/b)^{|\nu|}\big)\big((y/a)^{|\nu|} - (a/y)^{|\nu|}\big)}, \qquad a \le x \le y \le b$$

3.1.8 (1)
$$\mathbf{E}_x\Big\{e^{-\alpha H}; \sup_{0\le s\le H} V_s \ge y, V_H = a\Big\} \qquad a \le x \le y \le b$$

$$= \Big(\frac{a}{x}\Big)^{\nu} \frac{\big((b/y)^{\sqrt{\nu^2+2\alpha/\sigma^2}} - (y/b)^{\sqrt{\nu^2+2\alpha/\sigma^2}}\big)\big((x/a)^{\sqrt{\nu^2+2\alpha/\sigma^2}} - (a/x)^{\sqrt{\nu^2+2\alpha/\sigma^2}}\big)}{\big((b/a)^{\sqrt{\nu^2+2\alpha/\sigma^2}} - (a/b)^{\sqrt{\nu^2+2\alpha/\sigma^2}}\big)\big((y/a)^{\sqrt{\nu^2+2\alpha/\sigma^2}} - (a/y)^{\sqrt{\nu^2+2\alpha/\sigma^2}}\big)}$$

3.2.6
$$\mathbf{P}_x\Big(\inf_{0\le s\le H} V_s \le y, V_H = b\Big)$$

$$= \Big(\frac{b}{x}\Big)^{\nu} \frac{\big((b/x)^{|\nu|} - (x/b)^{|\nu|}\big)\big((y/a)^{|\nu|} - (a/y)^{|\nu|}\big)}{\big((b/y)^{|\nu|} - (y/b)^{|\nu|}\big)\big((b/a)^{|\nu|} - (a/b)^{|\nu|}\big)}, \qquad a \le y \le x \le b$$

9 $V_s = e^{\sigma^2\nu s+\sigma W_s}$ $\sigma > 0$ $H = H_{a,b} = \min\{s : V_s \notin (a,b)\}$

3.2.8 (1) $\mathbf{E}_x\{e^{-\alpha H}; \inf_{0\le s\le H} V_s \le y, V_H = b\}$ $a \le y \le x \le b$

$$= \Big(\frac{b}{x}\Big)^{\nu} \frac{\big((b/x)^{\sqrt{\nu^2+2\alpha/\sigma^2}} - (x/b)^{\sqrt{\nu^2+2\alpha/\sigma^2}}\big)\big((y/a)^{\sqrt{\nu^2+2\alpha/\sigma^2}} - (a/y)^{\sqrt{\nu^2+2\alpha/\sigma^2}}\big)}{\big((b/y)^{\sqrt{\nu^2+2\alpha/\sigma^2}} - (y/b)^{\sqrt{\nu^2+2\alpha/\sigma^2}}\big)\big((b/a)^{\sqrt{\nu^2+2\alpha/\sigma^2}} - (a/b)^{\sqrt{\nu^2+2\alpha/\sigma^2}}\big)}$$

3.3.1 $$\mathbf{E}_r e^{-\gamma\ell(H,r)} = \frac{|\nu|\big((b/a)^{|\nu|} - (a/b)^{|\nu|}\big)}{|\nu|\big((b/a)^{|\nu|} - (a/b)^{|\nu|}\big) + \frac{\gamma}{r\sigma^2}\big((b/r)^{|\nu|} - (r/b)^{|\nu|}\big)\big((r/a)^{|\nu|} - (a/r)^{|\nu|}\big)} =: L$$

$$\mathbf{E}_x e^{-\gamma\ell(H,r)} = \begin{cases} \Big(\frac{b}{x}\Big)^{\nu} \frac{(x/r)^{|\nu|} - (r/x)^{|\nu|}}{(b/r)^{|\nu|} - (r/b)^{|\nu|}} + \Big(\frac{r}{x}\Big)^{\nu} \frac{(b/x)^{|\nu|} - (x/b)^{|\nu|}}{(b/r)^{|\nu|} - (r/b)^{|\nu|}} L, & r \le x \\ & x \le b \\ \Big(\frac{a}{x}\Big)^{\nu} \frac{(r/x)^{|\nu|} - (x/r)^{|\nu|}}{(r/a)^{|\nu|} - (a/r)^{|\nu|}} + \Big(\frac{r}{x}\Big)^{\nu} \frac{(x/a)^{|\nu|} - (x/y)^{|\nu|}}{(r/a)^{|\nu|} - (a/r)^{|\nu|}} L, & a \le x \\ & x \le r \end{cases}$$

3.3.2 $$\mathbf{P}_x\big(\ell(H,r) \in dy\big) = \Big(\frac{r}{x}\Big)^{\nu} \exp\Big(-\frac{|\nu| y r\sigma^2\big((b/a)^{|\nu|} - (a/b)^{|\nu|}\big)}{\big((b/r)^{|\nu|} - (r/b)^{|\nu|}\big)\big((r/a)^{|\nu|} - (a/r)^{|\nu|}\big)}\Big)$$

$$\times \begin{cases} \frac{|\nu| r\sigma^2\big((b/x)^{|\nu|} - (x/b)^{|\nu|}\big)\big((b/a)^{|\nu|} - (a/b)^{|\nu|}\big)}{\big((b/r)^{|\nu|} - (r/b)^{|\nu|}\big)^2\big((r/a)^{|\nu|} - (a/r)^{|\nu|}\big)} dy, & r \le x \le b \\ \frac{|\nu| r\sigma^2\big((x/a)^{|\nu|} - (x/y)^{|\nu|}\big)\big((b/a)^{|\nu|} - (a/b)^{|\nu|}\big)}{\big((b/r)^{|\nu|} - (r/b)^{|\nu|}\big)\big((r/a)^{|\nu|} - (a/r)^{|\nu|}\big)^2} dy, & a \le x \le r \end{cases}$$

(1) $$\mathbf{P}_x\big(\ell(H,r) = 0\big) = \begin{cases} \Big(\frac{b}{x}\Big)^{\nu} \frac{(x/r)^{|\nu|} - (r/x)^{|\nu|}}{(b/r)^{|\nu|} - (r/b)^{|\nu|}}, & r \le x \le b \\ \Big(\frac{a}{x}\Big)^{\nu} \frac{(r/x)^{|\nu|} - (x/r)^{|\nu|}}{(r/a)^{|\nu|} - (a/r)^{|\nu|}}, & a \le x \le r \end{cases}$$

3.3.5 (a) $\mathbf{E}_x\{e^{-\gamma\ell(H,r)};\ V_H = a\}\Big(\frac{x}{a}\Big)^{\nu}$

$$= \begin{cases} \frac{|\nu|\big((b/x)^{|\nu|} - (x/b)^{|\nu|}\big)}{|\nu|\big((b/a)^{|\nu|} - (a/b)^{|\nu|}\big) + \frac{\gamma}{r\sigma^2}\big((b/r)^{|\nu|} - (r/b)^{|\nu|}\big)\big((r/a)^{|\nu|} - (a/r)^{|\nu|}\big)}, & r \le x \le b \\ \frac{|\nu|\big((b/x)^{|\nu|} - (x/b)^{|\nu|}\big) + \frac{\gamma}{r\sigma^2}\big((b/r)^{|\nu|} - (r/b)^{|\nu|}\big)\big((r/x)^{|\nu|} - (x/r)^{|\nu|}\big)}{|\nu|\big((b/a)^{|\nu|} - (a/b)^{|\nu|}\big) + \frac{\gamma}{r\sigma^2}\big((b/r)^{|\nu|} - (r/b)^{|\nu|}\big)\big((r/a)^{|\nu|} - (a/r)^{|\nu|}\big)}, & a \le x \le r \end{cases}$$

3.3.5 (b) $\mathbf{E}_x\{e^{-\gamma\ell(H,r)};\ V_H = b\}\Big(\frac{x}{b}\Big)^{\nu}$

9 $V_s = e^{\sigma^2 \nu s + \sigma W_s}$ $\sigma > 0$ $H = H_{a,b} = \min\{s : V_s \notin (a,b)\}$

$$=\begin{cases} \dfrac{|\nu|\big((x/a)^{|\nu|}-(a/x)^{|\nu|}\big)+\frac{\gamma}{r\sigma^2}\big((x/r)^{|\nu|}-(r/x)^{|\nu|}\big)\big((r/a)^{|\nu|}-(a/r)^{|\nu|}\big)}{|\nu|\big((b/a)^{|\nu|}-(a/b)^{|\nu|}\big)+\frac{\gamma}{r\sigma^2}\big((b/r)^{|\nu|}-(r/b)^{|\nu|}\big)\big((r/a)^{|\nu|}-(a/r)^{|\nu|}\big)}, & r\le x\le b\\ \dfrac{|\nu|\big((x/a)^{|\nu|}-(a/x)^{|\nu|}\big)}{|\nu|\big((b/a)^{|\nu|}-(a/b)^{|\nu|}\big)+\frac{\gamma}{r\sigma^2}\big((b/r)^{|\nu|}-(r/b)^{|\nu|}\big)\big((r/a)^{|\nu|}-(a/r)^{|\nu|}\big)}, & a\le x\le r \end{cases}$$

3.3.6 (a)

$$\mathbf{P}_x\big(\ell(H,r)\in dy, V_H=a\big)=\exp\Big(-\frac{|\nu|yr\sigma^2\big((b/a)^{|\nu|}-(a/b)^{|\nu|}\big)}{\big((b/r)^{|\nu|}-(r/b)^{|\nu|}\big)\big((r/a)^{|\nu|}-(a/r)^{|\nu|}\big)}\Big)$$

$$\times\begin{cases} \Big(\dfrac{a}{x}\Big)^{\nu}\dfrac{|\nu|r\sigma^2\big((b/x)^{|\nu|}-(x/b)^{|\nu|}\big)}{\big((b/r)^{|\nu|}-(r/b)^{|\nu|}\big)\big((r/a)^{|\nu|}-(a/r)^{|\nu|}\big)}dy, & r\le x\le b\\ \Big(\dfrac{a}{x}\Big)^{\nu}\dfrac{|\nu|r\sigma^2\big((x/a)^{|\nu|}-(x/y)^{|\nu|}\big)}{\big((r/a)^{|\nu|}-(a/r)^{|\nu|}\big)^2}dy, & a\le x\le r \end{cases}$$

(1)

$$\mathbf{P}_x\big(\ell(H,r)=0, V_H=a\big)=\begin{cases} 0, & r\le x\le b\\ \Big(\dfrac{a}{x}\Big)^{\nu}\dfrac{(r/x)^{|\nu|}-(x/r)^{|\nu|}}{(r/a)^{|\nu|}-(a/r)^{|\nu|}}, & a\le x\le r \end{cases}$$

3.3.6 (b)

$$\mathbf{P}_x\big(\ell(H,r)\in dy, V_H=b\big)=\exp\Big(-\frac{|\nu|yr\sigma^2\big((b/a)^{|\nu|}-(a/b)^{|\nu|}\big)}{\big((b/r)^{|\nu|}-(r/b)^{|\nu|}\big)\big((r/a)^{|\nu|}-(a/r)^{|\nu|}\big)}\Big)$$

$$\times\begin{cases} \Big(\dfrac{b}{x}\Big)^{\nu}\dfrac{|\nu|r\sigma^2\big((b/x)^{|\nu|}-(x/b)^{|\nu|}\big)}{\big((b/r)^{|\nu|}-(r/b)^{|\nu|}\big)^2}dy, & r\le x\le b\\ \Big(\dfrac{b}{x}\Big)^{\nu}\dfrac{|\nu|r\sigma^2\big((x/a)^{|\nu|}-(x/y)^{|\nu|}\big)}{\big((b/r)^{|\nu|}-(r/b)^{|\nu|}\big)\big((r/a)^{|\nu|}-(a/r)^{|\nu|}\big)}dy, & a\le x\le r \end{cases}$$

(1)

$$\mathbf{P}_x\big(\ell(H,r)=0, V_H=b\big)=\begin{cases} \Big(\dfrac{b}{x}\Big)^{\nu}\dfrac{(x/r)^{|\nu|}-(r/x)^{|\nu|}}{(b/r)^{|\nu|}-(r/b)^{|\nu|}}, & r\le x\le b\\ 0, & a\le x\le r \end{cases}$$

3.6.5 (a)

$$\mathbf{E}_x\Big\{\exp\Big(-\int_0^H\big(p\mathbb{I}_{[a,r]}(V_s)+q\mathbb{I}_{[r,b]}(V_s)\big)ds\Big); V_H=a\Big\}\Big(\frac{x}{a}\Big)^{\nu}$$

$a\le x$, $x\le r$

$$=\frac{\Upsilon_q\big((b/r)^{\Upsilon_q}+(r/b)^{\Upsilon_q}\big)\big((r/x)^{\Upsilon_p}-(x/r)^{\Upsilon_p}\big)+\Upsilon_p\big((b/r)^{\Upsilon_q}-(r/b)^{\Upsilon_q}\big)\big((r/x)^{\Upsilon_p}+(x/r)^{\Upsilon_p}\big)}{\Upsilon_q\big((b/r)^{\Upsilon_q}+(r/b)^{\Upsilon_q}\big)\big((r/a)^{\Upsilon_p}-(a/r)^{\Upsilon_p}\big)+\Upsilon_p\big((b/r)^{\Upsilon_q}-(r/b)^{\Upsilon_q}\big)\big((r/a)^{\Upsilon_p}+(a/r)^{\Upsilon_p}\big)}$$

$r\le x$, $x\le b$

$$=\frac{2\Upsilon_p\big((b/x)^{\Upsilon_q}-(x/b)^{\Upsilon_q}\big)}{\Upsilon_q\big((b/r)^{\Upsilon_q}+(r/b)^{\Upsilon_q}\big)\big((r/a)^{\Upsilon_p}-(a/r)^{\Upsilon_p}\big)+\Upsilon_p\big((b/r)^{\Upsilon_q}-(r/b)^{\Upsilon_q}\big)\big((r/a)^{\Upsilon_p}+(a/r)^{\Upsilon_p}\big)}$$

$$\Upsilon_s := \sqrt{\nu^2+2s/\sigma^2}$$

9 $V_s = e^{\sigma^2\nu s+\sigma W_s}$ $\sigma > 0$ $H = H_{a,b} = \min\{s : V_s \notin (a, b)\}$

3.6.5 (b)

$$\mathbf{E}_x\Big\{\exp\Big(-\int_0^H (p\mathbb{I}_{[a,r]}(V_s) + q\mathbb{I}_{[r,b]}(V_s))ds\Big); V_H = b\Big\}\Big(\frac{x}{b}\Big)^\nu$$

$a \le x$, $x \le r$

$$= \frac{2\Upsilon_q((x/a)^{\Upsilon_p} - (a/x)^{\Upsilon_p})}{\Upsilon_q((b/r)^{\Upsilon_q} + (r/b)^{\Upsilon_q})((r/a)^{\Upsilon_p} - (a/r)^{\Upsilon_p}) + \Upsilon_p((b/r)^{\Upsilon_q} - (r/b)^{\Upsilon_q})((r/a)^{\Upsilon_p} + (a/r)^{\Upsilon_p})}$$

$r \le x$, $x \le b$

$$= \frac{\Upsilon_q((x/r)^{\Upsilon_q} + (r/x)^{\Upsilon_q})((r/a)^{\Upsilon_p} - (a/r)^{\Upsilon_p}) + \Upsilon_p((x/r)^{\Upsilon_q} - (r/x)^{\Upsilon_q})((r/a)^{\Upsilon_p} + (a/r)^{\Upsilon_p})}{\Upsilon_q((b/r)^{\Upsilon_q} + (r/b)^{\Upsilon_q})((r/a)^{\Upsilon_p} - (a/r)^{\Upsilon_p}) + \Upsilon_p((b/r)^{\Upsilon_q} - (r/b)^{\Upsilon_q})((r/a)^{\Upsilon_p} + (a/r)^{\Upsilon_p})}$$

3.8.3

$$\mathbf{E}_x \exp\Big(-\alpha H - \gamma\int_0^H V_s ds\Big)$$

$$= \frac{a^{\nu-\Upsilon_\alpha} S_{\Upsilon_\alpha}\big(\frac{2\sqrt{2\gamma b}}{\sigma}, \frac{2\sqrt{2\gamma x}}{\sigma}\big) + b^{\nu-\Upsilon_\alpha} S_{\Upsilon_\alpha}\big(\frac{2\sqrt{2\gamma x}}{\sigma}, \frac{2\sqrt{2\gamma a}}{\sigma}\big)}{x^{\nu-\Upsilon_\alpha} S_{\Upsilon_\alpha}\big(\frac{2\sqrt{2\gamma b}}{\sigma}, \frac{2\sqrt{2\gamma a}}{\sigma}\big)}$$

3.8.7 (a)

$$\mathbf{E}_x\Big\{\exp\Big(-\alpha H - \gamma\int_0^H V_s ds\Big); V_H = a\Big\} = \frac{x^{\Upsilon_\alpha-\nu} S_{\Upsilon_\alpha}\big(\frac{2\sqrt{2\gamma b}}{\sigma}, \frac{2\sqrt{2\gamma x}}{\sigma}\big)}{a^{\Upsilon_\alpha-\nu} S_{\Upsilon_\alpha}\big(\frac{2\sqrt{2\gamma b}}{\sigma}, \frac{2\sqrt{2\gamma a}}{\sigma}\big)}$$

3.8.7 (b)

$$\mathbf{E}_x\Big\{\exp\Big(-\alpha H - \gamma\int_0^H V_s ds\Big); V_H = b\Big\} = \frac{x^{\Upsilon_\alpha-\nu} S_{\Upsilon_\alpha}\big(\frac{2\sqrt{2\gamma x}}{\sigma}, \frac{2\sqrt{2\gamma a}}{\sigma}\big)}{b^{\Upsilon_\alpha-\nu} S_{\Upsilon_\alpha}\big(\frac{2\sqrt{2\gamma b}}{\sigma}, \frac{2\sqrt{2\gamma a}}{\sigma}\big)}$$

3.12.5 (a)

$$\mathbf{E}_x\{e^{-\gamma\check{H}(H)}; V_H = a\}$$

$$= \int_x^b \frac{|\nu|(a/x)^\nu\big((x/a)^{\sqrt{\nu^2+2\gamma/\sigma^2}} - (a/x)^{\sqrt{\nu^2+2\gamma/\sigma^2}}\big)}{y\big((y/a)^{|\nu|} - (a/y)^{|\nu|}\big)\big((y/a)^{\sqrt{\nu^2+2\gamma/\sigma^2}} - (a/y)^{\sqrt{\nu^2+2\gamma/\sigma^2}}\big)}dy$$

3.12.5 (b)

$$\mathbf{E}_x\{e^{-\gamma\hat{H}(H)}; V_H = b\}$$

$$= \int_a^x \frac{|\nu|(b/x)^\nu\big((b/x)^{\sqrt{\nu^2+2\gamma/\sigma^2}} - (x/b)^{\sqrt{\nu^2+2\gamma/\sigma^2}}\big)}{y\big((b/y)^{|\nu|} - (y/b)^{|\nu|}\big)\big((b/y)^{\sqrt{\nu^2+2\gamma/\sigma^2}} - (y/b)^{\sqrt{\nu^2+2\gamma/\sigma^2}}\big)}dy$$

$$\Upsilon_s := \sqrt{\nu^2 + 2s/\sigma^2}$$

9 $V_s = e^{\sigma^2 \nu s + \sigma W_s}$ $\sigma > 0$ $H = H_{a,b} = \min\{s : V_s \notin (a,b)\}$

3.12.6 (a)
$$\mathbf{P}_x\big(\check{H}(H) \in du,\ V_H = a\big) = \Big(\frac{a}{x}\Big)^{\nu} e^{-\nu^2\sigma^2 u/2} \int_x^b \frac{|\nu|\,\mathrm{ss}_u\big(\frac{\ln(x/a)}{\sigma}, \frac{\ln(y/a)}{\sigma}\big)}{y\big((y/a)^{|\nu|} - (a/y)^{|\nu|}\big)} dy\,du$$

3.12.6 (b)
$$\mathbf{P}_x\big(\hat{H}(H) \in du,\ V_H = b\big) = \Big(\frac{b}{x}\Big)^{\nu} e^{-\nu^2\sigma^2 u/2} \int_a^x \frac{|\nu|\,\mathrm{ss}_u\big(\frac{\ln(b/x)}{\sigma}, \frac{\ln(b/y)}{\sigma}\big)}{y\big((b/y)^{|\nu|} - (y/b)^{|\nu|}\big)} dy\,du$$

3.12.8 (a)
$$\mathbf{P}_x\big(\check{H}(H) \in du,\ H \in dt,\ V_H = a\big) \qquad u < t$$
$$= \Big(\frac{a}{x}\Big)^{\nu} e^{-\nu^2\sigma^2 t/2} \int_x^b \frac{1}{y\sigma}\,\mathrm{ss}_u\Big(\frac{\ln(x/a)}{\sigma}, \frac{\ln(y/a)}{\sigma}\Big)\,\mathrm{s}_{t-u}\Big(\frac{\ln(y/a)}{\sigma}\Big) dy\,dudt$$

3.12.8 (b)
$$\mathbf{P}_x\big(\hat{H}(H) \in du,\ H \in dt,\ V_H = b\big) \qquad u < t$$
$$= \Big(\frac{b}{x}\Big)^{\nu} e^{-\nu^2\sigma^2 t/2} \int_a^x \frac{1}{y\sigma}\,\mathrm{ss}_u\Big(\frac{\ln(b/x)}{\sigma}, \frac{\ln(b/y)}{\sigma}\Big)\,\mathrm{s}_{t-u}\Big(\frac{\ln(b/y)}{\sigma}\Big) dy\,dudt$$

3.19.7 (a)
$$\mathbf{E}_x\Big\{\exp\Big(-\alpha H - \gamma\int_0^H \frac{ds}{V_s}\Big);\ V_H = a\Big\} = \frac{a^{\Upsilon_\alpha+\nu} S_{2\Upsilon_\alpha}\big(\frac{2\sqrt{2\gamma}}{\sigma\sqrt{b}}, \frac{2\sqrt{2\gamma}}{\sigma\sqrt{x}}\big)}{x^{\Upsilon_\alpha+\nu} S_{2\Upsilon_\alpha}\big(\frac{2\sqrt{2\gamma}}{\sigma\sqrt{b}}, \frac{2\sqrt{2\gamma}}{\sigma\sqrt{a}}\big)}$$

3.19.7 (b)
$$\mathbf{E}_x\Big\{\exp\Big(-\alpha H - \gamma\int_0^H \frac{ds}{V_s}\Big);\ V_H = b\Big\} = \frac{b^{\Upsilon_\alpha+\nu} S_{2\Upsilon_\alpha}\big(\frac{2\sqrt{2\gamma}}{\sigma\sqrt{x}}, \frac{2\sqrt{2\gamma}}{\sigma\sqrt{a}}\big)}{x^{-\Upsilon_\alpha+\nu} S_{2\Upsilon_\alpha}\big(\frac{2\sqrt{2\gamma}}{\sigma\sqrt{b}}, \frac{2\sqrt{2\gamma}}{\sigma\sqrt{a}}\big)}$$

3.20.7 (a)
$$\mathbf{E}_x\Big\{\exp\Big(-\alpha H - \frac{\gamma^2}{2}\int_0^H V_s^{2\beta} ds\Big); V_H = a\Big\} = \frac{x^{\Upsilon_\alpha \operatorname{sign}\beta - \nu} S_{\Upsilon_\alpha/|\beta|}\big(\frac{\gamma b^\beta}{\sigma|\beta|}, \frac{\gamma x^\beta}{\sigma|\beta|}\big)}{a^{\Upsilon_\alpha \operatorname{sign}\beta - \nu} S_{\Upsilon_\alpha/|\beta|}\big(\frac{\gamma b^\beta}{\sigma|\beta|}, \frac{\gamma a^\beta}{\sigma|\beta|}\big)}$$

3.20.7 (b)
$$\mathbf{E}_x\Big\{\exp\Big(-\alpha H - \frac{\gamma^2}{2}\int_0^H V_s^{2\beta} ds\Big); V_H = b\Big\} = \frac{x^{\Upsilon_\alpha \operatorname{sign}\beta - \nu} S_{\Upsilon_\alpha/|\beta|}\big(\frac{\gamma x^\beta}{\sigma|\beta|}, \frac{\gamma a^\beta}{\sigma|\beta|}\big)}{b^{\Upsilon_\alpha \operatorname{sign}\beta - \nu} S_{\Upsilon_\alpha/|\beta|}\big(\frac{\gamma b^\beta}{\sigma|\beta|}, \frac{\gamma a^\beta}{\sigma|\beta|}\big)}$$

$$\Upsilon_s := \sqrt{\nu^2 + 2s/\sigma^2}$$

9 $V_s = e^{\sigma^2 \nu s + \sigma W_s}$ $\sigma > 0$ $H = H_{a,b} = \min\{s : V_s \notin (a,b)\}$

3.21.7 (a)

$$\mathbf{E}_x\Big\{\exp\Big(-\alpha H - \int_0^H \big(\frac{p^2}{2}V_s^{2\beta} + qV_s^{\beta}\big)ds\Big);\ V_H = a\Big\}$$

$$= \frac{x^{\Upsilon_\alpha \operatorname{sign}\beta - \nu}\exp\big(-\frac{px^\beta}{\sigma|\beta|}\big)S\big(\frac{\Upsilon_\alpha}{|\beta|} + \frac{q}{p\sigma|\beta|} + \frac{1}{2}, \frac{2\Upsilon_\alpha}{|\beta|} + 1, \frac{2pb^\beta}{\sigma|\beta|}, \frac{2px^\beta}{\sigma|\beta|}\big)}{a^{\Upsilon_\alpha \operatorname{sign}\beta - \nu}\exp\big(-\frac{pa^\beta}{\sigma|\beta|}\big)S\big(\frac{\Upsilon_\alpha}{|\beta|} + \frac{q}{p\sigma|\beta|} + \frac{1}{2}, \frac{2\Upsilon_\alpha}{|\beta|} + 1, \frac{2pb^\beta}{\sigma|\beta|}, \frac{2pa^\beta}{\sigma|\beta|}\big)}$$

3.21.7 (b)

$$\mathbf{E}_x\Big\{\exp\Big(-\alpha H - \int_0^H \big(\frac{p^2}{2}V_s^{2\beta} + qV_s^{\beta}\big)ds\Big);\ V_H = b\Big\}$$

$$= \frac{x^{\Upsilon_\alpha \operatorname{sign}\beta - \nu}\exp\big(-\frac{px^\beta}{\sigma|\beta|}\big)S\big(\frac{\Upsilon_\alpha}{|\beta|} + \frac{q}{p\sigma|\beta|} + \frac{1}{2}, \frac{2\Upsilon_\alpha}{|\beta|} + 1, \frac{2px^\beta}{\sigma|\beta|}, \frac{2pa^\beta}{\sigma|\beta|}\big)}{b^{\Upsilon_\alpha \operatorname{sign}\beta - \nu}\exp\big(-\frac{pb^\beta}{\sigma|\beta|}\big)S\big(\frac{\Upsilon_\alpha}{|\beta|} + \frac{q}{p\sigma|\beta|} + \frac{1}{2}, \frac{2\Upsilon_\alpha}{|\beta|} + 1, \frac{2pb^\beta}{\sigma|\beta|}, \frac{2pa^\beta}{\sigma|\beta|}\big)}$$

$$\Upsilon_s := \sqrt{\nu^2 + 2s/\sigma^2}$$

9 $V_s = e^{\sigma^2\nu s+\sigma W_s}$ $\sigma > 0$ $\varrho = \varrho(v,z) = \min\{s : \ell(s,z) = v\}$

4. Stopping at inverse local time

4.0.1 $$\mathbf{E}_x e^{-\alpha\varrho} = \begin{cases} \left(\frac{x}{z}\right)^{\sqrt{\nu^2+2\alpha/\sigma^2}-\nu} e^{-vz\sigma^2\sqrt{\nu^2+2\alpha/\sigma^2}}, & x \le z \\ \left(\frac{z}{x}\right)^{\sqrt{\nu^2+2\alpha/\sigma^2}+\nu} e^{-vz\sigma^2\sqrt{\nu^2+2\alpha/\sigma^2}}, & z \le x \end{cases}$$

4.0.2 $$\mathbf{P}_x\big(\varrho \in dt\big) = \frac{|\ln(x/z)| + vz\sigma^2}{\sigma\sqrt{2\pi}t^{3/2}}\left(\frac{z}{x}\right)^{\nu}\exp\Big(-\frac{\nu^2\sigma^2 t}{2} - \frac{(|\ln(x/z)| + vz\sigma^2)^2}{2\sigma^2 t}\Big)dt$$

(1) $$\mathbf{P}_x\big(\varrho = \infty\big) = \begin{cases} 1 - \left(\frac{x}{z}\right)^{|\nu|-\nu} e^{-|\nu|vz\sigma^2}, & x \le z \\ 1 - \left(\frac{z}{x}\right)^{|\nu|+\nu} e^{-|\nu|vz\sigma^2}, & z \le x \end{cases}$$

4.1.2 $z < y$ $$\mathbf{P}_x\big(\sup_{0\le s\le\varrho} V_s < y,\ \varrho < \infty\big)$$

$$= \begin{cases} \left(\frac{x}{z}\right)^{|\nu|-\nu} \exp\Big(-\frac{|\nu|vz\sigma^2}{1-(z/y)^{2|\nu|}}\Big), & x \le z \\ \left(\frac{z}{x}\right)^{\nu}\frac{(y/x)^{|\nu|} - (x/y)^{|\nu|}}{(y/z)^{|\nu|} - (z/y)^{|\nu|}} \exp\Big(-\frac{|\nu|vz\sigma^2}{1-(z/y)^{2|\nu|}}\Big), & z \le x \le y \end{cases}$$

4.1.4 $$\mathbf{E}_x\big\{e^{-\alpha\varrho};\ \sup_{0\le s\le\varrho} V_s < y\big\}$$

$$= \begin{cases} \left(\frac{x}{z}\right)^{\sqrt{\nu^2+2\alpha/\sigma^2}-\nu} \exp\Big(-\frac{vz\sigma^2\sqrt{\nu^2+2\alpha/\sigma^2}}{1-(z/y)^{2\sqrt{\nu^2+2\alpha/\sigma^2}}}\Big), & x \le z,\ z \le y \\ \left(\frac{z}{x}\right)^{\nu}\frac{(y/x)^{\sqrt{\nu^2+2\alpha/\sigma^2}} - (x/y)^{\sqrt{\nu^2+2\alpha/\sigma^2}}}{(y/z)^{\sqrt{\nu^2+2\alpha/\sigma^2}} - (z/y)^{\sqrt{\nu^2+2\alpha/\sigma^2}}} \exp\Big(-\frac{vz\sigma^2\sqrt{\nu^2+2\alpha/\sigma^2}}{1-(z/y)^{2\sqrt{\nu^2+2\alpha/\sigma^2}}}\Big), & z \le x,\ x \le y \end{cases}$$

4.2.2 $y < z$ $$\mathbf{P}_x\big(\inf_{0\le s\le\varrho} V_s > y,\ \varrho < \infty\big)$$

$$= \begin{cases} \left(\frac{z}{x}\right)^{\nu}\frac{(x/y)^{|\nu|} - (y/x)^{|\nu|}}{(z/y)^{|\nu|} - (y/z)^{|\nu|}} \exp\Big(-\frac{|\nu|vz\sigma^2}{1-(y/z)^{2|\nu|}}\Big), & y \le x \le z \\ \left(\frac{z}{x}\right)^{|\nu|+\nu} \exp\Big(-\frac{|\nu|vz\sigma^2}{1-(y/z)^{2|\nu|}}\Big), & y \le z \le x \end{cases}$$

9 $V_s = e^{\sigma^2\nu s+\sigma W_s}$ $\sigma > 0$ $\varrho = \varrho(v,z) = \min\{s : \ell(s,z) = v\}$

4.2.4 $\mathbf{E}_x\{e^{-\alpha\varrho}; \inf_{0\le s\le\varrho} V_s > y\}$

$$= \begin{cases} \Big(\frac{z}{x}\Big)^{\nu} \dfrac{(x/y)^{\sqrt{\nu^2+2\alpha/\sigma^2}} - (y/x)^{\sqrt{\nu^2+2\alpha/\sigma^2}}}{(z/y)^{\sqrt{\nu^2+2\alpha/\sigma^2}} - (y/z)^{\sqrt{\nu^2+2\alpha/\sigma^2}}} \exp\Big(-\dfrac{vz\sigma^2\sqrt{\nu^2+2\alpha/\sigma^2}}{1-(y/z)^{2\sqrt{\nu^2+2\alpha/\sigma^2}}}\Big), & y \le x \\ & x \le z \\ \Big(\frac{z}{x}\Big)^{\sqrt{\nu^2+2\alpha/\sigma^2}+\nu} \exp\Big(-\dfrac{vz\sigma^2\sqrt{\nu^2+2\alpha/\sigma^2}}{1-(y/z)^{2\sqrt{\nu^2+2\alpha/\sigma^2}}}\Big), & y \le z \\ & z \le x \end{cases}$$

4.3.1 $\mathbf{E}_x\{e^{-\gamma\ell(\varrho,r)};\ \varrho < \infty\} = \exp\Big(-\dfrac{vz\sigma^2|\nu|(\gamma+|\nu|r\sigma^2)}{|\nu|r\sigma^2+\gamma(1-e^{-2|\ln(r/z)||\nu|})}\Big)$

$$= \Big(\frac{z}{x}\Big)^{\nu} e^{-|\nu||\ln(z/x)|} \begin{cases} 1, & r \le z \le x, \quad x \le z \le r \\ \dfrac{|\nu|r\sigma^2+\gamma(1-e^{-2|\nu||\ln(x/r)|})}{|\nu|r\sigma^2+\gamma(1-e^{-2|\nu||\ln(z/r)|})}, & z\wedge r \le x \le z \vee r \\ \dfrac{|\nu|r\sigma^2}{|\nu|r\sigma^2+\gamma(1-e^{-2|\nu||\ln(z/r)|})}, & z \le r \le x, \quad x \le r \le z \end{cases}$$

4.3.2 $\mathbf{P}_x\big(\ell(\varrho,r) \in dy,\ \varrho < \infty\big) = \Big(\frac{z}{x}\Big)^{\nu} \exp\Big(-\dfrac{(vz+yr)\sigma^2|\nu|}{1-e^{-2|\ln(r/z)||\nu|}}\Big)$

$$\times \begin{cases} \dfrac{|\nu|r\sigma^2 e^{-|\ln(z/x)||\nu|}}{2\,\mathrm{sh}(|\ln(r/z)||\nu|)} \dfrac{\sqrt{vz}}{\sqrt{yr}} I_1\Big(\dfrac{|\nu|\sigma^2\sqrt{vzyr}}{\mathrm{sh}(|\ln(r/z)||\nu|)}\Big)dy, & x \le z \le r, \quad r \le z \le x \\ \dfrac{|\nu|r\sigma^2}{2\,\mathrm{sh}^2(|\ln(r/z)||\nu|)} \Big[\mathrm{sh}(|\ln(r/x)||\nu|)\dfrac{\sqrt{vz}}{\sqrt{yr}} I_1\Big(\dfrac{|\nu|\sigma^2\sqrt{vzyr}}{\mathrm{sh}(|\ln(r/z)||\nu|)}\Big) & \\ \qquad + \mathrm{sh}(|\ln(z/x)||\nu|) I_0\Big(\dfrac{|\nu|\sigma^2\sqrt{vzyr}}{\mathrm{sh}(|\ln(r/z)||\nu|)}\Big)\Big]dy, & z\wedge r \le x \le r \vee z \\ \dfrac{|\nu|r\sigma^2 e^{-|\ln(r/x)||\nu|}}{\mathrm{sh}(|\ln(r/z)||\nu|)} I_0\Big(\dfrac{|\nu|\sigma^2\sqrt{vzyr}}{\mathrm{sh}(|\ln(r/z)||\nu|)}\Big)dy, & z \le r \le x, \quad x \le r \le z \end{cases}$$

(1) $\mathbf{P}_x\big(\ell(\varrho,r) = 0,\ \varrho < \infty\big)$

$$= \Big(\frac{z}{x}\Big)^{\nu} \exp\Big(-\dfrac{|\nu|vz\sigma^2}{1-e^{-2|\nu||\ln(z/r)|}}\Big) \begin{cases} e^{-|\nu||\ln(z/x)|}, & x \le z \le r, \quad r \le z \le x \\ \dfrac{\mathrm{sh}(|\nu||\ln(x/r)|)}{\mathrm{sh}(|\nu||\ln(z/r)|)}, & z\wedge r \le x \le r \vee z \\ 0, & z \le r \le x, \quad x \le r \le z \end{cases}$$

9 $V_s = e^{\sigma^2 \nu s + \sigma W_s}$ $\sigma > 0$ $\varrho = \varrho(v,z) = \min\{s : \ell(s,z) = v\}$

4.8.3 $$\mathbf{E}_x \exp\Big(-\alpha\varrho - \gamma\int_0^{\varrho} V_s ds\Big) = \exp\Big(-\frac{vz\sigma^2}{4I_{2\sqrt{\nu^2+2\alpha/\sigma^2}}\big(\frac{2\sqrt{2\gamma z}}{\sigma}\big)K_{2\sqrt{\nu^2+2\alpha/\sigma^2}}\big(\frac{2\sqrt{2\gamma z}}{\sigma}\big)}\Big)$$

$$\times\begin{cases}\Big(\dfrac{z}{x}\Big)^{\nu}\dfrac{I_{2\sqrt{\nu^2+2\alpha/\sigma^2}}(2\sqrt{2\gamma x}/\sigma)}{I_{2\sqrt{\nu^2+2\alpha/\sigma^2}}(2\sqrt{2\gamma z}/\sigma)}, & x \le z\\[2ex] \Big(\dfrac{z}{x}\Big)^{\nu}\dfrac{K_{2\sqrt{\nu^2+2\alpha/\sigma^2}}(2\sqrt{2\gamma x}/\sigma)}{K_{2\sqrt{\nu^2+2\alpha/\sigma^2}}(2\sqrt{2\gamma z}/\sigma)}, & z \le x\end{cases}$$

4.15.4 $$\mathbf{E}_x\{e^{-\alpha\varrho}; a < \inf_{0\le s\le\varrho} V_s, \sup_{0\le s\le\varrho} V_s < b\}$$

$$= \exp\Big(-\frac{vz\sigma^2\sqrt{\nu^2+2\alpha/\sigma^2}\big((b/a)^{\sqrt{\nu^2+2\alpha/\sigma^2}} - (a/b)^{\sqrt{\nu^2+2\alpha/\sigma^2}}\big)}{2\big((b/z)^{\sqrt{\nu^2+2\alpha/\sigma^2}} - (z/b)^{\sqrt{\nu^2+2\alpha/\sigma^2}}\big)\big((z/a)^{\sqrt{\nu^2+2\alpha/\sigma^2}} - (a/z)^{\sqrt{\nu^2+2\alpha/\sigma^2}}\big)}\Big)$$

$$\times\begin{cases}\Big(\dfrac{z}{x}\Big)^{\nu}\dfrac{(x/a)^{\sqrt{\nu^2+2\alpha/\sigma^2}} - (a/x)^{\sqrt{\nu^2+2\alpha/\sigma^2}}}{(z/a)^{\sqrt{\nu^2+2\alpha/\sigma^2}} - (a/z)^{\sqrt{\nu^2+2\alpha/\sigma^2}}}, & -b \le x \le z\\[2ex] \Big(\dfrac{z}{x}\Big)^{\nu}\dfrac{(b/x)^{\sqrt{\nu^2+2\alpha/\sigma^2}} - (x/b)^{\sqrt{\nu^2+2\alpha/\sigma^2}}}{(b/z)^{\sqrt{\nu^2+2\alpha/\sigma^2}} - (z/b)^{\sqrt{\nu^2+2\alpha/\sigma^2}}}, & z \le x \le b\end{cases}$$

4.19.3 $$\mathbf{E}_x \exp\Big(-\alpha\varrho - \gamma\int_0^{\varrho} \frac{ds}{V_s}\Big) = \exp\Big(-\frac{vz\sigma^2}{4I_{2\sqrt{\nu^2+2\alpha/\sigma^2}}\big(\frac{2\sqrt{2\gamma}}{\sigma\sqrt{z}}\big)K_{2\sqrt{\nu^2+2\alpha/\sigma^2}}\big(\frac{2\sqrt{2\gamma}}{\sigma\sqrt{z}}\big)}\Big)$$

$$\times\begin{cases}\Big(\dfrac{z}{x}\Big)^{\nu}\dfrac{K_{2\sqrt{\nu^2+2\alpha/\sigma^2}}(2\sqrt{2\gamma}/\sigma\sqrt{x})}{K_{2\sqrt{\nu^2+2\alpha/\sigma^2}}(2\sqrt{2\gamma}/\sigma\sqrt{z})}, & x \le z\\[2ex] \Big(\dfrac{z}{x}\Big)^{\nu}\dfrac{I_{2\sqrt{\nu^2+2\alpha/\sigma^2}}(2\sqrt{2\gamma}/\sigma\sqrt{x})}{I_{2\sqrt{\nu^2+2\alpha/\sigma^2}}(2\sqrt{2\gamma}/\sigma\sqrt{z})}, & z \le x\end{cases}$$

4.20.3 $$\mathbf{E}_x \exp\Big(-\alpha\varrho - \frac{\gamma^2}{2}\int_0^{\varrho} V_s^{2\beta}\, ds\Big) = \exp\Big(-\frac{vz\sigma^2|\beta|}{2I_{\frac{\sqrt{\nu^2+2\alpha/\sigma^2}}{|\beta|}}\big(\frac{\gamma z^{\beta}}{\sigma|\beta|}\big)K_{\frac{\sqrt{\nu^2+2\alpha/\sigma^2}}{|\beta|}}\big(\frac{\gamma z^{\beta}}{\sigma|\beta|}\big)}\Big)$$

$$\times\begin{cases}\Big(\dfrac{z}{x}\Big)^{\nu}\dfrac{I_{\sqrt{\nu^2+2\alpha/\sigma^2}/|\beta|}(\gamma x^{\beta}/\sigma|\beta|)}{I_{\sqrt{\nu^2+2\alpha/\sigma^2}/|\beta|}(\gamma z^{\beta}/\sigma|\beta|)}, & (z-x)\beta \ge 0\\[2ex] \Big(\dfrac{z}{x}\Big)^{\nu}\dfrac{K_{\sqrt{\nu^2+2\alpha/\sigma^2}/|\beta|}(\gamma x^{\beta}/\sigma|\beta|)}{K_{\sqrt{\nu^2+2\alpha/\sigma^2}/|\beta|}(\gamma z^{\beta}/\sigma|\beta|)}, & (z-x)\beta \le 0\end{cases}$$

9 $V_s = e^{\sigma^2 \nu s + \sigma W_s}$ $\sigma > 0$ $\varrho = \varrho(v, z) = \min\{s : \ell(s, z) = v\}$

4.21.3
$0 < x$

$$\mathbf{E}_x \exp\Big(-\alpha\varrho - \int_0^{\varrho} \big(\frac{p^2}{2} V_s^{2\beta} + qV_s^{\beta}\big) ds\Big)$$

$$= \exp\Big(-\frac{pvz^{\beta+1}\sigma\Gamma\big(\frac{2\sqrt{\nu^2\sigma^2+2\alpha}}{\sigma|\beta|}+1\big)\Gamma^{-1}\big(\frac{\sqrt{\nu^2\sigma^2+2\alpha}}{\sigma|\beta|}+\frac{1}{2}+\frac{q}{p\sigma|\beta|}\big)}{M_{-\frac{q}{p\sigma|\beta|},\frac{\sqrt{\nu^2\sigma^2+2\alpha}}{\sigma|\beta|}}\big(\frac{2pz^{\beta}}{\sigma|\beta|}\big) W_{-\frac{q}{p\sigma|\beta|},\frac{\sqrt{\nu^2\sigma^2+2\alpha}}{\sigma|\beta|}}\big(\frac{2pz^{\beta}}{\sigma|\beta|}\big)}\Big)$$

$$\times \begin{cases} \left(\dfrac{z}{x}\right)^{\nu} \dfrac{z^{\beta/2} M_{-q/p\sigma|\beta|,\sqrt{\nu^2+2\alpha/\sigma^2}/|\beta|}(2px^{\beta}/\sigma|\beta|)}{x^{\beta/2} M_{-q/p\sigma|\beta|,\sqrt{\nu^2+2\alpha/\sigma^2}/|\beta|}(2pz^{\beta}/\sigma|\beta|)}, & (z-x)\beta \geq 0 \\ \left(\dfrac{z}{x}\right)^{\nu} \dfrac{z^{\beta/2} W_{-q/p\sigma|\beta|,\sqrt{\nu^2+2\alpha/\sigma^2}/|\beta|}(2px^{\beta}/\sigma|\beta|)}{x^{\beta/2} W_{-q/p\sigma|\beta|,\sqrt{\nu^2+2\alpha/\sigma^2}/|\beta|}(2pz^{\beta}/\sigma|\beta|)}, & (z-x)\beta \leq 0 \end{cases}$$

APPENDIX 2

SPECIAL FUNCTIONS

1. Hyperbolic functions

$$\operatorname{sh} x := \frac{1}{2}(e^x - e^{-x}) = \sum_{k=0}^{\infty} \frac{x^{2k+1}}{(2k+1)!}$$

$$\operatorname{ch} x := \frac{1}{2}(e^x + e^{-x}) = \sum_{k=0}^{\infty} \frac{x^{2k}}{2k!}$$

$$\operatorname{th} x := \frac{\operatorname{sh} x}{\operatorname{ch} x}$$

$$\operatorname{cth} x := \frac{\operatorname{ch} x}{\operatorname{sh} x}$$

2. Gamma function

$$\Gamma(x) := \int_0^{\infty} u^{x-1} e^{-u}\, du, \quad \operatorname{Re} x > 0$$

$$\Gamma(x+1) = x\Gamma(x), \qquad \frac{1}{\Gamma(x)} \simeq x \quad \text{as } x \to 0$$

$$\Gamma(x)\Gamma(1-x) = \frac{\pi}{\sin(\pi x)}$$

$$\Gamma(2x) = \frac{2^{2x-1}}{\sqrt{\pi}} \Gamma(x)\Gamma\big(x + \tfrac{1}{2}\big)$$

$$\Gamma(ax+b) \simeq \sqrt{2\pi} e^{-ax} (ax)^{ax+b-1/2} \quad \text{as } x \to \infty, \quad a > 0$$

$$\Gamma(n+1) = n!, \qquad \Gamma(n+1/2) = 1 \cdot 3 \cdot \ldots \cdot (2n-1) 2^{-n} \sqrt{\pi}, \quad n = 1, 2, \ldots$$

$$\Gamma(1/2) = \sqrt{\pi}, \qquad \Gamma(3/2) = \sqrt{\pi}/2, \qquad \Gamma(-1/2) = -2\sqrt{\pi}$$

3. Bessel functions

$$J_\nu(x) := \sum_{k=0}^{\infty} \frac{(-1)^k (x/2)^{\nu+2k}}{k!\,\Gamma(\nu+k+1)}$$

$$Y_\nu(x) := \frac{1}{\sin(\nu\pi)} \big(J_\nu(x)\cos(\nu\pi) - J_{-\nu}(x)\big)$$

$$Y_\nu'(x) J_\nu(x) - Y_\nu(x) J_\nu'(x) = Y_\nu(x) J_{\nu+1}(x) - Y_{\nu+1}(x) J_\nu(x) = 2/(\pi x)$$

$0 < j_{\nu,1} < j_{\nu,2} < \ldots$ – positive zeros of $J_\nu(x)$ when $\nu \geq 0$

4. Modified Bessel functions

$$I_\nu(x) := \sum_{k=0}^{\infty} \frac{(x/2)^{\nu+2k}}{k!\,\Gamma(\nu+k+1)} = i^{-\nu} J_\nu(ix)$$

$$K_\nu(x) := \frac{\pi}{2\sin(\nu\pi)}\big(I_{-\nu}(x) - I_\nu(x)\big)$$

are linearly independent solutions of the Bessel equation.

$$x^2 Y''(x) + xY'(x) - (x^2+\nu^2)Y(x) = 0, \qquad x > 0$$

Integral representations: for $x > 0$

$$I_\nu(x) = \frac{(x/2)^\nu}{\sqrt{\pi}\Gamma(\nu+1/2)} \int_{-1}^{1} (1-t^2)^{\nu-1/2} e^{xt}\,dt, \qquad \operatorname{Re}\nu > -1/2$$

$$K_\nu(x) = \frac{\sqrt{\pi}(x/2)^\nu}{\Gamma(\nu+1/2)} \int_{1}^{\infty} (t^2-1)^{\nu-1/2} e^{-xt}\,dt, \qquad \operatorname{Re}\nu > -1/2$$

Properties:

$$I_{-\nu}(x) = I_\nu(x) \quad \nu = 1, 2, \ldots$$

$$K_{-\nu}(x) = K_\nu(x)$$

$$I_\nu'(x)K_\nu(x) - I_\nu(x)K_\nu'(x) = I_\nu(x)K_{\nu+1}(x) + I_{\nu+1}(x)K_\nu(x) = 1/x$$

$$\big(x^{-\nu} I_\nu(x)\big)' = x^{-\nu} I_{\nu+1}(x), \qquad \big(x^{\nu} I_\nu(x)\big)' = x^{\nu} I_{\nu-1}(x)$$

$$\big(x^{-\nu} K_\nu(x)\big)' = -x^{-\nu} K_{\nu+1}(x), \qquad \big(x^{\nu} K_\nu(x)\big)' = -x^{\nu} K_{\nu-1}(x)$$

$$I_{\nu-1}(x) - I_{\nu+1}(x) = \frac{2\nu}{x} I_\nu(x), \qquad K_{\nu+1}(x) - K_{\nu-1}(x) = \frac{2\nu}{x} K_\nu(x)$$

$$I_\nu(x) \simeq \frac{1}{\Gamma(\nu+1)}\left(\frac{x}{2}\right)^\nu \quad \text{as } x \to 0, \quad \nu \neq -1, -2, \ldots$$

$$K_\nu(x) \simeq \frac{\Gamma(\nu)}{2}\left(\frac{x}{2}\right)^{-\nu}, \quad \nu > 0, \qquad K_0(x) \simeq -\ln x \quad \text{as } x \to 0$$

$$I_\nu(x) \simeq \frac{1}{\sqrt{2\pi x}} e^x \quad \text{as } x \to \infty$$

$$K_\nu(x) \simeq \frac{\sqrt{\pi}}{\sqrt{2x}} e^{-x} \quad \text{as } x \to \infty$$

$$I_{\nu\sqrt{\lambda}}\big(\nu\sqrt{\gamma}\, e^{x/\nu}\big) \simeq \frac{1}{\sqrt{2\pi}(\lambda+\gamma)^{1/4}\sqrt{\nu}} \left(\frac{\sqrt{\gamma}}{\sqrt{\lambda}+\sqrt{\lambda+\gamma}}\right)^{\nu\sqrt{\lambda}} e^{(\nu+x)\sqrt{\lambda+\gamma}} \quad \text{as } \nu \to \infty$$

$$K_{\nu\sqrt{\lambda}}(\nu\sqrt{\gamma}\,e^{x/\nu}) \simeq \frac{\sqrt{\pi}}{(\lambda+\gamma)^{1/4}\sqrt{2\nu}}\Big(\frac{\sqrt{\gamma}}{\sqrt{\lambda}+\sqrt{\lambda+\gamma}}\Big)^{-\nu\sqrt{\lambda}} e^{-(\nu+x)\sqrt{\lambda+\gamma}} \quad \text{as } \nu \to \infty$$

Special cases:

$$I_{\frac{1}{2}}(x) = \frac{\sqrt{2}}{\sqrt{\pi x}}\operatorname{sh} x; \quad K_{\frac{1}{2}}(x) = K_{-\frac{1}{2}}(x) = \frac{\sqrt{\pi}}{\sqrt{2x}}e^{-x}; \quad I_{-\frac{1}{2}}(x) = \frac{\sqrt{2}}{\sqrt{\pi x}}\operatorname{ch} x;$$

$$I_{\frac{3}{2}}(x) = \frac{\sqrt{2}}{\sqrt{\pi}x^{3/2}}(x\operatorname{ch} x - \operatorname{sh} x); \qquad K_{\frac{3}{2}}(x) = \frac{\sqrt{\pi}}{\sqrt{2}x^{3/2}}e^{-x}(x+1).$$

5. Airy function

$$\operatorname{Ai}(x) := \frac{1}{3}\sqrt{x}\,\Big(I_{-1/3}\big(\tfrac{2}{3}x^{3/2}\big) - I_{1/3}\big(\tfrac{2}{3}x^{3/2}\big)\Big) = \frac{\sqrt{x}}{\pi\sqrt{3}}K_{1/3}\big(\tfrac{2}{3}x^{3/2}\big)$$

$\dots\alpha_k < \dots < \alpha_2 < \alpha_1 < 0$ – zeros of the Airy function $\operatorname{Ai}(x)$

$\dots\alpha'_k < \dots < \alpha'_2 < \alpha'_1 < 0$ – zeros of the derivative of the Airy function $\operatorname{Ai}'(x)$

6. Hermite polynomials and related functions

$$\operatorname{He}_n(x) := (-1)^n e^{x^2/2}\frac{d^n}{dx^n}\big(e^{-x^2/2}\big) = \sum_{0\le k\le n/2}\frac{(-1)^k 2^{-k} n!}{k!(n-2k)!}x^{n-2k}$$

$$\operatorname{h}_y(n,v) := \mathcal{L}_\gamma^{-1}\big((2\gamma)^{n/2-1/2}e^{-v\sqrt{2\gamma}}\big) = \frac{1}{\sqrt{2\pi}y^{(n+1)/2}}e^{-v^2/2y}\operatorname{He}_n\Big(\frac{v}{\sqrt{y}}\Big), \quad 0 < v$$

7. Binomial series

$$(1+x)^\mu = \sum_{k=0}^{\infty}\frac{\Gamma(\mu+1)}{\Gamma(\mu+1-k)\,k!}x^k = \sum_{k=0}^{\infty}\frac{(-1)^k\Gamma(-\mu+k)}{\Gamma(-\mu)\,k!}x^k, \qquad |x| < 1$$

8. Error functions

$$\operatorname{Erf}(x) := \frac{2}{\sqrt{\pi}}\int_0^x e^{-v^2}\,dv = \frac{2}{\sqrt{\pi}}\sum_{k=0}^{\infty}\frac{(-1)^k x^{2k+1}}{k!(2k+1)} = \frac{2}{\sqrt{\pi}}e^{-x^2}\sum_{k=0}^{\infty}\frac{2^k x^{2k+1}}{1\cdot 3\cdot 5\cdots(2k+1)}$$

$$\operatorname{Erfc}(x) := \frac{2}{\sqrt{\pi}}\int_x^{\infty} e^{-v^2}\,dv = 1 - \operatorname{Erf}(x)$$

$$\operatorname{Erfi}(x) := \frac{2}{\sqrt{\pi}}\int_0^x e^{v^2}\,dv = \frac{2}{\sqrt{\pi}}\sum_{k=0}^{\infty}\frac{x^{2k+1}}{k!(2k+1)}$$

$$\operatorname{Erfid}(x,y) := \operatorname{Erfi}\Big(\frac{x}{\sqrt{2}}\Big) - \operatorname{Erfi}\Big(\frac{y}{\sqrt{2}}\Big)$$

$$\operatorname{Erfc}(x) \simeq \frac{1}{\sqrt{\pi}x}e^{-x^2} \quad \text{as } x \to \infty$$

9. Parabolic cylinder functions

$$D_{-\nu}(x) := e^{-x^2/4} 2^{-\nu/2}\sqrt{\pi}\Big\{\frac{1}{\Gamma((\nu+1)/2)}\Big(1+\sum_{k=1}^{\infty}\frac{\nu(\nu+2)\cdots(\nu+2k-2)}{(2k)!}x^{2k}\Big)$$
$$-\frac{x\sqrt{2}}{\Gamma(\nu/2)}\Big(1+\sum_{k=1}^{\infty}\frac{(\nu+1)(\nu+3)\cdots(\nu+2k-1)}{(2k+1)!}x^{2k}\Big)\Big\}$$

$$\widetilde{D}_{-\nu}(x) := D_{-\nu}(-x)$$

are linearly independent solutions of the differential equation

$$Y''(x) - \Big(\frac{x^2}{4} + \frac{2\nu-1}{2}\Big)Y(x) = 0, \qquad x \in \mathbb{R}.$$

Integral representation: for $x \in \mathbb{R}$

$$D_{-\nu}(x) = \frac{1}{\Gamma(\nu)}e^{-x^2/4}\int_0^\infty t^{\nu-1}e^{-xt-t^2/2}dt, \qquad \operatorname{Re}\nu > 0$$

Properties:

$$D'_{-\nu}(x) = -\frac{x}{2}D_{-\nu}(x) - \nu D_{-\nu-1}(x) = \frac{x}{2}D_{-\nu}(x) - D_{-\nu+1}(x)$$

$$\big(e^{x^2/4}D_{-\nu}(x)\big)' = -\nu e^{x^2/4}D_{-\nu-1}(x), \qquad \big(e^{-x^2/4}D_{-\nu}(x)\big)' = -e^{-x^2/4}D_{-\nu+1}(x)$$

$$\big(x^{1-\nu}e^{-x^2/4}D_{-\nu}(x)\big)' = -x^{-\nu}e^{-x^2/4}D_{-\nu+2}(x)$$

$$D'_{-\nu}(-x)D_{-\nu}(x) + D_{-\nu}(-x)D'_{-\nu}(x) = -\frac{\sqrt{2\pi}}{\Gamma(\nu)}$$

$$\lim_{\theta\downarrow 0} 2^{\alpha/4\theta}\Gamma\Big(\frac{\alpha}{4\theta}+\frac{1}{2}\Big)D_{-\alpha/2\theta}(x\sqrt{2\theta}) = \sqrt{\pi}e^{-x\sqrt{\alpha}}, \quad x \in \mathbb{R}$$

Special cases:

$$D_n(x) = e^{-x^2/4}\operatorname{He}_n(x), \quad D_{-n-1}(x) = \frac{\sqrt{\pi}e^{-x^2/4}}{(-1)^n\sqrt{2}n!}\frac{d^n}{dx^n}\Big(e^{x^2/2}\operatorname{Erfc}\Big(\frac{x}{\sqrt{2}}\Big)\Big), \ n = 0, 1, \dots$$

$$D_{-1/2}(x) = \frac{\sqrt{x}}{\sqrt{2\pi}}K_{1/4}\Big(\frac{x^2}{4}\Big), \qquad D_{-1/2}(-x) - D_{-1/2}(x) = \sqrt{\pi x}I_{1/4}\Big(\frac{x^2}{4}\Big), \quad x \ge 0$$

$$D_{-3/2}(x) = \frac{2}{\sqrt{\pi}}\Big(\frac{x}{2}\Big)^{3/2}\Big(-K_{1/4}\Big(\frac{x^2}{4}\Big) + K_{3/4}\Big(\frac{x^2}{4}\Big)\Big), \quad x \ge 0$$

$$D_{3/2}(x) = \frac{1}{\sqrt{\pi}}\Big(\frac{x}{2}\Big)^{5/2}\Big(2K_{1/4}\Big(\frac{x^2}{4}\Big) + 3K_{3/4}\Big(\frac{x^2}{4}\Big) - K_{5/4}\Big(\frac{x^2}{4}\Big)\Big), \quad x \ge 0$$

$$D_{-1/2}(-x) + D_{-1/2}(x) = \sqrt{\pi x}I_{-1/4}\Big(\frac{x^2}{4}\Big), \quad x \ge 0$$

10. Kummer and Whittaker functions

$$M(a,b,x) := 1 + \sum_{k=1}^{\infty} \frac{a(a+1)\dots(a+k-1)x^k}{b(b+1)\dots(b+k-1)k!}$$

$$U(a,b,x) := \frac{\Gamma(1-b)}{\Gamma(1+a-b)} M(a,b,x) + x^{1-b}\frac{\Gamma(b-1)}{\Gamma(a)} M(1+a-b,2-b,x)$$

are linearly independent solutions of the Kummer equation

$$xY''(x) + (b-x)Y'(x) - aY(x) = 0, \qquad x > 0.$$

Kummer transformations:

$$M(a,b,x) = e^x M(b-a,b,-x), \qquad U(a,b,x) = x^{1-b}U(1+a-b,2-b,x)$$

$$x^{1-b}M(1+a-b,2-b,x) = x^{1-b}e^x M(1-a,2-b,-x)$$

$$e^{-x}U(b-a,b,x) = x^{1-b}e^{-x}U(1-a,2-b,x), \qquad x > 0$$

Whittaker functions:

$$M_{n,m}(x) := x^{m+1/2}e^{-x/2}M(m-n+1/2,2m+1,x)$$

$$W_{n,m}(x) := x^{m+1/2}e^{-x/2}U(m-n+1/2,2m+1,x)$$

Integral representations:

$$M(a,b,x) = \frac{\Gamma(b)}{\Gamma(a)\Gamma(b-a)} \int_0^1 e^{xt}t^{a-1}(1-t)^{b-a-1}dt, \qquad \operatorname{Re} b > \operatorname{Re} a > 0$$

$$U(a,b,x) = \frac{1}{\Gamma(a)} \int_0^\infty e^{-xt}t^{a-1}(1+t)^{b-a-1}dt, \quad \operatorname{Re} a > 0, \quad \operatorname{Re} x > 0$$

Properties:

$$W_{n,m}(x) = \frac{\Gamma(-2m)}{\Gamma(1/2-m-n)} M_{n,m}(x) + \frac{\Gamma(2m)}{\Gamma(1/2+m-n)} M_{n,-m}(x)$$

$$\frac{\partial}{\partial x} M(a,b,x)\, U(a,b,x) - M(a,b,x)\, \frac{\partial}{\partial x} U(a,b,x) = \frac{\Gamma(b)}{\Gamma(a)} x^{-b}e^x$$

$$M'_{n,m}(x)W_{n,m}(x) - M_{n,m}(x)W'_{n,m}(x) = \frac{\Gamma(2m+1)}{\Gamma(m-n+1/2)}$$

$$\frac{\partial}{\partial x} M(a,b,x) = \frac{a}{b} M(a+1,b+1,x), \qquad \frac{\partial}{\partial x} U(a,b,x) = -aU(a+1,b+1,x)$$

$$\frac{\partial}{\partial x}(x^b M(a,b+1,x)) = b\,x^{b-1}M(a,b,x)$$

$$\frac{\partial}{\partial x}(x^b U(a,b+1,x)) = (b-a)x^{b-1}U(a,b,x)$$

$$\frac{\partial}{\partial x}(e^{-x}M(a,b,x)) = \frac{a-b}{b}e^{-x}M(a,b+1,x)$$

$$\frac{\partial}{\partial x}(e^{-x}U(a,b,x)) = -e^{-x}U(a,b+1,x)$$

$$\frac{\partial}{\partial x}(x^b e^{-x}M(a+1,b+1,x)) = bx^{b-1}e^{-x}M(a,b,x)$$

$$\frac{\partial}{\partial x}(x^b e^{-x}U(a+1,b+1,x)) = -x^{b-1}e^{-x}U(a,b,x)$$

$$\lim_{\theta\downarrow 0} M\bigl(\tfrac{a}{4\theta}, b+1, \theta x\bigr) = 2^b\Gamma(b+1)(xa)^{-b/2}I_b(\sqrt{xa})$$

$$\lim_{\theta\downarrow 0} \theta^b\Gamma\bigl(\tfrac{a}{4\theta}\bigr)U\bigl(\tfrac{a}{4\theta}, b+1, \theta x\bigr) = 2^{1-b}(a/x)^{b/2}K_b(\sqrt{xa})$$

Special cases:

$$M_{0,m}(2x) = 2^{2m}\Gamma(m+1)\sqrt{2x}I_m(x), \quad x \ge 0$$

$$W_{0,m}(2x) = \sqrt{2x/\pi}K_m(x), \quad x \ge 0$$

$$M_{1/4-\nu/2,1/4}(x^2/2) = \frac{\Gamma(\nu/2)}{4\sqrt{\pi}}2^{\nu/2-1/4}\sqrt{x}\bigl(D_{-\nu}(-x) - D_{-\nu}(x)\bigr), \quad x \ge 0$$

$$M_{1/4-\nu/2,-1/4}(x^2/2) = \frac{\Gamma((\nu+1)/2)}{2\sqrt{\pi}}2^{\nu/2-1/4}\sqrt{x}\bigl(D_{-\nu}(-x) + D_{-\nu}(x)\bigr), \quad x \ge 0$$

$$W_{1/4-\nu/2,1/4}(x^2/2) = W_{1/4-\nu/2,-1/4}(x^2/2) = 2^{\nu/2-1/4}\sqrt{x}D_{-\nu}(x), \quad x \ge 0$$

$$M_{-1/4,1/4}(x^2) = \frac{\sqrt{\pi x}}{2}\, e^{x^2/2}\,\mathrm{Erf}(x), \qquad M_{1/4,1/4}(x^2) = \frac{\sqrt{\pi x}}{2}\, e^{-x^2/2}\,\mathrm{Erfi}(x)$$

$$W_{-1/4,1/4}(x^2) = \sqrt{\pi x}\, e^{x^2/2}\,\mathrm{Erfc}(x), \qquad W_{1/4,1/4}(x^2) = \sqrt{x}\, e^{-x^2/2}, \quad x \ge 0$$

11. Hypergeometric functions

For $-1 < x < 1$

$$F(\alpha,\beta,\gamma,x) := 1 + \sum_{k=1}^{\infty} \frac{\alpha(\alpha+1)\dots(\alpha+k-1)\beta(\beta+1)\dots(\beta+k-1)x^k}{\gamma(\gamma+1)\dots(\gamma+k-1)k!}$$

for $0 < x < 2$

$$G(\alpha,\beta,\gamma,x) := \frac{1}{\Gamma(\alpha+\beta+1-\gamma)}F(\alpha,\beta,\alpha+\beta+1-\gamma,1-x), \ \mathrm{Re}(\alpha+\beta+1-\gamma) > 0$$

are linearly independent solutions of the hypergeometric differential equation

$$x(1-x)Y''(x) + (\gamma - (\alpha+\beta+1)x)Y'(x) - \alpha\beta Y(x) = 0, \qquad 0 < x < 1.$$

Integral representations: for $-1 < x < 1$

$$F(\alpha,\beta,\gamma,x) = \frac{\Gamma(\gamma)}{\Gamma(\alpha)\Gamma(\gamma-\alpha)} \int_0^1 t^{\alpha-1}(1-t)^{\gamma-\alpha-1}(1-tx)^{-\beta} dt, \qquad \operatorname{Re}\gamma > \operatorname{Re}\alpha > 0$$

for $0 < x < \infty$

$$G(\alpha,\beta,\gamma,x) = \frac{1}{\Gamma(\alpha)\Gamma(\beta+1-\gamma)} \int_0^\infty t^{\alpha-1}(1+t)^{\gamma-\alpha-1}(1+tx)^{-\beta} dt, \quad \operatorname{Re}(\beta+1-\gamma) > 0, \ \operatorname{Re}\alpha > 0$$

Properties:

$$\frac{\partial}{\partial x}F(\alpha,\beta,\gamma,x)\,G(\alpha,\beta,\gamma,x) - F(\alpha,\beta,\gamma,x)\,\frac{\partial}{\partial x}G(\alpha,\beta,\gamma,x) = \frac{\Gamma(\gamma)x^{-\gamma}(1-x)^{\gamma-\alpha-\beta-1}}{\Gamma(\alpha)\Gamma(\beta)}$$

$$\frac{\partial}{\partial x}F(\alpha,\beta,\gamma,x) = \frac{\alpha\beta}{\gamma}F(\alpha+1,\beta+1,\gamma+1,x)$$

$$\frac{\partial}{\partial x}G(\alpha,\beta,\gamma,x) = -\alpha\beta G(\alpha+1,\beta+1,\gamma+1,x)$$

$$\frac{\partial}{\partial x}\big((1-x)^\alpha F(\alpha,\beta,\gamma,x)\big) = \frac{\alpha(\beta-\gamma)}{\gamma}(1-x)^{\alpha-1}F(\alpha+1,\beta,\gamma+1,x)$$

$$\frac{\partial}{\partial x}\big((1-x)^\alpha G(\alpha,\beta,\gamma,x)\big) = -\alpha(1-x)^{\alpha-1}G(\alpha+1,\beta,\gamma+1,x)$$

$$\frac{\partial}{\partial x}\big(x^\gamma(1-x)^{1+\alpha+\beta-\gamma}F(\alpha+1,\beta+1,\gamma+1,x)\big) = \gamma x^{\gamma-1}(1-x)^{\alpha+\beta-\gamma}F(\alpha,\beta,\gamma,x)$$

$$\frac{\partial}{\partial x}\big(x^\gamma(1-x)^{1+\alpha+\beta-\gamma}G(a+1,b+1,\gamma+1,x)\big) = -\frac{x^{\gamma-1}(1-x)^{\alpha+\beta-\gamma}}{1+\alpha+\beta-\gamma}G(\alpha,\beta,\gamma,x)$$

$$F(\alpha,\beta,\gamma,x) = (1-x)^{\gamma-\alpha-\beta}F(\gamma-\alpha,\gamma-\beta,\gamma,x), \qquad 0 \le x < 1$$

$$G(\alpha,\beta,\gamma,x) = x^{1-\gamma}G(\alpha+1-\gamma,\beta+1-\gamma,2-\gamma,x), \qquad 0 < x \le 1$$

$$F(\alpha,\beta,\gamma,1/x) = \frac{x^\beta}{\Gamma(\beta)}\int_0^\infty e^{-xt}t^{\beta-1}M(\alpha,\gamma,t)\,dt, \qquad 1 \le x, \quad \operatorname{Re}\beta > 0$$

$$G(\alpha,\beta,\gamma,1/x) = \frac{x^\beta}{\Gamma(\beta)\Gamma(\beta+1-\gamma)}\int_0^\infty e^{-xt}t^{\beta-1}U(\alpha,\gamma,t)\,dt, \qquad 0 \le x, \quad \operatorname{Re}\beta > 0$$

$$G(\alpha,\beta,\gamma,x) = \frac{\Gamma(1-\gamma)F(\alpha,\beta,\gamma,x)}{\Gamma(\alpha+1-\gamma)\Gamma(\beta+1-\gamma)} + \frac{\Gamma(\gamma-1)x^{1-\gamma}F(\alpha+1-\gamma,\beta+1-\gamma,2-\gamma,x)}{\Gamma(\alpha)\Gamma(\beta)} \qquad 0 < x < 1$$

$$\lim_{\beta\to\infty} F\big(\alpha,\beta,\gamma,\tfrac{x}{\beta}\big) = M(\alpha,\gamma,x), \qquad \lim_{\beta\to\infty}\Gamma(\beta+1-\gamma)G\big(\alpha,\beta,\gamma,\tfrac{x}{\beta}\big) = U(\alpha,\gamma,x)$$

Special cases:

$$F(\alpha,\beta,\gamma,1)=\frac{\Gamma(\gamma)\Gamma(\gamma-\alpha-\beta)}{\Gamma(\gamma-\alpha)\Gamma(\gamma-\beta)},\ F(\alpha,\beta,\beta,x)=(1-x)^{-\alpha},\ G(\alpha,\beta,\beta+1,x)=\frac{x^{-\beta}}{\Gamma(\alpha)}$$

$$G(\alpha,\beta,\gamma,0)=\frac{\Gamma(1-\gamma)}{\Gamma(\alpha+1-\gamma)\Gamma(\beta+1-\gamma)},\ G(\alpha,\beta,\gamma,1)=\frac{1}{\Gamma(\alpha+\beta+1-\gamma)}$$

12. Legendre functions

Legendre function of the first kind
for $-1<x<1$

$$\widetilde{P}_p^q(x):=\frac{1}{\Gamma(1-q)}\Big(\frac{1+x}{1-x}\Big)^{q/2}F\big(-p,1+p,1-q,\frac{1-x}{2}\big),\qquad \mathrm{Re}\,q<1$$

for $1<x$

$$P_p^q(x):=\frac{2^q x^{q-p-1}}{(x^2-1)^{q/2}\Gamma(1-q)}F\big(\frac{p-q+1}{2},\frac{p-q+2}{2},1-q,1-\frac{1}{x^2}\big),\qquad \mathrm{Re}\,q<1$$

Legendre function of the second kind, divided by $e^{iq\pi}$, for $1<x$

$$\widetilde{Q}_p^q(x):=\frac{\sqrt{\pi}\Gamma(p+q+1)(x^2-1)^{q/2}}{2^{p+1}\Gamma(p+3/2)\,x^{p+q+1}}F\big(\frac{p+q+1}{2},\frac{p+q+2}{2},p+\frac{3}{2},\frac{1}{x^2}\big)$$

are linearly independent solutions of the Legendre equation

$$(1-x^2)Y''(x)-2xY'(x)+\big(p(1+p)-\frac{q^2}{1-x^2}\big)Y(x)=0.$$

Integral representations: for $-1<x<1$

$$\widetilde{P}_p^q(x)=\frac{2^{-p}(1-x^2)^{-q/2}}{\Gamma(-p-q)\Gamma(1+p)}\int_0^\infty(x+\operatorname{ch}t)^{q-p-1}\operatorname{sh}^{1+2p}t\,dt,\qquad \mathrm{Re}\,q<\mathrm{Re}(-p)<1$$

for $1<x$

$$\widetilde{Q}_p^q(x)=\frac{2^{-q}(x^2-1)^{q/2}\sqrt{\pi}\Gamma(p+q+1)}{\Gamma(p-q+1)\Gamma(q+1/2)}\int_0^\infty(x+\sqrt{x^2-1}\operatorname{ch}t)^{-q-p-1}\operatorname{sh}^{2q}t\,dt$$

$$\mathrm{Re}(p\pm q+1)>0$$

Properties:

$$\frac{d}{dx}P_p^q(x)\widetilde{Q}_p^q(x)-P_p^q(x)\frac{d}{dx}\widetilde{Q}_p^q(x)=\frac{\Gamma(1+p+q)}{\Gamma(1+p-q)(x^2-1)},\qquad 1<x$$

$$\frac{d}{dx}\widetilde{P}_p^q(-x)\widetilde{P}_p^q(x)-\widetilde{P}_p^q(-x)\frac{d}{dx}\widetilde{P}_p^q(x)=\frac{2}{\Gamma(-p-q)\Gamma(1+p-q)(1-x^2)},\qquad -1<x<1$$

$$\widetilde{Q}_p^{-q}(x)=\frac{\Gamma(p-q+1)}{\Gamma(p+q+1)}\widetilde{Q}_p^q(x),\qquad P_p^q(x)=P_{-p-1}^q(x),\qquad \widetilde{P}_p^q(x)=\widetilde{P}_{-p-1}^q(x)$$

$$\lim_{x\downarrow 1}(x^2-1)^{q/2}P_p^q(x)=\frac{2^q}{\Gamma(1-q)},\qquad \widetilde{P}_p^{-p}(x)=\frac{(1-x^2)^{p/2}}{2^p\Gamma(p+1)}$$

$$\widetilde{P}_{-p}^{-p}(-x)+\widetilde{P}_{-p}^{-p}(x)=\frac{2^p\Gamma(p)}{\Gamma(2p)(1-x^2)^{p/2}},\qquad \widetilde{P}_{1-p}^{-p}(-x)-\widetilde{P}_{1-p}^{-p}(x)=\frac{2^p\Gamma(p)\,x}{\Gamma(2p-1)(1-x^2)^{p/2}}$$

$$\frac{d}{dx}\Big(\frac{P_p^q(x)}{(x^2-1)^{q/2}}\Big)=\frac{P_p^{q+1}(x)}{(x^2-1)^{(q+1)/2}},\qquad \frac{d}{dx}\Big(\frac{\widetilde{Q}_p^q(x)}{(x^2-1)^{q/2}}\Big)=-\frac{\widetilde{Q}_p^{q+1}(x)}{(x^2-1)^{(q+1)/2}},\qquad 1<x$$

$$\frac{d}{dx}\big((x^2-1)^{q/2}P_p^q(x)\big)=(p+q)(p-q+1)(x^2-1)^{(q-1)/2}P_p^{q-1}(x),\qquad 1<x$$

$$\frac{d}{dx}\big((1-x^2)^{q/2}\widetilde{P}_p^q(x)\big)=(p+q)(p-q+1)(1-x^2)^{(q-1)/2}\widetilde{P}_p^{q-1}(x),\qquad -1<x<1$$

$$\frac{d}{dx}\big((x^2-1)^{q/2}\widetilde{Q}_p^q(x)\big)=-(p+q)(p-q+1)(x^2-1)^{(q-1)/2}\widetilde{Q}_p^{q-1}(x),\qquad 1<x$$

$$\lim_{p\to\infty}p^qP_p^{-q}\big(\operatorname{ch}\big(\tfrac{z}{p}\big)\big)=I_q(z),\qquad \lim_{p\to\infty}p^{-q}\widetilde{Q}_p^q\big(\operatorname{ch}\big(\tfrac{z}{p}\big)\big)=K_q(z)$$

Special cases:

$$P_p^{-1/2}(\operatorname{ch}x)=\frac{2\sqrt{2}\operatorname{sh}(x(p+1/2))}{\sqrt{\pi}(2p+1)\operatorname{sh}^{1/2}x},\qquad P_p^{1/2}(\operatorname{ch}x)=\frac{\sqrt{2}\operatorname{ch}(x(p+1/2))}{\sqrt{\pi}\operatorname{sh}^{1/2}x}$$

$$Q_p^{-1/2}(\operatorname{ch}x)=\frac{\sqrt{2\pi}e^{-x(p+1/2)}}{(2p+1)\operatorname{sh}^{1/2}x},\qquad Q_p^{1/2}(\operatorname{ch}x)=\frac{\sqrt{\pi}e^{-x(p+1/2)}}{\sqrt{2}\operatorname{sh}^{1/2}x}$$

13. Theta functions of imaginary argument

$$\operatorname{cs}_y(v,t):=\mathcal{L}_\gamma^{-1}\Big(\frac{\operatorname{ch}(v\sqrt{2\gamma})}{\sqrt{2\gamma}\operatorname{sh}(t\sqrt{2\gamma})}\Big)=\frac{1}{\sqrt{2\pi y}}\sum_{k=-\infty}^{\infty}e^{-(v+t+2kt)^2/2y}$$

$$\operatorname{sc}_y(v,t):=\mathcal{L}_\gamma^{-1}\Big(\frac{\operatorname{sh}(v\sqrt{2\gamma})}{\sqrt{2\gamma}\operatorname{ch}(t\sqrt{2\gamma})}\Big)=\frac{1}{\sqrt{2\pi y}}\sum_{k=-\infty}^{\infty}(-1)^ke^{-(t-v+2kt)^2/2y}$$

$$\operatorname{ss}_y(v,t):=\mathcal{L}_\gamma^{-1}\Big(\frac{\operatorname{sh}(v\sqrt{2\gamma})}{\operatorname{sh}(t\sqrt{2\gamma})}\Big)=\sum_{k=-\infty}^{\infty}\frac{t-v+2kt}{\sqrt{2\pi}y^{3/2}}e^{-(t-v+2kt)^2/2y},\qquad v<t$$

$$\widetilde{\operatorname{ss}}_y(v,t):=\mathcal{L}_\gamma^{-1}\Big(\frac{\operatorname{sh}(v\sqrt{2\gamma})}{\gamma\operatorname{sh}(t\sqrt{2\gamma})}\Big)=\sum_{k=-\infty}^{\infty}\operatorname{sign}(t-v+2kt)\operatorname{Erfc}\Big(\frac{|t-v+2kt|}{\sqrt{2y}}\Big)$$

$$\operatorname{cc}_y(v,t):=\mathcal{L}_\gamma^{-1}\Big(\frac{\operatorname{ch}(v\sqrt{2\gamma})}{\operatorname{ch}(t\sqrt{2\gamma})}\Big)=\sum_{k=-\infty}^{\infty}(-1)^k\frac{v+t+2kt}{\sqrt{2\pi}y^{3/2}}e^{-(v+t+2kt)^2/2y},\qquad v<t$$

$$\widetilde{\operatorname{cc}}_y(v,t):=\mathcal{L}_\gamma^{-1}\Big(\frac{\operatorname{ch}(v\sqrt{2\gamma})}{\gamma\operatorname{ch}(t\sqrt{2\gamma})}\Big)=\sum_{k=-\infty}^{\infty}(-1)^k\operatorname{sign}(v+t+2kt)\operatorname{Erfc}\Big(\frac{|v+t+2kt|}{\sqrt{2y}}\Big)$$

$$\operatorname{s}_y(t):=L_\gamma^{-1}\Big(\frac{\sqrt{2\gamma}}{\operatorname{sh}(t\sqrt{2\gamma})}\Big)=\frac{\sqrt{2}}{\sqrt{\pi}y^{5/2}}\sum_{k=0}^{\infty}((2k+1)^2t^2-y)e^{-(2k+1)^2t^2/2y},\qquad 0<t$$

14. Special Inverse Laplace transforms

$$\mathrm{c}_y(\mu,\nu,t,z) := \mathcal{L}_\gamma^{-1}\Big(\frac{(2\gamma)^{\mu/2}}{\mathrm{ch}^\nu(t\sqrt{2\gamma})}e^{-z\sqrt{2\gamma}}\Big) \qquad t>0$$

$$= 2^\nu \sum_{k=0}^\infty \frac{(-1)^k\Gamma(\nu+k)e^{-(\nu t+z+2kt)^2/4y}}{\sqrt{2\pi}y^{1+\mu/2}\Gamma(\nu)k!}D_{\mu+1}\Big(\frac{\nu t+z+2kt}{\sqrt{y}}\Big), \quad \nu\ge 0, \quad \nu t+z>0$$

$$\mathrm{s}_y(\mu,\nu,t,z) := \mathcal{L}_\gamma^{-1}\Big(\frac{(2\gamma)^{\mu/2}}{\mathrm{sh}^\nu(t\sqrt{2\gamma})}e^{-z\sqrt{2\gamma}}\Big) \qquad t>0$$

$$= 2^\nu \sum_{k=0}^\infty \frac{\Gamma(\nu+k)e^{-(\nu t+z+2kt)^2/4y}}{\sqrt{2\pi}y^{1+\mu/2}\Gamma(\nu)k!}D_{\mu+1}\Big(\frac{\nu t+z+2kt}{\sqrt{y}}\Big), \qquad \nu\ge 0, \quad \nu t+z>0$$

$$\mathrm{r}_y(\mu,t,x,z) := \mathcal{L}_\gamma^{-1}\Big(\frac{(2\gamma)^{\mu/2}e^{-x\sqrt{2\gamma}}}{z\,\mathrm{sh}(t\sqrt{2\gamma})+\mathrm{ch}(t\sqrt{2\gamma})}\Big) \qquad z\ne -1, \quad t>0$$

$$= \sum_{k=0}^\infty \frac{\sqrt{2}(z-1)^k e^{-(x+t+2kt)^2/4y}}{\sqrt{\pi}(z+1)^{(k+1)}y^{1+\mu/2}}D_{\mu+1}\Big(\frac{x+t+2kt}{\sqrt{y}}\Big), \qquad x+t>0$$

$$\widetilde{\mathrm{rc}}_y(\mu,t,x,z) := \mathcal{L}_\gamma^{-1}\Big(\frac{(2\gamma)^{\mu/2}e^{-x\sqrt{2\gamma}}}{\mathrm{sh}(t\sqrt{2\gamma})+z\sqrt{2\gamma}\,\mathrm{ch}(t\sqrt{2\gamma})}\Big) \qquad z\ne 0, \quad t>0$$

$$= \sum_{k=0}^\infty \frac{(-1)^k k!}{2^k z^{k+1}}\sum_{l=0}^k \frac{(-1)^l}{(k-l)!l!}c_y(\mu-k-1,k+1,t,x+kt-2lt), \qquad x>-t$$

$$\mathrm{rc}_y(\mu,t,x,z) := \mathcal{L}_\gamma^{-1}\Big(\frac{(2\gamma)^{\mu/2}\,\mathrm{sh}(x\sqrt{2\gamma})}{\mathrm{sh}(t\sqrt{2\gamma})+z\sqrt{2\gamma}\,\mathrm{ch}(t\sqrt{2\gamma})}\Big) \qquad t>0$$

$$= \frac{1}{2}\widetilde{\mathrm{rc}}_y(\mu,t,-x,z) - \frac{1}{2}\widetilde{\mathrm{rc}}_y(\mu,t,x,z), \qquad x<t$$

$$\widetilde{\mathrm{rs}}_y(\mu,t,x,z) := \mathcal{L}_\gamma^{-1}\Big(\frac{(2\gamma)^{\mu/2}e^{-x\sqrt{2\gamma}}}{z\sqrt{2\gamma}\,\mathrm{sh}(t\sqrt{2\gamma})+\mathrm{ch}(t\sqrt{2\gamma})}\Big) \qquad z\ne 0, \quad t>0$$

$$= \sum_{k=0}^\infty \frac{(-1)^k k!}{2^k z^{k+1}}\sum_{l=0}^k \frac{1}{(k-l)!l!}s_y(\mu-k-1,k+1,t,x+kt-2lt), \qquad x>-t$$

$$\mathrm{rs}_y(\mu,t,x,z) := \mathcal{L}_\gamma^{-1}\Big(\frac{(2\gamma)^{\mu/2}\,\mathrm{ch}(x\sqrt{2\gamma})}{z\sqrt{2\gamma}\,\mathrm{sh}(t\sqrt{2\gamma})+\mathrm{ch}(t\sqrt{2\gamma})}\Big) \qquad t>0$$

$$= \frac{1}{2}\widetilde{\mathrm{rs}}_y(\mu,t,-x,z) + \frac{1}{2}\widetilde{\mathrm{rs}}_y(\mu,t,x,z), \qquad x<t$$

$$\mathrm{ec}_y(\mu,\nu,t,x,z) := \mathcal{L}_\gamma^{-1}\Big(\frac{(2\gamma)^{\mu/2}}{\mathrm{ch}^\nu(t\sqrt{2\gamma})}\exp\Big(-x\sqrt{2\gamma}-\frac{z\sqrt{2\gamma}\,\mathrm{sh}(t\sqrt{2\gamma})}{\mathrm{ch}(t\sqrt{2\gamma})}\Big)\Big) \qquad t>0$$

$$= \sum_{k=0}^\infty \frac{z^k}{k!}\,\mathrm{c}_y(\mu+k,\nu+k,t,x+z+kt), \qquad \nu\ge 0, \quad \nu t+x+z>0$$

$$\mathrm{es}_y(\mu,\nu,t,x,z) := \mathcal{L}_\gamma^{-1}\Big(\frac{(2\gamma)^{\mu/2}}{\mathrm{sh}^\nu(t\sqrt{2\gamma})}\exp\Big(-x\sqrt{2\gamma}-\frac{z\sqrt{2\gamma}\,\mathrm{ch}(t\sqrt{2\gamma})}{\mathrm{sh}(t\sqrt{2\gamma})}\Big)\Big) \qquad t>0$$

$$= \sum_{k=0}^{\infty}\frac{(-z)^k}{k!}\,\mathrm{s}_y(\mu+k,\nu+k,t,x+z+kt), \qquad \nu\ge 0, \quad \nu t+x+z>0$$

$$\mathrm{ee}_y(\nu,t,z,x) := \mathcal{L}_\gamma^{-1}\Big(\Big(\frac{\sqrt{2\gamma}}{\mathrm{sh}(t\sqrt{2\gamma})}\Big)^\nu\exp\Big(-\frac{z\sqrt{2\gamma}\,\mathrm{ch}(t\sqrt{2\gamma})}{\mathrm{sh}(t\sqrt{2\gamma})}-\frac{x\sqrt{2\gamma}}{\mathrm{sh}(t\sqrt{2\gamma})}\Big)\Big) \qquad t>0$$

$$= \sum_{k=0}^{\infty}\frac{(-z)^k}{k!}\sum_{l=0}^{\infty}\frac{(-x)^l}{l!}\,\mathrm{s}_y(\nu+k+l,\nu+k+l,t,z+kt), \qquad \nu\ge 0,\ \nu t+z>0$$

For $t+x+z-v>0, \quad t>0$

$$\mathrm{ecc}_y(v,t,x,z) := \mathcal{L}_\gamma^{-1}\Big(\frac{\mathrm{ch}(v\sqrt{2\gamma})}{\mathrm{ch}(t\sqrt{2\gamma})}\exp\Big(-x\sqrt{2\gamma}-\frac{z\sqrt{2\gamma}\,\mathrm{sh}(t\sqrt{2\gamma})}{\mathrm{ch}(t\sqrt{2\gamma})}\Big)\Big)$$

$$= \frac{1}{2}\,\mathrm{ec}_y(0,1,t,x-v,z)+\frac{1}{2}\,\mathrm{ec}_y(0,1,t,x+v,z)$$

$$\mathrm{ess}_y(v,t,x,z) := \mathcal{L}_\gamma^{-1}\Big(\frac{\mathrm{sh}(v\sqrt{2\gamma})}{\mathrm{sh}(t\sqrt{2\gamma})}\exp\Big(-x\sqrt{2\gamma}-\frac{z\sqrt{2\gamma}\,\mathrm{ch}(t\sqrt{2\gamma})}{\mathrm{sh}(t\sqrt{2\gamma})}\Big)\Big)$$

$$= \frac{1}{2}\,\mathrm{es}_y(0,1,t,x-v,z)-\frac{1}{2}\,\mathrm{es}_y(0,1,t,x+v,z)$$

$$\mathrm{esc}_y(v,t,x,z) := \mathcal{L}_\gamma^{-1}\Big(\frac{\mathrm{sh}(v\sqrt{2\gamma})}{\sqrt{2\gamma}\,\mathrm{ch}(t\sqrt{2\gamma})}\exp\Big(-x\sqrt{2\gamma}-\frac{z\sqrt{2\gamma}\,\mathrm{sh}(t\sqrt{2\gamma})}{\mathrm{ch}(t\sqrt{2\gamma})}\Big)\Big)$$

$$= \frac{1}{2}\,\mathrm{ec}_y(-1,1,t,x-v,z)-\frac{1}{2}\,\mathrm{ec}_y(-1,1,t,x+v,z)$$

$$\mathrm{ecs}_y(v,t,x,z) := \mathcal{L}_\gamma^{-1}\Big(\frac{\mathrm{ch}(v\sqrt{2\gamma})}{\sqrt{2\gamma}\,\mathrm{sh}(t\sqrt{2\gamma})}\exp\Big(-x\sqrt{2\gamma}-\frac{z\sqrt{2\gamma}\,\mathrm{ch}(t\sqrt{2\gamma})}{\mathrm{sh}(t\sqrt{2\gamma})}\Big)\Big)$$

$$= \frac{1}{2}\,\mathrm{es}_y(-1,1,t,x-v,z)+\frac{1}{2}\,\mathrm{es}_y(-1,1,t,x+v,z)$$

$$\widetilde{\mathrm{ecc}}_y(v,t,x,z) := \mathcal{L}_\gamma^{-1}\Big(\frac{\mathrm{ch}(v\sqrt{2\gamma})}{2\gamma\,\mathrm{ch}(t\sqrt{2\gamma})}\exp\Big(-x\sqrt{2\gamma}-\frac{z\sqrt{2\gamma}\,\mathrm{sh}(t\sqrt{2\gamma})}{\mathrm{ch}(t\sqrt{2\gamma})}\Big)\Big)$$

$$= \frac{1}{2}\,\mathrm{ec}_y(-2,1,t,x-v,z)+\frac{1}{2}\,\mathrm{ec}_y(-2,1,t,x+v,z)$$

$$\widetilde{\mathrm{ess}}_y(v,t,x,z) := \mathcal{L}_\gamma^{-1}\Big(\frac{\mathrm{sh}(v\sqrt{2\gamma})}{2\gamma\,\mathrm{sh}(t\sqrt{2\gamma})}\exp\Big(-x\sqrt{2\gamma}-\frac{z\sqrt{2\gamma}\,\mathrm{ch}(t\sqrt{2\gamma})}{\mathrm{sh}(t\sqrt{2\gamma})}\Big)\Big)$$

$$= \frac{1}{2}\,\mathrm{es}_y(-2,1,t,x-v,z)-\frac{1}{2}\,\mathrm{es}_y(-2,1,t,x+v,z)$$

For $2t+z-u-v>0, \quad t>0$

$$\mathrm{eccc}_y(u,v,t,z) := \mathcal{L}_\gamma^{-1}\Big(\frac{\mathrm{ch}(u\sqrt{2\gamma})\,\mathrm{ch}(v\sqrt{2\gamma})}{\mathrm{ch}^2(t\sqrt{2\gamma})}\exp\Big(-\frac{z\sqrt{2\gamma}\,\mathrm{sh}(t\sqrt{2\gamma})}{\mathrm{ch}(t\sqrt{2\gamma})}\Big)\Big)$$

$$= \frac{1}{4}\operatorname{ec}_y(0,2,t,-u-v,z) + \frac{1}{4}\operatorname{ec}_y(0,2,t,u+v,z)$$

$$+\frac{1}{4}\operatorname{ec}_y(0,2,t,-u+v,z) + \frac{1}{4}\operatorname{ec}_y(0,2,t,u-v,z)$$

$$\operatorname{esss}_y(u,v,t,z) := \mathcal{L}_\gamma^{-1}\Big(\frac{\operatorname{sh}(u\sqrt{2\gamma})\operatorname{sh}(v\sqrt{2\gamma})}{\operatorname{sh}^2(t\sqrt{2\gamma})}\exp\Big(-\frac{z\sqrt{2\gamma}\operatorname{ch}(t\sqrt{2\gamma})}{\operatorname{sh}(t\sqrt{2\gamma})}\Big)\Big)$$

$$= \frac{1}{4}\operatorname{es}_y(0,2,t,-u-v,z) + \frac{1}{4}\operatorname{es}_y(0,2,t,u+v,z)$$

$$-\frac{1}{4}\operatorname{es}_y(0,2,t,-u+v,z) - \frac{1}{4}\operatorname{es}_y(0,2,t,u-v,z)$$

$$\operatorname{ecsc}_y(u,v,t,z) := \mathcal{L}_\gamma^{-1}\Big(\frac{\operatorname{ch}(u\sqrt{2\gamma})\operatorname{sh}(v\sqrt{2\gamma})}{\sqrt{2\gamma}\operatorname{ch}^2(t\sqrt{2\gamma})}\exp\Big(-\frac{z\sqrt{2\gamma}\operatorname{sh}(t\sqrt{2\gamma})}{\operatorname{ch}(t\sqrt{2\gamma})}\Big)\Big)$$

$$= \frac{1}{4}\operatorname{ec}_y(-1,2,t,-u-v,z) - \frac{1}{4}\operatorname{ec}_y(-1,2,t,u+v,z)$$

$$-\frac{1}{4}\operatorname{ec}_y(-1,2,t,-u+v,z) + \frac{1}{4}\operatorname{ec}_y(-1,2,t,u-v,z)$$

$$\operatorname{escs}_y(u,v,t,z) := \mathcal{L}_\gamma^{-1}\Big(\frac{\sqrt{2\gamma}\operatorname{sh}(u\sqrt{2\gamma})\operatorname{ch}(v\sqrt{2\gamma})}{\operatorname{sh}^2(t\sqrt{2\gamma})}\exp\Big(-\frac{z\sqrt{2\gamma}\operatorname{ch}(t\sqrt{2\gamma})}{\operatorname{sh}(t\sqrt{2\gamma})}\Big)\Big)$$

$$= \frac{1}{4}\operatorname{es}_y(1,2,t,-u-v,z) - \frac{1}{4}\operatorname{es}_y(1,2,t,u+v,z)$$

$$+\frac{1}{4}\operatorname{es}_y(1,2,t,-u+v,z) - \frac{1}{4}\operatorname{es}_y(1,2,t,u-v,z)$$

For $\nu \ge -1, \quad t+\nu t+r+z>0, \quad t>0$

$$\operatorname{is}_y(\nu,t,r,z,x) := \mathcal{L}_\gamma^{-1}\Big(\frac{\sqrt{2\gamma}}{\operatorname{sh}(t\sqrt{2\gamma})}\exp\Big(-r\sqrt{2\gamma}-\frac{z\sqrt{2\gamma}\operatorname{ch}(t\sqrt{2\gamma})}{\operatorname{sh}(t\sqrt{2\gamma})}\Big)I_\nu\Big(\frac{2x\sqrt{2\gamma}}{\operatorname{sh}(t\sqrt{2\gamma})}\Big)\Big)$$

$$= \sum_{l=0}^{\infty}\frac{x^{\nu+2l}}{\Gamma(\nu+l+1)l!}\operatorname{es}_y(1+\nu+2l,1+\nu+2l,t,r,z)$$

For $\operatorname{Re} z>0$

$$\operatorname{i}_y(z) := \mathcal{L}_\gamma^{-1}\big(I_{\sqrt{\gamma}}(z)\big) = \frac{ze^{\pi^2/4y}}{\pi\sqrt{\pi y}}\int_0^\infty e^{-z\operatorname{ch}u-u^2/4y}\operatorname{sh}u\sin(\pi u/2y)du$$

For $\mu \ge \eta$

$$\operatorname{ki}_y(\mu,\eta) := \mathcal{L}_\gamma^{-1}\big(K_{\sqrt{\gamma}}(\mu)I_{\sqrt{\gamma}}(\eta)\big)$$

$$= \frac{\mu\eta e^{\pi^2/4y}}{\pi\sqrt{\pi y}}\int_0^\infty\frac{K_1\big((\mu^2+\eta^2+2\mu\eta\operatorname{ch}u)^{1/2}\big)}{(\mu^2+\eta^2+2\mu\eta\operatorname{ch}u)^{1/2}}e^{-u^2/4y}\operatorname{sh}u\sin(\pi u/2y)du$$

For $\mu \le \eta$ $\qquad \mathrm{ki}_y(\mu,\eta) := \mathrm{ki}_y(\eta,\mu)$

For $\operatorname{Re}\mu > -3/2,\ \operatorname{Re} z > 0$

$$\mathrm{m}_y(\mu,z) := \mathcal{L}_\gamma^{-1}\Big(\frac{\Gamma(\sqrt{\gamma}+\mu+1/2)}{\Gamma(2\sqrt{\gamma}+1)} M_{-\mu,\sqrt{\gamma}}(2z)\Big)$$

$$= \frac{8z^{3/2}\Gamma(\mu+\frac{3}{2})e^{\pi^2/4y}}{\pi\sqrt{2\pi y}} \int_0^\infty e^{-z\,\mathrm{ch}(2u)-u^2/y} M\big(-\mu,\frac{3}{2},2z\,\mathrm{sh}^2 u\big)\,\mathrm{sh}(2u)\sin\big(\frac{\pi u}{y}\big)du$$

For $\nu > 0, \quad t+z > 0$

$$\mathrm{ei}_y(\nu,t,z,x) := \mathcal{L}_\gamma^{-1}\Big(\frac{\sqrt{2\gamma}}{\mathrm{sh}(t\sqrt{2\gamma})}\exp\Big(-\frac{z\sqrt{2\gamma}\,\mathrm{ch}(t\sqrt{2\gamma})}{\mathrm{sh}(t\sqrt{2\gamma})}\Big)\,\mathrm{i}_\nu\Big(\frac{x\sqrt{2\gamma}}{\mathrm{sh}(t\sqrt{2\gamma})}\Big)\Big)$$

$$= \frac{xe^{\pi^2/4\nu}}{\pi\sqrt{\pi\nu}} \int_0^\infty e^{-u^2/4\nu}\,\mathrm{ee}_y(2,t,z,x\,\mathrm{ch}\,u)\,\mathrm{sh}\,u\sin(\pi u/2\nu)du$$

Properties:

$$\mathrm{m}_y(0,z) = \sqrt{2\pi z}\ \mathrm{i}_y(z)$$

$$\lim_{y\to\infty} e^{-(\mu+1/2)^2 y}\,\mathrm{m}_y(\mu,z) = \frac{(2z)^{-\mu}}{\Gamma(-2\mu-1)}\,e^{-z}, \qquad -3/2 < \mu < -1/2$$

$$\mathcal{L}_\gamma^{-1}\big(\mathrm{ki}_z(\mu\sqrt{\gamma},\eta\sqrt{\gamma})\big) = \frac{1}{2y}\exp\Big(-\frac{\mu^2+\eta^2}{4y}\Big)\,\mathrm{i}_z\Big(\frac{\mu\eta}{2y}\Big)$$

$$\mathcal{L}_\gamma^{-1}\big(\mathrm{is}_r(\sqrt{\gamma},t,0,z,x)\big) = \mathrm{ei}_r(y,t,z,2x)$$

15. Two parameter functions connected to the Bessel functions

$$S_\nu(x,y) := (xy)^{-\nu}\big(I_\nu(x)K_\nu(y) - K_\nu(x)I_\nu(y)\big)$$

$$C_\nu(x,y) := (xy)^{-\nu}\big(I_{\nu+1}(x)K_\nu(y) + K_{\nu+1}(x)I_\nu(y)\big)$$

$$F_\nu(x,z) := (xz)^{-\nu}I_\nu((x+z-|x-z|)/2)K_\nu((x+z+|x-z|)/2)$$

$$= \begin{cases} (xz)^{-\nu}I_\nu(x)K_\nu(z), & x \le z \\ (xz)^{-\nu}K_\nu(x)I_\nu(z), & z \le x \end{cases}$$

Properties:

$C_\nu(x,y)$ as a function of y and $S_\nu(x,y)$, $F_\nu(x,z)$ as functions of both variables satisfy the Bessel equation

$$\text{(B)} \qquad Z''(x) + \frac{2\nu+1}{x}Z'(x) - Z(x) = 0.$$

$$S_\nu(x,x) = 0, \quad C_\nu(x,x) = x^{-1-2\nu}, \quad S_\nu(x,y) = -S_\nu(y,x), \quad F_\nu(x,z) = F_\nu(z,x)$$

$$\frac{\partial}{\partial x}S_\nu(x,y) = C_\nu(x,y), \qquad \frac{\partial}{\partial y}S_\nu(x,y) = -C_\nu(y,x), \qquad \frac{\partial}{\partial y}C_\nu(x,y) = -xyS_{\nu+1}(x,y)$$

$$\frac{\partial}{\partial x}\big(x^{2\nu+1}C_\nu(x,y)\big) = x^{2\nu+1}S_\nu(x,y), \quad \frac{\partial}{\partial x}F_\nu(z+0,z) - \frac{\partial}{\partial x}F_\nu(z-0,z) = -z^{-1-2\nu}$$

$$S_\nu(q,p)S_\nu(r,z) + S_\nu(q,r)S_\nu(z,p) = S_\nu(q,z)S_\nu(r,p)$$

$$S_\nu(q,r)C_\nu(r,p) + C_\nu(r,q)S_\nu(r,p) = r^{-1-2\nu}S_\nu(q,p)$$

$$2\nu S_\nu(x,y) + xC_\nu(x,y) = y^{-1}C_{\nu-1}(y,x)$$

$$\lim_{\theta\downarrow 0} S_0(x\theta, y\theta) = \ln(x/y)$$

$$\lim_{\theta\downarrow 0}\lim_{\theta\downarrow 0} \theta^{2\nu}S_\nu(x\theta,y\theta) = \frac{1}{2\nu}(y^{-2\nu} - x^{-2\nu}), \qquad \nu \neq 0$$

$$\lim_{\nu\to\infty} \nu(\nu^2\gamma)^{\nu\sqrt{\lambda}}S_{\nu\sqrt{\lambda}}\big(\nu\sqrt{\gamma}\,e^{x/\nu}, \nu\sqrt{\gamma}\,e^{y/\nu}\big) = \frac{e^{-(x+y)\sqrt{\lambda}}}{\sqrt{\lambda+\gamma}}\,\mathrm{sh}((x-y)\sqrt{\lambda+\gamma})$$

$$\lim_{\nu\to\infty} (\nu^2\gamma)^{\nu\sqrt{\lambda}+1/2}C_{\nu\sqrt{\lambda}}\big(\nu\sqrt{\gamma}\,e^{x/\nu}, \nu\sqrt{\gamma}\,e^{y/\nu}\big)$$
$$= \frac{e^{-(x+y)\sqrt{\lambda}}}{\sqrt{\lambda+\gamma}}\big(\sqrt{\lambda+\gamma}\,\mathrm{ch}((x-y)\sqrt{\lambda+\gamma}) - \sqrt{\lambda}\,\mathrm{sh}((x-y)\sqrt{\lambda+\gamma})\big)$$

Special cases:

$$S_{-1/2}(x,y) = \mathrm{sh}(x-y), \qquad C_{-1/2}(x,y) = \mathrm{ch}(x-y)$$

$$S_{1/2}(x,y) = \frac{1}{xy}\,\mathrm{sh}(x-y), \qquad C_{1/2}(x,y) = \frac{1}{x^2y}(x\,\mathrm{ch}(x-y) - \mathrm{sh}(x-y))$$

$$F_{-1/2}(x,z) = \frac{1}{2}(e^{-|x-z|} + e^{-(x+z)}), \qquad F_{1/2}(x,z) = \frac{1}{2xz}(e^{-|x-z|} - e^{-(x+z)})$$

16. Two parameter functions connected to the parabolic cylinder functions

$$S(\nu,x,y) := \frac{\Gamma(\nu)}{\pi}e^{(x^2+y^2)/4}\big(D_{-\nu}(-x)D_{-\nu}(y) - D_{-\nu}(x)D_{-\nu}(-y)\big), \quad \nu > 0$$

$$C(\nu,x,y) := \frac{\Gamma(\nu+1)}{\pi}e^{(x^2+y^2)/4}\big(D_{-\nu-1}(-x)D_{-\nu}(y) + D_{-\nu-1}(x)D_{-\nu}(-y)\big)$$

$$F(\nu,x,z) := \frac{\Gamma(\nu)}{\pi}e^{(x^2+y^2)/4}D_{-\nu}(-(x+z-|x-z|)/2)D_{-\nu}((x+z+|x-z|)/2)$$
$$= \begin{cases} \dfrac{\Gamma(\nu)}{\pi}e^{(x^2+z^2)/4}D_{-\nu}(-x)D_{-\nu}(z), & x \le z \\ \dfrac{\Gamma(\nu)}{\pi}e^{(x^2+z^2)/4}D_{-\nu}(-z)D_{-\nu}(x), & z \le x \end{cases}$$

Properties:

$C(\nu,x,y)$ as a function of y and $S(\nu,x,y)$, $F(\nu,x,y)$ as functions of both variables x, y satisfy the Weber equation

(W) $$Z''(x) - xZ'(x) - \nu Z(x) = 0.$$

$$S(\nu,x,x) = 0, \quad S(\nu,x,y) = -S(\nu,y,x), \quad C(\nu,x,x) = \frac{\sqrt{2}}{\sqrt{\pi}}e^{x^2/2}$$

$$\frac{\partial}{\partial x}S(\nu,x,y)=C(\nu,x,y),\quad \frac{\partial}{\partial y}S(\nu,x,y)=-C(\nu,y,x)$$

$$\frac{\partial}{\partial y}C(\nu,x,y)=-\nu S(\nu+1,x,y),\quad \frac{\partial}{\partial x}\left(\nu^{-1}e^{-x^2/2}C(\nu,x,y)\right)=e^{-x^2/2}S(\nu,x,y)$$

$$\frac{\partial}{\partial x}F(\nu,z+0,z)-\frac{\partial}{\partial x}F(\nu,z-0,z)=-\frac{\sqrt{2}}{\sqrt{\pi}}e^{z^2/2}$$

$$S(\nu,q,p)S(\nu,r,z)+S(\nu,q,r)S(\nu,z,p)=S(\nu,q,z)S(\nu,r,p)$$

$$S(\nu,q,r)C(\nu,r,p)+C(\nu,r,q)S(\nu,r,p)=\frac{\sqrt{2}}{\sqrt{\pi}}e^{r^2/2}S(\nu,q,p)$$

$$\lim_{\theta\downarrow 0}\frac{\sqrt{\alpha}}{\sqrt{\theta}}S\left(\frac{\alpha}{2\theta},x\sqrt{2\theta},y\sqrt{2\theta}\right)=\frac{2}{\sqrt{\pi}}\operatorname{sh}((x-y)\sqrt{\alpha})$$

$$\lim_{\theta\downarrow 0}C\left(\frac{\alpha}{2\theta},x\sqrt{2\theta},y\sqrt{2\theta}\right)=\frac{\sqrt{2}}{\sqrt{\pi}}\operatorname{ch}((x-y)\sqrt{\alpha})$$

Special cases:

$$S(0,x,y):=\lim_{\nu\to 0}S(\nu,x,y)=\operatorname{Erfi}\left(\frac{x}{\sqrt{2}}\right)-\operatorname{Erfi}\left(\frac{y}{\sqrt{2}}\right)=:\operatorname{Erfid}(x,y)$$

$$S(1/2,x,y)=\frac{xy}{2\sqrt{2\pi}}e^{(x^2+y^2)/4}S_{1/4}\left(\frac{x^2}{4},\frac{y^2}{4}\right),\quad C(0,x,y)=\frac{\sqrt{2}}{\sqrt{\pi}}e^{x^2/2}$$

Properties:

$\operatorname{Erfid}(x,y)$ as a function of both variables satisfies the equation (W) when $\nu=0$.

17. Two parameter functions connected to the Kummer functions

$$S(a,b,x,y):=\frac{\Gamma(a)}{\Gamma(b)}\left(M(a,b,x)U(a,b,y)-U(a,b,x)M(a,b,y)\right)$$

$$C(a,b,x,y):=\frac{\Gamma(a+1)}{\Gamma(b+1)}\left(M(a+1,b+1,x)U(a,b,y)+bU(a+1,b+1,x)M(a,b,y)\right)$$

$$F(a,b,x,z):=\begin{cases}\frac{\Gamma(a)}{\Gamma(b)}M(a,b,x)U(a,b,z), & 0\le x\le z\\ \frac{\Gamma(a)}{\Gamma(b)}M(a,b,z)U(a,b,x), & z\le x\end{cases}$$

Properties:

$C(a,b,x,y)$ as a function of y and $S(a,b,x,y)$, $F(a,b,x,y)$ as functions of both variables x,y satisfy the Kummer equation

$$\text{(K)}\qquad xZ''(x)+(b-x)Z'(x)-aZ(x)=0.$$

$$S(a,b,x,x)=0,\quad S(a,b,x,y)=-S(a,b,y,x),\quad C(a,b,x,x)=x^{-b}e^{x}$$

$$\frac{\partial}{\partial x}S(a,b,x,y)=C(a,b,x,y),\quad \frac{\partial}{\partial y}S(a,b,x,y)=-C(a,b,y,x)$$

$$\frac{\partial}{\partial y}C(a,b,x,y)=-aS(a+1,b+1,x,y)$$

$$\frac{\partial}{\partial x}\left(a^{-1}x^be^{-x}C(a,b,x,y)\right)=x^{b-1}e^{-x}S(a,b,x,y)$$

$$\frac{\partial}{\partial x}F(a,b,z+0,z)-\frac{\partial}{\partial x}F(a,b,z-0,z)=-z^{-b}e^z$$

$$S(a,b,q,p)S(a,b,r,z)+S(a,b,q,r)S(a,b,z,p)=S(a,b,q,z)S(a,b,r,p)$$

$$S(a,b,q,r)C(a,b,r,p)+C(a,b,r,q)S(a,b,r,p)=r^{-b}e^rS(a,b,q,p)$$

$$\lim_{\theta\downarrow 0}(\theta/\alpha)^{\nu}S\left(\frac{\alpha}{4\theta},\nu+1,x\theta,y\theta\right)=2S_\nu(\sqrt{\alpha x},\sqrt{\alpha y})$$

$$\lim_{\theta\downarrow 0}(\theta/\alpha)^{\nu+1}C\left(\frac{\alpha}{4\theta},\nu+1,x\theta,y\theta\right)=\frac{1}{\sqrt{\alpha x}}C_\nu(\sqrt{\alpha x},\sqrt{\alpha y})$$

Special cases: for $x>0, y>0$

$$S\left(0,b,\frac{x^2}{2},\frac{y^2}{2}\right):=\lim_{a\to 0}S\left(a,b,\frac{x^2}{2},\frac{y^2}{2}\right)=2\int_{y/\sqrt{2}}^{x/\sqrt{2}}u^{1-2b}e^{u^2}du=:\mathrm{Ed}(b,x,y)$$

$$S\left(\frac{a}{2},\frac{1}{2},\frac{x^2}{2},\frac{y^2}{2}\right)=\sqrt{\pi}S(a,x,y),\quad S\left(\nu+\frac{1}{2},2\nu+1,2x,2y\right)=2^{-2\nu}e^{x+y}S_\nu(x,y)$$

$$\mathrm{Ed}(1/2,x,y)=\sqrt{\pi}\,\mathrm{Erfid}(x,y),\quad C(0,b,x,y)=x^{-b}e^x$$

Properties:

$\mathrm{Ed}(b,\sqrt{2x},\sqrt{2y})$ as a function of variables x,y satisfies the equation (K) when $a=0$.

APPENDIX 3

INVERSE LAPLACE TRANSFORMS

General formulae

$$\mathcal{L}_\gamma^{-1}(F(\gamma)) =: f(y), \quad \text{where} \quad F(\gamma) = \int_0^\infty e^{-\gamma y} f(y)dy, \qquad \operatorname{Re}\gamma \ge \sigma \ge 0$$

0. $f(y) = \frac{1}{2\pi i}\int_{c-i\infty}^{c+i\infty} e^{\gamma y} F(\gamma)d\gamma, \qquad c > \sigma$

a. $\mathcal{L}_\gamma^{-1}(F(\alpha\gamma+\beta)) = \frac{1}{\alpha} e^{-\beta y/\alpha} f\big(\frac{y}{\alpha}\big), \qquad \alpha > 0$

b. $\mathcal{L}_\gamma^{-1}(e^{-\beta\gamma}F(\gamma)) = f(y-\beta)\mathbb{I}_{[\beta,\infty)}(y), \qquad \beta > 0$

c. $\mathcal{L}_\gamma^{-1}\Big(\int_\gamma^\infty F(x)dx\Big) = \frac{1}{y} f(y)$

d. $\mathcal{L}_\gamma^{-1}\Big(\frac{F(\gamma)}{\gamma}\Big) = \int_0^y f(x)dx$

e. $\mathcal{L}_\gamma^{-1}(\gamma F(\gamma) - f(+0)) = f'(y)$

f. $\mathcal{L}_\gamma^{-1}(F_1(\gamma)F_2(\gamma)) = \int_0^y f_1(x)f_2(y-x)dx = \int_0^y f_1(y-x)f_2(x)dx = f_1(y) * f_2(y)$

g. $\mathcal{L}_\gamma^{-1}(F(\sqrt{\gamma})) = \frac{1}{2\sqrt{\pi}y^{3/2}}\int_0^\infty x e^{-x^2/4y} f(x)dx$

i. $\mathcal{L}_\gamma^{-1}(\gamma^\mu F(\sqrt{\gamma})) = \frac{\sqrt{2}}{\sqrt{\pi}(2y)^{\mu+1}}\int_0^\infty e^{-x^2/8y} D_{2\mu+1}\big(\frac{x}{\sqrt{2y}}\big) f(x)dx$

j. $\mathcal{L}_\gamma^{-1}\mathcal{L}_\eta^{-1}(F_1(\gamma+\eta)F_2(\eta)) = f_1(y)f_2(g-y)\mathbb{I}_{[0,g]}(y)$

k. $\mathcal{L}_\gamma^{-1}\mathcal{L}_\eta^{-1}(F_1(p\gamma+\eta)F_2(q\gamma+\eta)) = f_1\Big(\frac{|y-qg|}{|p-q|}\Big) f_2\Big(\frac{|y-pg|}{|q-p|}\Big)\frac{\mathbb{I}_{[(p\wedge q)g,(p\vee q)g]}(y)}{|q-p|}$

l. $\mathcal{L}_\gamma^{-1}\mathcal{L}_\eta^{-1}\mathcal{L}_\lambda^{-1}(F_1(\gamma+\eta+\lambda)F_2(\eta+\lambda)F_3(\lambda))$

$$= f_1(y)f_2(g-y)f_3(t-g)\mathbb{I}_{[0,g]}(y)\mathbb{I}_{[y,t]}(g)$$

1. $\mathcal{L}_\gamma^{-1}\Big(\frac{1}{\gamma+\beta}\Big) = e^{-\beta y}$

2. $\mathcal{L}_\gamma^{-1}\big(e^{-\sqrt{\alpha}\sqrt{\gamma}}\big) = \frac{\sqrt{\alpha}}{2\sqrt{\pi}y^{3/2}}\, e^{-\alpha/4y},$ $\qquad \alpha > 0$

3. $\mathcal{L}_\gamma^{-1}\big(\sqrt{\gamma}\, e^{-\sqrt{\alpha}\sqrt{\gamma}}\big) = \frac{1}{\sqrt{\pi}y^{5/2}}\Big(\frac{\alpha}{4} - \frac{y}{2}\Big)e^{-\alpha/4y},$ $\qquad \alpha > 0$

4. $\mathcal{L}_\gamma^{-1}\big(\gamma\, e^{-\sqrt{\alpha}\sqrt{\gamma}}\big) = \frac{\sqrt{\alpha}}{4\sqrt{\pi}\, y^{5/2}}\Big(\frac{\alpha}{2y} - 3\Big)e^{-\alpha/4y},$ $\qquad \alpha > 0$

5. $\mathcal{L}_\gamma^{-1}\Big(\frac{1}{\sqrt{\gamma}}\, e^{-\sqrt{\alpha}\sqrt{\gamma}}\Big) = \frac{1}{\sqrt{\pi y}}\, e^{-\alpha/4y},$ $\qquad \alpha \geq 0$

6. $\mathcal{L}_\gamma^{-1}\Big(\frac{1}{\gamma}\, e^{-\sqrt{\alpha}\sqrt{\gamma}}\Big) = \operatorname{Erfc}\Big(\frac{\sqrt{\alpha}}{2\sqrt{y}}\Big),$ $\qquad \alpha \geq 0$

7. $\mathcal{L}_\gamma^{-1}\Big(\frac{1}{\gamma^{3/2}}\, e^{-\sqrt{\alpha}\sqrt{\gamma}}\Big) = \frac{2\sqrt{y}}{\sqrt{\pi}}\, e^{-\alpha/4y} - \sqrt{\alpha}\operatorname{Erfc}\Big(\frac{\sqrt{\alpha}}{2\sqrt{y}}\Big),$ $\qquad \alpha \geq 0$

8. $\mathcal{L}_\gamma^{-1}\Big(\gamma^{\mu/2}e^{-\sqrt{\alpha}\sqrt{\gamma}}\Big) = \frac{\sqrt{2}}{\sqrt{\pi}\,(2y)^{\mu/2+1}}\, e^{-\alpha/8y} D_{\mu+1}\Big(\frac{\sqrt{\alpha}}{\sqrt{2y}}\Big),$ $\qquad \alpha > 0$

9. $\mathcal{L}_\gamma^{-1}\Big(\frac{e^{-\sqrt{\alpha}\sqrt{\gamma}}}{\gamma-\beta}\Big) = \frac{e^{\beta y+\sqrt{\alpha\beta}}}{2}\operatorname{Erfc}\Big(\frac{\sqrt{\alpha}}{2\sqrt{y}} + \sqrt{\beta y}\Big) + \frac{e^{\beta y-\sqrt{\alpha\beta}}}{2}\operatorname{Erfc}\Big(\frac{\sqrt{\alpha}}{2\sqrt{y}} - \sqrt{\beta y}\Big)$

10. $\mathcal{L}_\gamma^{-1}\Big(\frac{1}{\sqrt{\gamma}+\beta}e^{-\alpha\sqrt{\gamma}}\Big) = \frac{1}{\sqrt{\pi y}}\, e^{-\alpha^2/4y} - \beta e^{\alpha\beta+\beta^2 y}\operatorname{Erfc}\Big(\frac{\alpha}{2\sqrt{y}} + \beta\sqrt{y}\Big),$ $\qquad \alpha \geq 0$

11. $\mathcal{L}_\gamma^{-1}\Big(\frac{1}{\sqrt{\gamma}\,(\sqrt{\gamma}+\beta)}e^{-\alpha\sqrt{\gamma}}\Big) = e^{\alpha\beta+\beta^2 y}\operatorname{Erfc}\Big(\frac{\alpha}{2\sqrt{y}} + \beta\sqrt{y}\Big),$ $\qquad \alpha \geq 0$

12. $\mathcal{L}_\gamma^{-1}\Big(\frac{\beta}{\gamma(\sqrt{\gamma}+\beta)}e^{-\alpha\sqrt{\gamma}}\Big) = \operatorname{Erfc}\Big(\frac{\alpha}{2\sqrt{y}}\Big) - e^{\alpha\beta+\beta^2 y}\operatorname{Erfc}\Big(\frac{\alpha}{2\sqrt{y}} + \beta\sqrt{y}\Big),$ $\qquad \alpha \geq 0$

13. $\mathcal{L}_\gamma^{-1}\Big(\frac{1}{\sqrt{\gamma+\alpha}+\sqrt{\gamma+\beta}}\Big) = \frac{e^{-\beta y} - e^{-\alpha y}}{2\sqrt{\pi}(\alpha-\beta)y^{3/2}}$

14. $\mathcal{L}_\gamma^{-1}\Big(\frac{\sqrt{\gamma+\alpha}}{\sqrt{\gamma-\alpha}} - 1\Big) = \alpha\big(I_0(\alpha y) + I_1(\alpha y)\big),$ $\qquad \alpha > 0$

15. $\mathcal{L}_\gamma^{-1}\big(e^{\alpha/\gamma} - 1\big) = \frac{\sqrt{\alpha}}{\sqrt{y}} I_1(2\sqrt{\alpha y}),$ $\qquad \alpha > 0$

16. $\mathcal{L}_\gamma^{-1}\Big(\frac{1}{\gamma^{\mu+1}}e^{\alpha/\gamma}\Big) = \big(\frac{y}{\alpha}\big)^{\mu/2} I_\mu(2\sqrt{\alpha y}),$ $\qquad \alpha > 0 \quad \mu > -1$

17. $\mathcal{L}_\gamma^{-1}\Big(\frac{|\beta|^\mu e^{-\alpha\sqrt{\gamma^2-\beta^2}}}{\sqrt{\gamma^2-\beta^2}\,(\gamma+\sqrt{\gamma^2-\beta^2})^\mu}\Big) = \Big(\frac{y-\alpha}{y+\alpha}\Big)^{\mu/2} I_\mu(|\beta|\sqrt{y^2-\alpha^2}\,)\mathbb{I}_{(\alpha,\infty)}(y),$ $\qquad \alpha \geq 0$

18. $\mathcal{L}_\gamma^{-1}\big(D_{-2\nu}(2\sqrt{\alpha\gamma})\big) = \frac{2^{1/2-\nu}\sqrt{\alpha}(y-\alpha)^{\nu-1}}{\Gamma(\nu)(y+\alpha)^{\nu+1/2}}\mathbb{I}_{(\alpha,\infty)}(y),$ $\qquad \alpha > 0,\ \nu > 0$

19. $\mathcal{L}_\gamma^{-1}\big(\operatorname{Erfc}(\sqrt{\alpha\gamma})\big) = \frac{\sqrt{\alpha}}{\pi y\sqrt{y-\alpha}}\mathbb{I}_{(\alpha,\infty)}(y),$ $\qquad \alpha > 0$

20. $\mathcal{L}_\gamma^{-1}\big(e^{-\alpha\gamma} - \sqrt{\pi\alpha\gamma}\,\mathrm{Erfc}(\sqrt{\alpha\gamma})\big) = \frac{\sqrt{\alpha}}{2y^{3/2}}\mathbb{I}_{(\alpha,\infty)}(y),$ $\alpha > 0$

21. $\mathcal{L}_\gamma^{-1}\big(\sqrt{\gamma}e^{-2\alpha\sqrt{\gamma}}\,\mathrm{Erfc}(\beta+\sqrt{\gamma})\big)$ $\alpha+\beta>0$

$$= \Big\{\Big(\frac{\alpha+\beta}{\pi y(y-1)^{3/2}} + \frac{\alpha}{\pi y^2\sqrt{y-1}}\Big)e^{-\beta^2-(\alpha+\beta)^2/(y-1)}$$
$$+\Big(\frac{\alpha^2}{\sqrt{\pi}y^{5/2}} - \frac{1}{2\sqrt{\pi}y^{3/2}}\Big)e^{-\alpha^2/y}\,\mathrm{Erfc}\Big(\frac{\alpha}{\sqrt{y(y-1)}} + \frac{\beta\sqrt{y}}{\sqrt{y-1}}\Big)\Big\}\mathbb{I}_{[1,\infty)}(y)$$

22. $\mathcal{L}_\gamma^{-1}\big(\gamma^{\mu-1}K_{2\nu}(2\sqrt{\alpha\gamma})\big) = \frac{1}{2\sqrt{\alpha}}\,y^{1/2-\mu}e^{-\alpha/2y}W_{\mu-1/2,\nu}(\alpha/y),$ $\alpha>0$

23. $\mathcal{L}_\gamma^{-1}\big(\gamma^{-1/2}K_{2\nu}(2\sqrt{\alpha\gamma})\big) = \frac{1}{2\sqrt{\pi y}}\,e^{-\alpha/2y}K_\nu(\alpha/2y),$ $\alpha>0$

24. $\mathcal{L}_\gamma^{-1}\big(\gamma^{\nu/2}K_\nu(\sqrt{\alpha\gamma})\big) = \frac{\alpha^{\nu/2}}{(2y)^{\nu+1}}\,e^{-\alpha/4y},$ $\alpha>0,\ \nu\ge 0$

25. $\mathcal{L}_\gamma^{-1}\Big(\frac{e^{\alpha/2\gamma}}{\gamma^\mu}M_{-\mu,\nu}\big(\frac{\alpha}{\gamma}\big)\Big) = \frac{\sqrt{\alpha}\Gamma(2\nu+1)}{\Gamma(\mu+\nu+1/2)}\,y^{\mu-1/2}I_{2\nu}(2\sqrt{\alpha y}),$ $\mu+\nu>-1/2$

26. $\mathcal{L}_\gamma^{-1}\big(\Gamma(\gamma-\beta)W_{-\gamma,\beta+1/2}(4\alpha)\big) = \alpha^{-\beta}(\mathrm{sh}(y/2))^{2\beta}e^{-2\alpha\,\mathrm{cth}(y/2)},$ $\alpha>0\ge\beta$

27. $\mathcal{L}_\gamma^{-1}\Big(\frac{1}{\gamma}\exp\big(\frac{\alpha+\beta}{2\gamma}\big)I_\nu\big(\frac{\sqrt{\alpha\beta}}{\gamma}\big)\Big) = I_\nu(\sqrt{2\alpha y})I_\nu(\sqrt{2\beta y}),$ $\nu>-1$

28. $\mathcal{L}_\gamma^{-1}\big(I_\nu(x\sqrt{2\gamma})K_\nu(z\sqrt{2\gamma})\big) = \frac{1}{2y}\exp\Big(-\frac{x^2+z^2}{2y}\Big)I_\nu\Big(\frac{xz}{y}\Big),$ $0\vee x\le z$

29. $\mathcal{L}_\gamma^{-1}\big(\Gamma(\gamma)D_{-\gamma}(x)D_{-\gamma}(z)\big) = \frac{e^{y/2}}{\sqrt{2\,\mathrm{sh}\,y}}\exp\Big(-\frac{(x^2+z^2)\,\mathrm{ch}\,y+2xz}{4\,\mathrm{sh}\,y}\Big)$

30. $\mathcal{L}_\gamma^{-1}\big(\gamma^{-1}2^{\gamma/2}\Gamma(\tfrac{\gamma}{2}+\tfrac{1}{2})D_{-\gamma}(x)\big) = \sqrt{\pi}\,e^{-x^2/4}\,\mathrm{Erfc}\Big(\frac{x}{\sqrt{2(e^{2y}-1)}}\Big),$ $x\ge 0$

31. $\mathcal{L}_\gamma^{-1}\Big(\frac{D_{-\gamma}(x)}{D_{-\gamma}(0)}\Big) = \frac{xe^{y/2}}{2\sqrt{\pi}\,\mathrm{sh}^{3/2}y}\exp\Big(-\frac{x^2\,\mathrm{ch}\,y}{4\,\mathrm{sh}\,y}\Big),$ $x\ge 0$

32. $\mathcal{L}_\gamma^{-1}\big(\Gamma(1/2+\nu+\gamma)M_{-\gamma,\nu}(x^2)W_{-\gamma,\nu}(z^2)\big)$ $0\le x\le z$

$$= \frac{\Gamma(2\nu+1)xz}{2\,\mathrm{sh}(y/2)}\exp\Big(-\frac{(x^2+z^2)\,\mathrm{ch}(y/2)}{2\,\mathrm{sh}(y/2)}\Big)I_{2\nu}\Big(\frac{xz}{\mathrm{sh}(y/2)}\Big),$$ $\nu>-1/2$

33. $\mathcal{L}_\gamma^{-1}\big(\Gamma(\gamma)M(\gamma,\nu+1,x^2)U(\gamma,\nu+1,z^2)\big)$ $0\le x\le z$

$$= \frac{\Gamma(\nu+1)e^{(\nu+1)y/2}}{2x^\nu z^\nu\,\mathrm{sh}(y/2)}\exp\Big(\frac{x^2+z^2}{2} - \frac{(x^2+z^2)\,\mathrm{ch}(y/2)}{2\,\mathrm{sh}(y/2)}\Big)I_\nu\Big(\frac{xz}{\mathrm{sh}(y/2)}\Big),$$ $\nu>-1$

34. $\mathcal{L}_\gamma^{-1}\mathcal{L}_\eta^{-1}\Big(\frac{1}{\gamma\eta+a\gamma+b\eta+c}\Big) = e^{-by-ag}I_0(2\sqrt{(ab-c)yg})$

35. $\mathcal{L}_\gamma^{-1}\mathcal{L}_\eta^{-1}\Big(\frac{\gamma+b}{\gamma\eta+a\gamma+b\eta+c} - \frac{1}{\eta+a}\Big) = e^{-by-ag}\sqrt{(ab-c)g/y}\,I_1(2\sqrt{(ab-c)yg})$

DIFFERENTIAL EQUATIONS AND THEIR SOLUTIONS

Below ψ and φ denote a nonnegative increasing and decreasing, respectively, solutions of the differential equations, $w = \psi'\varphi - \psi\varphi'$ is the Wronskian, λ and q are given positive parameters.

1. $\frac{1}{2}Y''(x) - \lambda Y(x) = 0, \qquad x \in \mathbb{R}$

$\psi(x) = e^{x\sqrt{2\lambda}}, \qquad \varphi(x) = e^{-x\sqrt{2\lambda}}, \qquad w = 2\sqrt{2\lambda}$

2. $\frac{1}{2}Y''(x) - (\lambda + \gamma x)Y(x) = 0, \qquad x > 0$

$\psi(x) = \sqrt{\lambda + \gamma x}\, I_{1/3}\big(\frac{2\sqrt{2}}{3\gamma}(\lambda + \gamma x)^{3/2}\big), \qquad w = 3\gamma/2, \qquad \gamma > 0$

$\varphi(x) = \sqrt{\lambda + \gamma x}\, K_{1/3}\big(\frac{2\sqrt{2}}{3\gamma}(\lambda + \gamma x)^{3/2}\big) = \pi\sqrt{3}\,\mathrm{Ai}\big(2^{1/3}\gamma^{-2/3}(\lambda + \gamma x)\big)$

3. $\frac{1}{2}Y''(x) - \big(\lambda + \frac{\gamma^2}{2}x^2\big)Y(x) = 0, \qquad x \in \mathbb{R}$

$\psi(x) = D_{-1/2-\lambda/\gamma}(-x\sqrt{2\gamma}), \qquad \varphi(x) = D_{-1/2-\lambda/\gamma}(x\sqrt{2\gamma})$

$w = \frac{2\sqrt{\pi\gamma}}{\Gamma(1/2 + \lambda/\gamma)}, \qquad \gamma > 0$

4. $\frac{1}{2}Y''(x) - \big(\lambda + \frac{\gamma^2 - 2^{-2}}{2x^2}\big)Y(x) = 0, \qquad x > 0$

$\psi(x) = \sqrt{x}\, I_\gamma(x\sqrt{2\lambda}), \qquad \varphi(x) = \sqrt{x}\, K_\gamma(x\sqrt{2\lambda}), \qquad w = 1, \qquad \gamma \geq \frac{1}{2}$

a. $\frac{1}{2}Y''(x) - \frac{\gamma^2 - 2^{-2}}{2x^2}Y(x) = 0, \qquad x > 0$

$\psi(x) = x^{1/2+\gamma}, \qquad \varphi(x) = x^{1/2-\gamma}, \qquad w = 2\gamma, \qquad \gamma \geq \frac{1}{2}$

5. $\frac{1}{2}Y''(x) - \big(\lambda + \frac{p^2 - 2^{-2}}{2x^2} + \frac{q^2x^2}{2}\big)Y(x) = 0, \qquad x > 0$

$\psi(x) = x^{-1/2}M_{-\lambda/2q,p/2}(qx^2), \qquad \varphi(x) = x^{-1/2}W_{-\lambda/2q,p/2}(qx^2)$

$$w = \frac{2q\Gamma(1+p)}{\Gamma((1+p+\lambda/q)/2)}, \qquad p \geq \frac{1}{2}$$

a. $\frac{1}{2}Y''(x) - \big(\frac{p^2 - 2^{-2}}{2x^2} + \frac{q^2x^2}{2}\big)Y(x) = 0, \qquad x > 0$

$$\psi(x) = \sqrt{x}I_{p/2}(qx^2/2), \qquad \varphi(x) = \sqrt{x}K_{p/2}(qx^2/2), \qquad w = 2, \qquad p \geq \frac{1}{2}$$

6. $\frac{1}{2}Y''(x) - \big(\lambda + \frac{p^2 - 2^{-2}}{2x^2} + \frac{q}{x}\big)Y(x) = 0, \qquad x > 0$

$$\psi(x) = M_{-q/\sqrt{2\lambda},p}(2x\sqrt{2\lambda}), \qquad \varphi(x) = W_{-q/\sqrt{2\lambda},p}(2x\sqrt{2\lambda})$$

$$w = \frac{2\sqrt{2\lambda}\Gamma(2p+1)}{\Gamma(p+1/2+q/\sqrt{2\lambda})}, \qquad p \geq \frac{1}{2}$$

a. $\frac{1}{2}Y''(x) - \big(\frac{p^2 - 2^{-2}}{2x^2} + \frac{q}{x}\big)Y(x) = 0, \qquad x > 0$

$$\psi(x) = \sqrt{x}I_{2p}(2\sqrt{2qx}), \qquad \varphi(x) = \sqrt{x}K_{2p}(2\sqrt{2qx}), \qquad w = \frac{1}{2}, \qquad p \geq \frac{1}{2}$$

7. $\frac{1}{2}Y''(x) - (\lambda + \gamma e^{2\beta x})Y(x) = 0, \qquad x \in \mathbb{R}$

$$\psi(x) = I_{\sqrt{2\lambda}/|\beta|}\Big(\frac{\sqrt{2\gamma}}{|\beta|}e^{\beta x}\Big), \qquad \varphi(x) = K_{\sqrt{2\lambda}/|\beta|}\Big(\frac{\sqrt{2\gamma}}{|\beta|}e^{\beta x}\Big), \qquad w = \beta$$

8. $\frac{1}{2}Y''(x) - \big(\lambda + \frac{p^2}{2}e^{2\beta x} + qe^{\beta x}\big)Y(x) = 0, \qquad x \in \mathbb{R}$

$$\psi(x) = e^{-\beta x/2}M_{-q/p|\beta|,\sqrt{2\lambda}/|\beta|}\big(\tfrac{2p}{|\beta|}e^{\beta x}\big), \qquad w = \frac{2p\Gamma(2\sqrt{2\lambda}/|\beta|+1)}{\Gamma(\sqrt{2\lambda}/|\beta|+1/2+q/p|\beta|)}$$

$$\varphi(x) = e^{-\beta x/2}W_{-q/p|\beta|,\sqrt{2\lambda}/|\beta|}\big(\tfrac{2p}{|\beta|}e^{\beta x}\big), \qquad p > 0$$

9. $\frac{1}{2}Y''(x) + \mu Y'(x) - (\lambda + f(x))Y(x) = 0, \qquad x \in \mathbb{R}$

solution $Y_{\mu,\lambda}(x)$ of this equation satisfies the relation

$$Y_{\mu,\lambda}(x) = e^{-\mu x}Y_{0,\lambda+\mu^2/2}(x)$$

10. $\frac{1}{2}Y''(x) + \frac{1}{x}Y'(x) - (\lambda + f(x))Y(x) = 0, \qquad x > 0$

solution of this equation is $x^{-1}Y_{0,\lambda}(x)$, where $Y_{0,\lambda}(x)$ is the solution of the equation 9 for $\mu=0$

11. $Y''(x)+\frac{1}{x}Y'(x)-\Big(1+\frac{\nu^2}{x^2}\Big)Y(x)=0, \qquad x>0$

$\psi(x)=I_\nu(x), \qquad \varphi(x)=K_\nu(x), \qquad w=\frac{1}{x}, \qquad \nu>0$

12. $\frac{1}{2}Y''(x)+\frac{2\nu+1}{2x}Y'(x)-\lambda Y(x)=0, \qquad x>0$

$\psi(x)=x^{-\nu}I_\nu(x\sqrt{2\lambda}), \qquad \varphi(x)=x^{-\nu}K_\nu(x\sqrt{2\lambda}), \quad w=\frac{1}{x^{2\nu+1}}, \quad \nu>-1$

a. $\frac{1}{2}Y''(x)+\frac{2\nu+1}{2x}Y'(x)=0, \qquad x>0$

$\psi(x)=1, \qquad \varphi(x)=x^{-2\nu}, \qquad w=\frac{2\nu}{x^{2\nu+1}}, \qquad \nu>0$

b. $\frac{1}{2}Y''(x)+\frac{1}{2x}Y'(x)=0, \qquad x>1$

$\psi(x)=\ln x, \qquad \varphi(x)=1, \qquad w=\frac{1}{x}$

c. $\frac{1}{2}Y''(x)+\frac{1}{x}Y'(x)-\lambda Y(x)=0, \qquad x>0$

$\psi(x)=\frac{1}{x}\operatorname{sh}(x\sqrt{2\lambda}), \qquad \varphi(x)=\frac{1}{x}e^{-x\sqrt{2\lambda}}, \qquad w=\frac{2\sqrt{2\lambda}}{x^2}$

13. $\frac{1}{2}Y''(x)+\frac{2\nu+1}{2x}Y'(x)-\gamma xY(x)=0, \qquad x>0$

$\psi(x)=x^{-\nu}I_{2\nu/3}\big(\frac{2}{3}x^{3/2}\sqrt{2\gamma}\big), \qquad \varphi(x)=x^{-\nu}K_{2\nu/3}\big(\frac{2}{3}x^{3/2}\sqrt{2\gamma}\big)$

$w=\frac{3}{2}x^{-2\nu-1}, \qquad \nu>-1$

14. $\frac{1}{2}Y''(x)+\frac{2\nu+1}{2x}Y'(x)-\big(\lambda+\frac{\gamma^2}{2x^2}\big)Y(x)=0, \qquad x>0$

$\psi(x)=x^{-\nu}I_{\sqrt{\nu^2+\gamma^2}}(x\sqrt{2\lambda}), \qquad \varphi(x)=x^{-\nu}K_{\sqrt{\nu^2+\gamma^2}}(x\sqrt{2\lambda})$

$$w = x^{-2\nu-1}, \qquad \nu \ge 0$$

a. $$\frac{1}{2}Y''(x) + \frac{2\nu+1}{2x}Y'(x) - \frac{\gamma^2}{2x^2}Y(x) = 0, \qquad x > 0$$

$$\psi(x) = x^{\sqrt{\nu^2+\gamma^2}-\nu}, \qquad \varphi(x) = x^{-\sqrt{\nu^2+\gamma^2}-\nu}, \qquad w = \frac{2\sqrt{\nu^2+\gamma^2}}{x^{2\nu+1}}, \qquad \nu \ge 0$$

15. $$\frac{1}{2}Y''(x) + \frac{2\nu+1}{2x}Y'(x) - \big(\lambda + \frac{p^2}{2x^2} + \frac{q^2x^2}{2}\big)Y(x) = 0, \qquad x > 0$$

$$\psi(x) = x^{-\nu-1}M_{-\lambda/2q,\sqrt{p^2+\nu^2}/2}(qx^2), \qquad w = \frac{2q\Gamma(1+\sqrt{p^2+\nu^2})x^{-2\nu-1}}{\Gamma((1+\sqrt{p^2+\nu^2}+\lambda/q)/2)}$$

$$\varphi(x) = x^{-\nu-1}W_{-\lambda/2q,\sqrt{p^2+\nu^2}/2}(qx^2), \qquad \nu \ge 0$$

a. $$\frac{1}{2}Y''(x) + \frac{2\nu+1}{2x}Y'(x) - \big(\frac{p^2}{2x^2} + \frac{q^2x^2}{2}\big)Y(x) = 0, \qquad x > 0$$

$$\psi(x) = x^{-\nu}I_{\sqrt{\nu^2+p^2}/2}(qx^2/2), \qquad w = 2x^{-2\nu-1}$$

$$\varphi(x) = x^{-\nu}K_{\sqrt{\nu^2+p^2}/2}(qx^2/2), \qquad \nu \ge 0$$

16. $$\frac{1}{2}Y''(x) + \frac{2\nu+1}{2x}Y'(x) - \big(\lambda + \frac{p^2}{2x^2} + \frac{q}{x}\big)Y(x) = 0, \qquad x > 0$$

$$\psi(x) = \frac{1}{x^{\nu+1/2}}M_{-q/\sqrt{2\lambda},\sqrt{\nu^2+p^2}}(2x\sqrt{2\lambda}), \qquad w = \frac{2x^{1-2\nu}\sqrt{2\lambda}\Gamma(2\sqrt{\nu^2+p^2}+1)}{\Gamma(\sqrt{\nu^2+p^2}+1/2+q/\sqrt{2\lambda})}$$

$$\varphi(x) = \frac{1}{x^{\nu+1/2}}W_{-q/\sqrt{2\lambda},\sqrt{\nu^2+p^2}}(2x\sqrt{2\lambda}), \qquad \nu \ge 0$$

a. $$\frac{1}{2}Y''(x) + \frac{2\nu+1}{2x}Y'(x) - \big(\frac{p^2}{2x^2} + \frac{q}{x}\big)Y(x) = 0, \qquad x > 0$$

$$\psi(x) = x^{-\nu}I_{2\sqrt{\nu^2+p^2}}(2\sqrt{2qx}), \qquad w = 2^{-1}x^{-2\nu-1}$$

$$\varphi(x) = x^{-\nu}K_{2\sqrt{\nu^2+p^2}}(2\sqrt{2qx}), \qquad \nu \ge 0$$

17. $$Y''(x) - \Big(\frac{x^2}{4} + \frac{2\nu-1}{2}\Big)Y(x) = 0, \qquad x \in \mathbb{R}$$

$$\psi(x) = D_{-\nu}(-x), \qquad \varphi(x) = D_{-\nu}(x), \qquad w = \frac{\sqrt{2\pi}}{\Gamma(\nu)} \qquad \nu > 0$$

18. $$\sigma^2 Y''(x) - xY'(x) - \Big(\lambda + \frac{(\gamma^2-1)x^2}{4\sigma^2}\Big)Y(x) = 0, \qquad x \in \mathbb{R}$$

$$\psi(x) = e^{x^2/4\sigma^2} D_{-\frac{1}{2}+\frac{1}{2\gamma}-\frac{\lambda}{\gamma}}\Big(-\frac{x\sqrt{\gamma}}{\sigma}\Big), \qquad \varphi(x) = e^{x^2/4\sigma^2} D_{-\frac{1}{2}+\frac{1}{2\gamma}-\frac{\lambda}{\gamma}}\Big(\frac{x\sqrt{\gamma}}{\sigma}\Big)$$

$$w = \frac{\sqrt{2\pi\gamma}}{\sigma\Gamma(\frac{1}{2}-\frac{1}{2\gamma}+\frac{\lambda}{\gamma})} e^{x^2/2\sigma^2}, \qquad \sigma > 0, \qquad \gamma \ge 1$$

a. $$\sigma^2 Y''(x) - xY'(x) = 0, \qquad x \in \mathbb{R}$$

$$\psi(x) = \frac{2}{\sqrt{\pi}} \int_0^{x/\sigma\sqrt{2}} e^{v^2} dv, \qquad \varphi(x) = 1, \qquad w = \frac{\sqrt{2}}{\sqrt{\pi}\sigma} e^{x^2/2\sigma^2}, \qquad \sigma > 0$$

19. $$\sigma^2 Y''(x) - xY'(x) - \big(\lambda + \frac{\gamma x}{\sigma}\big)Y(x) = 0, \qquad x \in \mathbb{R}$$

$$\psi(x) = e^{x^2/4\sigma^2} D_{\gamma^2-\lambda}\big(-\frac{x}{\sigma} - 2\gamma\big), \qquad \varphi(x) = e^{x^2/4\sigma^2} D_{\gamma^2-\lambda}\big(\frac{x}{\sigma} + 2\gamma\big)$$

$$w = \frac{\sqrt{2\pi}}{\sigma\Gamma(\lambda-\gamma^2)} e^{x^2/2\sigma^2}, \qquad \sigma > 0, \qquad \sqrt{\lambda} > \gamma \ge 0$$

20. $$\sigma^2 Y''(x) - xY'(x) - \Big(\lambda + \frac{(p^2-2^{-2})\sigma^2}{x^2} + \frac{(q^2-1)x^2}{4\sigma^2}\Big)Y(x) = 0, \qquad x > 0$$

$$\psi(x) = x^{-1/2} e^{x^2/4\sigma^2} M_{\frac{1-2\lambda}{4q}, \frac{p}{2}}\Big(\frac{qx^2}{2\sigma^2}\Big), \qquad w = \frac{q\Gamma(p+1)e^{x^2/2\sigma^2}}{\sigma^2\Gamma\Big(\frac{p+1}{2} + \frac{2\lambda-1}{4q}\Big)}$$

$$\varphi(x) = x^{-1/2} e^{x^2/4\sigma^2} W_{\frac{1-2\lambda}{4q}, \frac{p}{2}}\Big(\frac{qx^2}{2\sigma^2}\Big), \qquad p \ge \frac{1}{2}, \qquad q \ge 1$$

21. $$\sigma^2 Y''(x) + \Big(\frac{\sigma^2(2\nu+1)}{x} - x\Big)Y'(x) - \Big(\lambda + \frac{p^2\sigma^2}{x^2} + \frac{(q^2-1)}{4\sigma^2}x^2\Big)Y(x) = 0, \qquad x > 0$$

$$\psi(x) = x^{-\nu-1} e^{x^2/4\sigma^2} M_{\frac{\nu+1-\lambda}{2q}, \frac{1}{2}\sqrt{\nu^2+p^2}}\Big(\frac{qx^2}{2\sigma^2}\Big)$$

$$\varphi(x) = x^{-\nu-1} e^{x^2/4\sigma^2} W_{\frac{\nu+1-\lambda}{2q}, \frac{1}{2}\sqrt{\nu^2+p^2}}\Big(\frac{qx^2}{2\sigma^2}\Big)$$

$$w = \frac{q e^{x^2/2\sigma^2}\Gamma\big(\sqrt{\nu^2+p^2}+1\big)}{\sigma^2 x^{2\nu+1}\Gamma\Big(\frac{1}{2}\sqrt{\nu^2+p^2}+\frac{1}{2}+\frac{\lambda-\nu-1}{2q}\Big)}, \qquad q \geq 1, \qquad \nu \geq 0$$

22. $\frac{1}{2}Y''(x) + \Big(\frac{2\nu+1}{2x} - \gamma x\Big)Y'(x) - \lambda Y(x) = 0, \qquad x > 0$

for $\gamma > 0$, $\nu > -1$

$$\psi(x) = M\big(\tfrac{\lambda}{2\gamma}, \nu+1, \gamma x^2\big), \qquad \varphi(x) = U\big(\tfrac{\lambda}{2\gamma}, \nu+1, \gamma x^2\big)$$

$$w = \frac{2\Gamma(\nu+1)x^{-2\nu-1}}{\Gamma(\lambda/2\gamma)\gamma^\nu}e^{\gamma x^2}$$

for $\gamma > 0$, $\nu < 0$

$$\psi(x) = x^{-2\nu}M\big(\tfrac{\lambda}{2\gamma}-\nu, 1-\nu, \gamma x^2\big), \qquad \varphi(x) = x^{-2\nu}U\big(\tfrac{\lambda}{2\gamma}-\nu, 1-\nu, \gamma x^2\big)$$

$$w = \frac{2\Gamma(1-\nu)x^{-2\nu-1}}{\Gamma(\lambda/2\gamma-\nu)\gamma^\nu}e^{\gamma x^2}$$

for $\gamma < 0$, $\nu > -1$

$$\psi(x) = e^{\gamma x^2}M\big(\nu+1-\tfrac{\lambda}{2\gamma}, \nu+1, -\gamma x^2\big), \quad \varphi(x) = e^{\gamma x^2}U\big(\nu+1-\tfrac{\lambda}{2\gamma}, \nu+1, -\gamma x^2\big)$$

$$w = \frac{2\Gamma(\nu+1)x^{-2\nu-1}}{\Gamma(\nu+1-\lambda/2\gamma)|\gamma|^\nu}e^{-\gamma x^2}$$

for $\gamma < 0$, $\nu < 0$

$$\psi(x) = x^{-2\nu}e^{\gamma x^2}M\big(1-\tfrac{\lambda}{2\gamma}, 1-\nu, -\gamma x^2\big), \ \varphi(x) = x^{-2\nu}e^{\gamma x^2}U\big(1-\tfrac{\lambda}{2\gamma}, 1-\nu, -\gamma x^2\big)$$

$$w = \frac{2\Gamma(1-\nu)x^{-2\nu-1}}{\Gamma(1-\lambda/2\gamma)|\gamma|^\nu}e^{-\gamma x^2}$$

23. $\frac{1}{2}Y''(x) + \rho\,\mathrm{cth}(x)Y'(x) - \frac{1}{2}(\mu^2-\rho^2)Y(x) = 0, \quad x > 0$

$$\psi(x) = \frac{P^{-\rho+1/2}_{\mu-1/2}(\mathrm{ch}\,x)}{\mathrm{sh}^{\rho-1/2}\,x}, \qquad \varphi(x) = \frac{\widetilde{Q}^{\rho-1/2}_{\mu-1/2}(\mathrm{ch}\,x)}{\mathrm{sh}^{\rho-1/2}\,x}, \quad \omega = \frac{1}{\mathrm{sh}^{2\rho}\,x}, \quad \mu \geq \rho > -\frac{1}{2}$$

24. $\frac{1}{2}Y''(x) - \rho\,\mathrm{th}(x)Y'(x) - 2\alpha(\rho+\alpha)Y(x) = 0, \qquad x > 0$

$$\psi(x) = \mathrm{ch}^\rho\,x\big(\widetilde{P}^{-2\alpha-\rho}_\rho(-\mathrm{th}\,x) + \widetilde{P}^{-2\alpha-\rho}_\rho(\mathrm{th}\,x)\big), \qquad \alpha > 0, \ \alpha+\rho \geq 0$$

$$\varphi(x) = \mathrm{ch}^\rho\,x\,\widetilde{P}^{-2\alpha-\rho}_\rho(\mathrm{th}\,x), \qquad \omega = \frac{2\,\mathrm{ch}^{2\rho}\,x}{\Gamma(2\alpha)\Gamma(2\alpha+2\rho+1)}$$

25. $\frac{1}{2}Y''(x) + \big(\mathrm{cth}\,x - \rho\,\mathrm{th}\,x\big)Y'(x) - 2\alpha(\rho+\alpha-1)Y(x) = 0, \qquad x > 0$

$$\psi(x) = \frac{\operatorname{ch}^\rho x}{\operatorname{sh} x}\big(\widetilde{P}_\rho^{1-2\alpha-\rho}(-\operatorname{th} x) - \widetilde{P}_\rho^{1-2\alpha-\rho}(\operatorname{th} x)\big), \qquad \alpha > 0,\ \alpha + \rho \geq 1$$

$$\varphi(x) = \frac{\operatorname{ch}^\rho x}{\operatorname{sh} x}\,\widetilde{P}_\rho^{1-2\alpha-\rho}(\operatorname{th} x), \qquad \omega = \frac{2\operatorname{ch}^{2\rho} x}{\operatorname{sh}^2 x\,\Gamma(2\alpha-1)\Gamma(2\alpha+2\rho)}$$

26. $\frac{1}{2}Y''(x) + \big((\gamma - \frac{1}{2})\operatorname{cth} x - (\beta - \alpha - \frac{1}{2})\operatorname{th} x\big)Y'(x) - 2\alpha(\beta-\gamma)Y(x) = 0$

$$\psi(x) = \frac{1}{\operatorname{ch}^{2\alpha} x}F\big(\alpha,\beta,\gamma,\operatorname{th}^2 x\big), \qquad \varphi(x) = \frac{1}{\operatorname{ch}^{2\alpha} x}G\big(\alpha,\beta,\gamma,\operatorname{th}^2 x\big)$$

$$w = \frac{2\Gamma(\gamma)\operatorname{ch}^{2(\beta-\alpha-1/2)} x}{\Gamma(\alpha)\Gamma(\beta)\operatorname{sh}^{2\gamma-1} x}, \qquad \alpha > 0,\ \beta \geq \gamma > 0$$

For the differential equations in 27-29 it is assumed that $\alpha\beta > 0$, $\alpha+\beta \geq 0$, $\alpha+\beta+1 > \gamma > 0$, and α, β can be complex conjugate.

27. $x(1-x)Y''(x) + (\gamma - (\alpha+\beta+1)x)Y'(x) - \alpha\beta Y(x) = 0, \qquad 0 < x < 1$

$$\psi(x) = F(\alpha,\beta,\gamma,x), \qquad \varphi(x) = G(\alpha,\beta,\gamma,x), \qquad \omega = \frac{\Gamma(\gamma)(1-x)^{\gamma-\alpha-\beta-1}}{x^\gamma\Gamma(\alpha)\Gamma(\beta)}$$

28. $Y''(x) + \Big(1 - \gamma + \frac{\alpha+\beta-1}{e^x+1}\Big)Y'(x) - \frac{\alpha\beta e^x}{(e^x+1)^2}Y(x) = 0, \qquad x \in \mathbb{R}$

$$\psi(x) = G\Big(\alpha,\beta,\gamma,\frac{1}{e^x+1}\Big), \qquad \varphi(x) = F\Big(\alpha,\beta,\gamma,\frac{1}{e^x+1}\Big)$$

$$w = \frac{\Gamma(\gamma)\,(e^x+1)^{\alpha+\beta-1}}{\Gamma(\alpha)\Gamma(\beta)\,e^{x(\alpha+\beta-\gamma)}}$$

29. $Y''(x) + \Big(\frac{\gamma}{1-e^{-x}} - \alpha - \beta\Big)Y'(x) - \frac{\alpha\beta e^{-x}}{1-e^{-x}}Y(x) = 0, \qquad x > 0$

$$\psi(x) = F\big(\alpha,\beta,\gamma,1-e^{-x}\big), \qquad \varphi(x) = G\big(\alpha,\beta,\gamma,1-e^{-x}\big)$$

$$w = \frac{\Gamma(\gamma)\,e^{x(\alpha+\beta-\gamma)}}{\Gamma(\alpha)\Gamma(\beta)\,(1-e^{-x})^\gamma}$$

FORMULAE FOR n-FOLD DIFFERENTIATION

For the n times differentiable real valued function f introduce

$$(f(x))^{(n)} := f^{(n)}(x) := \frac{d^n}{dx^n} f(x), \quad n = 1, 2, 3, \ldots, \quad f^{(0)}(x) := f(x).$$

1. Leibnitz's formula for the nth derivative of the product of two functions:

$$\big(f(x)g(x)\big)^{(n)} = \sum_{k=0}^{n} \frac{n\,!}{k\,!\,(n-k)\,!}\, f^{(k)}(x)\, g^{(n-k)}(x).$$

2. de Bruno's formula for the nth derivative of the composition of two functions:

$$\big(F(f(x))\big)^{(n)} = n\,! \sum_{m=1}^{n} F^{(m)}(f(x)) \sum_{\substack{k_1+2k_2+\cdots+nk_n=n\\ k_1+k_2+\cdots+k_n=m}} \prod_{j=1}^{n} \frac{1}{k_j\,!} \left(\frac{f^{(j)}(x)}{j\,!}\right)^{k_j}.$$

3. The nth derivative of the ratio of two functions:

$$\left(\frac{f(x)}{g(x)}\right)^{(n)} = \frac{f^{(n)}(x)}{g(x)} + \sum_{1\le m\le l\le n} \frac{(-1)^m\, m\,!\, n\,!}{(n-l)\,!}\, \frac{f^{(n-l)}(x)}{g^{m+1}(x)}$$

$$\times \sum_{\substack{k_1+2k_2+\cdots+lk_l=l\\ k_1+k_2+\cdots+k_l=m}} \prod_{j=1}^{l} \frac{1}{k_j\,!} \left(\frac{g^{(j)}(x)}{j\,!}\right)^{k_j}.$$

4. The $(n+1)$th derivative of the inverse function (see Bödewadt (1942)): Let F be a given smooth function with the inverse f. Then

$$f^{(n+1)}(x) = \sum_{m=1}^{n} \frac{(-1)^m\,(n+m)\,!}{(F'(f(x)))^{n+m+1}} \sum_{\substack{k_1+2k_2+\cdots+nk_n=n\\ k_1+k_2+\cdots+k_n=m}} \prod_{j=1}^{n} \frac{1}{k_j\,!} \left(\frac{F^{(j+1)}(f(x))}{(j+1)\,!}\right)^{k_j}.$$

The second sum in the formulae above is taken over all combinations of nonnegative integers $k_1, k_2, \ldots$ such that the indicated equalities hold.

5. The nth derivative of the composition of two functions (see Adams (1947)):

$$\big(F(f(x))\big)^{(n)} = \sum_{m=1}^{n} (-1)^m F^{(m)}(f(x)) \sum_{l=1}^{m} \frac{(-1)^l}{l\,!\,(m-l)\,!}\, \big(f^l(x)\big)^{(n)} f^{m-l}(x).$$

6. Examples:

$$\left(F\left(\tfrac{1}{x}\right)\right)^{(n)} = \sum_{k=0}^{n-1} \frac{(-1)^n (n-1)!\,n!}{k!\,(n-1-k)!\,(n-k)!} \; \frac{F^{(n-k)}(1/x)}{x^{2n-k}}$$

$$\left(e^{a/x}\right)^{(n)} = \sum_{k=0}^{n-1} \frac{(-1)^n (n-1)!\,n!}{k!\,(n-1-k)!\,(n-k)!} \; \frac{a^{n-k}\, e^{a/x}}{x^{2n-k}}$$

$$\left(F(x^2)\right)^{(n)} = \sum_{0\le k\le n/2} \frac{n!}{k!\,(n-2k)!} \; \frac{F^{(n-k)}(x^2)}{(2x)^{2k-n}}$$

$$\left(e^{ax^2}\right)^{(n)} = e^{ax^2} \sum_{0\le k\le n/2} \frac{n!}{k!\,(n-2k)!} \; a^{n-k}(2x)^{n-2k}$$

$$\left(F(\sqrt{x})\right)^{(n)} = \sum_{k=0}^{n-1} \frac{(-1)^k\,(n-1+k)!}{k!\,(n-1-k)!} \; \frac{F^{(n-k)}(\sqrt{x})}{(2\sqrt{x})^{n+k}}$$

$$\left(\frac{c+dx}{a+bx}\right)^{(n)} = (-1)^n n!\,\frac{(cb-ad)\,b^{n-1}}{(a+bx)^{n+1}}$$

$$\left(\frac{c+d\sqrt{x}}{a+b\sqrt{x}}\right)^{(n)} = \sum_{k=0}^{n-1} \frac{(-1)^n (n-k)\,(n-1+k)!}{k!} \; \frac{(cb-ad)\,b^{n-k-1}}{(a+b\sqrt{x})^{n-k+1}(2\sqrt{x})^{n+k}}.$$

BIBLIOGRAPHY

Abramowitz, M., and Stegun I.A., *Mathematical Functions*, Dover Publications, Inc., New York, 1970 (9th printing).

Adams, E.P., *Smithsonian Mathematical Formulae and Tables of Elliptic Functions*, Smithsonian Institution, Washington, 1947.

Alili, L., Dufresne, D., and Yor, M., *Sur l'identité de Bougerol pour les fonctionnelles exponentielles du mouvement brownien avec drift*, in "Exponential Functionals and Principal Values related to Brownian Motion; A collection of research papers", Revista Matemática Iberoamericana, Madrid, 1997, pp. 3–14.

Alili, L., and Gruet J-C., *An explanation of a generalized Bougerol's identity in terms of hyperbolic Brownian motion*, in "Exponential Functionals and Principal Values related to Brownian Motion; A collection of research papers", Revista Matemática Iberoamericana, Madrid, 1997, pp. 15–33.

Amir, M., *Sticky Brownian motion as the strong limit of a sequence of random walks*, Stoch. Proc. Appl. **39** (1991), 221–237.

Bachelier, L., *Théorie de la spéculation*, Ann. Sci. École Norm. Sup. **17** (1900), 21–86.

Balkema, A.A., and Chung, K.L., *Paul Lévy's way to his local time*, in "Seminar on Stochastic Processes, 1990", Birkhauser Verlag, Boston, Basel, and Berlin, 1991, pp. 5–14.

Barlow, M., *One-dimensional stochastic differential equation with no strong solution*, J. London Math. Soc. **26** (1982), 335–345.

Barlow, M., *Skew Brownian motion and a one-dimensional stochastic differential equation*, Stochastics **25** (1988), 1–2.

Bass, R.F., *Probabilistic Techniques in Analysis*, Springer Verlag, Berlin, Heidelberg, and New York, 1995.

Bass, R.F., and Khoshnevisan, D., *Laws of the iterated logarithm for local times of the empirical processes*, Ann. Probab. **23** (1995), 388–399.

Bertoin, J., *Excursions of a $BES_0(d)$ and its drift term* $(0 < d < 1)$, Probab. Th. Rel. Fields **84** (1990), 251–265.

Bertoin, J., *An extension of Pitman's theorem for spectrally positive Lévy processes*, Ann. Probab. **20** (1992), no. 3, 1464–1483.

Bertoin, J., and Pitman, J., *Path transformations connecting Brownian bridge, excursion and meander*, Bull. Sci. Math. **118** (1994), 147–166.

Biane, Ph., *Relations entre pont brownien et du mouvement Brownien réel*, Ann. Inst. H. Poincaré **22** (1986), 1–7.

Biane, Ph., *Decompositions of Brownian trajectories and some applications*, in "Probability and Statistics; Rencontres Franco–Chinoises en Probabilités et Statistiques. Proceedings of the Wuhan meeting", World Scientific, 1993, pp. 51–76.

Biane, Ph., and Yor, M., *Valeurs principales associées aux temps locaux browniens*, Bull. Sci. Math. **111** (1987), 23–101.

Biane, Ph., and Yor, M., *Quelques précisions sur le méandre Brownien*, Bull. Sci. Math. **112** (1988), 101–109.

Bismut, J.M., *Last exit decomposition and regularity at the boundary of transition probabilities*, Z. Wahrscheinlichkeitstheorie verw. Gebiete **69** (1985), 65–98.

Blumenthal, R.M., *Excursions of Markov Processes*, Birkhäuser, Boston, Basel, and Berlin, 1992.

Blumenthal, R.M., and Getoor, R.K., *Markov Processes and Potential Theory*, Academic Press, New York and London, 1968.

Bödewadt, U.T., *Die Kettenregel für Höhere Ableitungen*, Math. Zeitschrift **48** (1942), 735–746.

Borodin, A.N., *Distribution of integral functionals of Brownian motion*, Zapiski Nauchn. Semin. LOMI **119** (1982), 19–38; English transl. in J. Soviet Math.

Borodin, A.N., *On distribution of supremum of increments of Brownian local time*, Zapiski Nauchn. Semin. LOMI **142** (1985), 6–24; English transl. in J. Soviet Math.

Borodin, A.N., *Brownian local time*, Uspekhi Mat. Nauk **44** (1989a), no. 2, 7–48; English transl. in Russian Math. Surveys.

Borodin, A.N., *Distributions of functionals of the Brownian local time I and II*, Th. Probab. Appl. **34** (1989b), 385–401, 576–590.

Borodin, A.N., *On distribution of functionals of the Brownian motion stopped at the moment inverse to local time*, Zapiski Nauchn. Semin. LOMI **184** (1990), 37–61; English transl. in J. Soviet Math.

Borodin, A.N., *On distribution of Brownian motion stopped at the moment inverse to the range of the process*, Zapiski Nauchn. Semin. POMI **260** (1999), 50–72; English transl. in J. Math. Sciences.

Borodin, A.N., *On distributions of supremum of linear combinations of local times of random processes*, Trans. St. Petersburg Math. Soc. **8** (2000), 5–28; English transl. in Transl., Ser 2, AMS Russian.

Borodin, A.N., *Stochastic Processes*, Lan', St. Petersburg, 2013. (Russian)

Borodin, A.N., and Salminen, P., *Some exponential integral functionals of BM(μ) and BES(3)*, Zapiski Nauchn. Semin. LOMI **311** (2004), 51–78; English transl. in J. of Math. Sciences **133,** (2006), no. 3, 1231–1248.

Bougerol, Ph., *Exemple du théorèmes locaux sur les groupes résolubles*, Ann. Inst. H. Poincaré **19** (1983), 369–391.

Bouleau, N., and Yor, M., *Sur la variation quadratique des temps locaux de certaines semimartingales*, C.R. Acad. Sci. Paris, Série I **292** (1981), 491–494.

Breiman, L., *Probability*, Addison-Wesley Publ. Co, Reading, Mass., 1968.

Bruno de, F., *Note sur une nouvelle formule de calcul différentiel*, Quarterly J. Pure Appl. Math **1** (1857), 359–360.

Bulinski, A.V., and Shiryaev, A.N., *Theory of Stochastic Processes*, Fizmatlit, Moscow, 2005. (Russian)

Cameron, R.H., and Martin W.T., *Evaluations of various Wiener integrals by use of certain Sturm–Liouville differential equation*, Bull. Amer. Math. Soc. **51** (1945), 73–90.

Chaumont, L., Hobson, D., and Yor, M., *Some consequences of the cyclic exchangeability property for exponential functionals of Lévy processes*, in "Séminaire de Probabilités, XXXV", Lecture Notes in Mathematics, vol. 1755, Springer-Verlag, Berlin, Heidelberg, and New York, 2001, pp. 334–347.

Chitashvili, R., *On the nonexistence of a strong solution in the boundary problem for a sticky Brownian motion*, Proc. A. Razmadze Math. Inst. **115** (1997), 17–31.

Chong, K.S., Cowan, R., and Holst, L., *The ruin problem and cover times of asymmetric random walks and Brownian motions*, Adv. Appl. Probab. **32** (2000).

Chung, K.L., *Excursion in Brownian motion*, Arkiv för matematik **14** (1976), 155–177.

Chung, K.L., *A cluster of great formulas*, Acta Math. Acad. Sci. Hungar. **39** (1982a), 65–67.

Chung, K.L., *Lectures from Markov Processes to Brownian Motion*, Springer-Verlag, Berlin, Heidelberg, and New York, 1982b.

Chung, K.L., *Reminiscences of some of Paul Lévy's ideas in Brownian motion and in Markov chains*, in "Seminar on Stochastic Processes, 1988", Birkhauser Verlag, Boston, Basel, and Berlin, 1989, pp. 99–108.

Chung, K.L., *Green, Brown, and Probability*, World Scientific, Singapore, 1995.

Chung, K.L., *Probability and Doob*, Amer. Math. Monthly **105(1)** (1998), 28–35.

Chung, K.L., and Getoor, R., *The condenser problem*, Ann. Probab. **5(1)** (1977), 82–86.

Chung, K.L., and Walsh, J.B., *Markov processes, Brownian Motion, and Time Symmetry, 2nd edition*, Springer-Verlag, Berlin, Heidelberg, and New York, 2005.

Chung, K.L., and Williams, R., *Introduction to Stochastic Integration, 2nd edition*, Birkhäuser, Boston, 1990.

Comtet, A., Monthus, C., and Yor, M., *Exponential functionals of Brownian Motion and disordered systems*, J. Appl. Probab. **35** (1998), 255–271.

Cootner, P.H. (ed.), *The Random Character of Stock Market Prices*, The M.I.T. Press, Cambridge, MA., 1964.

Cox, J.C., *Notes on option pricing I: Constant elasticity of variance diffusions*, Working paper, Stanford University, 1975, reprinted in Journal of Portfolio Management **22** (1996), 15–17.

Cox, J.C., Ingersoll, J.E., and Ross, S., *A theory of the term structure of interest rates*, Econometrica **53** (1985), 385–407.

Csáki, E., and Földes, A., *How small are the increments of the local time of a Wiener process*, Ann. Probab. **14** (1986), 533–546.

Csáki, E., Földes, A., and Salminen, P., *On the joint distribution of the maximum and its location for a linear diffusion*, Ann. Inst. H. Poincaré **23** (1987), 179–194.

Csáki, E., and Salminen, P., *On additive functionals of diffusion processes*, Studia Sci. Math. Hungar. **31** (1996), 47–62.

Darling, D.A., and Siegert, A.J.F., *The first passage problem for a continuous Markov process*, Ann. Math. Statistics **24** (1953), 624–639.

Davydov, D., and Linetsky, V., *Pricing and hedging path-dependent options under the SEV process*, Management Science **47** (2001), no. 7, 949–965.

Decamps, M., Schepper de, A., Goovaerts, M. and Schoutens, W., *A note on some new perpetuities*, Scandinavian Actuarial J. **4** (2005), 261–270.

Dellacherie, C., and Meyer, P.A., *Probabilités et Potentiel I–IV*, Hermann, Paris, 1976, 1980, 1983, 1987.

Donati-Martin, C., Matsumoto, H., and Yor, M., *On positive and negative moments of the integral of geometric Brownian motion*, Stat. Probab. Letters **49** (2000a), 45–52.

Donati-Martin, C., Matsumoto, H., and Yor, M., *On striking identities about the exponential functionals of the Brownian bridge and Brownian motion*, Periodica Math. Hung. **41** (2000b), 103–119.

Donati-Martin, C., Matsumoto, H., and Yor, M., *The law of geometric Brownian motion and its integral, revisited; application to conditional moments*, in "Mathematical Finance, Bachelier

Congress, 2000(Paris),, Springer Verlag, Berlin, Heidelberg, and New York, 2002, pp. 221–243.

Doney, R., and Yor, M., *On the formula of Takàcs for Brownian motion with drift*, J. Appl. Probab. **35** (1998), 272–280.

Doob, J.L., *Heuristic approach to the Kolmogorov–Smirnov theorems*, Ann. Math. Stat. **20** (1949), 393–403.

Doob, J.L., *Stochastic Processes*, Wiley and Sons, New York, 1953.

Doob, J.L., *Wiener's work in probability theory*, Bull. Amer. Math. Soc. **72(1), part II** (1966), 69–72.

Doob, J.L., *Classical Potential Theory and Its Probabilistic Counterpart*, Springer-Verlag, Berlin, Heidelberg, and New York, 1984.

Dufresne, D., *The distribution of a perpetuity, with applications to risk theory and pension funding*, Scand. Actuarial J. **1-2** (1990), 39–79.

Dufresne, D., *Laguerre series for Asian and other options*, Report No. 69, Center for Actuarial Studies, University of Melbourne, Australia (1999).

Durrett, R., *Brownian Motion and Martingales in Analysis*, Wadsworth Inc., Belmont, California, 1984.

Durrett, R., *Stochastic Calculus: A Practical Introduction*, CRC Press, Boca Raton, Florida, 1996.

Durrett, R., and Iglehart, D.L., *Functionals of Brownian meander and Brownian excursion*, Ann. Probab. **5** (1977), 130–135.

Dynkin, E.B., *Markov Processes I, II*, Springer-Verlag, Berlin, Heidelberg, and New York, 1965.

Einstein, A., *On the movement of small particles suspended in a stationary liquid demanded by the molecular-kinetic theory of heat*, Ann. Physik **17** (1905), 549–560.

Eisenbaum, N., *Un théorème de Ray–Knight lié au supremum des temps locaux browniens*, Probab. Th. Rel. Fields **87** (1990), 79–95.

Emanuel, D., and MacBeth, J., *Further results on the constant elasticity of variance call option pricing model*, J. Financial and Quant. Anal. **17** (1982), 533–554.

Engelbert, H.J., and Schmidt, W., *On one-dimensional stochastic differential equations with generalized drift*, in "Stochastic Differential Systems", Lecture notes in Control and Information Sciences, Vol. 69, Springer-Verlag, Berlin, Heidelberg, and New York, 1984, pp. 143–155.

Engelbert, H.J., and Schmidt, W., *On solutions of stochastic differential equations without drift*, Z. Wahrscheinlichkeitstheorie verw. Gebiete **68** (1985), 287–317.

Engelbert, H.J., and Schmidt, W., *Strong Markov continuous local martingales and solutions of one-dimensional stochastic differential equations (Part III)*, Math. Nachr. **151** (1991), 149–197.

Engelbert, H.J., and Senf, T., *On functionals of Wiener process with drift and exponential local martingales*, Dozzi, M., Engelbert, H.J., and Nualart, D. (ed.), *Stochastic processes and related topics. Proc. Wintersch. Stochastic Processes, Optim. Control, Georgenthal/Ger.* 1990, N 61 in Math. Res., Academic Verlag, Berlin, 1991, pp. 45–58.

Erdelyi, A., et al., *Higher Transcendental Functions, I, II, III*, McGraw-Hill, New York, 1953.

Erdelyi, A., et al., *Tables of Integral Transforms, I, II*, McGraw-Hill, New York, 1954.

Ethier, S.N., and Kurtz, T.G., *Markov Processes. Characterization and Convergence*, Wiley and Sons, New York, 1986.

Feller, W., *Zur Theorie der stochastischen Prozesse*, Math. Ann. **113** (1936), 113–160.

Feller, W., *The asymptotic distribution of the range of sums of independent random variables*, Ann. Math. Stat. **22** (1951), no. 3, 427–432.

Fitzsimmons, P.J., and Pitman, J.W., *Kac's moment formula and the Feynman–Kac formula for additive functionals of a Markov process*, Stoch. Proc. Appl. **79** (1999), 117-134.

Föllmer, H., Protter, P., and Shiryaev, A.N., *Quadratic variation and an extension of Itô's formula*, Bernoulli **1** (1995), 149–169.

Föllmer, H., and Protter, P., *On Itô's formula for multidimensional Brownian motion*, Probab. Theory Relat. Fiedds **116** (2000), 1–20.

Freedman, D., *Brownian Motion and Diffusion*, Holden-Day, San Francisco, 1971.

Geman, D., and Horowitz, J., *Occupation densities*, Ann. Probab. **8** (1980), 1–67.

Geman, H., and Yor, M., *Bessel processes, Asian options and perpetuities*, Mathematical Finance **3** (1993), 349–375.

Getoor, R.K., *The Brownian escape process*, Ann. Probab. **7** (1979a), 864–867.

Getoor, R.K., *Excursions of a Markov process*, Ann. Probab. **8** (1979b), 244–266.

Getoor, R.K., *Transience and recurrence of Markov processes*, in "Séminaire de Probabilités, XIV, 1978/79", Lecture Notes in Mathematics, vol. 784, Springer Verlag, Berlin, Heidelberg, and New York, 1980, pp. 397–409.

Getoor, R.K., and Sharpe, M., *Excursions of Brownian motion and Bessel processes*, Z. Wahrscheinlichkeitstheorie verw. Gebiete. **47** (1979), 83–106.

Getoor, R.K., and Sharpe, M., *Excursions of dual processes*, Adv. in Math. **45** (1982), 259–309.

Gihman, I.I., and Skorohod, A.V., *Theory of Stochastic Processes I, II, III*, Springer-Verlag, Berlin, Heidelberg and New York, 1974, 1975, 1979.

Girsanov, I.V., *On transforming a certain class of stochastic processes by absolutely continuous substitution of measures*, Th. Probab. Appl. **5** (1960), 285–301.

Goldenberg, D., *A unified method for pricing options on diffusion processes*, J. Financial Econom. **29** (1991), 3–34.

Graversen, S.-E., and Shiryaev, A.N., *An extension of P. Lévy's distributional properties to the case of a Brownian motion*, Bernoulli **6** (2000), 615–620.

Gruet, J.-C., *Windings of hyperbolic Brownian motion*, Yor, M. (ed.), *Exponential Functionals and Principal Values related to Brownian Motion*, Revista Matemática Iberoamericana, Madrid, 1997, pp. 35–72.

Harrison, J.M., *Brownian Motion and Stochastic Flow Systems*, Wiley, New York, 1985.

Harrison, J.M., and Lemoine, A.J., *Sticky Brownian motion as the limit of storage processes*, J. Appl. Probab. **18** (1981), 216–226.

Harrison, J.M., and Shepp, L.A., *On skew Brownian motion*, Ann. Probab. **9** (1981), 309–313.

Ikeda, N., Kusuoka, S., and Manabe, S., *Lévy's stochastic area formula for Gaussian processes*, Comm. Pure App. Math. **XLVII** (1994), 329–360.

Ikeda, N., and Watanabe, S., *Stochastic Differential Equations and Diffusion Processes*, North-Holland Publ. Co. and Kodansha Ltd, Amsterdam, Oxford, New York, and Tokyo, 1981.

Imhof, J.-P., *Density factorization for Brownian motion, meander and the three-dimensional Bessel process, and applications*, J. Appl. Probab. **21** (1984), 500–510.

Imhof, J.-P., *On the time spent above a level by Brownian motion with negative drift*, Adv. Appl. Probab. **18** (1986), 1017–1018.

Itô, K., *Stochastic integral*, Proc. Imperial Acad. Tokyo **20** (1944), 519–524.

Itô, K., *Poisson point processes attached to Markov processes*, in "Proc. of the sixth Berkeley Symposium in Mathematical Statistics and Probability, vol. 3", University of California, Berkeley, 1970, pp. 225–239.

Itô, K., and McKean, Jr., H.P., *Diffusion Processes and Their Sample Paths*, Springer Verlag, Berlin, Heidelberg, and New York, 1974.

Jacobsen, M., *Splitting times for Markov processes and a generalized Markov property for diffusions*, Z. Wahrscheinlichkeitstheorie verw. Gebiete **30** (1974), 27–43.

Jacobsen, M., *Laplace and the origin of the Ornstein–Uhlenbeck process*, Bernoulli **2** (1996), 271–286.

Jeanblanc, M., Pitman, J., and Yor, M., *The Feynman–Kac formula and decomposition of Brownian paths*, Comput. Appl. Math. **16** (1997), 27–52.

Jeulin, T., *Semi-martingales et Grossissement d'une Filtration*, Lecture Notes in Mathematics, vol. 833, Springer-Verlag, Berlin, Heidelberg, and New York, 1980.

Jeulin, T., *Applications de la théorie du grossissement de filtrations á l'étude des temps locaux du mouvement Browniens*, in "Grossissement de filtrations: exemples et applications", Lecture Notes in Mathematics, vol. 1118, Springer Verlag, Berlin, Heidelberg, and New York, 1985, pp. 197–304.

Jeulin, T., and Yor, M., *Inègalitè de Hardy, semimartingales, et faux-amis*, in "Séminaire de Probabilités, XII, 1977/78", Lecture Notes in Mathematics, vol. 721, Springer-Verlag, Berlin, Heidelberg, and New York, 1979, pp. 332–359.

Jeulin, T., and Yor, M., *Sur les distributions de certaines fonctionnelles du mouvement brownien*, in "Séminaire de Probabilités, XV, 1979/80", Lecture Notes in Mathematics, vol. 850, Springer-Verlag, Berlin, Heidelberg, and New York, 1981, pp. 210–226.

Johnson, B., and Killeen, T., *An explicit formula for the c.d.f. of the L_1 norm of the Brownian bridge*, Ann. Probab. **11** (1983), 807–808.

Kac, M., *On distribution of certain Wiener functionals*, Trans. Amer. Math. Soc. **65** (1949), 1–13.

Kac, M., *On some connection between probability theory and differential and integral equations*, in "Proc. of the second Berkeley Symposium in Mathematical Statistics and Probability", University of California, Berkeley and Los Angeles, 1951, pp. 189–215.

Kac, M., *Wiener and integration in function space*, Bull. Amer. Math. Soc. **72(1), part II** (1966), 52–68.

Kac, M., Kiefer, J., and Wolfowitz, J., *On tests of normality and other tests of goodness of fit based on distance methods*, Ann. Math. Stat. **18** (1955), 189–211.

Kamke, E., *Differentialgleichungen, Lösungsmethoden und Lösungen*, Akademische Verlagsgesellschaft Becker & Erler Kom.-Ges., Leipzig, 1943.

Karatzas, I., and Shreve, S.E., *Trivariate density of Brownian motion, its local and occupation times, with application to stochastic control*, Ann. Probab. **12** (1984), 819–828.

Karatzas, I., and Shreve, S.E., *Brownian Motion and Stochastic Calculus, 2nd edition*, Springer-Verlag, Berlin, Heidelberg, and New York, 1991.

Karlin, S., and Taylor, H.M., *A second course in stochastic processes*, Academic Press, Boston, San Diego, and New York, 1981.

Kennedy, D., *The distribution of the maximum of Brownian excursion*, J. Appl. Probab. **13** (1976), 371–376.

Kesten, H., *An iterated logarithm law for local times*, Duke Math. J. **32** (1965), 447–456.

Kesten, H., *The influence of Mark Kac on probability theory*, Ann. Probab. **14** (1986), 1103–1128.

Khas'minskii, R. Z., *Probability distribution for functionals of the path of a random process of diffusion type*, Dokl. Akad. Nauk SSSR **104** (1955), no. 1, 22–25. (Russian)

Khoshnevisan, D., Salminen, P., and Yor, M., *A note on a.s. finiteness of perpetual integral functionals of diffusions*, Elect. Comm. in Probab. **11** (2006), 108–117.

Kinkladze, G.N., *A note on the structure of processes the measure of which is absolutely continuous with respect to the Wiener process modulus measure*, Stochastics **8** (1982), 39–84.

Knight, F.B., *Random walks and a sojourn density process of Brownian motion*, Trans. Amer. Math. Soc. **109** (1963), 56–86.

Knight, F.B., *Brownian local times and taboo processes*, Trans. Amer. Math. Soc. **143** (1969), 173–185.

Knight, F.B., *On the sojourn times of killed Brownian motion*, in "Séminaire de Probabilités, XII, 1976/77", Lecture Notes in Mathematics, vol. 649, Springer-Verlag, Berlin, Heidelberg, and New York, 1978, pp. 428–445.

Knight, F.B., *Essentials of Brownian Motion and Diffusion*, Math. Surv. No.18, Amer. Math. Soc., Providence and Rhode Island, 1981.

Kolmogorov, A.N., *Über die analytischen Methoden in der Wahrscheinlichkeitsrechnung*, Math. Ann. **104** (1931), 415–458.

Kramkov, D.O., and Mordecki, E., *Integral options*, Th. Probab. Appl. **39** (1994), 162–171.

Kunita, H., and Watanabe, T., *Markov processes and Martin boundaries, I*, Illinois J. Math. **9** (1965), 485–526.

Lai, T. L., *Space-time processes, parabolic functions and one-dimensional diffusions*, Trans. Amer. Math. Soc. **175** (1973), 409–438.

Lamperti, J., *Stochastic processes; a survey of the mathematical theory*, Springer-Verlag, New York, Heidelberg, Berlin, 1977.

Langer, H., and Schenk, W., *Generalised second-order differential operators, corresponding gap diffusions and superharmonic transformations*, Math. Nachr. **148** (1990), 7–45.

Lebedev, N.N., *Special functions and their applications*, Dover, New York, 1972.

Le Gall, J.F., *Applications du temps local aux équations différentielles stochastiques unidimensionelles*, in "Séminaire de Probabilités, XVII, 1981/82", Lecture Notes in Mathematics, vol. 986, Springer Verlag, Berlin, Heidelberg, and New York, 1983, pp. 15–31.

Lehoczky, J.P., *Formulas for stopped diffusion processes with stopping times based on the maximum*, Ann. Probab. **5** (1977), no. 4, 601–607.

Lenglart, E., Lépingle, D., and Pratelli, M., *Une présentation unifiée des inégalités en théorie des martingales*, in "Séminaire de Probabilités, XIV, 1978/79", Lecture Notes in Mathematics, vol. 784, Springer-Verlag, Berlin, Heidelberg, and New York, 1980, pp. 26–48.

Lévy, P., *Sur certains processus stochastiques homogènes*, Compositio Math. **7** (1939), 283–339.

Lévy, P., *Processus stochastiques et mouvement brownien*, Gauthier-Villars, Paris, 1948.

Lévy, P., *Construction du processus de W.Feller et H.P.McKean en partant du mouvement Brownien*, in "Probability and Statistics: The Harald Cramér Volume", Wiley and Sons, New York, 1959, pp. 162–174.

Lindvall, T., *On coupling of diffusion processes*, J. Appl. Probab. **20** (1983), 82–93.

Liptser, R.S. and Shiryaev, A.N., *Statistics of Random Processes I, II*, Springer-Verlag, Berlin, Heidelberg and New York, 1977, 1978.

Loève, M., *Paul Lévy, 1886-1971*, Ann. Probab. **1** (1973), 1–18.

Louchard, G., *Kac's formula, Lévy's local time and Brownian excursion*, J. Appl. Probab. **21** (1984), 479–499.

McGill, P., *Markov properties of diffusion local time: a martingale approach*, Adv. Appl. Probab. **14** (1982), 789–810.

McKean, H.P., *Stochastic Integrals*, Academic Press, New York, 1969.

McKean, H.P., *Brownian local time*, Adv. Math. **15** (1975), 91–111.

Mandl, P., *Analytical treatment of one-dimensional Markov processes*, Springer-Verlag, Berlin, Heidelberg, and New York, 1968.

Matsumoto, H., and Yor, M., *On Bougerol and Dufresne's identities for exponential Brownian functionals*, Proc. Japan Acad., Ser A **74** (1998), 152–155.

Matsumoto, H., and Yor, M., *A version of Pitman's 2M–X theorem for geometric Brownian motion*, C. R. Acad. Sc. Paris Séries I **328** (1999), 1067–1074.

Matsumoto, H., and Yor, M., *An analogue of Pitman's 2M–X theorem for exponential Wiener functionals, Part I: A time inversion approach*, Nagoya Math. J. **159** (2000), 1–42.

Matsumoto, H., and Yor, M., *An analogue of Pitman's 2M–X theorem for exponential Wiener functionals, Part II: The role of the generalized inverse Gaussian laws*, Nagoya Math. J. **162** (2001), 65–86.

Meyer, P.A., *Flot d'une equation differentielle stochastique*, in "Séminaire de Probabilités, XV, 1979/80", Lecture Notes in Mathematics, vol. 850, Springer-Verlag, Berlin, Heidelberg, and New York, 1981, pp. 103–117.

Meyer, P.A., Smythe, R.T., and Walsh, J.B., *Birth and death of Markov processes*, in "Proc. Sixth Berkeley Symposium III", University of California Press, 1971, pp. 295–305.

Molchanov, S.A., *Martin boundaries for invariant Markov process on a solvable group*, Th. Probab. Appl. **12** (1967), 310–314.

Nagasawa, M., *Time reversions of Markov processes*, Nagoya Math. J. **24** (1964), 177–204.

Nelson, E., *Dynamical theories of Brownian motion*, Princeton University Press, Princeton, NJ, 1967.

Neveu, J., *Processus aléatoires gaussiens*, Les Presses de l'Univ. de Montréal, Montréal, 1968.

Øksendal, B., *Stochastic Differential Equations, an introduction with applications, 5th edition*, Springer-Verlag, Berlin, Heidelberg, and New York, 1998.

Paulsen, J., *Risk theory in a stochastic economic environment*, Stoch. Proc. Appl. **46(2)** (1993), 327–362.

Perkins, E., *The exact Hausdorff measure of the level sets of Brownian motion*, Z. Wahrscheinlichkeitstheorie verw. Gebiete **58** (1981), 373–388.

Perkins, E., *Local time is a semimartingale*, Z. Wahrscheinlichkeitstheorie verw. Gebiete **60** (1982), 79–117.

Perkins, E., *Local time and pathwise uniqueness for stochastic differential equations*, in "Seminaire de probabilité XVI", Lecture notes in Mathematics., vol. 920, Springer Verlag, Berlin, Heidelberg, and New York, 1982b, pp. 201–208.

Pitman, J.W., *One-dimensional Brownian motion and three-dimensional Bessel process*, Adv. Appl. Probab. **7** (1975), 511–526.

Pitman, J.W., *The SDE solved by local times of a Brownian excursion or bridge derived from the height profile of a random tree or forest*, Ann. Probab. **27** (1999), 261–283.

Pitman, J.W., and Yor, M., *Bessel processes and infinitely divisible laws*, in "Stochastic integrals", Lecture notes in Mathematics., vol. 851, Springer-Verlag, Berlin, Heidelberg, and New York, 1981, pp. 285–370.

Pitman, J.W., and Yor, M., *A decomposition of Bessel bridges*, Z. Wahrscheinlichkeitstheorie verw. Gebiete **59** (1982), 425–457.

Pitman, J.W., and Yor, M., *Decomposition at the maximum for excursions and bridges of one-dimensional diffusions*, in "Itô's Stochastic Calculus and Probability Theory", Springer-Verlag, Berlin, Heidelberg, and New York, 1996.

Pitman, J.W., and Yor, M., *Laplace transforms related to excursions of a one-dimensional diffusions*, Bernoulli **5** (1999a), 249–255.

Pitman, J.W., and Yor, M., *The law of the maximum of a Bessel bridge*, Elect. J. Probab. **4** (1999b), 1–35.

Rauscher, B., *Some remarks on Pitman's theorem*, in "Seminaire de probabilité XXXI", Lecture notes in Mathematics., vol. 1655, Springer-Verlag, Berlin, Heidelberg, and New York, 1997, pp. 266–271.

Ray, D.B., *Sojourn times of a diffusion process*, Ill. J. Math. **7** (1963), 615–630.

Revuz, D., and Yor, M., *Continuous Martingales and Brownian Motion, 3rd edition*, Springer-Verlag, Berlin, Heidelberg, and New York, 2001.

Rice, S.O., *The integral of the absolute value of the pinned Wiener process – calculation of its probability density by numerical integration*, Ann. Probab. **10** (1982), no. 1, 240–243.

Roberts, A.W., and Varberg, D.E., *Convex Functions*, Academic Press, New York, 1973.

Rogers, L.C.G., *Characterizing all diffusions with the 2M–X property*, Ann. Probab. **9** (1981), no. 4, 561–572.

Rogers, L.C.G., *A guided tour through excursions*, Bull. London Math. Soc. **12** (1989), no. 4, 305–341.

Rogers, L.C.G., and Pitman, J., *Markov functions*, Ann. Probab. **9** (1981), no. 4, 573–582.

Rogers, L.C.J., and Williams, D., *Diffusions, Markov Processes, and Martingales, Volume 2: Itô Calculus, 2nd edition*, Cambridge University Press, Cambridge, 2000.

Rogers, L.C.J., and Williams, D., *Diffusions, Markov Processes, and Martingales, Volume 1: Foundations, 2nd edition*, Wiley and Sons, New York, 1994.

Rösler, U., *The tail σ-field of time-homogeneous one-dimensional diffusion processes*, Ann. Probab. **7** (1979), no. 5, 847–857.

Saisho, T., and Tanemura, H., *Pitman type theorem for one-dimensional diffusion process*, Tokyo J. Math. **13** (1990), no. 2, 429–440.

Salminen, P., *Mixing Markovian laws; with an application to path decompositions*, Stochastics **9** (1983), 223–231.

Salminen, P., *One-dimensional diffusions and their exit spaces*, Math. Scand. **54** (1984), 209–220.

Salminen, P., *Optimal stopping of one-dimensional diffusions*, Math. Nachr. **124** (1985), 85–101.

Salminen, P., *On the distribution of supremum of diffusion local time*, Statistics & Probability Letters **18** (1993), 219–225.

Salminen, P., *A pointwise limit theorem for the transition density of a linear diffusion*, in "Proc. of the fifth Finnish–Russian Symposium on Probability Theory and Mathematical Statistics", Frontiers in Pure and Applied Probability, VSP and TVP, Utrecht and Moscow, 1996.

Salminen, P., *On last exit decompositions of linear diffusions*, Studia Sci. Math. Hungar. **33** (1997), 251–262.

Salminen, P., and Vallois, P., *On first range times of linear diffusions*, J. Theor. Probab. **18** (2005), 567–593.

Salminen, P., and Yor, M., *Perpetual integral functionals as hitting and occupation times*, Elect. J. Probab. **10** (2005), 371–419.

Salminen, P., and Yor, M., *Properties of perpetual integral functionals of Brownian motion with drift*, Ann. I. H. Poincaré, PR **41** (2005), 335–347.

Schwartz, L., *Quelques réflexions et souvenirs sur Paul Lévy*, in "Colloque Paul Lévy sur les processus stochastiques", vol. 157–158, 1988, pp. 13–28.

Sharpe. M., *Some transformations of diffusions by time reversal*, Ann. Probab. **8** (1980), 1157–1162.

Sharpe, M.J., *General theory of Markov processes*, Academic Press, New York, 1988.

Shepp, L.A., *The joint density of the maximum and its location for a Wiener process with drift*, J. Appl. Probab. **16** (1979), 423–427.

Shepp, L.A., *On the integral of the absolute value of the pinned Wiener process*, Ann. Probab. **10** (1982), 234–239.

Shepp, L.A., and Shiryaev, A.N., *The Russian option: reduced regret*, Ann. Appl. Probab. **3** (1993), 631–640.

Shepp, L.A., and Shiryaev, A.N., *A new look at pricing of the "Russian" option*, Theory Probab. Appl. **39** (1994), 103–119.

Shiga, T., and Watanabe, S., *Bessel diffusions as a one-parameter family of diffusion processes*, Z. Wahrscheinlichkeitstheorie verw. Gebiete **27** (1973), 37–46.

Shiryaev, A.N., *Kolmogorov: life and creative activities*, Ann. Probab. **17** (1989), 866–944.

Stroock, D.W., and Varadhan, S.R.S., *Diffusion processes with continuous coefficients I, II*, Comm. Pure Appl. Math. **22** (1969), 345–400, 479–530..

Stroock, D.W., and Varadhan, S.R.S., *Multidimensional Diffusion Processes*, Springer-Verlag, Berlin, Heidelberg, and New York, 1979.

Takàcs, L., *On the generalization of the arc-sine law*, Ann. Appl. Probab. **6** (1996), 1035–1039.

Taqqu, M.S., *Bachelier and his times: A conversation with Bernard Bru*, Finance and Stochastics **5** (2001), 3–32.

Taylor, H.M., *A stopped Brownian motion formula*, Ann. Probab. **3** (1975), no. 2, 234–246.

Taylor, S.J., and Wendel, J.G., *The exact Hausdorff measure of the zero set of a stable process*, Z. Wahrscheinlichkeitstheorie verw. Gebiete **6** (1966), 170–180.

Trotter, H.F., *A property of Brownian motion paths*, Ill. J. Math. **2** (1958), 425–433.

Truman, A., and Williams, D., *A generalized arc-sine law and Nelson's stochastic mechanics of one-dimensional time-homogeneous diffusions*, in "Proc. Int. Conf., Evanston/IL 1989", Prog. Probab. vol. 22, Birkhauser Verlag, Boston/MA, 1990, pp. 117–135.

Vallois, P., *Diffusion arretée au premier instant où l'amplitude atteint un niveau donné*, Stoch. and Stoch. Reports **43** (1993), 93–115.

Vallois, P., *Decomposing the Brownian path via the range process*, Stoch. Proc. Appl. **55** (1995), 211–226.

Vervaat, W., *A relation between brownian bridge and brownian excursion*, Ann. Probab. **7(1)** (1979), 143–149.

Vincze, I., *Einige zweidimensionale Verteilungs- und Grenzverteilungssätze in der Theorie der geordneten Stichproben*, Publ. Math. Inst. Hungar. Acad. Sci. **11** (1957), 183–203.

Volkonskij, V.A., *Random time changes in strong Markov processes*, Th. Probab. Appl. **3** (1958), no. 2, 332–350.

Walsh, J.B., *A diffusion with discontinuous local time*, Astérisque **52-53** (1978), 37–46.

Walsh, J.B., *Downcrossings and the Markov property of local time*, Astérisque **52-53** (1978), 89–115.

Warren, J., *Branching processes, the Ray–Knight theorem, and sticky Brownian motion*, in "Séminaire de Probabilités, XXXI", Lecture Notes in Mathematics, vol. 1665, Springer-Verlag, Berlin, Heidelberg, and New York, 1997, pp. 1–15.

Wenocur, M., *Ornstein–Uhlenbeck process with quadratic killing*, J. Appl. Probab. **23** (1990), 707–712.

Wiener, N., *Differential space*, J. Math. Phys. **2** (1923), 131–174.

Williams, D., *Path decomposition and continuity of local time for one-dimensional diffusions I*, Proc. London Math. Soc. **28** (1974), 738–768.

Williams, D., *Lévy's downcrossing theorem*, Z. Wahrscheinlichkeitstheorie verw. Gebiete **40** (1977), 157–158.

Williams, D., *Diffusions, Markov Processes, and Martingales, I: Foundations.*, Wiley and Sons, New York, 1979.

Yamada, T., and Watanabe, S., *On the uniqueness of solutions of stochastic differential equations*, J. Math. Kyoto Univ. **11** (1971), 155–167.

Yor, M., *Un exemple de processus qui n'est pas une semi-martingale.*, Astérisque **52-53** (1978), 219–221.

Yor, M., *Loi de l'indice du lacet Brownien et distribution de Hartman–Watson.*, Z. Wahrscheinlichkeitstheorie verw. Gebiete **53** (1980), 71–95.

Yor, M., *Some Aspects of Brownian Motion. Part I: Some Special Functionals*, Birkhäuser, Basel, Boston, and Berlin, 1992a.

Yor, M., *On some exponential functionals of Brownian motion*, Adv. Appl. Probab. **24** (1992b), 509–531.

Yor, M., *On certain exponential functionals of real-valued Brownian motion*, J. Appl. Probab. **29** (1992c), 202–208.

Yor, M., *Local times and excursions for Brownian motion: a concise introduction*, Lecciones en Matemáticas, Universidad Central de Venezuela, 1995.

Yor, M., *Some Aspects of Brownian Motion. Part II: Some Recent Martingale Problems*, Birkhäuser, Basel, Boston, and Berlin, 1997a.

Yor, M. (ed.), *Exponential Functionals and Principal Values related to Brownian Motion; A collection of research papers*, Revista Matemática Iberoamericana, Madrid, 1997b.

Zvonkin, A. K., *A transformation of the phase space of a diffusion process that moves the drift*, Math. USSR Sbornik **22** (1974), 129–149. (Russian)

SUBJECT INDEX

Printed in the USA
CPSIA information can be obtained
at www.ICGtesting.com
LVHW020736231023
761793LV00006B/597